Polymers for Food Applications

Tomy J. Gutiérrez
Editor

Polymers for Food Applications

Springer

Editor
Tomy J. Gutiérrez
Thermoplastic Composite Materials (CoMP) Group
Faculty of Engineering
Institute of Research in Materials Science and Technology (INTEMA)
National University of Mar del Plata (UNMdP) and National Scientific
and Technical Research Council (CONICET)
Mar del Plata, Buenos Aires, Argentina

ISBN 978-3-030-06886-8 ISBN 978-3-319-94625-2 (eBook)
https://doi.org/10.1007/978-3-319-94625-2

This Springer imprint is published by the registered company Springer Nature Switzerland AG
The registered company address is: Gewerbestrasse 11, 6330 Cham, Switzerland

I would like to dedicate this book:
To my God and to my Guardian Angel,
Its energy stimulates me to enjoy the landscape we call life, and its peace encourages me to always continue towards the future, a place where we will all go and each one will be under the law of the creative father.

To my Mother (Dr. Mirian Arminda Carmona Rodríguez),
For forming my character and attitude towards life.

To my Grandmother (Mrs. Arminda Teresa Rodríguez Romero),
A person who unfortunately left before this world's time, but I am sure that she is up watching me and supporting me in all facets of my life. You are in my most beautiful memories.

To all the anonymous persons,
Those who give me their love, friendship, patience and support in various situations.

To Venezuela and Argentina,
The first for giving me my academic and professional training, and the second for welcoming me with love and friendship before the dictatorship that my country (Venezuela) is experiencing today.

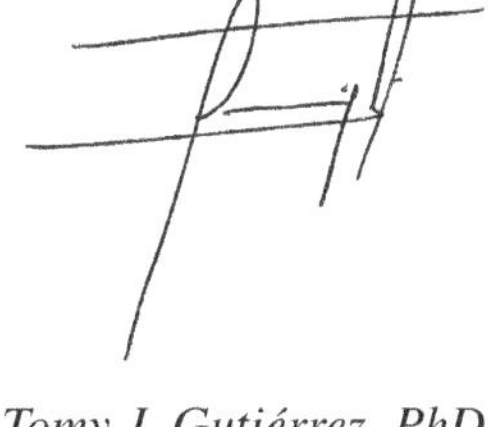

Tomy J. Gutiérrez, PhD
Editor

Contents

Contributors

Shakeel Ahmed Department of Chemistry, Government Degree College Mendhar, Poonch, Jammu and Kashmir, India

Priscila Alessio School of Technology and Applied Sciences, São Paulo State University (UNESP), Presidente Prudente, SP, Brazil

Akbar Ali Department of Chemistry, Jamia Millia Islamia, New Delhi, India

Vera A. Alvarez Thermoplastic Composite Materials (CoMP) Group, Institute of Research in Materials Science and Technology (INTEMA), Faculty of Engineering, National University of Mar del Plata (UNMdP) and National Scientific and Technical Research Council (CONICET), Buenos Aires, Argentina

C. Anandharamakrishnan Indian Institute of Food Processing Technology (IIFPT), Ministry of Food Processing Industries, Government of India, Thanjavur, India

K. K. Anas Biochemistry & Nutrition Division, Central Institute of Fisheries Technology (CIFT), ICAR, Cochin, Kerala, India

María R. Ansorena Food Engineering Group, Chemical Engineering Department, Engineering Faculty, National University of Mar del Plata (UNMdP), Mar del Plata, Argentina

Constantin Apetrei Department of Chemistry, Physics, and Environment, Faculty of Sciences and Environment, "Dunarea de Jos" University of Galati, Galati, Romania

Luis M. Araque Graduate Program in Materials Science and Engineering, Federal University of Piauí, Teresina, Piauí, Brazil

Teresita Arredondo-Ochoa DIPA, PROPAC, Facultad de Química, Universidad Autónoma de Querétaro, Querétaro, Qro, Mexico

Mohamed E. I. Badawy Department of Pesticide Chemistry and Technology, Faculty of Agriculture, Alexandria University, Alexandria, Egypt

Alicia Califano Centro de Investigación y Desarrollo en Criotecnología de Alimentos (CIDCA), Facultad de Cs. Exactas, UNLP—CICPBA—CONICET, La Plata, Argentina

M. A. Cerqueira International Iberian Nanotechnology Laboratory, Braga, Portugal

Amparo Chiralt Instituto de Ingeniería de Alimentos para el Desarrollo, Universitat Politècnica de València, Valencia, Spain

Ioannis S. Chronakis Nano-Bio Science Research Group, National Food Institute, Technical University of Denmark, Kgs. Lyngby, Denmark

Sofía Collazo-Bigliardi Instituto de Ingeniería de Alimentos para el Desarrollo, Universitat Politècnica de València, Valencia, Spain

Koro de la Caba BIOMAT Research Group, Department of Chemical and Environmental Engineering, Engineering College of Gipuzkoa, Donostia-San Sebastián, Spain

Paulo José do Amaral Sobral Department of Food Engineering, Faculty of Animal Science and Food Engineering, University of São Paulo, Pirassununga, São Paulo, Brazil

A. L. Ellis Chemical Engineering Department, University of Birmingham, Edgbaston, UK

Pedro J. García-Moreno Research Group for Bioactives—Analysis and Application, National Food Institute, Technical University of Denmark, Kgs. Lyngby, Denmark

Tania Garrido BIOMAT Research Group, Department of Chemical and Environmental Engineering, Engineering College of Gipuzkoa, Donostia-San Sebastián, Spain

John B. Golding Faculty of Science and Information Technology, School of Environmental and Life Sciences, University of Newcastle, Ourimbah, NSW, Australia

NSW Department of Primary Industries, Ourimbah, NSW, Australia

Manuel Gómez Food Technology Area, College of Agricultural Engineering, University of Valladolid, Palencia, Spain

Pedro Guerrero BIOMAT Research Group, Department of Chemical and Environmental Engineering, Engineering College of Gipuzkoa, Donostia-San Sebastián, Spain

Tomy J. Gutiérrez Thermoplastic Composite Materials (CoMP) Group, Faculty of Engineering, Institute of Research in Materials Science and Technology

(INTEMA), National University of Mar del Plata (UNMdP) and National Scientific and Technical Research Council (CONICET), Mar del Plata, Buenos Aires, Argentina

Dejian Huang National University of Singapore (Suzhou) Research Institute, Suzhou, People's Republic of China

Food Science and Technology Programme, c/o Department of Chemistry, National University of Singapore, Singapore, Singapore

Charlotte Jacobsen Research Group for Bioactives—Analysis and Application, National Food Institute, Technical University of Denmark, Kgs. Lyngby, Denmark

Lekshmi R. G. Kumar Biochemistry & Nutrition Division, Central Institute of Fisheries Technology (CIFT), ICAR, Cochin, Kerala, India

Biochemistry and Nutrition Division, Central Institute of Fisheries Technology (CIFT), Cochin, Kerala, India

A. Lazidis Chemical Engineering Department, University of Birmingham, Edgbaston, UK

Alex López-Córdoba Facultad Seccional Duitama, Escuela de Administración de Empresas Agropecuarias, Universidad Pedagógica y Tecnológica de Colombia, Boyacá, Colombia

Gabriel Lorenzo Centro de Investigación y Desarrollo en Criotecnología de Alimentos (CIDCA), Facultad de Cs. Exactas, UNLP—CICPBA—CONICET, La Plata, Argentina

Depto. Ingeniería Química, Facultad de Ingeniería, Universidad Nacional de La Plata (UNLP), La Plata, Argentina

Suzana Mali Department of Biochemistry and Biotechnology, State University of Londrina, Paraná, Brazil

Bianca Chieregato Maniglia Departamento de Química, Faculdade de Filosofia, Ciências e Letras de Ribeirão Preto, Universidade de São Paulo, Ribeirão Preto, SP, Brazil

Norma E. Marcovich Instituto de Investigaciones en Ciencia y Tecnología de Materiales (INTEMA), Universidad Nacional de Mar del Plata (UNMdP)–Consejo Nacional de Investigaciones Científicas y Técnicas (CONICET), Mar del Plata, Argentina

Cibely S. Martin School of Technology and Applied Sciences, São Paulo State University (UNESP), Presidente Prudente, SP, Brazil

Olga Martín-Belloso Department of Food Technology, University of Lleida–Agrotecnio Center, Lleida, Spain

A. J. Martins Centre of Biological Engineering, University of Minho, Braga, Portugal

Suseela Mathew Central Institute of Fisheries Technology (CIFT), ICAR, Cochin, Kerala, India

Mateus D. Maximino School of Technology and Applied Sciences, São Paulo State University (UNESP), Presidente Prudente, SP, Brazil

Ana C. Mendes Nano-Bio Science Research Group, National Food Institute, Technical University of Denmark, Kgs. Lyngby, Denmark

J. A. Moses Indian Institute of Food Processing Technology (IIFPT), Ministry of Food Processing Industries, Government of India, Thanjavur, India

Deepak Mudgil Department of Dairy and Food Technology, Mansinhbhai Institute of Dairy and Food Technology, Mehsana, India

Z. A. Nur Hanani Department of Food Technology, Faculty of Food Science and Technology, Universiti Putra Malaysia, Serdang, Selangor, Malaysia

Rodrigo Ortega-Toro Programa de Ingeniería de Alimentos, Facultad de Ingeniería, Universidad de Cartagena, Cartagena de Indias D.T y C, Colombia

L. M. Pastrana International Iberian Nanotechnology Laboratory, Braga, Portugal

Ashok R. Patel Guangdong Technion Israel Institute of Technology, Shantou, China

Mariana Pereda Instituto de Investigaciones en Ciencia y Tecnología de Materiales (INTEMA), Universidad Nacional de Mar del Plata (UNMdP)–Consejo Nacional de Investigaciones Científicas y Técnicas (CONICET), Mar del Plata, Argentina

Entsar I. Rabea Department of Plant Protection, Faculty of Agriculture, Damanhour University, Damanhour, Egypt

Carlos Regalado-González DIPA, PROPAC, Facultad de Química, Universidad Autónoma de Querétaro, Querétaro, Qro, Mexico

Laura Román Food Technology Area, College of Agricultural Engineering, University of Valladolid, Palencia, Spain

Bahareh Saberi Faculty of Science and Information Technology, School of Environmental and Life Sciences, University of Newcastle, Ourimbah, NSW, Australia

Jie Sun Xi'an Jiaotong-Liverpool University, Suzhou, People's Republic of China

National University of Singapore (Suzhou) Research Institute, Suzhou, People's Republic of China

Delia Rita Tapia-Blácido Departamento de Química, Faculdade de Filosofia, Ciências e Letras de Ribeirão Preto, Universidade de São Paulo, Ribeirão Preto, SP, Brazil

C. S. Tejpal Biochemistry & Nutrition Division, Central Institute of Fisheries Technology (CIFT), ICAR, Cochin, Kerala, India

Rahul Thakur Faculty of Science and Information Technology, School of Environmental and Life Sciences, University of Newcastle, Ourimbah, NSW, Australia

Milena Martelli Tosi Departamento de Química, Faculdade de Filosofia, Ciências e Letras de Ribeirão Preto, Universidade de São Paulo, Ribeirão Preto, SP, Brazil

Jone Uranga BIOMAT Research Group, Department of Chemical and Environmental Engineering, Engineering College of Gipuzkoa, Donostia-San Sebastián, Spain

Germán Ayala Valencia Department of Chemical and Food Engineering, Federal University of Santa Catarina, Florianópolis, SC, Brazil

A. A. Vicente Centre of Biological Engineering, University of Minho, Braga, Portugal

S. K. Vimala Bharathi Indian Institute of Food Processing Technology (IIFPT), Ministry of Food Processing Industries, Government of India, Thanjavur, India

Quan V. Vuong Faculty of Science and Information Technology, School of Environmental and Life Sciences, University of Newcastle, Ourimbah, NSW, Australia

Yixiang Xu Agricultural Research Station, Virginia State University, Petersburg, VA, USA

Liangkun Yan Keio-NUS CUTE Center, Smart System Institute (SSI), National University of Singapore, Singapore, Singapore

Jun Yang Physical Characterization Team, Measurement Science, PepsiCo Global R&D, Plano, TX, USA

Noemí Zaritzky Centro de Investigación y Desarrollo en Criotecnología de Alimentos (CIDCA), Facultad de Cs. Exactas, UNLP—CICPBA—CONICET, La Plata, Argentina

Depto. Ingeniería Química, Facultad de Ingeniería, Universidad Nacional de La Plata (UNLP), La Plata, Argentina

Haotian Zheng Animal Science Department, Leprino Dairy Innovation Institute, California Polytechnic State University, San Luis Obispo, CA, USA

Weibiao Zhou National University of Singapore (Suzhou) Research Institute, Suzhou, People's Republic of China

Food Science and Technology Programme, c/o Department of Chemistry, National University of Singapore, Singapore, Singapore

About the Editor

Tomy J. Gutiérrez has a degree in chemistry (geochemical option) from the Central University of Venezuela (UCV) (December, 2007), a degree in education (chemical mention) from the same university (UCV, July, 2008) and a specialization in International Negotiation of Hydrocarbons from the National Polytechnic Experimental University of the National Armed Force (UNEFA)—Venezuela (July, 2011). He also has a master's and PhD degree in Food Science and Technology obtained in October 2013, and April 2015, respectively, both from the UCV. He has also completed PhD studies in Metallurgy and Materials Science from the UCV and postdoctoral studies at the Research Institute in Materials Science and Technology (INTEMA). Dr. Gutiérrez has been professor-researcher both at the Institute of Food Science and Technology (ICTA) and at the School of Pharmacy at the UCV. He is currently an adjunct researcher in INTEMA—The National Scientific and Technical Research Council (CONICET), Argentina. Dr. Gutiérrez has at least 20 book chapters and 30 publications in international journals of high impact factor. Dr. Gutiérrez today is developing a line of research in nanostructured materials based on polymers (composite materials), which are obtained on a pilot scale to be transferred to the food, pharmaceutical and polymer industry. In addition, he is a collaborator of international projects between Argentina and France, Brazil, Venezuela and Colombia.

Chapter 1
Polymers for Food Applications: News

Tomy J. Gutiérrez

Abstract Polymers usually are found every day in a myriad of applications, but special importance has the polymers for food applications. In particular, edible polymers are of great importance for human subsistence. Edible polymers from the nutritional point of view have been classified as carbohydrates, proteins, fiber and lipids, i.e. they are considered as macronutrients. The study of edible polymers still booming because of the great demand for healthier and more convenient foods, as well as the development of new food products with better sensory properties, which may have a prolonged shelf life. Many edible polymers being modified have allowed the development of functional or medical foods. Obtaining new products from modified edible polymers has also led to the manufacture of more stable foods, and even that can be administered to people with special dietary regimens such as celiac, phenylketonuric, diabetic, lactose intolerant, among others. The edible polymers have had a positive impact on different sectors of the food industry, from food packaging to the detection of toxic food substances. The edible polymers in essence lead to the production of edible films and membranes, foamed foods, snack, micro- and nanoencapsulated, hydrogels, prebiotics and oligomers, as well as food colloids and emulsions. More recently, edible polymers have also given way to the development of printed and electrospinned foods. This chapter aims to be preamble to the study and analysis of polymers for food applications that will be addressed in the course of this book, which has the contribution of important researchers with extensive experience, which in some cases are editors of major international journals in the field of food science and technology.

Keywords Carbohydrate polymers · Edible polymers · Food hydrocolloids · Polysaccharide · Proteins

T. J. Gutiérrez (✉)
Thermoplastic Composite Materials (CoMP) Group, Institute of Research in Materials Science and Technology (INTEMA), Faculty of Engineering, National University of Mar del Plata (UNMdP) and National Scientific and Technical Research Council (CONICET), Colón 10850, Mar del Plata 7600, Buenos Aires, Argentina
e-mail: tomy.gutierrez@fi.mdp.edu.ar

T. J. Gutiérrez (ed.), *Polymers for Food Applications*,
https://doi.org/10.1007/978-3-319-94625-2_1

1.1 Present and Future in Perspective

The main foods elaborated by the food industry are based on edible polymers, which constitute the biomacromolecule structure of them (Gutiérrez et al. 2015a, 2016a). These edible polymers interact and react with each other, thus achieving the development of micro- or nanostructured systems with better properties and insurance (Gutiérrez and Álvarez 2017; Gutiérrez 2018a). Many polymeric structures are found for food applications such as composite films and membranes, layer-by-layer films, intelligent and active films, foams, micro- and nanoencapsules, hydrogels, emulsions, electrospun, printing, hyperbranched structures (dendrimers), among others (Suárez and Gutiérrez 2017; Gutiérrez and Alvarez 2018; Gutiérrez 2018b). The edible polymers can be classified into three categories: hydrocolloids (polysaccharides and proteins), lipids (fatty acids and waxes) and composites (hydrocolloids and lipids mixtures or combinations of components of the same group) (Álvarez et al. 2017; Gutiérrez 2017a). It can be highlighted among the edible polymers most used: starch, cellulose and derivatives, alginate, chitosan, collagen, gelatin, casein, whey protein, among others (Gutiérrez et al. 2014, 2015a, b, c, d). Foods based on edible polymers today are developed not only with the aim of nourishing, but are also directed for example to prevent diseases, or improve the quality of life of people with metabolic diseases (Gutiérrez and Álvarez 2016; Gutiérrez 2017b, 2018c; Gutiérrez et al. 2018a). Although many efforts have been made in the study of edible polymers, the trend in this field should be directed to the study of toxicity, compostability, surface properties, and the nutritional and molecular aspects of these materials, since are points keys for their application (Bracone et al. 2016; Gutiérrez et al. 2016b, 2018b; Gutiérrez and González 2016, 2017; Medina Jaramillo et al. 2016; Gutiérrez 2017c, 2018d). Likewise, studies of polymers for food applications must be designed on a large scale, since many investigations are evaluated on a laboratory scale but are not viable within the food industry for different technical and economic reasons (Gutiérrez and Alvarez 2017a, b, c, d, e; Gutiérrez et al. 2017). Finally, the polymers for food applications will continue to give firm footing not only for their consumption but also for the detection of toxic food substances through the development of biosensors, the manufacture of food packaging and the rise of food nanotechnology.

Acknowledgements The authors would like to thank the Consejo Nacional de Investigaciones Científicas y Técnicas (CONICET) (Postdoctoral fellowship internal PDTS-Resolution 2417), Universidad Nacional de Mar del Plata (UNMdP) for financial support, and Dr. Mirian Carmona-Rodríguez.

Conflicts of Interest: The author declares no conflict of interest.

References

Álvarez K, Famá L, Gutiérrez TJ (2017) Physicochemical, antimicrobial and mechanical properties of thermoplastic materials based on biopolymers with application in the food industry. In: Masuelli M, Renard D (eds) Advances in physicochemical properties of biopolymers: Part 1. Bentham Science, Sharjah. EE.UU. ISBN: 978-1-68108-454-1. eISBN: 978-1-68108-453-4, pp 358–400. https://doi.org/10.2174/97816810845341170100015

Bracone M, Merino D, González J, Alvarez VA, Gutiérrez TJ (2016) Nanopackaging from natural fillers and biopolymers for the development of active and intelligent films. In: Ikram S, Ahmed S (eds) Natural polymers: derivatives, blends and composites. Nova Science, New York. EE.UU. ISBN: 978-1-63485-831-1, pp 119–155

Gutiérrez TJ (2017a) Chitosan applications for the food industry. In: Ahmed S, Ikram S (eds) Chitosan: derivatives, composites and applications. WILEY-Scrivener, Beverly, MA. EE. UU. ISBN: 978-1-119-36350-7, pp 183–232. https://doi.org/10.1002/9781119364849.ch8

Gutiérrez TJ (2017b) Surface and nutraceutical properties of edible films made from starchy sources with and without added blackberry pulp. Carbohydr Polym 165:169–179. https://doi.org/10.1016/j.carbpol.2017.02.016

Gutiérrez TJ (2017c) Effects of exposure to pulsed light on molecular aspects of edible films made from cassava and taro starch. Innov Food Sci Emerg Technol 41:387–396. https://doi.org/10.1016/j.ifset.2017.04.014

Gutiérrez TJ (2018a) Processing nano- and microcapsules for industrial applications. In: Hussain CM (ed) Handbook of nanomaterials for industrial applications. Elsevier, Amsterdam EE.UU. pp 989-1011. ISBN: 978-0-12-813351-4. https://doi.org/10.1016/B978-0-12-813351-4.00057-2

Gutiérrez TJ (2018b) Active and intelligent films made from starchy sources/blackberry pulp. J Polym Environ 26:2374–2391. https://doi.org/10.1007/s10924-017-1134-y

Gutiérrez TJ (2018c) Characterization and *in vitro* digestibility of non-conventional starches from Guinea arrowroot and La Armuña lentils as potential food sources for special diet regimens. Starch-Stärke 70(1–2). https://doi.org/10.1002/star.201700124

Gutiérrez TJ (2018d) Biodegradability and compostability of food nanopackaging materials. In: Cirillo G, Kozlowski MA, Spizzirri UG (eds) Composite materials for food packaging. WILEY-Scrivener, Beverly, MA. EE.UU. ISBN: 978-1-119-16020-5, pp 269–296. https://doi.org/10.1002/9781119160243.ch9

Gutiérrez TJ, Álvarez K (2016) Physico-chemical properties and *in vitro* digestibility of edible films made from plantain flour with added *Aloe vera* gel. J Funct Foods 26:750–762. https://doi.org/10.1016/j.jff.2016.08.054

Gutiérrez TJ, Álvarez K (2017) Biopolymers as microencapsulation materials in the food industry. In: Masuelli M, Renard D (eds) Advances in physicochemical properties of biopolymers: Part 2. Bentham Science, Sharjah. EE.UU. ISBN: 978-1-68108-545-6. eISBN: 978-1-68108-544-9, pp 296–322. https://doi.org/10.2174/97816810854491170100009

Gutiérrez TJ, Alvarez VA (2017a) Properties of native and oxidized corn starch/polystyrene blends under conditions of reactive extrusion using zinc octanoate as a catalyst. React Funct Polym 112:33–44. https://doi.org/10.1016/j.reactfunctpolym.2017.01.002

Gutiérrez TJ, Alvarez VA (2017b) Cellulosic materials as natural fillers in starch-containing matrix-based films: a review. Polym Bull 74(6):2401–2430. https://doi.org/10.1007/s00289-016-1814-0

Gutiérrez TJ, Alvarez VA (2017c) Data on physicochemical properties of active films derived from plantain flour/PCL blends developed under reactive extrusion conditions. Data Brief 15:445–448. https://doi.org/10.1016/j.dib.2017.09.071

Gutiérrez TJ, Alvarez VA (2017d) Eco-friendly films prepared from plantain flour/PCL blends under reactive extrusion conditions using zirconium octanoate as a catalyst. Carbohydr Polym 178:260–269. https://doi.org/10.1016/j.carbpol.2017.09.026

Gutiérrez TJ, Alvarez VA (2017e) Films made by blending poly(ε-caprolactone) with starch and flour from Sagu rhizome grown at the Venezuelan Amazons. J Polym Environ 25(3):701–716. https://doi.org/10.1007/s10924-016-0861-9

Gutiérrez TJ, Alvarez VA (2018) Bionanocomposite films developed from corn starch and natural and modified nano-clays with or without added blueberry extract. Food Hydrocoll 77:407–420. https://doi.org/10.1016/j.foodhyd.2017.10.017

Gutiérrez TJ, González G (2016) Effects of exposure to pulsed light on surface and structural properties of edible films made from cassava and taro starch. Food Bioprocess Technol 9(11):1812–1824. https://doi.org/10.1007/s11947-016-1765-3

Gutiérrez TJ, González G (2017) Effect of cross-linking with *Aloe vera* gel on surface and physicochemical properties of edible films made from plantain flour. Food Biophys 12(1):11–22. https://doi.org/10.1007/s11483-016-9458-z

Gutiérrez TJ, Pérez E, Guzmán R, Tapia MS, Famá L (2014) Physicochemical and functional properties of native and modified by crosslinking, dark-cush-cush yam (*Dioscorea Trifida*) and cassava (*Manihot Esculenta*) starch. J Polym Biopolym Phys Chem 2(1):1–5. https://doi.org/10.12691/jpbpc-2-1-1

Gutiérrez TJ, Morales NJ, Pérez E, Tapia MS, Famá L (2015a) Physico-chemical study of edible films based on native and phosphating cush-cush yam and cassava starches. Food Packaging Shelf Life 3:1–8. https://doi.org/10.1016/j.fpsl.2014.09.002

Gutiérrez TJ, Morales NJ, Tapia MS, Pérez E, Famá L (2015b) Corn starch 80:20 "waxy":regular, "native" and phosphated, as bio-matrixes for edible films. Proc Mater Sci 8:304–310. https://doi.org/10.1016/j.mspro.2015.04.077

Gutiérrez TJ, Tapia MS, Pérez E, Famá L (2015c) Structural and mechanical properties of native and modified cush-cush yam and cassava starch edible films. Food Hydrocoll 45:211–217. https://doi.org/10.1016/j.foodhyd.2014.11.017

Gutiérrez TJ, Tapia MS, Pérez E, Famá L (2015d) Edible films based on native and phosphated 80:20 waxy:normal corn starch. Starch-Stärke 67(1–2):90–97. https://doi.org/10.1002/star.201400164

Gutiérrez TJ, Guzmán R, Medina Jaramillo C, Famá L (2016a) Effect of beet flour on films made from biological macromolecules: native and modified plantain flour. Int J Biol Macromol 82:395–403. https://doi.org/10.1016/j.ijbiomac.2015.10.020

Gutiérrez TJ, Suniaga J, Monsalve A, García NL (2016b) Influence of beet flour on the relationship surface-properties of edible and intelligent films made from native and modified plantain flour. Food Hydrocoll 54:234–244. https://doi.org/10.1016/j.foodhyd.2015.10.012

Gutiérrez TJ, Guarás MP, Alvarez VA (2017) Reactive extrusion for the production of starch-based biopackaging. In: Masuelli MA (ed) Biopackaging. CRC Press Taylor & Francis Group, Miami, FL. EE.UU. ISBN: 978-1-4987-4968-8, pp 287–315

Gutiérrez TJ, Herniou-Julien C, Álvarez K, Alvarez VA (2018a) Structural properties and *in vitro* digestibility of edible and pH-sensitive films made from Guinea arrowroot starch and wastes from wine manufacture. Carbohydr Polym 184:135–143. https://doi.org/10.1016/j.carbpol.2017.12.039

Gutiérrez TJ, Ollier R, Alvarez VA (2018b) Surface properties of thermoplastic starch materials reinforced with natural fillers. In: Thakur VK, Thakur MK (eds) Functional biopolymers. Springer International, Basel. EE.UU. ISBN: 978-3-319-66416-3. eISBN: 978-3-319-66417-0, pp 131–158. https://doi.org/10.1007/978-3-319-66417-0_5

Medina Jaramillo C, Gutiérrez TJ, Goyanes S, Bernal C, Famá L (2016) Biodegradability and plasticizing effect of *yerba mate* extract on cassava starch edible films. Carbohydr Polym 151:150–159. https://doi.org/10.1016/j.carbpol.2016.05.025

Suárez G, Gutiérrez TJ (2017) Recent advances in the development of biodegadable films and foams from cassava starch. In: Klein C (ed) Handbook on cassava: production, potential uses and recent advances. Nova Science, New York. EE.UU. ISBN: 978-1-53610-307-6, pp 297–312

Chapter 2
Edible Films

María R. Ansorena, Mariana Pereda, and Norma E. Marcovich

Abstract Self-supporting edible films are one of the emerging technologies used today to optimize food preservation. With an ever increasing demand of consumers for high quality food products in addition to environmental concerns regarding the adverse effect of plastic packaging, food industry drive to develop and implement new types of edible films. The use of edible films as food packaging can play an important role in the quality, safety, transportation, storage, and display of a wide range of fresh and processed foods. Edible films can provide replacement and/or fortification of natural layers preventing moisture losses, while selectively allowing for controlled exchange of important gases such as oxygen and carbon dioxide, therefore, extending shelf life by minimizing food quality deterioration. Moreover, edible films can act as carriers of food additives such as vitamins, antimicrobial and antioxidants agents, providing a highly localized functional effect and improving food organoleptic properties. In addition, edible films could add value to agricultural and food industries by-products, since they are formed from various renewable eco-friendly and edible substances such as proteins, lipids or carbohydrates. Lipid-based films have good water barrier properties but form brittle films. On the other hand, protein and polysaccharide based films generally have good mechanical properties and thus they may withstand handling. However, they are not good barriers to water vapor. The use of blends comprising such compounds or their combination with lipids is thus a way of developing composite edible films matching the requirements for use as food packaging. Accordingly, this chapter discusses the latest advances on edible films aimed for food packaging.

Keywords Active films · Bioactive films · Food packaging · Food preservation

---

M. R. Ansorena
Food Engineering Group, Chemical Engineering Department, Engineering Faculty, National University of Mar del Plata (UNMdP), Mar del Plata, Argentina

M. Pereda · N. E. Marcovich (✉)
Instituto de Investigaciones en Ciencia y Tecnología de Materiales (INTEMA), Universidad Nacional de Mar del Plata (UNMdP)–Consejo Nacional de Investigaciones Científicas y Técnicas (CONICET), Mar del Plata, Argentina
e-mail: marcovic@fi.mdp.edu.ar

T. J. Gutiérrez (ed.), *Polymers for Food Applications*,
https://doi.org/10.1007/978-3-319-94625-2_2

2.1 Introduction

Changes in the current lifestyles of modern consumers have led to an increase in the sales of ready-to-eat (RTE), minimally processed and preservative-free food products (Peelman et al. 2013; Raybaudi-Massilia et al. 2016; Nisar et al. 2017; Moghimi et al. 2017). Consumers are demanding microbiologically safe and fresh-like products that are healthy, shelf-stable, convenient, and produced using environmentally friendly technologies (Raybaudi-Massilia et al. 2016). Nevertheless, natural and preservative-free food products are more susceptible to be attacked by spoilage and pathogenic microorganisms due to their manipulation. The growth of spoilage microorganisms and food-borne pathogens is one of the most important causes for food degradation (Viuda-Martos et al. 2011; Atarés and Chiralt 2016; Giteru et al. 2017), since it can accelerate lipid and other oxidation processes, produce changes in the organoleptic properties of the foods (Saggiorato et al. 2012; Atarés and Chiralt 2016) and losses of nutritional quality (Ganiari et al. 2017). Moreover, food-borne pathogens are directly responsible for certain illnesses in the human organism, or can be indirectly responsible due to the production of toxins (Saggiorato et al. 2012; Atarés and Chiralt 2016).

Packaging materials play an important role in containing foods and preserving the quality and safety of a food product throughout the supply chain until consumption (Raybaudi-Massilia et al. 2016; López-Córdoba et al. 2017). The packaging material functions as a barrier to the migration of pollutants from the environment to the product (Raybaudi-Massilia et al. 2016). It is however also urged to meet aesthetic and mechanical requirements for food applications (Manrich et al. 2017). For more than 50 years, the food industry has come to use a wide range of synthetic petroleum-based polymers that have provided convenience and ease of use in different situations (Giteru et al. 2017). However, questions have been raised about their continued use: they are non-renewable and non-biodegradable and therefore they have become a serious environmental issue that concerns all the stakeholders of the food production chain (Raybaudi-Massilia et al. 2016; Vijayendra and Shamala 2014; Giteru et al. 2017; López-Córdoba et al. 2017). Therefore, innovative films derived from agro-food industry wastes and renewable low cost natural resources have received greater attention as effective and economical replacement for conventional plastics (Valdés et al. 2014; Balti et al. 2017; Nisar et al. 2017; Alzate et al. 2017). Some of these materials are edible, so they can be consumed together with the product or be in close contact with the food, while being eco-friendly and ensuring that the package meets the primary objective of protecting the food (Raybaudi-Massilia et al. 2016).

Edible films are preformed, thin layers, made of edible materials, which once formed can be placed on or between food components (McHugh 2000; Yuan et al. 2016). Besides being biodegradable and compostable, edible films are desirable because they also offer a lucrative outlet for surplus agricultural materials (Tomasula 2009). On the other hand, these films have similar functions as those of conventional packaging, including barriers against water vapor, gases, and flavor compounds and improving structural integrity and mechanical-handling properties of foods (Galus and Kadzińska 2015; Yuan et al. 2016). Smartly designed, they can also improve the quality, safety, shelf life, and functionality of food products, as well as increase food sensory attributes and convenience (Pascall and Lin 2013; Raybaudi-Massilia et al. 2016), while minimizing both spoilage and pathogenic microorganisms during storage, transportation, and handling (Arrieta et al. 2014; Raybaudi-Massilia et al. 2016; Gutierrez-Pacheco et al. 2016). Moreover, the edible films can be used to produce a soluble package for premixed food ingredients or additives and act as a separate layer of individual food portion (Harnkarnsujarit 2017).

2.2 Materials Used for Edible Films

Edible films can be made from materials such as lipids, proteins, and polysaccharides with the ability to form films (Gutierrez-Pacheco et al. 2016; Balti et al. 2017; Hashemi and Mousavi Khaneghah 2017; Ganiari et al. 2017; Dehghani et al. 2018). The polysaccharides used for making edible films include cellulose derivatives, chitosan, starch, starch hydrolysates (dextrins), konjac flour, gums, pullulan, alginate, carrageenan, pectin and others that should be chemically treated to increase water solubility like cellulose and chitin (Park et al. 2014; Desobry and Arab-Tehrany 2014; Dehghani et al. 2018; Ganiari et al. 2017). Polysaccharides are widely available and usually cost effective (Dehghani et al. 2018). Due to the presence of a large number of hydroxyl and other polar groups in their structure, hydrogen bonds have a crucial function in film formation and final characteristics (Dehghani et al. 2018; Harnkarnsujarit 2017). The major mechanism of film formation in polysaccharide films is the breaking apart of polymer segments and reforming of the polymer chain into a film matrix or gel. This is usually achieved by evaporation of a solvent creating hydrophilic and hydrogen bonding and/or electrolytic and ionic cross-linking (Park et al. 2014). Protein films originate from several sources including plant, meat, egg, and milk, for example, collagen, albumin, gelatin, casein, milk whey proteins, corn zein, ovalbumin, soy protein, peanut protein, pea protein, rice bran protein, cottonseed protein, keratin and wheat gluten (Harnkarnsujarit 2017; Kumari et al. 2017; Park et al. 2014; Desobry and Arab-Tehrany 2014; Ansorena et al. 2016). However, some considerations with respect to food intolerances, such as wheat gluten intolerance (celiac disease), or milk protein intolerance, allergies, or religious beliefs/banning, should be taken into account when protein-based films and coatings are used (Desobry and Arab-Tehrany 2014). The main mechanism of formation of protein films includes denaturation of the protein initiated by heat, solvents, or a change in pH, followed by association of peptide chains through new intermolecular interactions (Dehghani et al. 2018), being the protein-protein interactions, with disulfide, hydrogen, and hydrophobic bonds, the main associative forces in the film network (Park et al. 2014). Proteins have good film-forming properties and good adherence to hydrophilic surfaces (Dehghani et al. 2018).

Protein and polysaccharide-based films generally have good mechanical and sensory properties and are effective barriers to aroma compounds and gases such as oxygen and carbon dioxide, but due to their hydrophilic nature, they are not effective water vapor barriers (Desobry and Arab-Tehrany 2014; Otoni et al. 2016). Even so, protein or polysaccharide based films can be used as "sacrificing agents" that retard moisture loss from the food products by adding additional moisture on the surface that is lost first, as indicated by Kester and Fennema (1986). Contrastingly, lipid-based films have good water barrier properties due to their apolar nature, but possess undesirable sensory properties and form brittle or non-cohesive films (Desobry and Arab-Tehrany 2014; Otoni et al. 2016). Therefore, lipids are either used as coatings or incorporated into polysaccharides or proteins matrices to form composite films, giving a better water vapor barrier, due to their low polarity (Desobry and Arab-Tehrany 2014; Dehghani et al. 2018). Lipid incorporation into edible films and coatings can improve also cohesiveness and flexibility making better moisture barriers (Desobry and Arab-Tehrany 2014). Some of the lipids that have been used effectively in films or coating formulations are beeswax, mineral oil, vegetable oil, surfactants, acetylated monoglycerides, shellac, terpene, carnauba wax, and paraffin wax (Desobry and Arab-Tehrany 2014; Harnkarnsujarit 2017). Lipids offer limited oxygen barrier properties, due to the presence of microscopic pores and elevated solubility and diffusivity (Desobry and Arab-Tehrany 2014) and are mostly soft solids at room temperature (Harnkarnsujarit 2017). Distribution of chemical groups, the length of the aliphatic chains, and the presence and degree of unsaturation impact lipid polarity and thus, they could behave in different ways respect to moisture transfer. For example, waxes (esters of long-chain aliphatic acids with long-chain aliphatic alcohols) have very low content of polar groups and high content of long-chain fatty alcohols and alkanes and thus they are very resistant to water migration while the high polarity of the unsaturated fatty acids results in high moisture transfer (Dehghani et al. 2018).

Whichever material is used for films, it should form an unbroken film structure with good functional properties (Kumari et al. 2017), i.e. food packaging having poor mechanical properties may not withstand handling, whereas one with poor barrier properties may lead to food physical, chemical, and microbiological spoilage (Otoni et al. 2016). To achieve this generally protein and polysaccharide-based film formulations require the addition of a plasticizing agent above a minimum threshold to reduce film fragility, confer certain plastic properties (Kokoszka et al. 2010; Martins et al. 2012; Sánchez-Ortega et al. 2014; Costa et al. 2015; Pérez et al. 2016) and sometimes also to guarantee their processability (Martins et al. 2012). The plasticizer molecules modify the three-dimensional organization of the polymeric materials, leading to decreased intermolecular forces along the polymer chains, increased free volume and chain mobility, thus improving flexibility, extensibility, and toughness of the film (Kokoszka et al. 2010; Pérez et al. 2016). However, plasticizers also decrease film cohesion and thus the mechanical resistance and barrier properties of the films (Pérez et al. 2016). The most commonly used food-grade plasticizers are sorbitol, glycerol, mannitol, sugar and polyethylene glycol (Kokoszka et al. 2010; Pérez et al. 2016; Castro-Rosas et al. 2016). On the other hand, it was demonstrated that lipids sometimes can also plasticize edible films (Pereda et al. 2012; Rodrigues et al. 2014; Kowalczyk et al. 2016; Rocca-Smith et al. 2016; Otoni et al. 2016) possibly because the amphiphilic compounds (e.g. acetic acids of mono and di-glycerides, and glycerol monostereate) included in the lipid phase, could act as lubricants favoring sliding of molecular chains, leading therefore to a gentle and highly extensible polymer network, as pointed out by Rocca-Smith et al. (2016).

2.3 Composite Edible Films

As can be deduced from the previous section, one of the most used approach to tailor films' properties and thus to develop edible films matching the desired functional properties and requirements for use as food packaging, has been the obtaining of composite films based on the combination of different biopolymers, biopolymers and lipids, biopolymers and solid particles (i.e. nonsoluble substances such as fibers, hydrophobic proteins, organic, and/or inorganic nanoparticles (NPs)), and so on (Martins et al. 2012; Desobry and Arab-Tehrany 2014; Otoni et al. 2016; Balti et al. 2017; Hashemi and Mousavi Khaneghah 2017; Ganiari et al. 2017). As expected, the main objective is to take advantage of the properties of each compound but also to benefit from the synergy between them when possible (Ganiari et al. 2017). In this sense, composite films are those whose structure is heterogeneous, that is, composed of a continuous matrix with some added phases or composed of several layers (Buffo and Han 2005; Fortunati 2016; Ganiari et al. 2017). Usually, multilayered films have better mechanical and barrier efficiencies than emulsion-based films and coatings, but their manufacturing requires an additional step of spreading or lamination and drying for each layer that could lead to layer delamination (Otoni et al. 2016). Moreover, in an industrial plan they do not seem very practical because of too many steps in their manufacture. On the contrary, solution or emulsion-based edible films could provide nearly the same properties, while only requiring one operation in their preparation (Fortunati 2016). However, when dealing with film-forming emulsions usually the incorporation of an emulsifier to avoid phase separation during drying is required (Otoni et al. 2016), which adds complexity to the system. Furthermore, the mechanical and barrier properties of these films not only depend on the compounds used in the polymer matrix, but also on their compatibility (Ganiari et al. 2017). Nevertheless, there has been great scientific and commercial progress made in this area (Balti et al. 2017; Hashemi and Mousavi Khaneghah 2017) and thus, some recent examples of the obtained composite films are presented below. The following works were selected because their systems were able to exhibit synergistic contributions instead of a range of properties that go from those of one of the constituents to the other one, like the rule of mixtures could predict.

Kurt et al. (2017) optimized the concentrations of two polysaccharides, xanthan (XG) and locust bean gum (LBG), and glycerol as plasticizer in the film formulation using combined design (a statistical tool that consists of mixture design and response

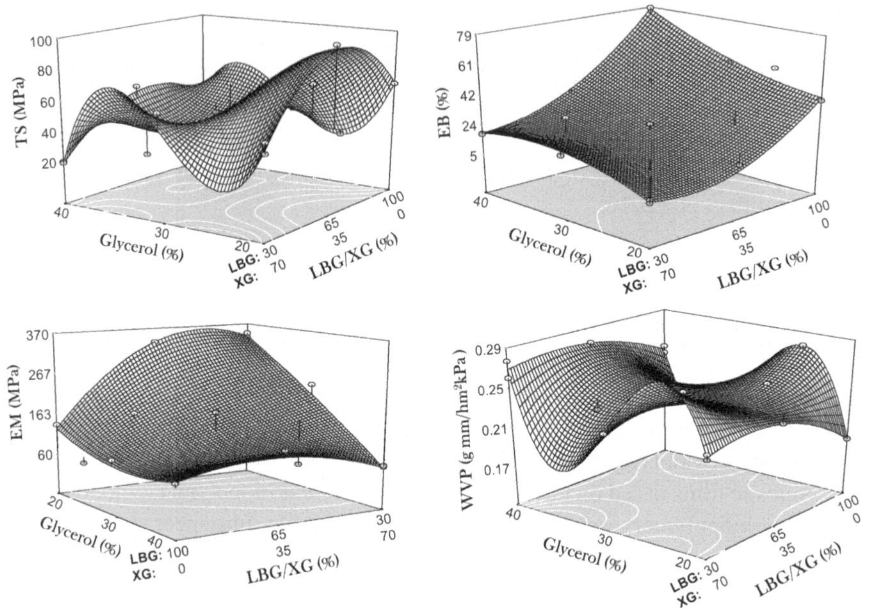

Fig. 2.1 Effect of different gum combinations and glycerol concentration on the mechanical and water vapor permeability properties of the edible films (*TS* tensile strength; *EB* percentage of elongation at break; *EM* elastic modulus; *WVP* water vapor permeability). From Kurt et al. (2017) with permission

surface methodology). The effect of all LBG/XG and glycerol concentrations used on the film's mechanical and water vapor permeability (WVP) properties is illustrated in Fig. 2.1. According to the authors, these two polymers had excellent blend miscibility and exhibited synergic interactions between them that allowed to maximize almost all the functional properties of the resulting edible film at 89.6%, 10.4% and 20% of LBG, XG and glycerol, respectively. At this optimum point, the WVP, tensile strength (TS), elongation at break (E%) and elasticity modulus (EM) values of film resulted 0.22 g mm h^{-1} m^2 kPa, 86.97 MPa, 33.34% and 177.25 MPa, respectively. Manrich et al. (2017) obtained water-resistant edible films derived from the tomato peel-extracted biopolyester cutin upon its casting in the presence of pectin (a water soluble polysaccharide). The obtained films were able to mimic tomato peel in terms of mechanical strength and thermal stability. Nevertheless, uniform surface hydrophobicity and greater stiffness were observed for cutin/pectin films as compared to that naturally occurring ultrastructure. Moreover, cutin-based edible films resulted slightly affected by moisture and thus, they were recommended to be used as water-resistant plastic wraps for short-time applications.

Silva et al. (2016) manufactured blended glycerol-plasticized films using whey protein isolate (WPI), four different concentrations of locust bean gum (LBG) and two different thermal treatments. Authors found that interactions between WPI and LBG strongly influence film properties and it is thus possible to tune the properties of WPI-based edible films to meet food packaging and edible coating needs adding different amounts of LBG and/or using different thermal treatments. The composites resulted more stable as longer was the thermal treatment due to the higher cross-linking effect and higher level of developed molecular interactions. This, consequently, resulted in the obtaining of stronger and less soluble films with improved barrier properties to carbon dioxide, oxygen and light. For example, films made from solutions with 5% WPI + 0.1% LBG + 2% glycerol, heated at 75 °C for 10 min presented the lowest oxygen permeability (more than 50% decrease as compared with the film with no LBG, with the same thermal treatment).

Kowalczyk et al. (2016) studied the effect of adding different lipids on the physicochemical and morphological properties of sorbitol-plasticized pea protein isolate (PPI) edible emulsion films. The chosen lipids were anhydrous milk fat (AMF), candelilla wax (CNW), lecithin (LEC) and oleic acid (OLA) and were incorporated into film-forming solutions at 0, 0.5, 1.0, 1.5, and 2.0%. Authors found that only AMF and CNW reduced WVP of the films, this last one in a dose dependent manner, leading to a WVP 2.5 times lower than that of the control for the films incorporated with 2.0% CNW. On the other hand, the incorporation of lipids into PPI films caused an increase in oxygen permeability and decreased the mechanical strength. Nevertheless, all the produced films were effective UV barriers. Surface microstructure of the emulsion films was influenced by the lipid type and lipid volume fraction, as shown in Fig. 2.2. Unlike the solid lipids, OLA did not reduce the film transparency and showed a plasticizing effect, making the films more extensible. Summarizing, this work showed that CNW was the most effective lipid for this system since it was able to improve the water vapor barrier properties and simultaneously provoked the lowest increase in the oxygen permeability and the lowest decrease in the mechanical strength of the films. Besides, CNW-added films had also higher transparency compared to the other films containing solid lipids (AMF, LEC).

Hosseini et al. (2015) successfully developed bio-nanocomposite films based on fish gelatin (FG) and spherical chitosan nanoparticles (CSNPs) with size range 40–80 nm. The incorporation of CSNPs to FG films improved their water vapor barrier, as well as TS and elastic modulus, which was associated to the evenly dispersion of the particles in the bio-polymeric matrix at lower loading levels (less than 8%, w/w) added to the interaction between CSNPs and FG through hydrogen bonding. Furthermore, addition of CSNPs contributed to the significant decrease of WVP, leading to a 50% reduction at 6% (w/w) filler. Composite films presented reduced values of transparency at 600 nm as compared to the control film (0% CSNPs) while they have excellent barrier properties against UV light. The results presented in this study show the feasibility of using bionanocomposite technology to improve the properties of biopolymer films based on fish gelatin. Antoniou et al. (2015) produced composite tara gum films with the inclusion of both, bulk chitosan or chitosan nanoparticles at various concentrations. The incorporation of bulk or chitosan nanoparticles resulted in improved mechanical (the TS of the control film films was 22.71 ± 2.98 MPa and increased, respectively to ~58 MPa or ~53 when 10 or 15 wt% of CSNPs or bulk chitosan were added, without reducing notably the

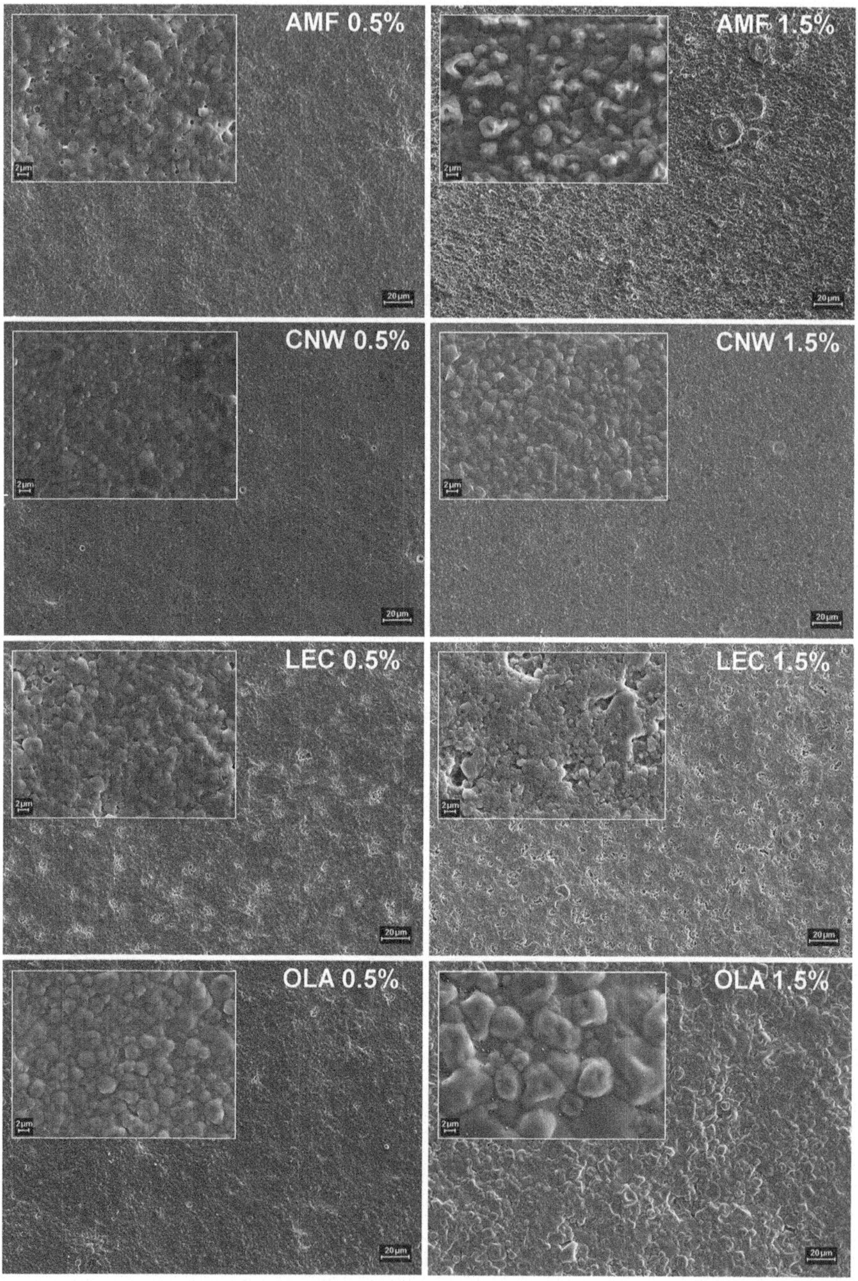

Fig. 2.2 SEM micrographs (×1000 and ×10 000 magnification) of surfaces of the emulsion PPI-based films. From Kowalczyk et al. (2016) with permission

Table 2.1 Antimicrobial activity of tara gum films containing bulk chitosan (CS) and chitosan nanoparticles (CSNPs) against food pathogenic bacteria of *E. coli* and *Staphylococcus aureus*[a]

		Inhibitory zone (mm^2)	
Film type	CS/CSNP content (% w/w)	*E. coli*	*S. aureus*
TG	0	0.0 ± 0.0^f	0.0 ± 0.0^d
TG + CS	5	65.43 ± 10.36^d	78.45 ± 10.13^c
	10	112.0 ± 16.06^b	117.96 ± 11.08^b
	15	138.87 ± 12.35^a	157.42 ± 10.95^a
TG + CSNPs	5	41.53 ± 7.60^e	85.30 ± 12.30c
	10	87.32 ± 6.72^c	111.71 ± 3.12^b
	15	85.80 ± 6.17^c	93.41 ± 8.59^c

From Antoniou et al. (2015), with permission
[a]Values are given as mean ± standard deviation. Different superscript letters in the same column indicate a statistically significant difference ($P < 0.05$)

EB. The incorporation of nanoparticles resulted in a more efficient strategy for decreasing water solubility and WVP leading to reductions of 74.3% and 22.7%, respectively, which was attributed to the compact structure of the chitosan nanoparticles that reduced the free volume of the polymer matrix more than bulk chitosan by obstructing the diffusion of water and thereby decreasing the moisture content of the films. However, tara gum films with bulk chitosan exhibited better antimicrobial activity, as shown in Table 2.1. Authors indicated that this behavior could be attributed to the size of the CSNPs that possible is not small enough to accumulate in the bacterial membrane; besides, when the CSNPs concentration increased, the particle size increased even further due to aggregation during the film formation. Moreover, since the formation of CSNPs is based on ionic gelation mechanisms the surface charge of CS was reduced thus making it less effective against Gram-negative bacteria (*Escherichia coli*).

Pereda et al. (2014) prepared edible composite films by casting film-forming emulsions based on chitosan/glycerol/olive oil containing dispersed cellulose nanocrystals (CNs). Due to cellulose–glycerol–chitosan interactions, these complex composite films appeared less opaque as the cellulose concentration increases (up to 7 wt% CN), which balanced the reduction of film transparency due to lipid addition. Moreover, both, nanocellulose and olive oil addition led to the reduction of the water vapor permeation and the total soluble matter of the films in a concentration-dependent manner, improving at the same time their tensile behavior. Results from dynamic mechanical tests revealed that all films present two main relaxations that could be ascribed to the glycerol- and chitosan-rich phases, respectively.

As an example of multilayers edible films Basiak et al. (2016) developed composites aimed to reduce the hygroscopic character of biodegradable starch-based films. In this case rapeseed oil was incorporated by lamination, leading to samples composed by 3-layers (starch-oil-starch). According to the authors, the lipid lamination followed by starch solution casting step induced an emulsion type structure of dried films. Thus, composite films resulted more opalescent and glossier, presenting also lower TS than fatty free starch films. On the other hand, lipid incorporation reduced the moisture absorption, particularly at higher RH, as well as the surface swelling index and the water vapor and oxygen permeability.

2.4 Active and Bioactive Edible Films

In addition, edible films could act as carriers not only of health-related compounds such as vitamins, minerals, nutraceuticals and other bioactive compounds such as colorants, flavor compounds or antibrowning agents but also of antimicrobial, antioxidant, and antisoftening agents (Raybaudi-Massilia et al. 2016; Fortunati 2016; Mellinas et al. 2016; Gutiérrez 2017; Harnkarnsujarit 2017; Moghimi et al. 2017; López-Córdoba et al. 2017). These substances can be naturally or synthetically derived, although consumers generally prefer those materials from natural sources (López-Córdoba et al. 2017; Moghimi et al. 2017). The antimicrobials extend product shelf life and reduce the risk of pathogen growth on food surfaces leading to safer food consumption (Gutiérrez 2018). The incorporation of antimicrobials to the films directly in contact to food surface, where the microbial growth and contamination are mostly found, can replace the addition of preservatives into food products (Harnkarnsujarit 2017; Nisar et al. 2017). In this line Xiao et al. (2015) also indicated that bioactive agents may not be applied directly to a food system due to possible evaporation losses, inactivation or rapid release into the food matrix. According to Nisar et al. (2017) the film approach can be more effective than applying antimicrobial substances directly to the food due to providing continuous migration of active substances into the food and thus remaining at high concentration for prolonged periods of time. Moreover, the inhibitory agents incorporated to a formulation can be specifically targeted to post-processing contaminants on the food surface (Raybaudi-Massilia et al. 2016). Diffusion of antimicrobials through an edible film or coating is influenced by the composition and manufacturing procedure of the film and coating, factors related to the food (pH, water activity), hydrophilic/hydrophobic balance of the compound, and storage conditions. Furthermore, the diffusion rates of the active compound into the product can be partially controlled by the compounds found in the coating matrix (Raybaudi-Massilia et al. 2016). All these factors need to be taken into account in order to design effective edible coating systems (Álvarez et al. 2017). Some examples of active edible films developed in the recent years are summarized below.

López et al. (2017) developed edible films with high antioxidant capacity based on salmon gelatin incorporating boldine, which is the major alkaloid obtained from Boldo tree. The concentration of both components was optimized by applying a Box-Behnken experimental design with the goal of maximizing radical scavenging capacity of film forming suspensions. Results showed synergistic effect between gelatin and boldine for both the antioxidant capacity and antimicrobial activity of gelatin against *E. coli* and *Listeria monocytogenes*. Boldine was released from the films into food simulant following the Fickian model, at a faster rate as lower was the concentration of gelatine in the film forming solution, which suggests that it is

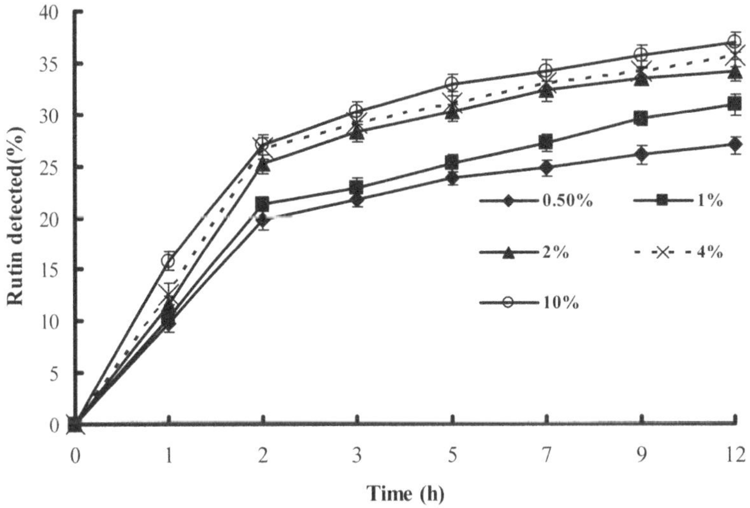

Fig. 2.3 Cumulative rutin release (%) as function of time in Milli-Q water. Samples studied were zein–rutin composite nanoparticle/corn starch films with zein–rutin composite nanoparticle contents of 0.5, 1, 2, 4, and 10% w/w. From Zhang and Zhao (2017), with permission

possible to design films to satisfy specific release criteria in order to have an active concentration of boldine in food with different shelf lives. The development of antioxidant-loaded edible films for preventing oxidation of lipid-containing foods has attracted also the attention of Zhang and Zhao (2017). These authors prepared edible active films based on zein–rutin composite nanoparticles (RNs) and corn starch. RNs (0, 0.5, 1, 2, 4, and 10%, w/w) were incorporated into the starch matrix, to act as a natural antioxidant. The RNs acted as a strong antioxidant in the corn starch films, as determined from three complementary radical-scavenging assays (ABTS, DPPH, and phosphomolybde-num), and thus the related values increased as the nanoparticle content increased. The active films also showed long-lasting antioxidant activity, which is important for the controlled release of rutin, as illustrated in Fig. 2.3. Moreover, the incorporation of RNs into the starch matrix led to the formation of a net-like structure, which decreased slightly the WVP and water solubility of nanocomposite films respect to the control film, while TS and EB increased from 1.19 to 2.42 MPa, and from 42.10 to 78.84%, respectively, with increasing RN loading. Thus, the incorporation of these complex nanoparticles into the edible films not only provided them with antioxidant properties but also improved their physical and mechanical properties.

Similar results, i.e. synergic effect of antioxidants on water barrier and mechanical properties were found by Benbettaïeb et al. (2016) for tyrosol and ferulic acid encapsulated in chitosan–gelatin films, even when this system is more complex than the previous one (i.e. a blend of polymers instead of a single polymer matrix). Moreover, these authors proved that applying radiation to the films, the retention and diffusivity of the antioxidants can be modulated. Piñeros-Hernandez et al.

(2017) incorporated polyphenols-rich rosemary extracts (RE) within cassava starch films in order to produce active food packaging with antioxidant properties. As the polyphenols content increased, the films showed an increase in their antioxidant activity. Moreover, the films containing the higher extract concentration (13.6 mg of gallic acid equivalents per gram) showed better barrier properties against UV light. However, in this case the WVP of the films increased and the elongation decreased as the extract concentration increased, which was explained considering that the presence of RE inhibited the bonding between glycerol and starch molecules. These authors also carried out migration tests using water and ethanol (95%) as food simulants for aqueous and fatty foods, respectively, finding that total polyphenols content loaded in the films migrated within the aqueous food simulant after 7 days of film exposition, while, only a negligible polyphenol amount was detected in the fatty food one.

Regarding antimicrobial active films, Arrieta et al. (2014) studied the effect of adding carvacrol into sodium caseinate (SC) and calcium caseinate (CC) matrices plasticized with two different glycerol concentrations (25 and 35 wt%). All films exhibited good performance in terms of optical properties, showing high transparency. Besides, the antimicrobial activity of SC and CC films containing carvacrol was clearly demonstrated against two indicator bacteria, *E. coli* (Gram negative) and *S. aureus* (Gram positive). Barrier properties to oxygen were excellent but diffusion of dyes through films was dependent on the caseinate type (SC or CC), since CC resulted in less permeable composites due the ability to promote cross-linking.

On the other hand, in terms of active ingredients that can be incorporated into films and coatings, essential oils (EOs) have received much attention in the last years (Calo et al. 2015; Vergis et al. 2015; Galus and Kadzińska 2015; Yuan et al. 2016; Gutierrez-Pacheco et al. 2016; Moghimi et al. 2017) due to their antimicrobial/antifungal activity and volatile nature, which facilitates the use of small concentrations that are safe for consumption (Sivakumar and Bautista-Banos 2014; Yuan et al. 2016). As the direct addition of essential oils to food may adversely impact sensory perception of applied foods, added to the fact that the essential oils lose effectiveness over time, the incorporation of these compounds into the formulation of edible films and coatings has been suggested as a better option (Yuan et al. 2016; Gutierrez-Pacheco et al. 2016). In this sense, antimicrobial edible films and coatings may provide increased inhibitory effects against spoilage and pathogenic bacteria by maintaining effective concentrations of the active compounds on the food surfaces (Gutierrez-Pacheco et al. 2016). The following are examples of recent developments in this line. Hashemi and Mousavi Khaneghah (2017) developed basil seed gum (BSG) edible films containing oregano essential oil (OEO) (1–6%). They found that WVP significantly decreased by the incorporation of OEO while moisture content, contact angle, transparency and swelling index of edible films increased. Films containing 2–6% OEO presented a significant antibacterial activity against *E. coli, S. Typhimurium, P. aeruginosa, S. aureus and B. cereus*. Moreover, the antioxidant activity (i.e. DPPH and ABTS radical scavenging activities and ferric reducing ability) of BSG films were enhanced considerably with increasing OEO concentration, as exemplified in Fig. 2.4. According to these results, the

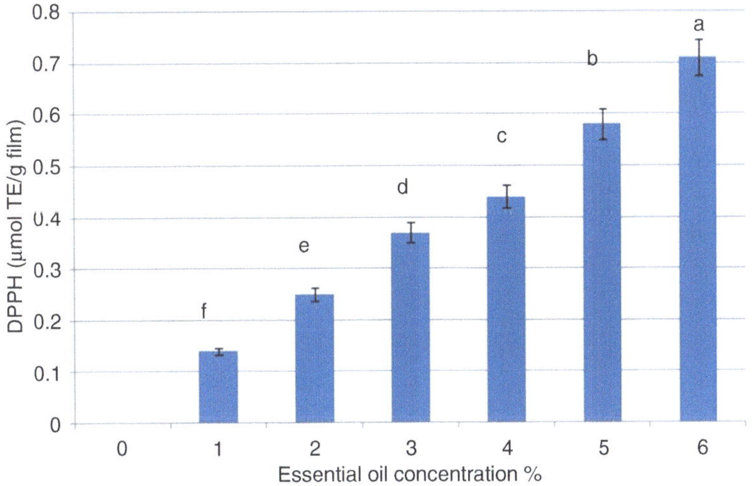

Fig. 2.4 DPPH radical scavenging activity of film at different essential oil concentration. Values represent means ± standard deviations. Each column with the same lowercase letters is not significantly different at $P < 0.05$. From Hashemi and Mousavi Khaneghah (2017), with permission

fabricated edible films with BSG and OEO can be considered for further approaching as an edible food packaging.

Moghimi et al. (2017), taking into account that EOs have low water solubility and could denature under light and heat condition, chose the nanoemulsion path for preparing antibacterial edible films based on hydroxypropyl methyl cellulose (HPMC) and Thymus daenensis (wild and cultivated) essential oil (TD EO). The authors showed the uniform incorporation of nanoemulsions into the edible film, but also that active edible films had less TS and Young's modulus compared to the control film, which was attributed to the plasticizing effect of the essential oil. However, both films exhibited a potent but differentiated antibacterial and antifungal activity, which was attributed to the variations of TD EOs components (i.e. wild plant had 53.28% thymol, while cultivated had 75.69%; the second main component of the cultivated *T. daenensis* was carvacrol (22.72%), whereas the wild plant had 25.26% of p-cymene (precursor of carvacrol)). Thus, films modified with EO extracted from the wild plant had more antibacterial activity against gram positive bacteria (including *S. aureus, S. epidermidis, B. subtilis, E. faecalis, E. faecium* and Methicillin resistant *S. aureus*) than the films based on cultivated TD. However, the formulation based on cultivated TD was more efficient against *Candida albicans* and gram negative bacteria (including *E. coli, S. typhi, S. dysenetriae, S. flexneri, A. baumannii and K. peneumoniae*). Therefore, both nanoemulsion based HPMC films could be used based on the microorganisms of interest for each packaging material.

Giteru et al. (2017) combined the previous approaches by incorporating both, citral, an antimicrobial monoterpene aldehyde extracted from the plant *Cymbopogon citratusis* and quercetin, an antioxidant compound naturally occurring in plants, into kafirin-based films. They compared the behavior of kafirin films consisting of 2.5%

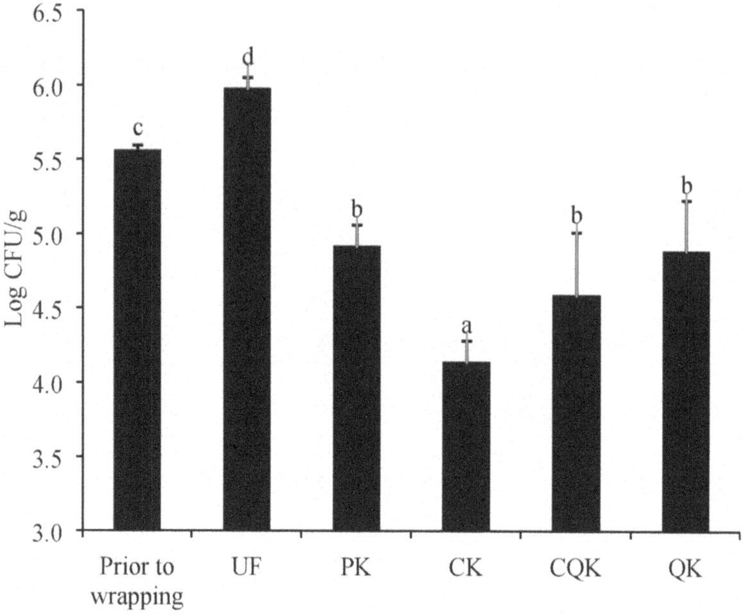

Fig. 2.5 Effect of kafirin films on the total viable count of chicken fillets stored at 2 ± 0.5 °C for 96 h. (UF) unwrapped fillet, (PK) fillets wrapped with plain film, (CK) fillet wrapped with citral film, (CQK) fillets wrapped with citral-quercetin film, (QK) fillets wrapped with quercetin film. Error bars represent ± SD (n = 9). Different letters indicate significantly different ($p < 0.05$). From Giteru et al. (2017), with permission

citral (CK), 2% quercetin (QK) or 1.25% citral +1% quercetin (CQK). It was found that incorporation of citral reduced the maximum stress and stiffness while imparting more flexibility to the kafirin bioactive films. Compared to the control kafirin (PK) and quercetin-containing films, those containing citral showed significant antimicrobial activity against the total viable count on chicken fillets stored at 2 ± 0.5 °C for 96 h, as shown in Fig. 2.5. The ability of bioactive kafirin films to inhibit oxidation of lipids on chicken fillets, which was evaluated by the thiobarbituric acid reactive substances assay, were lower for PK, CQK, and QK (0.18, 0.16 and 0.23 mg MDA/kg respectively), compared to unwrapped fillets and CK (0.41 and 0.59 mg MDA/kg respectively). These findings implied that quercetin and those polyphenolics likely co-extracted with kafirin possessed the ability to inhibit the development of lipid oxidative products. Regarding product appearance, authors found that wrapped fillets become more yellowish after the storage than unwrapped ones. This color transformation toward yellowness was attributed to the presence of citral and quercetin in the films since film swelling due to absorption of water by the hydroxyl groups in the film matrix could cause the release of originally encapsulated pigments, as indicated by Shojaee-Aliabadi et al. (2014). Thus, authors pointed out that those pigment-rich additives may have an influence on the organoleptic property of food products and that as these changes occurred on raw meat, their persistence during cooking still requires determination. Moreover, they emphasized that it is important to determine whether this pigmentation of raw meat would be perceived negatively by consumers with further investigations.

At this point it is important to introduce the concept of bioactive packaging. The concepts of active and bioactive packaging have been sometimes used indiscriminately, but there are differences (Espitia et al. 2016). Active food packaging systems are those previously exemplified, which go beyond the traditional passive role of food protection and include desirable interactions with the food, in a way that is relevant to extend food stability; bioactive food packaging systems, on the other hand, are those which may contribute to health benefits to the consumers (Espitia et al. 2016). Bioactive packaging materials would thus hold bioactive agents, which are eventually released into the food product (Lopez-Rubio et al. 2006). In the specific case of edible bioactive films and coatings, this release is not even required, since the film/coating itself is supposed to be eaten with the food (Espitia et al. 2016). Edible films enriched with probiotics (i.e. bacteria with beneficial effects on humans and animals) are the typical examples of these bioactive films. For probiotics to play the intended role in human health, it is essential that both viability and the metabolic activity are maintained throughout food processing and supply chain, as well as within human gastrointestinal tract (Espitia et al. 2016). Several methods have been tested to increase the quality of probiotic cultures (Nguyen et al. 2016), however, microencapsulation is by far the most exploited approach (Espitia et al. 2016; Nguyen et al. 2016). When combined, microencapsulation and bioactive edible packaging concepts denote a promising strategy for protecting and delivering probiotic species efficiently. In this line, Soukoulis et al. (2017) evaluated the inclusion of whey protein isolate in different bio-polymers with established good film forming properties, for their ability to stabilize live probiotic organisms. Edible films based on low (LSA) and high (HSA) viscosity sodium alginate, low esterified amidated pectin (PEC), kappa-carrageenan/locust bean gum (k-CAR/LBG) and gelatine (GEL) in the presence or absence of whey protein concentrate (WPC) were then shown to be feasible carriers for the delivery of *Lactobacillus rhamnosus* GG. While losses of *L. rhamnosus* GG throughout the drying process ranged from 0.87 to 3.06 log CFU/g for the systems without WPC, they were significantly reduced to 0 to 1.17 log CFU/g in the presence of WPC. Films fabricated with k-CAR/LBG or HSA were most effective at maintaining maximal biological activity of the probiotic cells (0.167 and 0.218 log CFU day^{-1} in average) compared to films made of PEC, GEL and LSA (0.251, 0.252 and 0.268 log CFU day^{-1}, respectively), which was explained by the low glass transition temperature (Tg), and low WVP of the binary system. Supplementation of the film forming solutions with WPC resulted in an enhanced *L. rhamnosus GG* storage stability (0.279 and 0.183 log average CFU day^{-1} for systems with and without the addition of WPC respectively), being the bioprotective role of WPC associated with its ability to reduce the osmolytic cell injuries arising throughout the dehydration process and their excellent cell adhesion properties. Furthermore, probiotic films based on HSA/WPC and k-CAR/LBG/WPC blends had both acceptable mechanical and barrier properties. Overall this work shows that the development of edible films as carriers for the delivery of probiotics is a plausible strategy. As pointed out by Soukoulis et al. (2017), the maintenance of the biological activity of the probiotic cells is the governing parameter for the selection of the substrate compositional aspects, however, other technological parameters such as the mechanical and barrier properties are essential to ensure adequate processability and shelf life. López de Lacey et al. (2014) over passed the previous findings by developing a bioactive film based on agar and incorporating both, green tea extract (antimicrobial and antioxidant) and probiotic strains (*Lactobacillus paracasei L26* and *Bifidobacterium lactis B94*), which was applied on hake fillets in order to investigate the effect during chilled storage. Hake was previously inoculated with *Shewanella putrefaciens* and *Photobacterium phosphoreum* (10^3–10^4 CFU/g) to simulate a spoilage process. It was found that green tea/probiotic film led to a reduction of the spoilage indicators, particularly of H_2S-producing bacteria counts and total viable bacteria throughout the 15 days of storage period. However probiotics alone had a smaller effect over reducing these chemical spoilage indicators. Authors pointed out that the probiotic strains added to the film were able to pass to the fish, producing an increment of lactic acid bacterial counts, even in the presence of green tea extract. They concluded that films with green tea and probiotic could extend shelf-life of hake at least for a week and, at the same time, it could be a way to incorporate beneficial probiotic bacteria to the fish.

2.5 Conclusions

Edibility, biodegradability, and increased food safety are the main benefits of edible films. Their environmental friendly aspects make them alternatives in packaging systems, without the ecological costs of synthetic non-biodegradable materials. Although edible films and coatings have not been enough developed to be able to replace conventional plastic packaging totally, new interesting approaches are being tested every day. Therefore, there is still a considerable requirement for the development of new edible packaging materials to fulfill the increasing consumer demands for more natural foods which need to be packed and also to meet the environmental concerns and regulations. Future work in this area should ensure that the new bioactive and biodegradable materials help improving mechanical handling of food and lessen the migration of gases, volatiles, vapors and lipids in synergy with conventional packaging, but also that the incorporation of active substances into the film lead to the enhancement of the shelf-life of food products.

References

Álvarez K, Famá L, Gutiérrez TJ (2017) Physicochemical, antimicrobial and mechanical properties of thermoplastic materials based on biopolymers with application in the food industry. In: Advances in physicochemical properties of biopolymers (part 1), pp 358–400. https://doi.org/10.2174/9781681084534117010015

Alzate P, Miramont S, Flores S, Gerschenson LN (2017) Effect of the potassium sorbate and carvacrol addition on the properties and antimicrobial activity of tapioca starch—hydroxypropyl methylcellulose edible films. Starch/Stärke 69:1600261
Ansorena MR, Zubeldía F, Marcovich NE (2016) Active wheat gluten films obtained by thermoplastic processing. LWT Food Sci Technol 69:47–54
Antoniou J, Liu F, Majeed H, Zhong F (2015) Characterization of tara gum edible films incorporated with bulk chitosan and chitosan nanoparticles: a comparative study. Food Hydrocoll 44:309–319
Arrieta MP, Peltzer MA, López J, Garrigós M d C, Valente AJM, Jiménez A (2014) Functional properties of sodium and calcium caseinate antimicrobial active films containing carvacrol. J Food Eng 121:94–101
Atarés L, Chiralt A (2016) Review. Essential oils as additives in biodegradable films and coatings for active food packaging. Trends Food Sci Technol 48:51–62
Balti R, Mansour MB, Sayari N, Yacoubi L, Rabaoui L, Brodu N, Massé A (2017) Development and characterization of bioactive edible films from spider crab (Maja crispata) chitosan incorporated with Spirulina extract. Int J Biol Macromol 105:1464–1472
Basiak E, Debeaufort F, Lenart A (2016) Effect of oil lamination between plasticized starch layers on film properties. Food Chem 195:56–63
Benbettaïeb N, Assifaoui A, Karbowiak T, Debeaufort F, Chambin O (2016) Controlled release of tyrosol and ferulic acid encapsulated in chitosan–gelatin films after electron beam irradiation. Radiat Phys Chem 118:81–86
Buffo RA, Han JH (2005) Chapter 17. Edible films and coatings from plant origin proteins. In: Food science and technology. Innovations in Food Packaging. Academic, London, pp 277–300. ISBN 9780123116321e. https://doi.org/10.1016/B978-012311632-1/50049-8
Calo JR, Crandall PG, O'Bryan CA, Ricke SC (2015) Essential oils as antimicrobials in food systems—a review. Food Control 54:111–119
Castro-Rosas J, Cruz-Galvez AM, Gomez-Aldapa CA, Falfan-Cortes RN, Guzman-Ortiz FA, Rodríguez-Marín ML (2016) Biopolymer films and the effects of added lipids, nanoparticles and antimicrobials on their mechanical and barrier properties: a review. Int J Food Sci Technol 51:1967–1978
Costa MJ, Cerqueira MA, Ruiz HA, Fougnies C, Richel A, Vicente AA, Teixeira JA, Aguedo M (2015) Use of wheat bran arabinoxylans in chitosan-based films: effect on physicochemical properties. Ind Crop Prod 66:305–311
Dehghani S, Hosseini SV, Regenstein JM (2018) Edible films and coatings in seafood preservation: a review. Food Chem 240:505–513
Desobry S, Arab-Tehrany E (2014) Diffusion barrier layers for edible food packaging. In: Hashmi S, Batalha GF, Van Tyne CJ, Yilbas B (eds) Comprehensive materials processing. Elsevier, Oxford, pp 499–518. ISBN 9780080965338. https://doi.org/10.1016/B978-0-08-096532-1.00419-2
Espitia PJP, Batista RA, Azeredo HMC, Otoni CG (2016) Review. Probiotics and their potential applications in active edible films and coatings. Food Res Int 90:42–52
Fortunati E (2016) Multifunctional films, blends, and nanocomposites based on chitosan: use in antimicrobial packaging. Chapter 38. In: Barros-Velazquez J (ed) Antimicrobial food packaging. Elsevier Inc, Amsterdam, pp 467–477. ISBN: 978-0-12-800723-5. https://doi.org/10.1016/B978-0-12-800723-5.09989-7
Galus S, Kadzińska J (2015) Food applications of emulsion-based edible filmsand coatings. Trends Food Sci Technol 45(2):273–283
Ganiari S, Choulitoudi E, Oreopoulou VR (2017) Edible and active films and coatings as carriers of natural antioxidants for lipid food. Trends Food Sci Technol 68:70–82
Giteru SG, Oey I, Ali MA, Johnson SK, Fang Z (2017) Effect of kafirin-based films incorporating citral and quercetin on storage of fresh chicken fillets. Food Control 80:37–44

Gutiérrez TJ (2017) Surface and nutraceutical properties of edible films made from starchy sources with and without added blackberry pulp. Carbohydr Polym 165:169–179. https://doi.org/10.1016/j.carbpol.2017.02.016

Gutiérrez TJ (2018) Active and intelligent films made from starchy sources/blackberry pulp. J Polym Environ 26:2374–2391. https://doi.org/10.1007/s10924-017-1134-y

Gutierrez-Pacheco MM, Ortega-Ramirez LA, Cruz-Valenzuela MR, Silva-Espinoza BA, Gonzalez-Aguilar GA, Ayala-Zavala JF (2016) Chapter 50. Combinational approaches for antimicrobial packaging: pectin and cinnamon leaf oil. In: Barros-Velazquez J (ed) Antimicrobial food packaging. Elsevier Inc, Amsterdam, pp 609–617. ISBN: 978-0-12-800723-5. https://doi.org/10.1016/B978-0-12-800723-5.09989-7

Harnkarnsujarit N (2017) Glass-transition and non-equilibrium states of edible films and barriers. Chapter 15. In: Bhandari B, Roos YH (eds) "Non-equilibrium states and glass transitions in foods processing effects and product-specific implications", a volume in Woodhead Publishing Series in Food Science, Technology and Nutrition. Elsevier, Amsterdam ISBN: 978–0–08-100309-1

Hashemi SMB, Mousavi Khaneghah A (2017) Characterization of novel basil-seed gum active edible films and coatings containing oregano essential oil. Prog Org Coat 110:35–41

Hosseini SF, Rezaei M, Zandi M, Farahmandghavi F (2015) Fabrication of bio-nanocomposite films based on fish gelatin reinforced with chitosan nanoparticles. Food Hydrocoll 44:172–182

Kester JJ, Fennema OR (1986) Edible films and coatings: a review. Food Technol 40(12):47–59

Kokoszka S, Debeaufort F, Hambleton A, Lenart A, Voilley A (2010) Protein and glycerol contents affect physico-chemical properties of soy protein isolate-based edible films. Innov Food Sci Emerg Technol 11:503–510

Kowalczyk D, Gustaw W, Zięba E, Lisiecki S, Stadnik J, Baraniak B (2016) Microstructure and functional properties of sorbitol-plasticized pea protein isolate emulsion films: effect of lipid type and concentration. Food Hydrocoll 60:353–363

Kumari M, Mahajan H, Joshi R, Gupta M (2017) Development and structural characterization of edible films for improving fruit quality. Food Packaging Shelf Life 12:42–50

Kurt A, Toker OS, Tornuk F (2017) Effect of xanthan and locust bean gum synergistic interaction on characteristics of biodegradable edible film. Int J Biol Macromol 102:1035–1044

López de Lacey AM, López-Caballero ME, Montero P (2014) Agar films containing green tea extract and probiotic bacteria for extending fish shelf-life. LWT Food Sci Technol 55:559–564

López D, Márquez A, Gutiérrez-Cutiño M, Venegas-Yazigi D, Bustos R, Matiacevich S (2017) Edible film with antioxidant capacity based on salmon gelatin and boldine. LWT Food Sci Technol 77:160–169

López-Córdoba A, Medina-Jaramillo C, Piñeros-Hernandez D, Goyanes S (2017) Cassava starch films containing rosemary nanoparticles produced by solvent displacement method. Food Hydrocoll 71:26–34

Lopez-Rubio A, Gavara R, Lagaron JM (2006) Bioactive packaging: turning foods into healthier foods through biomaterials. Trends Food Sci Technol 17:567–575

Manrich A, Moreira FKV, Otoni CG, Lorevice MV, Martins MA, Mattoso LHC (2017) Hydrophobic edible films made up of tomato cutin and pectin. Carbohydr Polym 164:83–91

Martins JT, Cerqueira MA, Bourbon AI, Pinheiro AC, Souza BWS, Vicente AA (2012) Synergistic effects between κ-carrageenan and locust bean gum on physicochemical properties of edible films made thereof. Food Hydrocoll 29:280–289

McHugh TH (2000) Protein-lipid interactions in edible films and coatings. Food Nahrung 44(3):148–151

Mellinas C, Valdés A, Ramos M, Burgos N, Garrigós MDC, Jiménez A (2016) Active edible films: current state and future trends. J Appl Polym Sci 133. https://doi.org/10.1002/APP.42631

Moghimi R, Aliahmadi A, Rafati H (2017) Antibacterial hydroxypropyl methyl cellulose edible films containing nanoemulsions of Thymus daenensis essential oil for food packaging. Carbohydr Polym 175:241–248

Nguyen H-T, Truong D-H, Kouhoundé S, Ly S, Razafindralambo H, Delvigne F (2016) Biochemical engineering approaches for increasing viability and functionality of probiotic bacteria. Int J Mol Sci 1:867–884
Nisar T, Wang Z-C, Yang X, Tian Y, Iqbal M, Guo Y (2017) Characterization of citrus pectin films integrated with clove bud essential oil: physical, thermal, barrier, antioxidant and antibacterial properties. Int J Biol Macromol 106:670–680. https://doi.org/10.1016/j.ijbiomac.2017.08.068
Otoni CG, Avena-Bustillos RJ, Olsen CW, Bilbao-Sáinz C, McHugh TH (2016) Mechanical and water barrier properties of isolated soy protein composite edible films as affected by carvacrol and cinnamaldehyde micro and nanoemulsions. Food Hydrocoll 57:72–79
Park HJ, Byun YJ, Kim YT, Whiteside WS, Bae HJ (2014) Processes and applications for edible coating and film materials from agropolymers, Chapter 10. In: Han JH (ed) Innovations in food packaging (second edition). A volume in food science and technology. Academic, London, pp 257–275. ISBN: 978-0-12-394601-0. https://doi.org/10.1016/B978-0-12-394601-0.00026-6
Pascall MA, Lin SJ (2013) The application of edible polymeric films and coatings in the food industry. J Food Process Technol 4(2):116–117
Peelman N, Ragaert P, De Meulenaer B, Adons D, Peeters R, Cardon L, Van Impe F, Devlieghere F (2013) Application of bioplastics for food packaging. F Trends Food Sci Technol 32:128–141
Pereda M, Amica G, Marcovich NE (2012) Development and characterization of edible chitosan/olive oil emulsion films. Carbohydr Polym 87:1318–1325
Pereda M, Dufresne A, Aranguren MI, Marcovich NE (2014) Polyelectrolyte films based on chitosan/olive oil and reinforced with cellulose nanocrystals. Carbohydr Polym 1011:1018–1026
Pérez LM, Piccirilli GN, Delorenzi NJ, Verdini RA (2016) Effect of different combinations of glycerol and/or trehalose on physical and structural properties of whey protein concentrate-based edible films. Food Hydrocoll 56:352–359
Piñeros-Hernandez D, Medina-Jaramillo C, López-Córdoba A, Goyanes S (2017) Edible cassava starch films carrying rosemary antioxidant extracts for potential use as active food packaging. Food Hydrocoll 63:488–495
Raybaudi-Massilia R, Mosqueda-Melgar J, Soliva-Fortuny R, Martín-Belloso O (2016) Chapter 52. Combinational edible antimicrobial films and coatings. In: Barros-Velázquez J (ed) Antimicrobial food packaging. Academic, San Diego, pp 633–646. ISBN 9780128007235. https://doi.org/10.1016/B978-0-12-800723-5.00052-8
Rocca-Smith JR, Marcuzzo E, Karbowiak T, Centa J, Giacometti M, Scapin F, Venir E, Sensidoni A, Debeaufort F (2016) Effect of lipid incorporation on functional properties of wheat gluten based edible films. J Cereal Sci 69:275–282
Rodrigues DC, Caceres CA, Ribeiro HL, de Abreu RFA, Cunha AP, Azeredo HMC (2014) Influence of cassava starch and carnauba wax on physical properties of cashew tree gum-based films. Food Hydrocoll 38:147–151
Saggiorato AG, Gaio I, Treichel H, De Oliveira D, Cichoski AJ, Cansian RL (2012) Antifungal activity of basil essential oil (Ocimum basilicum L.): evaluation in vitro and on an Italian-type sausage surface. Food Bioprocess Technol 5:378–384
Sánchez-Ortega I, García-Almendárez BE, Santos-López EM, Amaro-Reyes A, Barboza-Corona JE, Regalado C (2014) Antimicrobial edible films and coatings for meat and meat products preservation. Sci World J 2014:1–18
Shojaee-Aliabadi S, Hosseini H, Mohammadifar MA, Mohammadi A, Ghasemlou M, Hosseini SM et al (2014) Characterization of k-carrageenan films incorporated plant essential oils with improved antimicrobial activity. Carbohydr Polym 101:582–591
Silva KS, Mauro MA, Gonçalves MP, Rocha CMR (2016) Synergistic interactions of locust bean gum with whey proteins: effect on physicochemical and microstructural properties of whey protein-based films. Food Hydrocoll 54:179–188
Sivakumar D, Bautista-Banos S (2014) A review on the use of essential oils for postharvest decay control and maintenance of fruit quality during storage. Crop Prot 64:27–37

Soukoulis C, Behboudi-Jobbehdar S, Macnaughtan W, Parmenter C, Fisk ID (2017) Stability of lactobacillus rhamnosus GG incorporated in edible films: impact of anionic biopolymers and whey protein concentrate. Food Hydrocoll 70:345–355

Tomasula PM (2009) Chapter 23. Using dairy ingredients to produce edible films and biodegradable packaging materials. In: Corredig M (ed) Dairy-derived ingredients. Series in food science, technology and nutrition. Woodhead, Cambridge, pp 589–624. ISBN: 9781845694654. (http://www.sciencedirect.com/science/article/pii/B9781845694654500233). https://doi.org/10.1533/9781845697198.3.589

Valdés A, Mellinas AC, Ramos M, Garrigós MC, Jiménez A (2014) Natural additives and agricultural wastes in biopolymer formulations for food packaging. Front Chem 2(1). https://doi.org/10.3389/fchem.2014.00006

Vergis J, Gokulakrishnan P, Agarwal RK, Kumar A (2015) Essential oils as natural food antimicrobial agents: a review. Crit Rev Food Sci Nutr 55(10):1320–1323

Vijayendra SVN, Shamala TR (2014) Film forming microbial biopolymers for commercial applications - a review. Crit Rev Biotechnol 34(4):338–357

Viuda-Martos M, Mohamady MA, Fernández-López J, Abd ElRazik KA, Omer EA, Pérez-Álvarez JA et al (2011) In vitro antioxidant and antibacterial activities of essential oils obtained from Egyptian aromatic plants. Food Control 22:1715–1722

Xiao J, Li C, Huang Q (2015) Kafirin nanoparticle-stabilized Pickering emulsions as oral delivery vehicles: physicochemical stability and in vitro digestion profile. J Agric Food Chem 63(47):10263–10270

Yuan G, Chen X, Li DR (2016) Chitosan films and coatings containing essential oils: the antioxidant and antimicrobial activity, and application in food systems. Food Res Int 89:117–128

Zhang S, Zhao H (2017) Preparation and properties of zein–rutin composite nanoparticle/cornstarch films. Carbohydr Polym 169:385–392

Chapter 3
The Potential of Vegetal and Animal Proteins to Develop More Sustainable Food Packaging

Tania Garrido, Jone Uranga, Pedro Guerrero, and Koro de la Caba

Abstract Vegetal and animal proteins have been considered promising alternatives to develop new sustainable food packaging derived from bio-resources. Edible, easy to process, renewable and environmental friendly are just a few characteristics that turn proteins into excellent raw materials to develop edible or biodegradable films. Furthermore, the valorization of industrial wastes or by-products to develop more sustainable films could add value to these products. In this context, food processing industries generate wastes that cause economic and environmental problems; therefore, the valorization of these wastes to obtain proteins could facilitate the waste management as well as the development of value-added products. Despite these environmental benefits, vegetal and animal proteins have some drawbacks and different strategies are explored to overcome those limitations for their use as food packaging. Taking this into account, the aim of this chapter is to provide an overview of the currently developed films and coatings based on animal and vegetal proteins, including active packaging.

Keywords Active film · Animal protein · Food shelf life · Vegetal protein

3.1 Introduction

Food packaging technology is a research field that has shown a fast growing in recent years. In this context, a high amount of studies focused on the improvement of food safety and quality and the preservation and extension of the product shelf life are carried out (Pinheiro et al. 2016; Sousa-Gallagher et al. 2016; Ma et al. 2017). The largest part of the materials employed for food packaging finds their origin in the petrochemical industry, which causes serious environmental problems due to non-biodegradability and non-renewability; thus, research efforts are being

Tania Garrido and Jone Uranga contributed equally to this work.

T. Garrido · J. Uranga · P. Guerrero · K. de la Caba (✉)
BIOMAT Research Group, Department of Chemical and Environmental Engineering, Engineering College of Gipuzkoa, Donostia-San Sebastián, Spain
e-mail: koro.delacaba@ehu.eus

T. J. Gutiérrez (ed.), *Polymers for Food Applications*,
https://doi.org/10.1007/978-3-319-94625-2_3

driven to develop alternative materials derived from bio-resources (Wang et al. 2017a). In that way, unlikeable circumstances, such as hazardous microorganisms (including bacteria and fungi), external physical forces, chemical compounds, sunlight, permeable volatile compounds, oxygen and moisture, have to be controlled by the use of renewable and biodegradable packaging (Garavand et al. 2017).

Protein films and coatings could be an alternative to synthetic packaging since they can provide a range of barrier attributes that can benefit the packaged food, improving its aroma, taste, texture and stability. Thus, they play an important role in food preservation, as well as in food distribution and marketing (Falguera et al. 2011). Films can be defined as a performed thin layer, which can be placed surrounding the food product, between food components or even sealed into edible pouches; whereas a coating is directly formed onto food surface by dipping, spraying, brushing or panning (Junqueira-Gonçalves et al. 2017; Otoni et al. 2017). Proteins have been considered a promising alternative to develop food packaging films and coatings since they are edible, food compatible, renewable and able to increase the nutritional value of the coated product. Moreover, proteins are generally superior to polysaccharides in their ability to form three-dimensional macromolecular networks, stabilized and strengthened by hydrogen bonds, hydrophobic interactions and disulphide bonds, which contribute to enhance mechanical and barrier properties (Gupta and Nayak 2015). Apart of providing external protection for food, films and coatings can act as effective carriers of many types of compounds, including antimicrobial and antioxidant additives. The incorporation of antimicrobial agents into films can help to prevent or delay the growth of foodborne pathogens and thus, food spoilage (Aloui and Khwaldia 2016); while the incorporation of antioxidants can inhibit or retard the oxidation of food, thus, extending food shelf life and improving food safety and quality (Álvarez et al. 2017). Other additives can also be incorporated into films with the purpose of enhancing their functional properties, as well as the organoleptic properties of the packaged product. Some examples of these additives could be anti-browning agents, nutraceuticals, texture enhancers, flavor and color ingredients (Olivas et al. 2007; Gutiérrez 2017).

Regarding antioxidants, synthetic ones have been traditionally employed in packaging applications (Gutiérrez 2018). However, these additives are increasingly more challenged due to their potential risk resulting from their migration into food products as well as to the strict statutory controls currently existing in so many countries (Gómez-Estaca et al. 2014a; Álvarez et al. 2018). Fortunately, the use of natural antioxidants, such as tocopherol (Córdoba and Sobral 2017), plants extracts (Saberi et al. 2017) and essential oils (EOs) (Atarés and Chiralt 2016), among others, is being studied. In fact, obtaining bioactive compounds from vegetal sources, especially from inexpensive waste products from food, forest or agricultural industries, is taking importance in order to add value to an unavoidable amount of waste that grows day-by-day (Gutiérrez et al. 2018). High amounts of wastes are released during industrial manufacturing of apple, blueberry, olive, raspberry, grape or even citrus fruits, among many others. Hence, valuable compounds from wastes, such as anthocyanins, phenolic acids and flavonoids, could be successfully recovered and employed to develop active and intelligent packaging (Socaci et al. 2017). For instance, Luchese et al. (2018) employed blueberry pomace obtained from blueberry juice processing wastes, rich in phenolic compounds such as anthocyanins, to prepare films with the ability to change color when subjected to different pH values, which could be correlated with the pH changes in some food products. Prietto et al. (2017) and Ma and Wang (2016) also used this antioxidant extracted from other sources, such as black bean seeds, red cabbage leaves and grape skins. Olejar et al. (2017), studied the antioxidant activity of grape tannins extracted from an agro-waste stream from the wine industry and de Moraes Crizel et al. (2018) obtained flour and microparticles of olive pomace from olive oil manufacturing, which were considered suitable to provide antioxidant activity.

Concerning antimicrobials, the incorporation of natural antimicrobial agents into films has been also widely studied. EOs, chitosan, extracts of herbs, plants and species have been employed to control the microbial growth into the packaged product (Etxabide et al. 2017; Irkin and Esmer 2015). Arfat et al. (2014) studied the antimicrobial impact of basil leaf EO and ZnO nanoparticles into fish protein isolate and fish skin gelatin blend. The addition of basil leaf EO, especially in combination with ZnO nanoparticles, exhibited strong antibacterial activity against the evaluated foodborne pathogenic and spoilage bacteria. Other authors have analyzed the effectivity of the natural antimicrobial agents directly into food; Alparslan and Baygar (2017) studied the antimicrobial effect of chitosan combined with orange peel EO on the shelf life of deepwater pink shrimp, and Kakaei and Shahbazi (2016) analyzed the incorporation of ethanolic red grape seed extract and *Ziziphora clinopodioides* EO on fish fillets.

In recent years, a variety of bioactive compounds has been successfully incorporated into protein-based films since they are suitable for controlled release of different additives. Therefore, many research works have been conducted in order to develop films and coatings from various proteins sources, including collagen (Wang et al. 2017b), gelatin (Molinaro et al. 2015), soy (Galus et al. 2012), whey (Cecchini et al. 2017), zein (Pena-Serna et al. 2016), pea (Kowalczyk et al. 2016) or wheat gluten (Sharma et al. 2017), among others. Especially, soy protein and gelatin have received much attention since they are widely available at a relative low-cost, meet food grade standards and have interesting characteristics, such as film forming ability and biodegradability, to be employed in food packaging applications. Moreover, the use of soy protein and gelatin to develop films and coatings not only satisfies the needs of consumers, but also helps to add value to industrial by-products and waste materials from agricultural, meat, poultry and fish processing industries.

Taking the above into consideration, the aim of this chapter is to provide an overview of the current development of food packaging films and coatings based on animal and vegetal proteins, giving special emphasis to soy protein and gelatin-based films. Furthermore, the improvements carry out in protein-based packaging in terms of active materials are intended to be highlighted.

3.2 Protein Structure and Properties

A proper study of protein structure, folding and interactions is essential to understand their characteristic properties and functions. Proteins can be defined as biological macromolecules consisting of a linear chain of amino acids that fold into three-dimensional structures composed of different secondary structure elements (Yan et al. 2014); thus, α-amino acids are the basic structural units of proteins. Natural proteins contain up to 20 different primary amino acids, which are linked together by amide bonds. These amino acids are composed of a α-carbon atom covalently attached to a hydrogen atom, an amino group, a carboxyl group and to a side chain group. This side chain group is characteristic from each amino acid and have a direct relationship in the solubility, net charge, chemical reactivity and hydrogen bonding potential of the protein (Damodaran 2007). Each amino acid possesses a range of chemical properties, which collectively endows each protein molecule with a unique set of physicochemical characteristics. Amino acids can be classified according to the chemical properties of their side chain, which provides specific characteristics to each amino acid. In particular, the polarity of the amino acid side chain, can determine the ability the amino acid has to interact with other entities. For instance, polar amino acids are able to interact with other polar amino acids and even with water molecules surrounding the protein, improving its solubility. Moreover, these interactions have an important role in the protein folding (Kessel and Ben-Tal 2010).

Proteins possess different amino acid compositions, depending on protein source or origin, which influences their interactions with other compounds and their functional properties. As can be seen in Table 3.1, fish gelatins contain mostly glycine (31–37%), proline-hydroxiproline (14–20%) and alanine (10–12%); while bovine gelatin contains higher amounts of proline-hydroxiproline (22%) and alanine (12%). It is worth noting that hydroxyproline is a non-essential amino acid, present mainly in collagen but rarely in other proteins, which is produced by hydroxylation of proline. For most gelatins derived from type I collagen, cysteine and tryptophan are absent, and the content of tyrosine residues is below 1% (Gómez-Guillén et al. 2009; Lassoued et al. 2014). Regarding vegetal proteins, they have a higher amount of cysteine, which helps to promote the formation of disulphide bonds (Garrido et al. 2018). Soy protein contains mainly glutamic acid (20%) and aspartic acid (12%), as well as leucine (8%) and arginine (8%) (Kalman 2014). Zein is particularly rich in glutamic acid (27%), leucine (21%), proline (10%) and alanine (10%) and this high proportion of nonpolar amino acid residues is responsible for the limited solubility of zein, mainly restricted to aqueous alcohols (Shukla and Cheryan 2001). With regard to wheat gluten, glutamic acid (32%) and proline (14%) are the main amino acid residues (Rombouts et al. 2009).

Protein structure is stabilized through hydrogen bonding and hydrophobic and electrostatic interactions among the functional groups of amino acid residues in four protein levels (Verbeek and van den Berg 2010), as shown in Fig. 3.1. The primary structure refers to the sequence of amino acids linked by the α-carboxyl group of one amino acid to the α-amino group of another one through a peptide bond. The

Table 3.1 Amino acid concentration of some animal and vegetal proteins

	Composition (%)				
Amino acids	Fish gelatin	Bovine gelatin	Soy protein	Zein	Wheat gluten
Aspartic acid[a]	5.2	4.4	11.6	4.6	2.8
Threonine	2.5	1.7	3.6	3.5	2.8
Serine	6.4	2.9	5.2	7.05	5.7
Glutamic acid[b]	7.8	7.5	19.8	26.9	31.9
Proline	15.6[c]	21.9[c]	5.6	10.5	14.1
Glycine	34.4	34.5	4.1	–	5.4
Alanine	9.6	11.6	4.1	10.5	3.5
Cysteine	–	–	1.2	0.8	2.2
Valine	1.8	2.1	4.7	4.0	5.4
Methionine	1.7	0.5	1.3	2.4	1.3
Isoleucine	1.1	1.1	4.8	5.0	4.1
Leucine	2.2	2.5	7.7	21.1	7.2
Tyrosine	0.3	0.1	3.7	5.3	2.8
Phenylalanine	1.6	1.2	5.2	7.3	4.4
Histidine	0.8	0.5	2.6	1.3	1.7
Lysine	2.9	2.6	6.0	–	1.4
Arginine	5.6	4.8	7.6	4.7	3.2
Tryptophan	–	–	1.3	–	–

[a]Value for aspartic acid and asparagine
[b]Value for glutamic acid and glutamine
[c]Value for proline and hydroxyproline

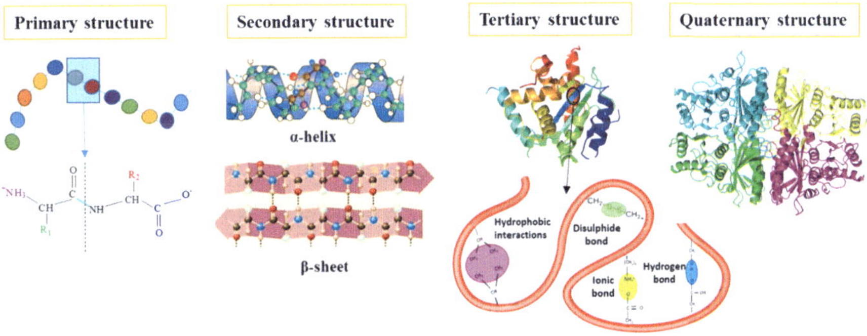

Fig. 3.1 The four protein structures: primary, secondary, tertiary and quaternary structures

chain size as well as the amino acid order define the physical, chemical, structural, biological and functional properties of the protein. The next level, the secondary structure, is conformed of regular arrangements of the backbone of the polypeptide chain. They can be found mainly in two secondary structures, α-helix and β-sheets, but also in unordered random coil structures. The folding of the secondary structure segments into a compact tridimensional form results in the tertiary structure. The formation of this structure assumes the optimization of hydrophobic, electrostatic and van der Walls interactions, and hydrogen bonds between the different available groups of the protein (Rodrigues et al. 2012). Finally, the disposition of more than one peptide chain constitutes the forth level of organization. The protein structure is especially important for film formation since it determines the ability of proteins to interact with themselves and other components.

Protein structure can be modified by physical (heating, shearing, hydrostatic pressure or irradiation), chemical (alkylation, acylation, acetylation or pH alteration) and biochemical methods (enzymes), which result in a structural or conformational change of the native structure, without altering the amino acid sequence (Zink et al. 2016). These structure modifications could improve functional properties of the proteins in order to adapt them to a specific final application. These modifications induce the protein denaturation, promoting protein unfolding and the exposure of functional groups followed by new chain associations through intermolecular interactions (Cordeiro de Azeredo 2012; Schmid et al. 2014). Depending on the protein source, the denaturation temperature can vary. As an example, thermal denaturation of fish collagen was found to be around 35 °C (Pati et al. 2010), whereas this is around 75 °C for soy protein isolate (Guerrero et al. 2010). Moreover, additives and processing conditions have also a direct influence on protein denaturation (de Graaf 2000; Gómez-Guillén et al. 2005). Solution pH value is a factor that conditions protein structure and therefore, the functional properties of proteins. Intermolecular interactions, such as hydrogen bonds, hydrophobic interactions and disulphide bonds, are especially influenced by a pH shift. Proteins exhibit a net negative charge at a pH higher than their isoelectric point (IP) and a net positive charge at a pH lower than their IP. Thus, when pH is adjusted away from the IP, electrostatic repulsion between protein molecules occurs, increasing protein solubility. In turn, at the IP, protein molecules have no net charge, which results in protein aggregation and precipitation (Wihodo and Moraru 2013).

In order to change protein-protein interactions, cross-linking have been explored as a viable method to improve the mechanical strength and barrier properties of protein films. In this context, chemicals like aldehydes are used to interact with the functional groups of proteins, such as the amino function in lysine and hydroxylysine or the carboxyl group in aspartic and glutamic acids (Etxabide et al. 2015a). However, aldehydes can be toxic, so natural cross-linkers are preferred for protein modifications. Aragui and Moslehi (2014) cross-linked fish gelatin with caffeic acid, a natural phenolic compound, to improve barrier and physicochemical properties, and Samsalee and Sothornvit (2017) employed rutin, caffeic acid and genipin to cross-link porcine plasma protein. Enzymes, such as transglutaminase, are also used to cross-link proteins (Al-Saadi et al. 2014; Song and Zhao 2014). In addition to physical, chemical and biochemical modifications, the addition of plasticizers is often used to modify the brittle behavior of proteins, improving their processability and those properties required to employ proteins for the development of films and coatings. Possible food grade plasticizers are glycerol, mannitol, sorbitol and sucrose, but water also acts as an effective plasticizer (Ustunol 2009). The effectiveness of a plasticizer depends on the size, shape and compatibility with the protein matrix. Rezaei and Motamedzadegan (2015) studied the effect of plasticizer type in protein films; results showed that glycerol was a better plasticizer than sorbitol since it increased percentage elongation to the breaking point besides imparting suitable tensile strength to the films.

Finally, regarding protein classification, these macromolecules can be classified depending on their shape as globular or fibrous proteins. The globular proteins fold into spherical or ellipsoidal-shaped structures, resulting from the enrolling on itself, holding mutually by an arrangement of hydrogen, ionic, hydrophobic and disulphide bonds; while the fibrous proteins are stem-like shaped, coupled in parallel constructions by hydrogen bonds to form fibers (Gupta and Nayak 2015). Vegetal proteins such as soy protein are an example of globular proteins. They are composed of a mixture of albumins and globulins, 90% of which are storage proteins with globular structure. According to the sedimentation rate, soy protein consists of four major fractions, 2S, 7S, 11S and 15S, being 7S (β-conglycinin) and 11S (glycinin) more than 80% of the total protein (Acosta-Dominguez et al. 2016). On the other hand, collagen can be an example of a fibrous structured protein. Collagen is composed of three cross-linked α-chains intertwined in the so-called collagen triple-helix. Likewise, this structure is mainly stabilized by intra- and inter-chain hydrogen bonding that varies according to animal species, age, tissue and other factors (Alfaro et al. 2015). Collagen fibers are hardly soluble, so they are treated by chemical denaturation to obtain gelatin. During this process, hydrogen and covalent bonds are cleaved, leading to a destabilization of the triple helix as a result of the helix-to coil transition and thus, the conversion into soluble gelatin. The structure of gelatin can change under the influence of some parameters, such as extraction temperature, pH, drying temperature and relative humidity (Duconseille et al. 2017). Overall, these structural differences reflect the specific characteristics and functionality of each protein. As shown in Table 3.1, fish gelatins show different amino acid profile than mammalian gelatins and, consequently, different thermal, rheological, viscoelastic and mechanical properties (Díaz-Calderón et al. 2017). Although the functional properties may vary, all proteins have in common some properties that make them suitable for food packaging. Their mayor advantages can be attributed to the excellent film forming ability, good transparency, excellent barrier properties against fat and oxygen, heat sealability, as well as odorless and tasteless characteristics (Song and Zheng 2014; Tongnuanchan et al. 2012, 2016). However, proteins possess great sensitivity to water and moderate mechanical properties and thus, the improvement of those properties is needed in order to develop new active packaging based on animal and vegetal proteins (Lin and Zhao 2007; Park et al. 2014).

3.3 Manufacture of Protein Films and Coatings

Different production techniques are used to manufacture protein films, mainly wet and dry processing. The wet process, also called solution casting, is the most widely used at laboratory scale owing to its simplicity (Kashiri et al. 2017a; Liu et al. 2017a), but at larger scale films can be produced by extrusion or compression molding (Ciannamea et al. 2017). Therefore, the technology used in the plastic industry can be also used for the production of renewable and biodegradable films. A scheme of these manufacturing methods is shown in Fig. 3.2.

Traditionally, protein films and coatings are manufactured by solution casting. Although water is the main solvent employed, ethanol-water mixtures can be also used for some proteins such as zein. Heating, stirring and pH adjustment are the parameters under control in order to obtain homogenous films and coatings, since those processing conditions affect protein denaturation, promoting or hindering interactions. Once the film forming solution is prepared, it is poured onto petri dishes and allowed to dry to form the film (Liu et al. 2016; Qazanfarzadeh and Kadivar 2016) or it is applied directly onto food surface by dipping or spraying (Zhong et al. 2014). In order to scale-up production of protein films, tape casting, also known as spread casting or knife-coating, has been used in plastic and paper industries. This technique consists in spreading the film forming solution on a tape and dry it by heat conduction, convection, infrared radiation or a combination of them. This technique requires more concentrated protein solutions than the traditional solution casting (Ortiz et al. 2017). Continuous processes can be carried out by this technique, although the measurement of rheological properties is needed in order to ensure appropriate flow conditions and minimize undesired sedimentation (de Moraes et al. 2013).

Regarding dry processes, when extrusion or compression are employed, proteins must be heated above their glass transition temperature to make them flow (Ghanbarzadeh and Oromiehi 2008). Additionally, pressure is also applied and plasticizers are added to decrease glass transition temperature and promote the thermoplasticity of the protein (Hernandez-Izquierdo and Krochta 2008; Visakh and Nazarenko 2017). Plasticizers such as polyols are usually used, but also water can act as a plasticizer (Bertuzzi and Slavutsky 2016; Nur Hanani et al. 2013). Concerning extrusion, this technique is a continuous, efficient, high-performance and low-cost process, advantageous for large-scale production (Gutiérrez and Alvarez 2018). Proteins are fed to the extruder where they are mixed with the incorporated additives (Koch et al. 2017). In the feeding zone, the mixture is slightly compressed, but compression increases in the kneading zone, where higher pressures are applied. In the heating zone, the highest shear rates, temperatures and pressures are achieved and, finally, the product exits from the die and it is pulled away at constant velocity (Bertuzzi and Slavutsky 2016). Single-screw extruders or twin-screw extruders can be used to push forward and mix the components of the film forming formulations (Bouvier and Campanella 2014). Extrusion is a complex multi-input-output process in which some parameters must be mastered (Emin et al. 2017), mainly extruder parameters (screw speed, barrel temperature, screw configuration, die dimension) and process parameters (moisture content, specific mechanical energy, residence time) (Guerrero et al. 2012). In the case of compression, the film forming components are mixed and a specific quantity of the mixture is placed inside the pre-heated press and pressure is applied to obtain the film (Tatara 2017; Türe et al. 2012). Protein films are successfully prepared in short times using this process.

3.4 Food Packaging Based on Animal Proteins

3.4.1 Collagen and Gelatin

Collagen is the main biopolymer in living organisms and the main component of connective tissues in vertebrates (around 30% of total body protein content) since it constitutes the main building blocks of many structural tissues, such as skin, bones, ligaments and tendons (Zuber et al. 2015). Twenty-eight different types of collagen can be found in nature, but type I collagen is the most common one (Zhang et al. 2018). This type of collagen is composed of three types of polypeptide chains arranged into a macromolecular fibrillar structure (Wang et al. 2017c; Tan and Chang 2018). These microfibrillar networks present in collagen contribute to its high strength, insolubility in water and durable nature (Berkowitz and Houde 2014). In spite of this resistance and durability, collagen can disintegrate and biodegrade in

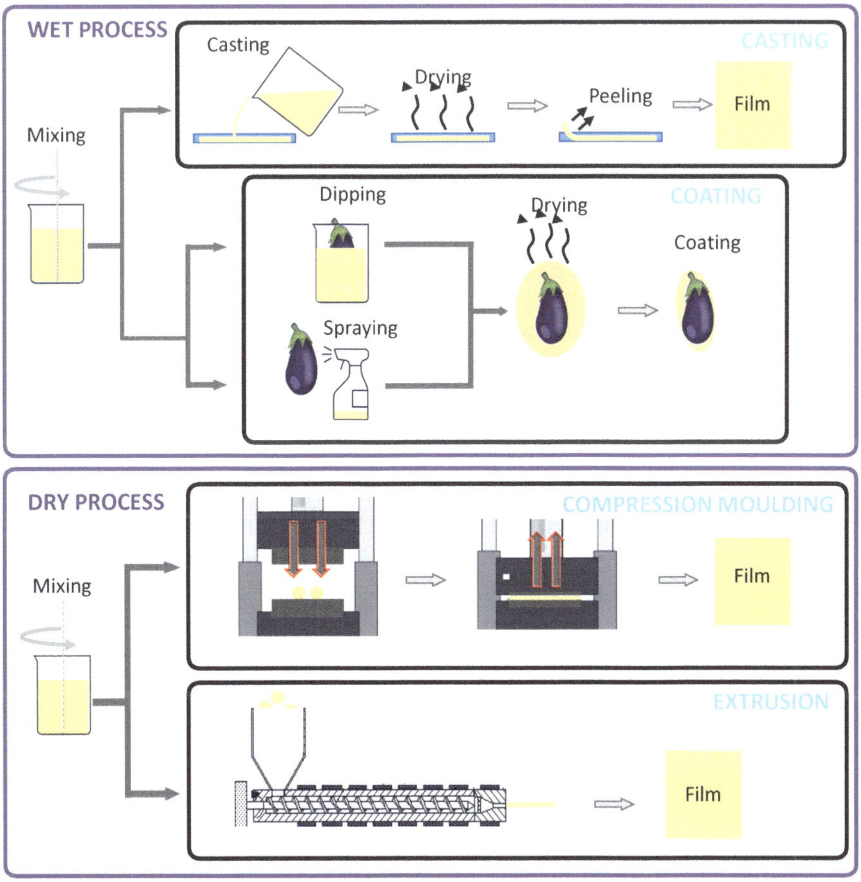

Fig. 3.2 Manufacture of films and coatings by wet and dry processes

certain conditions and, thus, it has been thoroughly studied for packaging applications (Liao et al. 2017). Collagen has been successfully commercialized as sausage casing (Wang et al. 2016a). Some synthetic cross-linkers, such as glutaraldehyde and formaldehyde have been used to improve collagen properties (Hu et al. 2014); however, since food safety concerns have increased, the employment of natural cross-linkers to replace synthetic ones is gaining importance. Wang et al. (2015) investigated the influence of some physical cross-linking procedures, believed to be safe and effective. Specifically, ultraviolet irradiation (UV), dehydrothermal treatment (DHT) and their combination (UV and DHT) were used to improve the properties of edible collagen casing. In that study, mechanical properties and thermal stability were enhanced without any significant impact on the film appearance. Other means of improving mechanical properties are also studied. Taking into account that upon gelatinization physicochemical interactions can be created between functional groups, Wang et al. (2017d) intended to promote hydrogen bonds between hydroxyl groups in starch and carbonyl or amine groups in the polypeptide chains of collagen, incorporating raw starch granules as a reinforcing agent in collagen matrix. As a result, interactions between collagen and starch molecules were promoted, improving not only mechanical strength but also the water resistance of the resultant composite films. In addition to films, collagen edible coatings have been widely applied onto various food products since these coatings can play an active role as protective barrier against oxidation and as carriers of active substances, such as antioxidants and antimicrobials (Hashim et al. 2015). In particular, lysozyme is a food grade antimicrobial enzyme, stable over a wide range of pH and temperature, which can be used with collagen. It has bacteriostatic, bacteriolytic and bactericidal activity and it is efficient in controlling the growth of a great number of food pathogens. It has been found that fresh salmon fillets coated with collagen-lysozyme maintain their quality longer during refrigerated storage (4 ± 1 °C) than non-coated fillets (Wang et al. 2017b).

Although collagen properties vary with the employed collagen source, the triple-helical structure and the insolubility of the native collagen are maintained, therefore, collagen can be hydrolyzed to obtain gelatin. Two types of gelatin are obtained depending on the pre-treatment procedure, commercially known as type A gelatin (acid pre-treatment conditions) and type B gelatin (alkaline pre-treatment conditions) (See et al. 2015). Gelatins are biodegradable and biocompatible and show swelling and gelling capacity as well as thermal stability (Hashim et al. 2015; Patel et al. 2018). Porcine skin was the first raw material used for the manufacture of gelatin in the 1930s (Gómez-Guillén et al. 2011). Nowadays, commercial gelatin is mainly produced from bovine and porcine skin and bones (Roy et al. 2017), although increased interest in other sources of gelatin, such as fish gelatin, has been shown (Table 3.2).

Wastes from fishery industries (fish head, viscera, skin, bones, scales) represent 20-60% of the initial raw material (Rebah and Miled 2013), so the valorization of these wastes to obtain fish gelatin could facilitate waste management and lead to the development of novel and more sustainable packaging. In recent years, gelatin derived from diverse aquatic fish species has been employed to prepare active films

Table 3.2 Bovine, porcine and fish gelatin-based active films and coatings

Gelatin	Active agent	Additive role	Reference
Bovine gelatin	α-tocopherol, cinnamaldehyde and garlic EOs	AO	Córdoba and Sobral (2017)
	Carrot residue fiber	AO	Iahnke et al. (2015)
	Brown seaweed extract	AO	Kadam et al. (2015)
	Oregano and lavender EOs	AMAO	Martucci et al. (2015)
Porcine gelatin	Ascorbic acid	AO	Kowalczyk (2016)
	Ethanol hop extract	AO	Kowalczyk and Biendl (2016)
	Curcuma ethanol extract	AO	Bitencourt et al. (2014)
Fish gelatin	Aqueous extracts of henna	AMAO	Jridi et al. (2018)
	Olive phenols	AM	Bermúdez-Oria et al. (2017)
	Aloe vera	AO	Chin et al. (2017)
	Tea polyphenol	AMAO	Feng et al. (2017)
	Esculine	AO	Liang et al. (2017)
	Boldine	AOAM	López et al. (2017)
	Coumarin	AO	Benbettaïeb et al. (2016)
	Chitosan nanoparticles	AM	Hosseini et al. (2016)
	Thyme EO	AM	Lee et al. (2016)
	Peppermint and citronella EOs	AM	Yanwong and Threepopnatkul (2015)

AO antioxidant; *AM* antimicrobial

with antioxidant and/or antimicrobial properties. The incorporation of polyphenols extracted from henna (Jridi et al. 2018), tea (Feng et al. 2017), or olives (Bermúdez-Oria et al. 2017) into fish gelatins has been found to extend the shelf life of meat, fish and fruit, respectively. Also EOs have been used to produce active packaging; in particular, thyme EO (Lee et al. 2016), oregano EO (Hosseini et al. 2016), and peppermint and citronella EOs (Yanwong and Threepopnatkul 2015), which have shown antimicrobial properties against some food spoilage bacteria such as *Staphylococcus aureus* and *Escherichia coli*. Beside natural antimicrobial and antioxidant additives, natural cross-linkers such as lactose (Etxabide et al. 2015b) and citric acid (Uranga et al. 2016) have been incorporated into fish gelatin film forming solutions to improve barrier properties and prevent food oxidation caused by light.

EOs have been also incorporated into bovine gelatin to inhibit the growth of microorganisms; specifically, oregano and lavender EOs have been found to be effective against food spoilage bacteria, especially gram-positive bacteria, being oregano the EO that exhibited the most effective antimicrobial and antioxidant effect (Martucci et al. 2015). The effect of other natural additives, such as brown seaweed extract, on bovine gelatin films has been also analyzed and results have shown that higher seaweed extract contents led to increased antioxidant activity (Kadam et al. 2015). Even waste from minimally processed carrots have been used to prepare bovine gelatin films that retard sunflower oil oxidation (Iahnke et al. 2015).

Some antioxidants, such as ascorbic acid (Kowalczyk 2016) or ethanolic hop extract (Kowalczyk and Biendl 2016) have been incorporated into porcine gelatin films and a controlled antioxidant release from gelatin film was observed. Also the ethanolic extract of curcuma has been added to porcine gelatin films (Bitencourt et al. 2014). Curcuma contains phenolic compounds responsible for its antioxidant and anti-inflammatory activities. The incorporation of this additive into porcine gelatin films resulted in interactions between the phenolic compounds and gelatin, improving the UV-vis light barrier properties of gelatin films, besides the improvement of their antioxidant capacity.

3.4.2 Milk Proteins: Casein and Whey Protein

Milk and dairy products, especially milk proteins such as caseins and whey proteins, contain numerous essential nutrients. Water makes up more than 80% of the total weight of milk, whereas protein composition varied with the source of milk (Massouras et al. 2017; Wang et al. 2016b). Casein is the main protein fraction of ruminant milk (Tsakalidou and Papadimitriou 2016), while whey protein constitutes the 18% of total cow milk protein content (Ganju and Gogate 2017).

Casein is responsible for the white and opaque appearance of milk and it comprises four different components, known as α_{S1}-, α_{S2}-, β- and κ-casein (Sarode et al. 2016; Zhang et al. 2017). It is present in the form of spherical complexes with an average radius of about 100 nm, called casein micelles, which contain nanoclusters of colloidal calcium phosphate (Balakrishnan et al. 2017). Casein-based films are attractive for food applications due to their high nutritional quality, excellent sensory properties and potential to protect food products from their surrounding environment (Ponce et al. 2016). Depending on the coagulation method, two different caseins can be obtained, known as rennet and acid caseins (Barbé et al. 2014). Acid casein can be solubilized by neutralization with a base as sodium, potassium or calcium hydroxide to get sodium, potassium or calcium caseinates (Early 2012). Rennet casein, acid casein and sodium caseinate films have been successfully manufactured by extrusion and a considerable range of properties was obtained, from water-soluble sodium caseinate films to insoluble and hydrophobic acid casein films (Chevalier et al. 2018). However, caseinates are more employed than acid and rennet caseins for food packaging, since they do not have colloidal calcium phosphates and are highly water-soluble. Sodium caseinate coatings containing different EOs, such as ginger EO (Noori et al. 2018) and *Zataria multiflora* EO (Emam-Djomeh et al. 2016) have been used to extend food shelf life. Although sodium caseinate films present higher flexibility than calcium caseinate films, calcium caseinate show higher thermal stability (Arrieta et al. 2013). Also Belyamani et al. (2014) prepared sodium and calcium caseinate films and observed that calcium caseinate films exhibited better mechanical properties than sodium caseinate films.

Whey protein is one of the most important by-products of dairy industry obtained during cheese and casein production. In fact, significant quantities of this protein are obtained in cheese production, 8–9 kg of whey are produced per each 1–2 kg of cheese (Corbatón-Báguena et al. 2015). Whey is a complex mixture of water-soluble globular proteins whose principal components are β-lactoglobulin, α-lactalbumin, serum albumin, lactoferrin, immunoglobulins, lactose and soluble mineral salts (Amaya-Farfan et al. 2016). Four major forms of whey protein are used: whey protein concentrate (WPC), whey protein isolate (WPI), whey protein hydrolysates (WPH) and native whey protein (NWP). In the field of food packaging, WPI (more than 90% protein content) and WPC (29–89% protein content) are mostly used (Li et al. 2018). WPI coatings have been applied onto peanuts (Riveros et al. 2013), pike-perch fillets (Shokri and Ehsani 2017) and cut apples, potatoes and carrots (Marquez et al. 2017). All of them contributed to delay oxidation and microbial growth, extending food shelf life. WPI films with different active additives have also been prepared and characterized. Xie et al. (2017) analyzed the effect of WPI-based films in modified atmosphere packaged (MAP) and vacuum packaged tuna, highlighting the effectiveness of WPI films to enhance physical and microbiological attributes. In another study, Boyaci et al. (2016) designed a novel activate-at-home-type antimicrobial packaging, based on lysozyme, oleic acid and WPI, to increase the safety of the remained food left in the package after opening the original MAP or vacuum packaging. In this study, the possibility of employing antimicrobial films with pH controlled release properties was tested and it was found that the films showed antimicrobial activity on cold stored smoked salmon slices. Overall, researches on whey protein films have been focused on the use of WPI; nevertheless, several WPCs with protein contents from 35 to 80% have been carried out. Ribeiro-Santos et al. (2017) assessed the optimal blend of cinnamon, rosemary and basil EOs to enhance the antimicrobial and antioxidant activities of WPC active packaging. Result showed that the optimal blend of EOs consisted of two species of cinnamon (51% of *Cinnamomum cassia* and 34% of *Cinnamomum zeylanicum*) and rosemary (15%). In addition, an active WPC coating containing *Origanum virens* EO was applied on sausages during industrial production and food was regularly monitored for four months (Catarino et al. 2017). This WPC coating effectively extended the sausages shelf life, delaying microbial spoilage and preventing lipid oxidation and color fading.

3.4.3 Other Animal Proteins

In addition to collagen, gelatin and milk proteins other animal proteins have been employed for food packaging. Among them, myofibrillar proteins, mainly composed of myosin (~200 kDa) and actin (~45 kDa), are soluble in moderate salt solutions and form films with good mechanical properties (Gómez-Estaca et al. 2014b). In general, myofibrillar proteins are derived from the muscle protein extracted from the fishery industry wastes. Different fish species can be used for the muscle protein extraction, such as tilapia (Kaewprachu et al. 2017, 2018), jumbo squid, the largest and most abundant squid species found in the pelagic zone of the eastern Pacific (Blanco-Pascual et al. 2013, 2014), and sliver carp, one of the main freshwater fish species in China (Nie et al. 2015); however, myofibrillar protein can be also extracted from other sources, such as chicken breast (Cercel et al. 2015).

Keratin, also a fibrous protein, is one of the most abundant proteins since it is the major constituent of fur, nails, wool, feathers, horns, and hooves (Holkar et al. 2018). Nevertheless, this protein is highly-structured due to disulphide linked polypeptides and, thus, few works are focused on turning this waste into films (Nassar et al. 2012; Song et al. 2014).

Finally, egg albumen is a mixture of eight globular proteins known as ovalbumin, ovotransferrin, ovomucoid, ovomucin, lysozyme, G2 globulin, G3 globulin and avidin. Lacroix and Vu (2014) reported that the utilization of egg albumen for clear and transparent films and coatings preparation is of particular nutritional interest due to its antioxidant effectiveness. As an example, gelatin-egg albumen films incorporating sepiolite and clove EO have shown antioxidant and antimicrobial activities (Giménez et al. 2012). It is also worth noting that some assessments related to the use of egg yolk as a film forming material have been recently published (Fuertes et al. 2017).

3.5 Food Packaging Based on Vegetal Proteins

3.5.1 Soy Protein

Soy protein is extracted from soybeans used to obtain soy oil. According to United States Department of Agriculture (USDA), the soybean production in 2016/2017 was around 350 million tons worldwide (USDA 2017), which make this resource abundant, accessible and low-cost. During soy oil extraction, secondary products such as soy flour (SF), soy protein concentrate (SPC) and soy protein isolate (SPI) are obtained (Preece et al. 2017). These three products differ in the protein concentration, SF has the less protein content, around 40–60% protein, combined with fats and carbohydrates; SPC contains around 60–70% protein, a polysaccharide fraction of around 8–15%, mainly composed of cellulose and pectic polysaccharides, and a minor content of fats (1%), fibers (1–3%) and ashes (3–5%); and SPI contains about 90% protein (Ciannamea et al. 2014). Due to this high amount of protein, its renewability, biodegradability and abundance, SPI has become a potential candidate in the polymer industry for food packaging applications (Garrido et al. 2014; Božič et al. 2015). These bio-based materials must fulfill certain conditions to maintain food sensory quality and safety by serving selective barriers to moisture transfer, oxygen uptake, lipid oxidation and losses of volatile aromas and flavors, without forgetting the requirement of having an adequate mechanical behavior (González et al. 2015). For that reason, a high amount of research works has been published focused on meeting all these requirements.

SPI films have inferior mechanical properties and lower resistance to moisture than synthetic plastics; thus, different methods have been studied in order to modify SPI with the aim of overcoming these drawbacks. Among these methods, blending with natural substances (Pan et al. 2014; Wang et al. 2014), chemical cross-linking (Xu et al. 2015; Jiang et al. 2017), employment of different processing methods (Garrido et al. 2016) and enzyme treatment (Meinlschmidt et al. 2015) can be highlighted.

The incorporation of additives, such as other proteins, polysaccharides, lipids or other natural substances, has been widely considered in order to take advantage of the properties of each compound and the synergy between them. In that way, the properties of SPI-based films and coatings can be enhanced and new functionalities can be provided. Bai et al. (2013) studied the incorporation of gelatin into SPI-based films. Results showed an increase of tensile strength, elongation at break and contact angle and a decrease of water vapor transmission rate. The addition of polysaccharides has also been considered. Sui et al. (2016) incorporated guar gum into SPI films and results indicated that guar gum induced increased network compactness, resulted from strong intermolecular interactions. Consequently, SPI-based films were more tensile-resistant, water-resistant and had better barrier properties to light. The addition of starch into SPI films has also been used. Starch nanocrystals improved the tensile strength and elastic modulus of SPI films, which became more rigid due to the interactions and the high cohesion between the two components; additionally, the solubility, swelling and water vapor permeability (WVP) decreased (González and Igarzabal 2015). Regarding mechanical properties, similar trend was observed when cassava starch (Chinma et al. 2012) or oxidized potato starch was incorporated into SPI films (Galus et al. 2013). On the other hand, taking advance of the good properties of lipids, Hopkins et al. (2015) combined SPI with flaxseed oil to perform films with lower moisture content and swelling but with higher strength.

The added compounds can also act as cross-linkers and contribute to the enhancement of the film network. Xia et al. (2015) employed an epoxidized soybean oil, an environmentally friendly cross-linking agent derived from soybean, which effectively improved tensile strength values in a 139.8% and the modulus in a 695.6% as compared to the untreated SPI-based films, whereas Friesen et al. (2015) determined that the incorporation of rutin into SPI films decreased the WVP and increased the film strengthen.

It is also well known that the processing methods also affect the film properties. In this context, Liu et al. (2017b) prepared soy protein isolate/propylene glycol alginate/lauric acid films by direct blending and by co-dried blending. The obtained results demonstrated that the films prepared by co-drying had better barrier and mechanical properties. Moreover, Garrido et al. (2018) carried out the assessment of hydrolyzed keratin/soy protein films processed by casting and compression molding by means of the analysis of physicochemical, thermal, mechanical, optical and surface properties. Films processed by compression achieved better mechanical properties in terms of tensile strength, which increased with the incorporation of hydrolyzed keratin.

On the other hand, the efficiency of soy protein films to work as carriers for antioxidants, such as chestnut bur extract (Wang et al. 2016c), tannins (Wang and Wang 2017) or catechin (Han et al. 2015), and antimicrobial agents, such as thymol (antibacterial) and natamycin (antifungal) (González and Igarzabal 2013), has been proved by many authors. Natural extracts or bioactives, such as anthocyanin-rich red raspberry (Wang et al. 2012), mango kernel extract (Adilah et al. 2018) or licorice residue extract (Han et al. 2018), can also have influence on the functional properties of films, apart from improving the antioxidant or antimicrobial activity inherent of each additive. For instance, it was determined that the incorporation of mango kernel extract increased the tensile strength of SPI-based films and reduced the water solubility and elongation at break; whereas, licorice residue extract improved the mechanical, water, oxygen and light barrier properties when its content was lower than 70 g per kg of protein. Echeverría et al. (2016) carried out the addition of clove EO into SPI films reinforced with montmorillonite. Besides the important antioxidant and antimicrobial properties provided by the EO, a plasticizing effect was exerted. Furthermore, the nanoclay caused a further strengthening effect in films containing clove EO; whilst nanocomposite films containing 10 g montmorillonite/100 g SPI reached an increase of 105 and 200% in tensile strength and Young's modulus, respectively, those that also contained clove reached higher variations (230 and 345%, respectively). Phenolic compounds, such as carvacrol, ferulic, caffeic and gallic acids, have been also successfully employed in SPI films. In general, the effect of phenolic acids incorporated into protein-based films is controlled by two different features: their ability to form hydrogen bonds or other interactions between the carboxyl groups and the amino groups of proteins, and their affinity towards water that results in high water absorption (Ganiari et al. 2017). Otoni et al. (2016) incorporated carvacrol and cinnamaldehyde and promoted an increase in the rigidity of films, while Insaward et al. (2015) studied the influence of ferulic, caffeic and gallic acids and their oxidized products. Results determined that gallic acid-containing films exhibited the highest tensile strength and elongation at break, and oxidized phenolic acids were shown to produce films with higher tensile strength and elongation at break than their unoxidized counterparts. Moreover, phenolic-containing films showed reduced WVP and water solubility and increased contact angles.

Some research studies have demonstrated that films and coatings based on soy protein have multiple benefits regarding the extension of food shelf life; in particular, reduction of oxidation and discoloration, retardation of rancidity processes, inhibition or reduction of the microbial contamination, prevention of moisture loss and diminution of the loss of flavor compounds, stand out as the most significant ones. Some studies that demonstrated the potential of SPI-based films and coatings for food products, such as meat, fish, fruits, vegetables and others, are outlined in Table 3.3.

Strategies to extend the quality of fruits and vegetables need to target several key challenges, such as minimizing dehydration, reducing or avoiding microbial growth and extending maturation and senescence periods. The incorporation of chitosan and stearic acid (Wu et al. 2017) and ferulic acid (Alves et al. 2017) into SPI films

Table 3.3 SPI-based films or coatings for food packaging

Active agent	Additives role	Test in food	Processing	Reference
Clove EO	AMAO	Tuna	Casting	Echeverría et al (2018)
NisinSodium lactate EDTA	AM	Pork	Casting	Liu et al. (2017c)
CinnamaldehydeEugenol	AM	Pork	Casting	Zhang et al. (2013)
Thyme EOOregano EO	AM	Beef	Casting	Yemiş and Candoğan (2017)
Thyme EOOregano EO	AO	Beef	Casting	Coşkun et al. (2014)
CitralLimonene	AM	Persian lime	Coating	González-Estrada et al. (2017)
Ferulic acid	AO	Apple	Casting	Alves et al. (2017)
Citronella EO	AM	Banana	Casting	Arancibia et al. (2014)
Cinnamon oil	AM	Dry tofu	Casting	Liu et al. (2014)
Catechin	AO	Walnut	Coating	Kang et al. (2013)

AO antioxidant; *AM* antimicrobial

extended the shelf life of apples and improved film properties, including gas permeability and tensile strength. In another study, Arancibia et al. (2014) showed that the addition of 3% w/v citronella EO to SPI films blended with lignin had good antifungal activity against pathogen microorganisms (*Fusariun oxysporum*) in bananas. Also the application of SPI coatings with citral and limonene preserved postharvest quality of lime and provided antifungal activity against *Penicillium italicum* in inoculated limes (González-Estrada et al. 2017). Other additives, such as cysteine (Ghidelli et al. 2014, 2015), lauric acid and propylenglycol alginate (Zeng et al. 2013), incorporated into SPI coatings have shown to reduce enzymatic browning and improve the quality of artichokes, eggplant and jujubes. The prevention of other effects indicative of quality loss, such as shrinkage, oxidative off-flavors, microbial contamination and discoloration in meat, fish and poultry products is also of great importance. Echeverría et al. (2018) applied soy protein/montmorillonite/clove EO films for the preservation of refrigerated of tuna fillets. It was observed that SPI-based films decreased *Pseudomonas* growth and lipid oxidation; furthermore, montmorillonite favored the release of clove EO, extending the tuna shelf life. Also SPI coatings have been found to be effective for meat preservation by delaying lipid oxidation and color deterioration and maintaining textural parameters in beef patties (Guerrero et al. 2015). Beef has also been coated employing SPI with thyme and oregano EOs. According to Yemiş and Candoğan (2017), these soy protein coatings exhibited antimicrobial activity against *E. coli* O157:H7, *Listeria monocytogenes* and *S. aureus*. Moreover, Coşkun et al. (2014) confirmed that the addition of these two EOs controlled lipid oxidation in SPI-coated beef patties. Other bioactives, such as cinnamaldehyde and eugenol (Zhang et al. 2013) or nisin, sodium lactate and EDTA (Liu et al. 2017c) have also showed significant inhibitory effect on the growth of *S. aureus*, *Pseudomonas* and *yeast*, being cinnamaldehyde-containing SPI films the most effective ones. The films containing 6% EOs showed the preservation effect on pork. Besides that, films showed a significant antimicrobial effect on three food pathogens, *E. coli*, *Salmonella* and *Bacillus cereus* when nisin, sodium lactate and EDTA were incorporated into SPI films, extending pork shelf life up to 3–6 days.

3.5.2 Zein

Zein is a protein used for food packaging applications, since it exerts better barrier against transmission of water vapor and volatile compounds as compared to other types of protein films (Ozcalik and Tihminlioglu 2013). Moreover, zein has excellent film forming ability since it is believed to involve development of hydrophobic, hydrogen and limited disulphide bonds between zein chains in the film matrix (Bourtoom 2008). This biopolymer can be defined as a water-insoluble hydrophobic storage protein found in corn. Based on solubility and sequence homology, it can be separated into α-zein (19 and 22 kDa), β-zein (14 kDa), γ-zein (16 and 27 kDa) and δ-zein (10 kDa). α-zein, which consists on highly homologous repeat units with a high content of α-helix (Zhang et al. 2015), composes around 70–85% of the total fraction of zein mass, while γ-zein is the second most abundant fraction (10–20%).

As happened with other proteins, the brittleness is the major disadvantage of zein. It is known that the incorporation of plasticizers helps to reduce the brittleness (Xu et al. 2012), but blending or mixing with other natural compounds can also be a solution due to the benefits that these components can provide to the film. Since zein and gliadin are both readily dissolved in aqueous ethanol and have a good film-forming property, Gu et al. (2013) prepared films with these two proteins. The results showed that the addition of gliadin enhanced the strain at break of zein films as a result of the increase in the content of α-helix and β-turn structures and the decrease in the level of β-sheet structure. Cheng et al. (2015) prepared active zein films with chitosan and phenolic compounds (ferulic acid or gallic acid) and dicarboxylic acids (adipic acid or succinic acid). The antimicrobial properties against *S. aureus* and *E. coli* were determined and the antioxidant activity was confirmed by DPPH and ABTS free radical scavenging tests. Additionally, zein glycosylated with chitosan by transglutaminase has resulted effective in retarding lipid oxidation of ground pork, indicating that enzymatic glycosylation might be a new approach to modify the functional properties of zein (Wang et al. 2017e).

Other authors have also employed the cross-linking strategy to obtain zein films with enhanced properties. Succinic anhydride, eugenol and citric acid were employed by Khalil et al. (2015) as natural cross-linking agents. All cross-linked films showed remarkable antibacterial activities against *B. cereus* ATCC 49064 and *Salmonella enterica* ATCC 25566. However, these cross-linkers were also added to modify zein chemically in order to improve the functional properties of the films; thus, the addition of these compounds resulted in two- to three-fold increases in tensile strength values. On the other hand, Santos et al. (2017) assessed the addition of tannic acid, especially in its oxidized form, to cross-link zein and concluded that higher tannic acid contents and pH values resulted in films with better physical properties.

In general, one of the advantages of working with zein comes from its hydrophobicity, which makes it compatible with natural antimicrobials (Yemenicioğlu 2016). Naturally occurring antimicrobial compounds are an alternative to synthetic preservatives and their use is increasingly growing. In this context, many natural antimicrobial agents have been employed to develop zein films with antimicrobial effects. *Z. multiflora* Boiss. is a thyme-like plant belonging to the *Lamiaceae* family, being carvacrol and thymol its main antimicrobial components. The addition of these two EOs into zein films was studied by Kashiri et al. (2017a) and Moradi et al. (2016). These authors confirmed the good antimicrobial properties of *Z. multiflora* Boiss. against *L. monocytogenes* and *E. coli*, as well as the good antioxidant properties of these substances. Moreover, Kashiri et al. (2017b) demonstrated that the addition of *Z. multiflora* Boiss. EO caused an increase in the percent of elongation at break of the films. Alkan and Yemenicioğlu (2016) not only evaluated the employment of carvacrol and thymol EOs into zein films, but also studied other EOs, such as eugenol and citral, some phenolic acids like gallic, vanillic and cinnamic acids, and phenolic extracts from clove, oregano, artichocke stem and walnut shells. The incorporation of these promising antimicrobial compounds into zein films show positive results against bacterial plant pathogens, such as *Erwinia amylovora*, *Erwinia carotovora*, *Xanthomonas vesicatoria* and *Pseudomonas syringae*. Moreover, these authors concluded that phenolic-containing zein coatings could provide an additional post-harvest benefit by delaying bacterial spoilage of coated fresh fruits and vegetables.

A limited number of studies have investigated the effectiveness of zein-based films in fresh products. Ünalan et al. (2013) determined the release profiles of lysozyme and mixtures of lysozyme and phenolic compounds (catechin and gallic acid) from zein and zein-wax composite films to cold-stored fresh Kashar cheese. They concluded that all lysozyme-containing films prevented the increase of *L. monocytogenes* counts in the cheese stored at 4 °C for 8 weeks. However, only zein-wax films with sustained lysozyme-release rates caused a significant reduction in initial microbial load of inoculated cheese samples. The mixture of catechin and gallic acid improved the *in vitro* antimicrobial effect of films against *L. monocytogenes*, but showed no considerable antimicrobial effect in cheese. In another study (Chen et al. 2016), grass carp fish balls were coated by zein containing a polymeric chelator based on hexadentate 3-hydroxypyridinones. The study demonstrated that this coating could effectively improve the sensory properties of fish balls and maintain their freshness due to the inhibition of microbial growth and the delay of protein decomposition and lipid oxidation during storage. Mehyar et al. (2014) coated Berhi date palm fruits, harvested at the khalal stage, with different edible materials, including zein protein. Results confirmed that zein coatings was one of the most effective coating in retarding fruit maturation by extending the khalal stage from 7 to more than 14 days.

3.5.3 Wheat Gluten

Wheat gluten is a storage protein obtained as a by-product of the isolation process of starch from wheat flour (Day 2011). This protein consists of a mixture of two main proteins, gliadins and glutenins, which differ in their solubility in aqueous alcohols and in their propensity to form intermolecular disulphide bonds (Zubeldía et al. 2015). Gliadins are monomers, while glutenins are composed of discrete polypeptide subunits linked together by interchain disulphide bonds to form high molecular weight polymers. Therefore, gluten proteins cover a broad range of molecular masses up to several million daltons. The amount of disulphide bonds as well as the hydrogen bonds present in wheat gluten play an important role in the structure and properties of this protein (Pommet et al. 2005). Zein and wheat gluten share many similarities, including the solubility in aqueous alcohols, the high amount of proline, low-cost and availability (Dahesh et al. 2016). Moreover, since these two proteins are insoluble in water, they produce insoluble coatings. However, wheat gluten as compared to zein, exhibits remarkable viscoelastic properties which promote largely the film forming ability. As other proteins, wheat gluten films have limited resistance to water vapor and weak mechanical properties, but the chief advantage of using gluten as a raw material is its low oxygen permeability rates (Tanada-Palmu et al. 2000; Mojumdar et al. 2011).

Regarding film forming techniques, many authors have used thermo-mechanical processing to manufacture wheat gluten films (Ansorena et al. 2016; Thammahiwes et al. 2017). Exposing wheat gluten to high temperatures results in important changes in the type and degree of covalent cross-linking within the molecular network, which leads to brittle wheat gluten films in the absence of plasticizers (Jansens et al. 2013). Thus, incorporation of plasticizers in wheat gluten film forming solutions is required to promote film flexibility, being glycerol the most used plasticizer (Duval et al. 2015; Sharma et al. 2017). As in the case of the previously mentioned proteins, blending or mixing with other biopolymers can be also a way to enhance wheat gluten film properties. In this regard, the influence of adding some polysaccharides, such as locust bean gum, methyl cellulose, carboxymethyl cellulose (Zárate-Ramírez et al. 2014), starch (Basiak et al. 2015, 2017) and chitosan (Chen et al. 2014), has been analyzed.

During the last few years, a number of studies have aimed to evaluate the effectiveness of wheat gluten films against food-contaminating fungi and microbes in order to provide evidence of their applicability in different food products. *E. coli*, *Salmonella Typhimurium*, *S. aureus*, *B. cereus* and *L. monocytogenes* are common pathogens that can damage the foodstuffs; thus, Barazi and Osman (2017) evaluated the antimicrobial activity of gluten films with different concentrations of *Origanum vulgare* EO against these pathogens. Other antimicrobial agents, such as clove, red thyme, carvacrol, cinnamaldehyde and white thyme also showed antimicrobial effectiveness against *Aspergillus niger*, *Candida albicans*, *E. coli* and *S. aureus* (Gómez-Heincke et al. 2016). Also the incorporation of formic acid and oregano EO into wheat gluten films has demonstrated antimicrobial activity against *A. niger*,

Candida kefyr, *B. cereus* and *E. coli*. (Martínez et al. 2013). Pomegranate peel and curry leaf extracts have also been employed in the development of wheat gluten-based films in order to extend the life of cherry tomatoes and mangoes (Kumari et al. 2017). It was observed that pomegranate peel extract showed significantly higher antibacterial activity than curry leaves powder against tested pathogens, *S. aureus* and *Micrococcus luteus*. The antimicrobial performance of wheat gluten films was tested on meat products (Massani et al. 2014). Wheat gluten-containing *Lactobacillus curvatus* CRL705 was employed to assess the release properties in contact with substances commonly used as food simulants (sunflower oil and water). According to the results, it should be expected that in the timescale of vacuum-packaged cooked sausages (approximately 30–40 days), the film could provide an efficient antimicrobial effect on these products.

Although the research interest has been focused on manufacturing active films for packaging applications, intelligent packaging development is growing gradually (Ghaani et al. 2016; Gutiérrez et al. 2016a, b). According to Biji et al. (2015), an intelligent packaging material can be defined as a material that monitor the condition of packaged food or the environment surrounding food to give information regarding the quality of the packaged food during transportation and storage. Therefore, monitoring carbon dioxide as well as controlling the variation of relative humidity in food packages could be interesting so as to give a better control on the evolution of food metabolism and to meet the consumers demand for high quality food products. With regard to wheat gluten, this is considered a polarizable material having dielectric properties (Sharma et al. 2010); thus, this protein could be suitable to manufacture intelligent packaging. In fact, Bibi et al. (2017) mentioned that the electric and dielectric properties of wheat gluten are known to be sensitive to carbon dioxide. For that reason, gluten-based films could be used for monitoring packaging headspace in intelligent packaging systems. In that context, the authors were able to determine the potential use of wheat gluten as a carbon dioxide sensor, showing interesting results particularly at high relative humidity values (90% RH). Moreover, Bibi et al. (2016) investigated wheat gluten protein to monitor relative humidity, confirming the good sensitivity of wheat gluten at high relative humidity values, ideal for foreseen applications related to the control of packed food products.

3.5.4 Other Vegetal Proteins

Although soy protein, zein and wheat gluten are widely studied and employed proteins, other vegetal proteins have also been considered to develop active films and coatings. Sunflower protein films with clove EO were used for the preservation of refrigerated sardine patties, retarding lipid oxidation and delaying the growth of total mesophiles (Salgado et al. 2013). The viability of canola for food packaging applications has also been studied. Canola is mostly obtained as a by-product from the oil industry, but it is underutilized in the marketplace since it is sold traditionally for its use as livestock feed. The protein content within the meal can be up to 50%

(on a dry weight basis) and has a well-balanced amino acid profile and thus, it has been used for the preparation of films plasticized with glycerol (Chang and Nickerson 2015), sorbitol or polyethylenglycol (Chang and Nickerson 2014). Other vegetal proteins such as bitter vetch seeds, which contain up to 25% of protein, are also an inexpensive source of protein and could thus be an affordable raw material to produce films for food applications; however, bitter vetch protein films do not satisfy mechanical requirements (Arabestani et al. 2013). Hence, different additives, such as oxidized ferulic acid (Arabestani et al. 2016a) or pomegranate juice (Arabestani et al. 2016b) have been incorporated into film forming formulations to enhance properties, obtaining markedly higher tensile strength and elongation at break values. Recently, edible oil production from hazelnuts has become increasingly important since hazelnut oil has similar fatty acid profile to olive oil, so due to its nutritive value and high protein content, the hazelnut meal obtained from oil extraction is gaining success. Aydemir et al. (2014) demonstrated that hazelnut protein isolate shows antioxidant, anticancerogenic and antihypertensive activity, but poor emulsifying, gelling and water absorption capacities. However, hazelnut proteins form transparent, light to brown colored and flexible films. Also sesame protein is obtained as a by-product of oil extraction. Sesame meal contains 35–40% protein and can be used for film formation. Sharma and Singh (2016) employed sesame protein isolate, extracted from defatted sesame meal, to prepare films. Although further research should be done to improve the mechanical and optical properties of those films, this protein can be used for food packaging of fruits and vegetables since it provides barrier to moisture.

3.6 Conclusions and Future Trends

Films and coatings based on animal and vegetal proteins are promising systems to be used for food packaging applications. However, the proper understanding of the role of each component of the packaging is a key issue in order to develop films and coatings with the suitable functional properties to control the mechanism of food deterioration in order to improve food quality and extend food shelf life, fulfilling all the specifications required for those materials in contact with food. Several studies have been focused on optimizing the composition, functional properties and processing methods of protein-based films and coatings in order to develop economically and environmentally sustainable materials able to compete against synthetic plastics and reduce their use as food packaging. Therefore, the employment of protein along with bioactive compounds, which can be extracted wastes or by-products from agriculture, meat, poultry and fish processing industries, could contribute to the cost-effective production of this environmentally friendly packaging, as well as to the promotion of new desirable functionalities. There is a growing number of research works that apply protein films and coatings onto fresh food, demonstrating that these proteins are able to improve the quality, safety, functionality and shelf life of food products, consequently, new market opportunities are opened for the commercial implementation of this protein-based food packaging.

References

Acosta-Dominguez L, Hernández-Sánchez H, Gutiérrez-López GF, Alamilla-Bletrán L, Azuara E (2016) Modification of the soy protein isolate surface at nanometric scale and its effect on physicochemical properties. J Food Eng 168:105–112

Adilah ZAM, Jamilah B, Nur Hanani ZA (2018) Functional and antioxidant properties of protein-based films incorporated with mango kernel extract for active packaging. Food Hydrocoll 74:207–218

Alfaro AT, Balbinot E, Weber CI, Tonial IB, Machado-Lunkes A (2015) Fish gelatin: characteristics, functional properties, applications and future potentials. Food Eng Rev 7(1):33–44

Alkan D, Yemenicioğlu A (2016) Potential application of natural phenolic antimicrobials and edible films technology against bacterial plant pathogens. Food Hydrocoll 55:1–10

Aloui H, Khwaldia K (2016) Natural antimicrobial edible coatings for microbial safety and food quality enhancement. Compr Rev Food Sci Food Saf 15(6):1080–1103

Alparslan Y, Baygar T (2017) Effect of chitosan film coating combined with orange peel essential oil on the shelf life of deepwater pink shrimp. Food Bioproc Tech 10(5):842–853

Al-Saadi JS, Shaker KA, Ustunol Z (2014) Effect of heat and transglutaminase on solubility of goat milk protein-based films. Int J Dairy Technol 67(3):420–426

Álvarez K, Famá L, Gutiérrez TJ (2017) Physicochemical, antimicrobial and mechanical properties of thermoplastic materials based on biopolymers with application in the food industry. In: Masuelli M, Renard D (eds) Advances in physicochemical properties of biopolymers (Part 1). Bentham Science, Sharjah, pp 358–400. https://doi.org/10.2174/9781681084534117010015

Álvarez K, Alvarez VA, Gutiérrez TJ (2018) Biopolymer composite materials with antimicrobial effects applied to the food Industry. In: Functional biopolymers. Springer, Cham, pp 57–96. https://doi.org/10.1007/978-3-319-66417-0_3

Alves MM, Gonçalves MP, Rocha CMR (2017) Effect of ferulic acid on the performance of soy protein isolate-based edible coatings applied to fresh-cut apples. LWT Food Sci Technol 80:409–415

Amaya-Farfan J, Moura CS, Morato PN, Lollo PCB (2016) Dietary whey protein and type 2 diabetes: molecular aspects. In: Mauricio D (ed) Molecular nutrition series, 1st edn. Academic, London, pp 211–220

Ansorena MR, Zubeldía F, Marcovich NE (2016) Active wheat gluten films obtained by thermoplastic processing. LWT Food Sci Technol 69:47–54

Arabestani A, Kadivar M, Shahedi M, Goli SAH, Porta R (2013) Properties of a new protein film from bitter vetch (*Vicia ervilia*) and effect of $CaCl_2$ on its hydrophobicity. Int J Biol Macromol 57:118–123

Arabestani A, Kadivar M, Shahedi M, Goli SAH, Porta R (2016a) The effect of oxidized ferulic acid on physicochemical properties of bitter vetch (*Vicia ervilia*) protein-based films. J Appl Polym Sci 133(2):42894–42901

Arabestani A, Kadivar M, Shahedi M, Goli SAH, Porta R (2016b) Characterization and antioxidant activity of bitter vetch protein-based films containing pomegranate juice. LWT Food Sci Technol 74:77–83

Aragui M, Moslehi Z (2014) Chemical mofification of cold water fish gelatin using phenolic cross-linker caffeic acid. Res J Pharm Biol Chem Sci 5(6):674–679

Arancibia MY, López-Caballero ME, Gómez-Guillén C, Montero P (2014) Release of volatile compounds and biodegradability of active soy protein lignin blend films with added citronella essential oil. Food Control 44:7–15

Arfat YA, Benjakul S, Prodpran T, Sumpavapol P, Songtipya P (2014) Properties and antimicrobial activity of fish protein isolate/fish skin gelatin film containing basil leaf essential oil and zinc oxide nanoparticles. Food Hydrocoll 41:265–273

Arrieta MP, Peltzer MA, del Carmen Garrigós M, Jiménez A (2013) Structure and mechanical properties of sodium and calcium caseinate edible active films with carvacrol. J Food Eng 114(4):486–494

Atarés L, Chiralt A (2016) Essential oils as additives in biodegradable films and coatings for active food packaging. Trends Food Sci Technol 48:51–62

Aydemir LY, Gökbulut AA, Baran Y, Yemenicioğlu A (2014) Bioactive, functional and edible film-forming properties of isolated hazelnut (*Corylus avellana* L.) meal proteins. Food Hydrocoll 36:130–142

Bai H, Xu J, Liao P, Liu X (2013) Mechanical and water properties of soy protein isolate film incorporated with gelatin. J Plast Film Sheeting 29(2):174–188

Balakrishnan G, Nguyen BT, Schmitt C, Nicolai T, Chassenieux C (2017) Heat-set emulsion gels of casein micelles in mixtures with whey protein isolate. Food Hydrocoll 73:213–221

Barazi A, Osman E (2017) Antimicrobial activity of a novel biodegradable edible film produced from Pistacia vera resin and *Origanum vulgare* essential oil. Res J BioTechnol 12(9):15–21

Barbé F, Ménard O, Le Gouar Y, Buffière C, Famelart MH, Laroche B, Le Feunteun S, Rémond D, Dupont D (2014) Acid and rennet gels exhibit strong differences in the kinetics of milk protein digestion and amino acid bioavailability. Food Chem 143:1–8

Basiak E, Galus S, Lenart A (2015) Characterisation of composite edible films based on wheat starch and whey-protein isolate. Int J Food Sci Technol 50(2):372–380

Basiak E, Lenart A, Debeaufort F (2017) Effects of carbohydrate/protein ratio on the microstructure and the barrier and sorption properties of wheat starch-whey protein blend edible films. J Sci Food Agric 97(3):858–867

Belyamani I, Prochazka F, Assezat G, Debeaufort F (2014) Mechanical and barrier properties of extruded film made from sodium and calcium caseinates. Food Packaging Shelf Life 2(2):65–72

Benbettaïeb N, Chambin O, Assifaoui A, Al-Assaf S, Karbowiak T, Debeaufort F (2016) Release of coumarin incorporated into chitosan-gelatin irradiated films. Food Hydrocoll 56:266–276

Berkowitz SA, Houde DJ (2014) The complexity of protein structure and the challenges it poses in developing biopharmaceuticals. In: Houde DJ, Berkowitz SA (eds) Biophysical characterisation of proteins in developing biopharmaceuticals, 1st edn. Elsevier, Dordrecht, pp 1–22

Bermúdez-Oria A, Rodríguez-Gutiérrez G, Vioque B, Rubio-Senent F, Fernández-Bolaños J (2017) Physical and functional properties of pectin-fish gelatin films containing the olive phenols hydroxytyrosol and 3,4-dihydroxyphenylglycol. Carbohydr Polym 178:368–377

Bertuzzi MA, Slavutsky AM (2016) Standard and new processing techniques used in the preparation of films and coatings at the lab level and scale-up. In: MPM G, Gómez-Guillén MC, López-Caballero ME, Barbosa-Cánovas GV (eds) Edible films and coatings: fundamentals and applications. CRC Press, Boca Raton, FL Chapter 1

Bibi F, Guillaume C, Vena A, Gontard N, Sorli B (2016) Wheat gluten, a bio-polymer layer to monitor relative humidity in food packaging: electric and dielectric characterization. Sens Actuators B 247:355–367

Bibi F, Guillaume C, Gontard N, Sorli B (2017) Wheat gluten, a bio-polymer to monitor carbon dioxide in food packaging: electric and dielectric characterization. Sens Actuators B 250:76–84

Biji KB, Ravishankar CN, Mohan CO, Gopal TKS (2015) Smart packaging systems for food applications: a review. J Food Sci Technol 52(10):6125–6135

Bitencourt CM, Fávaro-Trindade CS, Sobral PJA, Cavalho RA (2014) Gelatin-based films additivated with curcuma ethanol extract: Antioxidant activity and physical properties of films. Food Hydrocoll 40:145–152

Blanco-Pascual N, Fernández-Martín F, Montero MP (2013) Effect of different protein extracts from *Dosidicus gigas* muscle co-products on edible films development. Food Hydrocoll 33(1):118–131

Blanco-Pascual N, Fernández-Martín F, Montero P (2014) Jumbo squid (*Dosidicus gigas*) myofibrillar protein concentrate for edible packaging films and storage stability. LWT Food Sci Technol 55(2):543–550

Bourtoom T (2008) Edible films and coatings: characteristics and properties. Int Food Res J 15(3):237–248

Bouvier J, Campanella OH (2014) Extrusion processing technology: food and non-food biomaterials. Wiley-Blackwell, Hoboken, NJ

Boyaci D, Korel F, Yemenicioğlu A (2016) Development of activate-at-home-type edible antimicrobial films: an example pH-triggering mechanism formed for smoked salmon slices using lysozyme in whey protein films. Food Hydrocoll 60:170–178

Božič M, Majerič M, Denac M, Kokol V (2015) Mechanical and barrier properties of soy protein isolate films plasticized with a mixture of glycerol and dendritic polyglycerol. J Appl Polym Sci 132(17):41837–41849

Catarino MD, Alves-Silva JM, Fernandes RP, Gonçalves MJ, Salgueiro LR, Henriques MF, Cardoso SM (2017) Development and performance of whey protein active coatings with *Origanum virens* essential oils in the quality and shelf life improvement of processed meat products. Food Control 80:273–280

Cecchini JP, Spotti MJ, Piagentini AM, Milt VG, Carraca CR (2017) Development of edible films obtained from submicron emulsions based on whey protein concentrate, oil/beeswax and brea gum. Food Sci Technol Int 23(4):371–381

Cercel F, Stroiu M, Alexe P, Ianiţchi D (2015) Characterization of myofibrillar chicken breast proteins for obtain protein films and biodegradable coatings generation. Agric Agric Sci Proc 6:197–205

Chang C, Nickerson MT (2014) Effect of plasticizer type and genipin on the mechanical, optical, and water vapour barrier properties of canola protein isolate-based edible films. Eur Food Res Technol 238(1):35–46

Chang C, Nickerson MT (2015) Effect of protein and glycerol concentration on the mechanical, optical, and water vapour barrier properties of canola protein isolate-based edible films. Sci Technol Int 21(1):33–44

Chen F, Monnier X, Gällstedt M, Gedde UW, Hedenqvist MS (2014) Wheat gluten/chitosan blends: a new biobased material. Eur Polym J 60:186–197

Chen J, Li L, Yi R, Xu N, Gao R, Hong B (2016) Extraction and characterization of acid-soluble collagen from scales and skin of tilapia (*Oreochromis niloticus*). LWT Food Sci Technol 66:453–459

Cheng SY, Wang BJ, Weng YM (2015) Antioxidant and antimicrobial edible zein/chitosan composite films fabricated by incorporation of phenolic compounds and dicarboxylic acids. LWT Food Sci Technol 63(1):115–121

Chevalier E, Assezat G, Prochazka F, Oulahal N (2018) Development and characterization of a novel edible extruded sheet based on different casein sources and influence of the glycerol concentration. Food Hydrocoll 75:182–191

Chin SS, Lyn FH, Nur Hanani ZA (2017) Effect of Aloe vera (*Aloe barbadensis* Miller) gel on the physical and functional properties of fish gelatin films as active packaging. Food Packaging Shelf Life 12:128–134

Chinma CE, Ariahu CC, Abu JO (2012) Development and characterization of cassava starch and soy protein concentrate based edible films. Int J Food Sci Technol 47(2):383–389

Ciannamea EM, Stefani PM, Ruseckaite RA (2014) Physical and mechanical properties of compression molded and solution casting soybean protein concentrate based films. Food Hydrocoll 38:193–204

Ciannamea EM, Espinosa JP, Stefani PM, Ruseckaite RA (2017) Long-term stability of compression-molded soybean protein concentrate films stored under specific conditions. Food Chem 243:448–452

Corbatón-Báguena MJ, Álvarez-Blanco S, Vincent-Vela MC, Lora-García J (2015) Utilization of NaCl solutions to clean ultrafiltration membranes fouled by whey protein concentrates. Sep Purif Technol 150:95–101

Cordeiro de Azeredo HM (2012) Edible coatings. In: Rodrigues S, Fernandes FAN (eds) Advances in fruit processing technologies. CRC, Boca Raton, FL, pp 345–361

Córdoba LJP, Sobral PJA (2017) Physical and antioxidant properties of films based on gelatin, gelatin-chitosan or gelatin-sodium caseinate blends loaded with nanoemulsified active compounds. J Food Eng 213:47–53

Coşkun BK, Çalikoğlu E, Emiroğlu ZK, Candoğan K (2014) Antioxidant active packaging with soy edible films and oregano or thyme essential oils for oxidative stability of ground beef patties. J Food Qual 37(3):203–212
Dahesh M, Banc A, Duri A, Morel MH, Ramos L (2016) Spontaneous gelation of wheat gluten proteins in a food grade solvent. Food Hydrocoll 52:1–10
Damodaran S (2007) Amino acids, peptides and proteins. In: Damodaran S, Parkin KL, Fennema OR (eds) Fennema's food chemistry, 4th edn. CRC, Boca Raton, FL, pp 219–323
Day L (2011) Wheat gluten: production, properties and applications. In: Phillips GO, Williams PA (eds) Handbook of proteins. Woodhead, Cambridge, pp 267–288
de Graaf LA (2000) Denaturation of protein from a non-food perspective. J Biotechnol 79(3):299–306
de Moraes Crizel T, de Oliveira Rios A, Alves VD, Bandarra N, Moldão-Martins M, Hickmann Flôres S (2018) Active food packaging prepared with chitosan and olive pomace. Food Hydrocoll 74:139–150
de Moraes JO, Scheibe AS, Sereno A, Laurindo JB (2013) Scale-up of the production of cassava starch based films using tape-casting. J Food Eng 119(4):800–808
Díaz-Calderón P, Flores E, González-Muñoz A, Pepczynska M, Quero F, Enrione J (2017) Influence of extraction variables on the structure and physical properties of salmon gelatin. Food Hydrocoll 71:118–128
Duconseille A, Wien F, Audonnet F, Traore A, Refregiers M, Astruc T, Santé-Lhoutellier V (2017) The effect of origin of the gelatin and ageing on the secondary structure and water dissolution. Food Hydrocoll 66:378–388
Duval A, Molina-Boisseau S, Chirat C, Morel MH (2015) Dynamic mechanical analysis of the multiple glass transitions of plasticized wheat gluten biopolymer. J Appl Polym Sci 133(14):43254–43265
Early R (2012) Dairy products and milk-based food ingredients. In: Baines D, Seal R (eds) Natural food additives, ingredients and flavourings. Woodhead, Cambridge, pp 417–445
Echeverría I, López-Caballero ME, Gómez-Guillén MC, Mauri AN, Montero MP (2016) Structure, functionality and active release of nanoclay-soy protein films affected by clove essential oil. Food Bioproc Tech 9(11):1937–1950
Echeverría I, López-Caballero ME, Gómez-Guillén MC, Mauri AN, Montero MP (2018) Active nanocomposite films based on soy proteins-montmorillonite-clove essential oil for the preservation of refrigerated bluefin tuna (*Thunnus thynnus*) fillets. Int J Food Microbiol 266:142–149. https://doi.org/10.1016/j.ijfoodmicro.2017.10.003
Emam-Djomeh Z, Karami-Moghaddam A, Broumand A (2016) Casein-based *Zataria multiflora* boiss films: use in antimicrobial packaging. In: Barros-Velazquez J (ed) Antimicrobial food packaging, 1st edn. Academic, London, pp 515–526
Emin MA, Quevedo M, Wilhelm M, Karbstein HP (2017) Analysis of the reaction behavior of highly concentrated plant proteins in extrusion-like conditions. Innov Food Sci Emerg Technol 44:15–20
Etxabide A, Urdanpilleta M, Guerrero P, de la Caba K (2015a) Effect of cross-linking in nanostructure and physicochemical properties of fish gelatins for bio-applications. React Funct Polym 94:55–62
Etxabide A, Uranga J, Guerrero P, de la Caba K (2015b) Improvement of barrier properties of fish gelatin films promoted by gelatin glycation with lactose at high temperatures. LWT Food Sci Technol 63(1):315–321
Etxabide A, Uranga J, Guerrero P, de la Caba K (2017) Development of active films by means of valorisation of food processing waste: a review. Food Hydrocoll 68:192–198
Falguera V, Quintero JP, Jiménez A, Muñoz JA, Ibarz A (2011) Edible films and coatings: structures, active functions and trends in their use. Trends Food Sci Technol 22(6):292–303
Feng X, Ng VK, Mikš-Krajnik M, Yang H (2017) Effects of fish gelatin and tea polyphenol coating on the spoilage and degradation of myofibril in fish fillet during cold storage. Food Bioproc Tech 10(1):89–102

Friesen K, Chang C, Nickerson M (2015) Incorporation of phenolic compounds, rutin and epicatechin, into soy protein isolate films: mechanical, barrier and cross-linking properties. Food Chem 172:18–23

Fuertes S, Laca A, Oulego P, Paredes B, Rendueles M, Díaz M (2017) Development and characterization of egg yolk and egg yolk fractions edible films. Food Hydrocoll 70:229–239

Galus S, Mathieu H, Lenart A, Debeaufort F (2012) Effect of modified starch or maltodextrin incorporation on the barrier and mechanical properties, moisture sensitivity and appearance of soy protein isolate-based edible films. Innov Food Sci Emerg Technol 16:148–154

Galus S, Lenart A, Voilley A, Debeaufort F (2013) Effect of oxidized potato starch on the physicochemical properties of soy protein isolate-based edible films. Food Technol Biotechnol 51(3):403–409

Ganiari S, Choulitoudi E, Oreopoulou V (2017) Edible and active films and coatings as carriers of natural antioxidants for lipid food. Trends Food Sci Technol 68:70–82

Ganju S, Gogate PR (2017) A review on approaches for efficient recovery of whey proteins from dairy industry effluents. J Food Eng 215:84–96

Garavand F, Rouhi M, Hadi Razavi S, Cacciotti I, Mohammadi R (2017) Improving the integrity of natural biopolymer films used in food packaging by crosslinking approach: a review. Int J Biol Macromol 104:687–707

Garrido T, Etxabide A, Leceta I, Cabezudo S, de la Caba K, Guerrero P (2014) Valorization of soya by-products for sustainable packaging. J Clean Prod 64:228–233

Garrido T, Leceta I, Cabezudo S, Guerreo P, de la Caba K (2016) Tailoring soy protein film properties by selecting casting or compression as processing methods. Eur Polym J 85:499–507

Garrido T, Leceta I, de la Caba K, Guerrero P (2018) Chicken feathers as a natural source of sulphur to develop sustainable protein films with enhanced properties. Int J Biol Macromol 106:523–531

Ghaani M, Cozzolino CA, Castelli G, Farris S (2016) An overview of the intelligent packaging technologies in the food sector. Trends Food Sci Technol 51:1–11

Ghanbarzadeh B, Oromiehi AR (2008) Studies on glass transition temperature of mono and bilayer protein films plasticized by glycerol and olive oil. J Appl Polym Sci 109(5):2848–2854

Ghidelli C, Mateos M, Rojas-Argudo C, Pérez-Gago MB (2014) Extending the shelf life of fresh-cut eggplant with a soy protein-cysteine based edible coating and modified atmosphere packaging. Postharvest Biol Technol 95:81–87

Ghidelli C, Mateos M, Rojas-Argudo C, Pérez-Gago MB (2015) Novel approaches to control browning of fresh-cut artichoke: effect of soy protein-based coating and modifies atmosphere packaging. Postharvest Biol Technol 99:105–113

Giménez B, Gómez-Guillén MC, López-Caballero ME, Gómez-Estaca J, Montero P (2012) Role of sepiolite in the release of active compounds from gelatin–egg white films. Food Hydrocoll 27(2):475–486

Gómez-Estaca J, López de Dicastillo C, Hérnández-Muñoz P, Catalá R, Gavara R (2014a) Advances in antioxidant active food packaging. Trends Food Sci Technol 35(1):42–51

Gómez-Estaca J, Montero P, Gómez-Guillén MC (2014b) Shrimp (*Litopenaeus vannamei*) muscle proteins as source to develop edible films. Food Hydrocoll 41:86–94

Gómez-Guillén MC, Giménez B, Montero P (2005) Extraction of gelatin from fish skins by high pressure treatment. Food Hydrocoll 19(5):923–928

Gómez-Guillén MC, Pérez-Mateos M, Gómez-Estaca J, López-Caballero E, Giménez B, Montero P (2009) Fish gelatin: a renewable material for developing active biodegradable films. Trends Food Sci Technol 20(1):3–16

Gómez-Guillén MC, Gimenez B, Lopez-Caballero ME, Montero MP (2011) Functional and bioactive properties of collagen and gelatin from alternative sources: a review. Food Hydrocoll 25(8):1813–1827

Gómez-Heincke D, Martínez I, Partal P, Guerrero A, Gallegos C (2016) Development of antimicrobial active packaging materials based on gluten proteins. J Sci Food Agric 96(10):3432–3438

González A, Igarzabal CIA (2013) Soy protein-Poly (lactic acid) bilayer films as biodegradable material for active food packaging. Food Hydrocoll 33(2):289–296

González A, Igarzabal CIA (2015) Nanocrystal-reinforced soy protein films and their applications as active packaging. Food Hydrocoll 43:777–784

González A, Strumia MC, Igarzabal CIA (2015) Modification strategies of proteins for food packaging applications. In: Cirillo G, Spizzirri U, Iemma F (eds) Functional polymers in food science: from technology to biology. Wiley, New York, NY, pp 127–145

González-Estrada RR, Chalier P, Ragazzo-Sánchez JA, Konuk D, Calderón-Santoyo M (2017) Antimicrobial soy protein based coatings: application to Persian lime (*Citrus latifolia* Tanaka) for protection and preservation. Postharvest Biol Technol 132:138–144

Gu L, Wang M, Zhou J (2013) Effects of protein interactions on properties and microstructure of zein-gliadin composite films. J Food Eng 119(2):288–298

Guerrero P, Retegi A, Gabilondo N, de la Caba K (2010) Mechanical and thermal properties of soy protein films processed by casting and compression. J Food Eng 100(1):145–151

Guerrero P, Beatty E, Kerry JP, de la Caba K (2012) Extrusion of soy protein with gelatin and sugars at low moisture content. J Food Eng 110(1):53–59

Guerrero P, O'Sullivan MG, Kerry JP, de la Caba K (2015) Application of soy protein coatings and their effect on the quality and shelf life stability of beef patties. RSC Adv 5(11):8182–8189

Gupta P, Nayak KK (2015) Characteristics of protein-based biopolymers and its applications. Polym Eng Sci 55(3):485–498

Gutiérrez TJ (2017) Surface and nutraceutical properties of edible films made from starchy sources with and without added blackberry pulp. Carbohydr Polym 165:169–179. https://doi.org/10.1016/j.carbpol.2017.02.016

Gutiérrez TJ (2018) Active and intelligent films made from starchy sources/blackberry pulp. J Polym Environ 26:2374–2391. https://doi.org/10.1007/s10924-017-1134-y

Gutiérrez TJ, Alvarez VA (2018) Bionanocomposite films developed from corn starch and natural and modified nano-clays with or without added blueberry extract. Food Hydrocoll 77:407–420. https://doi.org/10.1016/j.foodhyd.2017.10.017

Gutiérrez TJ, Guzmán R, Jaramillo CM, Famá L (2016a) Effect of beet flour on films made from biological macromolecules: native and modified plantain flour. Int J Biol Macromol 82:395–403. https://doi.org/10.1016/j.ijbiomac.2015.10.020

Gutiérrez TJ, Suniaga J, Monsalve A, García NL (2016b) Influence of beet flour on the relationship surface-properties of edible and intelligent films made from native and modified plantain flour. Food Hydrocoll 54:234–244. https://doi.org/10.1016/j.foodhyd.2015.10.012

Gutiérrez TJ, Herniou-Julien C, Álvarez K, Alvarez VA (2018) Structural properties and in vitro digestibility of edible and pH-sensitive films made from guinea arrowroot starch and wastes from wine manufacture. Carbohydr Polym 184:135–143. https://doi.org/10.1016/j.carbpol.2017.12.039

Han J, Shin SH, Park KM, Kim KM (2015) Characterization of physical, mechanical and antioxidant properties of soy protein-based bioplastics films containing carboxymethylcellulose and catechin. Food Sci Biotechnol 24(3):939–945

Han Y, Yu M, Wang L (2018) Preparation and characterization of antioxidant soy protein isolate films incorporating licorice residue extract. Food Hydrocoll 75:13–21

Hashim P, Mohd Ridzwan MS, Bakar J, Mat Hashim D (2015) Collagen in food and beverage industries. Int Food Res J 22(1):1–8

Hernandez-Izquierdo VM, Krochta JM (2008) Thermoplastic processing of proteins for film formation—a review. J Food Sci 73(2):30–39

Holkar CR, Jain SS, Jadhav AJ, Pinjari DV (2018) Valorization of keratin based waste. Process Saf Environ Prot 115:85–98. https://doi.org/10.1016/j.psep.2017.08.045

Hopkins EJ, Chang C, Lam RSH, Nickerson MT (2015) Effect of flaxseed oil concentration on the performance of a soy protein isolated-based emulsion type film. Food Res Int 67:418–425

Hosseini SF, Rezaei M, Zandi M, Farahmandghavi F (2016) Development of bioactive fish gelatin/chitosan nanoparticles composite films with antimicrobial properties. Food Chem 194:1266–1274

Hu Y, Liu L, Gu Z, Dan W, Dan N, Yu X (2014) Modification of collagen with a natural derived cross-linker, alginate dialdehyde. Carbohydr Polym 102:324–332
Iahnke AOS, Costa TMH, Rios AO, Flôres SH (2015) Residues of minimally processed carrot and gelatin capsules: potential materials for packaging films. Ind Crop Prod 76:1071–1078
Insaward A, Duangmal K, Mahawanich T (2015) Mechanical, optical and barrier properties of soy protein film as affected by phenolic acid addition. J Agric Food Chem 63(43):9421–9426
Irkin R, Esmer OK (2015) Novel food packaging systems with natural antimicrobial agents. J Food Sci Technol 52(10):6095–6111
Jansens KJA, Vo Hong N, Telen L, Brijs K, Lagrain B, Van Vuure AW, Van Acker K, Verpoest I, Van Puyvelde P, Goderis B, Smet M, Delcour JA (2013) Effect of molding conditions and moisture content on the mechanical properties of compression molded glassy, wheat gluten bioplastics. Ind Crop Prod 44:480–487
Jiang P, Liu HF, Zhao XH, Ding Q (2017) Physicochemical properties of soybean protein isolate affected by the cross-linking with horseradish peroxidase, glucose oxidase and glucose. J Food Meas Charact 11(3):1196–1202
Jridi M, Mora L, Souissi N, Aristoy M, Nasri M, Toldrá F (2018) Effects of active gelatin coated with henna (*L. inermis*) extract on beef meat quality during chilled storage. Food Control 84:238–245
Junqueira-Gonçalves MP, Salinas GE, Bruna JE, Niranjan K (2017) An assessment of lactobiopolymer-montmorillonite composites for dip coating applications on fresh strawberries. J Sci Food Agric 97(6):1846–1853
Kadam SU, Pankaj SK, Tiwari BK, Cullen PJ, O'Donnell CP (2015) Development of biopolymer-based gelatin and casein films incorporating brown seaweed *Ascophyllum nodosum* extract. Food Packaging Shelf Life 6:68–74
Kaewprachu P, Rungraeng N, Osako K, Rawdkuen S (2017) Properties of fish myofibrillar protein film incorporated with catechin-Kradon extract. Food Packaging Shelf Life 13:56–65
Kaewprachu P, Osako K, Rungraeng N, Rawdkuen S (2018) Characterization of fish myofibrillar protein film incorporated with catechin-Kradon extract. Int J Biol Macromol 107(Pt B):1463–1473. https://doi.org/10.1016/j.ijbiomac.2017.10.011
Kakaei S, Shahbazi Y (2016) Effect of chitosan-gelatin film incorporated with ethanolic red grape seed extract and *Ziziphora clinopodioides* essential oil on survival of *Listeria monocytogenes* and chemical, microbial and sensory properties of minced trout fillet. LWT Food Sci Technol 72:432–438
Kalman DS (2014) Amino acid composition of an organic brown rice protein concentrate and isolate compared to soy and whey concentrates and isolates. Foods 3(3):394–402
Kang HJ, Kim SJ, You YS, Lacroix M, Han J (2013) Inhibitory effect of soy protein coating formulations on walnut (*Junglans regia* L.) kernels against lipid oxidation. LWT Food Sci Technol 51(1):393–396
Kashiri M, Cerisuelo JP, Domínguez I, López-Carballo G, Muriel-Gallet V, Gavara R, Hernández-Muñoz P (2017a) Zein films and coatings as carriers and release systems of *Zataria multiflora* Boiss. essential oil for antimicrobial food packaging. Food Hydrocoll 70:260–268
Kashiri M, Maghsoudlo Y, Khomeiri M (2017b) Incorporating *Zataria multiflora* Boiss. essential oil and sodium bentonite nano-clay open a new perspective to use zein films as bioactive packaging materials. Food Sci Technol Int 23(7):582–596
Kessel A, Ben-Tal N (2010) Introduction to proteins: structure, function and motion. CRC, Boca Raton, FL
Khalil AA, Deraz SF, Elrahman SA, El-Fawal G (2015) Enhancement of mechanical properties, microstructure, and antimicrobial activities of zein films cross-linked using succinic anhydride, eugenol, and citric acid. Prep Biochem Biotechnol 45(6):551–567
Koch L, Emin MA, Schuchmann HP (2017) Influence of processing conditions on the formation of whey protein-citrus pectin conjugates in extrusion. J Food Eng 193:1–9
Kowalczyk D (2016) Biopolymer/candelilla wax emulsion films as carriers of ascorbic acid–a comparative study. Food Hydrocoll 52:543–553

Kowalczyk D, Biendl M (2016) Physicochemical and antioxidant properties of biopolymer/candelilla wax emulsion films containing hop extract—a comparative study. Food Hydrocoll 60:384–392

Kowalczyk D, Gustaw W, Zieba E, Lisiecki S, Stadnik J, Baraniak B (2016) Microstructure and functional properties of sorbitol-plasticized pea protein isolate emulsion films: Effect of lipid type and concentration. Food Hydrocoll 60:353–363

Kumari M, Mahajan H, Joshi R, Gupta M (2017) Development and structural characterization of edible films for improving fruit quality. Food Packaging Shelf Life 12:42–50

Lacroix M, Vu KD (2014) Edible coating and film materials: proteins. In: Han JH (ed) Innovations in food packaging, 2nd edn. Academic, London, pp 277–304

Lassoued I, Jridi M, Nasri R, Dammak A, Hajji M, Nasri M, Barkia A (2014) Characteristics and functional properties of gelatin from thornback ray skin obtained by pepsin-aided process in comparison with commercial halal bovine gelatin. Food Hydrocoll 41:309–318

Lee KY, Lee JH, Yang HJ, Song KB (2016) Production and characterisation of skate skin gelatin films incorporated with thyme essential oil and their application in chicken tenderloin packaging. Int J Food Sci Technol 51(6):1465–1472

Li D, Chen E, Chen H, Zhou H, Li B, Li Y (2018) Impact of whey protein isolates and concentrates on the formation of protein nanoparticles-stabilised pickering emulsions. Int J Food Sci Technol 53(3). https://doi.org/10.1111/ijfs.13578

Liang C, Jia M, Tian D, Tang Y, Ju W, Ding S, Tian L, Ren X, Wang X (2017) Edible sturgeon skin gelatine films: tensile strength and UV light-barrier as enhanced by blending with esculine. J Funct Foods 37:19–228

Liao W, Liu Y, Xia G, Shen XR, Duan Z, Yang J (2017) Characterization of three-spot seahorse (*Hippocampus trimaculatus*) skin collagen and its fibrillar gel reinforced by EDC: a comparative study. J Food Biochem 41(5):12391–12400

Lin D, Zhao Y (2007) Innovations in the development and application of edible coatings for fresh and minimally processed fruits and vegetables. Compr Rev Food Sci Food Saf 6(3):60–75

Liu H, Li J, Zhu D, Wang Y, Zhao Y, Li J (2014) Preparation of soy protein isolate (SPI)-pectin complex film containing cinnamon oil and its effects on microbial growth of dehydrated soybean curd (dry tofu). J Food Process Preserv 38(3):1371–1376

Liu F, Majeed H, Antoniou J, Li Y, Ma Y, Yokoyama W, Ma J, Zhong F (2016) Tailoring physical properties of transglutaminase-modified gelatin films by varying drying temperature. Food Hydrocoll 58:20–28

Liu F, Chiou B, Avena-Bustillos RJ, Zhang Y, Li Y, McHugh TH, Zhong F (2017a) Study of combined effects of glycerol and transglutaminase on properties of gelatin films. Food Hydrocoll 65:1–9

Liu Y, Pan H, Liao T, Zhu B, Li Z, Wu J, Li Y, Cai J, Wang M (2017b) Combination of propylene glycol alginate and lauric acid on water retention and mechanical properties of soy protein isolate-based films. J Braz Chem Soc 28(12):2365–2374

Liu Y, Zhang H, Xu L, Chi Y, Wu Y, Cao W, Li T (2017c) Properties of soy protein isolate antimicrobial films and its application in preservation of meat. Emirates J Food Agric 29(8):589–600

López D, Márquez A, Gutiérrez-Cutiño M, Venegas-Yazigi D, Bustos R, Matiacevich S (2017) Edible film with antioxidant capacity based on salmon gelatin and boldine. LWT Food Sci Technol 77:160–169

Luchese CL, Garrido T, Spada JC, Tessaro IC, de la Caba K (2018) Development and characterization of cassava starch films incorporated with blueberry pomace. Int J Biol Macromol 106:834–839

Ma Q, Wang L (2016) Preparation of a visual pH-sensing film based on tara gum incorporating cellulose and extracts from grape skins. Sens Actuators B 235:401–407

Ma L, Zhang M, Bhandari B, Gao Z (2017) Recent developments in novel shelf life extension technologies of fresh-cut fruits and vegetables. Trends Food Sci Technol 64:23–38

Marquez GR, Di Pierro P, Mariniello L, Esposito M, Giosafatto CV, Porta R (2017) Fresh-cut fruit and vegetable coatings by transglutaminase-crosslinked whey protein/pectin edible films. LWT Food Sci Technol 75:124–130

Martínez I, Partal P, García-Morales M, Guerrero A, Gallegos C (2013) Development of protein-based bioplastics with antimicrobial activity by thermo-mechanical processing. J Food Eng 117(2):247–254
Martucci JF, Gende LB, Neira LM, Ruseckaite RA (2015) Oregano and lavender essential oils as antioxidant and antimicrobial additives of biogenic gelatin films. Ind Crop Prod 71:205–213
Massani MB, Botana A, Eisenberg P, Vignolo G (2014) Development of an active wheat gluten film with *Lactobacillus curvatus* CRL705 bacteriocins and a study of its antimicrobial performance during ageing. Food Addit Contam Part A 31(1):164–171
Massouras T, Triantaphyllopoulos KA, Theodossiou I (2017) Chemical composition, protein fraction and fatty acid profile of donkey milk during lactation. Int Dairy J 75:83–90
Mehyar GF, El Assi NM, Alsmairat NG, Holley RA (2014) Effect of edible coatings on fruit maturity and fungal growth on Berhi dates. Int J Food Sci Technol 49(11):2409–2417
Meinlschmidt P, Sussmann D, Schweiggert-Weisz U, Eisner P (2015) Enzymatic treatment of soy protein isolates: effects on the potential allergenicity, technofuncionality and sensory properties. Food Sci Nutr 4(1):11–23
Mojumdar SC, Moresoli C, Simon LC, Legge RL (2011) Edible wheat gluten protein films: preparation, thermal, mechanical and spectral properties. J Therm Anal Calorim 104(3):929–936
Molinaro S, Cruz-Romero M, Sensidoni A, Morris M, Lagazio C, Kerry JP (2015) Combination of high-pressure treatment, mild heating and holding time effects as a means of improving the barrier properties of gelatin-based packaging films using response surface modeling. Innov Food Sci Emerg Technol 30:15–23
Moradi M, Tajik H, Rohani SMR, Mahmoudian A (2016) Antioxidant and antimicrobial effects of zein edible film impregnated with *Zataria multiflora* Boiss. essential oil and monolaurin. LWT Food Sci Technol 72:37–43
Nassar MA, Sultan M, Mohamed S (2012) Preparation and characterization of PVA/starch blend composite film based on modified chicken feather protein and effect of plasticizer type on its properties. BioTechnol Indian J 6(12):375–381
Nie X, Gong Y, Wang N, Meng X (2015) Preparation and characterization of edible myofibrillar protein-based film incorporated with grape seed procyanidins and green tea polyphenol. LWT Food Sci Technol 64(2):1042–1046
Noori S, Zeynali F, Almasi H (2018) Antimicrobial and antioxidant efficiency of nanoemulsion-based edible coating containing ginger (*Zingiber officinale*) essential oil and its effect on safety and quality attributes of chicken breast fillets. Food Control 84:312–320
Nur Hanani ZA, McNamara J, Roos YH, Kerry JP (2013) Effect of plasticizer content on the functional properties of extruded gelatin-based composite films. Food Hydrocoll 31(2):264–269
Olejar KJ, Ray S, Kilmartin PA (2017) Utilisation of agro-waste extract in thermoplastics. Int J Nanotechnol 14(1-6):304–312
Olivas GI, Mattinson DS, Barbosa-Cánovas GV (2007) Alginate coatings for preservation of minimally processed 'Gala'apples. Postharvest Biol Technol 45(1):89–96
Ortiz CM, de Moraes JO, Vicente AR, Laurindo JB, Mauri AN (2017) Scale-up of the production of soy (*Glycine max* L.) protein films using tape casting: formulation of film-forming suspension and drying conditions. Food Hydrocoll 66:110–117
Otoni CG, Avena-Bustillos RJ, Olsen CW, Bilbao-Sainz C, McHugh TH (2016) Mechanical and water barrier properties of isolated soy protein composite edible films as affected by carvacrol and cinnamaldehyde micro and nanoemulsions. Food Hydrocoll 57:72–79
Otoni CG, Avena-Bustillos J, Azeredo HMC, Lorevide MV, Moura MR, Mattoso LHC, McHugh TH (2017) Recent advances on edible films based on fruits and vegetables—a review. Compr Rev Food Sci Food Saf 16(5):1151–1169
Ozcalik O, Tihminlioglu F (2013) Barrier properties of corn zein nanocomposite coated polypropylene films for food packaging applications. J Food Eng 114(4):505–513
Pan H, Jiang B, Chen J, Jin Z (2014) Blend-modification of soy protein/lauric acid edible films using polysaccharides. Food Chem 151:1–6

Park HJ, Byun YJ, Kim YT, Whiteside WS, Bae HJ (2014) Processes and applications for edible coating and film materials from agropolymers. In: Han JH (ed) Innovation in food packaging. Academic, London, pp 257–275

Patel S, Srivastava S, Singh MR, Singh D (2018) Preparation and optimization of chitosan-gelatin films for sustained delivery of lupeol for wound healing. Int J Biol Macromol 107(Pt B):1888–1897. https://doi.org/10.1016/j.ijbiomac.2017.10.056

Pati F, Adhikari B, Dhara S (2010) Isolation and characterization of fish scale collagen of higher thermal stability. Bioresour Technol 101(10):3737–3742

Pena-Serna C, Penna ALB, Lopes Filho JF (2016) Zein-based blend coating: Impact on the quality of a model cheese of short ripening period. J Food Eng 171:208–213

Pinheiro JC, Alegria CS, Abreu MM, Gonçalves EM, Silva CL (2016) Evaluation of alternative preservation treatments (water heat treatment, ultrasounds, thermosonication and uv-c radiation) to improve safety and quality of whole tomato. Food Bioproc Tech 9(6):924–935

Pommet M, Redl A, Guilbert S, Morel MH (2005) Intrinsic influence of various plasticizers on functional properties and reactivity of wheat gluten thermoplastic materials. J Cereal Sci 42(1):81–91

Ponce A, Roura SI, Moreira MR (2016) Casein and chitosan polymers: use in antimicrobial packaging. In: Barros-Velazquez J (ed) Antimicrobial food packaging, 1st edn. Academic, London, pp 455–466

Preece KE, Hooshyar N, Zuidam NJ (2017) Whole soybean protein extraction processes: a review. Innov Food Sci Emerg Technol 43:163–172

Prietto L, Mirapalhete TC, Pinto VZ, Hoffmann JF, Vanier NL, Lim LT, Dias ARG, da Rosa Zavareze E (2017) pH-sensitive films containing anthocyanins extracted from black bean seed coat and red cabbage. LWT Food Sci Technol 80:492–500

Qazanfarzadeh Z, Kadivar M (2016) Properties of whey protein isolate nanocomposite films reinforced with nanocellulose isolated from oat husk. Int J Biol Macromol 91:1134–1140

Rebah FB, Miled N (2013) Fish processing wastes for microbial enzyme production: a review. 3 Biotech 3(4):255–265

Rezaei M, Motamedzadegan A (2015) The effect of plasticizers on mechanical properties and water vapor permeability of gelatin-based edible films containing clay nanoparticles. World J Nano Sci Eng 5(04):178–193

Ribeiro-Santos R, Sanches-Silva A, Motta JFG, Andrade M, de Araújo Neves I, Teófilo RF, de Carvalho MG, de Melo NR (2017) Combined use of essential oils applied to protein base active food packaging: study in vitro and in a food simulant. Eur Polym J 93:75–86

Riveros CG, Mestrallet MG, Quiroga PR, Nepote V, Grosso NR (2013) Preserving sensory attributes of roasted peanuts using edible coatings. Int J Food Sci Technol 48(4):850–859

Rodrigues IM, Coelho JFJ, Carvalho MGVS (2012) Isolation and valorisation of vegetable proteins from oilseed plants: methods, limitations and potential. J Food Eng 109(3):337–346

Rombouts I, Lamberts L, Celus I, Lagrain B, Brijis K, Delcour JA (2009) Wheat gluten amino acid composition analysis by high-performance anion-exchange chromatography with integrated pulsed amperometric detection. J Chromatogr A 1216(29):5557–5562

Roy BC, Das C, Hong H, Betti M, Bruce HL (2017) Extraction and characterization of gelatin from bovine heart. Food Biosci 20:116–124

Saberi B, Vuong QV, Chockchaisawasdee S, Golding JB, Scarlett CJ, Stathopoulos CE (2017) Physical, barrier and antioxidant properties of pea starch-guar gum biocomposite edible films by incorporation of natural plant extracts. Food Bioproc Tech 10(12):2240–2250

Salgado PR, López-Caballero ME, Gómez-Guillén MC, Mauri AN, Montero MP (2013) Sunflower protein films incorporated with clove essential oil have potential application for the preservation of fish patties. Food Hydrocoll 33(1):74–84

Samsalee N, Sothornvit R (2017) Modification and characterization of porcine plasma protein with natural agents as potential cross-linkers. Int J Food Sci Technol 52(4):964–971

Santos TM, Souza Filho MDSM, Muniz CR, Morais JPS, Kotzebue LRV, Pereira ALS, Azeredo HMC (2017) Zein films with unoxidized or oxidized tannic acid. J Sci Food Agric 97(13):4580–4587

Sarode AR, Sawale PD, Khedkar CD, Kalyankar SD, Pawshe RD (2016) Casein and caseinate: methods of manufacture. In: Caballero B, Finglas P, Toldra F (eds) The encyclopedia of food and health. Academic, London, pp 676–682
Schmid M, Krimmel B, Grupa U, Noller K (2014) Effects of thermally induced denaturation on technological-functional properties of whey protein isolate-based films. J Dairy Sci 97(9):5315–5327
See SF, Ghassem M, Mamot S, Babji AS (2015) Effect of different pretreatments on functional properties of African catfish (*Clarias gariepinus*) skin gelatin. J Food Sci Technol 52(2):753–762
Sharma L, Singh C (2016) Sesame protein based edible films: development and characterization. Food Hydrocoll 61:139–147
Sharma K, Jain R, Bhargava N, Sharma R, Sharma KS (2010) Dielectric studies of wheat in powder from at microwave frequencies. Indian J Exp Biol 47:1002–1007
Sharma N, Khatkar BS, Kaushik R, Sharma P, Sharma R (2017) Isolation and development of wheat based gluten edible film and its physicochemical properties. Int Food Res J 24(1):94–101
Shokri S, Ehsani A (2017) Efficacy of whey protein coating incorporated with lactoperoxidase and α-tocopherol in shelf life extension of Pike-Perch fillets during refrigeration. LWT Food Sci Technol 85:225–231
Shukla R, Cheryan M (2001) Zein: the industrial protein from corn. Ind Crop Prod 13(3):171–192
Socaci SA, Rugină DO, Diaconeasa ZM, Pop OL, Fărcaş AC, Păucean A, Tofană M, Pintea A (2017) Antioxidant compounds recovered from food wastes. In: Chavarri M (ed) Functional food-improve health through adequate food. InTech, Rijeka, Croatia. https://doi.org/10.5772/intechopen.69124
Song CL, Zhao XH (2014) Structure and property modification of an oligochitosan-glycosylated and crosslinked soybean protein generated by microbial transglutaminase. Food Chem 163:114–119
Song Y, Zheng Q (2014) Ecomaterials based on food proteins and polysaccharides. Polym Rev 54:514–571
Song NB, Lee JH, Al Mijan M, Song KB (2014) Development of a chicken feather protein film containing clove oil and its application in smoked salmon packaging. LWT Food Sci Technol 57(2):453–460
Sousa-Gallagher MJ, Tank A, Sousa R (2016) Emerging technologies to extend the shelf life and stability. In: Subramaniam P, Wareing P (eds) The stability and shelf life of food. Woodhead, Cambridge, pp 399–424
Sui C, Zhang W, Ye F, Liu X, Yu G (2016) Preparation, physical and mechanical properties of soy protein isolate/guar gum composite films prepared by solution casting. J Appl Polym Sci 133(18):43375–43381
Tan Y, Chang SKC (2018) Isolation and characterization of collagen extracted from channel catfish (*Ictalurus punctatus*) skin. Food Chem 242:147–155
Tanada-Palmu P, Helén H, Hyvönen L (2000) Preparation, properties and applications of wheat gluten edible films. Agric Food Sci 9:23–35
Tatara RA (2017) Compression molding. In: Kutz M (ed) Applied plastics engineering handbook, processing, materials, and applications, 2nd edn. William Andrew, Delmar, NY, pp 291–320
Thammahiwes S, Riyajan SA, Kaewtatip K (2017) Preparation and properties of wheat gluten based bioplastics with fish scale. J Cereal Sci 75:186–191
Tongnuanchan P, Benjakul S, Prodpran T (2012) Properties and antioxidant activity of fish skin gelatin film incorporated with citrus essential oils. Food Chem 134(3):1571–1579
Tongnuanchan P, Benjakul S, Prodpran T, Pisuchpen S, Kazufumi O (2016) Mechanical, thermal and heat sealing properties of fish skin gelatin film containing palm oil and basil essential oil with different surfactants. Food Hydrocoll 56:93–107
Tsakalidou E, Papadimitriou K (eds) (2016) Non-bovine milk and milk products. Academic, London
Türe H, Gällstedt M, Hedenqvist MS (2012) Antimicrobial compression-moulded wheat gluten films containing potassium sorbate. Food Res Int 45(1):109–115

Ünalan İU, Arcan I, Korel F, Yemenicioğlu A (2013) Application of active zein-based films with controlled release properties to control *Listeria monocytogenes* growth and lipid oxidation in fresh Kashar cheese. Innov Food Sci Emerg Technol 20:208–214

Uranga J, Leceta I, Etxabide A, Guerrero P, de la Caba K (2016) Cross-linking of fish gelatins to develop sustainable films. Eur Polym J 78:82–90

USDA (2017) United States Department of Agriculture. World agricultural production. Available at https://www.fas.usda.gov/data/world-agricultural-production. Accessed 7 Dec 2017

Ustunol Z (2009) Edible films and coatings for meat and poultry. In: Huber KC, Embuscado ME (eds) Edible films and coatings for food applications. Springer, New York, NY, pp 245–268

Verbeek CJR, van den Berg LE (2010) Extrusion processing and properties of protein-based thermoplastics. Macromol Mater Eng 295(1):10–21

Visakh PM, Nazarenko O (eds) (2017) Soy protein-based blends, composites and nanocomposites. Wiley, New York, NY

Wang H, Wang L (2017) Developing a bio-based packaging film from soya by-products incorporated with valonea tannin. J Clean Prod 143:624–633

Wang S, Marcone M, Barbut S, Lim LT (2012) The impact of anthocyanin-rich red raspberry extract (ARRE) on the properties of edible soy protein isolate (SPI) film. J Food Sci 77(4):497–505

Wang Z, Zhou J, Wang X, Zhang N, Sun X, Ma Z (2014) The effects of ultrasonic/microwave assisted treatment on the water vapor barrier properties of soybean protein isolate-based oleic acid/stearic acid blend edible films. Food Hydrocoll 35:51–58

Wang W, Zhang Y, Ye R, Ni Y (2015) Physical crosslinkings of edible collagen casing. Int J Biol Macromol 81:920–925

Wang S, Liu Y, Liu A, Zhao Y, Chen X (2016a) Effect of *in situ* apatite on performance of collagen fiber film for food packaging applications. J Appl Polym Sci 133(44):44154–44163

Wang A, Duncan SE, Knowlton KF, Ray WK, Dietrich AM (2016b) Milk protein composition and stability changes affected by iron in water sources. J Dairy Sci 99(6):4206–4219

Wang H, Hu D, Ma Q, Wang L (2016c) Physical and antioxidant properties of flexible soy protein isolate films by incorporating chestnut (*Castanea mollissima*) bur extracts. LWT Food Sci Technol 71:33–39

Wang J, Qian W, He Y, Xiong Y, Song P, Wang R-M (2017a) Reutilization of discarded biomass for preparing functional polymer materials. Waste Manag 65:11–21

Wang Z, Hu S, Gao Y, Ye C, Wang H (2017b) Effect of collagen-lysozyme coating on fresh-salmon fillets preservation. LWT Food Sci Technol 75:59–64

Wang S, Sun X, Zhou D (2017c) Physicochemical characteristics and fibril-forming properties of collagen from paddlefish (*Polyodon spathula*) and globefish (*Fugu flavidus*) skin byproducts. Food Sci Technol Campinas 37(2):176–183

Wang K, Wang W, Ye R, Liu A, Xiao J, Liu Y, Zhao Y (2017d) Mechanical properties and solubility in water of corn starch-collagen composite films: effect of starch type and concentrations. Food Chem 216:209–216

Wang XJ, Zheng XQ, Liu XL, Kopparapu NK, Cong WS, Deng YP (2017e) Preparation of glycosylated zein and retarding effect on lipid oxidation of ground pork. Food Chem 227:335–341

Wihodo M, Moraru CI (2013) Physical and chemical methods used to enhance the structure and mechanical properties of protein films: a review. J Food Eng 114:292–302

Wu T, Dai S, Cong X, Liu R, Zhang M (2017) Succinylated soy protein film coating extended the shelf life of apple fruit. J Food Process Preserv 41(4):13024–13034

Xia C, Wang L, Dong Y, Zhang S, Shi SQ, Cai L, Li J (2015) Soy protein isolate-based films cross-linked by epoxidized soybean oil. RSC Adv 5(101):82765–82771

Xie J, Tang Y, Yang SP, Qian YF (2017) Effects of whey protein films on the quality of thawed bigeye tuna (*Thunnus obesus*) chunks under modified atmosphere packaging and vacuum packaging conditions. Food Sci Biotechnol 26(4):937–945

Xu H, Chai Y, Zhang G (2012) Synergistic effect of oleic acid and glycerol on zein film plasticization. J Agric Food Chem 60(40):10075–10081

Xu F, Dong Y, Zhang W, Zhang S, Li L, Li J (2015) Preparation of cross-linked soy protein isolate-based environmentally-friendly films enhanced by PTGE and PAM. Ind Crop Prod 67:373–380

Yan W, Zhou J, Sun M, Chen J, Hu J, Shen B (2014) The construction of an amino acid network for understanding protein structure and function. Amino Acids 46(6):1419–1439
Yanwong S, Threepopnatkul P (2015) Effect of peppermint and citronella essential oils on properties of fish skin gelatin edible films. IOP Conf Ser Mater Sci Eng 87(1):012064
Yemenicioğlu A (2016) Zein and its composites and blends with natural active compounds: development of antimicrobial films for food packaging. In: Barros-Velazquez J (ed) Antimicrobial food packaging. Elsevier, Dordrecht, pp 503–513
Yemiş GP, Candoğan K (2017) Antibacterial activity of soy edible coatings incorporated with thyme and oregano essential oils on beef against pathogenic bacteria. Food Sci Biotechnol 26(4):1113–1121
Zárate-Ramírez LS, Romero A, Bengoechea C, Partal P, Guerrero A (2014) Thermo-mechanical and hydrophilic properties of polysaccharide/gluten-based bioplastics. Carbohydr Polym 112:24–31
Zeng X, Duan Y, Zhe W, Jiang J, He L, Wang S, Wang M (2013) Edible coating based on soy protein to improve shelf life and overall quality of minimally processed jujubes. J Food Agric Environ 11(3-4):263–269
Zhang YB, Wang JW, Jiang J, Li YX (2013) Antimicrobial activity of soy protein isolate-essential pol monomers edible composite films for chilled pork preservation. Adv Mater Res 726:638–641
Zhang Y, Cui L, Che X, Zhang H, Shi N, Li C, Chen Y, Kong W (2015) Zein-based films and their usage for controlled delivery: origin, classes and current landscape. J Control Release 206:206–219
Zhang DD, Liu JL, Jiang TM, Li L, Fang GZ, Liu YP, Chen LJ (2017) Influence of Kluyveromyces marxianus on proteins, peptides, and amino acids in Lactobacillus-fermented milk. Food Sci Biotechnol 26(3):739–748
Zhang D, Wu X, Chen J, Lin K (2018) The development of collagen based composite scaffolds for bone regeneration. Bioact Mater 3:129–138. https://doi.org/10.1016/j.bioactmat.2017.08.004
Zhong Y, Cavender G, Zhao Y (2014) Investigation of different coating application methods on the performance of edible coatings on Mozzarella cheese. LWT Food Sci Technol 56(1):1–8
Zink J, Wyrobnik T, Prinz T, Schmid M (2016) Physical, chemical and biochemical modifications of protein-based films and coating: an extensive review. Int J Mol Sci 17(9):1376–1421
Zubeldía F, Ansorena MR, Marcovich NE (2015) Wheat gluten films obtained by compression molding. Polym Test 43:68–77
Zuber M, Zia F, Zia KM, Tabasum S, Salman M, Sultan N (2015) Collagen based polyurethanes—a review of recent advances and perspective. Int J Biol Macromol 80:366–374

Chapter 4
Properties of Micro- and Nano-Reinforced Biopolymers for Food Applications

Sofía Collazo-Bigliardi, Rodrigo Ortega-Toro, and Amparo Chiralt

Abstract Food packaging implies a significant consumption of different materials, of which plastics are the second most widely used. So, the development of biopolymers for food packaging applications is critically important. Although several biopolymers are available for different applications, they have some drawbacks and their functional properties need to be adapted for food packaging requirements. The incorporation of micro- and nano-fillers into the biopolymer matrix has proven to be an alternative means of improving their mechanical and barrier properties. In composites, the polymer forms the continuous matrix while the dispersed filler phase helps to positively modify the functional characteristics of the material. Different kinds of fillers have been used which modify the material characteristics as a function of their content and filler-matrix interactions. The particle size and shape, the amount and distribution and the chemical nature of the fillers are key factors in the final properties of the composite. In general, thermomechanical processes with high shearing forces and temperatures for the required time are needed to guarantee the convenient dispersion of the filler within the polymer matrix. In this chapter, the different kinds of fillers used in biopolymer composites have been summarized. The relevant surface properties and the changes induced by fillers on the mechanical, barrier and thermal properties of micro- and nano-composites have been discussed, with emphasis on food packaging applications. The processing techniques, formulation and final structure of materials have also been reviewed, as well as the influence of the fillers on the biodegradation behaviour of composites.

Keywords Biodegradability · Functional properties · Micro- and nanocomposites · Thermomechanical process

---

S. Collazo-Bigliardi · A. Chiralt
Instituto de Ingeniería de Alimentos para el Desarrollo, Universitat Politècnica de València, Valencia, Spain
e-mail: socol@alumni.upv.es; dchiralt@tal.upv.es

R. Ortega-Toro (✉)
Programa de Ingeniería de Alimentos, Facultad de Ingeniería, Universidad de Cartagena, Cartagena de Indias D.T y C, Colombia
e-mail: rortegap1@unicartagena.edu.co

T. J. Gutiérrez (ed.), *Polymers for Food Applications*,
https://doi.org/10.1007/978-3-319-94625-2_4

Abbreviations

AFM	Atomic Force Microscopy
Ag-NPs	Ag nanoparticles
ATBC	Acetyltributyl citrate
BCNW	Bacterial cellulose nanowhiskers
ChNC	Chitin nanocrystals
CMC	Carboxymethyl cellulose
CNC	Cellulose nanocrystals
CNF	Cellulose nanofibrils
DSC	Differential Scanning Calorimetry
FESEM	Field emission scanning electron microscopy
FTIR	Fourier-transform infrared spectroscopy
GTA	Glycerol triacetate
HPMC	(Hidroxypropil)metil cellulose
MC	Methylcellulose
MCC	Microcrystalline cellulose
Mnt	Montmorillonite
NCC	Nano-crystalline cellulose
PBS	Poly(butylene succinate)
PBTA	Poly(butylene adipate co-terephthalate)
PCL	Polycaprolactone
PEG	Polyethylen glycol
PHA	Polyhydroxyalcanoates
PHB	Polyhydroxybutyrate
PHBV	Polyhydroxyl-3-butyrate-co23-valerate
PHBV12	Polyhydroxybutyrate with 12 mol% of valerate and containing 10 wt% of the plasticizer citric ester
PLA	Poly(lactic) acid
PLLA	Poly(L-lactide)
PVA	Poly(vinyl alcohol)
SEM	Scanning Electron Microscopy
TPCS	Thermoplastic corn starch
TPS	Thermoplastic starch
WSNC	Waxy starch nanocrystals

4.1 Introduction

Of all the materials available for food packaging, plastics have increased exponentially over the past two decades, with an annual growth of approximately 5%. It is estimated that worldwide annual plastic production exceeds 300 million tonnes, and was about 59 million tonnes in Europe in 2014. In fact, nowadays, plastics represent almost 40% of the European packaging market (Muller et al. 2017a). Of the plastic materials, petroleum-based plastics, such as polyethylene (PE), polypropylene (PP), polyamide (PA), are widely used as packaging materials due to their ready availability at relatively low cost, good mechanical and barrier properties, thermo-processing ability and chemical characteristics, which make them suitable for food packaging. However, despite their good properties, their use and accumulation imply serious environmental problems and a dependence on fossil fuels. Around 63% of the current plastic waste comes from packaging applications, and it is estimated that less than 14% is recyclable. Taking this scenario into account, and bearing in mind the growing environmental awareness, research has focused on the development of alternative bio-packaging materials, derived from renewable sources, which are biodegradable or compostable.

Biopolymers can be used for food packaging applications or food coating purposes, reducing the environmental impact and oil-dependence (Rivero et al. 2017; Emadian et al. 2017). They can be divided into three main categories, on the basis of their origin and biodegradable nature. Together with the conventional, non-biodegradable, oil-based plastics, there are biobased-non-degradable bioplastics (e.g. polyethylene terephthalate: PET), biobased-biodegradable bioplastics (e.g. polylactic acid: PLA, starch and other polysaccharides, or proteins) or fossil-based biodegradable bioplastics (e.g. polycaprolactone: PCL, polyvinyl alcohol: PVA, or polybutylene succinate: PBS). So, biopolymers are biodegradable, biobased or both and can be classified as those directly obtained from biomass (polysaccharides and proteins), synthetic biopolymers from biomass or petrochemicals (e.g. PLA, PCL) or those obtained by microbial fermentation (polyhydroxyalcanoates: PHA and bacterial cellulose) (Nair et al. 2017). The former are directly extracted from biological and natural resources and they are hydrophilic and somewhat crystalline in nature, making an excellent gas barrier. Biodegradable polyesters (synthetic or biosynthesized) are more hydrophobic and constitute better barriers to water vapour. In general, the functional properties of biopolymer-based materials in terms of their mechanical and barrier properties need to be adapted to food requirements by using different strategies, such as physical or chemical modifications (crosslinking), blending with other components, fillers, plasticizers or compatibilizers (Ortega-Toro et al. 2017).

The industrial uses of biopolymers have been restricted because of their usually poor mechanical, barrier or thermal properties, and high price. The incorporation of micro- and nano-reinforcing agents into the matrix for the purposes of obtaining composites has been seen to improve their functional properties and so their competitiveness in the plastics market. Composites are made up of a continuous polymer matrix in which the filler particles are dispersed, thus contributing to a modification of the functional characteristics of the material (Azeredo 2009). Fillers differing in size, shape, amount, distribution and chemical nature have been used. Lignocellulosic or cellulosic materials obtained from agro-waste have been widely studied as organic micro-fillers (Gutiérrez and Alvarez 2017). Fibres from cotton (Ludueña et al. 2012), garlic straw (Kallel et al. 2016), rice husk (Johar et al. 2012), wheat straw (Berthet et al. 2015) or coffee silverskin (Sung et al. 2017), have been used as reinforcing agents in different biopolymer films. Micro-particles significantly improved the elastic modulus of composites while providing great thermal resistance to the matrices due to the presence of hydroxyl groups interacting with the biopolymer network (Ludueña et al. 2012; Berthet et al. 2015). Different organic nano-fillers can be obtained, mainly from cellulose (cellulose nanocrystals or nanofibres), chitin/chitosan nanocrystals from crustacean waste (Gutiérrez 2017) or starch nanoparticles. These nano-reinforcing agents improve the tensile strength and elastic modulus when they have a proper distribution, chemical affinity with the polymer and high aspect ratio. The crystalline structure of nanofillers enhances the tortuosity factor for the mass transport of gas molecules into the biopolymer matrix, contributing to the formation of a hydrogen-bonded network (Ng et al. 2015; Azeredo 2009; Azeredo et al. 2017). On the other hand, inorganic particles are relevant as filling agents in food packaging materials due to the enhancement of the mechanical and barrier properties (MgO, silicon carbide or nano-clays) Some of them also exhibited antimicrobial activity, such as Ag, TiO_2 and ZnO nanoparticles (Gutiérrez et al. 2017; Azeredo 2009).

It is remarkable that biodegradation behaviour is a crucial factor in the development of composites. The biodegradation process takes place in aerobic conditions by the action of a microorganism, which identifies the polymer as a source of energy to produce organic residues from the packaging material. The incorporation of nano-fillers can affect the biodegradability of composites (Gutiérrez 2018). In this sense, cellulose nanocrystals (CNC) promoted the material's water intake due to their hydrophilic nature, contributing to an acceleration of the biodegradation process (Ludueña et al. 2012; Luzi et al. 2016). Some inorganic nano-fillers could also affect the disintegration processes, such as what occurs with Ag nanoparticles (Ramos et al. 2014; Cano et al. 2016), or nano-clays, whose hydroxyl groups react with the chains of the polymer matrices (montmorillonite and fluorohectorite, Fukushima et al. 2013).

This chapter reviews the potential use of reinforcing agents of differing natures and sizes in biopolymer materials that are potentially useful for food packaging, analysing their effect on the mechanical and barrier properties and on the thermal resistance of the material. The surface properties and biodegradation behaviour were also analysed in different kinds of composites.

4.2 Bioplastics for Food Packaging

Over the last decade, several bioplastics, bio-based, biodegradable, or both, have been available as a suitable alternative to conventional plastics for food packaging applications (Fabra et al. 2014; Ortega-Toro et al. 2017). At least 90% of natural or synthetic biodegradable polymers decompose in less than 180 days (ASTM 2003). Figure 4.1 shows the main polymers of potential use in food packaging. Of the natural polymers, different polysaccharides and proteins and microbially-produced biopolymers have been extensively studied for food packaging applications. Starch is a promising polymer, suitable for processing by means of different techniques, such as the casting method (Moreno et al. 2017), compression moulding (Muller et al. 2017b), extrusion (Gutiérrez and Alvarez 2018) or injection moulding (López et al.

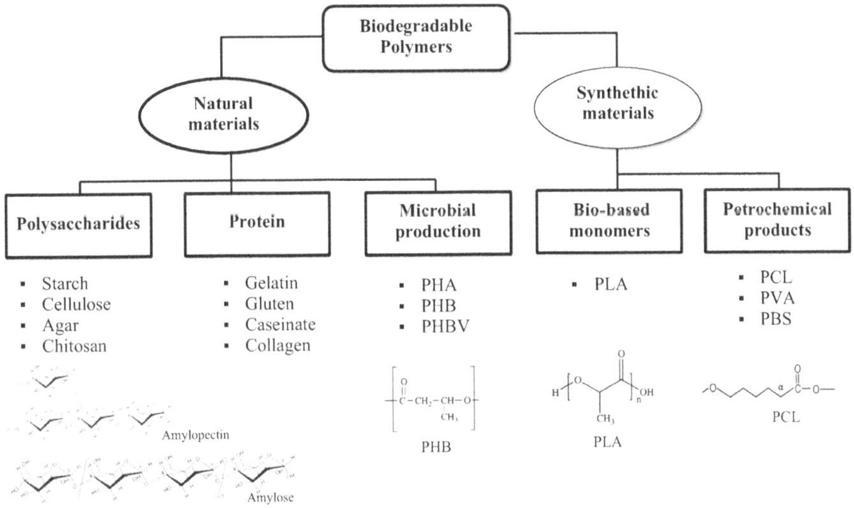

Fig. 4.1 Main biopolymers with potential application in food packaging. Some molecular structures are included

2015). Starch is naturally highly abundant, low cost and renewable. Cellulose, constituted by glucose units via β-1,4-glycosidic bonds, is the most naturally abundant carbohydrate (Xiao et al. 2014). This polymer is usually used as micro-filler or nano-filler (Shankar and Rhim 2016) in the composite formulation. These could be processed by compression moulding and injection moulding (Graupner et al. 2016) to obtain semi-rigid packaging (trays). On the other hand, agar is a fibrous polysaccharide obtained from marine algae, such as *Gelidium* sp. and *Gracilaria* sp., consisting of a mixture of agarose and agaropectin, which is slightly branched and sulphated. This is thermoplastic polysaccharide, biodegradable and biocompatible, which exhibits great mechanical strength with moderate water resistance (Giménez et al. 2013). Chitosan, the second most naturally abundant polysaccharide, has non-toxic, biodegradable, and antimicrobial characteristics, which are of great interest for packaging purposes (Leceta et al. 2013). Chitin, the precursor of chitosan, is a linear polymer of mainly β-(1 → 4)-2-acetamido-2-deoxy-D-glucopyranose units and low amounts of β-(1 → 4)-2-amino-2-deoxy-D-glucopyranose residues (Van den Broek et al. 2015). Others relevant natural polymers are proteins, such as gelatin or collagen, gluten proteins and dairy proteins. Gelatin is an animal protein obtained by the hydrolysis of the fibrous insoluble collagen from skins and bones of different animals. It is well known for its film-forming properties. It is abundantly available, low cost, and easily biodegradable and biocompatible (Kanmani and Rhim 2014). On the other hand, wheat gluten (constituted by gliadins and glutenin proteins) is an inexpensive protein from the milling process, which allows the production of membranes that are semipermeable to water vapour, oxygen and carbon dioxide molecules. This polymer can be applied as a food coating or edible film on different foods (Rocca-Smith et al. 2016), or processed by compression moulding (Zubeldía et al. 2015) and extrusion (Rombouts et al. 2013) for the purposes of developing flexible or semi-rigid packaging. Caseinates have been proposed as raw materials for food packaging development, since this protein exhibited good film-forming ability with good mechanical properties (Fabra et al. 2012; Arrieta et al. 2014a; Jiménez et al. 2013). The use of caseinates could be considered an alternative means of obtaining a high degree of protection from oxygen in modified atmosphere packaging (Arrieta et al. 2014a). Biopolymers obtained from several microorganisms, such as poly-hydroxyalkanoates -PHA-, poly-hydroxybutyrate -PHB-, or poly- hydroxybutyrate-co-hydroxyvalerate -PHBV-, are a family of biodegradable thermoplastic polymers. The polymer is produced in the microbial cells through a fermentation process and then collected by solvents, like chloroform. More than 100 PHA are identified, of which PHB is the most common (Peelman et al. 2013). PHB is a biopolymer produced from renewable sources and fermentation by certain micro-organisms, like *Halomonas hydrothermalis* and *Burkholderia sp.* and *Chelatococcus daeguensis*, inter alia (Bera et al. 2015). In addition to being biodegradable, PHB exhibits some properties similar to some synthetic polymers, especially polypropylene (PP) (Heitmann et al. 2016). PHBV is a copolymer of 3-hydroxybutanoic acid and 3-hydroxypentanoic acid produced directly by microorganisms. This polymer is less brittle and more stretchable than PHB (Requena et al. 2017).

Other biodegradable polymers are obtained by synthesis from biobased monomers (e.g. PLA) and non-biobased monomers (e.g. PCL, PVA or PBS). PLA is the most common synthetic polymer obtained from biobased monomers. The synthetic routes to obtain PLA are through the ring-opening polymerisation of the esters of the acid and the direct condensation of the free acid (Cheng et al. 2009). PLA is of great potential to the packaging industry because of its mechanical, barrier and optical properties. It can be processed using readily available production technologies, and exhibits good thermal behaviour and water vapour barrier properties, although it is brittle and only a moderate gas barrier (Bonilla et al. 2013).

In the category of biodegradable synthetic petroleum-based polymers, PCL, PVA and PBS are the most representative. PCL is a linear, semicrystalline hydrophobic polyester, highly flexible, tough and thermally stable (Correa et al. 2017). In contrast, PVA is a synthetic, water soluble polymer which forms translucent films with good tensile strength, elongation at break and barrier properties (Dominguez-Martinez et al. 2017). PVA has been used in polymeric blends, with glycerol as a proper plasticiser because of its chemical affinity. PBS is another biodegradable thermoplastic polymer which has desirable melt processability and good mechanical properties, which are closely comparable to those of widely-used polyethylene (PE) or polypropylene (PP) (Mizuno et al. 2015).

4.3 Micro and Nano-Reinforcing Agents

An alternative means of improving the properties of biopolymers for food packaging applications, and reducing some of their drawbacks, is by the incorporation of micro- or nano-fillers to the matrix for the purposes of obtaining micro- and nano-composites. Composites have immiscible phases constituted by the polymer continuous network in which the filler particles are dispersed, thus generating new structures with different properties to those of the original polymer matrix. The filler can positively modify the functional characteristics of the material, depending on the filler polymer network interactions. Different kinds of fillers have been used, which modify the material characteristics as a function of the filler-matrix interactions. The size, shape, amount, distribution and chemical nature of the fillers are crucial factors in the final properties of the composite.

Many studies reported the use of different kinds of fillers. As is shown in Table 4.1, cellulose has been frequently used as a reinforcing agent in different forms, such as cellulosic fibres (Martino et al. 2015; Moustafa et al. 2016; Ludueña et al. 2012), bacterial cellulose (Fabra et al. 2016), nano-cellulosic fibres (Abdul Khalil et al. 2016) or cellulose nanocrystals (El-Hadi 2017; Fortunati et al. 2013a; Follain et al. 2013). In the polymer matrices, the incorporation of cellulosic fillers directly affects the mechanical and barrier properties depending on the particle size (micro or nano), which is a significant factor. As has been observed by various authors, particle size has differing effects on the mechanical properties. Cho et al. (2006) studied the effect of the particle size on the mechanical properties of polymeric composites with spherical micro (0.5 mm) and nano (15 nm) particles. They observed that, at nano-scale, Young's modulus and tensile strength increased as the particle size decreased. As regards the barrier properties, it is expected that nano scale and a homogeneous distribution increases the tortuosity factor for the migration of molecules through the composite, decreasing the permeability of both water vapour and gases. On the other hand, the amount of filler included significantly affects the composite properties. At high concentrations, the polymer matrix could lose cohesiveness and continuity which could lead to a loss in functional properties (mechanical and barrier). In this sense, Maqsood et al. (2016) studied the reinforcing capacity of enzyme-hydrolysed longer jute micro-crystals in polylactic acid matrices. The elastic modulus and tensile strength increased by 40% and 28% respectively, once the filler loading rose to 5% with respect to neat PLA. However, a filler loading of 10% led to a decrease of 32% in the elastic modulus and 33% in tensile strength with respect to materials containing 5% of fillers. Other organic fillers are chitin nanocrystals (Herrera et al. 2016) from crustacean waste that improve the mechanical properties and transparency of neat PLA. Also, starch nanocrystals are frequently added to several biopolymers, improving the mechanical and barrier properties and decreasing the biodegradation time (Mukurubira et al. 2017; Le Corre and Angellier-Coussy 2014).

Some inorganic particles have been extensively studied as fillers in food packaging materials. Some of these, such as MgO, (Sanuja et al. 2014); silicon carbide, (Dash and Swain 2013) and nano-clays (Cavallaro et al. 2013; Majeed et al. 2013; Abdollahi et al. 2013), enhanced the mechanical and barrier properties of the films. Additionally, other ones can provide antimicrobial activity to the material, as occurs with Ag nanoparticles (Carbone et al. 2016; Gutiérrez

Table 4.1 Main inorganic and organic reinforcing agents used as fillers in biopolymer matrices and main provoked changes in the matrix

Micro- and nano-fillers	Filler information	Polymer matrix	Main provoked changes	Applications	Reference
Organic fillers					
Lignocellulosic fibres	Natural Lignocellulosic material	PHBV/wheat straw	– Decrease cost – Increase EM – Decrease deformation	– Food Packaging	Martino et al. (2015)
	Natural by-product	PBAT/coffee ground	– Compatibiliser not necessary – Improve mechanical properties	– Food packaging	Moustafa et al. (2016)
Cellulosic fibres	Natural Previously alkali treated agro-waste	PCL/cotton	– Improve mechanical properties and biodegradation in soil	– Food packaging	Ludueña et al. (2012)
Bacterial cellulose	Obtained from bacterial strain *Gluconacetobacter xylinus* 7351	TPS, PHB	– Improve mechanical and barrier properties	– Food packaging	Fabra et al. (2016)
Nanocellulosic fibres	Natural Glucose units (beta-1,4 glycosidic linkages) $(C_6H_{10}O_5)n$	TPS, PLA	– Improve mechanical, thermal and barrier properties	– Food packaging – Pharmaceutical and industrial packaging – Medical application	Abdul Khalil et al. (2016)
Cellulose nanocrystal	Natural Obtained from agro-wastes	PHB and PLA	– Reduce Tg, Tc and Tm – Improve mechanical properties	– Food packaging	El-Hadi (2017)
		PVA	– Increase tensile strength	– Food packaging	Fortunati et al. (2013a)
		PCL	– Decrease WVP	– Food packaging	Follain et al. (2013)

(continued)

Table 4.1 (continued)

Micro- and nano-fillers	Filler information	Polymer matrix	Main provoked changes	Applications	Reference
Chitin nanocrystals	Powder from crustacean waste	PLA	– Improve mechanical properties and transparency	– Packaging and food packaging	Herrera et al. (2016)
Starch nanocrystals	Amadumbe corms Botanical sources: normal maize, waxy maize, amylomaize, potato, rice, oat, peas or beans	TPS PLA, PVA, PBS, natural rubber, pullulan, carboximethyl chitosan	– Increase barrier properties – Improve mechanical properties – Increase the biodegradability	– Food packaging –Water treatment – Adhesive applications – Medical application	Mukurubira et al. (2017) and Le Corre and Angellier-Coussy (2014)
Inorganic fillers					
Ag nanoparticles	Metal	– Cellulose – HPMC – Pullulan – TPS – Alginate	– Provide antimicrobial and antiviral properties	– Fresh food packaging: meat, fruit and dairy products – Food packaging – Agricultural	Carbone et al. (2016) and Gutiérrez et al. (2017)
		– Agar	– Decrease WVP	– Food packaging	Rhim et al. (2014)
TiO_2 nanoparticles	Oxide	Several polymers TPS	– Provide antimicrobial properties – Change colour	– Food packaging: active and intelligent packaging – Food packaging – Agricultural	He et al. (2015) and Gutiérrez et al. (2017)
ZnO nanoparticles	Oxide	TPS	– Provide antimicrobial properties	– Food packaging – Agricultural	Gutiérrez et al. (2017)
MgO nanoparticles	Oxide	Chitosan	– Increase tensile strength and deformation – Decrease water solubility	– Food packaging	Sanuja et al. (2014)

Table 4.1 (continued)

Micro- and nano-fillers	Filler information	Polymer matrix	Main provoked changes	Applications	Reference
Silicon carbide	SiC Melting point: 2730 °C Molar mass: 40.11 g/mol	Starch	– Decrease OP	– Food packaging – Industrial packaging	Dash and Swain (2013)
Nano-clays	Phyllosilicates $Al_2Si_2O_5(OH)_4$ Specific gravity: 2–2.65	Pectin/PEG/hallosyte	– Decrease water uptake capacity – Increase EM	– Industrial packaging	Cavallaro et al. (2013)
	$M_x(A_{l4}\text{-}xMg_x)Si_8O_{20}(OH)_4$	Several polymers/Mnt	– Increase barrier properties – Improve mechanical properties	– Food packaging – Industrial packaging	Majeed et al. (2013)
		Alginate	– Reduced water solubility – Decrease elongation and increase tensile strength – Decrease WVP	– Food packaging	Abdollahi et al. (2013)

et al. 2017), TiO_2 nanoparticles (He et al. 2015; Gutiérrez et al. 2017) and ZnO nanoparticles (Gutiérrez et al. 2017).

4.4 The Effect of Reinforcing Agents on the Functional Properties of Biopolymers

In the following sections, the effect that micro- and nano-fillers have on the tensile behaviour, barrier properties, thermal resistance and biodegradability of biopolymer matrices is discussed, taking the processing method into account.

4.4.1 *Processing Methods*

In general, thermomechanical processes (melt compounding, extrusion and compression moulding) with high shearing forces, temperature and adequate time are necessary to guarantee the convenient dispersion of the filler in the polymer matrix. However, the casting of polymer-filler dispersions is an alternative method to obtain nano-composites, due to the high aggregation tendency of nanoparticles, which are better maintained in liquid dispersions. Table 4.2 shows some recent studies on composite materials, including micro- or nano- fillers of differing characteristics, using different processing methods.

Berthet et al. (2015) studied the properties of PHBV composites containing wheat straw micro-fibres (10, 20 and 30 wt%). Compounding was carried out with a lab-scale twin-screw extruder, using a temperature profile from the polymer feeding to the die of 180–160 °C. Composite films were obtained by compression moulding at 170 °C. The mechanical and barrier properties of composites were poorer than those of neat PHVB, although the authors point that the obtained materials could be applied as packaging for respiring fresh products. Melt compounding using a Plastograph mixer (16 cm^3) for 4 min at 160 °C was also applied to obtain PHBV composites with micro-particles of keratin at 0.5, 1, 3, 5, 10, 25 and 50 wt%. The composites exhibited improved mechanical and water vapour and oxygen barrier properties with only 1 wt% of micro-filler (Pardo-Ibáñez et al. 2014). Moustafa et al. (2016) studied the use of roasted coffee grounds as micro-reinforcing agent to produce high-quality biodegradable Polybutylene adipate co-terephthalate (PBAT) composites for food packaging applications. The composites were extruded at 160–165 °C with a screw speed of 100 rpm for 5 min mixing. The films were obtained by a special die attached to the mixing chamber. The authors observed an increase in the hydrophobicity and thermal stability compared to the control films without fillers. A compatibiliser was not necessary to obtain a filler-polymer matrix with suitable interfacial adhesion, especially at low filler content (<30%), which was attributed to the good grindability of roasted coffee, which improved compatibility and filler dispersion during processing.

Table 4.2 Recent studies on composite materials including micro- or nano- fillers of different nature, applying different processing methods

Composite material	Filler content	Processing method	Effect of filler	Possible application	Reference
Polymer/micro-filler					
PHBV/ wheat straw fibres	10, 20 and 30 wt% filler with different preparation process	Extrusion and compression moulding	– Increased the water vapour transmission – Decrease ultimate tensile strength	– Food packaging materials to respiring fresh products	Berthet et al. (2015)
PHBV/ keratin	0.5, 1, 3, 5, 10, 25 and 50 wt% filler	Melt compounding	– Reduction WVP and OP – Improve mechanical properties	– Packaging	Pardo-Ibáñez et al. (2014)
PBAT/ roasted coffee ground	10, 20 and 30 wt% filler with roating process at 250 and 270 °C	Extrusion	– Increase hydrophobicity – Increase thermal stability – Compatibiliser is not necessary	– Food packaging	Moustafa et al. (2016)
Polymer/nano-filler					
PCL-MC/ NCC	2 wt% NCC	Casting and compression moulding	– Increased the tightly of the matrix – Increased stability of active components – Increased the rough and density of the matrix	– Vegetable packages	Boumail et al. (2013)
PLA/CNC from MCC and Ag-NPs	1 and 5 wt% CNC with/ without surfactant and 1 wt% Ag nanoparticles	Casting	– Increase barrier effect – Antimicrobial effect	– Food active packaging	Fortunati et al. (2013b)
Agar/ Ag-NPs	20, 40 and 80 mg Ag-NPs	Casting	– Increase thermal stability – Barrier properties to water vapour increase slightly – Mechanical strength and stiffness decreased slightly – Antimicrobial activity against Listeria monocytogenes and Escherichia coli O157:H7	– Food packaging	Rhim et al. (2014)

(continued)

Table 4.2 (continued)

Composite material	Filler content	Processing method	Effect of filler	Possible application	Reference
Alginate/ nano-clays Mnt and CNC from MCC	1, 3 and 5 wt% fillers	Casting	– Decrease water solubility – Increase surface hydrophobicity with CNC and decrease of this parameter with nanoclays addition – Reduction in WVP – Tensile properties improved	– Food packaging	Abdollahi et al. (2013)
Chitosan/ MgO	0.1 g MgO/g chitosan	Casting	– Improve mechanical properties – Increase opacity – Decrease swelling, permeability and solubility – Antimicrobial properties	– Food active packaging	Sanuja et al. (2014)
Starch/ Silicon carbide	1, 2, 5, 8, 10 wt% filler	Casting	– Increase thermal stability – Reduce OP	– Adhesive application – Covering and protecting applications – Food packaging	Dash and Swain (2013)
Pectin-PEG/ Halloysite nanotubes	5, 10, 15, 20, 30 and 50 wt% filler	Casting	– Decrease wettability – Improve mechanical properties	– Coatings for food conservation	Cavallaro et al. (2013)

As concerns the incorporation of nano-fillers in composites, casting is the most commonly used method due to the better dispersion (more limited aggregation) of nano-particles in a liquid medium. Casting is suitable for the obtaining of coatings, mulch films and flexible films. In some cases, this technique has been used as a preliminary test to study the filler effect before thermomechanical processing with actual industrial applications. Boumail et al. (2013) characterized trilayer antimicrobial films based on methylcellulose and PCL composites with 2% of cellulose nanocrystals (CNC). These were prepared under stirring before sonication at room temperature for 30 min and the subsequent casting of the PCL-CNC dispersion. The trilayer films were obtained by compression moulding at 120 °C. An increase in the matrix toughness and greater stability of active components was found by the filler addition. Fortunati et al. (2013b) incorporated the CNCs and Ag-NPs PLA films obtained by casting, leading to improved barrier properties and antimicrobial activity in the composites. Ag-NPs (20, 40 and 80 mg Ag-NPs) have also been included in other polymers, such as agar matrices; although a significant increase in the thermal stability of the material was obtained, with improved water vapour barrier properties and antimicrobial activity against *Listeria monocytogenes* and *Escherichia coli* O157:H7, the mechanical strength and stiffness of the composites slightly decreased with respect to filler-free polymers (Rhim et al. 2014). Chitosan composites with MgO nano-filler, obtained by casting, also exhibited antimicrobial properties. In addition, the metallic oxide improved the mechanical properties and reduced the water swelling capacity and solubility and water vapour permeability of the films, which became less transparent.

Other nanocomposites with organic (CNC) or inorganic nano-clays:Mnt) obtained by casting, using an alginate matrix, exhibited improved functional properties with respect to the net polymer matrix (Abdollahi et al. 2013). The addition of both nano-fillers improved the tensile properties of the material and promoted a decrease in water solubility and water vapour permeability, whereas the surface hydrophobicity increased with the use of CNC but decreased with nano-clays. In the same way, nanoparticles of silicon carbide increased the thermal stability and reduced the oxygen permeability in starch composites (Dash and Swain 2013). This material could be used as adhesive and coating in food applications. Halloysite nanotubes promoted a decrease in film wettability and improved the mechanical properties in composites of pectin and polyethylene glycol blends (Cavallaro et al. 2013).

4.4.2 The Effect on Tensile Properties

In this section, the changes in the tensile properties of some biopolymers caused by incorporating nano- and micro-fillers of differing characteristics are analysed, as summarized in Table 4.3. The main mechanical properties characterized in plastic packaging materials are elastic modulus (EM) and the tensile strength (TS) and elongation (ε) at break, which provide information about the rigidity and resistance to deformation and break of the material, respectively.

The changes in the polymer's functional properties caused by filler addition are strongly associated with surface properties and polymer-filler interfacial interactions. In this sense, the effects caused by fillers on the mechanical properties of polymer are not always positive. The main reason is the interruption of the polymer matrix continuity, but this effect could be diminished if the polymer and fillers have chemical affinity or an interfacial agent is added into the composite formulation. As previously mentioned, Berthet et al. (2015) found a deterioration in the tensile properties of neat PHBV when wheat straw fibres were added to the matrix. Strain and stress at break decreased by 61% and 63%, respectively, with 30 wt% filler in the composite. However, Young's modulus increased by 13% with 20 wt% filler. Moustafa et al. (2016) found different trends in the tensile strength with the varia-

Table 4.3 Changes in tensile properties of some biopolymers by incorporating micro- or nano-fillers of different nature

Composite material	Filler content	Plasticizer/ equilibrium RH	Thickness (μm)	TS (MPa)	EM (MPa)	ε (%)	Reference
Polymer/micro-filler							
PHBV/wheat straw fibres	10, 20 and 30 wt% filler with different preparation process	–	–	2.2–3.13	14.4–39.2	0.89–2.3[a]	Berthet et al. (2015)
PBAT/ roasted coffee ground	10,20 and 30 wt% filler with roasting process at 250 and 270°°C	/50%	300–400	6.8–18.2	–	98–1545[a]	Moustafa et al. (2016)
PHBV / keratin	0.5, 1, 3, 5, 10, 25 and 50 wt% filler	–	100	–	540–600	2.9–5.5[a]	Pardo-Ibáñez et al. (2014)
Polymer/nano-filler							
PCL-MC/ NCC	7.7 wt% NCC	Glycerol/	225–280	20.3[a]–24.0	175.2–218.3[a]	–	Boumail et al. (2013)
PLA/CNC and ChNC	1 wt % nanocrystals	Triethyl citrate	100	15.8[a]–24.2	300[a]–1200	16–309	Herrera et al. (2016)
Corn starch-gelatin/CNC from eucalyptus wood pulp	0.44, 1.5, 2.56; 3% CNC and 20% plasticizer	Glycerol/50%	50–140	11–49	–	1.24–38	Alves et al. (2015)
Alginate/ nano-clays Mnt and CNC from MCC	1, 3 and 5 wt% fillers	Glycerol/53%	–	CNC: 18[a]–23	150[a]–270	9–12[a]	Abdollahi et al. (2013)
				Mnt: 17–19	150[a]–210	8–12[a]	
Pectin-PEG/ Halloysite nanotubes	5, 10, 15, 20, 30 and 50 wt% filler	/53%	60	25–26	2.6[a]–4.1	0.9–1.5[a]	Cavallaro et al. (2013)
Chitosan/ MgO	0.1 g MgO/g chitosan	–	220–470	30[a]–60	–	7.5[a]–15	Sanuja et al. (2014)
Agar/ Ag-NPs	20, 40 and 80 mg Ag-NPs	Glycerol/50%	62.2–65.8	45.2–49.6[a]	1290–1460	19.0–23.6[a]	Rhim et al. (2014)

TS tensile strength; *EM* young's modulus; *ε* elongation at break
Data show the range in the values of each property reported for the different formulations
[a]Identifies the value for the control sample (without filler), when it is in the edge of the range

tion in filler content, using differently roasted coffee ground (CG) in PBAT matrices. As compared with neat PHBV, the best tensile behaviour was obtained with 10% of filler roasted at the highest temperature (270 °C), while the worst behaviour was observed for 30% of non-roasted CG powder. As the authors mentioned in their study, roasted CG showed a better affinity with PBAT compared to untreated CG when obtaining green composites without the need for a compatibiliser. This is important because the greater chemical affinity among the components led to stronger structures and better tensile properties of the composites. The values of the strain at break were lower than in the net polymer films for every composite. Pardo-Ibáñez et al. (2014) also found a decrease in the elongation at break of PHBV composites in line with differing quantities of keratin microparticles, but the elastic modulus at low loading was significantly improved. The content of filler greatly affected the tensile behaviour of composites and, in general, low filler contents improved both tensile strength and elastic modulus.

As regards nano-fillers, different studies have shown the improvement in tensile strength and elastic modulus when nanoparticles are incorporated into the polymer matrices. This effect is enhanced by a good particle distribution, a chemical affinity between filler and polymer and a high aspect ratio of the particles and filler-polymer contact area (more interactions). However, the processing conditions and the filler-polymer ratio must be taken into account to optimize the composite properties. Excessively high filler content results in the polymer matrix interruption and formation of micro- nano-cracks. Herrera et al. (2016) studied the effect of the incorporation of CNCs and chitosan nano-crystals (ChNC) into PLA films obtained by extrusion and compression moulding using fast and slow cooling rates. The tensile properties of composites were affected by both the chemical nature of the filler and processing conditions. Strength at break was improved by CNC incorporation, both at fast and slow cooling rates, but ChNC only improved the film strength when processing at a slow cooling rate. Young's modulus increased and elongation at break decreased after the addition of both nanoparticles at both processing cooling rates. Alves et al. (2015) studied the effect of CNC and gelatin on corn starch plasticised films. Nanocrystals were added at 0.44, 1.5, 2.56 and 3% with respect to the polymers. They found a significant increase in the film resistance when the CNC ratio rose, with a slight fall in the elongation values. Similar results were obtained in nanocomposites based on pectin/polyethylene glycol blends containing halloysite nanotubes (Cavallaro et al. 2013). The incorporation of nanotubes led to significantly more rigid films, but reduced the elongation capacity of the material.

Metallic oxide nanoparticles have been used as a suitable option to improve the mechanical properties of biopolymers. Sanuja et al. (2014) reported a significant increase in both the tensile strength and elongation at break of chitosan composites with 10% magnesium oxide, obtained by casting. Abdollahi et al. (2013) compared fillers of differing chemical characteristics (Mnt and CNC) added to alginate matrices at 1, 3 and 5%. The tensile strength values exhibited a constant growth as the CNC content increased but this parameter decreased when the content of Mnt was higher than 1%. Young's modulus also behaved differently, depending on the filler. The composite stiffness was higher as the CNC content increased from 0 to 5%, but decreased when the Mnt content rose from 3 to 5%. However, the elongation values exhibited the same trend for both kinds of nanocomposites; decreasing when both CNC and Mnt contents increased.

4.4.3 The Effect on Barrier Properties

Some biodegradable materials need to improve the gas and water vapour permeability because it is a fundamental property in packaging (Ng et al. 2015). The incorporation of micro- and nano-fillers, of organic or inorganic nature, to biodegradable polymer matrices can modify the barrier properties (oxygen permeability: OP and water vapour permeability: WVP). The barrier properties of polymers containing fillers depend on their chemical nature, the particle size and shape of the particles, as well as other factors such as polarity, hydrogen bonding capacity and polydispersity (González et al. 2015; Pardo-Ibáñez et al. 2014). Table 4.4 summarizes the different effects several fillers were observed to have on the barrier properties of some biopolymer matrices. In every case, the addition of fillers can improve the oxygen permeability and the values of the control samples (without filler) were in the range of the corresponding composites or at the high end, especially when CNC fillers were used. Fabra et al. (2016) studied the incorporation of bacterial cellulose nanowhiskers (BCNW) into thermoplastic corn starch matrices (TPCS) and a 95% decrease in OP was obtained with 15% of filler with respect to the TPCS sample. The reinforced TPCS was assembled in multilayer PHB films to obtain more hydrophobic matrices and to improve the film performance. The best functional properties of multilayers were obtained with 15% BCNW-TPCS composite inner layer and PHB outer layers. Luzi et al. (2016) observed a reduction of 47% in OP with the incorporation of 3 wt% CNC from *Carmagnola* carded hemp fibres and a commercial surfactant into a PLA-PBS matrix. The addition of a surfactant could contribute to a better dispersion of the CNCs and PBS. They observed that the CNCs were well dispersed in the polymer matrices, through the Atomic Force Microscopy (AFM) analysis of surface roughness.

As regards values, analyses of WVP in composites provide similar tendencies to those OP values. In most cases, WVP was reduced after the incorporation of fillers, especially nano-size ones. Table 4.4 shows the different effect of micro- and nano-fillers on this barrier property for several biopolymer matrices. Raw lignocellulosic fibres (Berthet et al. 2015) or cellulose fibres obtained after chemical treatments (Ludueña et al. 2012) only slightly decreased, or did not affect, the WVP of some composites due to their hydrophilic nature. However, Pardo-Ibáñez et al. (2014) observed a 59% reduction in WPV of the PHVB matrices with 1 wt% of keratin micro-filler, while no significant differences were obtained with a filler load higher than 5 wt%. They detected a homogenous particle distribution in the matrix at 1 wt% filler, where the micro-particles are not aggregated, causing an increase in the tortuosity factor for the diffusion of gas molecules, which reduced the permeability values. On the other hand, the incorporation of nano-filler into different polymer matrices, especially CNCs, improved the WVP values. The crystalline structure of

Table 4.4 Changes in barrier properties (oxygen and water vapour permeability) of some biopolymers by incorporating micro- or nano-fillers of different nature

Composite material	Filler content	Processing method	Plasticizer/ equilibrium RH	OP ($m^3.m/m^2.Pa.s$)	WVP ($kg.m/Pa.s.m^2$)	RH (%)	T (°C)	Thickness (μm)	Reference
Polymer/micro-filler									
PHBV/wheat straw	10, 20 and 30 wt% filler with different preparation process	Extrusion and compression moulding	–	–	1.26 (10^{-7})[a]–1.27 (10^{-6}) ($kg/m^2.s$)	100	20		Berthet et al. (2015)
PHBV/keratin from poultry feathers	0.5, 1, 3, 5, 10, 25 and 50 wt% filler	Brabender Plastograph mixer and compression moulding	–	1.0 (10^{-18})–3.2 (10^{-18})	3.1 (10^{-15})–62.0 (10^{-15})	40	24	100	Pardo-Ibáñez et al. (2014)
PCL/cellulose fibres from cotton	5 and 15 wt% filler	Brabender Plastograph mixer and compression moulding	–	–	1.6 (10^{-14})[a]–1.7 (10^{-14})	68	–	–	Ludueña et al. (2012)
Polymer/nano-filler									
Corn starch-gelatin/CNC from eucalyptus wood pulp	0.44, −1.5, 2.56; 3% CNC	Casting	Glycerol/50%	–	5.21 (10^{-14})–6.99 (10^{-14})[a]	50	37.8	90–140	Alves et al. (2015)
Pea starch-PVA/CNC from MCC	1, 3 and 5 wt% CNC	Casting	/53%	–	2.32 (10^{-11})–2.43 (10^{-11})	100	25	1000	Cano et al. (2015)
TPS/CNC from gravata fibres	0.5, 1, 2 and 3 wt% CNC	Casting	Glycerol and lignin/	–	1.27 (10^{-13})–2.67 (10^{-13})[a]	100	25	–	Miranda et al. (2015)
Starch-CMC/CNC from sugarcane bagasse	0.5, 2.5 and 5 wt% CNC	Casting	Glycerol/	–	9.72 (10^{-14})–3.75 (10^{-13})[a]	50	32	–	El Miri et al. (2015)
TPS/WSNC	1, 2.5 and 5 wt% WSNC	Casting	Glycerol/	8.79 (10^{-16})–3.07 (10^{-15})	57.3 (10^{10})–66.7 (10^{10}) ($kg/m^2.s.Pa$)	50	23	–	González et al. (2015)

Composite material	Filler content	Processing method	Plasticizer/ equilibrium RH	OP ($m^3.m/m^2.Pa.s$)	WVP ($kg.m/Pa.s.m^2$)	RH (%)	T (°C)	Thickness (μm)	Reference
TPCS-PHB/ BCNW	2, 5, 10, 15, 20 and 25 wt% CNC	Brabender Plastograph internal Mixer, compression moulding and Electrospinning	Glycerol/0%	2.03 (10^{18})–41[a] ± 2.3 (10^{18}) (80% HR)	6.42 (10^{13})–15.52 (10^{13})[a]	0–100	25	–	Fabra et al. (2016)
Chitosan/ Commercial CNC	1, 3 and 5 wt% CNC	Casting	–	–	2.62 (10^{-7})–4.20 (10^{-7})[a] ($kg.d.m/m^2.Pa$)	75	25	50	Corsello et al. (2017)
Wheat gluten/CNC and CNF from sunflower stalks	1 and 3 wt% CNC or CNF	Casting	/53%	CNC 1 (10^7)–1.21 (10^7)[a]	CNC 1.40 (10^{-11})–1.55 (10^{-12})	53	25	–	Fortunati et al. (2016)
				CNF 1.08 (10^7)–1.37 (10^7)	CNF 1.39 (10^{-12}) 1.55 (10^{-12})				
PLA-PBS/CNC from *Carmagnola* carded hemp fibres	1 and 3 wt% CNC with/without surfactant	Casting	–	1.05 (10^{-6})–1.98 (10^{-6})[a]	1.52 (10^{-14})–2.41 (10^{-14})	53	25	–	Luzi et al. (2016)
PVA/CNC from potato peel waste	1 and 2 wt% CNC	Casting	Glycerol/50%	–	1.25 (10^{-5})–1.33 (10^{-5}) ($kg/m^2.s$)	50	23	150	Chen et al. (2012)
Alginate/ nano-clays Mnt and CNC from MCC	1, 3 and 5 wt% fillers	Casting	Glycerol/53%	–	1.6 (10^{-13})–1.99(10^{-13})[a]	20	1.5	–	Abdollahi et al. (2013)
Agar/Ag-NPs	20, 40 and 80 mg Ag-NPs	Casting	Glycerol/50%	–	1.38(10^{-12})–1.52 (10^{-12})[a]	–	–	–	Rhim et al. (2014)

OP oxygen permeability; *WVP* water vapour permeability; *RH* equilibrium relative humidity of samples; *T* temperature of the analysed samples
Data show the range in the values of each property reported for the different formulations
[a]Identifies the value of the control sample (without filler) when it is in the edge of the range

nanocrystals makes it difficult for certain molecules (O_2, CO_2, H_2O) to diffuse into the biopolymer matrix because of the formation of a hydrogen-bonded network (Brinchi et al. 2013; Ng et al. 2015), which favours the development of a percolation network, as reported by Miranda et al. (2015); the incorporation of 1 wt% of CNC from gravata fibres into thermoplastic corn starch matrices, provoked a 30% decrease in WVP. The addition of 2.5 wt% of CNC from sugarcane bagasse into matrices made up of corn starch and CMC decreased WVP by 50% due to the impermeable crystalline structure of CNC and a good dispersion of CNCs, creating a highly tortuous path for water vapour transfer (El Miri et al. 2015).

The use of inorganic fillers had similar effects on barrier properties to those of organic nano-fillers. The improvement in the barrier properties is due to the increased tortuosity factor for the gas molecule mass transport in the matrix and the impermeable nature of fillers, as reported by Abdollahi et al. (2013) in alginate films with 5% of Mnt nano-clays, where WVP decreased by about 20%. Studying agar films reinforced with Ag-NPs, Rhim et al. (2014) observed that the dispersed phase of Ag-NPs in the polymer impeded the mobility of its chains, inducing an improvement in WVP of the composites.

4.4.4 The Effect on Thermal Properties

The effects of micro- and nano-fillers of differing characteristics on the thermal properties of some biopolymers have been studied by several authors. Table 4.5 shows the main calorimetric parameters obtained from Differential Scanning Calorimetry (DSC) and the thermal stability of different materials obtained by Thermogravimetric Analysis (TGA) for different biopolymers and composites. Information about glass transition temperature (T_g), crystallisation temperature (T_c), melting temperature (T_m), melting enthalpy (ΔH_m), onset temperature (T_{onset}) and peak temperature (T_{peak}) of thermodegradation are given in the Table 4.5.

In general, the addition of micro- or nano-fillers can modify the Tg and crystallization/melting properties (T_c, T_m, ΔH_m) of polymer in line with the established interactions between particles and polymer chains. As expected, the addition of plasticizers to the filler-biopolymer blends decreases both the Tg and melting point (T_m). In this sense, Martino et al. (2015) analysed the effect of different plasticisers, such as ATBC (acetyltributyl citrate), GTA (glycerol triacetate) and PEG (polyethylen glycol) in PHVB films with 20 wt% of wheat straw fibres. Blends with ATBC showed the strongest Tg reduction due to its non-polar nature and great affinity with the polymer. Similar effects were observed in both polymer and composites. Cano et al. (2015) observed a ~5 °C reduction in Tg of pea starch-PVA (1:1) matrices after the addition of 3 wt% of CNC from MCC. This was related to the partial inhibition of the PVA crystallisation and to the lower mean molecular weight of the amorphous PVA fraction.

As regards the thermal degradation of materials, the addition of fillers generally improves the thermal stability of composites for both micro or nano fillers. The

Table 4.5 Changes in thermal behaviour (calorimetric parameters and thermal degradation) of some biopolymers by incorporating micro- or nano-fillers of different nature

Composite material	Filler content	Processing method	Plasticizer/ equilibrium RH	Calorimetric parameters				Thermal degradation		Reference
				T_g (°C)	T_c (°C)	T_m (°C)	ΔHm (J/g)	T_{peak} (°C)	T_{onset} (°C)	
Polymer/micro-filler										
PHBV/wheat straw fibres	10, 20 and 30 wt% filler with different preparation process	Extrusion and compression moulding	–	–	–	–	–	273–315[a]	227–276[a]	Berthet et al. (2015)
PHBV/wheat straw fibres	20 wt% fibres	Extrusion and compression moulding	ATBC, GTA and PEG/	−14–3[a]	–	161–170[a]	70–95	–	–	Martino et al. (2015)
PHBV/keratin from poultry feathers	0.5, 1, 3, 5, 10, 25 and 50 wt% filler	Brabender Plastograph blending and compression moulding	–	–	101.9–103.3	147.1–148.3	35–49.9	–	–	Pardo-Ibáñez et al. (2014)
PBAT/coffee ground	10,20 and 30 wt% filler with roasting process at 250 and 270 °C	Extrusion with die attached with the mixing chamber	–	–	80.7–88.1	95.8–115.1	6.3–6.7	334.7–404.8[a]	278.5–353.3[a]	Moustafa et al. (2016)
Mater Bi–KE/ cotton, kenaf, and hemps fibres	10% (w/w) fibres	Extrusion and compression moulding	–	–	41.85[a]–46.65	62.85[a]–67.95	61.1[a]–68.8	335[a]–356	319[a]–334	Moriana et al. (2011)
PCL/cellulose fibres from cotton	5 and 15 wt% filler	Brabender Plastograph blending and compression moulding	–	–	–	–	–	354–417[a]	–	Ludueña et al. (2012)

(continued)

Table 4.5 (continued)

Composite material	Filler content	Processing method	Plasticizer/ equilibrium RH	Calorimetric parameters				Thermal degradation		Reference
				T_g (°C)	T_c (°C)	T_m (°C)	ΔHm (J/g)	T_{peak} (°C)	T_{onset} (°C)	
Polymer/nano-filler										
Corn starch-gelatin/ CNC from eucalyptus wood pulp	0.44, 1.5, 2.56 and 3% CNC	Casting	Glycerol/	–	–	–	–	296.39–298.47[a]	248.34–304.13	Alves et al. (2015)
Pea starch-PVA/ CNC from MCC	1, 3 and 5 wt% CNC	Casting	Glycerol/53%	73.9–78.6[a]	200.7[a]–202.3	225.6–227.04[a]	61–108[a]	419[a]–431.9	–	Cano et al. (2015)
PLA-PHB/CNC from MCC	5% CNC	Extrusion and compression moulding	–	55.3–62.5		148.6–150.2	17.7[a]–28.6	267–357[a]	278[a]–280	Arrieta et al. (2014b)
PLA-PHB/CNC from MCC	1 and 5 wt% CNC 15 wt% plasticizer	Electrospinning Extrusion and compression moulding	ATBC/	27.1–51[a]	80.7–107.6[a]	147.5–155[a]	35–48[a]	332–340	76[a]–141.1	Arrieta et al. (2015)
PLA-PBS/CNC from *Carmagnola* carded hemp fibres	1 and 3 wt% CNC with/without surfactant	Casting	–	46.9–54.6	–	138.6–138.6[a]	21.5–30.3	344–364[a]	–	Luzi et al. (2016)
PLA/CNC from *Posidonia oceanica* plant waste	1 and 3 wt% CNC	Casting	–	40.6–54.3	91.5–118.6[a]	157.2–164.1	37.4[a]–40.4	313–332	240–270[a]	Fortunati et al. (2015)
PLA/CNC from *Phormium tenax* leaves	1 and 3 wt% CNC with or without 20 wt% plasticizer	Extrusion and compression moulding	Limonene	31.6–59.6	92.6–112.6[a]	143.2–147.9[a]	29.1[a]–44.0	352–357[a]	–	Fortunati et al. (2014)
PLLA/CNC from eucalyptus wood pulp	10 wt% CNC	Casting	–	59[a]	98–100[a]	160–162[a]	41[a]–48	–	–	De Paula et al. (2016)

Composite material	Filler content	Processing method	Plasticizer/ equilibrium RH	Calorimetric parameters				Thermal degradation		Reference
				T_g (°C)	T_c (°C)	T_m (°C)	ΔHm (J/g)	T_{peak} (°C)	T_{onset} (°C)	
PLA/CNC from MCC	1 and 3 wt% CNC-g-PLLA	Grafting Extrusion and compression moulding	–	57.7–60.4[a]	96.2–100.5[a]	166–168.8	42.4–47.9	300–310	–	Lizundia et al. (2016)
Poly(butylene/ triethylene succinate)/CNC from MCC	1 and 5 wt% CNC with surfactant	Extrusion and compression moulding	–	−40–33	50–89[a]	84–114[a]	27–63	402–407	–	Fortunati et al. (2017)

T_g glass transition temperature; T_c crystallisation temperature; T_m melting temperature; ΔHm melting enthalpy; T_{peak} peak temperature in DGTA; T_{onset} onset temperature in DTGA

Data show the range in the values of each property reported for the different formulations

[a]Identifies the value of the control sample (without filler) when it is in the edge of the range

network of the matrices becomes more resistant to heat based on the inherently high heat resistance of organic and inorganic fillers. Lignocellulosic fillers, such as wheat straw fibres (Berthet et al. 2015; Martino et al. 2015), kenaf fibres (Moriana et al. 2011), garlic straw (Kallel et al. 2016), rice husk (Johar et al. 2012), sisal fibres (Santos et al. 2015), pineapple leaf fibres (Shih et al. 2014), soy hull (Flauzino Neto et al. 2013), rice straw (Boonterm et al. 2015), coconut husk fibres (Rosa et al. 2010) or banana peel waste (Hossain et al. 2016) decompose in the temperature range of 150–500 °C: specifically, hemicellulose decomposes mainly from 150 to 350 °C, cellulose at between 275 and 350 °C and lignin undergoes gradual decomposition in the range of 250–500 °C. This high/wide range of decomposition temperatures promotes the greater thermal resistance of composites. Moriana et al. (2011), found a T_{peak} increase of 6% when natural micro-fibres (cotton, kenaf and hemp fibres) were incorporated into starch-based composites. The greatest increase was obtained with kenaf fibres, probably due to the better compatibility between this filler and the starch matrix. This was associated with the higher content of hemicellulose, which promotes the hydrogen bonding between the fibres and the matrix, improving the interfacial adhesion and thermal stability. However, with other biopolymer matrices, the addition of micro-fibres did not affect the thermal stability as described by Berthet et al. (2015) for PHBV-wheat straw micro-fibres blends. The presence of lignocellulosic micro-fibres could contribute to a reduction in the mean polymer molecular weight of the blend, reducing the overall thermal stability, as was also observed by Ludueña et al. (2012) in PCL-cotton micro-fibre films.

The particle size reduction from micro- to nano-scale of fibres (e.g. by means of alkali and bleaching treatments of lignocellulosic material and acid hydrolysis to obtain pure cellulose nanocrystals (Brinchi et al. 2013; Jonoobi et al. 2015; Zhou et al. 2016), implies a high yield in thermal resistance, as well as in the previously mentioned barrier and mechanical properties. The incorporation of CNCs into biopolymer composites improved their thermal stability due to the crystalline structure and compact chains present in the nanocrystals, which are not easily dissociated by heating, increasing the thermal stability (Ng et al. 2015). Arrieta et al. (2014a, b and 2015) reported greater thermal stability in PLA-PHB blends reinforced with 1 or 5 wt% of CNC, from commercial MCC, obtained by electrospinning or extrusion processes. Similar behaviour was observed by Cano et al. (2015) in PVA-starch matrices with 1, 3 and 5 wt% of the same reinforcing agent.

As concerns the influence of inorganic nano-fillers on the thermal properties of composites, they also enhanced the thermal stability of biopolymer matrices. Rhim et al. (2014) studied the use of Ag nanoparticles in glycerol plasticised agar matrices obtained by casting. The thermogravimetric analysis exhibited a high residual mass of the composite films due to the inclusion of the more thermically stable metallic nanoparticles. Cavallaro et al. (2013) obtained pectin-PEG blends with nano-clays, specifically hallosyte nanotubes, at 5, 10, 15, 20, 30 and 50 wt% by casting. The thermal degradation analyses reflected the fact that nano-composites had a high degree of thermal resistance in comparison with the control sample, which was attributed to the fact that the nano-clay lumen can encapsulate the pectin degradation products delaying the process. Moreover, the good dispersion of the nano-filler inside the polymer matrix improved the thermal stabilization of the biopolymer.

4.5 The Surface Properties of Micro- and Nano-Reinforced Polymers for Food Applications

In this section, recent studies into the effect of the addition of micro- and nano-fillers of differing characteristics on the surface properties of some biopolymers are analysed, and summarised in Table 4.6. The main changes in biopolymer functional properties caused by the addition of a filler are strongly associated with surface properties and the interfacial interactions between biopolymer and filler. Several methods have been used to characterise the morphology and the surface composition/structure of biomaterials, such as contact angle, electron spectroscopy for chemical analysis (ESCA) or X-ray photoelectron spectroscopy (XPS), secondary ion mass spectrometry (SIMS), scanning electron microscopy (SEM), atomic force microscopy (AFM) (Gutiérrez et al. 2018). The surface properties of the composites can directly impact on the macroscopic observation of the material gloss, which can have notable influence on their practical applications. In this section, recent studies into the effect of different fillers on the surface hydrophobicity (contact angle), topographic analysis (AFM) or sample gloss are discussed and summarised in Table 4.6.

The contact angle (θ) of a liquid in contact with a solid material mainly depends on the balance between the adhesive liquid-solid forces and the cohesive forces of the liquid (Gutiérrez et al. 2018). Aqueous or organic solvents can be used on a determined material in order to characterise the relative affinity of the material for polar or non-polar systems, thus obtaining information about its respective wettability properties according to the hydrophobic-hydrophilic nature of the surface. The θ values vary according to the type of biopolymer and nature of the filler (organic/inorganic). The inorganic fillers, such as nano-clays, could negatively affect the surface hydrophobicity of matrices due to their great water affinity, as reported by Abdollahi et al. (2013) for matrices of alginate-Mnt nano-clays at 1, 3 and 5 wt%. The film's surface was more hydrophilic than the control sample mainly due to the hydrophilic nature of the Mnt also present at surface level. However, the same authors observed an 87.5% increase in hydrophobicity when they used CNC from commercial MCC in the same matrices. Films with 5 wt% CNC exhibited the highest degree of hydrophobicity, which was associated with the high ratio of CNCs at surface level and their crystalline nature, with lower water affinity than the alginate matrix. Similar results are reported by Slavutsky and Bertuzzi (2014) for thermoplastic starch (TPS) matrices reinforced with CNCs from sugarcane bagasse. The water contact angle increased (rise in the surface hydrophobicity) when CNCs were incorporated into TPS, while strong interactions and the formation of hydrogen bonds between the starch chains and CNCs are expected. These strong internal bonds could also reduce the surface interactions between water molecules and the material. Cao et al. (2008) obtained nanocomposites with TPS and CNCs from hemp fibres and they also observed an increase in the water contact angle or surface hydrophobicity of the matrices. On the contrary, the incorporation of CNCs into hydrophobic polymer matrices, such as PBS, enhanced the water wettability of the films (decrease in contact angle). This could be expected from the surface presence

Table 4.6 Effect of fillers on surface properties of different composite films

Composite material	Formulation	Processing method	Surface analysis	Effect on polymer matrix	Reference
Polymer/micro-filler					
PHBV/wheat straw fibres (WSF)	20 wt% fibres 10 wt% plasticizer (ATBC, GTA or PEG)	Extrusion and compression moulding	– Surface hydrophobicity (contact angle reference liquids: distilled water, diiodomethane ethylene glycol and glycerol)	Contact angle values of the plasticized composite with WSFs were lower than that of the PHBV matrix	Martino et al. (2015)
Polymer/nano-filler					
Alginate/nano-clays Mnt and CNC from MCC	1, 3 and 5 wt% fillers	Casting	– Surface hydrophobicity (contact angle)	Composites with Mnt had more hydrophilic surface Composites with CNC had more hydrophobic surface due to their highly crystalline nature	Abdollahi et al. (2013)
Pea starch-PVA/CNC from MCC	1, 3 and 5 wt% CNC	Casting	– Gloss	The incorporation of filler do not affect gloss in composites	Cano et al. (2015)
TPS/CNC from sugarcane bagasse	Appropriate amount of CNC suspension and glycerol as plasticizer	Casting	– Surface hydrophobicity (contact angle)	Contact angle increased with CNC addition Strong interactions between starch chains and CNC, which reduced the water affinity of the film surface	Slavutsky and Bertuzzi (2014)

(continued)

Table 4.6 (continued)

Composite material	Formulation	Processing method	Surface analysis	Effect on polymer matrix	Reference
Wheat gluten/ CNC and CNF from sunflower stalks	1 and 3 wt% CNC or CNF	Casting	– Gloss – AFM topographic analysis	Good distribution of CNC into the matrix and some regions with aggregated CNF CNC promoted gloss a function of filler content CNF decreased gloss as a function of filler content	Fortunati et al. (2016)
Poly(butylene/ triethylene succinate)/ CNC from MCC	1 and 5 wt% CNC, with surfactant	Extrusion and compression moulding	– Gloss – Surface hydrophobicity (contact angle)	Decreased the gloss value as the amount of CNC increased Higher contact angle values for 1% filler	Fortunati et al. (2017)
PLA-PHB/ CNC from MCC	1 wt% and 5 wt% CNC 15 wt% plasticizer	Extrusion and compression moulding	– Topographic analysis by AFM	Presence of aggregated and individualized CNC The surfactant allowed for the polymer chain penetration between the cellulose structures	Arrieta et al. (2014b)

of the cellulose hydroxyl groups, which favour water affinity at the surface (Fortunati et al. 2017).

From the AFM analyses, the presence of nanoparticles on the composite surface and their aggregation/isolation state can be assessed, while their effect on the surface roughness can be verified. Arrieta et al. (2014b) studied the surface properties of PLA-PHB matrices containing CNCs from MCC. The AFM analysis showed the presence of some agglomerated and individualised CNCs in matrices. The aggregation of nanoparticles was reduced by the use of surfactants, which allowed for a better polymer chain interaction with the cellulose nanostructure. Nevertheless, an opposite effect was deduced by Fortunati et al. (2016) from the AFM images for CNCs in wheat gluten matrices, probably due to the different kinds of interactions between cellulose and the amphiphilic protein chains. The tendency of nanocrystals to aggregate has been widely found in numerous studies (Brinchi et al. 2013; Ng et al. 2015; Zhou et al. 2016) due to the spontaneous tendency to reduce the interfacial free energy of the system, accumulated at the contact surface area. The aggregated nanocrystals can be successfully dispersed and homogenised by strong mechanical shearing effects into a homogeneous suspension (Ng et al. 2015) or with the incorporation of surfactants to achieve a good dispersion in the matrices (Hu et al. 2015; Kaboorani and Riedl 2015); all of this is dependent on the nature of the polymer and filler and the processing conditions.

The influence of fillers on the material gloss is related with the surface topography achieved in the composite. Materials with aggregated fillers exhibit greater surface roughness so that they are less bright than other homogeneous material with a smoother surface. Fortunati et al. (2016) studied the homogeneity of CNC and cellulose nano-fibril (CNF) dispersion in wheat gluten composites and observed changes in the material gloss as a function of the filler. In the case of bionanocomposites that are reinforced with CNC, the values of gloss increase as a function of the filler percentage while the opposite behavior was observed in the CNF-reinforced materials. This could be related with the presence of CNF aggregates on the composite surface, evidenced by optical microscopy, whereas in CNC nanocomposites, nanoparticles were homogeneously distributed in the matrix. Cano et al. (2015) observed that the addition of CNCs to pea starch-PVA matrices did not affect the gloss values, as compared with the control samples, which was attributed to the good CNC dispersion in the biopolymer blends, with strong adhesion forces between the filler and the matrix.

4.6 The Effect of Reinforcing Agents on the Material Biodegradability

The disintegration and biodegradation behavior of the materials is analyzed through their composting under controlled aerobic processes, designed to produce organic residues from the biodegradable parts of the material, by the action of microorganisms. In this sense, ISO standards establish methodologies, where specific disposal pathways, specific time frames and criteria are indicated in order to unify a proper composting analysis (Cano et al. 2016). The biodegradation behavior is a crucial factor for the purposes of developing environmentally-friendly packaging materials. Biodegradable polymers are able to decompose in the medium by the enzymatic action of microorganisms in a defined period (Nair et al. 2017). In the disintegration and biodegradation processes produced by the action of microorganisms (bacteria, fungi and algae), these identify the polymer as a source of energy to produce organic residues from the biodegradable materials. These chemically react under the microbial enzymatic action and the polymer chains are fragmented (Cano et al. 2016; Nair et al. 2017).

Table 4.7 shows the effect of some reinforcing agents on the composite disintegration or biodegradationl, using different composting conditions. The degradation rate of the materials varies according to the type of polymer and reinforcing agent.

Table 4.7 Effect of reinforcing agents on the material biodegradability

Composite material	Formulation	Processing method	Biodegradation test	Control method	Effect of filler on biodegradation of polymer matrix	Reference
Polymer/micro-filler						
PCL/cellulose fibres from cotton	5 and 15 wt% filler	Brabender Plastograph blending and compression moulding	– Samples: 10 mm × 20 mm × 0.3–0.5 mm – Compost material: natural microflora present in soil (Pinocha type) – Incubation: 20 °C, 40% HR under aerobic conditions – Time tested: 6 months	– Average weight loss (%WL)	The high hydrophilicity of the natural fibres promoted the water intake and provides a rougher support for microbial growth	Ludueña et al. (2012)
Polymer/nano-filler						
PCL/CNC from cotton	5 and 15 wt% filler	Brabender Plastograph mixer and compression moulding	– Samples: 10 mm × 20 mm × 0.3–0.5 mm – Compost material: natural microflora present in soil (Pinocha type) – Incubation: 20 °C, 40% HR under aerobic conditions – Time tested: 6 months	– Average weight loss (%WL)	The crystalline structure of CNC promoted the water intake	Ludueña et al. (2012)
PLA-PHB/CNC from MCC	1 and 5 wt% CNC 15 wt% ATBC as plasticizer	Electrospinning	– Sample: 15 mm × 15 mm – Compost materials: 10% compost, 30% rabbit food, 10% starch, 5% sugar, 1% urea, 4% corn oil, 40% sawdust and 50 wt% of water content – Incubation: 58 °C, under aerobic conditions – Time tested: 28 days	– Photographs of physical changes – SEM analysis	The presence of nano-filler speeded up the disintegration process Matrices became breakable after 10 days of composting	Arrieta et al. (2015)

(continued)

Table 4.7 (continued)

Composite material	Formulation	Processing method	Biodegradation test	Control method	Effect of filler on biodegradation of polymer matrix	Reference
PLA/PVAc/CNC from Whatmann paper	3 wt% CNC and 10 wt% Glycidyl methacrylate resect to PVAc	Extrusion	– Samples: – – Compost material: sawdust, rabbit food, starch, sugar, oil and urea – Incubation: 58 °C, 50% humidity under aerobic conditions – Time tested: 60 days	– Disintegration value	Variation in terms of mass loss was limited because the water attack starts on the more susceptible component with hydroxyl groups available on the surface	Haque et al. (2017)
PLA/CNC from *Phormium tenax leaves*	1 and 3 wt% CNC with or without 20 wt% limonene as plasticizer	Extrusion and compression moulding	– Samples: 15 mm × 15 mm × 0.05 mm – Compost material: compost inoculum, sawdust, rabbit food, starch, sugar, oil, urea and 50 wt% water of content – Incubation: 58 °C, under aerobic conditions – Time tested: 14 days	– Photographs of physical changes – FESEM – FTIR – Disintegration value	The CNCs increased the crystallinity and inhibited water diffusion into the material, causing a lower disintegration rate	Fortunati et al. (2014)
PLA-PBS/CNC from *Carmagnola* carded hemp fibres	1 and 3 wt% CNC with/ without surfactant	Casting	– Samples: 15 mm × 15 mm × 0.03 mm – Compost material: sawdust, rabbit food, compost inoculum, starch, sugar, oil and urea – Incubation: 58 °C, 50%HR under aerobic conditions – Time tested: 90 days	– Photographs of physical changes – Degree of disintegration (*D*)	CNCs benefited disintegration process due to their hydrophilic nature Hydrophilic surfactant improved the CNC dispersion in the matrices and the *D* parameter	Luzi et al. (2016)

Composite material	Formulation	Processing method	Biodegradation test	Control method	Effect of filler on biodegradation of polymer matrix	Reference
PLA/Ag-NPs	1 wt% CNC and 2 ratio of antioxidant	Extrusion and compression moulding	– Samples: 15 × 5 × 2 mm^3 – Compost material: sawdust, rabbit food, starch, oil and urea-Incubation: 58 °C, under aerobic conditions – Time tested: 35 days	– DSC – FTIR – FESEM – Disintegration test	Ag-NPs and thymol (plasticizer) accelerated the PLA hydrolysis process Ag atoms could catalyse the disintegration process	Ramos et al. (2014)
Starch-PVA/Ag-NPs	0.6, 6, 16 and 32% respect to starch ratio	Casting	– Compost material: organic fraction of solid municipal waste and vermiculite – Incubation: 58 °C, under aerobic conditions – Time tested: 45 days	– CO_2 produced – Disintegration test	Ag-NPs enhanced film disintegration due to the incorporation of structural discontinuities in the composite network Low Ag-NPs concentrations are recommended to avoid alterations in the bio-degradation process	Cano et al. (2016)
PLA/Montmorillonite and Fluoro-hectorite nano-clays	5 and 10 wt% filler	Extrusion and compression moulding	– Samples: 75 mm × 0.5 mm – Compost material: pruning residues – Incubation: 40 °C, 50 70% RH under aerobics conditions – Time tested: 35 days	– Photographs of physical changes – DSC – Tensile properties	The hydroxyl groups of silicate layers and/or of their organic modifiers promoted biodegradation of the PLA matrix	Fukushima et al. (2013)

(continued)

Table 4.7 (continued)

Composite material	Formulation	Processing method	Biodegradation test	Control method	Effect of filler on biodegradation of polymer matrix	Reference
Methyl cellulose/ Montmorillonite	1, 2, 3, 5 and 10 wt% filler	Casting	– Samples: 5 g of fragmented samples – Compost material: 70% topsoil and 30% composted manure. 80–100% moisture content – Incubations: 25 °C in dark cabinet – Time tested: 42 days	– CO_2 produced	The filler improved the composite barrier properties, restricting the segmental movement at the interface The silicate layers on the film surface could hide part of the polymer chains, making biodegradation more difficult	Rimdusit et al. (2008)
Alginate/Chitin whiskers	0.1 and 2 wt% filler	Extrusion	– Medium: Tris-HCl buffer solution (pH =7.4) or buffer solution containing 0.4 mg/mL of lysozyme (pH = 7.4) – Incubation: Shaking incubator at 37 °C – Time tested: 5 days	– SEM – Mechanical integrity	Biodegradation process was improved by lysozyme Surface of the samples was smoother and weight loss was higher	Watthanaphanit et al. (2008)
Poly(butylene/ triethylene succinate)/CNC from MCC	1 and 5 wt% CNC, with surfactant	Extrusion and compression moulding	– Samples: 20 × 30 mm and ~35 mg of weight – Compost material: 22.08% organic carbon, 13.44% humic and fulvic carbon – Incubation: 58 °C – Time tested: 30 days	– Residual mass – Visual observations – SEM	The surface hydrophilicity increased when the CNC content rose. This provoked higher resistance to water uptake and diffusion, delaying disintegration process	Fortunati et al. (2017)

Ludueña et al. (2012) analysed the biodegradation behavior of PCL composites containing cotton fibres and CNC from cotton fibres at 20 °C and 40% relative humidity (RH), using compost material with the natural microflora present in soil (Pinocha type). The analysis was carried out throughout 6 months by controlling the average weight loss of the samples. They observed that the biodegradability of the reinforced material was enhanced with the addition of both kinds of fillers, which was attributed to the high hydrophilicity of the natural fibres, which promoted water transport and provided a rougher support for microbial growth. PCL is semicrystalline polyester and the reduction in the degree of crystallinity benefits the biodegradation process, since the amorphous regions are more quickly attacked by microorganisms. Similar conclusions were reported by Luzi et al. (2016), for PLA-PBS composites reinforced with CNC from *Carmagnola* carded hemp fibres submitted to composting in sawdust, rabbit food, compost inoculum, starch, sugar, oil and urea, at 58 °C and 50% HR throughout 90 days. The authors evaluated the degree of disintegration (D) and the physical changes and observed that the presence of CNC in the matrices benefited the biodegradation process. Likewise, the use of a hydrophilic surfactant improved the dispersion of cellulosic nano-fillers in the matrices and the D parameter. The biodegradationl in blends with PBS was retarded due to the more hydrophobic and semicrystalline nature of PBS. However, Fortunati et al. (2014) observed CNC from *Phormium tenax* leaves had the opposite effect on PLA composites. The presence of CNCs increased the crystallinity of the composites, limiting the water transport through the PLA matrices. When limonene was incorporated as plasticiser in the PLA composites, an increase in the biodegradation rate was observed due to an improvement in the chain mobility, which favoured the polymer erosion. Nevertheless, each sample was 90% disintegrated after 14 days of composting, which is within the limit defined by the ISO 20200. In PBS composites containing CNC from MCC, similar behavior was observed, but the hydrophobic nature of PBS and the degree of crystallinity slowed down the biodegradation process (Fortunati et al. 2017).

The addition of inorganic nano-fillers provoked similar effects to those brought about by organic fillers in the biodegradation process of composites. Ramos et al. (2014) studied the effect of Ag nanoparticles on the disintegrability of PLA composites at 58 °C, using a compost media made from sawdust, rabbit food, starch, oil and urea, throughout 35 days. They assumed that Ag atoms could catalyse the disintegration process and the synergies between the Ag-NPs and thymol could accelerate the hydrolysis process. The presence of homogeneously dispersed thymol in the PLA matrix could promote the polymer chain mobility and thus, diffuse the water molecules through the PLA structure. Cano et al. (2016) also observed an increase in the film disintegration rate when different ratios of Ag-NPs were incorporated in starch-PVA composites. Nevertheless, the generation of CO_2 as the result of total carbon conversion was notably reduced when the Ag ratio increased, probably due to its antimicrobial effect on the microorganisms responsible for the biodegradation process. Fukushima et al. (2013) also observed an increase in the disintegration rate of PLA composites with Mnt and fluorohectorite nano-clays at 5 and 10 wt%. The biodegradation of PLAI matrices was enhanced by the catalytic effect of the hydroxyl groups of silicate layers.

It is remarkable that many factors can affect the degradation rate of composites. The environmental conditions have a significant impact on microbial growth and parameters, such as humidity, temperature, pH, salinity, oxygen pressure, and microbial nutrients, have a great influence on the microbial degradation of polymers. The biodegradation process also depends on the chemical and physical characteristics of the biopolymer. Nair et al. (2017) report that the enzymatic degradation implies the binding of the enzyme to the bioplastic surface, followed by hydrolytic split; biopolymers are degraded into low-molecular-weight oligomers, dimers, monomers and finally mineralised to CO_2 and H_2O. For instance, the biodegradation of PLAI starts with the hydrolysis of the polymer chains promoted by the water diffusion in the matrices. When the molecular weight reaches about 10,000–20,000 g mol^{-1}, microorganisms, such as fungi and bacteria, can metabolise the macromolecules, converting them into carbon dioxide, water and humus (Luzi et al. 2016; Fortunati et al. 2014; Fukushima et al. 2013). Several microorganisms are able to decompose biomaterials, such as *Tritirachium album, Amycolatopsis* strain 41, *Amycolatopsis* sp. strain 3118, *Kibdelosporangium aridum* for PLA, *Penicillium* sp. strain 26–1 (ATCC 36507), *Aspergillus* sp. strain ST-01, *Clostridium* sp. for PCL or *Pseudomonas* sp., *Bacillus* sp., *Streptomyces* sp., *Aspergillus* sp. for PHB (Nair et al. 2017).

4.7 Final Remarks

The incorporation of reinforcing agents of different natures (organic or inorganic) and size (micro or nano-sized) represents a good strategy for the purposes of improving the functional properties of biopolymers. In general, improved barrier and mechanical properties can be achieved when compatible micro or nano particles are adequately dispersed in the polymer matrix. Nano-particles are generally more effective, but their natural tendency towards aggregation makes the dispersion process difficult, requiring carefully designed dispersion techniques. To a great extent, the surface interactions of the filler with the polymer matrix define the effectiveness of the reinforcement and the promotion of barrier properties. Therefore, if materials with optimal functionality are to be obtained, it is of relevance to make an adequate selection of both the filler for a determined polymer matrix and the processing conditions necessary to ensure high dispersion levels of the particles. Composite biodegradability is generally enhanced by the presence of the filler dispersed particles. However, the total conversion of carbon to CO_2 through the action of microorganisms could be limited when the filler exhibits antimicrobial action.

Acknowledgements The authors thank the Ministerio de Economía y Competitividad (Spain) for the financial support provided through Project AGL2016-76699-R.

References

Abdollahi M, Alboofetileh M, Rezaei M et al (2013) Comparing physico-mechanical and thermal properties of alginate nanocomposite films reinforced with organic and/or inorganic nanofillers. Food Hydrocoll 32:416–424

Abdul Khalil HP, Davoudpour Y, Saurabh C et al (2016) A review on nanocellulosic fibres as new material for sustainable packaging: process and applications. Renew Sust Energ Rev 64:823–836

Alves JS, dos Reis KC, Menezes EGT et al (2015) Effect of cellulose nanocrystals and gelatin in corn starch plasticized films. Carbohydr Polym 115:215–222

Arrieta M, Peltzer M, López J et al (2014a) Functional properties of sodium and calcium caseinate antimicrobial active films containing carvacrol. J Food Eng 121:94–101

Arrieta MP, Fortunati E, Dominici F et al (2014b) Multifunctional PLA–PHB/cellulose nanocrystal films: processing, structural and thermal properties. Carbohydr Polym 107:16–24

Arrieta MP, López J, Kenny JM et al (2015) Biodegradable electrospun bionanocomposite fibers based on plasticized PLA–PHB blends reinforced with cellulose nanocrystals. Ind Crop Prod 96:290–301

ASTM (2003) Standard test method for determining aerobic biodegradation of plastic materials under controlled composting conditions, incorporating thermophilic temperatures. Standards designations: D5338. In: Annual book of ASTM standards. American Society for Testing and Materials, Philadelphia, PA

Azeredo HMC (2009) Nanocomposites for food packaging applications. Food Res Int 42:1240–1253

Azeredo HMC, Rosa MF, Mattoso LHC (2017) Nanocellulose in bio-based food packaging applications. Ind Crop Prod 97:664–671

Bera A, Dubey S, Bhayani K et al (2015) Microbial synthesis of polyhydroxyalkanoate using seaweed-derived crude levulinic acid as co-nutrient. Int J Biol Macromol 72:487–494

Berthet MA, Angellier-Coussy H, Chea V et al (2015) Sustainable food packaging: valorising wheat straw fibres for tuning PHBV-based composites properties. Compos Part A Appl Sci Manuf 72:139–147

Bonilla J, Fortunati E, Vargas M (2013) Effects of chitosan on the physicochemical and antimicrobial properties of PLA films. J Food Eng 119(2):236–243

Boonterm M, Sunyadeth S, Dedpakdee S et al (2015) Characterization and comparison of cellulose fiber extraction from rice straw by chemical treatment and thermal stem explosion. J Clean Prod 134:592–599

Boumail A, Salmieri S, Klimas E et al (2013) Characterization of trilayer antimicrobial diffusion films (ADFs) based on methylcellulose–polycaprolactone composites. J Agric Food Chem 61:811–821

Brinchi L, Cotana F, Fortunati E et al (2013) Production of nanocrystalline cellulose from lignocellulosic biomass: technology and applications. Carbohydr Polym 94:154–169

Cano A, Fortunati E, Cháfer M et al (2015) Effect of cellulose nanocrystals on the properties of pea starch–poly(vinyl alcohol) blend films. J Mater Sci 50:6979–6992

Cano A, Cháfer M, Chiralt A et al (2016) Biodegradation behaviour of starch-PVA films as affected by the incorporation of different antimicrobials. Polym Degrad Stab 132:11–20

Cao X, Chen Y, Chang PR et al (2008) Green composites reinforced with hemp nanocrystals in plasticized starch. J Appl Polym Sci 109(6):3804–3810

Carbone M, Donia DM, Sabbatella G et al (2016) Silver nanoparticles in polymeric matrices for fresh food packaging. J King Saud Univ Sci 28:273–279

Cavallaro G, Lazzara G, Milioto S (2013) Sustainable nanocomposites based on halloysite nanotubes and pectin/polyethylene glycol blend. Polym Degrad Stab 98:2529–2536

Chen D, Lawton D, Thompson MR et al (2012) Biocomposites reinforced with cellulose nanocrystals derived from potato peel waste. Carbohydr Polym 90:709–716

Cheng Y, Deng S, Chen P et al (2009) Polylactic acid (PLA) synthesis and modifications: a review. Front Chem China 4(3):259–264
Cho J, Joshi MS, Sun CT (2006) Effect of inclusion size on mechanical properties of polymeric composites with micro and nano particles. Compos Sci Technol 66(13):1941–1952
Correa JP, Molina V, Sanchez M et al (2017) Improving ham shelf life with a polyhydroxybutyrate/polycaprolactone biodegradable film activated with nisin. Food Packaging Shelf Life 11:31–39
Corsello FA, Bolla PA, Anbinder PS et al (2017) Morphology and properties of neutralized chitosan-cellulose nanocrystals biocomposite films. Carbohydr Polym 156:452–459
Dash S, Swain SK (2013) Synthesis of thermal and chemical resistant oxygen barrier starch with reinforcement of nano silicon carbide. Carbohydr Polym 97:758–763
De Paula EL, Roig F, Mas A et al (2016) Effect of surface-grafted cellulose nanocrystals on the thermal and mechanical properties of PLLA based nanocomposites. Eur Polym J 84:173–187
Dominguez-Martinez B, Martínez-Flores H, Berrios J et al (2017) Physical characterization of biodegradable films based on chitosan, polyvinyl alcohol and opuntia mucilage. J Polym Environ 25(3):683–691
El Miri N, Abdelouahdi K, Barakar A et al (2015) Bio-nanocomposite films reinforced with cellulose nanocrystals: rheology of film-forming solutions, transparency, water vapor barrier and tensile properties of films. Carbohydr Polym 129:156–167
El-Hadi A (2017) Increase the elongation at break of poly (lactic acid) composites for use in food packaging films. Sci Rep 7:46767. https://doi.org/10.1038/srep46767
Emadian SM, Onay TT, Demirel B (2017) Biodegradation of bioplastics in natural environments. Waste Manag 59:526–536
Fabra MJ, Talens P, Gavara R et al (2012) Barrier properties of sodium caseinate films as affected by lipid composition and moisture content. J Food Eng 109(3):372–379
Fabra MJ, López-Rubio A, Lagaron JM (2014) Biopolymers for food packaging applications. In: Aguilar de Armas MR, Román JS (eds) Smart polymers and their applications. Elsevier, Amsterdam, pp 476–509
Fabra MJ, López-Rubio A, Ambrosio-Martín J et al (2016) Improving the barrier properties of thermoplastic corn starch-based films containing bacterial cellulose nanowhiskers by means of PHA electrospun coatings of interest in food packaging. Food Hydrocoll 621:261–268
Flauzino Neto WP, Silvério HA, Dantas NO et al (2013) Extraction and characterization of cellulose nanocrystals from agro-industrial residue soy hulls. Ind Crop Prod 42:480–488
Follain NG, Belbekhouche S, Bras J et al (2013) Water transport properties of bionanocomposites reinforced by Luffa cylindrica cellulose nanocrystals. J Membr Sci 427:218–229
Fortunati E, Puglia D, Luzi F et al (2013a) Binary PVA bio-nanocomposites containing cellulose nanocrystals extracted from different natural sources: part I. Carbohydr Polym 97:825–836
Fortunati E, Pelltzer M, Armentano I et al (2013b) Combined effects of cellulose nanocrystals and silver nanoparticles on the barrier and migration properties of PLA nano-biocomposites. J Food Eng 118:117–124
Fortunati E, Luzi F, Puglia D et al (2014) Investigation of thermo-mechanical, chemical and degradative properties of PLA-limonene films reinforced with cellulose nanocrystals extracted from Phormium tenax leaves. Eur Polym J 56:77–91
Fortunati E, Luzi F, Puglia D et al (2015) Processing of PLA nanocomposites with cellulose nanocrystals extracted from *Posidonia oceanica* waste: innovative reuse of coastal plant. Ind Crop Prod 67:439–447
Fortunati E, Luzi F, Jiménez A et al (2016) Revalorization of sunflower stalks as novel sources of cellulose nanofibrils and nanocrystals and their effect on wheat gluten bionanocomposite properties. Carbohydr Polym 149:357–368
Fortunati E, Gigli M, Luzi F et al (2017) Processing and characterization of nanocomposite based on poly(butylene/triethylene succinate) copolymers and cellulose nanocrystals. Carbohydr Polym 165:51–60
Fukushima K, Tabuani D, Arena M et al (2013) Effect of clay type and loading on thermal, mechanical properties and biodegradation of poly(lactic acid) nanocomposites. React Funct Polym 73:540–549

Giménez B, López de Lacey A, Pérez-Santín E et al (2013) Release of active compounds from agar and agar–gelatin films with green tea extract. Food Hydrocoll 30:264–271
González K, Retegi A, González A et al (2015) Starch and cellulose nanocrystals together into thermoplastic starch bionanocomposites. Carbohydr Polym 117:83–90
Graupner N, Ziegmann G, Wilde F et al (2016) Procedural influences on compression and injection moulded cellulose fibre-reinforced polylactide (PLA) composites: influence of fibre loading, fibre length, fibre orientation and voids. Compos Part A Appl Sci Manuf 81:158–171
Gutiérrez TJ (2017) Chitosan applications for the food industry. In: Ahmed S, Ikram S (eds) Chitosan: derivatives, composites and applications. Wiley-Scrivener, Beverly, MA, pp 183–232. EE.UU. ISBN: 978-1-119-36350-7. https://doi.org/10.1002/9781119364849.ch8
Gutiérrez TJ (2018) Biodegradability and compostability of food nanopackaging materials. In: Cirillo G, Kozlowski MA, Spizzirri UG (eds) Composite materials for food packaging. Wiley-Scrivener, Beverly, MA, pp 263–289 EE.UU. ISBN: 978-1-119-16020-5. https://doi.org/10.1002/9781119160243.ch9
Gutiérrez TJ, Alvarez VA (2017) Cellulosic materials as natural fillers in starch-containing matrix-based films: a review. Polym Bull 74(6):2401–2430. https://doi.org/10.1007/s00289-016-1814-0
Gutiérrez TJ, Alvarez VA (2018) Bionanocomposite films developed from corn starch and natural and modified nano-clays with or without added blueberry extract. Food Hydrocoll 77:407–420. https://doi.org/10.1016/j.foodhyd.2017.10.017
Gutiérrez T, González P, Medina C et al (2017) Effect of filler properties on the antioxidant response of thermoplastic starch composites. In: Thakur V, Thakur M, Kessler M (eds) Handbook of composites from renewable materials. Wiley, New York, pp 337–369. https://doi.org/10.1002/9781119441632.ch14
Gutiérrez TJ, Ollier R, Alvarez VA (2018) Surface properties of thermoplastic starch materials reinforced with natural fillers. In: Thakur V, Thakur M (eds) Functional biopolymers. Springer series on polymer and composite materials. Springer, Cham, pp 131–158. https://doi.org/10.1007/978-3-319-66417-0_5
Haque MM, Puglia D, Fortunati E et al (2017) Effect of reactive functionalization on properties and degradability of poly(lactic acid)/poly(vinyl acetate) nanocomposites with cellulose nanocrystals. React Funct Polym 110:1–9
He J, Chong Yap R, Wong S et al (2015) Polymer composites for intelligent food packaging. J Mol Eng Mater 3(1):1–12
Heitmann A, Patrício P, Coura I et al (2016) Nanostructured niobium oxyhydroxide dispersed poly (3-hydroxybutyrate) (PHB) films: highly efficient photocatalysts for degradation methylene blue dye. Appl Catal B Environ 189:141–150
Herrera N, Salaberria AM, Mathe A et al (2016) Plasticized polylactic acid nanocomposite films with cellulose and chitin nanocrystals prepared using extrusion and compression molding with two cooling rates: effects on mechanical, thermal and optical properties. Compos Part A 83:89–97
Hossain ABMS, Ibrahim N, AlEissa MS (2016) Nano-cellulose derived bioplastic biomaterial data for vehicle bio-bumper from banana peel waste biomass. Data Brief 8:286–294
Hu Z, Ballinger S, Pelton R et al (2015) Surfactant-enhanced cellulose nanocrystal pickering emulsions. J Coll Interface Sci 439:139–138
Jiménez A, Fabra MJ, Talens P et al (2013) Physical properties and antioxidant capacity of starch–sodium caseinate films containing lipids. J Food Eng 116(3):695–702
Johar N, Ahmad I, Dufresne A (2012) Extraction, preparation and characterization of cellulose fibres and nanocrystals from rice husk. Ind Crop Prod 37:93–99
Jonoobi M, Oladi R, Davaoudpour Y et al (2015) Different preparation methods and properties of nanostructured cellulose from various natural resources and residues: a review. Cellulose 22:935–969
Kaboorani A, Riedl B (2015) Surface modification of cellulose nanocrystals (CNC) by a cationic surfactant. Ind Crop Prod 65:45–55
Kallel F, Bettaieb F, Khiari R et al (2016) Isolation and structural characterization of cellulose nanocrystals extracted from garlic straw. Ind Crop Prod 87:287–296

Kanmani P, Rhim JW (2014) Physicochemical properties of gelatin/silver nanoparticle antimicrobial composite films. Food Chem 148:162–169
Le Corre D, Angellier-Coussy H (2014) Preparation and application of starch nanoparticles for nanocomposites: a review. React Funct Polym 85:97–120
Leceta I, Guerrero P, de la Caba K (2013) Functional properties of chitosan-based films. Carbohydr Polym 93(1):339–346
Lizundia E, Fortunati E, Dominici F et al (2016) PLLA-grafted cellulose nanocrystals: role of the CNC content and grafting on the PLA bionanocomposite film properties. Carbohydr Polym 142:105–113
López OV, Ninago MD, Soledad Lencina MM et al (2015) Thermoplastic starch plasticized with alginate–glycerol mixtures: melt-processing evaluation and film properties. Carbohydr Polym 126:83–90
Ludueña L, Vázquez A, Alvarez V (2012) Effect of lignocellulosic filler type and content on the behavior of polycaprolactone based eco-composites for packaging applications. Carbohydr Polym 87:411–421
Luzi F, Fortunati E, Jiménez A (2016) Production and characterization of PLA PBS biodegradable blends reinforced with cellulose nanocrystals extracted from hemp fibres. Ind Crop Prod 93:276–289
Majeed K, Jawaid M, Hassan A et al (2013) Potential materials for food packaging from nanoclay/natural fibres filled hybrid composites. Mater Des 46:391–410
Maqsood HS, Baheti V, Wiener J et al (2016) Reinforcement of enzyme hydrolyzed longer jute micro crystals in polylactic acid. Polym Compos 39:1089–1097. https://doi.org/10.1002/pc.24036
Martino L, Berthet MA, Angellier-Coussy H et al (2015) Understanding external plasticization of melt extruded PHBV–wheat straw fibers biodegradable composites for food packaging. J Appl Polym Sci 41611:2–11
Miranda CS, Ferreira MS, Magalhães MT, Santos WJ et al (2015) Mechanical, thermal and barrier properties of starch-based films plasticized with glycerol and lignin and reinforced with cellulose nanocrystals. Mater Today Proc 2:63–69
Mizuno S, Maeda T, Kanemura C et al (2015) Biodegradability, reprocessability, and mechanical properties of polybutylene succinate (PBS) photografted by hydrophilic or hydrophobic membranes. Polym Degrad Stab 117:58–65
Moreno O, Gil A, Atarés L et al (2017) Active starch-gelatin films for shelf-life extension of marinated salmon. LWT Food Sci Technol 84:189–195
Moriana R, Vilaplana F, Karlsson S et al (2011) Improved thermo-mechanical properties by the addition of natural fibres in starch-based sustainable biocomposites. Compos Part A 42:30–40
Moustafa H, Guizani C, Dupont C et al (2016) Utilization of Torrefied coffee grounds as reinforcing agent to produce high-quality biodegradable PBAT composites for food packaging applications. ACS Sustain Chem Eng 5(2):1906–1916
Mukurubira AR, Mellem JM, Amonsou EO (2017) Effects of amadumbe starch nanocrystals on the physicochemical properties of starch biocomposite films. Carbohydr Polym 165:142–148
Muller J, González-Martínez C, Chiralt A (2017a) Combination of poly(lactic) acid and starch for biodegradable food packaging. Materials 10. https://doi.org/10.3390/ma10080952
Muller J, González-Martínez C, Chiralt A (2017b) Poly(lactic) acid (PLA) and starch bilayer films, containing cinnamaldehyde, obtained by compression moulding. Eur Polym J 95:56–70
Nair NR, Sekhar VC, Nampoothir KM et al (2017) Biodegradation of biopolymers. In: Pandey A, Negi S, Soccol CR (eds) Current developments in biotechnology and bioengineering. Elsevier, Amsterdam, pp 739–755
Ng HM, Sin LT, Tee TT et al (2015) Extraction of cellulose nanocrystals from plant sources for application as reinforcing agent in polymers. Compos Part B 75:176–200
Ortega-Toro R, Bonilla J, Talens P et al (2017) Future of starch-based materials in food packaging. In: Vilar M (ed) Starch-based materials in food packaging: processing, characterization and application. Academic Press, London, pp 257–312

Pardo-Ibáñez P, López-Rubio A, Martínez-Sanz M et al (2014) Keratin–polyhydroxyalkanoate melt-compounded composites with improved barrier properties of interest in food packaging applications. J Appl Polym Sci 39947:1–10
Peelman N, Ragaert P, Meulenaer B et al (2013) Application of bioplastics for food packaging. Trends Food Sci Technol 32(2):128–141
Ramos M, Fortunati E, Peltzer M et al (2014) Influence of thymol and silver nanoparticles on the degradation of poly(lactic acid) based nanocomposites: thermal and morphological properties. Polym Degrad Stab 108:158–165
Requena R, Vargas M, Chiralt A (2017) Release kinetics of carvacrol and eugenol from poly(hydroxybutyrate-co-hydroxyvalerate) (PHBV) films for food packaging applications. Eur Polym J 92:185–193
Rhim JW, Wang LF, Lee Y et al (2014) Preparation and characterization of bio-nanocomposite films of agar and silver nanoparticles: laser ablation method. Carbohydr Polym 103:456–465
Rimdusit S, Jingjid S, Damrongsakkul S et al (2008) Biodegradability and property characterizations of methyl cellulose: effect of nanocompositing and chemical crosslinking. Carbohydr Polym 72:444–455
Rivero CP, Hu Y, Kwan TH et al (2017) Bioplastics from solid waste. In: Wong JWC, Tyagi RD, Pandey AR (eds) Current developments in biotechnology and bioengineering. Elsevier, Amsterdam, pp 1–26
Rocca-Smith J, Marcuzzo E, Karbowiak T et al (2016) Effect of lipid incorporation on functional properties of wheat gluten based edible films. J Cereal Sci 69:275–282
Rombouts I, Lagrain B, Delcour J et al (2013) Crosslinks in wheat gluten films with hexagonal close-packed protein structures. Ind Crop Prod 51:229–235
Rosa MF, Medeiros ES, Malmonge JA et al (2010) Cellulose nanowhiskers from coconut husk fibers: effect of preparation conditions on their thermal and morphological behavior. Carbohydr Polym 81:83–92
Santos RPO, Rodrigues BVM, Ramires EC et al (2015) Bio-based materials from the electrospinning of lignocellulosic sisal fibers and recycled PET. Ind Crop Prod 72:69–76
Sanuja S, Agalya A, Umapathy MJ (2014) Studies on magnesium oxide reinforced chitosan bionanocomposite incorporated with clove oil for active food packaging application. Int J Polym Mater Polym Biomater 63:733–740
Shankar S, Rhim JW (2016) Preparation of nanocellulose from micro-crystalline cellulose: the effect on the performance and properties of agar-based composite films. Carbohydr Polym 135:18–26
Shih YF, Chang WC, Liu WC et al (2014) Pineapple leaf/recycled disposable chopstick hybrid fiber-reinforced biodegradable composites. J Taiwan Inst Chem Eng 45(4):2039–2046
Slavutsky AM, Bertuzzi MA (2014) Water barrier properties of starch films reinforced with cellulose nanocrystals obtained from sugarcane bagasse. Carbohydr Polym 110:53–61
Sung SH, Chang Y, Han J (2017) Development of polylactic acid nanocomposite films reinforced with cellulose nanocrystals derived from coffee silverskin. Carbohydr Polym 169:495–503
Van den Broek L, Knoop R, Kappen F et al (2015) Chitosan films and blends for packaging material. Carbohydr Polym 116:237–242
Watthanaphanit A, Supaphol P, Tamura H et al (2008) Fabrication, structure, and properties of chitin whisker-reinforced alginate nanocomposite fibers. J Appl Polym Sci 110:890–899
Xiao S, Liu B, Wang Y et al (2014) Efficient conversion of cellulose into biofuel precursor 5-hydroxymethylfurfural in dimethyl sulfoxide–ionic liquid mixtures. Bioresour Technol 151:361–366
Zhou Y, Fan M, Chen L (2016) Interface and bonding mechanisms of plant fibre composites: an overview. Compos Part B 101:41–45
Zubeldía F, Ansorena M, Marcovich N (2015) Wheat gluten films obtained by compression molding. Polym Test 43:68–77

Chapter 5
Recent Trends on Nano-biocomposite Polymers for Food Packaging

Germán Ayala Valencia and Paulo José do Amaral Sobral

Abstract In recent years, much attention has been focused on research to replace petroleum-based polymers by biodegradable materials. Specify, polymer from natural sources have been considered as the most promising materials for this purpose. However, materials manufacture from natural polymers (e.g. polymeric films) generally present poor mechanical and high water sensibility. A recent alternative to improve the physical properties of natural polymeric films is a reinforcement with nanoparticles, producing nano-biocomposite polymers. The present chapter reviews the state-of-the-art with regard to the use of polymers obtained from biomass and their reinforced with nanoparticles, aiming food packaging applications. This chapter, especially include information about: (1) the use of casein, collagen/gelatin, chitin/chitosan, gluten, soya, starch, whey and zein as macromolecules to manufacture polymeric films, (2) the use of carbon nanotubes, chitin whiskers, metal nanoparticles, nanocellulose, nanoclays and starch nanocrystals to reinforced polymeric films, (3) the main physicochemical properties of nano-biocomposite polymers. The use of casting, tape casting, thermoforming and extrusion to manufacturing polymeric films; as well as the recent applications of nano-biocomposite polymers as food packaging and active/intelligent food packaging materials. Others topics such as nanoparticle migration, future prospects and limitations in nano-biocomposite polymers will be included in this chapter.

Keywords Biomass · Food applications · Nanoparticles · Nanotechnology · Polymers

---

G. A. Valencia (✉)
Department of Chemical and Food Engineering, Federal University of Santa Catarina, Florianópolis, SC 88040-900, Brazil
e-mail: g.ayala.valencia@ufsc.br

P. J. do Amaral Sobral
Department of Food Engineering, Faculty of Animal Science and Food Engineering, University of São Paulo, Pirassununga, São Paulo, Brazil

T. J. Gutiérrez (ed.), *Polymers for Food Applications*,
https://doi.org/10.1007/978-3-319-94625-2_5

5.1 Introduction

Petroleum-based polymers have been the most used materials for manufacture films[1] in food packaging industry due to their good mechanical, optical and gas barrier properties (Embuscado and Huber 2009). However, petroleum-based polymers are:

- Nonbiodegradable materials and may cause environmental and ecological problems due to the accumulation of waste polymers in living and non-living systems (Vieira et al. 2011; Benjakul et al. 2016).
- Produced from non-renewable resources, therefore, they are exhaustible raw materials that are subject to great price fluctuations (Gómez-Estaca et al. 2016).

New strategies are oriented to regarding the manufacture of packaging films using renewable polymers that could lead to crucial reductions of environmental and economic problems (Li et al. 2015a).

Polymers from natural sources or biopolymers can be classified in three categories, such as: polymers from biomass, polymers synthetized by means of microbial routes and polymers chemically synthesized using monomers from agro-resources (Vieira et al. 2011). Due to the abundant, renewable and inexpensive biomass characteristics (Ali and Ahmed 2016), most studies have been addressed to manufacture food packaging materials using polymers from biomass (Li et al. 2015a; Abreu et al. 2015; Montero et al. 2017; Belyamani et al. 2014; Colak et al. 2016; Balakrishnan et al. 2017; Huang et al. 2016; Oechsle et al. 2016; Etxabide et al. 2016; Matet et al. 2015; Mendes et al. 2016; Garrido et al. 2016; Wang and Padua 2003; Valencia et al. 2016). Polymers from biomass can be classified in two groups: polysaccharides and proteins, both isolated from animal or plant sources (Fig. 5.1) (Vieira et al. 2011).

Biopolymer-based films usually present some good functional or physical properties but are very sensitive to environmental conditions, as relative humidity, and are affected by several factors such as pH, heat treatment, addition of plasticizers, ion concentration, biopolymer concentration and its molecular conformation (Nisperos-Carriedo 1994; Arvanitoyannis 2002; Gutiérrez and Alvarez 2017a). An alternative to enhance the biopolymer-based films properties is their reinforcement with nanoparticles, producing a material often called as nano-biocomposite polymers or only nanocomposites (Lagaron et al. 2005). Thus, the nano-biocomposite polymer films are thin materials formed by a biopolymer matrix reinforced with a dispersed nanoscale filler (Gutiérrez et al. 2016).

The present chapter aims to reviews the recent trends about the use of polymers produced from biomass and used to manufacture nano-biocomposite polymer films; the main techniques used to manufacture nano-biocomposite polymer films (casting, tape casting, thermoforming and extrusion); the use of nanoparticles to manufacture nano-biocomposite and active nano-biocomposite polymer films; the nanoparticle

[1] Films are continuous polymeric materials with thickness less than 0.3 mm (Embuscado and Huber 2009).

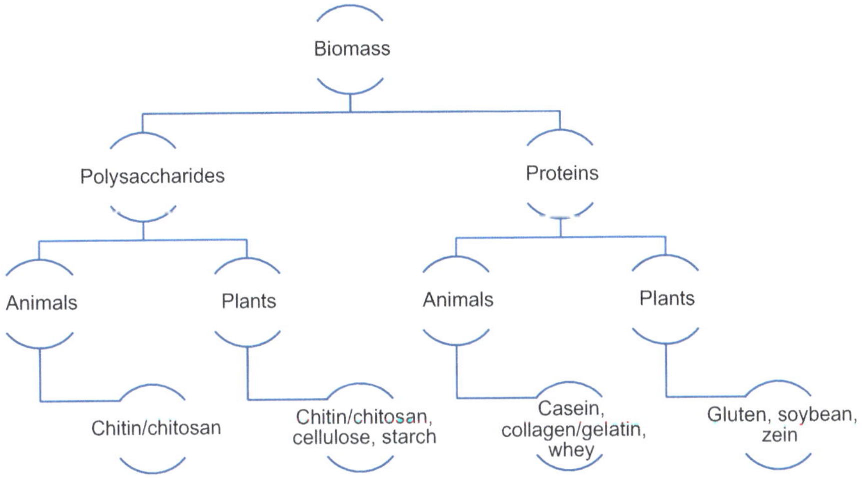

Fig. 5.1 Examples of polymers produced from biomass (adapted from Vieira et al. (2011))

migration from nano-biocomposite polymer films to food products; and the prospects of nano-biocomposite polymer films in food industry.

5.2 Main Polymers Isolated from Biomass

In this section, the main biopolymers produced from biomass and used to manufacture nano-biocomposite polymer films will be presented. These biopolymers are:

5.2.1 Chitin/Chitosan

Chitin is a N-acetylglucosamine mainly found in cell walls of fungi and exoskeleton of arthropods such as shrimps shell, carapace of crabs, epidermis and trachea of insects (Ali and Ahmed 2016; Kumar 2000; Usman et al. 2016; Merzendorfer and Zimoch 2003; Gutiérrez 2017). The chemical structure of chitin consists of 2-acetamido-2-deoxy-β-D-glucose through a β (1 → 4) linkage and its molecular weight (MW) varies between 1.030 and 2.500 kDa (Kumar 2000). Depending of the degree of hydration, size of unit cell, and number of chitin chains per unit cell, chitin can has a α, β, or γ crystalline structure. This polymer is broadly used in pharmaceutic and chemical industries (Ali and Ahmed 2016; Merzendorfer and Zimoch 2003).

Chitosan is a linear and partially acetylated (1–4)-2-amino-2-deoxy-β-D-glucan obtained by de-acetylation of chitin (Ali and Ahmed 2016). The MW of chitosan can varies between 100 and 500 kDa (Kumar 2000). Chitosan is a biodegradable polymer, non-toxic, biocompatible and displays antimicrobial properties against several microorganisms. Chitosan exhibits good filmogenic properties, allowing the manufacture of films with low barrier to oxygen and water vapor (Martínez-Camacho et al. 2013; Bonilla et al. 2014). More detailed characteristics and/or properties of chitin and chitosan can be found in the specialized literature (Ali and Ahmed 2016; Kumar 2000; Usman et al. 2016; Merzendorfer and Zimoch 2003).

5.2.2 Cellulose/Cellulose Derivatives

Cellulose is considered the polysaccharide most abundant in biomass (Gutiérrez and Alvarez 2017b). This is a crystalline biopolymer produced from plants, principally from cotton and wood, with molecular weight (MW) between 460 and 520 kDa (Ali and Ahmed 2016; Kathirgamanathan et al. 2017). Cellulose is composed of D-glucopyranose units linked by β(1 → 4)-glycosidic bonds (Ali and Ahmed 2016; Park et al. 2010). Hydrogen groups in cellulose can conform several intra- and intermolecular hydrogen bonds, resulting in different crystalline structures (I, II, III and IV) (Park et al. 2010).

Cellulose is a biopolymer insoluble in water due to its molecular and supramolecular structure (Mischnick and Momcilovic 2010). Modifications of cellulose are performed under controlled conditions aiming to destroy the crystalline structure of cellulose, hence, D-glucopyranose units should become equally accessible. In sequence, solvents are added to modify the swelling properties, followed by the addition of an alkyl halide or by an oxirane, obtaining cellulose derivatives (Mischnick and Momcilovic 2010; Doelker and Geneva 1993).

Cellulose derivatives have a MW less than 150 kDa, they may be water soluble and are used in a large number of pharmaceutical, cosmetic and food industrial applications (Kulicke et al. 2005; Kalyuzhnaya et al. 2015). The cellulose derivatives most used to manufacture biopolymer films are the methylcellulose (MC), carboximethylcellulose (CMC), hydroxypropylcellulose (HPC), hydroxypropylmethylcellulose (HMC) (Mischnick and Momcilovic 2010; Kulicke et al. 2005; Kalyuzhnaya et al. 2015). Cellulose derivatives have a good film-forming performance and its films are transparent and have a high mechanical strength and low barrier to oxygen and water vapor (Kalyuzhnaya et al. 2015; Deng et al. 2016; Yang et al. 2011). More detailed characteristics and/or properties of cellulose and cellulose derivatives can be found in several works (Kathirgamanathan et al. 2017; Park et al. 2010; Mischnick and Momcilovic 2010; Doelker and Geneva 1993; Kulicke et al. 2005; Kalyuzhnaya et al. 2015; Deng et al. 2016; Yang et al. 2011).

5.2.3 Starch

Starch is a polysaccharide conformed of amylose and amylopectin, in which amylose is the predominantly linear (1 → 4)-linked α-glucan, while, amylopectin has a linear (1 → 4)-linked α-glucan with many ramifications by means α-(1 → 6) branch points

(Álvarez et al. 2017; Gutiérrez et al. 2017a). The MW of amylose and amylopectin varies between 150 and 400 kDa; and between 10,000 and 15,000 kDa, respectively (Cornell 2004; Pérez et al. 2009). Amylose and amylopectin conform starch granules with different shapes such as spheres, ellipsoids, polygons, platelets, irregular tubules and sizes between 0.1 to at least 200 μm, depending on the botanical source (Cornell 2004; Pérez et al. 2009). Starch granules are partially crystalline with A, B or C crystalline structure (Gutiérrez et al. 2015; Gutiérrez and Álvarez 2016; Gutiérrez 2018). These granules are mainly found in roots, tubers and seeds and to a lesser extent in stems, leaves, fruits and pollen (Pérez et al. 2009). The amount of the amylose in starch granules oscillates between 15 and 40% (Tester et al. 2004). Starch has the ability to form a continuous matrix, allowing the manufacture of films where the mechanical, barrier and film formation properties are improved with the increasing in amylose content (Azevedo et al. 2017; Valencia et al. 2018). The main functional and physical properties of starch has been reviewed by many authors (Ali and Ahmed 2016; Cornell 2004; Pérez et al. 2009; Tester et al. 2004).

5.2.4 *Casein and Whey*

Casein and whey are milk proteins. Milk proteins constitute approximately 3.2% of bovine milk (Guo and Wang 2016). Casein constitute approximately 75% of the total proteins in milk, which precipitates at pH 4.6. The protein remaining in milk serum (~25%) are called whey protein (Guo and Wang 2016; Morgan et al. 2005).

In its native state in milk, casein is mostly a random coil with a high proline content, highly phosphorylated for calcium binding, existing in the form of a "micelle" structure with a MW of ~10^5 kDa, and comprising four different casein subunits (α-, β- and κ-casein) and calcium phosphate (Guo and Wang 2016; Morgan et al. 2005). The proportion of α-, β- and κ-casein varies among species. The MW of α-, β- and κ-casein are approximately 23, 25, 24 and 19 kDa, respectively (Gómez-Estaca et al. 2016; Morgan et al. 2005; Wang et al. 2013; Bonnaillie et al. 2014).

Whey proteins are ordered secondary structures, which is described as a "cup" or "calyx" shaped three-dimensional structure and with a low proline content, is non-phosphorylated, and has small soluble proteins (Guo and Wang 2016). The β-lactoglobulin (MW ≈ 18 kDa) is the dominant protein in whey, which accounts for about 50% of total whey proteins (Guo and Wang 2016). Other proteins such as α-lactalbumin (MW ≈ 14 kDa) and bovine serum albumin (MW ≈ 66 kDa) also are found in whey protein (Gómez Estaca et al. 2016; Schmid et al. 2012).

Casein and whey have good film-forming performance and their films have a good barrier to oxygen and they can form flavourless, flexible and transparent films (Bonnaillie et al. 2014; Schmid et al. 2012; Wagh et al. 2014). More detailed characteristics and/or properties of casein and whey proteins can be found in the specialized literature (Guo and Wang 2016; Morgan et al. 2005; Wang et al. 2013; Bonnaillie et al. 2014; Schmid et al. 2012).

5.2.5 Collagen/Gelatin

Collagen is comprised of homotrimers and heterotrimers that are formed by three polypeptide chains called α-chains (collagen triple helix), each one, with a MW between 345 and 360 kDa (Vieira et al. 2011; Gómez-Estaca et al. 2016; Mouw et al. 2014). To date, some 28 different types of collagen have been identified, classified as type I, II, II… XXVIII (Schrieber and Gareis 2007; Shoulders and Raines 2009). Type I collagen is formed by two α-1 chains and one α-2 chain, this protein exhibits primary, secondary, tertiary and quaternary structure and it has several industrial applications (Mouw et al. 2014; Schrieber and Gareis 2007).

Collagen is the primary fibrous protein present in the extracellular matrix of animals and is a major component of tendon, cartilage, bone and skin. Principally, this protein is produced from mammalian (bovine, goat, porcine and ovine) and non-mammalian (amphibian and fish) animals (Fauzi et al. 2016; Stylianou and Yova 2013). This protein is not water soluble, however, after a number of separations and hydrolysis, the collagen becomes a hydrocolloid, exhibiting filmogenic properties (Wolf et al. 2009).

Collagen can be hydrolyzed under thermal treatments, in acidic or alkaline conditions, obtaining type A or B gelatin, respectively (Schrieber and Gareis 2007). Depending of hydrolysis type, the isoelectric point (IEP) and MW of gelatins are altered. Hence, the IEP for type A and B gelatins are ranged between 8 and 9, and between 4.8 and 5.5, respectively (Schrieber and Gareis 2007), while MW of type A and B gelatins are 30–300 kDa, and ~100 kDa, respectively (Djabourov et al. 2013). The hydrolysis type of collagen results in gelatin with different amino acid composition, altering the viscosity and gelling properties of gelatin (Schrieber and Gareis 2007). Gelatin has an excellent filmogenic properties which allows the production of films with excellent optical properties (Valencia et al. 2016).

Collagen and gelatin films are good barrier to oxygen, carbon dioxide and oils (Chen et al. 2013). The functional and physical properties of collagen and gelatin have been reviewed by many authors (Mouw et al. 2014; Schrieber and Gareis 2007; Shoulders and Raines 2009; Djabourov et al. 2013).

5.2.6 Gluten

Gluten is a protein fraction from several cereal seeds such as wheat, rye, barley, oats or their crossbred varieties and derivatives (Rzychon et al. 2017). Gluten proteins can be divided in gliadins and glutenins, being that both fractions have high glutamine and proline contents. Gliadins are mainly monomeric proteins with a MW between 28 and 55 kDa and can be classified according to their different primary structures into the α/β-, γ- and ω-type. Glutenins are comprised of aggregated proteins linked by interchain disulphide bonds with MW varying between 500 and more than 10,000 kDa (Wieser 2007). Glutenin subunits can be obtained reducing disulphide bonds in glutenins. The resulting glutenin subunits are soluble in aqueous alcohols, similar to gliadins. Depending of primary structure, glutenins subunits can be classified in high-molecular-weight (HMW) and low-molecular-weight (LMW), with MW between 67 and 88 kDa; and between 32 and 35 kDa, respectively (Wieser 2007; Tilley et al. 2001).

Several researches have studied the gluten film-forming performance, particularly, its viscoelastic properties, ability to cross-link upon heating, low water solubility, low cost and availability as a co-product of the wheat starch industry (Tilley et al. 2001; Ansorena et al. 2016; Tanada-palmu and Grosso 2003). Gluten films are good barrier to oxygen, carbon dioxide and oils (Chen et al. 2013). Most information about the functional and physical properties of gluten can be found in the specialized literature (Wieser 2007; Tilley et al. 2001; Gras et al. 2001; Shewry et al. 2002).

5.2.7 Soybean/Soybean Derivatives and Zein

Soybean proteins are composed of a mixture of globular proteins such as albumins (10–50%) and globulins (50–90%) (Qi et al. 2011), with MW between 140 and 170 kDa; and between 340 and 375 kDa, respectively (Nehete et al. 2013). Globulin is the dominant storage protein in soybean (50–90%), this protein is composed of two major components: glycinin (11S) and β-conglycinin (7S). The 11S:7S proportion depend of the botanical origin, varying from 1:3 to 3:1 (Qi et al. 2011; Singh et al. 2015). Two proteins derivatives can be obtained from soybean protein: soy protein concentrate (SPC) and soy protein isolate (SPI). The difference between SPC and SPI is based on the protein content, hence SPC is defined as an protein product with a protein content ≥65%, whereas SPI is a product with a protein content ≥90%, both on dry weight basis (Wang et al. 2004). SPC and SPI are prepared in different process involving insolubilization of the protein to remove soluble sugars, whereas, SPI preparation involves solubilization of the protein to remove the insoluble fiber and then precipitation to remove soluble sugars (Wang et al. 2004; Tsumura 2009).

On another side, zein is a protein (prolamine) from maize with MW between 14 and 27 kDa (Wang and Padua 2003; Argos et al. 1982; Thompson and Larkins 1989). This protein displays significant hydrophobic properties due to the amino acid composition, having a large amount of strong or somewhat hydrophobic residues such as glutamine, proline, leucine and alanine (Matsushima et al. 1997). According to the solubility properties, zeins can be classified as α, β, γ and δ. Hence, α-zeins (MW between 19 and 22 kDa) can be extracted by aqueous alcohol alone, whereas β (MW ~ 14 kDa), γ (MW between 16 and 27 kDa) and δ-zeins (MW ~ 10 kDa) are solubilized by the addition of a reducing agent to aqueous alcohol. The β, γ and δ-zeins exhibit little charge heterogeneity and are probably single polypeptides (Thompson and Larkins 1989).

Films based on soybean/soybean derivatives proteins and zein have excellent water-resistance, as well as good mechanical and optical properties (Chen et al.

2013; Li et al. 2016). Most information about the functional and physical properties of soybean/soybean derivatives proteins (Qi et al. 2011; Singh et al. 2015; Wang et al. 2004; Tsumura 2009; Li et al. 2016), as well for zein (Wang and Padua 2003; Argos et al. 1982; Thompson and Larkins 1989; Matsushima et al. 1997) can be found in the specialized literature.

5.3 Main Techniques Used to Manufacture Nano-biocomposite Polymer Films

In general, films manufactured with biopolymers from biomass are sensitive to water vapor, thus limiting their food industrial applications (Valencia et al. 2016; Azevedo et al. 2017; Chen et al. 2013; Ansorena et al. 2016). Then, in the next sections we will discuss some new strategies to improve physicochemical properties (e.g. water vapor barrier, mechanical, optical and thermal properties) in films based on biopolymers from biomass. There are two common technologies to manufacture biopolymer films and nano-biocomposite polymer films: wet and dry processes.

5.3.1 Wet Process

In the wet process, solvents are used for the dispersion of biopolymer to obtain a film-forming solution (FFS). Then the desired additives, such as plasticizers, to allow the production of flexible films; cross-linkers, to produce films with modified biopolymers; antimicrobials/antioxidants, to produce active films; or, nanoparticles, when we are interested in to produce nano-biocomposite polymer films, are added to FFS. Finally, FFS is dried to remove the solvent and form a continuous film (Benjakul et al. 2016).

Wet process comprises casting and tape-casting techniques. In casting, FFS is poured onto the support, and then, the film thickness is a function of the weight of dry matter poured onto the support (Valencia et al. 2016, 2018). Casting is the most used technique to manufacture biopolymer films at laboratory-scale. This technique has been used for manufacture biopolymer films based on chitosan (Bonilla et al. 2014; Ma et al. 2016), cellulose (Yang et al. 2011), starch (Abreu et al. 2015; Valencia et al. 2018), casein (Bonnaillie et al. 2014; Wagh et al. 2014), whey (Wagh et al. 2014), collagen (Wolf et al. 2009), gelatin (Valencia et al. 2016), gluten (Fakhouri et al. 2018), soybean (Ma et al. 2016) and zein (Matthews et al. 2011), among others.

Tape-casting, or spread casting or knife-coating, is a well-known technique to manufacture paper, plastic and ceramics (Richard and Twiname 2000). In this technique, FFS is placed in a reservoir with a blade, whose height can be adjusted with micrometric screws (Fig. 5.2). The FFS is cast as a film on a tape or support, due to the movement of the doctor blade, in batch process; or the movement of the carrier

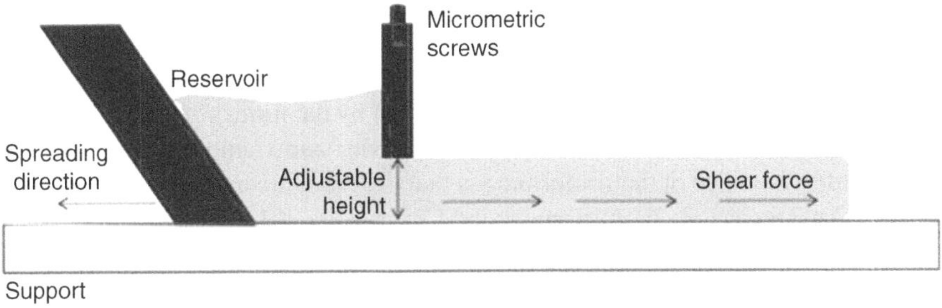

Fig. 5.2 A schematic illustration of the tape-casting process (adapted from De Moraes et al. (2013))

tape, in continuous process (De Moraes et al. 2013). FFS is dried on the tape, resulting in a reduction of its thickness, hence, the film thickness is a function of the weight of dry matter poured onto the tape, which can be controlled knowing the dry matter of FFS and the height of the blade (Flaker et al. 2015). Principally, tape-casting has been used to manufacture biopolymer films based on starch (De Moraes et al. 2013, 2015) and gelatin (Flaker et al. 2015). The disadvantage of tape-casting technique is that FFS must have a shear-thinning behavior with viscosity values that allow the flow of FFS to high- or low-shear stresses (De Moraes et al. 2013).

In the wet process (casting and tape-casting), the selection of solvents is one of the most important factors. Normally, solvents are water and mixtures of water/ethanol or water/acetic acid (Benjakul et al. 2016; Bonilla et al. 2014). The type of solvent, its pH and temperature can affect the biopolymer network during its dispersion. For example, chitosan, collagen and some proteins are dispersed in acid conditions (Bonilla et al. 2014; Wolf et al. 2009; Li et al. 2012); whereas, starch and gelatin can be dispersed using only hot water (Valencia et al. 2016, 2018).

By means of wet process, it is possible to manufacture biopolymer films with better barrier properties (e.g., oxygen, water, light), however, other physical properties such as mechanical properties, thermal sealebility and printability are not necessarily improved (Gómez-Estaca et al. 2016).

5.3.2 Dry Process

In the dry process or thermal process, solvents are not necessary needed, thus, from economic and environmental viewpoints, this process is interesting to manufacture biopolymer films in an industrial-scale (Bouvier and Campanella 2014). In dry process, heat is applied to increase the temperature above the glass transition (T_g) and/or the melting (T_m) temperatures of biopolymer and then promoting the material flow (Benjakul et al. 2016; Gómez-Estaca et al. 2016). Normally, biopolymer film formation can be carried out using thermoforming and extrusion (Bouvier and Campanella 2014; Gutiérrez and Alvarez 2017c, d).

Thermoforming is a high-temperature process where the biopolymer is forced to form a specific shape in a mold. Normally, this process is done under high pressure and high temperature, promoting important changes in the molecular network of biopolymer due to the thermal transitions and even by the formation of covalent and cross-linking bonds (Ansorena et al. 2016; Bouvier and Campanella 2014). The principal disadvantage of thermoforming is that materials are manufactured in batch process (Ansorena et al. 2016; Bouvier and Campanella 2014).

The principal variables to manufacture biopolymer films by thermoforming are temperature and pressure molding, as well process time, depending of the biopolymer to be manufactured (Table 5.1). Sothornvit et al. (2007) observed that to produce compression-molded whey protein films is necessary a minimum temperature of 104 °C. However, a temperature higher than 140 °C lead to the polymer degradation. In this same research, these authors observed that continuous polymeric material were obtained using pressures between 0.8 and 2.2 MPa.

In the last years, most researchers have studied the extrusion because materials can be manufactured in a continuous process (Bouvier and Campanella 2014). Extrusion is widely used in polymer industry to manufacture plastic packaging and recently used to manufacture biopolymer films (Hanani et al. 2013).

Extrusion is a thermomechanical process based on combination of multiple unit operations such as the transport, mixing, shearing, plasticizing, melting, polymerization, and fragmentation (Emin and Schuchmann 2017). Extrusion uses one or two co-rotating twin screws fitted in a barrel in order to gradually increase the pressure and push forward and mix the ingredients required to manufacture the product through a die section where expansion may take place (Fig. 5.3) (Hanani et al. 2013; Emin and Schuchmann 2017). Actually, extruders allow extensive variation in thermal and mechanical energy inputs, control of the residence time and efficient of mixing ingredients (Bouvier and Campanella 2014). The principal disadvantage of extrusion is that several parameters control the process, hence slight variations in processing conditions can have a considerable effect on physicochemical properties of biopolymer films (Bouvier and Campanella 2014).

The principal variables to manufacture biopolymer films and nano-biocomposite polymer films by extrusion are screw speed, temperature profile in screw section (e.g. temperature 1/temperature 2, etc) and temperature in die section (temperature at the end of extruder) (Table 5.2).

Extrusion can cause chemical reactions or reorganization of the chain structure in biopolymers, allowing to improve some physicochemical properties in biopolymer films (Garrido et al. 2016; Hanani et al. 2012). Hanani et al. (Hanani et al. 2012) developed films based on gelatins from beef, pork and fish sources using twin-screw extrusion (Table 5.2). These authors observed that as the screw speed increased up to 300 rpm, the tensile strength (TS) of films based on gelatin from beef and pork increased, however, at 400 rpm, the TS values of both films decreased. Also, as the screw speed of the extruder was increased, from 100 to 300 rpm, an increase in elongation at break (EB) for fish gelatin films was observed. These authors observed that higher screw speed provoked the gelatin melting by friction and lead to most homogeneous films due to the better compound mixture. Also,

Table 5.1 Principal thermoforming variables used to manufacture biopolymer films and nano-biocomposite polymer films

Biopolymer	Thermoforming variables Temperature (°C)	Time (min)	Pressure (kPa)	Film thickness (mm)	Reference
Cellulose derivatives	100	8	6863	0.08	Shih et al. (2009)
	160	2	3000	0.3	Ortega-toro et al. (2014)
Starch	110 to 170		29 to 69	Not informed	Guimarães et al. (2010)
	110	3	7400	0.3	Hietala et al. (2013)
	140	6	14,706	0.1	López et al. (2015)
	160	2 and 6	5000 and 15,000	0.2 to 0.3	Ortega-Toro et al. (2015)
	180	15	10,000	0.7	Gutiérrez and Alvarez (2017)
Whey	150	5	20,000	Not informed	Sharma and Luzinov (2013)
Gelatin	80	2	64	1.0	Krishna et al. (2012)
Casein	70 to 140	7	50,000	0.5	Colak et al. (2016)
Gluten	80	10	1	Not informed	Zubeldía et al. (2015)
	100	10	1	Not informed	Ansorena et al. (2016)
Soybean derivatives	150	2	12,000	0.7	Guerrero et al. (2011)
	150	2	13,000	0.4	Garrido et al. (2016)

higher screw speed increased extruder shear rate and decreased the viscosity of gelatin solutions. These authors concluded that the mechanical properties of films based on fish gelatin decreasing when the processing temperature increased from 90 to 120 °C, due to thermomechanical process that disrupt the structures of the protein from fish gelatin. Water vapor permeability (WVP) values in gelatin films decreased with the increasing in extrusion screw speeds due to the decreasing in the molecular size of gelatin. Finally, the water solubility in gelatin films increased with the extrusion screw speeds and with the processing temperature due to the modifications in secondary and tertiary structures of proteins into films.

Hanani et al. (2014) studied the effect of the processing temperature on the mechanical and water vapor barrier properties of extruded gelatin films (Table 5.2) and observed that temperature had a significant effect on the mechanical and WVP properties in composite films. Hence, TS values for gelatin films increased, from 1.4 to 5.4 MPa, as temperature increased until 120 °C. This could be attributed to the melting protein at higher temperature, which also contributed to tighter compact protein networks. Also, EB was not altered with temperature in screw section (EB = 2.1%). WVP values decreased from $9.3x10^{-10}$ to $4.6x10^{-10}$ $g{\cdot}m/m^2{\cdot}s{\cdot}Pa$, as

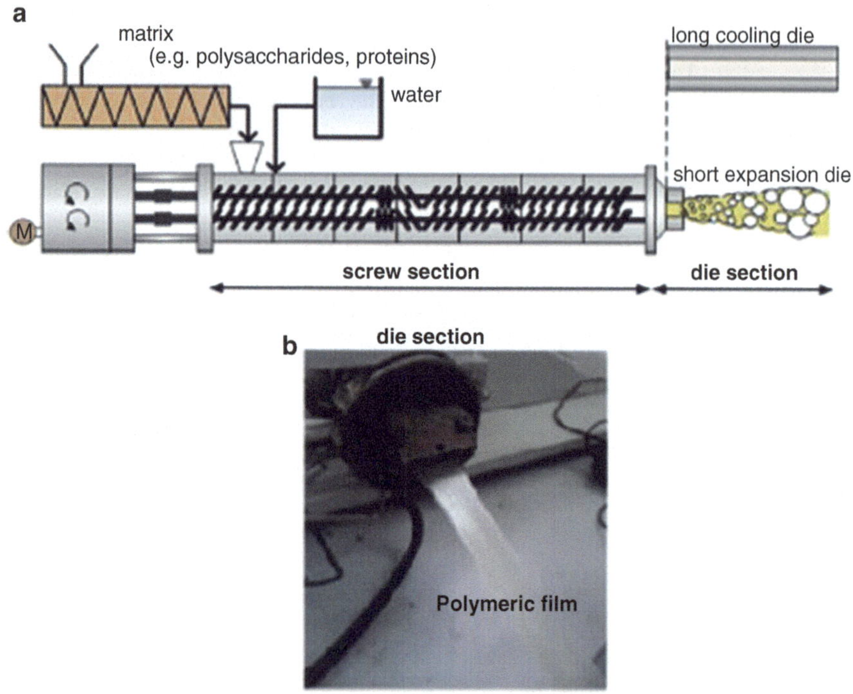

Fig. 5.3 (**a**) A schematic illustration of twin screw extrusion with the principal sections (adapted from Emin and Schuchmann (2017)). (**b**) A photography of a polymeric extruded film (adapted from Camacho et al. (2013))

temperature increased, from 90 °C until 120 °C, in screw section. Similar results were observed by Krishna et al. (2012), working with extruded gelatin films, which also observed that the T_g values in extruded gelatin films increased with increase in extrusion temperature, from 110 °C (T_g = 11.6 °C) to 120 °C (T_g = 26.3 °C). This behavior was due to the decreasing in films moisture content with the increasing of processing temperature (Krishna et al. 2012).

Andreuccetti et al. (2012) compared the casting and extrusion process to manufacture gelatin films. These authors observed that extruded films showed much better elongation at break and lower WVP that those manufactured by casting. Similar results were observed by Guerrero et al. (2010) and Yan et al. (2012), both, comparing the casting and extrusion process to manufacture soybean and corn starch films, respectively.

However, others researchers have been observed that extrusion process not necessarily improve the physicochemical properties of biopolymer films when compared with the same films manufactured by means of wet process. These behavior could be due to the biopolymer degradation during extrusion process (Krishna et al. 2012; Yan et al. 2012; Andreuccetti et al. 2012; Park et al. 2008).

Table 5.2 Principal extrusion variables used to manufacture biopolymer films and nano-biocomposite polymer films

Biopolymer	Extruder	Extrusion variables			Film thickness (mm)	Reference
		Screw speed (rpm)	Temperature profile (°C)			
			Screw section	Die section		
Chitosan	Single-screw	120	150/160/160	160	0.3	Martínez-Camacho et al. (2013)
	Single-screw	40	130	140	1.5	Quiroz-Castillo et al. (2014)
	Twin-screw	170	90/95/105/110/115/120/120/120/120	125	Not informed	Shih et al. (2009)
	Single-screw	35 to 45	130–140	150	0.3	Dang and Yoksan (2016)
Cellulose derivatives	Twin-screw	150	185	185	Not informed	Quintana et al. (2016)
Starch	Twin-screw	30	95 to 125	Not informed	0.1	Yan et al. (2012)
	Twin-screw	300	90/90/90/100/110/110	100	0.2	Hietala et al. (2014)
	Twin-screw	200	80/90/90/100/100/110	110	0.3	Hietala et al. (2013)
	Single-screw	Not informed	120	125 to 130	0.1	López et al. (2013)
	Twin-screw	35	100/140/140	140	0.3	Ayumi et al. (2013)
	Single-screw	40 to 80	160 to 190	Not informed	Not informed	Zepon et al. (2013)
	Single-screw	35 to 45	130/140/140/140	150	Not informed	Dang and Yoksan (2015)
Casein	Twin-screw	45	80	80	0.3	Belyamani et al. (2014)
	Twin-screw	Not informed	65	80	0.5	Colak et al. (2016)
Collagen	Single-screw	65	5	5	0.6	Oechsle et al. (2016)

(continued)

Table 5.2 (continued)

Biopolymer	Extruder	Extrusion variables			Film thickness (mm)	Reference
		Screw speed (rpm)	Temperature profile (°C)			
			Screw section	Die section		
Gelatin	Twin-screw	100 to 400	90 (all sections)	Not informed	0.04	Hanani et al. (2012)
	Twin-screw	300	90/120/90/90	Not informed	0.022	Hanani et al. (2013)
	Twin-screw	300	90/90–130/90/90	Not informed	0.025	Hanani et al. (2014)
	Twin-screw	70	90	90	3.8	Etxabide et al. (2016)
Zein	Twin-screw	100	80	80	Not informed	Chen et al. (2013)
	Single-screw	12 to 115	105/132/145	145	No informed	Selling and Utt (2013)

5.4 Main Nanoparticles Used to Manufacture Nano–biocomposite Polymer Films

The use of nanoparticles is a recent alternative to improve the physicochemical properties of biopolymer films (Valencia et al. 2016; Flaker et al. 2015; Perotti et al. 2014), and even to manufacture active packaging[2] (Krepker et al. 2017). Nanoparticles are characterized by having at least one dimension in nanometric scale, i.e. between 1 and 100 nm (Aouada et al. 2011). Hence, when the particle size is equivalent to the dimension of a molecule, the atomic and molecular interactions can have a significant influence on the macroscopic properties of liquid and solid systems (Valencia et al. 2015). It is important that nanoparticles are well-dispersed in the biopolymer network to enhance the contact area between nanoparticle and biopolymer chain and then improve the physicochemical properties of nano-biocomposite polymer films (Valencia et al. 2016; Flaker et al. 2015; Yamakawa et al. 2017; Ghelejlu et al. 2016).

The principal nanoparticles used in studies about nano-biocomposite polymer films are:

5.4.1 Carbon Nanotubes

Carbon nanotubes (CNTs) are sheets of graphite that has been rolled into a tube, forming cylinders in nanometric scale (Belin and Epron 2005). There are two types of CNTs: single-walled carbon nanotubes (SWCNTs) and multi-walled carbon nanotubes (MWCNTs) (Zhanjun et al. 2011). SWCNTs have a well-defined atomic structure, with high length to diameter ratio, having a diameter between 0.8 and 1.2 nm and length between 100 and 1000 nm. MWCNTs are elongated cylindrical nanoparticles made of sp^2 carbon, having a diameter between 3 and 30 nm and length of several cm long, thus their aspect ratio can vary between 10 and ten million. CNTs have electronic properties and can be either metallic or semiconducting depending on their geometry (Belin and Epron 2005).

CNTs have been broadly used in electrical applications due to its high electrical conductivity (Yamakawa et al. 2017). Most recently, CNTs have been used as reinforcement load for biopolymer films due to its high mechanical properties (elastic modulus, EM = 0.91 TPa and tensile strength, TS = 0.15 TPa) and thermal stability (Ortiz-Zarama et al. 2014). Zhanjun et al. (2011) manufactured nano-biocomposite polymer films based on corn starch containing until 3% of CNTs (based on the mass of biopolymer) and observed that the weight loss of nano-biocomposite polymer films, analyzed by thermogravimetric analysis (TGA), decreased with CNTs con-

[2] Active packaging: active packaging are materials that contain deliberately incorporated components intended to release (controlled) or absorb substances into or from the packaged food or from the environment surrounding the food (Krepker et al. 2017).

centration, in the temperature between 260 and 315 °C. However, the CNTs decreased the compatibility between corn starch with other compounds such as polyols (e.g. glycerol) used to plasticizer the biopolymer.

Famá et al. (2011) developed nano-biocomposite polymer films based on tapioca starch containing until 0.055% of MWCNTs (based on the mass of biopolymer) and observed that nano-biocomposite polymer films exhibited highly improved tensile and impact properties, with increments of approximately 70% in stiffness, 35% in ultimate tensile strength and 50% in tensile toughness, keeping deformations higher than 80% without break. Similar results were observed by Jose et al. (2015) and Ortiz-Zarama et al. (2014), working on nano-biocomposite polymer films based on corn starch or bovine gelatin, both containing until 2% of CNTs (based on the mass of biopolymer), respectively.

However, the high price and toxicity concerns are the principals disadvantages of CNTs, thus limiting its applications in nano-biocomposite polymer films for food industry (Famá et al. 2011).

5.4.2 Chitosan Nanoparticle

Chitosan nanoparticles (CNPs) have spherical shape with diameters less than 100 nm (Hosseini et al. 2015), with are produced from chitosan by means of several methods such as emulsion crosslinking, reverse micelles, precipitation, ionotropic gelation, among others (Gomathi et al. 2017). CNPs are broadly used for various applications due to their biodegradability, high permeability through biological membranes, non-toxicity to human, cost effectiveness and broad antifungal activities (Dananjaya et al. 2017; Ma et al. 2017a).

In the nano-biocomposite polymer film domain using CNPs. Hosseini et al. (2015) developed nano-biocomposite polymer films based on fish gelatin containing until 8% of CNPs (based on the mass of biopolymer) and observed that the addition of CNPs caused remarkable increase in TS and EM in nano-biocomposite polymer films when compared with control films (gelatin films without CNPs). Also, the addition of CNPs decreased in approximately 50% the WVP and the transparency at 600 nm of values of nano-biocomposite polymer films when compared with control films (without CNPs). The improvement of mechanical and barrier (WVP) properties of nano-biocomposite polymer films with the CNPs concentration was due to the decreasing in the free volume between gelatin chains caused by an increase in intermolecular attractive forces, making the gelatin network highly dense and thus less permeable. CNPs were not dissolved into biopolymer matrix, hence, nano-biocomposite polymer films were most opaque with CNPs, reflecting the UV light.

Barreras et al. (2016) observed that CNPs enhanced the antibacterial activity of nano-biocomposite polymer films based on collagen, suggesting that CNPs could be used to develop active packaging with food applications (Krepker et al. 2017).

5.4.3 Metal Nanoparticles

The most common nanoparticles from noble metals are silver and gold. Silver (AgNPs) and gold (AuNPs) nanoparticles are synthetized by chemical reactions from silver nitrate and chloroauric acid, respectively. Normally, AgNPs and AuNPs have a spherical form with diameters less than 100 nm (Valencia et al. 2013, 2014a; Cheviron et al. 2014; Ji et al. 2016).

AgNPs are known to have inhibitory and antimicrobial properties and low toxicity, receiving special attention in food packaging industry. These nanoparticles have an absorbent effect, being applied in food packaging to absorb moisture and fluids exuded from foods such as meat and fish (Abreu et al. 2015). AgNPs are broadly used to develop active food packaging (Carbone et al. 2016).

Regarding nano-biocomposite polymer film loaded with AgNPs, Abreu et al. (2015) developed nano-biocomposite polymer films based on corn starch containing until 1.7% of AgNPs (based on the mass of biopolymer) and observed that this material had antimicrobial activity against *Staphylococcus aureus*, *Escherichia coli* and *Candida albicans* without significant differences between AgNPs concentrations in nano-biocomposite polymer films. Similar results were observed in nano-biocomposite polymer films based on chitosan (Gabriel et al. 2017), starch (Valencia et al. 2013; Cheviron et al. 2014; Ji et al. 2016; Ortega et al. 2017) and gelatin (Kanmani and Rhim 2014) loaded with AgNPs.

Ortega et al. (2017) developed also nano-biocomposite polymer films based on corn starch containing until 0.02% of AgNPs (based on the mass of biopolymer) and observed that the nano-biocomposite polymer films extended the shelf-life of fresh cheese samples for 21 days due to the antimicrobial activity of AgNPs.

Moreover, AgNPs can cause an increasing in opacity and thermal stability, as well as decreasing in mechanical properties and WVP in nano-biocomposite polymer films (Ji et al. 2016; Ortega et al. 2017; Kanmani and Rhim 2014).

Regarding gold nanoparticles (AuNPs), Abdel-Raouf et al. (2017) and Mohamed et al. (2017) recently observed that this nanoparticle had antibacterial activity against Gram-positive and Gram-negative bacteria. In addition, Bumbudsanpharoke and Ko (2018) observed also antioxidant properties in nano-biocomposite polymer films charged with AuNPs and explained this behavior by the electron configuration of AuNP, allowing accept or donate an electron to quench radicals. However, AuNPs are less investigate in food packaging research due to its high price. Most research about AuNPs are focused on the development of drug delivery systems (Pooja et al. 2015) and biosensors (Valencia et al. 2014b).

5.4.4 Cellulose Nanowhiskers or Nanocrystals

Cellulose nanowhiskers (CNWs) or cellulose nanocrystals (CNCs) can be obtained by strong acid hydrolysis of cellulose fibres from cotton, hemp, flax, microcrystalline cellulose and bacterial cellulose, using sulfuric acid under controlled conditions, such as temperature, agitation and time, producing highly crystalline rod-like nanostructures (Alves et al. 2015; Cano et al. 2015). Thus, CNWs are from renewable, sustainable and abundant resources. CNWs have a needle shaped with the length between 10 and 400 nm and the diameter between 3 and 30 nm, depending on the chemical reaction (Wang et al. 2017; Soni et al. 2016).

Cano et al. (2015) developed nano-biocomposite polymer films based on pea starch containing until 5% of CNWs (based on the mass of biopolymer) and observed that the incorporation of CNWs lead to phase separation between CNWs and starch. These authors also observed that CNWs did not improved water vapor barrier in nano-biocomposite polymer films.

Noshirvani et al. (2018) developed nano-biocomposite polymer films based on potato starch containing until 20% of CNWs (based on the mass of biopolymer) and observed that nano-biocomposite polymer films showed a decreasing in solubility, water absorption, WVP and EB values with CNW concentration, and that contact angle, TS, T_g and T_m values increased with CNW concentration. The physicochemical modifications in nano-biocomposite polymer films with CNWs was due to the three-dimensional networks of intermolecular hydrogen-bonding interactions between CNW and potato starch matrix. CNW has also a crystalline nature with high aspect ratio, altering the films crystallinity and then their physical properties (Noshirvani et al. 2018). Similar results have been obtained in nano-biocomposite polymer films based on chitosan containing CNWs (Soni et al. 2016; Mujtaba et al. 2017; Ma et al. 2017b; Llanos and Tadini 2018).

In the same sense, Soni et al. (2016) observed that CNWs improved the thermal stability in nano-biocomposite polymer films based on chitosan due to the presence of the crystalline structure and great compactness between both chitosan and CNWs. Ma et al. (2017b) observed that CNWs improved barrier properties against UV light in nano-biocomposite polymer films based on chitosan as a consequence that CNWs hinder light transmission in the near ultraviolet range (between 300 and 400 nm). These results may be due to the interaction between the CNWs and chitosan matrix. Mujtaba et al. (2017) observed that CNWs improved the antimicrobial properties in nano-biocomposite polymer films based on chitosan against *Escherichia coli* ATCC 35218, *Pseudomonas aeruginosa* ATCC 27853, *Enterococcus faecalis* ATCC 29212, *Staphylococcus aureus* ATCC 25923 and *Listeria monocytogenes* ATCC 7644.

5.4.5 Nanoclays

Nanoclays are nanoparticles of layered mineral silicates, that can be classified as natural and synthetic nanoclays (Bracone et al. 2016; Gutiérrez et al. 2017b). The most common natural and synthetic nanoclays used to manufacture nano-biocomposite polymer films are montmorillonite (Mnt) and laponite (Lap), respectively (Valencia et al. 2016, 2018; Flaker et al. 2015; Llanos and Tadini 2018). Mnt and Lap have a disk-shape with a thickness of approximately 10 nm for Mnt and 1 nm for Lap, and a diameter of approximately 1000 nm for Mnt and 25 nm for Lap (Cummins 2007; Cadene et al. 2005).

Chung et al. (2010) developed nano-biocomposite polymer films based on corn starch containing until 7% of Mnt or Lap (based on the mass of biopolymer) and observed that nanoclays improved EM and TS in nano-biocomposite polymer films. Nanoclays can improve others physical properties such as thermal stability (Perotti et al. 2014; Aouada et al. 2011; Gutiérrez and Alvarez 2018), water vapor barrier and water solubility (Flaker et al. 2015; Barreras et al. 2016; Tang and Alavi 2012) in nano-biocomposite polymer films. Optical properties of nano-biocomposite polymer films were not altered by nanoclays as a consequence of well dispersed nanoclays (exfoliated) into biopolymer matrix (Valencia et al. 2016, 2018).

About the antibacterial activity, Hong and Rhim (2008) observed that natural Mnt (Cloisite Na^+) did not show any antibacterial activity. However, modified Mnt (Cloisite 20A and Cloisite 30B) displayed antimicrobial effect against Gram-positive and Gram-negative bacteria. Liu et al. (2014) observed that films based on synthetic polymer containing modified Mnt (I.34TCN) displayed antimicrobial activity against Gram-positive bacteria (i.e., *Listeria monocytogenes* and *Staphylococcus aureus*). Nanoclays can also have antioxidant effect. Echeverría et al. (2018) observed that the presence of Mnt in polyethylene films was able to decrease microbial growth and also the lipid autoxidation of tuna fillets during the storage period studied (17 days of storage at 2 °C). Li et al. (2015a) developed nano-biocomposite polymer films based on gelatin containing until 20% of Lap (based on the mass of biopolymer) and observed that Lap avoided lipid oxidation and protein decomposition of meat during storage due to the barrier capability of Lap. In this way, nanoclays are important candidates to manufacture active food packaging, however, the main disadvantage is that nanoclays are not from renewable sources, neither biodegradables (Abdollahi et al. 2013).

5.4.6 Starch Nanocrystals

Starch nanocrystals (SNC) can be obtained from native starch by strong acid hydrolysis, using HCl or H_2SO_4, under controlled conditions such as temperature, agitation and time. SNC has crystalline platelets with a length between 20 and 40 nm, a width between 15 and 30 nm, and a thickness between 5 and 7 nm (Gong et al. 2016). The main advantage is that SCN are from renewable sources and are biodegradables (González and Igarzabal 2015).

SNC has been used to improve the physicochemical properties of nano-biocomposite polymer films. Li et al. (2015b) developed nano-biocomposite polymer films based on pea starch containing until 9% of SNC (based on the mass of biopolymer) and observed that SCN improved some physicochemical properties in nano-biocomposite polymer films such as mechanical properties, water vapor barrier and thermal stability. González & Igarzabal (2015) also developed nano-biocomposite polymer films based on soybean protein isolate containing until 40% of SNC (based on the mass of biopolymer). These authors observed that the opacity, degree of crystallinity and mechanical properties in nano-biocomposite polymer films increased with SNC. The moisture content, total soluble matter and swelling in water showed also a marked effect on SNC additions. As the amount of SNC increased, the nano-biocomposite polymer films exhibited less affinity for water. These authors observed that nano-biocomposite polymer films containing SNC sequester cholesterol when brought into contact with cholesterol-rich food such as milk. Hence, SNC can also be used to develop active food packaging.

5.5 Nanoparticle Migration

As previously described, nanoparticles can improve several physicochemical properties and provide active properties in biopolymer films. However, the use of nanoparticles to manufacture nano-biocomposite polymer films for food packaging applications is a recent technology and can have several risks associated with the potential ingestion of nanoparticles migrated from films to drinks and foods (Carbone et al. 2016). To date, the toxicity, genotoxicity and carcinogenicity effect of nanoparticles in humans is not clear (Dimitrijevic et al. 2015).

The European Union (EU) and Switzerland are the only world region where nano-specific provisions have been incorporated in legislation. In other regions, nanomaterials are regulated more implicitly by mainly building on guidance for industry (Amenta et al. 2015). European Food Safety Authority (EFSA) did not approve the use of nanomaterials in food packaging (Carbone et al. 2016; Amenta et al. 2015).

Actually, nanoparticle migration into drinks and foods is studied based on overall migration limit (OML) proposed by European directives on food packaging normative (Commission Regulation No. 10/2011) (European Food Contact Materials Legislation 2002). For that, the OML in plastic materials due be less than 60 mg (of substances)/kg (of foodstuff or food simulant) for all substances. However, this normative is not specify for nanoparticles migration.

There are a limited number of studies about nanoparticle migration from nano-biocomposite polymer films. In some studies, the AgNPs and Mnt migration into foods have been studied. Abreu et al. (2015) studied the AgNPs migration from nano-biocomposite polymer films based on corn starch to food simulants and observed that the OML was less (42.9 mg/kg) than that OML proposed by European directives on food packaging normative. Similar results were observed by Metak et al. (2015) who studied the AgNPs migration from polyethylene packaging to different food simulants and real foods samples. These authors concluded that polymer films containing nanoparticles appears to be safe for food packaging since insignificant levels of AgNPs were released.

Echeverría et al. (2018) studied the Mnt migration from polyethylene films to tuna fillets during the storage period studied (17 days of storage at 2 °C) and did not observe significant migration of Mnt compounds (Mg, Al, Si) to tuna fillets, proving that polymer films containing Mnt can be used as food packaging.

However, it is necessary to study the nanoparticle migration in extreme conditions of temperature, pH and time, as well as in most liquid and solid foods. Several nanoparticles have been proposed as possible nanomaterials to manufacture nano-biocomposite polymer films for future food packaging applications (e.g. carbon nanotubes, chitosan nanoparticles, cellulose nanowhiskers, starch nanocrystals and some nanoclays such as laponite and modified Mnt), however, any study about the nanoparticle migration into food systems were reported to date. Finally, it is necessary to quantify the OML permissible of each nanoparticle into foods. For that, is necessary to understand the nanoparticle accumulation in biosystems and its toxicity, genotoxicity and carcinogenicity effects (Carbone et al. 2016).

5.6 Prospects in Nano-biocomposite Polymer Films

Aiming the improvement of physicochemical properties in nano-biocomposite polymer films, some authors have charged more than one nanoparticle in the same biopolymer matrix. Thus, Abreu et al. (2015) and Kanmani and Rhim (2014) developed nano-biocomposite polymer films based on starch or gelatin, both loaded with Mnt and AgNPs. In these research, AgNPs provided antimicrobial properties, whereas Mnt improved some mechanical and water barrier properties in the nano-biocomposite polymer films. Hence, these authors developed active nano-biocomposite polymer films with better physicochemical properties when compared with control films (without nanoparticles).

Other mixtures of nanoparticles such as CNTs and CNWs (Yamakawa et al. 2017), CNPs and AgNPs (Dananjaya et al. 2017) and SNC and Mnt (Orsuwan and Sothornvit 2017) have been charged in nano-biocomposite polymer films with better physicochemical and antimicrobial properties. Nevertheless, according to Yamakaka et al. (2017), these nanomaterials should have a similar diameter and similar aspect ratio to improve the nanoparticle compatibility into biopolymer matrix.

Another approach to improve physicochemical properties of nano-biocomposite polymer films is to blending biopolymers or biopolymers/synthetic polymer in presence of nanoparticles (Ortega-Toro et al. 2015; Jose et al. 2015; Alves et al. 2015; Noshirvani et al. 2018; Tang and Alavi 2012; Taghizadeh and Favis 2013). However, the phase separation between nanoparticle and biopolymer/polymer matrixes is a problem frequently observed (Li et al. 2015a; Valencia et al. 2018; Flaker et al. 2015; Cano et al. 2015; Tang and Alavi 2012). A recent alternative to improve compatibility between biopolymer/polymer and nanoparticles is the use of reactive extrusion (Gutiérrez and Alvarez 2017; Quintana et al. 2016; Dhar et al. 2016). Reactive extrusion is a process where chemical reactions are controlled during extrusion process (Bouvier and Campanella 2014). To blending biopolymer by reactive extrusion it is necessary that biopolymers or biopolymer/synthetic polymer have similar melting temperatures (Quintana et al. 2016).

5.7 Conclusions

Based on the economic and environmental viewpoints, biopolymer films are probably the future materials to be used in food industry. In nature, there are several biopolymers able to be used in manufacture process of biopolymer films. However, it is necessary to understand as biopolymer films can become materials with important characteristics to food industry. The use of news manufacture process and/or nanomaterials can improve the mechanical, thermal, optical and barrier (to water, oxygen and UV/light) properties of biopolymer films. Particularly, the use of nanoparticles can introduce new properties such as antimicrobial and antioxidant activity as well as sequester effects in biopolymer films, opening a windows of news applications as active packaging films for food industry. However, the toxicity, genotoxicity and carcinogenicity effects of nanoparticles are not clear to date. Hence, it is necessary most studies about the nanoparticle migration into drinks and foods as well as its risks associated with the potential ingestion, before the nanoparticle application in industrial food packaging.

References

Abdel-Raouf N, Al-Enazi NM, Ibraheem IBM (2017) Green biosynthesis of gold nanoparticles using Galaxaura elongata and characterization of their antibacterial activity. Arab J Chem 10:S3029–S3039

Abdollahi M, Alboofetileh M, Rezaei M, Behrooz R (2013) Comparing physico-mechanical and thermal properties of alginate nanocomposite fi lms reinforced with organic and/or inorganic nano fillers. Food Hydrocoll 32(2):416–424

Abreu AS, Oliveira M, De SA, Rodrigues RM, Cerqueira MA, Vicente AA et al (2015) Antimicrobial nanostructured starch based films for packaging. Carbohydr Polym 129:127–134

Ali A, Ahmed S (2016) Natural polymers: an overview. In: Ikram S, Ahmed S (eds) Natural polymers: derivatives, blends and composites. Nova, New York

Álvarez K, Famá L, Gutiérrez TJ (2017) Physicochemical, antimicrobial and mechanical properties of thermoplastic materials based on biopolymers with application in the food industry. In: Masuelli M, Renard D (eds) Advances in physicochemical properties of biopolymers: Part 1. Bentham Science, Sharjah, pp 358–400. EE.UU. ISBN: 978-1-68108-454-1. eISBN: 978-1-68108-453-4. https://doi.org/10.2174/97816810845341170100015

Alves JS, Reis KC, Menezes EGT, Pereira FV, Pereira J (2015) Effect of cellulose nanocrystals and gelatin in corn starch plasticized films. Carbohydr Polym 115:215–222

Amenta V, Aschberger K, Arena M, Bouwmeester H, Botelho F, Brandhoff P et al (2015) Regulatory aspects of nanotechnology in the agri/feed/food sector in EU and non-EU countries. Regul Toxicol Pharmacol 73(1):463–476

Andreuccetti C, Carvalho RA, Galicia-garcía T, Martinez-bustos F, González-nuñez R, Grosso CRF (2012) Functional properties of gelatin-based films containing Yucca schidigera extract produced via castin, extrusion and blown extrusion processes: a preliminary study. J Food Eng 113(1):33–40

Ansorena MR, Zubeldía F, Marcovich NE (2016) Active wheat gluten films obtained by thermoplastic processing. LWT Food Sci Technol. 69:47–54

Aouada FA, Mattoso LHC, Longo E (2011) A simple procedure for the preparation of lapoonite and thermoplastic starch nanocomposites: structural, mechanical, and thermal characterizations. J Thermoplast Compos 26(1):109–124

Argos P, Pedersenfl K, Marksl MD, Larkinsflll BA (1982) Structural model for maize Zein proteins. J Biol Chem 257(1):9984–9990
Arvanitoyannis IS (2002) Formation and properties of collagen and gelatin films and coatings. In: Protein-based films and coatings. CRC, Boca Raton, FL, pp 275–304
Ayumi M, Bonametti J, Salomão P, Maria C, Müller O, Victória M et al (2013) Thermoplastic starch/polyester films: effects of extrusion process and poly lactic acid addition. Mater Sci Eng C 33(7):4112–4117
Azevedo VM, Borges SV, Marconcini JM, Yoshida MI, Neto ARS, Pereira TC et al (2017) Effect of replacement of corn starch by whey protein isolate in biodegradable film blends obtained by extrusion. Carbohydr Polym 157:971–980
Balakrishnan P, Sreekala MS, Kunaver M, Huskic M, Thomas S (2017) Morphology, transport characteristics and viscoelastic polymer chain confinement in nanocomposites based on thermoplastic potato starch and cellulose nanofibers from pineapple leaf. Carbohydr Polym 169:176–188
Barreras US, Méndez FT, Martínez REM, Valencia CS, Rodríguez PRM, Rodríguez JPL (2016) Chitosan nanoparticles enhance the antibacterial activity of chlorhexidine in collagen membranes used for periapical guided tissue regeneration. Mater Sci Eng C 58:1182–1187
Belin T, Epron F (2005) Characterization methods of carbon nanotubes: a review. Mater Sci Eng B 119:105–118
Belyamani I, Prochazka F, Assezat G (2014) Production and characterization of sodium caseinate edible films made by blown-film extrusion. J Food Eng 121:39–47
Benjakul S, Nagarajan M, Prodpran T (2016) Films and coatings from collagen and gelatin. In: Montero GMP, Gómez-Guillén M, Carmen LCE, B-C V (eds) Edible films and coatings: fundamentals and applications, 1st edn. Taylor & Francis Group, CRC, Boca Raton, FL, pp 103–124
Bonilla J, Fortunati E, Atarés L, Chiralt A, Kenny JM (2014) Physical, structural and antimicrobial properties of poly vinyl alcohol e chitosan biodegradable films. Food Hydrocoll 35:463–470
Bonnaillie LM, Zhang H, Akkurt S, Yam KL, Tomasula PM (2014) Casein films: the effects of formulation, environmental conditions and the addition of citric pectin on the structure and mechanical properties. Polymers (Basel) 6:2018–2036
Bouvier J-M, Campanella OH (2014) Extrusion processing technology: food and non-food biomaterials, 1st edn. Wiley, New York
Bracone M, Merino D, Gonzalez JS, Alvarez VA, Gutiérrez TJ (2016) Nanopackaging from natural fillers and biopolymers for the development of active and intelligent films. In: Ikram S, Ahmed S (eds) Natural polymers: derivatives, blends and composites. Nova Science, New York, pp 119–155 EE.UU. ISBN: 978-1-63485-831-1
Bumbudsanpharoke N, Ko S (2018) The green fabrication, characterization and evaluation of catalytic antioxidation of gold nanoparticle-lignocellulose composite papers for active packaging. Int J Biol Macromol 107:1782–1791
Cadene A, Durand-vidal S, Turq P, Brendle J (2005) Study of individual Na-montmorillonite particles size, morphology, and apparent charge. J Colloid Interface Sci 285:719–730
Cano A, Fortuna E, Cháfer M, Gonzáles-Martínez C, Chiralt A, Kenny J (2015) Effect of cellulose nanocrystals on the properties of pea starch-poly (vinyl alcohol) blend films. J Mater Sci 50:6979–6992
Carbone M, Tommasa D, Sabbatella G (2016) Silver nanoparticles in polymeric matrices for fresh food packaging. J King Saud Univ Sci 28(4):273–279
Chen Y, Ye R, Li X, Wang J (2013) Preparation and characterization of extruded thermoplastic zein-poly(propylene carbonate) film. Ind Crop Prod 49:81–87
Cheviron P, Gouanvé F, Espuche E (2014) Green synthesis of colloid silver nanoparticles and resulting biodegradable starch/silver nanocomposites. Carbohydr Polym 108:291–298
Chung Y, Ansari S, Estevez L, Hayrapetyan S, Giannelis EP, Lai H (2010) Preparation and properties of biodegradable starch-clay nanocomposites. Carbohydr Polym 79(2):391–396
Colak BY, Peynichou P, Galland S, Oulahal N, Degraeve P (2016) Antimicrobial activity of Nisin and Natamycin incorporated Sodium Caseinate extrusion-blown films: a comparative study with heat-pressed/solution cast films. J Food Sci 81(5):E1141–E1150

Cornell H (2004) The functionality of wheat starch. In: Eliasson A-C (ed) Starch in food: structure, function and applications, 1st edn. CRC, Cambridge
Cummins HZ (2007) Liquid, glass, gel: the phases of colloidal Laponite. J Non-Cryst Solids 353:3891–3905
Dananjaya SHS, Erandani WKCU, Kim C, Nikapitiya C, Lee J, De ZM (2017) Comparative study on antifungal activities of chitosan nanoparticles and chitosan silver nano composites against Fusarium oxysporum species complex. Int J Biol Macromol 105:478–488
Dang KM, Yoksan R (2015) Development of thermoplastic starch blown film by incorporating plasticized chitosan. Carbohydr Polym 115:575–581
Dang KM, Yoksan R (2016) Morphological characteristics and barrier properties of thermoplastic starch/chitosan blown film. Carbohydr Polym 150:40–47
De Moraes JO, Scheibe AS, Sereno A, Laurindo JB (2013) Scale-up of the production of cassava starch based films using tape-casting. J Food Eng 119(4):800–808
De Moraes JO, Scheibe AS, Augusto B, Carcio M (2015) Conductive drying of starch-fiber films prepared by tape casting: drying rates and film properties. LWT Food Sci Technol. 64:356–366
Deng S, Huang R, Zhou M, Chen F, Fu Q (2016) Hydrophobic cellulose films with excellent strength and toughness via ball milling activated acylation of microfibrillated cellulose. Carbohydr Polym 154:129–138
Dhar P, Tarafder D, Kumar A, Katiyar V (2016) Thermally recyclable polylactic acid/cellulose nanocrystal films through reactive extrusion process. Polymer (Guildf) 87:268–282
Dimitrijevic M, Karabasil N, Boskovic M, Teodorovic V (2015) Safety aspects of nanotechnology applications in food packaging. Procedia Food Sci 5:57–60
Djabourov M, Nishinari K, Ross-Murphy SB (2013) Physical gels from biological and synthetic polymers., 1st edn. Cambridge University Press, Cambridge
Doelker E, Geneva C (1993) Cellulose derivatives. In: Advances in polymer science, pp 199–265
Echeverría I, López-caballero ME, Carmen M, Mauri AN, Montero MP (2018) Active nanocomposite films based on soy proteins- montmorillonite- clove essential oil for the preservation of refrigerated bluefin tuna (Thunnus thynnus) fillets. Int J Food Microbiol 266:142–149
Embuscado M, Huber K. Edible films and coatings for food applications. Embuscado M, Huber K, editors. New York: Springer; 2009
Emin MA, Schuchmann HP (2017) A mechanistic approach to analyze extrusion processing of biopolymers by numerical, rheological, and optical methods. Trends Food Sci Technol 60:88–95
Etxabide A, De Caba K, Guerrero P (2016) A novel approach to manufacture porous biocomposites using extrusion and injection moulding. Eur Polym J 82:324–333
European Food Contact Materials Legislation. European directives on food packaging. Materials and articles in contact with foodstuffs—plastics. 2002
Fakhouri FM, Martelli SM, Caon T, Velasco I, Buontempo RC, Bilck AP et al (2018) The effect of fatty acids on the physicochemical properties of edible films composed of gelatin and gluten proteins. LWT Food Sci Technol. 87:293–300
Famá LM, Pettarin V, Goyanes SN, Bernal CR (2011) Starch/multi-walled carbon nanotubes composites with improved mechanical properties. Carbohydr Polym 83:1226–1231
Fauzi MB, Lokanathan Y, Aminuddin BS, Ruszymah BHI, Chowdhury SR (2016) Ovine tendon collagen: extraction, characterisation and fabrication of thin films for tissue engineering applications. Mater Sci Eng C 68:163–171
Flaker CHC, Lourenço RV, Bittante AMQB, Sobral PJA (2015) Gelatin-based nanocomposite films: a study on montmorillonite dispersion methods and concentration. J Food Eng 167:65–70
Gabriel JS, Gonzaga VAM, Poli AL, Schmitt CC (2017) Photochemical synthesis of silver nanoparticles on chitosans/montmorillonite nanocomposite films and antibacterial activity. Carbohydr Polym 171:202–210
Garrido T, Etxabide A, Guerrero P, De CK (2016) Characterization of agar/soy protein biocomposite films: effect of agar on the extruded pellets and compression moulded films. Carbohydr Polym 151:408–416
Ghelejlu SB, Esmaiili M, Almasi H (2016) Characterization of chitosan-nanoclay bionanocomposite active films containing milk thistle extract. Int J Biol Macromol 86:613–621

Gomathi T, Prasad PS, Sudha PN, Anil S (2017) Size optimization and in vitro biocompatibility studies of chitosan nanoparticles. Int J Biol Macromol 104:1794–1806

Gómez-Estaca J, Gavara R, Catalá R, Hernández-muñoz P (2016) The potential of proteins for producing food packaging materials: a review. Packag Technol Sci 29:203–224

Gong B, Liu W, Tan H, Yu D, Song Z, Lucia LA (2016) Understanding shape and morphology of unusual tubular starch nanocrystals. Carbohydr Polym 151:666–675

González A, Igarzabal CIA (2015) Nanocrystal-reinforced soy protein films and their application as active packaging. Food Hydrocoll 43:777–784

Gras P, Anderssen R, Keentok M, Békés F, Appels R (2001) Gluten protein functionality in wheat flour processing: a review. Aust J Agric Res 52:1311–1323

Guerrero P, Retegi A, Gabilondo N, De CK (2010) Mechanical and thermal properties of soy protein films processed by casting and compression. J Food Eng 100(1):145–151

Guerrero P, Stefani PM, Ruseckaite RA, De CK (2011) Functional properties of films based on soy protein isolate and gelatin processed by compression molding. J Food Eng 105(1):65–72

Guimarães JL, Wypych F, Saul CK, Ramos LP, Satyanarayana KG (2010) Studies of the processing and characterization of corn starch and its composites with banana and sugarcane fibers from Brazil. Carbohydr Polym 80(1):130–138

Guo M, Wang G (2016) Milk protein polymer and its application in environmentally safe adhesives. Polymers (Basel) 8:1–12

Gutiérrez TJ (2017) Chitosan applications for the food industry. In: Ahmed S, Ikram S (eds) Chitosan: derivatives, composites and applications. Wiley-Scrivener, Beverly, MA, pp 185–232 EE.UU. ISBN: 978-1-119-36350-7. https://doi.org/10.1002/9781119364849.ch8

Gutiérrez TJ (2018) Characterization and *in vitro* digestibility of non-conventional starches from guinea arrowroot and La Armuña lentils as potential food sources for special diet regimens. Starch-Stärke 70(1–2):1700124. https://doi.org/10.1002/star.201700124

Gutiérrez TJ, Álvarez K (2016) Physico-chemical properties and *in vitro* digestibility of edible films made from plantain flour with added *Aloe vera gel*. J Funct Foods 26:750–762. https://doi.org/10.1016/j.jff.2016.08.054

Gutiérrez TJ, Alvarez VA (2017) Properties of native and oxidized corn starch/polystyrene blends under conditions of reactive extrusion using zinc octanoate as a catalyst. React Funct Polym 112:33–44. https://doi.org/10.1016/j.reactfunctpolym.2017.01.002

Gutiérrez TJ, Alvarez VA (2017a) Films made by blending poly (ε-caprolactone) with starch and flour from sagu rhizome grown at the Venezuelan amazons. J Polym Environ 25(3):701–716. https://doi.org/10.1007/s10924-016-0861-9

Gutiérrez TJ, Alvarez VA (2017b) Cellulosic materials as natural fillers in starch-containing matrix-based films: a review. Polym Bull 74(6):2401–2430.2430. https://doi.org/10.1007/s00289-016-1814-0

Gutiérrez TJ, Alvarez VA (2017c) Data on physicochemical properties of active films derived from plantain flour/PCL blends developed under reactive extrusion conditions. Data Brief 15:445–448. https://doi.org/10.1016/j.dib.2017.09.071

Gutiérrez TJ, Alvarez VA (2017d) Eco-friendly films prepared from plantain flour/PCL blends under reactive extrusion conditions using zirconium octanoate as a catalyst. Carbohydr Polym 178:260–269. https://doi.org/10.1016/j.carbpol.2017.09.026

Gutiérrez TJ, Alvarez VA (2018) Bionanocomposite films developed from corn starch and natural and modified nano-clays with or without added blueberry extract. Food Hydrocoll 77:407–420. https://doi.org/10.1016/j.foodhyd.2017.10.017

Gutiérrez TJ, Morales NJ, Pérez E, Tapia MS, Famá L (2015) Physico-chemical properties of edible films derived from native and phosphated cush-cush yam and cassava starches. Food Packaging Shelf Life 3:1–8. https://doi.org/10.1016/j.fpsl.2014.09.002

Gutiérrez TJ, Seligra PG, Medina Jaramillo C, Famá L, Goyanes S (2016) Effect of filler properties on the antioxidant response of thermoplastic starch composites. In: Thakur VK, Thakur MK, Kessler MR (eds) Handbook of composites from renewable materials. Wiley-Scrivener, Beverly, MA, pp 337–370 EE.UU. ISBN: 978-1-119-22362-7. https://doi.org/10.1002/9781119441632.ch14

Gutiérrez TJ, Guarás MP, Alvarez VA (2017a) Reactive extrusion for the production of starch-based biopackaging. In: Masuelli MA (ed) Biopackaging. CRC, Taylor & Francis Group, Miami, FL, pp 287–315 EE.UU. ISBN: 978-1-4987-4968-8
Gutiérrez TJ, Ponce AG, Alvarez VA (2017b) Nano-clays from natural and modified montmorillonite with and without added blueberry extract for active and intelligent food nanopackaging materials. Mater Chem Phys 194:283–292. https://doi.org/10.1016/j.matchemphys.2017.03.052
Hanani ZAN, Beatty E, Roos YH, Morris MA, Kerry JP (2012) Manufacture and characterization of gelatin films derived from beef, pork and fish sources using twin screw extrusion. J Food Eng 113(4):606–614
Hanani ZAN, Mcnamara J, Roos YH, Kerry JP (2013) Effect of plasticizer content on the functional properties of extruded gelatin-based composite films. Food Hydrocoll 31(2):264–269
Hanani ZAN, Mahony JAO, Roos YH, Oliveira PM, Kerry JP (2014) Extrusion of gelatin-based composite films: effects of processing temperature and pH of film forming solution on mechanical and barrier properties of manufactured films. Food Packag Shelf Life 2(2):91–101
Hietala M, Mathew AP, Oksman K (2013) Bionanocomposites of thermoplastic starch and cellulose nanofibers manufactured using twin-screw extrusion. Eur Polym J 49(4):950–956
Hietala M, Rollo P, Kekäläinen K, Oksman K (2014) Extrusion processing of green biocomposites: compounding, fibrillation efficiency, and fiber dispersion. J Appl Polym Sci 39981:1–9
Hong S, Rhim J (2008) Antimicrobial activity of organically modified nano-clays. J Nanosci Nanotechnol 8:5818–5824
Hosseini SF, Rezaei M, Zandi M, Farahmandghavi F (2015) Fabrication of bio-nanocomposite films based on fish gelatin reinforced with chitosan nanoparticles. Food Hydrocoll 44:172–182
Huang C, Kuo J, Wu S, Tsai H (2016) Isolation and characterization of fish scale collagen from tilapia (Oreochromis sp.) by a novel extrusion-hydro-extraction process. Food Chem 190:997–1006
Ji N, Liu C, Zhang S, Xiong L, Sun Q (2016) Elaboration and characterization of corn starch fi lms incorporating silver nanoparticles obtained using short glucan chains. LWT Food Sci Technol. 74:311–318
Jose J, Alma MA, Dakua JB, Sreekumar PA, Sougrat R, Al-harthi MA (2015) Compatibilizing role of carbon nanotubes in poly (vinyl alcohol)/starch blend. Starch Stärke 30:147–153
Kalyuzhnaya LM, Bochek AM, Shevchuk IL (2015) Compatibility of carboxymethyl cellulose with hydroxypropyl cellulose in composite films based on them. Polym Syst Technol 88(6):1062–1069
Kanmani P, Rhim J (2014) Physicochemical properties of gelatin/silver nanoparticle antimicrobial composite films. Food Chem 148:162–169
Kathirgamanathan K, Grigsby W, Edmonds NR, Al HJ (2017) Molecular weight fractionation of high polydispersity native celluloses. Cellulose 24:5261–5265
Krepker M, Shemesh R, Danin Y, Kashi Y, Vaxman A, Segal E (2017) Active food packaging films with synergistic antimicrobial activity. Food Control Ltd 76:117–126
Krishna M, Nindo CI, Min SC (2012) Development of fish gelatin edible films using extrusion and compression molding. J Food Eng 108(2):337–344
Kulicke W-M, Clasen C, Lohman C (2005) Characterization of water-soluble cellulose derivatives in terms of the molar mass and particle size as well as their distribution. Macromol Symp 223:151–174
Kumar MNVR (2000) A review of chitin and chitosan applications. React Funct Polym 46:1–27
Lagaron JM, Cabedo L, Cava D, Feijoo JL, Gavara R, Gimenez E (2005) Improving packaged food quality and safety. Part 2: nanocomposites. Food Addit Contam 22:37–41
Li Y, Li J, Xia Q, Zhang B, Wang Q, Huang Q (2012) Understanding the dissolution of α-Zein in aqueous ethanol and acetic acid solutions. J Phys Chem B 116:12057–12064
Li X, Liu A, Ye R, Wang Y, Wang W (2015a) Fabrication of gelatin-laponite composite films: effect of the concentration of laponite on physical properties and the freshness of meat during storage. Food Hydrocoll 44:390–398
Li X, Qiu C, Ji N, Sun C, Xiong L, Sun Q (2015b) Mechanical, barrier and morphological properties of starch nanocrystals-reinforced pea starch films. Carbohydr Polym 121:155–162

Li S, Donner E, Xiao H, Thompson M, Zhang Y, Rempel C et al (2016) Preparation and characterization of soy protein films with a durable water resistance-adjustable and antimicrobial surface. Mater Sci Eng C 69:947–955

Liu G, Song Y, Wang J, Zhuang H, Ma L, Li C et al (2014) Effects of nanoclay type on the physical and antimicrobial properties of PVOH-based nanocomposite films. LWT Food Sci Technol. 57(2):562–568

Llanos JH, Tadini C (2018) Preparation and characterization of bio-nanocomposite films based on cassava starch or chitosan, reinforced with montmorillonite or bamboo nanofibers. Int J Biol Macromol 107:371–382

López OV, Zaritzky NE, Grossmann MVE, García MA (2013) Acetylated and native corn starch blend films produced by blown extrusion. J Food Eng 116:286–297

López O, Versino F, Villar M, García M (2015) Agro-industrial residue from starch extraction of Pachyrhizus ahipa as filler of thermoplastic corn starch films. Carbohydr Polym 134:324–332

Ma Q, Zhang Y, Critzer F, Davidson PM, Zivanovic S (2016) Physical, mechanical, and antimicrobial properties of chitosan films with microemulsions of cinnamon bark oil and soybean oil. Food Hydrocoll 52:533–542

Ma Z, Garrido-maestu A, Casey K (2017a) Application, mode of action, and in vivo activity of chitosan and its micro- and nanoparticles as antimicrobial agents: a review. Carbohydr Polym 176:257–265

Ma X, Lv M, Anderson DP, Chang PR (2017b) Natural polysaccharide composites based on modified cellulose spheres and plasticized chitosan matrix. Food Hydrocoll 66:276–285

Martínez-Camacho AP, Cortez-Rocha MO, Graciano-Verdugo AZ, Rodríguez-Félix F (2013) Extruded films of blended chitosan, low density polyethylene and ethylene acrylic acid. Carbohydr Polym 91(2):666–674

Matet M, Heuzey M, Ajji A, Sarazin P (2015) Plasticized chitosan/polyolefin films produced by extrusion. Carbohydr Polym 117:177–184

Matsushima N, Danno G, Takezana H, Izumi Y (1997) Three-dimensional structure of maize alpha-zein proteins studied by small-angle X-ray scattering. Biochim Biophys Acta 1339:14–22

Matthews LB, Kunkel ME, Acton JC, Ogale AA, Dawson PL (2011) Bioavailability of soy protein and corn zein films. Food Nutr Sci 2:1105–1113

Mendes JF, Paschoalin RT, Carmona VB, Sena AR, Marques ACP, Marconcini JM et al (2016) Biodegradable polymer blends based on corn starch and thermoplastic chitosan processed by extrusion. Carbohydr Polym 137:452–458

Merzendorfer H, Zimoch L (2003) Chitin metabolism in insects: structure, function and regulation of chitin synthases and chitinases. J Exp Biol 206:4393–4412

Metak AM, Nabhani F, Connolly SN (2015) Migration of engineered nanoparticles from packaging into food products. LWT Food Sci Technol 64:781–789

Mischnick P, Momcilovic D (2010) Advances in carbohydrate chemistry and biochemistry. Chem Struct Anal Starch Cellulose Derivatives 64:117–210

Mohamed MM, Fouad SA, Elshoky HA, Mohammed GM, Salaheldin TA (2017) Antibacterial effect of gold nanoparticles against Corynebacterium pseudotuberculosis. Int J Vet Sci Med 5(1):23–29

Montero B, Rico M, Rodríguez-llamazares S, Barral L (2017) Effect of nanocellulose as a filler on biodegradable thermoplastic starch films from tuber, cereal and legume. Carbohydr Polym 157:1094–1104

Morgan PE, Treweek TM, Linder RA, Price WE, Carver JA (2005) Casein proteins as molecular chaperones. J Agric Food Chem 53:2670–2683

Mouw JK, Ou G, Weaver VM (2014) Extracellular matrix assembly: a multiscale deconstruction. Nat Rev Mol Cell Biol 15:771–785

Mujtaba M, Salaberria AM, Andres MA, Kaya M, Gunyakti A, Labidi J (2017) Utilization of flax (Linum usitatissimum) cellulose nanocrystals as reinforcing material for chitosan films. Int J Biol Macromol 104:944–952

Nehete JY, Bhambar RS, Narkhede MR, Gawali SR (2013) Natural proteins: sources, isolation, characterization and applications. Pharmacogn Rev 7(14):107–116
Nisperos-Carriedo MO (1994) Edible coatings and films based on polysaccharides. In: Edible coatings and films to improve food quality. CRC, Boca Raton, FL, pp 305–336
Noshirvani N, Hong W, Ghanbarzadeh B, Fasihi H, Montazami R (2018) Study of cellulose nanocrystal doped starch-polyvinyl alcohol bionanocomposite films. Int J Biol Macromol 107:2065–2074
Oechsle AM, Häupler M, Weigel F, Gibis M, Kohlus R, Weiss J (2016) Modulation of extruded collagen films by the addition of co-gelling proteins. J Food Eng 171:164–173
Orsuwan A, Sothornvit R (2017) Development and characterization of banana flour film incorporated with montmorillonite and banana starch nanoparticles. Carbohydr Polym 174:235–242
Ortega F, Giannuzzi L, Arce VB, García MA (2017) Active composite starch films containing green synthetized silver nanoparticles. Food Hydrocoll 70:152–162
Ortega-toro R, Jiménez A, Talens P, Chiralt A (2014) Properties of starch-hydroxypropyl methylcellulose based films obtained by compression molding. Carbohydr Polym 109:155–165
Ortega-Toro R, Morey I, Talens P, Chiralt A (2015) Active bilayer films of thermoplastic starch and polycaprolactone obtained by compression molding. Carbohydr Polym 127:282–290
Ortiz-Zarama MA, Jiménez-Aparicio A, Perea-Flores MJ, Solorza-Feria J (2014) Barrier, mechanical and morpho-structural properties of gelatin films with carbon nanotubes addition. J Food Eng 120:223–232
Park JW, Whiteside S, Cho SY (2008) Mechanical and water vapor barrier properties of extruded and heat-pressed gelatin films. LWT Food Sci Technol. 41:692–700
Park S, Baker JO, Himmel ME, Parilla PA, Johnson DK (2010) Cellulose crystallinity index: measurement techniques and their impact on interpreting cellulase performance. Biotechnol Biofuels 3(1):10
Pérez S, Baldwin PM, Gallant DJ (2009) Structural features of starch granules I. In: BeMiller J, Whistler R (eds) Starch: chemistry and technology, 3rd edn. Elsevier B.V., Amsterdam
Perotti GF, Tronto J, Bizeto MA, Izumi CMS, Temperini MLA, Lugão AB et al (2014) Biopolymer-clay nanocomposites: cassava starch and synthetic clay cast films. J Braz Chem Soc 25(2):320–330
Pooja D, Panyaram S, Kulhari H, Reddy B (2015) Natural polysaccharide functionalized gold nanoparticles as biocompatible drug delivery carrier. Int J Biol Macromol 80:48–56
Qi G, Venkateshan K, Mo X, Zhang L, Sun XS (2011) Physicochemical properties of soy protein: effects of subunit composition. J Agric Food Chem 59:9958–9964
Quintana R, Persenaire O, Lemmouchi Y, Bonnaud L, Dubois P (2016) Compatibilization of co-plasticized cellulose acetate/water soluble polymers blends by reactive extrusion. Polym Degrad Stab 126:31–38
Quiroz-Castillo JM, Rodríguez-Félix DE, Grijalva-Monteverde H (2014) Preparation of extruded polyethylene/chitosan blends compatibilized with polyethylene-graft-maleic anhydride. Carbohydr Polym 101:1094–1100
Richard ME, Twiname ER (2000) Tape casting: theory and practice. Wiley, New York, p 293
Rzychon M, Brohée M, Cordeiro F, Haraszi R, Ulberth F, O'Connor G (2017) The feasibility of harmonizing gluten ELISA measurements. Food Chem 234:144–154
Schmid M, Dallmann K, Bugnicourt E, Cordoni D, Wild F, Lazzeri A et al (2012) Properties of whey-protein-coated films and laminates as novel recyclable food packaging materials with excellent barrier properties. Int J Polym Sci 562381:1–7
Schrieber R, Gareis H (2007) Gelatine handbook, 1st edn. Wiley, Weinheim
Selling GW, Utt KD (2013) Effect of multiple extrusion passes on zein. Polym Degrad Stab 98(1):184–189
Sharma S, Luzinov I (2013) Whey based binary bioplastics. J Food Eng 119(3):404–410
Shewry PR, Halford NG, Belton PS, Tatham AS (2002) The structure and properties of gluten: an elastic protein from wheat grain. R Soc 357:133–142

Shih C, Shieh Y, Twu Y (2009) Preparation and characterization of cellulose/chitosan blend films. Carbohydr Polym 78(1):169–174

Shoulders MD, Raines RT (2009) Collagen structure and stability. Annu Rev Biochem 78:929–958

Singh A, Meena M, Kumar D, Dubey AK, Hassan I (2015) Structural and functional analysis of various globulin proteins from soy seed. Crit Rev Food Sci Nutr 8398:1491–1502

Soni B, Barbary E, Schilling MW, Mahmoud B (2016) Transparent bionanocomposite films based on chitosan and TEMPO-oxidized cellulose nanofibers with enhanced mechanical and barrier properties. Carbohydr Polym 151:779–789

Sothornvit R, Olsen C, McHuhg T, Krochta J (2007) Tensile properties of compression-molded whey protein sheets: determination of molding condition and glycerol-content effects and comparison with solution-cast films. J Food Eng 78:855–860

Stylianou A, Yova D (2013) Surface nanoscale imaging of collagen thin films by atomic force microscopy. Mater Sci Eng C 33(5):2947–2957

Taghizadeh A, Favis BD (2013) Carbon nanotubes in blends of polycaprolactone/thermoplastic starch. Carbohydr Polym 98(1):189–198

Tanada-palmu PS, Grosso CRF (2003) Development and characterization of edible films based on gluten from semi-hard and soft brazilian wheat flours (development of films based on gluten from wheat flours). Food Sci Technol 23(2):264–269

Tang X, Alavi S (2012) Structure and physical properties of starch/poly vinyl alcohol/laponite RD nanocomposite films. J Agric Food Chem 60:1954–1962

Tester RF, Karkalas J, Qi X (2004) Starch-composition, fine structure and architecture. J Cereal Sci 39:151–165

Thompson G, Larkins B (1989) Structural elements regulating Zein gene expression. BioEssays 10(4):108–113

Tilley KA, Benjamin RE, Bagorogoza KE, Okot-kotber BM (2001) Tyrosine cross-links: molecular basis of gluten structure and function. J Agric Food Chem 49:2627–2632

Tsumura K (2009) Improvement of the physicochemical properties of soybean proteins by enzymatic hydrolysis. Food Sci Technol Res 15(4):381–388

Usman A, Zia KM, Zuber M, Tabasum S, Rehman S, Zia F (2016) Chitin and chitosan based polyurethanes: a review of recent advances and prospective biomedical applications. Int J Biol Macromol 86:630–645

Valencia GA, Vercik LC de O, Ferrari R, Vercik A (2013) Synthesis and characterization of silver nanoparticles using water-soluble starch and its antibacterial activity on Staphylococcus aureus. Starch Stärke 65:931–937

Valencia GA, Ferreira LG, Llanos JHR, Vercik A (2014a) Synthesis and characterisation of gold nanoparticles using Mentha piperita leaf extract: a green, non-toxic and rapid method. Int J Nano Biomater 5(2/3):181–192

Valencia GA, de Vercik LC, Vercik A (2014b) A new conductometric biosensor based on horseradish peroxidase immobilized on chitosan and chitosan/gold nanoparticle films. J Polym Eng 34(7):633–638

Valencia GA, Moraes ICF, Hilliou LHG, Lourenço RV, Sobral PJA (2015) Nanocomposite-forming solutions based on cassava starch and laponite: viscoelastic and rheological characterization. J Food Eng 166:174–181

Valencia GA, Lourenço RV, Bittante AMQB, Sobral PJA (2016) Physical and morphological properties of nanocomposite films based on gelatin and Laponite. Appl Clay Sci 124–125:260–266

Valencia GA, Luciano CG, Lourenço RV, Sobral PJA (2018) Microstructure and physical properties of nano-biocomposite films based on cassava starch and laponite. Int J Biol Macromol 107:1576–1583

Vieira MGA, Silva MA, Santos LO, Beppu MM (2011) Natural-based plasticizers and biopolymer films: a review. Eur Polym J 47(3):254–263

Wagh YR, Pushpadass HA, Emerald FM, Nath B (2014) Preparation and characterization of milk protein films and their application for packaging of Cheddar cheese. J Food Sci Technol 51(12):3767–3775

Wang Y, Padua GW (2003) Tensile properties of extruded zein sheets and extrusion blown films. Macromol Mater Eng 288:886–893

Wang H, Johnson L, Wang T (2004) Preparation of soy protein concentrate and isolate from extruded-expelled soybean meals. J Am Oil Chem Soc 81:713–714

Wang J, Su Y, Jia F, Jin H (2013) Characterization of casein hydrolysates derived from enzymatic hydrolysis. Chem Cent J 7:1–8

Wang L, Shankar S, Rhim J (2017) Properties of alginate-based films reinforced with cellulose fibers and cellulose nanowhiskers isolated from mulberry pulp. Food Hydrocoll 63:201–208

Wieser H (2007) Chemistry of gluten proteins. Food Microbiol 24:115–119

Wolf K, Sobral PJA, Telis VR (2009) Physicochemical characterization of collagen fibers and collagen powder for self-composite film production. Food Hydrocoll 23:1886–1894

Yamakawa A, Suzuki S, Oku T, Enomoto K, Ikeda M, Rodrigue J et al (2017) Nanostructure and physical properties of cellulose nanofiber-carbon nanotube composite films. Carbohydr Polym 171:129–135

Yan Q, Hou H, Guo P, Dong H (2012) Effects of extrusion and glycerol content on properties of oxidized and acetylated corn starch-based films. Carbohydr Polym 87(1):707–712

Yang Q, Fukuzumi H, Saito T, Isogai A, Zhang L (2011) Transparent cellulose films with high gas barrier properties fabricated from aqueous alkali/urea solutions. Biomacromolecules 12:2766–2771

Zepon KM, Vieira LF, Soldi V, Salmoria GV, Kanis LA (2013) Influence of process parameters on microstructure and mechanical properties of starch-cellulose acetate/silver sulfadiazine matrices prepared by melt extrusion. Polym Test 32(6):1123–1127

Zhanjun L, Lei Z, Minnan C, Jiugao Y (2011) Effect of carboxylate multi-walled carbon nanotubes on the performance of thermoplastic starch nanocomposites. Carbohydr Polym 83(2):447–451

Zubeldía F, Ansorena MR, Marcovich NE (2015) Wheat gluten films obtained by compression molding. Polym Test 43:68–77

Chapter 6
Surface Properties of Biodegradable Polymers for Food Packaging

Z. A. Nur Hanani

Abstract Biodegradable polymers derived from biomass such as polysaccharides (starches, chitosan, and gums) and proteins (gelatin, soy, and zein) have been explored tremendously as potential food packaging materials. Their unique characteristics, for example, edible, abundance, renewable and low-cost allow these materials to be utilized in many forms such as films and coatings. However, biodegradable polymers exhibit high water vapour permeability and solubility. Functional properties of biodegradable polymers can be enhanced by blending with other polymers, lipids, surfactants, emulsifiers or other additives. Combining some polymers and additives will change the microstructure, mechanical, barrier and surface properties of films. Therefore, surface properties can influence the final applications of films and coatings. Interestingly, surface properties of polymers can be tailored using some treatment. Lack of discussion on surface properties of biodegradable films is noticeable. This chapter presents the surface properties of biodegradable films and coatings from various sources and their characterizations. Some surface treatments on films aiming to improve their characteristics and effect of the surface on active packaging are also discussed.

Keywords Surface treatment

6.1 Biodegradable Polymers for Food Applications

Nowadays, food processing involves some other treatments to prolong the shelf life of foods in the package such as freezing, irradiation, high pressure and ozone treatment. Also, the modern lifestyle and convenient options on the usage of ready to eat (RTE) meals require some further steps before consuming the foods such as microwave and oven and this indirectly may change the film surface properties and affect foods specifically.

Z. A. Nur Hanani (✉)
Department of Food Technology, Faculty of Food Science and Technology, Universiti Putra Malaysia, Serdang, Selangor, Malaysia
e-mail: hanani@upm.edu.my

T. J. Gutiérrez (ed.), *Polymers for Food Applications*,
https://doi.org/10.1007/978-3-319-94625-2_6

Synthetic polymers such as polypropylene (PP), low-density polyethylene (LDPE), high-density polyethylene (HDPE), polyethylene terephthalate (PET) and nylon have been widely used in food packaging to protect food products from physical and mechanical damages and microbiological spoilage. This scenario is due to their good performance properties such as stable, water resistance and strong. However, these materials are being utilized as food packages for short term usage despite their properties and durability that remained for a longer period. Furthermore, the recycling of these materials is impractical due to food contamination. Thus, due to environmental awareness, biodegradable materials are seen as alternatives to replace synthetic plastics which are non-biodegrade; causing some negative issues to nature. Biodegradable polymers are claimed as green materials, renewable and environmental-friendly causing this area to be tremendously exploited today.

6.2 Biodegradable Polymers

Biodegradable polymers are gaining more popularity due to the eco-friendly effect they offered, renewable, abundance and cheap. These polymers will break down and produce natural by-products such as gases (CO_2 and N_2), water, biomass and inorganic salts. Biodegradable polymers are classified into few categories depending on the sources obtained; natural biopolymers and synthetic biopolymers. Natural polymers are produced from agro-based materials such as polysaccharides (starch, cellulose derivatives, chitosan, alginate, etc.) and proteins (soy, corn, wheat, gelatin, keratins, etc.) whereas synthetic biopolymers are obtained from synthetic or natural monomers and microorganism. Biodegradable polymers can also be in combination with synthetic biodegradable polyesters (Gutiérrez and Alvarez 2017a).

Natural polymers derived from polysaccharides- and protein-based are unique since they are also edible (can be consumed together with the food) and can provide additional nutrients; giving an advantage to applying with food products (Álvarez et al. 2017). Hence, this type of polymers can be developed as films or/ and coatings. The films produced also perform good gas barrier, good mechanical properties, transparent and good carriers for active compounds such as antimicrobials, antioxidants, etc.. The main issue related to this group of polymers is the high permeability towards moisture or water (Gutiérrez and Alvarez 2017b). Great efforts are still in progress to improve the water barrier so that these materials have wider applications and can be utilised for several food types. One of the alternatives is adding hydrophobic compounds like lipids or fats to produce composites or blend films. However, incorporation of lipids to the solutions will alter the properties of films produced. Also, surface properties of the single polymer are different compared to the blends polymers (Sionkowska and Płanecka 2013).

Currently, there are high interest on developing biodegradable films, manly focused on improving mechanical and barrier properties of the materials for further applications. Only few studies focused on the improvement of surface properties of biodegradable films to enhance the characteristics of films or other surface such as paper. Surface properties are also important because they will influence the structure of films (Gutiérrez et al. 2018). However, surface properties of biodegradable polymers can be modified depending on the final applications either as coatings or films since both applications may have different purpose. Biodegradable polymers can also be used in the surface treatment of cellulose-based materials, either by coating or by extrusion/lamination (Andersson 2008).

6.3 Surface Properties of Biodegradable Polymers

Biodegradable films can be tailored based on the combination and composition of materials used which consequently, will affect the mechanical properties, water transmission rate, optical properties and also surface properties of films. Like other polymers, surface properties of food packaging polymers include wettability, sealability, printability, dye uptake, resistance to glazing and adhesion to food surfaces or other polymers (Sengupta and Han 2014). These are some of the important elements to food packaging manufacturers for maintaining the shelf life of products while ensuring good quality appearance. Furthermore, surface roughness, polarity, wettability and modification of the surface will also determine the biocompatibility of polymeric materials (Sionkowska and Płanecka 2013).

6.3.1 *Wettability*

Wetting is the ability of a liquid to interact with the solid surface or another fluid either maintains the interaction or penetrates the surface. Wettability is the degree of wetting due to the interaction between the fluid and solid phases. It determines the properties of polymers' surface in allowing liquid seep through and plays a crucial role in liquid coating and printing. Wettability can be determined using contact angle measurement whereby small contact angles ($\leq 90^\circ$) indicate high wettability where as large contact angles ($\geq 90^\circ$) indicate low wettability.

6.3.2 *Sealability*

Sealing involved melting of the polymer by supplying heat, followed by linking of the surfaces under pressure. Sealability determines film's ability to turn as its own bond-forming agent without requiring extra hot melt adhesive. Some materials may need less energy to be sealed, depending on the degree of materials crystallinity. Thus, temperature, time and pressure should be identified so that sufficient energy is supplied to ensure good sealability. A polymer that displays low-temperature sealability and maintains seal integrity over a broad seal temperature, dwell time and seal pressure can increase packaging line speeds, improve efficiencies and reduce seal failures (Butler and Morris 2012). However, the presence of fillers, additives or waxes may cause weaker seal due to the migration of these additives to the surface (Andersson 2008; Bracone et al. 2016; Gutiérrez et al. 2017). The–OH groups are also capable of forming hydrogen bonds and thereby provide good adhesion between different surfaces (Andersson 2008). Defects or weak spots during and after sealing may produce pathways for transportation of gases or liquids either in or out of the package (Andersson 2008).

6.3.3 Printability

Printability of polymer is the ability of the polymer to be printed without displacement of the ink. Surface properties such as smoothness, levelness, ink absorbency, gloss, etc. influence this property. Printability test can be done by using tape which will be used for lifting ink off the printed samples according to a pre-established number of peels (López-García et al. 2013). The tape is lifted off with consistent force at an angle 90° before the samples are recorded using UV-VIS spectrophotometer.

6.4 Surface Properties of Polysaccharides-Based Films

Polysaccharides have a wide range of structures, depending on the types such as starch, cellulose derivatives, chitosan, and alginate. Polysaccharides-based films possess high gas barrier due to their well-ordered hydrogen-bonded network shape (Hassan et al. 2017). However, these materials are very hydrophilic causing in high water vapour permeability. This drawback limits the usage of polysaccharides films for high moisture or semi-solid food products. To optimum the applications, they can be applied as thick films or coatings on the surface of food to absorb water, giving temporary protections to foods from moisture loss (Cazón et al. 2017). Polysaccharides films and coatings are colourless, however can change the film's colour depending on the additives or active compounds added. Mechanical property, i.e. tensile strength values of some polysaccharides based films are comparable to those values obtained in high density polyethylene (HDPE) films (Cazón et al. 2017). Highly structured polysaccharide attributes to the homogeneity of films and smoothness of the film's surface (Caro et al. 2016).

6.4.1 Starch

Starch is an agricultural biopolymer that composed of anhydroglucose. It is abundant, low price and unique. This is because starch granules vary in shape, size, structure and chemical composition depending on the botanical source (Molavi et al. 2015). The starch granules comprise two main polysaccharides; amylose and amylopectin, apart from other components such as proteins and lipids. Amylose which responsible for film-forming properties is a linear chain polymer of a-1,4 anhydroglucose units with a molecular size ranging from 20 to 800 kg/mol (Cazón et al. 2017). Meanwhile, amylopectin is a highly branched polymer of short a-1,4 chains linked by a-1,6 glycosidic branching points occurring every 25–30 glucose units and with a very high molecular weight (5000–30,000 kg/mol) (Cazón et al. 2017; Jiménez et al. 2012). Starch films are generally tasteless, odorless and transparent. However, at higher concentration, films produced tend to become whitish. Starch films that compose higher crystalline structure are less sensitive to moisture and the environmental relative humidity (Cazón et al. 2017; Molavi et al. 2015; Jonhed et al. 2008; Mali et al. 2004). Regarding surface properties, starch application increases both the roughness and the hydrophilicity of the coated surface (Andersson 2008). Films contain higher amylose content exhibits greater surface roughness (Gutiérrez and González 2016). Films with different starches have shown that glutinous rice starch and normal rice starch-based films possessed higher contact angle values than cassava starch due to higher lipid content (Phan et al. 2005). Study on the surface properties of starchy films with blackberry pulp revealed that the pulp increased the contact angle and lower surface roughness (Gutiérrez 2017a).

6.4.2 Chitosan

Chitosan is a natural polymer derived by deacetylation of chitin, the second most abundant natural polymer after cellulose (Gutiérrez 2017b). Chitosan has various applications because of its functional properties such as antibacterial activity, non-toxicity, ease of modification and biodegradability (Muxika et al. 2017). Furthermore, chitosan films are transparent and flexible and have semicrystalline structure. Chitosan addition in cassava films helps to increase the contact angle, which means improving the surface hydrophobicity of films. This is due to the hydrophobic acetyl groups present in chitosan chain, suggesting that chitosan is more hydrophobic than starch (Kampeerapappun et al. 2007). Kurek et al. (2014) also observed that larger contact angles on the chitosan surface were found compared to whey protein (support surface) of bilayer films. However, an addition of plasticizer on the chitosan films lessens water contact angle due to the water binding capacity (hygroscopicity) of plasticizers. As chitosan has highly structured polysaccharides; films produced are smooth and flat with no cracks and pores (Caro et al. 2016). However, hybrid chitosan films may have surface irregularities. Higher ferulic incorporation in the chitosan films had caused phase separation (from AFM analysis) and might be responsible for the reduction in tensile strength (Mathew and Abraham 2008). However, the authors indicated that the cross section images of SEM showed films were more compact due to the networking introduced by the acid than those control films (chitosan-starch blend) which were having discontinuous zones.

6.4.3 Pectin

Pectin is a plant cell wall polysaccharide rich in D-galacturonic acid and mostly obtained from citrus fruits or fruit processing industry waste. Films from pectin possess good hardness and adhesiveness. However, they can become rigid and brittle. The SEM images of the surface and cross-sectional of pectin films added with clove bud essential oil (CEO) showed that oil produced a dense sheet-like structure, whereas the cross-sectional images had the sheets stacked in compact layers demonstrating that CEO added uniformly in the film matrix (Nisar et al. 2018). The authors had found out that at a low level of oil (0.5%), smoother surfaces without any phase separation was observed. However, higher oil (1 and 1.5%) had caused the surface to become rough and looser texture. According to Nisar et al. (2018), the different surface morphology of films could be due to the structural changes of components of micro-emulsions during the drying process. The effect of pectin surface density on the high methoxyl pectin-based films was also investigated (Giancone et al. (2011). It was revealed that the surface density did not affect the film structure, yet, it increased the WVP.

6.4.4 Galactomannans

Galactomannans are heterogeneous polysaccharides composed of linear chains of β-(1–4)-D-mannan backbone with a single D-galactose branch linked α-(1–6) (Cerqueira and Bourbon 2011). There are three major gallactomannans used for food industry which are guar gum, tara gum and locust bean gum, mainly vary because of the different ratio of mannose and galactose. Almost similar with other common biodegradable films, galactomannans films are essentially hydrophilic. Irregular surface of galactomannans films caused the films to have higher WVP due to the presence of voids (Albuquerque et al. 2017). Furthermore, the addition of bioactive compound contributed to rougher surface and increased in the hydrophobicity.

6.4.5 Fiber

Fiber refers to edible parts of carbohydrates that cannot be digested. Natural fiber is cheap, has low specific weight, recyclable and competitive mechanical properties (Gutiérrez and Alvarez 2017c). However, fiber has poor adhesion with the matrix with the creation of voids at the interface and non-uniform dispersion (John and Thomas 2008). Modifying surface of fiber had enhanced the fiber-matrix interaction by improving the mechanical properties (Geogiopoulos et al. 2016).

6.5 Surface Properties of Protein-Based Films

Proteins are linear polymers constructed by monomer unit called amino acids through a covalent peptide bond. There are 20 types of amino acids in protein which having different chemical properties and roles (hydrophobic, polar or charged). Proteins can be classified into two types; plant proteins and animal proteins. Soy protein, corn protein and wheat protein are among the plant proteins whereas casein, collagen, gelatin and keratin are the types of extensively used animal proteins. Lactate dehydrogenase, chymotrypsin, and fumarase are the main bacterial proteins. Compared to polysaccharides and lipids, films obtained from proteins exhibit valuable characteristics for the production of food packaging as these films have good film-forming ability, mechanical properties and transparency. In fact, protein-based films have excellent oxygen and carbon dioxide barrier properties than polysaccharides-based films.

6.5.1 Soy Protein Isolate

Soy protein isolate is obtained from soy protein, a by-product from soy oil production. Hydrolyzed keratin produced from chicken feather had been used into soy protein films (Garrido et al. 2018). The findings discovered that an additional of keratin decreased the gloss significantly and enhanced the hydrophobicity. Surface hydrophobicity of protein films such as soy protein isolate, whey protein concentrate, gelatin, peanut protein isolate and sodium caseinate were also improved with the treatment of the enzyme crosslinking (transglutaminase) (Tang and Jiang 2007).

6.5.2 Gelatin

Gelatin is an insoluble protein gained by hydrolysis of collagen, a fundamental structure of animal bodies (Nur Hanani 2016). Gelatin from fish, pork and bovine have been studied greatly. The surface properties of bilayer and blend films based on bovine gelatin showed that all films have hydrophobic surface, based on the contact angle (Abdelhedi et al. 2018). Chemical reaction occured between gelatin and lactose also influenced the structure of gelatin films with tetrahydrocurcumin, causing the films to become less glossy and rougher surfaces (Etxabide et al. 2017). Deng et al. (2018) have discovered that the gelatin/zein nanofibrous film had a hydrophobic surface with 118.0°, whereas casted gelatin/zein film had a hydrophilic surface (53.5°). Addition of chitin in gelatin film also had decreased the hydrophilicity of film with film containing the highest oil had contact angle higher than 90° (Sahraee et al. 2017).

6.5.3 Zein

Zein is a hydrophobic and thermoplastic material derived from corn. Its high hydrophobicity is attributed by its high content of nonpolar amino acids. The incorporation of sugar plasticizers in zein films reduced the contact angle with higher hygroscopicity of plasticizer contributed to the higher hydrophilic due to the higher water binding capacity (Ghanbarzadeh et al. 2007).

6.6 Lipids and Other Additives

Due to the limitation on water barrier properties of starch- and protein-based films, different components mainly lipids (waxes, oils and fats) and other additives (surfactants, emulsifiers, etc.) are being added to produce composite films. Lipids exhibit excellent barriers against moisture migration. Extensive studies are ongoing considering these lipids help to enhance the water barrier of biodegradable films. Polysaccharides- and proteins-based films incorporated with lipids generally have higher mechanical properties. Nonetheless, the composite films may have higher moisture permeability than that of pure lipids (Hassan et al. 2017; Bravin et al. 2004). Higher amount of lipids used can increase gloss and decrease transparency. Meanwhile, the production of lipid based films and coating are believed to be highly effective to block the delivery of moisture due to their low polarity (Hassan et al. 2017; 104). Their hydrophobicity also causes the films to become brittle and thicker.

In general, plasticizers contribute to the decrease of contact angle of hydrocolloid films due to hydrophilicity of these materials. Higher concentration of glycerol decreases the contact angle due to the increases in the surface tension (Caro et al. 2016). In contrast, additional of plasticizers such as glycerol, sorbitol and polyethylene (glycol) in sage seed gum films had caused higher contact angle than the control films (Razavi et al. 2015). Meanwhile, calcium chloride has been used as a firming agent in mesquite gum films. The surface morphology of the films was influenced by the agent concentration, whereby the film surface became gradually rough at higher concentration (Bosquez-Molina et al. 2010).

6.7 Surface Characteristics

There are some techniques to characterize the surface properties of biodegradable polymers such as contact angle measurement, atomic force microscopy (AFM) and scanning electron microscopy (SEM). This section is not going to discuss details on each instrument. However, brief information is delivered to relate with the surface properties of biodegradable polymers.

6.7.1 Contact Angle

Contact angle is the wetting angle between the surface of the liquid and the outline of the contact surface. The analysis determines the surface hydrophobicity of polymers. Small contact angles (≤90°) occur due to spreading of the drop (molecular attraction). Meanwhile, greater angles (≥90°) occur due to the liquid becomes bead or shrink away from the solid surface. Lower contact angle indicates the films have high polarity and better bonding of adhesive. This process is a crucial index to determine the wettability of the solid phase by the liquid and establish the formation of a good bonding interface. The increase of the contact angle with water in biodegradable polymers could be due to a strong hydrogen bond inter-molecular by below of the surface of the film, i.e. the more polar sites (Lewis sites) would be affected, thus generating a decrease in the surface polarity of biopolymer-based films (Gutiérrez et al. 2016a, b; Karbowiak et al. 2006). Storage also can increase the contact angle due to the loss of moisture content and plasticizer (Suyatma et al. 2005). On the other hand, dynamic contact can reflect the degree of difficulty of coating the solid phase with the liquid in the real wetting process (Zhang et al. 2017).

Surface free energy is another parameter to discuss about the surface properties. Surface free energy (interfacial free energy) is work required to increase the size of the solid surface (work per unit area). In contrast, the term 'surface tension' is used for a liquid phase. Surface tension can be measured using some techniques depending on the nature of the liquid, the condition during the measurement and the stability of the surface. However, the contact angle is normally used to measure the surface free energy indirectly. According to Young's equation, the surface free energy can be determined using equation below:

$$\sigma_s = \sigma_{sl} + \sigma_l \cdot \cos\theta \tag{6.1}$$

where:
σ_s = surface free energy
σ_{sl} = interfacial tension
σ_l = surface tension of the liquid
θ = contact angle

In conventional plastics, polymer with high surface energy can be used as the first surface of few layers structures and requires an adhesive layer to bond to the other different layer (Butler and Morris 2012).

6.7.2 Atomic Force Microscopy (AFM)

Atomic Force Microscopy (AFM) offers a 3D profile on a nanoscale, by measuring forces between a sharp probe (with radius less than 10 nm) and surface at very short distance (0.2–10 nm probe-sample preparation) (De Oliviera et al. 2012). The probe is supported on a flexible cantilever and the AFM tip softly touches the surface and records the small force between the probe and the surface (De Oliviera et al. 2012). The images obtained give some information about the surface roughness of films such as roughness average (R_a) and root mean square roughness (R_q). R_a, is the arithmetic mean of the absolute values of the height of the surface profile (De Oliviera et al. 2012). It is used widely because of easy to obtain. R_q is similar to roughness average, except it is the mean squared absolute values of surface roughness profile. The information delivered by AFM is beneficial to determine structural changes of the film matrix. Despite of providing high resolution at the nanoscale, this analysis is restricted due to time-consuming measurement. Smoother surface of films can also be related to the increased of the transparency as revealed by Gutiérrez et al. (Gutiérrez et al. 2016a, b).

6.7.3 Scanning Electron Microscopy (SEM)

Scanning Electron Microscopy (SEM) is used to determine the surface structure of polymers. The knowledge of the morphology is an important parameter to determine structural changes in the films and to predict their porosity, permeability, flexibility and resistance (Giosafatto et al. 2014; de Paula Herrmann et al. 2004). Smooth surface indicates good compatibility between compounds and the plasticizers. For cross section analysis, film with compact structure indicates the networking is developed. However, film with some porosity allows the distribution of active compound to a greater depth which in turn caused the films to release the compounds in slower rate.

6.7.4 X-Ray Photoelectron Spectroscopy (XPS)

X-ray Photoelectron Spectroscopy (XPS) can also be used to analyse the surface chemistry of biodegradable polymers. It is employed to obtain quantitative insight into the elemental composition of the surface that does not extend beyond certain depth (López-García et al. 2013). In this analysis, X-rays irradiates onto the surface of polymers in vacuum environment. The energy from the photoelectrons radiated from the surface is measured. XPS can be applied for various type of materials (conducting and non-conducting samples) by providing about the surface layer or structures.

6.7.5 Gloss Analysis

Gloss, transparency, clarity, haze, and colour are some optical properties of films. Gloss analysis is a simple method, yet its usage is not widely emphasized. Gloss analysis can directly relate to the surface roughness, with lower values indicate the surface is rougher (Garrido et al. 2018; Ward and Nussinovitch 2017). Different techniques of films manufacturing produce different surface properties. Films produced by compression moulding exhibited higher gloss values than those prepared by casting contributing to the smoother surface, as supported by SEM (Garrido et al. 2018). Also, gloss values may decrease due to the immiscibility of the polymers.

6.8 Surface Treatment

As different polymers possess various surface characteristics, therefore, it is important to have another step to alter the surface of biodegradable polymers and improve the functionality for fulfil their application. There are some techniques used to modify the surface properties such as utraviolet (UV)-light irradiation, incorporation of nano-materials, plasma surface treatment and lamination process.

6.8.1 Ultraviolet (UV)-Light

Ultraviolet (UV) radiation has been used to modify films surface properties. In the case of protein films, the radiation is absorbed by double bonds and aromatic rings of some amino acids, producing free radicals and causes intermolecular covalent bonding (Díaz et al. 2017; Rhim et al. 1999). UV treatment on film-forming solution has contributed to significant effect on the mechanical properties, colour and stability of whey protein films (Díaz et al. 2016). UV treated whey protein films also improved their puncture properties than the control (Díaz et al. 2017). UV radiation is normally applied on materials based on natural polymers for sterilisation process, whereby blended polymers may change differently than single components (Sionkowska and Płanecka 2013). Tarek et al. (2015) have studied the effect of UV-light treatment on the surface properties of plastic films for beef by determining the surface-free energy. They have found that UV-C light treatment had decreased the polymer surface roughness; however, it did not affect the surface free energy of films. UV-irradiation may cause the reduction of surface roughness of chitosan films and chitosan with silk fibroin (Sionkowska and Płanecka 2013). Surface modification using UV irradiation reduced the surface hydrophilicity and enhanced the water resistance and tensile strength of blended starch films (Zhou et al. 2009).

6.8.2 Nano Sized Materials

Packaging materials utilizing nanotechnology is also being explored and become one of the emerging areas today. Nanomaterials have high surface area and charge density. Clays, silica, nanocellulose, organic and inorganic fillers, etc. are some nanomaterials that are used to improve the mechanical and barrier properties of films. Nanofillers possess good interfacial interactions on polymer branches since they have large specific surface area and high surface energy (Nafchi et al. 2012; Kovačević et al. 2008). Chitosan nanoparticles have been added in tara gum edible films revealing that no significant difference of surface structure between films with bulk chitosan and chitosan nanoparticles (Antoniou et al. 2015). However, from cross-sectional profiles, the appearance of these two films were different whereby, nanoparticles caused the surface became rougher. Increasing the roughness had caused the contact angle of the surface also higher. In general, addition of nanoparticles had improved the hydrophobicity of film's surface by lowering the water vapour permeability and solubility. This result is in agreement with Abreu et al. (2015) whereby the incorporation of silver nanoparticles increased the contact angle significantly. Nanorod-rich zinc oxide also increased the contact angle, showing the tendency of films to absorb water decreased (Nafchi et al. 2012). The introduction of calcium montmorillonite into carboxymethyl starch films had increased the contact angle due to the clay platelets presented on the film surface (Wilpiszewska et al. 2015). In contrast, Shankar et al. (2016) claimed that adding silver nanoparticles in pectin films had caused a decrease in contact angle due to increase in roughness.

6.8.3 Plasma Surface Treatment

Plasma, the fourth state of matter is an ionised gaseous substance which if in contact with the material surface, will modify its surface properties due to the additional energy transferred from the plasma. This mechanism enables the surface to have a treatment or an alteration process to fulfil further applications such as printing, painting or laminating. It improves the adhesion properties, wettability and surface chemistry of polymers. During the treatment of polymers, energetic particles and photons generated in the plasma interact strongly with the polymer surface. Consequently, treated surface may have additional functional groups that increase the surface free energy of the polymer, enhance the printability and improve hydrophobicity through surface-chemical changes (Liston et al. 1994). This technique is an effective tool, which is convenient, quick technology, environmentally friendly and only requires low-cost processing devices (López-García et al. 2013). The polymer layers activated by cold plasma have controlled surface wetting properties, varying from superhydrophilic to superhydrophobic (Dowling and Stallard 2015). The air plasma applied on the whey protein gels had a greater effect on the surface wettability than roughness, due to polar groups deposited on the surface (Terpiłowski et al. 2017). However, a study on the effect of plasma treatment on chitosan films indicated that there were more water molecules surrounding the plasma treated sample (greater hydrophilic) and gave rougher surface compared to unmodified (Chang and Chain 2013).

6.8.4 Others

Other surface modification techniques such as silylation, acetylation, esterification and polymer grafting are also being used to improve nanocellulose composites (Zhang et al. 2018). However, most surface modification of nanocellulose use the hazardous solvents which are not preferable. Lamination of biodegradable polymers with lipids can cause to lower of WVP. Biodegradable materials can also be used in the surface treatment of cellulose-based substrates, either by coating or lamination (Andersson 2008).

6.9 Surface Properties in Active Packaging

Biodegradable polymers can act as good carriers for active compounds such as antimicrobial and antioxidant agents producing a system called active packaging. Active packaging helps to alter the package system in a positive approach as it contributes to extending the shelf life of food by inhibiting the microbial growth and delaying the lipid oxidation of some food products. Active packaging materials may have active compounds on the surface of the packaging inside the package, contributing to surface modifications of polymers.

In general, there are some research on active packaging indirectly investigated the surface properties of films. However, lack of emphasis occurred because most of the studies are keen to observe the efficiency of the active compounds in performing their role particularly for food systems. Caro et al. (2016) had developed active packaging films based on chitosan using thermal inkjet printing and found out that the efficiency of thymol as active agents is depending on few factors such as the number of oriented layers, the contact angle, the amount of glycerol used and the film type. The efficiency of thymol improved proportionally with the contact angles. However, increasing the concentration of glycerol had lowered the contact angle due to the increase in the surface tension. Meanwhile, adding active compounds in galactomannans films also improved the hydrophobicity of films despite no effect of the concentration used (Albuquerque et al. 2017).

6.10 Future Trends

Analyses on the surface properties of biodegradable polymers are essential due to their significant effects on the physical and mechanical properties. However, research on this aspect is still narrow. Some treatments to improve the surface properties can be discovered further to establish the area and can be a platform and database for future applications.

References

Abdelhedi O, Nasri R, Jridi M, Kchaou H, Nasreddine B, Karbowiak T, Debeaufort F, Nasri M (2018) Composite bioactive films based on smooth-hound viscera proteins and gelatin: physicochemical characterization and antioxidant properties. Food Hydrocoll 74:176–186

Abreu AS, Oliveira M, de Sá A, Rodrigues RM, Cerqueira MA, Vicente AA, Machado AV (2015) Antimicrobial nanostructured starch based films for packaging. Carbohydr Polym 129:127–134

Albuquerque PBS, Cerqueira MA, Vicente AA, Teixeira JA, da Cunha C (2017) Immobilization of bioactive compounds in Cassia grandis galactomannan-based films: influence on physicochemical properties. Int J Biol Macromol 96:727–735

Álvarez K, Famá L, Gutiérrez TJ (2017) Physicochemical, antimicrobial and mechanical properties of thermoplastic materials based on biopolymers with application in the food industry. In: Masuelli M, Renard D (eds) Advances in physicochemical properties of biopolymers: Part 1. Bentham Science, Sharjah, pp 358–400. EE.UU. ISBN: 978-1-68108-454-1. eISBN: 978-1-68108-453-4. https://doi.org/10.2174/9781681084534117010015

Andersson C (2008) New ways to enhance the functionality of paperboard b surface treatment- a review. Packag Technol Sci 21(6):339–373

Antoniou J, Liu F, Majeed H, Zhong F (2015) Characterization of tara gum edible films incorporated with bulk chitosan and chitosan nanoparticles: a comparative study. Food Hydrocoll 44:309–319

Bosquez-Molina E, Tomás SA, Rodríguez-Hueza ME (2010) Influence of CaCl2 on the water vapor permeability and the surface morphology of mesquite gum based edible films. LWT Food Sci Technol 43:1419–1425

Bracone M, Merino D, González J, Alvarez VA, Gutiérrez TJ (2016) Nanopackaging from natural fillers and biopolymers for the development of active and intelligent films. In: Ikram S, Ahmed S (eds) Natural polymers: derivatives, blends and composites. Nova Science, New York, pp 119–155 EE.UU. ISBN: 978-1-63485-831-1

Bravin B, Peressini D, Sensidoni A (2004) Influence of emulsifier type and content on functional properties of polysaccharide lipid-based edible films. J Agric Food Chem 52(21):6448–6455

Butler TI, Morris BA (2012) PE-based multilayer film structures in plastic films in food packaging- materials, technology, and applications. Ebnesajjad, S. Elsevier, Oxford, pp 21–52

Caro N, Medina E, Díaz-Dosque M, López L, Abugoch L, Tapia C (2016) Novel active packaging based on films of chitosan and chitosan/quinoa protein printed with chitosan-tripolyphosphate-thymol nanoparticles via thermal ink-jet printing. Food Hydrocoll 52:520–532

Cazón P, Velazquez G, Ramírez JA, Vázquez M (2017) Polysaccharide-based films and coatings for food packaging: a review. Food Hydrocoll 68:136–148

Cerqueira MA, Bourbon AI (2011) Galactomannans use in the development of edible films/ coatings for food applications. Trends Food Sci Technol 22:662–671

Chang SH, Chain C (2013) Plasma surface modification effects on biodegradability and protein adsorption properties of chitosan films. Appl Surf Sci 282:735–740

De Oliviera RRL, Albuquerque DAC, Cruz TGS, Yamaji FM, Leite FL (2012) In: Bellitto V (ed) Measurement of the nanoscale roughness by atomic force microscopy: basic principles and applications in atomic force microscopy—imaging, measuring and manipulating surfaces at the atomic scale. InTech, Rijeka, pp 147–175

de Paula Herrmann PS, Cristiana M, Pedroso Yoshida CM, Antunes AJ, Marcondes JA (2004) Surface evaluation of whey protein films by atomic force microscopy and water vapour permeability analysis. Packag Technol Sci 17:267–273

Deng L, Kang X, Liu Y, Feng F, Zhang H (2018) Characterization of gelatin/zein films fabricated by electrospinning vs solvent casting. Food Hydrocoll 74:324–332

Díaz O, Candia D, Cobos A (2016) Effects of ultraviolet radiation on properties of films from whey protein concentrate treated before or after film formation. Food Hydrocoll 55:189–199

Díaz O, Candia D, Cobos A (2017) Whey protein film properties as affected by ultraviolet treatment under alkaline conditions. Int Dairy J 73:84–91

Dowling DP, Stallard CP (2015) Achieving enhanced material finishing using cold plasma treatments. Trans Inst Met Finish 93(3):119–125

Etxabide A, Coma V, Guerrero P, Gardrat C, de la Caba K (2017) Effect of cross-linking in surface properties and antioxidant activity of gelatin films incorporated with a curcumin derivative. Food Hydroll 66:168–175

Garrido T, Leceta I, de la Caba K, Guerrero P (2018) Chicken feathers as a natural source of Sulphur to develop sustainable protein films with enhanced properties. Int J Biol Macromol 106:523–531

Geogiopoulos P, Christopoulos A, Koutsoumpis S, Kontou E (2016) The effect of surface treatment on the performance of flax/biodegradable composites. Compos Part B 106:88–98

Ghanbarzadeh B, Musavi M, Oromiehie AR, Rezayi K, Rad ER, Milani J (2007) Effect of plasticizing sugars on water vapor permeability, surface energy and microstructure properties of zein films. LWT 40:1191–1197

Giancone T, Torrieri E, Pierro PD, Cavella S, Giosafatto CVL, Masi P (2011) Effect of surface density on the engineering properties of high methoxyl pectin-based edible films. Food Bioprocess Technol 4:1228–1236

Giosafatto CVL, Pierro PD, Gunning P, Mackie A, Porta R, Mariniello L (2014) Characterization of *citrus* pectin edible films containing transglutaminase-modified phaseolin. Carbohydr Polym 106:200–208

Gutiérrez TJ (2017a) Surface and nutraceutical properties of edible films made from starchy sources with and without added blackberry pulp. Carbohydr Polym 165:169–179. https://doi.org/10.1016/j.carbpol.2017.02.016

Gutiérrez TJ (2017b) Chitosan applications for the food industry. In: Ahmed S, Ikram S (eds) Chitosan: derivatives, composites and applications. Wiley-Scrivener, Beverly, MA, pp 183–232. EE.UU. ISBN: 978-1-119-36350-7. https://doi.org/10.1002/9781119364849.ch8

Gutiérrez TJ, Alvarez VA (2017a) Films made by blending poly(ε-caprolactone) with starch and flour from Sagu rhizome grown at the Venezuelan amazons. J Polym Environ 25(3):701–716. https://doi.org/10.1007/s10924-016-0861-9

Gutiérrez TJ, Alvarez VA (2017b) Eco-friendly films prepared from plantain flour/PCL blends under reactive extrusion conditions using zirconium octanoate as a catalyst. Carbohydr Polym 178:260–269. https://doi.org/10.1016/j.carbpol.2017.09.026

Gutiérrez TJ, Alvarez VA (2017c) Cellulosic materials as natural fillers in starch-containing matrix-based films: a review. Polym Bull 74(6):2401–2430. https://doi.org/10.1007/s00289-016-1814-0

Gutiérrez TJ, González G (2016) Effects of exposure to pulsed light on surface and structural properties of edible films made from cassava and taro starch. Food Bioprocess Technol 9(11):1812–1824. https://doi.org/10.1007/s11947-016-1765-3

Gutiérrez TJ, Guzmán R, Medina Jaramillo C, Famá L (2016a) Effect of beet flour on films made from biological macromolecules: native and modified plantain flour. Int J Biol Macromol 82:395–403. https://doi.org/10.1016/j.ijbiomac.2015.10.020

Gutiérrez TJ, Suniaga J, Monsave A, García NL (2016b) Influence of beet flour on the relationship surface-properties of edible and intelligent films made from native and modified plantain flour. Food Hydrocoll 54:234–244. https://doi.org/10.1016/j.foodhyd.2015.10.012

Gutiérrez TJ, González Seligra P, Medina Jaramillo C, Famá L, Goyanes S (2017) Effect of filler properties on the antioxidant response of thermoplastic starch composites. In: Thakur VK, Thakur MK, Kessler MR (eds) Handbook of composites from renewable materials. Wiley-Scrivener, Beverly, MA, pp 337–370. EE.UU. ISBN: 978-1-119-22362-7. https://doi.org/10.1002/9781119441632.ch14

Gutiérrez TJ, Ollier R, Alvarez VA (2018) Surface properties of thermoplastic starch materials reinforced with natural fillers. In: Thakur VK, Thakur MK (eds) Functional biopolymers. Springer International, Basel, pp 131–158. EE.UU. ISBN: 978-3-319-66416-3. eISBN: 978-3-319-66417-0. https://doi.org/10.1007/978-3-319-66417-0_5

Hassan B, Chatha SAS, Hussain AI, Zia KM, Akhtar N (2017) Recent advances on polysaccharides, lipids and protein based edible films and coatings: a review. Int J Biol Macromol 109:1095–1107. https://doi.org/10.1016/j.ijbiomac.2017.11.097

Jiménez A, Fabra MJ, Talens P, Chiralt A (2012) Edible and biodegradable starch films: a review. Food Bioprocess Technol 5(6):2058–2076

John MJ, Thomas S (2008) Biofibres and biocomposites. Carbohydr Polym 71(3):343–364

Jonhed A, Andersson C, Järnström L (2008) Effects of film forming and hy- drophobic properties of starches on surface sized packaging paper. Packag Technol Sci 21(3):123–135

Kampeerapappun P, Aht-ong D, Pentrakoon D, Srikulkit K (2007) Preparation of cassava starch/montmorillonite composite film. Carbohydr Polym 67:155–163

Karbowiak T, Debeaufort F, Champion D, Voilley A (2006) Wetting properties at the surface of iota-carrageenan-based edible films. J Colloid Interface Sci 294:400–410

Kovačević V, Vrsaljko D, Lučić Blagojević S, Leskovac M (2008) Adhesion parameters at the interface in nanoparticulate filled polymer systems. Polym Eng Sci 48(10):1994–2002

Kurek M, Galus S, Debeaufort F (2014) Surface, mechanical and barrier properties of bio-based composite films based on chitosan and whey protein. Food Packaging Shelf Life 1:56–67

Liston EM, Mrtinu L, Wertheimer MR (1994) In: Strobel M, Lyons CS, Mittal KL (eds) Plasma surface modification of polymers for improved adhesion: a critical review in plasma surface modification of polymers: relevance to adhesion. VSP BV, Zeist, pp 3–42

López-García J, Bílek F, Lehocký M, Junkar I, Mozetic M, Sowe M (2013) Enhanced printability of polyethylene through air plasma treatment. Vacuum 95:43–49

Mali S, Grossmann MVE, Garcıa MA, Martino MN, Zaritzky NE (2004) Barrier, mechanical and optical properties of plasticized yam starch films. Carbohydr Polym 56(2):129–135

Mathew S, Abraham TE (2008) Characterisation of ferulic acid incorporated starch–chitosan blend films. Food Hydrocoll 22:826–835

Molavi H, Behfar S, Shariati MA, Kaviani M, Atarod S (2015) A review on biodegradable starch based film. J Microbiol Biotechnol Food Sci 4(5):456–461

Muxika A, Extabide A, Uranga J, Guerrero P, de la Caba K (2017) Chitosan as a bioactive polymer: processing, properties and applications. Int J Biol Macromol 105:1358–1368

Nafchi AM, Alias AK, Mahmud S, Robal M (2012) Antimicrobial, rheological, and physicochemical properties of sago starch films filled with nanorod-rich zinc oxide. J Food Eng 113:511–519

Nisar T, Wang Z, Yang X, Tian Y, Iqbal M, Guo Y (2018) Characterization of citrus pectin films integrated with clove bud essential oil: physical, thermal, barrier, antioxidant and antibacterial properties. Int J Biol Macromol 106:670–680

Nur Hanani ZA (2016) In: Caballero B, Finglas PM, Toldrá F (eds) Gelatin in encyclopedia of food and health. Academic Press, Oxford, pp 191–195

Phan TD, Debeaufort F, Luu D, Voilley A (2005) Functional properties of edible agar-based and starch-based films for food quality preservation. J Agric Food Chem 53:973–981

Razavi SMA, Amini AM, Zahedi Y (2015) Characterisation of a new biodegradable edible film based on sage seed gum: influence of plasticiser type and concentration. Food Hydrocoll 43:290–298

Rhim JW, Gennadios A, Fu D, Weller CL, Hanna MA (1999) Properties of ultraviolet irradiated protein films. Lebensm Wiss Technol 32:129–133

Sahraee S, Milani JM, Ghanbarzadeh B, Hamishehkar H (2017) Effect of corn oil on physical, thermal, and antifungal properties of gelatin-based nanocomposite films containing nano chitin. LWT Food Sci Technol 76:33–39

Sengupta T, Han JH (2014) In: Han JH (ed) Surface chemistry of food, packaging, and biopolymer materials in innovations in food packaging, 2nd edn. Academic Press, Tokyo, pp 52–86

Shankar S, Tanomrod N, Rawdkuen S, Rhim JW (2016) Preparation of pectin/silver nanoparticles composite films with UV-light barrier and properties. Int J Biol Macromol 92:842–849

Sionkowska A, Płanecka A (2013) Surface properties of thin films based on the mixtures of chitosan and silk fibroin. J Mol Liq 186:157–162

Suyatma NE, Tighzert L, Copinet A (2005) Effects of hydrophilic plasticizers on mechanical, thermal, and surface properties of chitosan films. J Agric Food Chem 53:3950–3957
Tang C, Jiang Y (2007) Modulation of mechanical and surface hydrophobic properties of food protein films by transglutaminase treatment. Food Res Int 40(4):504–509
Tarek AR, Rasco BA, Sablani SS (2015) Ultraviolet-C light inactivation kinetics of *E. coli* on bologna beef packaged in plastic films. Food Bioprocess Technol 8(6):1267–1280
Terpiłowski K, Tomczyńska-Mleko M, Nishinari K, Mleko S (2017) Surface properties of ion-inducted whey protein gels deposited on cold plasma treated support. Food Hydrocoll 71:17–25
Ward G, Nussinovitch A (2017) Characterizing the gloss properties of hydrocolloid films. Food Hydrocoll 11:357–365
Wilpiszewska K, Antosik AK, Spychaj T (2015) Novel hydrophilic carboxymethyl starch/montmorillonite nano composite films. Carbohydr Polym 128:82–89
Zhang X, Zhang Y, Ma QY, Dai Y, Hu FP, Wei GB, Xu TC, Zeng QW, Wang SZ, Xie WD (2017) Effect of surface treatment on the corrosion properties of magnesium-based fibre metal laminate. Appl Surf Sci 396:1264–1272
Zhang Y, Jiang Y, Han L, Wang B, Xu H, Zhong Y, Zhang L, Mao Z, Sui X (2018) Biodegradable regenerated cellulose-dispersed composites with improved properties *via* a Pickering emulsion process. Carbohydr Polym 179:86–92
Zhou J, Ma Y, Ren L, Tong J, Liu Z, Xie L (2009) Preparation and characterization of surface crosslinked TPS/PVA blend films. Carbohydr Polym 76:632–638

Chapter 7
Transport Phenomena in Edible Films

Delia Rita Tapia-Blácido, Bianca Chieregato Maniglia, and Milena Martelli Tosi

Abstract Edible films and coatings help to control transfer of water vapor, oxygen, CO_2, and active compounds between the food product and the environment providing additional protection during storage of fresh and processed food. Mass transfer phenomena are involved in these processes because edible films can act as functional interfaces between the food product and the environment. Edible films and coatings can also modify the heat transfer mechanism that takes place during food drying and frying, as well. In addition, they can function as controlled release packaging or active packaging—such packaging can be effectively impregnated with antimicrobial or antioxidant compounds, to deliver them over a stipulated period. Release and delivery of active compounds by these materials depend on the type of biopolymer that composes the film matrix and on the environmental conditions during storage. In a particular study, a turmeric dye extraction residue previously submitted to mechanical and chemical treatments was employed as coating in bananas. The treated turmeric residue coating effectively extended the coated banana shelf life by 4 days as compared to uncoated bananas.

Keywords Coating · Diffusion · Turmeric

7.1 Introduction

Edible protective films or coatings have been generally defined as thin layers of materials that provide a barrier to mass transfer (moisture, oxygen, and solute movement) in the food itself or between the food and its environment (Gutiérrez and Álvarez 2017). In this sense, applying edible films and coatings to food products can prevent moisture loss, aroma loss, solute transport, water absorption, and oxygen penetration (Cazón et al. 2017). Whereas films consist of stand-alone sheets of material, coatings form directly on the product (Guilbert 1986; Shellhammer and

D. R. Tapia-Blácido (✉) · B. C. Maniglia · M. M. Tosi
Departamento de Química, Faculdade de Filosofia, Ciências e Letras de Ribeirão Preto, Universidade de São Paulo, Ribeirão Preto, SP, Brazil
e-mail: delia@ffclrp.usp.br

T. J. Gutiérrez (ed.), *Polymers for Food Applications*,
https://doi.org/10.1007/978-3-319-94625-2_7

Krochta 1997). Edible coatings also help to reduce excessive oil uptake by fried foodstuff and water migration to and from such foodstuff. In this case, coatings operate by forming thin films that require drying before frying; these coatings can also operate through their gelation properties during heating (Kulp 2011). Two transport mechanisms underlie the frying process: mass and heat (Garcia and Zaritzky 2017).

Nowadays, using edible films as "active packaging" is one of the most challenging research areas. Active packaging comprises a package system that deliberately incorporates components and releases substances into or absorbs substances from the packaged food or the environment surrounding the food. Active packaging aims to extend the food product shelf life and to keep or to improve the packaged food conditions while maintaining their mechanical integrity and handling characteristics (Bracone et al. 2016; Ganiari et al. 2017). Antioxidant/antimicrobial agent incorporation into edible films or coatings reduces the need to introduce larger quantities of such additives into the food bulk (Van Long et al. 2016). Thus, active materials can act as a source of antioxidants/antimicrobials that are released into the food at controlled rates (Gutiérrez 2018). Consequently, a predetermined active compound concentration is maintained in the food, which compensates for continuous additive consumption during food storage (Ganiari et al. 2017; LaCoste et al. 2005; Mastromatteo et al. 2010).

Although research into active edible films has been extensive release mechanisms are still poorly understood. Mass transfer mathematical modeling is necessary to achieve deeper understanding of this process and to optimize active systems by considering not only the polymer matrix kinetic and thermodynamic parameters but also its structural and interfacial characteristics (Mastromatteo et al. 2010; Voilley et al. 2011). Pure diffusion, polymer matrix swelling, and polymer erosion and degradation are mechanisms that lead polymeric devices to release compounds (Faisant et al. 2002; Jain 2000; Pinheiro et al. 2013; Polakovic et al. 1999).

In this chapter, we depict the fundamental mechanisms underlying the transfer of mass, oxygen, and active compounds in edible films. We also describe recent advances in the development of delivery devices. Finally, we present some results regarding active films produced from turmeric dye residue and their use in post-harvest banana preservation.

7.2 Fundamentals of Transport Phenomena in Polymeric Films: Heat and Mass

If we consider a biodegradable film, which represents a simple geometry for a thin layer, the main transport phenomena concern mass exchange by diffusion and energy exchange by conduction (Fig. 7.1), with the concentration of a certain component (C) on sides 1 and 2 of the layer being $C_1 > C_2$ or the temperature (T) on sides 1 and 2 of the layer being $T_1 > T_2$. Equation (7.1) shows that the magnitude of

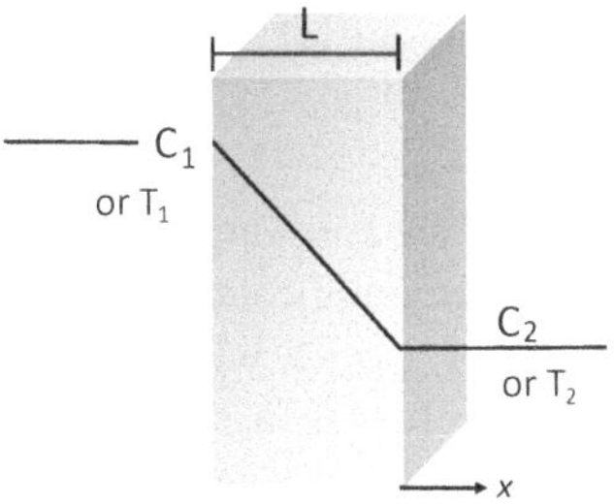

Fig. 7.1 Schematic representation of mass transfer at the interface (boundary layers). Permeation of a component through thin layers (interfaces) or difference between temperatures on the two sides of the layer. C1 and C2 are the concentration of a component in compartments 1 and 2, separated by the film, and T1 and T2 are the respective temperatures (adapted from Voilley et al. 2011)

transport phenomenon rates (χ) is directly proportional (constant K) to a driving force and to a concentration or temperature difference ($\Delta\omega$).

$$\chi = \mathrm{K} \cdot \Delta\omega \tag{7.1}$$

Heat flow density or heat flux (q) is proportional to the negative temperature gradient and is the unidimensional form of Fourier's law of heat conduction (Eq. 7.2) (Ibarz and Barbosa-Cánovas 2002):

$$q = -k\frac{dT}{dx} \tag{7.2}$$

In this case, only the increase in temperature as a function of the distance **x** is taken into account because the temperature differences in the other axes (dT/dy), (dT/dz) are not as significant as compared to the temperature difference in the x axis. This proportionality constant **k** is called thermal conductivity of the solid and is normally expressed in W/(m K) or kcal/(h m °C). The medium is assumed to be isotropic, i.e. **k** has the same value in all the directions of the material, which is only true for lower ΔT.

Thermal diffusivity, defined according to Eq. (7.3), can also be used:

$$\alpha = \frac{k}{\rho \cdot \hat{C}_{\mathrm{P}}} \tag{7.3}$$

where

ρ = density of the material;

$\hat{C}_p$ =specific heat of the material.

Therefore, Fourier's law expression for one direction is (Eq. 7.4):

$$q = -\alpha \frac{d\left(\rho \cdot \hat{C}_P \cdot T\right)}{dx} \tag{7.4}$$

for an isotropic material; ρ and $\hat{C}_P$ are constants (Ibarz and Barbosa-Cánovas 2002).

Studying mass diffusion is more complicated than studying heat transfer because diffusion involves movement of a species within a mixture by natural or forced means. Fick's law of diffusion refers to a substance moving through a binary mixture of components A and B due to a concentration gradient (Eq. 7.5) (Ibarz and Barbosa-Cánovas 2002). Fickian diffusion involves a substantially stochastic phenomenon (related to Brownian motion) that is analogous to heat transfer, where thermal conductivities are defined as proportionality factors between heat flux and temperature gradient (Fourier's law of heat conduction).

$$\bar{J}_A^* = -C \cdot D_{AB} \cdot \bar{\nabla} X_A \tag{7.5}$$

where

$\bar{J}_A^*$ = diffusion molar flux (mol/m s);

C = molar concentration (mol/m^3);

D_{AB} = effective diffusion coefficient (m^2/s);

$\bar{\nabla} X_A$ = concentration gradient;

X = species i molar fraction divided by the mixture total molar density: $X_i = C_i/C$.

The negative sign expresses that diffusion takes place from higher to lower concentration zones.

If total molar density is constant, and if there is no chemical reaction or if the number of moles does not change, Eq. (7.5) can be transformed into Eq. (7.6):

$$\bar{J}_A^* = -D_{AB} \cdot \bar{\nabla} C_A \tag{7.6}$$

In a binary mixture, diffusivity of A to B,D_{AB}, is equal to diffusivity of B to A, D_{BA}, and ordinary diffusion is defined as the proportionality factor between mass flux and concentration gradient. The sum of diffusions in a binary mixture is zero $\left(\bar{J}_A^* + \bar{J}_B^* = 0\right)$ because one component diffuses in one direction, while the other component diffuses in the opposite direction.

When the solid has some porosity, D_{AB} corresponds to an effective diffusion coefficient (D_{ef}) (Cremasco 1998). This coefficient depends on variables that influence diffusion, such as temperature and pressure, and on the porous matrix properties (porosity—ε, sphericity—ϕ, and tortuosity—τ), according to Eq. 7.7:

$$D_{ef} = D_{AB} \cdot \frac{\varepsilon}{\tau} \tag{7.7}$$

Other types of diffusion exist depending on the property that confers movement to the mixture component. If movement is due to a pressure gradient, it is called pressure diffusion; if movement is due to a thermal gradient, it is called thermal diffusion (Ibarz and Barbosa-Cánovas 2002).

Hydrophilic polymer matrixes can undergo gradual hydration when they are immersed in liquid media. Polymer chain relaxation causes volume expansion or swelling and affects the mechanism through which active agents are transported within the polymer matrix (Pinheiro et al. 2013). The active agent diffusion coefficient in the swollen part of the matrix increases, so the active agent diffuses out. Mastromatteo et al. (2010) mentioned that mass transport mechanism in these systems is generally classified into three different types: ideal Fickian diffusion, anomalous behavior, and Case II transport. If the geometry is planar, the quantity of active agent, Mt., released at time **t** is given by (Crank 1955; Peppas 1984):

$$\frac{M_t}{M_\infty} = k_a t^n \tag{7.8}$$

where

M_∞ = initial polymer loading with the active agent;

k_a and n = system parameters that depend on both the nature of the polymer-penetrant-active agent interactions and on the release device geometry.

The parameter **n** can take a range of values that indicate the type of transport: n = 0.5 means that the active agent is released by simple Fickian diffusion; 0.5 < n < 1.0 means that the diffusion process is a combination of Fickian and non-Fickian diffusion and is known as "anomalous diffusion"; n = 1.0 corresponds to diffusion described as "Case II diffusion". In this case, the rate of solvent uptake by the polymer is largely determined by the polymer chain swelling and relaxation rates; and n > 1.0 corresponds to the region known as "Super Case II transport".

Several authors consider that diffusion of active compounds released from a hydrophilic polymeric matrix occurs by Fickian diffusion and the relaxation phenomenon, so they have used a linear superimposition of both mechanisms to obtain a mathematical model (Flores et al. 2007; Mastromatteo et al. 2010; Pinheiro et al. 2013). The linear superimposition approach assumes that molecule transport observed within the polymer can be described as the sum of molecules transported due to Brownian motion and of molecules transported due to polymer relaxation (Berens and Hopfenberg 1978; Pinheiro et al. 2013):

$$M_t = M_{t,F} + M_{t,R} \tag{7.9}$$

where

$M_{t,F}$ and $M_{t,R}$ = Fickian and relaxation process contributions at time **t**, respectively.

In the specific case of diffusion through a planar sheet with constant boundary conditions, the solution of Fick's second law is (Langer and Peppas 1983; Mastromatteo et al. 2010):

$$M_{F,t} = M_F^{eq.} \left\{ 1 - \frac{8}{\pi^2} \sum_{n=0}^{n=\infty} \frac{1}{(2.n+1)^2} \exp\left[-D.(2.n+1)^2 .\pi^2 . \frac{t}{l^2} \right] \right\} \quad (7.10)$$

where

$\mathbf{M_F^{eq}}$ = amount of compound with low molecular mass released at equilibrium as a consequence of stochastic phenomena;

D = diffusion coefficient through the swollen polymer matrix;

l = film thickness.

Relaxation is related to stress dissipation induced by permeant entry. The relaxation process can be seen as a distribution of relaxation times and can be quantitatively described through a first-order kinetics equation:

$$M_{R,t} = M_R^{eq.} \left[1 - \exp\left(-\frac{t}{\tau} \right) \right] \quad (7.11)$$

where

$M_R^{eq.}$ = amount of compound with low molecular weight released at equilibrium as a consequence of polymer relaxation;

τ = relaxation time associated with polymer relaxation.

Buonocore et al. (2003) and (2003b) developed a mathematical model that can predict the kinetics of compound release from crosslinked polyvinylalcohol (PVOH) into aqueous solution. Their model considers water molecule diffusion into the polymeric film and incorporated antimicrobial agent counter diffusion from the film to the aqueous solution. In particular, active agent release from a swelling polymer can be regarded as "anomalous diffusion" with moving boundary conditions. A study has been conducted by developing two models: one that describes water uptake kinetics and another that describes enzyme release, as shown in Eqs. (7.12) and (7.13), respectively (Crank 1955):

$$\frac{d}{dt} \int_{V(t)} (\rho \cdot Cw) \cdot dV = \int_{S(t)} \left[D_F^W . \frac{\partial}{\partial x} (\rho.Cw) \cdot \vec{n} \right] \cdot dS \quad (7.12)$$

$$\frac{d}{dt} \int_{V(t)} (\rho \cdot C_L) \cdot \mathrm{d}V = \int_{S(t)} \left[D_F^L \cdot \frac{\partial}{\partial x} (\rho \cdot C_L) \cdot \vec{n} \right] \cdot dS \quad (7.13)$$

where

D_F^W = water diffusion coefficient;

D_F^L = active agent diffusion coefficient;

C_W = local water concentration;

C_L = active agent concentration;

ρ = polymer matrix density;

S(t) = volume V(t) surface;

$\vec{n}$ = vector normal to the surface;

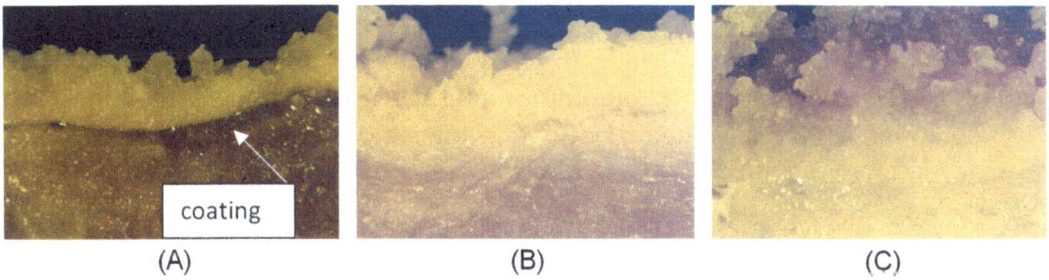

Fig. 7.2 Micrographs of the transversal section of chicken nuggets with gelatin coating (10% of gelatin and 25% of glycerol) between the meat and the crust in the following conditions: (**a**) before frying (in natura), (**b**) after pre-frying (30s at 180 °C), and (**c**) after frying (3 min at 180 °C) in hydrogenated vegetable oil

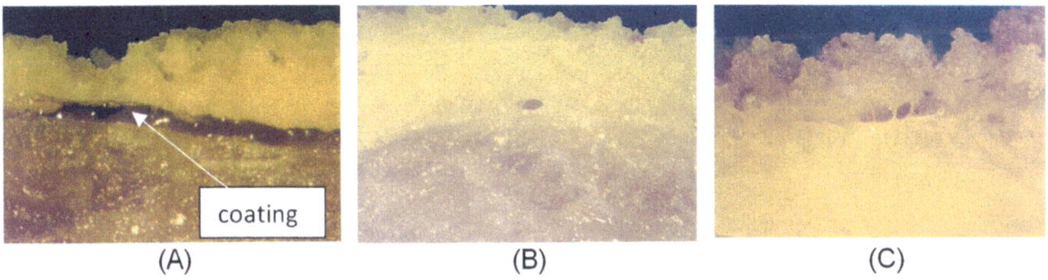

Fig. 7.3 Micrographs of the transversal section of chicken nuggets with gelatin coating (10%of gelatin and 55%of glycerol) between the meat and the crust in the following conditions: (**a**) before frying (in natura), (**b**) after pre-frying (30s at 200 °C), and (**c**) after frying (3 min at 200 °C) in hydrogenated vegetable oil

x = axial coordinate.

Instead of considering elements with a fixed volume (Eulerian approach), the authors assumed that the volume of the elements changed during hydration (Lagrangian approach). They fitted the first model to the water sorption data and used the obtained parameters to fit the release data in the second model. These models consider that the water diffusion coefficient depends on the degree of cross-linking. Mass balance for the permeant and the active substance are numerically solved by means of the Finite Elements Method (Mastromatteo et al. 2010).

Frying is another important process that involves two simultaneous transport mechanisms—heat and mass transport. The product must be heated for proper cooking to be achieved; meanwhile, oil is transported into the food. High temperatures dehydrate the product by evaporating water, which is replaced with oil (Pinthus et al. 1993; Saguy et al. 1998). Because foodstuffs absorb part of the oil they are fried in, edible coatings can be used to decrease oil absorption.

Methylcellulose, hydroxypropyl cellulose, hydroxypropyl methyl cellulose, and starch are the most studied biopolymers (Mallikarjunan et al. 1997; Albert and Mittal 2002; Garcia et al. 2002; Martelli et al. 2008; Gutiérrez and Alvarez 2017a; Suárez and Gutiérrez 2017). Some modification of the surface structure formed during frying reduces moisture loss and oil uptake (Gutiérrez 2017a). Martelli et al.

(2005) studied oil absorption during chicken nugget deep-frying. Figures 7.2 and 7.3 illustrate the optical micrographs of the transversal section of the samples. The nuggets had been coated with gelatin film solution (10% of gelatin) containing 25% (Fig. 7.2) or 55% glycerol (Fig. 7.3) in the following conditions: (A) before frying (in natura), (B) after pre-frying (30s at 200 °C), and (C) after frying (3 min at 200 °C) in fat.

In condition A, the arrow in Figs. 7.2a and 7.3a indicates the gelatin coating. After pre-frying (Figs. 7.2b and 7.3b) and frying (Figs. 7.2c and 7.3c), the coating is no longer visible in the micrographs. Although the coating may have lost its integrity during frying, it helps to preserve the samples against dehydration and to decrease oil uptake after pre-frying. It is worth noting that oil uptake is higher after pre-frying. Besides that, nuggets containing the coating are the softest, which is desirable from a technological viewpoint.

7.3 Water Vapor and Gas Transport Mechanism in Edible Films

Knowledge of barrier properties in edible films has become increasingly important in recent years, especially because this property is related to food quality (Gutiérrez et al. 2015a; Álvarez et al. 2018). Studying these properties is a large and growing segment of commercial manufacturing: this sector seeks extended food shelf life through moisture, oxygen, carbon dioxide, lipid, flavor, and aroma control, which depends on interaction between food components and the surrounding atmosphere (Jongjareonrak et al. 2006). Food quality may decrease due to (i) oxidation of aroma components by interaction with oxygen or (ii) loss of specific aroma compounds to the packaging material or to the environment (Miller and Krochta 1997).

Selecting the best packaging material is crucial for the food market. Food packaging materials need to be versatile enough to with stand handling process forces while maintaining their physical and chemical integrity, and they must display suitable barrier properties to several gases (e.g., O_2, N_2, and CO_2) (Gutiérrez et al. 2017). Knowledge of the solubility/diffusion/permeation of these molecules through the polymeric film gives insight into the barrier properties of a given polymeric material. Furthermore, the packed food intrinsic composition (e.g. pH, fat content, and aroma compound) may influence the packaging material sorption characteristics, whereas environmental factors like temperature and, for some polymers, relative humidity may affect b 'from one side to the other side of the edible film, and permeant desorption from arrier characteristics (Johansson and Leufven 1995).

In general, gas molecule transport through an edible film comprises three main steps (Fig. 7.4): permeant **adsorption** onto the edible film surface, permeant **diffusion** from one side to the other side of the edible film, and permeant **desorption** from the edible film (Skurtys et al. 2011).

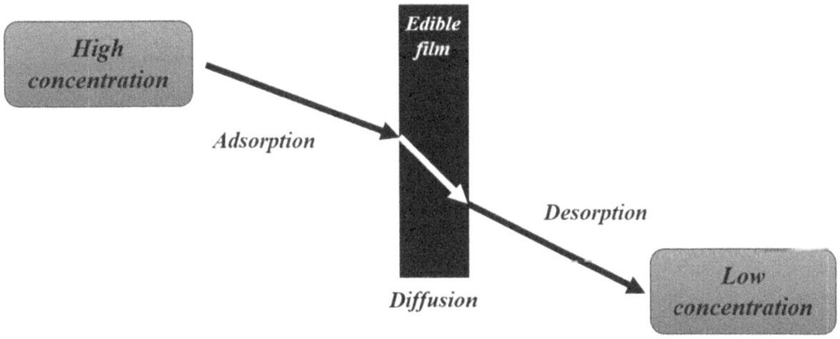

Fig. 7.4 Gas molecule transport in edible films

The diffusion mechanism of simple gases follows Fick's First law:

$$J = -D\frac{dC}{dX} = D\frac{(C_1 - C_2)}{l} \tag{7.14}$$

where
l = edible film thickness;
D = diffusion coefficient;
C = concentration.

D reflects the rate at which the permeant diffuses through the polymeric film, and the negative signal indicates that migration occurs toward lower concentration.

Moreover, according to Henry's law of solubility, gas concentration can be expressed as:

$$C = SP \tag{7.15}$$

where
S = gas solubility coefficient in the membrane;
P = gas pressure.

The combination of Fick's First law of diffusion with Henry's law of solubility is used to express the permeant steady-state permeability through a nonporous barrier without significant imperfections.

$$J = DS\frac{(p_1 - p_2)}{l} \tag{7.16}$$

The product of the diffusion coefficient and the solubility coefficient is equal to the permeability coefficient, Π = DS, which characterizes the edible film intrinsic permeability. If we rearrange Eq. (7.16), the permeability coefficient can be expressed as:

$$\Pi = DS = \frac{J\,l}{(p_1 - p_2)} \tag{7.17}$$

Diffusion takes place in one direction only, through the film and not along or across it; coefficients D and S do not depend on permeant concentration. Molecular diffusion in polymers usually follows a Fickian behavior. When the system takes long to reach the steady state (like glassy polymer) or when coefficients D and S correlate with interaction between the permeant and the polymer, such as interaction between water or solvent vapor and a hydrophilic film, and their diffusion therein, the system follows a non-Fickian behavior. Because film and food are in contact, studying the film barrier characteristics under realistic conditions is essential as these characteristics may alter the polymeric film performance.

Polymer type also impacts film barrier properties. For example, gas molecules cannot permeate through polymer crystallites because they are insoluble in this material. Therefore, gas permeation into semi-crystalline polymers is restricted to amorphous regions (Kofinas et al. 1994) because crystalline zones have small volume available for gas penetration and the path between the crystallites is tortuous. Reduction in permeability is proportional to the crystalline phase volume fraction (Gutiérrez and Álvarez 2017).

7.3.1 *Water Vapor Permeability (WVP)*

Water vapor permeability (WVP) in thin, nonporous, non-swelling, highly hydrophobic films can be determined by combining Fick's First Law of diffusion (Eq. 7.14) with Henry's Law of solubility (Eq. 7.15) as shown in Eq. (7.18).

$$\mathrm{WVP} = \mathrm{D.S} = \frac{\mathrm{J.x}}{\mathrm{A}.\Delta p.\mathrm{M}} \tag{7.18}$$

where

WVP = water vapor permeability [mol/(m.s.Pa)];

D = permeant diffusivity corresponding to the rate at which the concentration gradient is dissipated (m^2/s);

S = solubility coefficient defined as the maximum migrating molecule mass that dissolves in a unit volume of the material at equilibrium [(mol/m^3.Pa)];

p = permeant partial pressure in adjacent air [Pa];

A = barrier surface [m^2];

M = water molar weight [g/mol].

On the other hand, in the case of moisture permeation through hydrophilic, composite, and moderately hydrophobic films, coefficients D and S vary due to plasticization/swelling after moisture absorption. These films present nonlinear water sorption isotherms and water content-dependent diffusivities, and RH conditions during testing greatly influence WVP (Mchugh et al. 1993). Hydrophilic materials such as proteins and polysaccharides present high a_w range and much higher moisture sorption isotherms than hydrophobic materials (Gutiérrez et al. 2015, 2015b).

Water sorption in these materials is usually non-ideal and leads to plasticization and/or clustering phenomena, to result in complex type II or sigmoidal isotherms (Despond et al. 2005; Bourlieu et al. 2009). According to Bourlieu et al. (2009), hydrophilic films cannot be classified on the basis of moisture sorption isotherms due to (i) influence of formulation (macromolecule polymerization degree, presence of lateral groups, addition of plasticizer, and addition of non-lipid components), (ii) impact of film-forming conditions (pH, thermal treatment, or any cross-linking treatment), and (iii) shift in the relative positions of moisture sorption isotherms over the full a_w range.

On the basis of the Flory equation, Buonocore et al. (2005) proposed a model that describes WVP dependence on water activity with good accuracy for a_w ranging from 0.3 to 0.8 at 20 °C (Eq. 7.19). This model integrates nonlinear water sorption in hydrophilic polymer-based films (alginate, casein, chitosan, and zein) and moisture concentration dependence on effective water diffusivity. Variations in moisture effective diffusivity are an exponential function of the polymer moisture content:

$$WVP_{(aw_1\_aw_2)} = \frac{1}{po.\left(aw_1\_aw_2\right)} \cdot \int_{C_w(aw_2)}^{C_w(aw_1)} D.dC_w$$

$$= \frac{1}{po.\left(aw_1\_aw_2\right)} \cdot \int_{C_w(aw_2)}^{C_w(aw_1)} (D_0.\exp\left(\propto .C_w\right).dC_w \tag{7.19}$$

where

$WVP_{(aw1,aw2)}$ = water vapor permeability coefficient for a water activity difference between the upstream and downstream side of the film equal to aw_1 and aw_2respectively;

p_0 = water vapor pressure;

D = water diffusion at zero moisture concentration;

α = constant that accounts for the water ability to plasticize the polymeric matrix;

C_w = polymer moisture content.

The positive slope relationship between a hydrophilic film WVP and thickness has been reported. Some materials present an exponential relationship (McHugh et al. 1993). The way thickness affects WVP has been explained on the basis of structural aspects: (i) film thickness influences film structure and its homogeneity; (ii) air gap between the solution and the film leads to equilibrium moisture relationships at the film/air interface, but these relationships differ from the test cup solution equilibrium conditions. Increased thickness lowers the effect of such limit layers. Generally accepted explanations are related to nonlinear moisture sorption in the film; (iii) increased thickness results in higher quantity of hydrophilic component, which may interact with water molecules and cause swelling and apparent thickness effect (McHugh et al. 1993; Bourlieu et al. 2009).

Hydrophilic film swelling is an especially important characteristic regarding film thickness during moisture transport. Roca et al. (2008) used a Fickian model to analyze moisture sorption in hydrophobic (acetylated monoglyceride), hydrophilic (wheat gluten), and solid dispersion (dark chocolate) barrier matrixes. The authors aimed to compare moisture D_{eff} values by using either a numerical solution (taking material deformation after sorption into account) or an analytical solution; verifying that the deformation hypothesis did not affect the D_{eff} value in the case of the lipid matrix, but it strongly impacted D_{eff} for the hydrophilic and solid dispersion matrixes. Assuming that the solid matrix does not swell, the analytical solution to Fick's second law underestimates product thickness when water activity increases. Magnetic Resonance Imaging (MRI) aids diffusion coefficient calculation by means of a unidirectional Fickian model applied to a rectangular polymer sample. This model considers an exponential D_{eff} dependence on moisture concentration in the film (Eq. 7.20):

$$D_{eff} = D_0.e^{A(C_w/C_{w0})} \tag{7.20}$$

where

C_w = moisture concentration at a point;

C_{w0} = moisture concentration at the surface;

D_0 and A = constants.

Polymer swelling, defined as the increase in sample thickness relative to the sample initial thickness, can be accurately fitted by using the following equation:

$$S = S_{max}\left(1 - e^{-kt}\right) \tag{7.21}$$

where

S = swelling at time t;

S_{max} = maximum swelling as time approaches ∞;

k = swelling rate.

Temperature also impacts the film moisture barrier property—higher temperatures increase molecular mobility and diffusivity, inducing accelerated water movement through barrier matrixes. Provided the coating structure remains unaltered, the dependence of the diffusion, sorption, and permeability coefficients on temperature can be expressed by the Arrhenius Law (Rogers 1985):

$$P = P_o \exp\left(-\frac{E_{a.p}}{RT}\right) \tag{7.22}$$

$$D = D^* \exp\left(-\frac{E_{a.D}}{RT}\right) \tag{7.23}$$

$$S = S_o \exp\left(-\Delta H_s / RT\right) \tag{7.24}$$

where

$E_{a,p}$ and $E_{a,D}$ = permeation and diffusion process activation energy(kJ/mol);

ΔH_s = heat of sorption (kJ/mol);
R = perfect gas constant (8.314 J/mol/K);
T = temperature (K);
D^*, P_0, S_0 = pre-exponential factors for the three processes.

If the moisture diffusion coefficient and the solubility are constant over the investigated water activity range, the permeability energy of activation can be obtained from the relation:

$$E_{a,p} = E_{a,D} + \Delta H_s \tag{7.25}$$

Water vapor diffusion is always a thermally activated process with positive $E_{a,D}$. However, $E_{a,P}$ can be negative or positive for both hydrophilic and hydrophobic films. $E_{a,P}$ depends on the mechanism that predominates in moisture transport through films and on the result of the sum of ΔH_s and $E_{a,D}$.

7.3.2 Oxygen Permeability

Oxygen permeability is the second most commonly studied transport property of edible polymeric films. Edible film oxygen permeability depends on many factors, such as temperature and relative humidity. High temperature promotes exponential gas transfer across the film (Mate and Krochta 1996). Higher relative humidity increases interaction between water molecules and the polymer, making the film more plasticized (Hong and Krochta 2006). These conditions favor mobility and extensive mass transfer across the film. For this reason, the edible film antioxidant ability should always be tested under controlled relative humidity conditions.

The apparatus that is generally used to determine oxygen permeability in films is based on a colorimetric sensor, like Oxtran (Mocon, Modern Controls Inc., Minneapolis, MN, U.S.A), or on manometric methods, including the Lyssy L100 series (Lyssy, Zurich, Switzerland) for dry and noncorrosive permanent gases, both certified by the ASTM F1927 standard (ASTM 2014). However, these apparatus and techniques do not allow gas permeability measurements for different relative humidity values, which is necessary during food product simulation. Nevertheless, Oxtran is the most suitable apparatus available to determine gas permeability through edible films.

Analysis with the OXTRAN equipment is used to measure the amount of oxygen that passes through a unit of area parallel to the packaging material surface during a given period. This equipment consists of two test cells, which hold the samples (Fig. 7.5). During the process, oxygen gas is released between the two cells, while the entrainment gas, which is composed of a mixture of nitrogen (major component) and hydrogen, passes outside the cells. As oxygen permeates the film, it mixes with the drag and passes through a colorimetric sensor that measures the electric current resulting from the cathodic and anodic reactions. This current is proportional to the amount of oxygen (reactions represented in Fig. 7.5).

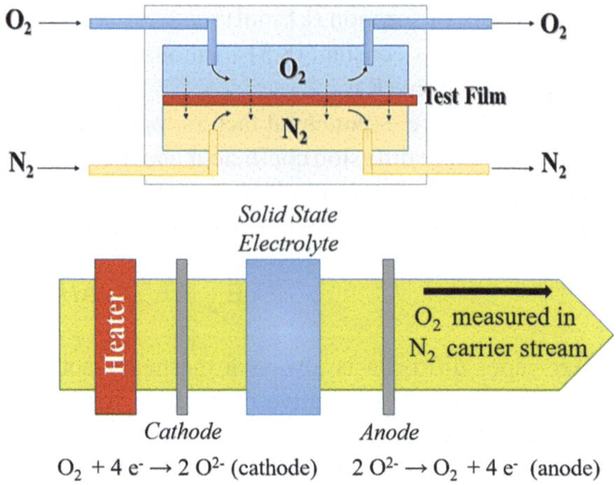

Fig. 7.5 System used to measure oxygen permeability in films

The equipment is connected to a microcomputer operating with software that provides all the results. Each reading yields a permeability oxygen rate (TPO_2). The experiment ends when the TPO_2 values stabilize; that is, when the $TPO_2 \times$ time curve is constant. Oxygen permeability (PO_2) is the highest value obtained throughout the analysis and can be calculated by Eq. (7.26):

$$PO_2 = \frac{TPO_2}{\Delta P} \tag{7.26}$$

where

TPO_2 = oxygen permeability rate;

ΔP = difference between oxygen partial pressure between the two sides of the film, which corresponds to the atmospheric pressure (101.3 kPa) when the samples are subjected to pure oxygen gas (100%) on one side of the film and to entrainment gas containing 98% nitrogen and 2% hydrogen on the other side of the film.

The oxygen permeability coefficient ($P'O_2$) is calculated by multiplying oxygen permeability (PO_2) by sample thickness.

7.3.3 Aroma Permeability

Aroma transport through edible films can result in flavor loss, scalping, and/or contamination, to culminate in unacceptable product quality. Therefore, aroma barrier is an important parameter when selecting polymeric films. However, there is limited information on the aroma barrier property of many polymeric materials mainly because several aroma compounds exist, and interaction between aroma compounds and polymers can be very complex. Additionally, permeation measurement methods and equipment to detect aroma compounds are more complicated than methods and

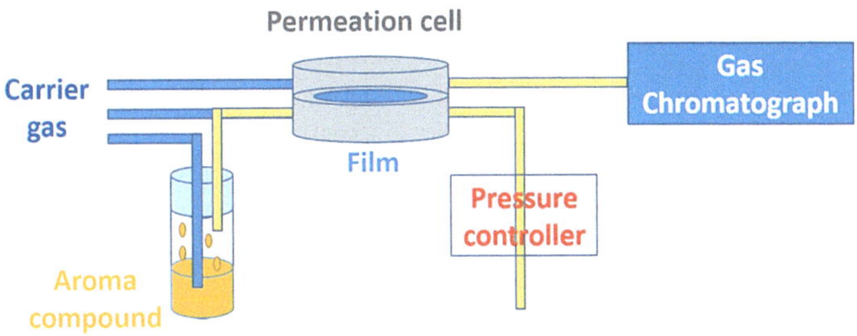

Fig. 7.6 System used to determine aroma permeability in films

equipment available for gas detection (oxygen, carbon dioxide, and water vapor) (Leelaphiwat et al. 2017).

Debeaufort et al. (1998) described a dynamic method to measure aroma vapor fluxes through edible films (Fig. 7.6). The permeation cell consists of two chambers divided by the studied film. The two chambers are continuously swept with a nitrogen flow. Aroma concentrations in the vapor phase on the upper side of the cell are obtained by mixing two flows: one containing the volatile compound, and the other containing dry nitrogen. Flows containing vapors are obtained by bubbling dry nitrogen through pure compounds. The volatile compounds passing across the film are swept by the carrier gas (N_2) and carried to an automatic injection valve through a transfer line that is heated to prevent adsorption. A carrier gas is automatically injected in the gas chromatograph at regular intervals.

Leelaphiwat et al. (2017) used gas chromatography techniques to determine eucalyptol and estragol permeability in low-density polyethylene (LDPE), polypropylene (PP), nylon (Nylon), polyethylene terephthalate (PET), metalized polyethylene terephthalate (MPET), and poly(lactic acid) (PLA) films at 15 and 25 °C.

7.4 Edible Film Barrier Properties and Attempts to Improve this Property

7.4.1 Edible Film Barrier Properties

Edible films and coatings consist of hydrocolloids (such as polysaccharides or proteins) or hydrophobic compounds (e.g., lipids or waxes) (Álvarez et al. 2017). Films made from polysaccharides are expected to be good oxygen and carbon dioxide barriers due to their tightly packed and ordered hydrogen-bonded network structure. Nevertheless, the relatively lower water resistance and poorer vapor barrier of polysaccharide films, which are a consequence of their hydrophilic nature, limit their use in food packaging (Yang and Paulson 2000). In addition, polysaccharide films only present good gas barrier properties if they are not plasticized with water or other plasticizers. Gas permeability increases significantly with rising water/

Table 7.1 Barrier properties of films produced from different raw materials

Edible films		Water vapor permeability ($\times 10^{-10}$ g.m^{-1}.s^{-1}.Pa^{-1})	Oxygen permeability coefficient (cm^3 µm.m^{-2}.d^{-1}.kPa^{-1})	References
Starch	Oat	0.042	–	Galdeano et al. (2009)
	Babassu	3.52–5.52	–	Maniglia et al. (2017)
	Achira	3.2	–	Andrade-Mahecha et al. (2012a)
	Amaranth	2.7	–	Tapia-Blácido (2006)
	Quinoa	0.6	–	Araújo-Farro et al. (2010)
	Rice	0.5	–	Dias et al. (2010)
	Banana	2.0	–	Pelissari et al. (2013)
	Corn	0.9	–	Luchese et al. (2017)
	Cassava	1.17	–	
	Wheat	–	57% RH = >0.12 to 1.23	Gaudin et al. (2000)
	Sweet potato	2.02	50% RH = >3.52	Shen et al. (2010)
Chitosan	Glycerol	–	0% RH = >0.01–0.04	Butler et al. (1996)
	Sorbitol	–	0% RH = >0.04–0.08	
Celullose	Microfibrillated cellulose	–	50% RH = >3.52 to 5.03	Syverud and Stenius (2009)
	Carboxymethylated microfibrillated cellulose	–	0% RH = >0.0006 50% RH = >0.85	Aulin et al. (2010)
Protein	Fish	1.45	–	Romani et al. (2017)
	Collagen	–	0% RH = >1.20	Mchugh and Krochta (1994a, b, c)
		0.21	–	He et al. (2011)
	Sesame	0.09–0.16	–	Sharma and Singh (2016)
	Whey protein	–	Glycerol: 50% RH = >40 to 330	Sothornvit and Krochta (2000)
		–	Sorbitol: 30% RH = >1.03	Mchugh and Krochta (1994a, b, c)
	Soy	12.2	–	Martelli-Tosi et al. (2017)

(continued)

Table 7.1 (continued)

Edible films		Water vapor permeability ($\times 10^{-10}$ g.m^{-1}.s^{-1}.Pa^{-1})	Oxygen permeability coefficient (cm^3 µm.m^{-2}.d^{-1}.kPa^{-1})	References
Flour	Turmeric	Glycerol: 0.823	–	Maniglia et al. (2015)
		Sorbitol: 0.461	–	Maniglia et al. (2014)
	Babassu	9.30	–	Maniglia et al. (2017)
	Achira	5.3	–	Andrade-Mahecha et al. (2012a)
	Amaranth	0.7	–	Tapia-Blácido (2006)
	Quinoa	0.6	–	Araújo-Farro et al. (2010)
	Rice	1.1	–	Dias et al. (2010)
	Banana	2.1	–	Pelissari et al. (2013)
		–	50% RH = >23	Sothornvit and Pitak (2007)
Synthetic plastics	Cellophane	0.84	57% RH = >159.48	Embuscado and Huber (2009)
	LDPE (low density polyethylene)	0.009	50% RH = >1870	
	HDPE (high density polyethylene)	0.002	50% RH = >427	

plasticizer content in the films as observed by Dole et al. (2004) and Gontard et al. (1996). This restricts their fields of application, namely in food packaging.

Table 7.1 lists the water vapor and gas barrier properties of some edible films made from polysaccharides (starch, chitosan, and cellulose). Starch is an abundant, inexpensive polysaccharide obtained from cereals, legumes, and tubers. It consists of two macromolecules: amylose, which is essentially linear, and amylopectin, which is highly branched. The proportion of these polymers in starch depends on the starch source, and this proportion provides the film with particular properties (Han et al. 2006). Starch films are hydrophilic, so their properties change when relative humidity fluctuates; for example, the starch film barrier properties decrease with increasing relative humidity (Bai et al. 2002). Therefore, starch is not the best option when it comes to working with minimally processed high-water activity commodities.

Chitosan is a derivative of chitin, which is obtained from marine invertebrates. After cellulose, chitosan is the second most abundant polysaccharide resource on Earth (Tuil et al. 2000). Chitosan films have lower oxygen and carbon dioxide permeability than polyethylene films (Hosokawa et al. 1990; Butler et al. 1996).

Table 7.1 shows the oxygen permeability of chitosan films with different plasticizers (sorbitol and glycerol).

Cellulose is an inexpensive abundant natural polymer with crystalline structure. It is insoluble in water, which makes its use as coating difficult. However, some commercially produced cellulose derivatives like carboxymethyl cellulose (CMC), methylcellulose (MC), hydroxypropyl cellulose (HPC), and hydroxypropyl methylcellulose (HPMC) can overcome limitations associated with native cellulose. Films based on cellulose derivatives tend to have moderate oxygen and poor water vapor barrier properties because of their inherent hydrophilic nature.

Proteins can form edible films because their side chains can establish intermolecular crosslinks (Álvarez et al. 2017). Film properties will depend on the nature of these linkages. In general, protein films are better barriers than polysaccharides and are considered to have good gas barrier properties. Nevertheless, their water barrier property is generally poor (Gennadios et al. 1994) because it depends on the environment RH and/or on the food water activity. This may be related to the protein film with a more polar nature and linear (non-ring) structure, which culminates in higher cohesive energy density and smaller free volume (Miller and Krochta 1997).

Myosin is the most abundant protein in myofibrils (60–70% of muscle protein), and it accounts for the myofibril functional properties, including gelation (Bourtoom et al. 2006; Xiong 1997). However, fish protein films have poor water vapor barrier performance, which could jeopardize food quality and safety, thus improving these materials is an important matter (Romani et al. 2017).

Gluten films have good oxygen and carbon dioxide barrier properties, but they exhibit relatively high WVP (Gennadios and Weller 1990). Whey protein produces a translucent and flexible film with excellent oxygen and aroma barrier properties at low RH (McHugh and Krochta 1994a, b, c; Miller and Krochta 1997). Soy protein isolate films also presented high WVP as compared to other protein sources.

Nowadays, collagen is one of the most widely investigated proteins—it not only represents the main structural protein accounting for approximately one-third of all vertebrate body proteins (Lee and Mooney 2001), but also has commercial and industrial significance, as exemplified by its traditional use in the leather industry and its current biomedical applications.

Flour is another source of compounds to produce films. Flour materials have a complex structure that contains starch, protein, fibers, and lipids (Gutiérrez et al. 2016a, b; Gutiérrez and Alvarez 2017b, c). Interest in these materials lies on the favorable and natural thermodynamic compatibility of its biopolymers, which prevents phase separation (Grinberg and Tolstoguzov 1997) and is in contrast to what often occurs when biopolymers are mixed during film processing (Arvanitoyannis and Kassaveti 2009). Flour film properties depend on the interaction established by the mixture of biopolymers (starch, protein, and fibers) and lipid, on the distribution of these interactions within the film matrix, on the balance between hydrophilic and hydrophobic interactions, and on the concentration of each component within the film (Andrade-Mahecha et al. 2012; Tapia-Blácido et al. 2007). Table 7.1 depicts the barrier properties of some films based on flour (turmeric, babassu, achira, amaranth, quinoa, rice, and banana).

Comparison between edible and synthetic films (cellophane, low density polyethylene - LDPE, and high polyethylene - HDPE) reveals that the latter films have higher oxygen permeability. On the other hand, some edible films and cellophane have similar WVP. LDPE and HDPE films display the lowest WVP.

7.4.2 Research into Improved Edible Film Barrier Properties

Several strategies have been tested to improve the edible film barrier properties of bio-based materials for sustainable food packaging applications (Table 7.2). One of these strategies is to use additives; e.g., to include inorganic impermeable particles like mica flakes, to decrease CO_2, O_2, and water vapor permeability by 73%, 27%, and 40%, respectively (Alves et al. 2010). Another strategy is to include essential oils (EOs) like lemon EO (2%), which can reduce film WVP by 16.08% as compared to the control (Song et al. 2018). Plasticizer type also influences barrier properties. Al-Hassan and Norziah (2012) noted that starch-gelatin blends plasticized with sorbitol have higher WVP (~ 50%) as compared to the same blends plasticized with glycerol. Fibers are another type of additive. Ma et al. (2017) added cellulose nanocrystals obtained from sweet potato to cassava starch film, to diminish WVP by 50%.

Another way to enhance barrier properties is to control the film-forming process conditions. For example, Angellier-Coussy et al. (2011) prepared wheat gluten films and evaluated process temperature. Increasing the temperature from 80 to 120 °C reduced the gluten film WVP by 28%. Hernández-Izquierdo and Krochta (2008) produced films from whey protein isolated by compressing molding (336 g mm/d kPa m^2), to obtain more water-permeable films than films prepared by solution casting (120 g mm/d kPa m^2).

Finally, chemical modifications such as crosslinking are another way to improve barrier properties. Balaguer et al. (2013) produced wheat gluten film crosslinked with cinnamaldehyde, to achieve 64%, 75%, and 79% reduction in water vapor, oxygen, and carbon dioxide permeability, respectively.

7.5 Release Mechanism of Active Compounds

Traditionally, packaging materials are selected so that they interact minimally with the food they contain and can therefore be considered as inert barriers. However, in the last decades, several packaging systems that interact with the food product in a desirable way, the so-called active packaging, have been developed. Active packaging is defined as a system where product, packaging, and free space interact, to result in improved product quality and safety (Suppakul et al. 2003; Vermeiren et al. 2002). This packaging can modify the environment inside the food package, thereby altering the state of the packaged food system and its headspace. Food quality is

Table 7.2 Strategies to improve edible film barrier properties

	Films	Strategies	Results	References
Additives	Kappa-carrageenan/pectin blends	Mica flakes	Films with mica flakes have significantly lower CO_2 (73%), O_2 (27%), and water vapor (40%) permeability	Alves et al. (2010)
	Corn starch/wheat gluten blends	Lemon essential oil	Compared to the control, the film with 2% lemon essential oil has 16.08% lower WVP	Song et al. (2018)
	Starch-gelatin blends	Plasticizer type	Sorbitol furnishes more water vapor-permeable films than glycerol (~ 50%)	Al-Hassan and Norziah (2012)
	Cassava starch films	Without cellulose nanocrystals	WVP: 2 g mm/m^2.d.kPa	Ma et al. (2017)
		0.4 g cellulose crystals of sweet potato/100 ml solution	WVP: 1.1 g mm/m^2.d.kPa	
Process conditions	Wheat gluten	Processing temperature	Increasing temperature from80 °C to 120 °C caused reduction of the water vapor (28%) permeability	Angellier-Coussy et al. (2011)
	Whey protein isolate	Compression molding	336 g mm/d kPa m^2	Hernández-Izquierdo and Krochta (2008)
		Solution-casting	120 g mm/d kPa m^2	
Chemical treatment	Wheat gluten	Crosslinking with cinnamaldehyde	5% cinnamaldehyde reduces water vapor, oxygen, and carbon dioxide permeability by 64%, 75%, and 79%, respectively	Balaguer et al. (2013)

improved as judged from shelf life extension, sensory quality enhancement, and microbial safety protecting food from external conditions (Debeaufort et al. 2000; Quintavalla and Vicini 2002; Suppakul et al. 2003;Cha and Chinnan 2004; Ozdemir and Floros 2004).

In active packaging technologies, the final coated film should fulfill three main requirements (Gómez-Estaca et al. 2014): (i) the active coating should adhere well to the film substrate and should be suitable for direct contact with the food; (ii) the active agent release should be adjusted so as to produce efficient antioxidant activity; and (iii) the final active coated structure should fulfill the food product functional packaging requirements, which are basically the same as the functional packaging requirements of conventional passive packaging.

Active packaging with antimicrobial and antioxidant action has been the most often applied because lipid oxidation and microbial growth are the main factors underlying food degradation and hence food sensorial and nutritional quality loss

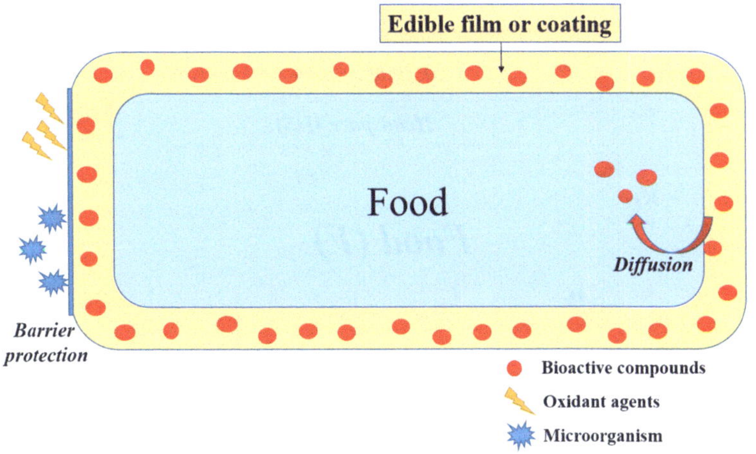

Fig. 7.7 Mass transfer process by active compound migration to the food and the barrier protection formed by the presence of the active compound

(Gómez-Estaca et al. 2010; Pereira Jr et al. 2015). Active compound migration to food is slow, which allows for incorporation of a lower content of preservatives into the food product.

Mass transfer; i.e., diffusion, governs the release of active compounds within the food and the film or coating used as packaging (Han and Scanlon 2014). Through diffusion, bioactive compounds included in the polymer matrix can be transferred to the product or headspace, where they exert a protective action (antioxidants, antimicrobials, etc.) (Fig. 7.7).

When the active material acts as an antioxidant/antimicrobial-releasing system, active agent release is controlled by combined mass transport processes involving the active agent partition equilibrium at the interphases and kinetic processes in the food, headspace, coating, and substrate phases (López-Carballo et al. 2012). Gómez-Estaca et al. (2014) explained that active substances are partitioned in all phases constituting the food/package/environment system as observed in Fig. 7.8.

In the case of non-volatile active agents and at equilibrium, concentration ratios at the interphases are given by the partition coefficients describing equilibrium between the film substrate (FS) and the film coating (FC), and between the FC and the food product (F):

$$K_{\frac{FC}{FS}} = \frac{C_{FC}}{C_{FS}}; \quad K_{\frac{FC}{F}} = \frac{C_{FC}}{C_F} \tag{7.27}$$

To obtain high active agent release into the food, the coating matrix should have low $K_{FC/F}$. The package must be designed with a substrate material that has low affinity for the active agent, so that $K_{FC/FS}$ is high. This should prevent the substrate from retaining the active agent. In the case of volatile active agents, the package

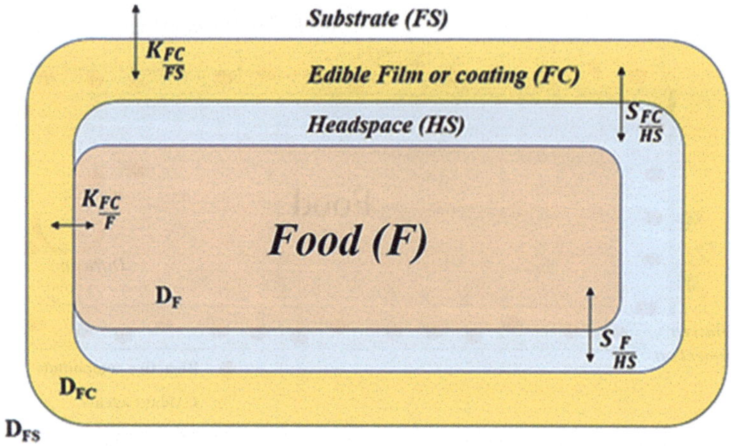

Fig. 7.8 Diagram of the mass transport parameters (partition K, solubility S, and diffusion coefficients D) involved in the various parts of a food stored in a bag consisting of a coated substrate: substrate (FS), edible film or coating (FC), food (F), and headspace (HS)

headspace (HS) also participates in the mass transport process, and the active agent concentration in the HS is related to its concentration in the food (F) and coating (FC), obtained by the corresponding partition coefficients

$$K_{\frac{FC}{HS}} = \frac{C_{FC}}{C_{HS}};\quad K_{\frac{F}{HS}} = \frac{C_F}{C_{HS}} \tag{7.28}$$

or by the solubility coefficients as described by Henry's law, which expresses the amount of agent in the HS as partial pressure (p_{HS}):

Low solubility coefficients should provide high active agent concentration in the vapor phase, whereas low $S_{FC/HS}$ and high $S_{F/HS}$ coefficients ensure high active compound concentration on the food surface.

$$S_{\frac{FC}{HS}} = \frac{C_{FC}}{P_{HS}};\quad S_{\frac{F}{HS}} = \frac{C_F}{P_{HS}} \tag{7.29}$$

Active agent diffusion in the coating is commonly the slowest process, which means that D_{FC} (edible film or coating diffusion coefficient) controls active agent mass transport. The above-mentioned partition and solubility coefficients characterize the extent of the mass transport process and denote the final active agent concentration throughout the various phases that constitute the food packaging system, assuming that equilibrium is achieved. However, the pace at which the system advances towards equilibrium depends on the active agent diffusion kinetics in the various system phases. According to Fick's laws, substance (J) flow in a phase is proportional to the established concentration gradient ($\delta c/\delta x$):

$$J = -D\alpha \frac{\delta c}{\delta x}; \quad \alpha = FC, FS, F, HS \tag{7.30}$$

Active packaging could also modify the product (dyes, peptides, vitamins, etc.) organoleptic and nutritional characteristics. Diffusion of these components—initially present in the film or coating—should not pose toxicity risks or alter the properties of the resulting polymeric materials (Salgado et al. 2015).

Salgado et al. (2015) pointed out that the active compound transfer rate depends on the polymeric material chemical nature and crosslinking degree, on the active compound concentration, on the active compound affinity for the polymer matrix and food, and on environmental conditions such as relative humidity, temperature, and contact time. Therefore, knowing how these variables affect active packaging effectiveness is essential.

Although edible films and coatings are promising systems for use as active ingredient carriers, application of these materials as food packaging must be evaluated. Knowledge of food deterioration mechanisms, of package mode of action, and of their relations is crucial. Studies in real, pragmatic, everyday systems are important to prove that the developed package is truly effective. Sometimes, the *in vitro* properties of a material are not evident in food systems, which is often due to difficult active compound release or active compound inactivation during processing (Silva-Weiss et al. 2013).

The concentration of a given active compound added to biopolymer films should be analyzed because high concentrations of this active agent could generate undesirable odors, turbidity, and/or compound precipitation in the films and even affect their structure, which could consequently impact their functionality (Wambura et al. 2011). According to Cosgrove (2008), the active compound dose needs to be relatively low; in general, the filmstrip can include 30% of active ingredient. The nature of the interaction between biopolymers and additives for preservation depends on the nature, chemical characteristics, concentration, and transport mechanism of the compounds and on the pH of both the biopolymer and additives as well as on the structural parameters of the active compounds (stereochemistry, conformational flexibility, and molecular weight).

7.5.1 Films with Antimicrobial Activity

Antimicrobial packaging is one of the most promising active packaging systems. Because antimicrobial packaging contains antimicrobial agents, it can effectively kill pathogenic microorganisms or at least inhibit food product spoilage by microorganisms that contaminate food (Salleh et al. 2007; Dutta et al. 2009). This type of packaging prevents microbial growth on the food surface by direct contact of the packaging material with said surface. Controlled bacteriocin release from packaging film toward the food surface is advantageous over food dipping into bacteriocins or food spraying with bacteriocins: in the latter situations, antimicrobial activity may be lost or reduced due to bacteriocin inactivation by food components or to considerable bacteriocin dilution below active concentration after they migrate into the food product(Malhotra et al. 2015). Application of antimicrobial packaging systems is currently limited by lack of suitable antimicrobials, scarcity of new polymer materials, regulatory concerns, and need for appropriate testing methods (Jin and Zhang 2008).

Although antimicrobial mechanisms are still not fully understood, active compounds are thought to exert their effect during the lag phase (Valero and Salmeron 2003; Valero and Ginger 2006), which last several hours for bacteria. Materials have different antimicrobial activities probably due to distinct antimicrobial compound diffusion through the polymer matrix. The diffusion coefficient of organic species in polymers is mainly a function of the organic species molecular weight and polymer type (Reynier et al. 2002). Because the molecular weight range covered by active compounds is not very large, their diffusion coefficients are quite similar. Consequently, effective diffusion depends mainly on polymer type and specifically on polymeric material polarity (Hernandez-Muñoz et al. 2002).

Finding an efficient method to deliver antimicrobials within food packaging materials is necessary. The Food and Drug Administration (FDA) regulates direct addition of antimicrobials to food formulation and food wrapping films and specifies safety levels for antimicrobial substances in food. Instant addition of antimicrobials to food packaging film formulations often results in instant inhibition of undesired microorganisms (Table 7.3). However, the surviving population will continue to grow as soon as the added antimicrobial is depleted. Antimicrobials are primarily depleted due to complex interactions with the food matrix and to their natural degradation along time, which decreases shelf life (Kester and Fennema 1986; Ouattara et al. 2000; Chi-Zhang et al. 2004).

Chitosan is a natural food preservative, but its antimicrobial mechanism has not been elucidated yet (Gutiérrez 2017b). Positively charged chitosan molecules are believed to interact with negatively charged microbial cell membranes, to modify microbial cell permeability and consequent leakage of cell constituents (No et al. 2007). Chitosan films or coatings have been satisfactorily used on fruits and vegetables; antimicrobial activity against *Bacillus cereus*, *Brochothrix thermosphact*, *Lactobacillus curvatus*, *Lactobacillus sakey*, *Listeria monocytogenes*, *Pediococcus acidilactici*, *Photobacterium phosphoreum*, *Pseudomona fluorescens*, *Candida lambica*, *Cryptococcus humiculus*, and *Botrytis cinerea* has been detected (Devlieghere et al. 2004; Romanazzi et al. 2002).

Benzoates, proprionates, sorbates, parabens, acidifying agents (e.g., acetic and lactic acids), curing agents (e.g., sodium chloride and sodium nitrite), bacteriocins, and natural preservatives (e.g., natural oils, lysozyme, and liquid smoke) are some of the preservatives and antimicrobials that are more commonly used in edible films and coatings (Cagri et al. 2004). Antifungal compounds, organic acids, potassium sorbate, and the bacteriocin nisin reduce foodborne microorganism levels more effectively when they are immobilized on or incorporated into edible gels (i.e., starch, carrageenan, waxes, cellulose ethers, or alginate) and applied to meat surfaces as compared to these agents alone (Cutter and Sumner 2002; Ustunol 2009).

Table 7.3 Active films with antimicrobial activity

Biopolymers	Antimicrobial compounds	Effect of active compounds	References
Gelatin and chitosan	Clove, fennel cypress, lavender, thyme, herb-of-the-cross, pine, and rosemary essential oils	Clove essential oil showed the highest inhibitory effect, followed by rosemary and lavender essential oils Clove-containing films inhibited the microorganisms *Pseudomonas fluorescens*, *Shewanella putrefaciens*, *Photobacterium phosphoreum*, *Listeria innocua*, *Escherichia coli*, and *Lactobacillus acidophilus*	Gómez-Estaca et al. (2010)
Soy protein	1, 2, 3, 4, and 5% oregano (OR) or thyme (TH) essential oils	*E. coli* and *S. aureus* were significantly inhibited by antimicrobial films. *L. plantarum* and *P. aeruginosa* were the most resistant bacteria Essential oil in soy protein films reduced coliform and *Pseudomonas spp.* counts in beef patties	Emiroğlu et al. (2010)
Carragen	*Satureja hortensis* essential oil (SEO)	Films containing SEO inhibited selected bacteria (*Staphylococcus aureus ATCC 25923*, *Bacillus cereus PTCC 1154*, *Escherichia coli ATCC 25922*, *Pseudomonas aeruginosa ATCC 27853*, and *Salmonella typhimurium ATCC 14028*)	Shojaee-Aliabadi et al. (2013)
Chitosan	*Thymus moroderi* and *Thymus piperella* essential oils	In chitosan films, *Thymus piperella* essential oil was more effective against *Serratia marcenscens* and *Listeria innocua* than *Thymus moroderi* essential oil	Ruiz-Navajas et al. (2013)
Sweetpotato starch (SPSF)	*Origanum* essential oil (OG)	Antimicrobial activity of SPSF films containing OG increased with increasing OG concentration. SPSF/OG films exhibited greater inhibitory effects against gram-negative bacteria such as *S. enteritidis* and *E. coli* O157:H7 than against the gram-positive L. *monocytogenes*	Ehivet et al. (2011)

(continued)

Table 7.3 (continued)

Biopolymers	Antimicrobial compounds	Effect of active compounds	References
Cassava starch	Cinnamon essential oil	Effective antimicrobial activity against the fungi *P. commune* and *E. amstelodami*	Souza et al. (2013)
Chitosan	Microemulsions formed from Corn-bio-fiber gum (C-BFG) as emulsifier additive with allyl isothiocyanate (AIT) and lauric arginate ester (LAE) as antimicrobials	Micro emulsions create micro pores (100 to 300 nm) and micro channels that hold antimicrobials effectively and facilitate antimicrobial release from the center to the surface of films or coatings, thus enhancing their antimicrobial efficacy. Films with 1% AIT reduced *Listeria innocua* populations in ready-to-eat meat and strawberries. Films with 1% LAE reduced *Escherichia coli* and *Salmonella spp*. populations in strawberries	Guo et al. (2008)
Carboxymethylcellulose	Potassium sorbate	Pistachios were coated with this edible antimicrobial film containing 1, 0.5, or 0.25 g of sorbate/100 mL of film solution. Potassium sorbate inhibited *Aspergillus flavus* and *Aspergillus Parasiticus* growth	Sayanjali et al. (2011)
Corn starch	Silver nanoparticles (ag-NPs)	Films presented antimicrobial activity *against Staphylococcus aureus*, *Escherichia coli*, and *Candida albicans* without significant differences between ag-NPs concentrations. Migration of components from nanostructured starch films, assessed by food contact tests, was minor and under the legal limits	Abreu et al. (2015)
Chitosan	Propionic acid	Propionic acid incorporation into chitosan films inhibited *Candida spp* and *Penicillium spp* growth and extended food shelf life by maintaining microbial growth in the latency period	Rivero et al. (2013)

The most common test used to determine film antimicrobial activity is based on inhibition zone evaluation in the disc diffusion method (Barry and Meyer 1979). This method consists in using Petri dishes containing the medium with inoculated microorganisms. Circular film pieces with diameter of approximately 1.5 cm are placed on the surface of previously inoculated culture medium. The Petri dishes are stored at 32–37 °C inside a circulation oven overnight to allow bacteria to grow. Inhibition zones are directly measured on the dishes on the basis of the clear zone average diameter.

7.5.2 Films with Antioxidant Activity

Oxygen underlies many degradation processes in foods such as lipid oxidation, microorganism growth, enzymatic browning, and vitamin loss (Ayranci and Tunc 2003). Fat oxidation results in altered flavor and color and nutrient loss (Hong and Krochta 2006). Oxidative processes cause degradation of meat proteins, pigments, and lipids, which reduces the food product shelf life (Liu et al. 2010). Although some oxygen availability is necessary for living tissue respiration, this process accelerates consumption of sugars and other compounds, thus increasing ethylene production and eliciting senescence in fruits and vegetables (Oms-Oliu et al. 2008; Rojas-Graü et al. 2007).

The antioxidant effect of stand-alone edible films is strongly linked to their oxygen permeability (OP), which can be directly measured by the oxygen permeation tests shown in item 3.2. Application of antioxidant and anti-browning agents, such as ascorbic and citric acids, can delay deleterious oxygen effects on food.

Nowadays, researchers have tended to include strongly flavored antioxidants (e.g. essential oils, N-acetylcysteine, and glutathione) in edible films and coatings to reduce strong aroma. Table 7.4 summarizes some films added with antioxidant agents like protocatechuic acid, hydroxybenzoic acids, epigallocatechin gallate nanocapsules, mango kernel extracts, polysorbate-thymol micelles, free lycopene and lycopene nanocapsules, rosemary extracts, husk powder, montmorillonite, murta leaves extract, and clove essential oil. Recent studies have also dealt with renewable sources that already contain active compounds associated with macromolecules and which can be used as active packaging for food. This natural system of renewable sources makes phase separation between the bioactive substance and the polymer matrix like babassu, turmeric, and red rice more difficult (Muñoz-Bonilla and Fernández-García 2012). The addition of vitamin E has been also evaluated in whey protein isolate edible coatings (Lee et al. 2002). Duan et al. (2010) observed that chitosan–fish oil coating increased total lipid and omega-3 fatty acid contents in fish by about threefold, reduced TBARS (thiobarbituric reactive substances) values in both fresh and frozen samples, and decreased drip loss of frozen samples by 14.1–27.6%. The chitosan antioxidant property is attributed to its ability to chelate free iron released by myoglobin degradation during meat storage (Kamil et al. 2002).

Table 7.4 Edible films with antioxidant activity

Biopolymer	Active compound	Effect of active compounds	References
Chitosan	Protocatechuic acid (PA)	Chitosan/PA composite films presented more total phenolic content and antioxidant activity than chitosan films.	Liu et al. (2017a)
Chitosan	Hydroxybenzoic acids: Gallic acid (GLA), gentisic acid (GTA), protocatechuic acid (PA), syringic acid (SA), and vanillic acid (VA)	Antioxidant activity assays showed that chitosan films with hydroxybenzoic acid had higher DPPH scavenging activity than films consisting of chitosan only. Gallic acid provided higher antioxidant activity.	Liu et al. (2017b)
Chitosan/zein	Epigallocatechin gallate nanocapsules (EGCG)	Film DPPH scavenging activity increased as the nanocapsule suspension concentration increased.	Liang et al. (2017)
Soy protein isolate (SPI) and fish gelatin (FG)	Mango kernel extracts (MKE) from 1 to 5%	Antioxidant activity increased at high MKE concentration in both films, with more impact in the case of SPI films. DPPH analysis of SPI films revealed that these films had the greatest antioxidant activity (89%) upon inclusion of 5% MKE extract.	Adilah et al. (2018)
Potato starch	Polysorbate-thymol micelle	Starch/polysorbate/thymol exhibited ABTS and DPPH radical scavenging activity, which was lower than polysorbate/thymol. Thymol micelle incorporation into starch significantly decreased thymol antioxidant capacity. This reduced antioxidant activity might be attributed to thymol encapsulation in the starch chain as confirmed by atomic microscope analysis.	Davoodi et al. (2017)
Cassava starch	Free lycopene and lycopene nanocapsules	Lycopene nanocapsules provided greater protection to sunflower oil stored under accelerated oxidation conditions, which attested to their potential application as antioxidant packaging to prevent high-fat food oxidation.	Assis et al. (2017)

(continued)

Table 7.4 (continued)

Biopolymer	Active compound	Effect of active compounds	References
Babassu flour and starch	–	Films produced from babassu flour or starch showed antioxidant activity as evaluated by the DPPH method.	Maniglia et al. (2017)
Turmeric flour	–	Films produced from turmeric flour showed antioxidant activity as evaluated by the DPPH method	Maniglia et al. (2015)
Cassava starch	Rosemary extracts	Rosemary extracts provided cassava starch films with antioxidant activity (DPPH method). Polyphenols migrated from edible films to food simulants after 7 days of film exposure.	Piñeros-Hernandez et al. (2017)
Brazilian pine seed (*Araucaria angustifolia*) flour	Husk powder	The highest antioxidant activity (as measured by the DPPH method) of Brazilian pine seed husk powder films was due to the presence of high amounts of flavonoids, mainly quercetin and apigenin, and tanins, mainly catechin and epicatechin, in A. *angustifolia* husk	Daudt et al. (2017)
Red rice flour and starch	–	Red rice flour and starch displayed antioxidant activity due to the presence of antioxidant compounds such as phenolic compounds and procyanidins in red rice. These compounds should exert a protective action against lipid oxidation in sunflower oil.	Vargas et al. (2017)
Carboxymethylcellulose (CMC)	Montmorillonite(MMT) activated with murta (*Ugni molinae Turcz*) leaf extract	The antioxidant capacity (ABTS method) of films added with murta extract increased over 18-fold as compared to the CMC control film due to the presence of gallic acid, myricetin, and quercetin in the murta extract. MMT addition to the CMC–murta extract formulations significantly increased the antioxidant activity.	Gutiérrez et al. (2012)

(continued)

Table 7.4 (continued)

Biopolymer	Active compound	Effect of active compounds	References
Carboxymethylcellulose (CMC)	Clove essential oil (CEO)	CMC films had lower antioxidant activity (0.32%) than CMC/CEO films. Addition of 3% CEO yielded films with high antioxidant activity (71.76%).	Dashipour et al. (2014)

7.6 Case Study: Application of Turmeric Coating Produced from Turmeric Dye Residue for Post-harvest Banana Preservation

Coatings produced from renewable sources increase food product durability, preserve foodstuff texture and nutritional value, decrease water loss or gain, and allow greater flexibility during fruit and vegetable handling and trade along post-harvest storage (Vargas et al. 2008; Baldwin et al. 2011).

Curcuma longa L. is a plant native to India. It is cultivated throughout the tropical world for medical uses and flour production. Recently, Maniglia et al. (2014, 2015) evaluated the residue from the pigment extracted from turmeric rhizomes as raw material for biodegradable film processing, which will be referred to as turmeric residue. This residue contains starch, proteins, fibers, and lipids and has a residual content of curcuminoids, such as curcumin, demethoxycurcumin, and bisdemethoxycurcumin. These phenolic compounds have antioxidant (Maniglia et al. 2014, 2015), anti-inflammatory (Menon and Sudheer 2007), antimicrobial (Arutselvi et al. 2012), and anticarcinogenic activities(Jiang et al. 2012). Curcumin is considered to be the main bioactive compound in this residue (Paramasivam et al. 2009). The great advantage of preparing coatings with turmeric residue is its intrinsic antioxidant activity, which dismisses the need to incorporate additional active substances in the polymer matrix.

In this part of the Chapter, we will show results concerning bananas coated with treated turmeric residue. This research was developed in the Chemistry Department of the University of São Paulo (USP) campus located in the city of Ribeirão Preto.

We failed to obtain films from turmeric residue (Maniglia et al. 2014, 2015): its closely packed structure retained starch granules and prevented water access for film formation. For this reason, we submitted the turmeric residue to mechanical and chemical treatments. Mechanical treatment reduced the turmeric residue particle diameter, to increase the superficial area of this material for future chemical treatment. Particle size reduction increased surface area by significantly raising the number of surface atoms as compared to the total particle volume, thus altering chemical reactivity (Martinez and Alves 2013). Mechanical treatment was performed by ball milling at 5 Hz for 24 h. To evaluate the mechanical treatment effect, we determined the D_{80} diameter (the maximum diameter of 80% of the particles).

The mechanically treated turmeric residue (TRM) had lower D_{80} diameter (0.265 ± 0.007) than the initial material (0.685 ± 0.007).

We carried out chemical treatment to break the fibers in the turmeric residue, to solubilize the protein, to remove ashes and lipids, and to release starch without damaging the granules. Chemical treatment comprised two stages: alkaline treatment and bleaching. For the alkaline treatment, the ground sample (20 g) was dispersed in 2.5% sodium hydroxide solution (400 mL) with subsequent shaking at ambient temperature for 4 h in a shaker (SL222, Solab, Brazil). Subsequently, the suspension was centrifuged in Quimis centrifuge (Q222RM, Brazil) at 1500 rpm and 10 °C for 10 min. The material was washed and centrifuged under the same conditions, until pH 7 was achieved, and dispersed in a solution containing 3.3% sodium chlorite and 0.7% acetic acid (350 mL). This procedure was performed in a jacketed beaker at room temperature for 4 h; a magnetic stirrer was used. The suspension was centrifuged in Quimis centrifuge (Q222RM, Brazil) at 1500 rpm and 10 °C for 10 min. The material was washed and centrifuged under the same conditions, until pH 7 was reached. The chemically treated turmeric residue (TRC) was dried in an oven at 40 °C (Q314M, CHEMIS).

Table 7.5 shows that the turmeric residue, the mechanically treated turmeric residue (TRM), and the chemically treated turmeric residue (TRC) consisted mainly of starch and contained significant fiber content, low humidity, and small fractions of proteins and lipids. Mechanical treatment only significantly affected the moisture and ash contents. Chemical treatment reduced the lipid, protein, ash, hemicellulose, and lignin contents. The cellulose content was higher in TRM and TRC probably because cellulose resisted to chemical hydrolysis because it contained fewer amorphous regions susceptible to attack by chemical reagents (Behera et al. 2014). The starch content was also higher in TRM and TRC, but the amylose content was lower in the treated residues (turmeric residue: 53.18% and treated turmeric residues: 37.17%).

Because turmeric residue is advantageous not only for its high starch content, but also for the presence of curcuminoids, we determined the antioxidant activity and the contents of curcuminoids and total phenolic compounds in the treated turmeric residues (Table 7.6). The treated residues still presented antioxidant activity and significant content of curcuminoids. TRC could be an interesting food coating, so we applied it to coat bananas as an attempt to extend the fruit (*Musa acuminata*) shelf life. The coating solution was prepared by following the same methodology used to obtain turmeric films (Maniglia et al. 2014). The turmeric coating formulation consisted of 6 g of TRC/100 g of solution and of 25 g of glycerol/100 g of TRC and was applied to the fruit by immersion. The bananas were dipped into the turmeric coating formulation for 1 min, and the excess gel was allowed to drain away. The samples were then hung inside an oven with forced circulation (MA Q314M, Quimis, Brazil) and stored at 25 °C and 65% RH for 12 days. The way the turmeric coating affected banana weight loss, firmness, pH, titratable acidity, contents of soluble solids and reducing sugars, and peel color was evaluated along 12 days of storage. Analyses were performed on days 0, 2, 4, 6, 8, 10, and 12 after the turmeric coating was applied. To analyze soluble solids, pH, titratable acidity, and reducing sugars, bananas were peeled, and pulps were removed and centrifuged at 10,000 rpm

Table 7.5 Turmeric residue, mechanically treated turmeric residue, and chemically treated turmeric residue chemical compositions (g/100 g of material on dry basis)

Sample	Moisture*	Lipid	Protein	Ash	Celullose	Hemicellulose	Lignin	Starch
Turmeric residue	11.33 ± 0.19^b	4.93 ± 0.21^a	4.17 ± 0.18^a	5.76 ± 0.12^b	8.27 ± 0.05^a	5.90 ± 0.40^a	$7,13 \pm 0,90^a$	63.84 ± 0.98^b
TRM	9.26 ± 0.52^c	4.68 ± 0.79^a	4.00 ± 0.19^a	6.56 ± 0.20^a	8.24 ± 0.15^a	5.94 ± 0.45^a	$8,09 \pm 0,97^a$	62.49 ± 1.27^b
TRC	13.24 ± 0.32^a	1.60 ± 0.10^b	1.93 ± 0.26^b	3.46 ± 0.42^c	12.66 ± 0.41^a	2.02 ± 0.19^b	2.79 ± 0.06^b	75.54 ± 0.69^a

TRM mechanically treated turmeric residue; *TRC* chemically treated turmeric residue

[a–c] Means with different superscript letters in the same column are statistically different (Tukey test, $p < 0.05$)

*Expressed on moisture basis; S: starch (obtained by difference) = 100 – (Ashes (%) + Lipid (%) + Protein (%) + Food Fiber (%))

Table 7.6 Antioxidant activity and contents of curcuminoids and total phenolic compounds as measured in the turmeric residue, mechanically treated turmeric residue, and chemically treated turmeric residue

Sample	Total phenolic compounds (mg of GAE/g of sample)	Curcumin (mg. L^{-1})	Demethoxycurcumin (mg.L^{-1})	Bisdemethoxycurcumin (mg.L^{-1})	Antioxidant activity (μM Trolox/g of sample)
Turmeric residue	*7.10 ± 0.05*^{a}	*401.75 ± 4.70*^{a}	*118.30 ± 8.10*^{a}	*171.70 ± 15.96*^{a}	*202.74 ± 7.14*^{a}
TRM	*3.42 ± 0.09*^{b}	*121.65 ± 9.76*^{b}	*41.53 ± 8.10*^{b}	*87.70 ± 4.90*^{b}	*79.84 ± 5.94*^{b}
TRC	*1.72 ± 0.20*^{c}	*65.53 ± 8.90*^{c}	*21.43 ± 3.11*^{c}	*46.12 ± 2.54*^{c}	*46.76 ± 4.23*^{c}

TRM mechanically treated turmeric residue; *TRC* chemically treated turmeric residue

[a–c]Different small caps in the same column indicate significant difference between the turmeric materials, as revealed by Tukey test, $p < 0.05$

Calibration curves for: total phenolics (galic acid) = >45 a 650 mg/L ($y = 0.0002\times + 0.2565$, $R^2 = 0.9960$), curcumin = >2.00 to 34.00 mg/L ($y = 7.10^7x$ - 12,546, $R^2 = 0.9998$); demethoxycurcumin = > 0.40 to 60.80 mg/L ($y = 9.10^7$ x + 45,744, $R^2 = 0,9969$), bisdemethoxycurcumin = > 0.30 to 24.20 mg/L ($y = 9.10^7$ x + 110,501, $R^2 = 0.9920$, and antioxidant activity ABTS (Trolox) = > 10 to 2000 μM ($y = -0.0003x + 0.6949$, $R^2 = 0.9964$)

and 10 °C for 15 min (MA Q222RM, Quimis, Brazil). The supernatant was analyzed. Weight loss was determined as the average of individual sample weights measured with a digital balance (Sartorius BL210S, New Jersey, USA). Results are expressed as the percentage loss of initial weight (day 0). pH was measured in 6 g of centrifuged solution diluted in 50 mL of water. Reading was performed in a pH meter (MA522, Marconi, Brazil) with automatic correction values, as a function of temperature. Titratable acidity was expressed as mg of malic acid per 100 g of sample because malic acid is the prevalent organic acid in banana fruit (Cano et al. 1997). Banana pulp firmness was determined with a texture analyzer TA TX Plus (TA Instrument, England). Soluble solids were directly measured with a digital refractometer (HI 96801, Hanna, Brazil); results are expressed as degree Brix (°Brix). Reducing sugars were determined by the DNS (3,5-dinitrosalicylate) method according to Miller (1959). Skin color was visually scored by adopting the standard banana color chart, which ranged from 1 to 8, where 1 = green; 2 = green with yellow stains; 3 = more green than yellow; 4 = more yellow than green; 5 = yellow with green tinge; 6 = completely yellow; 7 = yellow slightly mottled brown; and 8 = yellow samples with big brown areas(Alves 1999).

Figure 7.9 shows the firmness and weight loss evolution of uncoated and coated bananas cv. 'Maçã' stored at 25 °C for 12 days. Weight loss variation was similar for uncoated and coated bananas. It was possible to adjust the data to a simple linear regression model "$y = a + bx$". The coefficients of determination (R^2) were higher than 0.9, which confirmed the excellent fit and indicated that mass loss followed the same mechanism in both samples. However, uncoated bananas had higher weight loss than coated bananas along storage for 12 days, approximately 28% and 23% of the initial mass, respectively.

Both samples showed intensive firmness decay from day 2. In general, the turmeric coating delayed softening only slightly. At the end of the storage period, coated bananas showed higher firmness (1.20 N) than uncoated samples (0.98 N), which represented an overall decrease of 91% and 93%, respectively.

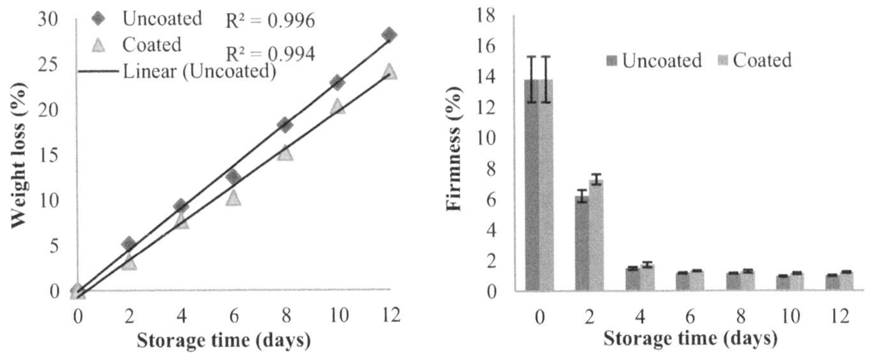

Fig. 7.9 Weight loss and firmness evolution for uncoated and coated bananas stored at 25 °C for 12 days. Each data point is the mean of three replicate samples

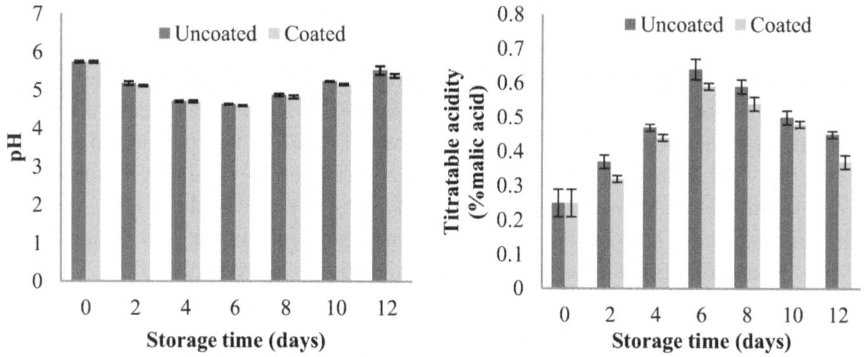

Fig. 7.10 pH and titratable acidity of uncoated and coated bananas stored at 25 °C for 12 days. Each data point is the mean of three replicate samples

Figure 7.10 depicts the pH and titratable acidity of uncoated and coated bananas cv. 'Maçã' stored at 25 °C for 12 days.

The pH value varied slightly along storage: from 5.7 (day 0) to 5.5 (day 12) and from 5.7 (day 0) to 5.4 (day 12) for uncoated and coated bananas, respectively. Higher decay occurred on day 6 for both uncoated and coated bananas (pH ~4.6), which indicated that this was the fruit climacteric peak. Figure 7.10 also shows variations in pulp titratable acidities. The turmeric coating effectively reduced acidity along all the studied period as compared to uncoated bananas. The percentage of titratable acidity increased simultaneously for both uncoated and coated samples up to day 6, with maximum values of 0.64 and 0.59%, respectively. After this period, titratable acidity decreased for both samples (0.45% for uncoated and 0.37% for coated samples).

Figure 7.11 reveals that both uncoated and coated samples had significantly higher content of soluble solids after day 2. Values stabilized at 24.9% and 23.4% for uncoated and coated bananas, respectively. The peak value of soluble solids was on 6 day. On the other hand, coated bananas presented lower content of reducing sugars than uncoated samples along all the storage. This difference was more evident from day 6. At the end of the storage period, the contents of reducing sugars were 29.3 and 21.5 g/L for uncoated and coated bananas, respectively. These contents suggested that the turmeric coating delayed maturation—the presence of reducing sugars indicated that starch hydrolysis and inversion of sucrose into glucose plus fructose occurred, which is directly related to fruit ripening (Kays 1997).

Figure 7.12 illustrates the color and the visual aspect of uncoated and coated bananas during storage determined by visual inspection. Banana skin color varies according to the degree of maturation. During ripening, chlorophyll is degraded, to expose carotenoids, which are the main pigments accounting for the yellow color of banana skin (Prill et al. 2012). Banana peel also contains phenolic compounds, which can be oxidized by polyphenoloxidase, to produce quinine. This raises the levels of macromolecules, to intensify brown pigmentation (Siriphanich 2006).

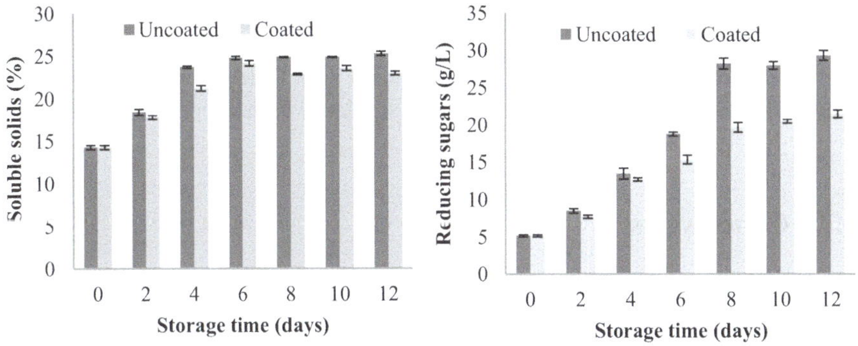

Fig. 7.11 Contents of soluble solids and reducing sugars in coated and uncoated bananas stored at 25 °C for 12 days. Each data point is the mean of three replicate samples

Samples	Day 0	Day 2	Day 4	Day 6	Day 8	Day 10	Day 12
Uncoated							
Skin color (level 1-8)	3	6	-	-	8	8	8
Coated							
Skin color (level 1-8)	3	6	6	6	6	-	-

Fig. 7.12 Visual aspect and skin color of uncoated and coated banana peels stored at 25 °C for 12 days

Figure 7.12 shows a significant difference in the colors of uncoated and coated fruits along storage. Distinctions were noted on day 2 for both samples. In the final storage period (days 8 and 12), uncoated fruits exhibited brown areas, which contrasted with the coated banana color. Development of darker shades and loss of yellow tones could indicate banana over-ripeness and quality loss (Yap et al. 2017). Coated bananas maintained their color up to day 10, whilst uncoated bananas had shelf life of around 6 days in the storage conditions adopted herein. The turmeric coating delayed banana peel color deterioration because it increased CO_2 concentration and decreased O_2 concentration, thereby reducing metabolic rates and inducing slow chlorophyll degradation in the peel. Hence, the turmeric coating created a modified atmosphere around the fruit, to reduce fruit respiration by restricting O_2 access to the tissue and consequently diminishing the enzymatic browning rate

(Jiang and Li 2001). In addition, curcuminoids and phenolic compounds in the turmeric coating acted as enzymatic inhibitors because they contained aromatic acids such as carboxylic, benzoic, and cinnamic acids, which are competitive inhibitors of polyphenoloxidase due to their structural similarity with phenolic substrates.

7.7 Conclusions

Edible films and coatings can are effective barriers for gases, water vapor, and/or aromatic substances, and can act as an active packaging for foodstuff preventing the oxidative and microbial degradation, which consequently extends food product shelf life. Knowledge of transport phenomena within a specific edible film or coating is important when choosing the best packaging for each food type. For this, the physicochemical and structural characteristics of polymer matrixes must be considered in order to use mathematical modeling that accounts for the mass transfer processes that take place in this system. This modeling is necessary for deeper understanding and optimization of edible packaging. Studies in real food cases, pragmatic, everyday systems are essential and should be performed to prove whether a developed package is truly effective. In this chapter was demonstrated that the turmeric residue coating effectively extended the coated banana (*Musa acuminata*) shelf life by 4 days as compared to uncoated bananas. This coating can be considered as an active packaging due to its antioxidant activity.

References

Abreu AS et al (2015) Antimicrobial nanostructured starch based films for packaging. Carbohydr Polym 129:127–134

Adilah ZM et al (2018) Functional and antioxidant properties of protein-based films incorporated with mango kernel extract for active packaging. Food Hydrocoll 74:207–218

Albert S, Mittal GS (2002) Comparative evaluation of edible coatings to reduce fat uptake in a deep-fried cereal product. Food Res Int 35:445–458

Al-Hassan AA, Norziah MH (2012) Starch–gelatin edible films: water vapor permeability and mechanical properties as affected by plasticizers. Food Hydrocoll 26(1):108–117

Álvarez K, Famá L, Gutiérrez TJ (2017) Physicochemical, antimicrobial and mechanical properties of thermoplastic materials based on biopolymers with application in the food industry. In: Masuelli M, Renard D (eds) Advances in physicochemical properties of biopolymers: Part 1. Bentham Science, Sharjah. EE.UU. ISBN: 978–1–68108-454-1. eISBN: 978–1–68108-453-4, pp 358–400. https://doi.org/10.2174/9781681084534117010015

Álvarez K, Alvarez VA, Gutiérrez TJ (2018) Biopolymer composite materials with antimicrobial effects applied to the food industry. In: Thakur VK, Thakur MK (eds) Functional biopolymers. Springer International, Basel, pp 57–96. EE.UU. ISBN: 978-3-319-66416-3. eISBN: 978-3-319-66417-0. https://doi.org/10.1007/978-3-319-66417-0_3

Alves EJ (1999) The banana cultivation: technical, socio-economic and agro-industrial aspects. Embrapa-SPI; Cruz das Almas: Embrapa-CNPMF

Alves VD et al (2010) Barrier properties of biodegradable composite films based on kappa-carrageenan/pectin blends and mica flakes. Carbohydr Polym 79(2):269–276
American Society for Testing and Materials—ASTM (2014) ASTM F1927—Standard test method for determination of oxygen gas transmission rate, permeability and permeance at controlled relative humidity through barrier materials using a coulometric detector. ASTM International, West Conshohocken, PA
Andrade-Mahecha MM et al (2012) Physical–chemical, thermal, and functional properties of achira (*Canna indica L.*) flour and starch from different geographical origin. Starch-Starke 64:348–358
Angellier-Coussy H et al (2011) Influence of processing temperature on the water vapor transport properties of wheat gluten based agromaterials. Ind Crop Prod 33(2):457–461
Araújo-Farro PC et al (2010) Development of films based on quinoa (*Chenopodium quinoa, Willdenow*) starch. Carbohydr Polym 81(4):839–848
Arutselvi R et al (2012) Phytochemical screening and comparative study of anti microbial activity of leaves and rhizomes of turmeric varieties. Asian J Plant Sci Res 2(2):212–219
Arvanitoyannis S, Kassaveti A (2009) HACCP and ISO 22000—a comparison of the two systems. In: Arvanitoyannis IS (ed) HACCP and ISO 22000: application to foods of animal origin. Wiley-Blackwell, Oxford, pp 3–45
Assis RQ et al (2017) Active biodegradable cassava starch films incorporated lycopene nanocapsules. Ind Crop Prod 109:818–827
Aulin C et al (2010) Oxygen and oil barrier properties of microfibrillated cellulose films and coatings. Cellulose 17(3):559–574
Ayranci E, Tunc S (2003) A method for the measurement of the oxygen permeability and the development of edible films to reduce the rate of oxidative reactions in fresh foods. Food Chem 80(3):423–431
Bai J et al (2002) Alternatives to shellac coatings provide comparable gloss, internal gas modification, and quality for 'Delicious' apple fruit. Hort Sci 37(3):559–563
Balaguer MP et al (2013) Antifungal properties of gliadin films incorporating cinnamaldehyde and application in active food packaging of bread and cheese spread foodstuffs. Int J Food Microbiol 166(3):369–377
Baldwin EA et al (2011) Edible coatings and films to improve food quality. CRC, Boca Raton, FL
Barry BW, Meyer MC (1979) The rheological properties of carbopol gels II. Oscillatory properties of carbopol gels. Int J Pharm 2(1):27–40
Behera S et al (2014) Importance of chemical pretreatment for bioconversion of lignocellulosic biomass. Renew Sust Energ Rev 36:91–106
Berens AR, Hopfenberg HB (1978) Diffusion and relaxation in glassy polymer powders. Separation of diffusion and relaxation parameters. Polymer 19(5):489–496
Bourlieu C et al (2009) Edible moisture barriers: how to assess of their potential and limits in food products shelf-life extension? Crit Rev Food Sci Nutr 49:474–499
Bourtoom T et al (2006) Effect of plasticizer type and concentration on the properties of edible films from water-soluble fish proteins in surimi wash-water. Food Sci Technol Int 12(2):119–126
Bracone M, Merino D, González J, Alvarez VA, Gutiérrez TJ (2016) Nanopackaging from natural fillers and biopolymers for the development of active and intelligent films. In: Ikram S, Ahmed S (eds) Natural polymers: derivatives, blends and composites. Nova Science, New York, pp 119–155 EE.UU. ISBN: 978-1-63485-831-1
Buonocore GG et al (2003) Modeling the lysozyme release kinetics from antimicrobial films intended for food packaging applications. J Food Sci 68(4):1365–1370
Buonocore GG et al (2003b) A general approach to describe the antimicrobial agent release from highly swell able films intended for food packaging applications. J Control Release 90:97–107
Buonocore GG et al (2005) Mono-and multilayer active films containing lysozyme as antimicrobial agent. Innov Food Sci Emerg Technol 6(4):459–464
Butler BL et al (1996) Mechanical and barrier properties of edible chitosan films as affected by composition and storage. J Food Sci 61:953–955

Cagri A et al (2004) Antimicrobial edible films and coatings. J Food Prot 67(4):833–848

Cano MP et al (1997) Differences among Spanish and Latin-American banana cultivars: morphological, chemical and sensory characteristics. Food Chem 59:411–419

Cazón P et al (2017) Polysaccharide-based films and coatings for food packaging: a review. Food Hydrocoll 68:136–148

Cha DS, Chinnan MS (2004) Biopolymer-based antimicrobial packaging: a review. Crit Rev Food Sci Nutr 44(4):223–237

Chi-Zhang YD et al (2004) Effective control of *Listeria monocytogenes* by combination of nisin formulated and slowly released into a broth system. Int J Food Microbiol 90:15–22

Cosgrove J (2008) Emerging edible films: dissolving strips have made minor supplement inroads, but advancing technologies point to progress. Available online at: http://www.nutraceuticals-world.com/contents/view_online-exclusives/2008-01-01/emerging708 edible-films/. Accessed 10 Jan 2018

Crank J (1955) The mathematics of diffusion. Clarendon Press, Oxford, pp 56–60

Cremasco, MA (1998) Fundamentos de transferência de massa. Editora da UNICAMP

Cutter CN, Sumner SS (2002) Application of edible coatings on muscle foods. In: Gennedios A (ed) Protein-based films and coatings. CRC, Boca Raton, FL, pp 467–484

Dashipour A et al (2014) Physical, antioxidant and antimicrobial characteristics of carboxymethylcellulose edible film cooperated with clove essential oil. Zahedan J Res Med Sci 16:34–42

Daudt RM et al (2017) Development of edible films based on Brazilian pine seed (*Araucaria angustifolia*) flour reinforced with husk powder. Food Hydrocoll 71:60–67

Davoodi M et al (2017) Preparation and characterization of potato starch-thymol dispersion and film as potential antioxidant and antibacterial materials. Int J Biol Macromol 104:173–179

Debeaufort FJ et al (1998) Edible films and coatings: tomorrow's packaging: a review. Crit Rev Food Sci Nutr 38:299–313

Debeaufort F et al (2000) Lipid hydrophobicity and physical state effects on the properties of bilayer edible films. J Membr Sci 180(1):47–55

Despond S et al (2005) Barrier properties of paper-chitosan and paper-chitosan-carnauba wax films. J Appl Polym Sci 98:704–710

Devlieghere F et al (2004) Chitosan: antimicrobial activity, interactions with food components and applicability as a coating on fruit and vegetables. Food Microbiol 21(6):703–714

Dias AB et al (2010) Biodegradable films based on rice starch and rice flour. J Cereal Sci 51(2):213–219

Dole P et al (2004) Gas transport properties of starch based films. Carbohydr Polym 58:335–343

Duan J et al (2010) Quality enhancement in fresh and frozen lingcod (*Ophiodon elongates*) fillets by employment of fish oil incorporated chitosan coatings. Food Chem 119(2):524–532

Dutta PK et al (2009) Perspectives for chitosan based antimicrobial films in food applications. Food Chem 114(4):1173–1182

Ehivet FE et al (2011) Characterization and antimicrobial activity of Sweetpotato starch-based edible film containing Origanum (Thymus capitatus) oil. J Food Sci 76(1):C178–C184

Embuscado ME, Huber KC (2009) Edible films and coatings for food applications. Springer, New York, NY, pp 213–214

Emiroğlu ZK et al (2010) Antimicrobial activity of soy edible films incorporated with thyme and oregano essential oils on fresh ground beef patties. Meat Sci 86(2):283–288

Faisant N et al (2002) PLGA-based microparticles: elucidation of mechanisms and a new, simple mathematical model quantifying drug release. Eur J Pharm Sci 15(4):355–366

Flores S et al (2007) Mass transport properties of tapioca-based active edible films. J Food Eng 81(3):580–586

Galdeano MC et al (2009) Effects of plasticizers on the properties of oat starch films. Mater Sci Eng C 9:532–538

Ganiari S et al (2017) Edible and active films and coatings as carriers of natural antioxidants for lipid food. Trends Food Sci Technol 68:70–82

Garcia MA, Zaritzky N (2017) Transport phenomena in films and coatings including their mathematical modeling. In: Montero García MP, Gómez-Guillén MC, López-Caballero ME, Barbosa-Canóvas GV (eds) Edible films and coatings. CRC, Boca Raton, pp 25–53

Garcia MA et al (2002) Edible coatings from cellulose derivatives to reduce oil uptake in fried products. Innov Food Sci Emerg Technol 3:391–397

Gaudin S et al (2000) Antiplasticisation and oxygen permeability of starch–sorbitol films. Carbohydr Polym 43:33–37

Gennadios A, Weller CL (1990) Edible films and coatings from wheat and corn proteins. Food Technol 44(10):63–69

Gennadios A et al (1994) Edible coatings and films based on proteins. In: Krochta JM, Baldwin EA, Nisperos-Carriedo M (eds) Edible coatings and films to improve food quality. Technomic, Lancaster, pp 210–278

Gómez-Estaca J et al (2010) Biodegradable gelatin–chitosan films incorporated with essential oils as antimicrobial agents for fish preservation. Food Microbiol 27(7):889–896

Gómez-Estaca J et al (2014) Advances in antioxidant active food packaging. Trends Food Sci Technol 35(1):42–51

Gontard N et al (1996) Influence of relative humidity and film composition on oxygen and carbon dioxide permeability's edible films. J Agric Food Chem 44:1064–1069

Grinberg VY, Tolstoguzov VB (1997) Thermodynamic incompatibility of proteins and polysaccharides in solutions. Food Hydrocoll 11:145–158

Guilbert S (1986) Technology and application of edible protective films. In: Mathlouthi M (ed) Food packaging and preservation: theory and practice. Elsevier, Applied Science, London, pp 371–394

Guo Z et al (2008) The influence of molecular weight of quaternized chitosan on antifungal activity. Carbohydr Polym 71(4):694–697

Gutiérrez TJ (2017a) Surface and nutraceutical properties of edible films made from starchy sources with and without added blackberry pulp. Carbohydr Polym 165:169–179. https://doi.org/10.1016/j.carbpol.2017.02.016

Gutiérrez TJ (2017b) Chitosan applications for the food industry. In: Ahmed S, Ikram S (eds) Chitosan: derivatives, composites and applications. Wiley-Scrivener, Beverly, MA, pp 183–232. EE.UU. ISBN: 978-1-119-36350-7. https://doi.org/10.1002/9781119364849.ch8

Gutiérrez TJ (2018) Active and intelligent films made from starchy sources/blackberry pulp. J Polym Environ 15:445–448. https://doi.org/10.1007/s10924-017-1134-y

Gutiérrez TJ, Álvarez K (2017) Transport phenomena in biodegradable and edible films. In: Masuelli M (ed) Biopackaging. CRC, Boca Raton, FL, pp 58–89

Gutiérrez TJ, Alvarez VA (2017a) Cellulosic materials as natural fillers in starch-containing matrix-based films: a review. Polym Bull 74(6):2401–2430. https://doi.org/10.1007/s00289-016-1814-0

Gutiérrez TJ, Alvarez VA (2017b) Eco-friendly films prepared from plantain flour/PCL blends under reactive extrusion conditions using zirconium octanoate as a catalyst. Carbohydr Polym 178:260–269. https://doi.org/10.1016/j.carbpol.2017.09.026

Gutiérrez TJ, Alvarez VA (2017c) Films made by blending poly(ε-caprolactone) with starch and flour from Sagu rhizome grown at the Venezuelan amazons. J Polym Environ 25(3):701–716. https://doi.org/10.1007/s10924-016-0861-9

Gutiérrez MQ et al (2012) Carboxymethylcellulose-montmorillonite nanocomposite films activated with murta (*Ugni molinae Turcz*) leaves extract. Carbohydr Polym 87:1495–1502

Gutiérrez TJ, Morales NJ, Pérez E, Tapia MS, Famá L (2015) Physico-chemical study of edible films based on native and phosphating cush-cush yam and cassava starches. Food Packaging Shelf Life 3:1–8. https://doi.org/10.1016/j.fpsl.2014.09.002

Gutiérrez TJ, Tapia MS, Pérez E, Famá L (2015a) Structural and mechanical properties of native and modified cush-cush yam and cassava starch edible films. Food Hydrocoll 45:211–217. https://doi.org/10.1016/j.foodhyd.2014.11.017

Gutiérrez TJ, Tapia MS, Pérez E, Famá L (2015b) Edible films based on native and phosphated 80:20 waxy:normal corn starch. Starch Stärke 67(1–2):90–97. https://doi.org/10.1002/star.201400164

Gutiérrez TJ, Guzmán R, Medina Jaramillo C, Famá L (2016a) Effect of beet flour on films made from biological macromolecules: native and modified plantain flour. Int J Biol Macromol 82:395–403. https://doi.org/10.1016/j.ijbiomac.2015.10.020

Gutiérrez TJ, Suniaga J, Monsalve A, García NL (2016b) Influence of beet flour on the relationship surface-properties of edible and intelligent films made from native and modified plantain flour. Food Hydrocoll 54:234–244. https://doi.org/10.1016/j.foodhyd.2015.10.012

Gutiérrez TJ, González Seligra P, Medina Jaramillo C, Famá L, Goyanes S (2017) Effect of filler properties on the antioxidant response of thermoplastic starch composites. In: Thakur VK, Thakur MK, Kessler MR (eds) Handbook of composites from renewable materials. Wiley-Scrivener, Beverly, MA, pp 337–370. EE.UU. ISBN: 978-1-119-22362-7. https://doi.org/10.1002/9781119441632.ch14

Han JH, Scanlon MG (2014) Mass transfer of gas and solute through packaging materials. In: Han JH (ed) Innovations in food packaging, 2nd edn. Academic Press, London, pp 37–49

Han SM et al (2006) Determining hardness of thin films in elastically mismatched film-on-substrate systems using nanoindentation. Acta Mater 54(6):1571–1581

He L et al (2011) Modification of collagen with a natural cross-linker, procyanidin. Int J Biol Macromol 48(2):354–359

Hernández-Izquierdo VM, Krochta JM (2008) Thermoplastic processing of proteins for film formation—a review. J Food Sci 73(2)

Hernandez-Muñoz P et al (2002) Simple method for the selection of the appropriate food simulant for the evaluation of a specific food/packaging interaction. Food Addit Contam 19(S1):192–200

Hong SI, Krochta JM (2006) Oxygen barrier performance of whey protein coated plastic films as affectes by temperature, relative humidity, base film and protein type. J Food Eng 77:736–745

Hosokawa J et al (1990) Biodegradable film derived from chitosan and homogeneized cellulose. Ind Eng Chem Res 29:800–805

Ibarz A, Barbosa-Cánovas GV (2002) Unit operations in food engineering. CRC Press, Boca Raton 889 p

Jain RA (2000) The manufacturing techniques of various drug loaded biodegradable poly(lactide-co-glycolide) (PLGA) devices. Biomaterials 21(23):2475–2490

Jiang Y, Li Y (2001) Effects of chitosan coating on postharvest life and quality of longan fruit. Food Chem 73:139–143

Jiang T et al (2012) Novel disease-modifying therapies for Alzheimer's disease. J Alzheimers Dis 31:475–492

Jin T, Zhang H (2008) Biodegradable polylactic acid polymer with nisin for use in antimicrobial food packaging. J Food Sci 73(3):M127–M134

Johansson F, Leufven A (1995) Food packaging polymer as barrier against aroma vapor and oxygen in fat or humid environments. In: Ackermann P, Tagerstand M, Ohesson T (eds) Food and packaging materials: chemical interactions. The Royal Society of Chemistry, Cambrige

Jongjareonrak A et al (2006) Characterization of edible films from skin gelatin of brown stripe red snapper and big eye snapper. Food Hydrocoll 20:492–501

Kamil JY et al (2002) Antioxidative activity of chitosans of different viscosity in cooked comminuted flesh of herring (*Clupea harengus*). Food Chem 79(1):69–77

Kays SJ (1997) Stress in harvested products. In: Kays SJ (ed) Postharvest physiology of perishable plant products. Exon Press, Athens, GA, pp 335–407

Kester JJ, Fennema OR (1986) Edible films and coatings: a review. Food Technol 40(12):47–59

Kofinas P et al (1994) Gas permeability of polyethylene/poly (ethylene-propylene) semicrystalline diblock copolymers. Polymer 35:1229–1235

Kulp K (2011) Batters and breadings in food processing. Academic Press, New York

Lacoste A et al (2005) Advancing controlled release packaging through smart blending. Food Packaging Technol Sci 18:77–87

Langer R, Peppas NA (1983) Chemical and physical structure of polymers as carriers for controlled release of bioactive agents: a review. Macromol Chem Phys C23:61–126

Lee KY, Mooney DJ (2001) Hydrogels for tissue engineering. Chem Rev 101(7):1869–1880

Lee SY et al (2002) Whey-protein-coated peanuts assessed by sensory evaluation and static headspace gas chromatography. J Food Sci 67:1212–1218

Leelaphiwat P et al (2017) Effects of packaging materials on the aroma stability of Thai 'tom yam' seasoning powder as determined by descriptive sensory analysis and gas chromatography–mass spectrometry. J Sci Food Agric 97:1854–1860

Liang J et al (2017) Encapsulation of epigallocatechin gallate in zein/chitosan nanoparticles for controlled applications in food systems. Food Chem 231:19–24
Liu F et al (2010) Optimizing color and lipid stability of beef patties with a mixture design incorporating with tea catechins, carnosine and atocopherol. J Food Eng 98:170–177
Liu J et al (2017a) Effect of protocatechuic acid incorporation on the physical, mechanical, structural and antioxidant properties of chitosan film. Food Hydrocoll 73:90–100
Liu J et al (2017b) Physical, mechanical and antioxidant properties of chitosan films grafted with different hydroxybenzoic acids. Food Hydrocoll 71:176–186
López-Carballo G et al (2012) Active antimicrobial food and beverage packaging. In: Emerging food packaging technologies, pp 27–54
Luchese CL et al (2017) Starch content affects physicochemical properties of corn and cassava starch-based films. Ind Crop Prod 109:619–626
Ma X et al (2017) Hydrophilic modification of cellulose nanocrystals improves the physicochemical properties of cassava starch-based nanocomposite films. LWT Food Sci Technol 86:318–326
Malhotra B et al (2015) Antimicrobial food packaging: potential and pitfalls. Front Microbiol 6:611
Mallikarjunan P et al (1997) Edible coatings for deep-fat frying of starchy products. Lebensm Wiss u-Tecnol 30:709–714
Maniglia BC et al (2014) Development of bioactive edible film from turmeric dye solvent extraction residue. LWT Food Sci Technol 56(2):269–277
Maniglia BC et al (2015) Turmeric dye extraction residue for use in bioactive film production: optimization of turmeric film plasticized with glycerol. LWT Food Sci Technol 64(2):1187–1195
Maniglia BC et al (2017) Bioactive films based on babassu mesocarp flour and starch. Food Hydrocoll 70:383–391
Martelli MR et al (2005) Efeito de Coberturas a Base de Gelatina na Fritura de Nuggets In: 5° Congreso IberoAmericano de Ingeniería de Alimentos Puerto Vallarta. Anales del 5° Congreso IberoAmericano de Ingeniería de Alimentos
Martelli MR et al (2008) Reduction of oil uptake in deep fat fried chicken nuggets using edible coatings based on cassava starch and methylcellulose. Ital J Food Sci 20(1):111–117
Martelli-Tosi M et al (2017) Chemical treatment and characterization of soybean straw and soybean protein isolate/straw composite films. Carbohydr Polym 157:512–520
Martinez DST, Alves OL (2013) Interação de nanomateriais com biossistemas e a nanotoxicologia: na direção de uma regulamentação. Ciência e Cultura 65(3):32–36
Mastromatteo M et al (2010) Advances in controlled release devices for food packaging applications. Trends Food Sci Technol 21:591–598
Mate JI, Krochta JM (1996) Comparison of oxygen and water vapor permeabilities of whey protein isolate and -lactoglobulin edible films. J Agric Food Chem 44(10):3001–3004
Mchugh TH, Krochta JM (1994a) Milk-protein based films and coatings. Food Technol 48(1):97–103
Mchugh TH, Krochta JM (1994b) Sorbitol vs glycerol whey protein edible films: integrated oxygen permeability and tensile property evaluation. J Agric Food Chem 42:841–845
Mchugh TH, Krochta JM (1994c) Water vapor permeability properties of edible whey protein-lipid emulsion films. J Am Oil Chem Soc 71(3):307–312
Mchugh TH et al (1993) Hydrophilic edible films: modified procedure for water vapor permeability and explanation of thickness effects. J Food Sci 58:899–903
Menon VP, Sudheer AR (2007) Antioxidant and anti-inflammatory properties of curcumin. Adv Exp Med Biol 595:105–125
Miller GL (1959) Use of dinitrosalicylic acid reagent for determination of reducing sugar. Anal Chem 31(3):426–428
Miller KS, Krochta JM (1997) Oxygen and aroma barrier properties of edible films: a review. Trends Food Sci Technol 8:228–237
Muñoz-Bonilla A, Fernández-García M (2012) Polymeric materials with antimicrobial activity. Prog Polym Sci 37(2):281–339

No HK et al (2007) Applications of chitosan for improvement of quality and shelf life of foods: a review. J Food Sci 72(5):R87–R100
Oms-Oliu GO et al (2008) Edible coatings with antibrowning agents to maintain sensory quality and antioxidant properties of fresh-cut pears. Postharvest Biol Technol 50(1):87–94
Ouattara B et al (2000) Inhibition of surface spoilage bacteria in processed meats by application of antimicrobial films prepared with chitosan. Int J Food Microbiol 62:139–148
Ozdemir M, Floros JD (2004) Active food packaging technologies. Crit Rev Food Sci Nutr 44(3):185–193
Paramasivam M et al (2009) High-performance thin layer chromatographic method for quantitative determination of curcuminoids in Curcuma longa germplasm. Food Chem 113:640–644
Pelissari FM et al (2013) Comparative study on the properties of flour and starch films of plantain bananas (*Musa paradisiaca*). Food Hydrocoll 30(2):681–690
Peppas N (1984) In: Anderson JM, Kim SW (eds) Recent advances in drug delivery systems. Plenum Press, New York, pp 279–298
Pereira VA Jr et al (2015) Active chitosan/PVA films with anthocyanins from brassica oleraceae (red cabbage) as time–temperature indicators for application in intelligent food packaging. Food Hydrocoll 43:180–188
Piñeros-Hernandez D et al (2017) Edible cassava starch films carrying rosemary antioxidant extracts for potential use as active food packaging. Food Hydrocoll 63:488–495
Pinheiro AC et al (2013) Transport mechanism of macromolecules on hydrophilic bio-polymeric matrices—diffusion of protein-based compounds from chitosan films. J Food Eng 116:633–638
Pinthus EJ et al (1993) Criterion for oil uptake during deep fat frying. J Food Sci 58(1):204–205
Polakovic M et al (1999) Lidocaine loaded biodegradable nanospheres: II. Modelling of drug release. J Control Release 60(2–3):169–177
Prill MAS et al (2012) Atmosfera modificada e controle de etileno para bananas 'Prata-Anã' cultivadas na Amazônia Setentrional Brasileira. Rev Bras Frutic 34(4):990–1003
Quintavalla S, Vicini L (2002) Antimicrobial food packaging in meat industry. Meat Sci 62(3):373–380
Reynier A et al (2002) Migration of additives from polymers into food simulants: numerical solution of a mathematical model taking into account food and polymer interactions. Food Addit Contam 1(1):89–102
Rivero S et al (2013) Controlled delivery of propionic acid from chitosan films for pastry dough conservation. J Food Eng 116(2):524–531
Roca E et al (2008) Effective moisture diffusivity modelling versus food structure and hygroscopicity. Food Chem 106:1428–1437
Rogers CE (1985) Permeation of gases and vapours in polymers. Chapter 2. In: Comyn J (ed) Polymer permeability. Elsevier Applied Science, London, pp 11–73
Rojas-Graü MA et al (2007) Alginate and gellan-based edible coatings as carriers of antibrowning agents applied on fresh-cut Fuji apples. Food Hydrocoll 21(1):118–127
Romanazzi G et al (2002) Effects of pre-and postharvest chitosan treatments to control storage grey mold of table grapes. J Food Sci 67(5):1862–1867
Romani VP et al (2017) Active and sustainable materials from rice starch, fish protein and oregano essential oil for food packaging. Ind Crops Prod 97:268–274
Ruiz-Navajas Y et al (2013) In vitro antibacterial and antioxidant properties of chitosan edible films incorporated with *Thymus moroderi* or *Thymus piperella* essential oils. Food Control 30(2):386–392
Saguy IS et al (1998) Oil uptake in deep fat frying: review. Ocl-Oleagineux corps Gras. Lipids 5(1):30–35
Salgado PR et al (2015) Edible films and coatings containing bioactives. Curr Opin Food Sci 5:86–92
Salleh E et al (2007) Preparation, characterization and antimicrobial analysis of antimicrobial starch-based film incorporated with chitosan and lauric acid. Asian Chitin J 3:55–68
Sayanjali S et al (2011) Evaluation of antimicrobial and physical properties of edible film based on carboxymethyl cellulose containing potassium sorbate on some mycotoxigenic aspergillus species in fresh pistachios. LWT Food Sci Technol 44(4):1133–1138

Sharma L, Singh C (2016) Sesame protein based edible films: development and characterization. Food Hydrocoll 61:139–147
Shellhammer TH, Krochta JM (1997) Water vapor barrier and rheological properties of simulated and industrial milk fat fractions. Trans ASAE 40:1119–1127
Shen XL et al (2010) Antimicrobial and physical properties of sweet potato starch films incorporated with potassium sorbate or chitosan. Food Hydrocoll 24:285–290
Shojaee-Aliabadi S et al (2013) Characterization of antioxidant-antimicrobial κ-carrageenan films containing *Satureja hortensis* essential oil. Int J Biol Macromol 52:116–124
Silva-Weiss A et al (2013) Natural additives in bioactive edible films and coatings: functionality and applications in foods. Food Eng Rev 5(4):200–216
Siriphanich J (2006) Physiology and postharvest Technology in vegetable and fruit. Office of Extension and Training Kamphaeng Saen, Nakhon Pathom, Thailand
Skurtys O et al. (2011) Food hydrocolloid edible films and coatings, Series: food science and technology
Song X et al (2018) Effect of essential oil and surfactant on the physical and antimicrobial properties of corn and wheat starch films. Int J Biol Macromol 107:1302–1309
Sothornvit R, Krochta JM (2000) Oxygen permeability and mechanical properties of films from hydrolyzed whey protein. J Agric Food Chem 48(9):3913–3916
Sothornvit R, Pitak N (2007) Oxygen permeability and mechanical properties of banana films. Food Res Int 40(3):365–370
Souza AC et al (2013) Cassava starch composite films incorporated with cinnamon essential oil: antimicrobial activity, microstructure, mechanical and barrier properties. LWT Food Sci Technol 54(2):346–352
Suárez G, Gutiérrez TJ (2017) Recent advances in the development of biodegadable films and foams from cassava starch. In: Klein C (ed) Handbook on Cassava: production, potential uses and recent advances. Nova Science, New York, pp 297–312 EE.UU. ISBN: 978-1-53610-307-6
Suppakul P et al (2003) Active packaging technologies with an emphasis on antimicrobial packaging and its applications. J Food Sci 68(2):408–420
Syverud K, Stenius P (2009) Strength and barrier properties of MFC films. Cellulose 16(1):75–85
Tapia-Blácido DR (2006) Films based on derivates of amaranth for use in foods PhD thesis, Unicamp, Brazil
Tapia-Blácido DR et al (2007) Contribution of the starch, protein, and lipid fractions to the physical, thermal, and structural properties of amaranth (*Amaranthus caudatus*) flour films. J Food Sci 72(5):E293–E300
Tuil R et al (2000) Converting biobased polymers into food packagings. Proceedings of the Food Biopack Conference, Copenhagen 27–29 August 2000, Copenhagen, Denmark, pp 28–30
Ustunol Z (2009) Edible films and coatings for meat and poultry. In: Edible films and coatings for food applications. Springer, New York, NY, pp 245–268
Valero M, Ginger MJ (2006) Effects of antimicrobial components of essential oils on growth of Bacillus cereus INRA L2104 in and the sensory qualities of carrot broth. Int J Food Microbiol 106(1):90–94
Valero M, Salmeron MC (2003) Antibacterial activity of 11 essential oils against Bacillus cereus in tyndallized carrot broth. Int J Food Microbiol 85(1–2):73–81
Van Long NN et al (2016) Active packaging with antifungal activities. Int J Food Microbiol 220:73–90
Vargas M et al (2008) Recent advances in edible coatings for fresh and minimally processed fruits. Crit Rev Food Sci Nutr 48(6):496–511
Vargas CG et al (2017) Comparative study on the properties of films based on red rice (*Oryza glaberrima*) flour and starch. Food Hydrocoll 65:96–106
Vermeiren L et al (2002) Effectiveness of some recent antimicrobial packaging concepts. Food Addit Contam 19(S1):163–171
Voilley A et al (2011) Transfer of water and active molecules at the interfaces in complex food systems: theoretical and practical aspects. Proc Food Sci 1:879–885

Wambura P et al (2011) Effects of sonication and edible coating containing rosemary and tea extracts on reduction of peanut lipid oxidative rancidity. Food Bioprocess Technol 4(1):107–115
Xiong L (1997) Structure-functionality relationships of muscle proteins. In: Damodaran S, Paraf A (eds) Food proteins and their applications. Marcel Dekker, New York, pp 341–392
Yang L, Paulson A (2000) Mechanical and water vapor barrier properties of edible gellan films. Food Res Int 33(7):563–570
Yap M et al (2017) The effects of banana ripeness on quality indices for puree production. LWT Food Sci Technol 80:10–18

Chapter 8
Antimicrobial Films and Coatings Incorporated with Food Preservatives of Microbial Origin

Alex López-Córdoba

Abstract Food quality and safety constitute main issues for the food industry. However, in spite of several efforts that have been carried out, food preservation is still challenging. Synthetic substances have been widely applied in the food industry as preservatives, however some of them have been associated with harmful effects to human health. This fact has prompted the quest of new methods for food preservation using natural and safer agents. Biopreservatives such as lactic acid bacteria and their bacteriocins have been widely recognized as potent natural compounds able to inhibit or prevent the growth of spoilage and pathogenic microorganisms in food systems. Therefore, the incorporation of these biopreservatives into polymeric films and coatings constitutes a promising strategy to develop new antimicrobial packaging materials to ensure food safety and extend the food shelf-life. This chapter presents the main developments regarding active packaging intended for food biopreservation. Different strategies for the incorporation of biopreservatives into food packaging materials are analyzed. Finally, the challenges against the large-scale production and successful commercialization of these materials containing biopreservatives are also addressed.

Keywords Antimicrobial food packaging · Bacteriocin · Biopreservation · Lactic acid bacteria

8.1 Introduction

Global initiatives on food loss and waste reduction have gained growing attention in the last years. According to Food and Agriculture Organization of the United Nations (FAO), roughly one third of the food produced in the world for human consumption every year (i.e., approximately 1.3 billion tonnes) gets lost or wasted (FAO 2011). Food losses and waste amounts to roughly US$ 680 billion in industrialized countries and US$ 310 billion in developing countries (FAO 2011). In

A. López-Córdoba (✉)
Facultad Seccional Duitama, Escuela de Administración de Empresas Agropecuarias, Universidad Pedagógica y Tecnológica de Colombia, Boyacá, Colombia

T. J. Gutiérrez (ed.), *Polymers for Food Applications*,
https://doi.org/10.1007/978-3-319-94625-2_8

addition, food losses represent a waste of resources used in production such as land, water, energy and inputs. Producing food that will not be consumed leads to unnecessary CO_2 emissions in addition to loss of economic value of the food produced.

Active packaging's constitute an actual choice for reducing food waste along the value chain (FAO 2014). Food packaging can be termed active when it performs a certain desirable role other than providing an inert barrier to external conditions (Anu Bhushani and Anandharamakrishnan 2014). Active packaging systems could include antioxidants, antimicrobial agents or oxygen scavengers (Chang-Bravo et al. 2014; López-Córdoba et al. 2017; Piñeros-Hernandez et al. 2017; Gutiérrez 2017, 2018). Other than these, moisture absorbing, flavor or odor absorbing active packaging systems are also being developed for food applications (Anu Bhushani and Anandharamakrishnan 2014). In particular, the fabrication of antimicrobial packaging's has received increasing importance in the past years because they offer slow and continuous migration of antimicrobial agents from packaging material to food surfaces, increasing their shelf-life (Blanco Massani et al. 2014a, b; Garcia et al. 2012; Woraprayote et al. 2018). Packaging's containing antimicrobial agents from natural sources are preferred, instead synthetic additives, since the latest alternatives have been associated with harmful effects on human health (Gutiérrez and Álvarez 2016, 2018a). In this context, lactic acid bacteria (LAB) have been proposed as natural preservatives to inhibit or prevent the growth of spoilage and pathogenic microorganisms in food systems and, consequently, to enhance their safety and prolong their shelf life (Aloui and Khwaldia 2016). LAB have ability to produce various types of antimicrobial compounds, the most important being bacteriocins. Bacteriocins and bacteriocin-producing cultures have the potential to increase the shelf-life of foods and contribute towards decreasing the incidence of foodborne diseases.

This chapter provides an overview of the current applications of lactic acid bacteria and their bacteriocins in food packaging (film and coatings) and highlight useful applications for these materials to extend shelf life of different food products such as meat, fish, dairy fruit, vegetables or other food products.

8.2 Lactic Acid Bacteria and Their Bacteriocins as Food Biopreservatives

Lactic acid bacteria (LAB) comprise a group of Gram-positive bacteria, nonsporulating, cocci or rods, and catalase-negative organisms with high tolerance for low pH (Calo-Mata et al. 2008). LAB are characterized by the production of lactic acid as the major end product during the fermentation of carbohydrates, lowering the pH of the food and also directly inhibiting the growth of many microorganisms.

LAB are categorized into homofermentative and heterofermentative microorganisms, based on the products of the fermented carbohydrates. Homofermentative

LAB degrade hexoses to lactate, whereas heterofermentative LAB degrade hexoses to lactate and additional products such as acetate, ethanol, CO_2, formate, or succinate (Calo-Mata et al. 2008). LAB are widely used as starter cultures in the food industry for the production of fermented foods, including dairy (e.g. yogurt and cheese), meat (e.g. sausages), fish, cereals (e.g. bread and beverages such as beer), fruit (malolactic fermentation processes in wine production), and vegetables (e.g. sauerkraut, kimchi and silage) (Chelule et al. 2010). Most LAB are considered GRAS (generally recognized as safe) by the US Food and Drug Administration. As probiotics, LAB are increasingly being used owing to their contribution to the healthy microflora of human mucosal surfaces (Mokoena and Paul 2017; Porto et al. 2017).

The LAB group is currently classified in the phylum *Firmicutes*, class *Bacilli*, and order *Lactobacillales*, Families *Aerococcaceae*, *Carnobacteriaceae*, *Enterococcaceae*, *Lactobacillaceae*, *Leuconostocaceae*, and *Streptococcaceae*. LAB are classified based on cellular morphology, mode of glucose fermentation, range of growth temperature, and sugar utilization patterns. LAB genera include *Lactobacillus*, *Lactococcus*, *Leuconostoc*, *Pediococcus*, *Streptococcus*, *Aerococcus*, *Alloiococcus*, *Carnobacterium*, *Dolosigranulum*, *Enterococcus*, *Oenococcus*, *Tetragenococcus*, *Vagococcus* and *Weissella* (Mokoena and Paul 2017). Among LAB, *Lactobacillus* is the genus including a high number of GRAS species and many strains are among the most important bacteria in food microbiology and human nutrition, due to their contribution to fermented food production or their use as probiotics (Salvetti et al. 2012).

Biopreservation refers to the use of natural or controlled microbiota or its antibacterial metabolites to extend the shelf life and enhance the safety of foods (Hugas 1998; Stiles 1996). This strategy can help to reduce the addition of synthetic preservatives as well as the intensity of heat treatments, resulting in foods which are more naturally preserved and richer in organoleptic and nutritional properties (Gálvez et al. 2007; García et al. 2010).

LAB have widely recognized as biopreservatives because can protect foods from microbial spoilage by the lowering their pH, by competitive growth against spoilage and pathogenic bacteria and by the production of antagonistic metabolic products such as organic acids (e.g., lactic acid), diacetyl, fatty acids, CO_2, peroxide, and bacteriocins (Calo-Mata et al. 2008).

Bacteriocins are ribosomally-synthesized peptides or proteins with antimicrobial activity, produced by many Gram-positive and Gram-negative microorganisms (Abbasiliasi et al. 2017; Woraprayote et al. 2016). However, bacteriocins produced by Gram-positive microorganisms such as LAB are more frequently used in the food industry (Cotter et al. 2005; Gálvez et al. 2007; García et al. 2010).

The bacteriocins produced by LAB offer several desirable properties that make them suitable for food preservation: (i) are generally recognized as safe substances (GRAS), (ii) are not active and nontoxic on eukaryotic cells, (iii) become inactivated by digestive proteases, having little influence on the gut microbiota, (iv) are usually pH and heat-tolerant, (v) they have a relatively broad antimicrobial spectrum, against many food-borne pathogenic and spoilage bacteria, (vi) they show a

Table 8.1 Classification of bacteriocins from lactic acid bacteria

Class	Properties	Producer strains	Examples
I	Small peptides (<5 kDa) that possess the eponymous lanthionine or β-methyllanthionine residues	*Lactobacillus lactis; Streptococcus mutans*	Nisin, mersacidin, lacticin 481; lacticin 3147; cytolysin
IIa	Small (<10 kDa) heat-stable peptides, which do not undergo extensive posttranslational modification	*Lactobacillus sakei; enterococcus faecium*	Pediocin PA1, leucocin A
IIb	Consist of two different individual peptide molecules that require equal peptide ratio of each peptide to exert its optimal antimicrobial activity	*Lactobacillus plantarum; Lactococcus lactis*	Lactacin F, lactococcin G
IIc	Circular LAB bacteriocins consist of N-to-C-terminally linked antimicrobial peptides, produced by gram-positive bacteria of the phylum *Firmicutes*	*Enterococcus faecalis; enterococcus faecium*	Enteriocin AS48, reuterin 6
IId	Include the remaining well-characterized bacteriocins, combined as miscellaneous, which are now including nonpediocin like single linear peptides	*Lactococcus lactis*	Lactococcin A, divergecin A

Adapted from Cotter et al. (2005) and Woraprayote et al. (2016)

bactericidal mode of action, usually acting on the bacterial cytoplasmic membrane: no cross resistance with antibiotics, and (vii) their genetic determinants are usually plasmid-encoded, facilitating genetic manipulation (Cotter et al. 2005; Gálvez et al. 2007; Woraprayote et al. 2016).

Bacteriocins are classified according to their chemical structure, molecular mass, enzymatic susceptibility, genetics, mechanism of microbial destruction, thermostability, producing strains, antimicrobial activities and the presence of post-translational modified amino acid residues (Kaškonienė et al. 2017; Mokoena and Paul 2017). Therefore, one only classification is not currently available and some authors distinguish two, three or four classes with subclasses/subcategories (Cotter et al. 2005; Kaškonienė et al. 2017; Mokoena and Paul 2017; Woraprayote et al. 2016). Between them, the classification proposed by Cotter et al. (2005) seems better for LAB bacteriocins at this moment. Accordingly, LAB bacteriocins can be classified into two major classes: class I lantibiotics (lanthionine-containing antibiotics) and class II which can further be grouped into four subclasses: IIa, IIb, IIc, and IId, respectively (Table 8.1). Bacteriocins from class I and IIa (pediocin-like bacteriocins) are among the best biochemically and genetically characterized antimicrobial peptides and the most likely to be used in food applications due to their target specificity. Among hundreds of bacteriocins, nisin is the most popular and extensively investigated bacteriocin, probably because it is approved for use as additive (code E234) in food products and is now available commercially (Woraprayote et al. 2016). Nisin is produced by many strains of *Lactococcus lactis*, a species widely used for cheese manufacture. It has a broad antimicrobial spectrum against a wide range of Gram-positive genera, including *Staphylococci*, *Streptococci*,

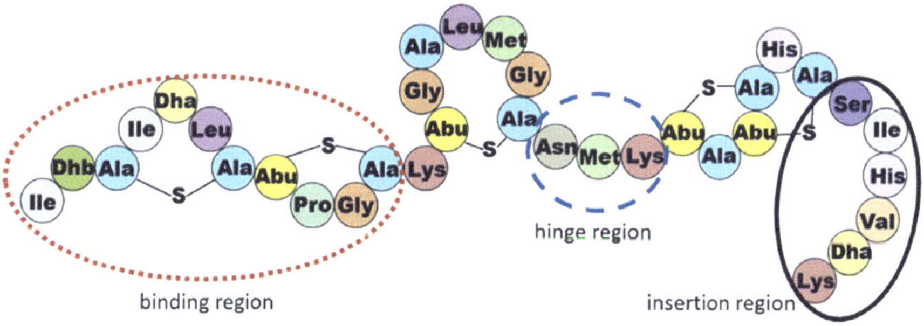

Fig. 8.1 Nisin molecular structure showing key regions responsible for its antimicrobial activity: binding region to the cell, hinge region to create cage and inhibit cell wall synthesis, and insertion region into the pore. Reprinted with permission from Han et al. (2017)

Listeria spp., *Bacilli*, and *Enterococci*, with a minimal inhibitory concentration in the nanomolar range (Woraprayote et al. 2016). To date, eight types of natural nisin were discovered: nisins A, Z, F and Q produced by *Lactococci* and nisins U, U2, P and H by some *Streptococcus* strains (Kaškonienė et al. 2017; O'Connor et al. 2015; Woraprayote et al. 2016).

Nisin acts on target bacteria by two major steps: (1) passage through the cell wall; (2) interaction with lipid II (e.g. binding to lipid II, pore formation on cell membrane), which is essential for the biosynthesis on the cell wall. Nisin's mechanism of action (see Fig. 8.1) involves first binding of the N-terminus to the lipid II complex and forming a pyrophosphate cage that inhibits cell wall synthesis. The C terminal is then responsible for pore-formation. A three amino acid hinge region exists between the N and C domains and allows conformational changes to occur upon contacting a microbe (Han et al. 2017).

Pediocins are another widely studied biopreservatives which are produced by *Pediococcus spp*. These class IIa bacteriocins are recognized because present a broad spectrum of antimicrobial activity against Gram-positive bacteria, with highlights efficient bactericidal effects against pathogenic bacteria, such as *Listeria monocytogenes* (Porto et al. 2017; Ríos Colombo et al. 2017; Rodríguez et al. 2002). It has been demonstrated that class IIa bacteriocins act on the cytoplasmic membrane of Gram-positive cells dissipating the transmembrane electrical potential, which results in an intracellular ATP depletion. These peptides induce the exit of ions, amino acids and other essential molecules by forming hydrophilic pores in the target (Ríos Colombo et al. 2017).

Among the pediocins isolated from different strains, only pediocin PA1 (*P. acidilactici PAC 1.0*) and pediocin AcH (*P. acidilactici LB42–923*) have been well characterized. Despite that pediocins have been widely studied, they have no official approved use in foods (Woraprayote et al. 2016).

In addition to nisin and pediocin, many other LAB bacteriocins have characteristics that make them ideal candidates for the preservation of food products including Lactococcin G, lactacin F, lactocin 705, enteriocin AS-48, lacticin Q and others.

More detailed information about LAB bacteriocins can be found in the recently published works by Garsa et al. (2014), Kaškonienė et al. (2017) and Mokoena and Paul (2017).

8.3 Incorporation of Lactic Acid Bacteria and Their Bacteriocins into Films and Coatings

Lactic acid bacteria and their bacteriocins have been applied in food products using mainly two different strategies: (i) direct inoculation of bacteriocin producting LAB culture into food products (*in situ* production) and (ii) direct application of purified or semi-purified bacteriocin as a food additive (*ex situ* production) (Castro et al. 2017). *Ex situ* preparations are obtained by growing the producer strain at industrial scale and then concentration and purification processes are needed to obtain a pure form of the bacteriocin. Both *in situ* and *ex situ* strategies have been reported to have limitations associated with interactions of biopreservatives with other food components (e.g., lipids and proteins) and to the loss of activity due to enzymatic degradation (Aloui and Khwaldia 2016). In order to overcome these disadvantages, biopreservatives-containing films and coatings has been evaluated obtaining feasible results. This strategy allows to combine the preservative function of antimicrobials with the protective function of packaging. For this purpose, a wide range of non-edible polypropylene- and polyethylene-based packaging materials and several biodegradable protein- and polysaccharide-based edible films have been used (Garcia et al. 2012; Muriel-Galet et al. 2015; Woraprayote et al. 2018).

Antimicrobial packaging offers slow and continuous migration of antimicrobial agent from packaging material to food surfaces which enables antimicrobial agents to maintain at high concentration over a long period. Antimicrobial coatings can be obtained by incorporation of the antimicrobials into an edible polymer blend that is then applied by dipping, brushing or spraying onto the food (Guo et al. 2014). Compared to direct application, edible coatings containing LAB and/or their bacteriocins may impart a highly localized functional effect without affecting the food organoleptic properties (Campos and others 2011). Moreover, edible coatings may act as a semipermeable barrier providing an additional protection for foods against moisture loss, solute migration, gas exchange, respiration, and oxidative reactions (Aloui and Khwaldia 2016).

Biopreservatives containing polymer films have been also used as antimicrobial packaging. Several film-forming methods have been evaluated, including casting and heat-pressing. It has been found that bacteriocin activity of cast film (solvent compounding) retained three times greater than that of heat-pressed films (Dawson et al. 2003).

In this context, several studies have investigated the application of films and coatings incorporating LAB and their bacteriocins as antimicrobial packagings to extend the shelf life of different food systems, including meat, fish, dairy fruit, vegetables or other food products.

8.4 Antimicrobial Films and Coatings for Meat and Meat Products

Meat and meat products are consumed extensively throughout the world because they are an important source of nutrients including fats, proteins, vitamin B12, zinc and iron. However, these products are perishable and susceptible to microbial contamination, leading to an increased health risk to consumers as well as to the economic loss in the meat industry (Woraprayote et al. 2016). Microorganisms commonly involved in spoilage of meat and meat products include *Pseudomonas (P. fragi, P. uorescens, P. putida and P. lundensis)*, *Shewanella putrefaciens*, *Photobacterium phosphereum*, *Brochothrix thermosphacta*, cold-tolerant *Enterobacteriaceae* (e.g., *Hafnia alvei, Serratia liquefaciens* and *Enterobacter agglomerans*), *Acinetobacter* spp., *Alcaligenes* spp., *Moraxella* spp., *Flavobacterium* spp., *Staphylococcus* spp., *Micrococcus* spp., coryneforms, fecal streptococci, lactic acid bacteria (LAB), among others (Sofos 2014). In addition, meat and meat products are also susceptible to contamination by pathogenic microorganisms such as *Salmonella* spp., thermophilic *Campylobacter jejuni*, enterohemorrhagic *Escherichia coli* O157:H7, *Clostridium perfringens*, anaerobic *Clostridium botulinum*, *Listeria monocytogenes*, *Staphylococcus aureus*, *Bacillus cereus*, and *Yesinia enterocolitica*. Among the meat-borne pathogens, *Listeria monocytogenes* consider one of the main causes of food borne illness and has been associated with cooked, ready-to-eat (RTE) meat and poultry product (Buchanan et al. 2017; Guo et al. 2014).

The application of LAB and their bacteriocins in meat and meat products has received a considerable attention in the last years. As above mentioned, several LAB produce microbial antagonists, such as bacteriocins, which are active against pathogens microorganism, including *Listeria monocytogenes*. The most-studied bacteriocins in meat and meat products are nisin, enterocin AS-48, enterocins A and B, sakacin, leucocin A and especially pediocin PA-l/AcH. These biopreservatives have been used alone or in combination with other hurdle treatments such as modified atmosphere packaging, high hydrostatic pressure (HHP), heat and chemical food preservatives (Castro et al. 2017; Cleveland et al. 2001; Delves-Broughton et al. 1996; Paul Ross et al. 2002).

The incorporation of LAB and/or their bacteriocins in food packaging's have been proposed as a useful strategy to prevent their degradation and achieved their controlled release towards the food products. Woraprayote et al. (Woraprayote et al. 2013) developed a poly(lactic acid) (PLA)/sawdust particle biocomposite film with anti-listeria activity by incorporation of pediocin PA-1/AcH. The addition of sawdust particle promoted the embedding of pediocin into the hydrophobic PLA film.

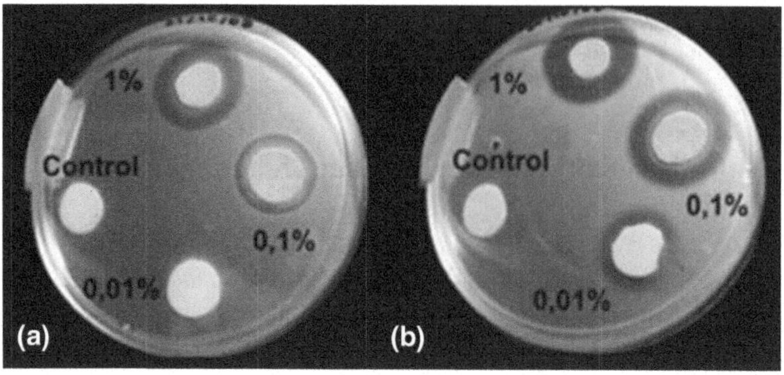

Fig. 8.2 Lactocin 705 (**a**) and AL705 (**b**) antimicrobial activity on wheat gluten films doped with bacteriocin crude extract (0.01%, 0.1% and 1%) against *L. plantarum CRL691* and *L. innocua 7*. Control film had no antimicrobials in its formulation. Reprinted with permission from Blanco Massani et al. (2014a, b).

The films significantly reduced the listerial population by about 1.5–2 log cycles from 1 to 14 days. This effect was enhanced when a film pre-conditioned by dry-heat treatment was carried out. Blanco Massani et al. (2014a, b) developed antimicrobial wheat gluten film containing *Lactobacillus curvatus* CRL705 bacteriocins (lactocin 705 and lactocin AL705). In order to determine the antimicrobial minimum inhibitory concentration, different bacteriocin crude extract concentrations were added to the film-forming solution (0.01%, 0.1% and 1% v/v) and the obtained films were assayed for antimicrobial activity by the agar well diffusion method (Fig. 8.2). *Lactobacillus plantarum* CRL691 and *Listeria innocua* 7 were used as indicators of lactocin 705 and lactocin AL705, respectively. It was found that wheat gluten films containing a bacteriocin crude extract concentration above 0.1% were active against both *L. innocua 7* and *L. plantarum* CRL691 bacteria (Fig. 8.2).

In other work, Blanco Massaniet al. (2014a, b) assayed the antimicrobial effectiveness of gluten film using Wieners inoculated with *Lactobacillus plantarum* CRL691 and *Listeria innocua* 7 (10^4 CFU/g) stored at 5 °C during 45 days. The wieners were separately inoculated under sterile conditions by immersion (30 s) in a solution containing *L. innocua* 7 (10^4 CFU/g) and *L. plantarum* CRL691 (10^4 CFU/g). After drying, three Wieners (42 g) were placed into each active and control packagings previously prepared. In parallel, control (without bacteriocins) uninoculated Wieners packages were included. Typical growth of both inoculated microorganisms was observed in control packages which reached 10^6–10^7 CFU/g at the end of storage period. In the active packages, *L.innocua* 7 was effectively inhibited (2.5 log cycles reduction at day 45), while *L. plantarum* CRL691 was only slightly inhibited (0.5 log cycles) up to the second week of storage, then counts around 10^6–10^7 CFU/g were reached. More recently, Correa et al. 2017 worked on the development of polyhydroxybutyrate/polycaprolactone (PHB/PCL) nisin activated films with and without the addition of organo-clays (Cloisite1 30 B and 10A).

Organo-clays were able to act as a filler, increasing the thermal stability and the barrier and mechanical properties of the nanocomposites. PHB/PCL nisin activated films were effective against *L. plantarum* CRL691 (used as processed meat spoilage bacterium model) inoculated on sliced ham. Lag phase was extended from 7.03 to 22.39 days due to the nisin effect in the active packages, avoiding LAB counts to reach more than 6 log units, thus extending the ham shelf life up to 28 days at 5 °C.

Several authors have suggested promising ways to enhance antimicrobial effectiveness of bacteriocin in meat products by the combination with another hurdle technology. For example, Guo et al. 2014 suggesting edible antimicrobial coating solutions incorporating chitosan, lauric arginate ester and nisin to reduce foodborne pathogen contamination on ready-to-eat (RTE) meats. Two different approaches were evaluated: RTE deli meat samples were directly coated with the solutions, or treated with a solution-coated polylactic acid (PLA) films. The antimicrobial efficacy of the coatings and films against *Listeria innocua* inoculated onto the surface of RTE meat samples was investigated. It was found that the addition of nisin to chitosan coating solutions significantly reduced more *Listeria* than the chitosan coating solutions without nisin. However, chitosan coatings with nisin exhibited less anti-listerial activity than lauric arginate ester and the combination of nisin with this agent did not contribute to a synergistic or additional anti-listerial effect (Guo et al. 2014).

Huq et al. (2015) evaluated the synergistic effect of gamma (γ)-irradiation and microencapsulated antimicrobials (nisin and oregano and cinnamon essential oils), alone or in combination, against *Listeria monocytogenes* on ready-to-eat (RTE) ham. Microencapsulation of essential oils and nisin showed a synergistic anti-listerial effect with γ-irradiation on RTE meat products. These combinations led to a lag phase of bacterial growth and provoked a reduction in the bacterial growth rate of 32%, compared to microencapsulated combined antimicrobials without irradiation.

8.5 Antimicrobial Films and Coatings for Fish and Fish Products

Fishery products have a high economic importance and they are one of the most important protein sources in human nutrition. However, these products are perishable and, if left unpreserved, spoil rapidly (Calo-Mata et al. 2008). Main spoilage causes of fishery products include enzymatic, microbial and chemical action.

Films and coatings incorporating LAB and/or their bacteriocins have been used to extend the shelf life and to maintain the quality of fresh and processed fish. Recently, Woraprayote et al. (2018) developed an antimicrobial biodegradable food packaging for control of pathogens in pangasius fish fillets. Bacteriocin 7293, a new found antimicrobial peptide produced by *W. hellenica* BCC 7293, was chosen as a biopreservative because its broad antimicrobial spectrum against both Gram-positive

Table 8.2 Application of bacteriocin and bacteriogenic strains in dairy products

Bacteriocin	Bacteriocin-producing culture	Application	Pathogen	Product
Lacticin 3147	*Lc. lactis* DPC 3147	Spray-dried powder	*L. monocytogenes*	Cottage cheese
Pediocin	*P. acidilactici* PAC1.0	Dry powder	*L. monocytogenes*	Cottage cheese and yogurt
Piscicolin 126	*C. piscicola* JG 126	Concentrated supernatant	*L. monocytogenes*	Camembert cheese
Enterocin CRL35	*E. faecium* CRL 35	Concentrated supernatant	*L. monocytogenes*	Goat milk cheese
Nisin	*Lc. lactis* CNRZ 150	Starter culture	*L. monocytogenes*	Camembert cheese
Nisin	*Lc. lactis* TAB 50	Starter culture	*L. monocytogenes*	Semihard cheese
Lacticin 481	*Lc. lactis* TAB 24	Starter culture	*L. monocytogenes*	Semihard cheese
Lacticin 3147	*Lc. lactis* DPC 4275	Starter culture	*L. monocytogenes*	Cottage cheese
Enterocin AS-48	*E. faecalis* TAB 28	Starter culture	*L. monocytogenes*	Semihard cheese
Enterocin AS-48	*E. faecalis* INIA 4	Starter or adjunct culture	*L. monocytogenes*	Manchego cheese
Pediocin	*Lc. lactis* MM 217	Starter culture	*L. monocytogenes*	Cheddar cheese
Pediocin	*Lb. plantarum* WHW 92	Surface sprayed cell suspension	*L. monocytogenes*	Munster cheese
Pediocin	*Lc. lactis* CL1	Adjunct culture	*L. monocytogenes*	Semihard cheese
Pediocin	*Lc. lactis* CL1	Adjunct culture	*S. aureus*	Semihard cheese
Nisin	*Lc. lactis* ESI 515	Adjunct culture	*S. aureus*	Semihard cheese

Reprinted from Arqués et al. (2015)

and Gram-negative bacteria (Woraprayote et al. 2015). The antimicrobial effectiveness of the produced PLA/sawdust particle films impregnated with Bac7293 on raw pangasius fish fillet was evaluated and it was found that these films effectively inhibited both Gram-positive (*Listeria monocytogenes* and *Staphylococcus aureus*) and Gram-negative bacteria (*Pseudomonas aeruginosa, Aeromonas hydrophila, Escherichia coli* and *Salmonella Typhimurium*) which have been considered as a reason for the rejection of pangasius fish fillets in worldwide markets.

8.6 Antimicrobial Films and Coatings for Dairy Products

Milk and dairy products have been an important part of the human diet for some 8000 years and are part of the official nutritional recommendations in many countries worldwide. However, it is well known, that these products provide a potential growth medium for the development of spoilage and pathogen microorganisms such as *Listeria monocytogenes*, *Salmonella* spp., *Staphylococcus aureus*, and pathogenic *Escherichia coli*. Several bacteriocin and bactioricin-producing strain have been used in dairy foods (Table 8.2). Moreover, different strategies of incorporation of biopreservatives into dairy foods have been considered in order to prevent the loss of antimicrobial activity of these agents.

LAB and their bacteriocins have been included in polymers for the production of active packaging for dairy foods. Recently, Marques et al. (2017) evaluated the effectiveness of a biodegradable film, with antimicrobial metabolites produced by *Lactobacillus curvatus* P99 incorporated, targeting the control of *Listeria monocytogenes* in sliced "Prato" cheese. *Lactobacillus curvatus* is part of the microbiota of many fermented products and stands out for its bacteriocinogenic activity, due to the production of different antimicrobial metabolites, especially bacteriocins, known as curvacins and sakacins and characterized by their antilisterial activity (de Souza Barbosa et al. 2015). Starch films incorporating cell-free supernatant containing bacteriocins from *Lactobacillus curvatus* P99 were prepared by casting. Films with added minimum bactericidal concentration (62.5 μL/mL) showed activity against different indicator microorganisms and were able to control *L. monocytogenes* Scott A when used in sliced "Prato" cheese. During 10 days of storage at 4 °C, the target microorganism count remained below the limit of detection (2.7 Log CFU/g).

Ollé Resa et al. (2016) proposed an innovative approach to prevent the contamination of an Argentinian Port Salut cheese with mixed cultures (bacteria, molds and yeast). Nisin was used as an antibacterial agent while natamycin was employed to prevent yeasts and moulds contamination. Both active compounds were incorporated together within tapioca starch edible films and the effectiveness of the active films was evaluated, at 7 ± 1 °C, in relation to the improvement of the microbiological stability of Argentinian Port Salut cheese. The films inhibited the growth of yeasts and moulds and controlled the growth of psychrotrophic bacteria originally present in the Port Salut cheese stored at refrigeration temperature. It also inhibited the development of a mixed culture (*Saccharomyces cerevisiae* and *Listeria innocua*) present in the cheese due to a superficial contamination, along a storage of 8 days at 7 ± 1 C.

8.7 Antimicrobial Films and Coatings for Fruits and Vegetables

The promotion of greater consumption of fruits and vegetables constitutes a worldwide challenge. This is due to that the low fruit and vegetable intake is estimated to cause some 5.2 million deaths each year, and was among the top 10 risk factors contributing to mortality (FAO 2017). Fruits and vegetables can be consumed either fresh or processed. Production and consumption of minimally processed foods, such as fresh-cut fruits and vegetables, is gaining popularity due to consumer preferences towards healthier foods. However, challenge for fresh-cut industry is to maintain fresh like characteristics of fresh-cut produce for a prolonged storage time. Fresh-cut products have much larger cut surface and consequently much shorter shelf-life. Loss of quality parameters such as color, firmness, juiciness, flavor and excessive moisture loss results in limited shelf-life and increased chances of rejection of the produce by the consumers (Yousuf et al. 2018).

Different approaches have been used to preserve the quality of fresh-cut fruits and vegetables (Barbosa et al. 2017; Yousuf et al. 2018). Between them, the application of antimicrobial packaging's containing plant extracts, antimicrobial polymers (e.g. chitosan), enzymes (e.g. lysozyme) and other agents has been proposed as a useful strategy to extend the shelf life of these products (Álvarez et al. 2018; Gutiérrez et al. 2018b). However, despite that LAB and their bacteriocins have been extensively studied to preserve foods of animal origin, little information is available for their use in vegetable products, especially in minimally processed ready-to-eat fruits (Barbosa et al. 2017). Some studies deal with the application of bacteriocins to extend the shelf life of pineapple pulps, apple juice and minimally processed fruits and vegetables (e.g., sliced apples and lamb's lettuce) have been reported (Leite et al. 2016; Pei et al. 2017; Siroli et al. 2015). Antimicrobial packagings containing LAB and/or their bacteriocins also have been developed. Narsaiah et al. (2015) developed pediocin-containing calcium alginate coating for preservation of minimally processed papaya fruit. Fruit quality parameters such as firmness, weight loss, color, head space gas composition, acidity, total soluble solids and microbial load were evaluated for 21 days of refrigerated storage. It was found that the pediocin-containing alginate coating prolonged the shelf-life of minimally processed papaya by maintaining physicochemical properties and microbial safety along 21 days of refrigerated storage.

Barbosa et al. (2013) studied the effects of nisin-incorporated films on the microbiological and physicochemical quality of minimally processed mangoes. Films were produced using a blend of cellulose acetate and nisin by the casting method and the antimicrobial activity of the films was tested against *S. aureus* ATCC 8095, *B. cereus* ATCC 4504, *A. acidoterrestris* DSMZ 2498 and *L. monocytogenes* ATCC 7644 using the diffusion method. In addition, the antimicrobial activity of the films on *S. aureus* or *L. monocytogenes* inoculated mango slices was evaluated. Antimicrobial tests using the diffusion method showed that the antimicrobial film inhibited *S. aureus*, *L. monocytogenes*, *B. cereus* and *A. acidoterrestris* strains *in*

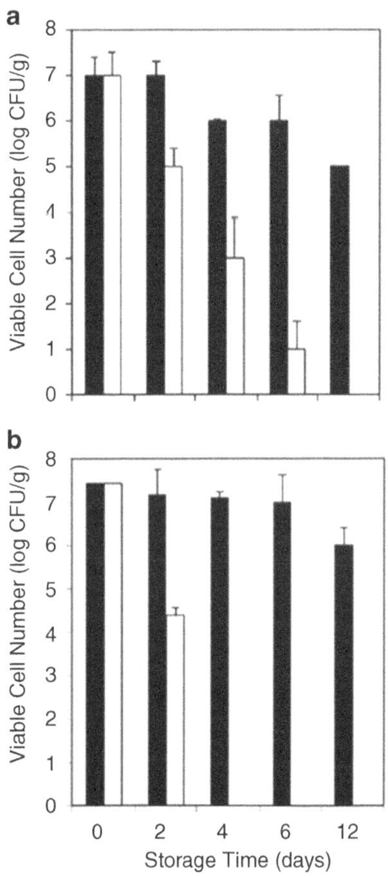

Fig. 8.3 Effects of nisin on the viability of *S. aureus* (**a**) and *L. monocytogenes* (**b**) on minimally processed mangoes. A total of 25 g of mango slices were inoculated with 10^7 CFU/g of each microorganism and packed with (white bars) and without (black bars) nisin. Reprinted with permission from Barbosa et al. (2013)

vitro. Moreover, cellulose films incorporated with nisin were efficient in eliminating *S. aureus* and *L. monocytogenes* contamination from minimally processed mango slices (Fig. 8.3), without interfering in the organoleptic characteristics of mangoes.

8.8 Conclusion

Lactic acid bacteria and their bacteriocins constitute an actual choice to decrease the use of synthetic additives in foods and also the application of thermal treatments, allowing to obtain more healthier foods. Despite that, these biopreservatives can be direct incorporated in foods, some disadvantages have reported regarding its loss of antimicrobial activity due to the interaction with food components and to its enzymatic degradation.

The incorporation of LAB and/or their bacteriocins in polymeric packaging (films or coatings) constitute a useful strategy to overcome these difficulties, improving the effectiveness of the biopreservatives. Moreover, these delivery systems allow to increase the shelf-life of foods and contribute towards decreasing the incidence of food-borne diseases, as it has been substantiated by the diversity of researches described in the current chapter.

The application of bacteriocinogenic LAB strains or their bacteriocins combined with other hurdle methods such as chemical compounds or physical processes can make use of synergies to increase microbial inactivation, without altering nutritional value and organoleptic properties of food.

Prior to successful commercialization of any of the effective films and coatings containing biopreservatives described in this study, large scale-production assays, shelf-life studies and, quality and sensory analyses will be needed. Moreover, close linkages between the scientific community and the industrial sector are highly required.

Acknowledgement The author would like to thank the Universidad Pedagógica y Tecnológica de Colombia (UPTC) and the Escuela de Administracion de Empresas Agropecuarias (UPTC Facultad Seccional Duitama) for their support.

References

Abbasiliasi S, Tan JS, Tengku Ibrahim TA, Bashokouh F, Ramakrishnan NR, Mustafa S, Ariff AB (2017) Fermentation factors influencing the production of bacteriocins by lactic acid bacteria: a review. RSC Adv 7(47):29395–29420. https://doi.org/10.1039/C6RA24579J

Aloui H, Khwaldia K (2016) Natural antimicrobial edible coatings for microbial safety and food quality enhancement. Compr Rev Food Sci Food Saf 15(6):1080–1103. https://doi.org/10.1111/1541-4337.12226

Álvarez K, Alvarez VA, Gutiérrez TJ (2018) Biopolymer composite materials with antimicrobial effects applied to the food industry. In: Thakur VK, Thakur MK (eds) Functional biopolymers. Springer International, Basel, pp 57–96. EE.UU. ISBN: 978-3-319-66416-3. eISBN: 978-3-319-66417-0. https://doi.org/10.1007/978-3-319-66417-0_3

Anu Bhushani J, Anandharamakrishnan C (2014) Electrospinning and electrospraying techniques: potential food based applications. Trends Food Sci Technol 38(1):21–33. https://doi.org/10.1016/j.tifs.2014.03.004

Arqués JL, Rodríguez E, Langa S, Landete JM, Medina M (2015) Antimicrobial activity of lactic acid Bacteria in dairy products and gut: effect on pathogens. Biomed Res Int 2015:9

Barbosa AAT, Silva de Araújo HG, Matos PN, Carnelossi MAG, Almeida de Castro A (2013) Effects of nisin-incorporated films on the microbiological and physicochemical quality of minimally processed mangoes. Int J Food Microbiol 164(2):135–140. https://doi.org/10.1016/j.ijfoodmicro.2013.04.004

Barbosa AAT, Mantovani HC, Jain S (2017) Bacteriocins from lactic acid bacteria and their potential in the preservation of fruit products. Crit Rev Biotechnol 37(7):852–864. https://doi.org/10.1080/07388551.2016.1262323

Blanco Massani M, Botana A, Eisenberg P, Vignolo G (2014a) Development of an active wheat gluten film with lactobacillus curvatus CRL705 bacteriocins and a study of its antimicrobial performance during ageing. Food Addit Contam Part A 31(1):164–171. https://doi.org/10.1080/19440049.2013.859398

Blanco Massani M, Molina V, Sanchez M, Renaud V, Eisenberg P, Vignolo G (2014b) Active polymers containing lactobacillus curvatus CRL705 bacteriocins: effectiveness assessment in wieners. Int J Food Microbiol 178:7–12. https://doi.org/10.1016/j.ijfoodmicro.2014.02.013

Buchanan RL, Gorris LGM, Hayman MM, Jackson TC, Whiting RC (2017) A review of Listeria monocytogenes: an update on outbreaks, virulence, dose-response, ecology, and risk assessments. Food Control 75:1–13. https://doi.org/10.1016/j.foodcont.2016.12.016

Calo-Mata P, Arlindo S, Boehme K, de Miguel T, Pascoal A, Barros-Velazquez J (2008) Current applications and future trends of lactic acid bacteria and their Bacteriocins for the biopreservation of aquatic food products. Food Bioprocess Technol 1(1):43–63. https://doi.org/10.1007/s11947-007-0021-2

Castro SM, Kolomeytseva M, Casquete R, Silva J, Queirós R, Saraiva JA, Teixeira P (2017) Biopreservation strategies in combination with mild high pressure treatments in traditional Portuguese ready-to-eat meat sausage. Food Biosci 19:65–72. https://doi.org/10.1016/j.fbio.2017.05.008

Chang-Bravo L, López-Córdoba A, Martino M (2014) Biopolymeric matrices made of carrageenan and corn starch for the antioxidant extracts delivery of Cuban red propolis and yerba mate. React Funct Polym 85:11–19. https://doi.org/10.1016/j.reactfunctpolym.2014.09.025

Chelule PK, Mbongwa HP, Carries S, Gqaleni N (2010) Lactic acid fermentation improves the quality of amahewu, a traditional south African maize-based porridge. Food Chem 122(3):656–661. https://doi.org/10.1016/j.foodchem.2010.03.026

Cleveland J, Montville TJ, Nes IF, Chikindas ML (2001) Bacteriocins: safe, natural antimicrobials for food preservation. Int J Food Microbiol 71(1):1–20. https://doi.org/10.1016/S0168-1605(01)00560-8

Correa JP, Molina V, Sanchez M, Kainz C, Eisenberg P, Massani MB (2017) Improving ham shelf life with a polyhydroxybutyrate/polycaprolactone biodegradable film activated with nisin. Food Packaging Shelf Life 11:31–39. https://doi.org/10.1016/j.fpsl.2016.11.004

Cotter PD, Hill C, Ross RP (2005) Bacteriocins: developing innate immunity for food. Nat Rev Microbiol 3:777–788. https://doi.org/10.1038/nrmicro1273

Dawson PL, Hirt DE, Rieck JR, Acton JC, Sotthibandhu A (2003) Nisin release from films is affected by both protein type and film-forming method. Food Res Int 36(9):959–968. https://doi.org/10.1016/S0963-9969(03)00116-9

de Souza Barbosa M, Todorov SD, Ivanova I, Chobert J-M, Haertlé T, de Melo Franco BDG (2015) Improving safety of salami by application of bacteriocins produced by an autochthonous lactobacillus curvatus isolate. Food Microbiol 46:254–262. https://doi.org/10.1016/j.fm.2014.08.004

Delves-Broughton J, Blackburn P, Evans RJ, Hugenholtz J (1996) Applications of the bacteriocin, nisin. Antonie Van Leeuwenhoek 69(2):193–202. https://doi.org/10.1007/BF00399424

FAO (2011) Global food losses and food waste. http://www.fao.org/docrep/014/mb060e/mb060e00.pdf. Accessed 19 Mar 2018

FAO (2014) Appropriate food packaging solutions for developing countries. Rome

FAO (2017) Fruit and vegetables for health initiative

Gálvez A, Abriouel H, López RL, Omar NB (2007) Bacteriocin-based strategies for food biopreservation. Int J Food Microbiol 120(1):51–70. https://doi.org/10.1016/j.ijfoodmicro.2007.06.001

García P, Rodríguez L, Rodríguez A, Martínez B (2010) Food biopreservation: promising strategies using bacteriocins, bacteriophages and endolysins. Trends Food Sci Technol 21(8):373–382. https://doi.org/10.1016/j.tifs.2010.04.010

Garcia LC, Pereira LM, de Luca Sarantópoulos CIG, Hubinger MD (2012) Effect of antimicrobial starch edible coating on shelf-life of fresh strawberries. Packag Technol Sci 25(7):413–425. https://doi.org/10.1002/pts.987

Garsa AK, Kumariya R, Sood SK, Kumar A, Kapila S (2014) Bacteriocin production and different strategies for their recovery and purification. Probiotics Antimicrob Proteins 6(1):47–58. https://doi.org/10.1007/s12602-013-9153-z

Guo M, Jin TZ, Wang L, Scullen OJ, Sommers CH (2014) Antimicrobial films and coatings for inactivation of Listeria innocua on ready-to-eat deli Turkey meat. Food Control 40:64–70. https://doi.org/10.1016/j.foodcont.2013.11.018

Gutiérrez TJ (2017) Surface and nutraceutical properties of edible films made from starchy sources with and without added blackberry pulp. Carbohydr Polym 165:169–179. https://doi.org/10.1016/j.carbpol.2017.02.016

Gutiérrez TJ (2018) Active and intelligent films made from starchy sources/blackberry pulp. J Polym Environ 15:445–448. https://doi.org/10.1007/s10924-017-1134-y

Gutiérrez TJ, Álvarez K (2016) Physico-chemical properties and *in vitro* digestibility of edible films made from plantain flour with added Aloe vera gel. J Funct Foods 26:750–762. https://doi.org/10.1016/j.jff.2016.08.054

Gutiérrez TJ, Herniou-Julien C, Álvarez K, Alvarez V (2018a) Structural properties and *in vitro* digestibility of edible and pH-sensitive films made from Guinea arrowroot starch and wastes from wine manufacture. Carbohydr Polym 184:135–143. https://doi.org/10.1016/j.carbpol.2017.12.039

Gutiérrez TJ, Ollier R, Alvarez VA (2018b) Surface properties of thermoplastic starch materials reinforced with natural fillers. In: Functional biopolymers. Vijay Kumar Thakur, and Manju Kumari Thakur (Eds). Editorial Springer International, Basel, 131-158. EE.UU. ISBN: 978-3-319-66416-3. eISBN: 978-3-319-66417-0. https://doi.org/10.1007/978-3-319-66417-0_5

Han D, Sherman S, Filocamo S, Steckl AJ (2017) Long-term antimicrobial effect of nisin released from electrospun triaxial fiber membranes. Acta Biomater 53:242–249. https://doi.org/10.1016/j.actbio.2017.02.029

Hugas M (1998) Bacteriocinogenic lactic acid bacteria for the biopreservation of meat and meat products. Meat Sci 49:S139–S150. https://doi.org/10.1016/S0309-1740(98)90044-4

Huq T, Vu KD, Riedl B, Bouchard J, Lacroix M (2015) Synergistic effect of gamma (γ)-irradiation and microencapsulated antimicrobials against Listeria monocytogenes on ready-to-eat (RTE) meat. Food Microbiol 46:507–514. https://doi.org/10.1016/j.fm.2014.09.013

Kaškonienė V, Stankevičius M, Bimbiraitė-Survilienė K, Naujokaitytė G, Šernienė L, Mulkytė K et al (2017) Current state of purification, isolation and analysis of bacteriocins produced by lactic acid bacteria. Appl Microbiol Biotechnol 101(4):1323–1335. https://doi.org/10.1007/s00253-017-8088-9

Leite JA, Tulini FL, Reis-Teixeira FBD, Rabinovitch L, Chaves JQ, Rosa NG, De Martinis ECP (2016) Bacteriocin-like inhibitory substances (BLIS) produced by Bacillus cereus: preliminary characterization and application of partially purified extract containing BLIS for inhibiting Listeria monocytogenes in pineapple pulp. LWT Food Sci Technol 72:261–266. https://doi.org/10.1016/j.lwt.2016.04.058

López-Córdoba A, Medina-Jaramillo C, Piñeros-Hernandez D, Goyanes S (2017) Cassava starch films containing rosemary nanoparticles produced by solvent displacement method. Food Hydrocoll 71:26–34. https://doi.org/10.1016/j.foodhyd.2017.04.028

Marques J d L, Funck GD, Dannenberg G d S, Cruxen CE d S, Halal SLM, Dias ARG, da Silva WP (2017) Bacteriocin-like substances of lactobacillus curvatus P99: characterization and application in biodegradable films for control of Listeria monocytogenes in cheese. Food Microbiol 63:159–163. https://doi.org/10.1016/j.fm.2016.11.008

Mokoena MP, Paul M (2017) Lactic acid Bacteria and their Bacteriocins: classification, biosynthesis and applications against Uropathogens: a mini-review. Molecules 22(12):1255. https://doi.org/10.3390/molecules22081255

Muriel-Galet V, Cran MJ, Bigger SW, Hernández-Muñoz P, Gavara R (2015) Antioxidant and antimicrobial properties of ethylene vinyl alcohol copolymer films based on the release of oregano essential oil and green tea extract components. J Food Eng 149:9–16. https://doi.org/10.1016/j.jfoodeng.2014.10.007

Narsaiah K, Wilson RA, Gokul K, Mandge HM, Jha SN, Bhadwal S et al (2015) Effect of bacteriocin-incorporated alginate coating on shelf-life of minimally processed papaya (Carica papaya L.). Postharvest Biol Technol 100:212–218. https://doi.org/10.1016/j.postharvbio.2014.10.003

O'Connor PM, O'Shea EF, Guinane CM, O'Sullivan O, Cotter PD, Ross RP, Hill C (2015) Nisin H is a new Nisin variant produced by the gut-derived strain Streptococcus hyointestinalis DPC6484. Appl Environ Microbiol 81(12):3953–3960. https://doi.org/10.1128/AEM.00212-15

Ollé Resa CP, Gerschenson LN, Jagus RJ (2016) Starch edible film supporting natamycin and nisin for improving microbiological stability of refrigerated argentinian port Salut cheese. Food Control 59:737–742. https://doi.org/10.1016/j.foodcont.2015.06.056

Paul Ross R, Morgan S, Hill C (2002) Preservation and fermentation: past, present and future. Int J Food Microbiol 79(1):3–16. https://doi.org/10.1016/S0168-1605(02)00174-5

Pei J, Yue T, Jin W (2017) Application of bacteriocin RC20975 in apple juice. Food Sci Technol Int 23(2):166–173. https://doi.org/10.1177/1082013216668691

Piñeros-Hernandez D, Medina-Jaramillo C, López-Córdoba A, Goyanes S (2017) Edible cassava starch films carrying rosemary antioxidant extracts for potential use as active food packaging. Food Hydrocoll 63(Supplement C):488–495. https://doi.org/10.1016/j.foodhyd.2016.09.034

Porto MCW, Kuniyoshi TM, Azevedo POS, Vitolo M, Oliveira RPS (2017) Pediococcus spp.: an important genus of lactic acid bacteria and pediocin producers. Biotechnol Adv 35(3):361–374. https://doi.org/10.1016/j.biotechadv.2017.03.004

Ríos Colombo NS, Chalón MC, Navarro SA, Bellomio A (2017) Pediocin-like bacteriocins: new perspectives on mechanism of action and immunity. Curr Genet 64:345–351. https://doi.org/10.1007/s00294-017-0757-9

Rodríguez JM, Martínez MI, Kok J (2002) Pediocin PA-1, a wide-Spectrum Bacteriocin from lactic acid Bacteria. Crit Rev Food Sci Nutr 42(2):91–121. https://doi.org/10.1080/10408690290825475

Salvetti E, Torriani S, Felis GE (2012) The genus lactobacillus: a taxonomic update. Probiotics Antimicrob Proteins 4(4):217–226. https://doi.org/10.1007/s12602-012-9117-8

Siroli L, Patrignani F, Serrazanetti DI, Tabanelli G, Montanari C, Gardini F, Lanciotti R (2015) Lactic acid bacteria and natural antimicrobials to improve the safety and shelf-life of minimally processed sliced apples and lamb's lettuce. Food Microbiol 47:74–84. https://doi.org/10.1016/j.fm.2014.11.008

Sofos JN (2014) Chapter 6. Meat and meat products A2 - Motarjemi, Yasmine. In: Lelieveld HBT-FSM (ed). Academic Press, San Diego, pp 119–162. https://doi.org/10.1016/B978-0-12-381504-0.00006-8

Stiles ME (1996) Biopreservation by lactic acid bacteria. Antonie Van Leeuwenhoek 70(2):331–345. https://doi.org/10.1007/BF00395940

Woraprayote W, Kingcha Y, Amonphanpokin P, Kruenate J, Zendo T, Sonomoto K et al (2013) Anti-listeria activity of poly(lactic acid)/sawdust particle biocomposite film impregnated with pediocin PA-1/AcH and its use in raw sliced pork. Int J Food Microbiol 167(2):229–235. https://doi.org/10.1016/j.ijfoodmicro.2013.09.009

Woraprayote W, Pumpuang L, Tosukhowong A, Roytrakul S, Perez RH, Zendo T, Visessanguan W (2015) Two putatively novel bacteriocins active against gram-negative food borne pathogens produced by Weissella hellenica BCC 7293. Food Control 55:176–184. https://doi.org/10.1016/j.foodcont.2015.02.036

Woraprayote W, Malila Y, Sorapukdee S, Swetwiwathana A, Benjakul S, Visessanguan W (2016) Bacteriocins from lactic acid bacteria and their applications in meat and meat products. Meat Sci 120:118–132. https://doi.org/10.1016/j.meatsci.2016.04.004

Woraprayote W, Pumpuang L, Tosukhowong A, Zendo T, Sonomoto K, Benjakul S, Visessanguan W (2018) Antimicrobial biodegradable food packaging impregnated with Bacteriocin 7293 for control of pathogenic bacteria in pangasius fish fillets. LWT Food Sci Technol 89:427–433. https://doi.org/10.1016/j.lwt.2017.10.026

Yousuf B, Qadri OS, Srivastava AK (2018) Recent developments in shelf-life extension of fresh-cut fruits and vegetables by application of different edible coatings: a review. LWT 89:198–209. https://doi.org/10.1016/j.lwt.2017.10.051

Chapter 9
Postharvest Application of Biopolymer-Based Edible Coatings to Improve the Quality of Fresh Horticultural Produce

Bahareh Saberi and John B. Golding

Abstract Fresh fruit and vegetables are essential part of the human diet and with increasing growth in the world population and globalization of the supply chain, there is high consumer demand for fresh high quality fruit and vegetables with satisfactory shelf-life. Fruit and vegetables are living perishable commodities and maintaining their eating quality requires delaying and slowing the loss of quality and microbial spoilage in the supply chain. To actively manage produce quality, it is essential to understand the physiology and pathology associated with the storage and handling of fresh fruit and vegetables, for example changes in internal gas and volatiles composition within the produce affect the respiration and senescence rates of stored produce and hence final eating quality. Biopolymers including polysaccharides, proteins and lipids used alone or in combination, have been widely studied as edible coatings for fresh produce due to their widespread availability, ability to control gaseous exchange and to delay senescence, as well as their capability as a medium to carry various additives which can assist with the maintenance of produce quality. This chapter describes and discusses on selected biopolymer based edible coatings and their various applications on fresh fruit and vegetables to extend their shelf life and to improve final eating and nutritional quality through the supply chain.

Keywords Shelf life

B. Saberi (✉)
Faculty of Science and Information Technology, School of Environmental and Life Sciences, University of Newcastle, Ourimbah, NSW, Australia
e-mail: bahareh.saberi@uon.edu.au

J. B. Golding
Faculty of Science and Information Technology, School of Environmental and Life Sciences, University of Newcastle, Ourimbah, NSW, Australia

NSW Department of Primary Industries, Ourimbah, NSW, Australia
e-mail: john.golding@dpi.nsw.gov.au

T. J. Gutiérrez (ed.), *Polymers for Food Applications*,
https://doi.org/10.1007/978-3-319-94625-2_9

9.1 Introduction

Moisture loss from horticultural produce is mainly due to the gradient of water pressure (vapor pressure deficit) between the less saturated ambient atmosphere and the fruit and can be one of the major losses in quality and quantity of fresh fruit and vegetables. Moisture loss after harvest has detrimental impact on the quality of fruit and vegetables and results in commercial losses as the price of produce depends on its weight (Tesfay and Magwaza 2017). The preliminary signs of moisture loss from fresh fruit are shriveling of the peel/skin and can affect fruit shine (glossiness), firmness and apparent 'freshness' (Tesfay and Magwaza 2017). The extension of fruit shelf life and eating quality is optimized by reducing and managing water loss, ripening and senescence, and development of decay (Vanoli et al. 2015). Many postharvest techniques such as maintaining the optimum storage temperatures and atmosphere during storage and other postharvest treatments including some approved chemical treatments are commonly applied in many horticulture supply chains.

However due to increasing consumer demands for improved health and fruit quality, along with the industry desire to use ecologically friendly coatings for fresh horticultural products, a substantial increase in awareness has been grown among researchers and technologists to develop natural, edible, biodegradable and food safe coatings for fruit and vegetables. Edible coatings have been used for a long time as a method with pronounced capability to increase safety of fresh produces by reducing the effect of external environmental factors and increasing the shelf life (Carneiro-da-Cunha et al. 2009; Cerqueira et al. 2011; Álvarez et al. 2017). Application of a thin layer of edible coatings to the surface of the fresh produce creates a semi-permeable barrier to respiratory gases and water vapor between the surrounding atmosphere and the fruit and also develops a modified atmosphere around the product, decelerating respiration, senescence, and enzymatic processes (Vanoli et al. 2015). Edible coatings need to be water-resistant, stable, and semi-permeable to water vapor during storage. Moreover, they should not excessively reduce oxygen (O_2) or accumulate carbon dioxide (CO_2), and should not cause off-flavors or reduce glossiness, aroma, texture, taste and appearance in order to efficiently preserve food quality. Edible coatings also must have low viscosity, be transparent and cost-effective (Dhall 2013; Mahajan et al. 2014). It has been shown that the incorporation of functional components including antibrowning, antimicrobial agents, nutraceuticals, volatile precursors, and colors can also improve the characteristics of the coatings (Olivas and Barbosa-Cánovas 2005; Gutiérrez 2017a). Besides the microbial stability, appearance, and texture of coated products can be improved by addition of other constituents, such as preservatives, antioxidants, and firming agents (Bai and Plotto 2012; Cerqueira et al. 2009; Gutiérrez 2018). Nevertheless, it is crucial to select appropriate components in order to enhance the effectiveness and stability of edible coatings. An extensive range of different components involving polysaccharides, proteins, lipids or resins, alone or more commonly in combination have been used to produce edible coatings (Flores-López et al. 2016).

Natural biopolymers have excellent potential to improve the safety, quality and functionality of fresh and fresh-cut fruit and vegetables. While many researches have been conducted and reviewed on application of edible coatings for fresh and fresh-cut products, this chapter summarizes and highlights the application of biopolymer based edible coatings on fresh horticultural produce for the past 5 years.

9.2 Polysaccharide–Based Coatings

The application of polysaccharides edible coatings to fresh and fresh-cut fruit and vegetables has been of increasing interest over the last few years due to the wide availability, low cost, and nontoxicity of polysaccharides, where they can be applied as film-forming solution to improve and control the texture, flavor, and shelf-life of produce (Williams and Phillips 2000). Water-soluble polysaccharides are long-chain polymers that give a thickening or viscosity-building effect by dissolving or dispersing in water (Nussinovitch 1997). These compounds are commercially available for use in food and non-food industries such as adhesive, mouth feeling, stabilizing, thickening and gelling agents, crystallization inhibiting, and encapsulating agents (Whistler and Daniel 1990; Izydorczyk et al. 2005; Gutiérrez and Álvarez 2017). They can also act as emulsifier because their stabilizing function on emulsions originates from an increase in viscosity of the aqueous phase of the edible coating (Nussinovitch 1997), therefore the kinetic motion of the oil droplets is decreased, leading to a lower rate of flocculation and coalescence in the coating (Nisperos-Carriedo 1994). Polysaccharides consist of many monosaccharide residues that are linked one to the other by O-glycosidic linkages. The physical attributes of polysaccharides including solubility, flow behavior, gelling potential, and/or surface and interfacial characteristics are dictated by their great diversity of structural features deriving from variations in the monosaccharide composition, linkage types and patterns, chain shapes, and degree of polymerization (Izydorczyk et al. 2005).

Polysaccharide based coatings usually have less moisture barriers in comparison with those of commercial ones. They also have reasonably low oxygen permeability but have selective permeability to O_2 and CO_2 (Lacroix and Le Tien 2005). Consequently, polysaccharide based coatings have been applied to, either fresh or minimally processed fruit and vegetables to protect from dehydration, decrease their respiration rate by producing modified atmosphere conditions inside the product, provide a moderate barrier to moisture, improve mechanical handling characteristics, and to deliver additives along with contributing to the preservation and even the creation of volatile compounds (Olivas and Barbosa-Cánovas 2005). Polysaccharide coatings can act as a sacrificial moisture barrier to the atmosphere by reducing moisture loss of the coated produce (Kester and Fennema 1986). The major advantage of application of polysaccharides is, acting a gas barrier rather than postponing water loss due to their hydrophilic nature (Lin and Zhao 2007).

Polysaccharides based edible coatings made of starch and non-starch polysaccharides (hydrocolloids or gums), which can be made from a variety of sources including cellulose derivatives (carboxymethylcellulose (CMC), methylcellulose (MC), hydroxypropyl cellulose (HPC), hydroxypropyl methyl cellulose (HPMC), microcrystalline cellulose), seaweed extracts (agar, alginates, carrageenans), various plant and microbial gums (arabic, ghatti, tragacanth, guar, locust bean, xanthan, gellan, pullulan), connective tissue extracts of crustaceans (chitosan) (Gutiérrez 2017b), and some mucilage compounds (Lacroix and Le Tien 2005), which have been extensively investigated by means of prolonging the shelf life of fresh and minimally processed fruit and vegetables.

9.2.1 Starch-Based Coatings

Starch is widely found in nature and has been extensively used as an alternative for producing edible coatings as it is abundant, inexpensive, biodegradable and edible (Gutiérrez et al. 2015; Cazón et al. 2017). Starch granules constitute two main polysaccharides: amylose and amylopectin. Amylose is a linear chain polymer of α-1,4 anhydroglucose units with a molecular size varying from 20 to 800 kg/mol and constitutes about 20–25% of the granular starches (Cazón et al. 2017). Amylopectin is a branched polymer of short α-1,4 chains connected by α-1,6 glycosidic branching points which occurs every 25–30 glucose units and normally has a very high molecular weight (5000–30,000 kg/mol) (Peressini et al. 2003; Jiménez et al. 2012). The changes of structure and molecular weight between amylose and amylopectin result in various molecular properties, which can be utilized to develop coatings (Cazón et al. 2017). The starch based coatings characterized by forming hydrogen bonds between hydroxyl groups during drying of a gelatinized dispersion (Jiménez et al. 2012). Starch-based coatings are often very fragile and have poor mechanical characteristics due to amorphous regions shaped by amylose causing extensive intermolecular forces which consequently impart brittleness of starch coatings (Peressini et al. 2003). It is therefore essential to add a plasticizer or combine starch with other compounds or chemically modify the starch to manage this problem (Bertuzzi et al. 2007; Xiong et al. 2008).

A key benefit with the use of these coatings on fresh fruit and vegetables is to lessen the fruit's natural high respiration and senescence rate, however it is critical that the coating allows some gas exchange for the fruit to continue to respire. Starch coatings have good oxygen barrier properties due to the high-ordered hydrogen-bonded configuration in which the amylose and amylopectin establish crystalline and non-crystalline districts in irregular layers (Lin and Zhao 2007). Starch based coatings can reduce the fruit deterioration rate by reduction of respiratory metabolism and retardation of some enzymatic reactions related to ripening, and accordingly preserve the appearance, texture and nutritional composition of coated fruit (Franco et al. 2017; Álvarez et al. 2018).

Nawab et al. (2017) showed that coatings made of mango kernel starch were effective at maintaining overall postharvest quality of tomato during storage at 25 °C and 60% RH for 20 days. The strawberry fruit coated with oxidized and acetylated cassava starch had lower weight loss, better texture, lighter and brighter appearance, lower soluble solids and higher total acidity compared with native cassava starch coated fruit, indicating that the modifications in the starch coating possibly improved its function in retarding some enzymatic reactions related to ripening (Franco et al. 2017).

Starch based coatings have also been used as a medium for carrying additives for preservation fresh and minimally processed (fresh-cut) fruit and vegetables. Ojeda et al. (2014) found that cassava starch coatings with ascorbic acid reduced browning, changes in hue and the activity of the polyphenol oxidase, phenylalanine ammonia lyase and polyphenol oxidase in minimally processed sweet potato. Whilst Pająk et al. (2017) showed that potato starch with white and green tea extracts prevented weight loss, darkening, and reduction in antioxidant activity and total phenolic content in apple slices during storage at 5 °C and 84% RH for 6 days.

9.2.2 Non-Starch Polysaccharide-Based Coatings

The application of non-starch polysaccharides edible coatings on fresh and fresh-cut fruit and vegetables, has been significant interest in recent years due to their low cost and nontoxicity (Williams and Phillips 2000).

Water-soluble non-starch polysaccharides are heterogeneous groups of long chain polymers categorized by their capability of developing viscous dispersions and/or gels when dispersed in water (Saha and Bhattacharya 2010). These compounds are commercially used in the food industry as agents for improving mouth feel, stability, thickening and gelling inhibiting crystalization, and encapsulating agents (Stephen et al. 2006; Izydorczyk et al. 2005). Non-starch polysaccharides can also act as emulsifiers due to their stabilizing functions, which increase the viscosity of the aqueous phase of the edible coating (Nussinovitch 1997). This reduces the kinetic motion of the oil droplets, leading to a low rate of flocculation and coalescence in the coating (Nisperos-Carriedo 1994).

Typically, non-starch polysaccharide coatings have poor moisture barriers, but low oxygen permeability with selective permeability to O_2 and CO_2 (Lacroix and Le Tien 2005). Consequently, non-starch polysaccharide based coatings have been examined on fresh and minimally processed fresh fruit and vegetables, to reduce water loss, decrease the respiration rates by modifying the atmospheric conditions inside the product, improve mechanical handling characteristics, deliver additives, and to maintain and improve volatile profiles (Olivas and Barbosa-Cánovas 2005). Table 9.1 summarizes several studies on application of various non-starch polysaccharides edible coatings and gums in extending shelf life of fruit and vegetables.

Table 9.1 Several recent studies on application of non-starch polysaccharide-based coatings on fresh and fresh-cut fruit and vegetables

Coatings	Additional compounds	Fruit/ vegetable	Results	References
Cellulose		Honeydew winter melon	Coating reduced internal O_2 levels, leading to a considerable increase in ethane, acetaldehyde, ethanol, and ethyl acetate amounts. The cellulose coating caused anaerobic respiration in fruit	Vanoli et al. (2015)
CMC	Garlic essential oil	Strawberry	Coatings had a positive effect on the WL, DR, TSS, TA, and AAC, and on maintaining higher concentrations of TPC and anthocyanins of strawberries	Dong and Wang (2017)
	Moringa leaf extract	Avocado	The combination of CMC (1%) and 2% moringa reduced WL, electrical conductivity, and RR and had higher values for F and phytochemical characteristics	Tesfay and Magwaza (2017)
	Nano-ZnO	Fresh cut pomegranate aril	Coatings reduced total yeast + mold during 12 days of storage while total mesophilic bacteria were decreased during 6 days of storage. Coatings decreased WL, suppressed TSS and TPC changes and also the greatest juice percent was in coated arils. Total anthocyanin, AAC, and antioxidant capacity were higher in coated arils	Saba and Amini (2017)
	Extract of *Impatiens balsamina L.* stems	Tangerine and navel orange	Coatings had an inhibitory influence on mold growth, decreased DR and WL, maintained commercial quality and enhanced the activities of antioxidant and defense-related enzymes	Zeng et al. (2013) and Chen et al. (2017)

(continued)

Table 9.1 (continued)

Coatings	Additional compounds	Fruit/ vegetable	Results	References
CMC	Calcium chloride and ascorbic acid	Fresh-cut apple	Coatings suppressed browning, retained F, AAC and antioxidant capacity and decreased PPO and POX activity, TSS, TA and pH changes of the slices	Saba and Sogvar (2016)
HPMC	Oregano and bergamot essential oils	Plum	Coatings decreased RR, EP, WL, TMC, C change, and fruit softening	Choi et al. (2016)
MC		Strawberry	Coating did not affect TA, anthocyanin and antioxidant activity of coated strawberry compared to the control one	Nadim et al. (2015)
Pectin	Ascorbic acid, citric acid and sodium chlorite	Fresh-cut apple	Coatings reduced microbial spoilage while did not significantly influence sensory and nutritional qualities. The anti-browning agents further enhanced this ability	Guerreiro et al. (2017)
	Orange peel essential oil	Fresh-cut orange	Coatings reduced the quality loss and improved the sensory scores during storage. The nanoemulsion pectin-based coatings containing 1% essential oil were the most effective in bacterial and fungal inactivation	Radi et al. (2017)
		Blueberry	Coating increased the F and the blue hue color and decreased the growth kinetics of yeasts and mesophilic aerobic bacteria	Mannozzi et al. (2017)

(continued)

Table 9.1 (continued)

Coatings	Additional compounds	Fruit/ vegetable	Results	References
Pectin	Citral and eugenol	Raspberry	Coatings were not cytotoxic and did not considerably change the general physicochemical and nutritional characteristics of raspberries. Their impact was mainly on decreasing food spoilage microorganisms and accordingly extending shelf-life	Guerreiro et al. (2016)
	Oregano essential oil (OEO)	Tomato	Coatings with OEO exhibited antifungal influence on inoculated tomatoes, and increased TPC and antioxidant activity. The sensorial acceptability of the coated tomatoes was well accepted by panelists	Rodriguez-Garcia et al. (2016)
Alginate	Lemon essential oil or orange essential oil	Red raspberry	The less red color verified in coated samples was coincident with the lower concentration of anthocyanins as well as the lower capacity for scavenging ABTS free radicals or quenching singlet oxygen. The coatings with the essential oil of orange were very efficient for controlling yeast and mold growth after 15 days of storage	Gomes et al. (2017)
	Olive oil	Ber fruit	Coatings decreased DR, WL, TSS and total sugars and increased the level of antioxidants. The delayed activity of PG, PL and PME was noticed in coated fruit representing the reduced softening and ripening process	Rao et al. (2016)

(continued)

Table 9.1 (continued)

Coatings	Additional compounds	Fruit/ vegetable	Results	References
Alginate	Tea polyphenols	Chinese winter jujube	Coatings decreased red indices, TCC, RR, electrolyte leakage and malonaldehyde content and maintained the AAC, TPC and the activities of antioxidant enzymes while had no significant effect on F	Zhang et al. (2016)
	Ficus hirta fruit extract	Nanfeng mandarin	The DR, WL, RR and MDA content were much lower in the coated samples. The coating treatment enhanced the activities of antioxidant and defense-related enzymes such as SOD, CAT, POD, CHI, GLU and PAL and the accumulation of phenolic compounds	Chen et al. (2016)
	Bacteriocin	Minimally processed papaya	The alginate coating performed as a barrier to WVT and gas exchange, which delayed changes in TSS values, F, WL and ripening in coated samples. Coating preserved minimally processed papaya for 3 weeks without reducing physico-chemical qualities or microbial safety	Narsaiah et al. (2015)
	Grapefruit seed extract (GSE) or grapefruit essential oil (GEO)	Table grape	Coatings reduced WL, maintained F during storage, preserved the antioxidant activity of treated grapes and decreased DR in inoculated fruit	Aloui et al. (2014)

(continued)

Table 9.1 (continued)

Coatings	Additional compounds	Fruit/ vegetable	Results	References
Xanthan	β-Carotene nanocapsules	Fresh-cut melon	The minor changes observed in the whiteness index and F. Incorporating of β-carotene nanocapsules into xanthan gum increased fruit shelf life	Zambrano-Zaragoza et al. (2017)
	α-Tocopherol nanocapsules	Fresh-cut apple	Firmness changes were reduced and the quality of fresh-cut apples was maintained by coating	Zambrano-Zaragoza et al. (2014a)
	Candeuba® wax nanoparticles	Guava	Coated fruit showed the lowest range of WL. High contents of nanoparticles caused physiological damage and also delayed the fruit maturation	Zambrano-Zaragoza et al. (2013)
	Cinnamic acid	Fresh-cut pear	Coating caused significant retardation of the oxidative browning, decline of AAC, degradation of TPC and reduction in antioxidant capacity	Sharma and Rao (2015)
Pullulan	Caraway essential oil	Fresh baby carrot	Coatings were active against all tested microorganisms and maintained better visual acceptability in comparison with control	Gniewosz et al. (2013)
	Sweet basil extract	Apple	Coating showed low antibacterial activity against mesophilic bacteria and good antifungal protection against *Rhizopus arrhizus* on apple surfaces They also contributed to a reduction in WL and lower changes in the C and TSS of apples. Coated fruit presented better overall preference parameters	Synowiec et al. (2014)

(continued)

Table 9.1 (continued)

Coatings	Additional compounds	Fruit/ vegetable	Results	References
Pullulan	Sodium benzoate and potassium sorbate	Strawberry	Coating decreased WL, fruit softening and microbial growth and delayed changes of C and TSS content. In contrast, pH and TA were not affected. Sensory quality (color, flavor, texture, and acceptance) improved and DR decreased in coated strawberries	Treviño-Garza et al. (2015)
	n-Octenyl succinic anhydride	Sapota (*Manilkara zapota*)	Coating reduced flesh F loss and WL and delayed the ripening and senescence	Shah et al. (2016a)
		Jujube	Coating treatment inhibited fruit softening, increase in redness index scores, WL rates, and reduced the loss of AA and freshness. Defensive enzymes and antioxidant activity and antioxidant compounds increased in coating-treated fruit	Kou et al. (2017)
Gum Arabic	Oregano and rosemary essential oils	Plum	Coatings delayed the occurrence of soft rot and decreased the rotted plums at the end of storage. Coated fruit exhibited greater F, decreased WL and lower decrease of sugars and phenolics at the end of storage	Andrade et al. (2017)
	Ginger oil and ginger extract	Papaya	Coating delayed the ripening and showed antifungal activity. Quality of papaya fruit in terms of F, C, TSS and TA was maintained	Ali et al. (2016)

(continued)

Table 9.1 (continued)

Coatings	Additional compounds	Fruit/ vegetable	Results	References
Gum Arabic	Calcium chloride	Mango	Coating reduced chilling injury, MDA content and electrolyte leakage. This treatment increased antioxidant activity and effectively inhibited the loss of TPC and AA. The treated fruit maintained cell membrane integrity	Khaliq et al. (2016)
		Tomato	Coatings delayed the ripening process by slowing down RR and EP and also maintained total antioxidant capacity, lycopene content, TPC and total carotenoids during storage	Ali et al. (2013)
Chitosan	*Cymbopogon citratus* (Dc. Ex Nees) (lemongrass) essential oil	Tomato	Coating decreased the severity of *Rhizopus* soft rot and more strongly delayed the infection when the fruit were artificially contaminated after coating application. The application of the coating preserved the general quality of tomato fruit	Athayde et al. (2016)
	Natamycin, nisin, pomegranate and grape seed extract	Strawberry	Coating reduced the O_2consumption of the fruit and showed better effects on delaying changes of pH, TSS, water activity and TMC. The incorporation of different antimicrobial agents into chitosan matrix did not reveal any significant effect on C of strawberry	Duran et al. (2016)

(continued)

Table 9.1 (continued)

Coatings	Additional compounds	Fruit/ vegetable	Results	References
Chitosan		Red kiwifruit berries	Coating did not affect the changes in weight during the first 12 days of storage. No significant differences in terms of F were determined at the end of storage. Coated berries showed higher amounts of biologically active substances and TSS	Kaya et al. (2016)
	Lemongrass oil	Grape berry	Coating with nano-droplet of oil showed higher initial inhibition of *Salmonella typhimurium*; greater growth inhibition of microorganisms and higher retention of C, TSS, antioxidant activity and better SE during storage	Oh et al. (2017)
	Salvia fruticosa Mill. extract	Table grapes	The efficacy of the coating against grey mold was statistically equal to the synthetic fungicide thiabendazole. Coating decreased the rate of fruit WL during cold storage, while preserved TSS and TA. Coatings did not affect quality attributes and the bioactive compounds in table grapes	Kanetis et al. (2017)
	Thyme essential oil nanoparticles	Avocado	The coating reduced the incidence of *C. gloeosporioides* on avocado. Coating did not affect the quality of avocado; moreover, fruit F was better maintained than untreated fruit	Correa-Pacheco et al. (2017)

(continued)

Table 9.1 (continued)

Coatings	Additional compounds	Fruit/ vegetable	Results	References
Chitosan		Mango	Chitosan delayed the climacteric peak, WL and F. Further, few changes in TSS, TA, pH of the pulp as well as in sugar content and decreased starch degradation were observed	Silva et al. (2017)
		Pomegranate	Coating effectively reduced rot incidence of *Botrytis sp*	Munhuweyi et al. (2017)
Carrageenan	Bacteriocin	Apple cubes	Reduced viable *Listeria* counts compared to the untreated apple cubes	Aguayo et al. (2016)
		Papaya	Increased firmness and delayed ripening and C changes	Hamzah et al. (2013)
Tragacanth gum	*Satureja khuzistanica essential oil*	Button mushroom (*Agaricus bisporus*)	Increased F, reduced TMC and browning index. Higher levels of TPC and AAC were observed in coated samples. SE demonstrated the capability of coating for preserving the quality of mushroom during the storage	Nasiri et al. (2017)
Locust bean gum	Biocontrol agent	Mandarin	Coating provided excellent control of postharvest decays caused by *P. digitatum* and *P. italicum* on mandarins	Parafati et al. (2016)
Tara gum	Ascorbic acid, citric acid and $CaCl_2$	Minimally processed peach	Reduced WL, C alteration and growth of molds and yeasts and maintained F	Pizato et al. (2013)
Guar gum	Silver nanoparticle	Kinnow	Coating preserved the fruit aroma and sensory quality	Shah et al. (2016b)

(continued)

Table 9.1 (continued)

Coatings	Additional compounds	Fruit/ vegetable	Results	References
Gellan	Sunflower oil, ascorbic acid, citric acid and $CaCl_2$	Fresh-cut pineapple	Coating reduced RR and WL and maintained the F and C. Coating did not change pH, TA and TSS and did not show any antimicrobial effect	Azarakhsh et al. (2014)
	Calcium gluconolactate	Ready-to-eat mango	Coating improved mango bars sensory characteristics (appearance and F) and stability in terms of syneresis, C and volatiles content during storage	Danalache et al. (2016)
	Potassium sorbate, calcium chloride, 1-methylcyclopropene (1-MCP)	Pre-cut jackfruit	Coating decreased the ripening rate; WL, and RR and preserved F, C, TSS, TA as well as pH until 12 d of storage. Microbial growth was hindered until the12th d of storage in coated fruit	Vargas-Torres et al. (2017)
Lignin		Lime	Maintained WL and C change. This coating formula also exhibited higher antifungal activities	Jonglertjunya et al. (2014)
Basil-seed gum	*Origanum vulgare subsp. viride* essential oil	Fresh cut apricot	Coatings reduced the TMC and significantly enhanced TPC and antioxidant activity of coated samples at the end of cold storage	Hashemi et al. (2017)
Flaxseed gum	Lemongrass essential oil	Ready-to-eat pomegranate arils	Coatings were effective in reducing TMC and reduced WL, ripening index, and changes in TSS, pH, TA and C	Yousuf and Srivastava (2017)
Almond gum		Sweet cherry	Fruit coated with almond gum showed a significant decrease in RR as well as EP. Moreover, coatings were able to delay changes in weight, F, TA, TSS and C development	Mahfoudhi and Hamdi (2015)

(continued)

Table 9.1 (continued)

Coatings	Additional compounds	Fruit/ vegetable	Results	References
Persian gum		Mandarin	Coatings reduced the weight loss. Persian gum caused glossiness in coated fruit; however, they were less effective than the commercial wax coating	Khorram et al. (2017a)
		Orange	Coating increased fruit gloss and was effective at improving postharvest quality of fruit during storage. Coating was not stable and showed visible cracks during storage	Khorram et al. (2017b)
Nopal mucilage	Nanoemulsion of α-tocopherol	Fresh-cut apple	PME activity in the coated apples was lower helping maintain the F of the coated fruit. At 21 days of storage, PPO activity decreased by 65% in the coated apples with nanoemulsion, as reflected in the lower browning indexes	Zambrano-Zaragoza et al. (2014b)
Opuntia cladodes mucilage		Fig	Coating improved the quality of fig during storage by maintaining fruit fresh weight, visual score values, F and total carotenoid content. Coated fruit showed a significantly lower development of *Enterobacteriaceae* than control ones	Allegra et al. (2017)
	TWEEN® 20	Kiwifruit slices	Coated samples showed a significant higher F and a lower WL until 5 d of shelf life. The treatment with Tween 20 did not affect the flavor of the kiwifruit slices, while increased microbial growth at the end of the monitoring period	Allegra et al. (2016)

(continued)

Table 9.1 (continued)

Coatings	Additional compounds	Fruit/ vegetable	Results	References
Aloe vera		Strawberry	Coating reduced rot incidence, RR, WL, and DR and preserved F, AAC, and other quality parameters. Furthermore, aloe vera gel delayed the changes in external C and retained all other postharvest quality	Nasrin et al. (2017)
		Tomato	Coating reduced fruit EP, ripening index (TSS/TA) and maintained the overall quality of the tomato fruit. Lycopene and β-carotene content were reduced while AAC, TPC and antioxidative status were increased. Fruit F, TA, WL, RR and fruit C did not differ among treatments	Chrysargyris et al. (2016)
		Kiwifruit slice	Aloe vera coating maintained the F of the fruit, prevented the AA losses and yellowing due to ripening and reduced microbial proliferation. The sensory panel preferred the kiwifruit slices treated with Aloe vera compared to the other coatings	Benítez et al. (2015)
	Calcium chloride, ascorbic acid, and vanillin	Fresh-cut papaya	Coating preserved TPC and AA and reduced microbial load and relatively low PPO and POX activity during storage	Kuwar et al. (2015)
		Raspberry	Coated samples maintained higher levels of antioxidant capacity, TPC, total anthocyanin and antioxidant enzymes during storage periods	Hassanpour (2015)

(continued)

Table 9.1 (continued)

Coatings	Additional compounds	Fruit/ vegetable	Results	References
Aloe vera	Rosehip oil	Peach, plum, nectarine and sweet cherry	Coating reduced RR in all fruit, and EP in the climacteric ones (peaches, plums and nectarine). In addition, all the parameters related with fruit ripening and quality, such as WL, softening, C change and ripening index, were also delayed in treated compared with control fruit	Paladines et al. (2014)

AAC ascorbic acid content, *C* color, *CAT* catalase, *CHI* chitinase, *DR* decay rate, *EP* ethylene production, *F* firmness, *GLU* β-1,3-glucanase, *MDA* maleicdialdehyde. *PAL* phenylalanine ammonia lyase, *PG* polygalacturonase, *PL* pectate lyase, *PME* pectin methyl esterase, *POX* peroxidase, *PPO* polyphenol oxidase, *RR* respiration rate, *SE* sensory evaluation, *SOD* superoxide dismutase, *TCC* total chlorophylls content, *TFC* total flavonoid content, *TMC* total microbial count, *TPC* total phenolic content, *WL* weight loss, *WVT* water vapor transmission

9.3 Protein–Based Coatings

Coatings may also be developed from both animal and plant protein sources (Cerqueira et al. 2016). Proteins are macromolecules with amino acid sequences and molecular conformation structures. The unique conformational structure, electrostatic charges, and amphiphilic nature are the most exclusive characteristics of proteins compared with other coating-forming materials (Hun and Cennadios 2005). The presence of different charged, polar and non-polar amino acids in the protein provides multiple sites for chemical interaction to improve and tailor the protein functional properties. Proteins are not entirely hydrophobic and encompass a range of hydrophilic amino acid residues restricting their moisture-barrier characteristics (Shit and Shah 2014). However protein coatings have a good oxygen barrier and be used to create an efficient semi-permeable barrier to respiratory gases resulting in development of a modified atmosphere (MA) within the fresh produce, in which O_2 is not freely replaced, nor can the CO_2 produced freely escape (Baldwin and Baker 2002). The MA within the produce reduces respiration and ethylene production rate, delays ripening and senescence, and ultimately extends the produce shelf life. Proteins which have been used in coatings include zein, soy protein, wheat gluten, whey protein, casein, gelatin, and collagen (Hun and Cennadios 2005). Protein based coating films serving as a vehicle to incorporate various active agents have been investigated on different fresh produce.

Soy protein isolate (SPI)-coatings incorporated with the essential component citral and limonene at 10% reduced weight loss of Persian lime (*Citrus latifolia Tanaka*) after 9 days storage at 13 °C and 95% RH, demonstrating their effectiveness in producing a physical barrier to moisture loss and hence postponing dehydration and fruit shriveling (González-Estrada et al. 2017). It was deduced that storage of coated fruits at 25 °C and 75% RH caused a darkening of the rind and/or desiccation of the fruit due to creation of a modified atmosphere around the coated product, which influenced the respiration rate, during this process that comprises the production of CO_2, water and heat through oxidation carbohydrates, negative changes in color of the fruit occurred (Galus and Kadzińska 2015). At the end of storage, coatings started to remove easily from surface of the limes demonstrating that the biodegradation of the coatings such as hydrolysis of polymer chain due to the acidity of fruit or respiration and gas exchanges was dependent on the storage conditions (González-Estrada et al. 2017).

The cereal protein gluten has been also examined as a fruit coating where gluten-based coatings incorporating pomegranate peel and curry leaf extracts did not cause any undesirable color changes in cherry tomatoes and mangoes, but they considerably enhanced the glossiness of the fruit (Kumari et al. 2017). However the use of gluten protein as a coating ingredient maybe limited with widespread gluten intolerance in the community.

Baraiya et al. (2015) evaluated the efficiency of zein-based coatings including cysteine (0.2%), ascorbic acid (0.2%), and jamun leaves extract (JLE) (0.2%) on shelf life and quality of fresh jamun fruit at 10 °C for 2 weeks storage. The results demonstrated that coated fruit had lower weight loss and decay incidence, low accumulation of sugars, and reduced softening and ripening compared to that of uncoated fruit. The levels of antioxidants were found to higher in zein coated fruit which also had higher antioxidant activity. Furthermore the coated fruit had higher firmness and textural qualities with lower activities of polygalacturonase (PG) and pectate lyase (PL). Yun et al. (2015) also showed that zein-based coatings with the addition of cinnamon essential oils decreased the population of *Salmonella enterica serovar Typhimurium* in cherry tomato during 7 days of storage at 10 °C. These coatings decreased the loss of firmness and weight and preserved or improved the quality of fruit (Yun et al. 2015).

Butt et al. (2015) reported that coating with a mixture of 2% sodium caseinate fortified with 40 ppm ZnO was the most effective coating in maintaining the quality characteristics of fresh apricots at 4–6 °C and 85% RH for 6 weeks.

Gelatin coatings with low content of lactic acid (below 0.2%) have been shown to delay ripening in plums after 12 days of storage at 20 °C (Peter et al. 2017). Yang et al. (2017) also studied the potential of application of gelatin extracted from dried Alaska pollock with addition of pine needle extract on quality maintenance of grape berries. They showed that this coating reduced the populations of total aerobic bacteria and yeast and molds and retarded weight loss and changes in anthocyanin content of the grape berries after 20 days of storage.

9.4 Lipid-Based Coatings

Edible lipids including neutral lipids (esters of glycerol and fatty acids), fatty acids, waxes (esters of long-chain monohydric alcohols and fatty acids), and resins (a group of acidic substances that are usually secreted by special plant cells into long resin ducts or canals in response to injury or infection in many trees and shrubs) are traditional coating materials for fresh produce, which provide a good moisture barrier and improve surface appearance (Hagenmaier and Baker 1994, 1995; Morillon et al. 2002). Lipids can be added as formulations of edible coatings as a single layer of lipid dispersed in a hydrocolloid network or in a secondary layer (a lipid layer over a hydrocolloid layer) (Olivas and Barbosa-Cánovas 2009). Emulsion-based coatings in which the lipid is dispersed in the biopolymer matrix are prepared during only one film-forming casting and one drying process (Galus and Kadzińska 2015). Their characteristics are dependent on preparation methods, type and number of components (hydrocolloid and lipid) and their compatibility, as well as the microstructural heterogeneity (Fabra et al. 2011). These emulsion-based coatings are more water permeable than bilayer coatings due to the lack of the homogeneous distribution of lipids in coating matrix, but they have desirable mechanical strength and are relatively easy to manufacture and apply, while multilayer coatings require a complex set of procedures depending on the number of coatings (Galus and Kadzińska 2015). However, it should be noted that the small size of lipid particles and their homogeneous distribution reduce emulsion instability and water vapor permeability (Pérez-Gago and Krochta 2001).

The compatibility of using lipid-based coatings with other coating-forming agents with high gas-barrier and water vapor properties is a major advantage as compared to the use of polysaccharides- and protein-based coatings (Greener and Fennema 1989), since lipids exhibit a very low tendency for water absorption, coatings comprising of lipids generally have good moisture barrier characteristics. However, the differential properties of the lipid component such as its physical state, degree of saturation and chain length of fatty acids all affect the physical and mechanical properties of lipid-based coatings (Olivas and Barbosa-Cánovas 2009). Saturated long-chain fatty acids have a more densely packed structure and less mobility than unsaturated short-chain fatty acids, and therefore have been shown to have the best water vapor barrier properties (Morillon et al. 2002). Lipids which are in the liquid at the desired storage temperature, will have lower water vapor barrier properties than lipids which are solid under same conditions, since the solubility of water vapor in lipids is principally higher in coatings having less ordered molecular organization (Galus and Kadzińska 2015). The gas barrier characteristics of lipids are also strongly determined by their chemical composition. Hydrophobic compounds are generally more permeable to gases owing to their greater chemical affinity and solubility (Miller and Krochta 1997), causing an increase in oxygen permeability of the coating (Navarro-Tarazaga et al. 2011; Jiménez et al. 2013). The shorter the hydrocarbon chain length of the fatty acids, the weaker the attraction forces between molecules, which consequently increases the CO_2 transmission rate of these coatings (Ayranci and Tunc 2001).

The candelilla wax-based edible coating with fermented extract of tarbush (as natural antioxidants source) and carnauba-shellac wax containing lemon essential oil, significantly improved the quality and shelf life of apples stored at room temperature and cold storage (De León-Zapata et al. 2015; Jo et al. 2014). Ochoa-Reyes et al. (2013) also described that an edible coating of candelilla wax with extract of tarbush inhibited weight and firmness loss, and preserved the appearance of green bell peppers.

An increase in the quality and shelf life was also observed for strawberry coated with candelilla wax-biocontrol bacteria (*B. subtilis*) (Oregel-Zamudio et al. 2017), pomegranate coated with carnauba wax-putrescine (Barman et al. 2014) and plum with carnauba wax-lemongrass oil coatings (Kim et al. 2013). Rice bran wax coating has been reported as an efficient alternative for preservation cherry tomato at 0 °C for 20 days (Zhang et al. 2017).

The activity of the antioxidant enzymes has also been shown to increase in the flesh of Ponkan (*Citrus reticulate* Blanco) fruit coated with wax incorporated with cinnamaldehyde (Duan et al. 2017). Lipid-based coatings are thought to protect the cell membrane structure and the fruit tissue by hindering the buildup of reactive oxygen species, leading to less oxidative stress and destruction to fruit (Duan et al. 2017).

9.5 Biocomposite Coatings

Coatings based on polysaccharides commonly exhibit efficient O_2 and CO_2 barriers, but have poor mechanical strength and high moisture sensitivity (Al-Hassan and Norziah 2012). To overcome these problems, their physical and functional properties can be modified by combining with other biopolymers, hydrophobic constituents, and antimicrobial/antioxidant compounds (Saberi et al. 2017; Álvarez et al. 2018). The blending of biopolymers has been demonstrated to improve the mechanical characteristics of the resultant coating (Veiga-Santos et al. 2005) depending on the compatibility/incompatibility of binary polymeric blends, their molecular weight, chemical structures, conformations, and hydration behaviors (Phan The et al. 2009; Gutiérrez and Alvarez 2017).

Forato et al. (2015) applied a composite edible coating made of cashew gum (CG) and carboxymethylcellulose (CMC) on fresh and fresh-cut red guavas. The results showed that the uncoated samples had the fastest degradation rate (around 2.5% of their mass daily), while the coating formulations performed appropriately as a conservative agent, decreasing the mass loss to approximately 2% a day for both intact and sliced fruit. Textural softening in coated guavas in both intact and in sliced forms was delayed as a consequence of reduction of the level of ethylene and enzymatic activity, as well as respiratory activity. Magnetic Resonance Imaging (MRI) confirmed that the addition of CMC presented a relevant role in forming coatings which decreased the free water content in coated fruit resulting from the hydrolysis of starch into sugar plus water and preserved the appearance by reducing color change as the maturation proceeded.

Lai et al. (2013) concluded that tapioca starch/decolorized hsian-tsao leaf gum (dHG) composite coating was suitable for prolonging the shelf life of fresh-cut carrots. The application of tapioca starch/dHG coating containing ascorbic acid and calcium chloride was also suggested for fresh-cut apples, as it could protect qualities in terms of color and firmness, and extended the shelf life up to 5–7 days by conferring preferable microbial quality (Pan et al. 2013). The application of biocomposite edible coating based on sodium alginate and pectin reduced the loss of firmness and microbial growth on blueberries (Mannozzi et al. 2017).

Proteins have the capability to develop extensive intermolecular hydrogen bonds with other biopolymers owing to the existence of a large number of polar (-OH and -NH) groups in their structure; electrostatic and hydrophobic bonds due to their random coil nature. Moreover, the moisture barrier, mechanical strength, O_2 and CO_2 barrier properties of protein based coatings can be improved by incorporating polysaccharide materials in their matrix (Siew et al. 1999).

Murmu and Mishra (2017) noted that an edible coating containing Arabic gum and sodium caseinate had a significant influence on changing the O_2, CO_2, and water vapor transmission rate of the coated guava in comparison with the uncoated control. Coating formulation with low concentration of Arabic gum (5 g/100 mL) was too thin to decelerate rate of respiration, ripening, senescence, and mold growth, while those with high concentration (12 g/100 mL) produced too thick coating leading to high rate of O_2 consumption, CO_2 evolution, mass loss, lower softening and overall acceptability of guava following 7 days of storage at 28 °C. It is thus crucial to optimize the solid content in the coating formulation so that coating may not have an unnecessary constraint of gas exchange through the skin, causing anaerobic and further development of off-flavors (Vargas et al. 2008).

Improved appearance and lower weight loss in Red Crimson grapes treated with starch-gelatin coating was observed after 21 days storage under refrigerated conditions. Sensory evaluation also presented that the coatings did not influence acceptability scores (Fakhouri et al. 2015). A composite chitosan–gelatin coating was applied to peppers and its effectiveness on fruit quality and storability was analyzed (Poverenov et al. 2014b). It was concluded that the composite coating decreased microbial decay, noticeably improved fruit texture and extended the possible cold storage period up to 21 days and fruit shelf-life up to 14 days, without changing the respiration or nutritional content of the fruit.

Fresh cut apples, potatoes and carrots were coated by a composite whey protein-pectin coating by addition of transglutaminase (Marquez et al. 2017). Coating not only prohibited microbial growth in all samples analyzed, but also preserved the phenolic content and carotenoid in carrots. Finally, an obvious reduction of hardness and chewiness loss was noticed after 10 days of storage in all the coated samples.

Lipid materials are also commonly examined in combination with polysaccharide- or protein-based coating materials to make composite coatings. Because lipid-based coatings alone can produce a greasy surface with adverse organoleptic characteristics such as a waxy taste and lipid rancidity of the coated produce (Olivas and Barbosa-Cánovas 2009). In addition, some waxes can result in a lower gas exchange of O_2 and CO_2 between atmosphere which causes the internal O_2 level reaching too low to provide aerobic respiration (Alleyne and Hagenmaier 2000). The subsequent anaerobic respiration results in the formation of ethanol, acetaldehyde which are responsive for many of the off-flavors in fruit (Dhall 2013; Porat et al. 2005). Furthermore, some lipid materials such as shellac, are unstable when exposed to variations of temperature, where a white waxy layer usually forms when transferring fruit from cold storage to the market (Lin and Zhao 2007). This is known as 'chalking' and is not acceptable in many markets.

Moreover, high water activity ($a_w > 0.94$) and high biochemical activity because of mechanical damages during peeling or slicing are two main features of minimally processed fruit and vegetables (Galus and Kadzińska 2015). The high solubility of hydrocolloid coatings at high water activity restricts their application due to disintegration and loss of their properties. Lipid based coatings can be a good substitute owing to their stability. Nonetheless, they can adversely affect the sensory characteristics of produce, for example, resulting in a waxy sensation. That confirms the need for more studies which aim to produce composite edible coatings preserving fruit and vegetables effectively and at the same time having no undesirable effect on product properties.

The application of lipid based coatings (carnauba wax emulsion coating formulated with poly ethylene glycol and sodium alginate) have been shown to reduce the loss of firmness, moisture, weight, lightness, TPC and antioxidants activity in eggplants packaged in 35 μ polypropylene pouches during ambient storage (Singh et al. 2016).

Pérez-Gallardo et al. (2015) investigated the effect of a starch-beeswax dispersion comprising 2% (w/v) modified tapioca starch added with either 0.5 or 1.0% (w/v) beeswax for spray coating on freshly harvested blackberries (*Rubus spp.*) during 16 days storage at 4 °C and 88% RH. The micrographs of fruit presented a smooth and continuous surface without cracks or pores and more surface area was protected by a thicker coating as the beeswax concentration enhanced. Both coatings caused a considerable increase in volatiles such as 1-octanol, and ethanol and aldehydes after 9 days, which have been associated with fermentative metabolism. The authors explained that this behavior was related to an enhanced ripening process generated by coatings, resulting in higher production of ethylene and CO_2 and accumulation of some volatiles in the intracellular tissue (Amarante and Banks 2010). Coating application reduced the anthocyanins and total phenols in blackberries due to the stress prompted by coatings and increased ethylene production, as well as accumulation of CO_2 destructing internal tissues and inducing oxidation of phenolic compounds by enzymatic reactions, including polyphenoloxidase and peroxidase (Duan et al. 2011).

It has been reported that incorporation of oleic acid and palm oil into MC-based coating improved shelf life of green chili (*Pusa jwala*) and Sapota (*Manilkara zapota* L. var. *Kalipatti*), respectively (Chaple et al. 2017; Vishwasrao and

Ananthanarayan 2017). The efficiency of edible composite coatings based on HPMC, beeswax, and various food preservatives with antifungal activities on preservation of cherry tomatoes during cold storage was also observed (Fagundes et al. 2013, 2014, 2015).

The use of quinoa protein-chitosan-sunflower oil coating presented a substantial lower amount of mold and yeast growth in coated strawberries and blueberries during storage (Abugoch et al. 2016; Valenzuela et al. 2015). Kowalczyk et al. (2017) demonstrated that the coating composed of CMC, candelilla wax (CnW) and potassium sorbate (KS) caused anaerobic respiration and the signs of superficial scald in pears after 9 days storage at 22 ± 1 °C, 50 ± 5% RH. It was explained that superficial scald was probably due to accumulation of CO_2 (up to a critical level) in the internal atmosphere of coated fruit (Lurie and Watkins 2012). High CO_2 level may trigger the creation of reactive oxygen species (Larrigaudière et al. 2004), which are greatly active and may extensively induce lipid peroxidation, leading to the more development of free radicals. However, the coating treatment retained green color of pears owing to decreasing chlorophyllase activity and creation of the modified atmosphere within the fruit tissues. (Guevara et al. 2001). It was also observed that uncoated fruit had common symptoms of contamination including softening, browning, or necrosis on the 2nd or the 3rd/4rd day of storage.

9.6 Layer–By–Layer Coatings

The layer-by-layer (LBL) electrostatic deposition technique as an approach to improve the functional and mechanical properties of coatings has extensive range of applications (Arnon et al. 2015). Breaks in the fruit skin and wax layers increase the movement of water from the fruit, initiating mass flow and quality reduction (Liu et al. 2017). Therefore, the LBL technique is a useful and simple approach to develop thin coating that prevents these surface breaks (Wang et al. 2011). LBL technique is based on the different deposition of oppositely charged polyelectrolytes (Arnon et al. 2015), which can be originated by many weak interactions, such as electrostatic interactions, hydrogen-bonds, coordination bonds, charge transfer interactions, and guest-host interactions (Jia and Li 2015). A LBL approach based on a combination of two polysaccharides, CMC as an internal layer and chitosan as an external layer, was applied on 'Rishon' and 'Michal' mandarins as shown in Fig. 9.1 (Arnon et al. 2015). It was found that the mandarins coated by LBL formulation were firmer than the fruits coated by commercial wax. Regarding weight loss, bi-layered coating was more effective than was the single coating and less efficient than the commercial wax. The bi-layered coating slowed down the ripening progress, as was detected by restriction of mandarin color change. There was not significant differences in CO_2 concentration in the internal atmosphere of mandarins coated by LBL coating and the ones coated by commercial wax. Whereas, the

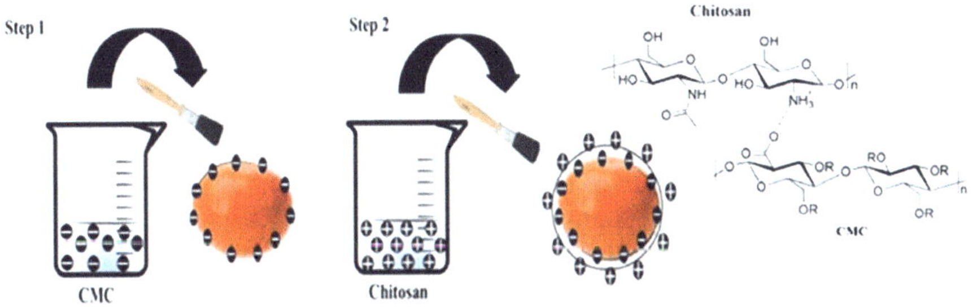

Fig. 9.1 Schematic presentation of the electrostatic deposition method used to form LBL edible coatings of mandarin fruit. (Reprinted from Arnon et al. (2015), Food Chem 166, 465–472. With permission)

ethanol concentration in the juice of LBL-coated fruit was considerably higher than the ethanol concentration in the juice of commercial wax-coated fruit. A similar observation of higher levels of ethanol causing increased perception of off-flavors upon application of the LBL method (pea starch-guar gum as an internal layer and shellac as an external layer) was determined in 'Valencia' oranges (Saberi et al. 2018).

Reyes-Avalos et al. (2016) treated figs (*Ficus carica*) with an alginate-chitosan bilayer edible coating and stored fruit at 6 °C. They showed that the application of the bilayer coating resulted in lower CO_2 production and weight loss and increased firmness and ethylene emission in figs. Moreover, coated figs had better retention of color as well as lower visual fungal infection than uncoated fruit during storage. A bilayer CMC-chitosan coating was also shown to inhibit weight loss, softening and slow surface browning of fresh-cut apples (Liu et al. 2017). A multilayer coating with the following elements: calcium chloride, chitosan with encapsulated trans-cinnamaldehyde, pectin, and calcium chloride, maintained the physicochemical and sensory quality of fresh-cut cantaloupe for 3–5 additional days than the uncoated control slices during storage at 4 °C (Martiñon et al. 2014).

The LBL formulations based on gelatin-chitosan for fresh cut melon, CMC-chitosan, alginate-chitosan for fresh cut mango, and cactus mucilage/pullulan/lin-seed/aloe mucilage-chitosan on fresh cut pineapple demonstrated improved quality during storage (Poverenov et al. 2014a; Arnon et al. 2014; Souza et al. 2015; Treviño-Garza et al. 2017). Sipahi et al. (2013) showed that the color, odor, and flavor attributes of fresh-cut watermelon coated with a multilayered antimicrobial coating containing sodium alginate, beta-cyclodextrin and microencapsulated tran-scinnamaldehyde (natural antimicrobial agent), pectin, and calcium lactate was acceptable to consumers after 13 days storage, whereas the control sample showed signs of decay. Three layer beeswax-chitosan-beeswax coating considerably reduced the senescence rate and weight loss of strawberries and resulted in better overall quality of the strawberries (Velickova et al. 2013), however the sensory evaluation showed that the three-layer coating had lower visual appearance and taste scores, and was less preferable by the consumer panel.

9.7 Combination of Biopolymer–Based Coatings with Other Physical Storage Technologies

Physical-based storage technologies (such as temperature, humidity, pressure and gas composition) are the primary techniques to maintain quality and shelf life of horticultural products (Krasaekoopt and Bhandari 2011). Cold storage, modified atmosphere packaging (MAP), and gamma irradiation are some physical-based methods to maintain quality and shelf-life of intact and fresh-cut fruit and vegetables (Ma et al. 2017). Various studies presented in Table 9.2 have shown that biopolymer-based coatings can act in synergy with physical-based preservation techniques to improve the physicochemical, nutritional and microbiological quality of fruit and vegetables during storage.

9.8 Conclusion

The proper combination of product and coating is essential for the establishment of correct storage conditions to optimize fruit and vegetable quality after storage. The results presented in this chapter focused on the potential applications of biopolymer-based edible coating and packaging for fresh and minimally processed fruit and vegetables and showed the different effects of biopolymer-based edible coatings on the physiological and quality of fruit and vegetables. Biopolymers have been shown to maintain fruit and vegetable quality in the laboratory but commercial development of application edible coatings on various fruit and vegetables is limited. Numerous factors including inadequate understanding and accessibility of suitable coating substances, poor moisture-barrier characteristics, weak surface adhesion of some coating components, possible allergenicity to coating materials, unpleasant sensory quality of some coating materials, and viability of scale-up to an industrial setting have been restricting the application of edible films and coatings for fruit and vegetables. New application in biopolymer research should highlight the production of tailor-made coatings, comprising the most compatible coating formulation ingredients and active compounds for intact and minimally processed fruit and vegetables, in relation to particular industrial requirements. It is critical to work with industry to ensure that its application is practical and cost-effective. To support this, there is also a need to investigate contribution of coatings on biochemical and secondary metabolites to manage the association of the internal atmosphere induced by the coating with the rate of physiological ripening processes.

Table 9.2 Several examples of combination biopolymer-based coatings with physical-based preservation techniques on fresh and fresh-cut fruit and vegetables

Fruit	Physical-based preservation	Coatings	Additives	Storage condition	Results	References
Broccoli	γ-Irradiation	Alginate	Lemongrass essential oil, sodium diacetate, natamycin	4 °C for 14 d	Active coating acted in synergy with γ-irradiation on broccoli floret to demolish pathogens by increasing the lag phase to 12 days and prolonging the shelf-life.	Ben-Fadhel et al. (2017)
Fresh-cut cantaloupe	Pulsed light	Pectin, alginate, chitosan, and gellan		4 °C for 28 d	Alginate and RPL treatment was the most efficient treatment condition to prolong the shelf-life of fresh-cut cantaloupes by preserving physicochemical, nutritional and microbiological quality up to 28 d with reducing fluid loss and enhancing firmness compared to samples treated with RPL alone.	Koh et al. (2017)
Plum	Gamma irradiation	CMC		25 °C for 20 d and 3 °C for 35 d	CMC at 1.0% w/v and 1.5 kGy irradiation was superior in retaining the chlorophyll and retarding the decay rate, resulting in retention of storage quality.	Hussain et al. (2015)
Wolfberry	Hot water dip	Chitosan		2 °C for 28 d	The synergistically treated fruit had higher ascorbic acid and total phenolic contents, antioxidant capacity and lower decay along with higher acceptability achieved by sensory analysis.	Ban et al. (2015)
Cauliflower florets	γ-radiation or negative air ionization (NAI) with ozone	MC + maltodextrin	Lactic acid, citrus extract, lemongrass essential oil	4 °C for 7 d	The bioactive coating performed in synergy with γ-radiation, leading to no bacterial growth of *L. innocua* and *E. coli*, in addition to the inhibition of the growth of mesophilic bacteria during 7 d.	Boumail et al. (2016)

(continued)

Table 9.2 (continued)

Fruit	Physical-based preservation	Coatings	Additives	Storage condition	Results	References
Fresh-cut mango	Pulsed light	Alginate	Malic acid	4 °C for 14 d	The combination contributed to preservation of the color and firmness of fruit and reduction of microbial population for 14 d.	Salinas-Roca et al. (2016)
Fresh-cut carrots	MAP	Cassava starch	Montmorillonite (MMT) nanoparticles	4 °C for 4 w	The combined application of coating and MAP led to the protection of the total antioxidant activity, the volatile and organic acids of fresh-cut carrots.	Guimarães et al. (2016)
Fresh-cut apple	Pulsed light	Pectin	Apple fiber	4 °C for 14 d	Coated and PL-treated fruit significantly presented higher antioxidant activity values than fresh and PL control samples. At the end of storage, the combination of both treatments caused microbial count reduction.	Moreira et al. (2017)
Fresh-cut eggplant	MAP	Soy protein isolate	Cysteine	5 °C for 8 d	MAP packaging conditions (low O_2 and high CO_2) were not suitable for storage of fresh-cut eggplants, because it caused damage of the tissue. The coating under air atmospheric conditions was the best and cheapest method for preserving fresh-cut eggplant	Ghidelli et al. (2014)
Green bean	High hydrostatic pressure (HHP) or pulsed light	Chitosan	Nanoemulsion of mandarin essential oil	4 °C for 14 d	The combination induced a significant reduction of *L. innocua* and firmness retention during storage, due to an antimicrobial synergism effect. However, the combination of the coating application with PL had a slight antagonistic impact, and had a slight unfavorable effect on color attributes	Donsì et al. (2015)

Fruit	Physical-based preservation	Coatings	Additives	Storage condition	Results	References
Fresh-cut persimmon	MAP	Pectin	Nisin, citric acid, and calcium chloride	5 °C for 9 d	The combined methods decreased the growth of mesophilic aerobic bacteria, browning index, and the CO_2 emission and O_2 consumption in the package	Sanchís et al. (2017)
Fresh-cut artichoke	MAP	Soy protein isolate (SPI) + beeswax	L-cysteine	5 °C for 7 d	The combination of the coating with MAP did not prolong the shelf-life of artichoke slices, but maintained the antioxidant capacity as compared with the control packaging conditions	Ghidelli et al. (2015)
Peach	γ -radiation	CMC		25 °C for 15 d and 3 °C for 35 d	Combination of CMC at 1.0% (w/v) and 1.2 kGy irradiation prohibited disease incidence of peach up to 7 days during ambient storage at 25 ± 2 °C, RH 70% following 30 d of refrigeration	Hussain et al. (2016b)
Cherry	γ -radiation	CMC		25 °C for 9 d and 3 °C for 28 d	Combinatory treatments demonstrated positive influence in preserving the storage quality as well as retarding the decay rate of cherry fruit	Hussain et al. (2016a)
Fresh-cut apple	Pulsed light	Gellan gum	Apple fiber	4 °C for 14 d	The combined application of coating and PL treatment delayed the microbiological contamination of fresh-cut apples and maintained the sensory attributes during storage	Moreira et al. (2015)
Green chili	MAP	Shellac		8 °C for 48 d	Coated and MA packed chilies showed 48 d shelf life compared with uncoated and MA packed (28 d), control (15 d) ones, and shellac coated chilies (30 d)	Chitravathi et al. (2016)

Acknowledgement This chapter was supported by NSW Department of Primary Industries and the Australian Citrus Postharvest Science Program (CT15010) with Horticulture Innovation Australia Ltd.

References

Abugoch L, Tapia C, Plasencia D, Pastor A, Castro-Mandujano O, López L et al (2016) Shelf-life of fresh blueberries coated with quinoa protein/chitosan/sunflower oil edible film. J Sci Food Agric 96(2):619–626

Aguayo M d CL, Burgos MJG, Pulido RP, Gálvez A, López RL (2016) Effect of different activated coatings containing enterocin AS-48 against *Listeria monocytogenes* on apple cubes. Innovative Food Sci Emerg Technol 35:177–183

Al-Hassan AA, Norziah MH (2012) Starch–gelatin edible films: water vapor permeability and mechanical properties as affected by plasticizers. Food Hydrocoll 26(1):108–117

Ali A, Maqbool M, Alderson PG, Zahid N (2013) Effect of gum arabic as an edible coating on antioxidant capacity of tomato (*Solanum lycopersicum L.*) fruit during storage. Postharvest Biol Technol 76:119–124

Ali A, Hei GK, Keat YW (2016) Efficacy of ginger oil and extract combined with gum arabic on anthracnose and quality of papaya fruit during cold storage. J Food Sci Technol 53(3):1435–1444

Allegra A, Inglese P, Sortino G, Settanni L, Todaro A, Liguori G (2016) The influence of Opuntia ficus-indica mucilage edible coating on the quality of 'Hayward'kiwifruit slices. Postharvest Biol Technol 120:45–51

Allegra A, Sortino G, Inglese P, Settanni L, Todaro A, Gallotta A (2017) The effectiveness of Opuntia ficus-indica mucilage edible coating on post-harvest maintenance of 'Dottato'fig (Ficus carica L.) fruit. Food Packaging Shelf Life 12:135–141

Alleyne V, Hagenmaier RD (2000) Candelilla-shellac: an alternative formulation for coating apples. Hortscience 35(4):691–693

Aloui H, Khwaldia K, Sánchez-González L, Muneret L, Jeandel C, Hamdi M et al (2014) Alginate coatings containing grapefruit essential oil or grapefruit seed extract for grapes preservation. Int J Food Sci Technol 49(4):952–959

Álvarez K, Famá L, Gutiérrez TJ (2017) Physicochemical, antimicrobial and mechanical properties of thermoplastic materials based on biopolymers with application in the food industry. In: Masuelli M, Renard D (eds) Advances in physicochemical properties of biopolymers: Part 1. Bentham Science, Sharjah, pp 358–400. EE.UU. ISBN: 978-1-68108-454-1. eISBN: 978-1-68108-453-4. https://doi.org/10.2174/9781681084534117010015

Álvarez K, Alvarez VA, Gutiérrez TJ (2018) Biopolymer composite materials with antimicrobial effects applied to the food industry. In: Thakur VK, Thakur MK (eds) Functional biopolymers. Springer International, Basel, pp 57–96. EE.UU. ISBN: 978-3-319-66416-3. eISBN: 978-3-319-66417-0. https://doi.org/10.1007/978-3-319-66417-0_3

Amarante C, Banks NH (2010) Postharvest physiology and quality of coated fruits and vegetables. Hortic Rev 26:161–238

Andrade SC, Baretto TA, Arcanjo NM, Madruga MS, Meireles B, Cordeiro ÂM et al (2017) Control of Rhizopus soft rot and quality responses in plums (Prunus domestica L.) coated with gum arabic, oregano and rosemary essential oils. J Food Process Preserv 41

Arnon H, Zaitsev Y, Porat R, Poverenov E (2014) Effects of carboxymethyl cellulose and chitosan bilayer edible coating on postharvest quality of citrus fruit. Postharvest Biol Technol 87:21–26

Arnon H, Granit R, Porat R, Poverenov E (2015) Development of polysaccharides-based edible coatings for citrus fruits: a layer-by-layer approach. Food Chem 166(Supplement C):465–472

Athayde A, De Oliveira P, Guerra I, Da Conceicao M, De Lima M, Arcanjo N et al (2016) A coating composed of chitosan and *Cymbopogon citratus* (dc. Ex Nees) essential oil to con-

trol Rhizopus soft rot and quality in tomato fruit stored at room temperature. J Hortic Sci Biotechnol 91(6):582–591
Ayranci E, Tunc S (2001) The effect of fatty acid content on water vapour and carbon dioxide transmissions of cellulose-based edible films. Food Chem 72(2):231–236
Azarakhsh N, Osman A, Ghazali HM, Tan CP, Adzahan NM (2014) Effects of gellan-based edible coating on the quality of fresh-cut pineapple during cold storage. Food Bioprocess Technol 7(7):2144–2151
Bai J, Plotto A (2012) Coatings for fresh fruits and vegetables. In: Baldwin EA, Hagenmaier R, Bai J (eds) Edible coatings and films to improve food quality, 2nd edn. CRC, Taylor & Francis Group, Boca Raton, FL, pp 185–242
Baldwin EA, Baker RA (2002) Use of proteins in edible coatings for whole and minimally processed fruits and vegetables. In: Gennadios A (ed) Protein-based films and coatings. CRC, Boca Raton, FL
Ban Z, Wei W, Yang X, Feng J, Guan J, Li L (2015) Combination of heat treatment and chitosan coating to improve postharvest quality of wolfberry (*Lycium barbarum*). Int J Food Sci Technol 50(4):1019–1025
Baraiya NS, Rao TVR, Thakkar VR (2015) Improvement of postharvest quality and storability of jamun fruit (*Syzygium cumini L. Var. Paras*) by zein coating enriched with antioxidants. Food Bioprocess Technol 8(11):2225–2234
Barman K, Asrey R, Pal R, Kaur C, Jha S (2014) Influence of putrescine and carnauba wax on functional and sensory quality of pomegranate (*Punica granatum L.*) fruits during storage. J Food Sci Technol 51(1):111–117
Ben-Fadhel Y, Saltaji S, Khlifi MA, Salmieri S, Vu KD, Lacroix M (2017) Active edible coating and γ-irradiation as cold combined treatments to assure the safety of broccoli florets (*Brassica oleracea L.*). Int J Food Microbiol 241:30–38
Benítez S, Achaerandio I, Pujolà M, Sepulcre F (2015) Aloe vera as an alternative to traditional edible coatings used in fresh-cut fruits: a case of study with kiwifruit slices. LWT Food Sci Technol 61(1):184–193
Bertuzzi M, Vidaurre EC, Armada M, Gottifredi J (2007) Water vapor permeability of edible starch based films. J Food Eng 80(3):972–978
Boumail A, Salmieri S, Lacroix M (2016) Combined effect of antimicrobial coatings, gamma radiation and negative air ionization with ozone on *Listeria innocua*, *Escherichia coli* and mesophilic bacteria on ready-to-eat cauliflower florets. Postharvest Biol Technol 118(Supplement C):134–140
Butt MS, Suleria HAR, Ahmed W (2015) Physicochemical behavior of zinc-fortified, sodium caseinate-based, edible-coated apricots during storage in controlled atmosphere. J Food Process Preserv 39(6):2431–2441
Carneiro-da-Cunha MG, Cerqueira MA, Souza BW, Souza MP, Teixeira JA, Vicente AA (2009) Physical properties of edible coatings and films made with a polysaccharide from *Anacardium occidentale L.* J Food Eng 95(3):379–385
Cazón P, Velazquez G, Ramírez JA, Vázquez M (2017) Polysaccharide-based films and coatings for food packaging: a review. Food Hydrocoll 68:136–148
Cerqueira MA, Lima ÁM, Teixeira JA, Moreira RA, Vicente AA (2009) Suitability of novel galactomannans as edible coatings for tropical fruits. J Food Eng 94(3):372–378
Cerqueira M, Bourbon A, Pinheiro A, Martins J, Souza B, Teixeira J et al (2011) Galactomannans use in the development of edible films/coatings for food applications. Trends Food Sci Technol 22(12):662–671
Cerqueira MAPR, Pereira RNC, da Silva Ramos OL, Teixeira JAC, Vicente AA (2016) Edible food packaging: materials and processing technologies, vol. 36. CRC, Boca Raton, FL
Chaple S, Vishwasrao C, Ananthanarayan L (2017) Edible composite coating of methyl cellulose for post-harvest extension of shelf-life of finger hot Indian pepper (*Pusa jwala*). J Food Process Preserv 41(2)

Chen C, Peng X, Zeng R, Chen M, Wan C, Chen J (2016) Ficus hirta fruits extract incorporated into an alginate-based edible coating for Nanfeng mandarin preservation. Sci Hortic 202:41–48

Chen C, Peng X, Zeng R, Wan C, Chen M, Chen J (2017) Physiological and biochemical responses in cold-stored citrus fruits to carboxymethyl cellulose coating containing ethanol extract of *Impatiens balsamina L.* stems. J Food Process Preserv 41(4)

Chitravathi K, Chauhan O, Raju P (2016) Shelf life extension of green chillies (*Capsicum annuum L.*) using shellac-based surface coating in combination with modified atmosphere packaging. J Food Sci Technol 53(8):3320–3328

Choi WS, Singh S, Lee YS (2016) Characterization of edible film containing essential oils in hydroxypropyl methylcellulose and its effect on quality attributes of 'Formosa'plum (*Prunus salicina L.*). LWT Food Sci Technol 70:213–222

Chrysargyris A, Nikou A, Tzortzakis N (2016) Effectiveness of Aloe vera gel coating for maintaining tomato fruit quality. N Z J Crop Hortic Sci 44(3):203–217

Correa-Pacheco ZN, Bautista-Baños S, Valle-Marquina MÁ, Hernández-López M (2017) The effect of nanostructured chitosan and chitosan-thyme essential oil coatings on *Colletotrichum gloeosporioides* growth in vitro and on cv Hass avocado and fruit quality. J Phytopathol 165(5):297–305

Danalache F, Carvalho CY, Alves VD, Moldão-Martins M, Mata P (2016) Optimisation of gellan gum edible coating for ready-to-eat mango (*Mangifera indica L.*) bars. Int J Biol Macromol 84:43–53. https://doi.org/10.1016/j.ijbiomac.2015.11.079

De León-Zapata MA, Sáenz-Galindo A, Rojas-Molina R, Rodríguez-Herrera R, Jasso-Cantú D, Aguilar CN (2015) Edible candelilla wax coating with fermented extract of tarbush improves the shelf life and quality of apples. Food Packaging Shelf Life 3:70–75

Dhall R (2013) Advances in edible coatings for fresh fruits and vegetables: a review. Crit Rev Food Sci Nutr 53(5):435–450

Dong F, Wang X (2017) Effects of carboxymethyl cellulose incorporated with garlic essential oil composite coatings for improving quality of strawberries. Int J Biol Macromol 104:821–826

Donsì F, Marchese E, Maresca P, Pataro G, Vu KD, Salmieri S et al (2015) Green beans preservation by combination of a modified chitosan based-coating containing nanoemulsion of mandarin essential oil with high pressure or pulsed light processing. Postharvest Biol Technol 106:21–32

Duan J, Wu R, Strik BC, Zhao Y (2011) Effect of edible coatings on the quality of fresh blueberries (*Duke and Elliott*) under commercial storage conditions. Postharvest Biol Technol 59(1):71–79

Duan X, OuYang Q, Tao N (2017) Effect of applying cinnamaldehyde incorporated in wax on green mold decay in citrus fruits. J Sci Food Agric 98(2):527–533

Duran M, Aday MS, Zorba NND, Temizkan R, Büyükcan MB, Caner C (2016) Potential of antimicrobial active packaging 'containing natamycin, nisin, pomegranate and grape seed extract in chitosan coating'to extend shelf life of fresh strawberry. Food Bioprod Process 98:354–363

Fabra M, Pérez-Masiá R, Talens P, Chiralt A (2011) Influence of the homogenization conditions and lipid self-association on properties of sodium caseinate based films containing oleic and stearic acids. Food Hydrocoll 25(5):1112–1121

Fagundes C, Pérez-Gago MB, Monteiro AR, Palou L (2013) Antifungal activity of food additives in vitro and as ingredients of hydroxypropyl methylcellulose-lipid edible coatings against *Botrytis cinerea* and *Alternaria alternata* on cherry tomato fruit. Int J Food Microbiol 166(3):391–398

Fagundes C, Palou L, Monteiro AR, Pérez-Gago MB (2014) Effect of antifungal hydroxypropyl methylcellulose-beeswax edible coatings on gray mold development and quality attributes of cold-stored cherry tomato fruit. Postharvest Biol Technol 92:1–8

Fagundes C, Palou L, Monteiro AR, Pérez-Gago MB (2015) Hydroxypropyl methylcellulose-beeswax edible coatings formulated with antifungal food additives to reduce alternaria black spot and maintain postharvest quality of cold-stored cherry tomatoes. Sci Hortic 193:249–257

Fakhouri FM, Martelli SM, Caon T, Velasco JI, Mei LHI (2015) Edible films and coatings based on starch/gelatin: film properties and effect of coatings on quality of refrigerated red crimson grapes. Postharvest Biol Technol 109:57–64
Flores-López ML, Cerqueira MA, de Rodríguez DJ, Vicente AA (2016) Perspectives on utilization of edible coatings and nano-laminate coatings for extension of postharvest storage of fruits and vegetables. Food Eng Rev 8(3):292–305
Forato LA, de Britto D, de Rizzo JS, Gastaldi TA, Assis OB (2015) Effect of cashew gum-carboxymethylcellulose edible coatings in extending the shelf-life of fresh and cut guavas. Food Packaging Shelf Life 5:68–74
Franco MJ, Martin AA, Bonfim LF, Caetano J, Linde GA, Dragunski DC (2017) Effect of plasticizer and modified starch on biodegradable films for strawberry protection. J Food Process Preserv 41(4)
Galus S, Kadzińska J (2015) Food applications of emulsion-based edible films and coatings. Trends Food Sci Technol 45(2):273–283
Ghidelli C, Mateos M, Rojas-Argudo C, Pérez-Gago MB (2014) Extending the shelf life of fresh-cut eggplant with a soy protein–cysteine based edible coating and modified atmosphere packaging. Postharvest Biol Technol 95:81–87
Ghidelli C, Mateos M, Rojas-Argudo C, Pérez-Gago MB (2015) Novel approaches to control browning of fresh-cut artichoke: effect of a soy protein-based coating and modified atmosphere packaging. Postharvest Biol Technol 99:105–113
Gniewosz M, Kraśniewska K, Woreta M, Kosakowska O (2013) Antimicrobial activity of a pullulan–caraway essential oil coating on reduction of food microorganisms and quality in fresh baby carrot. J Food Sci 78(8):M1242–M1248
Gomes M d S, Cardoso M d G, Guimarães ACG, Guerreiro AC, Gago CML, Boas V et al (2017) Effect of edible coatings with essential oils on the quality of red raspberries over shelf-life. J Sci Food Agric 97(3):929–938
González-Estrada RR, Chalier P, Ragazzo-Sánchez JA, Konuk D, Calderón-Santoyo M (2017) Antimicrobial soy protein based coatings: application to Persian lime (*Citrus latifolia Tanaka*) for protection and preservation. Postharvest Biol Technol 132:138–144
Greener IK, Fennema O (1989) Barrier properties and surface characteristics of edible, bilayer films. J Food Sci 54(6):1393–1399
Guerreiro AC, Gago CM, Miguel MG, Faleiro ML, Antunes MD (2016) The influence of edible coatings enriched with citral and eugenol on the raspberry storage ability, nutritional and sensory quality. Food Packaging Shelf Life 9:20–28
Guerreiro AC, Gago CM, Faleiro ML, Miguel MG, Antunes MD (2017) The effect of edible coatings on the nutritional quality of 'Bravo de Esmolfe'fresh-cut apple through shelf-life. LWT Food Sci Technol 75:210–219
Guevara JC, Yahia EM, Brito de la Fuente E (2001) Modified atmosphere packaging of prickly pear cactus stems (*Opuntia spp.*). LWT Food Sci Technol 34(7):445–451
Guimarães IC, dos Reis KC, Menezes EGT, Borges PRS, Rodrigues AC, Leal R et al (2016) Combined effect of starch/montmorillonite coating and passive MAP in antioxidant activity, total phenolics, organic acids and volatile of fresh-cut carrots. Int J Food Sci Nutr 67(2):141–152
Gutiérrez TJ (2017a) Surface and nutraceutical properties of edible films made from starchy sources with and without added blackberry pulp. Carbohydr Polym 165:169–179. https://doi.org/10.1016/j.carbpol.2017.02.016
Gutiérrez TJ (2017b) Chitosan applications for the food industry. In: Ahmed S, Ikram S (eds) Chitosan: Derivatives, Composites and Applications. Wiley-Scrivener, Beverly, MA, pp 185–232 EE.UU. ISBN: 978-1-119-36350-7. https://doi.org/10.1002/9781119364849.ch8
Gutiérrez TJ (2018) Active and intelligent films made from starchy sources/blackberry pulp. J Polym Environ 26(6):2374–2391. https://doi.org/10.1007/s10924-017-1134-y
Gutiérrez TJ, Alvarez VA (2017) Films made by blending poly (ε-caprolactone) with starch and flour from Sagu rhizome grown at the venezuelan amazons. J Polym Environ 25(3):701–716 25. https://doi.org/10.1007/s10924-016-0861-9

Gutiérrez TJ, Álvarez K (2017) Biopolymers as microencapsulation materials in the food industry. In: Masuelli M, Renard D (eds) Advances in physicochemical properties of biopolymers: Part 2. Bentham Science, Sharjah, pp 296–322. EE.UU. ISBN: 978–1–68108-545-6. eISBN: 978–1–68108-544-9. https://doi.org/10.2174/9781681085449117010009

Gutiérrez TJ, Morales NJ, Pérez E, Tapia MS, Famá L (2015) Physico-chemical properties of edible films derived from native and phosphated cush-cush yam and cassava starches. Food Packaging Shelf Life 3:1–8. https://doi.org/10.1016/j.fpsl.2014.09.002

Hagenmaier RD, Baker RA (1994) Wax microemulsions and emulsions as citrus coatings. J Agric Food Chem 42(4):899–902

Hagenmaier RD, Baker RA (1995) Layered coatings to control weight loss and preserve gloss of citrus fruit. Hortscience 30(2):296–298

Hamzah HM, Osman A, Tan CP, Ghazali FM (2013) Carrageenan as an alternative coating for papaya (*Carica papaya L. cv. Eksotika*). Postharvest Biol Technol 75:142–146

Hashemi SMB, Khaneghah AM, Ghahfarrokhi MG, Eş I (2017) Basil-seed gum containing origanum vulgare subsp. Viride essential oil as edible coating for fresh cut apricots. Postharvest Biol Technol 125:26–34

Hassanpour H (2015) Effect of Aloe vera gel coating on antioxidant capacity, antioxidant enzyme activities and decay in raspberry fruit. LWT Food Sci Technol 60(1):495–501

Hun JH, Cennadios A (2005) Edible films and coatings: a review. In: Han J (ed) Innovations in food packaging. Elsevier Science & Technology, London, pp 239–262

Hussain PR, Suradkar PP, Wani AM, Dar MA (2015) Retention of storage quality and post-refrigeration shelf-life extension of plum (*Prunus domestica L.*) cv. Santa Rosa using combination of carboxymethyl cellulose (CMC) coating and gamma irradiation. Radiat Phys Chem 107(Supplement C):136–148

Hussain P, Rather S, Suradkar P, Parveen S, Mir M, Shafi F (2016a) Potential of carboxymethyl cellulose coating and low dose gamma irradiation to maintain storage quality, inhibit fungal growth and extend shelf-life of cherry fruit. J Food Sci Technol 53(7):2966–2986

Hussain PR, Suradkar PP, Wani AM, Dar MA (2016b) Potential of carboxymethyl cellulose and γ-irradiation to maintain quality and control disease of peach fruit. Int J Biol Macromol 82:114–126

Izydorczyk M, Cui SW, Wang Q (2005) Polysaccharide gums: structures, functional properties, and applications. In: Cui SW (ed) Food carbohydrates: chemistry, physical properties, and applications. CRC, Taylor & Francis Group, Boca Raton, FL, pp 269–313

Jia Y, Li J (2015) Molecular assembly of schiff base interactions: construction and application. Chem Rev 115(3):1597–1621

Jiménez A, Fabra MJ, Talens P, Chiralt A (2012) Edible and biodegradable starch films: a review. Food Bioprocess Technol 5(6):2058–2076

Jiménez A, Fabra MJ, Talens P, Chiralt A (2013) Phase transitions in starch based films containing fatty acids. Effect on water sorption and mechanical behaviour. Food Hydrocoll 30(1):408–418

Jo W-S, Song H-Y, Song N-B, Lee J-H, Min SC, Song KB (2014) Quality and microbial safety of 'Fuji' apples coated with carnauba-shellac wax containing lemongrass oil. LWT Food Sci Technol 55(2):490–497

Jonglertjunya W, Juntong T, Pakkang N, Srimarut N, Sakdaronnarong C (2014) Properties of lignin extracted from sugarcane bagasse and its efficacy in maintaining postharvest quality of limes during storage. LWT Food Sci Technol 57(1):116–125

Kanetis L, Exarchou V, Charalambous Z, Goulas V (2017) Edible coating composed of chitosan and *Salvia fruticosa Mill.* extract for the control of grey mould of table grapes. J Sci Food Agric 97(2):452–460

Kaya M, Česonienė L, Daubaras R, Leskauskaitė D, Zabulionė D (2016) Chitosan coating of red kiwifruit (*Actinidia melanandra*) for extending of the shelf life. Int J Biol Macromol 85:355–360

Kester JJ, Fennema O (1986) Edible films and coatings: a review. Food Technol 40:47–59

Khaliq G, Mohamed MTM, Ghazali HM, Ding P, Ali A (2016) Influence of gum arabic coating enriched with calcium chloride on physiological, biochemical and quality responses of mango (*Mangifera indica L.*) fruit stored under low temperature stress. Postharvest Biol Technol 111:362–369

Khorram F, Ramezanian A, Hosseini SMH (2017a) Effect of different edible coatings on postharvest quality of 'Kinnow'mandarin. J Food Measurement Character 11:1827–1833

Khorram F, Ramezanian A, Hosseini SMH (2017b) Shellac, gelatin and Persian gum as alternative coating for orange fruit. Sci Hortic 225:22–28

Kim IH, Lee H, Kim JE, Song KB, Lee YS, Chung DS et al (2013) Plum coatings of lemongrass oil-incorporating carnauba wax-based nanoemulsion. J Food Sci 78(10):E1551–E1559

Koh PC, Noranizan MA, Hanani ZAN, Karim R, Rosli SZ (2017) Application of edible coatings and repetitive pulsed light for shelf life extension of fresh-cut cantaloupe (*Cucumis melo L. reticulatus cv. Glamour*). Postharvest Biol Technol 129:64–78

Kou X, Li Y, Wu J, Chen Q, Xue Z (2017) Effects of edible coatings on quality and antioxidant activity of *Zizyphus Jujuba* miller cv. Dongzao during storage. Trans Tianjin Univ 23(1):51–61

Kowalczyk D, Kordowska-Wiater M, Zięba E, Baraniak B (2017) Effect of carboxymethylcellulose/candelilla wax coating containing potassium sorbate on microbiological and physicochemical attributes of pears. Sci Hortic 218:326–333

Krasaekoopt W, Bhandari B (2011) Fresh-cut vegetables. Handbook of vegetables and vegetable processing. Wiley-Blackwell, Hoboken, NJ

Kumari M, Mahajan H, Joshi R, Gupta M (2017) Development and structural characterization of edible films for improving fruit quality. Food Packaging Shelf Life 12:42–50

Kuwar U, Sharma S, Tadapaneni VRR (2015) Aloe vera gel and honey-based edible coatings combined with chemical dip as a safe means for quality maintenance and shelf life extension of fresh-cut papaya. J Food Qual 38(5):347–358

Lacroix M, Le Tien C (2005) 20—Edible films and coatings from nonstarch polysaccharides. In: Han JH (ed) Innovations in food packaging. Academic Press, London, pp 338–361

Lai T-Y, Chen C-H, Lai L-S (2013) Effects of tapioca starch/decolorized hsian-tsao leaf gum-based active coatings on the quality of minimally processed carrots. Food Bioprocess Technol 6(1):249–258

Larrigaudière C, Lentheric I, Puy J, Pintó E (2004) Biochemical characterisation of core browning and brown heart disorders in pear by multivariate analysis. Postharvest Biol Technol 31(1):29–39

Lin D, Zhao Y (2007) Innovations in the development and application of edible coatings for fresh and minimally processed fruits and vegetables. Compr Rev Food Sci Food Saf 6(3):60–75

Liu X, Tang C, Han W, Xuan H, Ren J, Zhang J et al (2017) Characterization and preservation effect of polyelectrolyte multilayer coating fabricated by carboxymethyl cellulose and chitosan. Colloids Surf A Physicochem Eng Asp 529:1016–1023

Lurie S, Watkins CB (2012) Superficial scald, its etiology and control. Postharvest Biol Technol 65(Supplement C):44–60

Ma L, Zhang M, Bhandari B, Gao Z (2017) Recent developments in novel shelf life extension technologies of fresh-cut fruits and vegetables. In: Trends in Food Science & Technology, vol 64, pp 23–38

Mahajan PV, Caleb OJ, Singh Z, Watkins CB, Geyer M (2014) Postharvest treatments of fresh produce. Phil Trans R Soc A 372(2017):20130309

Mahfoudhi N, Hamdi S (2015) Use of almond gum and gum arabic as novel edible coating to delay postharvest ripening and to maintain sweet cherry (*Prunus avium*) quality during storage. J Food Process Preserv 39(6):1499–1508

Mannozzi C, Cecchini J, Tylewicz U, Siroli L, Patrignani F, Lanciotti R et al (2017) Study on the efficacy of edible coatings on quality of blueberry fruits during shelf-life. LWT Food Sci Technol 85:440–444

Marquez GR, Di Pierro P, Mariniello L, Esposito M, Giosafatto CV, Porta R (2017) Fresh-cut fruit and vegetable coatings by transglutaminase-crosslinked whey protein/pectin edible films. LWT Food Sci Technol 75:124–130

Martiñon ME, Moreira RG, Castell-Perez ME, Gomes C (2014) Development of a multilayered antimicrobial edible coating for shelf-life extension of fresh-cut cantaloupe (*Cucumis melo L.*) stored at 4 C. LWT Food Sci Technol 56(2):341–350

Miller KS, Krochta JM (1997) Oxygen and aroma barrier properties of edible films: a review. Trends Food Sci Technol 8(7):228–237

Moreira MR, Tomadoni B, Martín-Belloso O, Soliva-Fortuny R (2015) Preservation of fresh-cut apple quality attributes by pulsed light in combination with gellan gum-based prebiotic edible coatings. LWT Food Sci Technol 64(2):1130–1137

Moreira MR, Álvarez MV, Martín-Belloso O, Soliva-Fortuny R (2017) Effects of pulsed light treatments and pectin edible coatings on the quality of fresh-cut apples: a hurdle technology approach. J Sci Food Agric 97(1):261–268

Morillon V, Debeaufort F, Blond G, Capelle M, Voilley A (2002) Factors affecting the moisture permeability of lipid-based edible films: a review. Crit Rev Food Sci Nutr 42(1):67–89. https://doi.org/10.1080/10408690290825466

Munhuweyi K, Lennox CL, Meitz-Hopkins JC, Caleb OJ, Sigge GO, Opara UL (2017) Investigating the effects of crab shell chitosan on fungal mycelial growth and postharvest quality attributes of pomegranate whole fruit and arils. Sci Hortic 220:78–89

Murmu SB, Mishra HN (2017) Optimization of the arabic gum based edible coating formulations with sodium caseinate and tulsi extract for guava. LWT Food Sci Technol 80:271–279

Nadim Z, Ahmadi E, Sarikhani H, Amiri Chayjan R (2015) Effect of methylcellulose-based edible coating on strawberry fruit's quality maintenance during storage. J Food Process Preserv 39(1):80–90

Narsaiah K, Wilson RA, Gokul K, Mandge H, Jha S, Bhadwal S et al (2015) Effect of bacteriocin-incorporated alginate coating on shelf-life of minimally processed papaya (*Carica papaya L.*). Postharvest Biol Technol 100:212–218

Nasiri M, Barzegar M, Sahari M, Niakousari M (2017) Application of Tragacanth gum impregnated with *Satureja khuzistanica* essential oil as a natural coating for enhancement of postharvest quality and shelf life of button mushroom (*Agaricus bisporus*). Int J Biol Macromol 106, 218–226

Nasrin T, Rahman M, Hossain M, Islam M, Arfin M (2017) Postharvest quality response of strawberries with aloe vera coating during refrigerated storage. J Hortic Sci Biotechnol:1–8

Navarro-Tarazaga ML, Massa A, Pérez-Gago MB (2011) Effect of beeswax content on hydroxypropyl methylcellulose-based edible film properties and postharvest quality of coated plums (*Cv. Angeleno*). LWT Food Sci Technol 44(10):2328–2334

Nawab A, Alam F, Hasnain A (2017) Mango kernel starch as a novel edible coating for enhancing shelf-life of tomato (*Solanum lycopersicum*) fruit. Int J Biol Macromol 103:581–586

Nisperos-Carriedo M (1994) Edible coating and films based on polysaccharide. In: Krochta JM, Baldwin EA, Nisperos-Carriedo M (eds) Edible coatings and films to improve food quality. Technomic Publishing, Lancaster, PA, pp 305–335

Nussinovitch A (1997) Hydrocolloid applications: gum technology in the food and other industries. Springer, Berlin

Ochoa-Reyes E, Martínez-Vazquez G, Saucedo-Pompa S, Montañez J, Rojas-Molina R, de Leon-Zapata MA et al (2013) Improvement of shelf life quality of green bell peppers using edible coating formulations. J Microbiol Biotechnol Food Sci 2(6):2448

Oh YA, Oh YJ, Song AY, Won JS, Song KB, Min SC (2017) Comparison of effectiveness of edible coatings using emulsions containing lemongrass oil of different size droplets on grape berry safety and preservation. LWT Food Sci Technol 75:742–750

Ojeda GA, Sgroppo SC, Zaritzky NE (2014) Application of edible coatings in minimally processed sweet potatoes (*Ipomoea batatas L.*) to prevent enzymatic browning. Int J Food Sci Technol 49(3):876–883

Olivas G, Barbosa-Cánovas G (2005) Edible coatings for fresh-cut fruits. Crit Rev Food Sci Nutr 45(7–8):657–670

Olivas GI, Barbosa-Cánovas G (2009) Edible films and coatings for fruits and vegetables. In: Embuscado ME, Huber KC (eds) Edible films and coatings for food applications. Springer, New York, NY, pp 211–244

Oregel-Zamudio E, Angoa-Pérez MV, Oyoque-Salcedo G, Aguilar-González CN, Mena-Violante HG (2017) Effect of candelilla wax edible coatings combined with biocontrol bacteria on strawberry quality during the shelf-life. Sci Hortic 214:273–279

Pająk P, Socha R, Łakoma P, Fortuna T (2017). Antioxidant properties of apple slices stored in starch-based films. International Journal of Food Properties 20(5), 1117–1128

Paladines D, Valero D, Valverde JM, Díaz-Mula H, Serrano M, Martínez-Romero D (2014) The addition of rosehip oil improves the beneficial effect of Aloe vera gel on delaying ripening and maintaining postharvest quality of several stonefruit. Postharvest Biol Technol 92:23–28

Pan S-Y, Chen C-H, Lai L-S (2013) Effect of tapioca starch/decolorized hsian-tsao leaf gum-based active coatings on the qualities of fresh-cut apples. Food Bioprocess Technol 6(8):2059–2069

Parafati L, Vitale A, Restuccia C, Cirvilleri G (2016) The effect of locust bean gum (LBG)-based edible coatings carrying biocontrol yeasts against *Penicillium digitatum* and *Penicillium italicum* causal agents of postharvest decay of mandarin fruit. Food Microbiol 58:87–94

Peressini D, Bravin B, Lapasin R, Rizzotti C, Sensidoni A (2003) Starch–methylcellulose based edible films: rheological properties of film-forming dispersions. J Food Eng 59(1):25–32

Pérez-Gago MB, Krochta JM (2001) Lipid particle size effect on water vapor permeability and mechanical properties of whey protein/beeswax emulsion films. J Agric Food Chem 49(2):996–1002

Pérez-Gallardo A, García-Almendárez B, Barbosa-Cánovas G, Pimentel-González D, Reyes-González L, Regalado C (2015) Effect of starch-beeswax coatings on quality parameters of blackberries (*Rubus spp.*). J Food Sci Technol 52(9):5601–5610

Peter A, Nicula C, Mihaly-Cozmuta L, Mihaly-Cozmuta A (2017) An efficient and innovative method to preserve the harvested plums during storage. J Food Process Preserv 42:e13398

Phan The D, Debeaufort F, Voilley A, Luu D (2009) Biopolymer interactions affect the functional properties of edible films based on agar, cassava starch and arabinoxylan blends. J Food Eng 90(4):548–558

Pizato S, Cortez-Vega WR, Souza JTA, Prentice-Hernández C, Borges CD (2013) Effects of different edible coatings in physical, chemical and microbiological characteristics of minimally processed peaches (*prunus persica l. batsch*). J Food Saf 33(1):30–39

Porat R, Weiss B, Cohen L, Daus A, Biton A (2005) Effects of polyethylene wax content and composition on taste, quality, and emission of off-flavor volatiles in 'Mor'mandarins. Postharvest Biol Technol 38(3):262–268

Poverenov E, Rutenberg R, Danino S, Horev B, Rodov V (2014a) Gelatin-chitosan composite films and edible coatings to enhance the quality of food products: layer-by-layer vs. blended formulations. Food Bioprocess Technol 7(11):3319–3327

Poverenov E, Zaitsev Y, Arnon H, Granit R, Alkalai-Tuvia S, Perzelan Y et al (2014b) Effects of a composite chitosan–gelatin edible coating on postharvest quality and storability of red bell peppers. Postharvest Biol Technol 96:106–109

Radi M, Akhavan-Darabi S, Akhavan HR, Amiri S (2017) The use of orange peel essential oil microemulsion and nanoemulsion in pectin-based coating to extend the shelf life of fresh-cut orange. J Food Process Preserv 42:e13441

Rao TR, Baraiya NS, Vyas PB, Patel DM (2016) Composite coating of alginate-olive oil enriched with antioxidants enhances postharvest quality and shelf life of Ber fruit (*Ziziphus mauritiana Lamk. Var. Gola*). J Food Sci Technol 53(1):748–756

Reyes-Avalos M, Femenia A, Minjares-Fuentes R, Contreras-Esquivel J, Aguilar-González C, Esparza-Rivera J et al (2016) Improvement of the quality and the shelf life of figs (*Ficus carica*) using an alginate–chitosan edible film. Food Bioprocess Technol 9(12):2114–2124

Rodriguez-Garcia I, Cruz-Valenzuela MR, Silva-Espinoza BA, Gonzalez-Aguilar GA, Moctezuma E, Gutierrez-Pacheco MM et al (2016) Oregano (*Lippia graveolens*) essential oil added within pectin edible coatings prevents fungal decay and increases the antioxidant capacity of treated tomatoes. J Sci Food Agric 96(11):3772–3778

Saba MK, Amini R (2017) Nano-ZnO/carboxymethyl cellulose-based active coating impact on ready-to-use pomegranate during cold storage. Food Chem 232:721–726

Saba MK, Sogvar OB (2016) Combination of carboxymethyl cellulose-based coatings with calcium and ascorbic acid impacts in browning and quality of fresh-cut apples. LWT Food Sci Technol 66:165–171

Saberi B, Thakur R, Bhuyan DJ, Vuong QV, Chockchaisawasdee S, Golding JB et al (2017) Development of edible blend films with good mechanical and barrier properties from pea starch and guar gum. Starch Stärke 69(1–2)

Saberi B, Golding JB, Marques JR, Pristijono P, Chockchaisawasdee S, Scarlett CJ et al (2018) Application of biocomposite edible coatings based on pea starch and guar gum on quality, storability and shelf life of 'Valencia' oranges. Postharvest Biol Technol 137:9–20

Saha D, Bhattacharya S (2010) Hydrocolloids as thickening and gelling agents in food: a critical review. J Food Sci Technol 47(6):587–597

Salinas-Roca B, Soliva-Fortuny R, Welti-Chanes J, Martín-Belloso O (2016) Combined effect of pulsed light, edible coating and malic acid dipping to improve fresh-cut mango safety and quality. Food Control 66:190–197

Sanchís E, Ghidelli C, Sheth CC, Mateos M, Palou L, Pérez-Gago MB (2017) Integration of antimicrobial pectin-based edible coating and active modified atmosphere packaging to preserve the quality and microbial safety of fresh-cut persimmon (*Diospyros kaki Thunb. cv. Rojo Brillante*). J Sci Food Agric 97(1):252–260

Shah NN, Vishwasrao C, Singhal RS, Ananthanarayan L (2016a) n-Octenyl succinylation of pullulan: effect on its physico-mechanical and thermal properties and application as an edible coating on fruits. Food Hydrocoll 55:179–188

Shah SWA, Qaisar M, Jahangir M, Abbasi KS, Khan SU, Ali N et al (2016b) Influence of CMC- and guar gum-based silver nanoparticle coatings combined with low temperature on major aroma volatile components and the sensory quality of kinnow (*Citrus reticulata*). Int J Food Sci Technol 51(11):2345–2352

Sharma S, Rao TR (2015) Xanthan gum based edible coating enriched with cinnamic acid prevents browning and extends the shelf-life of fresh-cut pears. LWT Food Sci Technol 62(1):791–800

Shit SC, Shah PM (2014) Edible polymers: challenges and opportunities. J Polym 2014:1–14

Siew DC, Heilmann C, Easteal AJ, Cooney RP (1999) Solution and film properties of sodium caseinate/glycerol and sodium caseinate/polyethylene glycol edible coating systems. J Agric Food Chem 47(8):3432–3440

Silva GMC, Silva WB, Medeiros DB, Salvador AR, Cordeiro MHM, da Silva NM et al. (2017) The chitosan affects severely the carbon metabolism in mango (*Mangifera indica L.* cv. Palmer) fruit during storage. Food Chem 237, 372–378

Singh S, Khemariya P, Rai A, Rai AC, Koley TK, Singh B (2016) Carnauba wax-based edible coating enhances shelf-life and retain quality of eggplant (*Solanum melongena*) fruits. LWT Food Sci Technol 74:420–426

Sipahi R, Castell-Perez M, Moreira R, Gomes C, Castillo A (2013) Improved multilayered antimicrobial alginate-based edible coating extends the shelf life of fresh-cut watermelon (*Citrullus lanatus*). LWT Food Sci Technol 51(1):9–15

Souza MP, Vaz AF, Cerqueira MA, Texeira JA, Vicente AA, Carneiro-da-Cunha MG (2015) Effect of an edible nanomultilayer coating by electrostatic self-assembly on the shelf life of fresh-cut mangoes. Food Bioprocess Technol 8(3):647–654

Stephen A, Phillips GO, Williams PA (2006) Food polysaccharides and their applications. CRC, Taylor & Francis Group, Boca Raton, FL

Synowiec A, Gniewosz M, Kraśniewska K, Przybył JL, Bączek K, Węglarz Z (2014) Antimicrobial and antioxidant properties of pullulan film containing sweet basil extract and an evaluation of coating effectiveness in the prolongation of the shelf life of apples stored in refrigeration conditions. Innov Food Sci Emerg Technol 23:171–181

Tesfay SZ, Magwaza LS (2017) Evaluating the efficacy of moringa leaf extract, chitosan and carboxymethyl cellulose as edible coatings for enhancing quality and extending postharvest life of avocado (*Persea americana Mill.*) fruit. Food Packaging Shelf Life 11:40–48

Treviño-Garza MZ, García S, del Socorro Flores-González M, Arévalo-Niño K (2015) Edible active coatings based on pectin, pullulan, and chitosan increase quality and shelf life of strawberries (*Fragaria ananassa*). J Food Sci 80(8):M1823–M1830

Treviño-Garza MZ, García S, Heredia N, Alanís-Guzmán MG, Arévalo-Niño K (2017) Layer-by-layer edible coatings based on mucilages, pullulan and chitosan and its effect on quality and preservation of fresh-cut pineapple (*Ananas comosus*). Postharvest Biol Technol 128:63–75

Valenzuela C, Tapia C, López L, Bunger A, Escalona V, Abugoch L (2015) Effect of edible quinoa protein-chitosan based films on refrigerated strawberry (*Fragaria× ananassa*) quality. Electron J Biotechnol 18(6):406–411

Vanoli M, Grassi M, Buccheri M, Rizzolo A (2015) Influence of edible coatings on postharvest physiology and quality of Honeydew melon fruit (*Cucumis melo L. inodorus*). Adv Horticul Sci 29:65–74

Vargas M, Pastor C, Albors A, Chiralt A, González-Martínez C (2008) Development of edible coatings for fresh fruits and vegetables: possibilities and limitations. Fresh Produce 2(2):32–40

Vargas-Torres A, Becerra-Loza AS, Sayago-Ayerdi SG, Palma-Rodríguez HM, de Lourdes García-Magaña M, Montalvo-González E (2017) Combined effect of the application of 1-MCP and different edible coatings on the fruit quality of jackfruit bulbs (*Artocarpus heterophyllus Lam*) during cold storage. Sci Hortic 214:221–227

Veiga-Santos P, Oliveira LM, Cereda MP, Alves AJ, Scamparini ARP (2005) Mechanical properties, hydrophilicity and water activity of starch-gum films: effect of additives and deacetylated xanthan gum. Food Hydrocoll 19(2):341–349

Velickova E, Winkelhausen E, Kuzmanova S, Alves VD, Moldão-Martins M (2013) Impact of chitosan-beeswax edible coatings on the quality of fresh strawberries (*Fragaria ananassa* cv Camarosa) under commercial storage conditions. LWT Food Sci Technol 52(2):80–92

Vishwasrao C, Ananthanarayan L (2017) Delayed post-harvest ripening-associated changes in *Manilkara zapota L.* var. Kalipatti with composite edible coating. J Sci Food Agric 97(2):536–542

Wang X, Liu F, Zheng X, Sun J (2011) Water-enabled self-healing of polyelectrolyte multilayer coatings. Angew Chem Int Ed 50(48):11378–11381

Whistler RL, Daniel JR (1990) Functions of polysaccharides in foods. In: Branen AL, Davidson PM, Salminen S (eds) Food additives. Marcel Dekker, New York, NY, pp 395–424

Williams PA, Phillips GO (2000) In: Phillips G, Williams P (eds) Handbook of hydrocolloids. CRC, Cambridge, pp 1–19

Xiong H, Tang S, Tang H, Zou P (2008) The structure and properties of a starch-based biodegradable film. Carbohydr Polym 71(2):263–268

Yang HJ, Lee JH, Lee KY, Bin Song K (2017) Application of gelatin film and coating prepared from dried Alaska Pollock by-product in quality maintanance of grape berries. J Food Process Preserv 41:e13228

Yousuf B, Srivastava AK (2017) Flaxseed gum in combination with lemongrass essential oil as an effective edible coating for ready-to-eat pomegranate arils. Int J Biol Macromol 104:1030–1038

Yun J, Fan X, Li X, Jin TZ, Jia X, Mattheis JP (2015) Natural surface coating to inactivate *Salmonella enterica serovar Typhimurium* and maintain quality of cherry tomatoes. Int J Food Microbiol 193:59–67

Zambrano-Zaragoza M, Mercado-Silva E, Ramirez-Zamorano P, Cornejo-Villegas M, Gutiérrez-Cortez E, Quintanar-Guerrero D (2013) Use of solid lipid nanoparticles (SLNs) in edible coatings to increase guava (*Psidium guajava L.*) shelf-life. Food Res Int 51(2):946–953

Zambrano-Zaragoza M, Mercado-Silva E, Gutiérrez-Cortez E, Cornejo-Villegas M, Quintanar-Guerrero D (2014a) The effect of nano-coatings with α-tocopherol and xanthan gum on shelf-life and browning index of fresh-cut "Red Delicious" apples. Innov Food Sci Emerg Technol 22:188–196

Zambrano-Zaragoza ML, Gutiérrez-Cortez E, Del Real A, González-Reza RM, Galindo-Pérez MJ, Quintanar-Guerrero D (2014b) Fresh-cut red delicious apples coating using tocopherol/mucilage nanoemulsion: effect of coating on polyphenol oxidase and pectin methylesterase activities. Food Res Int 62:974–983

Zambrano-Zaragoza ML, Quintanar-Guerrero D, Del Real A, Piñon-Segundo E, Zambrano-Zaragoza JF (2017) The release kinetics of β-carotene nanocapsules/xanthan gum coating and quality changes in fresh-cut melon (cantaloupe). Carbohydr Polym 157:1874–1882

Zeng R, Zhang A, Chen J, Fu Y (2013) Impact of carboxymethyl cellulose coating enriched with extract of Impatiens balsamina stems on preservation of 'Newhall' navel orange. Sci Hortic 160:44–48

Zhang L, Li S, Dong Y, Zhi H, Zong W (2016) Tea polyphenols incorporated into alginate-based edible coating for quality maintenance of Chinese winter jujube under ambient temperature. LWT Food Sci Technol 70:155–161

Zhang L, Chen F, Zhang P, Lai S, Yang H (2017) Influence of rice bran wax coating on the physicochemical properties and pectin nanostructure of cherry tomatoes. Food Bioprocess Technol 10(2):349–357

Chapter 10
Edible Foams Stabilized by Food-Grade Polymers

Ashok R. Patel

Abstract Many food products which are consumed on a regular basis, can be categorized as complex colloidal systems that are created by close contact of dispersed and continuous phases. Foams are one such colloidal systems where dispersed air bubbles are incorporated and stabilized in a continuous phase. In principle, foams can be stabilized by use of either low molecular weight surfactants, amphiphilic polymers or surface-active rigid or soft particles. In food products, foams have traditionally being stabilized by biopolymers such as proteins and certain polysaccharides but lately, there has been a tremendous interest in improving the foaming functionality of biopolymers either through physical or chemical modifications and utilizing polymer-based colloidal particles for foam stabilization. The recent understanding and advances in this highly active area of foam stabilization are reviewed in this chapter with the help of illustrative examples.

Keywords Aggregates · Air-water interfaces · Biopolymers · Colloidal particles

10.1 Introduction

Foams form an integral part of many edible products ranging from frothy beer head to dehydrated foam of meringues. These air-filled systems are either stabilized by molecular layers (of macromolecules or low molecular weight surfactants, LMWS) adsorbed at the energy-rich interfaces or through physical entrapment of bubbles in solid matrices. In some specific cases, the interfaces are stabilized by colloidal particles such as casein micelles and fat crystals or partially aggregated emulsion droplets. Regardless of their final state, all foams are prepared in liquid form (i.e. air bubbles dispersed in continuous liquid phase). The microstructure of bubbles dispersed in a continuous phase of these liquid foams varies from spherical (wet foam) to polyhedral (dry foam) depending on the volume fraction of gas phase. As shown in Fig. 10.1, microstructure of liquid foam evolves from spherical closely packed

A. R. Patel (✉)
Guangdong Technion Israel Institute of Technology, Shantou, China

T. J. Gutiérrez (ed.), *Polymers for Food Applications*,
https://doi.org/10.1007/978-3-319-94625-2_10

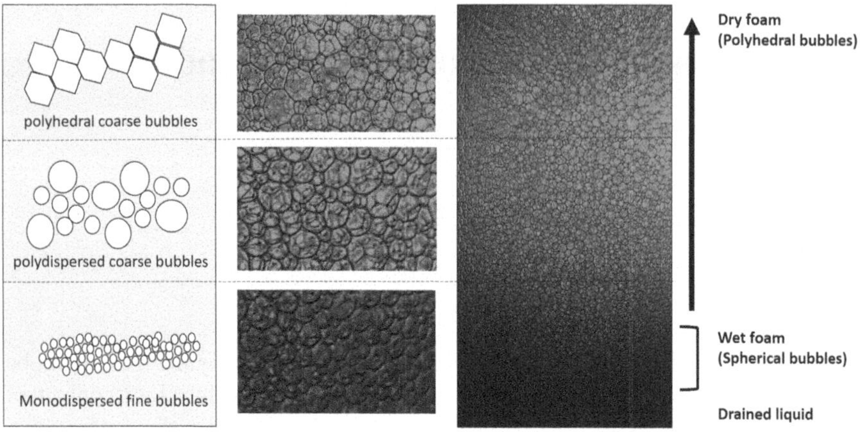

Fig. 10.1 Schematic representation of a foam sample displaying different microstructures

fine bubbles to large polyhedral shapes through an intermediate stage where the fine spherical bubbles grow into polydispersed coarse units. This non-equilibrium transition is driven both by gravitational and colloidal instabilities. Under the influence of gravity the liquid drains, while the buoyancy of large gas bubbles pushes them up. The wide distribution in the bubble size is due to the bubble growth induced both by coalescence (caused by film rupture) and Ostwald ripening or disproportionation (caused by diffusion of gas molecules through the liquid phase from smaller to larger bubbles). Figure 10.1 also shows the distinction between wet and dry foam which is characterized by the gas volume fraction. Although there is no strict criterion to distinguish wet from dry foam, it is generally accepted that a system having a gas volume fraction near 0.63 is a very wet foam, while gas fractions exceeding 0.8 constitute dry foam (Bergeron and Walstra 2005). As seen in the figure, at the very bottom sitting on the drained liquid is a zone of very wet foam. The typical length of this zone is $\xi = \gamma/\rho g D$, and the liquid fraction in this zone varies from 0.36 to 0.18 (ρ is the surfactant solution density, γ is the surface tension, D is bubble diameter) (Saint-Jalmes 2006). Most of the aerated food products are wet foams with spherical bubbles that remain separated from each other. Gravitational instability (creaming or drainage) is controlled in these systems by tuning the yielding properties of the continuous phase. This is usually achieved by either gelation of continuous phase with protein/polysaccharides in aqueous foams and structuring of continuous phase with fat particles in non-aqueous foams. However, the continuous phase viscosity only has a small influence in slowing down the colloidal instability (mainly coalescence), it is majorly controlled by the properties of foam-stabilizing components. In case of small molecular weight surfactants, an easy exchange of surfactant molecule can occur between the bulk and the interface due to the low desorption energy of soluble surfactant molecules. This leads to a low resistance of surfactant monolayer to compression making the foam less resistant to coalescence. When the surfactant monomers desorb from the interface, the surface coverage may in some cases become too low to avoid the coalescence phenomenon (Fameau 2018). In case of polymers (including proteins), the adsorption energies are much higher than those for surfactants which slows down the molecular exchange at the interface. In addition, the interfaces are relatively more viscoelastic than those made up of surfactants. Together this leads to a comparatively higher stability against coalescence. When particles are used for foam stabilisation the adsorption energy increases even further and can lead to even more elastic interfaces and in some cases, particles could also lead to a complete arrest of the ageing of foams (Stocco et al. 2011).

As far as food polymers are concerned, proteins are the most widely used foaming agents, used either as such or in particulate form such as aggregates, micelles and colloidal complexes together with polysaccharides (Wierenga and Gruppen 2010). In addition to proteins, food polymers such as cellulose derivatives have also been explored for foam stabilization in form of colloidal particles (Jin et al. 2012; Patel et al. 2012, 2013b).

10.2 Protein Stabilized Foams

Proteins are macromolecules consisting of a chain of amino acids (AAs) linked by peptide bonds. There are a total of 20 AAs which can be distinguished by their side groups that can be ionic, non-ionic or hydrophobic. The primary structure of a protein refers to the sequence of AAs in the polypeptide chain. The AA chain is folded into secondary structural elements, called e.g. α-helix, β-sheet or random coil. These structural elements are further ordered into a tertiary structure stabilized by covalent (e.g. S-S bridges) and/or non-covalent bonds. Proteins can be distinguished as globular, random coil and fibrillar proteins. Globular proteins such as β-lactoglobulin and bovine serum albumin have defined structural elements in a ‘fixed’ arrangement, while flexible ‘random coil’ proteins consist predominantly of random coil elements without a fixed structure, e.g. β-casein. Fibrillar proteins (e.g. collagen or myosin) are typically insoluble in water in their ‘native’ state.

When protein is used in non-particulate form, the adsorption of protein at an air-water interface takes place in three main stages: (i) diffusion of the native protein molecules from the bulk phase to the interface and their subsequent adsorption; (ii) surface denaturation, i.e. uncoiling of the polypeptide chains at the interface (surface denaturation); and (iii) aggregation of the surface denatured protein into a coagulum which is largely devoid of surface activity.

Although proteins lower the surface tension (due to the presence of hydrophilic and hydrophobic groups), the main factor determining the stability of a foam is the cohesion and elasticity of the protein monolayer film formed at the interface, this is controlled by the denatured but uncoagulated protein at the surface and more specifically by the rate at which the protein denatures at the interface and the rate at which it is removed by coagulation. It has been found that the rate of surface denaturation does not vary greatly from one protein to another, although it does depend upon the pH of the solution (Neurath and Bull 1938), but their rates of coagulation differ markedly and is affected by the mechanical agitation to which the protein film is subjected and by the electrostatic forces between the polypeptide chains (Cumper 1953).

10.3 Foams Stabilized by Protein Aggregates

Proteins are liable to undergo conformational changes and form supramolecular structures depending on the micro environment (pH, ionic strength etc.) and the external factors such as temperature. Protein aggregates could be made by heat-treatment induced fibrillization, denaturation of proteins and gelation (Amagliani and Schmitt 2017). It can be speculated that the aggregates may have a negative influence on the foaming behaviour of proteins. First of all, they are large structures (few hundred nanometers to 10s of microns) so their diffusion to the air-water interface is limited. Secondly, the nature of protein-protein interactions in aggregates may influence their cohesiveness and consequently their spreading behaviour at the interface (Rullier et al. 2008).

However, it has been shown in some cases that the presence of aggregates increases the foaming properties of proteins. The mechanism by which aggregates contribute to foam stabilization is not completely elucidated and different hypotheses have been proposed: the protein aggregates can adsorb at the interface and increase the viscoelasticity of the interface, thus providing a better foam stabilization or if aggregates do not adsorb to the interface, they remain in the aqueous phase and can become confined into foam films and lead to the formation of a gel-like network by undergoing a percolation process (Davis and Foegeding 2004). They can further prevent destabilization of foam by reducing the drainage by acting as cork in the Plateau borders (Saint-Jalmes et al. 2005).

In addition to aggregates, protein-based fluid gels have also be explored for foam stabilization (Lazidis et al. 2016). Fluid gels are formed when a separation process is applied to a gelling polymer undergoing its sol–gel transition. More precisely, fluid gels are formed when the (bio)polymers are introduced into a gelling environment (usually by cooling) while undergoing physical disruption (shear). Gelled particles are created which, depending on the shear field experienced, are gels on the microns rather than macro length scale. The resultant properties of these systems are, therefore, fluid or paste-like rather than those of typical gels (Norton et al. 1999). Whey protein isolate (WPI) fluid gels which were produced by thermally treating WPI under shear showed interesting foaming properties that were dependent not only on the pH at which these structures were originally formed but also the pH at which they were aerated (Lazidis et al. 2016). The small aggregates formed at pH 5 (close to its iso-electric point) were able to fill effectively the spaces between the bubbles and reduce the rate of liquid drainage by the particles acting as corks anchored on the interface and by increasing the local viscosity of film around the bubbles. While, fluid gels formed at pH 8 resulted in less viscoelastic films probably due to the larger size of the aggregates, which limited their ability to pack and leading to their exclusion from the spaces around the air bubbles (Lazidis et al. 2016).

It is also worth mentioning in this section, the stabilization of foams provided by unique class of proteins called hydrophobins. These are small proteins produced by filamentous fungi, they display a strong tendency to self-assemble at the air–water interface and form highly viscoelastic layers (Green et al. 2013). A striking structural feature of the hydrophobin molecule is that it is rigid like a small solid particle; it is also amphiphilic in nature, having a hydrophobic patch on one side of the molecule. An individual hydrophobin molecule might thus be represented like a nanosized Janus particle (Walther and Muller 2008). Due to their irreversible adsorption at the interface, they are known to produce foams at concentration as low as 0.1%wt (Green et al. 2013).

10.4 Foams Stabilized by Protein–Surfactant Mixtures

Proteins and low molecular weight surfactants (LMWS) are commonly used as foaming agents in food products. They stabilize air-water interfaces by two different mechanisms, proteins by forming a visco-elastic network and LMWS by fast adsorption and desorption and Gibbs-Marangoni mechanism. It has been suggested that for effective stabilization of foam by proteins a high dilatational interfacial elasticity is required while for foam stabilization by LMWS, a rapid increase in equilibrium interfacial pressure is needed (Damodaran 2005; Acharya et al. 2005; Saint-Jalmes et al. 2005). Because of these opposing concepts, it is difficult to predict and understand the foam stabilization behaviour of protein-LMWS mixtures. In addition, the competitive adsorption and displacement of proteins by LMWS also needs to be taken into consideration (Maldonado-Valderrama and Patino 2010). It is hypothesized that due to the relatively low adsorption energies of LMWS, they can adsorb into the protein-adsorb layer, probably at first into the packing defects. As more molecules of LMWS adsorb into the defects, LMWS-rich domains are formed that makes the interface weaker and consequently decreases the stability of foam due to increased drainage and coalescence rates (Wilde et al. 2004). However, this concept is valid when LMWS are added to a system where proteins are already adsorbed at the interface. In case where aeration is carried out with an aqueous dispersion which contains a protein-LMWS mixture, a completely different scenario is anticipated (Rodrıguez Patino et al. 2003).

The foaming stabilizing behaviour of protein-LMWS mixtures strongly depends on the charge of protein and LMWS as well as the mixing ratios of the two components. Depending on the conditions (pH, ionic strength, protein:LMWS ratio etc.), the mixtures may contain free proteins, free LMWS and/or protein-LMWS complexes. The relative proportions of these states and the surface activity of complexes will determine their overall contribution to foam stability (Lech et al. 2014).

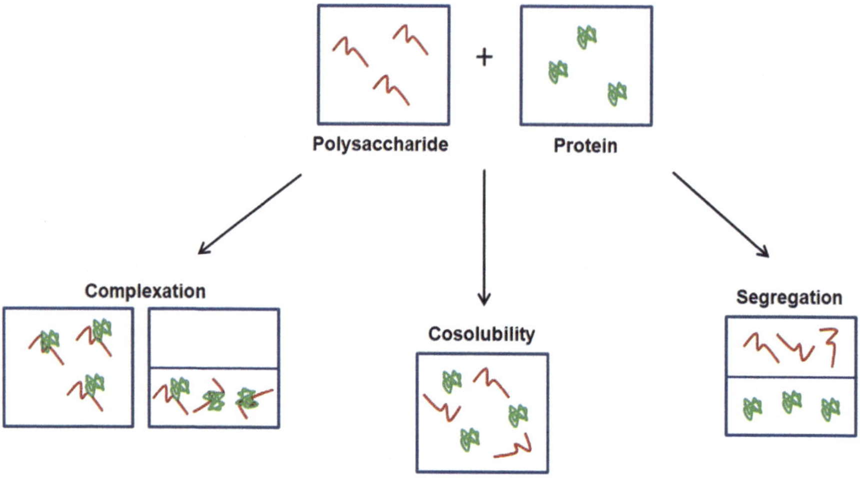

Fig. 10.2 Protein-polysaccharide mixing behaviour: complexation, co-solubility and segregation. Reprinted from Wijaya et al. Functional colloids from proteins and polysaccharides for food applications. Trends Food Sci Technol 68:56–69. Copyright (2017), with permission from Elsevier

10.5 Foams Stabilized by Protein-Polysaccharide Colloidal Complexes

When proteins and polysaccharides are mixed together in an aqueous environment, three different scenarios (Wijaya et al. 2017) are possible (Fig. 10.2): (i) complexation, which depending on the strength of interactions, may lead to the formation of either soluble complexes or coacervation (associative phase separation) where a colloid rich phase (of protein-polysaccharide complexes) separates from the solvent phase; (ii) cosolubility; protein-polysaccharide combination is compatible with each other and the solution is stable; and (iii) segregative phase separation; due to the incompatibility among protein and polysaccharide two phases appear, one rich in polysaccharide and poor in protein and one rich in protein and poor in polysaccharide.

Polysaccharides are generally used to enhance the colloid (emulsions and foams) stabilizing properties of proteins in foods. Most often than not, they are used simply as viscosifying agents to increase the bulk viscosity which can slow down the gravitational and colloidal instabilities in these biphasic systems. However, in some cases polysaccharide is specifically used to interact with the protein adsorbed at the interface. A classical application of such protein-polysaccharide interaction in the food industry is the use of pectin to stabilize casein micelles in acidified milk drinks. The negatively charged pectin molecules adsorb to the positively charged casein micelles due to electrostatic interactions. As a consequence, the electrostatic and steric repulsion prevent the micelles from acid-induced aggregation (Syrbe et al. 1998).

As for foaming properties, protein-polysaccharide complexes can lead to positive effect such as increased overrun and yield stress by improving the elasticity of the foam film or by altering the gelation behaviour of protein at the air-water interface (Wang et al. 2015). The mechanism of foam stabilization depends on the type of complex formed by protein-polysaccharide interaction, a soluble complex can increase the viscosity of the foam film and in the plateau region to limit the drainage and inhibit the bubble coalescence (Schmidt et al. 2010; Sadahira et al. 2016). Larger size insoluble complexes can help build an interfacial viscoelastic network at the air-water interface with reduced gas permeability, leading to greater stability concerning the disproportionation (Sadahira et al. 2014).

10.6 Foams Stabilized by Rigid Colloidal Particles

Rigid particles that have the required surface activity, wettability, shape and size can be utilized to create the so-called super stable foams lasting for months or even years. Stabilization of interfaces by rigid particles is also known as the Pickering stabilization mechanism and it has recently drawn a lot of scientific and industrial interest (Lam et al. 2014). Much of the research into particle-stabilized biphasic dispersions has been focused on using inorganic particles which have limited relevance for applications in foods. However, in the past few years, there has been a shift toward studying materials of biological origin for stabilization of biphasic dispersions with the goal of utilizing them for edible applications (Dickinson 2010; Lam et al. 2014). As far as Pickering stabilization of foams is concerned, most of the particles are either made from non-food grade materials or processes or their surfaces have been modified using non-edible materials or a process that involves a chemical modification (Du et al. 2003; Cervantes Martinez et al. 2008; Park et al. 2009). Only few studies have been reported with food-grade materials such as shellac, ethylcellulose (EC), calcium carbonate, cellulose nanocrystals, chitin nanocrystals, hydrophobic cellulose and some natural origin components such as bacteria, spores, cells and viruses (Lam et al. 2014; Binks et al. 2005; Campbell et al. 2008, 2009; Zhou et al. 2009; Hu et al. 2016; Wege et al. 2008; Tzoumaki et al. 2015; Yucel Falco et al. 2017).

Rigid particles from hydrophobic polymer such as EC can be prepared using a rather facile method based on controlled precipitation by altering the solvent quality. Such process is known by several names in the literature including liquid-liquid dispersion, anti-solvent precipitation, nano-precipitation, drowning-out etc. A stock solution of EC is first prepared in an organic solvent such as acetone, followed by dilution in a non-solvent such as water to trigger instant precipitation as a result of supersaturation. The resultant colloidal particles formed via nucleation and growth mechanism are in the nanoscale range (Fig. 10.3). The morphology of particles can be controlled from spherical to anisotropic shapes such as rods and fibre-like. Although EC is a neutral polymer, foam stabilizing properties of EC particles is found to be strongly influenced by pH (Jin et al. 2012). As suggested by Jin et al., at

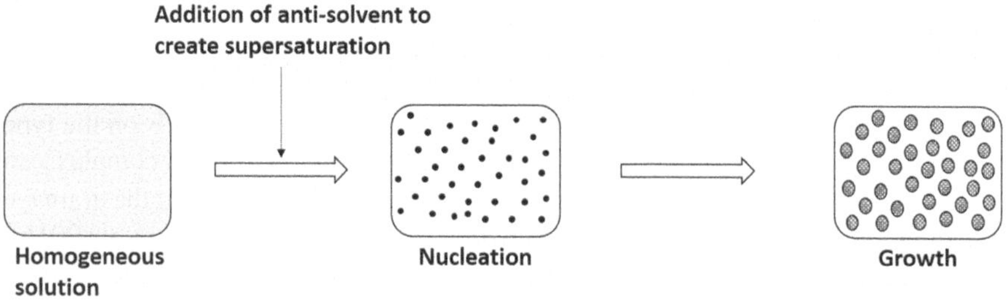

Fig. 10.3 Schematic representation of anti-solvent precipitation process

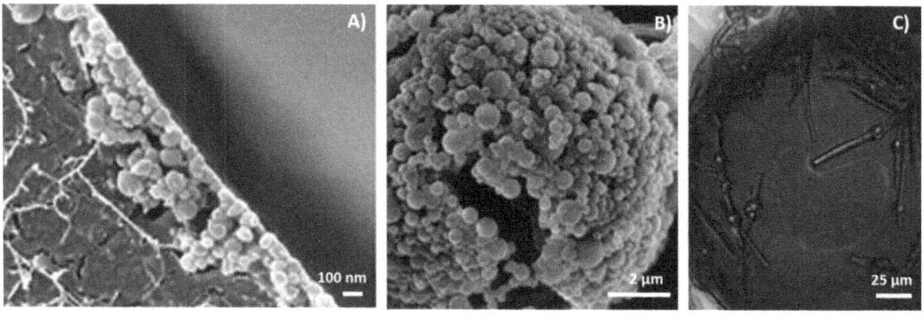

Fig. 10.4 Images of interfacial (air-water interface) accumulation of spherical (**a**, **b**) and anisotropic particles (**c**) created using anti-solvent precipitation. Figures **a** and **c** are adapted from (Jin et al. 2012; Campbell et al. 2009) with permission of the Royal Society of Chemistry, Figure **b** is reproduced with permission from Peng et al. (2017). Copyright (2017) American Chemical Society

low and high pH, the electrostatic repulsion among particles is reduced by lowering of surface potential and screening of surface charges respectively. This reduction in repulsion results in efficient loading of particles at the interfaces and consequently leads to enhanced stabilization of foam.

In addition as reported by Peng et al., such rigid particles prepared from other food polymers such as natural proteins (gliadin) also display good adsorption at the air-water interface (Fig. 10.4b) and leads to an improved foam ability and foam stability of the protein (Peng et al. 2017). These discrete adsorbed particles were further found to fuse into a continuous film over storage time.

In some cases, a combination of rigid particles and polymers have also been utilized to stabilize foams, e.g. cellulose nanocrystals + methylcellulose (CNC + MCE) and crystalline α-cellulose + EC (Fig. 10.5).

Nano materials derived from cellulose such as nanocrystals (CNC) and nanofibers (CNF) have recently gained a lot of popularity as 'green' components with multiple functionalities. With respect to foam stabilization, CNC on their own does not show lowering of surface tension and hence are not capable of stabilizing air-water interfaces. In some cases, hydrophobic modification of CNF and CNC have been explored to make them more suitable for foam stabilization (Cervin et al. 2013, 2015). Hu et al. observed that when unmodified CNC was used together with

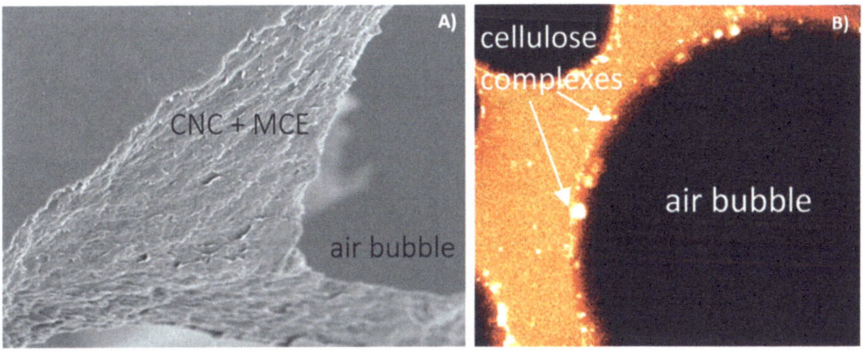

Fig. 10.5 Microscopy images of aqueous foams stabilized by combination of particles and polymers, (**a**) CNC + MCE and (**b**) α cellulose particles + EC (cellulose complexes). Reprinted with permission from Murray et al. (2011) and Hu et al. (2016). Copyright (2011 and 2016) American Chemical Society

a interfacial active component – MCE, the nanocrystal particles did not drain in the serum phase and were retained in the foam phase (Fig. 10.5a). The authors of the study speculated two possible mechanisms: (i) MCE bound CNC particles mechanically stabilized the interface and inhibit MCE desorption and (ii) in the aqueous phase, the CNC-MCE mixture forms a weak gel that inhibit bubble coalescence and slow down the drainage (Hu et al. 2016).

In case of cellulose complexes, the crystalline α cellulose particles were made hydrophobic by precipitating EC on their surface through anti-solvent precipitation. These surface active particles were found to display a good foam stabilization properties (Fig. 10.5b). Further, by combining these particles with proteins, a synergistic enhancement in foam stabilization could be obtained by just using a 0.1%wt concentration of particles in combination with proteins such as sodium caseinate and whey protein isolate (Murray et al. 2011).

10.7 Foams Stabilized by Gelled Particles of Cellulose–Polyphenol Complexes

Hydrophilic cellulose derivatives such as methylcellulose (MCE) and hydroxyl propyl methylcellulose (HPMC) display interesting associative behaviour in aqueous environment. They are also surface active and are known to stabilize oil-water and air-water interfaces (Patel et al. 2013a, 2014). However, when used as foaming agents, although they show good foamability (high air incorporation), the stability of foam is quite limited. For instance, lower viscosity grades of HPMC can even give an overrun of over 300% but these foams have a very short life span (Patel et al. 2013a). The foam stabilization property of cellulose derivatives is attributed to the gelled film formation of polymer at the air-water interface and therefore the higher

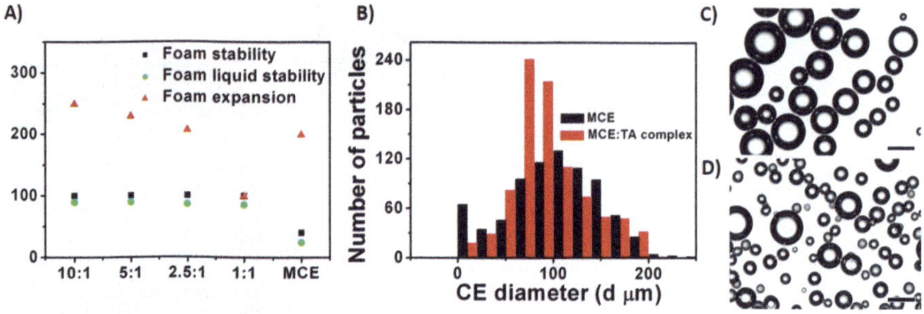

Fig. 10.6 (**a**) Foaming parameters (foam expansion, foam stability and foam liquid stability) for foams prepared using MCE (2%wt) and MCE:TA complexes (at MCE:TA ratios of 10:1, 5:1, 2.5:1 and 1:1 wt/wt); (**b**) Histograms showing the distribution of the mean CE diameter of bubbles in foams prepared from MCE and MCE:TA (10:1 wt/wt); (**c, d**) Microscopy images of diluted foams prepared from MCE and MCE:TA (10:1 wt/wt) respectively (scale bars = 100 μm). Reproduced from Patel et al. (2013b) with permission from The Royal Society of Chemistry

viscosity grade of these polymers provide a comparatively better stabilization as compared to lower viscosity grades due to the formation of the interfacial film of relatively higher stiffness (Patel et al. 2013a). In one of our previously reported work we found that the functionality of cellulose derivatives (including foaming properties) could be significantly improved by complexing them with natural polyphenols. For our experiments we selected MCE with molecular weight of 40,000 $gmol^{-1}$ (viscosity grade 2%, 20 °C = 400 cP) and degree of substitution of 1.8. Complexes of MCE and tannic acid, TA (one of the most researched natural polyphenol) were prepared at varying proportion of MCE: TA (10:1, 5:1, 2.5:1 and 1:1 wt/wt) while keeping the concentration of MCE constant at 2%wt. The MCE-TA complexes were compared to un-complexed MCE in terms of foam properties (Fig. 10.6a) such as foam expansion (or overrun = $[(V_t - V_0)/V_0] \times 100$), foam stability ($= 100 \times [F_2/F_1]$), and foam liquid stability ($= 100 - [100 \times V_{60}/V_0]$), where V_t and V_0 are the total foam and liquid volumes respectively; F_1 and F_2 are gas volume at t = 0 and t = 60 min respectively and V_{60} is liquid volume at 60 min.

As seen from Fig. 10.6a, MCE was a weak foaming agent showing a good foam expansion but a quick drop in foam volume within 60 min along with a high drainage which is reflected in poor foam liquid stability. In contrast, complexes prepared at all MCE: TA ratios showed improvement in both foamability as well as foam stability. The enhanced foam stabilization could be attributed to a combination of multiple factors such as an increased bulk viscosity due to the gelation of MCE: TA complexes, enhancement in interfacial gelation, and adsorption of colloidal particles at the air–water interface. Further, the increased proportion of TA resulted in a decreased foam expansion, with MCE: TA 1: 1 wt/wt showing a drop of over 100%, the most efficient system was MCE: TA 10: 1 wt/wt and was thus selected for further comparative studies. The average bubble size of foam was obtained by measuring the mean circular equivalent (CE) diameter on an automated microscope (Fig. 10.6b). The average bubble sizes for foams prepared using MCE and MCE: TA

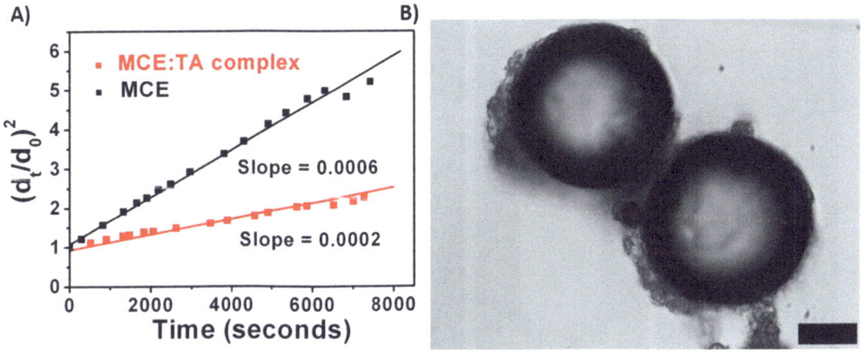

Fig. 10.7 (**a**) Comparative coarsening data for foams prepared using MCE and MCE: TA (10:1 wt/wt) complexes. MCE concentration in all cases was kept constant at 2 wt%; and (**b**) Microscopy image showing the air bubbles surrounded by a microscopic layer of MCE: TA complexes gelled at the air–water interface (scale bar = 20 μm). Reproduced from Patel et al. (2013b) with permission from The Royal Society of Chemistry

complexes were 118 and 96 μm, respectively. For visual comparison the representative microscopy images of foams stabilized by MCE and MCE: TA complexes are also shown in Fig. 10.6c, d. The stability of foams was also followed by measuring the coarsening time using the backscattering and transmission of light. A comparative graph of coarsening time, τ (from model $d^2(t)/d^2(0) = 1 + t/\tau$) is shown in Fig. 10.7a. The average coarsening times calculated from the plots (27.8 and 83.3 min for MCE and MCE: TA foams respectively) confirms the significant enhancement in the foam stabilizing property of MCE due to its interaction with TA. As the study was done on diluted foams, the increased stability can be at least partly attributed to the enhancement of the interfacial stiffness due to the surface gelation invoked by the MCE: TA complexes. The microscopy image of the MCE: TA foam is presented in Fig. 10.7b; as seen from the image, the increased interfacial stiffness could be attributed to the layer of MCE: TA complexes gelled at the air–water interface.

10.8 Some Innovative Applications of Polymer Stabilized Food Foams

10.8.1 Polymer Foams as Templates for Edible Oil Structuring

As a research domain, edible oil structuring is still in its infancy but it has already received a great deal of interest from academic researchers and industrial scientists. The goal of oil structuring is to replace solid fats in food products with components that can immobilize and structure liquid oil to provide a fat-mimicking functionality. Although this strategy has a huge potential, it is faced with a major drawback of

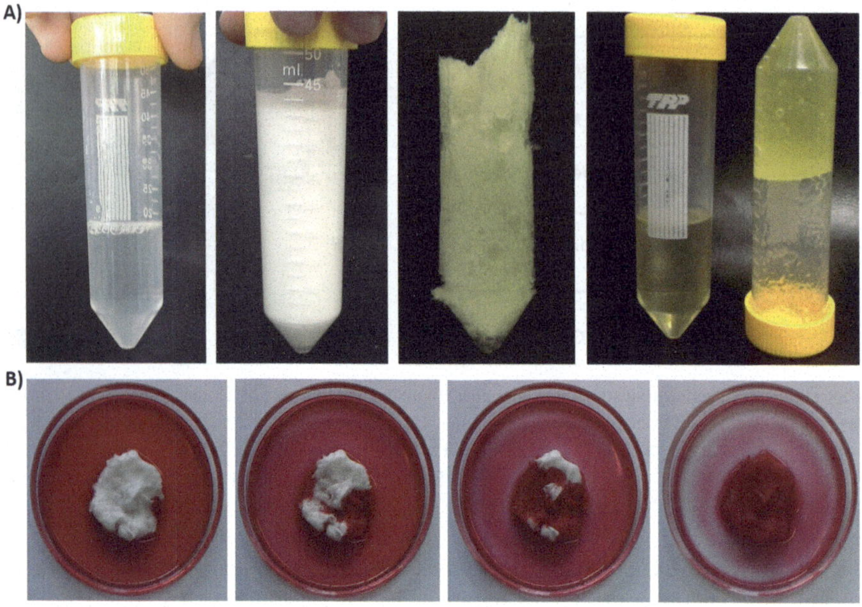

Fig. 10.8 (**a**) Photographic representation of steps involved in formulating oleogels using aqueous foams of HPMC as templates and (**b**) A series of photographs displaying the quick oil absorption by porous dried foam. Sunflower oil with Nile red dye was used for clear representation and images were captured every 60 s. Reproduced with permission from Royal Society of Chemistry (Patel et al. 2013a)

identifying food-grade components which could be used in edible products. Among several components tried for oil structuring, polymers appear to be the most promising candidates. However, most food-approved polymers are inherently hydrophilic in nature and therefore cannot be dispersed easily in oil to achieve the necessary structure/network formation which is required for gelation. Hydrophilic polymers play an important role in providing structural framework to water-based gels and the functionality of polymers to form structural framework in aqueous solvent is attributed to their hydration into an extended conformation which result in stronger molecular interactions with the solvent. Therefore, in order to use hydrophilic polymers for oil structuring, it is important to first prehydrate them in water phase and arrest these hydrated conformations in dehydrated form such that they can be used for physical entrapment of oils (Patel 2015). Working on this principle, conformational framework of some food polymers such as proteins and modified polysaccharides have been created from their water dispersions by first promoting their adsorption to air–water interfaces followed by stripping-off the water to obtain dried microstructures. These microstructures can then be used for oil structuring to create oleogels (Patel et al. 2013a). The strength and rheological properties of these oleogels can be controlled by choosing the suitable viscosity grade and the by altering the concentration of polymer in the oleogels (Fig. 10.8).

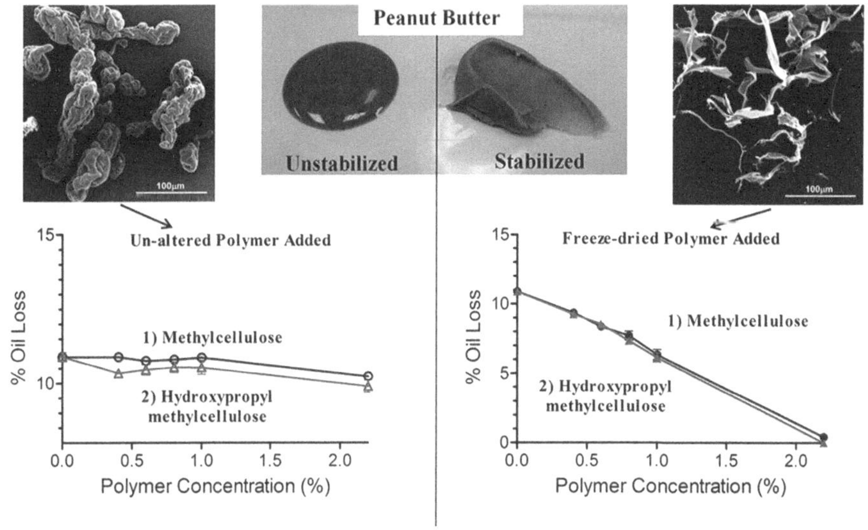

Fig. 10.9 Comparative representation of peanut butter structured using un-altered and freeze dried porous polymer. Reprinted from Tanti et al. Oil stabilization of natural peanut butter using food grade polymers. Food Hydrocoll 61:399–408. Copyright (2016), with permission from Elsevier

The approach of using polymer foam as templates for structure oil has also been explored for practical applications in real food systems such as peanut butter and sandwich cookie cream (Tanti et al. 2016a, b). The authors found that up to 75% of saturated fats could be replaced with polymer structured liquid oil without affecting the texture and stability of cookie creams (Tanti et al. 2016a). In case of peanut butter (Fig. 10.9), freeze dried polymer foam of HPMC and MCE was used as a replacement of fully hydrogenated oil stabilizer. It was observed that desirable oil stability and textural properties of commercial products could be achieved at just 1 and 2% level of addition of freeze dried foams (Tanti et al. 2016b).

10.8.2 Intensely Coloured Edible Foams Stabilized by Polymer-Polyphenol Complexes

Intensely coloured foams are of a large industrial interest as they have huge potential for applications in food products. However, it is difficult to impart intense colour to aqueous foams because foaming in the presence of water-soluble colorants typically results in weak colour intensity in the bubble phase as the majority of colorants are retained in the bulk solution or the serum phase. The colour intensity of the very thin foam films is insufficient to impart intense colour to the foam phase (Kim et al. 2009).

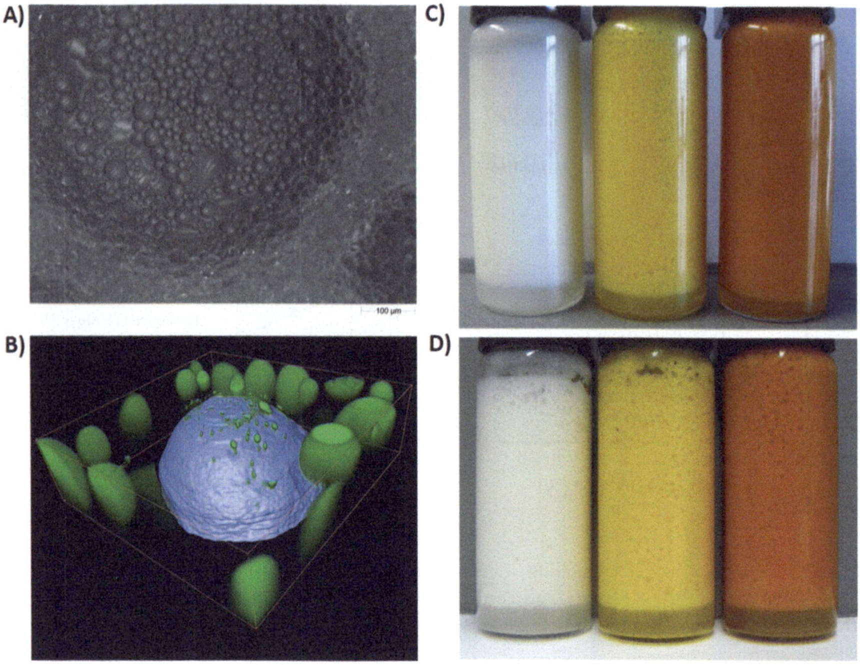

Fig. 10.10 (**a**) Optical microscope image of foamulsion showing the closely packed oil droplets around the air bubble (Scale bar = 100 µm); (**b**) 3D construction based on confocal microscopy images where air bubble (shown in blue) is surrounded by oil droplets (shown in green); (**c**, **d**) Photographs of foams (from left to right) containing no pigment, curcumin and β-carotene taken on day one and after 4 weeks of storage respectively. Reproduced with permission from Patel et al. (2012)

In one of our earlier published work (Patel et al. 2012), it was found that MCE-TA complexes (discussed in previous section) could stabilize foamulsions (foamed emulsions) at high oil volume fraction ($\varphi_{oil} > 0.5$). These foamulsions which were found to be stable against collapse for several days, showed a close packing of oil droplets around the air bubbles in the foam phase (Fig. 10.10a, b). This microstructure arrangement was further exploited to load oil-soluble pigments to generate intensely coloured foams (Fig. 10.10c) due to the preferential localization of the hydrophobic colorants in the foam phase and absence of any soluble colour in the serum phase. The foamulsions showed excellent stability without showing any significant macroscopic collapse after 4 weeks of storage.

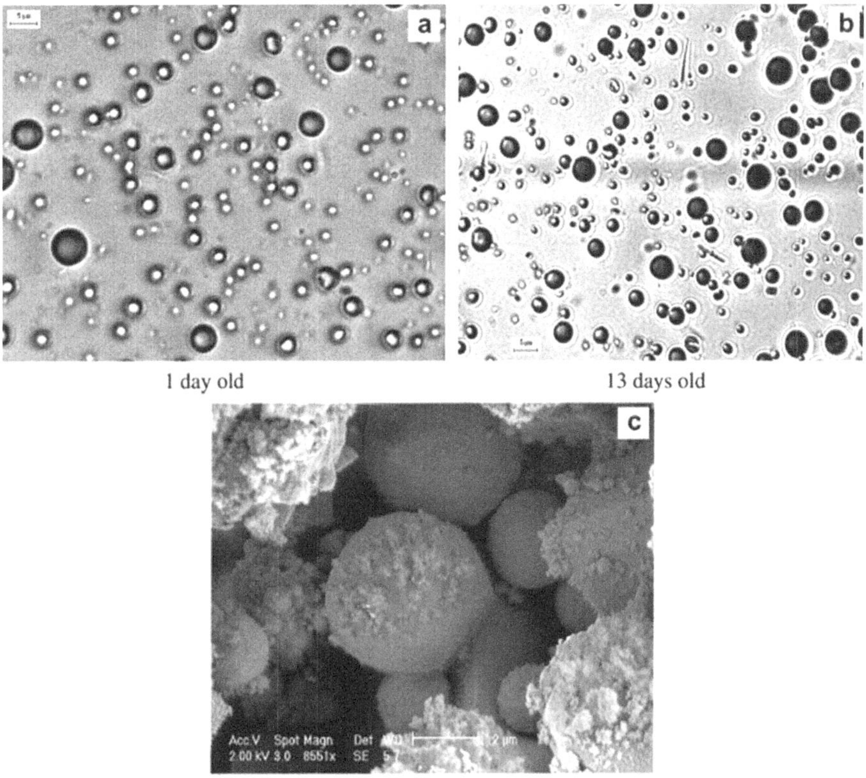

Fig. 10.11 Phase contrast optical micrograph of A/W emulsion showing bubbles size (**a**) is a brightfield image of a 1 day old sample; (**b**) a phase contrast image of a 13 day old sample and (**c**) etched Cryo-SEM micrograph of a 5 day old sample. Reprinted from Tchuenbou-Magaia et al. Hydrophobins stabilised air-filled emulsions for the food industry, Food Hydrocoll 23(7):1877–1885. Copyright (2009), with permission from Elsevier

10.8.3 Air Filled Emulsions as Fat Replacers in Food Products

Fat reduction is a growing area of interest to food manufacturers as consumers are increasingly turning to low-fat, low-calorie versions of their favourite foods. The key challenge in formulating these product types is retaining the sensorial properties provided by fat globules/oil droplets in lipid containing food products. It has been proposed that a new generation of food products can be developed by designing air bubbles that resemble oil droplets or fat globules which could be subsequently used to replace a part of lipid with air in the food products (Tchuenbou-Magaia et al. 2009). However, it is quite a challenge to produce an air based fat mimic with acceptable shelf-life for soft solid or liquid foods. Researchers from Birmingham were successful in fabricating air filled emulsions comprising of fat globule mimics constructed by coating air cells with hydrophobin proteins and subsequently used them in prototype triphasic air/oil/water emulsions (Tchuenbou-Magaia et al. 2009).

Air filled emulsion or air cells with hydrophobin coats were first prepared by ultrasonic irradiation of hydrophobin rich solution. Air cells ranging in the size from 1 to 100 μm (with approximately 40% of air cells falling in the range of 1–2 μm) were first prepared (Fig. 10.11). Triphasic air/oil/water emulsions were then created by mixing the air filled emulsion with oil in water emulsion with different level of total included phase volumes. The tri-phasic emulsions allowed the fat content of the model food to be reduced by more than 50% and they were found to be stable in terms of both volume and air content for up to 45 days.

10.9 Conclusion

Proteins are the commonest polymers used for stabilization of foam in food formulations. Some traditional food proteins such as milk, egg, plant and meat proteins have been extensively used in food formulations. Due to their inherent amphiphilicity, proteins and certain polysaccharides (cellulose derivatives) are able to reduce the surface tension and adsorb at the air-water interface to form an immobile viscoelastic film. However, these foams usually have a rather short life span due to the limited resistance provided by interfacial films which makes them susceptible to instabilities such as coalescence and disproportionation. This poor foaming functionality of proteins and polysaccharides can be improved by physical modifications such as formation of aggregates or by chemical modification through formation of complexes with LMWS or other reactive molecules such as polyphenols. A more bold approach is to stabilize foams by rigid particles created from these polymers. The advantage of using rigid particles over soluble polymers is that fluid interfaces can be effectively stabilized against disproportionation and coalescence on a longer time scale. The disadvantage however, is that a lot of energy is required to create a foam. Furthermore, rigid particles are very specific with regard to the type of interface to which they can adsorb. In contrast, soft gelled particles fabricated from polymers, have been found to spontaneously adsorb to a variety of fluid interfaces and are effective at resisting colloidal instabilities. In some cases, rigid particles and proteins can also be used in combination where the proteins can influence and promote the accumulation of particles at the interface, giving rise to increased jamming of the particles at the interface.

In my opinion, long term objectives of research in this area should focus on: (i) improvement of the foaming functionality of food proteins (especially the lesser explored plant proteins); (ii) understanding the foaming behaviour of binary mixtures (protein-LMWS, protein-polysaccharides etc.); (iii) studying the influence of product matrix on foaming behaviour of polymer and complex mixtures of polymers with other components; (iv) identifying new (industrially-feasible) methods to create rigid and soft particles from a range of food polymers and understanding their interfacial adsorption properties; and (v) exploring innovative applications of polymer stabilized foams to solve long standing and emerging industrial challenges in food formulation.

References

Acharya DP et al (2005) Interfacial properties and foam stability effect of novel gemini-type surfactants in aqueous solutions. J Colloid Interface Sci 291(1):236–243 Available at: http://www.sciencedirect.com/science/article/pii/S0021979705004996

Amagliani L, Schmitt C (2017) Globular plant protein aggregates for stabilization of food foams and emulsions. Trends Food Sci Technol 67(Supplement C):248–259 Available at http://www.sciencedirect.com/science/article/pii/S0924224417302649

Bergeron V, Walstra P (2005) Foams. Fundamentals of interface and colloid science, 5(C), pp 1–38. Available at http://www.sciencedirect.com/science/article/pii/S187456790580011X. Accessed 11 Oct 2017

Binks BP et al (2005) Naturally occurring spore particles at planar fluid interfaces and in emulsions. Langmuir 21(18):8161–8167. https://doi.org/10.1021/la0513858

Campbell AL et al (2008) Scalable fabrication of anisotropic micro-rods from food-grade materials using an in shear flow dispersion-solvent attrition technique. J Mater Chem 18(34):4074–4078. https://doi.org/10.1039/B807738J

Campbell AL, Stoyanov SD, Paunov VN (2009) Fabrication of functional anisotropic food-grade micro-rods with micro-particle inclusions with potential application for enhanced stability of food foams. Soft Matter 5(5):1019–1023. https://doi.org/10.1039/B812706A

Cervantes Martinez A et al (2008) On the origin of the remarkable stability of aqueous foams stabilised by nanoparticles: link with microscopic surface properties. Soft Matter 4(7):1531–1535. https://doi.org/10.1039/B804177F

Cervin NT et al (2013) Lightweight and strong cellulose materials made from aqueous foams stabilized by nanofibrillated cellulose. Biomacromolecules 14(2):503–511. https://doi.org/10.1021/bm301755u

Cervin NT et al (2015) Mechanisms behind the stabilizing action of cellulose nanofibrils in wet-stable cellulose foams. Biomacromolecules 16(3):822–831. https://doi.org/10.1021/bm5017173

Cumper CWN (1953) The stabilization of foams by proteins. Trans Faraday Soc 49(0):1360–1369. https://doi.org/10.1039/TF9534901360

Damodaran S (2005) Protein stabilization of emulsions and foams. J Food Sci 70(3):R54–R66. https://doi.org/10.1111/j.1365-2621.2005.tb07150.x

Davis JP, Foegeding EA (2004) Foaming and interfacial properties of polymerized whey protein isolate. J Food Sci 69(5):C404–C410. https://doi.org/10.1111/j.1365-2621.2004.tb10706.x

Dickinson E (2010) Food emulsions and foams: stabilization by particles. Curr Opin Colloid Interface Sci 15(1):40–49 http://www.sciencedirect.com/science/article/pii/S1359029409001010

Du Z et al (2003) Outstanding stability of particle-stabilized bubbles. Langmuir 19(8):3106–3108. https://doi.org/10.1021/la034042n

Fameau A-L (2018) Non-aqueous foams based on edible oils. In: Edible oil structuring: concepts, methods and applications. The Royal Society of Chemistry, London, pp 275–307. https://doi.org/10.1039/9781788010184-00275

Green AJ et al (2013) Formation and stability of food foams and aerated emulsions: hydrophobins as novel functional ingredients. Curr Opin Colloid Interface Sci 18(4):292–301 http://www.sciencedirect.com/science/article/pii/S1359029413000563

Hu Z et al (2016) Stable aqueous foams from cellulose nanocrystals and methyl cellulose. Biomacromolecules 17(12):4095–4099. https://doi.org/10.1021/acs.biomac.6b01641

Jin H et al (2012) Super stable foams stabilized by colloidal ethyl cellulose particles. Soft Matter 8(7):2194–2205. https://doi.org/10.1039/C1SM06518A

Kim S, Barraza H, Velev OD (2009) Intense and selective coloration of foams stabilized with functionalized particles. J Mater Chem 19(38):7043–7049. https://doi.org/10.1039/B908054F

Lam S, Velikov KP, Velev OD (2014) Pickering stabilization of foams and emulsions with particles of biological origin. Curr Opin Colloid Interface Sci 19(5):490–500 http://www.sciencedirect.com/science/article/pii/S1359029414000806

Lazidis A et al (2016) Whey protein fluid gels for the stabilisation of foams. Food Hydrocoll 53(Supplement C):209–217 http://www.sciencedirect.com/science/article/pii/S0268005X15000831

Lech FJ et al (2014) Identifying changes in chemical, interfacial and foam properties of β-lactoglobulin–sodium dodecyl sulphate mixtures. Colloids Surf A Physicochem Eng Asp 462(Supplement C):34–44 http://www.sciencedirect.com/science/article/pii/S0927775714006931

Maldonado-Valderrama J, Patino JMR (2010) Interfacial rheology of protein–surfactant mixtures. Curr Opin Colloid Interface Sci 15(4):271–282 http://www.sciencedirect.com/science/article/pii/S1359029409001150

Murray BS et al (2011) Preparation and characterization of the foam-stabilizing properties of cellulose–ethyl cellulose complexes for use in foods. J Agric Food Chem 59(24):13277–13288. https://doi.org/10.1021/jf203501p

Neurath H, Bull HB (1938) The surface activity of proteins. Chem Rev 23(3):391–435 http://pubs.acs.org/doi/abs/10.1021/cr60076a001

Norton IT, Jarvis DA, Foster TJ (1999) A molecular model for the formation and properties of fluid gels. Int J Biol Macromol 26(4):255–261 http://www.sciencedirect.com/science/article/pii/S0141813099000914

Park JI et al (2009) A microfluidic approach to chemically driven assembly of colloidal particles at gas–liquid interfaces. Angew Chem Int Ed 48(29):5300–5304. https://doi.org/10.1002/anie.200805204

Patel AR (2015) In: Patel AR (ed) Polymer-based oleogels created using indirect methods. Springer International, Cham, pp 29–39. https://doi.org/10.1007/978-3-319-19138-6_3

Patel AR et al (2012) Stable and temperature-responsive surfactant-free foamulsions with high oil-volume fraction. ChemPhysChem 13(17):3777–3781. https://doi.org/10.1002/cphc.201200564

Patel AR, Schatteman D et al (2013a) A foam-templated approach for fabricating organogels using a water-soluble polymer. RSC Adv 3(45):22900–22903. https://doi.org/10.1039/C3RA44763D

Patel AR, Seijen ten-Hoorn J et al (2013b) Colloidal complexation of a macromolecule with a small molecular weight natural polyphenol: implications in modulating polymer functionalities. Soft Matter 9(5):1428–1436. https://doi.org/10.1039/C2SM27200H

Patel AR et al (2014) Edible oleogels based on water soluble food polymers: preparation, characterization and potential application. Food Funct 5(11):2833–2841. https://doi.org/10.1039/C4FO00624K

Peng D et al (2017) Adsorption and distribution of edible gliadin nanoparticles at the air/water interface. J Agric Food Chem 65(11):2454–2460. https://doi.org/10.1021/acs.jafc.6b05757

Rodrıguez Patino JM, Rodrıguez Niño MR, Sánchez CC (2003) Protein–emulsifier interactions at the air–water interface. Curr Opin Colloid Interface Sci 8(4):387–395 http://www.sciencedirect.com/science/article/pii/S1359029403000955

Rullier B, Novales B, Axelos MAV (2008) Effect of protein aggregates on foaming properties of β-lactoglobulin. Colloids Surf A Physicochem Eng Asp 330(2):96–102 http://www.sciencedirect.com/science/article/pii/S0927775708004925

Sadahira MS et al (2014) Influence of protein–pectin electrostatic interaction on the foam stability mechanism. Carbohydr Polym 103(Supplement C):55–61 http://www.sciencedirect.com/science/article/pii/S0144861713012101

Sadahira MS et al (2016) Effect of egg white protein-pectin electrostatic interactions in a high sugar content system on foaming and foam rheological properties. Food Hydrocoll 58(Supplement C):1–10 http://www.sciencedirect.com/science/article/pii/S0268005X16300376

Saint-Jalmes A (2006) Physical chemistry in foam drainage and coarsening. Soft Matter 2(10):836–849. https://doi.org/10.1039/B606780H

Saint-Jalmes A et al (2005) Differences between protein and surfactant foams: microscopic properties, stability and coarsening. Colloids Surf A Physicochem Eng Aspects 263(1):219–225 http://www.sciencedirect.com/science/article/pii/S0927775705001226

Schmidt I et al (2010) Foaming properties of protein/pectin electrostatic complexes and foam structure at nanoscale. J Colloid Interface Sci 345(2):316–324 http://www.sciencedirect.com/science/article/pii/S0021979710000482

Stocco A et al (2011) Aqueous foams stabilized solely by particles. Soft Matter 7(4):1260–1267. https://doi.org/10.1039/C0SM01290D

Syrbe A, Bauer WJ, Klostermeyer H (1998) Polymer science concepts in dairy systems—an overview of milk protein and food hydrocolloid interaction. Int Dairy J 8(3):179–193 http://www.sciencedirect.com/science/article/pii/S0958694698000417

Tanti R, Barbut S, Marangoni AG (2016a) Hydroxypropyl methylcellulose and methylcellulose structured oil as a replacement for shortening in sandwich cookie creams. Food Hydrocoll 61:329–337 http://www.sciencedirect.com/science/article/pii/S0268005X16302387

Tanti R, Barbut S, Marangoni AG (2016b) Oil stabilization of natural peanut butter using food grade polymers. Food Hydrocoll 61(Supplement C):399–408 http://www.sciencedirect.com/science/article/pii/S0268005X16302429

Tchuenbou-Magaia FL, Norton IT, Cox PW (2009) Hydrophobins stabilised air-filled emulsions for the food industry. Food Hydrocoll 23(7):1877–1885 http://www.sciencedirect.com/science/article/pii/S0268005X09000605

Tzoumaki MV et al (2015) Aqueous foams stabilized by chitin nanocrystals. Soft Matter 11(31):6245–6253. https://doi.org/10.1039/C5SM00720H

Walther A, Muller AHE (2008) Janus particles. Soft Matter 4(4):663–668. https://doi.org/10.1039/B718131K

Wang Z, Zhang S, Vardhanabhuti B (2015) Foaming properties of whey protein isolate and λ-carrageenan mixed systems. J Food Sci 80(8):N1893–N1902. https://doi.org/10.1111/1750-3841.12940

Wege HA et al (2008) Long-term stabilization of foams and emulsions with in-situ formed microparticles from hydrophobic cellulose. Langmuir 24(17):9245–9253. https://doi.org/10.1021/la801634j

Wierenga PA, Gruppen H (2010) New views on foams from protein solutions. Curr Opin Colloid Interface Sci 15(5):365–373 http://www.sciencedirect.com/science/article/pii/S1359029410000592

Wijaya W et al (2017) Functional colloids from proteins and polysaccharides for food applications. Trends Food Sci Technol 68(Supplement C):56–69 http://www.sciencedirect.com/science/article/pii/S0924224417301577

Wilde P et al (2004) Proteins and emulsifiers at liquid interfaces. Adv Colloid Interf Sci 108–109(Supplement C):63–71 http://www.sciencedirect.com/science/article/pii/S0001868603001386

Yucel Falco C et al (2017) Edible foam based on pickering effect of probiotic bacteria and milk proteins. Food Hydrocoll 70(Supplement C):211–218 http://www.sciencedirect.com/science/article/pii/S0268005X17302278

Zhou W et al (2009) How rigid rods self-assemble at curved surfaces. Angew Chem Int Ed 48(2):378–381. https://doi.org/10.1002/anie.200804194

Chapter 11
Foams for Food Applications

A. L. Ellis and A. Lazidis

Abstract Numerous food products contain bubbles within their structure which in most cases play an important role either by contributing to the texture of the product or its visual appearance or both. Bubbles, therefore define the texture of commercial products that are sold aerated like bread, wafers, ice cream, chocolate mousse and meringues but are also important elements of products that are foamed in the point of consumption such as cappuccinos, whipped cream, milk shakes and more. At the same time, an increasing interest in incorporating bubbles in foods as a way to reduce their energy density is emerging. Furthermore, foams are structures that can potentially deliver new experiences to consumers from aroma release to visually enhancing traditional products. Consequently, finding ways to further understand how to construct aqueous systems with enhanced foaming properties is of great interest. This chapter will first introduce the basic concepts of foams in terms of structure, production and stability, before exploring the different biopolymer sources and their uses in aqueous food foams in particular, their role in stabilisation, with relevant examples from the literature.

Keywords Aeration · Hydrocolloids · Particles · Pickering · Proteins

11.1 Introduction

The aeration of foods was first explored 6000 years ago when the Egyptians first started baking bread. Aeration is now utilised in a wide range of products and applications, from aerated chocolate to vegan meringue, as our increased understanding has helped to exploit the novelty and versatility of bubbles as food ingredients. The incorporation of air into food products not only reduces the cost and calorific content of products but provides a luxurious texture desirable to consumers. For example, whipped cream and mousses acquire a creaminess, whilst the carbonation of soft drinks delivers the desired "fizziness" and solid products such as crisps, develop

A. L. Ellis · A. Lazidis (✉)
Chemical Engineering Department, University of Birmingham, Edgbaston B15 2TT, UK
e-mail: a.lazidis@bham.ac.uk

T. J. Gutiérrez (ed.), *Polymers for Food Applications*,
https://doi.org/10.1007/978-3-319-94625-2_11

a light and crisp texture. Introducing a gas phase into a food matrix not only affects texture but also appearance, colour and physical properties. In addition, air cells in liquid foam precursors can be used to manipulate the structural properties of products such as bread and meringue during the cooking process. As well as advances in foam formulation, processing methods have continued to be developed, for example, microfluidic devices are being developed currently for the production of microbubbles, which are thought to have exciting practical applications in the food industry as fat-replacers and texture modifiers (Rovers et al. 2016).

11.1.1 Foam Structure

Foams are two-phase systems that consist of gas cells dispersed throughout a continuous phase (Walstra 1989). This continuous phase can be either solid or liquid and the gas cells which constitute the dispersed phase are commonly called bubbles (Campbell and Mougeot 1999). The packing of gas bubbles in foams gives this state its distinctive mechanical and physical properties, which are notably different from that of both the liquid and the gas from which they are made (Suárez and Gutiérrez 2017). For example, foams can behave as viscoplastic solids that can be shaped as desired, much like rosettes of whipped cream on the top of a cake (Cantat et al. 2013).

A newly formed foam is considered wet due to the large content of water present around the bubbles. While the foam matures and the liquid drains away, the film between the bubbles thins considerably and becomes drier leaving behind polyhedral bubbles (Bergeron and Walstra 2005). The transition from a wet foam with spherical bubbles to a dry one with polyhedral bubbles (Fig. 11.1) is dependent on the ratio of liquid to gas, also known as the gas volume fraction (φ_g). Although there is no clear cut-off value that distinguishes the foam to be dry or wet, it is generally believed that a foam with a gas fraction around 0.63 (which is the limit for randomly packed spheres) is a very wet foam, while when the gas fraction exceeds 0.8 the foam is dry (Bergeron and Walstra 2005). For the majority of liquid foams present in foods most of the foaming functionality is exploited while being in the wet regime. Even foams that are found in solid form across the food industry (e.g. bread, ice cream and foam packaging) begin their life in the wet regime before they are solidified, usually by either cooling, heating or curing (Campbell and Mougeot 1999).

11.1.2 The Role of Surfactants

11.1.2.1 Surface Tension

Molecules within a fluid attract each other due to cohesion forces (hydrogen bonds, van der Waals, etc.). At an interface, these interactions result in a force with inward direction. Surface-active molecules are used to lower this surface tension and

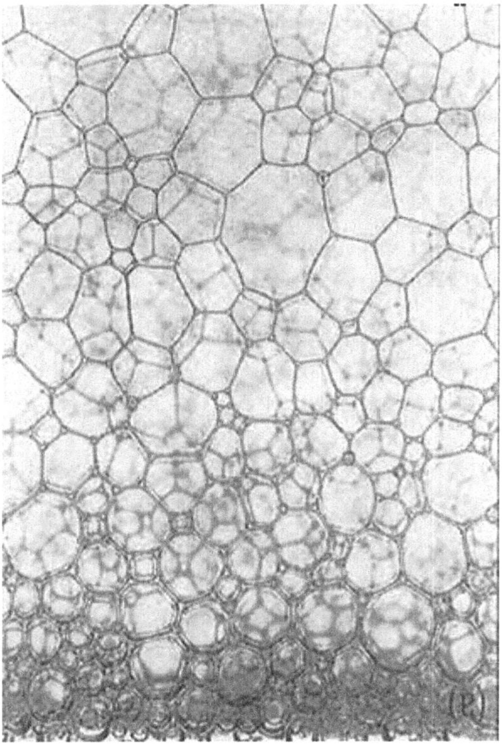

Fig. 11.1 Photograph of bubbles in an aqueous foam. Lower down, the foam is wet and bubbles appear spherical while towards the top, the bubbles are drier and polyhedral in shape. Reproduced with permission from Larson (1999)

provide greater stability. Surfactants can adsorb at an interface by positioning themselves partly in the hydrophilic phase (liquid) and partly in the hydrophobic phase (air) due to their amphiphilic nature. Their adsorption is in dynamic equilibrium where surfactant molecules are constantly replacing each other at the interface and is expressed by the Gibbs adsorption isotherm. SFT reduces proportionally to the concentration of the surfactants up to the critical micelle concentration (CMC). Above the CMC, the interface is covered with an adsorbed monolayer of surfactant molecules whilst micelles and free monomers reside in the bulk (Fig. 11.2) where all three phases are in equilibrium (Eastoe 2005).

Surfactants are used to extend the lifetime of bubbles in a liquid foam by reducing surface tension. Without surfactants, the SFT of pure water is 72 mN m^{-1} and bubbles collapse within a few milliseconds. In a solution that contains surfactants, the equilibrium SFT is not achieved instantly but is a relatively lengthy process taking milliseconds to days depending on the type of surfactant and its concentration. Surfactants fall into two main categories depending on their size; these are polymers and low molecular weight (LMW) surfactants.

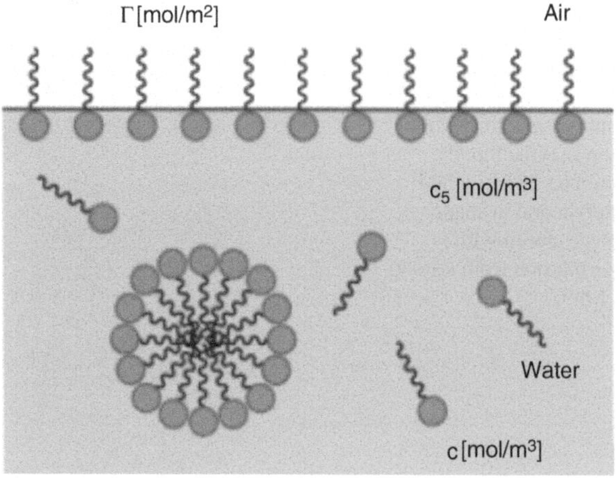

Fig. 11.2 Surfactant molecules on the interface and in the bulk at equilibrium conditions. When the surfactant concentration exceeds the CMC, surfactant molecules form micellar structures. Reproduced with permission from Cantat et al. (2013)

11.1.2.2 Low Molecular Weight Surfactants

Low molecular weight (LMW) surfactants are soap-like substances where the hydrophobic (lipophilic) part is typically an aliphatic chain and the hydrophilic part can be more diverse (Fennema 1996). The majority of the amphiphilic substances are not highly soluble in either water or oil/air, but attract the least repulsive interactions from these media when they reside partly in a hydrophilic environment (e.g. water) and partly in a hydrophobic one (e.g. air). The most commonly used LMW surfactants in foods are phospholipids (Lecithin), mono- and di-glycerides (glycerol monostearate); polysorbates (Tweens), sorbitan monostearate, sorbitan monooleate (Span 80), polyoxyethylene sorbitan monosterate (Tween 20) and sucrose esters (Bos and Vliet 2001).

LMW surfactants can be further categorised, depending on the charge of the head group, into non-ionic or ionic surfactants. When ionic LMW surfactants are used for foaming, the electrostatic repulsion between the two interfaces of the lamella increases the kinetic stability of the system. However, these systems are considerably unstable to environmental conditions such as pH and salt concentration. This can be a challenge in food foams where pH and salt concentrations can typically only be altered within a certain range. However, the use of LMW surfactants compared to polymers has a few main advantages. Firstly, they have the ability to lower the interfacial tension more effectively than amphiphilic polymers, like proteins and hydrocolloids, due to their higher adsorption energies per m^2. Secondly, as they are smaller in size, they can adsorb faster to the newly formed interface due to better mobility.

11.1.2.3 Polymeric Surfactants

Polymeric surfactants are macromolecules which contain both hydrophilic and hydrophobic parts much alike LMW surfactants. Compared to the latter, they have a more complex structure and therefore significantly different behaviour when at an interface. This difference in behaviour is mostly due to changes in their structural conformation while adsorbing to an interface caused by the need to expose and spread their hydrophobic regions (Bos and Vliet 2001).

The most notable example of polymeric surfactants used in foods are proteins, which will be discussed extensively in Sect. 11.2. Other examples of amphiphilic polymeric surfactants used in food applications are polysaccharides like chitosan (Gutiérrez 2017), hydrophobically modified cellulose (Gutiérrez and Alvarez 2017), propylene glycol alginate, gum arabic, pectin and certain galactomanans (e.g. guar gum, fenugreek gum) (Dickinson 2003; Raffa et al. 2015) as later discussed in Sect. 11. 3. While adsorbing to the a/w interface, unfolding of surface active polymers creates "trains", "loops" and "tails" in the continuous phase (Fennema 1996), which help to create a physical barrier at the interface. The unfolding of polymers on the interfaces also results in a change in conformation, which can affect the properties of the polymer. Some polymers go through such a structural change that they denature and loose some of their chemical and physical properties once desorbed (e.g. viscosity, -SH group reactivity and more) while others manage to retain their structure (Fennema 1996). Surface active polymers seem to be less sensitive to the presence of high electrolyte concentrations and heat when compared to LMW surfactants rendering them more versatile (Tadros et al. 2004). However, the performance of polymeric surfactants as interface stabilisers can be limited when compared to LMW surfactants due to their increased size, their poor solubility and insufficient amphiphilic character (Dickinson 2003).

11.1.3 Particle Stabilisation

Colloidal particles can have the ability to accumulate at the interface of two immiscible fluids in a similar manner to surfactants and can therefore be surface active (Binks 2002). This phenomenon was first studied by Ramsden (1903) and Pickering (1907) who noticed that colloidal particles could adsorb on water interfaces with oil or air and provide similar results to surfactants like saponin. The mechanism, especially when it is applied on solid particles adsorbing on the oil-water interface of emulsion droplets is commonly known as Pickering stabilisation (Dickinson 2010). The behaviour of particles in mixed aerated systems in the presence of surfactants depends on parameters such as their hydrophobicity and size. If the particles are reasonably hydrophilic, the foams produced can be further stabilised from the particles present in the aqueous phase within the foam lamella that accumulate in the Plateau borders significantly slowing down drainage (Binks 2002). If particles are too hydrophobic, they can protrude the a/w interface causing instability and

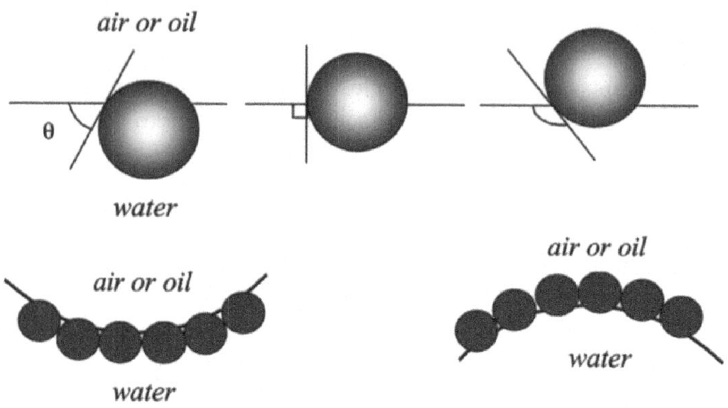

Fig. 11.3 (Above) Position of small spherical particles at the planar fluid water interface, measured in aqueous phase <90°(left), =90°(middle) and >90°(right). (Below) Probable positioning of particles on curved interfaces. For θ<90°, particle stabilised aqueous foams may form (left) and for θ >90°, particle stabilised aerosols may form (right). Reproduced with permission from (Binks 2002)

coalescence leading to foam instability via the bridging-dewetting mechanism. Finally, if particles have an intermediate hydrophobicity and they are partially wetted, they can accumulate on the bubble surface providing a strong barrier preventing shrinkage and coalescence (Binks 2002). This mechanism is commonly known as Pickering stabilisation (Dickinson 2010).

The key parameter that determines the ability of solid particles to attach to an a/w interface is their wettability as defined by the three-phase contact angle θ (Fig. 11.3). When the contact angle of the particles is around 90°, the particles have the tendency to reside on the interface and reduce the surface free energy (Binks 2002). The monolayers or particles that form on the interface curve in such a way that a larger area of the particle surface remains on the external side forming air in water when $\theta < 90°$ and water in air when $\theta > 90°$ (Binks and Horozov 2005). The energy of desorption is heavily influenced by particle size, for example, very small particles (<0.5 nm) similar to the majority of surfactant molecules, can be easily detached and therefore do not make very effective stabilisers (Binks 2002).

Although adsorption of particles on interfaces in pure particle systems is much slower compared to surfactants (Binks 2002), they have been able to form close-packed layers at the gas-liquid interface which generates a strong barrier that can retard or even eliminate the destabilisation of foams (Dickinson 2010). For example, it has been shown that particle-stabilised foams can endure disproportionation (the main source of foam instability, explained further in Sect. 11.1.5.2) for days or weeks compared to minutes or hours of the equivalent foams stabilised by LMW or polymers such as proteins (Kostakis et al. 2006).

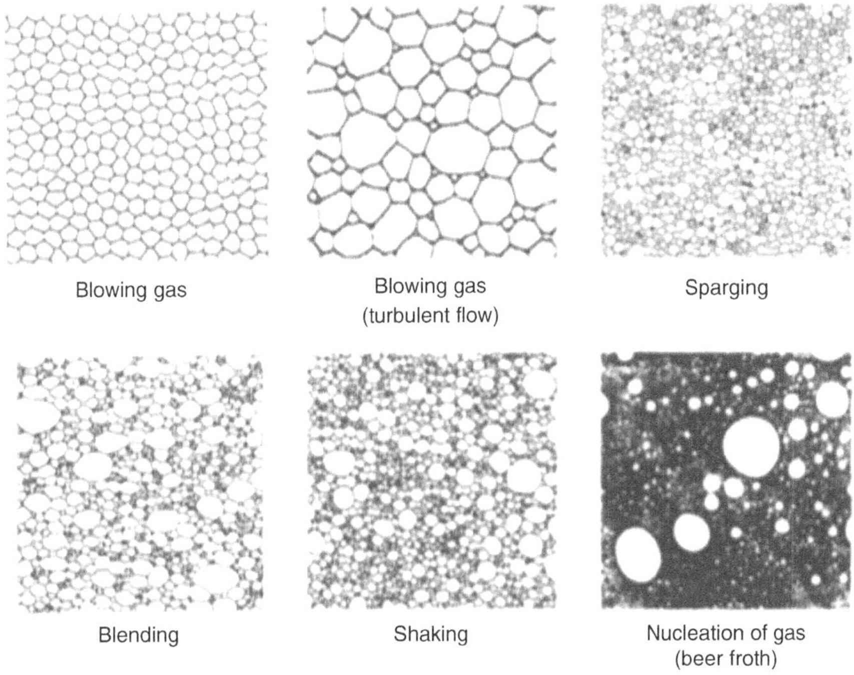

Fig. 11.4 Representation of bubble size distributions for foams created by different foaming methods. Reproduced with permission from Weaire and Hutzler (1999)

11.1.4 Foam Formation

There are three main routes to producing a foam: agitation/whipping, gas sparging and in-situ gas generation. When comparing different aeration methods, the formation of bubbles by agitation is a faster process than supersaturation or injection (Bergeron and Walstra 2005) due to the higher energy input involved. Generally speaking, gas sparging can produce bubbles that are mono-disperse especially when the flow rates are low and the capillaries used are of similar sizes. Supersaturation can also produce fairly homogeneous bubbles while agitation usually provides a wide distribution of bubble sizes (Weaire and Hutzler 1999). An illustration of the representative bubbles created by the different methods can be seen in Fig. 11.4.

11.1.4.1 Mechanical Agitation

This method involves actively forcing liquid around an external gas that protrudes the surface of the liquid (e.g. whipping, shaking) (Campbell and Mougeot 1999). It usually relies on mechanical energy being applied to liquid and according to

Bergeron and Walstra (2005) the following phenomena take place during this type of aeration:

1. Bubbles form by the entrapment of gas.
2. Surfactant molecules diffuse and adsorb to the newly formed interface.
3. Bubbles break up and reduce in size.
4. Bubbles start to coalesce due to poor coverage by surfactant until the system reaches a steady state where the surface of bubbles is covered with enough surfactant to become stable.
5. The liquid that is entrapped within the lamella and the Plateau borders starts draining (even within the turbulent flow). This causes the separation of the drier foam at the top of the liquid but is a phenomenon that becomes more prominent after agitation stops.

All these phenomena take place several times during this aeration process and occur concurrently. In industrial and professional application of aeration via mixing the main processing parameters that can affect both the air uptake and the bubble size distribution are the type and geometry of the whisk, the mixing speed and the mixing time (Campbell and Mougeot 1999).

11.1.4.2 Gas Sparging

Gas sparging includes processes where gas is being forced through the liquid (Campbell and Mougeot 1999). Often gas is injected through small capillaries inside the liquid and bubbles form and detach from the end of the capillary due to buoyancy. Sparging is extensively used in laboratory techniques for the study of foams where a gas is sparged in a liquid through a sintered glass (Waniska and Kinsella 1979). In this case the processing parameters are the size and the distance between the capillaries and the gas flow rate. If the capillaries are too close or if the gas flow rate is too high then the bubbles will coalesce immediately (Bergeron and Walstra 2005).

A similar method of producing foam is employed in modern coffee shops where milk is being frothed by injecting steam through a nozzle. This hot jet of steam drags air from the atmosphere into the mass of the milk due to the Bernoulli effect and aerates it (Jimenez-Junca et al. 2011). Apart from bubbling the milk, steam also increases the temperature which has a series of effects mostly on the protein conformation.

11.1.4.3 In Situ Generation

This mechanism involves the creation of bubbles inside the volume of the liquid either due to some type of reaction (chemical or biological), or due to a phase transition (Campbell and Mougeot 1999). At elevated pressure a gas can be dissolved within a liquid, upon release of this pressure the gas becomes supersaturated and gas bubbles form. The gases used for this type of foam formation need to be easily soluble in water like CO_2 and N_2O.

A common example of the formation of bubbles through a chemical reaction in the food industry is the production of CO_2 within the matrix of cake dough. Sodium bicarbonate (baking soda) reacts with cream of tartar (acid) when in contact with water, producing small bubbles (Pugh 2016). Another common example is the metabolism cycle of yeast in bread dough; CO_2 is produced from the fermentation of sugars when starch is broken down, creating small bubbles throughout the dough (Campbell and Mougeot 1999).

11.1.5 Foam (In)Stability

Foams are thermodynamically unstable systems that generally have a lifetime of a number of hours (Pieter Walstra 2003a). There are three main mechanisms by which the structure of foam changes over time resulting in its overall collapse:

1. The flow of liquid through the volume of the foam due to gravity known as **foam drainage**.
2. Rupturing of the film between neighbouring bubbles leading to **coalescence**.
3. Gas diffusion between adjacent bubbles which causes bubbles to become larger and is known as **disproportionation or coarsening**.

These mechanisms of instability are all occurring simultaneously and can therefore reinforce each other (Dickinson 2010). For example, coarsening results in larger bubbles and therefore larger films, which in turn accelerates drainage (Pieter Walstra 2003a).

11.1.5.1 Drainage

Drainage can be defined as the flow of liquid through a volume of foam (Cantat et al. 2013). It is observed in a plethora of every-day examples such as the liquid draining out of a soap froth or from the head of a pint of beer. Liquid begins to drain from the top of the foam, flowing from the lamella to the Plateau borders, following the direction of gravitational force.

All general models and approaches to describing liquid drainage indicate that both density and viscosity are important factors in controlling the rate at which it takes place (Cantat et al. 2013). Whilst the density of liquid cannot be altered significantly in most food applications, the viscosity can be significantly enhanced with the addition of biopolymers and/or suspended particles of polymeric or inorganic origin. Furthermore, drainage is not only influenced by the viscosity of the bulk phase but also by the viscosity of the lamellar film (Kinsella 1981). It has previously been shown that in the presence of particles, drainage can be significantly contained without an increase of the bulk viscosity but instead affecting the

Table 11.1 Equilibrium solubility mole fraction of commonly used gases in food aeration at 20 °C and atmospheric pressure (101.325 kPa), data from Gevantman (2013)

Gas	Solubility
CO_2	7.07×10^{-4}
N_2O	5.068×10^{-4}
O_2	2.501×10^{-5}
N_2	1.274×10^{-5}

permeability parameters of the foam (Britan et al. 2009). Moreover, the study of liquid drainage in the presence of nanoparticles has shown a traffic jamming mechanism comparable to the formation of cork-like structures. These structures were formed by aggregates of particles in the Plateau borders that significantly affected drainage (Carn et al. 2009; Lazidis et al. 2017).

In addition, the type of surfactant can also considerably affect the rate of liquid drainage.

When liquid drains through the lamella, a shear stress is applied on the layer of surfactants that are adsorbed on the interfaces, which results in a surface tension gradient. If the system contains surface active molecules that are less mobile such as polymers, as the surfactant layer moves against the direction of drainage (from high to low surface tension regions) drainage can be significantly reduced. LMW surfactants that can adsorb fast to the newly formed regions do not affect the rate of drainage as significantly.

11.1.5.2 Coarsening

Coarsening of the bubble size distribution can occur via disproportionation, that is, the movement of gas from regions of high concentrations to low concentrations via diffusion. The higher concentration is provided by a higher local curvature and hence a higher Laplace pressure. Unless a foam is completely mono-disperse, it contains a variety of bubbles sizes and therefore regions of different pressure and solubility of gas (Garrett 1993). This gradient of solubility gives rise to the driving force for directional diffusion as described by Fick's Law (Kilcast and Subramaniam 2011) and as a result, smaller bubbles have the tendency to shrink, whilst larger bubbles grow in size. Coarsening first occurs at the top of foam where bubbles are exposed to the atmosphere and then continues within the volume of the foam (Fennema 1996).

Coarsening in foams depends heavily on the solubility of the gas that is dispersed within the liquid. In the case of aqueous foams, the solubility of different gases commonly used in the food industry can be viewed in Table 11.1. It is evident that the more soluble a gas is in the water, the faster it can diffuse through a bubble if the circumstances permit it (a pressure gradient is in place). For example, a bubble of nitrogen with a radius of 1 mm in water will disappear in approximately 3 min while the same size bubble of carbon dioxide will need only about 4 s (Fennema 1996).

This has been exploited in the production of Guinness, where the gas used for carbonation is primarily composed of nitrogen resulting in a longer lasting foam.

Apart from controlling the rate of gas diffusion by altering the type of gas that the foam is made of, the type of surfactant can also play an important role. The permeability of the film around the bubbles can be reduced, to an extent, by the adsorption polymeric surfactant molecules, which provide a thick adsorbed layer (Bergeron and Walstra 2005). Increasing the interfacial elasticity of the films can also reduce coarsening. Most LMW surfactants do not provide a sufficient increase in the interfacial elasticity modulus and can be readily displaced or dissolved in the bulk phase during the shrinkage of bubbles (Damodaran 2005). In contrast, polymeric surfactants and strongly adsorbed particles can create a strong film with a high surface dilatational modulus (E), which can resist bubble shrinkage and therefore coarsening. The fact that many proteins have a high energy of desorption, sometimes due to denaturation at the interface, causes a significant change in surface tension while the bubble shrinks which translates to a high dilatational modulus. In fact, coarsening due to gas diffusion can be theoretically stopped if $E > \gamma/2$ (Garrett 1993).

11.1.5.3 Coalescence

The rupture of the film between two neighbouring bubbles leads to immediate coalescence of these bubbles into a larger one (Bergeron and Walstra 2005). Coalescence reduces the number of bubbles in a foam, which decreases the total surface area and therefore the interfacial Gibbs energy. This is the main driving force for film rupture since it is thermodynamically unfavourable to have such a high surface area (Álvarez et al. 2017).

The mechanism which leads to coalescence depends on the morphology of the film and the type of surfactants present on the surface of the bubbles. In thick films where colloidal interactions become less evident (Fennema 1996), the surface dilatational modulus is usually substantial enough to prevent film rupture (Bergeron and Walstra 2005). Coalescence typically only takes place if the surfactant concentration is low and the bubbles are unstable during creation when the films are rapidly stretched and deformed. When films are thin enough for colloidal interactions to be important (usually only observed at the top of the foam where water evaporation occurs), vdW attractions cause the films to come into contact resulting in bubble coalescence (Fennema 1996). When ionic surfactants are used, the repulsion provided between the two interfaces due to the electrostatic forces prevents this. Non-ionic surfactants on the other hand, usually provide stability against coalescence by steric repulsions between the interfaces (Bergeron and Walstra 2005).

Particles present in the bulk phase of foam systems can also have a detrimental effect on the longevity of the foam by causing film rupture (Fennema 1996). Firstly, if hydrophobic particles are present in the system ($\theta > 90°$), particles can potentially bridge the surfaces of two bubbles causing film rupture (Wilson 1981). Secondly, if hydrophobic particles, e.g. proteins, are present, they can spread at the a/w interface

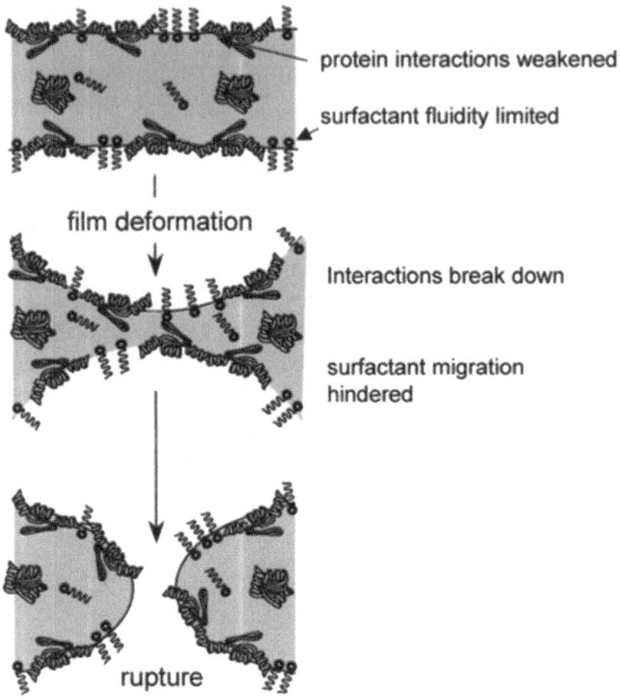

Fig. 11.5 Schematic visualising the mechanism of competitive adsorption on the lamella that can lead to destabilisation. Reproduced with permission from Wilde et al. (2004)

causing film rupture. This is why the head of a beer collapses quicker when the consumer has lipstick on or the glass used is dirty (Weaire and Hutzler 1999).

Finally, while polymeric and LMW surfactants can respectively stabilise foams alone, when used in combination this can have a detrimental effect (Wilde et al. 2004). The process is known as competitive destabilisation, where LMW surfactants displace proteins from the a/w interface reducing the interfacial elasticity and deforming the film to the point that it ruptures (Fig. 11.5).

11.1.6 Effect of Biopolymers on Foam Stability

Biopolymer ingredients including structurally diverse proteins and polysaccharides are capable of behaving as stabilising agents in multi-phase food systems. Most commonly, biopolymers are structurally modified during extraction and purification through several stages of processing (mechanical, biochemical and thermal). However, memory of their parent structure results in favouring towards specific types of aggregation or self-assembly (Dickinson 2017; Phillips and Williams 2009). Proteins and polysaccharides therefore exist in a variety of aggregated states,

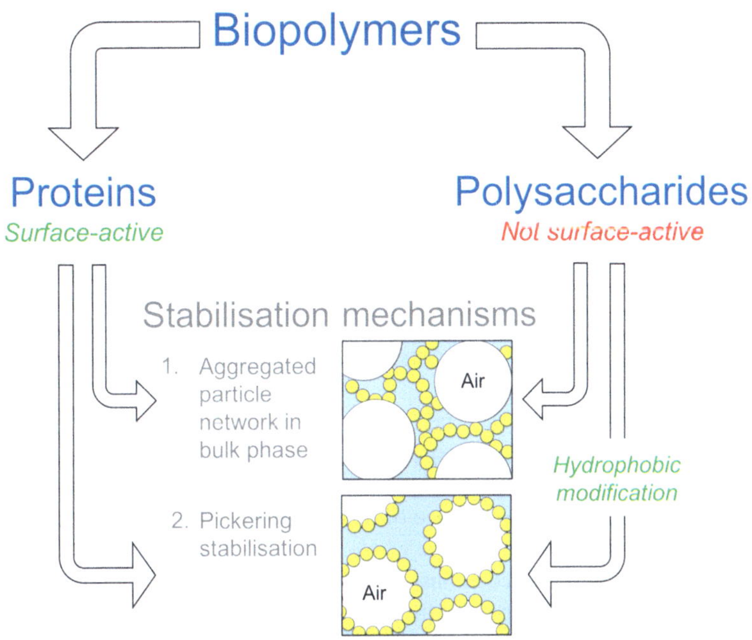

Fig. 11.6 A scheme representing the two foam stabilisation mechanisms available to biopolymer particles and the pathways for protein and polysaccharides to reach them based on their surface activity

providing food technologists with a diverse range of available structures and properties for the stabilisation of foams (Dickinson 2017). In their simplest form, biopolymers increase the viscosity of the aerated system providing stability during the foaming process e.g. gelatin stabilises marshmallows (Phillips and Williams 2009). Alternatively, biopolymers can stabilise foams in the form of particulate entities of nano- or micro-scale dimensions. This occurs through two main mechanisms, which are dependent on the surface activity of the biopolymer (Fig. 11.6). Firstly, when present in sufficiently high concentrations, particles can aggregate together in foam channels (Plateau borders), which provides a barrier to liquid drainage (Guillermic et al. 2009). Secondly, providing the particles are suitably hydrophobic, they can adsorb at the air-water interface, forming a protective layer against bubble coalescence and coarsening. In reality, protein stabilised foams will often experience a combination of these two mechanisms (Dickinson 2017). In contrast, polysaccharides are inherently hydrophilic and therefore need modifying in order to act as Pickering stabilisers. In food systems, this can be manipulated chemically or using thermal treatment. For example, starch particles can be heat treated to increase their surface hydrophobicity (Seguchi 2001). In addition, interaction with proteins such as soy, egg or milk to form polysaccharide-protein complexes can also increase polysaccharide surface activity (Binks et al. 2017; Ghosh and Bandyopadhyay 2012).

11.2 Protein-Based Biopolymers

Proteins are natural biopolymers with low toxicity that are cheap and readily available (Wilde et al. 2004). Once adsorbed on an interface, they unfold and rearrange their secondary and tertiary structure exposing the hydrophobic regions to the hydrophobic phase (Fang and Dalgleish 1997). The inherent properties of proteins are dependent on the content and composition of amino acids, the molecular size, the shape, the conformation, net charge and charge distribution, the hydrophobic/hydrophilic character of their surface and inter-protein interactions (Kinsella 1981). Moreover, their functional properties rely on extrinsic factors like pH, ionic strength and temperature, together with their interaction with other components present (Zhu and Damodaran 1994b).

Hydrophobicity, as mentioned before, plays an important role in the ability of proteins to stabilise interfaces, although a distinction should be made between the surface and the average hydrophobicity. Surface hydrophobicity relates to the extent of hydrophobic patches on the surface of the protein while average hydrophobicity refers to the amino acid residues in the whole structure of the protein (Hettiarachchy and Ziegler 1994). It has been demonstrated that the average hydrophobicity is more important when proteins are being utilised to stabilise foams, whilst the opposite is true for emulsions (Kato et al. 1983). This is due to the fact that at the high free energy of an a/w interface, the adsorbed proteins denature to a greater extent and so whilst the hydrophobic patches on their surface play an important role in the initial anchoring to the a/w interface, it is the unfolded protein that controls the behaviour on the interface (Damodaran 2005).

11.2.1 Comparison to LMW Surfactants

LMW surfactants are smaller and therefore more mobile than proteins, resulting in fast diffusion and subsequent adsorption to the newly formed interface. The adsorbed monolayers are in a dynamic equilibrium with the molecules that remain in the bulk (Pugh 2016). On the contrary, proteins adsorb on the interface while they unfold, aggregate and denature forming layers that resemble 3D structures similar to gels. These layers cannot completely block the passage of gas molecules but provide an additional steric barrier to foam instability. The main benefit of this added barrier is the resistance towards bubble shrinkage during disproportionation (Murray and Ettelaie 2004). Foam bubbles with an elastic interface or a purely elastic continuous phase (solid foams) can be completely resistant to shrinkage once the Gibbs stability criterion is satisfied (Kloek et al. 2001). The network formed by proteins at the interface is almost always viscoelastic and it is the viscous element which permits a reduced but constant shrinkage (Murray 2002). In regards to drainage, the highly viscoelastic layers that proteins form on the interface as opposed to

LMW surfactants, provide high surface shear viscosities which in turn result in rigid Plateau borders that hinder the flow of the liquid (Saint-Jalmes et al. 2005).

According to Murray (2007), the reasons that proteins are good foaming agents are:

1. They strongly adsorb to the gas-water interface.
2. They provide steric stabilisation and also electrostatic stabilisation when the environmental conditions (pH and salt concentration) allow it.
3. Films with adsorbed proteins have a certain structural consistency due to the inter- actions between the adsorbed molecules which translates as high surface rheological moduli.

Furthermore, proteins are already present in many food systems and there are several food grade protein sources that seem appealing to consumers and provide final products with clean labels. Commonly used proteins in foods such as whey, caseins, egg and meat proteins possess very good functional properties for several applications. It has been nevertheless pointed out that each of these proteins is actually a mixture of proteins and their desirable functionality might be due to the contribution of the individual components in different functions during the stabilisation of foams (Damodaran 2005). Moreover, protein mixtures that are regularly used for their foaming properties contain not only several species but also non-protein species (e.g. minerals). Generally, linear proteins are more mobile and flexible than proteins with more compact globular structure and can adsorb faster to the a/w interface (Zhang et al. 2004).

Proteins are effective foaming agents both in their native but also in particular form, which comprises of aggregates of the original polymers. The latter, can stabilise foams via a Pickering mechanism with enhanced stability (Lazidis et al. 2015; Schmitt et al. 2011). These particles, contain interfacial properties of their individual building blocks as they can adsorb or anchor on the interface providing a steric effect and an interface with enhanced viscoelasticity. Moreover, they increase the viscosity of the bulk phase and constrain the flow of the liquid through the plateau borders significantly decreasing drainage (Lazidis et al. 2017). Nevertheless, native proteins themselves can also be seen as colloidal particles which adsorb on the interface and follow rules that describe colloidal systems instead of molecular ones such as net charge, surface hydrophobicity and particle size (Wierenga and Gruppen 2010).

11.2.2 Factors Affecting the Foaming Properties of Proteins

11.2.2.1 pH

Electrostatic interactions play an important role in both the rate of adsorption and the interfacial rheology of films stabilised by proteins (Foegeding et al. 2006). Several studies looking at a range of proteins have demonstrated that foaming properties are optimal near the isoelectric point (pI) as surface pressure reaches a maximum (Davis et al. 2004; Zhu and Damodaran 1994a). These studies argue that at the pI, protein adsorption is faster due to the lack of electrostatic repulsions. Furthermore, the viscoelasticity of interfacial films from several proteins is higher near the pI. According to this, the culinary concept of adding acid (in the form of lemon juice or vinegar) to egg whites in order to improve its foaming properties, is justified by the shift of the pH closer to the pI. This is due to the absence of repulsions between the protein residues they can pack more densely and create a rigid film (Kinsella 1981). Moreover, the reduction in solubility close to the pI can increase stability of bubbles due to the enhanced steric mechanisms of the adsorbed species (P. Walstra and Roos 1993). On the contrary, a number of studies looking at vegetable proteins have shown the opposite, that when the pH of the system is close to the pI the lack of interactions and therefore the aggregation of the proteins at the a/w interface has a negative effect on the foaming properties (Rodríguez Niño et al. 2005; Rodríguez Patino et al. 2008). Additionally, Fujioka and Matsumoto (1995) observed an increase of the viscoelastic properties of albumen foam when the pH was higher than the pI. This behaviour was attributed to the competing effects of the interfacial properties of the adsorbed protein layers and the electrostatic repulsion affecting the drainage.

11.2.2.2 Ionic Strength

Increase of ionic strength has been demonstrated to decrease both the foam stability and foaming capacity of milk proteins and more specifically whey. Addition of NaCl in whey proteins solutions, at pH below or above the pI resulted in a higher protein adsorption in terms of surface tension and a more viscoelastic film (Davis et al. 2004). This was attributed to the salt counter ions screening the charge of the proteins. This is linked to neutralisation of their charge which leads to protein aggregation and exposure of the protein interfacial film which in turn accelerates disproportionation (Damodaran 2005). However, in the presence of the whole range of milk proteins, addition of NaCl (up to 0.8 M) has shown to have a positive effect in the foaming properties of the proteins. This is proposed to be associated with the antagonistic effect of sodium with the calcium ions in the core of the casein micelles which leads to their dissociation. Moreover, as mentioned before, the dissociated caseins are more surface active and adsorb faster than their micelles (Zhang et al. 2004).

Divalent cations (e.g. Ca^{2+} and Mg^{2+}) have a great effect on the foaming properties of whey protein isolate (WPI) due to the slow aggregation of the proteins in the presence of low concentrations of these ions (Zhu and Damodaran 1994a). The mechanism of this aggregation relies on electrostatic bridging interactions that are promoted by the presence of these multivalent cations (Foegeding et al. 2006). Since the aggregation is heavily time-dependant it usually takes place in situ on the a/w interface which facilitates in the production of a more viscoelastic film that slows down drainage (Damodaran 2005). Faster aggregation reduces foaming ability by decreasing the rate of adsorption of the protein to the a/w interface during aeration (Zhu and Damodaran 1994a).

Sagis et al. (2001) have reported that addition of copper ions in egg whites before aeration leads to further denaturation of the proteins on the a/w interface that in turn increases the interfacial elasticity and makes foam more resistant to drainage compared to the control without the copper ions.

11.2.2.3 Temperature

Heat can induce conformational changes to proteins that lead to what is known as heat induced denaturation (Bals and Kulozik 2003). Several proteins such as soy globulins, κ-casein, ovalbumin, β-lg and BSA have their surface hydrophobicity increased during heat denaturation (Kato et al. 1983). In the same study, whilst the increase in surface hydrophobicity has shown a significant increase in the foaming ability and foam stability of κ-casein, ovalbumin and soy globulins, it did not show significant improvement in the foaming performance of β-lg and BSA. The increase in the foaming properties was accredited to the decrease in SFT and since proteins that were already very hydrophobic did not demonstrate a significant drop in SFT to justify an enhancement of their foaming properties. Nevertheless, other studies have indicated that heating globular proteins, like α-la and β-lg, induces their partial unfolding which can facilitate foam formation (Kinsella 1981).

It has been shown that whilst controlled heating can improve foaming properties of proteins, extensive heating has an adverse effect. Indeed, Zhu and Damodaran (1994b) have indicated that in the case of whey proteins heating does not necessarily increase the surface hydrophobicity but has an impact on the overall hydrophobicity which becomes more important when proteins adsorb on the interface and unfold exposing additional hydrophobic groups. The structural changes in the proteins present in whey do not only affect their hydrophobicity but also promote polymerisation of the species to gel like structures known as aggregates. These polymers are believed to form due to disulphide bonds amongst the monomers of β-lg and α-lac (Monahan et al. 1995). In turn, these changes affect the foaming proteins of the whey solutions and the extent of heating (both in time and temperature) has been shown to be an important parameter (Patel et al. 1990). Whilst moderate heat denaturation seems to improve the foaming ability and foam stability, extended heating in both temperature or time seems to have an adverse effect. This observation has been connected to the ratio of monomeric to polymeric proteins present in the system after the heat treatment.

11.2.2.4 Protein Modification

The properties of proteins and therefore their behaviour and functionality can be altered with a series of modification methods that have been developed over the last years. The key factor when choosing the modification route of any food system is retaining the formulation food grade and safe for consumption whilst adhering to the enforced legislation. Some of the modifications are more pervasive than others and some require the use of additional molecules either in the form of enzymes or carbohydrate chains. Modifications on proteins that aim to improve their functional properties and the foaming behaviour in particular, usually aim to either trigger the controlled aggregation in order to create particles that will increase the stability of foams or expose the hydrophobic patches of proteins and increase their interfacial activity. Ways to achieve the first are either enzymatic crosslinking or high-pressure processing (HPP). While the latter can be accomplished by either partially unfolding proteins, cleaving peptide bonds or covalently binding carbohydrate chains on their surface with methods such as enzymatic hydrolysis, sonication, oxidation and glycosylation.

Aside from heating and pH regulation protein aggregation can be initiated by either enzymatic cross-linking or high-pressure processing with beneficial effect on foam stability. The foam stability of α-lactalbumin have been significantly improved by modification of the native protein by cross-linking using transglutaminase or peroxidase which allowed the fabrication of α-lac nanoparticles (Dhayal et al. 2015). HPP technology has also been successfully utilised in improving the foaming capacity and stability of bovine lactoferrin and soy protein isolate by partially denaturing the proteins, exposing the hydrophobic groups and subsequently causing aggregation (He et al. 2016; Martínez et al. 2011).

Enhancement of foaming properties of proteins can arise from making the secondary and tertiary structure less complicated. This allows the exposure of functional groups which increases the hydrophobicity, the reduction of the size which increases the molecular mobility and the rate of diffusion to the a/w interface and finally improves the solubility by increasing the available ionising groups (Panyam and Kilara 1996). Proteins with limited or low functionality, such as plant proteins, can be modified this way into very functional ingredients. The most common methods that can achieve this are enzymatic hydrolysis or sonication. Enzymatically hydrolysed wheat gluten for example, has shown superior ability to form and stabilise foams (Wouters et al. 2017). Similarly, hydrolysed bean protein isolate has shown higher foaming capacity and foam stability at a wider range of pH compared to the non-hydrolysed isolate (Betancur-Ancona et al. 2009; Mune Mune 2015). The parameters that are important when hydrolysing proteins for upgrading their functionality, are the degree of hydrolysis (DH) which is directly affected by the treatment time, temperature, pH, ionic strength and enzyme chosen. Another treatment that has drawn increasing attention the last years because it doesn't involve the use of additional substances is sonication. During the treatment of biopolymer suspensions with ultrasounds, physicochemical changes take place that relate to the cavitation, heating, dynamic agitation, shear stresses and turbulence (Knorr et al. 2004). The application of ultrasounds on proteins seem to have an effect on the viscosity of their solutions, the gelling properties and the hydrophobicity which is mainly attributed to molecular modifications (Arzeni et al. 2012). Foaming capacity of soy protein isolate seem to have been enhanced by the application of ultrasounds due to the reduction in particle size and increase of the molecular mobility (Morales et al. 2015). On systems with gluten and whey proteins sonication has a positive effect on both the foaming ability and foam stability which has been accredited to the partial denaturation of the proteins (Jambrak et al. 2008; Zhang et al. 2004).

Another modification method that has gained attention due to the utilisation of food grade biopolymers that provide potential "clean label" solutions is glycosylation, where usually a long chain carbohydrate is covalently bonded to the surface of the proteins through the Maillard reaction. During this reaction, conjugation of the carbonyl group of reducing sugar with an available unprotonated amino group, mainly the ε-amino group of lysine, takes placing with heat being the main trigger (Oliver et al. 2006). Controlling the extend and the conditions (molar ratio, pH, humidity, temperature and time) of this reaction, allows the formation of carbohydrate-protein complexes with improved functional properties (de Oliveira et al. 2016). A range of carbohydrates with various chain lengths and degree of branching such as maltodextrins, dextrans, galactose have been studied in conjugation with many animal and plant proteins. Maillard conjugates offer the possibility of an ideal steric foam stabiliser which combines the ability of proteins to strongly attach to the a/w interface with strong solubility to the aqueous phase of polysaccharides. Indeed the effect of glycosylation on the foaming properties of lysozyme, β-lactoglobulin and bovine serum albumin has been profoundly improved (Corzo-Martínez et al. 2012; Corzo-Martínez et al. 2017; Dickinson and Izgi 1996; Medrano et al. 2009). Similar results can be obtained by taking advantage of the inherent charge of proteins and electrostatically binding charged polysaccharides. Identifying charged polysaccharides is rather challenging but a good example is pectin which can result electrostatic complexes with whey proteins with promising foaming properties (Schmidt et al. 2010).

11.2.3 Animal Proteins

11.2.3.1 Milk/Dairy

Milk and the ingredients that derive from it are one of the most used foodstuffs in the food industry and home-based cooking. Dairy as a whole, is a staple food category for many diets around the wold. Many of the dairy formulations include the formation of air bubbles that are important for the final application either for aesthetics or functional purposes. These bubbles are predominantly stabilised by the proteins present in milk. Understanding therefore the mechanisms associated with stabilising air bubbles with milk proteins has and always be a field of great interest.

There are two main categories of proteins in milk and the way that they are distinguished is based on their solubility at pH 4.6 at 20 °C (Fox and Kelly 2004). The proteins that precipitate at these conditions are the caseins and the ones that remain soluble are the serum or whey proteins.

Table 11.2 Composition of whey proteins in bovine milk (data from de Wit (1998))

Whey protein fraction	Weight contribution (g/L of milk)
β-lg	3.2
α-lac	1.2
BSA	0.4
IgG	0.8
LF	0.2
LP	0.03
Enzymes	0.03
Proteose-peptones	≥1

Whey

Whey proteins, is the milk fraction that is stable to pH and enzymatic precipitation and usually produced as a by-product from cheese manufacturing (Kilara and Vaghela 2004). Whey was and still is in many cases treated as a waste stream and does not get fully utilised although its nutritional value and the functional properties of its proteins is well known. Much attention has focused on investigating the full potential of whey proteins and looking into possible pathways of improving their functional proteins. Most of the work carried out in this study focuses on looking at ways to exploit the potential of whey proteins as an ingredient used for stabilising foams.

Whey is composed by a group protein fractions, β-lactoglobulin (β-lg), α-lactalbumin (α-la), bovine serum albumin (BSA), Immunoglobulins (IgG), lactoferrin (LF), lactoperoxidase (LP), proteose-peptones and some enzymes. The contribution of each fraction to the total whey can be found in Table 11.2. The largest fraction of whey (68%) is made of the two main proteins β-lg and α-la that determine most of the properties of whey.

β-Lactoglobulin

β-lactoglobulin (β-lg) consists the higher fraction (about 58%) of the whey proteins in bovine milk (Kilara and Vaghela 2004). It consists of 162 amino acids and has a molecular weight of 18.3 kDa in its monomeric form but it is usually present as a dimer in the pH range of 5.5 and 7.5 (Mulvihill and Donovan 1987). The dimers are spherical and have diameters of around 18 Å. Whilst at pH 3.1–5.1 and at low temperatures it associates to an octamer (Kilara and Vaghela 2004). β-lg has two cysteine and one cysteine residues per monomer. The thiol group is equally distributed between position 119 and 121. When that is present at residue 119, a disulphide bridge forms between residues 106 and 121, while when it is at residue 121 the same bridge forms between residues 106 and 121. Another disulphide bridge is invariably present between residues 66 and 160. The presence of these disulphides and the single sulfhydryl is important not only to the structure but also to the properties of this protein (Mulvihill and Donovan 1987). In terms of its foaming ability, it

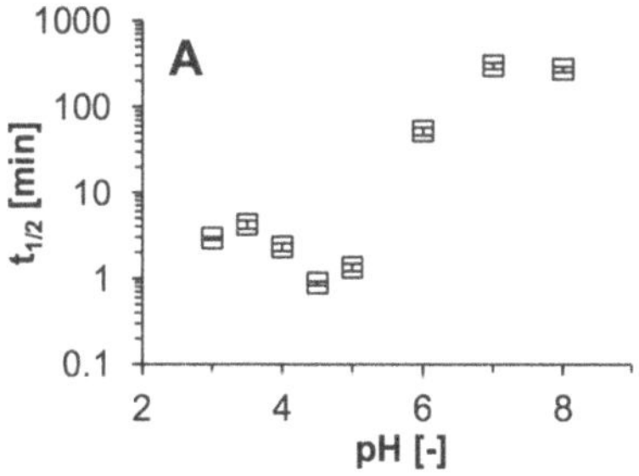

Fig. 11.7 Foam stability in terms of half-life as a function of the pH at 0.5 mg/mL β-lg. Reproduced with permission from Lech et al. (2016)

is considered one of the most surface active dairy proteins both in its native but also it is aggregated particle form, mostly due to its distinct tertiary and quaternary structure (Dombrowski et al. 2016b).

The dependence of the structure and the charge amongst the protein moieties on the pH affects significantly the foaming properties of its solutions. The octamers that form close to pH values of 3 have a positive effect in the stability of the foams formed. Whilst at pH values around 5 (close to the pI) the charge of the proteins is close to zero, the interfacial packing on the interface is high forming more rigid films. Nevertheless, at pH values higher than 7 the dimers present have significant charge to inhibit electrostatic interactions which cause the films around the bubbles to be thicker and the viscosity in the plateau borders to be higher which has a positive effect on foam stability as seen in Fig. 11.7 (Lazidis et al. 2015; Lech et al. 2016).

α-Lactalbumin

α-lactalbumin (α-lac) is the second higher in content protein present in bovine whey (13% of the whey) (Kilara and Vaghela 2004). It consists of 123 amino acids, it is compact and spherical and has a molecular weight of 14.0 kDa. It is high in tryptophan and aspartate whilst the presence of single arginine and methionine residues, four disulphide bonds and the absence of phosphoryl and sulfhydryl groups are notable and affect its structure and properties (Mulvihill and Donovan 1987). Its ability to adsorb in ordered fashion forming α-helixes makes it an efficient foaming stabiliser (Cheung 2017).

Zhu and Damodaran (1994b) have found that increasing the ratio of native to denatured WPI up to 40:60 increased the stability of the produced foams. Further increase of the amount of denatured protein had a negative effect on foam stability. Foaming ability, on the other hand, was optimal at lower concentrations of denatured WPI (60:40 native to denatured ratio). The same behaviour was also documented by Bals and Kulozik (2003) who observed that SFT of whey protein solutions decreased with increasing the degree of denaturation up to 34% while further increase caused an in- crease of the SFT. This indicates that native proteins contribute significantly to the foam generation because they are smaller and more flexible and can diffuse fast and adsorb at the surface during the creation of the foam. However, the denatured polymeric protein species have a significant contribution to the stability of the foams by adsorbing on the surface at a later stage increasing the viscoelasticity of the films.

Caseins

Caseins consist the main protein fraction of milk with a share of approximately 80% of the total protein content and significantly the properties of milk. In their native state they are in a micellar form with a calcium phosphate linked network of nanoclusters (Dalgleish and Corredig 2012). Caseinates are the water soluble form of casein, usually in the form of a sodium salt, they are linear in disordered coil arrangement and consist of a mixture of four individual fractions, α_{s1}, α_{s2}, β and κ caseins with β and α_{s1} being the most abundant fractions (Abascal and Gracia-Fadrique 2009). Those two fractions show a competitive adsorption on the interface with β being the most surface active (Fang and Dalgleish 1993). At concentrations over 0.01–0.1 wt%. caseinates self-associate into spherical molecules. In the surface of caseinate suspensions at concentrations above the CMC, where the surface tension reaches the minimum of 40 mN m^{-1}, amino acid segments are aligned on the interface but also extend to the suspension and form loops and trains.

Casein micelles separated from milk via ultrafiltration are close equivalent to the colloidal micelles natively present in milk. They can also be used in their dry form and dispersed in water to create casein micelle dispersions (CMD) which can also adsorb to the air/water interface and stabilise foams. Their size can be tailored and affects their foaming properties with larger micelles (400 nm) forming more stable foams than smaller ones (Chen et al. 2016). Another ways to affect the functionality of the casein micelles and increase their foam stability is by elevating the pH to alkaline regions (pH~9) which apart from increasing the electrostatic interactions between the micelles increases their size and the amount of monomers dissociated from the micelles causing the produced films to be more viscoelastic as seen in Fig. 11.8 (Dombrowski et al. 2016a).

BSA

Bovine serum albumin (BSA) is a protein from the bovine whey family and accounts for approximately 10% of the total whey fraction. It precipitates from milk along with the rest of the whey proteins and has an isoelectric point of 4.7. It is globular and large in size (66 kDa) and consists of 580 amino acid residues with 17 disulphide bonds and one sulfhydryl group at residue 34 (Kinsella and Whitehead 1989). Its secondary structure consists of ~54% α-helix and ~40% β-structure. Although it has the ability of stabilising the air/water interface, mostly by forming gel-like structures upon adsorption that consist of aggregates that develop due to the interaction of the hydrophilic groups being exposed upon unfolding on the interface (Li et al. 2017). Foams prepared with 1% wt. BSA solutions at neutral pH show narrow bubble size distributions, which depend a lot on the foaming method used, and

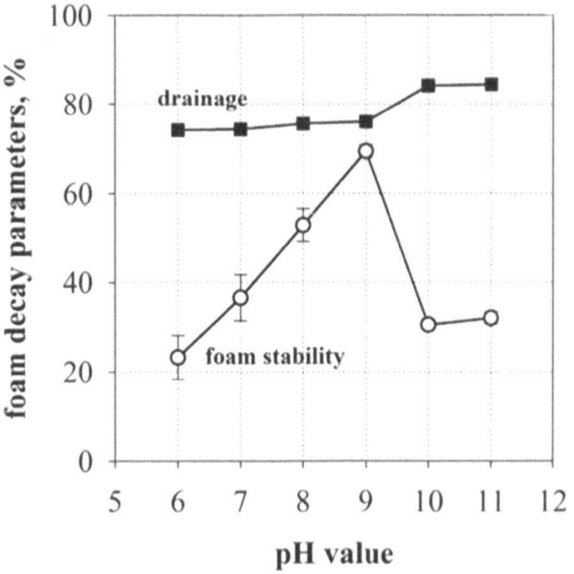

Fig. 11.8 Impact of pH on foam stability and drainage of micellar casein suspensions at 10 g/L concentration. Reproduced with permission from Dombrowski et al. (2016a, b)

exhibit low stability (Fig. 11.9). A way to improve the ability of BSA to create stable foams is by using mixed systems with other substances that perform synergistically. Such examples are mixed systems with low molecular weight surfactants such as Tween 20 (Zhang et al. 2013), electrostatic complexes with protamine (Glaser et al. 2007), aggregates with other proteins present in whey such as β-lg and α-lac (Gezimati et al. 1996; Havea et al. 2001) and acetylated BSA (Berthold et al. 2007).

Lactoferrin

Bovine lactoferrin is a cationic glycoprotein protein with a high isoelectric point at pH 8.9 and a molecular weight of 88 kDa. It has a special iron binding capability which makes it useful for iron fortification and also explains its biological functionality as antibacterial and anti-inflammatory agent (Ward et al. 2005). Its opposite charge with the rest of the milk proteins means that in milk lactoferrin is bound on the surface of the negatively charged casein micelles. Lactoferrin, at its native state denatures at temperatures as low as 70 °C making it prone to heat processing which is common to foods. The extent of this sensitivity to heat depends on the degree of iron saturation which is rather low at the native state (15–20%). The ability of lactoferrin to readily create electrostatic complexes with other proteins and negatively charged polysaccharides has been utilised in several research papers in order to increase its sensitivity to pH and temperature and create controlled sized aggregates with improved ability to stabilise a/w interfaces. Complexation of lactoferrin sodium caseinate has shown an improved ability to lower the surface tension of solutions further than the two constituents alone when in a lactoferrin:sodium caseinate ratio of 2:1 (Li and Zhao 2017). Apart from mixing it with other surface-active species,

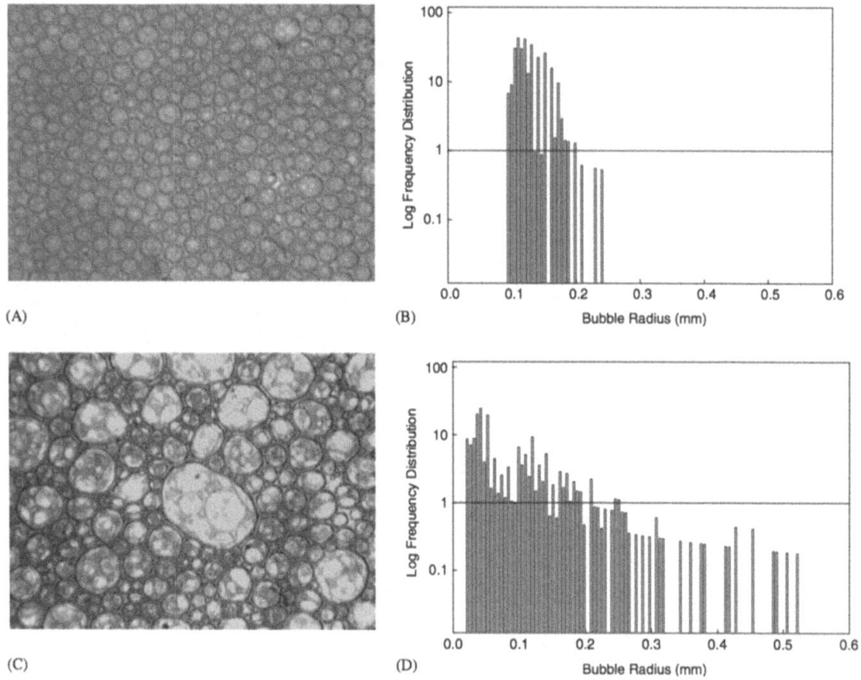

Fig. 11.9 Bubble size distribution of BSA foam at pH 7. (**A**) Foam structure at t = 0 min (immediately following foam formation); (**B**) bubble size distribution at t = 0 min; (**C**) foam structure at t = 16 min; (**D**) bubble size distribution at t = 16 min. Reproduced with permission from Glaser et al. (2007)

processing can improve its ability to adsorb on the a/w interface. On of this processes is high pressure treatment which has shown improve the solubility and foaming due to changes in the tertiary structure (He et al. 2016).

11.2.3.2 Egg Proteins

Egg proteins are considered the reference for proteins with enhanced foaming properties, which is explained by their wide spread application throughout the culinary world with products such as meringues, cakes and sоufflés. A commonly known trick amongst chefs in order to obtain the best foam while beating egg whites is the use of copper bowls (This 2010). This is due to the ability of copper ions to promote disulphide bonds and enhance the ability of the proteins to form gel like structures on the air water interface. This affects the viscoelasticity of the films (Fig. 11.10) making them more robust and shielding them against coalescence and coarsening (Sagis et al. 2001). Egg white proteins like every natural protein system consist of different fractions the main of which are ovalbumin and lysozyme.

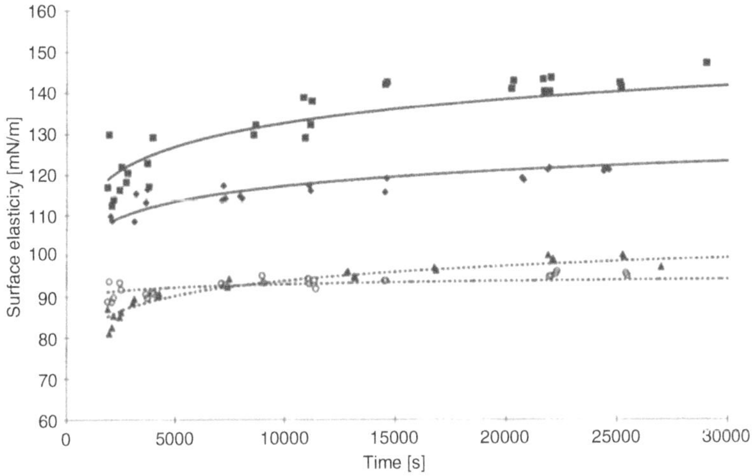

Fig. 11.10 Surface dilatational elasticity (surface storage modulus) as a function of time of the a/w interface of a diluted egg white solution 1:7. Diamonds and squares are samples with copper ions (measured on different days), triangles and open circles are samples without copper ions. Reproduced with permission from Sagis et al. (2001)

Ovalbumin

Egg albumin (ovalbumin), the protein predominantly present in egg whites, when used in foams seems to be forming particulate structures at the a/w interface even before the bubbles begin to shrink (Lechevalier et al. 2003). It has been demonstrated that the interactions between the 5 major components of ovalbumin are responsible for forming aggregates, induced by the formation of disulphide bonds amongst the molecules, on the lamella (Damodaran 2005). This ability of forming a continuous network on the interface, can explain the enhanced capacity of egg whites to form stable foams and possibly justifies the choice of egg whites as a foaming agent in numerous culinary products (Murray and Ettelaie 2004). For a protein therefore, to form an ultra-stable form it needs to be able to coagulate on the interface and provide a rigid network (Martin et al. 2002; Walstra 2003b). This behaviour of ovalbumin can explain the fact that egg white foams are shown to have higher yield stress (τ) than whey protein foams for the same air fractions and bubble size distributions (Pernell et al. 2002).

Lysozyme

Lysozyme, a protein also present both in egg white but also milk and secretions such as saliva, tear and mucus. It has lots of similarities with α-lac both in size (129 residues and 14.5 kDa) and hydropathy (Cheung 2017). Moreover, they both have four disulphide bonds at similar locations and over one-third of their residues identical (Dickinson 2013). Nevertheless, despite their resemblance they behave significantly differently in terms of foaming ability with lysozyme being inferior, making this a good example of how minor changes in structure of proteins can affect their functionality considerably. The lack of the α-helix secondary structure in lysozyme and its tendency to adsorb without a specific manner on interfaces makes it a less efficient foam stabiliser. Unlike ovalbumin, when lysozyme adsorbs in the air/water interface it does not undergo any structural modification (Lechevalier et al. 2003).

11.2.4 Plant Based Proteins

The majority of the proteins within conventional human food is of animal origin. Whilst plant proteins are cheaper and easier to produce relatively to animal proteins their majority is used as feed for the production of functional animal proteins form milk, eggs and meat (Day 2013). Nevertheless, the production of animal proteins requires 100 times more water than producing the equal amount of plant protein. It has been calculated that on average, for every 1 kg of meat protein approximately 6 kg of plant protein are needed as a result of animal metabolism (Pimentel and Pimentel 2003). This gives an idea of the toll that animal protein production has on land, water and energy use which heavily contribute to the loss of biodiversity, depletion of freshwater and climate change (Aiking 2011). Plant proteins originate mostly from the seeds and grains with the highest nitrogen content. The majority of the globulins present in plants are storage proteins that serve as nitrogen sources for the new embryos after germination (Tzitzikas et al. 2006). Plant proteins are admittedly less utilised than animal ones. The main reasons for this is that they are nutritionally inferior (on a single source basis), they pose difficulties in order to unravel similar functionality which is associated to their large molecular weight and poor solubility in water and it is costly to isolate and recover them (Day 2013). Nevertheless, extensive work from the research community has already revealed ways to expose and increase the functionality of plant-based proteins in terms of providing structure and also stabilising interfaces.

11.2.4.1 Soy Proteins

Soy bean proteins while having high functionality in their native state, after their denaturation during heat treatment as part of pasteurisation or drying processes loose this functionality in stabilising emulsion and foams (McSweeney 2008). The major fractions of soy proteins are the globular conglycinin (11S) and glycinin (7S). 11S is a trimeric glycoprotein with a size of 141–170 kDa which contains three subunits: α, α' and β which associate with each other by hydrophobic interactions. Glycinin is made of two identical hexamer rings each made of three pairs of acidic and basic disulphide-linked subunits which in turn are hydrophobically associated

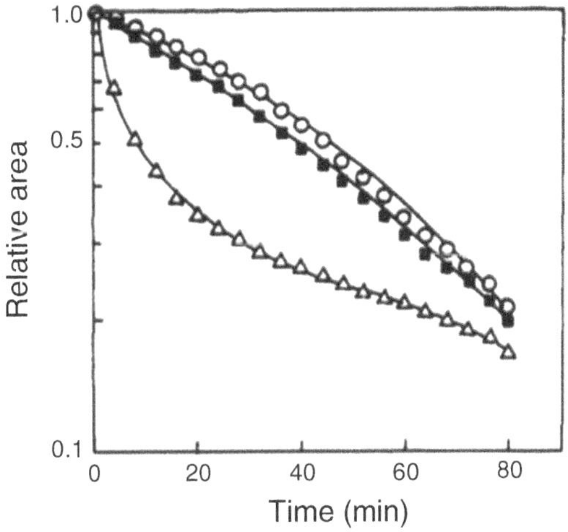

Fig. 11.11 Surface area decay of soy protein foams (2%, pH 7.0) at 25 °C: (open circle) 11s globulin; (open triangle) 7s globulin; (filled square) soy isolate. Each curve is an average of three experiments. Reproduced with permission from Yu and Damodaran (1991)

to each other (Utsumi and Kinsella 1985). The quaternary structure of both 7S and 11S is affected by environmental factors such as pH, ionic strength and temperature (Yu and Damodaran 1991). Optimum functionality occurs at pH <5 and ionic strength higher than 0.5 M (Pizones Ruíz-Henestrosa et al. 2007). Upon heating those subunits dissociate and rearrange in various manners forming gels.

In terms of foaming properties, 11S appears to be able to make foams at lower concentrations than 7S. Moreover, bubbles stabilised with 11S are more stable against drainage (Fig. 11.11) which suggests that the films formed are more viscoelastic compared to the ones made by 7S even if the bulk viscosity of solutions is higher for the ones of 7S (Yu and Damodaran 1991). Soy globulins whilst are able to form films with increased viscoelasticity, do not undergo a denaturation to the extent that they form gel structures on the air water interface such as other proteins (e.g. egg) (Sánchez et al. 2004).

11.2.4.2 Wheat Gluten

The insoluble protein fraction of wheat flour (80–85% of total protein content), known as gluten is the constituent that provides most of the functional properties of flour, an ingredient heavily used in the diet of several countries or cultures in a wide range of applications that includes breads, pasta, noodles and cakes (Veraverbeke and Delcour 2002). Gluten proteins contain comparable levels of two main fractions, the monomeric gliadins (30–80 kDa) and the heterogeneous mixture of glutenin polymers (<80 kDa). Both fractions are rich in proline and glutamine but it is the cysteine residues that play the most important role in their functionality due to their involvement in the formation of disulphide bonds within the same peptide and between different ones. Wheat gluten proteins have a unique ability of forming a viscoelastic network upon hydration which is capable of trapping gas which is

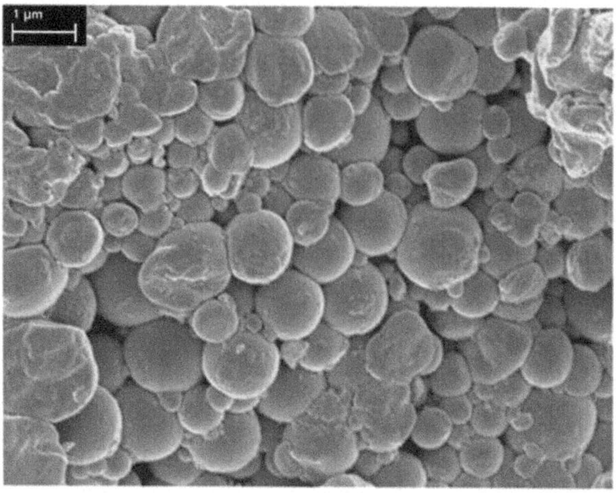

Fig. 11.12 SEM image of gliadin particles. Reproduced with permission from Quester et al. (2014)

either formed *in situ* during fermentation or incorporated by mixing. Upon heating, gluten proteins undergo a transition from a viscoelastic liquid to a solid and the initial foam structure becomes indefinitely stable giving the characteristic open foam structure to bread. The elastic properties of dough are attributed to the glutenin polymers while gliadins are believed to act as plasticiser that weaken the interactions between the glutenin chains increasing the viscosity of the dough (Thewissen et al. 2011). A suitable balance between viscosity and elasticity is necessary for efficient entrapment of CO_2 and adequate bread making results. Low elasticity leads to low bread volume while overly increased elasticity hinders the expansion of the cells. Sufficient elasticity is needed in order for bubbles to be entrapped and later expanded upon baking. Gliadins are the most surface active and are therefore responsible for the initial stabilisation of the gas bubbles (Mita et al. 1978).

Gliadins are in turn classified into ω-, α- and γ-gliadins. The α- and γ- type, have short amino acid terminal domain and consist of few amino acid residues and a large central domain and carboxyl-terminal domain with 6 and 8 cysteine residues respectively. The ω- type have a large central domain and short terminal ones with no cysteine residues. The existence of hydrophilic and hydrophobic parts in the α- and γ-gliadins makes them the most surface-active ones (Banc et al. 2007). Isolated gliadins from wheat flour through ethanol extraction show a spherical particulate shape varying from 100 nm to 1 μm (Fig. 11.12). The rheological properties of gluten proteins are also affected from the other components of flour (lipids, arabinoxylans and soluble proteins) and can be further modified by the addition of oxidants, reducing agents, proteases, lipid based surfactants or hemicellulases (Veraverbeke and Delcour 2002).

What also affects the ability of gluten proteins to stabilise gas bubbles is like in any other protein system, the environmental conditions such as the pH and ionic strength. With a pI at ~7.8, Gliadins, the proteins responsible of the initial stabilisation of gas bubbles in gluten matrixes, both the foaming ability (here as foam volume) and foam stability is higher at pH values around the pI and lower at the acidic

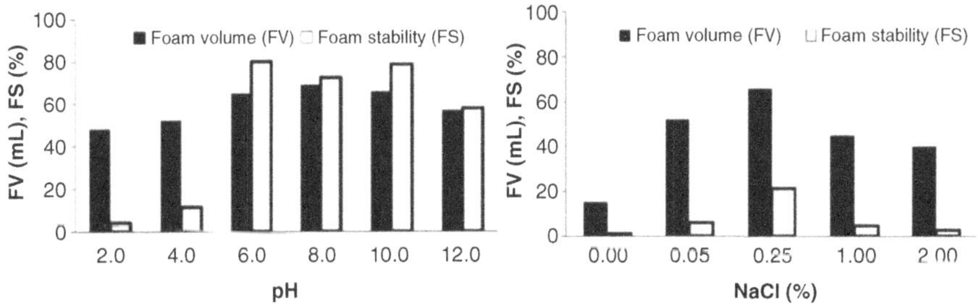

Fig. 11.13 Foam volume (FV) and foam stability (FS) measured 60 min after the start of whipping of: (left) gliadin solutions with a protein concentration of 0.10%, w/v at pH 2.0, 4.0, 6.0, 8.0, 10.0, and 12.0, respectively, (right) gliadin at pH 2 and 0.02%, w/v concentration in the presence of 0.00, 0.05, 0.25, 1.00, and 2.00% (w/v) NaCl. Reproduced with permission from Thewissen et al. (2011)

and alkaline regions (Fig. 11.13). The presence of electrostatic interactions between the moieties reduces the ability to diffuse fast to the interface which affects their foaming ability. This, coupled with the abilities of the proteins to have a more compact conformation when the net charge is zero and therefore pack more closely on the interface, makes the foams more stable when made at the neutral pH regions (Thewissen et al. 2011). Similarly adding a salt like NaCl, especially at pH where electrostatic interaction between the proteins are present (away from the pI), screens the available charges with the ions introduced masking them. As seen in Fig. 11.13, adding NaCl in gliadin solutions at pH 2 improves both foaming ability and foam stability in the ionic condition where there are only enough ions to mask the charges of the proteins. At pH close to the pI, the addition of salts has a negative effect on the foaming properties (Thewissen et al. 2011). The ability of wheat gluten proteins to stabilise and trap gas bubbles during the making of bread and other foods depends strongly not only on the gluten content (flour strength), but also in other factors such as the levels of the individual fractions (affected by genotype and cultivation conditions), the environmental conditions and the presence of other substances with synergistic effects.

11.2.4.3 Legume Proteins

Legumes and their dried derivatives (known as pulses) are an integral part of the human diet. Pulse crops include peas, chickpeas, lentils, beans and lupins and are produced in many continents worldwide, with the majority being produced in North America (Canada) and areas within Asia and the Middle East (Roy et al. 2010). They are an excellent source of protein, carbohydrates and fibre and contain many essential vitamins and minerals. Pulse seeds accumulate large amounts of protein during their development in the form of storage proteins which provide a nitrogen source for the seed during the germination process. Chickpea, lentil and pea contain approximately 22%, 28.6% and 23.3% of protein respectively on dry basis (Sotelo and Adsule 1996). These concentrations can vary significantly depending on factors such as plant species, variety, maturity and the growing conditions. From a nutritional point of view, pulse storage proteins are relatively low in sulphur-containing amino acids such methionine, cysteine and tryptophan but have higher lysine content compared to cereals (Duranti 2006). The main storage proteins present in pulses are albumins, globulins and glutenins and they are classified according to their solubility behaviour. The fraction soluble in salt-water solutions, globulins, make approximately the 70% of the total protein in pulses and one or two types are present and classified according to the sedimentation coefficient in vicilin (7S) and legumin (11S). The water-soluble albumins, represent 10–20% of the total protein in pulses. The soluble in dilute acids and bases are finally the glutenins which comprise the 10–20% of total proteins present in legumes. Alongside the storage proteins, there are some other protein species present such as various enzymes, inhibitors and lectins which are known as anti-nutritional compounds and the majority of which is within the water-soluble albumin fraction. These compounds are molecules which disrupt the digestion process of monogastric species and have been evolved within the seed in order to act as protective mechanism. They need to be removed or inactivated by soaking and/or hear processing before the consumption of legumes (Carbonaro et al. 2000). Legume proteins are usually isolated from flours or dried pulses by alkaline extraction followed by isoelectric precipitation but also membrane extraction such as ultraciltration or air classification is possible (Kiosseoglou and Paraskevopoulou 2011). The method of extraction often affects the functionality of the isolates produced (Boye et al. 2010).

Pea Proteins

Amongst legumes, peas (*Pisum sativum L.*) are the ones with the most widespread use due to their availability around the world, their hull is easily removable and their fractionation is economically viable (Day 2013). The protein content of pea is 22–23% on dry basis with globulins and albumin accounting for 55–65% and 18–25% of the total protein content respectively. Pea globulins mostly consist of vicilin and convicilin (7S) and legumin (11S) fractions and have a pI at pH 4.5 (Barać et al. 2015). Legumin in peas is similar in structure with the one in soy, it exists in a hexameric form with each monomer consisting of an acidic subunit (~40 kDa) and a basic subunit (~20 kDa) linked together by a disulphide bond. The 7S proteins, viciln and convicilin, have a size of 47–50 kDa and ~70 kDa respectively and can form trimers with a size of ~150 kDa and ~210 kDa respectively (Tzitzikas et al. 2006). Whilst legumins denature and aggregate upon heating (77 °C) forming large non-reversible aggregates via disulphide bonding and hydrophobic interactions, vicilin forms reversible aggregates at lower temperatures (69 °C) based on non-covalent interactions due to the lack of cysteine residues within their structure. Vicilin has also shown higher surface hydrophobicity than legumin which maybe indicates its potential to adsorb on the a/w interface (Kimura et al. 2008). Pea proteins have shown a surface activity that allows the increase of the surface pressure of their solutions (Tsoukala et al. 2006).

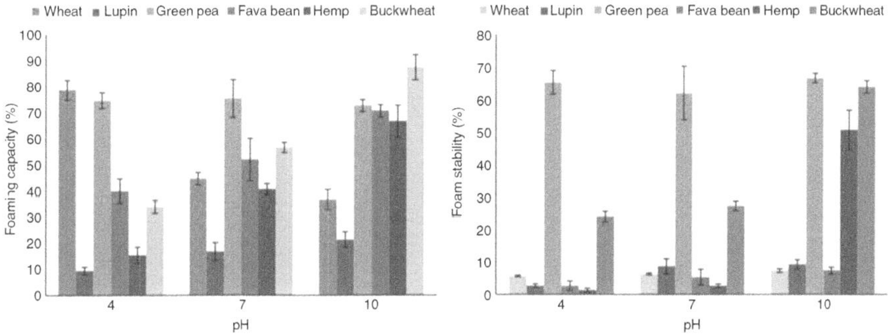

Fig. 11.14 Foaming capacity (left) and foam stability (right) of flours from different seeds over a range of pH (4, 7, 10). Reproduced with permission from Raikos et al. (2014)

The foaming properties of the proteins in pea flours have been reported in a comparative study amongst other types of flours (wheat, lupin, fava beans, hemp and buckwheat) (Raikos et al. 2014). Pea flour seemed to have consistently good foaming ability over the whole range of the pH studied (4–10) and a higher foam stability compared to the other flours (Fig. 11.14).

When looking at the foaming properties of pea protein isolates, studies have shown contradictory results when comparing the foaming properties of pea proteins against the ones from soy. Whilst some show that pea proteins have more flexible conformation at pH 3 and 7 and are more effective foam agents than soy proteins (Aluko et al. 2009; Sosulski and McCurdy 1987), other studies showed less enhanced foaming ability from pea proteins compared to soy (Tömösközi et al. 2001). Finally another study showed that whereas soy proteins, at pH 3–8, have higher foaming capacity, pea proteins demonstrate higher foam stability (Barac et al. 2015). The ability of pea proteins to stabilise foams can allow the replacement of other protein sources with pea. An example of that is sponge cakes where the eggs have been replaced with pea protein isolate along with xanthan gum with resulting cakes with very similar physical properties in terms of crump pore size and specific gravity to egg containing cakes (Lin et al. 2017). The variety of peas seems to be significant factor that affects the protein fraction profile of the isolate used (Barac et al. 2010) although there are factors such as the method of extraction used that has more significant effect on the functional properties of the pea isolate end foaming specifically (Stone et al. 2015).

Lentil Proteins

Lentil is a leguminous plant *(Lens culinaris)* that was originally grown in the Middle East and is now mainly produced in India, Turkey and Canada. It is high in fibre, low in fat and cholesterol free. Lentils contain 20.6–31.4% of protein with a legumin-like protein being the major (~50%) globulin fraction which is comprised of a number of 6–19 polypeptides with a molecular weight of 18–43 kDa (Barbana

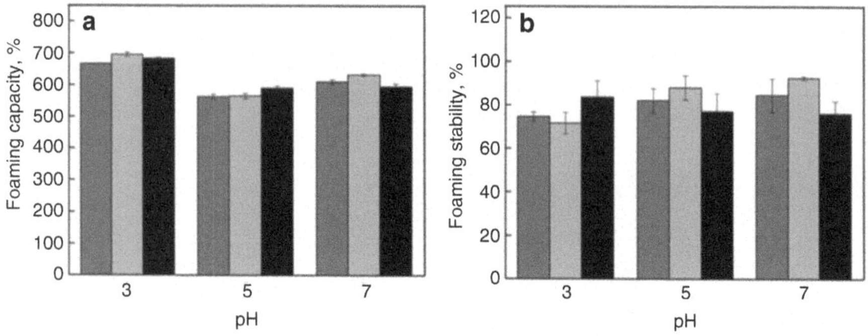

Fig. 11.15 Effect of environmental pH on the foaming capacity in terms of (%) overrun (**a**) and foam stability (**b**) in terms of % foam volume fraction after 0.5 h of lentil protein concentrates extracted at pH 8 (grey), 9 (light grey) and 10 (black). Reproduced with permission from Jarpa-Parra et al. (2014)

and Boye 2011). Legumins in lentils are similar to the ones in peas, present in hexamers with a Mw of 320–380 kDa that consist of six polypeptide pairs that interact non-covalently. Each pair, as in pea proteins is comprised of an acidic and a basic subunit that are linked via a single disulphide bond (Barbana and Boye 2011). Lentil proteins are often isolated via alkaline extraction that is followed by isoelectric precipitation which can take place at a range of pH values (7.2–11). The proteins obtained have significant functionality in terms of foaming, emulsifying and gelling (Bora 2002). Looking at the foaming properties of lentil proteins (Fig. 11.15) it can be noted that there is a dependence on the environmental pH mainly on the foaming capacity with proteins being able to incorporate more air when in solutions at acidic pH (3), far from the pI (4.5).

This was attributed to the higher solubility and charge at low pH which contributes to the ability of the proteins in lentils to be flexible and diffuse more rapidly on the a/w interface and orient their hydrophilic and hydrophobic segments accordingly while creating films that are resistant and elastic due to the repulsion amongst the protein species (Jarpa-Parra et al. 2014). At pH 7, foams are dense and have strong networks at the interface due to a combination of the α-helix secondary structure, medium hydrodynamic molecular size and balance between solubility and hydrophobicity (Jarpa-Parra et al. 2015). Closer to the pI at pH 5, lentil proteins formed dense and thick films that were composed of randomly aggregated particles that occur due to the lack of charge. Finally, the stability of the lentil protein foams seemed not to be affected by the pH of protein suspensions which was higher compared to other plant proteins such as pea (35–40%) and barley (40–70%) (Boye et al. 2010; Wang et al. 2010).

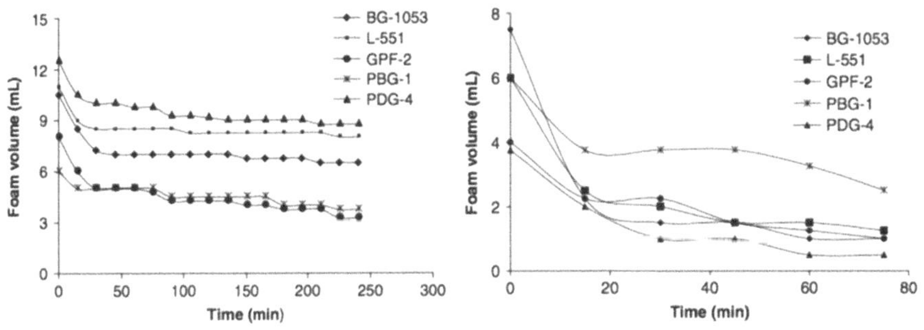

Fig. 11.16 Foaming properties of protein concentrates from different chicpea cultivars of the Kabuli and Desi types at pH 7 (left) and pH 4.5 (right). Reproduced with permission from Deep Singh et al. (2008)

Chickpea Proteins

Chickpea (*Cicer arietinum L.*), one of the first vegetables to be ever cultivated, is believed to have originated in the Middle East 7500 years ago, is now grown around the world in semi-arid regions of the Asia, Europe, Australia and North America making it second crop in growing area 15.3% of total legume area) and third in production (14.6% of total legume production) around the world (Frimpong et al. 2009; Roy et al. 2010). Chickpea proteins, similarly to the other pulses, are classified in two major fractions, globulins and albumins. Globulins, being the main seed proteins, mainly contain legumin and vicilin and represent the 60–80% of the total proteins. The albumin fraction accounts for the 15–25% of the total proteins on the seed and plays an essential role since it includes most of the enzymatic and metabolic proteins (Deep Singh et al. 2008).

The foaming properties of chickpea proteins have been studied and results have shown that close to the pI (pH 4.5) foaming capacity and foam stability is lower than when pH is higher of lower (Deep Singh et al. 2008; Tontul et al. 2017) (Fig. 11.16).

The poor foam stability at pH 4.5 improved significantly with the addition of NaCl and sucrose because of the introduction of electrostatic repulsion and increase in viscosity which significantly slows down drainage (Deep Singh et al. 2008). The presence of small amounts of fat that remains during the protein extraction process can have an adverse effect on the stability of the foams due to the anti-foaming mechanism of small quantities of liquid oils spreading on the a/w interface and displacing the proteins. Removal of the fat traces by defatting with a solvent can therefore significantly increase the stability of the foams stabilised by chickpea proteins (Toews and Wang 2013).

The introduction of new eating trends driven by conscious eating habits, sustainability and religious dietary restrictions has underlined the need of identifying alternative plant materials that provide the same functionality of animal derived ones. This exploration is motivated not only by the effort to address a consumer need by the industry and the research community but also by groups of consumers that like to experiment with food and discover new uses for food materials. In a similar fashion, in the quest for finding vegan alternatives to egg white, members of groups on social media have started exchanging their recipes at first using the mucilage created when soaking linseeds with interesting results. This until the discovery that the cooking water of chickpeas, often found in cans containing the particular pulse, can be whipped to a foam that reportedly has the same visual and organoleptic properties to whipped egg whites and can be equally baked to a meringue. The ingredient gained lots of attention amongst nutrition aficionados on social media and was named "aquafaba" from the Latin words for water (aqua) and bean (faba) (Twine 2018). The ability of cooking water of various legumes to effectively stabilise foams has been attributed to its composition in water soluble carbohydrates (sugars, soluble and insoluble fibre), small fractions of proteins and saponins that leach to the water from within the seeds. The foaming ability seems to be particularly accredited to the amount of proteins presence, although saponins are hypothesised to play a crucial role as well (Stantiall et al. 2018). Saponins, phytochemicals present in plants, are glycosides that contain a steroidal or triterpenoid aglycone and one or more sugar chains, structure that allows them to be surface active (Güçlü-Üstündağ and Mazza 2007). Nevertheless, they are considered anti-nutrients and the content in raw chickpeas is around 0.9 mg/kg and during cooking the original amount is reduced in half (Alajaji and El-Adawy 2006).

Lupin Proteins

Lupins (*Lupinus*) are legumes that have been cultivated in ancient Greece and Egypt before 2000 BC as grains for human and animal consumption alike. Lupin contains about 34% of protein from which the 87% are globulins and 13% albumins (Van de Noort 2017). The main isolation method is isoelectric precipitation after an alkaline treatment but also ultrafiltration provides satisfactory yields. Studies suggest that during the isolation of the lupin proteins some protein-polysaccharide complexes form which enhance the foaming properties of these systems due to the steric repulsion effects amongst the adsorbed complexes which increase the viscoelasticity of the interface (Alamanou and Doxastakis 1997; Karapantsios et al. 2011). Similar to the other plant proteins, lupin proteins exhibit better foaming proteins at pH values away from the isoelectric point where the solubility is higher and the diffusion and the unfolding on a/w is facilitated.

11.2.4.4 Potato Proteins

Potato protein extracted from the by-products of potato starch manufacturing is an ingredient with increasing popularity. The created side stream, called potato fruit juice, contains 1.5% of nitrogenous compounds (Knorr et al. 1977). The protein fractions present are the patatins (40% of total), protease inhibitors (50%) and other high Mw proteins (10%). Patatins are glycoproteins with a Mw of 39–43 kDa in a dimer conformation with different pIs at pH 4.45–5.17. Protease inhibitors (PI)

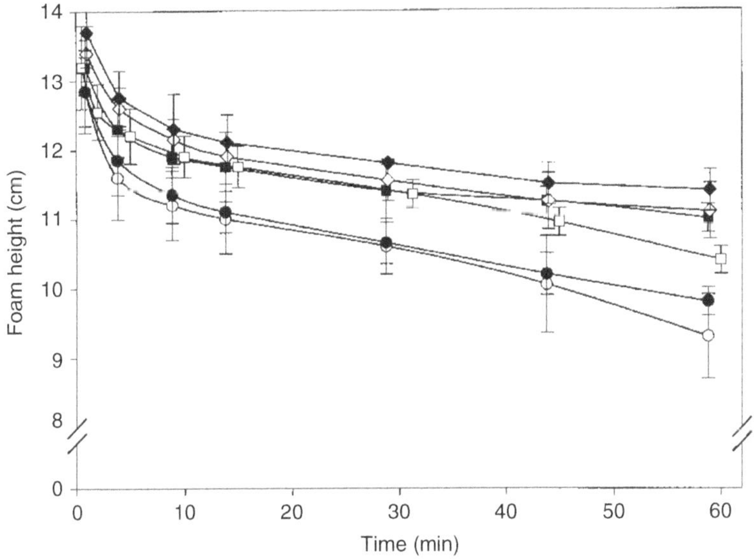

Fig. 11.17 Foam height as a function of time for foam prepared with whipping for 70 and 360 s at pH 7 and I = 50 mM with 10 mg/mL solutions of β-casein (70 s), (open circle); β-lactoglobulin (70 s), (filled circle); potato protein isolate (PPI) (70 s), (filled square); PPI (360 s), (filled diamond); ammonium sulfate precipitate (ASP) (70 s), (open square); and ASP (360 s), (open triangle). Reproduced with permission from van Koningsveld et al. (2002)

further consist of 7 subgroups with a wide range or Mw 4.3 to 20.6 kDa with pI values of pH 5.1–9.0 and they include oxidative enzymes like lipoxygenase, polyphenol oxidase and enzymes which are associated with starch synthesis. The amino acid sequence of patatin, with 366 amino acids, does not show specific clusters with either enhanced hydrophobicity or hydrophilicity and the positive and negative charges are randomly spread (Lomolino et al. 2015). Patatin contains 17 tyrosine and 2 tryptophan residues, the latter being in close proximity on the sequence at residues 256 and 260. It is also glycosylated at the 60 and 90 asparagine residues (Sonnewald et al. 1989). Patatin is believed to serve both as storage protein but also exhibits enzymatic activities which indicate its role in the plants defence mechanism (Andrews et al. 1988).

Whilst native potato proteins obtained via isoelectric precipitation have shown foaming properties inferior to those of whey proteins when either isolated with more gentle methods such as ultrafiltration and anion-exchange chromatography or precipitation with solvents such as ethanol or ammonium sulphate, have exhibited foaming properties similar to or inferior of β-casein and β-lac as seen on Fig. 11.17 (van Koningsveld et al. 2002). The patatin fraction has shown to stabilise foams for longer than the potato proteins in the 16–25 kDa range which include the protease inhibitors (Ralet and Guéguen 2001) which has been attributed to the ability of the patatins to adsorb faster to the a/w interface which is also indicated by their lower equilibrium surface tension and faster diffusion rates (Schmidt et al. 2018).

11.2.4.5 Hydrophobins

They are surface-active proteins produced by filamentous fungi (Wessels 1996). They are small molecules with a molecular size of 7–9 kDa and have eight cysteine residues that produce four disulphide linkages in unique symmetry (Cox et al. 2007). They are classified as classes I and II based on the occurrence of hydrophilic and hydrophobic amino acid residues in their sequences (Linder 2009). Class I (HFBI) forms aggregates that are highly insoluble in aqueous solutions while the ones of Class II (HFBII) are easier to dissolve. Cox et al. (2007) studied thoroughly the ability of hydrophobins to create highly viscoelastic layers on the a/w interface. They reported surface elasticities for films stabilised by adsorbed hydrophobins, both HBFI and HBFII, far higher than any other know protein and described them as nature-designed "Janus" particles. It was concluded that the bubbles stabilised by these proteins were completely stable to disproportionation. The highly viscoelastic films that are able to form around the air bubbles makes them resistant to high shear processing allowing the production of three phase systems where air droplets coexist with emulsion droplets giving way to partly replacing fat by air in emulsion-based formulations (Tchuenbou-Magaia et al. 2009).

11.3 Polysaccharide Based Bio-Polymers

Polysaccharides are polymeric carbohydrates most commonly obtained from plant sources. Their molecular structures are well defined as chains of monosaccharide units bound together by glyosidic linkages, although they generally occur in complex mixtures. Polysaccharides are isolated from their plant material typically via hot water extraction followed by purification. Molecular size, shape and conformation influence their water solubility, viscosity and gelling behaviour along with external factors such as pH, ions, complexing agents and temperature (Stephen 1995). The most abundant polysaccharide in nature is cellulose, followed by starch which is the main source of carbohydrate in our diets. The primary use of polysaccharides in food foams is to increase the viscosity of the solution through thickening or gelling mechanisms, which helps to stabilise the foam structure. In particulate form, they can stabile foams through aggregation in the bulk phase which reduces liquid drainage or by adsorbing on the air-water interface when hydrophobically modified providing a protective barrier around the air cells.

11.3.1 Botanical Polysaccharides

11.3.1.1 Cellulose

Cellulose, originating in plant cell walls, is the most abundant natural biopolymer. It is an attractive material for use in foods, due it's low-expense, health benefits and biodegradability. Cellulose is found in cell walls as macroscopic fibres and microfibrillar aggregates (Tingaut et al. 2012). The long, flexible fibres contain both amorphous and crystalline regions. The amorphous regions can be removed via hydrolysis leaving charged rod-like particles of microcrystalline cellulose and, upon further hydrolysis, nanocrystalline cellulose particles (Dickinson 2017). Both fibres and crystalline particles hold the potential to stabilise foam systems (Lam et al. 2014). Much of the research into this area focuses on the modification of cellulose to enable Pickering stabilisation. The surface of cellulose particles need to be hydrophobically modified in order to adsorb at the air-water interface. Common derivatives include methylcellulose, ethylcellulose and hydroxypropyl methylcellulose. Cervin et al. (2013) reported the stabilisation of foams using nanofibrillated cellulose particles modified through adsorption of cationic octylamine. The particles had a high aspect ratio, which facilitated the formation of a gel-like network between bubbles, as well as a large viscoelastic modulus at the particle-saturated surface. This increased the efficiency of foam stabilisation compared to shorter cellulose nanocrystals with the same charge density (Cervin et al. 2015).

Similarly, Wege et al. (2008) stabilised foams for long periods of time by in-situ formation of modified cellulose particles through solvent attrition. The cellulose derivative, hypromellose phthalate (HP), was dissolved in water-miscible solvents such as acetone and subsequently sheared in aqueous media to produce micron sized particles. These particles adsorbed onto the air-water interface during aeration resulting in highly stable foams. Dense shells were observed around the air cells, which kept the bubbles far apart preventing their interaction and therefore coalescence (Fig. 11.18). The particles efficiency at stabilising foams was controlled by HP concentration in the stock solution and the solvent chemistry. Jin et al. (2012) used the same method to produce ethyl cellulose particles capable of Pickering stabilisation. Scanning electron microscopy confirmed the adsorption of particles at

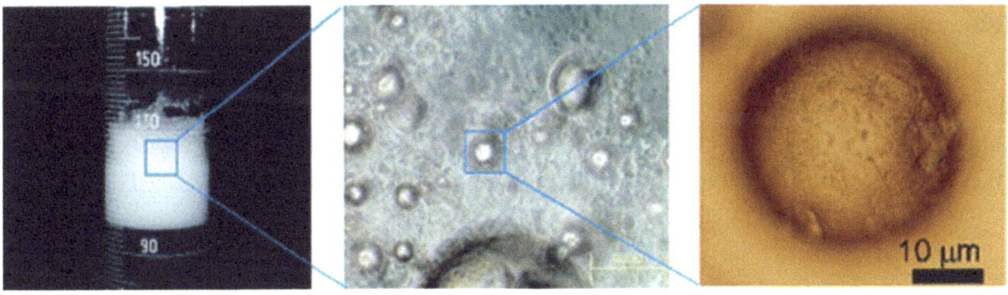

Fig. 11.18 Foams stabilised by cellulose derivative, hypromellulose phthalate, micron-sized particles. Reproduced with permission from Wege et al. (2008)

the air-water interface when exposed to acidic conditions. Current research in this area is focusing on the potential use of cellulose nanocrystals in systems of appetite control and obesity treatments, due to their resistance to digestion enzymes and gastric acids (Madadlou et al. 2016).

11.3.1.2 Starch

Starch is produced by most green plants to store energy and is a very popular carbohydrate in our diets. It occurs as semi-crystalline hydrophilic granules in many different shapes and sizes ranging from 0.5 to 100 μm (Dickinson 2017). Most of the recent research focuses on hydrophobically modifying starch particles for use as Pickering stabilisers, in particular quinoa starch particles, due to their relatively homogenous small size (approx. 2.5 μm). Starch particles can be modified chemically, through the addition of ocentyl succinic anhydride (OSA) or alternatively, using heat treatment. Dry-heating of starch alters the nature of the surface proteins resulting in a change in surface character from hydrophilic to hydrophobic (Seguchi 2001). Asghari et al. (2016) compared the performance of OSA-modified starch to heat-treated particles for use in protein-starch mixed systems. A dependence on particle size and hydrophobicity was observed, where the most effective system was that of OSA-modified starch particles and egg white protein, which enhanced foam stability 12-fold.

A key role of starch in food foams is its function in bread making. Texture and sensory properties of bread are heavily influenced by the gas phase (Wilde 2012) and so it is essential to understand the formation and stability of the wet foam structure. Bubbles are created during the mixing stage and are dependent on the rheological properties of the dough (Wilde 2012). During proving, the gluten-starch matrix is very important in controlling the structure and stabilising air cells (described in Fig. 11.19); the starch-gluten matrix needs to increase the viscoelastic modulus of the mixture sufficiently to prevent drainage whilst at the same time, allow bubble movement. At the latter stage of proofing and the early stages of cooking, bubbles are stabilised by thin liquid lamellae until the gluten and starch matrix gels, locking the structure in place (Gan et al. 1995). The bubbles then rupture converting the foam to a stable, open sponge structure. The most recent challenge in bread making was the production of gluten-free bread. To make a successful gluten-free product, the gas-binding capacity and the stabilisation of the starch gel during baking needs to be controlled (Schober 2009). Several gluten-free starches as well as modified starches were found to alter these dough parameters sufficiently to produce a quality product (Houben et al. 2012). Hydrocolloids are also good alternatives to gluten due to their high water-binding capacity (Houben et al. 2012).

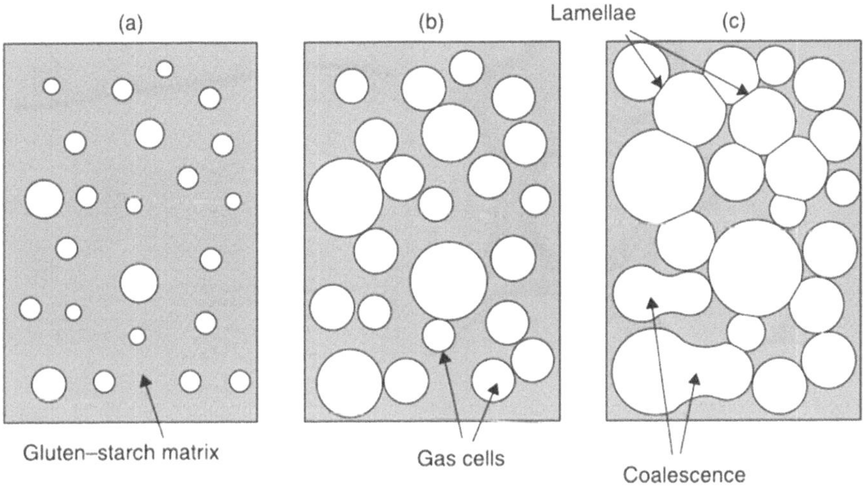

Fig. 11.19 Bubble evolution during the proof stage of bread making. (**a**) Post-mixing, the gas cells are supported by the gluten-starch matrix. (**b**) Mid-way through the proof, the gas cells are growing and beginning to make contact. (**c**) At the end of the proof, gluten-starch matrix is excluded from the space between some air cells, leaving thing liquid lamellae, which can cause coalescence. Reproduced with permission from Wilde (2012)

11.3.2 Animal Polysaccharides

11.3.2.1 Chitin

Chitin, a polysaccharide found in the shells of crustaceans and insects, is another effective material for the Pickering stabilisation of foams. The semi-crystalline material can be hydrolysed to form chitin nanocrystals, which can further provide stability through gel-like structuring (Dickinson 2017). Tzoumaki et al. (2015) used rod-like chitin nanocrystals to stabilise aqueous foams. Scanning electron microscopy (SEM) images revealed the presence of particles at the air-water interface and in the continuous phase, after drying. Packing of the nanocrystals at the air-water interface was optimised through increasing concentration and adjusting the pH close to chitin's pK_a, which minimised their electrostatic repulsion. A valuable derivative of chitin is chitosan; chitosan shows pH-dependent solubility in water due to the deprotonated chitin $-NH_3^+$ groups (Ho et al. 2016). Chitosan hydrogels have also been used in foaming applications; solidified foam structures have been produced from the gelling of chitosan hydrogel precursors using no additional surfactants (Testouri et al. 2010). The cationic nature of chitosan also increases its functionality through the ability to form electrostatic complexes with negatively charged hydrocolloids and proteins. This has been exploited in the Pickering stabilisation of emulsions (Nan et al. 2014; Shah et al. 2016; Wang et al. 2015). For example, Wang et al. (2015) used chitosan/zein complex particles as a potential pathway for producing antioxidant emulsions. However, chitosan complexes have not been as widely reported in foam systems, especially in foods.

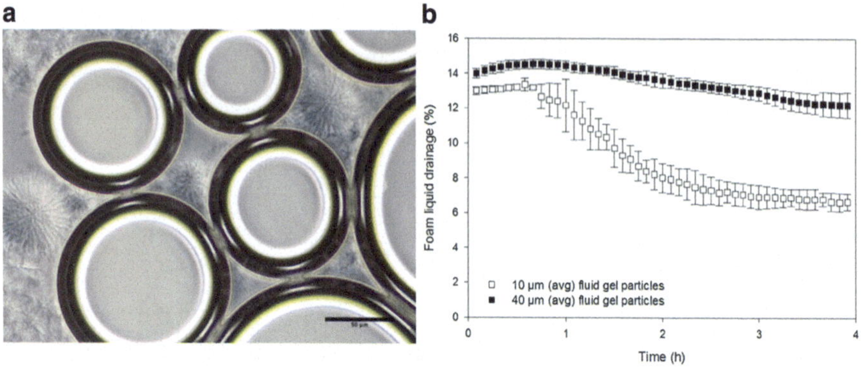

Fig. 11.20 (**a**) Optical microscopy of an aerated agar fluid gel revealing the confinement of particles to foam channels. (**b**) Liquid content measurements revealed the ability to considerably reduce foam drainage of aerated agar fluid gels by increasing the size of the particles. Reproduced with permission from Ellis et al. (2017)

11.3.3 Algal Polysaccharides

11.3.3.1 Agar

Agar is a polysaccharide originating from red seaweed which is not surface-active in its native form. It cannot easily be modified and so its main use in foam systems is through increasing the viscosity of the continuous phase when gelled. Recent work has investigated the use of agar fluid gels, which provide further advantages (Ellis et al. 2017). In combination with Tween 20, foams were stabilised by agar fluid gels at various concentrations. Tween 20 was used to establish short term stability and therefore creation of the foam through quick adsorption to the interface, which reduced the surface tension. Once the foam had been generated, the continuous phase quickly started to drain and a large number of agar particles remained in the foam channels and their nodes (Fig. 11.20a). This accumulation of particles increased their concentration at the Plateau borders which in turn caused a significant increase in the local micro-rheology. Drainage was therefore reduced, which had a large impact on the overall foam stability. In addition, the effect of particle size on this mechanism was studied. This revealed that drainage was considerably reduced when the continuous phase consisted of larger particles, due to more effective plugging of the channels (Fig. 11.20b). Whilst drainage of the continuous phase was occurring in these foams, the dispersed gas phase was coarsening (Saint-Jalmes 2006). Eventually, the size of the foam channels started to increase, resulting in the loss of confinement. The gravitational stress became larger than the yield stress and the foam collapsed (Guillermic et al. 2009). The yield stress of the fluid gel therefore gave a strong indication of foam half-life and could be manipulated through altering particle interaction and elasticity. The advantage of this particulate system over weak gels is the ability of particles to recover quickly upon shear as well as their potential to mimic fat droplets and therefore be used in reduced fat foam systems.

11.3.3.2 Alginate

Originating from brown seaweed, the polysaccharide alginate is primarily used as a stabiliser, thickener and gelling agent in food. Alginate undergoes gelation in the presence of multivalent ions such as Ca^{2+} resulting in heat-stable gels. In its native form, it primarily provides stabilisation in foams through increasing the viscosity of the continuous phase, for example, in ice-cream and whipped products. A derivate of alginate, propylene glycol alginate (PGA), is used as a popular foam stabiliser in the brewing industry. The presence of a stable and attractive head of foam is an important aspect of beer quality (Bamforth 1985). Bubbles are typically formed through the production of carbon dioxide during fermentation and are stabilised by proteins and other surface-active agents in the beer. However, lipids in the beer can displace these proteins at the interface resulting in instability, which causes the foam to collapse (Cooper et al. 2002). The electrostatic interaction of anionic alginate and the adsorbed protein increases local viscosity by thickening the adsorbed layer on the bubble surface, which prevents this from occurring and therefore increases foam stability (Jackson et al. 1980).

Additionally, alginate foams have been prepared using novel foaming techniques for application in the food industry; Ahmad et al. (2012) prepared highly porous calcium alginate foams using a microfluidic T-junction system. Monodispersed microbubbles were generated where size could be controlled by altering bubbling conditions such as solution flow rate and viscosity. A uniform size of air cells is advantageous as the pressure gradient between bubbles and therefore disproportionation will be significantly reduced, which will greatly improve overall foam stability. Furthermore, microbubbles are an exciting new trend in the food industry as it has been widely speculated that they will have novel applications as fat-replacers and texture modifiers (Rovers et al. 2016).

11.3.3.3 Carrageenan

Carrageenan is an anionic sulphated polysaccharide, originating from the same seaweed family as agar. There are three types of carrageenan, iota (ι), lambda (λ) and kappa (κ), which differ in the proportion and location of ester sulphate groups providing a wide range of rheological properties (Phillips and Williams 2009). Carrageenan is commonly used as a foam stabiliser in foods such as ice-cream, whipped toppings and meringues (Panda 2010). κ-carrageenan fluid gels can also provide stabilisation in foams through the same mechanism described for agar (Sect. 11.3.3.1). Lazidis et al. (2017) investigated kappa carrageenan fluid gels in addition with whey protein (WPI). Whey protein provided short-term stability through adsorption at the bubble interface and after initial drainage of the bulk phase, kappa carrageenan particles accumulated in the space between the bubbles and their nodes. The concentration of κ-carrageenan was the most significant factor affecting the viscosity of the fluid gels (Fig. 11.21a), which in turn affected foam half-life. Upon increasing the viscosity of the fluid gel, the foam half-life rose

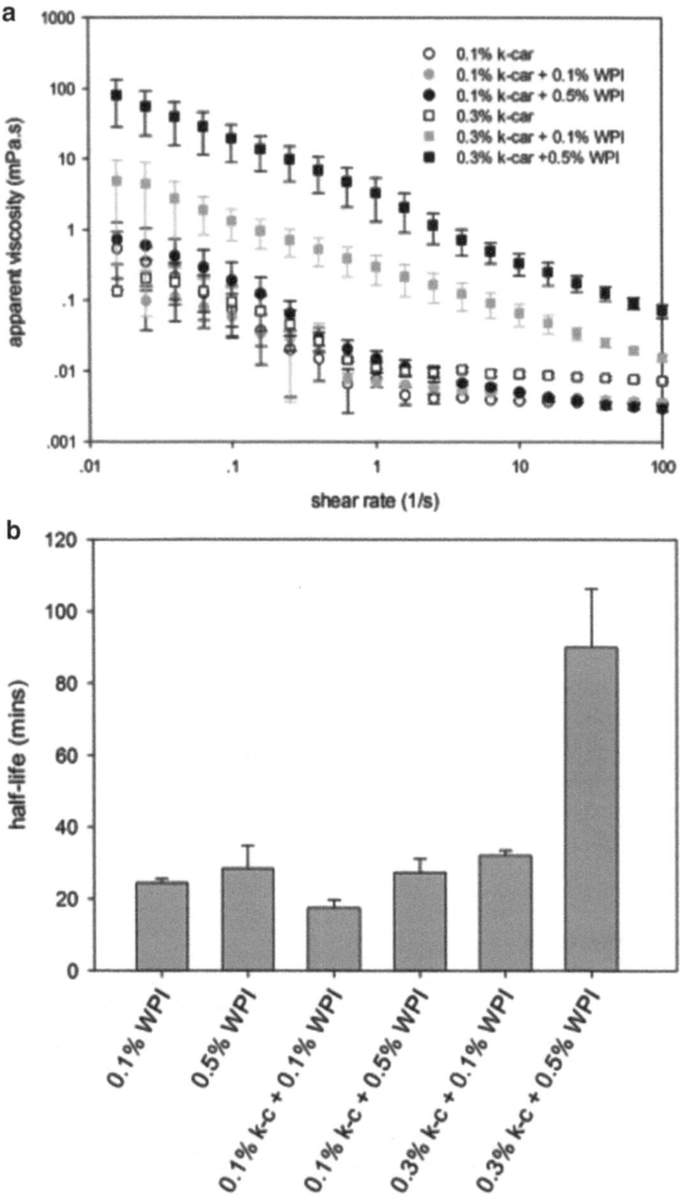

Fig. 11.21 (**a**) Flow curves of κ-carrageenan fluid gels containing WPI at different concentrations of both. (**b**) Foam half-lives of these aerated systems. Reproduced with permission from Lazidis et al. (2017)

(Fig. 11.21b) suggesting that the mechanism of stability relies on the viscosity of the fluid gel as well as the presence of the particles in the foam channels, which have a greater effect on the flowing properties of the bulk phase. κ-carrageenan also has the potential to interact with positively charged entities, such as proteins, due to its high negative charge. This would provide further foam stability by increasing the thickness of the adsorbed viscoelastic layer at the air-water interface.

11.4 Protein-Polysaccharide Mixtures

Proteins and polysaccharides are rarely found without one another in most food systems. Their interaction is therefore essential to understand in order to control the properties of products and to utilise in new food formulations. Their complex structural arrangements due, in part, to aggregation and gelation behaviour alters the rheological properties of the product which may in turn affect texture and stability (Benichou et al. 2002). It is also the reason that despite knowledge in this area increasing over the last few decades (Baeza et al. 2005; Bos and Vliet 2001; Dickinson 2003; Ghosh and Bandyopadhyay 2012; Krägel et al. 2008; Turgeon et al. 2007) there is still much more to understand and utilise. Due to their hydrophilic nature, polysaccharides usually remain in the aqueous phase, whilst proteins being surface active are typically found at the air-water interface. Protein-hydrocolloid interactions therefore generally increase foam stability by affecting both the rheology of the bulk phase and the thickness of the adsorbed viscoelastic layer at the air-water interface.

The overall interaction between two biopolymers is a contribution of a number of different intermolecular forces between their various segments and side chains (Dickinson 1998). It can be net attractive or repulsive depending on these interactions, which are controlled by solution conditions, such as pH. This has been illustrated in Fig. 11.22; a net repulsive interaction usually results in phase separation of the two biopolymers due to electrostatic repulsion or high steric exclusion, whilst an attractive interaction leads to either soluble complexes or insoluble precipitates. The formation and solubility of such complexes are affected by a number of factors, such as the charge and nature of the biopolymers, ionic strength, pH, the presence of external surfactants and temperature (Ghosh and Bandyopadhyay 2012). Their formation occurs primarily through non-covalent interactions such as hydrogen bonding, electrostatic, hydrophobic and steric interactions. Below the isoelectric point (pI) of a protein, -NH_3 groups become protonated and positively charged. Above the pK_a of a polysaccharide, -CO_2H or –OSO_3H groups (depending on the polysaccharide) become deprotonated contributing to an overall negative charge. Interactions between proteins and polysaccharides are largely reported as a function of the electrostatic interaction between these oppositely charged groups, where a stronger attraction is reported for sulphated polysaccharides (Dickinson 1998). Hydrogen bonding and hydrophobic interactions play an important secondary role in the formation of these complexes, which are heavily influenced by solution tem-

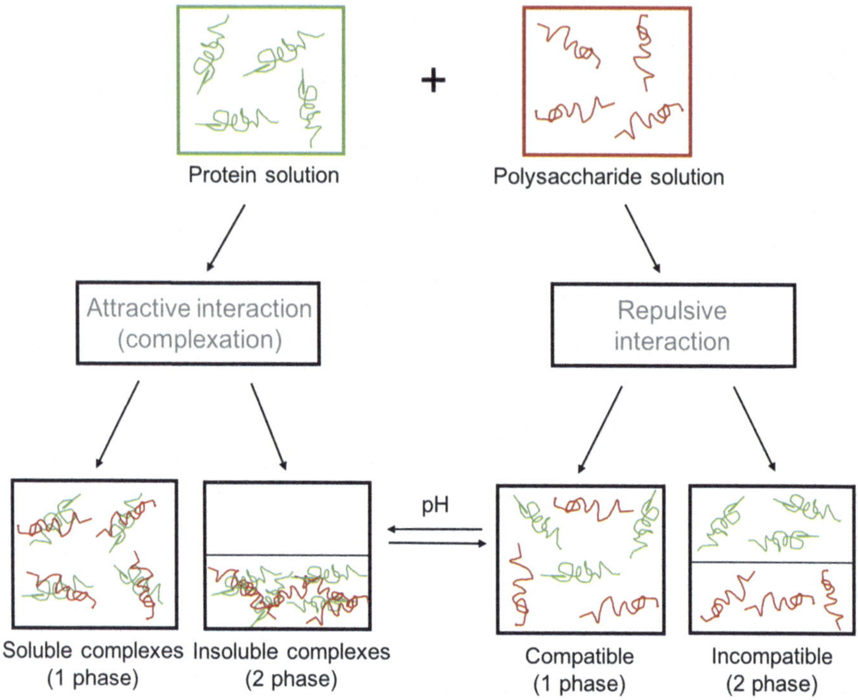

Fig. 11.22 Protein and polysaccharide solutions can interact either attractively resulting in soluble or insoluble complexes (left) or interact repulsively leading to either a one-phase or two-phase system (right). This can be controlled by altering the pH

perature (McClements 2006). The complexes formed can be both soluble and insoluble depending primarily on their concentration (McClements 2006); initial binding may cause charge neutralisation and therefore precipitation of the complexes but by increasing the concentration of one biopolymer, an overall charge should increase solubility.

11.4.1 Pectin-Protein Systems

Pectin is a polysaccharide present in citrus and apple fruits. It is widely used in the food industry for its gelling and stabilising properties, most commonly in jams and marmalades. As a result of its negative charge, pectin is often used to form electrostatic complexes with proteins. For example, Ganzevles et al. (2006) investigated foam stabilisation using β-lactoglobulin/pectin complexes. Soluble complexes of β-lactoglobulin/low methoxyl pectin were produced at pH 4.5 (below the pI of β-lactoglobulin and at the pK_a of pectin), which adsorbed at the air-water interface. The behaviour of adsorbed layers were compared when prepared by either sequential adsorption of pectin to a previously formed protein layer or adsorption of the two biopolymers simultaneously. The surface shear modulus was measured and observed to be up to a factor of 6 higher when layers were formed sequentially compared to those produced via simultaneous (complex) adsorption. In addition, work by Kudryashova et al. (2007) investigated the role of egg white ovalbumin/pectin complexes at the air-water interface. Complexes were produced in the pH range 4.5–7 with conformations dependent on the protein/pectin ratio. The complexes adsorbed at the air-water interface at a slower rate than pure ovalbumin alone, but did not dissociate when there. The protein therefore remained in its aqueous microenvironment, which was not the case when adsorbed without pectin. This demonstrates the ability to better control surface activity of a biopolymer with use of a second.

11.4.2 Other Polysaccharide-Protein Systems

Dickinson and Izgi (1996) investigated the foaming properties of complexes of dextran with three different proteins (bovine serum albumin (BSA), lysozyme and β-casein). The complexes were prepared by dry heating mixtures, which were subsequently aerated by sparging with nitrogen. Complexation with lysozyme led to a substantial enhancement in foaming properties whereas complexation with BSA and β-casein did not. The effects of complexation were enhanced with increasing molecular weight of the polysaccharide.

More recently, Schmitt et al. (2005) studied the use of β-lactoglobulin/acacia gum complexes to stabilise ice-cream. Complexes were formed at pH 4.2 and were observed to have a similar surface activity to that of the pure protein, however complexes formed a much stronger viscoelastic interfacial film. As a result, gas permeability of thin films stabilised by the complexes were considerably reduced (0.021 cm s^{-1}) compared to pure protein (0.521 cm s^{-1}), suggesting that foam stability would consequently be higher. Schmitt and co-workers argued that this difference was due to the ability of complexes to re-organise at the interface after coalescence forming an interfacial microgel. These findings were utilised in ice-cream where it was observed that increased stability due to complexes was correlated to perceived creaminess (Schmitt and Kolodziejczky 2009).

11.5 Conclusions and Future Research Challenges

The literature reviewed in this chapter has showcased the diverse and adaptable behaviour of biopolymers which makes them very effective functional ingredients for the generation and stabilisation of foams. Despite the difference in inherent surface-activity of proteins and polysaccharide-based biopolymers, both can be exploited by food engineers to generate and stabilise foams, whether this is by providing Pickering stabilisation through adsorption at the a/w interface or through the formation of an aggregated particle network in the bulk phase or a combination of the two.

Proteins naturally have excellent foaming capabilities due to their inherent interfacial properties. Such properties and therefore the functionality of proteins can be altered by a variety of different modification methods. Modifications aim to improve foaming behaviour by either triggering controlled aggregation in order to create particles (enzymatic crosslinking or high-pressure processing (HPP)) or by exposing the hydrophobic patches of proteins to increase their interfacial activity using methods such as enzymatic hydrolysis, sonication, oxidation and glycosylation. Polysaccharide-based biopolymers are typically modified to increase their hydrophobicity and therefore enable their ability to adsorb at the a/w interface. They can be modified chemically, using thermal treatment or through the formation of electrostatic complexes with proteins.

An important consideration when modifying biopolymers is whether their properties are altered to the extent that ingredients are no longer food grade. Nevertheless, in some cases the use of model systems can help to increase knowledge, which can aid in the search for alternative ingredients. It is also important to consider that ingredients in real food systems are not pure chemical compounds, which means that the interfaces in foods are complex (Sagis and Scholten 2014). For example, a common issue with measuring interfacial properties of biopolymer particles is whether they are free from contaminates such as surface-active lipids. These particles will also compete with other ingredients in a real food system for occupation of the interfaces (Dickinson 2017). Of course, there is a need to investigate mixtures of biopolymer ingredients to understand their interactions and the mechanisms in real food products but there is also the need to continue studying well-defined purified ingredients, in order to provide fundamental understanding of their behaviour, which can help to explain what is happening in real food systems.

Further future challenges in this evolving area of research involve better understanding particle-laden a/w interfaces and in particular, the mechanisms which lead to their collapse (Dickinson 2017). It is known that long-term stability of foams can be provided by the formation of an interfacial layer at the a/w interface through adsorption of particles, which provides a barrier to the destabilising processes of bubble shrinkage and coalescence. Future progress aims to further understand the time-dependent behaviour of these layers. This can be studied by measuring both surface rheology and microstructure simultaneously (Barman and Christopher 2014). Theoretical modelling of complex particle-based layers is also being developed to compliment experimental advances (Knoche and Kierfeld 2015; Sagis and Scholten 2014). It is hoped that they can address the lack of understanding as to how particle properties and particle-particle interactions can affect the collapse of particle-laden interfaces.

As well as continuing research into the surface properties of both pure ingredients and mixtures of biopolymers as found in real food systems, the search for new functional ingredients as well as novel sources is always ongoing. For example, insects are a novel source of protein that are not yet being taken advantage of in developed countries. It is reported that they have a protein content as high as 35–61% (Rumpold and Schlüter 2013), which is higher than that of beans and lentils. However, in order to exploit this source of protein consumer acceptance of eating insects needs to increase. Consumer trends continue to lead the direction of food research, for example the rise in veganism has led to increased research into alternative, non-dairy foaming ingredients.

There are numerous, exciting emerging areas of research, from applications in controlled release and digestion management to new methods of foam generation to create microbubbles, which will continue to see the research into biopolymer foams in the food industry thrive.

References

Abascal DM, Gracia-Fadrique J (2009) Surface tension and foam stability of commercial calcium and sodium caseinates. Food Hydrocoll 23(7):1848–1852 https://doi.org/10.1016/j.foodhyd.2009.02.012

Ahmad B, Stride E, Edirisinghe M (2012) Calcium alginate foams prepared by a microfluidic T-junction system: stability and food applications. Food Bioproc Tech 5(7):2848–2857. https://doi.org/10.1007/s11947-011-0650-3

Aiking H (2011) Future protein supply. Trends Food Sci Technol 22(2):112–120 https://doi.org/10.1016/j.tifs.2010.04.005

Alajaji SA, El-Adawy TA (2006) Nutritional composition of chickpea (Cicer arietinum L.) as affected by microwave cooking and other traditional cooking methods. J Food Compos Anal 19(8):806–812 https://doi.org/10.1016/j.jfca.2006.03.015

Alamanou S, Doxastakis G (1997) Effect of wet extraction methods on the emulsifying and foaming properties of lupin seed protein isolates (Lupinus albus ssp. Graecus). Food Hydrocoll 11(4):409–413 https://doi.org/10.1016/S0268-005X(97)80038-0

Aluko RE, Mofolasayo OA, Watts BM (2009) Emulsifying and foaming properties of commercial Yellow Pea (Pisum sativum L.) seed flours. J Agric Food Chem 57(20):9793–9800 https://doi.org/10.1021/jf902199x

Álvarez K, Famá L, Gutiérrez TJ (2017) Physicochemical, antimicrobial and mechanical properties of thermoplastic materials based on biopolymers with application in the food industry. In: Masuelli M, Renard D (eds) Advances in physicochemical properties of biopolymers: Part 1. Bentham Science, Sharjah, pp 358–400. EE.UU. ISBN: 978-1-68108-454-1. eISBN: 978-1-68108-453-4. https://doi.org/10.2174/9781681084534117010015

Andrews DL, Beames B, Summers MD, Park WD (1988) Characterization of the lipid acyl hydrolase activity of the major potato (Solanum tuberosum) tuber protein, patatin, by cloning and abundant expression in a baculovirus vector. Biochem J 252(1):199–206 https://doi.org/10.1042/bj2520199

Arzeni C, Martínez K, Zema P, Arias A, Pérez OE, Pilosof AMR (2012) Comparative study of high intensity ultrasound effects on food proteins functionality. J Food Eng 108(3):463–472 https://doi.org/10.1016/j.jfoodeng.2011.08.018

Asghari AK, Norton I, Mills T, Sadd P, Spyropoulos F (2016) Interfacial and foaming characterisation of mixed protein-starch particle systems for food-foam applications. Food Hydrocoll 53:311–319. https://doi.org/10.1016/j.foodhyd.2015.09.007

Baeza R, Sanchez CC, Pilosof AM, Patino JMR (2005) Interactions of polysaccharides with β-lactoglobulin adsorbed films at the air–water interface. Food Hydrocoll 19(2):239–248

Bals A, Kulozik U (2003) Effect of pre-heating on the foaming properties of whey protein isolate using a membrane foaming apparatus. Int Dairy J 13(11):903–908. https://doi.org/10.1016/S0958-6946(03)00111-0

Bamforth C (1985) The foaming properties of beer. J Inst Brewing 91(6):370–383

Banc A, Desbat B, Renard D, Popineau Y, Mangavel C, Navailles L (2007) Structure and orientation changes of ω- and γ-Gliadins at the air-water interface: a PM–IRRAS Spectroscopy and Brewster Angle Microscopy Study. Langmuir 23(26):13066–13075. https://doi.org/10.1021/la702037k

Barac M, Cabrilo S, Pesic M, Stanojevic S, Zilic S, Macej O, Ristic N (2010) Profile and functional properties of seed proteins from Six Pea (Pisum sativum) genotypes. Int J Mol Sci 11(12):4973–4990. https://doi.org/10.3390/ijms11124973

Barać MB, Pešić MB, Stanojević SP, Kostić AŽ, Čabrilo SB (2015) Techno-functional properties of pea (Pisum sativum) protein isolates: a review. Acta Periodica Technol 46:1–18

Barac M, Pesic M, Stanojevic SP, Kostic AZ, Bivolarevic V (2015) Comparative study of the functional properties of three legume seed isolates: adzuki, pea and soy bean. J Food Sci Technol 52(5):2779–2787. https://doi.org/10.1007/s13197-014-1298-6

Barbana C, Boye JI (2011) Angiotensin I-converting enzyme inhibitory properties of lentil protein hydrolysates: determination of the kinetics of inhibition. Food Chem 127(1):94–101. https://doi.org/10.1016/j.foodchem.2010.12.093

Barman S, Christopher GF (2014) Simultaneous interfacial rheology and microstructure measurement of densely aggregated particle Laden interfaces using a modified double wall ring interfacial rheometer. Langmuir 30(32):9752–9760. https://doi.org/10.1021/la502329s

Benichou A, Aserin A, Garti N (2002) Protein-polysaccharide interactions for stabilization of food emulsions. J Dispers Sci Technol 23(1-3):93–123

Bergeron V, Walstra P (2005) Foams. In: Soft colloids, vol. 5. Academic, San Diego, pp 1–38

Berthold A, Schubert H, Brandes N, Kroh L, Miller R (2007) Behaviour of BSA and of BSA-derivatives at the air/water interface. Colloids Surf A Physicochem Eng Asp 301(1):16–22. https://doi.org/10.1016/j.colsurfa.2006.11.054

Betancur-Ancona D, Martínez-Rosado R, Corona-Cruz A, Castellanos-Ruelas A, Jaramillo-Flores ME, Chel-Guerrero L (2009) Functional properties of hydrolysates from Phaseolus lunatus seeds. Int J Food Sci Technol 44(1):128–137. https://doi.org/10.1111/j.1365-2621.2007.01690.x

Binks BP (2002) Particles as surfactants—similarities and differences. Curr Opin Colloid Interface Sci 7(1–2):21–41. https://doi.org/10.1016/S1359-0294(02)00008-0

Binks BP, Horozov TS (2005) Aqueous foams stabilized solely by silica nanoparticles. Angew Chem Int Ed 44(24):3722–3725. https://doi.org/10.1002/anie.200462470

Binks BP, Muijlwijk K, Koman H, Poortinga AT (2017) Food-grade Pickering stabilisation of foams by in situ hydrophobisation of calcium carbonate particles. Food Hydrocoll 63:585–592. https://doi.org/10.1016/j.foodhyd.2016.10.002

Bora PS (2002) Functional properties of native and succinylated lentil (Lens culinaris) globulins. Food Chem 77(2):171–176. https://doi.org/10.1016/S0308-8146(01)00332-6

Bos MA, van Vliet T (2001) Interfacial rheological properties of adsorbed protein layers and surfactants: a review. Adv Colloid Interface Sci 91(3):437–471. https://doi.org/10.1016/S0001-8686(00)00077-4

Boye JI, Aksay S, Roufik S, Ribéreau S, Mondor M, Farnworth E, Rajamohamed SH (2010) Comparison of the functional properties of pea, chickpea and lentil protein concentrates processed using ultrafiltration and isoelectric precipitation techniques. Food Res Int 43(2):537–546. https://doi.org/10.1016/j.foodres.2009.07.021

Britan A, Liverts M, Ben-Dor G, Koehler SA, Bennani N (2009) The effect of fine particles on the drainage and coarsening of foam. Colloids Surf A Physicochem Eng Asp 344(1–3):15–23. https://doi.org/10.1016/j.colsurfa.2009.03.011

Campbell GM, Mougeot E (1999) Creation and characterisation of aerated food products. Trends Food Sci Technol 10(9):283–296. https://doi.org/10.1016/S0924-2244(00)00008-X

Cantat I, Cohen-Addad S, Elias F, Graner F, Höhler R, Pitois O et al (2013) Foams: structure and dynamics (First English edition). Oxford University Press, Oxford

Carbonaro M, Grant G, Cappelloni M, Pusztai A (2000) Perspectives into factors limiting in vivo digestion of legume proteins: antinutritional compounds or storage proteins? J Agric Food Chem 48(3):742–749. https://doi.org/10.1021/jf991005m

Carn F, Colin A, Pitois O, Vignes-Adler M, Backov R (2009) Foam drainage in the presence of nanoparticle−surfactant mixtures. Langmuir 25(14):7847–7856. https://doi.org/10.1021/la900414q

Cervin NT, Andersson L, Ng JBS, Olin P, Bergström L, Wågberg L (2013) Lightweight and strong cellulose materials made from aqueous foams stabilized by nanofibrillated cellulose. Biomacromolecules 14(2):503–511

Cervin NT, Johansson E, Benjamins J, Wågberg L (2015) Mechanisms behind the stabilizing action of cellulose nanofibrils in wet-stable cellulose foams. Biomacromolecules 16(3):822–831

Chen M, Bleeker R, Sala G, Meinders MBJ, van Valenberg HJF, van Hooijdonk ACM, van der Linden E (2016) Particle size determines foam stability of casein micelle dispersions. Int Dairy J 56:151–158. https://doi.org/10.1016/j.idairyj.2016.01.020

Cheung DL (2017) Adsorption and conformations of lysozyme and α-lactalbumin at a water-octane interface. J Chem Phys 147(19):195101. https://doi.org/10.1063/1.4994561

Cooper D, Husband F, Mills E, Wilde P (2002) Role of beer lipid-binding proteins in preventing lipid destabilization of foam. J Agric Food Chem 50(26):7645–7650

Corzo-Martínez M, Carrera-Sánchez C, Villamiel M, Rodríguez-Patino JM, Moreno FJ (2012) Assessment of interfacial and foaming properties of bovine sodium caseinate glycated with galactose. J Food Eng 113(3):461–470. https://doi.org/10.1016/j.jfoodeng.2012.06.025

Corzo-Martínez M, Moreno FJ, Villamiel M, Rodríguez Patino JM, Carrera Sánchez C (2017) Effect of glycation and limited hydrolysis on interfacial and foaming properties of bovine β-lactoglobulin. Food Hydrocoll 66:16–26. https://doi.org/10.1016/j.foodhyd.2016.12.008

Cox AR, Cagnol F, Russell AB, Izzard MJ (2007) Surface properties of Class II Hydrophobins from Trichoderma reesei and influence on bubble stability. Langmuir 23(15):7995–8002. https://doi.org/10.1021/la700451g

Dalgleish DG, Corredig M (2012) The structure of the casein micelle of milk and its changes during processing. Annu Rev Food Sci Technol 3(1):449–467. https://doi.org/10.1146/annurev-food-022811-101214

Damodaran S (2005) Protein stabilization of emulsions and foams. J Food Sci 70(3):R54–R66. https://doi.org/10.1111/j.1365-2621.2005.tb07150.x

Davis JP, Foegeding EA, Hansen FK (2004) Electrostatic effects on the yield stress of whey protein isolate foams. Colloids Surf B Biointerfaces 34(1):13–23. https://doi.org/10.1016/j.colsurfb.2003.10.014

Day L (2013) Proteins from land plants—potential resources for human nutrition and food security. Trends Food Sci Technol 32(1):25–42. https://doi.org/10.1016/j.tifs.2013.05.005

de Oliveira FC, Coimbra JS d R, de Oliveira EB, Zuñiga ADG, Rojas EEG (2016) Food protein-polysaccharide conjugates obtained via the Maillard reaction: a review. Crit Rev Food Sci Nutr 56(7):1108–1125. https://doi.org/10.1080/10408398.2012.755669

Deep Singh G, Wani AA, Kaur D, Sogi DS (2008) Characterisation and functional properties of proteins of some Indian chickpea (Cicer arietinum) cultivars. J Sci Food Agric 88(5):778–786. https://doi.org/10.1002/jsfa.3144

Dhayal SK, Delahaije RJBM, Vries RJ, de Gruppen H, Wierenga PA (2015) Enzymatic cross-linking of α-lactalbumin to produce nanoparticles with increased foam stability. Soft Matter 11(40):7888–7898. https://doi.org/10.1039/C5SM01112D

Dickinson E (1998) Stability and rheological implications of electrostatic milk protein–polysaccharide interactions. Trends Food Sci Technol 9(10):347–354

Dickinson E (2003) Hydrocolloids at interfaces and the influence on the properties of dispersed systems. Food Hydrocoll 17(1):25–39. https://doi.org/10.1016/S0268-005X(01)00120-5

Dickinson E (2010) Food emulsions and foams: stabilization by particles. Curr Opin Colloid Interface Sci 15(1–2):40–49. https://doi.org/10.1016/j.cocis.2009.11.001
Dickinson E (2013) In: Wilson AJ (ed) Foams: physics, chemistry and structure. Springer Science & Business Media, Berlin
Dickinson E (2017) Biopolymer-based particles as stabilizing agents for emulsions and foams. Food Hydrocoll 68:219–231. https://doi.org/10.1016/j.foodhyd.2016.06.024
Dickinson E, Izgi E (1996) Foam stabilization by protein-polysaccharide complexes. Colloids Surf A Physicochem Eng Asp 113(1):191–201. https://doi.org/10.1016/0927-7757(96)03647-3
Dombrowski J, Dechau J, Kulozik U (2016a) Multiscale approach to characterize bulk, surface and foaming behavior of casein micelles as a function of alkalinisation. Food Hydrocoll 57:92–102. https://doi.org/10.1016/j.foodhyd.2015.12.022
Dombrowski J, Johler F, Warncke M, Kulozik U (2016b) Correlation between bulk characteristics of aggregated β-lactoglobulin and its surface and foaming properties. Food Hydrocoll 61:318–328. https://doi.org/10.1016/j.foodhyd.2016.05.027
Duranti M (2006) Grain legume proteins and nutraceutical properties. Fitoterapia 77(2):67–82. https://doi.org/10.1016/j.fitote.2005.11.008
Eastoe J (2005) Surfactant chemistry. Wuhan University Press, Wuhan
Ellis AL, Norton AB, Mills TB, Norton IT (2017) Stabilisation of foams by agar gel particles. Food Hydrocoll 73:222–228
Fang Y, Dalgleish DG (1993) Dimensions of the adsorbed layers in oil-in-water emulsions stabilized by caseins. J Colloid Interface Sci 156(2):329–334. https://doi.org/10.1006/jcis.1993.1120
Fang Y, Dalgleish DG (1997) Conformation of β-lactoglobulin studied by FTIR: effect of pH, temperature, and adsorption to the oil–water interface. J Colloid Interface Sci 196(2):292–298. https://doi.org/10.1006/jcis.1997.5191
Fennema OR (1996) Food chemistry. M. Dekker, New York
Foegeding EA, Luck PJ, Davis JP (2006) Factors determining the physical properties of protein foams. Food Hydrocoll 20(2–3):284–292. https://doi.org/10.1016/j.foodhyd.2005.03.014
Fox PF, Kelly AL (2004) In: Yada RY (ed) Proteins in food processing. CRC; Woodhead, Boca Raton, Cambridge
Frimpong A, Sinha A, Tar'an B, Warkentin TD, Gossen BD, Chibbar RN (2009) Genotype and growing environment influence chickpea (Cicer arietinum L.) seed composition. J Sci Food Agric 89(12):2052–2063. https://doi.org/10.1002/jsfa.3690
Fujioka T, Matsumoto S (1995) Factors affecting the viscoelastic properties of albumen foams. J Texture Stud 26(4):411–419. https://doi.org/10.1111/j.1745-4603.1995.tb00981.x
Gan Z, Ellis PR, Schofield JD (1995) Gas cell stabilisation and gas retention in wheat bread dough. J Cereal Sci 21(3):215–230. https://doi.org/10.1006/jcrs.1995.0025
Ganzevles RA, Zinoviadou K, van Vliet T, Cohen Stuart MA, de Jongh HH (2006) Modulating surface rheology by electrostatic protein/polysaccharide interactions. Langmuir 22(24):10089–10096
Garrett PR (1993) Recent developments in the understanding of foam generation and stability. Chem Eng Sci 48:367–392. https://doi.org/10.1016/0009-2509(93)80023-J
Gezimati J, Singh H, Creamer LK (1996) Aggregation and gelation of bovine beta-lactoglobulin, alpha-lactalbumin, and serum albumin. In: Macromolecular interactions in food technology http://agris.fao.org/agris-search/search.do?recordID=US201302863036
Ghosh AK, Bandyopadhyay P (2012) Polysaccharide-protein interactions and their relevance in food colloids. INTECH Open Access, Rijeka, Croatia
Glaser LA, Paulson AT, Speers RA, Yada RY, Rousseau D (2007) Foaming behavior of mixed bovine serum albumin–protamine systems. Food Hydrocoll 21(4):495–506. https://doi.org/10.1016/j.foodhyd.2006.05.008
Güçlü-Üstündağ Ö, Mazza G (2007) Saponins: properties, applications and processing. Crit Rev Food Sci Nutr 47(3):231–258. https://doi.org/10.1080/10408390600698197
Guillermic R-M, Salonen A, Emile J, Saint-Jalmes A (2009) Surfactant foams doped with laponite: unusual behaviors induced by aging and confinement. Soft Matter 5(24):4975–4982

Gutiérrez TJ (2017) Chitosan applications for the food industry. In: Ahmed S, Ikram S (eds) Chitosan: derivatives, composites and applications. Wiley-Scrivener, Beverly, MA, pp 183–232. EE.UU. ISBN: 978-1-119-36350-7. https://doi.org/10.1002/9781119364849.ch8

Gutiérrez TJ, Alvarez VA (2017) Cellulosic materials as natural fillers in starch-containing matrix-based films: a review. Polym Bull 74(6):2401–2430. https://doi.org/10.1007/s00289-016-1814-0

Havea P, Singh H, Creamer LK (2001) Characterization of heat-induced aggregates of beta-lactoglobulin, alpha-lactalbumin and bovine serum albumin in a whey protein concentrate environment. J Dairy Res 68(3):483–497

He X, Mao L, Gao Y, Yuan F (2016) Effects of high pressure processing on the structural and functional properties of bovine lactoferrin. Innov Food Sci Emerg Technol 38:221–230. https://doi.org/10.1016/j.ifset.2016.10.014

Hettiarachchy NS, Ziegler GR (1994) Protein functionality in food systems. CRC, Boca Raton, FL

Ho KW, Ooi CW, Mwangi WW, Leong WF, Tey BT, Chan E-S (2016) Comparison of self-aggregated chitosan particles prepared with and without ultrasonication pretreatment as Pickering emulsifier. Food Hydrocoll 52:827–837

Houben A, Höchstötter A, Becker T (2012) Possibilities to increase the quality in gluten-free bread production: an overview. Eur Food Res Technol 235(2):195–208

Jackson G, Roberts RT, Wainwright T (1980) Mechanism of beer foam stabilisation by prolylene glycol alginate. J Inst Brewing 86(1):34–37. https://doi.org/10.1002/j.2050-0416.1980.tb03953.x

Jambrak AR, Mason TJ, Lelas V, Herceg Z, Herceg IL (2008) Effect of ultrasound treatment on solubility and foaming properties of whey protein suspensions. J Food Eng 86(2):281–287. https://doi.org/10.1016/j.jfoodeng.2007.10.004

Jarpa-Parra M, Bamdad F, Wang Y, Tian Z, Temelli F, Han J, Chen L (2014) Optimization of lentil protein extraction and the influence of process pH on protein structure and functionality. LWT Food Sci Technol 57(2):461–469. https://doi.org/10.1016/j.lwt.2014.02.035

Jarpa-Parra M, Bamdad F, Tian Z, Zeng H, Temelli F, Chen L (2015) Impact of pH on molecular structure and surface properties of lentil legumin-like protein and its application as foam stabilizer. Colloids Surf B Biointerfaces 132:45–53. https://doi.org/10.1016/j.colsurfb.2015.04.065

Jimenez-Junca CA, Gumy JC, Sher A, Niranjan K (2011) Rheology of milk foams produced by steam injection. J Food Sci 76(9):E569–E575. https://doi.org/10.1111/j.1750-3841.2011.02387.x

Jin H, Zhou W, Cao J, Stoyanov SD, Blijdenstein TB, de Groot PW, Pelan EG et al (2012) Super stable foams stabilized by colloidal ethyl cellulose particles. Soft Matter 8(7):2194–2205

Karapantsios TD, Papoti VT, Doxastakis G (2011) Foaming activity of lupin protein isolates in the absence of insoluble protein aggregates. Colloids Surf A Physicochem Eng Asp 382(1):74–80. https://doi.org/10.1016/j.colsurfa.2010.11.025

Kato A, Osako Y, Matsudomi N, Kobayashi K (1983) Changes in the emulsifying and foaming properties of proteins during heat denaturation. Agric Biol Chem 47(1):33–37

Kilara A, Vaghela MN (2004) Whey proteins. In: Yada RY (ed) Proteins in food processing. Woodhead, Cambridge, pp 72–100

Kilcast D, Subramaniam P (eds) (2011) Food and beverage stability and shelf life. Woodhead, Cambrdige

Kimura A, Fukuda T, Zhang M, Motoyama S, Maruyama N, Utsumi S (2008) Comparison of physicochemical properties of 7S and 11S globulins from pea, fava bean, cowpea, and french bean with those of soybean—French bean 7S globulin exhibits excellent properties. J Agric Food Chem 56(21):10273–10279. https://doi.org/10.1021/jf801721b

Kinsella JE (1981) Functional properties of proteins: possible relationships between structure and function in foams. Food Chem 7(4):273–288

Kinsella JE, Whitehead DM (1989) Proteins in whey: chemical, physical, and functional properties. In: Kinsella JE (ed) Advances in food and nutrition research (vol. 33). Academic Press, San Diego, pp 343–438. https://doi.org/10.1016/S1043-4526(08)60130-8

Kiosseoglou V, Paraskevopoulou A (2011) 3—Functional and physicochemical properties of pulse proteins. In: Tiwari BK, Gowen A, McKenna B (eds) Pulse foods. Academic Press, San Diego, pp 57–90. https://doi.org/10.1016/B978-0-12-382018-1.00003-4

Kloek W, van Vliet T, Meinders M (2001) Effect of bulk and interfacial rheological properties on bubble dissolution. J Colloid Interface Sci 237(2):158–166. https://doi.org/10.1006/jcis.2001.7454

Knoche S, Kierfeld J (2015) Elasticity of interfacial rafts of hard particles with soft shells. Langmuir 31(19):5364–5376

Knorr D, Kohler GO, Betschart AA (1977) Potato protein concentrates: the influence of various methods of recovery upon yield, compositional and functional characteristics. J Food Process Preserv 1(3):235–247. https://doi.org/10.1111/j.1745-4549.1977.tb00326.x

Knorr D, Zenker M, Heinz V, Lee D-U (2004) Applications and potential of ultrasonics in food processing. Trends Food Sci Technol 15(5):261–266. https://doi.org/10.1016/j.tifs.2003.12.001

Kostakis T, Ettelaie R, Murray BS (2006) Effect of high salt concentrations on the stabilization of bubbles by silica particles. Langmuir 22(3):1273–1280. https://doi.org/10.1021/la052193f

Krägel J, Derkatch S, Miller R (2008) Interfacial shear rheology of protein–surfactant layers. Adv Colloid Interface Sci 144(1-2):38–53

Kudryashova EV, Visser AJWG, van Hoek A, de Jongh HHJ (2007) Molecular details of ovalbumin−pectin complexes at the air/water interface: a spectroscopic study. Langmuir 23(15):7942–7950. https://doi.org/10.1021/la700379m

Lam S, Velikov KP, Velev OD (2014) Pickering stabilization of foams and emulsions with particles of biological origin. Curr Opin Colloid Interface Sci 19(5):490–500. https://doi.org/10.1016/j.cocis.2014.07.003

Larson RG (1999) The structure and rheology of complex fluids. Books24x7.com, Norwood, MA http://www.books24x7.com/marc.asp?bookid=9316

Lazidis A, Hancocks RD, Spyropoulos F, Kreuß M, Berrocal R, Norton IT (2015) Whey protein fluid gels for the stabilisation of foams. Food Hydrocoll. https://doi.org/10.1016/j.foodhyd.2015.02.022

Lazidis A, de Almeida Parizotto L, Spyropoulos F, Norton IT (2017) Microstructural design of aerated food systems by soft-solid materials. Food Hydrocoll 73:110–119. https://doi.org/10.1016/j.foodhyd.2017.06.032

Lech FJ, Delahaije RJBM, Meinders MBJ, Gruppen H, Wierenga PA (2016) Identification of critical concentrations determining foam ability and stability of β-lactoglobulin. Food Hydrocoll 57:46–54. https://doi.org/10.1016/j.foodhyd.2016.01.005

Lechevalier V, Croguennec T, Pezennec S, Guérin-Dubiard C, Pasco M, Nau F (2003) Ovalbumin, ovotransferrin, lysozyme: three model proteins for structural modifications at the air−water interface. J Agric Food Chem 51(21):6354–6361. https://doi.org/10.1021/jf034184n

Li Q, Zhao Z (2017) Formation of lactoferrin/sodium caseinate complexes and their adsorption behaviour at the air/water interface. Food Chem 232:697–703. https://doi.org/10.1016/j.foodchem.2017.04.072

Li R, Chen X e, Chang Y, Zhang L, Zhang Y, Zhu Y, Wang T (2017) Increase of bubble size playing a critical role in foam-induced protein aggregation: aggregation of BSA in foam fractionation. Chem Eng Sci 174:387–395. https://doi.org/10.1016/j.ces.2017.09.036

Lin M, Tay SH, Yang H, Yang B, Li H (2017) Development of eggless cakes suitable for lacto-vegetarians using isolated pea proteins. Food Hydrocoll 69:440–449. https://doi.org/10.1016/j.foodhyd.2017.03.014

Linder MB (2009) Hydrophobins: proteins that self assemble at interfaces. Curr Opin Colloid Interface Sci 14(5):356–363. https://doi.org/10.1016/j.cocis.2009.04.001

Lomolino G, Vincenzi S, Gazzola D, Crapisi A, Curioni A (2015) Foaming properties of potato (Solanum tuberosum) proteins: a study by the gas sparging method. Colloids Surf A Physicochem Eng Asp 475:75–83. https://doi.org/10.1016/j.colsurfa.2015.01.093

Madadlou A, Rakhshi E, Abbaspourrad A (2016) Engineered emulsions for obesity treatment. Trends Food Sci Technol 52:90–97

Martin AH, Grolle K, Bos MA, Stuart MAC, van Vliet T (2002) Network forming properties of various proteins adsorbed at the air/water interface in relation to foam stability. J Colloid Interface Sci 254(1):175–183. https://doi.org/10.1006/jcis.2002.8592

Martínez KD, Ganesan V, Pilosof AMR, Harte FM (2011) Effect of dynamic high-pressure treatment on the interfacial and foaming properties of soy protein isolate–hydroxypropylmethylcelluloses systems. Food Hydrocoll 25(6):1640–1645. https://doi.org/10.1016/j.foodhyd.2011.02.013

McClements DJ (2006) Non-covalent interactions between proteins and polysaccharides. Biotechnol Adv 24(6):621–625

McSweeney SL (2008) Emulsifiers in infant nutritional products. In: Food emulsifiers and their applications. Springer, New York, NY, pp 233–261. https://doi.org/10.1007/978-0-387-75284-6_8

Medrano A, Abirached C, Panizzolo L, Moyna P, Añón MC (2009) The effect of glycation on foam and structural properties of β-lactoglobulin. Food Chem 113(1):127–133. https://doi.org/10.1016/j.foodchem.2008.07.036

Mita T, Ishida E, Matsumoto H (1978) Physicochemical studies on wheat protein foams. II Relationship between bubble size and stability of foams prepared with gluten and gluten components. J Colloid Interface Sci 64(1):143–153. https://doi.org/10.1016/0021-9797(78)90344-2

Monahan FJ, German JB, Kinsella JE (1995) Effect of pH and temperature on protein unfolding and thiol/disulfide interchange reactions during heat-induced gelation of whey proteins. J Agric Food Chem 43(1):46–52. https://doi.org/10.1021/jf00049a010

Morales R, Martínez KD, Pizones Ruiz-Henestrosa VM, Pilosof AMR (2015) Modification of foaming properties of soy protein isolate by high ultrasound intensity: particle size effect. Ultrason Sonochem 26:48–55. https://doi.org/10.1016/j.ultsonch.2015.01.011

Mulvihill DM, Donovan M (1987) Whey proteins and their thermal denaturation—a review. Irish J Food Sci Technol 11(1):43–75

Mune Mune MA (2015) Optimizing functional properties of bambara bean protein concentrate by enzymatic hydrolysis using Pancreatin. J Food Process Preserv 39(6):2572–2580. https://doi.org/10.1111/jfpp.12507

Murray BS (2002) Interfacial rheology of food emulsifiers and proteins. Curr Opin Colloid Interface Sci 7(5–6):426–431. https://doi.org/10.1016/S1359-0294(02)00077-8

Murray BS, Ettelaie R (2004) Foam stability: proteins and nanoparticles. Curr Opin Colloid Interface Sci 9:314–320. https://doi.org/10.1016/j.cocis.2004.09.004

Nan F, Wu J, Qi F, Liu Y, Ngai T, Ma G (2014) Uniform chitosan-coated alginate particles as emulsifiers for preparation of stable pickering emulsions with stimulus dependence. Colloids Surf A Physicochem Eng Asp 456:246–252. https://doi.org/10.1016/j.colsurfa.2014.05.017

Oliver CM, Melton LD, Stanley RA (2006) Creating proteins with novel functionality via the Maillard reaction: a review. Crit Rev Food Sci Nutr 46(4):337–350. https://doi.org/10.1080/10408690590957250

Panda H (2010) The complete book on gums and stabilizers for food industry. Asia Pacific Business Press Inc, Delhi

Panyam D, Kilara A (1996) Enhancing the functionality of food proteins by enzymatic modification. Trends Food Sci Technol 7(4):120–125. https://doi.org/10.1016/0924-2244(96)10012-1

Patel MT, Kilara A, Huffman LM, Hewitt SA, Houlihan AV (1990) Studies on whey protein concentrates. 1. Compositional and thermal properties. J Dairy Sci 73(6):1439–1449. https://doi.org/10.3168/jds.S0022-0302(90)78808-X

Pernell CW, Foegeding EA, Luck PJ, Davis JP (2002) Properties of whey and egg white protein foams. Colloids Surf A Physicochem Eng Asp 204(1–3):9–21. https://doi.org/10.1016/S0927-7757(01)01061-5

Phillips GO, Williams PA (2009) Handbook of hydrocolloids, 2nd edn. Woodhead, Cambridge

Pickering SU (1907) CXCVI—emulsions. J Chem Soc Trans 91:2001–2021. https://doi.org/10.1039/CT9079102001

Pimentel D, Pimentel M (2003) Sustainability of meat-based and plant-based diets and the environment. Am J Clin Nutr 78(3):660S–663S. https://doi.org/10.1093/ajcn/78.3.660S

Pizones Ruíz-Henestrosa V, Carrera Sánchez C, Rodríguez Patino JM (2007) Formulation engineering can improve the interfacial and foaming properties of soy globulins. J Agric Food Chem 55(15):6339–6348. https://doi.org/10.1021/jf070918a

Pugh RJ (2016) Bubble and foam chemistry. Cambridge University Press, Cambridge

Quester S, Dahesh M, Strey R (2014) Microcellular foams made from gliadin. Colloid Polym Sci 292(9):2385–2389. https://doi.org/10.1007/s00396-014-3317-6

Raffa P, Wever DAZ, Picchioni F, Broekhuis AA (2015) Polymeric surfactants: synthesis, properties, and links to applications. Chem Rev 115(16):8504–8563. https://doi.org/10.1021/cr500129h

Raikos V, Neacsu M, Russell W, Duthie G (2014) Comparative study of the functional properties of lupin, green pea, fava bean, hemp, and buckwheat flours as affected by pH. Food Sci Nutr 2(6):802–810. https://doi.org/10.1002/fsn3.143

Ralet M-C, Guéguen J (2001) Foaming properties of potato raw proteins and isolated fractions. LWT Food Sci Technol 34(4):266–269. https://doi.org/10.1006/fstl.2000.0684

Ramsden W (1903) Separation of solids in the surface-layers of solutions and "Suspensions" (observations on surface-membranes, bubbles, emulsions, and mechanical coagulation)—preliminary account. Proc Roy Soc Lond A 72:156–164

Rodríguez Niño MR, Sánchez CC, Ruíz-Henestrosa VP, Patino JMR (2005) Milk and soy protein films at the air–water interface. Food Hydrocoll 19(3):417–428. https://doi.org/10.1016/j.foodhyd.2004.10.008

Rodríguez Patino JM, Carrera Sánchez C, Rodríguez Niño MR (2008) Implications of interfacial characteristics of food foaming agents in foam formulations. Adv Colloid Interface Sci 140(2):95–113. https://doi.org/10.1016/j.cis.2007.12.007

Rovers TA, Sala G, Van der Linden E, Meinders MB (2016) Potential of microbubbles as fat replacer: effect on rheological, tribological and sensorial properties of model food systems. J Texture Stud 47(3):220–230

Roy F, Boye JI, Simpson BK (2010) Bioactive proteins and peptides in pulse crops: pea, chickpea and lentil. Food Res Int 43(2):432–442. https://doi.org/10.1016/j.foodres.2009.09.002

Rumpold BA, Schlüter OK (2013) Nutritional composition and safety aspects of edible insects. Mol Nutr Food Res 57(5):802–823

Sagis LM, Scholten E (2014) Complex interfaces in food: structure and mechanical properties. Trends Food Sci Technol 37(1):59–71

Sagis LM, de Groot-Mostert AE, Prins A, van der Linden E (2001) Effect of copper ions on the drainage stability of foams prepared from egg white. Colloids Surf A Physicochem Eng Asp 180(1-2):163–172. https://doi.org/10.1016/S0927-7757(00)00734-2

Saint-Jalmes A (2006) Physical chemistry in foam drainage and coarsening. Soft Matter 2(10):836–849

Saint-Jalmes A, Peugeot M-L, Ferraz H, Langevin D (2005) Differences between protein and surfactant foams: microscopic properties, stability and coarsening. Colloids Surf A Physicochem Eng Asp 263(1–3):219–225. https://doi.org/10.1016/j.colsurfa.2005.02.002

Sánchez CC, Rodriguez Nino MR, Molina Ortiz SE, Anon MC, Rodriguez Patino JM (2004) Soy globulin spread films at the air–water interface. Food Hydrocoll 18(2):335–347. https://doi.org/10.1016/S0268-005X(03)00089-4

Schmidt I, Novales B, Boué F, Axelos MAV (2010) Foaming properties of protein/pectin electrostatic complexes and foam structure at nanoscale. J Colloid Interface Sci 345(2):316–324. https://doi.org/10.1016/j.jcis.2010.01.016

Schmidt JM, Damgaard H, Greve-Poulsen M, Larsen LB, Hammershøj M (2018) Foam and emulsion properties of potato protein isolate and purified fractions. Food Hydrocoll 74:367–378. https://doi.org/10.1016/j.foodhyd.2017.07.032

Schmitt C, Kolodziejczky E (2009) Protein-polysaccharide complexes: from basics to food applications gums and stabilisers for the food industry, vol. 15. The Royal Society of Chemistry, London, pp 211–222

Schmitt C, Kolodziejczyk E, Leser ME (2005) Interfacial and foam stabilization properties of b-lactoglobulin e Acacia gum electrostatic complexes. Food colloids: interactions, microstructure and processing, pp 289–300

Schmitt C, Bovay C, Vuilliomenet A-M, Rouvet M, Bovetto L (2011) Influence of protein and mineral composition on the formation of whey protein heat-induced microgels. Food Hydrocoll 25(4):558–567. https://doi.org/10.1016/j.foodhyd.2010.05.010

Schober TJ (2009) Manufacture of gluten-free specialty breads and confectionery products. In: Gluten free food science and technology, pp 130–180

Seguchi M (2001) Oil binding ability of chlorinated and heated wheat starch granules and their use in breadmaking and pancake baking. Starch Stärke 53(9):408–413

Shah BR, Li Y, Jin W, An Y, He L, Li Z, Li B et al (2016) Preparation and optimization of pickering emulsion stabilized by chitosan-tripolyphosphate nanoparticles for curcumin encapsulation. Food Hydrocoll 52:369–377. https://doi.org/10.1016/j.foodhyd.2015.07.015

Sonnewald U, Sturm A, Chrispeels MJ, Willmitzer L (1989) Targeting and glycosylation of patatin the major potato tuber protein in leaves of transgenic tobacco. Planta 179(2):171–180. https://doi.org/10.1007/BF00393687

Sosulski FW, McCurdy A (1987) Functionality of flours, protein fractions and isolates from field peas and faba bean. J Food Sci 52(4):1010–1014. https://doi.org/10.1111/j.1365-2621.1987.tb14263.x

Sotelo A, Adsule RN (1996) Chickpea. In: Smartt J, Nwokolo E (eds) Food and feed from legumes and oilseeds. Springer US, Basel, pp 82–89 www.springer.com/gp/book/9780412459306

Stantiall SE, Dale KJ, Calizo FS, Serventi L (2018) Application of pulses cooking water as functional ingredients: the foaming and gelling abilities. Eur Food Res Technol 244(1):97–104. https://doi.org/10.1007/s00217-017-2943-x

Stephen AM (1995) Food polysaccharides and their applications (vol. 67). CRC, Boca Raton, FL

Stone AK, Karalash A, Tyler RT, Warkentin TD, Nickerson MT (2015) Functional attributes of pea protein isolates prepared using different extraction methods and cultivars. Food Res Int 76:31–38. https://doi.org/10.1016/j.foodres.2014.11.017

Suárez G, Gutiérrez TJ (2017) Recent advances in the development of biodegadable films and foams from cassava starch. In: Klein C (ed) Handbook on Cassava: production, potential uses and recent advances. Nova Science, New York, pp 297–312 EE.UU. ISBN: 978-1-53610-307-6

Tadros TF, Vandamme A, Levecke B, Booten K, Stevens CV (2004) Stabilization of emulsions using polymeric surfactants based on inulin. Adv Colloid Interface Sci 108–109:207–226. https://doi.org/10.1016/j.cis.2003.10.024

Tchuenbou-Magaia FL, Norton IT, Cox PW (2009) Hydrophobins stabilised air-filled emulsions for the food industry. Food Hydrocoll 23(7):1877–1885. https://doi.org/10.1016/j.foodhyd.2009.03.005

Testouri A, Honorez C, Barillec A, Langevin D, Drenckhan W (2010) Highly structured foams from chitosan gels. Macromolecules 43(14):6166–6173

Thewissen BG, Celus I, Brijs K, Delcour JA (2011) Foaming properties of wheat gliadin. J Agric Food Chem 59(4):1370–1375. https://doi.org/10.1021/jf103473d

This H (2010) Kitchen mysteries: revealing the science of cooking = Les secrets de la casserole. Columbia University Press, Chichester

Tingaut P, Zimmermann T, Sèbe G (2012) Cellulose nanocrystals and microfibrillated cellulose as building blocks for the design of hierarchical functional materials. J Mater Chem 22(38):20105–20111

Toews R, Wang N (2013) Physicochemical and functional properties of protein concentrates from pulses. Food Res Int 52(2):445–451. https://doi.org/10.1016/j.foodres.2012.12.009

Tömösközi S, Lásztity R, Haraszi R, Baticz O (2001) Isolation and study of the functional properties of pea proteins. Food Nahrung 45(6):399–401 10.1002/1521-3803(20011001)45:6<399::AID-FOOD399>3.0.CO;2-0

Tontul İ, Kasimoglu Z, Asik S, Atbakan T, Topuz A (2017) Functional properties of chickpea protein isolates dried by refractance window drying. Int J Biol Macromol. https://doi.org/10.1016/j.ijbiomac.2017.11.135

Tsoukala A, Papalamprou E, Makri E, Doxastakis G, Braudo EE (2006) Adsorption at the air–water interface and emulsification properties of grain legume protein derivatives from pea and broad bean. Colloids Surf B Biointerfaces 53(2):203–208. https://doi.org/10.1016/j.colsurfb.2006.08.019

Turgeon SL, Schmitt C, Sanchez C (2007) Protein–polysaccharide complexes and coacervates. Curr Opin Colloid Interface Sci 12(4–5):166–178. https://doi.org/10.1016/j.cocis.2007.07.007

Twine R (2018) Materially constituting a sustainable food transition: the case of Vegan eating practice. Sociology 52(1):166–181. https://doi.org/10.1177/0038038517726647

Tzitzikas EN, Vincken J-P, de Groot J, Gruppen H, Visser RGF (2006) Genetic variation in pea seed globulin composition. J Agric Food Chem 54(2):425–433. https://doi.org/10.1021/jf0519008

Tzoumaki MV, Karefyllakis D, Moschakis T, Biliaderis CG, Scholten E (2015) Aqueous foams stabilized by chitin nanocrystals. Soft Matter 11(31):6245–6253

Utsumi S, Kinsella JE (1985) Structure-function relationships in food proteins: subunit interactions in heat-induced gelation of 7S, 11S, and soy isolate proteins. J Agric Food Chem 33(2):297–303. https://doi.org/10.1021/jf00062a035

Van de Noort M (2017) Chapter 10. Lupin: an important protein and nutrient source. In: Nadathur SR, Wanasundara JPD, Scanlin L (eds) Sustainable protein sources. Academic Press, San Diego, pp 165–183. https://doi.org/10.1016/B978-0-12-802778-3.00010-X

van Koningsveld GA, Walstra P, Gruppen H, Wijngaards G, van Boekel MAJS, Voragen AGJ (2002) Formation and stability of foam made with various potato protein preparations. J Agric Food Chem 50(26):7651–7659. https://doi.org/10.1021/jf025587a

Veraverbeke WS, Delcour JA (2002) Wheat protein composition and properties of wheat glutenin in relation to breadmaking functionality. Crit Rev Food Sci Nutr 42(3):179–208. https://doi.org/10.1080/10408690290825510

Walstra P (1989) Principles of foams formation and stability. In: Wilson AJ (ed) Foams: physics, chemistry and structure. Springer, London, pp 1–15

Walstra P (2003a) Physical chemistry of foods. CRC, Boca Raton, FL

Walstra P (2003b) Studying food colloids: past, present and future. In: Dickinson E, Van Vliet T (eds) Food colloids, biopolymers and materials. Royal Society of Chemistry, Cambridge, pp 391–400 http://ebook.rsc.org/?DOI=10.1039/9781847550835-00391

Walstra P, Roos ALD (1993) Proteins at air-water and oil-water interfaces: static and dynamic aspects. Food Rev Intl 9(4):503–525. https://doi.org/10.1080/87559129309540976

Wang C, Tian Z, Chen L, Temelli F, Liu H, Wang Y (2010) Functionality of barley proteins extracted and fractionated by alkaline and alcohol methods. Cereal Chem 87(6):597–606. https://doi.org/10.1094/CCHEM-06-10-0097

Wang L-J, Hu Y-Q, Yin S-W, Yang X-Q, Lai F-R, Wang S-Q (2015) Fabrication and characterization of antioxidant pickering emulsions stabilized by zein/chitosan complex particles (ZCPs). J Agric Food Chem 63(9):2514–2524

Waniska RD, Kinsella JE (1979) Foaming properties of proteins: evaluation of a column aeration apparatus using ovalbumin. J Food Sci 44(5):1398–1402. https://doi.org/10.1111/j.1365-2621.1979.tb06447.x

Ward PP, Paz E, Conneely OM (2005) Lactoferrin. Cell Mol Life Sci 62(22):2540. https://doi.org/10.1007/s00018-005-5369-8

Weaire DL, Hutzler S (1999) The physics of foams. Oxford University Press, Oxford

Wege HA, Kim S, Paunov VN, Zhong Q, Velev OD (2008) Long-term stabilization of foams and emulsions with in-situ formed microparticles from hydrophobic cellulose. Langmuir 24(17):9245–9253

Wessels JGH (1996) Fungal hydrophobins: proteins that function at an interface. Trends Plant Sci 1(1):9–15. https://doi.org/10.1016/S1360-1385(96)80017-3

Wierenga PA, Gruppen H (2010) New views on foams from protein solutions. Curr Opin Colloid Interface Sci 15(5):365–373. https://doi.org/10.1016/j.cocis.2010.05.017

Wilde P (2012) Foam formation in dough and bread quality. In: Breadmaking: improving quality, pp 370–399

Wilde P, Mackie A, Husband F, Gunning P, Morris V (2004) Proteins and emulsifiers at liquid interfaces. Adv Colloid Interface Sci 108–109:63–71. https://doi.org/10.1016/j.cis.2003.10.011

Wilson JC (1981) A study of particulate foams. University of Bristol, Bristol

Wouters AGB, Rombouts I, Fierens E, Brijs K, Blecker C, Delcour JA, Murray BS (2017) Foaming and air-water interfacial characteristics of solutions containing both gluten hydrolysate and egg white protein. Food Hydrocoll. https://doi.org/10.1016/j.foodhyd.2017.09.033

Yu MA, Damodaran S (1991) Kinetics of destabilization of soy protein foams. J Agric Food Chem 39(9):1563–1567. https://doi.org/10.1021/jf00009a005

Zhang Z, Dalgleish D, Goff H (2004) Effect of pH and ionic strength on competitive protein adsorption to air/water interfaces in aqueous foams made with mixed milk proteins. Colloids Surf B Biointerfaces 34(2):113–121. https://doi.org/10.1016/j.colsurfb.2003.11.009

Zhang H, Xu G, Liu T, Xu L, Zhou Y (2013) Foam and interfacial properties of Tween 20–bovine serum albumin systems. Colloids Surf A Physicochem Eng Asp 416:23–31. https://doi.org/10.1016/j.colsurfa.2012.10.028

Zhu H, Damodaran S (1994a) Effects of calcium and magnesium ions on aggregation of whey protein isolate and its effect on foaming properties. J Agric Food Chem 42(4):856–862. https://doi.org/10.1021/jf00040a003

Zhu H, Damodaran S (1994b) Heat-induced conformational changes in whey protein isolate and its relation to foaming properties. J Agric Food Chem 42(4):846–855. https://doi.org/10.1021/jf00040a002

Chapter 12
Biodegradable Foams in the Development of Food Packaging

Suzana Mali

Abstract The use of biodegradable and renewable-source polymers as substitutes for petroleum-based plastic packaging has been increasingly discussed by the academic community over the last few decades, particularly application in food packaging. Between the conventional plastic materials that have been used for food packaging, expanded polystyrene (EPS) is not degradable and is difficult to recycle, creating environmental issues. The development of biodegradable materials is an alternative to replace EPS. This chapter discusses recent advances in the development of biodegradable foams for food packaging, the main processes employed to produce foams with different cellular structures and properties, such as extrusion and baking, polymers that have been used to produce these foams, and their resulting water sorption capacity, morphological, mechanical and barrier properties.

Keywords Baking · Biodegradability · Extrusion

12.1 Introduction

The use of biodegradable materials for food packaging has attracted the attention of researchers all over the world, and several authors have reported on production of biodegradable foams for this purpose (Vercelheze et al. 2012; Matsuda et al. 2013; Debiagi et al. 2014, 2015; Mello and Mali 2014; Carvalho et al. 2017; Suárez and Gutiérrez 2017).

Biodegradable foams can be employed as substitutes for expanded polystyrene (EPS) products, which are largely used for food storage but represent an important disposal problem for companies and municipalities; EPS products are lightweight and bulky and are not a viable economic or environmentally responsible method due to expensive handling and transportation costs, leading to serious environmental problems (Nabar et al. 2005; Machado et al. 2017), especially in short-term applications (Kaseem et al. 2012).

S. Mali (✉)
Department of Biochemistry and Biotechnology, State University of Londrina, Paraná, Brazil
e-mail: smali@uel.br

T. J. Gutiérrez (ed.), *Polymers for Food Applications*,
https://doi.org/10.1007/978-3-319-94625-2_12

Among the biodegradable polymers employed for foam production, starch is a promising material and has been studied in combination with other polymers, additives and fillers to improve the properties of the final product. Polyvinyl alcohol (PVOH) and other biodegradable polyesters, such as poly (lactic acid) (PLA), have been employed to obtain biodegradable foams, mostly as part of starch blends.

Food packaging materials must meet the storage needs of the food product, possessing mechanical stability throughout the storage time and under the storage conditions (relative humidity and temperature), preserving the physical and chemical characteristics inherent to the food, as well as maintaining its microbiological safety.

This chapter discusses recent advances in the development of biodegradable foams for food packaging, the processes employed for foam production, such as extrusion and baking, polymers that have been used to produce these foams, and their resulting water sorption capacity, morphological, mechanical and barrier properties.

12.2 Processes to Produce Biodegradable Foams

Biodegradable foams can be produced through several processes described in the literature, such as extrusion, baking (thermoforming), microwave, freeze-drying/solvent exchange and supercritical fluid extrusion (Soykeabkaew et al. 2015); however, here, we focus on the use of extrusion and baking processes, which are the most promising for large scale use.

The formation of a foam structure involves nucleation, growth and stabilization of gas bubbles within a polymeric matrix (Bonin 2010). According to Hepburn and Alam (1991), foam structure can be divided into two classes: open-cell foams, in which the various cells are interconnected and closed-cell foams, in which cells are not connected to one another.

Foam properties vary extensively depending on processing conditions, polymer type and materials formulations. Density, cell size and shape, mechanical properties and water sensitivity are important properties that need to be monitored during development and use of these materials for food packaging.

Density is one of the most important characteristics of a foam and is defined as the ratio between mass and volume of a material. For foams obtained by extrusion, bulk density has been reported by several authors, and it is measured (calculated) by packing many small foam pieces together into a container with a known volume. This technique is useful for extruded materials because is very difficult to measure the volume of an individual piece. Several investigators have also reported measuring the expansion ratio for extruded samples, which is defined as the ratio of the cross-sectional area of the extruded foam to that of the extruder die. Thus, the higher expansion ratio, the lower the density (Bhatnagar and Hanna 1995, 1996; Cha et al. 2001; Nabar et al. 2005, 2006; Xu and Hanna 2005; Zhang and Sun 2007a, b; Bonin

2010; Mali et al. 2010a; Debiagi et al. 2011). According to Tatarka and Cunningham (1998), the bulk density of extruded starch foams ranges from 0.017 to 0.023 g/cm^3.

For baked foams, several authors (Vercelheze et al. 2012; Kaisangsri et al. 2012, 2014; Matsuda et al. 2013; Marengo et al. 2013; Mello and Mali 2014; Debiagi et al. 2015; Carvalho et al. 2017) reported measuring density employing a method described by Shogren et al. (1998) in which density is calculated from foams dimensions by the ratio between weight and volume of each sample specimen. According to Glenn et al. (2001), the density of baked starch foams is approximately 0.063 g/cm^3.

The morphology of extruded and baked foams, including cell size and shape and obtaining open or closed cells structures, has been reported by several authors using scanning electron microscopy (Cinelli et al. 2006; Vercelheze et al. 2012; Matsuda et al. 2013; Marengo et al. 2013; Mello and Mali 2014; Zhang and Sun 2007a, b; Wang et al. 2017). Tatarka and Cunningham (1998) compared properties of EPS-based foams and starch-based foams manufactured by extrusion, concluding that all starch-based foams have higher open-cell content than EPS-based foams, a finding related to the poor melt strength of starches that results in rupture of the foam cell walls. These authors also reported that the foam density of starch-based products was much higher than for EPS-based foams, and the open cells created during expansion prevent the foam from continuing to expand thereafter.

With respect to mechanical properties, most studies have concentrated on different approaches to increase resistance and decrease brittleness, an inherent characteristic of foams based on starch and other biodegradable polymers.

12.2.1 Use of Extrusion to Produce Biodegradable Foams

Starch was the first material employed to obtain fully biodegradable foams by extrusion processing technology, which has been used since the late 1980s to develop starch-based plastic foams as alternative to EPS (Cha et al. 2001). Starches from different botanical sources have been used for the production of biodegradable packaging for multiple reasons, including their biodegradability, renewable source, low cost and wide availability (Kaseem et al. 2012).

Native starches are homopolysaccharides formed by units of glucose, mostly composed of linear amylose and highly branched amylopectin. Native starches are present in granules (2–100 μm) that vary in their amylose/amylopectin ratio, composition, size, shape, and functionality depending upon their botanical source. Native starches are semi-crystalline structures consisting of concentric alternating amorphous and semi-crystalline growth rings (120–400 nm) composed of blocklets (20–50 nm) formed by a cluster structure of alternating crystalline and amorphous lamellae (9–10 nm) containing amylopectin and amylose chains (0.1–1 nm) (Le Corre and Angellier-Coussy 2014).

Native starches cannot be processed for use as thermoplastic materials due to their strong intermolecular and intramolecular hydrogen bonds, meaning that producing a thermoplastic material from starch requires plasticizers, such as water or glycerol. To obtain a thermoplastic starch-based material, its semi-crystalline granular structure must be destroyed, giving rise to a homogeneous and essentially amorphous polymer matrix. Starch granules are destructurized in the presence of water and plasticizers, giving rise to a continuous phase in the form of a viscous melt that can be processed using the traditional plastic processing techniques, such as extrusion. This type of starch material is called thermoplastic starch (Aichholzer and Fritz 1998; Kaseem et al. 2012).

The phenomena that allow destruction of starch granule organization are gelatinization and melting. Gelatinization is the irreversible transformation of the granular starch into a viscoelastic paste, a process that occurs in the presence of excess water under the action of heat and shear, leading to destruction of the crystallinity and the molecular order of the granule through rupture of the hydrogen bonds maintaining its integrity. On the other hand, when the starch is heated in the presence of small amounts of water, melting occurs, which requires much higher temperatures than for gelatinization, and requires presence of plasticizers (Mali et al. 2010b). After gelatinization and melting, the starch molecules begin to re-associate through hydrogen bonds, favoring formation of a more ordered structure, which, under favorable conditions, can form a new crystalline structure; this set of changes is called retrogradation or recrystallization (Zobel 1964; Van Soest et al. 1996).

According to Shogren (1992), the melting temperature of native corn starch with low water content (11–30%) ranged from 190 to 200 °C, and gelatinization temperature with moisture content >30% was approximately 70 °C. Fujita and Fujiyama (1993) reported that gelatinization temperatures of starches from different botanical sources range between 57 and 80 °C.

Biodegradable foams produced exclusively with thermoplastic starch are brittle, yielding materials with poor processability and moldability, which is a limiting factor for production. Starch foams are also water sensitive and disintegrate when exposed to moisture, thus limiting their use in humid environments (Bhatnagar and Hanna 1996; Soykeabkaew et al. 2015). Novel approaches have been reported in the literature to improve mechanical properties (brittleness), water resistance and other properties of starch-based foams, including chemical modification of starches, blending with additional biodegradable polymers, incorporation of natural fibers, and addition of fillers and nanofillers (Soykeabkaew et al. 2015).

Extrusion technology is a high-temperature, short-duration, continuous process with the advantages of high versatility and absence of effluents (Harper 1981). This technology combines temperature, pressure and shear rate, and it has been employed to produce several types of biodegradable materials since the 1970s.

In a typical extrusion process to obtain starch foams, starch is mixed with water, which acts as a blowing agent and plasticizer, as well as other plasticizers, polymers and additives. These mixtures contain approximately 12–30% moisture, and they are melted, employing temperatures approximately 120–180 °C and screw speeds from 70 to 400 rpm, allowing the blowing agent to diffuse within the melted polymeric matrix. When the melted polymer begins emerging from the extruder, it is subjected to rapid decompression, resulting in nucleation or bubble formation.

These bubbles expand, grow and stabilize as the blowing agent diffuses into them (Bhatnagar and Hanna 1995, 1996; Cha et al. 2001; Willett and Shogren 2002; Nabar et al. 2005, 2006; Xu and Hanna 2005; Bonin 2010; Mali et al. 2010a; Debiagi et al. 2011; Bénézet et al. 2012).

According to Zhang and Sun (2007a), an increase in water content during extrusion of starch implies results in a decrease of its melting point, and also in a decrease in foam expansion because the low viscosity of melted polymeric matrix, which favors the water loss and the foam cell collapse.

When other polymers, such as PVOH or PLA were processed by extrusion, the extrusion conditions have to be sufficient to melt these polymers, and depending on variations in their structure their melting points can change. According to Wang et al. (2017), PLA melting temperature ranges from 150 to 220 °C, while the melting temperature of PVOH ranges from 180 to 240 °C (Tang and Alavi 2011), and also the use of plasticizers affects the melting temperatures of these polymers.

Extrusion conditions (melt pressure, temperature, and screw speed) and material compositions (polymer type, feed moisture, blowing and nucleating agents) can significantly affect the properties of extruded foams. Several investigators have reported the use of blowing agents, including water and other compounds, to obtain foam materials. Blowing agents directly affect the density of foams, and the use of nucleating agents has been reported by several authors (Boehmer and Hanlon 1993; Bhatnagar and Hanna 1993, 1995, 1996; Cha et al. 2001; Guan and Hanna 2004, 2006; Nabar et al. 2005, 2006; Zhang and Sun 2007a, b).

Blowing agents can be divided into two main subgroups: physical and chemical. Physical blowing agents consist of compressed gasses, such as nitrogen, carbon dioxide and oxygen, that expand when pressure is released (Bonin 2010). Chemical blowing agents are solids or liquids at room temperature that, upon heating, release a gas. The main advantage of chemical agents over gaseous agents is that the former can be added to the solid plastic material prior to thermal processing, while gaseous agents must be injected into the already fluidized plastic material (Fisk 2002). Water has typically been used as a blowing agent, as well as a plasticizer, for extruded starch-based foams (Nabar et al. 2005, 2006; Zhang and Sun 2007a, b), with the advantages of being inexpensive and environmentally friendly. Other chemical solvents have been reported to obtain starch foams, including methanol (Cha et al. 2001), ethanol (Guan and Hanna 2004, 2006), azidocarbonamide (Bhatnagar and Hanna 1996; Zimmermann et al. 2013), and a mixture of sodium bicarbonate and citric acid (Zimmermann et al. 2013).

Addition of nucleating agents aids in the formation of a larger number of smaller cells in foams by providing nucleation sites for water vapor, resulting in improved uniformity of extruded foam structures (Soykeabkaew et al. 2015). Magnesium silicate (talc) is generally used in starch-based materials as a nucleating agent (Bhatnagar and Hanna 1996; Guan and Hanna 2004, 2006; Nabar et al. 2005, 2006; Zhang and Sun 2007a, b).

12.2.2 Use of Baking to Produce Biodegradable Foams

Until the 1990s, the most efficient process for producing starch foams was extrusion; however, resulting extruded foams were brittle and difficult to mold into shaped products. Shogren et al. (1998) and Glenn and Orts (2001) described a method for producing molded starch foams with a cellular structure based on baking, a similar process to that used in making waffles (Tiefenbacher 1993).

For the development of expanded foams by baking, specific conditions are required during processing, such as raising the temperature above that which would gelatinize the starch (or melt other biodegradable polymers), as well as boiling the water (Shogren et al. 1998). With baking, foam shape and thickness are controlled by choosing the geometry of the mold (Soykeabkaew et al. 2015).

The manufacturing process consists of preparing a suspension of starch, additives and water that will be mixed in a mechanical stirrer until a homogeneous mass is obtained. This mass will be added to a preheated mold and subjected to the baking process, standardizing the molding time, temperature and pressure (Schmidt and Laurindo 2010). Several conditions for preparation have been reported, and generally, when the temperature increases, the time decreases. Vercelheze et al. (2012) reported the production of baked starch foams employing 130 °C for 20 min, while Soykeabkaew et al. (2014) reported the preparation of starch foams by baking using 220 °C for only 30 s.

In the baking process, gelatinization of the starch occurs during heating, and excess water is evaporated, causing mass expansion and filling of the mold. Solidification occurs with water evaporation, forming a molded material with a porous structure (Hofmann et al. 1998). Starch pastes must have certain rheological characteristics to prevent collapse as the water evaporates, which is controlled by the water content of starch pastes. Pastes with low water content are very viscous and result in less expandable and higher density foams. In addition, the presence of fibers and other solids in the formulation can be responsible for increasing viscosity of the mixture, which can decrease foaming ability (Shogren et al. 1998; Cinelli et al. 2006; Vercelheze et al. 2012).

Baked foams have a sandwich-type structure with dense outer skins that contain small cells on the surface of the foam. The interior of the foam has large cells with thin walls, a type of structure that has been reported by Cinelli et al. (2006), Vercelheze et al. (2012), and Mello and Mali (2014).

Vercelheze et al. (2012) also reported another parameter of baking concerns the amount of batter used to fill the mold and form complete trays. According to Chiellini et al. (2009), the batter volume must be adjusted when starch is mixed with other components that do not promote the foaming process. Lawton et al. (1999) reported that large amounts of batter increase the quantity of released vapor, leading to broken trays. Some additives have been reported to prevent the starch foam sticking to the mold, such as magnesium stearate; in addition, guar gum is added to prevent solid separation, and glycerol is added as a plasticizer (Shogren et al. 1998; Salgado et al. 2008; Vercelheze et al. 2012; Marengo et al. 2013; Matsuda et al. 2013; Mello and Mali 2014; Carvalho et al. 2017).

Different additives can affect the chemical and physical stability of starch foams. For example, the water resistance of these materials can be increased by coating them with other less hydrophilic polymers such as PVOH (Mali et al. 2016; Carvalho et al. 2017), or by adding proteins and fibers in the formulations (Salgado et al. 2008). Fibers from agroindustrial residues (Marengo et al. 2013; Mello and Mali 2014) or nanoclays (Vercelheze et al. 2012; Matsuda et al. 2013) can also be used to increase water resistance.

12.3 Foams Based on Starch, Starch Blends and Composites

In the 1990s, several studies reported the use of starch blends with synthetic polymers, such as polystyrene, aimed at reducing use of petroleum-based plastic and disintegration of the material, while providing water resistance (Lacourse and Altieri 1991a, b; Chinnaswamy and Hanna 1996; Bhatnagar and Hanna 1993, 1995, 1996; Cha et al. 2001).

US Patent 5,035,930 (Lacourse and Altieri 1991a) and US Patent 5,043,196 (Lacourse and Altieri 1991b) disclosed the use of extrusion for the production of starch foams based on high amylose starch including the addition of synthetics, such as PVOH, polyvinyl acetate and polyurethane, to improve product properties. To impart hydrophobicity to the starch foams, Chinnaswamy and Hanna (1996) also reported the use of starch (70%) and polystyrene (30%) in their patents to obtain extruded foams.

Bhatnagar and Hanna (1993, 1995, 1996) studied production of starch foams by extrusion based on blends of polystyrene and starches from different sources. Despite these foams not being entirely biodegradable, authors reported that the use of 70% of starch in these blends did not affect their properties, and its water solubility indexes were much lower compared to the solubility of starch foams. Density of these materials ranged from 0.030 to 0.132 g/cm^3, higher than commercial polystyrene foams (Bhatnagar and Hanna 1995).

Density is an important physical property of foams. Low density is ideal for these products because it reduces production costs. Several studies reported higher density values for starch-based foams obtained by extrusion or baking compared to polystyrene foams (Shogren et al. 1998; Cha et al. 2001; Salgado et al. 2008; Mali et al. 2010a; Debiagi and Mali 2012; Vercelheze et al. 2012; Marengo et al. 2013; Mello and Mali 2014; Debiagi et al. 2014, 2015).

Cha et al. (2001) reported the use of blends that were not completely biodegradable containing wheat and corn starch (49%), synthetic polymers—polystyrene and poly (styrene/maleic anhydride) (33%), water (10.5%), blowing agents (7%) and nucleating agents (0.5%) to obtain starch-based foams by extrusion. Starch-foams were four to eight times denser than commercial polystyrene foam, likely due to its moisture content and volume being unaffected by relative humidity, while starch-based foams properties were strongly affected by moisture content employed in the extrusion process and by relative humidity during storage.

Beginning in the late 1990s, some researchers began to prioritize the use of biodegradable polymers in blends with starches from different botanical sources to obtain fully biodegradable foams. Altieri and Tessler (1999) patented biodegradable starch foams with improved water and/or humidity resistant properties obtained by extrusion from blends of starch and starch esters. Bastioli et al. (1994, 1998) patented biodegradable foams produced by extrusion from blends of starch and other biodegradable polymers, such as PVOH, poly (caprolactone) (PCL) and cellulose acetate.

Shogren et al. (1998) employed baking to produce biodegradable foams based on corn, wheat, potato and cassava starches and compared to polystyrene foams, their starch foams had higher densities, lower strengths and a hydrophilic character. This study also suggested that the use of coatings based on more hydrophobic biodegradable polymers might be an alternative method to improve water sensitivity.

Willett and Shogren (2002) obtained extruded foams by blending native corn starch, high amylose corn starch, wheat starch, and potato starch with PVOH, cellulose acetate, and biodegradable polyesters such as poly (lactic acid) (PLA), PCL, poly(hydroxyester ether) (PHEE), poly(ester amide) (PEA), poly(hydroxybutyrate-*co*-valerate) (PHBV), and poly(butylene adipate-*co*-terephthalate) (PBAT) to improve hydrophobicity of starch foams. The density of foams produced with 20% polymers ranged from 0.019 to 0.042 g/cm^3, lower than the density of pure corn starch foam (0.061 g/cm^3), suggesting that the polymer particles might also act as a nucleating agent during extrusion of the blend foams.

Guan and Hanna (2004) reported that one possible alternative to increase the hydrophobicity of starch is to substitute some of the hydroxyl groups on the starch backbone with acetyl groups to prevent hydrogen bond formation with water. However, this approach results in higher costs, and thus, they used blends of both acetylated and native starches to obtain biodegradable foams by extrusion with lower water absorption indexes. Additionally, Xu and Hanna (2005) produced biodegradable foams by extruding starch acetate with poly (tetramethylene adipate-*co*-terephthalate) (EBC). Low EBC content favored miscibility of the two polymers, resulting in formation of a homogeneous morphology observed by scanning electron microscopy. These foams exhibited low densities; however, biodegradation rates of the foams decreased with the addition of EBC to starch acetate.

Nabar et al. (2006) studied production of starch foams with hydrophobic character and were able to improve mechanical properties using other biodegradable polymers, such as PCL, poly (butylene adipate-*co*-terephthalate) (PBAT), cellulose acetate, methylated pectin and PVOH, and cross-linkers such as glyoxal. Formulations with PVOH, polyesters such as PCL and PBAT, and glyoxal with PVOH resulted in materials with lower density values (<0.025 g/cm^3) than control starch films (0.030 g/cm^3). These authors reported that starch foams exposed to a relative humidity of 95% and a temperature of 38 °C lost approximately 50–55% radial and longitudinal dimensions; however, under the same conditions, starch foams with PCL or PBAT resulted in losses of only 12–20%. Nabar et al. (2005) also reported that the hygroscopic character of starch-based foams makes these foams likely to gain weight when exposed to a humid environment. These observations were accompanied by losses in radial and longitudinal dimensions as well.

Utilizing the same strategy of biodegradable polyesters in mixtures with starch with the goal of obtaining superior foams, Zhang and Sun (2007a) produced starch-PLA foams. These foams were less hydrophilic than starch foams and exhibited both open and closed cellular structures observed by scanning electron microscopy. The findings suggested that the cell structure tends toward closed as the starch ratio of the matrix is reduced, whereas the cell wall was constituted primarily of PLA.

Stagner and Narayan (2011) produced biodegradable foams by extrusion of a biodegradable polyester, PBAT, and blends with malleated thermoplastic starch using a chemical blowing agent. In general, increasing levels of PBAT resulted in lower density and higher expansion ratios. Greater concentrations of PBAT in the samples decreased hydrophilicity, and when 5–7% chemical blowing agent is used, the samples exhibit cells throughout the foams matrices.

Another approach to improve the properties of starch foams is incorporation of cellulosic fibers, resulting in composite foams. Cellulosic fibers can be added to composites as reinforcing agents that improve mechanical properties by increasing resistance and flexibility. These fibers also increase degradability and lower the cost of production. Several fiber sources have been employed to reinforce starch foams, including fibers from cellulosic pulps (Salgado et al. 2008; Kaisangsri et al. 2012, 2014), pure cellulose (Guan and Hanna 2006), or natural fibers (Soykeabkaew et al. 2004). Fibers from lignocellulosic residues originating from agriculture or agroindustry have even been utilized (Cinelli et al. 2006; Mali et al. 2010a; Debiagi et al. 2011; Marengo et al. 2013; Mello and Mali 2014; Machado et al. 2017).

Soykeabkaew et al. (2004) reported production of starch foams incorporated with jute or flax fibers. Mechanical properties improved with addition of 5–10% fibers, which was related to the strong interaction between fibers and the starch matrix. Jute fibers had a greater reinforcing effect than did flax fibers, and increases in fiber content and size increased foam density, which ranged from 0.322 to 0.360 g/cm^3 for composite foams compared to 0.214 g/cm^3 for pure starch foams.

Cinelli et al. (2006) produced potato starch-based foam trays with a relatively high content of corn fibers (28.9 to 54.7%); however, the corn fiber did not act as a reinforcing agent for starch-based foams, resulting in a decrease of tensile strength and flexibility of trays with increased fiber content. The addition of PVOH to the starch-corn fiber matrix mitigates the reduction in mechanical properties, and the addition of corn fibers and PVOH together decreases water absorption of the foams.

Biodegradable foam trays based on cassava starch, sunflower proteins and cellulose fibers have been produced employing a baking strategy reported by Salgado et al. (2008). These authors reported that cellulose fibers act as reinforcing agents, with fiber concentrations from 10 to 20%, improving the mechanical properties but increasing water absorption capacity of the material by at least 15%. A formulation containing 20% fiber and 10% protein isolate exhibited the best combination of properties, with a maximal resistance of 6.57 MPa and a 38% reduction in water absorption capacity, corresponding to a more compact, homogeneous and dense microstructure. Density values of these foams ranged from 0.46 to 0.59 g/cm^3, with the highest density values obtained for fiber content up to 15%.

Guan and Hanna (2006) prepared acetylated starch–cellulose foams by extrusion, and the thermal stability of the foams increased with increasing starch acetylation. Mali et al. (2010a) employed extrusion to obtain biodegradable foams made from cassava starch, sugarcane bagasse fibers, and PVOH. The addition of PVOH in high proportions (40%) increased the expansion index and resulted in lower hydrophilicity of starch foams. Furthermore, addition of fibers at low levels (<20%) improved water resistance. The density values of composite foams ranged from 0.20 to 0.33 g/cm^3, and higher levels of fibers (40%) resulted in denser foams. The mechanical properties were affected by both relative humidity of storage and foam formulation. When the relative humidity increased, the foam strength decreased. At high relative humidity (90%), addition of fibers at high concentrations (40%) resulted in a reinforcing effect on starch foams, improving their strength.

Biodegradable composite foams produced by extrusion from a mixture of cassava starch, glycerol (plasticizer) and two different types of lignocellulosic fibers (sugarcane bagasse or oat hulls) were affected by fiber addition, resulting in materials with lower water solubility index when fiber content reached 10% in the polymeric matrix (Debiagi et al. 2010).

Schmidt and Laurindo (2010) reported the use of baking to obtain foam trays using cassava starch, dolomitic limestone and eucalypt cellulose fibers that could be used to pack foodstuffs. The increase in cellulose fibers resulted in a decrease in density and tensile strength of the foams. These results showed that tensile strength of starch foams with 5% cellulose (3.03 MPa) were higher or comparable to expanded polystyrene trays used to pack foods in supermarkets (1.49 MPa); however, elongation of the baked foams was approximately 20% lower than for those made with polystyrene.

Debiagi et al. (2011) studied the effects of cassava starch, PVOH, sugarcane bagasse fibers and chitosan on properties of extruded foams. The addition of starch/PVOH in high proportions decreased density and increased mechanical resistance of studied foams. Fiber addition decreased the density and improved mechanical properties of the foams. All formulations were resistant to moisture content increase up to 75% relative humidity of storage.

Bénézet et al. (2012) employed extrusion to produce foams from potato starch with addition of natural fibers, resulting in decreased density and water adsorption of starch foams. The mechanical properties of these foams were affected by both relative humidity of storage and foam formulation. When relative humidity increased, foam strength decreased. The formulation with the best combination of properties contained 10% hemp fiber and had a maximal resistance of 4.14 MPa, corresponding to a more compact and dense microstructure.

Biodegradable foam trays have also been produced by Kaisangsri et al. (2012), using cassava starch blended with kraft fibers and chitosan by baking. Foam produced from cassava starch with 30% kraft fiber and 4% chitosan exhibited mechanical properties similar to polystyrene foam; however, the water absorption index and the water solubility index were greater for this foam than for polystyrene foam.

Marengo et al. (2013) obtained starch foams using a baking process employing three different agroindustrial lignocellulosic residues (coconut fiber, soy bran and sugarcane bagasse) as reinforcing agents. Their results illustrated that foam properties were significantly affected by the type of residue employed. Starch-soy bran foams exhibited the highest density, due to the higher protein content in this residue. The addition of sugarcane bagasse resulted in less dense and resistant samples, and coconut fiber composites showed the highest tensile strength. The samples obtained from this work could be used as an alternative packaging material with low water content.

Mello and Mali (2014) reported that malt bagasse, a byproduct of the brewing industry, could be used as a reinforcing agent to produce foam trays based on cassava starch by baking. The foams trays formulated with 5 and 10% malt bagasse resulted in the highest production yields (100%), showing a homogeneous distribution of the malt bagasse throughout the polymeric matrix. Variation in relative humidity from 33 to 58% did not affect the mechanical properties of these foams, but tensile strength decreased and elongation increased when the trays were stored at 90% relative humidity. The addition of 10% malt bagasse decreased the hygroscopicity of starch foams, and the addition of bagasse at concentrations up to 15% decreased initial water adsorption rates of the trays.

Kaisangsri et al. (2014) reported the improvement of the mechanical properties and water resistance of baked cassava starch foams due to the addition of bio-based materials such as proteins (gluten and zein), kraft fibers and palm oil, which could be used as an alternative to EPS foam trays for oily and less moist foods. These authors reported that increases in protein content increase the density of starch foams.

Machado et al. (2017) produced baked foams based on cassava starch with the addition of sesame cake, a byproduct of the oil industry. The addition of sesame cake improved mechanical properties and reduced density and water capacity absorption of starch foams; however, the authors conceded that more improvements are needed to increase flexibility and decrease moisture sensibility in these foams. Overall, these foams could be considered an alternative for packing dry foods and foods with low moisture content, reducing the use of EPS.

The addition of fillers at the nanoscale level, such as layered silicates (nanoclays) and nanocelluloses, have also been implemented to produce nanocomposite foams (Lee et al. 2008; Lee and Hanna 2008; Debiagi and Mali 2012; Vercelheze et al. 2012; Matsuda et al. 2013; Silva et al. 2013). Lee et al. (2008) used cassava (tapioca) starch, PLA and four different organic nanoclays (Cloisite® 10A, Cloisite® 25A, Cloisite® 93A and Cloisite® 15A) to produce nanocomposite foams by melt-intercalation, which was observed by the shift of X-ray diffraction peaks to lower angles. The addition of different organoclays into the polymeric matrix of starch and PLA resulted in foams with higher cell density, smaller cell size and more uniformity. Lee and Hanna (2008) produced foams with the same polymeric matrix of starch and PLA and addition of another nanoclay (Cloisite® 30B) and concluded that the thermal stability of foams increased with addition of these nanofillers. Debiagi and Mali (2012) produced expanded nanocomposites based on cassava starch, PVOH and sodium montmorillonite (Cloisite® Na) by extrusion obtaining an intercalated structure. The addition of nanoclays resulted in increased tensile strength and decreased water absorption of the foams.

Vercelheze et al. (2012) developed biodegradable trays from cassava starch, sugarcane fibers and sodium montmorillonite (Cloisite® Na) using a baking process, which allowed a good nanoclay dispersion, leading to the formation of an exfoliated structure. The authors obtained well-shaped trays with densities between 0.1941 and 0.2966 g/cm^3, and incorporation of fibers and nanoclays resulted in less dense and less rigid trays, representing an alternative for packaging of foods with low water content.

Matsuda et al. (2013) produced biodegradable trays of cassava starch and organically modified montmorillonite (Cloisite®10A and 30B) using a baking process, resulting in materials with density values ranging from 0.2809 to 0.3075 g/cm^3. The addition of nanoclays to the starch matrix resulted in foams with lower water absorption capacities without dimensional changes during storage at all relative humidity conditions tested. The nanoclays acted as a reinforcement, increasing tensile strength of the trays.

Silva et al. (2013) produced composite foams trays based on cassava starch reinforced with nanocellulose from bacterial cellulose using a baking process. Bacterial cellulose was incorporated into starch composite foams by two methods: direct incorporation of bacterial cellulose powder into the starch matrix during baking (method 1) or coating the tray surface with bacterial cellulose films (method 2) after they were produced. The addition of bacterial cellulose by method 1 resulted in improved foaming ability of starch, generating trays with lower density. In addition, water absorption capacity was reduced by the incorporation of bacterial cellulose, independent of the method of incorporation.

Considering the use of coatings to improve hydrophilicity of starch foams, Mali et al. (2016) described in a Brazilian patent the developing of cassava starch foam trays with an antimicrobial coating based on PVOH and oregano and clove essential oils. Carvalho et al. (2017) also produced biodegradable foam trays based on cassava starch coated with PVOH with a higher degree of hydrolysis (98%), and they observed an approximately 50% decrease in water absorption capacity of the coated trays compared with the uncoated trays.

12.4 Foams Based on PVOH and Other Biodegradable Polymers

Polyvinyl alcohol (PVOH) is a synthetic, non-toxic, biodegradable, biocompatible promising polymer for the production of biodegradable packaging materials. PVOH is as strong as conventional plastics, showing high mechanical strength, adhesion, high tensile strength and compression, good flexibility, barrier to oxygen, flavors, oils and solvents (Sudhamani et al. 2003; Moraes et al. 2008; Faria et al. 2012; Wang et al. 2014).

Polyvinyl alcohol (PVA) is obtained by partial or complete hydrolysis of polyvinyl acetate to remove acetate groups, and partially hydrolyzed grades contain residual acetate groups (Tang and Alavi 2011). Considering its degree of hydrolysis (DH), PVAs are classified as partially hydrolyzed (above 88%) or highly hydrolyzed (98–99%). Based on their degree of polymerization (DP), PVAs are further classified as very low viscosity (viscosity = 3–4 cP; DP = 150–300), low viscosity (viscosity = 5–6 cP; DP = 350–650), medium viscosity (viscosity = 22–30 cP; DP = 1000–1500) and high viscosity (viscosity = 45–72 cP; DP = 1600–2200) (Maria et al. 2008; Goodship 2009; Tang and Alavi 2011).

Several studies have reported employing PVOH as an additive to improve properties of starch foams as discussed above, and some studies have used the production of biodegradable foams based on PVOH. Avella et al. (2012) reported preparation of PVOH-based foams containing pulp cellulose fibers and microfibrillated cellulose. These foams were obtained with an eco-friendly preparation method able to generate a pore structure by trapping air in the polymer/filler aqueous dispersion during a high-speed mixing followed by lyophilization. The addition of small amounts of microfibrillated cellulose (1–5%) resulted in a decrease of average cell diameter and an increase in cell density, and the addition of pulp cellulose fibers affected cell shape and regularity of PVOH-based foams. Both pulp cellulose fibers and microfibrillated cellulose additions resulted in foams with lower water vapor absorption when compared to the PVOH foam.

Additionally, Liu et al. (2014) reported production of biodegradable composite foams based on PVOH supported by cellulose nanofibrils using a unidirectional freeze-drying technology. Incorporation of cellulose nanofibrils enhanced the tensile strength, water resistance and dimensional stability of composite foams. Furthermore, these improvements were related to the rigid and semicrystalline nature of the nanofibrils, as well as its regular and compact pore structure.

PLA is an aliphatic, biodegradable polyester synthesized from lactic acid, which can be obtained from renewable sources such as corn, potatoes, sugar beets and sugarcane, and it is used to create foams for food packaging employing extrusion (Zimmermann et al. 2013; Wang et al. 2017). Processing conditions have to be adjusted to provide efficient melting and foaming of the polymer; however, its intrinsic brittleness, which is the result of a relatively short and semi-rigid molecular backbone, greatly limits its range of use in large-scale commercial applications (Wang et al. 2017).

Some studies have reported the use of PLA to obtain extruded biodegradable foams; however, the use of PLA for this purpose is limited by its low heat resistance, poor melt strength, slow crystallization kinetics and intrinsic brittleness (Xu et al. 2014; Wang et al. 2017). Mihai et al. (2007) described production of low-density, open-cell foams from PLA and PLA and starch blends using CO_2 as a blowing agent, and they observed a significant increase in crystallization rate of PLA foams extruded with the CO_2, with these PLA foams exhibiting up to 15% crystallinity.

PLA was further employed by Zimmermann et al. (2013) to obtain extruded foams in the context of studying the effects of incorporating different blowing agents (endothermic and exothermic) on foams properties. The density of pure PLA foams was 1.26 g/cm^3, and this value decreased with addition of endothermic blowing agents (0.50 g/cm^3) and exothermic blowing agents (0.44 g/cm^3), exhibiting foams with smaller and more numerous cells than foams obtained from addition of endothermic blowing agents.

Wang et al. (2017) reported large-scale production of well-defined cell structures of PLA foams using a series of asymmetric biodegradable poly(L-lactide) (PLLA) and poly(D-lactide) (PDLA) blends containing small amounts of PDLA using a continuous extrusion process. They stressed an existing potential to obtain fully bio-based, superior PLA foams for food packaging applications.

12.5 Final Considerations

Several approaches used in the last few years to improve the properties of biodegradable foams were presented in this chapter, and in most cases, mechanical properties and processability were improved. Hydrophilicity, however, while able to be decreased, remains an issue to be solved. Thus, the greatest challenge for researchers working with starch materials and other biodegradable polymers is to obtain biodegradable foams with lower hydrophilicity compared to synthetic conventional polymers. Currently, water sensitivity restricts the use of these materials for food packaging applications, especially for foods with high moisture contents or foods that are stored in high relative humidity. It is thus necessary the development of new technologies and materials that can extend the application of biodegradable foams for food packaging.

References

Aichholzer W, Fritz HG (1998) Rheological characterization of thermoplastic starch materials. Starch/Stärke 50(2–3):77–83

Altieri PA, Tessler MM (1999) Water/humidity resistant starch foams. US Patent CA 2171774 C, 16 Nov 1999

Avella M, Cocca M, Errico ME et al (2012) Polyvinyl alcohol biodegradable foams containing cellulose fibres. J Cell Plast 48:459–470

Bastioli C, Bellotti V, Del Giudice L et al (1994) Expanded articles of biodegradable plastic materials. US Patent 5,360,830, 1 Nov 1994

Bastioli C, Bellotti V, Del Tredici G et al (1998) Biodegradable foamed articles and process for preparation thereof. US Patent 5,801,207, 1 Sept 1998

Bénézet JC, Stanojlovic-Davidovic A, Bergeret A et al (2012) Mechanical and physical properties of expanded starch, reinforced by natural fibres. Ind Crop Prod 37:435–440

Bhatnagar S, Hanna MA (1993) Physical, mechanical and thermal properties of starch based plastic foams. ASAE, St. Joseph

Bhatnagar S, Hanna MA (1995) Properties of extruded starch based plastic foams. Ind Crop Prod 4:71–77

Bhatnagar S, Hanna MA (1996) Starch-based plastic foams from various starch sources. Cereal Chem 73:601–604

Boehmer EW, Hanlon DL (1993) Biodegradable expanded foam material. US Patent 5,272,181, 21 Dec 1993

Bonin M (2010) An Investigation into the properties of starch-based foams. Thesis, Brunel University

Carvalho F, Bilck AP, Yamashita F et al (2017) Baked foams based on cassava starch coated with polyvinyl alcohol with a higher degree of hydrolysis. J Polym Environ 26(4):1445–1452. https://doi.org/10.1007/s10924-017-1046-x

Cha JY, Chung DS, Seib PA et al (2001) Physical properties of starch-based foams as affected by extrusion temperature and moisture content. Ind Crop Prod 14:23–30

Chiellini E, Cinelli P, Ilieva VI, Imam SH et al (2009) Environmentally compatible foamed articles based on potato starch, corn fiber, and poly (vinyl alcohol). J Cell Plast 45:17–32

Chinnaswamy R, Hanna MA (1996) Biodegradable polymers. US Patent 5,496,895, 5 Mar 1996

Cinelli P, Chiellini E, Lawton JW et al (2006) Foamed articles based on potato starch, corn fibers and poly (vinyl alcohol). Polym Degrad Stab 91:1147–1155

Debiagi F, Mali S (2012) Functional properties of extruded nanocomposites based on cassava starch, polyvinyl alcohol and montmorillonite. Macromol Symp 319:235–239

Debiagi F, Mali S, Grossmann MVE et al (2010) Efeito de fibras vegetais nas propriedades de compósitos biodegradáveis de amido de mandioca produzidos via extrusão. Cienc Agrotec 34:1522–1529

Debiagi F, Mali S, Grossmann MVE et al (2011) Biodegradable foams based on starch, polyvinyl alcohol, chitosan and sugarcane fibers obtained by extrusion. Braz Arch Biol Technol 54:1043–1052

Debiagi F, Kobayashi RKT, Nakazato G et al (2014) Biodegradable active packaging based on cassava bagasse, polyvinyl alcohol and essential oils. Ind Crop Prod 52:666–670

Debiagi F, Marim BM, Mali S (2015) Properties of cassava bagasse and polyvinyl alcohol biodegradable foams. J Polym Environ 23:269–276

Faria FO, Vercelheze AES, Mali S (2012) Propriedades físicas de filmes biodegradáveis à base de amido de mandioca, álcool polivinílico e montmorilonita. Quim Nova 35:487–492

Fisk D (2002) Method for forming biodegradable foamed product from starch. US Patent 6,406,649 B1, 18 June 2002

Fujita S, Fujiyama G (1993) The study of melting temperature and enthalpy of starch from rice, barley, wheat, foxtail and proso-millets. Starch/Starke 45:436–441

Glenn GM, Orts WJ (2001) Properties of starch-based foam formed by compression/explosion processing. Ind Crop Prod 13:135–143

Glenn GM, Orts WJ, Nobes GAR (2001) Starch, fiber and $CaCO_3$ effects on the physical properties of foams made by a baking process. Ind Crop Prod 14:201–212

Goodship V (2009) Polyvinyl alcohol: materials, processing and applications. Smithers Rapra Press, Shrewsbury

Guan J, Hanna MA (2004) Functional properties of extruded foam composites of starch acetate and corn cob fiber. Ind Crop Prod 19:255–269

Guan J, Hanna MA (2006) Selected morphological and functional properties of extruded acetylated starch–cellulose foams. Bioresour Technol 97:1716–1726

Harper JM (ed) (1981) Extrusion of foods. CRC Press, Boca Raton

Hepburn C, Alam N (1991) Influence of fillers on rubber foam properties. Cell Polym 10:99–116

Hofmann T, Linke L, Tsiapouris A et al (1998) Porous materials made from starch. Chem Eng Technol 27:580–584

Kaisangsri N, Kerdchoechuen O, Laohakunjit N (2012) Biodegradable foam tray from cassava starch blended with natural fiber and chitosan. Ind Crop Prod 37:542–546

Kaisangsri N, Kerdchoechuen O, Laohakunjit N (2014) Characterization of cassava starch based foam blended with plant proteins, kraft fiber, and palm oil. Carbohydr Polym 110:70–77

Kaseem M, Hamad K, Deri F (2012) Thermoplastic starch blends: a review of recent works. Polym Sci 54:165–176

Lacourse NL, Altieri PA (1991a) Biodegradable shaped products and the method of preparation thereof. US Patent, 5,035,930, 30 July 1991

Lacourse NL, Altieri PA (1991b) Biodegradable shaped products and the method of preparation thereof. US Patent, 5,043,196, 27 Aug 1991

Lawton JW, Shogren RL, Tiefenbacher KF (1999) Effect of batter solids and starch type on the structure of baked starch foams. Cereal Chem 75:682–687

Le Corre D, Angellier-Coussy H (2014) Preparation and application of starch nanoparticles for starch nanocomposites: a review. React Funct Polym 85:97–120

Lee SY, Hanna MA (2008) Tapioca starch-poly (lactic acid)-cloisite 30B nanocomposite foams. Polym Compos 30:665–672

Lee SY, Chen H, Hanna MA (2008) Preparation and characterization of tapioca starch–poly (lactic acid) nanocomposite foams by melt intercalation based on clay type. Ind Crop Prod 28:95–106

Liu D, Ma Z, Wang Z et al (2014) Biodegradable poly(vinyl alcohol) foams supported by cellulose nanofibrils: processing, structure, and properties. Langmuir 30:9544–9550

Machado CM, Benelli P, Tessaro IC (2017) Sesame cake incorporation on cassava starch foams for packaging use. Ind Crop Prod 102:115–121

Mali S, Debiagi F, Grossmann MVE et al (2010a) Starch, sugarcane bagasse fibre and polyvinyl alcohol effects on extruded foam properties: a mixture design approach. Ind Crop Prod 32:353–359

Mali S, Grossmann MVE, Yamashita F (2010b) Filmes de amido: produção, propriedades e potencial de utilização. Semin-Cienc Agrar 31:137–156

Mali S, Bilck AP, Yamashita F et al (2016) Processo de produção de bandeja biodegradável a base de amido de mandioca e álcool polivinílico com ação antimicrobiana. BR Patent 1020160012759, 20 Jan 2016

Marengo VA, Vercelheze AES, Mali S (2013) Compósitos biodegradáveis de amido de mandioca e resíduos da agroindústria. Quim Nova 36:680–685

Maria TCM, Carvalho RA, Sobral PJA et al (2008) The effect of the degree of hydrolysis of the PVA and the plasticizer concentration on the color, opacity, and thermal and mechanical properties of films based on PVA and gelatin blends. J Food Eng 87:191–199

Matsuda DNK, Vercelheze AES, Carvalho GM et al (2013) Baked foams of cassava starch and organically modified nanoclays. Ind Crop Prod 44:705–711

Mello LRPF, Mali S (2014) Use of malt bagasse to produce biodegradable baked foams made from cassava starch. Ind Crop Prod 55:187–193

Mihai M, Huneault MA, Favis BD et al (2007) Extrusion foaming of semi-crystalline PLA and PLA/thermoplastic starch blends. Macromol Biosci 7:907–920

Moraes IC, Silva GGD, Carvalho RA et al (2008) Influência do grau de hidrólise do poli(vinil álcool) nas propriedades físicas de filmes à base de blendas de gelatina e poli(vinil álcool) plastificados com glicerol. Cienc Tecnol Alime 28:738–745

Nabar Y, Raquez JM, Dubois P et al (2005) Production of starch foams by twin-screw extrusion: effect of maleated poly (butylene adipate-co-terephthalate) as a compatibilizer. Biomacromolecules 6:807–817

Nabar Y, Narayan R, Schindler M (2006) Twin-screw extrusion production and characterization of starch foam products for use in cushioning and insulation applications. Polym Eng Sci 46:438–451

Salgado PR, Schmidt VC, Ortiz SE et al (2008) Biodegradable foams based on cassava starch, sunflower proteins and cellulose fibers obtained by a baking process. J Food Eng 85:435–443

Schmidt VC, Laurindo JB (2010) Characterization of foams obtained from cassava starch, cellulose fibres and dolomitic limestone by a thermopressing process. Braz Arch Biol Technol 53:185–192

Shogren RL (1992) Effect of moisture content on the melting and subsequent physical aging of cornstarch. Carbohydr Polym 19:83–90

Shogren RL, Lawton JW, Doanne WM et al (1998) Structure and morphology of baked foams. Polymer 39:6649–6655

Silva A, Nievola LM, Tischer CA et al (2013) Cassava starch-based foams reinforced with bacterial cellulose. J Appl Polym Sci 130:3043–3049

Soykeabkaew N, Supaphol P, Rujiravanit R (2004) Preparation and characterization of jute and flax reinforced starch-based composite foams. Carbohydr Polym 58:53–63

Soykeabkaew N, Thanomsilp C, Suwantong O (2015) A review: starch-based composite foams. Compos Part A Appl Sci Manuf 78:246–263

Stagner J, Narayan R (2011) Preparation and properties of biodegradable foams. J Polym Environ 19:598–606

Suárez G, Gutiérrez TJ (2017) Recent advances in the development of biodegadable films and foams from cassava starch. In: Klein C (ed) Handbook on Cassava: production, potential uses and recent advances. Editorial Nova Science Publishers, Inc., New York, pp 297–312. EE.UU. ISBN: 978-1-53610-307-6

Sudhamani SR, Prasad MS, Sankar KU (2003) DSC and FTIR studies on gellan and polyvinyl alcohol (PVA) blend films. Food Hydrocoll 17:245–250

Tang S, Alavi S (2011) Recent advances in starch, polyvinyl alcohol based polymer blends, nanocomposites and their biodegradability. Carbohydr Polym 85:7–16

Tatarka PD, Cunningham RL (1998) Properties of protective loose-fill foams. J Appl Polym Sci 67:1157–1176

Tiefenbacher KF (1993) Starch-based foamed materials—use and degradation properties. JMS Pure Appl Chem A 30(9/10):727–731

Van Soest JJG, Hulleman SHD, de Wit D et al (1996) Crystallinity in starch bioplastics. Ind Crop Prod 5:11–22

Vercelheze AES, Fakhouri FM, Dall'antônia LH et al (2012) Properties of baked foams based on cassava starch, sugarcane bagasse fibers and montmorillonite. Carbohydr Polym 87:1302–1310

Wang S, Ren J, Li W et al (2014) Properties of polyvinyl alcohol/xylan composite films with citric acid. Carbohydr Polym 103:94–99

Wang L, Lee RE, Wang G (2017) Use of stereocomplex crystallites for fully-biobased microcellular low-density poly(lactic acid) foams for green packaging. Chem Eng J 327:1151–1162

Willett JL, Shogren RL (2002) Processing and properties of extruded starch/Polymer foams. Polymer 43:5935–5947

Xu Y, Hanna MA (2005) Preparation and properties of biodegradable foams from starch acetate and poly(tetramethylene adipate-co-terephthalate). Carbohydr Polym 59:521–529

Xu H, Xie L, Jiang X et al (2014) Structural basis for unique hierarchical cylindrites induced by ultrahigh shear gradient in single natural fiber reinforced poly(lactic acid) green composites. Biomacromolecules 15:1676–1686

Zhang JF, Sun X (2007a) Biodegradable foams of poly(lactic acid)/starch I. Extrusion condition and cellular size distribution. J Appl Polym Sci 106:857–862

Zhang JF, Sun X (2007b) Biodegradable foams of poly(lactic acid)/starch II. Cellular structure and water resistance. J Appl Polym Sci 106:3058–3062

Zimmermann MVG, Brambilla VC, Brandalise RN (2013) Observations of the effects of different chemical blowing agents on the degradation of poly(lactic acid) foams in simulated soil. Mater Res 16:1266–1273

Zobel HF (ed) (1964) Methods in carbohydrate chemistry. Academic, New York

Chapter 13
Composite Foams Made from Biodegradable Polymers for Food Packaging Applications

Luis M. Araque, Vera A. Alvarez, and Tomy J. Gutiérrez

Abstract Polymeric foams are cell structures (porous microstructures) that have been frequently made from synthetic polymers for use in the development of food packaging. Due to the problems concerning the environmental impact caused by polymers from the petrochemical industry, the foams have been more recently studied from biodegradable polymers. However, the polymer materials obtained are usually susceptible to moisture, thus conditioning the collapse of the porous structure of the material. As an alternative, the composite foams have been investigated from nano-fillers such as clays, cellulose, nanoparticles, among others. This chapter aims to analyze the recent advances in the studies of composite foams.

Keywords Biopolymers · Composite materials · Polymer composites

13.1 Introduction

Foams produced from petroleum-based conventional polymers, such as polystyrene (PS) and polyurethane (PU), are widely used in the food packaging industry, because of their excellent properties and low production costs. However, these polymers do not undergo biodegradation and its recycling is unprofitable (Li et al. 2017). Hence, the increased customer awareness about the environment and the creation of government laws on environmental protection and waste disposal has encouraged scientists to develop new eco-friendly materials. Biodegradable polymers are an alternative to conventional ones, especially those produced from renewable

L. M. Araque
Graduate Program in Materials Science and Engineering, Federal University of Piauí, Ininga 64049-550, Teresina, Piauí, Brazil

V. A. Alvarez · T. J. Gutiérrez (✉)
Thermoplastic Composite Materials (CoMP) Group, Institute of Research in Materials Science and Technology (INTEMA), Faculty of Engineering, National University of Mar del Plata (UNMdP) and National Scientific and Technical Research Council (CONICET), Colón 10850, Mar del Plata 7600, Buenos Aires, Argentina
e-mail: tomy.gutierrez@fi.mdp.edu.ar

T. J. Gutiérrez (ed.), *Polymers for Food Applications*,
https://doi.org/10.1007/978-3-319-94625-2_13

resources since they can be mineralized by microorganism without negative environmental impact (Suárez and Gutiérrez 2017). Nevertheless, biodegradable polymer produce foams with poorer morphology, water and moisture resistance, and mechanical, thermal and barrier properties as compared to conventional polymer foams (Luzi et al. 2016; Mitrus et al. 2016).

13.2 Composite Foams

It has been reported that the formation of composite foams offers a solution to improve the properties of biodegradable polymers, especially when the filler has a dimension on the range of nanometers. The addition of a low nanofiller content as a reinforcement can provide a significant improvement of several properties, i.e. dimensional, mechanical and thermal stability. In addition, nanofiller may serve as nucleation agent for bubble generation and as a stabilizer of the nucleating bubbles by decreasing the cell coalescence with an increase of melt strength (Soykeabkaew et al. 2015; Zhao et al. 2015). In order to not alter the biodegradability and biocompatibility of biodegradable composite foams are added natural reinforcements, i.e. cellulose and clays. Cellulose nanocrystals (CNC) is a highly-crystalline and rod-like nanoparticle extracted from cellulose, which possesses several advantages such as sustainability, recyclability, non-toxicity, high surface area, low density, etc. When it is used as a filler of composite foams, CNC improves the thermal and mechanical properties and affect the polymer nucleating process (Srithep et al. 2012; Liu et al. 2014; Mi et al. 2014). On the other hand, nanoclays are layered silicate minerals, such as montmorillonite has high availability, versatility, low cost, and minimal adverse effects on the environment and human health. The individual clay particle presents a platelet structure, with a thickness of approximately 1 nm and lateral dimensions up to 1 μm. In nanocomposites, the nanoclays reinforcing efficiency, good barrier properties, and improved dimensional and thermal stability are strongly related to their aspect ratio and large surface area (Keshtkar et al. 2014; Kumar and Maiti 2015; Li et al. 2015). The most investigated biodegradable polymers for the developed of biodegradable composite foams are polybutylene succinate (PBS), polycaprolactone (PCL), polylactic acid (PLA), polyvinyl alcohol (PVA), and starch. In Table 13.1 are listed several works about edible nanocomposite foams based on the most used biodegradable polymers and CNC or nanoclays as the filler.

PBS is a biodegradable aliphatic polyester, which is obtained by polycondensation reaction of 1,4-butanediol and succinic acid, or by fermentation of agricultural crops containing cellulose, glucose, and lactose. For its reasonable production cost, outstanding biodegradability, melt processability, and chemical and thermal resistance PBS is a promising material. However, PBS presents some insufficient properties such as its linear molecular structure, low melt viscosity and strength, and its barrier properties (Chen et al. 2015; Hu et al. 2015; Luzi et al. 2016). Lin et al. (2015) developed biodegradable microcellular foams based on PBS and CNC by

Table 13.1 Edible nanocomposite foams with different biodegradable polymers as matrix and cellulose nanocrystals or nanoclays as filler

Polymer	Reinforcement	Foam processing	Reference
PBS	Cellulose nanocrystals (CNC)	Compressing method	Lin et al. (2015)
	Surface acetylated cellulose nanocrystals (ACNA)	Compressing method	Hu et al. (2015)
	Organically modified Closite®15A, Closite®20A, and Closite®30A	Baking method	Lim et al. (2011)
PCL	Cellulose nanocrystals	Supercritical CO_2 injection	Mi et al. (2014)
	Organically modified Closite®30B	Pressure vessel	Di et al. (2003)
	Organically modified Closite®25A	Supercritical CO_2 and ethanol atmosphere pressure quench	Tsimpliaraki et al. (2013)
PVA	Microfibrillated cellulose (MFC)	Supercritical CO_2 extrusion	Zhao et al. (2014)
	Cellulose nanofibrils (CNF)	Freeze-drying method	Liu et al. (2014)
	Cellulose nanocrystals (CNC)	Baking process	Song et al. (2016)
	Pristine Na^+ montmorillonite	Extrusion	Li et al. (2015)
PLA	Cellulose nanocrystals (CNC)	Casting and leaching method	Borkotoky et al. (2017)
	Organically modified I.34TCN nanoclay	Supercritical CO_2 extrusion	Liu et al. (2013)
	Organically modified Closite®30B	Supercritical CO_2 extrusion	Keshtkar et al. (2014)
Starch	Cellulose nanofibrils (CNF)	Freeze-drying method	Svagan et al. (2008)
	Cellulose nanofibrils (CNF)	Freeze-drying method	Yildirim et al. (2014)
	Organically modified Closite®30B and Closite10A	Baking method	Matsuda et al. (2013)
	Wizkay TU-90 kaolin	Compressing method	Kaewtatip et al. (2013)

compressing method. Azodicarbonamide (AC) was the blowing agent and zinc oxide (ZnO) the blowing promoter. The flexural strength and modulus of the nanocomposite foam with 5 wt% of CNC was increased by 50% and 62.9%, respectively, as compared with the neat PBS foam. Behavior attributed to the stress transfer between the matrix and nanofiller. The cells morphology, structure and stability were also modified by the filler due to the nucleation effect of CNC.

Hu et al. (2015) studied the effect of the chemical acetylation modification of CNC on the properties of PBS biodegradable foams. The foaming process was made by compressing method, and AC and ZnO were the blowing agent and the blowing promoter. The chemical treatment to obtain acetylated cellulose nanocrystals (ACNC) promoted the compatibility with the matrix. Nanocomposite foams with 5 wt% of ACNC showed flexural strength and modulus improved by 50% and 31.4%, respectively, as compared to that the neat PBS foam. The good dispersion of

ACNC along the PBS matrix, and to the compatibility between nanocomposite foam components improved those properties. Additionally, it was observed that the microstructure was affected by the filler, increasing the density and average cell size. Clays are other nanofiller used as the reinforcement of nanocomposite foams. Lim et al. (2011) prepared PBS and organoclay nanocomposite foams by the baking method, employing AC, *N*, *N′*-dinitroso pentatetramine, and urea activator as the blowing agent. The nanocomposite with 2 wt% of clay displayed the better mechanical and thermal properties. The thermal degradation temperature, tensile strength, and elongation at break of composites with 2 wt% of the filler increased by 5, 64 and 470%, respectively, as compared to the neat PSB foam. The improved on the mechanical properties was due to the strong interaction between the PBS matrix and the nanoclay, which resulted in the transfer of stress to the nanofiller. Furthermore, it was observed that the blowing ratio depended upon the blowing temperature and time, and that nanocomposite foams had closed cells, which were oval-shaped.

Another biodegradable promising polymer is the PCL that is a hydrophobic and semicrystalline polymer, which presents low melting temperature at approximately 60 °C and exceptional blend-compatibility. PCL is synthesized by ring opening polymerization of ε-caprolactone using a variety of catalyst, and by ring opening polymerization of 2-methylene-1-3-dioxepane. PCL properties can be modified by adding functional groups. However, PCL has high degraded times and poor mechanical performance (Di et al. 2003; Mi et al. 2014). Mi et al. (2014) prepared CNC and PCL composites by injection molding with supercritical CO_2 as the blowing agent CNC acted as a nucleating agent and had a strong interface with the PCL matrix improving the tensile modulus, complex viscosity, and storage modulus. It was also found that nanocomposite foams presented fine cell structure.

Studying the effect of organically modified clay upon the crystallization and foaming process of PCL. Di et al. (2003) found that the crystallization kinetics increase by 81% with the addition of 5 wt% of the filler, because of the nucleating effect of the silicate layers, as compared to the neat PCL foam. Furthermore, the complex viscosity and density of the nanocomposite foams were higher than that of neat PCL foam. Likewise, Tsimpliaraki et al. (2013) prepared PCL and organically modified clay nanocomposite foams by one-step or two-step press quench method, with supercritical CO_2 or CO_2–ethanol atmosphere as a blowing agent. The dispersion of the filler and the cell density and structure were better when the blowing agent was the mixture CO_2–ethanol. Besides, it was not found a significant difference between nanocomposite foams structure when the processing method was either one-step or two-step.

PVA is another biodegradable polymer widely studied as a matrix of edible nanocomposite foams, as a matter of fact, PVA foams has been extensively applied in the biomedical and pharmaceutical fields. PVA is a semicrystalline and water-soluble polymer with several advantages, for instance, its compatibility with organic and inorganic materials, good ion exchange, physical adsorption, chelation, polarity, biocompatibility, and versatility. However, PVA has narrow process window, which makes its melting temperature very close to its decomposition temperature (Mali et al. 2010; Debiagi et al. 2014; Li et al. 2015). Liu et al. (2014) developed

PVA and cellulose nanofibrils nanocomposite foams by the freeze-drying method. The addition of 30 wt% of CNF increased the Young's modulus, compressive stress and energy absorption by 75, 60.58 and 14.36%, respectively, as compared with the neat PVA foam. Those results were attributed to the effect of CNF on the foam morphology since composite foams with 30 wt% or lower CNF content presented smaller pore sizes with the larger surface area, which could support stronger compression forces. Those pores were composed of fibrils with amorphous PVA playing the role of cross-linker or glue. Comparatively, Zhao et al. (2014) prepared PVA and microfibrillated cellulose (MFC) by extrusion with water and CO_2 as the co-blowing agents and studied the effect of water, CO_2 and MFC content on the properties of the composite foam. It was observed that the addition of 0.5 wt% of MFC increased the crystallinity by 37.48% as compared to the neat PVA foam. This behavior was due to nucleation effect of the filler that contributed to cell nucleation and prevented large crystal formation. Moreover, it was perceived that the composite with 0.05 wt% of MFC prepared using 12.5 wt% of water and 9 wt% of CO_2 displayed the best properties. The cell density was increased by almost an order of magnitude and the cell size was decreased by 6.66%.

Likewise, the morphology, density, water uptake and mechanical properties of PVA and CNC nanocomposite foams, developed by Song et al. (2016) through the baking process, were modified with the addition of the filler. The foaming process was carried out with initial reaction times of 10 and 120 s, and calcium carbonate and formaldehyde were employed as the blowing and cross-linking agent, respectively. It was observed that the nanocomposite foam with 1.5 wt% of CNC and the initial time of 120 s displayed tensile strength and compressive modulus increased by 737.31 and 8.75%, respectively, as compared to the neat PVA foam and nanocomposite foams with 10 s initial reaction time. Those results were attributed to longer initial reaction time, allowing the formation of high density with small pore size foams, and to the strong interaction between the low CNC content that presented uniformly dispersion and the matrix. On the other hand, there have been also developed PVA nanocomposite foams with nanolayered silicates as the filler. Li et al. (2015) studied the effect of pristine montmorillonite on the properties of PVA nanocomposites foams prepared by extrusion and using AC as the blowing agent. The clay was intercalated during the extrusion forming the exfoliated and intercalated structures. Due to the strong interaction between the nanocomposite components, the melt elasticity and strength were improved. As a consequence of the increased melt strength and reduced crystallization time, the density of nanocomposite foam decreased.

Edible nanocomposite foams have also been prepared using PLA as a matrix. PLA is polyester synthesized by ring opening of lactide or L- and D-lactide acid, which can be produced from renewable agricultural materials. PLA is a biocompatible polymer that has low toxicity and produces foams with high surface area, high toughness and low thermal conductivity (Lee et al. 2008; Ding et al. 2016). However, due to its low crystallinity, low melt strength, low deflection temperature, and slow crystallization, PLA presents limited applicability in different manufacturing processes (Bocz et al. 2016). Borkotoky et al. (2017) developed PLA and CNC nanocomposite foams by a novel and economic casting and leaching method and studied the filler effect on the morphology and thermal properties of the foams. It was observed that the addition of 3 wt% of CNC increased the foam density by 38.09% and decreased the pore size by 47.14%, as compared with the neat PLA foam, as a consequence of the nucleating effect of the nanoparticles, which formed a greater number of smaller size cells. The crystallinity of the nanocomposite with 3 wt% of CNC increased by 12.9%, reaffirming the nucleating effect of the filler.

Additionally, clays are also used as a filler of PLA nanocomposite foams. Liu et al. (2013) found that organically modified I.34TCN clay and the chain-extender BL 10069 N affected the properties of PLA foams prepared by the supercritical CO_2 extrusion method. The nanocomposite with 3 wt% of the clay presented the exfoliated and intercalated structures. It was also perceived that the addition of 3 wt% of the clay increased the complex viscosity, storage modulus, melt strength and cell density, besides it decreased the loss factor and pore size of the nanocomposite foam. The behavior of the complex viscosity was attributed to the intercalation and exfoliation of the nanoclay that impede polymer chains to move and the increased of the melt strength was related to the increase of the storage modulus, which was ascribed to the intercalation of polymer chains into the clay layers. Similarly, Keshtkar et al. (2014) studied the effect of organically modified clay and CO_2 content on the morphology, crystallization and rheological properties of nanocomposite foams preparer by extrusion with supercritical CO_2 as the blowing agent. The nanocomposite foam with 5 wt% of the filler, 9 wt% of CO_2 and prepared with die temperature at 115 °C displayed the best crystallization, cell density, expansion rate and complex viscosity. Those results were attributed to the crystallization and nucleation effects of the filler that were pronounced by the presence of plasticizing CO_2. Besides, the filler increases the melt strength, and thereby the foaming behavior.

Finally, the most used biodegradable polymer to prepared composite foams is the starch, which is a renewable and low-cost polymer produced by a vast variety of sources. Starch is the main form in which carbohydrates are stored in plants, cereal grains, tubers, and roots. The major starch polymer components are amylose and amylopectin (Soykeabkaew et al. 2015). Since it is a renewable resource, abundant, inexpensive and non-toxic, starch is considered the most used polysaccharide for foam packaging applications (Palma-Rodríguez et al. 2016). However, starch-based foams have poor water resistance and mechanical properties (Kaisangsri et al. 2014; Pornsuksomboon et al. 2016). Aiming to study the mechanical performance of nanofibrillated cellulose (NFC) and amylopectin potato starch nanocomposite foams, prepared by the freeze-drying method, Svagan et al. (2008) found that the best mechanical properties were for obtained adding 40 wt% of the filler. The Young's and yield strength of the nanocomposite foam with 40 wt% of CNF increased by 42.85 and 200%, respectively, as compared to that the neat starch foam. Furthermore, the density and the water content decreased by 7.69 and 23.63%, respectively. Besides the reinforcement effect of the nanofiller, the cell structure of the nanocomposite foams was a formed by a mixture of closed and open cells, which influence the mechanical behavior in the linear-elastic and cell collapse regions. The decreased in water content was due to the hydrophobic nature of the nanofiller. Likewise, Yildirim et al. (2014) developed corn starch and CNF nanocomposite foams by the freeze-drying method, with different starch and CNF contents, and studied the mechanical flexural and compressive behavior of nanocomposite foams. The best mechanical performance was displayed by the nanocomposite foam with 6 wt% of corn starch and 1.5 wt% of NFC. The elastic modulus, modulus of rupture, compression modulus and compressive resistance of nanocomposites with 6 wt% of starch and 1.5 wt% of CNF increased by 109,869; 16,375; 5946 and 2212% as compared to the nanocomposite foam with 0.5 wt% of starch and 0.5 wt% of CNF.

Furthermore, starch-based nanocomposite foams with nanoclays as filler have been prepared. Matsuda et al. (2013) developed cassava starch and two organically modified clay (Closite®30B and Closite®10A) nanocomposite foams by the baking method and evaluated the influence of the content and the kind of filler upon the microstructural and mechanical properties. After 30 min of immersion, all nanocomposite foams had lower capacities of water absorption than the neat starch foam, due to the clays hydrophobicity, which is higher to Closite®10A. Both organoclays presented good dispersion along the matrix and formed the exfoliated structure. Also, both clays improved the stress at break, due to the strong interaction between the nanoclays and the matrix, which enhanced the mechanical reinforcement. Correspondingly, evaluating the effect of kaolin upon the morphology, water absorption and mechanical properties of starch foams that were prepared by compressing method. Kaewtatip et al. (2013) found that the addition of 15% of kaolin increased the Izod impact strength by 368%, as compared to the neat starch foam. Result attributed to the good dispersion and distribution of kaolin along the starch matrix, and to the kaolin efficiency absorbing and diluting impact energy. Additionally, the addition of kaolin the water absorption decreased by 6 and 8% after the nanocomposite foams were exposed at 55% of relative humidity for 7 and 45 days, respectively.

Acknowledgements The authors would like to thank the Consejo Nacional de Investigaciones Científicas y Técnicas (CONICET) (Postdoctoral fellowship internal PDTS-Resolution 2417), Universidad Nacional de Mar del Plata (UNMdP) for financial support, and Dr. Mirian Carmona-Rodríguez.

Conflicts of Interest: The authors declare no conflict of interest.

References

Bocz K, Tábi T, Vadas D, Sauceau M, Fages J, Marosi G (2016) Characterisation of natural fibre reinforced PLA foams prepared by supercritical CO2 assisted extrusion. Express Polym Lett 10(9):771–779. https://doi.org/10.3144/expresspolymlett.2016.71

Borkotoky SS, Dhar P, Katiyar V (2017) Biodegradable poly (lactic acid)/Cellulose nanocrystals (CNCs) composite microcellular foam: effect of nanofillers on foam cellular morphology, thermal and wettability behavior. Int J Biol Macromol 106:433–446. https://doi.org/10.1016/j.ijbiomac.2017.08.036

Chen RY, Zou W, Zhang HC, Zhang GZ, Yang ZT, Jin G, Qu JP (2015) Thermal behavior, dynamic mechanical properties and rheological properties of poly(butylene succinate) composites filled with nanometer calcium carbonate. Polym Test 42:160–167. https://doi.org/10.1016/j.polymertesting.2015.01.015

Debiagi F, Kobayashi RKT, Nakazato G, Panagio LA, Mali S (2014) Biodegradable active packaging based on cassava bagasse, polyvinyl alcohol and essential oils. Ind Crop Prod 52:664–670. https://doi.org/10.1016/j.indcrop.2013.11.032

Di ME, Iannace S, Di Y, Del GE, Nicolais L (2003) Heterogeneous bubble nucleation in PCL/clay nanocomposite foams. Plast Rubber Compos 32:313–317. https://doi.org/10.1179/146580103225003479. Copyright (C) 2010 American Chemical Society (ACS). All Rights Reserved

Ding W, Jahani D, Chang E, Alemdar A, Park CB, Sain M (2016) Development of PLA/cellulosic fiber composite foams using injection molding: crystallization and foaming behaviors. Compos A: Appl Sci Manuf 83:130–139. https://doi.org/10.1016/j.compositesa.2015.10.003

Hu F, Lin N, Chang PR, Huang J (2015) Reinforcement and nucleation of acetylated cellulose nanocrystals in foamed polyester composites. Carbohydr Polym 129:208–215. https://doi.org/10.1016/j.carbpol.2015.04.061

Kaewtatip K, Tanrattanakul V, Phetrat W (2013) Preparation and characterization of kaolin/starch foam. Appl Clay Sci 80–81:413–416. https://doi.org/10.1016/j.clay.2013.07.011

Kaisangsri N, Kerdchoechuen O, Laohakunjit N (2014) Characterization of cassava starch based foam blended with plant proteins, kraft fiber, and palm oil. Carbohydr Polym 110:70–77. https://doi.org/10.1016/j.carbpol.2014.03.067

Keshtkar M, Nofar M, Park CB, Carreau PJ (2014) Extruded PLA/clay nanocomposite foams blown with supercritical CO_2. Polymer (United Kingdom) 55(16):4077–4090. https://doi.org/10.1016/j.polymer.2014.06.059

Kumar S, Maiti P (2015) Understanding the controlled biodegradation of polymers using nanoclays. Polymer (United Kingdom) 76:25–33. https://doi.org/10.1016/j.polymer.2015.08.044

Lee SY, Chen H, Hanna MA (2008) Preparation and characterization of tapioca starch-poly(lactic acid) nanocomposite foams by melt intercalation based on clay type. Ind Crop Prod 28(1):95–106. https://doi.org/10.1016/j.indcrop.2008.01.009

Li Y, Tian H, Jia Q, Niu P, Xiang A, Liu D, Qin Y (2015) Development of polyvinyl alcohol/intercalated MMT composite foams fabricated by melt extrusion. J Appl Polym Sci 132(43):1–7. https://doi.org/10.1002/app.42706

Li R, Du J, Zheng Y, Wen Y, Zhang X, Yang W et al (2017) Ultra-lightweight cellulose foam material: preparation and properties. Cellulose 24(3):1417–1426. https://doi.org/10.1007/s10570-017-1196-y

Lim S-K, Lee J-J, Jang S-G, Lee S-I, Lee K-H, Choi HJ, Chin I-J (2011) Synthetic aliphatic biodegradable poly(butylene succinate)/clay nanocomposite foams with high blowing ratio and their physical characteristics. Polym Eng Sci 51(7):1316–1324. https://doi.org/10.1002/pen.21927

Lin N, Chen Y, Hu F, Huang J (2015) Mechanical reinforcement of cellulose nanocrystals on biodegradable microcellular foams with melt-compounding process. Cellulose 22(4):2629–2639. https://doi.org/10.1007/s10570-015-0684-1

Liu W, Wang X, Li H, Du Z, Zhang C (2013) Study on rheological and extrusion foaming behaviors of chain-extended poly (lactic acid)/clay nanocomposites. J Cell Plast 49(6):535–554. https://doi.org/10.1177/0021955X13503845

Liu D, Ma Z, Wang Z, Tian H, Gu M (2014) Biodegradable poly (vinyl alcohol) foams supported by cellulose nanofibrils: processing, structure, and properties. Langmuir 30:9544–9550. https://doi.org/10.1021/la502723d

Luzi F, Fortunati E, Jiménez A, Puglia D, Pezzolla D, Gigliotti G et al (2016) Production and characterization of PLA_PBS biodegradable blends reinforced with cellulose nanocrystals extracted from hemp fibres. Ind Crop Prod 93:276–289. https://doi.org/10.1016/j.indcrop.2016.01.045

Mali S, Debiagi F, Grossmann MVE, Yamashita F (2010) Starch, sugarcane bagasse fibre, and polyvinyl alcohol effects on extruded foam properties: a mixture design approach. Ind Crop Prod 32(3):353–359. https://doi.org/10.1016/j.indcrop.2010.05.014

Matsuda DKM, Verceheze AES, Carvalho GM, Yamashita F, Mali S (2013) Baked foams of cassava starch and organically modified nanoclays. Ind Crop Prod 44:705–711. https://doi.org/10.1016/j.indcrop.2012.08.032

Mi HY, Jing X, Peng J, Salick MR, Peng XF, Turng LS (2014) Poly(ε-caprolactone) (PCL)/cellulose nano-crystal (CNC) nanocomposites and foams. Cellulose 21(4):2727–2741. https://doi.org/10.1007/s10570-014-0327-y

Mitrus M, Combrzyński M, Kupryaniuk K, Wójtowicz A, Oniszczuk T, Kręcisz M et al (2016) A study of the solubility of biodegradable foams of thermoplastic starch. J Ecol Eng 17(4):184–189. https://doi.org/10.12911/22998993/64554

Palma-Rodríguez HM, Berrios JDJ, Glenn G, Salgado-Delgado R, Aparicio-Saguilán A, Rodríguez-Hernández AI, Vargas-Torres A (2016) Effect of the storage conditions on mechanical properties and microstructure of biodegradable baked starch foams. CYTA J Food 14(3):415–422. https://doi.org/10.1080/19476337.2015.1117142

Pornsuksomboon K, Holló BB, Szécsényi KM, Kaewtatip K (2016) Properties of baked foams from citric acid modified cassava starch and native cassava starch blends. Carbohydr Polym 136:107–112. https://doi.org/10.1016/j.carbpol.2015.09.019

Song T, Tanpichai S, Oksman K (2016) Cross-linked polyvinyl alcohol (PVA) foams reinforced with cellulose nanocrystals (CNCs). Cellulose 23(3):1925–1938. https://doi.org/10.1007/s10570-016-0925-y

Soykeabkaew N, Thanomsilp C, Suwantong O (2015) A review: starch-based composite foams. Compos A: Appl Sci Manuf 78:246–263. https://doi.org/10.1016/j.compositesa.2015.08.014

Srithep Y, Turng LS, Sabo R, Clemons C (2012) Nanofibrillated cellulose (NFC) reinforced polyvinyl alcohol (PVOH) nanocomposites: properties, solubility of carbon dioxide, and foaming. Cellulose 19(4):1209–1223. https://doi.org/10.1007/s10570-012-9726-0

Suárez G, Gutiérrez T (2017) Recent advances in the development of biodegadable films and foams from cassava starch. In: Klein C (ed) Handbook on Cassava: production, potential uses and recent advances. Nova Science Publishers, Inc., New York, pp 297–312

Svagan AJ, Samir MASA, Berglund LA (2008) Biomimetic foams of high mechanical performance based on nanostructured cell walls reinforced by native cellulose nanofibrils. Adv Mater 20(7):1263–1269. https://doi.org/10.1002/adma.200701215

Tsimpliaraki A, Tsivintzelis I, Marras SI, Zuburtikudis I, Panayiotou C (2013) Foaming of PCL/clay nanocomposites with supercritical CO_2 mixtures: the effect of nanocomposite fabrication route on the clay dispersion and the final porous structure. J Supercrit Fluids 81:86–91. https://doi.org/10.1016/j.supflu.2013.05.003

Yildirim N, Shaler SM, Gardner DJ, Rice R, Bousfield DW (2014) Cellulose nanofibril (CNF) reinforced starch insulating foams. Cellulose 21(6):4337–4347. https://doi.org/10.1007/s10570-014-0450-9

Zhao N, Mark LH, Zhu C, Park CB, Li Q, Glenn R, Thompson TR (2014) Foaming poly(vinyl alcohol)/microfibrillated cellulose composites with CO_2 and water as co-blowing agents. Ind Eng Chem Res 53(30):11962–11972. https://doi.org/10.1021/ie502018v

Zhao H, Zhao G, Turng LS, Peng X (2015) Enhancing nanofiller dispersion through prefoaming and its effect on the microstructure of microcellular injection molded polylactic acid/clay nanocomposites. Ind Eng Chem Res 54(28):7122–7130. https://doi.org/10.1021/acs.iecr.5b01130

Chapter 14
Nano and Microencapsulation Using Food Grade Polymers

S. K. Vimala Bharathi, J. A. Moses, and C. Anandharamakrishnan

Abstract Encapsulation is the technique in which an active component (core material) is entrapped in a polymeric (wall) material. This can be done at the macro, micro or nano-level. In the food industry, apart from providing protection to the core material, encapsulation has found several applications in terms of controlled release, targeted delivery, enhanced bioavailability, improved storage stability and control over unpleasant flavours. These attributes are highly linked to surface-volume ratios and hence differentiates nano and microencapsulation from macroencapsulation. Common wall materials include lipids and food grade polymers such as polysaccharides and proteins or their combinations. This chapter presents a detailed note on various approaches for nano and microencapsulation, emphasising on the types and potential of using different types of food grade polymers. A summary of research on such aspects and the various core materials of interest are also presented. Methods to characterize encapsulated materials and challenges faced in these practices are included to provide an in-depth understanding on the subject.

Keywords Bioactive components · Encapsulation techniques · Nanoencapsulation · Polymeric wall material

14.1 Introduction

Encapsulation is one of the efficient ways of protecting bioactive components through entrapment of active component (core material) within a polymeric wall material (Gutiérrez and Álvarez 2017). The major functions of encapsulation include controlled release, targeted delivery, control over unpleasant flavours and odours, ease of handling and storage, enhanced bioavailability and absorption through cells. The core material can be flavours, colours, preservatives, sweeteners, proteins, minerals, lipids, probiotics, acids, bases, buffers, antioxidants,

S. K. Vimala Bharathi · J. A. Moses · C. Anandharamakrishnan (✉)
Indian Institute of Food Processing Technology (IIFPT), Ministry of Food Processing Industries, Government of India, Thanjavur, India
e-mail: director@iifpt.edu.in

T. J. Gutiérrez (ed.), *Polymers for Food Applications*,
https://doi.org/10.1007/978-3-319-94625-2_14

polyphenols and so on (Anandharamakrishnan and Ishwarya 2015). The wall materials include lipids and food grade polymers such as polysaccharides, proteins and some of the synthetic polymers or their combinations (Lakkis 2016). Among these wall materials, food grade polymers are most commonly used because of properties like biocompatibility, biodegradability, low viscosity at high solid content, solubility, retention of volatiles, suitability for high temperature application, possibility of modification and ability to provide structure (Fathi et al. 2014; Gharsallaoui et al. 2007; Shahidi and Han 1993; Gómez-Mascaraque et al. 2016). Further, polymers can act as versatile wall material to bind or encapsulate wide variety of bioactive core material.

Depending on the size of the resultant encapsulated particle obtained, encapsulation process can be termed as macro, micro and nanoencapsulation. In general, the properties of the encapsulated materials improve with increase in surface-volume ratio and hence, micro and nanoencapsulation are preferred than macroencapsulation. This chapter deals with various techniques of nano and microencapsulation, emphasizing on the characterization of the encapsulated particles. Also, provides a detailed note on food grade polymers as wall materials and some of the related core materials highlighting the researches on the respective areas.

14.2 Wall Materials for Encapsulation

Wall materials are the encapsulant materials that provide protection to core material and hence, the selection of these materials is based on characteristics of material to be encapsulated, as well as physico-chemical properties of wall material including solubility, film forming properties, emulsifying properties, molecular weight, and core and wall material interaction (Gharsallaoui et al. 2007). On the other hand, the wall materials used in food application must be generally regarded as safe (GRAS), biodegradable and should not affect the taste or flavour of the food (Augustin and Hemar 2009; Fathi et al. 2014).

Food grade polymers used as wall materials can be natural or synthetic. Commonly used natural polymers include polysaccharides (starch and their derivatives, cellulose and their derivatives, plant extracts and exudates, microbial and animal extracts, and marine extracts) (Wandrey et al. 2010) and proteins (plant and animal protein); whereas synthetic polymers for encapsulation include polypropylene, polyvinylacetate, polystyrene, polybutadiene, polylactic acid, polyglycolic acid, polycaprolactic acid and so on (Lakkis 2016). Different kinds of food grade polymers and their application in encapsulation is discussed in this section.

14.2.1 Natural Polymers

Natural polymers are materials obtained either directly or indirectly from natural sources. They are mainly polysaccharides and proteins from plant and animal sources. Polysaccharides from plant sources include maltodextrins, corn syrup, cyclodextrin, methylcellulose, ethylcellulose, carboxymethyl cellulose, gum tragacanth, gum Arabic, locust bean gum, guar gum, pectin and so on; whereas carrageenans and alginates are polysaccharides from marine source. And, polysaccharides from microbiological and animal extracts includes xanthan, gellan, dextran, chitosan, etc., Similarly, proteins from plant sources include gluten, soy isolate, pea isolate; whereas milk proteins (caseins, whey protein) and gelatin are from animal sources (Wandrey et al. 2010).

In food applications, these polymers either alone or in combination are used as wall materials. For instance, Xue et al. (2013) encapsulated tomato oleoresin using zein, obtained from corn gluten meal through ultrasound assisted extraction. A 18 days study on lycopene content found that at 35 °C storage, maximum 81.91% of lycopene content was lost; whereas at 4 °C storage temperature, only 23.63% lycopene content was lost. The study concludes that by decreasing the storage time and temperature, lycopene content loss can be reduced. A similar study was conducted by Li et al. (2015) by encapsulating tomato oleoresin with Soy protein isolate-gum acacia conjugate and during 30 days of lycopene retention study, it was found that loss % of lycopene content was 8.96% and 30.94% at 4 °C and 35 °C storage temperature, respectively. Both these studies found that encapsulation reduced the release of lycopene content in stomach and improved the rate of release in intestine. This proves that encapsulation is better suitable for lycopene retention because lycopene must be protected at the stomach and must be released only in the intestine. Also, on comparing the above two studies, it can be seen that the conjugates retained more lycopene content for longer storage period. At times, single wall material can be beneficial and effective than that of combinations. For instance, sulforaphane stability was higher when encapsulated with maltodextrin than that of maltodextrin and gum arabic (25:75) (Wu et al. 2014). Hence the choice of wall material chosen has greater impact on the final encapsulated product. The impact of different combinations of wall material on final product is explained in Table 14.1.

14.2.2 Synthetic Polymers

Synthetic polymers are the polymers synthesized by chemical reaction, through polymerization of repeating units of monomers. Though synthetic polymers are widely used in drug encapsulation, not all polymers used in drug encapsulation are accepted in food applications (Kailasapathy 2016). Lemma et al. (2015) improved the shelf life and thermal stability of retinyl acetate by encapsulating with poly (vinyl alcohol) (PVA) and β-cyclodextrin. The study also showed that retinyl acetate

Table 14.1 Effect of combined wall materials on the properties of final encapsulated product

Wall materials				
Primary	Combined with	Core material	Changes	References
Gum arabic	Maltodextrin	Spent coffee grounds extract	Improved retention of phenolic compounds, flavonoids and reduced antioxidant activity (by total antioxidant activity assay)	Ballesteros et al. (2017)
Maltodextrin	Pectin	Anthocyanin rich extract from Jaboticaba pomace	Decrease in solubility with slight increase in water activity and hygroscopicity; Increased retention of phenolics	Souza et al. (2017)
	Soy protein isolate		Decrease in solubility and hygroscopicity; Insignificant decrease in water activity; Increased retention of phenolics	
	Pectin and soy protein isolate		Water activity increased significantly with decrease in hygroscopicity and solubility; Increased retention of phenolics	
Lactose and whey protein isolate	Maltodextrin	Ethyl butyrate	Delayed lactose crystallization and sustained flavour release during storage	Li et al. (2016)
Maltodextrin	Gum arabic	Anthocyanins from *Berberis vulgari*	Highest encapsulation efficiency; Reduced porosity and hygroscopicity, and increased solubility and degree of caking	Mahdavi et al. (2016)
	Gelatin		Higher encapsulation efficiency; Insignificant increase in hygroscopicity, solubility and degree of caking, and reduced porosity	
Fructooligosaccharide	Whey protein isolate	*Lactobacillus plantarum*	Reduced stickiness and aggregation	Rajam and Anandharamakrishnan (2015a)
	Denatured whey protein isolate		Reduced stickiness and aggregation, and improved encapsulation efficiency	
Whey protein concentrate and sodium caseinate	Glucose syrup	Shrimp oil	Improved encapsulation efficiency; With appropriate antioxidants enhanced oxidative stability	Takeungwongtrakul et al. (2015)

Wall materials				
Primary	Combined with	Core material	Changes	References
Whey protein isolate	Maltodextrin/ inulin	Fish oil	Improved wettability and reduced surface oil; Decreased glass transition temperature	Botrel et al. (2014a)
Fructooligosaccharide	Denatured whey protein isolate	*Lactobacillus plantarum*	Less reduction in the relative cell viability during gastric and bile juice exposure	Rajam et al. (2014)
Gum arabic	β-cyclodextrin	Sulforaphane	Improved storage stability	Wu et al. (2014)
	Maltodextrin		Slight decrease in encapsulation efficiency and encapsulation yield; Improved storage stability	
Maltodextrin	Gum arabic		Increased encapsulation yield, whereas slight decrease in encapsulation efficiency and storage stability	
Whey protein isolate	Sodium caseinate	Fish oil	Reduced particle size with corresponding increase in bulk density; Slight decrease in encapsulation efficiency	Aghbashlo et al. (2012a)

with PVA/β-cyclodextrin exhibited better thermal stability than that with only PVA nanofibers. Similarly, to improve the stability of the curcumin, Rachmawati et al. (2016) encapsulated curcumin with poly (lactic acid) and dichloromethane by using vitamin E and polyethylene glycol succinate as solvent and surfactant, respectively. The maximum encapsulation efficiency was found to be 89.42% with 5% curcumin in it. A study on nanoencapsulation of roasted coffee oil using poly (L-Lactic acid) (PLLA) and poly (hydroxyl butyrate-*co*-hydroxyvalerate) (PHBV) as wall materials, revealed that volatile concentration and aroma constituents can be retained effectively using encapsulation (Freiberger et al. 2015). Several patents were filed on the encapsulation of active food ingredients using synthetic polymers (Cherukuri et al. 1991; Cherukuri and Mansukhani 1989, 1990; Wei et al. 1986; Yatka et al. 1992a). For instance, polyvinyl acetate blended with plasticizers, when used as encapsulant material for sweeteners, forms a film that can be retain the moisture and enhance the controlled release of the core material. This when applied in chewing gum composition with proper manipulation, the sweetness perception can be prolonged during mastication (Yang 1988). A similar patent was filed for the method to produce crystalline sucralose encapsulated with polyvinyl acetate, to retain the sweetness for longer time with controlled release (Yatka et al. 1992b). Recently, gum based chewing gum was replaced with 70% by weight of polyvinyl acetate with improved release of flavours and sweeteners. The obtained chewing gum was similar to toffee in terms of texture (Wittorff and Lund 2017).

14.3 Core Materials for Encapsulation

Core materials are the active components (solid, liquid or gas) that are encapsulated within the wall material. The properties of the core materials define the nature of the final product. Thus, based on the arrangement of core materials, final encapsulates obtained can be mononuclear (single continuous phase of core material within the wall material), polynuclear (multiple units of core material inside the wall material) or matrix-type capsule (core material is homogeneously distributed in the wall material) (Anandharamakrishnan and Ishwarya 2015).

14.4 Techniques for Micro and Nanoencapsulation

Apart from the properties of core materials and wall materials, encapsulation techniques adapted can alter the physicochemical properties of the final encapsulated material (Table 14.2). For instance, Ezhilarasi et al. (2013a) microencapsulated hydroxycitric acid, a thermally sensitive compound with maltodextrin, whey protein isolate, and their combination, using freeze drying. Upon incorporation in bread, the whey protein isolate encapsulated microcapsules were better in terms of

Table 14.2 Particle size and encapsulation efficiency different encapsulated products with various encapsulation techniques

Wall material	Core material	Encapsulation technique	Particle size	Encapsulation efficiency	References
Cashew tree gum	Fish oil	Spray drying	29.9 μm (D_{43})	75.80%	Botrel et al. (2017)
Modified starch			10.2 μm (D_{43})	75.80%	
Arabic gum			8.83 μm (D_{43})	59.80%	
Zein	Green tea catechins	Electrospraying	174.3–188.1 nm	97.45–86.84%	Bhushani et al. (2017)
Collagen and alginate	Avocado peel extract	Complex coacervation and spray drying	37.95 μm (D_{43}) 9.94 μm (D_{32})	64.34%	Calderón-Oliver et al. (2017)
		Complex coacervation and freeze drying		67%	
	Nisin	Complex coacervation and spray drying		63.70%	
		Complex coacervation and freeze drying		66.50%	
Collagen and pectin	Avocado peel extract	Complex coacervation and spray drying	39.39 μm (D_{43}) 10.17 μm (D_{32})	67.08%	
		Complex coacervation and freeze drying		69.64%	
	Nisin	Complex coacervation and spray drying		68.28%	
		Complex coacervation and freeze drying		69.88%	

(continued)

Table 14.2 (continued)

Wall material	Core material	Encapsulation technique	Particle size	Encapsulation efficiency	References
Poly (3-hydroxybutyrate-*co*-3-hydroxyvalerate)	Resveratrol	Solution enhanced dispersion by supercritical fluids	390–570 nm	99.54%	Dal magro et al. (2017)
Gelatin	Curcumin	Electrohydrodynamic atomization	200–400 nm	97%	Gómez-Estaca et al. (2017)
Hydroxypropyl methyl cellulose	Fish oil	Supercritical antisolvent process	58.35 μm	69.55–81.75%	Karim et al. (2017)
Chitosan	Squalene	Spray drying	344.6–489.9 nm	26%	Kumar et al. (2017)
Sodium caseinate and maltodextrin	Nigella sativa oil	Spray drying	17.48–31 μm	33.33–92.71%	Mohammed et al. (2017)
Alginate and Shellac	Sunflower oil	External gelation	2.13–2.80 mm	63.20–98.70%	Morales et al. (2017)
Poly (DL-lactide-*co*-glycolide)	Passion fruit seed and cake extracts	Emulsion/solvent evaporation	355–470 nm	23.80–79%	Oliveira et al. (2017a)
Poly(lactic-*co*-glycolic) acid	Passion fruit seed oil	Supercritical antisolvent process	721–1498 nm	67.80–91%	Oliveira et al. (2017b)
Alginate	Linseed oil	Ionotropic gelation	1.77 μm	98.30%	Piornos et al. (2017)
Maltodextrin and gum arabic	Drumstick (*Moringa oleifera*) oil	Spray drying	22.56–28.03 μm (D_{43})	82.67–91.05%	Premi and Sharma (2017)
Maltodextrin and whey protein isolate			11.43–18.48 μm (D_{43})	66.23–73.43%	
Poly caprolactone	Fish oil	Supercritical fluid extraction of emulsions	6–73 nm	40%	Prieto and Calvo (2017b)

Wall material	Core material	Encapsulation technique	Particle size	Encapsulation efficiency	References
Poly vinylpyrrolidone	α-tocopherol	Supercritical antisolvent coprecipitation	1.69–4.08 μm	53%	Prosapio et al. (2017)
	Menadione		2.64–5.09 μm	50%	
Octenyl succinic anhydride modified starches (Hi-Cap 100™)	β-carotene, eugenol and flax seed oil	Spray drying	–	92.08%	Sharif et al. (2017)
Octenyl succinic anhydride modified starches (Gum 2000)				88.71%	
Maltodextrin	Anthocyanin rich extract from Jaboticaba pomace	Freeze drying	370.89 μm (D_{43})	–	Souza et al. (2017)
Maltodextrin and pectin			341.60 μm (D_{43})		
Maltodextrin and whey protein isolate			311.66 μm (D_{43})		
Maltodextrin, whey protein isolate and pectin			319.77 μm (D_{43})		
Vanillic acid grafted chitosan	Sardine oil	Spray drying	2.3 μm	84%	Vishnu et al. (2017)
Alginate and whey proteins	Dandelion (*Taraxacum officinale* L.) polyphenols and β-carotene	Emulsification/internal gelation	318.48 μm	77.35%	Belščak-Cvitanović et al. (2016)
Lactoferrin-glycomacropeptide	Curcumin	Thermal gelation	112 nm	95.10%	Bourbon et al. (2016)
	Caffeine		126 nm	90%	

(continued)

Table 14.2 (continued)

Wall material	Core material	Encapsulation technique	Particle size	Encapsulation efficiency	References
Lentil protein isolate and maltodextrin	Canola oil	Spray drying	9.15 μm (D_{43})	75.30%	Chang et al. (2016)
Lentil protein isolate, maltodextrin and sodium alginate			9.04 μm (D_{43})	87.90%	
Sodium alginate and citric pectin	*Lactobacillus plantarum*	Electrospraying	80–170 μm	–	Coghetto et al. (2016)
Gelatin and cashew gum	Astaxanthin	Complex coacervation	32.7 μm	59.90%	Gomez-Estaca et al. (2016)
Brown rice flour and *β*-cyclodextrin	curcumin (1,7-bis(4-hydroxy-3-methoxyphenyl)-1,6-heptadien-3,5-dione)	Spray drying		7.89–97.11%	Laokuldilok et al. (2016)
	Demethoxycurcumin			7.49–86.66%	
	Bis-demethoxycurcumin			3.30–92.86%	
Lactose and whey protein isolate	Ethyl butyrate	Spray drying	23.73 μm	48.80%	Li et al. (2016)
Lactose, whey protein isolate and maltodextrin			21.57–26.50 μm	36.81–46.52%	
Maltodextrin and gum arabic	Anthocyanins from *Berberis vulgari*	Spray drying	–	89.09–96.21%	Mahdavi et al. (2016)
Maltodextrin and gelatin				87.57–94.97%	
Maltodextrin				86.07–93.09%	
Whey protein isolate and inulin	Annatto seed oil	Freeze drying	124 μm (D_{43})	88%	Silva et al. (2016)
Amaranth protein isolate and pullulan	Quercetin	Electrospinning	260.8 nm	93.60%	Aceituno-Medina et al. (2015b)
	Ferulic acid		362.6 nm	83.70%	

Wall material	Core material	Encapsulation technique	Particle size	Encapsulation efficiency	References
Gelatin	*Saccharomyces cerevisiae*	Spray drying	4.08 μm (D_{32}) 21.38 μm (D_{43})	–	Arslan et al. (2015)
Whey protein concentrate			3.31 μm (D_{32}) 10.64 μm (D_{43})		
Modified starch			3.53 μm (D_{32}) 13.03 μm (D_{43})		
Maltodextrin			3.47 μm (D_{32}) 10.91 μm (D_{43})		
Pea protein isolate			3.12 μm (D_{32}) 8.56 μm (D_{43})		
Gum arabic			3.31 μm (D_{32}) 9.94 μm (D_{43})		
Poly caprolactone	Indomethacin	Nanoprecipitation	290–359 nm	65–75%	Badri et al. (2015)

(continued)

Table 14.2 (continued)

Wall material	Core material	Encapsulation technique	Particle size	Encapsulation efficiency	References
Ethyl cellulose	Quercetin	Supercritical anti-solvent process	150–350 nm	99%	Fernández-Ponce et al. (2015)
Whey protein isolate	Vanillin	Spray freeze drying	24.76 μm	72%	Hundre et al. (2015)
		Freeze drying	36.91 μm	76.8%	
		Spray drying	14.18 μm	86.2%	
β-cyclodextrin		Spray freeze drying	165.40 μm	71%[a]	
		Freeze drying	120.10 μm	80%[a]	
		Spray drying	42.58 μm	78%[a]	
Whey protein isolate and *β*-cyclodextrin		Spray drying	41.18 μm	77%[a]	
Soy protein isolate-gum acacia conjugate	Tomato oleoresin	Spray drying	–	69.25–84.69%	Li et al. (2015)
Fructooligosaccharide	*Lactobacillus plantarum*	Spray drying	15.44–23.89 μm (D_{43})	70.77–72.82%	Rajam and Anandharamakrishnan (2015a)
Fructooligosaccharide and whey protein isolate			7.34–8.97 μm (D_{43})	88–96%[a]	
Fructooligosaccharide and denatured whey protein isolate			6.68–13.62 μm (D_{43})	98.63%	
Maltodextrin	*Cinnamomum zeylanicum*	Spray drying	10.01–12.51 μm (D_{43})	85%	Santiago-Adame et al. (2015)

Wall material	Core material	Encapsulation technique	Particle size	Encapsulation efficiency	References
Whey protein isolate and sodium caseinate	Shrimp oil	Spray drying	10.09 μm (D_{43})	51.71%	Takeungwongtrakul et al. (2015)
Whey protein isolate, sodium caseinate and gum arabic			10.32 μm (D_{43})	51.78%	
Whey protein isolate, sodium caseinate and glucose syrup			10.91 μm (D_{43})	86.31%	
Whey protein isolate, sodium caseinate and maltodextrin			10.04 μm (D_{43})	59.81%	
Whey protein isolate and inulin	Fish oil	Spray drying	9.6–15.6 μm (D_{43})	–	Botrel et al. (2014a)
Whey protein isolate	Fish oil	Spray drying	18.6 μm (D_{43})	–	Botrel et al. (2014b)
Whey protein isolate and maltodextrin			17.3 μm (D_{43})		
Whey protein isolate and inulin			15.6 μm (D_{43})		
Starch	Rosemary essential oil	Spray drying	13.4 μm (D_{43})	26.31–61.81%	De Barros Fernandes et al. (2014a)
Gum arabic			13.5 μm (D_{43})		
Whey protein isolate and inulin	Rosemary essential oil	Spray drying	11.5–11.9 μm	28.97–38.34%	De Barros Fernandes et al. (2014b)
Poly (3-hydroxybutyrate-*co*-3-hydroxyvalerate)	astaxanthin from *Haematococcus pluvialis*	Solution enhanced dispersion by supercritical fluids	128 nm	48.25%	Machado Jr et al. (2014)

(continued)

Table 14.2 (continued)

Wall material	Core material	Encapsulation technique	Particle size	Encapsulation efficiency	References
Maltodextrin	Polyphenol from *Orthosiphon stamineus*	Spray drying	4.87–6.93 μm (D_{43})	–	Pang et al. (2014)
Whey protein isolate			4.09–9.30 μm (D_{43})		
Whey protein isolate and soluble corn fiber	Astaxanthin	Spray drying	2–25 μm	63.20–92.69%	Shen and Quek (2014)
Sodium caseinate and Soluble corn fiber				65.40–95.19%	
Maltodextrin	Sulforaphane	Spray drying	2–4 μm	39.10%	Wu et al. (2014)
Gum Arabic				39.80%	
κ-carrageenan				12.60%	
Maltodextrin and gum arabic				34%	
Gum arabic and β-cyclodextrin				29.40%	
Poly(hydroxybutyrate-*co*-hydroxyvalerate)	Grape seed extract	Solution enhanced dispersion by supercritical fluids	0.70 μm	66.01%	Boschetto et al. (2013)
Maltodextrin and modified starch (Hi-Cap 100™)	Flax seed oil	Spray drying	19.79 μm (D_{43})	95.7%	Carneiro et al. (2013)
Maltodextrin and whey protein concentrate			17.98 μm (D_{43})	62.3%	
Maltodextrin and gum Arabic			23.03 μm (D_{43})	80%[a]	
Maltodextrin and modified starch (Capsul TA®)			15.32 μm (D_{43})	88%[a]	

Wall material	Core material	Encapsulation technique	Particle size	Encapsulation efficiency	References
Zein	Gallic acid	Electrospinning	327–387 nm	–	Neo et al. (2013)
Poly ε-caprolactone	α-tocopherol	Nanoprecipitation	184.4–219.1 nm	65.55–99.97%	Noronha et al. (2013)
Zein	Tomato oleoresin	Spray drying	–	74.19–90.46%	Xue et al. (2013)
Whey protein concentrate	Fish oil	Spray drying	2.26–2.62 μm	40.59–49.60%	Aghbashlo et al. (2012a)
Whey protein isolate			2.38–3.10 μm	68.36–75.65%	
Skim milk powder			3.09–4.59 μm	71.26–81.94%	
Whey protein isolate and sodium caseinate			1.37–1.53 μm	65.33–72.19%	
Whey protein isolate and milk protein concentrate			1.68–2.41 μm	66.28–75.72%	

(continued)

Table 14.2 (continued)

Wall material	Core material	Encapsulation technique	Particle size	Encapsulation efficiency	References
Skim milk powder	Fish oil	Spray drying	4.25 μm	80.75%	Aghbashlo et al. (2012b)
Skim milk powder and maltodextrin			3.21 μm	76.22%	
Skim milk powder and lactose			5.21 μm	84.96%	
Skim milk powder and sucrose			5.37 μm	84.92%	
Skim milk powder and tween			3.96 μm	81.42%	
Skim milk powder, maltodextrin and tween			3.07 μm	77.21%	
Skim milk powder, lactose and tween			5.11 μm	84.94%	
Skim milk powder, sucrose and tween			5.33 μm	85.12%	
Soybean lecithin	*β*-carotene	Particles from Gas Saturated Solutions-drying	10–500 μm	60%	de Paz et al. (2012)

[a]Approximate values based on graph given in the reference
[b]Surface mean diameter (D_{32}) and volume mean diameter (D_{43}) (Arslan *et al. 2015)*

free hydroxycitric acid concentration, loaf volume, texture, color and other sensory characteristics. Similarly, on encapsulating the same compound with same set of wall materials using spray drying, the better quality bread were obtained for capsules encapsulated with maltodextrin (Ezhilarasi et al. 2014). It is, thus, crucial to select the appropriate encapsulation technique. The selection of encapsulation techniques depend on nature of the core material and wall material, final particle size required, encapsulation efficiency, release mechanism and so on. Microencapsulation techniques can be divided into chemical and physical processes. Chemical microencapsulation techniques include: polymerization (*in situ* and interfacial) and polycondensation; And, physical microencapsulation techniques include: physico-chemical (coacervation, sol-gel and supercritical CO_2 assisted microencapsulation) and physico-mechanical (spray drying, fluid bed coating, centrifugal encapsulation, vacuum encapsulation and electrostatic encapsulation) (Ghosh 2006). Further, microencapsulation technique can also be divided on the basis of medium used for suspension. The techniques that use liquid as suspending medium are coacervation, emulsification-solvent evaporation, *in situ* and interfacial polymerization, and the techniques that use gas as suspending medium are spray drying, spray chilling/cooling, fluidized bed coating (Schrooyen et al. 2001).

Nanoencapsulation techniques adopt either bottom-up or top-down approach for synthesis of nanoparticles. Bottom-up approach techniques like coacervation, inclusion complexation, nanoprecipitation and super critical fluid technique, synthesize nanomaterials by self-assembly and self-organisation of individual molecules. pH, temperature, concentration, ionic strength, stresses due to mechanical, electrical, and magnetic forces influences the particle formed. And, top-down approach techniques like emulsification and emulsification-solvent evaporation, involves size reduction of bulk materials to get nanoparticles (Augustin and Sanguansri 2009; Ezhilarasi et al. 2013b). Moreover, nanoencapsulation techniques can also be divided into liquid-based (coacervation, emulsification-solvent evaporation, inclusion complexation, nanoprecipitation and supercritical fluid technique), electro-hydrodynamic processes (electrospinning and electrospraying) and drying techniques (spray drying and freeze drying) (Anandharamakrishnan 2014). In this section, some of the most commonly used encapsulation techniques are explained by highlighting some of the recent researches explored in particular areas (Gutiérrez 2018).

14.4.1 Coacervation

Coacervation technique involves suspension of the core material in polymeric (wall material) solution and then deposition of the wall material around core material after phase separation. Phase separation can either be induced by de-solvating (coacervating) agents (simple coacervation) or by oppositely charged polymers (complex coacervation). Finally, the stabilization of particles can be done using suitable cross-linking agent or by heat treatment (Jyothi et al. 2010). The nature of the polymers and their interaction plays a major role in size of the final capsule formed. For example, formation of nano-sized particles is restricted for strongly negative charged polysaccharides, as their coacervation leads to insoluble or precipitated complexes (Lv et al. 2014). Hence, Lv et al. (2014) used gelatin and gum arabic as wall material for preparation of nanocapsules of jasmine essential oil with transglutaminase as a cross-linking agent. The resulted encapsulated oil possessed a good heat-resistant characteristic against humid heat (80 °C).

Coacervation technique is advantageous in many aspects like very high achievable payloads (max 99%), high encapsulation efficiency (approx. 90%), requires neither high temperature nor organic/toxic solvents, capsules obtained are heat-resistant, controlled and sustained release on mechanical stress and temperature, but the major restriction is in food applications is the use of gluteraldehyde as cross-linking agent. Yet, this can be overcome by the application of enzymes or natural cross-linking agent such as genipin, anionic polysaccharides and proteins (Dima et al. 2014; Ezhilarasi et al. 2013b; Gouin 2004; Yin and Li 2012). For instance, Jain et al. (2016) microencapsulated β-carotene using casein and gum tragacanth as wall material and genipin as cross-linking agent. The study found that the intensity of electrostatic attraction of complex coacervation was optimum at pH of 4.35 with casein and gum tragacanth ratio as 2:1. The stability and antioxidant activity of microencapsulated β-carotene was found to be improved. *In vitro* studies revealed the release pattern of β-carotene as biphasic (burst release initially and then sustained release). Several other studies also used genipin as cross-linking agent that includes microencapsulation of vanilla oil with chitosan and gum arabic (Yang et al. 2014) and mustard seed oil with gelatin and gum arabic (Peng et al. 2014).

Another limitation is that this technique is more suitable only for hydrophobic compounds. Still, Santos et al. (2015) encapsulated xylitol (a hydrophilic compound) through coacervation technique by using double emulsion as the intermediate process, followed by freeze drying. Here, W/O and W/O/W emulsion was prepared by adding xylitol into corn oil and gelatin respectively. The obtained microcapsules possessed characteristics like optimum particle size (78.45–109.31 μm), low solubility (5.31–13.90%) and good encapsulation efficiency (31.42–71.93%); those are suitable for food applications in terms of sweetness and freshness. The *in vitro* study conducted for the obtained microcapsule revealed that over 70% xylitol was released in artificial saliva within 20 min. Thus, coacervation technique can be efficiently used to micro and/or nanoencapsulate heat-sensitive compounds with appropriate use of cross-linking agents and wall materials.

14.4.2 Fluid Bed Coating

Fluid bed coating can be used to encapsulate solid particles like powders and granules uniformly (Gupta et al. 2016). It involves spray coating of the wall material on the fluidized core material. Initial works on fluid bed coating was initiated by Dr. Dale E. Wurster in 1950s (Battista et al. 1965; Lindlof and Wurster 1964, 1965; Wurster 1957, 1963, 1966; Wurster and Lindlof 1966) and hence the process is also known as 'Wurster process'. Fluid bed coating can be bottom spray, top spray and tangential spray (rotary and static), depending on whether the direction of spray is concurrent, counter-current or tangential to the fluidized particles respectively. Here, the major difference between rotary and static tangential spray type lies in the base plate provided for fluidizing the core material. The rotary type has non-perforated rotating plate and provides air in the outer edge, whereas static type has diagonally cut perforations and provides air angularly (Frey 2014). For food industry that deals with powders of wide particle size distribution, tapered/conical fluid bed with top-spray type is preferred, because of relatively higher versatility and batch size, as well as simplicity of cleaning and disassembly (Depypere et al. 2009). The efficiency of fluidized bed coating is governed by factors like process temperature and humidity, inlet air and atomization air temperature, liquid feed rate, fluidization air and atomization air volume (Meiners 2012). Several studies were conducted and appropriate models were validated, especially computational fluid dynamics, related to the above mentioned factors (Duangkhamchan et al. 2012, 2015). A study was conducted by Palamanit et al. (2013) on the effect of inlet air temperature and spray rate of coating solution on quality attributes of rice with turmeric extract during top-spray fluidized bed coating and the study revealed that the percentage of fissured rice kernel and its moisture content depends on both spray rate of coating solution and temperature of inlet air. Further, the spray rate of coating solution had effect on the colour, total antioxidant capacity and total phenolics content of turmeric coated rice kernel.

Fluidized bed coating is advantageous over freeze drying in terms of cost (Schell and Beermann 2014). And, at times, spray dried encapsulated particles may leave some portion of the active core material exposed because of structural collapse of wall material on exposure to heat. To protect those active components, fluid bed coating technique can be adopted for spray dried particles. Coronel-Aguilera and San Martín-González (2015) encapsulated spray dried β-carotene emulsion using fluid bed coating techniques and hydroxypropyl cellulose as wall material. The obtained microcapsules had particles size and moisture contents in the range of 48.6–69.5 μm and 1.69–8.53 g/100 g, respectively. Furthermore, the particles obtained at lower coating temperatures (60 and 70 °C) and lower feed rate of wall material solution showed better encapsulation than that at 80 °C. And, the particles exhibited stable colour in acidic media (yogurt) for a period of 4 weeks of refrigerated storage. Thus, fluidized bed coating with proper understanding of the parameters involved, can be used to overcome some of the disadvantages of spray and freeze drying to obtain uniform coated powders with high encapsulation efficiency.

14.4.3 Emulsification-Solvent Evaporation

Encapsulation by emulsification-solvent evaporation is done by dissolving the wall material and the core material in water immiscible, low boiling point solvents and then emulsifying the solution in aqueous emulsifier solution. Finally, microspheres or nanospheres are produced by removal of solvent through evaporation (Jafari et al.

2017). The evaporation may be through the application of temperature, pressure or continuous stirring (Soppimath et al. 2001). The size of capsules formed depends on type and amount of dispersing agent, stirring rate, viscosity and quantity of the organic and aqueous phase, and temperature (Tice and Gilley 1985). Also, the rate of solvent evaporated has a considerable effect on the characteristics of the capsule formed. Rapid solid removal encourages the formation of porous capsules with amorphous and hardened polymer (Tewes et al. 2006). The x-ray diffraction and differential scanning calorimetry studies on nanocapsule of β-carotene encapsulated with whey protein isolate, sodium caseinate and soybean protein isolate found that crystalline β-carotene was transformed into amorphous form in nanosystems. The reason predicted in the study was it may be due to precipitation during solvent evaporation or residual lipids extraction from the proteins (Yi et al. 2015).

In general, this technique is employed for encapsulation of lipophilic compounds (Ezhilarasi et al. 2013b). But, double emulsion solvent evaporation facilitates to adopt this technique for both lipophilic and hydrophilic compounds (Zafar et al. 2017). The most commonly used polymer wall materials are polylactic acid, poly(lactic-*co*-glycolic acid), ethylcellulose, cellulose acetate phthalate, poly(ε-caprolactone) and poly(β-hydroxybutyrate) (Reis et al. 2006). Recently, Teo et al. (2017) conducted a comparative study on the kinetic stability and cellular uptake of lutein in whey protein isolate-stabilized nanoemulsions, prepared by solvent evaporation method and conventional emulsions. The particle sizes of nanoemulsions and emulsions were 68.8 nm and 147.3 nm, respectively. And, the encapsulation efficiencies were 80.7% and 86.3%, respectively. The results of the studies revealed that though nanoemulsions showed higher lutein loss than conventional emulsions, the Caco-2 cellular uptake of lutein was almost 2.5 times higher in nanoemulsions than that of conventional emulsions.

Considering the usage of solvent, this technique is not preferred for food application (Lee and Wong 2014), but this technique is suited for active components that are susceptible to heat and oxygen. Recently, Iron salt ($FeSO_4$. $7H_2O$) was encapsulated in gum tragacanth using solvent evaporation technique with alcohol as the solvent. The optimized contents of 5% iron salt, 22% gum tragacanth and 11:1 alcohol: mixture ratio resulted in maximum encapsulation efficiency of 60.13% with 55.31 μm sized microcapsules. Also, these capsules exhibited complete release and upto 80% release in simulated gastric fluid and water, respectively (Asghari-Varzaneh et al. 2017). A similar study was conducted using maltodextrin, gum Arabic and modified starch as wall material. The microcapsules prepared with maltodextrin, gum arabic and modified starch in the proportion of 1:4:1 and 10:1 absolute alcohol: mixture ratio resulted in maximum encapsulation efficiency of 91.58% and 15.54 μm average sized particles. These microcapsules when fortified in milk showed remarkable difference in sensory scores and higher *in vitro* bioavailability of iron as compared to those of iron salt fortified milk. And, TBA values were also comparatively lower upto fifth day of storage (Gupta et al. 2015). Thus, emulsification-solvent evaporation technique helps in better encapsulation of lipophilic components and high temperature and oxygen susceptible components with particles of required size.

14.4.4 Inclusion Complexation

Inclusion complexation refers to encapsulation through various intermolecular interactions like hydrophobic interaction, hydrogen bonds, van der Waals interactions, stearic interactions between hydrophopic molecules (core material) and amphiphilic structured molecular compounds like β-cyclodextrin and β-lactoglobulin (wall material) (Ezhilarasi et al. 2013b; Kfoury et al. 2016). Shpigelman et al. (2014) formed complex between naringenin and β-lactoglobulin to improve the solubility. The complex prevented crystallization of naringenin up to 3 times its solubility limit, with the particle size of 10 nm (approx.). These two factors make the complex suitable for application in clear beverages. Similarly, limonene and β-cyclodextrin complexes were spray dried and incorporated into lemon juice. The encapsulated particles size ranged from 1 to 3 μm and accelerated aged studies shown that after 10 days, which mimics 9 months of storage, 40% of complexed limonene remained in the model beverage (Do Carmo et al. 2017). The encapsulation of certain essential oils like eugenol, thymol, carvacrol, *Citrus sinensis, Hyptis martiusii,* cinnamon bark and clove bud extract in β-cyclodextrin using inclusion complexation technique were proved to improve/prolong the antimicrobial activity against *Peronophythora litchii* (Gong et al. 2016), *Escherichia coli* (Tao et al. 2014), *Aspergillus niger* (Bernardos et al. 2015), *Aedes aegypti.* larvae (Galvão et al. 2015), *Salmonella enterica* serovar Typhimurium LT2 and *Listeria innocua* (Hill et al. 2013).

In food packaging applications, the inclusion complexes formed were incorporated in electrospun fibers and factors like thermal stability, durability and antimicrobial activities were studied (Aytac et al. 2014; Kayaci et al. 2013, 2014; Mallardo et al. 2016; Wen et al. 2016a). Recently, Aytac et al. (2017) fabricated zein electrospun nanofibers incorporating thymol and γ-cyclodextrin inclusion complex and the formed material was most effective at reducing the bacterial count in meat stored up to 5 days at 4 °C. Another study was conducted by Celebioglu et al. (2016) to produce polymer-free nanofibers using vanillin/cyclodextrin inclusion complexes. The study revealed that maximum achievable vanillin loading was higher in the polymer-free fiber than that of the polymeric nanofiber with improved thermal stability and solubility. Deng et al. (2016) prepared clove essential oil- β-cyclodextrin inclusion complex based active packaging film using polyvinyl alcohol as film forming material and studied the migration behaviour with three kinds of solvents: 3% acetic acid, 10% ethanol and olive oil. The migration behaviour was different for different solvents with olive oil exhibiting the fastest migration. The concentration of clove oil that reached equilibrium in olive oil, ethanol and acetic acid was after 32 h, 72 h and 80 h, respectively. Thus, inclusion complexation can be used to encapsulate essential oils, vitamins, minerals, polyphenols and other volatile components and convert them into free flowing powders. And, encapsulated components can exhibit improved thermal stability, solubility, and antimicrobial and antioxidant activity, reduce volatility and controlled release, without the use of organic solvents.

14.4.5 Nanoprecipitation

Both nanospheres and nanocapsules can be formed through nanoprecipitation (also known as solvent displacement/interfacial deposition) technique of incorporating the organic phase containing dissolved wall material and core material in organic solvent into the aqueous phase containing a stabilizer (surfactant), followed by vaporization of solvent (Ezhilarasi et al. 2013b). Nanoprecipitation can produce particles of size between 100 and 300 nm (Bareras-Urbina et al. 2016). A comparative study on nanoprecipitation and emulsification-diffusion of diclofenac revealed that the lower particle size and higher zeta potential were obtained for nanocapsules obtained via nanoprecipitation method (Mora-Huertas et al. 2012). The commonly used polymers for encapsulation are poly(lactic acid), poly(lactide-*co*-glycolide), poly(alkyl cyanoacrylate) and poly(ε-caprolactone), and the corresponding copolymers with poly(ethylene-glycol) (Lepeltier et al. 2014). Noronha et al. (2014) incorporated α-tocopherol nanocapsules encapsulated with poly(ε-caprolactone) via nanoprecipitation method into methylcellulose. The film exhibited potential antioxidant ability because of the presence of α-tocopherol. Further, the study found that films acted as a barrier against UV and visible light that prevents photooxidation of foods and concluded that α-tocopherol incorporated methylcellulose films can be used as an efficient tool for food preservation.

Nanoprecipitation technique includes: flash precipitation and two step nanoprecipitation. Flash nanoprecipitation is used to obtain uniformly sized nanoparticles with high encapsulation efficiency. It involves rapid mixing with the help of high-efficiency mixers to obtain particles in the range of 50–150 nm (Bareras-Urbina et al. 2016). A study was conducted on encapsulation of curcumin using flash nanoprecipitation with an auxiliary stabilizer, followed by freeze drying, in order to improve the stability of nanoparticles prepared. The study revealed that stability and particle size of the generated nanosuspensions is influenced by molecular weight of polymeric stabilizer (polyethylene glycol-*b*-poly(DL-lactide) di-block copolymer), mixing rate and drug-to-copolymer ratio. The nanosuspensions were stable only for about 2 h. But, poly (vinylpyrrolidone) (as auxiliary stabilizer) addition extended the stability of the nanosuspensions to 5 days and 2 weeks at ambient conditions and 4 °C, respectively (Chow et al. 2015).

Two step nanoprecipitation involves two steps: In step 1, the core material is precipitated using a solvent and nanoparticles are created; whereas in step 2, the formed nanoparticles are mixed with the mixture of polymer and solvent. Finally, the polymer precipitates out encapsulating the core material nanoparticle (Bareras-Urbina et al. 2016). Morales-Cruz et al. (2012) encapsulated cytochrome *c* (Cyt-c) in poly (lactic-*co*-glycolic) acid. The enzymes, lysozyme and α-chymotrypsin, were used for the optimization. They concluded that two-step nanoprecipitation helped in obtaining high encapsulation efficiency and improve protein protein stability, which is not possible through one-step nanoprecipitation as protein gets denatured when dissolved in organic solvent. Nanoprecipitation technique can be efficiently used to obtain uniform sized particles with high efficiency and high drug loading capacity without the use of high temperature (Bazylińska et al. 2014). Further, the desired particle size rang can be obtained through control of the nucleation and growth of particles in solution (Chow et al. 2015).

14.4.6 Supercritical Fluid Technique

Supercritical fluids possess the properties of both liquids and gases, when exposed above its critical temperature and pressure. In general, supercritical fluids like alkanes (C_2 to C_4), supercritical carbon-dioxide and nitrous oxide (N_2O) are used (Ghosh 2006). Among these, carbon-dioxide is commonly used in food applications. The spotlighted properties of supercritical fluids include high diffusivities, high solvating power, high mass transfer rates, low density and low viscosity (Gouin 2004). Supercritical fluid techniques involve dispersion of core material and dissolved (hydrophobic) or swollen (hydrophilic) wall material in supercritical fluids and then the fluid is released to form nanoencapsulates (Gouin 2004; Jafari 2017). Some of the supercritical fluid techniques are aerosol solvent extraction system (ASES), rapid expansion of supercritical solutions (RESS), particles from gas-saturated solutions (PGSS), gas or supercritical fluid antisolvent (GAS/SAS), solution-enhanced dispersion by supercritical fluid (SEDS) and rapid expansion of supercritical solutions into a liquid solvent chamber (RESOLVE) (Augustin and Sanguansri 2009).

Giufrida et al. (2016) encapsulated medroxyprogesterone in poly (3-hydroxybut yrate-*co*-3-hydroxyvalerate) (PHBV) via supercritical fluid extraction of emulsion (SFEE) technique and compared with conventional emulsion solvent evaporation (ESE) technique. The particle sizes obtained using SFEE and ESE were 850–183 nm and 932–215 nm, respectively, with varying PHBV molecular weight. It can be seen that encapsulates produced using SFEE were comparatively smaller than those of ESE. Further, ESE consumed more time for solvent removal than that required for SFEE. Another study encapsulated resveratrol in PHBV via SEDS technique using CO_2 and dichloromethane as anti-solvent and organic solvent, respectively, and revealed that resveratrol encapsulation either increased or maintained the antioxidant activity. Also, the particles have the encapsulation efficiency and particles sizes of 99.54% and 390–570 nm, respectively and are free of organic solvent (Dal Magro et al. 2017). Lévai et al. (2017) encapsulated quercetin using SFEE and then compared with PGSS-drying and lyophilisation. Antioxidant activity and quercetin encapsulation efficiency were similar for both PGSS-dried and lyophilised products, but PGSS-dried particles are more homogenous and less crystalline than that of lyophilised products, proving that microsized quercetin loaded particles can be produced efficiently through PGSS-drying.

Supercritical fluid techniques can be performed at low temperature to facilitate encapsulation of heat sensitive bioactive components. Further, the final product can be made solvent free by controlling the temperature, pressure and flow rate of CO_2 to favour food application (Prieto and Calvo 2017a). The major drawback of this technique is its initial cost owing to high-pressure equipment (Gouin 2004).

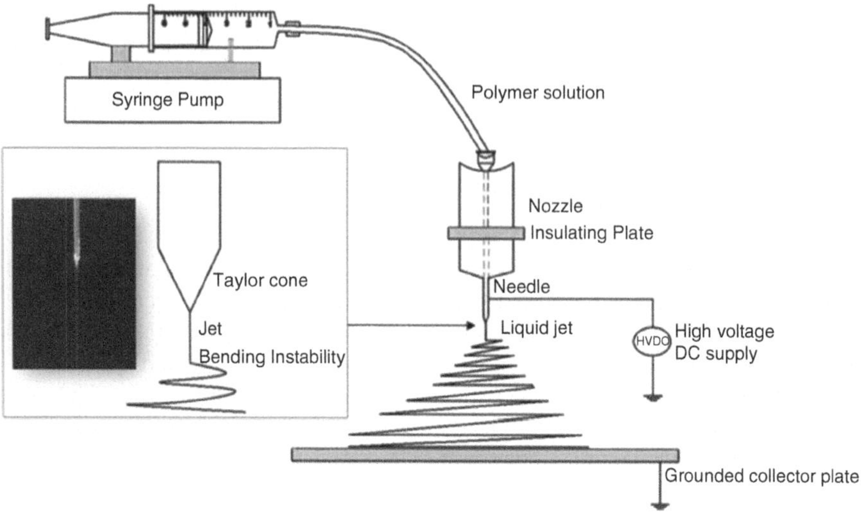

Fig. 14.1 Electrospinning setup (model) (Bhushani and Anandharamakrishnan 2014)

14.4.7 Electrospinning

This technique involves production of polymer fibers of diameters less than 1 μm by applying high voltage to the polymer droplets (Ghorani et al. 2017). The method uses a set-up (Fig. 14.1) that consists of two electrodes, one of which is connected to the solution (containing core material in organic polymer solution) in the needle and the other is connected to the collector at a pre-set distance. At the tip of the needle, the droplet initially forms Taylor cone and then transformed into jet, when the electric charge approaches and overcomes the droplet's surface tension. Further, in the space between the electrodes the solvent gets evaporated, transforming the jet into nanofibers. Voltage applied, flow rate of solution, electrode distance, molecular weight and weight concentration of polymer, and volatility and conductivity of solvent used governs the morphology and properties of the fiber formed (Anandharamakrishnan 2014). Aceituno-Medina et al. (2015a) studied the photo-protection of amaranth protein isolate: pullulan based electrospun fibers with folic acid as core material. Electrospinning of vitamin with protein matrix increased thermal stability of folic acid with very high encapsulation efficiency (>95%). Further, after 2 h of UV exposure, no degradation of folic acid was witnessed. Thus, electrospinning amaranth protein isolate: pullulan structures can protect photosensitive components and has great potential in food application. Another application of electrospinning in food was explored by fabricating electrospun polyvinyl alcohol/cinnamon essential oil/β-cyclodextrin (PVA/CEO/β-CD) nanofibrous film. The film showed improved thermal stability and excellent antimicrobial activity against *Staphylococcus aureus* and *Escherichia coli* with average diameter of 240 ± 40 nm. Also, these films were also able to prolong the shelf life of strawberry and pork with

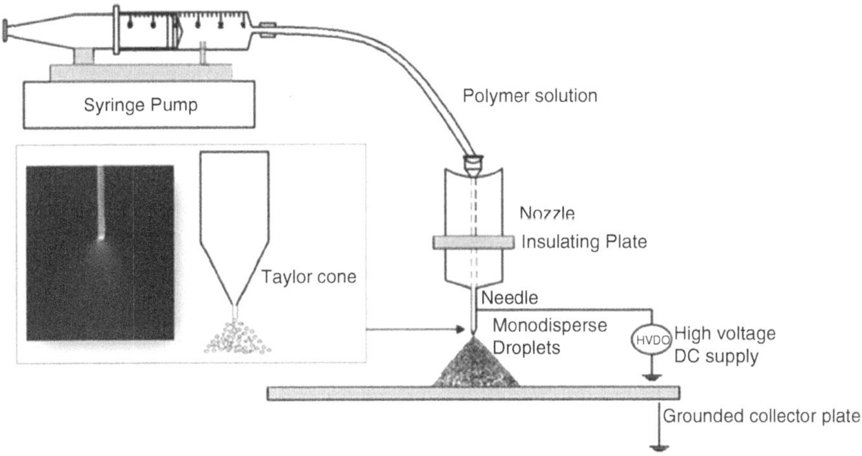

Fig. 14.2 Electrospraying setup (model) (Bhushani and Anandharamakrishnan 2014)

polyvinyl alcohol and polylactic acid, respectively (Wen et al. 2016a, b). A study was conducted on release of Plantaricin 423, from *Lactobacillus plantarum,* and bacteriocin ST4SA, from *Enterococcus mundtii,* those were electrospun into nanofibers blends of poly(D,L-lactide) (PDLLA) and poly(ethylene oxide) (PEO) combinations. The study found that 88% of the original antimicrobial activity was retained by both the peptides at 37 °C. Hence, electrospinning technique can be used for sustained release with appropriate combination of polymers (Heunis et al. 2011).

14.4.8 Electrospraying

Electrospraying is similar to electrospinning, with differences like use of low viscous solutions and hence degree of cohesion is lower, so as to form fine droplets of liquid (Costa et al. 2010; Ghorani et al. 2017). The self-dispersing characteristic of these charged droplets prevents agglomeration (Fig. 14.2) (Bhushani and Anandharamakrishnan 2014). Similar to electrospinning, the nature of the final particle formed depends on various parametes like electric potential applied, flow rate of the solution, distance between the electrodes, concentration, viscosity, density and electrical conductivity of the solution, and relative humidity of the surroundings. For instance, spherical and small sized particle can be obtained with higher electrical potential, low flow rate and large electrode distance (Tapia-Hernández et al. 2017). On encapsulating the antioxidant molecule, epigallocatechin gallate using gelatin as wall material via electrospraying, the encapsulation efficiency was found to be 96 ± 3% (Gómez-Mascaraque et al. 2015). This explains the possibility of incorporating almost all the core material into the wall material using these electro hydrodynamic techniques. Further, electrospraying can be effectively used to preserve heat sensitive ingredients. Encapsulation of β-carotene using whey protein concentrate as wall material, via electrospraying provided stability for the core material against photo-oxidation (López-Rubio and Lagaron 2012).

Electrospinning and electrospraying are the areas recently being explored, especially in food applications because of varied reasons like their ability to better preserve oxidation and heat sensitive ingredients, high encapsulation efficiency, different tactile sensation and mouth feel to food, high specific surface, processing at room temperature and ability to obtain various sized particle with narrow size distribution (Drosou et al. 2017; Echegoyen et al. 2017; Ghorani and Tucker 2015; Gómez-Estaca et al. 2015; Nieuwland et al. 2013).

14.4.9 Spray Drying

Spray drying technique not only encapsulates the particle but also facilitates drying. The process involves dissolving, dispersing or emulsifying the core material in an aqueous solution containing wall material, followed by encapsulation through spraying into hot drying chamber (Arpagaus et al. 2017). Briefly, four stages involved during transformation of liquid/suspension into powder includes atomization of the solution, contact of spray with the hot gas, evaporation of moisture and particle separation (Anandharamakrishnan and Ishwarya 2015). In general, commercially available spray drying results in micron-sized particles. Upon some modifications like replacing cyclone separator with electrostatic particle collector, atomizer with vibrating mesh (piezoelectric actuator) and turbulent air flow with laminar air flow, spray drying can be used to produce nano-sized particles (Arpagaus et al. 2017).

Several studies have been conducted on the encapsulation of food bioactive components using spray drying either as a single encapsulation technique (Arslan et al. 2015; Aghbashlo et al. 2012a, b; Botrel et al. 2014a, 2017; Li et al. 2015, 2016; Premi and Sharma 2017; Rajam and Anandharamakrishnan 2015a; Sharif et al. 2017; Shen and Quek 2014; Takeungwongtrakul et al. 2015; Xue et al. 2013) or as a post encapsulation treatment (Calderón-Oliver et al. 2017; Oliveira et al. 2007) for any of the techniques discussed above. A study was conducted on the effect of inlet temperature (140, 160 and 180 °C) and feed rate (8 and 10 mL/min) on spray dried cinnamon infusions using maltodextrin as wall material. The results revealed that best encapsulation yield was obtained at inlet temperature of 160 and 180 °C and feed rate of 10 mL/min. Under these conditions, both phenolic content and antioxidant capacity of cinnamon infusions were protected, whereas the particle size remained almost same for all the treatments (Santiago-Adame et al. 2015). A similar study was conducted by Arslan et al. (2015) with different wall materials (gelatin, whey protein concentrate, modified starch, maltodextrin, pea protein isolate and gum arabic) at two different inlet temperatures (80 and 125 °C) for spray dried encapsulation of *Saccharomyces cerevisiae var. boulardii.* The study found that gelatin and gum arabic are the most suitable wall materials for the encapsulation of

S. boulardii based on the tested temperatures and simulated gastric solution (pH—1, 1.5 and 2 and exposure time—1, 2 and 3 h). Also, the microcapsules produced at 125 °C were more resistance to gastric solution than those at 80 °C. During encapsulation of vanillin with whey protein isolate, spray drying showed better encapsulation efficiency (86.2%) than that of spray freeze drying (72%) and freeze drying (76.8%). Also, mean particles size distribution and mean particle diameter of spray freeze dried and spray dried samples were superior to that of freeze drying (Hundre et al. 2015). Similarly, a comparative study was conducted by Alvim et al. (2016) on spray drying (wall material: arabic gum) and spray chilling (wall material: stearic acid and hydrogenated vegetable fat) encapsulation of ascorbic acid. The average diameters and encapsulation efficiencies of particles obtained from spray drying and spray chilling were 9.3 and 31.2 μm, and 100.8% and 97.8%, respectively. It can be seen that comparatively smaller sized particles and higher encapsulation efficiency were obtained for spray dried sample than that of spray chilled samples. Also, spray drying provided greater protection for encapsulated ascorbic acid than spray chilling on incorporation into biscuits. Thus, spray drying is better than spray chilling for encapsulation of ascorbic acid.

Spray drying is one the efficient and widely used method for encapsulation of heat-sensitive components. This may because of various reasons like wide range of size and shape of the particle can be obtained by optimizing the process parameters, the properties of the material used and method of preparation (Balassa et al. 1971; Nandiyanto and Okuyama 2011), cost effective as the production cost is 30–50 times cheaper than freeze drying (Desobry et al. 1997), short exposure time to the hot air facilitates minimum loss and degradation (Augustin and Hemar 2009).

Spray drying, spray cooling, spray chilling and freeze drying are the techniques used to recover product after encapsulation by converting nano and micro suspensions into capsules, to maintain stability, to facilitate easier handling and storage, and to promote controlled release (Anandharamakrishnan 2014; Vishwakarma et al. 2016).

14.4.10 Freeze Drying

Similar to spray drying, freeze drying also aids in encapsulation as well as drying. Freeze drying (also known as lyophilization) involves freezing the solution containing core material, wall material and solvent, and then vaporizing the solvent from the solid phase to vapour phase directly through sublimation (Anandharamakrishnan 2014). Freeze drying produces particles that are stable in dry form, rapidly soluble in aqueous solution, retains color with minimum possible degradation (Gharibzahedi and Jafari 2017; Reis et al. 2006).

Chranioti and Tzia (2014) encapsulated fennel oleoresin using freezing drying. Among different combinations of wall material (binary and ternary mixtures of gum arabic combined with modified starch, maltodextrin, and chitosan) studied, gum arabic with modified starch proved to be the best in terms of storage stability, encapsulation efficiency (74.88%) and redispersibility (volume mean diameter = 2.74 μm).

The study found also that the moisture content and redispersibility of obtained final freeze-dried encapsulates were influenced by the properties like emulsion mean diameter and stability of initial formed emulsions. Similarly, Dianawati et al. (2013) encapsulated a probiotic (*Bifidobacterium longum*) with five types of proteins (skim milk 12%, whey protein concentrate 12%, sodium caseinate 12%, sodium caseinate:whey protein concentrate 6%:6%, and soy protein isolate 12%) combined with three types of sugars (glycerol (3% w/v), mannitol (3% w/v) and maltodextrin (3% w/v)) using freeze drying. The study showed that milk proteins and sugar alcohols (glycerol and mannitol) were to superior soy protein isolate and maltodextrin, respectively, in terms of surface hydrophobicity, acid and bile tolerance, retention of β-glucosidase, adenosine triphosphate and lactate dehydrogenase. Several comparative studies were done encapsulation of bioactive components with spray drying and freeze drying (Chen et al. 2013; Laokuldilok and Kanha 2015; Wilkowska et al. 2016). During microencapsulation of flax oil with zein, higher particle yield were obtained by freeze drying than by spray drying using the same core:wall ratio. But, Hausner ratio and Carr's index value showed that the powders obtained using freeze drying had very poor flowability (Quispe-Condori et al. 2011). Similarly, a study on encapsulation of black carrot juice with different wall materials (gum arabic, maltodextrin 20 DE and tapioca starch) using spray and freeze drying, showed that the most satisfactory product was freeze dried product with higher anthocyanin content, antioxidant activity, encapsulation efficiency, water solubility index and color (Murali et al. 2015). Saikia et al. (2015) encapsulated polyphenols extracted from star fruit *(Averrhoa carambola)* pomace using maltodextrin as wall material by spray and freeze drying techniques. It was found that freeze dried encapsulates was superior to spray dried encapsulates in terms of solubility, color, encapsulation efficiency.

Freeze drying is a highly stabilizing process, helps in improving the physicochemical stability of the encapsulated particles (Reis et al. 2006). It is more suitable for heat-sensitive compounds like flavour, colour, aroma, sensitive natural oil, anthocyanins, probiotics and so on (Dianawati et al. 2013; Jafari et al. 2016). But, the main limitations of freeze drying is energy intensiveness, long processing time, open porous structure leading to exposure of core materials to the surrounding environment, final product with poor flowing property (Anandharamakrishnan and Ishwarya 2015; Semalty 2014). To overcome the drawbacks of freeze drying, spray freeze drying can be used as an alternative to reduce the pore size and drying time (Anandharamakrishnan et al. 2010).

14.4.11 Spray Freeze Drying

Spray freeze drying is nothing but the combination of spray drying and freeze drying, involving three processes: atomization, freezing and drying. Thus, the advantages of both spray drying and freeze drying can be obtained, simultaneously in the form spherical shaped free flowing powder with porous structure (Hundre et al.

2015). Spray freezing can be performed in any of the three ways: spray freezing into vapour, spray freezing into liquid and spray freezing into vapour over liquid (Anandharamakrishnan and Ishwarya 2015). Parthasarathi and Anandharamakrishnan (2016) performed encapsulation of vitamin E with whey protein isolate using spray freeze drying technique and compared with spray drying and freeze drying techniques. The study revealed that spray freeze drying, is best suited for encapsulation of poorly water soluble compounds, with 1.13 and 1.19 fold increase in oral bioavailability of spray freeze dried microcapsules than that of spray dried and freeze dried microcapsules, respectively. Spray freeze drying is also used to encapsulate heat sensitive compounds like probiotics (Rajam and Anandharamakrishnan 2015b; Semyonov et al. 2010). In a study of encapsulation of *Lactobacillus plantarum* with various mixtures of food grade polymers showed that the processing time required for spray freeze drying reduced by 12 h than that of conventional freeze drying (20 h) (Rajam and Anandharamakrishnan 2015b). Hundre et al. (2015) microencapsulated vanillin by spray freeze drying into vapour over liquid technique and found that thermal stability of spray freeze dried microcapsules were better than that of spray dried and freeze dried particles. Similarly, Karthik and Anandharamakrishnan (2013) demonstrated microencapsulation of docosahexaenoic acid by spray freezing into vapour over liquid, followed by conventional freeze drying process. Though the encapsulation efficiency of spray freeze dried microcapsules (70.77%) was lower than that of spray dried (82.16%) and freeze dried (73.08%) microcapsules, the percentage of oxidation values reduced for spray freeze dried samples (14%) than those of spray dried and freeze dried (around 33%) samples. Further, the porous spherical shaped powder of spray freeze drying showed good rehydration behaviour. Thus, spray freeze drying is better suited for producing good quality capsules.

Though the use of spray freeze drying can facilitate reduced drying time and particle size control, the drawbacks like energy intensiveness, high fixed and operating cost, and complexity of process limits its application to high value products (Ishwarya et al. 2015; Rajam and Anandharamakrishnan 2015b).

14.5 Characterization of Micro and Nanocapsules

Characterization of capsules formed is essential for the optimization and validation of process as well as the product and the application or incorporation of the formed capsules. Moreover, in food applications, characteristics of the capsules are of major concern in food safety and regulatory aspects (Anandharamakrishnan and Ishwarya 2015). The most commonly used techniques for characterization of micro and nanocapsules (Fig. 14.3) are reviewed in this section.

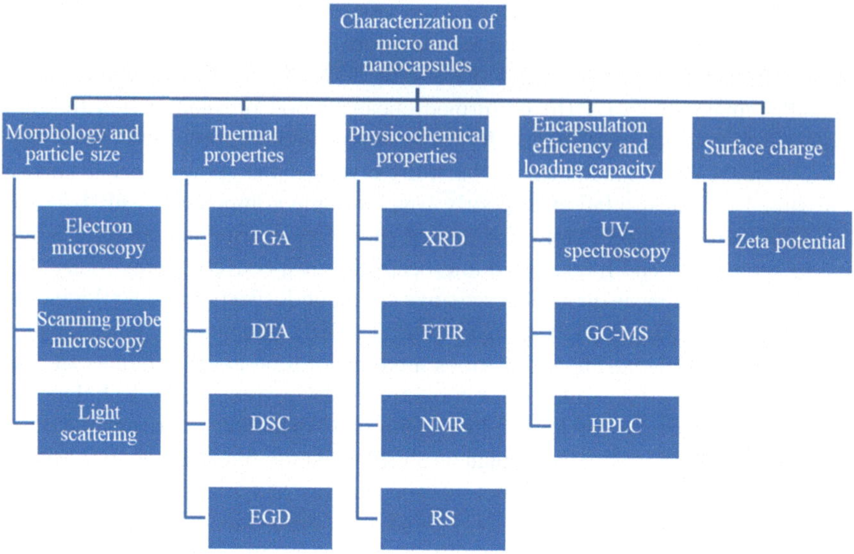

Fig. 14.3 Commonly used techniques for characterization of micro and nanocapsules

14.5.1 Morphology and Particle Size

Morphology and particle size analyses provide information about structure, size and range of particle, state of agglomeration, particle size distribution, shape and so on, that are useful for understanding the physicochemical and functional properties of the encapsulated particle (Jafari and Esfanjani 2017). Generally, these characteristics can be analysed by electron microscopy (scanning electron microscopy (SEM) and transmission electron microscopy (TEM)), scanning probe microscopy (atomic force microscopy (AFM) and confocal laser scanning microscopy (CLSM)) and light scattering (dynamic light scattering (DLS) and laser diffraction (LD)) (Cocero et al. 2009; Jafari and Esfanjani 2017).

Electron microscopy is similar to light microscopy, except that beam of electrons strikes the sample instead of light, thus providing high resolution images. SEM helps to explore the surface morphology through back scattered and secondary electrons from the surface of the sample, whereas TEM helps in study of micro-structural features as well as thickness of wall material (Anandharamakrishnan 2014; Tiede et al. 2008; Vishwakarma et al. 2016). In SEM, electrons are emitted when a tungsten source or lanthanum hexaboride thermionic emitter sources are heated in the temperature range of 2000–2700 K, through the electron gun with the energy of 0.1–30 keV (Anandharamakrishnan 2014). TEM works on the principle of de Broglie wavelength with the energy of electron beam in the range of 100–400 keV (Williams and Carter 1996). Thus, TEM provides better resolution (smaller than 1 nm) than that of SEM (1–20 nm or lesser) (Anandharamakrishnan 2014; Jafari and Esfanjani 2017). These electron microscopic imaging techniques can be used only for conducting surfaces. This can be overcome by the use of scanning probe microscopies (Anandharamakrishnan 2014).

Scanning probe microscopy uses a physical probe to scan the specimen. CLSM combines high-resolution imaging with selectivity in depth and facilitates construction of 3D structured image (Jafari and Esfanjani 2017). AFM uses a movable sharp tip to measure the structure of the nuclear, sub-atomic and molecular level capsules. AFM facilitates the study of surface morphology at x, y (resolutions 1 nm) and z (resolution less than 0.1 nm) directions (Anandharamakrishnan 2014; Jafari and Esfanjani 2017). DLS (also known as correlation spectroscopy) uses Brownian movement and Doppler shift to identify the structure and motion of the particles (Goldburg 1999). Doppler shift occurs when the particle in Brownian movement are exposed to laser beam. This Doppler shift relates to the particle size, which is larger particles exhibit smaller shift, whereas smaller particles exhibit larger shift. Laser diffraction (also known as static light scattering (SLS)) works by analysing the diffracted laser beam from the particles. Thus, DLS measures the change in scattering over time, whereas SLS measures the intensity.

14.5.2 Thermal Properties

Characterization of thermal properties of the capsules is crucial to understand the thermal stability, heat energy storage, decomposition, melting temperature, freezing temperature, glass transition temperature and so on. Thermo-gravimetric analysis (TGA), differential thermal analysis (DTA), differential scanning calorimetry (DSC) and evolved gas detection (EGA) techniques are generally used in characterization of thermal properties of the micro and nano encapsulated materials (Vishwakarma et al. 2016).

TGA monitors the change in weight of the sample during the application of controlled heat to determine the thermal stability of the capsules (Anandharamakrishnan and Ishwarya 2015). In case of DTA, sample and a reference material are subjected to similar heat treatments and the difference in temperature is recorded. While, DSC measures the heat flow rate of the sample and the reference material at same temperature (Hohne et al. 2013). Thus, DSC helps in analysing and identifying the melting point, glass transition temperature, heat capacity and recrystallization times of capsules (Jafari and Esfanjani 2017). EGD is used in detection of volatile substances from the sample during thermal degradation. Furthermore, these techniques can be coupled with techniques like Fourier transform infrared spectroscopy (FTIR) or mass spectrometry (MS) to facilitate structural identification of materials obtained during thermal degradation (Evolved gas analysis) (Xie and Pan 2001).

14.5.3 Physicochemical Properties

Physicochemical nature of the micro and nano encapsulated materials can be characterized through techniques like X ray diffraction (XRD), FTIR, nuclear magnetic resonance and Raman spectroscopy (RS). XRD and NMR are used to characterize the degree of crystallinity of the capsules, in particular wall materials. When a fine beam of collimated X-rays are passed through the sample, a diffraction grating is obtained depending on the spatial arrangement of the crystals present in the sample (Anandharamakrishnan and Ishwarya 2015). The application of NMR in characterizing the crystalline degree is based on the concept that different NMR spectra is obtained for crystalline (broad component) and amorphous (narrow component) domains of nuclei of same functional group due to difference in magnetic moment experienced (De Oca et al. 2004; Hronsky et al. 2014). FTIR aids in measuring the molecular fingerprint of the sample through spectra obtained when IR radiation is passed through the sample (Boydston-White et al. 1999). FTIR is mainly used to assess the chemical interaction between the molecules (Khadiran et al. 2015).

14.5.4 Encapsulation Efficiency and Loading Capacity

Encapsulation efficiency is the percentage of the core material encapsulated within the wall material with respect to the total core material used, whereas loading capacity is the percentage mass ratio between the core material and the total mass of the capsules (Prieto and Calvo 2017b). Characterization of these parameters can be performed using UV spectroscopy, high performance liquid chromatography (HPLC) and gas chromatography-mass spectrometry (GC-MS). UV spectroscopy involves measurement of a beam of radiation after passing through the sample. HPLC helps in identification and quantification of the components, based on the interaction between the stationary phase and the sample mixture. GC-MS combines GC and MS for identification and quantification of the sample components. The major difference between HPLC and GC-MS is that HPLC is for components from low to high molecular weight, whereas GC-MS is limited to high molecular weight components. All these methods require extraction of core materials from the sample using solvents and preparation of calibration curve (Vishwakarma et al. 2016).

14.5.5 Surface Charge

Surface charge is described in terms of zeta potential (Jafari and Esfanjani 2017). Zeta potential measures the overall charge of a particle in a given medium and the value describes the potential stability, electrostatic interaction and mobility of the suspension or colloids (Anandharamakrishnan 2014; Tiede et al. 2008). Zeta potential value ranging from −10 to +10 mV, implies a neutral system, whereas value lesser than −30 mV and greater than +30 mV represent strongly anionic and cationic systems, respectively (Clogston and Patri 2011). Moreover, cationic and anionic systems are more stable and have lesser or no tendency to aggregate (Peres 2011).

14.6 Application of Encapsulation in Food

Recently, encapsulation is one of the areas that is being explored in food applications, because it facilitates controlled and/or targeted release, provides stability to the bioactive components, masking unpleasant colour, flavour or odour, enhances the viability of probiotics, facilitates easy handling, extends the shelf life, protects the sensitive ingredients from heat, light, oxygen or moisture and enhancing bioavailability (Anal and Singh 2007; Huang et al. 2010). Addition of encapsulated products in food can enhance the properties of the food material. For instance, incorporation of microencapsulated pomegranate peel phenolics in ice cream, showed increase in antioxidant and α-glucosidase inhibitory activities with better sensory scores than that of control (Çam et al. 2014). Similarly, microencapsulated hydroxycitric acid, a thermally sensitive and hygroscopic compound was incorporated in bread and pasta to improve the sensory and nutritional characteristics (Ezhilarasi et al. 2013a, 2014; Pillai et al. 2012). Table 14.3 illustrates some of the application of encapsulated materials in food packaging and their effect on the product and/or packaging material.

14.7 Conclusion

Though encapsulation is used in cosmetics and pharmaceutical applications since long time, it is not completely accepted in food applications. Micro and nano encapsulation in food applications is still in research areas and commercial application is in slow pace. This may be due to reasons like unavailability of suitable food grade wall materials, high production cost and processing time, reluctance of people/food industries to accept new products and technologies (Patel and Bhandari 2014). Also, to obtain required product of good quality and high encapsulation efficiency, appropriate techniques must be selected with reduced cost. Further, food grade polymers possess several advantages as wall material, mainly because of their ability to be modified to attain the required characteristics (Fathi et al. 2014). Hence, their application in protecting the active food ingredients must be explored.

Table 14.3 Effect of incorporation of encapsulated materials in food packaging application

Nano/microcapsules				
Core material	Wall material	Matrix material	Property	References
Thymol	γ-cyclodextrin	Zein	Reduced the bacterial count in meat stored at 4 °C for 5 days and hence can be used as an antimicrobial food packaging material for meat and meat products	Aytac et al. (2017)
Green tea	Aluminosilicates	Polyethylene	Extended the shelf life fresh meat by 3 days	Wrona et al. (2017)
Essential oils (clove, cinnamon bark and lemongrass)	*B*-chitosan	Cellulose nanocrystals	Enhanced mechanical and antibacterial property of film	Zhang et al. (2017)
Peppermint oil	Halloysite	Pectin	Exhibited antimicrobial property against *Escherichia coli* and *Staphylococcus aureus*	Biddeci et al. (2016)
Clove oil	β-cyclodextrin	Polyvinyl alcohol	Exhibited inhibitory effect on *Aspergillus niger*	Deng et al. (2016)
Carvacrol	Halloysite	Polyamide	Showed fungicidal effect on postharvest pathogens and extended the shelf lige of fresh produce	Shemesh et al. (2016)
Cinnamon oil	β-cyclodextrin	Poly lactic acid	Exhibited better antimicrobial property against *Escherichia coli* and *Staphylococcus aureus*	Wen et al. (2016a)
Cinnamon oil	β-cyclodextrin	Polyvinyl alcohol	Exhibited antimicrobial property against *Escherichia coli* and *Staphylococcus aureus,* and extended the shelf life of strawberry	Wen et al. (2016b)
α-tocopherol	poly(ε-caprolactone)	Methylcellulose	Exhibited potential antioxidant ability Excellent barrier against UV and visible light that prevents photooxidation of foods	Noronha et al. (2014)

References

Aceituno-Medina M, Mendoza S, Lagaron JM, López-Rubio A (2015a) Photoprotection of folic acid upon encapsulation in food-grade amaranth (*Amaranthus hypochondriacus* L.) protein isolate–Pullulan electrospun fibers. LWT Food Sci Technol 62(2):970–975

Aceituno-Medina M, Mendoza S, Rodríguez BA, Lagaron JM, López-Rubio A (2015b) Improved antioxidant capacity of quercetin and ferulic acid during in-vitro digestion through encapsulation within food-grade electrospun fibers. J Funct Foods 12:332–341

Aghbashlo M, Mobli H, Madadlou A, Rafiee S (2012a) Influence of wall material and inlet drying air temperature on the microencapsulation of fish oil by spray drying. Food Bioprocess Technol 6(6):1561–1569

Aghbashlo M, Mobli H, Madadlou A, Rafiee S (2012b) The correlation of wall material composition with flow characteristics and encapsulation behavior of fish oil emulsion. Food Res Int 49(1):379–388

Alvim ID, Stein MA, Koury IP, Dantas FB, Cruz CL (2016) Comparison between the spray drying and spray chilling microparticles contain ascorbic acid in a baked product application. LWT Food Sci Technol 65:689–694

Anal AK, Singh H (2007) Recent advances in microencapsulation of probiotics for industrial applications and targeted delivery. Trends Food Sci Technol 18(5):240–251

Anandharamakrishnan C (2014) Techniques for nanoencapsulation of food ingredients. Springer, New York

Anandharamakrishnan C, Ishwarya SP (2015) Spray drying techniques for food ingredient encapsulation. Wiley, Hoboken

Anandharamakrishnan C, Rielly CD, Stapley AG (2010) Spray-freeze-drying of whey proteins at sub-atmospheric pressures. Dairy Sci Technol 90(2–3):321–334

Arpagaus C, John P, Collenberg A, Rutti D (2017) Nanocapsules formation by nano spray drying. In: Jafari SM (ed) Nanoencapsulation technologies for the food and nutraceutical industries. Academic, New York, pp 1–34

Arslan S, Erbas M, Tontul I, Topuz A (2015) Microencapsulation of probiotic *Saccharomyces cerevisiae* var. boulardii with different wall materials by spray drying. LWT Food Sci Technol 63(1):685–690

Asghari-Varzaneh E, Shahedi M, Shekarchizadeh H (2017) Iron microencapsulation in gum tragacanth using solvent evaporation method. Int J Biol Macromol 103:640–647

Augustin MA, Hemar Y (2009) Nano-and micro-structured assemblies for encapsulation of food ingredients. Chem Soc Rev 38(4):902–912

Augustin MA, Sanguansri P (2009) Nanostructured materials in the food industry. Adv Food Nutr Res 58:183–213

Aytac Z, Dogan SY, Tekinay T, Uyar T (2014) Release and antibacterial activity of allyl isothiocyanate/β-cyclodextrin complex encapsulated in electrospun nanofibers. Colloids Surf B: Biointerfaces 120:125–131

Aytac Z, Ipek S, Durgun E, Tekinay T, Uyar T (2017) Antibacterial electrospun zein nanofibrous web encapsulating thymol/cyclodextrin-inclusion complex for food packaging. Food Chem 233:117–124

Badri W, Miladi K, Eddabra R, Fessi H, Elaissari A (2015) Elaboration of nanoparticles containing indomethacin: argan oil for transdermal local and cosmetic application. J Nanomater 16(1):113

Balassa LL, Fanger GO, Wurzburg OB (1971) Microencapsulation in the food industry. Crit Rev Food Sci Nutr 2(2):245–265

Ballesteros LF, Ramirez MJ, Orrego CE, Teixeira JA, Mussatto SI (2017) Encapsulation of antioxidant phenolic compounds extracted from spent coffee grounds by freeze-drying and spray-drying using different coating materials. Food Chem 237:623–631

Bareras-Urbina CG, Ramírez-Wong B, López-Ahumada GA, Burruel-Ibarra SE, Martínez-Cruz O, Tapia-Hernández JA, Rodriguez Felix F (2016) Nano-and micro-particles by nanoprecipitation: possible application in the food and agricultural industries. Int J Food Prop 19(9):1912–1923

Battista JV, Lindlof JA, Wurster DE (1965) Process for preparing agglomerates. US Patent 3207824

Bazylińska U, Lewińska A, Lamch Ł, Wilk KA (2014) Polymeric nanocapsules and nanospheres for encapsulation and long sustained release of hydrophobic cyanine-type photosensitizer. Colloids Surf A Physicochem Eng Asp 442:42–49

Belščak-Cvitanović A, Bušić A, Barišić L, Vrsaljko D, Karlović S, Špoljarić I, Vojvodic A, Mrsic G, Komes D (2016) Emulsion templated microencapsulation of dandelion (*Taraxacum officinale* L.) polyphenols and β-carotene by ionotropic gelation of alginate and pectin. Food Hydrocoll 57:139–152

Bernardos A, Marina T, Žáček P, Pérez-Esteve É, Martínez-Mañez R, Lhotka M, Kouřimská L, Pulkrábek J, Klouček P (2015) Antifungal effect of essential oil components against *Aspergillus niger* when loaded into silica mesoporous supports. J Sci Food Agric 95(14):2824–2831

Bhushani JA, Anandharamakrishnan C (2014) Electrospinning and electrospraying techniques: potential food based applications. Trends Food Sci Technol 38(1):21–33

Bhushani JA, Kurrey NK, Anandharamakrishnan C (2017) Nanoencapsulation of green tea catechins by electrospraying technique and its effect on controlled release and in-vitro permeability. J Food Eng 199:82–92

Biddeci G, Cavallaro G, Di Blasi F, Lazzara G, Massaro M, Milioto S, Parisi F, Riela S, Spinelli G (2016) Halloysite nanotubes loaded with peppermint essential oil as filler for functional biopolymer film. Carbohydr Polym 152:548–557

Boschetto DL, Dalmolin I, de Cesaro AM, Rigo AA, Ferreira SR, Meireles MA, Batista EA, Oliveira JV (2013) Phase behavior and process parameters effect on grape seed extract encapsulation by SEDS technique. Ind Crop Prod 50:352–360

Botrel DA, Borges SV, de Barros Fernandes RV, Lourenço Do Carmo E (2014a) Optimization of fish oil spray drying using a protein: inulin system. Dry Technol 32(3):279–290

Botrel DA, de Barros Fernandes RV, Borges SV, Yoshida MI (2014b) Influence of wall matrix systems on the properties of spray-dried microparticles containing fish oil. Food Res Int 62:344–352

Botrel DA, Borges SV, de Barros Fernandes RV, Antoniassi R, de Faria-Machado AF, de Andrade Feitosa JP, de Paula RCM (2017) Application of cashew tree gum on the production and stability of spray-dried fish oil. Food Chem 221:1522–1529

Bourbon AI, Cerqueira MA, Vicente AA (2016) Encapsulation and controlled release of bioactive compounds in lactoferrin-glycomacropeptide nanohydrogels: curcumin and caffeine as model compounds. J Food Eng 180:110–119

Boydston-White S, Gopen T, Houser S, Bargonetti J, Diem M (1999) Infrared spectroscopy of human tissue. V. Infrared spectroscopic studies of myeloid leukemia (ML-1) cells at different phases of the cell cycle. Biospectroscopy 5(4):219–227

Calderón-Oliver M, Pedroza-Islas R, Escalona-Buendía HB, Pedraza-Chaverri J, Ponce-Alquicira E (2017) Comparative study of the microencapsulation by complex coacervation of nisin in combination with an avocado antioxidant extract. Food Hydrocoll 62:49–57

Çam M, İçyer NC, Erdoğan F (2014) Pomegranate peel phenolics: microencapsulation, storage stability and potential ingredient for functional food development. LWT Food Sci Technol 55(1):117–123

Carneiro HC, Tonon RV, Grosso CR, Hubinger MD (2013) Encapsulation efficiency and oxidative stability of flaxseed oil microencapsulated by spray drying using different combinations of wall materials. J Food Eng 115(4):443–451

Celebioglu A, Kayaci-Senirmak F, İpek S, Durgun E, Uyar T (2016) Polymer-free nanofibers from vanillin/cyclodextrin inclusion complexes: high thermal stability, enhanced solubility and antioxidant property. Food Funct 7(7):3141–3153

Chang C, Varankovich N, Nickerson MT (2016) Microencapsulation of canola oil by lentil protein isolate-based wall materials. Food Chem 212:264–273

Chen Q, Zhong F, Wen J, McGillivray D, Quek SY (2013) Properties and stability of spray-dried and freeze-dried microcapsules co-encapsulated with fish oil, phytosterol esters, and limonene. Dry Technol 31(6):707–716

Cherukuri SR, Mansukhani G (1989) Sweetener delivery systems containing polyvinyl acetate. US Patent 4816265

Cherukuri SR, Mansukhani G (1990) Multiple encapsulated sweetener delivery system. US Patent 4933190

Cherukuri SR, Mansukhani G, Faust SM (1991) Food acid delivery systems containing polyvinyl acetate. US Patent 5057328

Chow SF, Wan KY, Cheng KK, Wong KW, Sun CC, Baum L, Chow AH (2015) Development of highly stabilized curcumin nanoparticles by flash nanoprecipitation and lyophilization. Eur J Pharm Biopharm 94:436–449

Chranioti C, Tzia C (2014) Arabic gum mixtures as encapsulating agents of freeze-dried fennel oleoresin products. Food Bioprocess Technol 7(4):1057–1065

Clogston JD, Patri AK (2011) Zeta potential measurement. In: McNeil SE (ed) Characterization of nanoparticles intended for drug delivery, vol 697. Humana Press, New York, pp 63–70

Cocero MJ, Martín Á, Mattea F, Varona S (2009) Encapsulation and co-precipitation processes with supercritical fluids: fundamentals and applications. J Supercrit Fluids 47(3):546–555

Coghetto CC, Brinques GB, Siqueira NM, Pletsch J, Soares RMD, Ayub MAZ (2016) Electrospraying microencapsulation of *Lactobacillus plantarum* enhances cell viability under refrigeration storage and simulated gastric and intestinal fluids. J Funct Foods 24:316–326

Coronel-Aguilera CP, San Martín-González MF (2015) Encapsulation of spray dried β-carotene emulsion by fluidized bed coating technology. LWT Food Sci Technol 62(1):187–193

Costa LM, Bretas RE, Gregorio R (2010) Effect of solution concentration on the electrospray/electrospinning transition and on the crystalline phase of PVDF. Mater Sci Appl 1(04):247

Dal Magro C, Aguiar GP, Veneral JG, dos Santos AE, de Chaves LM, Oliveira JV, Lanza M (2017) Co-precipitation of trans-resveratrol in PHBV using Solution Enhanced Dispersion by Supercritical Fluids technique. J Supercrit Fluids 127:182–190

De Barros Fernandes RV, Borges SV, Botrel DA (2014a) Gum arabic/starch/maltodextrin/inulin as wall materials on the microencapsulation of rosemary essential oil. Carbohydr Polym 101:524–532

De Barros Fernandes RV, Borges SV, Botrel DA, Oliveira CRD (2014b) Physical and chemical properties of encapsulated rosemary essential oil by spray drying using whey protein–inulin blends as carriers. Int J Food Sci Technol 49(6):1522–1529

De Oca HM, Ward IM, Klein PG, Ries ME, Rose J, Farrar D (2004) Solid state nuclear magnetic resonance study of highly oriented poly (glycolic acid). Polymer 45(21):7261–7272

De Paz E, Martín Á, Cocero MJ (2012) Formulation of β-carotene with soybean lecithin by PGSS (Particles from Gas Saturated Solutions)-drying. J Supercrit Fluids 72:125–133

Deng J, Li W, Lin Q (2016) Study on preparation and migration behavior of polyvinyl alcohol active packaging film based on clove essential oil/β-cyclodextrin inclusion complex. J Nanosci Nanotechnol 16(12):12617–12620

Depypere F, Pieters JG, Dewettinck K (2009) PEPT visualisation of particle motion in a tapered fluidised bed coater. J Food Eng 93(3):324–336

Desobry SA, Netto FM, Labuza TP (1997) Comparison of spray-drying, drum-drying and freeze-drying for β-carotene encapsulation and preservation. J Food Sci 62(6):1158–1162

Dianawati D, Mishra V, Shah NP (2013) Survival of *Bifidobacterium longum* 1941 microencapsulated with proteins and sugars after freezing and freeze drying. Food Res Int 51(2):503–509

Dima C, Cotârlet M, Alexe P, Dima S (2014) Microencapsulation of essential oil of pimento [*Pimenta dioica* (L) Merr.] by chitosan/k-carrageenan complex coacervation method. Innovative Food Sci Emerg Technol 22:203–211

Do Carmo CS, Pais R, Simplício AL, Mateus M, Duarte CM (2017) Improvement of aroma and shelf-life of non-alcoholic beverages through cyclodextrins-limonene inclusion complexes. Food Bioprocess Technol 10(7):1297–1309

Drosou CG, Krokida MK, Biliaderis CG (2017) Encapsulation of bioactive compounds through electrospinning/electrospraying and spray drying: a comparative assessment of food-related applications. Dry Technol 35(2):139–162

Duangkhamchan W, Ronsse F, Depypere F, Dewettinck K, Pieters JG (2012) CFD study of droplet atomisation using a binary nozzle in fluidised bed coating. Chem Eng Sci 68(1):555–566

Duangkhamchan W, Ronsse F, Siriamornpun S, Pieters JG (2015) Numerical study of air humidity and temperature distribution in a top-spray fluidised bed coating process. J Food Eng 146:81–91

Echegoyen Y, Fabra MJ, Castro-Mayorga JL, Cherpinski A, Lagaron JM (2017) High throughput electro-hydrodynamic processing in food encapsulation and food packaging applications. Trends Food Sci Technol 60:71–79

Ezhilarasi PN, Indrani D, Jena BS, Anandharamakrishnan C (2013a) Freeze drying technique for microencapsulation of Garcinia fruit extract and its effect on bread quality. J Food Eng 117(4):513–520

Ezhilarasi PN, Karthik P, Chhanwal N, Anandharamakrishnan C (2013b) Nanoencapsulation techniques for food bioactive components: a review. Food Bioprocess Technol 6(3):628–647
Ezhilarasi PN, Indrani D, Jena BS, Anandharamakrishnan C (2014) Microencapsulation of Garcinia fruit extract by spray drying and its effect on bread quality. J Sci Food Agric 94(6):1116–1123
Fathi M, Martín Á, McClements DJ (2014) Nanoencapsulation of food ingredients using carbohydrate based delivery systems. Trends Food Sci Technol 39(1):18–39
Fernández-Ponce MT, Masmoudi Y, Djerafi R, Casas L, Mantell C, de La Ossa EM, Badens E (2015) Particle design applied to quercetin using supercritical anti-solvent techniques. J Supercrit Fluids 105:119–127
Freiberger EB, Kaufmann KC, Bona E, de Araújo PH, Sayer C, Leimann FV, Gonçalves OH (2015) Encapsulation of roasted coffee oil in biocompatible nanoparticles. LWT Food Sci Technol 64(1):381–389
Frey C (2014) Fluid bed coating-based microencapsulation. In: Gaonkar AG, Vasisht N, Khare AR, Sobel R (eds) Microencapsulation in the food industry: a practical implementation guide. Elsevier, pp 65–79
Galvão JG, Silva VF, Ferreira SG, França FR, Santos DA, Freitas LS, Alves PB, Araújo AA, Cavalcanti SC, Nunes RS (2015) β-cyclodextrin inclusion complexes containing *Citrus sinensis* (L.) Osbeck essential oil: an alternative to control *Aedes aegypti* larvae. Thermochim Acta 608:14–19
Gharibzahedi SMT, Jafari SM (2017) Nanocapsule formation by cyclodextrin. In: Jafari SM (ed) Nanoencapsulation technologies for the food and nutraceutical industries. Academic, pp 187–261
Gharsallaoui A, Roudaut G, Chambin O, Voilley A, Saurel R (2007) Applications of spray-drying in microencapsulation of food ingredients: an overview. Food Res Int 40(9):1107–1121
Ghorani B, Tucker N (2015) Fundamentals of electrospinning as a novel delivery vehicle for bioactive compounds in food nanotechnology. Food Hydrocoll 51:227–240
Ghorani B, Alehosseini A, Tucker N (2017) Nanocapsule formation by electrospinning. In: Jafari SM (ed) Nanoencapsulation technologies for the food and nutraceutical industries. Academic, pp 264–319
Ghosh SK (ed) (2006) Functional coatings: by polymer microencapsulation. Wiley, Hoboken
Giufrida WM, Cabral VF, Cardoso-Filho L, dos Santos Conti D, de Campos VE, da Rocha SR (2016) Medroxyprogesterone-encapsulated poly (3-hydroxybutirate-co-3-hydroxyvalerate) nanoparticles using supercritical fluid extraction of emulsions. J Supercrit Fluids 118:79–88
Goldburg WI (1999) Dynamic light scattering. Am J Phys 67(12):1152–1160
Gómez-Estaca J, Gavara R, Hernández-Muñoz P (2015) Encapsulation of curcumin in electrosprayed gelatin microspheres enhances its bioaccessibility and widens its uses in food applications. Innovative Food Sci Emerg Technol 29:302–307
Gomez-Estaca J, Comunian TA, Montero P, Ferro-Furtado R, Favaro-Trindade CS (2016) Encapsulation of an astaxanthin-containing lipid extract from shrimp waste by complex coacervation using a novel gelatin–cashew gum complex. Food Hydrocoll 61:155–162
Gómez-Estaca J, Balaguer MP, López-Carballo G, Gavara R, Hernández-Muñoz P (2017) Improving antioxidant and antimicrobial properties of curcumin by means of encapsulation in gelatin through electrohydrodynamic atomization. Food Hydrocoll 70:313–320
Gómez-Mascaraque LG, Lagarón JM, López-Rubio A (2015) Electrosprayed gelatin submicroparticles as edible carriers for the encapsulation of polyphenols of interest in functional foods. Food Hydrocoll 49:42–52
Gómez-Mascaraque LG, Ambrosio-Martín J, Fabra MJ, Pérez-Masiá R, López-Rubio A (2016) Novel nanoencapsulation structures for functional foods and nutraceutical applications. In: Sen S, Pathak Y (eds) Nanotechnology in nutraceuticals: production to consumption. CRC Press, Taylor & Francis Group, New York, pp 373–395
Gong L, Li T, Chen F, Duan X, Yuan Y, Zhang D, Jiang Y (2016) An inclusion complex of eugenol into β-cyclodextrin: preparation, and physicochemical and antifungal characterization. Food Chem 196:324–330

Gouin S (2004) Microencapsulation: industrial appraisal of existing technologies and trends. Trends Food Sci Technol 15(7):330–347
Gupta C, Chawla P, Arora S, Tomar SK, Singh AK (2015) Iron microencapsulation with blend of gum arabic, maltodextrin and modified starch using modified solvent evaporation method–milk fortification. Food Hydrocoll 43:622–628
Gupta S, Khan S, Muzafar M, Kushwaha M, Yadav AK, Gupta AP (2016) Encapsulation: entrapping essential oil/flavors/aromas in food. In: Grumezescu A (ed) Encapsulations, vol 2. Academic, pp 229–268
Gutiérrez TJ (2018) Processing nano- and microcapsules for industrial applications. In: Chaudhery Mustansar Hussain (ed) Handbook of nanomaterials for industrial applications. Elsevier, pp. 989–1011. https://doi.org/10.1016/B978-0-12-813351-4.00057-2. EE.UU. ISBN: 978-0-12-813351-4
Gutiérrez TJ, Álvarez K (2017) Biopolymers as microencapsulation materials in the food industry. In: Masuelli M, Renard D (eds) Advances in physicochemical properties of biopolymers: part 2. Bentham Science Publishers, pp 296–322. https://doi.org/10.2174/97816810854491170100 09. EE.UU. ISBN: 978-1-68108-545-6. eISBN: 978-1-68108-544-9
Heunis T, Bshena O, Klumperman B, Dicks L (2011) Release of bacteriocins from nanofibers prepared with combinations of poly (D, L-lactide) (PDLLA) and poly (ethylene oxide) (PEO). Int J Mol Sci 12(4):2158–2173
Hill LE, Gomes C, Taylor TM (2013) Characterization of beta-cyclodextrin inclusion complexes containing essential oils (trans-cinnamaldehyde, eugenol, cinnamon bark, and clove bud extracts) for antimicrobial delivery applications. LWT Food Sci Technol 51(1):86–93
Hohne G, Hemminger WF, Flammersheim HJ (2013) Differential scanning calorimetry, 2nd edn. Springer, New York
Hronsky V, Kovalakova M, Vrabel P, Uhrinova M, Olcak D (2014) Estimation of the degree of crystallinity of partially crystalline polypropylenes using 13C NMR. In: Proceedings of the 15th Czech and Slovak conference on magnetism, Kosice, Slovakia, vol 126
Huang Q, Yu H, Ru Q (2010) Bioavailability and delivery of nutraceuticals using nanotechnology. J Food Sci 75(1):R50–R57
Hundre SY, Karthik P, Anandharamakrishnan C (2015) Effect of whey protein isolate and b-cyclodextrin wall systems on stability of microencapsulated vanillin by spray–freeze drying method. Food Chem 174:16–24
Ishwarya SP, Anandharamakrishnan C, Stapley AG (2015) Spray-freeze-drying: a novel process for the drying of foods and bioproducts. Trends Food Sci Technol 41(2):161–181
Jafari SM (2017) An overview of nanoencapsulation techniques and their classification. In: Jafari SM (ed) Nanoencapsulation technologies for the food and nutraceutical industries. Academic, pp 1–34
Jafari SM, Esfanjani AF (2017) Instrumental analysis and characterization of nanocapsules. In: Jafari SM (ed) Nanoencapsulation technologies for the food and nutraceutical industries. Academic, pp 524–544
Jafari SM, Mahdavi-Khazaei K, Hemmati-Kakhki A (2016) Microencapsulation of saffron petal anthocyanins with cress seed gum compared with Arabic gum through freeze drying. Carbohydr Polym 140:20–25
Jafari SM, Paximada P, Mandala I, Assadpour E, Mehrnia MA (2017) Encapsulation by nano-emulsions. In: Jafari SM (ed) Nanoencapsulation technologies for the food and nutraceutical industries. Academic, pp 36–73
Jain A, Thakur D, Ghoshal G, Katare OP, Shivhare US (2016) Characterization of microcapsulated β-carotene formed by complex coacervation using casein and gum tragacanth. Int J Biol Macromol 87:101–113
Jyothi NVN, Prasanna PM, Sakarkar SN, Prabha KS, Ramaiah PS, Srawan GY (2010) Microencapsulation techniques, factors influencing encapsulation efficiency. J Microencapsul 27(3):187–197

Kailasapathy K (2016) Bioencapsulation technologies for incorporating bioactive components into functional foods. In: Rai VR (ed) Advances in food biotechnology. Wiley, Hoboken, pp 313–333
Karim FT, Ghafoor K, Ferdosh S, Al-Juhaimi F, Ali E, Yunus KB, Hamed MH, Islam A, Asif M, Sarker MZI (2017) Microencapsulation of fish oil using supercritical antisolvent process. J Food Drug Anal 25(3):654–666
Karthik P, Anandharamakrishnan C (2013) Microencapsulation of docosahexaenoic acid by spray-freeze-drying method and comparison of its stability with spray-drying and freeze-drying methods. Food Bioprocess Technol 6(10):2780–2790
Kayaci F, Ertas Y, Uyar T (2013) Enhanced thermal stability of eugenol by cyclodextrin inclusion complex encapsulated in electrospun polymeric nanofibers. J Agric Food Chem 61(34):8156–8165
Kayaci F, Sen HS, Durgun E, Uyar T (2014) Functional electrospun polymeric nanofibers incorporating geraniol–cyclodextrin inclusion complexes: high thermal stability and enhanced durability of geraniol. Food Res Int 62:424–431
Kfoury M, Hadaruga NG, Hadaruga DI, Fourmentin S (2016) Cyclodextrins as encapsulation material for flavors and aroma. In: Grumezescu A (ed) Encapsulations, vol 2. Academic, pp 127–192
Khadiran T, Hussein MZ, Zainal Z, Rusli R (2015) Encapsulation techniques for organic phase change materials as thermal energy storage medium: a review. Sol Energy Mater Sol Cells 143:78–98
Kumar LR, Chatterjee NS, Tejpal CS, Vishnu KV, Anas KK, Asha KK, Anandan R, Mathew S (2017) Evaluation of chitosan as a wall material for microencapsulation of squalene by spray drying: characterization and oxidative stability studies. Int J Biol Macromol 104(Pt B):1986–1995
Lakkis JM (2016) Introduction. In: Lakkis JM (ed) Encapsulation and controlled release technologies in food systems. Wiley, Hoboken, pp 1–15
Laokuldilok T, Kanha N (2015) Effects of processing conditions on powder properties of black glutinous rice (*Oryza sativa* L.) bran anthocyanins produced by spray drying and freeze drying. LWT Food Sci Technol 64(1):405–411
Laokuldilok N, Thakeow P, Kopermsub P, Utama-ang N (2016) Optimisation of microencapsulation of turmeric extract for masking flavour. Food Chem 194:695–704
Lee SJ, Wong M (2014) Nano-and microencapsulation of phytochemicals. In: Kwak HS (ed) Nano-and microencapsulation for foods. Wiley, pp 117–165
Lemma SM, Scampicchio M, Mahon PJ, Sbarski I, Wang J, Kingshott P (2015) Controlled release of retinyl acetate from β-cyclodextrin functionalized poly (vinyl alcohol) electrospun nanofibers. J Agric Food Chem 63(13):3481–3488
Lepeltier E, Bourgaux C, Couvreur P (2014) Nanoprecipitation and the "Ouzo effect": application to drug delivery devices. Adv Drug Deliv Rev 71:86–97
Lévai G, Martín Á, Moro A, Matias AA, Gonçalves VS, Bronze MR, Duarte CM, Rodríguez-Rojo S, Cocero MJ (2017) Production of encapsulated quercetin particles using supercritical fluid technologies. Powder Technol 317:142–153
Li C, Wang J, Shi J, Huang X, Peng Q, Xue F (2015) Encapsulation of tomato oleoresin using soy protein isolate-gum acacia conjugates as emulsifier and coating materials. Food Hydrocoll 45:301–308
Li R, Roos YH, Miao S (2016) Characterization of mechanical and encapsulation properties of lactose/maltodextrin/WPI matrix. Food Hydrocoll 63:149–159
Lindlof JA, Wurster DE (1964) Apparatus for coating particles in a fluidized bed. US Patent 3117027
Lindlof JA, Wurster DE (1965) Apparatus for the encapsulation of discrete particles. US Patent 3196827
López-Rubio A, Lagaron JM (2012) Whey protein capsules obtained through electrospraying for the encapsulation of bioactives. Innovative Food Sci Emerg Technol 13:200–206

Lv Y, Yang F, Li X, Zhang X, Abbas S (2014) Formation of heat-resistant nanocapsules of jasmine essential oil via gelatin/gum arabic based complex coacervation. Food Hydrocoll 35:305–314
Machado FR Jr, Reis DF, Boschetto DL, Burkert JF, Ferreira SR, Oliveira JV, Burkert CA (2014) Encapsulation of astaxanthin from Haematococcus pluvialis in PHBV by means of SEDS technique using supercritical CO2. Ind Crop Prod 54:17–21
Mahdavi SA, Jafari SM, Assadpoor E, Dehnad D (2016) Microencapsulation optimization of natural anthocyanins with maltodextrin, gum Arabic and gelatin. Int J Biol Macromol 85:379–385
Mallardo S, De Vito V, Malinconico M, Volpe MG, Santagata G, Di Lorenzo ML (2016) Poly (butylene succinate) based composites containing β-cyclodextrin/d-limonene inclusion complex. Eur Polym J 79:82–96
Meiners JA (2012) Fluid bed microencapsulation and other coating methods for food ingredient and nutraceutical bioactive compounds. In: Garti N, McClements DJ (eds) Encapsulation technologies and delivery systems for food ingredients and nutraceuticals. Elsevier, Woodhead Publishing Limited, Cambridge, pp 150–176
Mohammed NK, Tan CP, Manap YA, Alhelli AM, Hussin ASM (2017) Process conditions of spray drying microencapsulation of Nigella sativa oil. Powder Technol 315:1–14
Mora-Huertas CE, Garrigues O, Fessi H, Elaissari A (2012) Nanocapsules prepared via nanoprecipitation and emulsification–diffusion methods: comparative study. Eur J Pharm Biopharm 80(1):235–239
Morales E, Rubilar M, Burgos-Díaz C, Acevedo F, Penning M, Shene C (2017) Alginate/Shellac beads developed by external gelation as a highly efficient model system for oil encapsulation with intestinal delivery. Food Hydrocoll 70:321–328
Morales-Cruz M, Flores-Fernández GM, Morales-Cruz M, Orellano EA, Rodriguez-Martinez JA, Ruiz M, Griebenow K (2012) Two-step nanoprecipitation for the production of protein-loaded PLGA nanospheres. Results Pharma Sci 2:79–85
Murali S, Kar A, Mohapatra D, Kalia P (2015) Encapsulation of black carrot juice using spray and freeze drying. Revista de Agaroquimica y Tecnologia de Alimentos 21(8):604–612
Nandiyanto AB, Okuyama K (2011) Progress in developing spray-drying methods for the production of controlled morphology particles: from the nanometer to submicrometer size ranges. Adv Powder Technol 22(1):1–9
Neo YP, Ray S, Jin J, Gizdavic-Nikolaidis M, Nieuwoudt MK, Liu D, Quek SY (2013) Encapsulation of food grade antioxidant in natural biopolymer by electrospinning technique: a physicochemical study based on zein–gallic acid system. Food Chem 136(2):1013–1021
Nieuwland M, Geerdink P, Brier P, Van Den Eijnden P, Henket JT, Langelaan ML, Stroeks N, van Deventer HC, Martin AH (2013) Food-grade electrospinning of proteins. Innovative Food Sci Emerg Technol 20:269–275
Noronha CM, Granada AF, de Carvalho SM, Lino RC, de OB Maciel MV, Barreto PL (2013) Optimization of α-tocopherol loaded nanocapsules by the nanoprecipitation method. Ind Crop Prod 50:896–903
Noronha CM, de Carvalho SM, Lino RC, Barreto PL (2014) Characterization of antioxidant methylcellulose film incorporated with α-tocopherol nanocapsules. Food Chem 159:529–535
Oliveira AC, Moretti TS, Boschini C, Baliero JC, Freitas O, Favaro-Trindade CS (2007) Stability of microencapsulated *B. lactis* (BI 01) and *L. acidophilus* (LAC 4) by complex coacervation followed by spray drying. J Microencapsul 24(7):685–693
Oliveira DA, Angonese M, Ferreira SR, Gomes CL (2017a) Nanoencapsulation of passion fruit by-products extracts for enhanced antimicrobial activity. Food Bioprod Process 104:137–146
Oliveira DA, Mezzomo N, Gomes C, Ferreira SRS (2017b) Encapsulation of passion fruit seed oil by means of supercritical antisolvent process. J Supercrit Fluids 129:96–105
Palamanit A, Soponronnarit S, Prachayawarakorn S, Tungtrakul P (2013) Effects of inlet air temperature and spray rate of coating solution on quality attributes of turmeric extract coated rice using top-spray fluidized bed coating technique. J Food Eng 114(1):132–138
Pang SF, Yusoff MM, Gimbun J (2014) Assessment of phenolic compounds stability and retention during spray drying of *Orthosiphon stamineus* extracts. Food Hydrocoll 37:159–165

Parthasarathi S, Anandharamakrishnan C (2016) Enhancement of oral bioavailability of vitamin E by spray-freeze drying of whey protein microcapsules. Food Bioprod Process 100:469–476

Patel AR, Bhandari B (2014) Nano-and microencapsulation of vitamins. In: Kwak HS (ed) Nano- and microencapsulation for foods. Wiley, pp 223–248

Peng C, Zhao SQ, Zhang J, Huang GY, Chen LY, Zhao FY (2014) Chemical composition, antimicrobial property and microencapsulation of Mustard (*Sinapis alba*) seed essential oil by complex coacervation. Food Chem 165:560–568

Peres IMNFV (2011) Encapsulation of active compounds: particle characterization, loading efficiency and stability. Doctoral dissertation, Universidade do Porto (Portugal)

Pillai DS, Prabhasankar P, Jena BS, Anandharamakrishnan C (2012) Microencapsulation of Garcinia cowa fruit extract and effect of its use on pasta process and quality. Int J Food Prop 15(3):590–604

Piornos JA, Burgos-Díaz C, Morales E, Rubilar M, Acevedo F (2017) Highly efficient encapsulation of linseed oil into alginate/lupin protein beads: optimization of the emulsion formulation. Food Hydrocoll 63:139–148

Premi M, Sharma HK (2017) Effect of different combinations of maltodextrin, gum arabic and whey protein concentrate on the encapsulation behaviour and oxidative stability of spray dried drumstick (*Moringa oleifera*) oil. Int J Biol Macromol 105(Pt 1):1232–1240

Prieto C, Calvo L (2017a) Supercritical fluid extraction of emulsions to nanoencapsulate vitamin E in polycaprolactone. J Supercrit Fluids 119:274–282

Prieto C, Calvo L (2017b) The encapsulation of low viscosity omega-3 rich fish oil in polycaprolactone by supercritical fluid extraction of emulsions. J Supercrit Fluids 128:227–234

Prosapio V, Reverchon E, De Marco I (2017) Incorporation of liposoluble vitamins within PVP microparticles using supercritical antisolvent precipitation. J CO2 Util 19:230–237

Quispe-Condori S, Saldaña MD, Temelli F (2011) Microencapsulation of flax oil with zein using spray and freeze drying. LWT Food Sci Technol 44(9):1880–1887

Rachmawati H, Yanda YL, Rahma A, Mase N (2016) Curcumin-loaded PLA nanoparticles: formulation and physical evaluation. Sci Pharm 84(1):191–202

Rajam R, Anandharamakrishnan C (2015a) Microencapsulation of *Lactobacillus plantarum* (MTCC 5422) with fructooligosaccharide as wall material by spray drying. LWT Food Sci Technol 60(2):773–780

Rajam R, Anandharamakrishnan C (2015b) Spray freeze drying method for microencapsulation of *Lactobacillus plantarum*. J Food Eng 166:95–103

Rajam R, Kumar SB, Prabhasankar P, Anandharamakrishnan C (2014) Microencapsulation of *Lactobacillus plantarum* MTCC 5422 in fructooligosaccharide and whey protein wall systems and its impact on noodle quality. J Food Sci Technol 52(7):4029–4041

Reis CP, Neufeld RJ, Ribeiro AJ, Veiga F (2006) Nanoencapsulation I. Methods for preparation of drug-loaded polymeric nanoparticles. Nanomedicine 2(1):8–21

Saikia S, Mahnot NK, Mahanta CL (2015) Optimisation of phenolic extraction from *Averrhoa carambola* pomace by response surface methodology and its microencapsulation by spray and freeze drying. Food Chem 171:144–152

Santiago-Adame R, Medina-Torres L, Gallegos-Infante JA, Calderas F, González-Laredo RF, Rocha-Guzmán NE, Ochoa-Martinez LA, Bernad-Bernad MJ (2015) Spray drying-microencapsulation of cinnamon infusions (*Cinnamomum zeylanicum*) with maltodextrin. LWT Food Sci Technol 64(2):571–577

Santos MG, Bozza FT, Thomazini M, Favaro-Trindade CS (2015) Microencapsulation of xylitol by double emulsion followed by complex coacervation. Food Chem 171:32–39

Schell D, Beermann C (2014) Fluidized bed microencapsulation of *Lactobacillus reuteri* with sweet whey and shellac for improved acid resistance and in-vitro gastro-intestinal survival. Food Res Int 62:308–314

Schrooyen PM, van der Meer R, De Kruif CG (2001) Microencapsulation: its application in nutrition. Proc Nutr Soc 60(4):475–479

Semalty A (2014) Cyclodextrin and phospholipid complexation in solubility and dissolution enhancement: a critical and meta-analysis. Expert Opin Drug Deliv 11(8):1255–1272

Semyonov D, Ramon O, Kaplun Z, Levin-Brener L, Gurevich N, Shimoni E (2010) Microencapsulation of *Lactobacillus paracasei* by spray freeze drying. Food Res Int 43(1):193–202

Shahidi F, Han XQ (1993) Encapsulation of food ingredients. Crit Rev Food Sci Nutr 33(6):501–547

Sharif HR, Goff HD, Majeed H, Shamoon M, Liu F, Nsor-Atindana J, Haider J, Liang R, Zhong F (2017) Physicochemical properties of β-carotene and eugenol co-encapsulated flax seed oil powders using OSA starches as wall material. Food Hydrocoll 73:274–283

Shemesh R, Krepker M, Nitzan N, Vaxman A, Segal E (2016) Active packaging containing encapsulated carvacrol for control of postharvest decay. Postharvest Biol Technol 118:175–182

Shen Q, Quek SY (2014) Microencapsulation of astaxanthin with blends of milk protein and fiber by spray drying. J Food Eng 123:165–171

Shpigelman A, Shoham Y, Israeli-Lev G, Livney YD (2014) β-Lactoglobulin–naringenin complexes: nano-vehicles for the delivery of a hydrophobic nutraceutical. Food Hydrocoll 40:214–224

Silva EK, Zabot GL, Cazarin CB, Maróstica MR, Meireles MAA (2016) Biopolymer-prebiotic carbohydrate blends and their effects on the retention of bioactive compounds and maintenance of antioxidant activity. Carbohydr Polym 144:149–158

Soppimath KS, Aminabhavi TM, Kulkarni AR, Rudzinski WE (2001) Biodegradable polymeric nanoparticles as drug delivery devices. J Control Release 70(1):1–20

Souza ACP, Gurak PD, Marczak LDF (2017) Maltodextrin, pectin and soy protein isolate as carrier agents in the encapsulation of anthocyanins-rich extract from jaboticaba pomace. Food Bioprod Process 102:186–194

Takeungwongtrakul S, Benjakul S, H-kittikun A. (2015) Wall materials and the presence of antioxidants influence encapsulation efficiency and oxidative stability of micro-encapsulated shrimp oil. Eur J Lipid Sci Technol 117(4):450–459. https://doi.org/10.1002/ejlt.201400235

Tao F, Hill LE, Peng Y, Gomes CL (2014) Synthesis and characterization of β-cyclodextrin inclusion complexes of thymol and thyme oil for antimicrobial delivery applications. LWT Food Sci Technol 59(1):247–255

Tapia-Hernández JA, Rodríguez-Félix F, Katouzian I (2017) Nanocapsule formation by electrospraying. In: Jafari SM (ed) Nanoencapsulation technologies for the food and nutraceutical industries. Academic, pp 320–345

Teo A, Lee SJ, Goh KK, Wolber FM (2017) Kinetic stability and cellular uptake of lutein in WPI-stabilised nanoemulsions and emulsions prepared by emulsification and solvent evaporation method. Food Chem 221:1269–1276

Tewes F, Boury F, Benoit JP (2006) Biodegradable microspheres: advances in production technology. In: Benita S (ed) Microencapsulation: methods and industrial applications. CRC Press: Taylor and Francis Group, pp 1–53

Tice TR, Gilley RM (1985) Preparation of injectable controlled-release microcapsules by a solvent-evaporation process. J Control Release 2:343–352

Tiede K, Boxall AB, Tear SP, Lewis J, David H, Hassellöv M (2008) Detection and characterization of engineered nanoparticles in food and the environment. Food Addit Contam 25(7):795–821

Vishnu KV, Chatterjee NS, Ajeeshkumar KK, Lekshmi RGK, Tejpal CS, Mathew S, Ravishankar CN (2017) Microencapsulation of sardine oil: application of vanillic acid grafted chitosan as a bio-functional wall material. Carbohydr Polym 174:540–548

Vishwakarma GS, Gautam N, Babu JN, Mittal S, Jaitak V (2016) Polymeric encapsulates of essential oils and their constituents: a review of preparation techniques, characterization, and sustainable release mechanisms. Polym Rev 56(4):668–701

Wandrey C, Bartkowiak A, Harding SE (2010) Materials for encapsulation. In: Zuidam NJ, Nedovic VA (eds) Encapsulation technologies for active food ingredients and food processing. Springer, New York, pp 31–100

Wei YC, Cherukuri SR, Hriscisce F, Piccolo DJ, Bilka KP (1986) Elastomer encapsulation of flavors and sweeteners, long lasting flavored chewing gum compositions based thereon and process of preparation. US Patent 4590075
Wen P, Zhu DH, Feng K, Liu FJ, Lou WY, Li N, Zong MH, Wu H (2016a) Fabrication of electrospun polylactic acid nanofilm incorporating cinnamon essential oil/β-cyclodextrin inclusion complex for antimicrobial packaging. Food Chem 196:996–1004
Wen P, Zhu DH, Wu H, Zong MH, Jing YR, Han SY (2016b) Encapsulation of cinnamon essential oil in electrospun nanofibrous film for active food packaging. Food Control 59:366–376
Wilkowska A, Ambroziak W, Czyżowska A, Adamiec J (2016) Effect of microencapsulation by spray-drying and freeze-drying technique on the antioxidant properties of blueberry *(Vaccinium myrtillus)* juice polyphenolic compounds. Polish J Food Nutr Sci 66(1):11–16
Williams DB, Carter CB (1996) Transmission electron microscopy. Springer, New York
Wittorff H, Lund K (2017) Confectionery product comprising low molecular weight polyvinyl acetate. US Patent 9565867
Wrona M, Nerín C, Alfonso MJ, Caballero MÁ (2017) Antioxidant packaging with encapsulated green tea for fresh minced meat. Innovative Food Sci Emerg Technol 41:307–313
Wu Y, Zou L, Mao J, Huang J, Liu S (2014) Stability and encapsulation efficiency of sulforaphane microencapsulated by spray drying. Carbohydr Polym 102:497–503
Wurster DE (1957) Means for applying coatings to tablets or the like. US Patent 2799241
Wurster DE (1963) Granulating and coating process for uniform granules. US Patent 3089824
Wurster DE (1966) Particle coating process. US Patent 3253944
Wurster DE, Lindlof JA (1966) Particle coating apparatus. US Patent 3241520
Xie W, Pan WP (2001) Thermal characterization of materials using evolved gas analysis. J Therm Anal Calorim 65(3):669–685
Xue F, Li C, Liu Y, Zhu X, Pan S, Wang L (2013) Encapsulation of tomato oleoresin with zein prepared from corn gluten meal. J Food Eng 119(3):439–445
Yang RK (1988) Encapsulation composition for use with chewing gum and edible products. US Patent 4740376
Yang Z, Peng Z, Li J, Li S, Kong L, Li P, Wang Q (2014) Development and evaluation of novel flavour microcapsules containing vanilla oil using complex coacervation approach. Food Chem 145:272–277
Yatka RJ, Broderick KB, Song JH, Zibell SE, Meyers MA, Campbell AA (1992a) Polyvinyl acetate encapsulation of codried sucralose for use in chewing gum. US Patent 5139798
Yatka RJ, Broderick KB, Song JH, Zibell SE, Record DW (1992b) Polyvinyl acetate encapsulation of crystalline sucralose for use in chewing gum. US Patent 5169658
Yi J, Lam TI, Yokoyama W, Cheng LW, Zhong F (2015) Beta-carotene encapsulated in food protein nanoparticles reduces peroxyl radical oxidation in Caco-2 cells. Food Hydrocoll 43:31–40
Yin Y, Li J (2012) Formation of chitosan-based hydrogels network. In: Yao K, Li J, Yao F, Yin Y (eds) Chitosan-based hydrogels. Functions and applications. CRC Press, Taylor & Francis Group, Boca Raton, London, pp 183–197
Zafar N, Agusti G, Fessi H, Elaissari A (2017) Elaboration of sponge-like biodegradable cationic particles via double-emulsion solvent evaporation. J Dispers Sci Technol 38(4):577–583
Zhang H, Jung J, Zhao Y (2017) Preparation and characterization of cellulose nanocrystals films incorporated with essential oil loaded β-chitosan beads. Food Hydrocoll 69:164–172

Chapter 15
Food-Grade Biopolymers as Efficient Delivery Systems for Nutrients: An Overview

Lekshmi R. G. Kumar, K. K. Anas, C. S. Tejpal, and Suseela Mathew

Abstract The technological advancements in the field of food and nutrition have made the development of functional foods/nutraceuticals an easier task. Globally, researchers are in the process of isolating biomolecules of potential health significance from natural sources and subjecting to various in vitro and in vivo assays to investigate its feasibility as a functional food ingredient. However, incorporation of many such biomolecules into food systems often faces difficulties owing to the issues associated with its stability, bioavailability and sustained release. Such problems can be easily addressed through the use of efficient delivery systems that guarantees the safety, stability and sustained release of the nutrients. Such delivery systems are often referred to as 'encapsulation systems'. Till date, a wide variety of biopolymers such as proteins, polysaccharides, protein-polysaccharide conjugates, maillard products, natural gums, structurally modified polysaccharides etc. are being employed for delivery of nutrients. Reports says that the success of encapsulation to a greater extent depends on the prudent selection of food grade biopolymers which can deliver the nutrients effectively. In the present review, a comprehensive list of food grade biopolymers used as delivery systems for nutrients is discussed in detail.

Keywords Coacervates · Electrospraying · Encapsulation efficiency · Functional foods · Hydrogels · Nanoliposomes

15.1 Introduction

The role of foods has undergone a major transformation from alleviating hunger to reducing the risk rate of life style associated diseases. This along with the increase in consumer awareness about the importance of health and nutrition and the ever growing world population has put much pressure on the global food industry.

L. R. G. Kumar (✉) · K. K. Anas · C. S. Tejpal · S. Mathew
Biochemistry & Nutrition Division, Central Institute of Fisheries Technology (CIFT), ICAR, Cochin, Kerala, India

T. J. Gutiérrez (ed.), *Polymers for Food Applications*,
https://doi.org/10.1007/978-3-319-94625-2_15

Scientific evidences proposing the ability of certain foods to combat the life style associated ailments such as neurodegenerative, cardiovascular, diabetes, cancer etc. has been accumulating over the past decades (Dickinson and MacKay 2014; Luo 2014). This has spawned the development of new area within the food industry known as functional foods/nutraceuticals. Functional foods are those foods which provide a long term beneficial physiological or health effects beyond the nutritional properties. Such foods are often made by identifying, isolating and incorporating bioactive molecules of potential health significance (Blasco and Pico 2011). Globally, the nutraceutical market was valued at around $250 billion in 2014 and is expected to reach around $385 billion by 2020 (Suleria et al. 2015). Presently, the global functional food market is dominated by the USA, Europe and Japan contributing to more than 85% of the market.

Owing to the reported side effects of synthetic additives, the food industry has shifted their focus towards natural bioactive compounds which are considered as safer and viable substitutes for the former ones. Consequently, a number of substances of natural origin has been explored for isolation of bioactive molecules from them. Several analytical techniques such as high performance liquid chromatography (HPLC), gas chromatography (GC), fourier transform infrared spectroscopy (FTIR), thermogravimetric analysis (TGA), X-ray diffraction studies (XRD), nuclear magnetic resonance spectroscopy (NMR), mass spectroscopy (MS) etc. are being employed to determine the purity and physico-chemical attributes of the newly isolated biomolecules. The developed biomolecules, if meant for functional food application, need to further undergo safety assessment studies to establish the safe limits before their market release.

15.2 Challenges in Development of Functional Foods

The development of functional foods often involves the process of identifying, isolating, purifying, characterizing biomolecules and their subsequent incorporation in food systems. However, the development of functional foods is not an easier task; it involves several challenges. They are listed as follows:

15.2.1 Loss of Physico-chemical Stability

The important characteristic of most of the bioactive compounds is that they are subjected to rapid degradation processes, either during the processing or in the storage step. Several factors can affect the stability of biomolecules in a food system such as the presence of oxygen, moisture, pressure, endogenous enzymes, trace metals, pH etc. However, not all the bioactive molecules are equally affected; some may be highly sensitive (phenolic compounds and vitamins) and some may be least sensitive (minerals, fibres). Similarly, the chemical stability of many lipophilic compounds such as ω-3 rich oils, carotenoids, curcumin, squalene etc. are reported to be affected by the presence of oxygen, moisture, trace metals etc. (McClements and Decker 2000; Boon et al. 2010; Waraho et al. 2011; Heger et al. 2014; Lekshmi et al. 2017). These degradation processes often lead to the loss of bioactivity and stability of the product. The physico-chemical stability of such compounds has to be retained by developing certain intervention strategies such as encapsulating them in suitable polymeric matrices.

15.2.2 *Solubility*

An important pre-requisite for the development of bioactive molecules for functional food application is the solubility aspect. For instance, bioactive of hydrophobic nature such as ω-3 fatty acids, carotenoids, phytosterols etc. exhibits poor solubility in aqueous systems which hinder its application in many aqueous based food systems. As a consequence, these bioactive substances will remain underutilized. Hence such issues have to be addressed, while developing a hydrophobic biomolecule based functional food product by employing cost-effective technologies.

15.2.3 *Bioavailability*

The bioavailability of many bioactive molecules will be affected by several intrinsic and extrinsic factors such as its structure, chemical state, interaction with the gut microflora, interaction with other components in the food matrix. This can be in turn correlated with the loss of physico-chemical stability due to the transformation or degradation processes happening in the gastrointestinal fluids. Sealing those compounds in suitable delivery systems can either improve or increase the bioavailability.

15.2.4 *Melting Point*

The crystalline nature of certain biomolecules such as phytosterols, carotenoids, curcumin, fatty alcohols etc. makes their incorporation into food products a difficult task. When introduced into food systems, they may adversely affect the bioavailability, appearance, odour, taste and even shelf life of the final product. This can in turn affect the market potential and consumer demand of the product. In case of such crystalline compounds, the incorporation can be made successful by either dissolving them in suitable solvents or by introducing in the form of nanocrystals (McClements 2012).

15.3 Delivery Systems

Realising the major scientific and technological challenges associated with the formulation of functional foods, various efforts have been taken by researchers and nutritionists to battle the issue. Development of suitable delivery systems can be a promising solution for designing nutraceuticals by playing a major role in the encapsulation, protection and sustained release of biomolecules (Mezzenga et al. 2005; Sanguansri and Augustin 2006; McClements et al. 2007; Gutiérrez and Álvarez 2017). Micro and nanotechnology can be employed for the development of delivery systems. Materials of macro scale when converted to micro or nano size range are reported to exhibit different physico-chemical and biological properties than the former ones. The larger surface area to volume ratio achieved through size reduction strategy can help to improve the solubility, bioavailability, delivery, sensory attributes and total functionality of the biomolecules (Acosta 2009; Duran and Marcato 2013; Cerqueira et al. 2014). Moreover, size reduction is also reported to improve the bioadhesive properties, thereby prolonging the gastrointestinal residence time which will finally lead to a higher bioavailability at the target site. However, delivery systems must have certain characteristics to be used in development of functional foods/nutraceuticals which are listed below:

1. Delivery system intended for food application should be fabricated only from biomaterials which are of food-grade quality, safe and biodegradable in nature. It should also have Generally Recognized As Safe (GRAS) status (Augustin and Hemar 2009).
2. The material used for the development of delivery systems should be easy available, economically feasible and the benefits gained out of the encapsulation should outweigh the additional cost incurred in the process (McClements et al. 2007; Gutiérrez 2018).
3. The incorporation of the delivery systems should not adversely affect the physical, chemical, textural and sensory quality of the final product. That is, it should be compatible with other ingredients of the food matrix (Joye et al. 2014).
4. The delivery system developed should be robust. It should be physically and chemically stable and can offer the biomolecule considerable protection from any sort of degradation processes.
5. Ideally, the delivery system should have high loading capacity and retention of the biomolecule. Loading capacity refers to the amount of bioactive substance present per unit mass of the encapsulation material (McClements 2015).
6. Delivery system developed should ensure the sustained and targeted release of the biomolecule in response to a specific environmental stimulus such as pH, enzymatic action, temperature or ionic strength (Shegokar and Müller 2010).

15.3.1 Types of Delivery Systems

Delivery systems for functional food formulation can be developed from a variety of food-grade biopolymers such as proteins, lipids or carbohydrates etc. The biopolymers will be used either singly or in combination to increase the functionality. The properties of biomolecule to be encapsulated and the nature of the surrounding food matrix has to be taken into account while selecting biopolymers. Adequate knowledge about the molecular structure of the biopolymer is also essential as it in turn determines the functionality of food systems. The different types of delivery systems used in food industry are discussed below:

15.3.1.1 Protein Based Delivery Systems

Proteins are generally preferred for the development of delivery systems owing to their high nutritional value, biodegradability, economical and GRAS nature (Elzoghby et al. 2012; Benshitrit et al. 2012). They are also reported to have excellent functional properties including emulsification, gelation, water binding capacity, foaming etc. (Chen et al. 2006; Elzoghby et al. 2012). Its structural diversity attributed due to the multiple functional groups present in the primary sequence of polypeptides makes it an excellent candidate for the delivery of bioactives over a wide range of platforms such as hydrogels, micro and nano particles, molecular coacervates, emulsion droplet stabilization, films etc. (Chen et al. 2006). Among these, protein hydrogels are the most convenient and commonly employed matrix in food applications. In case of solid and semisolid foods, it is advantageous to decrease the matrix size from micrometers to nanometers to develop new protein vehicles of improved functional properties and facilitating their incorporation in food products without affecting the sensory (Augustin 2003). Two approaches are being used for size reduction; top down approach and bottom up approach.

Protein based delivery systems are relatively simple to prepare, economical and deliver both hydrophobic and hydrophilic bioactives (Luykx et al. 2008). Proteins can be obtained from various sources such as bacterial, fungal, plant and animals and among this, the latter two are commonly employed for food applications (Elzoghby et al. 2012; Kimura et al. 2014). Gelatin, collagen, elastin casein, albumin and whey proteins are some of the commonly used proteins of animal origin for functional food applications. Similarly, soy glycinin, zein and wheat gliadin are some of the plant proteins used as food delivery systems (Song et al. 2007; Xiao et al. 2011). Compared to animal proteins, plant proteins are more of hydrophobic in nature and can be easily self-assembled to stable structures without employing any severe processing conditions (Elzoghby et al. 2012). Some of the most commonly used proteins for the development of delivery systems are discussed below:

Whey protein which is obtained as a by-product of dairy industry accounts approximately 20% of the total milk protein (Brandelli et al. 2015). The main components of whey proteins are β-lactoglobulin, α-lactalbumin, and bovine serum albumin (Loveday et al. 2012). The essential amino acids such as leucine, valine, isoleucine and cysteine which are important agents in metabolism, neural function and homeostasis are present in rich amounts in whey protein (Patel 2015). The high nutritional content, GRAS status, binding properties with various bioactive molecules and other functional properties made it as a promising option in development of functional foods. It has been successfully used for encapsulation of blueberry pomace extract (Flores et al. 2014), anthocyanins and flavonoids from elderberry extract (Stănciuc et al. 2018), vitamin D3 (Abbasi et al. 2014), riboflavin and peptides (O'Neill et al. 2014) etc. By employing techniques such as gelation, emulsification, spray drying, coacervation etc. hydrogels, hydrocolloids, micelles, and micro- and nanoparticles of whey protein can be prepared and utilized as carriers of nutraceuticals. Studies show that properties of whey protein can be improved when combined with other biopolymers than when it is used alone.

Gunasekaran et al. (2007) have reported the release profile of nutraceuticals from whey protein hydrogels can be modulated by coating with oppositely charged polysaccharides such as alginate and chitosan. Paraskevopoulou et al. (2014) have reported that the incorporation of egg yolk or its plasma into WPI solution increased the storage and elastic modulus of mixed gels because of interaction between egg lipoproteins and whey proteins especially β-lactoglobulin. The successful encapsulation of black carrot extract using polysaccharide blended heat-set whey protein isolate hydrogels has been also reported (Ozel et al. 2017). López-Rubio and Lagaron (2012) have reported the potential of the electrospinning technique to generate whey protein capsules obtained through electrospraying for the encapsulation of β-carotene. Drosou et al. (2018) designed novel composite pullulan-whey protein isolate fibers using electrospinning which can find innovative applications in the food and pharmaceutical industries for formulation of products with specific material properties. Similarly, reports have shown that whey protein/sodium alginate gel microparticles prepared by cold gelation can be used to formulate a soft gel which will be soft to chew and swallow by elderly people (Leon et al. 2016). Complex coacervate of whey proteins was proven to protect and improve the oral delivery of folic acid (Chapeau et al. 2017).

Gelatin, a biodegradable protein derived from collagen by acidic (type-A) or alkaline (type-B) pre-treatment conditions, consists of repeating sequences of glycine, proline and hydroxyproline (Gómez-Guillén et al. 2009; Lai 2013). It has extensive applications as a carrier system for nutrients due to its high stabilizing activity, excellent emulsifying capacity, thickening ability, water-solubility, low cost and commercial availability (Shu et al. 2006). It can form thermoreversible hydrogels in water due to the formation of collagen-like triple helices below the helix–coil transition temperature and thereby leading to chain entanglement and formation of network (Peña et al. 2010). It has been used in combination with other biopolymers for encapsulation of curcumin (Wang et al. 2009a, b), sulforaphane of broccoli seed extract (García-Saldaña et al. 2016), epigallocatechin gallate (Shutava et al. 2009) etc. Gelatin sub microparticles prepared by electrospraying was also found to have promising potential as edible carriers for polyphenols of interest in functional foods (Gómez-Mascaraque et al. 2015). The gelatin–curcumin microparticles obtained by electrodynamic atomization was found to improve the antioxidant activity, water solubility and bioaccessibility of curcumin and even improved its dispersion in a fish gellified product (Gómez-Estaca et al. 2015). Gelatin nanofibres loaded with curcumin was found to have considerable antimicrobial activity with effective inhibition against *Staphylococcus aureus* (Deng et al. 2017).

Zein which is the main storage protein of corn has garnered much attention in the recent past as an attractive delivery system for functional ingredients because of its ability to entrap a large number of hydrophobic compounds (Davidov-Pardo et al. 2015; Patel et al. 2010). It is one among the few hydrophobic water-insoluble materials approved by Food and Drug Administration (FDA) for food use (Donsì et al. 2017). Apart from this, it is food grade, biodegradable, non-toxic and have amphiphilic properties. It helps in the controlled release of bioactive core material and in the prolongation of shelf life (Wu et al. 2011). This has been successfully used to encapsulate bioactive compounds such as vitamin E, grape seed extract, essential oils etc. (Joye et al. 2015). Zein-chitosan based nanoparticles were found successful for the encapsulation of epigallocatechin gallate, the major polyphenol in tea, resulting in its slow and sustained release (Liang et al. 2017). Zein-carboxymethyl chitosan-tea polyphenol ternary complexes when used for the encapsulation of β-carotene showed good redispersibility in distilled water, good DPPH scavenging activity due to stronger hydrophobic interaction, controlled release profile and improved encapsulation efficiency (Wang et al. 2018). Veneranda et al. (2018) developed zein-caseinate-pectin complex nanoparticles for encapsulation of eugenol and found that spray dried eugenol powders exhibited exceptional redispersibility property while maintaining the original nanoscale size range. Liang et al. (2018) studied the feasibility of ultrasound treatments on the formation of resveratrol-loaded zein particles and reported that triple-frequency simultaneous ultrasound with 20/28/40 kHz resulted in the highest encapsulation efficiency (84.75%) and loading capacity (84.78%) of resveratrol. They have concluded that ultrasound can be a potential process to produce zein nanoparticles. Similarly, Dai et al. (2017) evaluated the stability of curcumin in zein-lecithin composite nanoparticles and suggested it as a potential delivery system for water-insoluble bioactive compounds with enhanced encapsulation efficiency and chemical stability.

15.3.1.2 Carbohydrate Based Delivery Systems

Carbohydrates which accounts for calorific value, sensory and textural properties form a significant component of many food systems. It is considered as suitable carrier for many nutraceuticals owing to its biocompatibility, biodegradability, structural versatility, site digestion properties (Sinha and Kumria 2001; Baldwin and Kiick 2010). The presence of functional groups makes it an excellent candidate for the development of delivery systems as it can interact with a wide range of bioactive compounds of both hydrophobic and hydrophilic nature. Among the various forms of carbohydrates, polysaccharides are often considered as better carrier agents because of their massive molecular structure which enables in the efficient entrapment of biomolecules (Fathi et al. 2014). Polysaccharides which are natural polymers of monosaccharides vary in type, distribution, number and bonding of the monomers in the chain (Fathi et al. 2014). Carbohydrate based delivery systems can be categorized into four main groups based on their source such as plant origin (e.g. starch, gum Arabic, guar gum, pectin), animal origin (chitin, chitosan), algal (agar, carrageenan, alginate) and microbial origin (xanthan gum, dextran, cyclodextrin etc.) (Kosaraju 2005). Polysaccharides can be further categorized based on their charges such as neutral, anionic and cationic (Daniel-da-Silva and Trindade 2011). However, the magnitude of electrical charge is often dependent on pH relative to the pKa value (Liu 2008). For instance, anionic polysaccharides behave as neutral polysaccharides at a pH below their pKa value, whereas cationic polysaccharides tend to be neutral at pH values above their pKa value (Kumari et al. 2010). The physicochemical properties and application of some of the commonly used carbohydrate based delivery systems are reviewed below:

Chitosan, a natural nontoxic biopolymer derived by deacetylation of chitin, has received much attention in the past few decades due to its numerous biological activities such as antitumor, immunoenhancing, antimicrobial, hypocholesterolemic properties etc. (Sudarshan et al. 1992; Sekiguchi et al. 1994; Gutiérrez 2017). These immense bioactivities made it as suitable candidate in the food, pharmaceutical and even in the cosmetic industries. However, many times its application is restricted due to its insolubility in water, high viscosity and molecular weight (Lodhi et al. 2014). This has geared up the research for possible alternatives which will make the application of chitosan easier. As a result, many chitosan derivatives such as oligomers and grafted chitosan were prepared by subjecting to various enzymatic or chemical reactions (Hasegawa et al. 1993; Won-Seok et al. 2002; Ilyina et al. 2000). Chitosan with degrees of polymerization (DPs) <20 and an average molecular weight less than 3900 Da are called chitosan oligomers, chitooligomers, or chitooligosaccharides. Chitosan oligomer is found to be readily soluble in water because of their shorter chain lengths (Jeon et al. 2000). In addition, the low viscosity of chitosan oligomer makes it application easier. Akin to chitosan, the chitosan derivatives are also reported to possess numerous bioactivities such as antibacterial, antifungal, antitumor, radical scavenging, antimicrobial, immunomodulatory and wound healing effect (Muzzarelli 2009; Aam 2010; Xia et al. 2011; Li et al. 2012). Apart from this, the absorption rate of chitosan oligomers in human body is reported to be nearly 100% and having improved properties than the unmodified chitosan.

Grafting chitosan with appropriate functional moieties to improve the solubility and other properties are well documented. Grafting of chitosan with phenolic acid such as coumaric acid, gallic acid and vanillic acid has showed promising improvement in the bio-active properties (Cho et al. 2011; Aytekin et al. 2011; Tai et al. 2012). Grafted chitosan derivatives are also finding applications in food industry as a functional food supplement or functional food ingredient. Hu et al. (2015) reported that gallic acid grafted chitosan could be used for the delivery of bioactive compounds. Chatterjee et al. (2016) also reported the application of ferulic acid grafted chitosan for microencapsulation and controlled release of vitamins. Tejpal et al. (2017) showed that dietary supplementation of thiamine and pyridoxine loaded vanillic acid grafted chitosan could effectively enhance the growth performance, metabolic and immune responses and can be used as a functional food ingredient.

Pectin which is a linear anionic polysaccharide comprises of galacturonic acid-rich polysaccharides such as rhamnogalacturonan I, homogalacturonan, substituted galacturonan rhamnogalacturonan and xylogalacturonan (Mohnen 2008). Based on the degree of esterification, pectin is classified into two: low methoxyl pectin (25–50% methoxylation) and high methoxyl pectin (50–80% methoxylation) (Liang et al. 2012). Studies have demonstrated the potential of pectin to form nano-scale particles suitable for delivery of bioactive ingredients. However, it is found that relatively large particles (r > 1000 nm) are formed when pectin alone is used as delivery system, whereas smaller particles are formed when it is used combination with other biopolymers such as chitosan (Dutta and Sahu 2012; Fathi and Varshosaz 2013). Badrul et al. (2018) suggested a viable method for production of low molecular weight pectin and its nanoparticles by subjecting pectin to a combination treatment of microwave and inorganic salts.

Though it is resistant to enzymatic digestion in the mouth and stomach it can be degraded by colon microflora and hence can be used for the delivery of acid sensitive food bioactives (Sinha and Kumria 2001). Hence, pectin has been mainly employed for probiotic encapsulation in food products. Sun et al. (2014) reported that high methoxy pectin-soy protein isolate complex can enhance the viability of *Lactobacillus delbrueckii* strains by 3 log units compared with the control. Similarly, Dafe et al. (2017) formulated novel food-grade hydrogel particle based on pectin-starch for probiotic colon delivery and was found to be resistant against adverse conditions of the gastro–intestinal tract and bile salt solution compared to non-encapsulated cells.

Alginate, an anionically charged polysaccharide, is composed of 1,4-linked-β-D-mannuronic acid and α-L-guluronic acid residues (Rehm 2009). It is mainly extracted from red algae and recent reports have shown that can also be obtained by bacterial biosynthesis (Lee and Mooney 2012). It is non-toxic, biocompatible, economical, biodegradable and hence finds wide application in many industries (Josef et al. 2010; Zeeb et al. 2015). Its hydrophilic nature is utilized for the efficient entrapment of hydrophilic food bioactives. By interacting with certain cations such as Cu^{2+}, Zn^{2+}, Ca^{2+}, Ba^{2+}, Al^{3+} it can undergo ionotropic gelation and form egg-box dimers (Fang et al. 2007; Nayak et al. 2012). The gelled alginate thus formed is a pH-sensitive polymer that is insoluble in low pH and swells in a high-pH environment (Wang et al. 2009a, b). This characteristic makes it a favourable candidate for delivery of acid sensitive bioactive ingredients in the intestine and to protect or retard its release in acidic environment (George and Abraham 2006).

Starch, which is the most abundant storage polysaccharide in plants, is composed of two main structural components: amylose and amylopectin (Barsby et al. 2001; Gutiérrez et al. 2015, 2018). The hydrophilic nature of starch limits its use as delivery system for hydrophobic bioactives and hence structural modification is often employed to widen its application. Hydrophobic starch derivatives such as dialdehyde starch (Yu et al. 2007), propyl starch nanoparticles (Jain et al. 2011; Santander-Ortega et al. 2010) have been designed and used for encapsulation and delivery of lipophilic pharmaceutical agents. Various attempts have been carried out for the encapsulation of nutraceuticals using either modified starches or a combination of several other biopolymers. The use of modified starches with other biopolymers for the encapsulation of cardamom oleoresin (Krishnan et al. 2005), cumin oleoresin (Kanakdande et al. 2007), monoterpenoids (Mourtzinos et al. 2008), β-carotene (Loksuwan 2007) etc. has been already reported. Apart from this, modified starches such as n-octenylsuccinylated (n-OSA) and phosphorylated waxy maize starches, modified sago and tapioca starch has been used for encapsulation of flavouring agents (Murúa-Pagola et al. 2009; Baranauskienė et al. 2007; Varavinit et al. 2001).

Gum Arabic is a heterogeneous mixture of 90–99% of arabinogalactan oligosaccharides and 1% glycoproteins (Gulão et al. 2014). It is mainly extracted from trunks and branches of *Acacia senegal* and *Acacia seyal*; it can be also extracted from other species of Acacia such as *Acacia karroo*, *Acacia polyacantha*, *Acacia sieberiana* etc. (Masuelli 2013). Some of the unique properties of gum Arabic includes colourless nature, emulsification, good retention of volatile compounds, low cost, high solubility, low viscosity anionic properties and inhibition of oxidation reactions It is also reported to have the ability to create a strong protective film around oil droplets due to the presence of protein and polysaccharide moieties in it and hence can be used for protection of oil from oxidation (Krishnan et al. 2005). It is widely being used in the food industry as a stabilizer, encapsulant, thickener and emulsifier. Recently, the modified form also finding applications as useful hydrocolloids in a variety of applications. Chemically modified gum Arabic with sodium trimetaphosphate crosslinker are finding applications in encapsulation of essential oils (Aguilar et al. 2010), octenyl succinic anhydride gum arabic as food additive (Sarkar and Singhal 2011), acetylated gum Arabic for preparing iodine complexes (Ahmad et al. 2013), blend of gum arabic with other biopolymers for encapsulation of grape phenols (Kuck and Noreña 2016). Tsai et al. (2017) reported that gum Arabic modified alginate helped in preparation of liquid hydrogel beads of higher encapsulation efficiency, better protection from degradation and followed a fickian diffusion pattern. Similarly, gum Arabic-whey protein complex was found to be very efficient in encapsulation of citrus flavonoids resulted in high powder yield (72.74%), encapsulation efficiency (97.60 ± 0.99%) with better retention of antioxidant activity (Hu et al. 2018).

Dextran, a polysaccharide of bacterial origin, is mainly linear neutral polymers of α-D-glucose linked by α (1 → 6) glycosidic bonds and a small percentage of α-(1 → 3) linkages. The presence of hydroxyl groups is mainly utilized for the covalent attachment of various organic functional groups especially hydrophobic compounds (Kaewprapan et al. 2012). Modified dextran made by varying degree of substitution makes it either water soluble or insoluble. Aumelas et al. (2007) reported that substitution with hydrophobic group resulted in dramatic decrease of dextran biodegradability. Falco et al. (2017) developed novel chitosan-dextran sulfate hydrogels for probiotic bacteria and found that the method was successful in retaining the viability of bacterial cells. Dextran - Bovine serum albumin conjugate nanoparticles was found effective in loading of curcumin by offering high stability and cellular antioxidant activity (Fan et al. 2018).

15.3.1.3 Lipid Based Delivery Systems

Lipid based delivery systems are reported to have better encapsulation efficiency and low toxicity than other delivery systems. In general, four major lipid based delivery systems are being used, namely, nanoemulsions, nanoliposomes, solid lipid nanoparticles (SLN) and nanostructure lipid carriers (NLC). They are discussed below:

Nanoemulsions, also known as submicron emulsions, are formed by dispersing of one liquid in another immiscible liquid by physical share-induced rupturing (Mukherjee et al. 2009; Liu et al. 2006). When compared to their microsized counterparts, nanoemulsions have certain advantages such as (1) it can remain stable in diluted water without any change in its relative size distribution (Gutiérrez et al. 2008) (2) appear optically transparent as its size is much smaller than visible wavelengths; this makes its application in easier in beverages (3) demonstrated better against the gravity due to Brownian motion of nano sized droplets caused by entropic driving forces (Mason et al. 2006) (4) nanoemulsions are kinetically stable whereas microemulsions are thermodynamically stable systems (Henry et al. 2010). Nanoemulsions can be prepared using two methods: mechanical and non-mechanical techniques. Mechanical methods include high-pressure homogenization, microfluidization and ultrasonication (Anton et al. 2007). Non-mechanical methods include solvent diffusion technique can be used for preparation of nanoemulsions (Anton et al. 2007; Tadros et al. 2004; Unger et al. 2004).

Another lipid based delivery system which is widely being used in the food research industry is liposomes. They are reported to possess wide range of benefits such as (1) can be produced from materials of natural origin (2) used for production, entrapment, release of compounds having wide range of solubility such as water-soluble, lipid-soluble, and amphiphilic materials (Mozafari et al. 2008; Thompson et al. 2006) (3) deliver and release their load in the target site inside and outside the body (Mozafari 2006). They are widely used in food industry for making formulations of antimicrobials (Malheiros et al. 2010), enzymes (Dufour et al. 1996), lipophilic vitamins (Gonnet et al. 2010) and minerals (Arnaud 1995). Mechanical (extrusion, sonification, high pressure homogenization, microfluidization and colloid mill) and non-mechanical methods (reversed-phase evaporation and depletion of mixed detergent-lipid micelles) are generally being employed for the production of nanosized liposomes.

Liposomes are classified into unilamellar vesicles, multilamellar vesicles and multivesicular vesicles based on their lamellar structure. Based on size, liposomes can be further classified into small unilamellar vesicle (size between 20 and 100 nm) and large unilamellar vesicle (size up to few micrometers). Among these, small unilamellar vesicles (SUV) have a low aqueous core volume to lipid ratio and hence can be considered as efficient encapsulants of large functional foods and nutraceutical compounds (Sharma and Sharma 1997); whereas, large unilamellar vesicle (LUV) and multivesicular vesicles (MVV) have a large aqueous core volume to lipid ratio. These two are reported to carry large loads of encapsulated water-soluble compounds in their internal core and hence considered appropriate for encapsulation of large hydrophilic compounds. As of now, several methods are being employed for the stabilization of liposomes such as lyophilization, freezing, spray-drying, supercritical fluid (SCF) technology etc. and among these SCF is considered as better option (Fathi et al. 2012).

Solid lipid nanoparticle (SNP) is another category of lipid-based nutrient delivery system which has gained considerable attention in the recent past (Harde et al. 2011; Almeida and Souto 2007). SNP generally exists as spherical particles of size 10–1000 nm and possess a solid or semisolid lipid core matrix. They can solubilize lipophilic molecules and their stability is usually maintained by addition of surfactants (Kalepu et al. 2013). When compared to liposomes and nanoemulsions, SNP have some distinct advantages such as protection against GI tract degradation, high encapsulation efficiency, minimum use of organic solvents in their preparation, feasibility of large scale production and controlled release (Mader and Mehnert 2005; Saupe and Rades 2006; Ting 2014; Onoue 2012). Two basic production methods used for large-scale production of SNP in food processing are hot and cold homogenization (Muller et al. 2000). SNP was found useful in the successful encapsulation of nutrients such as curcumin (Wang 2013) and resveratrol, a bioactive polyphenolic phytoalexin (Pandita 2014).

Though SNP is reported to have several advantages, it has certain inherent problems such as low encapsulation load and possibility of explosion during storage. The lower encapsulation efficiency is attributed due to the less space available to accommodate the drug and nutraceutical molecules and explosion due to the formation of α and β into a perfect β transition form (Westesen et al. 1997). Hence a novel carrier known as the nanostructure lipid carrier (NLC) was developed by mixing different lipid molecules to overcome the limitation (Radtke and Müller 2001). By giving nanostructure to the lipid matrix, the encapsulation efficiency of biomolecules is found to be enhanced and risk of explosion is also reduced by preventing the formation of perfect crystals (Chen et al. 2010). The use of NLC for chemical stability enhancement of ascorbyl palmitate has been reported by Teeranachaideekul et al. (2007). The utility of NLC for the encapsulation of lipophilic nutrients such vitamin E and omega 3 fatty acids was reported by Zhuang et al. (2010). The bioavailability of quercetin in different lipid delivery systems showed the loading efficiency of quercetin was higher when incorporated in NLC and nanoemulsions than in SNPs (Aditya et al. 2014). In spite of its listed advantages, it is reported to have certain limitations when used in the food industry such as relatively low purity of the bio-based materials and undergo flocculation and agglomeration under different pH conditions, osmotic pressure and ionic strength (Liu et al. 2012).

15.3.1.4 Mixed Delivery Systems

Of late, there is an increasing trend in application of mixtures of polymers instead of individual ones with a view to improve and broaden the application range and functionality. Among the various biopolymer mixture studied so far, protein–polysaccharide complex is considered more advantageous, because of their higher chemical and colloidal protection. The formation and stability of protein–polysaccharide complex is found to be affected by a number of factors such as pH, ionic strength, and biopolymer concentration, charge distribution, molecular weight, temperature, pressure, etc. (Ye 2008). In such cases, the sequence of biopolymers adsorption onto the interface determines the structure and stabilizing properties of the mixed protein-polysaccharide complex (Dickinson 2008). Two phenomena can occur during the mixing of proteins and polysaccharides in a liquid medium depending on the pH and isoelectric point (1) the attractive interactions can lead to the formation of soluble and insoluble complexes and (2) the repulsive interactions can separation of the two biopolymers from each other (Weinbreck et al. 2004).

The potential of protein-polysaccharide complex nanoparticles to encapsulate lipophilic bioactive compounds, such as curcumin (Wang et al. 2016), rutin (Luo et al. 2015), essential oils (Zhang et al. 2014), chia seed oil (Timilsena et al. 2016), turmeric oleoresin (Zuanon et al. 2013), lutein (Qv et al. 2011), vitamin E (Alencastre et al. 2006) and lycopene (Silva et al. 2012) has been reported. Complexation of β-lactoglobulin and pectin has been employed for the encapsulation of both hydrophilic and hydrophobic nutraceuticals including different vitamins, flavors, essential oils, fatty acids, etc. (Zimet and Livney 2009). Shaddel et al. (2018) have reported that complex coacervates of gelatin and gum Arabic when used for encapsulation of black raspberry anthocyanins was found to increase the stability of anthocyanins up to 23.66% even after 2 months of storage. The potential of zein-caseinate-pectin complex nanoparticles as an efficient delivery system for curcumin owing to their high encapsulation efficiency, slow release, enhanced antioxidant activity, redispersibility has been reported (Chang et al. 2017).

15.4 Future Perspectives

Food delivery systems has emerged as a cost effective and potential tool to combat malnutrition and ill effects of many life style associated diseases. As a result, diverse range of nutraceutical incorporated food delivery systems are being designed and tested for their physico-chemical characteristics and other stability aspects. Innovations are happening on these arenas to develop newer and efficient unique biopolymer complexes by either structurally modifying or by conjugating with functional moieties for improving its functionality. While developing new delivery systems certain aspects has to be kept in mind such as its redispersibility, bioavailability, controlled release, loading efficiency, behaviour within the gastrointestinal fluids, sensory attributes, stability and safety. Most importantly, the technology should be commercial viable and hence costs incurred until the production, packaging and storage needs to be considered.

Nanotechnology is being extensively employed for development of newer delivery systems. Undoubtedly, nanotechnology has become one of the most promising technology in the functional food industry; however, the aftermath of the long term exposure of human health to nanomaterials is still unknown and has to be studied. The challenges within the ethical, safety and regulatory aspects also needs to be considered and assessed while developing nanosystems or nanoparticles for nutrient delivery. By adopting proper regulatory measures and addressing the possible human and environmental concerns, nanobiomaterials can be used. The future holds ample prospects in improvement of human health and prevention of life style associated diseases by rational designing of tailor made foods by application of such newer and innovative technologies.

References

Aam BB (2010) Production of chitooligosaccharides and their potential applications in medicine. Mar Drugs 8:1482–1517

Abbasi A, Emam-Djomeh Z, Mousavi MAE, Davoodi D (2014) Stability of vitamin D 3 encapsulated in nanoparticles of whey protein isolate. Food Chem 143:379–383

Acosta E (2009) Bioavailability of nanoparticles in nutrient and nutraceutical delivery. Curr Opin Colloid Interface Sci 14(1):3–15

Aditya NP, Macedo AS, Doktorovova S, Souto EB, Kim S, Chang PS, Ko S (2014) Development and evaluation of lipid nanocarriers for quercetin delivery: a comparative study of solid lipid nanoparticles (SLN), nanostructured lipid carriers (NLC), and lipid nanoemulsions (LNE). LWT Food Sci Technol 59(1):115–121

Aguilar F, Charrondiere UR, Dusemund B, Galtier P, Gilbert J, Gott DM (2010) Scientific opinion on the use of Gum Acacia modified with Octenyl Succinic Anhydride (OSA) as a food additive. EFSA J 8(3):1–23

Ahmad SI, Mazumdar N, Kumar S (2013) Functionalization of natural gum: an effective method to prepare iodine complex. Carbohydr Polym 92(1):497–502

Alencastre JB, Bentley MVLB, Garcia FS, Moragas MD, Viladot JL, Marchetti JM (2006) A study of the characteristics and in vitro permeation properties of CMC/chitosan microparticles as a skin delivery system for vitamin E. Rev Bras Ciênc Farm 42(1):69–76

Almeida AJ, Souto E (2007) Solid lipid nanoparticles as a drug delivery system for peptides and proteins. Adv Drug Deliv Rev 59(6):478–490

Anton N, Saulnier P, Beduneau A, Benoit JP (2007) Salting-out effect induced by temperature cycling on a water/nonionic surfactant/oil system. J Phys Chem B 111(14):3651–3657

Arnaud JP (1995) Pro-liposomes for the food industry. Food Technol Europe 2:30–34

Augustin MA (2003) The role of microencapsulation in the development of functional dairy foods. Aust J Dairy Technol 58(2):156

Augustin MA, Hemar Y (2009) Nano-and micro-structured assemblies for encapsulation of food ingredients. Chem Soc Rev 38(4):902–912

Aumelas A, Serrero A, Durand A, Dellacherie E, Leonard M (2007) Nanoparticles of hydrophobically modified dextrans as potential drug carrier systems. Colloids Surf B: Biointerfaces 59(1):74–80

Aytekin AO, Morimura S, Kida K (2011) Synthesis of chitosan–caffeic acid derivatives and evaluation of their antioxidant activities. J Biosci Bioeng 111:212–216

Badrul HZ, Wong TW, Halimaton H (2018) Design of low molecular weight pectin and its nanoparticles through combination treatment of pectin by microwave and inorganic salts. Polym Degrad Stab 147:35–40

Baldwin AD, Kiick KL (2010) Polysaccharide modified synthetic polymeric biomaterials. J Pept Sci 94(1):128–140

Baranauskienė R, Bylaitė E, Žukauskaitė J, Venskutonis RP (2007) Flavor retention of peppermint (Mentha piperita L.) essential oil spray-dried in modified starches during encapsulation and storage. J Agric Food Chem 55(8):3027–3036

Barsby TL, Donald AM, Frazier PJ (2001) Starch: advances in structure and function. Royal Society of Chemistry, Cambridge

Benshitrit RC, Levi CS, Tal SL, Shimoni E, Lesmes U (2012) Development of oral food-grade delivery systems: current knowledge and future challenges. Food Funct 3(1):10–21

Blasco C, Pico Y (2011) Determining nanomaterials in food. TrAC Trends Anal Chem 30(1):84–99

Boon CS, McClements DJ, Weiss J, Decker EA (2010) Factors influencing the chemical stability of carotenoids in foods. Crit Rev Food Sci Nutr 50(6):515–532

Brandelli A, Daroit DJ, Corrêa APF (2015) Whey as a source of peptides with remarkable biological activities. Food Res Int 73:149–161

Cerqueira MA, Pinheiro AC, Silva HD, Ramos PE, Azevedo MA, Flores-López ML, Rivera MC, Bourbon AI, Ramos OL, Vicente AA (2014) Design of bio-nanosystems for oral delivery of functional compounds. Food Eng Rev 6(1–2):1–19

Chang C, Wang T, Hu Q, Luo Y (2017) Zein/caseinate/pectin complex nanoparticles: formation and characterization. Int J Biol Macromol 104(Pt A):117–124

Chapeau AL, Bertrand N, Briard-Bion V, Hamon P, Poncelet D, Bouhallab S (2017) Coacervates of whey proteins to protect and improve the oral delivery of a bioactive molecule. J Funct Foods 38;197–204

Chatterjee NS, Anandan R, Navitha M, Asha KK, Kumar KA, Mathew S, Ravishankar CN (2016) Development of thiamine and pyridoxine loaded ferulic acid-grafted chitosan microspheres for dietary supplementation. J Food Sci Technol 53(1):551–560

Chen L, Remondetto GE, Subirade M (2006) Food protein-based materials as nutraceutical delivery systems. Trends Food Sci Technol 17(5):272–283

Chen CC, Tsai TH, Huang ZR, Fang JY (2010) Effects of lipophilic emulsifiers on the oral administration of lovastatin from nanostructured lipid carriers: physicochemical characterization and pharmacokinetics. Eur J Pharm Biopharm 74(3):474–482

Cho YS, Kim SB, Ahn CB, Je JY (2011) Preparation characterization, and antioxidant properties of gallic acid-grafted-chitosans. Carbohydr Polym 83:1617–1622

Dafe A, Etemadi H, Dilmaghani A, Mahdavinia GR (2017) Investigation of pectin/starch hydrogel as a carrier for oral delivery of probiotic bacteria. Int J Biol Macromol 97:536–543

Dai L, Sun C, Li R, Mao L, Liu F, Gao Y (2017) Structural characterization, formation mechanism and stability of curcumin in zein-lecithin composite nanoparticles fabricated by antisolvent co-precipitation. Food Chem 237:1163–1171

Daniel-da-Silva AL, Trindade T (2011) Biofunctional composites of polysaccharides containing inorganic nanoparticles. Advances in nanocomposite technology. InTech, Sheffield, pp 275–298

Davidov-Pardo G, Joye IJ, McClements DJ (2015) Encapsulation of resveratrol in biopolymer particles produced using liquid antisolvent precipitation. Part 1: preparation and characterization. Food Hydrocoll 45:309–316

Deng L, Kang X, Liu Y, Feng F, Zhang H (2017) Effects of surfactants on the formation of gelatin nanofibres for controlled release of curcumin. Food Chem 231:70–77

Dickinson E (2008) Interfacial structure and stability of food emulsions as affected by protein–polysaccharide interactions. Soft Matter 4(5):932–942

Dickinson A, MacKay D (2014) Health habits and other characteristics of dietary supplement users: a review. Nutr J 13(1):14

Donsì F, Voudouris P, Veen SJ, Velikov KP (2017) Zein-based colloidal particles for encapsulation and delivery of epigallocatechin gallate. Food Hydrocoll 63:508–517

Drosou C, Krokida M, Biliaderis CG (2018) Composite pullulan-whey protein nanofibers made by electrospinning: impact of process parameters on fiber morphology and physical properties. Food Hydrocoll 77:726–735

Dufour P, Laloy E, Vuillemard JC, Simard R (1996) Liposomes in cheesemaking. In: Lasic D, Barenholz Y (eds) Handbook of nonmedical applications of liposomes. CRC Press, Boca Raton

Duran N, Marcato PD (2013) Nanobiotechnology perspectives. Role of nanotechnology in the food industry: a review. Int J Food Sci Technol 48(6):1127–1113

Dutta RK, Sahu S (2012) Development of diclofenac sodium loaded magnetic nanocarriers of pectin interacted with chitosan for targeted and sustained drug delivery. Colloids Surf B: Biointerfaces 97:19–26
Elzoghby AO, Samy WM, Elgindy NA (2012) Albumin-based nanoparticles as potential controlled release drug delivery systems. J Control Release 157(2):168–182
Falco CY, Falkman P, Risbo J, Cárdenas M, Medronho B (2017) Chitosan-dextran sulfate hydrogels as a potential carrier for probiotics. Carbohydr Polym 172:175–183
Fan Y, Yi J, Zhang Y, Yokoyama W (2018) Fabrication of curcumin-loaded bovine serum albumin (BSA)-dextran nanoparticles and the cellular antioxidant activity. Food Chem 239:1210–1218
Fang Y, Al-Assaf S, Phillips GO, Nishinari K, Funami T, Williams PA, Li L (2007) Multiple steps and critical behaviors of the binding of calcium to alginate. J Phys Chem B 111(10):2456–2462
Fathi M, Varshosaz J (2013) Novel hesperetin loaded nanostructure lipid carriers coated by different carbohydrate for food fortification: production and characterization. J Funct Foods 5:1382–1391
Fathi M, Mozafari MR, Mohebbi M (2012) Nanoencapsulation of food ingredients using lipid based delivery systems. Trends Food Sci Technol 23(1):13–27
Fathi M, Martín Á, McClements DJ (2014) Nanoencapsulation of food ingredients using carbohydrate based delivery systems. Trends Food Sci Technol 39(1):18–39
Flores FP, Singh RK, Kong F (2014) Physical and storage properties of spray-dried blueberry pomace extract with whey protein isolate as wall material. J Food Eng 137:1–6
García-Saldaña JS, Campas-Baypoli ON, López-Cervantes J, Sánchez-Machado DI, Cantú-Soto EU, Rodríguez-Ramírez R (2016) Microencapsulation of sulforaphane from broccoli seed extracts by gelatin/gum arabic and gelatin/pectin complexes. Food Chem 201:94–100
George M, Abraham TE (2006) Polyionic hydrocolloids for the intestinal delivery of protein drugs: alginate and chitosan—a review. J Control Release 114(1):1–14
Gómez-Estaca J, Gavara R, Hernández-Muñoz P (2015) Encapsulation of curcumin in electrosprayed gelatin microspheres enhances its bioaccessibility and widens its uses in food applications. Innov Food Sci Emerg Technol 29:302–307
Gómez-Guillén MC, Pérez-Mateos M, Gómez-Estaca J, López-Caballero E, Giménez B, Montero P (2009) Fish gelatin: a renewable material for developing active biodegradable films. Trends Food Sci Technol 20(1):3–16
Gómez-Mascaraque LG, Lagarón JM, López-Rubio A (2015) Electro sprayed gelatine submicroparticles as edible carriers for the encapsulation of polyphenols of interest in functional foods. Food Hydrocoll 49:42–52
Gonnet M, Lethuaut L, Boury F (2010) New trends in encapsulation of liposoluble vitamins. J Control Release 146(3):276–290
Gulão EDS, de Souza CJ, da Silva FA, Coimbra JS, Garcia-Rojas EE (2014) Complex coacervates obtained from lactoferrin and gum arabic: formation and characterization. Food Res Int 65:367–374
Gunasekaran S, Ko S, Xiao L (2007) Use of whey proteins for encapsulation and controlled delivery applications. J Food Eng 83(1):31–40
Gutiérrez TJ (2017) Chitosan applications for the food industry. In: Ahmed S, Ikram S (eds) Chitosan: derivatives, composites and applications. Wiley-Scrivener Publisher, pp 185–232. https://doi.org/10.1002/9781119364849.ch8. EE.UU. ISBN: 978-1-119-36350-7
Gutiérrez TJ (2018) Processing of nano- and microcapsules for industrial applications. In: Hussain CM (ed) Handbook of nanomaterials for industrial applications. Elsevier, pp 989-1011. https://doi.org/10.1016/B978-0-12-813351-4.00057-2. EE.UU. ISBN: 978-0-12-813351-4
Gutiérrez TJ, Álvarez K (2017) Biopolymers as microencapsulation materials in the food industry. In: Masuelli M, Renard D (eds) Advances in physicochemical properties of biopolymers: part 2. Bentham Science Publishers, pp 296–322. https://doi.org/10.2174/978168108544911701000 9. EE.UU. ISBN: 978-1-68108-545-6. eISBN: 978-1-68108-544-9
Gutiérrez JM, González C, Maestro A, Sole I, Pey CM, Nolla J (2008) Nano-emulsions: new applications and optimization of their preparation. Curr Opin Colloid Interface Sci 13(4):245–251

Gutiérrez TJ, Morales NJ, Pérez E, Tapia MS, Famá L (2015) Physico-chemical properties of edible films derived from native and phosphated cush-cush yam and cassava starches. Food Packag Shelf Life 3:1–8. https://doi.org/10.1016/j.fpsl.2014.09.002

Gutiérrez TJ, Ollier R, Alvarez VA (2018) Surface properties of thermoplastic starch materials reinforced with natural fillers. In: Thakur VK, Thakur MK (eds) Functional biopolymers. Editorial Springer International Publishing, pp 131–158. https://doi.org/10.1007/978-3-319-66417-0_5. EE.UU. ISBN: 978-3-319-66416-3. eISBN: 978-3-319-66417-0

Harde H, Das M, Jain S (2011) Solid lipid nanoparticles: an oral bioavailability enhancer vehicle. Expert Opin Drug Deliv 8(11):1407–1424

Hasegawa M, Isogai A, Onabe F (1993) Preparation of low molecular weight chitosan using phosphoric acid. Carbohydr Polym 20:279–283

Heger M, van Golen RF, Broekgaarden M, Michel MC (2014) The molecular basis for the pharmacokinetics and pharmacodynamics of curcumin and its metabolites in relation to cancer. Pharmacol Rev 66(1):222–307

Henry JV, Fryer PJ, Frith WJ, Norton IT (2010) The influence of phospholipids and food proteins on the size and stability of model sub-micron emulsions. Food Hydrocoll 24(1):66–71

Hu B, Wang Y, Xie M, Hu G, Ma F, Zeng X (2015) Polymer nanoparticles composed with gallic acid grafted chitosan and bioactive peptides combined antioxidant, anticancer activities and improved delivery property for labile polyphenols. J Funct Foods 15:593–603

Hu Y, Li Y, Zhang W, Kou G, Zhou Z (2018) Physical stability and antioxidant activity of citrus flavonoids in arabic gum-stabilized microcapsules: modulation of whey protein concentrate. Food Hydrocoll 77:588–597

Ilyina AV, Tikhonov VE, Albulov AI, Varlamov VP (2000) Enzymic preparation of acidfree-water-soluble chitosan. Process Biochem 35:563–568

Jain R, Dandekar P, Loretz B, Melero A, Stauner T, Wenz G (2011) Enhanced cellular delivery of idarubicin by surface modification of propyl starch nanoparticles employing pteroic acid conjugated polyvinyl alcohol. Int J Pharm 420:147–155

Jeon YJ, Shahidi F, Kim SK (2000) Preparation of chitin and chitosan oligomers and their applications in physiological functional foods. Food Rev Int 16:159–176

Josef E, Zilberman M, Bianco-Peled H (2010) Composite alginate hydrogels: an innovative approach for the controlled release of hydrophobic drugs. Acta Biomater 6(12):4642–4649

Joye IJ, Davidov-Pardo G, McClements DJ (2014) Nanotechnology for increased micronutrient bioavailability. Trends Food Sci Technol 40(2):168–182

Joye IJ, Davidov-Pardo G, Ludescher RD, McClements DJ (2015) Fluorescence quenching study of resveratrol binding to zein and gliadin: towards a more rational approach to resveratrol encapsulation using water-insoluble proteins. Food Chem 185:261–267

Kaewprapan K, Inprakhon P, Marie E, Durand A (2012) Enzymatically degradable nanoparticles of dextran esters as potential drug delivery systems. Carbohydr Polym 88(3):875–881

Kalepu S, Manthina M, Padavala V (2013) Oral lipid-based drug delivery systems–an overview. Acta Pharm Sin B 3(6):361–372

Kanakdande D, Bhosale R, Singhal RS (2007) Stability of cumin oleoresin microencapsulated in different combination of gum arabic, maltodextrin and modified starch. Carbohydr Polym 67(4):536–541

Kimura A, Kabasawa Y, Tabata Y, Aoki K, Ohya K, Omura K (2014) Gelatin hydrogel as a carrier of recombinant human fibroblast growth factor-2 during rat mandibular distraction. J Oral Maxillofac Surg 72(10):2015–2031

Kosaraju SL (2005) Colon targeted delivery systems: review of polysaccharides for encapsulation and delivery. Crit Rev Food Sci Nutr 45(4):251–258

Krishnan S, Bhosale R, Singhal RS (2005) Microencapsulation of cardamom oleoresin: evaluation of blends of gum arabic, maltodextrin and a modified starch as wall materials. Carbohydr Polym 61(1):95–102

Kuck LS, Noreña CPZ (2016) Microencapsulation of grape (Vitislabrusca var. Bordo) skin phenolic extract using gum Arabic, polydextrose, and partially hydrolyzed guar gum as encapsulating agents. Food Chem 194:569–576

Kumari A, Yadav SK, Yadav SC (2010) Biodegradable polymeric nanoparticles based drug delivery systems. Colloids Surf B Biointerfaces 75(1):1–18

Lai JY (2013) Influence of solvent composition on the performance of carbodiimide cross-linked gelatin carriers for retinal sheet delivery. J Mater Sci Mater Med 24(9):2201–2210

Lee KY, Mooney DJ (2012) Alginate: properties and biomedical applications. Prog Polym Sci 37(1):106–126

Lekshmi RGK, Chatterjee NS, Tejpal CS, Vishnu KV, Anas KK, Asha KK, Anandan R, Mathew S (2017) Evaluation of chitosan as a wall material for microencapsulation of squalene by spray drying: characterization and oxidative stability studies. Int J Biol Macromol 104:1986–1995

Leon AM, Medina WT, Park DJ, Aguilera JM (2016) Mechanical properties of whey protein/Na alginate gel microparticles. J Food Eng 188:1–7

Li K, Xing R, Liu S, Qin Y, Li B, Wang X, Li P (2012) Separation and scavenging superoxide radical activity of chitooligomers with degree of polymerization 6–16. Int J Biol Macromol 51:826–830

Liang RH, Chen J, Liu W, Liu CM, Yu W, Yuan M, Zhou XQ (2012) Extraction, characterization and spontaneous gel-forming property of pectin from creeping fig (Ficus pumila Linn.) seeds. Carbohydr Polym 87(1):76–83

Liang J, Yan H, Wang X, Zhou Y, Gao X, Puligundla P, Wan X (2017) Encapsulation of epigallocatechin gallate in zein/chitosan nanoparticles for controlled applications in food systems. Food Chem 231:19–24

Liang Q, Ren X, Zhang X, Hou T, Chalamaiah M, Ma H, Xu B (2018) Effect of ultrasound on the preparation of resveratrol-loaded zein particles. J Food Eng, 221:88–94

Liu Z (2008) Polysaccharides-based nanoparticles as drug delivery systems. Adv Drug Deliv Rev 60(15):1650–1662

Liu W, Sun D, Li C, Liu Q, Xu J (2006) Formation and stability of paraffin oil-in-water nanoemulsions prepared by the emulsion inversion point method. J Colloid Interface Sci 303(2):557–563

Liu GY, Wang JM, Xia Q (2012) Application of nanostructured lipid carrier in food for the improved bioavailability. Eur Food Res Technol 234(3):391–398

Lodhi G, Kim YS, Hwang JW, Kim SK, Jeon YJ, Je JY, Ahn CB, Moon SH, Jeon BT, Park PJ (2014) Chitooligosaccharide and its derivatives: preparation and biological applications. Biomed Res Int 2014:654913

Loksuwan J (2007) Characteristics of microencapsulated β-carotene formed by spray drying with modified tapioca starch, native tapioca starch and maltodextrin. Food Hydrocoll 21(5):928–935

López-Rubio A, Lagaron JM (2012) Whey protein capsules obtained through electrospraying for the encapsulation of bioactives. Innovative Food Sci Emerg Technol 13:200–206

Loveday SM, Su J, Rao MA, Anema SG, Singh H (2012) Whey protein nanofibrils: kinetic, rheological and morphological effects of group IA and IIA cations. Int Dairy J 26(2):133–140

Luo Y (2014) Nutrient delivery systems: the future strategy for chronic diseases. Austin J Nutr Food Sci 2(9):1049

Luo Y, Pan K, Zhong Q (2015) Casein/pectin nanocomplexes as potential oral delivery vehicles. Int J Pharm 486(1):59–68

Luykx DM, Peters RJ, van Ruth SM, Bouwmeester H (2008) A review of analytical methods for the identification and characterization of nano delivery systems in food. J Agric Food Chem 56(18):8231–8247

Mader K, Mehnert W (2005) Solid Lipid nanoparticles—concepts, procedures, and physicochemical aspects. In: Lipospheres in drug targets and delivery. CRC Press

Malheiros PS, Daroit DJ, Silveira NP, Brandelli A (2010) Effect of nanovesicle-encapsulated nisin on growth of Listeria monocytogenes in milk. Food Microbiol 27:175–178

Mason TG, Wilking JN, Meleson K, Chang CB, Graves SM (2006) Nanoemulsions: formation, structure, and physical properties. J Phys Condens Matter 18:R635–R666

Masuelli MA (2013) Hydrodynamic properties of whole arabic gum. Am J Food Sci Technol 1(3):60–66

McClements DJ (2015) Nanoscale nutrient delivery systems for food applications: improving bioactive dispersibility, stability, and bioavailability. J Food Sci 80(7):N1602–N1611
McClements DJ, Decker EA (2000) Lipid oxidation in oil in water emulsions: impact of molecular environment on chemical reactions in heterogeneous food systems. J Food Sci 65(8):1270–1282
McClements DJ, Decker EA, Weiss J (2007) Emulsion-based delivery systems for lipophilic bioactive components. J Food Sci 72(8):R109–R124
McClements DJ (2012) Crystals and crystallization in oil-in-water emulsions: implications for emulsion-based delivery systems. Adv Colloid Interface Sci 174:1–30.
Mezzenga R, Schurtenberger P, Burbidge A, Michel M (2005) Understanding foods as soft materials. Nat Mater 4(10):729–740
Mohnen D (2008) Pectin structure and biosynthesis. Curr Opin Plant Biol 11(3):266–277
Mourtzinos I, Kalogeropoulos N, Papadakis SE, Konstantinou K, Karathanos VT (2008) Encapsulation of nutraceutical monoterpenes in β-cyclodextrin and modified starch. J Food Sci 73(1):S89–S94
Mozafari MR (2006) Nanocarrier technologies: frontiers of nanotherapy. Springer, Dordrecht, p 225
Mozafari MR, Johnson C, Hatziantoniou S, Demetzos C (2008) Nanoliposomes and their applications in food nanotechnology. J Liposome Res 18:309–327
Mukherjee S, Ray S, Thakur RS (2009) Solid lipid nanoparticles: a modern formulation approach in drug delivery system. Indian J Pharm Sci 71(4):349
Muller RH, Dingler A, Schneppe T, Gohla S (2000) Largescale production of solid lipid nanoparticles (SLN) and nanosuspensions. In: Wise DL (ed) Handbook of pharmaceutical controlled release technology. CRC Press, New York, pp 377–392
Murúa-Pagola B, Beristain-Guevara CI, Martínez-Bustos F (2009) Preparation of starch derivatives using reactive extrusion and evaluation of modified starches as shell materials for encapsulation of flavoring agents by spray drying. J Food Eng 91(3):380–386
Muzzarelli RAA (2009) Chitins and chitosans for the repair of wounded skin, nerve, cartilage and bone. Carbohydr Polym 76:167–182
Nayak AK, Das B, Maji R (2012) Calcium alginate/gum Arabic beads containing glibenclamide: development and in vitro characterization. Int J Biol Macromol 51(5):1070–1078
O'Neill GJ, Egan T, Jacquier JC, O'Sullivan M, O'Riordan ED (2014) Whey microbeads as a matrix for the encapsulation and immobilisation of riboflavin and peptides. Food Chem 160:46–52
Onoue S (2012) Novel solid self-emulsifying drug delivery system of coenzyme Q10 with improved photochemical and pharmacokinetic behaviors. Eur J Pharm Sci 46(5):492–499
Ozel B, Cikrikci S, Aydin O, Oztop MH (2017) Polysaccharide blended whey protein isolate-(WPI) hydrogels: a physicochemical and controlled release study. Food Hydrocoll 71:35–46
Pandita D (2014) Solid lipid nanoparticles enhance oral bioavailability of resveratrol, a natural polyphenol. Food Res Int 62:1165–1174
Paraskevopoulou A, Amvrosiadou S, Biliaderis CG, Kiosseoglou V (2014) Mixed whey protein isolate-egg yolk or yolk plasma heat-set gels: rheological and volatile compounds characterisation. Food Res Int 62:492–499
Patel S (2015) Functional food relevance of whey protein: a review of recent findings and scopes ahead. J Funct Foods 19:308–319
Patel A, Hu Y, Tiwari JK, Velikov KP (2010) Synthesis and characterisation of zein–curcumin colloidal particles. Soft Matter 6(24):6192–6199
Peña C, De La Caba KORO, Eceiza A, Ruseckaite R, Mondragon I (2010) Enhancing water repellence and mechanical properties of gelatin films by tannin addition. Bioresour Technol 101(17):6836–6842
Qv XY, Zeng ZP, Jiang JG (2011) Preparation of lutein microencapsulation by complex coacervation method and its physicochemical properties and stability. Food Hydrocoll 25(6):1596–1603
Radtke M, Müller RH (2001) Nanostructured lipid drug carriers. New Drugs 2:48–52
Rehm BH (2009) Alginates: biology and applications, vol 13. Springer Science & Business Media

Sanguansri P, Augustin MA (2006) Nanoscale materials development–a food industry perspective. Trends Food Sci Technol 17(10):547–556

Santander-Ortega MJ, Stauner T, Loretz B, Ortega-Vinuesa JL, Bastos-Gonzalez D, Wenz G (2010) Nanoparticles made from novel starch derivatives for transdermal drug delivery. J Control Release 141:85–92

Sarkar S, Singhal RS (2011) Esterification of guar gum hydrolysate and gum Arabic with n-octenyl succinic anhydride and oleic acid and its evaluation as wall material in microencapsulation. Carbohydr Polym 86(4):1723–1731

Saupe A, Rades T (2006) Solid lipid nanoparticles. In: Nanocarrier technologies. Springer, Netherlands, pp 41–50

Sekiguchi S, Miura Y, Kaneko H, Nishimura SI, Nishi N, Iwase M, Tokura S (1994) Molecular weight dependency of antimicrobial activity by chitosan oligomers. In: Nishinari K, Doi E (eds) Food hydrocolloids: structures, properties, and functions. Plenum, New York, pp 71–76

Shaddel R, Hesari J, Azadmard-Damirchi S, Hamishehkar H, Fathi-Achachlouei B, Huang Q (2018) Use of gelatin and gum Arabic for encapsulation of black raspberry anthocyanins by complex coacervation. Int J Biol Macromol 107(Pt B):1800–1810

Sharma A, Sharma US (1997) Liposomes in drug delivery: progress and limitations. Int J Pharm 154:123–140

Shegokar R, Müller RH (2010) Nanocrystals: industrially feasible multifunctional formulation technology for poorly soluble actives. Int J Pharm 399(1):129–139

Shu B, Yu W, Zhao Y, Liu X (2006) Study on microencapsulation of lycopene by spray-drying. J Food Eng 76(4):664–669

Shutava TG, Balkundi SS, Lvov YM (2009) Epigallocatechin gallate/gelatin layer-by-layer assembled films and microcapsules. J Colloid Interface Sci 330(2):276–283

Silva DF, Favaro Trindade CS, Rocha GA, Thomazini M (2012) Microencapsulation of lycopene by gelatin–pectin complex coacervation. J Food Process Preserv 36(2):185–190

Sinha VR, Kumria R (2001) Polysaccharides in colon-specific drug delivery. Int J Pharm 224(1):19–38

Song WO, Chun OK, Hwang I, Shin HS, Kim BG, Kim KS, Lee SY, Shin D, Lee SG (2007) Soy isoflavones as safe functional ingredients. J Med Food 10(4):571–580

Stănciuc N, Oancea AM, Aprodu I, Turturică M, Barbu V, Ioniţă E, Râpeanu G, Bahrim G (2018) Investigations on binding mechanism of bioactives from elderberry (Sambucus nigra L.) by whey proteins for efficient microencapsulation. J Food Eng 223:197–207

Sudarshan NR, Hoover DG, Knorr D (1992) Antibacterial action of chitosan. Food Biotechnol 6:257–272

Suleria HAR, Osborne S, Masci P, Gobe G (2015) Marine-based nutraceuticals: an innovative trend in the food and supplement industries. Mar Drugs 13(10):6336–6351

Sun Q, Wang F, Han D, Zhao Y, Liu Z, Lei H, Song Y, Huang X, Li X, Ma A, Yuan G (2014) Preparation and optimization of soy protein isolate–high methoxy pectin microcapsules loaded with Lactobacillus delbrueckii. Int J Food Sci Technol 49(5):1287–1293

Tadros T, Izquierdo P, Esquena J, Solans C (2004) Formation and stability of nano-emulsions. Adv Colloid Interf Sci 108:303–318

Tai A, Sawano T, Ito H (2012) Antioxidative properties of vanillic acid esters in multiple antioxidant assays. Biosci Biotechnol Biochem 76:314–318

Teeranachaideekul V, Müller RH, Junyaprasert VB (2007) Encapsulation of ascorbyl palmitate in nanostructured lipid carriers (NLC)—effects of formulation parameters on physicochemical stability. Int J Pharm 340(1):198–206

Tejpal CS, Chatterjee NS, Elavarasan K, Lekshmi RGK, Anandan R, Asha KK, Ganesan B, Mathew S, Ravishankar CN (2017) Dietary supplementation of thiamine and pyridoxine-loaded vanillic acid-grafted chitosan microspheres enhances growth performance, metabolic and immune responses in experimental rats. Int J Biol Macromol 104:1874–1881

Thompson AK, Hindmarsh JP, Haisman D, Rades T, Singh H (2006) Comparison of the structure and properties of liposomes prepared from milk fat globule membrane and soy phospholipids. J Agric Food Chem 54(10):3704–3711

Timilsena YP, Adhikari R, Barrow CJ, Adhikari B (2016) Microencapsulation of chia seed oil using chia seed protein isolate-chia seed gum complex coacervates. Int J Biol Macromol 91:347–357
Ting Y (2014) Common delivery systems for enhancing in vivo bioavailability and biological efficacy of nutraceuticals. J Funct Foods 7:112–128
Tsai FH, Kitamura Y, Kokawa M (2017) Effect of gum arabic-modified alginate on physicochemical properties, release kinetics, and storage stability of liquid-core hydrogel beads. Carbohydr Polym 174:1069–1077
Unger EC, Porter T, Culp W, Labell R, Matsunaga T, Zutshi R (2004) Therapeutic applications of lipid-coated microbubbles. Adv Drug Deliv Rev 56(9):1291–1314
Varavinit S, Chaokasem N, Shobsngob S (2001) Studies of flavor encapsulation by agents produced from modified sago and tapioca starches. Starch Stärke 53(6):281–287
Veneranda M, Hu Q, Wang T, Luo Y, Castro K, Madariaga JM (2018) Formation and characterization of zein-caseinate-pectin complex nanoparticles for encapsulation of eugenol. LWT Food Sci Technol 89:596–603
Wang P (2013) The formulation and delivery of curcumin with solid lipid nanoparticles for the treatment of on non-small cell lung cancer both in vitro and in vivo. Mater Sci Eng C 33(8):4802–4808
Wang Q, Zhang J, Wang A (2009a) Preparation and characterization of a novel pH-sensitive chitosan-g-poly (acrylic acid)/attapulgite/sodium alginate composite hydrogel bead for controlled release of diclofenac sodium. Carbohydr Polym 78(4):731–737
Wang Y, Lu Z, Lv F, Bie X (2009b) Study on microencapsulation of curcumin pigments by spray drying. Eur Food Res Technol 229(3):391–396
Wang T, Hu Q, Zhou M, Xia Y, Nieh MP, Luo Y (2016) Development of "all natural" layer-by-layer redispersible solid lipid nanoparticles by nano spray drying technology. Eur J Pharm Biopharm 107:273–285
Wang M, Fu Y, Chen G, Shi Y, Li X, Zhang H, Shen Y (2018) Fabrication and characterization of carboxymethyl chitosan and tea polyphenols coating on zein nanoparticles to encapsulate β-carotene by anti-solvent precipitation method. Food Hydrocoll 77:577–587
Waraho T, McClements DJ, Decker EA (2011) Mechanisms of lipid oxidation in food dispersions. Trends Food Sci Technol 22(1):3–13
Weinbreck F, Nieuwenhuijse H, Robijn GW, de Kruif CG (2004) Complexation of whey proteins with carrageenan. J Agric Food Chem 52(11):3550–3555
Westesen K, Bunjes H, Koch MHJ (1997) Physicochemical characterization of lipid nanoparticles and evaluation of their drug loading capacity and sustained release potential. J Control Release 48(2):223–236
Won-Seok C, Kil-Jin A, Dong-Wook L, Myung-Woo B, Hyun-Jin P (2002) Preparation of chitosan oligomers by irradiation. Polym Degrad Stab 78:533–538
Wu LY, Wen QB, Yang XQ, Xu MS, Yin SW (2011) Wettability, surface microstructure and mechanical properties of films based on phosphorus oxychloride-treated zein. J Sci Food Agric 91(7):1222–1229
Xia W, Liu P, Zhang J, Chen J (2011) Biological activities of chitosan and chitooligosaccharides. Food Hydrocoll 25:170–179
Xiao D, Davidson PM, Zhong Q (2011) Spray-dried zein capsules with coencapsulated nisin and thymol as antimicrobial delivery system for enhanced antilisterial properties. J Agric Food Chem 59(13):7393–7404
Ye A (2008) Complexation between milk proteins and polysaccharides via electrostatic interaction: principles and applications–a review. Int J Food Sci Technol 43(3):406–415
Yu D, Xiao S, Tong C, Chen L, Liu X (2007) Dialdehyde starch nanoparticles: preparation and application in drug carrier. Chin Sci Bull 52:2913–2918
Zeeb B, Saberi AH, Weiss J, McClements DJ (2015) Formation and characterization of filled hydrogel beads based on calcium alginate: factors influencing nanoemulsion retention and release. Food Hydrocoll 50:27–36

Zhang Y, Niu Y, Luo Y, Ge M, Yang T, Yu LL, Wang Q (2014) Fabrication, characterization and antimicrobial activities of thymol-loaded zein nanoparticles stabilized by sodium caseinate–chitosan hydrochloride double layers. Food Chem 142:269–275

Zhuang CY, Li N, Wang M, Zhang XN, Pan WS, Peng JJ, Pan YS, Tang X (2010) Preparation and characterization of vinpocetine loaded nanostructured lipid carriers (NLC) for improved oral bioavailability. Int J Pharm 394(1):179–185

Zimet P, Livney YD (2009) Beta-lactoglobulin and its nanocomplexes with pectin as vehicles for ω-3 polyunsaturated fatty acids. Food Hydrocoll 23(4):1120–1126

Zuanon LAC, Malacrida CR, Telis VRN (2013) Production of turmeric oleoresin microcapsules by complex coacervation with gelatin–gum Arabic. J Food Process Eng 36(3):364–373

Chapter 16
Current Processing Methods in the Development of Micro- and Nanoencapsulation from Edible Polymers

Teresita Arredondo-Ochoa, Carlos Regalado-González, and Olga Martín-Belloso

Abstract Micro- and nanoencapsulation are processes that enclose a substance within a wall material at the micro and nano scale levels. This technology shows an important role in the food industry due to the many advantages and a variety of new properties that it can offer to the encapsulated material. Among them it can be highlighted the improvement of stability by protecting a food ingredient from an adverse environment, masking organoleptic properties, facilitation of handling equipment, enhanced bioavailability of bioactive compounds, and controlled release, which could reduce doses and potential toxicity of the encapsulated compound. Edible polymers such as polysaccharides and proteins have been proposed as wall materials in micro- and nanoencapsulation, due to the benefits that they offer over synthetic polymers. In addition, edible polymers are highly available, safe, convenient, and can increase the quality of the final product. Hence, the techniques used to successfully achieve these processes depend on the carrier wall materials used. This chapter will focus on describing the characteristics of the different processing methods used for the production of micro- and nanoencapsulated compounds, their advantages, disadvantages and applications. It will also provide insights about recent advances in this area.

Keywords Biopolymers · Lipids · Polysaccharides · Proteins

T. Arredondo-Ochoa · C. Regalado-González
DIPA, PROPAC, Facultad de Química, Universidad Autónoma de Querétaro, Querétaro, Querétaro, Mexico

O. Martín-Belloso (✉)
Department of Food Technology, University of Lleida–Agrotecnio Center, Lleida, Spain
e-mail: omartin@tecal.udl.cat

T. J. Gutiérrez (ed.), *Polymers for Food Applications*,
https://doi.org/10.1007/978-3-319-94625-2_16

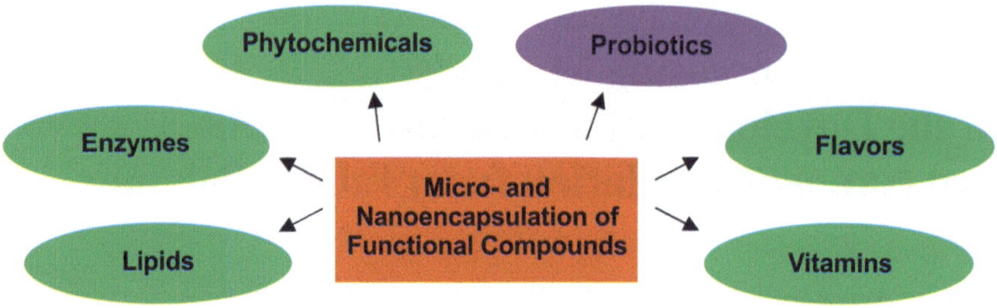

Fig. 16.1 Schematic representation of functional compounds encapsulation. Adapted from Kwak (2014)

16.1 Introduction

The development of functional foods involving the addition of bioactive compounds is being considered as a promising field in the food industry due to the health promoting benefits for the consumer (Olaiya et al. 2016). In addition, appearance, product safety, storage conditions, and sensory properties of the food product should not be compromised by the incorporation of the functional ingredient (Dordevic et al. 2015). However, it has been proved that many of these compounds may be unstable when facing environmental and processing conditions, which causes either a loss of functionality, chemical degradation or a premature or incomplete release (Nedovic et al. 2011). Based on the previous statements, the development of new technologies to incorporate health-promoting compounds into food products is becoming increasingly important (Alexander et al. 2012).

Micro- and nanoencapsulation processes have great potential in the food industry since they allow the incorporation of different components such as lipids, enzymes, phytochemicals, vitamins, flavors and probiotics in food products providing many advantages in their preservation and contributes to the development of functional foods (Fig. 16.1) (Kwak 2014; Quirós-Sauceda et al. 2014; Gutiérrez and Álvarez 2017).

The main function of micro- and nanoencapsulation technologies is to improve the physical, chemical, and biological properties of the functional ingredients entrapped in a variety of ways: (1) to mask the organoleptic properties such as taste, color and odor of the functional component; (2) to protect the encapsulated substance from external conditions during processing and storage in the food system; (3) to generate sustained, targeted, or controlled release of the bioactive component; (4) to provide better uptake, absorption, and bioavailability of specific nutrients in the body (Ray et al. 2016).

These technologies provide protection to labile compounds, by covering them with a resistant layer to preserve their availability, stability and functional properties (Nagavarma et al. 2012). In addition, several encapsulation procedures categorized into chemical, physical or mechanical methods, are widely used to promote the micro- and nanoencapsulation of bioactive compounds, but none of them can be considered as a universally applicable technique for a bioactive substance or food component (Vincekovic et al. 2017; Gutiérrez 2018).

The challenge lies in selecting a convenient encapsulation material and the most suitable method which mainly depend on the properties of the functional compound to be encapsulate (Dias et al. 2017). This chapter presents an overview of the most common edible polymers used to protect bioactive compounds and the methods currently used to produce micro-and nanoencapsulation systems.

16.2 Micro- and Nanoencapsulation

Micro- and nanoencapsulation techniques are inclusion processes of small particles of a functional compound, also known as the core material, the active agent, the fill or the internal phase, within a secondary material, also known as the wall material, the shell, the external phase, or the matrix, to form small capsules (Fang and Bhandari 2010).

The micro-and nanoencapsulation technologies have the ability to preserve a substance in a finely divided state and to be slowly released for its intended function. However, there are a number of requirements for the encapsulated compound to be used in a food system, which are listed below (McClements 2012):

- *Food grade:* the encapsulated compound must be fabricated entirely from food ingredients and processing operations that have regulatory approval.
- *Food matrix compatibility:* the encapsulated product should be compatible with the surrounding food matrix, i.e. should not adversely affect the appearance, texture, flavor or shelf life of the product.
- *Protection against chemical degradation:* the encapsulated compound has to stay stable against factors like oxygen, pH and temperature, which can trigger chemical degradation reactions such as oxidation and hydrolysis, among others.
- *Loading capacity and retention:* the encapsulated product should be capable of enclosing a large amount of food components per unit mass of carrier material and should efficiently retain them until the functional compound needs to be delivered.
- *Delivery capacity:* the encapsulated compound may have to be designed to release the food component at a particular site-of-action, at a controlled rate or in response to a specific environmental stimulus such as pH, ionic strength, enzyme activity or temperature, which may occur during food storage or within the human body.
- *Bioavailability/bioactivity:* the encapsulated component should enhance (or at least not adversely affect) the bioavailability/bioactivity of the food ingredient.
- *Cost-benefits balance:* the encapsulated product should be capable of being economically manufactured from inexpensive ingredients. In this sense, the benefits gained from encapsulating a food ingredient within a micro- and nano system such as shelf life improvement, marketability enhanced, novel functionality and better bioavailability, should outweigh the additional costs associated with encapsulation.

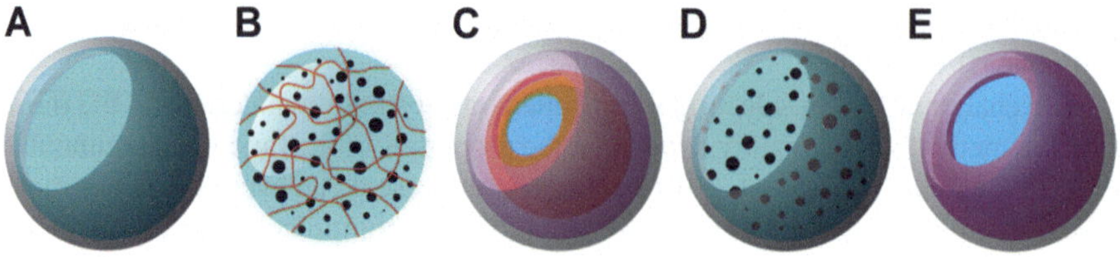

Fig. 16.2 Morphologies of capsules: (**a**) single core capsule; (**b**) dispersed core in polymer network; (**c**) multi-layered capsule; (**d**) dual-core capsule; (**e**) dual-layer capsule. Adapted from Augustin and Hemar (2009)

16.2.1 Classification

The morphology of a capsule depends mainly on the inner core and the position of the wall material. In this sense, the structure of the encapsulation system can be classified into capsules with (a) a core that is surrounded by a wall of the polymer material or (b) a core that is entrapped within a continuous polymer network. Variations of these morphologies include multi-layered capsules and dual-core or dual-layer capsules (Fig. 16.2) (Augustin and Hemar 2009). On the other hand, capsules can be classified according to their size in macrocapsules (>5000 μm), microcapsules (0.2–5000 μm) and nanocapsules (<0.2 μm) (Da Silva et al. 2014).

16.2.2 Core Material

The core material nature, determines the selection of the wall material, and the choice of the most suitable method used for encapsulate (Dias et al. 2017). In this sense, core material properties such as thermal stability, viscosity if it is a liquid, particle size and shape if it is a solid, density, reactivity, and solubility will determine the effective design and development of the preferred micro- and nanocapsule (Oxley 2016).

16.2.3 Wall Material

The wall or shell material, should be capable of a film-forming that is cohesive with the core material. In this sense, the wall material determines the degree of protection for the inner core, influences on stability of micro- and nanocapsules and efficiency of the encapsulation process (Bakry et al. 2016).

The wall material should have the following characteristics (Kwak 2014):

- Compatible or not reactive with the core material.
- Able to seal and maintain the core inside the micro or nanocapsuled.
- Capable to provide maximum protection to the core against adverse conditions.
- Lack an unpleasant taste in the case of food application.
- Economically viable.

Different types of edible polymers such as polysaccharides, proteins, and lipids are commonly used to encapsulate functional components, but their selection depends on the physical and chemical properties of the functional compounds to be encapsulated (Shit and Shah 2014).

16.3 Materials Used for Encapsulation

Micro- and nanoencapsulation involves the incorporation of functional compounds with specific molecular characteristics such as conformation, weight, and polarity (Chen et al. 2017). Consequently, an adequate encapsulating material greatly improves the effectiveness of these bioactive compounds (Suganya and Anuradha 2017).

It is important to note that the materials used to encapsulate functional ingredients for food applications should be biodegradable and biocompatible, which means they have to be certified for food applications as "generally recognized as safe" (GRAS) to facilitate their incorporation in the food matrix (Nazzaro et al. 2012). Materials that can fulfill these requirements are edible polymers, which are divided in three main categories (Fig. 16.3):

One of the limitations that can arise when working with natural ingredients such as edible polymers is the variation in composition and quality that exist in the original source from which they are isolated. Consequently, when these edible polymers are used as wall material, there is a need to combine them with the most suitable method to produce high-quality micro- and nanoencapsulation products (Gbassi and Vandamme 2012).

16.3.1 Polysaccharides

Polysaccharides are commercially available for use in the food industry as stabilizers, thickening and gelling agents, crystallization inhibitors, and encapsulation support as the wall material or carrier (De Vos et al. 2014). Polysaccharides are conveniently categorized according to their biological origin. Starch, cellulose, pectin, Arabic gum, and guar gum from plant origin; algal origin such as agar, alginate, and carrageenan; chitosan and xanthan from animal and microbial origin (Fig. 16.3) (Liu et al. 2015). Polysaccharides are excellent candidates for micro-and

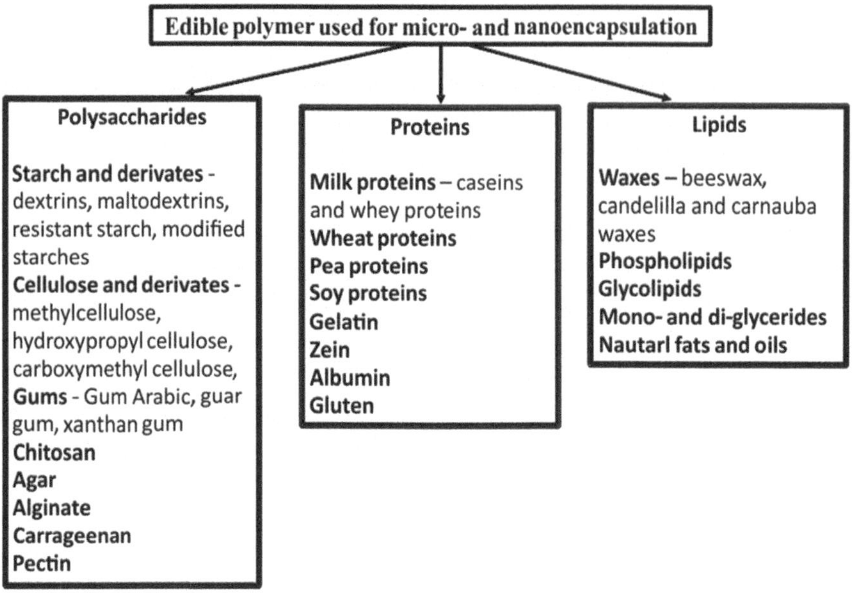

Fig. 16.3 Edible polymers used for micro and nanoencapsulation. Adapted from Vincekovic et al. (2017)

nanoencapsulation due to several attributes: (1) integral part of many food systems, (2) inexpensive, (3) wide range of polymer sizes, and (4) desirable physicochemical properties such as solubility, melting and phase change, low viscosity, wall-forming ability, and efficient drying properties (Fathi et al. 2014). The mechanisms by which polysaccharides retain a functional component during micro- and nanoencapsulation process involves physicochemical interactions depending on molecular association by hydrogen bonding (Cook et al. 2012). The use of κ-carrageenan as wall material have been reported to encapsulate curcumin (Xu et al. 2014). Also, Ferrari et al. (2012) used maltodextrin or gum Arabic to encapsulate anthocyanins from blackberry powder. On the other hand, polysaccharide blends from alginate and chitosan have been used to encapsulate vitamin B_2 (Azevedo et al. 2014). The mixture of xanthan and guar gum was used to encapsulate β-carotene (Toniazzo et al. 2014), while the use of resistant starch with guar gum was used to encapsulate folic acid (Pérez-Masiá et al. 2015).

16.3.2 Proteins

Proteins are natural macromolecules composed of linear chains of amino acids that can be used in a number of applications including micro- and nanoencapsulation processes (Lam and Nickerson 2013). Proteins show several advantages such as biocompatibility, biodegradability, good water solubility, film-forming and barrier properties, ability for adhesion, and acting as organic solvents (Riaz and Masud 2013; Wang et al. 2015). The use of animal origin proteins (caseins, whey proteins and gelatin), vegetable origin proteins (pea and soy proteins) and cereal proteins (zein and wheat proteins) have been proposed to micro- and nanoencapsulate bioactive compounds (Fig. 16.3). Haham et al. (2012) reported the use of casein micelles to encapsulate vitamin D. Liang et al. (2011) encapsulated α-tocopherol using β-lactoglobulin as shell material, while Gómez-Estaca et al. (2015) encapsulated curcumin incorporating gelatin as wall material. On the other hand, vegetable origin sources such as soy protein have been used to encapsulate α-tocopherol and ascorbic acid (Nesterenko et al. 2014). Also, Bajaj et al. (2017) used pea protein to encapsulate flaxseed oil. Likewise, cereal proteins such as zein was studied as wall material to encapsulate quercetin, while wheat protein was used to encapsulate resveratrol (Patel et al. 2012; Qiu et al. 2017). Often polysaccharides such as carboxy methyl cellulose and pectin are mixed with whey proteins to improve the emulsifying and film-forming properties during microencapsulation of antioxidants from Roselle extract (Serrano-Cruz et al. 2013). Pea protein isolate mixed with sodium alginate, carrageenan and gellan gum was used to encapsulate probiotics (Varankovich et al. 2015). Also, maltodextrin, gum Arabic and gelatin blends were used to encapsulate saffron (Rajabi et al. 2015).

16.3.3 Lipids

The use of lipids, such as phospholipids, glycolipids, natural fats and oils, mono and di-glycerides, beeswax, candelilla and carnauba waxes (Fig. 16.3) as wall materials in micro- and nanoencapsulation relies on their advantages to reduce off-flavors, to control release of nutrients, to improve the stability under extreme temperatures or moisture, and to reduce the interaction between a functional compounds with other ingredients in the food product (Fathi et al. 2012). Lipophilic micronutrients, such as resveratrol (Sessa et al. 2014), and quercetin (Pool et al. 2013) have been encapsulated within the lipid droplets in nanoemulsions to increase their bioavailability.

16.4 Methods to Obtain Micro- and Nanoencapsulation of Functional Compounds

The selection of an encapsulation technique for a specific application is based on parameters such as mean particle size required, physical/chemical properties of both core and wall materials, targeted applications of the encapsulated material, desired release mechanisms, manufacturing scale-up potential and acceptable processing cost (Ezhilarasi et al. 2013).

Many encapsulation methods have been proposed but none of them can be considered as a universally applicable procedure for functional food components. This is due to the fact that individual functional food components have their own characteristic molecular properties (Nazzaro et al. 2012). In general, the selected method needs to involve wall formation around the functional compound and ensure that leakage does not occur, while the undesired materials are kept out (Kumar et al. 2013).

Nanoencapsulation has emerged as the most attractive field for bioactive compounds encapsulation. However, the encapsulation methods currently being used by food industries, are based on microencapsulation, which is commonly achieved using techniques such as spray drying, spray cooling, simple extrusion, ionic gelation, fluidized bed coating, and emulsification (Table 16.1).

16.4.1 Spray Drying

Spray drying method involves the transformation of a liquid fluid to dry particles, in where the functional compound dispersed in aqueous coating material, known as the feed solution, is atomized into dryer chamber with a hot air flow producing a fine powder (Ray et al. 2016). Polysaccharides and proteins are the main wall materials used with this method (Kavitake et al. 2018). Moreover, the addition of surfactants such as polysorbates, sulfates, fatty acids, among others, are often required to stabilize the polymer particles (Di Battista et al. 2015).

Spray drying process consists of three fundamental steps (Arpagaus et al. 2017):

1. Atomization: The nozzle atomizes the feed solution into droplets in which the reduction in particle size leads to a large increase in the surface area.
2. Dry particles formation: In the drying chamber, the solvent, which can be water, organic solvents, or mixtures thereof, is quickly removed by the continuous flow of the hot drying gas (usually air or inert gas) allowed the dry particles formation.
3. Particle collection: The dried solid particles, which are entrained in the gas outlet stream from the drying chamber, are separated from the gas stream by a cyclone separator and collected in a collection vessel.

Spray drying offers several advantages over other micro- and nanoencapsulation techniques, such as low cost and high flexibility to control particle size and morphology by optimizing the process parameters and feed formulation (Nandiyanto and Okuyama 2011). The sizes of the dry particles formed through encapsulation are in a micro range (10–400 μm), but in some cases, they can reach a nanometric sizes up to 0.2 μm (Fang and Bhandari 2012). On the other hand, the particles morphology depends on the mass and heat transfer, in which the cohesion force of the solid particles, including the diffusion and wall flexibility, tends to equilibrate with compression stresses resulting from evaporation and capillary forces. There are different granule morphologies that can be produced by spray drying, for example,

Table 16.1 Micro- and nanoencapsulation methods currently used in food products

Method	Particle size	Functional encapsulated compound	Advantages	Disadvantages	References
Spray drying	3–100 μm	Flavors, colors, vitamins, minerals, oils, antioxidants, enzymes and probiotics	– For hydrophilic and hydrophobic functional compounds – Continuous operation – Simple, fast and easy to scale up – Low operational cost	– Loss of the material – Micro-cracks on the capsule surface – Installation costs – Low thermal efficiency	Arpagaus et al. (2017) Stunda-Zujeva et al. (2017)
Spray cooling	20–1000 μm	Volatiles (flavors and aromatic compounds), vitamins, antioxidants, enzymes and probiotics	– For heat sensitive compounds – Continuous operation – Easy of scaling up – Low operational cost	– Lipophilic coating (water-insoluble particles) – Expulsion of functional compound during storage	Okuro et al. (2013) Oxley (2016)
Extrusion (Melt injection, melt extrusion and co-extrusion)	150–8000 μm	Flavors and probiotics	– For heat and oxygen sensitive compounds – Decrease gas diffusivity – Low water content – Low energy consumption – Does not involve deleterious solvents	– Laborious technology – Large particles sizes – Limited choice of wall materials – Susceptibility of polysaccharide towards damage and structural defect – Non-scalable	Repka et al. (2012) Oxley (2016) Zuidam and Shimoni (2010)
Ionic gelation (external and internal)	200–500 μm	Vitamins, antioxidants probiotics	– For hydrophobic functional compounds – Easy to perform – No require specialized equipment – Does not involve high temperatures and deleterious solvents – Low operational cost	– Higher diffusion by heterogeneous gelation – Uneven capsule formation – Limitation batch production	Kurozawa and Hubinger (2017) Kim et al. (2016)

(continued)

Table 16.1 (continued)

Method	Particle size	Functional encapsulated compound	Advantages	Disadvantages	References
Fluidized bed coating	5–5000 μm	Volatiles (flavors and aromatic compounds), vitamins, and antioxidants	– For heat sensitive functional compounds – Total control over temperature – Easy to performance – Low particle sizes – Low operational cost	– Longer duration process – Uneven coating particles	Onwulata (2012) Srivastava et al. (2013)
Emulsification	0.1–100 μm	Antioxidants, vitamins, oils, fats	– For hydrophilic and hydrophobic functional compounds – Simply operation – Low particle sizes – Low operational cost	– Aggregation – Flocculation	Iqbal et al. (2015) Lu et al. (2016)

spheres, donuts, capsules, porous, hollow and hairy granules and raspberry-shaped granules (Stunda-Zujeva et al. 2017).

Due to the low moisture powder production, spray drying is the most frequently method used to micro- and nanoencapsulate functional compounds such as flavors, dyes, vitamins, minerals, oils, antioxidants, enzymes and probiotics. Luna-Guevara et al. (2017) reported an increase in antioxidant capacity during storage of walnuts, peanuts and pecans oils microencapsulated with mixtures of gum Arabic, maltodextrin and gelatin. Also, Berg et al. (2012) improved the anthocyanin retention in simulated gastric fluid in blueberry extract microcapsules coated with shellac using maltodextrin and pectins of lower degree of esterification as wall materials. Eckert et al. (2017) demonstrated a survivability to conditions simulating the gastrointestinal tract of *Lactobacillus plantarum* microencapsulated with whey and whey retentate.

On the other hand, nanoencapsulation by spray drying depends on other nanoencapsulation techniques prior to spray drying. In this sense, De Paz et al. (2012) encapsulated β-carotene through solvent emulsification-evaporation from ethyl acetate-in-water emulsion using modified n-octenyl succinate starch along with spray drying technique. Dahili and Feczkó (2015) reported the use of ethyl cellulose and poly(lactic *co*-glycolic acid) carrier particles obtained by spray drying to immobilize horseradish peroxidase enzyme by cross-linking achieving to prevent aggregation and agglomeration of nanoparticles. Besides, nanoencapsulation of vitamin B_{12} in Arabic gum, cashew nut gum, sodium alginate, and carboxymethyl cellulose and folic acid in whey protein and resistant starch by spray drying was obtained by Oliveira et al. (2013) and Pérez-Masiá et al. (2015).

The main limitation of spray-drying method is the high temperatures generated during the process, which can affect the capsule surface due to the pores or cracks formation, decreasing the protection and retention of the bioactive encapsulated material (Dordevic et al. 2015). In this sense, Laohasongkrama et al. (2011) obtained macadamia oil spherical microcapsules with some wrinkles or dimples using sodium caseinate and maltodextrin blends as wall materials. Also, Yang et al. (2012) reported the presence of pores in microcapsules of whey protein hydrolysate encapsulated with maltodextrin or maltodextrin/β-cyclodextrin blends.

On the other hand, Ray et al. (2016) reported that an adequate wall material selection can prevent the loss of bacterial viability caused by the high temperatures reached during spray drying process. Yonekura et al. (2014) compared the viability of spray dried *Lactobacillus acidophilus* using sodium alginate, chitosan and hydroxypropyl methylcellulose and found that sodium alginate and HPMC did not affect cell viability. Also, spray drying of *Bifidobacterium lactis* Bb-12 in whey protein enhanced cell viability under bile conditions (De Castro-Cislaghi et al. 2012).

16.4.2 Spray Congealing

Spray congealing is another method that involves the solidification of an atomized liquid, in which the initial set-up is quite similar to spray drying but here there is no water evaporation (Alvim et al. 2016). This encapsulation process involves the dispersion of the active compound in a coating material melt, commonly fatty acids, triacylglycerols, and waxes. Then, droplets of molten solution are atomized into a chilled chamber which are solidified and finally recovered as fine particles (Zuidam and Shimoni 2010). There is no agreement in the literature regarding the nomenclature of spray congealing encapsulation technique which can be considered synonymous with spray chilling and spray cooling. The terms spray chilling is the process that uses wall materials with a melting point between 32 and 42 °C for the production of spray particles, and spray cooling is the encapsulation process in which the carrier has a melting point between 45 and 122 °C (Celli et al. 2015). The term spray congealing is employed indiscriminately when molten wall materials are used regardless of their melting temperature (Okuro et al. 2013). In terms of particle size, this tends to increase as atomization moves from spray drying to spray congealing, producing particles from a few microns (20 μm) up to several millimeters (1000 μm) (Oxley 2016).

The main advantage of spray congealing is the encapsulation of highly heat-sensitive materials such as volatiles (flavors and aromatic compounds), vitamins, antioxidants, enzymes and probiotics since the dehydration process is carried out after freezing and at low temperature under vacuum, unlike the spray drying technique at high temperatures (Arslan-Tontul and Erbas 2017). Consoli et al. (2016) evaluated the implementation of spray chilling technique to microencapsulate gallic acid as a model phenolic compound, showed an increased encapsulation efficiency using blends of soybean oil with higher concentration of fully hydrogenated soybean oil used as wall materials. Gamboa et al. (2011) also microencapsulated α-tocopherol by spray chilling using interesterified fats with fully hydrogenated soybean oil as wall material, showing high efficiency of microencapsulation and high levels of retention of the active product. Engelmann and Kragl (2018) reported the microencapsulation of laccase N51003 using cetyl alcohol, palmitic acid and palmityl palmitate as wall materials, showing a good enzyme immobilization inside the particles as a results of spraying congealing method. Pedroso et al. (2012) produced also solid lipid microparticles containing *Bifidobacterium animalis* subsp. *lactis and L. acidophilus* by spray chilling using an interesterified fat produced with fully hydrogenated palm oil and kernel oil. The microparticles showed greater resistance to simulated gastric and intestinal fluids as compared with free microorganisms. Others authors have reported the application of some acidulants and textural ingredients microencapsulated by spray chilling directly in a food system. Alvim et al. (2016) reported the application in biscuits of ascorbic acid microencapsulated by spray drying using Arabic gums as wall material, and spray chilling using stearic acid and hydrogenated vegetable fat as coating material. These authors reported the inhibition of dark spots formation on the biscuits that were associated with the thermal degradation ascorbic acid during baking and a reduction of close to half of the loss observed for the free active substance after baking.

The spray congealing method has a drawback related to fast cooling rates that can crystallize the lipid wall material into an unstable arrangement, influencing on the barrier properties of encapsulated compound with the possibility of the active ingredient can be expulsed during storage, and thereby decreasing the encapsulation capacity. Zaky et al. (2010) encapsulated a protein (BSA-albumin) using hydrogenated palm oil by spray chilling. These authors observed that the size and morphology of the particles presented a significant effect on the kinetics of protein release, suggesting that water entered into it and dissolved the incorporated protein during the encapsulation process, leading to the formation of pores that were filled with water, and enabling the diffusion of the protein to the outside of the matrix.

Other possible disadvantages include the effect of particles on food texture, which depends on the particle size, as well as the possibility of particles to float in liquid systems (Okuro et al. 2013). Finally, the hydrophobic character of the microcapsules obtained by spray congealing due to the lipid wall material can limit or not allow some applications (Gadkari and Balaraman 2015).

16.4.3 Extrusion

Extrusion is the most common microencapsulation approach in which the functional compound is embedded into thermoplastic polymers carriers, usually molten polysaccharides such as starch, maltodextrin, sugar alcohols and low meting waxes (Repka et al. 2012).

Generally, extrusion methods include three processes; melt injection, melt extrusion or centrifugal extrusion (co-extrusion):

(a) **Melt injection**: Droplets of molten polysaccharide in which the functional compound was added, are extruding by a syringe needle into a bath consisting of a dehydrating liquid such as liquid nitrogen for gelation. The coating material hardens on contacting the liquid, forming gelled droplets, also known as "beads", which entrap the bioactive compound. Finally, the granules are recovered by filtration or centrifugation and the residual solvent is removed by air drying or vacuum drying (Zuidam and Shimoni 2010; Mishra 2016).

(b) **Melt extrusion**: An extruder equipped with one or two rotating screws (either corotating or counter rotating) inside a stationary cylindrical barrel was used in this extrusion process. Firstly, the functional compound is added to the molten polysaccharide carrier at the inlet port of the extruder or injected under pressure within the extruder to be mixed with carrier system before being extruder. The feed formulation is transported along the length of the extruder barrel, where it gets melted, plasticized, mixed and compressed, while thermal energy is generated by shearing, imposed by the rotating screw and from conduction from the barrel via electrical heating bands. Finally, an end-plate die is connected to the end of the barrel which determine the shape (e.g. sheets, ropes or threads) of the microencapsulate product (Zuidam and Shimoni 2010; Repka et al. 2012).

(c) **Centrifugal extrusion (co-extrusion)**: The core and wall materials are pumped separately into a concentric and multifluid nozzle extruder, which may be stationary, rotating or vibrating. While the core material flows through the center tube, wall material flows through the outer tube. The entire device is attached to a rotating shaft so that the head rotates around its vertical axis. As the head rotates, the core and coating materials are co-extruded through the concentric orifices of the nozzles as a fluid rod of the core sheathed in coating material. Centrifugal force impels the rod outward, causing it to break into tiny particles. By the action of surface tension, the coating material envelops the core material, thus accomplishing encapsulation (Mishra 2016; Oxley 2016).

The extrusion method enables the encapsulation of volatile and unstable compounds, because the polysaccharide matrices in gelled state provide and almost impermeable barrier against gases such as oxygen providing protection to normally oxidation prone compounds. Chang et al. (2010) prevented an oxidative degradation of ascorbic acid microencapsulated by melt extrusion using a glassy low-dextrose equivalent maltodextrin. These authors suggested that ascorbic acid extrudates can thus be used to impart or modify the sensorial and nutrimental properties of a product without being affected by other ingredients of the food system or by the typical food processing conditions. Sun-Waterhouse et al. (2012) improved the oxidative stability and suppressed hydrolytic rancidity of phenolic-fortified avocado oil microencapsulated by co-extrusion using an alginate polymer matrix. Also, extrusion by melt injection and co-extrusion allows the encapsulation of probiotics which cannot be encapsulated under much higher temperatures Also, the low water content of the microparticles could prevent the deterioration by other microorganisms. Silva et al. (2018) compared the microencapsulation of *L. acidophilus* LA3 by melt injection using alginate or blend of alginate–shellac as wall materials and the microencapsulation of sunflower oil loaded with the probiotic by co-extrusion with the same encapsulate materials. As results, the probiotic *L. acidophilus* LA3 incorporated into sunflower oil and loaded into capsules produced with a blend of alginate–shellac by coextrusion presented reduced porosity and an improved structure. Also, the addition of drying step in a fluidized bed after the extrusion process, extended the probiotic viability during storage, since the water activity was reduced and, consequently, the microorganism metabolism.

On the other hand, during the extrusion processes, the selection of the correct wall material is critical for the generation of gelling matrices for the achievement of an adequate bioactive compound microencapsulation. In this sense, Belščak-Cvitanović et al. (2011) reported the microencapsulation of water-soluble polyphenol antioxidant extracts of medicinal herbal plants (raspberry leaf, hawthorn, ground ivy, yarrow, nettle, and olive leaf) in alginate/chitosan hydrogel by an extrusion method. The results showed that the encapsulation efficiency, antioxidant activity of polyphenols, and size of the microbeads were affected by the microelements content of extracts from different plants (e.g. divalent cations, such as calcium, copper, strontium, and zinc) which affected the gelling properties of alginate.

Finally, the diameter of the dropping tool and viscoelasticity of the solution in the extrusion method are the most important factors that can impact on the size of microencapsulated material (Celli et al. 2015). In this sense, particle size from 20 to 200 μm are obtained by melt injection, 300 to 5000 μm by melt extrusion, and 150 to 8000 μm using co-extrusion (Mishra 2016).

16.4.4 Ionic Gelation

Ion gelation is the recommended method for entrapping hydrophobic functional compounds into a gel network based on alginate, low methoxyl pectin, chitin or chitosan (Mishra 2016). This encapsulation technique is classified into two groups depending on the method used to form the beads: extrusion (droplet method) and emulsion (two-phase system), also called external ionic gelation and internal ionic gelation techniques, respectively (Chun et al. 2014).

(a) **External gelation**: Microbeads made of gel-type polymers are produced by dissolving the polymer in an aqueous solution then, suspending the bioactive ingredient in the blend followed by extruding through a precision device (a pipette, syringe, vibrating nozzle, spraying nozzle, jet cutter, atomizing disk, coaxial air-flow, or electric field). The microdroplets obtained are hardened by cross-linking the wall material or polymer chain by using di- or multivalent metal ion (such as Ca^{+2}) aqueous solutions (Kurozawa and Hubinger 2017). It is a simple and easy procedure, which does not require specialized equipment, high temperature or organic solvents, and can be considered inexpensive. On the other hand, one of external gelation disadvantages is the occurrence of heterogeneous gelation of particles showed problems of easy diffusion and fast release through the ionic gel network regardless of pH (Moura et al. 2018). Also, the low production capacity, limitation of size reduction, and uneven capsule formation are one of the many challenges for scaling-up of the process and application on an industrial scale (Dong et al. 2013).

(b) **Internal gelation:** There is a loss of bioactive compounds during external ionic gelation. Internal gelation could be an alternative to minimize or to avoid active compounds diffusion toward the cross-linking solution, as encapsulation process occurs in an oil phase, enhancing encapsulation efficiency (Kim et al. 2016). In this sense, the production of capsules by internal ionic gelation was carried out dispersing Ca^{+2} into an ionic polymer solution, such as carrageenan, alginate and gelan gum, containing the functional compound. This solution is emulsified by adding into oil, and Ca^{+2} is released form insoluble calcium carbonate by pH reduction to promote cross-linking with the polymer (Kurozawa and Hubinger 2017). Internal gelation method has been suggested as an alternative for the formation of spherically structured capsules with sizes of less than 1000 μm and it is easy to scale up (Rosas-Flores et al. 2013). However, this technique is limited to batch production and oil removal is difficult, which requires chemicals and further separation.

Belščak-Cvitanović et al. (2016) reported a higher encapsulation efficiency of dandelion (*Taraxacum officinalis* L.) polyphenols and ß-carotene active compounds in alginate-based hydrogel beads obtained by internal gelation when compared with external encapsulation. In this work, the authors also used whey proteins in combination with alginate to improve encapsulation efficiency. As expected, these microspheres presented higher total polyphenols and chlorogenic acids retention than samples obtained by external gelation. Apart from that, an enhancement of functional compounds entrapment in hydrogels could be due to differences observed in microstructure of beads. Lupo et al. (2015) observed larger porous in the microstructure of alginate beads with encapsulated cocoa extract produced by external gelation. On the other hand, beads obtained by internal gelation presented a more compact structure. According to the authors, during external gelation, there was an initial formation of a superficial gel layer, retarding the calcium ion diffusion into the beads, as consequence a heterogeneous matrix was formed with a low cross-linked core due to the lack of the calcium ion. The compact microstructure of beads obtained by internal gelation reduced polyphenol release when compared with those obtained by external gelation. In order to improve the efficiency of external gelation microencapsulation, Belščak-Cvitanović et al. (2015) evaluated the addition of calcium caseinate or whey protein for producing alginate particles with higher retention of green tea polyphenols. Also, Wichchukit et al. (2013) reported riboflavin microencapsulation in whey protein/alginate gel beads. These authors observed that pure alginate bead presented rough and cracked surface, allowing higher water penetration into the sample. In contrast whey proteins incorporation can decrease diffusion of Ca^{+2} from the hydrogels, leading to lower interaction between alginate molecules with water and, as a consequence, the ability of the gel to swell and suffer erosion.

16.4.5 Fluidized Bed Drying

Fluidized bed drying or coating is similar to spray drying in which a coating of liquid material is sprayed to cover a functional compound as a core material in high-pressure aerosol. The difference between spray drying and fluidized bed coating is that the core material is already in powdered form (Zuidam and Shimoni 2010). In this technique, a cyclone of inlet hot air forces the fluid coating into a chamber contacting the powdered particles. After each contact of the core particle with the hot atmosphere, a minute layering occurs; over time, the hot air then evaporates the mixtures and every particle has a thin film of coating (Onwulata 2012). Different types of fluid-bed coating techniques include top spray, bottom spray, and tangential spray which are used for encapsulating solid or liquids absorbed into porous particles (Bandhavi 2013).

Fluidized bed coating is ideal for a wide range of heat sensitive products as a result of efficient heat and mass transport during drying of the coated particles at lower temperatures than spray drying, which results in greater protection of bioactive components (Augustin and Hemar 2009). Other advantage includes the lower capital cost and easy control of the method (Srivastava et al. 2013). Also, this method allows the micro- and nanoencapsulation of very fine particles ranging from 0.01 to 0.04 μm and from 100 μm to several millimeters, respectively (Tolve et al. 2016). A β-carotene microencapsulated was obtained by Coronel-Aguilera and Martín-González (2015) using spray dried method followed by fluidized bed coating with a hydroxypropyl cellulose solution. This carotenoid is sometimes limited because of its susceptibility to oxidation and thermal degradation which is related directly with color parameters. These authors observed a higher protection of β-carotene and vivid color during storage of coated powder at different temperatures. Also, the coated powders showed color stability when applied to yogurt for up to 4 weeks of storage since low values of total difference color were observed. Schell and Beermann (2014) developed a microencapsulation process for *Lactobacillus reuteri* by a two-step fluidized bed granulation with sweet whey followed by top spray coating with and aqueous shellac solution. The microencapsulate product offered resistance to acid conditions and enabling improved survival during gastro-intestinal.

On the other hand, a good core material for fluid bed coating need to be smooth and spherical to minimize the amount of fluid needed to coat it and to reduce the likelihood of uneven coating along the jagged edges, which limits coating efficiency and functionality. It was reported that 0.2–1.5% of particle remains uncoated during the fluidized bed technique (Srivastava et al. 2013).

16.4.6 *Emulsification*

This micro- and nanoencapsulation method by emulsification is a process widely used in the industry to combine immiscible solution in a mixture, through dispersions stabilized using emulsifying agents, also known as surfactants (Ye and Chi 2018). Hence, according to which solutions are dispersed in a continuous phase, simple water-in-oil or oil-in-water (W/O, O/W) emulsions, or even double emulsions of water-in-oil-in-water and oil-in-water-in-oil ($W_1/O/W_2$, $O_1/W/O_2$) could be achieved (Iqbal et al. 2015).

Droplet sizes of micro- and nanoemulsion obtained that range from 0.1 to 100 μm are highly dependent on hydrophobicity or hydrophilicity of active component (Lu et al. 2016). Hydrophobic active agents such as carotenoids, fats, oil-soluble vitamins and flavor oils can be encapsulated and delivered by O/W, while W/O was used to encapsulate water-soluble food active agents such as polyphenols (Zuidam and Shimoni 2010). In this sense, Ziani et al. (2012) designed a food grade colloidal delivery system to encapsulate vitamin E, vitamin D, and lemon oil. The study was focused on the influence of oil type and physicochemical properties on the formation and stability of colloidal dispersions using Tween 20 as emulsifier agent. The authors found that lemon oil was capable of forming stable microemulsions that were optically transparent at high emulsifier-to-oil ratios, whereas vitamin D and E were not able to form microemulsions, and emulsions stability strongly depended on oil type. Vitamin D and vitamin E acetate emulsions were relatively stable to droplet growth once they have been formed, but lemon oil emulsions were highly unstable. Ydjedd et al. (2017) found that encapsulation of phenolic compounds from carob (*Ceratonia silique L.*) pulp into $W_1/O/W_2$ emulsions protected them from chemical conditions of gastrointestinal tract and mainly released the bioactive compounds during intestinal phase, probably enhancing their absorption to systemic circulation. Absorption of tea catechins was improved when they were incorporated into nanoemulsions; particularly, epigallocatechin gallate increased its concentration in plasma to 28.6% compared with non-encapsulated solution (Peng et al. 2018). At the same time, it has been reported that encapsulation in coated-emulsions could enhance the antimicrobial activity of essential oils due to the interaction between positively charged surface coating with the negative surface potential of bacteria by attractive electrostatic forces (Krogsgård Nielsen et al. 2016).

On the other hand, a balance must be found between droplet size and concentration of emulsifying agent. Small droplets have a larger surface area that must be covered with the emulsifier. Therefore, a high concentration of emulsifier is required to avoid flocculation and aggregation (Torres et al. 2016).

16.5 Conclusions

Micro- and nanoencapsulation are promising techniques that can enhance the effect of functional compounds in a food system, broadens their application range, and thereby being cost-effective for the industry's survival in a competitive marketplace. The selection of a proper micro- and nanoencapsulation technology requires an adequate good wall material, and suitable micro- and nanoencapsulation method. Finally, the design of micro- and nanoencapsulated systems should fulfill these characteristics: (1) The physicochemical properties and functionality of the bioactive compound in the final product. (2) Appropriate food processing conditions to keep functional properties of the active micro- and nanoencapsulated ingredient. (3) The events that trigger the release of the bioactive ingredient from micro- and nanocapsules, and the mechanism(s) involved. (4) The cost constraints.

References

Alexander M, Acero Lopez A, Fang Y, Corredig M (2012) Incorporation of phytosterols in soy phospholipids nanoliposomes: encapsulation efficiency and stability. LWT Food Sci Technol 47(2):427–436

Alvim ID, Stein MA, Koury IP, Dantas FBH, Cruz CLCV (2016) Comparison between the spray drying and spray chilling microparticles contain ascorbic acid in a baked product application. LWT Food Sci Technol 65:689–694

Arpagaus C, John P, Collenberg A, Rütti D (2017) Nanocapsules formation by nano spray drying. In: Jafari SM (ed) Nanoencapsulation technologies for the food and nutraceutical industries. Academic, Cambridge, pp 346–401

Arslan-Tontul S, Erbas M (2017) Single and double layered microencapsulation of probiotics by spray drying and spray chilling. LWT Food Sci Technol 81:160–169

Augustin MA, Hemar Y (2009) Nano- and micro-structured assemblies for encapsulation of food ingredients. Chem Soc Rev 38:902–912

Azevedo MA, Bourbon AI, Vicente AA, Cerqueira MA (2014) Alginate/chitosan nanoparticles for encapsulation and controlled release of vitamin B_2. Int J Biol Macromol 71:141–146

Bajaj PR, Bhunia K, Kleiner L, Melito HSJ, Smith D, Ganjyal G, Sablani SS (2017) Improving functional properties of pea protein isolate for microencapsulation of flaxseed oil. J Microencapsul 34(2):218–230

Bakry AM, Abbas S, Ali B, Majeed H, Abouelwafa MY, Mousa A, Liang L (2016) Microencapsulation of oils: a comprehensive review of benefits, techniques, and applications. Compr Rev Food Sci Food Saf 15:143–182

Bandhavi P (2013) Microencapsulation process materials, manufacturing methods and applications: a review. Int J Pharm Res Dev 5(03):008–020

Belščak-Cvitanović A, Stojanović R, Manojlović V, Komes D, Cindrić IJ, Nedović V, Bugarski B (2011) Encapsulation of polyphenolic antioxidants from medicinal plant extracts in alginate–chitosan system enhanced with ascorbic acid by electrostatic extrusion. Food Res Int 44:1094–1101

Belščak-Cvitanović A, Đorđević V, Karlović S, Pavlović V, Komes D, Ježek D, Bugarski B, Nedović V (2015) Protein-reinforced and chitosan-pectin coated alginate microparticles for delivery of flavan-3-ol antioxidants and caffeine from green tea extract. Food Hydrocoll 51:361–374

Belščak-Cvitanović A, Bušić A, Barišić L, Vrsaljko D, Karlović S, Špoljarić I, Vojvodić A, Mršić G, Komes D (2016) Emulsion templated microencapsulation of dandelion (*Taraxacum officinale* L.) polyphenols and β-carotene by ionotropic gelation of alginate and pectin. Food Hydrocoll 57:139–152
Berg S, Bretz M, Hubbermann EM, Schwarz K (2012) Influence of different pectins on powder characteristics of microencapsulated anthocyanins and their impact on drug retention of shellac coated granulate. J Food Eng 108:158–165
Celli GB, Ghanem A, Brooks MSL (2015) Bioactive encapsulated powders for functional foods—a review of methods and current limitations. Food Bioprocess Technol 8(9):1825–1837
Chang D, Abbas S, Hayat K, Xia S, Zhang X, Xie M, Kim J (2010) Encapsulation of ascorbic acid in amorphous maltodextrin employing extrusion as affected by matrix core ratio and water content. Int J Food Sci Technol 45:1895–1901
Chen J, Wang Q, Liu CM, Gong J (2017) Issues deserve attention in encapsulating probiotics: critical review of existing literature. Crit Rev Food Sci Nutr 57(6):1228–1238
Chun H, Kim CH, Cho YH (2014) Microencapsulation of *Lactobacillus plantarum* DKL 109 using external ionic gelation method. Korean J Food Sci Anim Resour 34(5):692–699
Consoli L, Grimaldi R, Sartori T, Menegalli FC, Hubinger MD (2016) Gallic acid microparticles produced by spray chilling technique: production and characterization. LWT Food Sci Technol 65:79–87
Cook MT, Tzortzis G, Charalampopoulos D, Khutoryanskiy VV (2012) Microencapsulation of probiotics for gastrointestinal delivery. J Control Release 162(1):56–67
Coronel-Aguilera C, Martín-González M (2015) Encapsulation of spray dried β-carotene emulsion by fluidized bed coating technology. LWT Food Sci Technol 62:187–193
Da Silva PT, Fries LLM, De Menezes CR, Holkem AT, Schwan CL, Wigmann EF, Bastos JO, Da Silva CB (2014) Microencapsulation: concepts, mechanisms, methods and some applications in food technology. Cienc Rural 44(7):1304–1311
Dahili LA, Feczkó T (2015) Cross-linking of horseradish peroxidase enzyme to fine particles generated by Nano Spray Dryer B-90. Period Polytech Chem Eng 59:209–214
De Castro-Cislaghi FP, Silva CRE, Fritzen-Freire CB, Lorenz JG, Sant'Anna ES (2012) *Bifidobacterium* Bb-12 microencapsulated by spray drying with whey: survival under simulated gastrointestinal conditions, tolerance to NaCl, and viability during storage. J Food Eng 113(2):186–193
De Paz E, Martín A, Estrella A, Rodríguez-Rojo S, Matias AA, Duarte CMM, Cocero MJ (2012) Formulation of β-carotene by precipitation from pressurized ethyl acetate-on water emulsions for application as natural colorant. Food Hydrocoll 26:17–27
De Vos P, Lazarjani HA, Poncelet D, Faas MM (2014) Polymers in cell encapsulation from an enveloped cell perspective. Adv Drug Deliv Rev 67–68:15–34
Di Battista CA, Constenla D, Ramírez-Rigo MV, Piña J (2015) The use of arabic gum, maltodextrin and surfactants in the microencapsulation of phytosterols by spray drying. Powder Technol 286:193–201
Dias DR, Botrel DA, Fernandes RVB, Borges SV (2017) Encapsulation as a tool for bioprocessing of functional foods. Curr Opin Food Sci 13:31–37
Dong QY, Chen MY, Xin Y, Qin XY, Cheng Z, Shi L, Tang ZX (2013) Alginate-based and protein-based materials for probiotics encapsulation: a review. Int J Food Sci Technol 48(7):1339–1351
Dordevic V, Balanc B, Cvitanovic AB, Levic S, Trifkovic K, Kalusevic A, Kostic I, Komes D, Bugarski B, Nedovic V (2015) Trends in encapsulation technologies for delivery of food bioactive compounds. Food Eng Rev 7(4):452–490
Eckert C, Garcia-Serpa V, Dos Santos ACF, Da Costa SM, Dalpubel V, Lehn DN, Volken de Souza CF (2017) Microencapsulation of *Lactobacillus plantarum* ATCC 8014 through spray drying and using dairy whey as wall materials. LWT Food Sci Technol 82:176–183
Engelmann C, Kragl U (2018) Spray congealing as innovative technique for enzyme encapsulation. J Chem Technol Biotechnol 93:191–197
Ezhilarasi PN, Karthik P, Chhanwal N, Anandharamakrishnan C (2013) Nanoencapsulation techniques for food bioactive components: a review. Food Bioprocess Technol 6(3):628–647

Fang Z, Bhandari B (2010) Encapsulation of polyphenols—a review. Trends Food Sci Technol 21(10):510–523
Fang Z, Bhandari B (2012) Encapsulation techniques for food ingredient systems. In: Bhandari B, Roos YH (eds) Food materials science and engineering. Blackwell Publishing Ltd, Oxford, pp 320–348
Fathi M, Mozafari MR, Mohebbi M (2012) Nanoencapsulation of food ingredients using lipid based delivery systems. Trends Food Sci Technol 23(1):13–27
Fathi M, Martín A, McClements DJ (2014) Nanoencapsulation of food ingredients using carbohydrate based delivery systems. Trends Food Sci Technol 39(1):18–39
Ferrari CC, Germer SPM, Alvim ID, Vissotto FZ, De Aguirre JM (2012) Influence of carrier agents on the physicochemical properties of blackberry powder produced by spray drying. Int J Food Sci Technol 47(6):1237–1245
Gadkari P, Balaraman M (2015) Catechins: sources, extraction and encapsulation: a review. Food Bioprod Process 93:122–138
Gamboa OD, Gonçalves LG, Grosso CF (2011) Microencapsulation of tocopherols in lipid matrix by spray chilling method. Proc Food Sci 1:1732–1739
Gbassi GK, Vandamme T (2012) Probiotic encapsulation technology: from microencapsulation to release into the gut. Pharmaceutics 4(1):149–163
Gómez-Estaca J, Gavara R, Hernández-Muñoz P (2015) Encapsulation of curcumin in electrosprayed gelatin microspheres enhances its bioaccessibility and widens its uses in food applications. Innov Food Sci Emerg Technol 29:302–307
Gutiérrez TJ (2018) Processing nano- and microcapsules for industrial applications. In: Hussain CM (ed) Handbook of nanomaterials for industrial applications. Editorial Elsevier, pp 989–1011. https://doi.org/10.1016/B978-0-12-813351-4.00057-2. EE.UU. ISBN: 978-0-12-813351-4
Gutiérrez TJ, Álvarez K (2017) Biopolymers as microencapsulation materials in the food industry. In: Masuelli M, Renard D (eds) Advances in physicochemical properties of biopolymers: part 2. Bentham Science Publishers, pp 296–322. https://doi.org/10.2174/978168108544911701000 09. EE.UU. ISBN: 978-1-68108-545-6. eISBN: 978-1-68108-544-9
Haham M, Ish-Shalom S, Nodelman M, Duek I, Segal E, Kustanovich M, Livney YD (2012) Stability and bioavailability of vitamin D nanoencapsulated in casein micelles. Food Funct 3(7):737–744
Iqbal M, Zafar N, Fessi H, Elaissari A (2015) Double emulsion solvent evaporation techniques used for drug encapsulation. Int J Pharm 496(2):173–190
Kavitake D, Kandasamy S, Devi P, Shetty PH (2018) Recent developments on encapsulation of lactic acid bacteria as potential starter culture in fermented foods-a review. Food Biosci 21:34–44
Kim ES, Lee JS, Lee HG (2016) Calcium-alginate microparticles for sustained release of catechin prepared via an emulsion gelation technique. Food Sci Biotechnol 25(5):1337–1343
Krogsgård Nielsen C, Kjems J, Mygind T, Snabe T, Schwarz K, Serfert Y, Meyer RL (2016) Enhancing the antibacterial efficacy of isoeugenol by emulsion encapsulation. Int J Food Microbiol 229:7–14
Kumar KPS, Sk T, Banu S, Lakshmi PN, Bhowmik D (2013) Microencapsulation technology. Indian J Res Pharm Biotechnol 1(3):324–328
Kurozawa L, Hubinger M (2017) Hydrophilic food compounds encapsulation by ionic gelation. Curr Opin Food Sci 15:50–55
Kwak HS (2014) Overview of nano- and microencapsulation for foods. In: Kwak HS (ed) Nano- and microencapsulation for foods. Wiley Blackwell, Chichester, pp 1–14
Lam RSH, Nickerson MT (2013) Food proteins: a review on their emulsifying properties using a structure-function approach. Food Chem 141(2):975–984
Laohasongkrama K, Mahamaktudsanee T, Chaiwanichsiri S (2011) Microencapsulation of Macadamia oil by spray drying. Proc Food Sci 1:1660–1665
Liang L, Tremblay-Hébert V, Subirade M (2011) Characterisation of the b-lactoglobulin/a--tocopherol complex and its impact on α-tocopherol stability. Food Chem 126:821–826
Liu J, Willför S, Xun C (2015) A review of bioactive plant polysaccharides: biological activities, functionalization, and biomedical applications. Bioact Carbohydr Diet Fibre 5(1):31–61

Lu W, Kelly A, Miao S (2016) Emulsion-based encapsulation and delivery systems for polyphenols. Trends Food Sci Technol 47:1–9
Luna-Guevara JJ, Ochoa-Velasco CE, Hernández-Carranza P, Guerrero-Beltrán JA (2017) Microencapsulation of walnut, peanut and pecan oils by spray drying. Food Struct 12:26–32
Lupo B, Maestro A, Gutiérrez JM, González C (2015) Characterization of alginate beads with encapsulated cocoa extract to prepare functional food: comparison of two gelation mechanisms. Food Hydrocoll 49:24–34
McClements DJ (2012) Requirements for food ingredient and nutraceutical delivery systems. In: Garti N, McClements DJ (eds) Encapsulation technologies and delivery systems for food ingredients and nutraceuticals. Woodhead, Cambridge, pp 3–18
Mishra MK (2016) Overview of encapsulation and controlled release. In: Mishra M (ed) Handbook of encapsulation and controlled release. CRC Press, Taylor & Francis Group, Boca Raton, pp 3–22
Moura S, Berling C, Germer S, Alvim I, Hubinger M (2018) Encapsulating anthocyanins from *Hibiscus sabdariffa* L. calyces by ionic gelation: pigment stability during storage of microparticles. Food Chem 241:317–327
Nagavarma BVN, Hemant KSY, Ayaz AVLS, Shivakumar HG (2012) Different techniques for preparation of polymeric nanoparticles—a review. Asian J Pharm Clin Res 5(3):16–23
Nandiyanto ABD, Okuyama K (2011) Progress in developing spray-drying methods for the production of controlled morphology particles: from the nanometer to submicrometer size ranges. Adv Powder Technol 22(1):1–19
Nazzaro F, Orlando P, Fratianni F, Coppola R (2012) Microencapsulation in food science and biotechnology. Curr Opin Biotechnol 23(2):182–186
Nedovic V, Kalusevic A, Manojlovic V, Levic S, Bugarski B (2011) An overview of encapsulation technologies for food applications. Proc Food Sci 1:1806–1815
Nesterenko A, Alric I, Silvestre F, Durrieu V (2014) Comparative study of encapsulation of vitamins with native and modified soy protein. Food Hydrocoll 38:172–179
Okuro PK, Junior FEM, Silvia Favaro-Trindade CS (2013) Technological challenges for spray chilling encapsulation of functional food ingredients. Food Technol Biotechnol 51(2):171–182
Olaiya C, Soetan K, Esan A (2016) The role of nutraceuticals, functional foods and value added food products in the prevention and treatment of chronic diseases. Afr J Food Sci 10(10):185–193
Oliveira A, Guimarães K, Cerize N, Tunussi A, Poço J (2013) Nano spray drying as an innovative technology for encapsulating hydrophilic active pharmaceutical ingredients (API). Nanomed Nanotechnol 4(6):1–6
Onwulata C (2012) Encapsulation of new active ingredients. Annu Rev Food Sci Technol 3:183–202
Oxley J (2016) Process-selection criteria. In: Mishra M (ed) Handbook of encapsulation and controlled release. CRC Press, Taylor & Francis Group, Boca Raton, pp 23–34
Patel AR, Heussen PCM, Hazekamp J, Drost E, Velikov KP (2012) Quercetin loaded biopolymeric colloidal particles prepared by simultaneous precipitation of quercetin with hydrophobic protein in aqueous medium. Food Chem 133(2):423–429
Pedroso DL, Thomazini M, Heinemann RJB, Favaro-Trindade CS (2012) Protection of *Bifidobacterium lactis* and *Lactobacillus acidophilus* by microencapsulation using spray chilling. Int Dairy J 26(2):127–132
Peng Y, Meng Q, Zhou J, Chen B, Xi J, Long P, Zhang L, Hou R (2018) Nanoemulsion delivery system of tea polyphenols enhanced the bioavailability of catechins in rats. Food Chem 242:527–532
Pérez-Masiá R, López-Nicolás R, Periago MJ, Ros G, Lagaron JM, López-Rubio A (2015) Encapsulation of folic acid in food hydrocolloids through nanospray drying and electrospraying for nutraceutical applications. Food Chem 168:124–133
Pool H, Mendoza S, Xiao H, McClements DJ (2013) Encapsulation and release of hydrophobic bioactive components in nanoemulsion based delivery systems: impact of physical form on quercetin bioaccessibility. Food Funct 4:162–174

Qiu C, Wang B, Wang Y, Teng Y (2017) Effects of colloidal complexes formation between resveratrol and deamidated gliadin on the bioaccessibility and lipid oxidative stability. Food Hydrocoll 69:466–472

Quirós-Sauceda AE, Ayala-Zavala JF, Olivas GI, González-Aguilar GA (2014) Edible coatings as encapsulating matrices for bioactive compounds: a review. J Food Sci Technol 51(9):1674–1685

Rajabi H, Ghorbani M, Jafari SM, Sadeghi A, Rajabzadeh Q (2015) Retention of saffron bioactive components by spray drying encapsulation using maltodextrin, gum Arabic and gelatin as wall materials. Food Hydrocoll 51:327–337

Ray S, Raychaudhuri U, Chakraborty R (2016) An overview of encapsulation of active compounds used in food products by drying technology. Food Biosci 13:76–83

Repka MA, Shah S, Lu J, Maddineni S, Morott J, Patwardhan K, Mohammed NN (2012) Melt extrusion: process to product. Expert Opin Drug Deliv 9(1):105–125

Riaz QUA, Masud T (2013) Recent trends and applications of encapsulating materials for probiotic stability. Crit Rev Food Sci Nutr 53(3):231–244

Rosas-Flores W, Ramos-Ramirez EG, Salazar-Montoya JA (2013) Microencapsulation of *Lactobacillus helveticus* and *Lactobacillus delbrueckii* using alginate and gellan gum. Carbohydr Polym 98:1011–1017

Schell D, Beermann C (2014) Fluidized bed microencapsulation of *Lactobacillus reuteri* with sweet whey and shellac for improved acid resistance and in-vitro gastro-intestinal survival. Food Res Int 62:308–314

Serrano-Cruz MR, Villanueva-Carvajal A, Morales-Rosales EJ, Ramírez-Dávila JF, Dominguez-Lopez A (2013) Controlled release and antioxidant activity of Roselle (Hibiscus sabdariffa L.) extract encapsulated in mixtures of carboxymethyl cellulose, whey protein, and pectin. LWT Food Sci Technol 50(2):554–561

Sessa M, Balestrieri ML, Ferrari G, Servillo L, Castaldo D, D'Onofrio N, Donsi F, Tsao R (2014) Bioavailability of encapsulated resveratrol into nanoemulsion-based delivery systems. Food Chem 147:42–50

Shit SC, Shah PM (2014) Edible polymers: challenges and opportunities. J Polym Hindawi 2014:1–13

Silva MP, Tulini FL, Martins E, Penning M, Fávaro-Trindade CS, Poncelet D (2018) Comparison of extrusion and co-extrusion encapsulation techniques to protect *Lactobacillus acidophilus* LA3 in simulated gastrointestinal fluids. LWT Food Sci Technol 89:392–399

Srivastava Y, Semwal AD, Sharma GK (2013) Application of various chemical and mechanical microencapsulation techniques in food sector—a review. Int J Food Ferment Technol 3(1):1–13

Stunda-Zujeva A, Irbe Z, Berzina-Cimdina L (2017) Controlling the morphology of ceramic and composite powders obtained via spray drying—a review. Ceram Int 43(15):11543–11551

Suganya V, Anuradha V (2017) Microencapsulation and nanoencapsulation: a review. Int J Pharm Clin Res 9(3):233–239

Sun-Waterhouse D, Penin-Peyta L, Wadhwa SS, Waterhouse GIN (2012) Storage stability of phenolic-fortified avocado oil encapsulated using different polymer formulations and co-extrusion technology. Food Bioprocess Technol 5(8):3090–3102

Tolve R, Galgano F, Caruso MC, Tchuenbou-Magaia FL, Condelli N, Favati F, Zhang Z (2016) Encapsulation of health-promoting ingredients: applications in foodstuffs. Int J Food Sci Nutr 67(8):888–918

Toniazzo T, Berbel IF, Cho S, Fávaro-Trindade CS, Moraes ICF, Pinho SC (2014) β-carotene-loaded liposome dispersions stabilized with xanthan and guar gums: physico-chemical stability and feasibility of application in yogurt. LWT Food Sci Technol 59(2):1265–1273

Torres O, Murray B, Sarkar A (2016) Emulsion microgel particles: novel encapsulation strategy for lipophilic molecules. Trends Food Sci Technol 55:98–108

Varankovich NV, Khan NH, Nickerson MT, Kalmokoff M, Korber DR (2015) Evaluation of pea protein-polysaccharide matrices for encapsulation of acid-sensitive bacteria. Food Res Int 70:118–124

Vincekovic M, Viskic M, Juric S, Giacometti J, Kovacevic DB, Putnik P, Donsi F, Barba FJ, Jambrak AR (2017) Innovative technologies for encapsulation of Mediterranean plants extracts. Trends Food Sci Technol 69(Part A):1–12

Wang Z, Ju X, He R, Yuan J, Wang L (2015) The effect of rapeseed protein structural modification on microstructural properties of peptide microcapsules. Food Bioprocess Technol 8(6):1305–1318

Wichchukit S, Oztop MH, McCarthy MJ, McCarthy KL (2013) Whey protein/alginate beads as carriers of a bioactive component. Food Hydrocoll 33:66–73

Xu W, Jin W, Zhang C, Li Z, Lin L, Huang Q, Ye S, Li B (2014) Curcumin loaded and protective system based on complex of κ-carrageenan and lysozyme. Food Res Int 59:61–66

Yang S, Mao XY, Li FF, Zhang D, Leng XJ, Ren FZ, Teng GX (2012) The improving effect of spray-drying encapsulation process on the bitter taste and stability of whey protein hydrolysate. Eur Food Res Technol 235:91–97

Ydjedd S, Bouriche S, López-Nicolás R, Sánchez-Moya T, Frontela-Saseta C, Ros-Berruezo G, Rezqui F, Louaileche H, Kati DE (2017) Effect of in vitro Gastrointestinal digestion on encapsulated and nonencapsulated phenolic compounds of carob (*Ceratonia siliqua L.*) pulp extracts and their antioxidant capacity. J Agric Food Chem 65(4):827–835

Ye C, Chi H (2018) A review of recent progress in drug and protein encapsulation: approaches, applications and challenges. Mater Sci Eng C Mater Biol Appl 83:233–246

Yonekura L, Sun H, Soukoulis C, Fisk I (2014) Microencapsulation of *Lactobacillus acidophilus* NCIMB 701748 in matrices containing soluble fibre by spray drying: technological characterization, storage stability and survival after in vitro digestion. J Funct Food 6:205–214

Zaky A, Elbakry A, Ehmer A, Breuning M, Goepferich A (2010) The mechanism of protein release from triglyceride microspheres. J Control Release 147:202–210

Ziani K, Fang Y, McClements DJ (2012) Encapsulation of functional lipophilic components in surfactant-based colloidal delivery systems: vitamin E, vitamin D, and lemon oil. Food Chem 134:1106–1112

Zuidam NJ, Shimoni E (2010) Overview of microencapsulates for use in food products or processes and methods to make them. In: Zuidam NJ, Nedović N (eds) Encapsulation technologies for active food. Sciences and Business, Israel, pp 3–29

Chapter 17
Biopolymers for the Nano-microencapsulation of Bioactive Ingredients by Electrohydrodynamic Processing

Pedro J. García-Moreno, Ana C. Mendes, Charlotte Jacobsen, and Ioannis S. Chronakis

Abstract Electrohydrodynamic processing, including electrospinning and electrospraying, is an emerging technique for the encapsulation of bioactive ingredients (e.g. omega-3, vitamins, antioxidants, probiotics) with interest for the functional food industry. This chapter presents the fundamentals of electrohydrodynamic processes for the production of nano-microstructures (fibers or capsules) loaded with bioactive compounds. Particularly, it focuses on the properties as well as electrospinning and electrospray processing of food-grade polymers. The physicochemical characteristics of the resulting nano-microencapsulates will also be discussed. Electrospun and electrospray food-grade polymers include biopolymers such as proteins (e.g. zein, gelatin, whey, casein, amaranth, soy, egg and fish protein) and polysaccharides (e.g. pullulan, dextran, chitosan, starch, alginate, cellulose, cyclodextrin, xanthan gum), as well as blends of biopolymers with biocompatible synthetic polymers (e.g. poly-vinyl alcohol).

Keywords Electrospinning · Electrospraying · Encapsulation · Polysaccharides · Proteins

Pedro J. García-Moreno and Ana C. Mendes contributed equally to this work.

P. J. García-Moreno (✉) · C. Jacobsen
Research Group for Bioactives—Analysis and Application, National Food Institute, Technical University of Denmark, Kgs. Lyngby, Denmark
e-mail: pejeg@food.dtu.dk

A. C. Mendes (✉) · I. S. Chronakis
Nano-Bio Science Research Group, National Food Institute, Technical University of Denmark, Kgs. Lyngby, Denmark
e-mail: anac@food.dtu.dk

T. J. Gutiérrez (ed.), *Polymers for Food Applications*,
https://doi.org/10.1007/978-3-319-94625-2_17

17.1 Introduction

Consumers have raising awareness about nutrition and consumption of healthy food (e.g. fortified food products). As a consequence, food industry has an increasing interest in the enrichment of traditional food matrices with bioactive compounds such as omega-3 fatty acids, vitamins, antioxidants and probiotics, which have several benefits on human health. However, the incorporation of these active agents into the food matrix and its delivery into the right place in the gastrointestinal tract (GI) is challenging due to the high instability and absorption of these ingredients. Therefore, bioactive compounds are commonly encapsulated in order to prevent their degradation during food processing and digestion. Encapsulation, which is defined as the process of entrapping active agents within a layer of coating material, protects the unstable ingredient from several factors such as oxygen, light, heat, water, pH and enzymes; whereby undesirable reactions between the bioactive compound and other food components can be avoided (Gutiérrez and Álvarez 2017). Moreover, encapsulation enables the masking of potential unpleasant odor, taste or color of the active compounds and facilitates the handling of liquid products when dried to free-flowing solids (Sobel et al. 2014).

Several techniques are available for the encapsulation of liquids, solids or gases within a solid wall material (e.g. spray-drying, spray-cooling, fluid-bed coating, extrusion, and complex coacervation followed by drying) (Gutiérrez 2018). Among them, spray-drying is the most common process used in the food industry for microencapsulation of active ingredients. In spray-drying, the bioactive ingredients are emulsified in a solution containing the shell material. Subsequently, the solution is atomized and dried by contact with a hot fluid (e.g. air) (Jacobs 2014). In the last years, electrohydrodynamic processes such as electrospinning and electrospraying are emerging as alternative encapsulation techniques. The resulting nano-microstructures (e.g. fibers or capsules) present high-performance and functionalities due to their ultra-thin size and large surface-to-volume ratio. Moreover, these processes do not require heat to dry the emulsions (e.g. by solvent evaporation), which makes them extremely adequate for encapsulating heat-sensitive compounds such as antioxidants, vitamins, probiotics or omega-3 fatty acids (Jacobsen et al. 2018).

The development of encapsulates by electrohydrodynamic processing intended for food applications requires the use of food-grade polymers and solvents (e.g. preferably water or water/ethanol mixtures). Biopolymers such as proteins and polysaccharides, which are macronutrients normally ingested through the diet, are the most commonly employed shell materials (García-Moreno et al. 2017a). Moreover, biocompatible synthetic polymers such as poly-vinyl alcohol (PVA) and poly-ethylene oxide (PEO) are also used (e.g. blended with biopolymers) (Lim 2015). In addition, lipids such as phospholipids and cocoa butter have also been employed as wall materials for the production of encapsulates by electrospinning and electrospraying, respectively (Bocanegra et al. 2005; Shekarforoush et al. 2017a).

This book chapter aims to provide an overview of the biopolymers used for the encapsulation of bioactive compounds by electrospinning and electrospraying processes. Food-grade biopolymers including proteins, polysaccharides as well as their blends with biocompatible polymers are discussed. Moreover, the physicochemical characteristics of the resulting micro-nanoencapsulates are also presented.

17.2 Electrohydrodynamic Processes

Electrospinning was first observed in 1897 by Rayleigh, whereas electrospraying was first studied in 1914 by Zeleny (Zeleny 1914). Nevertheless, they have gained an increasing attention in the last decades for the production of nano- and microsized materials with different applications (e.g. filtration, immobilization of enzymes, encapsulation, drug-delivery, scaffolds and others) (Bhardwaj and Kundu 2010).

17.2.1 Fundamentals

Electrohydrodynamic processes use a high-voltage electrostatic field to charge the surface of a polymer solution or melt droplet at the end of a capillary tube. Polymer solutions containing the bioactive compound (e.g. in the form of an emulsion) are normally used since the high temperature required to melt the polymer will degrade the heat-sensitive active ingredient. In these processes, the polymer solution is pumped through a capillary spinneret (e.g. a conductive needle) and an electrical potential is applied between the needle and a collector, which is normally grounded. The resulting electrostatic repulsion forces in the droplet surface cause the droplet to elongate from a semispherical to a conical shape known as the Taylor cone. Further increase in the electric potential strengthens repulsion forces, which eventually overcomes the surface tension of the droplet. Then, a charged jet is ejected from the tip of the Taylor cone to the collector. In electrospinning (Fig. 17.1a), the jet is stable due to high polymer chain entanglements. On the way to the collector, the jet elongates as a consequence of shearing of the viscoelastic solution and of bending and whipping motions. The latter is crucial for the extensive reduction in size of the jet and the evaporation of the solvent, allowing deposition of solidified nano-microfibers in the collector (Fig. 17.1b) (Chronakis 2005). In electrospraying (Fig. 17.2a), the jet is destabilized due to varicose instability as a consequence of the low viscoelasticity of the polymer solution. Fine highly charged droplets are formed, which subsequently break up due to electrostatic repulsion. This favors solvent evaporation and deposition of solids nano-microcapsules on the collector (Fig. 17.2b) (Bhardwaj and Kundu 2010).

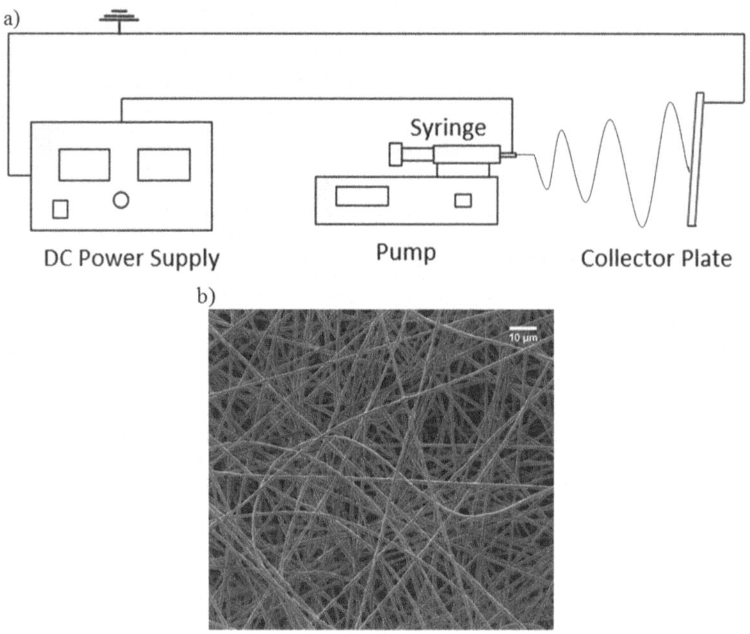

Fig. 17.1 (**a**) Schematic illustration of the electrospinning setup, and (**b**) SEM image of electrospun nano-microfibers. Figure (**a**) was modified from Pelayo (2017)

17.2.2 Influencing Factors

Several factors such as solution properties (e.g. viscosity, surface tension and conductivity), process variables (e.g. voltage, flow rate, spinneret-collector distance) and environmental conditions (temperature and relative humidity) affect the electrospinning and electrospray processing of biopolymers (Chronakis 2010).

17.2.2.1 Solutions Properties

Solutions properties are determined by the type and the concentration of the biopolymer used as well as the type of solvent. The viscoelasticity of the biopolymer solution determines the polymer chain entanglements, which control the stability of the jet, leading to fiber or capsule formation. Biopolymers with high molecular weight or high biopolymer concentration are required to have sufficient chain entanglements in the solution, suppressing the action of the surface tension, and stresses of whipping and bending motions, which tend to break the jet and favor the formation of beads (Chronakis 2010). Conductive solutions, which allow to build-up charge on the surface of the droplet, are necessary to initiate jetting. Increases in conductivity led to production of more uniform fibers because of more intense charge repulsion within the jet resulting in further elongation. However, excessive

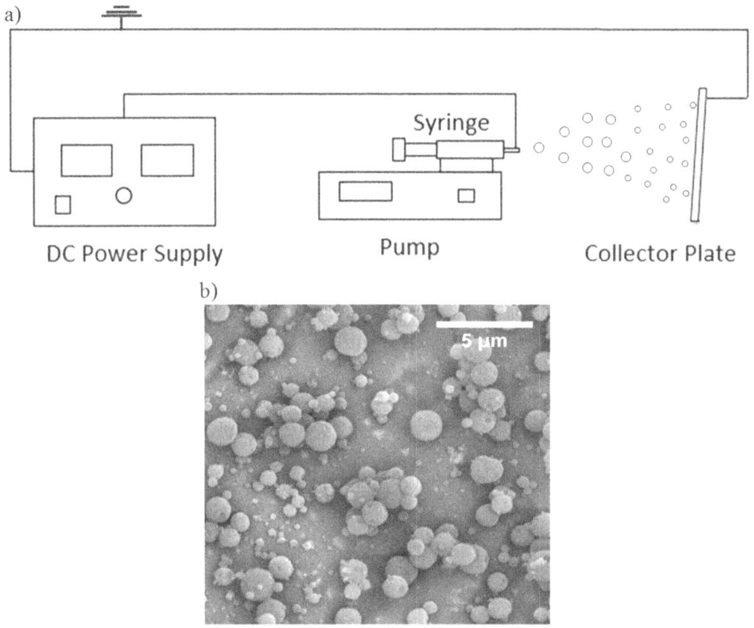

Fig. 17.2 (**a**) Schematic illustration of electrospraying process, and (**b**) SEM image of electrosprayed nano-microcapsules. Figure (**a**) was modified from Pelayo (2017)

charges can slide along the surface of the polymer jet in highly conductive solutions, which is detrimental for jet stretching (Lim 2015). Low surface tension is preferred for initiating jetting in both electrospinning and electrospraying, and to avoid the formation of beads in electrospinning. Both conductivity and surface tension of the polymer solution can be altered by adding ionic salts (e.g. NaCl) or surfactants (e.g. Tween20), respectively (Ghorani and Tucker 2015). The vapor pressure of the solvent also plays an important role in the electrohydrodynamic processes. It has to be low enough to allow evaporation of the solvent during the transportation of the charged jet/droplets from the needle to the collector, avoiding the deposition of wet nano-microstructures; but also high enough to not evaporate in the Taylor cone leading to clogging (Weiss et al. 2012).

17.2.2.2 Process Variables and Environmental Conditions

The main variables of electrohydrodynamic processes are voltage, flow rate and spinneret-collector distance. It is extremely important to find an optimum combination of voltage and flow rate which allows to maximize throughput while maintaining a stable Taylor cone (e.g. avoiding dripping or wet droplets in the collector). In general, a minimum voltage (e.g. 6–15 kV) is required to form the Taylor cone and to induce the jet formation (Ghorani and Tucker 2015). Increasing the voltage results in higher repulsion of the charges, which favors stretching of the electrospinning jet or breaking-up of the electrospray droplets leading to fibers and capsules with lower diameter. Nevertheless, too high voltage will destabilize the Taylor cone. Increasing spinneret-collector distance rises the time of the jet/droplet to reach the collector, favoring solvent evaporation (if dry conditions are used), which results in the formation of nano-microstructures with reduced diameter. In addition, decreasing spinneret diameter also favors the formation of capsules and fibers with lower average diameters (Weiss et al. 2012). Apart from processing variables, the temperature and the humidity also affect the electrospinning and electrospraying processes. Temperature influences solvent evaporation as well as viscosity, conductivity and surface tension of the polymer solution. For instance, Mit-Uppatham et al. (2004) reported the production of fibers with lower diameter by increasing the temperature. It has also been reported that increasing humidity led to the appearance of pores in the surface of electrospun fibers (Bhardwaj and Kundu 2010).

17.2.3 Advantages

Low temperature (e.g. room temperature) can be used for the evaporation of the solvent(s) and the production of nano-microencapsulates during the electrohydrodynamic processes. This is of great importance to avoid the degradation of thermolabile compounds during encapsulation processing, including active ingredients and wall materials (e.g. proteins). Furthermore, electrospun/electrosprayed encapsulates present a reduced size when compared to encapsulates produced by other traditional techniques (e.g. spray-drying). This facilitates their incorporation into food matrices without affecting organoleptic properties and favors the release of the encapsulated bioactive compounds as a consequence of the increased surface-to-volume ratio (Jacobsen et al. 2018). Nevertheless, it should be noted that electrospun fibers are continuous in length and exhibit a high tensile strength as a consequence of a high alignment of polymer chains produced in the stretching process (Weiss et al. 2012). Therefore, their incorporation into foods needs to be properly optimized. In electrospraying (where capsules aggregation is avoided due to the electrostatic repulsion forces) smaller capsules are produced, with a narrow droplet size distribution, in comparison to traditional encapsulation techniques (e.g. spray-drying). Moreover, electrospraying processing results in higher production yields than spray-drying because the collection of the capsules is favored by electrostatic attraction forces (García-Moreno et al. 2016). In addition, nano-microencapulates (e.g. fibers or capsules) with enhanced functionalities can be obtained by co-processing of polymer mixtures, utilization of co-axial processes which result in core/shell structures, or by tuning their surface properties (e.g. crosslinking) (Mendes et al. 2017). Finally, electrohydrodynamic processes are cost-effective and possible to scale-up, which makes them suitable for the production of encapsulates intended for food applications (Jacobsen et al. 2018).

17.3 Food Grade Polymers Used in Electrohydrodynamic Processes

Food grade polymers include biopolymers such as proteins and polysaccharides. Proteins are a class of natural polymers with unique properties/functionalities that can interact with a broad range of hydrophobic and hydrophilic bioactive compounds, and through changes in pH, temperature and ionic strength their conformation can change to self-assemble into different nano-micro structures such as particles, fibers, gels (El-Salam and El-Shibiny 2016). In addition, proteins have low resistance to high acidic conditions (such as stomach pH), which can degrade them into potential bioactive peptides with nutritional value. On the other hand, the consumption of proteins (particularly of animal origin) can bring allergenic side effects. Proteins consist of long chains of amino acids residues linked by peptide bonds. Thus, the type, sequence and number of amino acids determines the properties of a particular protein such as molecular weight, structural organization (e.g. globular or random coil), electrical charge, hydrophobicity, interactions and chemical reactivity (e.g. cross-linking) (McClements 2015).

Common proteins used in food industry are classified in globular, Prolamins and Phosphoproteins. Globular proteins usually adopt naturally an ordered structure, which can change into non-ordered when denatured by exposure to heat under specific pH and ionic strength leading to gel formation. Examples of globular proteins include β-lactoglobulin, whey proteins, soy proteins, ovalbumin, lactoferrin (El-Salam and El-Shibiny 2016). Prolamins include zein and gliadin and are known for their hydrophobicity. Phosphoproteins exhibit an open structure with distinct hydrophilic and randomly distributed hydrophobic domains over the polypeptide chain and caseins are the main examples.

The employment of proteins as shell materials in the production of nano- and microencapsulates by electrohydrodynamic processing is challenging due to: (1) high surface tension (in water solution) which hinders the formation of stable Taylor cones, (2) low chain entanglements (particularly required for electrospinning) as a consequence of the low molecular weight, chain flexibility and rigid or globular structure, and (3) their polyelectrolytic nature, which weakens the charge build-up and reduces stability (Weiss et al. 2012; Lim et al. 2015). For fiber formation, proteins need to be unfolded in order to increase polymer chain entanglements. For particular proteins, this is achieved by changing protein structure/aggregation and intra/inter-molecular disulfide bonds by: (a) using specific solvents, although organic solvents (e.g. hexafluoro-2-propanol or trifluoroethanol) cannot be used for food applications, (b) adding a denaturing agent (e.g. β-mercaptonol, which is not food grade) capable of reducing the disulphide bonds, and (c) applying pH changes or heat (Lim 2015; Mendes et al. 2017). Electrospinnability/sprayability of aqueous protein solutions is also commonly enhanced by using surfactants in order to reduce surface tension or by using mixtures with aided polymers such as polysaccharides (e.g. guar guam or pullulan) or biocompatible polymers (PVA or PEO) (Pérez-Masiá et al. 2014a; Nieuwland et al. 2013).

Polysaccharides are a natural class of biopolymers consisting of repeating monomeric units of monosaccharides. They are often classified according to their charge into anionic, neutral and cationic, but also according to their origin (bacterial, plants, sea weeds). Common examples of anionic biopolymers used in food include xanthan, gellan, arabic gums and also alginates, carrageenan's and pectin's. The most popular neutral polysaccharides are pullulan, guar gum, starches, cellulose and dextran. Positively charged polysaccharides are very limited and so far only chitosan has been used as cationic biopolymer is some food applications.

Similar to proteins, the electrospray/electrospinning of polysaccharides comes along with some challenges (Mendes et al. 2017) that depend on many factors, including chemical properties (e.g. molecular weight, functional groups, charges) and the selection of the solvent. Those characteristics will lead to different conductivities, extensional and intrinsic viscosities, surface tension and vapor pressure (Stijnman et al. 2011) necessary to produce capsules/fibers. Typically, the formation of electrospun polysaccharide fibers is dependent on the degree of their chain entanglements (Stijnman et al. 2011). The concentration needed for significant entanglements to electrospun polysaccharides is the concentration at which the molecular hydrodynamic radius start to overlap. Compact globular-like polysaccharides give rise to fewer entanglements than random walk-coil-like chains at the same concentration. A shear thinning behavior implies a decreased extensional viscosity, which prevents jet and fiber formation, thus a weak shear thinning is essential to electrospin polysaccharide into fibers (Stijnman et al. 2011).

In this section, the processing and properties of proteins and polysaccharides as shell materials for the production of electrospun/electrosprayed encapsulates are reviewed. The physicochemical characteristics of the resulting encapsulates are also outlined.

17.3.1 Proteins

17.3.1.1 Plant Proteins

Zein

Zein is a highly hydrophobic protein extracted from corn, which is not soluble in water, but in concentrated ethanol/aqueous solutions. Zein exhibits interesting characteristic as shell material such as low moisture absorption, high thermal resistance, oxygen barrier and antioxidant properties (Shukla and Cheryan 2001; Moomand and Lim 2015). Although zein structure is globular, it has a good electrospinnability due to the formation of globules aggregates with rod shape that lead to an extended network stabilized by disulfide bonds (Weiss et al. 2012). Thus, zein has been widely employed to encapsulate an extensive number of bioactive ingredients (omega-3 fatty acids, antioxidants and vitamin-E) in nano-microencapsulates produced by both electrospinning and electrospraying processes (Table 17.1).

Table 17.1 Overview of studies using proteins to encapsulate bioactive ingredients by electrohydrodynamic processes

Biopolymer		Morphology	Bioactive ingredient	Solvent	Reference
Plant proteins	Zein	Fibers	Epigallocatechin gallate	Ethanol/water	Li et al. (2009)
			β-carotene		Fernandez et al. (2009)
			Gallic acid		Neo et al. (2013)
			Fish oil	Ethanol/water Isopropanol/water	Moomand and Lim (2014)
		Capsules	DHA	Ethanol/water	Torres-Giner et al. (2010)
			Fish oil		Moomand and Lim (2015)
			Green tea extract		Gómez-Mascaraque et al. (2017b)
			β-carotene		Gómez-Mascaraque et al. (2017a)
	Zein + PEO (blend)	Fibers	Vitamin E		Li et al. (2016a)
	Zein + PVP (co-axial)		Fish oil		Yang et al. (2017)
	SPI + PEO	Fibers	Allyl isothiocyanate	1% NaOH solution (and Titron X-100)	Vega-Lugo and Lim (2009)
	SPI	Capsules	α-linolenic acid	Water (and Tween-20)	Gómez-Mascaraque and Lopez-Rubio (2016)
	API + pullulan	Fibers	Folic acid	95% formic acid (and Tween-80)	Aceituno-Medina et al. (2015a)
			Quercetin or ferulic acid		Aceituno-Medina et al. (2015b)
			Curcumin		Blanco-Padilla et al. (2015)

(continued)

Table 17.1 (continued)

Biopolymer		Morphology	Bioactive ingredient	Solvent	Reference
Milk proteins	WPI + PEO BLG + PEO	Fibers	Rhodamine B	Water	Sullivan et al. (2014)
	WPC	Capsules	Bifidobacterium strains	PBS or skimmed milk	López-Rubio et al. (2012)
			Lactobacillus plantarum	Water or skimmed milk (and Tween-20)	Gómez-Mascaraque et al. (2016a)
			Bifidobacterium longum subsp. infantis CECT 4552	Skimmed milk (and Span 20)	Librán et al. (2016)
			β-carotene	Water	Lopez-Rubio and Lagaron (2012)
			Lycopene	Water	Pérez-Masiá et al. (2014b)
			Folic acid	Water (and Span 20)	Pérez-Masiá et al. (2015)
			α-linolenic acid	Water (and Tween-20)	Gómez-Mascaraque and Lopez-Rubio (2016)
	WPC + Pullulan + dextran or glucose syrup	Capsules	Fish oil	Water (and Citrem)	García-Moreno et al. (2018)
	Casein + PEO	Fibers	Lipase (type VII, from *Candida rugosa*)	Aqueous triethanolamine	Xie and Hsieh (2003)

Biopolymer		Morphology	Bioactive ingredient	Solvent	Reference
Meat and marine proteins	Gelatin	Fibers	Vitamin A palmitate and vitamin E TPGS	Acetic acid	Li et al. (2016a, b)
			Moringa oleifera extract	Acetic acid (30% v/v)	Hani et al. (2017)
			Curcumin	Acetic acid (40% v/v) (and Tween80 or SDS or CTAB)	Deng et al. (2017)
		Capsules	Curcumin	Ethanol/Acetic acid/water	Gómez-Estaca et al. (2015)
			Epigallocatechin gallate	Acetic acid (20% v/v)	Gómez-Mascaraque et al. (2015)
			α-linolenic acid		Gómez-Mascaraque and Lopez-Rubio (2016)
			Green tea extract		Gómez-Mascaraque et al. (2017b)
Egg proteins	EMP	Fibers	Catechin	Water	Kang et al. (2010)

PEO poly-ethylene oxide, *PVP* poly-vinylpyrrolidone, *SPI* soy protein isolate, *API* amaranth protein isolate, *WPI* whey protein isolate, *BLG* β-lactoglobulin, *WPC* whey protein concentrate, *PBS* phosphate-buffered saline, *SDS* anionic sodium dodecyl sulfonate, *CTAB* cationic cetyltrimethyl ammonium bromide, *EMP* eggshell membrane proteins

For instance, Torres-Giner et al. (2010) reported lower oxidation of DHA encapsulated in zein nano-microcapsules (~490 nm) when compared to neat DHA, as indicated by a lower formation of secondary volatile oxidation products after 1 week storage at 18 °C. Likewise, Moomand and Lim (2014) stated higher oxidative stability for fish oil encapsulated in zein nano-microfibers (diameter of ~500 nm) than neat fish oil. Fernandez et al. (2009) indicated that encapsulation of β-carotene in zein electrospun fibers (~1140 nm) prevented photooxidation compared to the non-encapsulated compound. Recently, Gómez-Mascaraque et al. (2017a) reported an increase in the *in vitro* bioaccessibility of β-carotene encapsulated in zein capsules (~750–880 nm) by emulsion electrospraying. Nonetheless, it is worth mentioning that, besides its excellent encapsulation properties, aqueous zein solutions quickly gel and cause clogging of the spinneret during electrospinning/spraying (if the processes are not carried out at high relative humidity conditions) (Weiss et al. 2012).

Soy Protein

Soy protein is a widely used plant protein consisting of a complex mixture of globular proteins (e.g. 2S, 7S, 11S, and 15S fractions) (McClements 2015). Soy protein is an adequate shell material due to its superior functional properties (e.g. emulsification, fat absorption, film forming and oxygen barrier property) and it has been extensively studied in spray-drying (Meng and Cloutier 2014). Nevertheless, its use as wall material in electrospun/sprayed encapsulates is limited. This may be a consequence of its globular structure and reduced chain entanglements. Soy protein needs to be denatured (e.g. by applying heat) and use of surfactants (e.g. Triton X-100, Tween-20) to facilitate its electrospin/electrospray ability is required (Table 17.1). Heating soy protein solutions (e.g. neutral or alkaline solutions) unfolds the protein, exposing sulfhydryl and hydrophobic groups, which interact with each other, and this will increase chain entanglements (Meng and Cloutier 2014). Among the few examples reported in literature, Vega-Lugo and Lim (2009) incorporated allyl isothiocyanate in β-cyclodextrin that was encapsulated into soy protein isolate (SPI)/PEO electrospun fibers, which was aimed to be used as antimicrobial packaging material. Lately, Gómez-Mascaraque and Lopez-Rubio (2016) encapsulated α-linolenic acid (ALA) in SPI capsules (0.75–3.25 μm) by electrospraying. The authors observed that the degradation of encapsulated ALA within SPI capsules was significantly delayed with respect to free ALA.

Amaranth Protein

Amaranth protein isolate (API), extracted from the grain of the plant *Amaranthus hypochondriacus*, has recently been investigated as coating material in electrospun/sprayed encapsulates. Aceituno-Medina et al. (2013a) concluded that capsules of API were obtained when dissolving the protein in formic acid, and that its electrosprayability was improved when adding both Tween-80 as surfactant and

2-mercaptoethanol to reduce disulfide linkages in the protein. Nevertheless, these authors reported that API electrospun fibers could only be obtained when using hexafluoro-2-propanol as solvent, since it promotes the formation of random coil structures increasing polymer entanglements. Subsequently, API electrospun fibers were developed by the same authors by using blends of API with a spinable carbohydrate polymer such as pullulan (Aceituno-Medina et al. 2013b). Lately, the protective effect of API and pullulan blends as shell material in ultra-thin fibers produced by electrospinning has been demonstrated for several bioactive ingredients such as folic acid (Aceituno-Medina et al. 2015a), quercetin and ferulic acid (Aceituno-Medina et al. 2015b) as well as curcumin (Blanco-Padilla et al. 2015) (Table 17.1). Folic acid was encapsulated within API/pullulan nanofibers increasing its thermal stability and protection from degradation after UV light exposure (Aceituno-Medina et al. 2015a). This confirmed the great potential of electrospun fibers for the encapsulation and protection of photosensitive vitamins. Electrospun fibers of API/pullulan were produced to encapsulate and release quercetin or ferulic acid (Aceituno-Medina et al. 2015b). *In vitro* digestion studies showed a more sustained release of quercetin comparatively to ferulic acid over 250 min due to the differences in fiber composition as well as interactions between the bioactives and the encapsulating materials. Moreover, the fibers preserved the antioxidant capacity of both bioactives and showed inhibition percentages that doubled those of the non-encapsulated compounds.

17.3.1.2 Milk Proteins

Milk proteins (whey and casein) have excellent functional (e.g. solubility, emulsifying, viscosity and film-forming) as well as antioxidant properties (e.g. radical scavenging and metal chelating activities) which makes them effective encapsulating material (Augustin and Oliver 2014).

Whey Protein

Whey protein consists of mainly globular proteins, with the most abundant being β-lactoglobulin (~55%) and α-lactalbumin (~24%) (McClements 2015). Similarly to most proteins (apart from zein and gelatin), electrospinning of whey protein alone (native or denatured) in a food grade solvent (e.g. water) is not feasible due to its globular nature, lack of intermolecular entanglements and high conductivity (Drosou et al. 2018). As a consequence, electrospinning of whey protein has been facilitated by using carrier polymer such as PEO (Sullivan et al. 2014), gelatin (at 40 °C) (Nieuwland et al. 2013) and pullulan (Drosou et al. 2018). From these studies, only Sullivan et al. (2014) investigated the encapsulation of a bioactive compound such as Rhodamine B (RhB) (as a model flavonoid). The authors reported that RhB was uniformly distributed in whey protein/PEO fibers and that 90% of the RhB was released in water at room temperature after 10 min. Moreover, crosslinking of the protein by heating for 24–44 h at 100 °C made the fiber insoluble in water after several days of soaking. Conversely, electrospraying of whey protein has been studied for the encapsulation of numerous bioactive ingredients, including omega-3 fatty acids, antioxidants vitamins and probiotics (Table 17.1). For example, Pérez-Masiá et al. (2015) produced WPC electrosprayed capsules loaded with folic acid having an encapsulation efficiency of 80.8%. The authors observed that folic acid encapsulated in WPC nano-microcapsules was protected against degradation during storage in both aqueous solution and dryness. Gómez-Mascaraque et al. (2016a) reported an improved stability for a fresh culture of *Lactobacillus plantarum* during storage (53% of relative humidity) when encapsulated in WPC electrosprayed capsules. It is worth mentioning that sprayability of whey protein aqueous solutions, which have high conductivity and surface tension, is commonly enhanced by the addition of surfactants (Table 17.1).

Caseins

Caseins (αs1-casein, αs2-casein, β-casein, and κ-casein) are the major milk proteins accounting for 80% of the total protein in milk (Augustin and Oliver 2014). Although caseins have some secondary and tertiary structure, they have relatively flexible and disordered structures when dissolved in water (McClements 2015). Nevertheless, electrospinning of caseinates alone from water solutions is not possible, mainly due to the clustering of caseinate molecules, which prevents polymer chain entanglements (Nieuwland et al. 2013). Hence, caseinates have been electrospun in combination with carrier polymers such as PEO or PVA (Xie and Hsieh 2003), gelatin (Nieuwland et al. 2013) and pullulan (Tomasula et al. 2016). Xie and Hsieh (2003) reported the immobilization of lipase type VII from *Candida rugosa* in electrospun fibers produced from blends of casein and PEO (Table 17.1). Nonetheless, the authors indicated that the enzyme immobilized in electrospun fibers produced from a blend of PEO/casein had a lower catalytic activity than the enzyme immobilized in PVA electrospun fibers. This was attributed to chemical crosslinking effects and different polymer/enzyme interaction, which were not further detailed. To the best of the authors' knowledge, bioactive ingredients have not been encapsulated by electrospraying using casein alone or in combination with carriers as shell material.

17.3.1.3 Meat and Marine Proteins

Gelatin

Gelatin is one of the most widely used proteins in food applications due to its ability to form thermally reversible gels. Although gelatin is not as good emulsifier as casein or whey protein, it is also commonly used as coating material in encapsulates because of its excellent film-forming properties (Meng and Cloutier 2014). Gelatin is derived from animal collagen (e.g. pig, fish, cow) by acidic (type A gelatin) or alkaline (type B gelatin) treatments. Gelatin adopts a helix conformation below its gelling temperature (e.g. 10–25 °C for pig and cow gelatin, and 0–5 °C for fish gelatin) (McClements 2015). Thus, electrospinning/spraying of gelatin needs to be carried out from gelatin solutions where the triple-helix structure has been destabilized to random coil conformation (Mendes et al. 2017). This has been favored by using specific organic solvents such as acetic (Okutan et al. 2014) or formic acids (Songchotikunpan et al. 2008) and also by using aqueous solutions at 40 °C (Li et al. 2006). However, the electrospinnability of gelatin in aqueous solution is very poor due to extensive polymer-polymer interaction and high surface tension (Weiss et al. 2012). As a consequence, fiber formation using aqueous solutions as solvent has been facilitated by using another carrier polymer such as hyaluronic acid (Li et al. 2006) or PVA (Yang et al. 2007). Bioactive compounds such as vitamins and antioxidants have been encapsulated within gelatin electrospun fibers (Table 17.1). Hani et al. (2017) encapsulated an ethanol extract from Moringa oleifera (rich in polyphenolic compounds) in gelatin electrospun fibers. The fibers, with a dimeter lower than 100 nm, were loaded with 3–5 wt.% of extract resulting in an encapsulation efficiency of 80–85%. Deng et al. (2017) evaluated the protection and release of curcumin by encapsulation in gelatin nano-microfibers containing different surfactants. The authors indicated that adding CTAB (cationic cetyltrimethyl ammonium bromide) enhanced the release of curcumin in aqueous solvents leading to higher radical scavenging activity and reducing power as well as stronger antimicrobial activity. Electrosprayed gelatin capsules have also been developed for the encapsulation of antioxidants and omega-3 fatty acids (Table 17.1). For instance, Gómez-Mascaraque et al. (2017b) encapsulated green tea extract within gelatin nano-microcapsules. They observed that the encapsulation efficiency was lower when using gelatin when compared to zein protein, probably because of the instability of catechins in aqueous media. Although the encapsulated green tea extract showed antioxidant activity in enriched biscuits, the activity was not significantly higher than that obtained by the addition of free green tea extract. The same authors reported the encapsulation of α-linolenic acid within gelatin electrosprayed capsules (Gómez-Mascaraque and Lopez-Rubio 2016). Contrary to whey protein or soy protein capsules, they found that the oxidative stability of gelatin capsules loaded with α-linolenic acid was lower than that of unprotected α-linolenic acid.

Other Marine Proteins

Other marine proteins have been investigated as potential encapsulating materials for bioactive compounds. Verdugo et al. (2014) studied the electrospinning of a protein concentrate from microalgae *Botryococcus braunii* residual biomass (MPC). The authors reported that MPC solutions (distilled water, aqueous sodium hydroxide 1% solution, or glacial acetic acid) could not be electrospun without PEO due to the impossibility to form a jet from the Taylor cone. On the other hand, Stephansen et al. (2014) reported the feasibility to produce electrospun fibers from fish sarcoplasmic proteins (FSP) from cod (*Gadus morhua*) when dissolved in HFIP

(1,1,1,3,3,3-Hexafluoro-2-propanol). The resulting fibers were insoluble in water, despite the water-soluble nature of FSP. The FSP fibers showed potential as carrier system for delivery of drugs, bioactive agents, and nutraceuticals. The dipeptide Ala-Trp, rhodamine B, or insulin was encapsulated into the fibers, and the release was studied in biorelevant media (Stephansen et al. 2014, 2015, 2016).

17.3.1.4 Egg Proteins

Eggshell membrane proteins (EMP), consisting of collagen and phosphoproteins (e.g. osteopontin and sialoprotein), have been investigated as biomaterial for the production of ultra-thin fibers by electrospinning (Table 17.1) (Kang et al. 2010). PEO and PVA were used as carrier polymer in order to electrospin aqueous solutions of EMP. The authors reported that the EMP/PEO and EMP/PVA fibers became less soluble in water when treated with a catechin/ethanol solution.

17.3.2 Polysaccharides

17.3.2.1 Anionic Polysaccharides

Alginate

Alginate is an anionic linear polysaccharide extracted from many species of brown seaweed that consists of (1–4) linked β-D-mannuronic acid (M) and α-L-glucuronic acid (G) units in various sequences and compositions (Nie et al. 2008; Bonino et al. 2011; Reddy and Yang 2015). Several alginate nano-microstructures loaded with bioactive ingredients have been reported (Table 17.2). Nie et al. fabricated alginate electrospun fibers by using a strong polar co-solvent such as glycerol (Nie et al. 2008). Fibers with diameters ranging from 120 to 300 nm were produced and the diameter was observed to be dependent on the concentration of alginate and the content of glycerol/water. The presence of glycerol in alginate solutions was observed to improve the flexibility and the entanglement of alginate chains by disrupting the strong inter- and intramolecular hydrogen bonding among alginate chains. Due to the rigid and extended structure of alginate, the electrospinning/electrospay of this biopolymer is challenging. Therefore blending alginate with other polymers (e.g. PVA, PEO) is a common process to produce alginate based particles/fibers by electrohydrodynamic processes. Bonino et al. (2011) produced alginate fibers by electrospinning blends of alginate/PEO. The collected fibers were further cross-linked with calcium ions to preserve their structure in aqueous environment and subsequent the removal of the PEO was achieved by soaking them in water. Electrospun alginate-pectin-PEO fibers were also used to encapsulate folic acid (Alborzi et al. 2010, 2012, 2014). The recovery of folic acid loaded in electrospun fibers was about 100% when stored at pH 3 in dark conditions for 31 days of

Table 17.2 Overview of studies using polysaccharides to encapsulate bioactive ingredients by electrohydrodynamic processes

Biopolymer		Morphology	Bioactive ingredient	Solvent	Reference
Anionic	Alginate-pectin-poly(ethylene oxide)	Fibers	Folic acid	Sodium hydroxide solution (pH 11.5)	Alborzi et al. (2010, 2012, 2014)
	Sodium alginate/PVA		Moxifloxacin hydrochloride	Water	Fu et al. (2015)
	Alginate-zein (core-shell)	Capsules	*Lactobacillus acidophilus*	Glycerol (core); citric acid, ethanol (shell)	Laelorspoen et al. (2014)
	Alginate Egg albumen		*Lactobacillus acidophilus*	Glycerol; stearic acid	Pitigraisorn et al. (2017)
	Xanthan gum	Fibers	Antioxidants	Formic acid	Shekarforoush et al. (2017b)
Neutral	Pullulan	Fibers	Omega-3 polyunsaturated fatty acids; Tocopherol, rosemary extract	Water	García-Moreno et al. (2017a, b)
	Pullulan/β-cyclodextrin		R-(+)-limonene	Water	Fuenmayor et al. (2013)
	API/pullulan		Quercetin, Ferulic acid	Formic acid/water	Aceituno-Medina et al. (2015b)
		Capsules	*Bifidobacterium animalis* subsp. *lactis* Bb12	Water	López-Rubio et al. (2012)

(continued)

Table 17.2 (continued)

Biopolymer		Morphology	Bioactive ingredient	Solvent	Reference
	Dextran	Capsules	Omega-3 polyunsaturated fatty acids	Water	García-Moreno et al. (2017a)
			Lycopene	Water	Pérez-Masiá et al. (2014b)
	Starch	Fibers	*L. paracasei*	Formic acid (shell) Glycerol (core)	Lancuški et al. (2017)
	Starch acetate		Diclofenac	Formic acid/water	Xu et al. (2009)
	Resistant starch/guar gum	Capsules	Folic acid	Water	Pérez-Masiá et al. (2015)
	Cellulose acetate	Fibers	Vitamins A and E	Acetone/*N*,*N*-dimethylacetamide	Taepaiboon et al. (2007)
			Candida rugosa lipase	Acetone/*N*,*N*-dimethylacetamide	Huang et al. (2011)
	PVA/cyclodextrin	Fibers	Vanillin	Water	Kayaci and Uyar (2012)
	PVA/cyclodextrin		Eugenol	Water	Kayaci et al. (2013a)
	Cyclodextrins		Vanillin	Water, DMF and DMAc	Celebioglu et al. (2016)
Cationic	Chitosan/phospholipids	Fibers	Vitamin B12 diclofenac curcumin	TFA/DCM	Mendes et al. (2016)
	Chitosan	Capsules	(−)-Epigallocatechin gallate (EGCG)	Acetic acid	Gómez-Mascaraque et al. (2016b)

API amaranth protein isolate, *PVA* polyvinyl alcohol, *DMF* dimethylformamide, *DMAc* dimethylacetamide, *TFA* trifluoroacetic acid, *DCM* dichloromethane storage, while the recovery of non-encapsulated folic acid was 8% within the first day at the same conditions (pH and light (Alborzi et al. 2012). The release of folic acid, in aqueous solution at different pHs (1.2, 3 and 7.8) from alginate-pectin/PEO electrospun fibers (Alborzi et al. 2010, 2014) showed that the highest release took place at pH 7.8 due to the swelling and partial dissolution of the fibers at this pH. Thus, electrospun alginate based fibers can be utilized to encapsulate, stabilize and release micronutrients such as folic acid in the small intestine. In a different study, sodium alginate was blended with PVA to produce nanofibers to encapsulate and release moxifloxacin hydrochloride for antibacterial applications (Fu et al. 2015).

Core-shell alginate–zein microcapsules were also produced to encapsulate *Lactobacillus acidophilus* (Laelorspoen et al. 2014). The core was made of electrosprayed alginate/glycerol solution collected in the shell solution composed of acidic zein (zein, citric acid, ethanol) with $CaCl_2$. The diameter and morphology of the microcapsules was observed to decrease with the increase of the voltages from 4 to 6 kV. The viability of the encapsulated cells suffered only 1-log reduction, while the number of the non-encapsulated bacteria decreased to 5-log reduction, attesting the potential of this system to preserve the viability of encapsulated cells. Alginate/glycerol solutions were electrosprayed directly into solutions of egg albumen (EA) and stearic acid (SA) to encapsulate *L. acidophilus* (Pitigraisorn et al. 2017). Further these capsules were coated with cassava starch granules and dried in a fluidized bed dryer using cassava pearls as drying help. The encapsulation efficiencies were higher than 90%, and by increasing the amount of SA the cell viability was improved. These results confirm the potential of these matrices to encapsulate thermo-sensitive compounds that can be used for food fortification purposes.

Xanthan Gum

Xanthan gum is an extracellular anionic biopolymer produced by the bacterium *Xanthomonas campestris*. The primary structure of xanthan consists of β(1,4)-linked glucose units, similar to the structure of cellulose backbone, substituted at O-3 of alternate glucose residues, with a tri-saccharide. The tri-saccharide consists of one glucuronic acid unit between two mannose units (Rosalam and England 2006; Faria et al. 2011). For specific concentrations, temperature and pH, xanthan gum can assume a helical conformation, with the side branches positioned almost parallel to the helix axis that stabilizes the structure. Xanthan gum can form very viscous aqueous solutions, and at sufficiently high polymer concentration, it exhibits weak gel-like properties. Therefore, it has been widely used and investigated in areas such as food (e.g. as thickener), pharmaceutical, cosmetics, biomedical, tissue engineering and oil industries (Ungeheuer et al. 1989; Zirnsak et al. 1999; Lachke 2004; Mendes et al. 2012). The production of electrospun xanthan fibers has been challenging due to the insufficient entanglements required for electrospinning using water as solvent (Stijnman et al. 2011). However, recently, electrospun xanthan polysaccharide nanofibers with an average diameter of 128 ± 36.7 to 240 ± 80.7 nm were prepared using formic acid as a solvent (Shekarforoush et al. 2017b). At xanthan concentrations above 1 wt/vol% chain entanglements led to an increase of its elastic modulus, apparent viscosity, and first normal stress differences (N1), resulting in the stabilization of the xanthan jet during electrospinning and formation of electrospun fibers. In a further study from the same authors, it was investigated and demonstrated the potential of these nanofibers to encapsulate and release anti-oxidants.

17.3.2.2 Neutral Polysaccharides

Pullulan

Pullulan is a linear polysaccharide, of microbial origin, consisting of repeating units of α-(1→6)-linked maltotriose, with three glucopyranoses units linked by α-(1→4) glycosidic bonds. Electrospun pullulan nanofibers with average diameters ranging from hundreds of nanometers to micrometer scale, were produced using water (Sun et al. 2013), mixtures of DMSO/water (Kong and Ziegler 2014a) and 95% formic acid (Aceituno-Medina et al. 2013b) as solvents. To increase the tensile strength and the thermal stability of the pullulan matrix, montmorillonite/Pullulan composite fibers, were fabricated using 20 wt.% aqueous Pullulan concentrations with different contents of clay (1–10 wt.%), which resulted in fibers with diameters ranging from 50 to 500 nm (Karim et al. 2009). Pullulan has been used as encapsulating material for the production of bioactive-loaded nano-microstructures (Table 17.2). Pullulan and β-cyclodextrin emulsions in water were also electrospun to encapsulate R-(+)-limonene (Fuenmayor et al. 2013). The release of limonene from the fibers was modulated by the relative humidity changes, which could be used as an active packaging device. Pullulan was investigated as wall material for the production of electrospun omega-3 polyunsaturated fatty acids encapsulates (García-Moreno et al. 2017a) Electrospun fibers containing neat fish oil exhibited a good oxidative stability after production. To improve oxidative stability during storage, natural antioxidants were added to fish oil-loaded pullulan nano-microfibers (García-Moreno et al. 2017b). The oxidation of these electrospun encapsulates was reduced when tocopherol (500 ppm) was combined with rosemary extract (500 ppm), while a prooxidant effect was observed when higher concentration of the same antioxidants (2000 ppm tocopherol and 1000 ppm rosemary extract) were added (García-Moreno et al. 2017b). The encapsulation of *Bifidobacterium animalis* subsp. *lactis* Bb12 within electrosprayed WPC and pullulan capsules was observed to increase the viability of the probiotics (particularly at 20 °C) comparatively to non-encapsulated in freeze-dried form (López-Rubio et al. 2012). Pullulan capsules were observed to prolong the survival of the cells even at high humidity, compared to WPC capsules.

Dextran

Dextran is a water soluble, bacterial polysaccharide consisting of α-1,6 linked d-glucopyranose residues with α-1,2, α-1,3 or α-1,4 linked side chains. Uniform electrospun dextran fibers with diameters ranging from micron to several hundred of nanometers were produced using water, DMSO/water, and DMSO/DMF mixtures as co-solvents (Hongliang Jiang et al. 2004). To avoid the solubility of electrospun dextran fiber mats in aqueous medium, dextran fibers can be cross-linked with glutaraldehyde (Ritcharoen et al. 2008). Dextran electrospun fibers were also prepared using a mixture of 78% deionized water, 20% ethanol and 2% PBS as solvents resulting in fibers with average diameters of 200 nm (Spano and Massaro 2012). In another study, dextran, and WPC were used to produce particles via either electrospraying or spray-drying to encapsulate lycopene (Pérez-Masiá et al. 2014b). Overall spray drying led to lower encapsulation efficiencies. WPC and dextran were also used for co-axial or emulsion electrospraying. Emulsion electrospraying of dextran based solutions led to more homogenous capsule sizes than coaxial electrospraying, which produced more aggregate capsules. García-Moreno et al. (2017a) evaluated the encapsulation of omega-3 fatty acids in dextran nano-microcapsules. The encapsulates showed a poor oxidative stability due to emulsion instability which led to a low encapsulation efficiency.

Starch

Starch is one of the most abundant polysaccharides, from plant origin and it is constituted by mixtures of amylose and amylopectin (Lancuški et al. 2015; Suárez and Gutiérrez 2017; Gutiérrez et al. 2017). The linearity of amylose and its ability to associate intra-molecularly has led to the development of amylose electrospun fibers (Kong and Ziegler 2014b). In addition, the electrospinning of starch molecules dissolved in 95% DMSO and mixtures of DMSO/water has been attempted by Kong and Ziegler (Kong and Ziegler 2012, 2013, 2014b, c). These studies verified that the conformation of amylose helices into random coil structures promote the molecular entanglements of starch that enable electrospinning and nanofibers formation. The content of amylopectin also played a critical role in the electrospinning of starch and it has to be below 65% (Kong and Ziegler 2012). Examples of starches used in electrospinning studies are: Gelose 80 (Kong and Ziegler 2014b, c), with amylose content about 80%; Hylon VII, Hylon V, Melojel, and Amioca starches with amylose content of approximately 70, 55, 25, and 0–1%, respectively (Kong and Ziegler 2012). Starch electrospun fibers with diameters ranging from 3.35 to 22.35 μm were produced using Hylon VII dissolved in DMSO 95% (Kong and Ziegler 2012, 2014b, c). However, to increase the stability of these fibers in aqueous media, a post-spinning treatment (immersion in ethanol and heat at 70 °C for 1 h) and crosslinking with 25% (v/v) aqueous glutaraldehyde had to be executed (Kong and Ziegler 2014b).

The electrospinning of high-amylose content starch from aqueous formic acid dispersions was also reported (Lancuški et al. 2015). Herein Hylon VII maize starch (17 wt.%) was dissolved in solutions of formic acid (60–100%) to promote starch gelatinization and molecular entanglements essential for the electrospinning processing. The presence of water (80 and 90 vol%) significantly delayed gelatinization and dissolution of starch, and consequently weakened fiber quality (Lancuški et al. 2015). Later, the same authors produced core-shell electrospun fibers made of glycerol/starch formate (shell) to encapsulate *L. paracasei* bacteria (Lancuški et al. 2017) (Table 17.2). These fibers were observed to be stable with retained bacterial activity at 4 °C and room temperature up to 21 days. Due to the low water stability and weak mechanical properties of starch, Xu et al. (2009) used starch acetate dissolved in formic acid/water to produce nanofibers with diameters of 50 μm with highest tenacity of 17.9 MPa. These fibers were further tested as drug delivery carrier, where diclofenac was used as a model drug.

Resistant starch and WPC matrix were used in electrospray and nanospray drying methodologies to encapsulate folic acid (Pérez-Masiá et al. 2015) (Table 17.2). Electrospray led to capsules with lower diameters than nanospray. The encapsulation efficiencies were similar when comparing both encapsulation technologies. However when comparing the polymers, WPC led to higher encapsulation efficiencies around 80.8–83.9% due to the interaction between the protein matrix and the folic acid. Starch provided encapsulation efficiencies ranging from 44 to 52%. In addition, WPC has shown to provide higher protection against degradation of folic acid in both aqueous and dry conditions.

Cellulose

Cellulose is one of the most abundant polysaccharides that consists of (1,4)-linked β-D-glucose units (Gutiérrez and Alvarez 2017). Cellulose is almost undissolved in a broad range of solvents (Lindman et al. 2010) which therefore limits its electrospinning processing. To produce electrospun cellulose fibers, solvents such as *N*-methylmorpholine *N*-oxide/water (nNMMO/H_2O), lithium chloride/dimethylacetamide (LiCl/DMAc), ionic liquids and ethylene diamine/salt can be used (Frey 2008) and the remained solvent can be removed from fresh fibers by the use of coagulant baths and temperature control at the spinneret (Frey 2008). In addition, the synthesis of cellulose derivatives with enhanced solubility and the possibility to be converted back to cellulose have been considered as possibilities to create pure cellulose electrospun fibers (Frey 2008). Examples of cellulose derivatives for electrospinning are cellulose acetate, cellulose triacetate (Lee et al. 2009), ethyl cellulose (Wu et al. 2005; Lee et al. 2009), methyl cellulose (Lee et al. 2009; Guo et al. 2013), hydroxypropyl cellulose (Lee et al. 2009; Guo et al. 2013) and ethylcyanoethyl cellulose (Frey 2008). However, it is worth noting that these nano-microstructures have not been designed for encapsulation of food bioactives.

Electrospun cellulose based nanofibers have been used for enzyme immobilization, for drug/vitamin encapsulation, and as antimicrobial and biosensor systems

(Rezaei et al. 2015) (Table 17.2). Cellulose acetate nanofibers with diameters ranging from 247 to 265 nm, (produced using acetone/*N,N*-dimethylacetamide as solvents), were utilized for the delivery of vitamins A and E in a sustained manner (Taepaiboon et al. 2007). In addition, cellulose acetate nanofibers with diameters of 200 nm were used to immobilize *Candida rugosa* lipase, which showed significantly higher thermal stability and durability comparatively to the equivalent free enzyme (Huang et al. 2011).

Cyclodextrin

Cyclodextrins (CDs) are truncated cone shape oligosaccharides consisting of 6, 7, or 8 (α, β or γ-CDs) glucose units linked by α-(1→4) glicosidic bonds (Celebioglu and Uyar 2010; Fathi et al. 2014). The electrospinning of methyl-β-cyclodextrin using water and DMF as a solvent, produced fibers with diameters ranging from 20 to 1200 nm (Celebioglu and Uyar 2010). Hydroxypropyl-β-cyclodextrin (HPβCD) and hydroxypropyl-γ-cyclodextrin (HPγCD) were also electrospun using water, dimethylformamide and dimethylacetamide as solvents producing fibers with diameters ranging from 250 to 1860 nm (Celebioglu and Uyar 2013). The electrospinning of these oligosaccharides was observed to be dependent on the type of solvent, concentration of the CD and intermolecular interactions between the CD molecules.

CD based electrospun fibers have been used to incorporate anti-microbial agents. As example, electrospun composite polyacrylonitrile nanofibers containing CD and Ag nanoparticles with diameters ranging from 180 to 260 nm, were shown to have improved antibacterial activity against (*S. aureus*) and Gram-negative *Escherichia coli* bacteria (Wang et al. 2012). In addition, electrospun HPβCD inclusion complexes, with the anti-bacterial agent triclosan, produced fibers with diameters ranging from 50 to 900 nm (Celebioglu and Uyar 2011). The inclusion of triclosan/CDs within poly(lactic acid) (PLA) nanofibers showed higher antibacterial activity against *Staphylococcus aureus* and *Escherichia col*i bacteria compared to PLA nanofibers containing only triclosan without CD. Thus the high surface area, nanoporous structure, as well as efficient antibacterial properties of CD/PLA/triclosan, suggests their great potential to be used in active food packaging applications (Kayaci et al. 2013b).

In addition, CD based fibers have been also used in applications that required antioxidant properties (Table 17.2). Vanillin/cyclodextrin inclusion complex (vanillin/CD-IC) was incorporated within electrospun PVA nanofibers (Kayaci and Uyar 2012). These fibers were shown to prolong shelf-life and high temperature stability to vanillin. A similar approach was investigated by loading eugenol/cyclodextrin within electrospun PVA solutions to produce electrospun fibers, which improved the thermal stability of eugenol (Kayaci et al. 2013a). Later, vanillin/cyclodextrins inclusion complexes were successfully electrospun into nanofibers without using a polymer carrier to allow the loading of higher amounts of vanillin while improving its antioxidant property (Celebioglu et al. 2016).

17.3.2.3 Cationic Polysaccharides

Chitosan

Chitosan is a chitin-derived polysaccharide made of glucosamine and *N*-acetylglucosamine (Gutiérrez 2017). Chitosan is one of the few positively charged polysaccharides that exhibits several biological properties, including non-toxicity, biocompatibility, biodegradability, hemostatic activity, and antibacterial and antimycotic (Balan and Verestiuc 2014; Jayakumar et al. 2010a; Luo and Wang 2014). Chitosan nanofibers were initially produced by using acetic acid as solvent (Geng et al. 2005). Concentration of acetic acid higher than 30% is necessary to lower the surface tension and increase the charge density, needed for the formation of nanofibers without significant effect on the viscosity of the solution (Geng et al. 2005). In another study, electrospun chitosan fibers were produced using TFA as a solvent (Ohkawa et al. 2004). However, these fibers can easily be dissolved in neutral and weak basic aqueous solvents (Sun and Li 2011; Pakravan et al. 2012) due to the high solubility of the TFA-chitosan salt residues. Efforts to enhance chitosan nanofiber stability have been reported through the neutralization of electrospun chitosan fiber mats with saturated solutions of Na_2CO_3 (Sangsanoh and Supaphol 2006; Gudjónsdóttir et al. 2015), and by crosslinking chitosan electrospun fibers with food-approved crosslinkers such as genipin (Austero et al. 2012), tripolyphosphate, glycerol phosphate and tannic acid, (Kiechel and Schauer 2013). Other crosslinkers (non-food approved) used to stabilize chitosan electrospun fibers include glutaraldehyde (Schiffman and Schauer 2007), hexamethylene-1,6-diaminocarboxysulphonate (HDACS) and epichlorohydrin (ECH) (Austero et al. 2012).

Alternatively, water stable and biocompatible hybrid chitosan/phospholipids electrospun nanofibers with diameters ranging from 250 to 600 nm were developed and used to encapsulate vitamin B12, diclofenac and curcumin (Mendes et al. 2016) (Table 17.2). The solubility of the bioactives used in the present study in PBS follow the subsequent order: vitamin B12 > diclofenac > curcumin. Thus, the release of vitamin B12 was observed to be higher than the other two model bioactives. Likewise, curcumin, a hydrophobic bioactive, exhibits the slowest release profile in comparison with diclofenac and vitamin B12. The effect of the phospholipid content on the nanofibers was also observed to play a major role on the release properties of vitamin B12. For longer release periods (after day 1), the increase in the phospholipid content slightly increased the hydrophilicity of the matrix and this further facilitated the diffusion of vitamin B12 molecules to the release media.

Chitosan blends with other polymers such as polyvinyl alcohol (PVA) (Li and Hsieh 2006; Jeannie Tan and Zhang 2011), polycaprolactone (PCL) (Shalumon et al. 2010), polylactic acid (PLA) (Ignatova et al. 2009), and polyethylene oxide (PEO) (Bhattarai et al. 2005; Desai et al. 2009; Pakravan et al. 2012), have been used to prepare composite electrospun fibers. Moreover chitosan has also been blended with other biopolymers including polysaccharides (Ma et al. 2012; Devarayan et al. 2013) and proteins (e.g. collagen (Chen et al. 2007), silk fibroin (Park et al. 2004) and zein (Torres-Giner et al. 2009)) to produce composite chitosan fibers for a broad range of applications, including encapsulation and delivery of bioactives (Jayakumar et al. 2010b; Sun and Li 2011; Hu et al. 2014; Reddy and Yang 2015). Electrospun chitosan fibers have also been used as a wrapping material for dry-ageing of meat (Gudjónsdóttir et al. 2015), as well as for anti-microbial (Torres-Giner et al. 2008; Ignatova et al. 2013), filtering (Jayakumar et al. 2010b; Sun and Li 2011) applications.

Studies of electrosprayed chitosan particles have not been so well reported as fibers. The interest on chitosan electrosprayed particles was first explored in drug delivery applications through the encapsulation and release of anticancerous drugs like doxorubicin (Songsurang et al. 2011), indomethacin (Thien et al. 2012) and antibiotics (ampicillin) (Arya et al. 2009) or genetic material like deoxyribonucleic acid (DNA) (Sreekumar et al. 2017). Later the production of chitosan electrosprayed particles started to be explored for the encapsulation of food ingredients such as antioxidants, nutraceuticals or probiotics, that can be beneficial to the food industry (Ghorani and Tucker 2015). Sreekumar et al. (2017) concluded that chitosan's with low degree of acetylation, (below 10%), and degree of polymerization ranging from 500 to 1500 are more favorable for the production of electrosprayed nanoparticles due to the reduced solution conductivity, when using acetic acid/water/ethanol as solvents.

Gómez-Mascaraque et al. (2016b) investigated the effect of the molecular weight on the electrospraying of chitosan, and tested the potential of these particles to encapsulate and release (−)-epigallocatechin gallate (EGCG) a polyphenol very abundant in green tea with additional healthy benefits, e.g. anti-inflammatory, cardiovascular, and neurodegenerative. The combination of different molecular weights (lowest 25 kDa) with different concentrations (highest 5% w/v) allowed electrospraying of chitosan into capsules that encapsulated about 80% of EGCG. The antiviral activity against the murine norovirus in simulated physiological conditions was demonstrated, thus suggesting the potential of these particles for the encapsulation and release of bioactives.

17.4 Conclusions and Future Perspectives

Electrospinning and electrospraying processes facilitate the encapsulation of a broad range of bioactives using a variety of biopolymers (e.g. proteins and polysaccharides), at room temperature with high encapsulation efficiency. In addition, the encapsulation of bioactives within electrohydrodynamically produced biopolymeric nano- and microcapsules/fibers with controllable diameters, morphologies, porosities and functionality facilitates their controlled delivery, bioactivity and bioavailability.

Zein, WPC and gelatin are among the most popular proteins used for the processing and encapsulation of bioactives, due to their suitable physicochemical properties for electrohydrodynamic processing when dissolved in food-grade solvents. Dextran, pullulan and alginates are some of the most studied polysaccharides to encapsulate food bioactives via electrohydrodynamic processes, most probably due to the utilization of aqueous solutions. Although water-soluble biopolymers are the most desirable biopolymers to be processed electrohydrodynamically into capsular or fibrillar structures for the encapsulation of food bioactives, the low processing throughput using water as a solvent, and the lack of mass production facilities, limits their commercial applications at industrial scale. The use of organic solvents considered as GRAS can be utilized for the electrohydrodynamic processing of a higher variability of biopolymers leading to higher yield of production. However, the selection of the food biopolymers and solvent has to be done in respect to the final food application and considering the acceptance of the consumers in the market. Therefore, the investment from academia, industry and regulatory agencies is essential not only to improve the implementation and scaling up of electrohydrodynamic processes, but also to create high value bioactive functional ingredients and foods that can be accepted by the consumers.

References

Aceituno-Medina M, Lopez-Rubio A, Mendoza S, Lagaron JM (2013a) Development of novel ultrathin structures based in amaranth (amaranthus hypochondriacus) protein isolate through electrospinning. Food Hydrocoll 31:289–298

Aceituno-Medina M, Mendoza S, Lagaron JM, López-Rubio A (2013b) Development and characterization of food-grade electrospun fibers from amaranth protein and pullulan blends. Food Res Int 54:667–674

Aceituno-Medina M, Mendoza S, Lagaron JM, Lopez-Rubio A (2015a) Photoprotection of folic acid upon encapsulation in food-grade amaranth (Amaranthus hypochondriacus L.) protein isolate—Pullulan electrospun fibers. LWT Food Sci Technol 62:970–975

Aceituno-Medina M, Mendoza S, Rodríguez BA et al (2015b) Improved antioxidant capacity of quercetin and ferulic acid during in-vitro digestion through encapsulation within food-grade electrospun fibers. J Funct Foods 12:332–341

Alborzi S, Lim LT, Kakuda Y (2010) Electrospinning of sodium alginate-pectin ultrafine fibers. J Food Sci 75:100–107

Alborzi S, Lim L-T, Kakuda Y (2012) Encapsulation of folic acid and its stability in sodium alginate-pectin-poly(ethylene oxide) electrospun fibres. J Microencapsul 30:1–8

Alborzi S, Lim L-T, Kakuda Y (2014) Release of folic acid from sodium alginate-pectin-poly(ethylene oxide) electrospun fibers under in vitro conditions. LWT Food Sci Technol 59:383–388

Arya N, Chakraborty S, Dube N, Katti DS (2009) Electrospraying: a facile technique for synthesis of chitosan-based micro/nanospheres for drug delivery applications. J Biomed Mater Res B Appl Biomater 88B:17–31

Augustin MA, Oliver CM (2014) Use of milk proteins for encapsulation of food ingredients. In: Gaonkar AG, Vasisht N, Khare AR (eds) Microencapsulation in the food industry. Academic Press, San Diego, pp 211–226

Austero MS, Donius AE, Wegst UGK, Schauer CL (2012) New crosslinkers for electrospun chitosan fibre mats. I. Chemical analysis. J R Soc Interface 9:2551–2562

Balan V, Verestiuc L (2014) Strategies to improve chitosan hemocompatibility: a review. Eur Polym J 53:171–188

Bhardwaj N, Kundu SC (2010) Electrospinning: a fascinating fiber fabrication technique. Biotechnol Adv 28:325–347

Bhattarai N, Edmondson D, Veiseh O et al (2005) Electrospun chitosan-based nanofibers and their cellular compatibility. Biomaterials 26:6176–6184

Blanco-Padilla A, López-Rubio A, Loarca-Piña G, Gómez-Mascaraque LG, Mendoza S (2015) Characterization, release and antioxidant activity of curcumin-loaded amaranth-pullulan electrospun fibers. LWT Food Sci Technol 63:1137–1144

Bocanegra R, Gaonkar AG, Barrero A, Loscertales IG, Pechack D, Marquez M (2005) Production of cocoa butter microcapsules using an electrospray process. J Food Sci 70:492–497

Bonino CA, Krebs MD, Saquing CD et al (2011) Electrospinning alginate-based nanofibers: from blends to crosslinked low molecular weight alginate only systems. Carbohydr Polym 85:111–119

Celebioglu A, Uyar T (2010) Cyclodextrin nanofibers by electrospinning. Chem Commun (Camb) 46:6903–6905

Celebioglu A, Uyar T (2011) Electrospinning of polymer-free nanofibers from cyclodextrin inclusion complexes. Langmuir 27:6218–6226

Celebioglu A, Uyar T (2013) Electrospinning of nanofibers from non-polymeric systems: electrospun nanofibers from native cyclodextrins. J Colloid Interface Sci 404:1–7

Celebioglu A, Kayaci-Senirmak F, Kusku Sİ et al (2016) Polymer-free nanofibers from vanillin/cyclodextrin inclusion complexes: high thermal stability, enhanced solubility and antioxidant property. Food Funct 7:3141–3153

Chen Z, Mo X, Qing F (2007) Electrospinning of collagen-chitosan complex. Mater Lett 61:3490–3494

Chronakis IS (2005) Novel nanocomposites and nanoceramics based on polymer nanofibers using electrospinning process: a review. J Mater Process Technol 167:283–293

Chronakis IS (2010) Nano-microfibers by electrospinning technology: processing, properties and applications. In: Quin Y (ed) Micromanufacturing engineering and technology. Elsevier, Oxford, pp 264–286

Deng L, Kang X, Liu Y, Feng F, Zhang H (2017) Effects of surfactants on the formation of gelatin nanofibres for controlled release of curcumin. Food Chem 231:70–77

Desai K, Kit K, Li J et al (2009) Nanofibrous chitosan non-wovens for filtration applications. Polymer (Guildf) 50:3661–3669

Devarayan K, Hanaoka H, Hachisu M et al (2013) Direct electrospinning of cellulose-chitosan composite nanofiber. Macromol Mater Eng 298:1059–1064

Drosou C, Krokida M, Biliaderis CG (2018) Composite pullulan-whey protein nanofibers made by electrospinning: Impact of process parameters on fiber morphology and physical properties. Food Hydrocoll 77:726–735

El-Salam MHA, El-Shibiny S (2016) Natural biopolymers as. Elsevier

Faria S, De Oliveira Petkowicz CL, De Morais SAL et al (2011) Characterization of xanthan gum produced from sugar cane broth. Carbohydr Polym 86:469–476

Fathi M, Martín Á, McClements DJ (2014) Nanoencapsulation of food ingredients using carbohydrate based delivery systems. Trends Food Sci Technol 39:18–39

Fernandez A, Torres-Giner S, Lagaron JM (2009) Novel route to stabilization of bioactive antioxidants by encapsulation in electrospun fibers of zein prolamine. Food Hydrocoll 23:1427–1432

Frey MW (2008) Electrospinning cellulose and cellulose derivatives. Polym Rev 48:378–391

Fu R, Li C, Yu C et al (2015) A novel electrospun membrane based on moxifloxacin hydrochloride/poly(vinyl alcohol)/sodium alginate for antibacterial wound dressings in practical application. Drug Deliv 7544:1–12

Fuenmayor CA, Mascheroni E, Cosio MS et al (2013) Encapsulation of R-(+)-limonene in edible electrospun nanofibers. Chem Eng Trans 32:1771–1776

García-Moreno PJ, Stephansen K, Van Der Kruijs J, Guadix A, Guadix EM, Chronakis IS, Jacobsen C (2016) Encapsulation of fish oil in nanofibers by emulsion electrospinning: physical characterization and oxidative stability. J Food Eng 183:39–49

García-Moreno PJ, Özdemir N, Stephansen K, Mateiu RV, Echegoyen Y, Lagaron JM, Chronakis IS, Jacobsen C (2017a) Development of carbohydrate-based nano-microstructures loaded with fish oil by using electrohydrodynamic processing. Food Hydrocoll 69:273–285

García-Moreno PJ, Damberg C, Chronakis IS, Jacobsen C (2017b) Oxidative stability of pullulan electrospun fibers containing fish oil: effect of oil content and natural antioxidants addition. Eur J Lipid Sci Technol 1600305:1–11

García-Moreno PJ, Pelayo A, Yu S, Busolo M, Lagaron JM, Chronakis IS, Jacobsen C (2018) Physicochemical characterization and oxidative stability of fish oil-loaded electrosprayed capsules: combined use of whey protein and carbohydrates as wall materials. Food Hydrocoll 231:42–53

Geng X, Kwon O-H, Jang J (2005) Electrospinning of chitosan dissolved in concentrated acetic acid solution. Biomaterials 26:5427–5432

Ghorani B, Tucker N (2015) Fundamentals of electrospinning as a novel delivery vehicle for bioactive compounds in food nanotechnology. Food Hydrocoll 51:227–240

Gómez-Estaca J, Gavara R, Hernández-Muñoz P (2015) Encapsulation of curcumin in electrosprayed gelatin microspheres enhances its bioaccessibility and widens its uses in food applications. Innovative Food Sci Emerg Technol 29:302–307

Gómez-Mascaraque LG, Lopez-Rubio A (2016) Protein-based emulsion electrosprayed micro- and submicroparticles for the encapsulation and stabilization of thermosensitive hydrophobic bioactives. J Colloid Interface Sci 465:259–270

Gómez-Mascaraque LG, Lagarón JM, López-Rubio A (2015) Electrosprayed gelatin submicrocapsules as edible carriers for the encapsulation of polyphenols of interest in functional foods. Food Hydrocoll 49:42–52

Gómez-Mascaraque LG, Morfin RC, Pérez-Masiá R, Sanchez G, Lopez-Rubio A (2016a) Optimization of electrospraying conditions for the microencapsulation of probiotics and evaluation of their resistance during storage and in-vitro digestion. LWT Food Sci Technol 69:438–446

Gómez-Mascaraque LG, Sanchez G, López-Rubio A (2016b) Impact of molecular weight on the formation of electrosprayed chitosan microcapsules as delivery vehicles for bioactive compounds. Carbohydr Polym 150:121–130

Gómez-Mascaraque LG, Perez-Masia R, Gonzalez-Barrio R, Jesus Periago M, Lopez-Rubio A (2017a) Potential of microencapsulation through emulsion-electrospraying to improve the bioaccesibility of beta-carotene. Food Hydrocoll 73:1–12

Gómez-Mascaraque LG, Hernández-Rojas M, Tarancón P, Tenon M, Feuillére N et al (2017b) Impact of microencapsulation within electrosprayed proteins on the formulation of green tea extract—enriched biscuits. LWT Food Sci Technol 81:77–86

Gudjónsdóttir M, Gacutan MD, Mendes AC et al (2015) Effects of electrospun chitosan wrapping for dry-ageing of beef, as studied by microbiological, physicochemical and low-field nuclear magnetic resonance analysis. Food Chem 184:167–175

Guo C, Zhou L, Lv J (2013) Effects of expandable graphite and modified ammonium polyphosphate on the flame-retardant and mechanical properties of wood flour-polypropylene composites. Polym Polym Compos 21:449–456

Gutiérrez TJ (2017) Chitosan applications for the food industry. In: Ahmed S, Ikram S (eds) Chitosan: derivatives, composites and applications. Wiley-Scrivener Publisher, pp 185–232. 232. https://doi.org/10.1002/9781119364849.ch8. ISBN: 978-1-119-36350-7

Gutiérrez TJ (2018) Processing nano- and microcapsules for industrial applications. In: Hussain CM (ed) Handbook of nanomaterials for industrial applications. Elsevier, pp. 989-1011. https://doi.org/10.1016/B978-0-12-813351-4.00057-2. ISBN: 978-0-12-813351-4

Gutiérrez TJ, Álvarez K (2017) Biopolymers as microencapsulation materials in the food industry. In: Masuelli M, Renard D (eds) Advances in physicochemical properties of biopolymers: Part 2. Bentham Science Publishers, pp 296–322. https://doi.org/10.2174/97816810854491170100 09. ISBN: 978-1-68108-545-6. eISBN: 978-1-68108-544-9

Gutiérrez TJ, Alvarez VA (2017) Cellulosic materials as natural fillers in starch-containing matrix-based films: a review. Polym Bull 74(6):2401–2430. https://doi.org/10.1007/s00289-016-1814-0

Gutiérrez TJ, Guarás MP, Alvarez VA (2017) Reactive extrusion for the production of starch-based biopackaging. In: Masuelli MA (ed) Biopackaging. CRC Press, Taylor & Francis Group, Miami, pp 287–315. ISBN: 978-1-4987-4968-8
Hani NM, Torkamani AE, Azarian MH, Mahmood KWA, Ngalim SH (2017) Characterisation of electrospun gelatine nanofibres encapsulated with Moringa oleifera bioactive extract. J Sci Food Agric 97:3348–3358
Hu X, Liu S, Zhou G et al (2014) Electrospinning of polymeric nanofibers for drug delivery applications. J Control Release 185:12–21
Huang XJ, Chen PC, Huang F et al (2011) Immobilization of Candida rugosa lipase on electrospun cellulose nanofiber membrane. J Mol Catal B Enzym 70:95–100
Ignatova M, Manolova N, Markova N, Rashkov I (2009) Electrospun non-woven nanofibrous hybrid mats based on chitosan and PLA for wound-dressing applications. Macromol Biosci 9:102–111
Ignatova M, Manolova N, Rashkov I (2013) Electrospun antibacterial chitosan-based fibers. Macromol Biosci 13:860–872
Jacobs IC (2014) Atomization and spray-drying processes. In: Gaonkar AG, Vasisht N, Khare AR (eds) Microencapsulation in the food industry. Academic Press, San Diego, pp 47–56
Jacobsen C, García-Moreno PJ, Mendes AC, Mateiu RV, Chronakis IS (2018) Use of electrohydrodynamic processes for encapsulation of sensitive bioactive compounds and applications in food. Annu Rev Food Sci Technol 9:525–549
Jayakumar R, Menon D, Manzoor K et al (2010a) Biomedical applications of chitin and chitosan based nanomaterials—a short review. Carbohydr Polym 82:227–232
Jayakumar R, Prabaharan M, Nair SV, Tamura H (2010b) Novel chitin and chitosan nanofibers in biomedical applications. Biotechnol Adv 28:142–150
Jeannie Tan ZY, Zhang XW (2011) Influence of chitosan on electrospun PVA nanofiber mat. Adv Mater Res 311–313:1763–1768
Jiang HL, Fang DF, Hsiao BS, Chu B, Chen WL (2004) Optimization and characterization of dextran membranes prepared by electrospinning. Biomacromolecules 5:326–333
Kang J, Kotaki M, Okubayashi S, Sukigara S (2010) Fabrication of electrospun eggshell membrane nanofibers by treatment with catechin. J Appl Polym Sci 117:2042–2049
Karim MR, Lee HW, Kim R et al (2009) Preparation and characterization of electrospun pullulan/montmorillonite nanofiber mats in aqueous solution. Carbohydr Polym 78:336–342
Kayaci F, Uyar T (2012) Encapsulation of vanillin/cyclodextrin inclusion complex in electrospun polyvinyl alcohol (PVA) nanowebs: prolonged shelf-life and high temperature stability of vanillin. Food Chem 133:641–649
Kayaci F, Ertas Y, Uyar T (2013a) Enhanced thermal stability of eugenol by cyclodextrin inclusion complex encapsulated in electrospun polymeric nanofibers. J Agric Food Chem 61:8156–8165
Kayaci F, Umu OCO, Tekinay T, Uyar T (2013b) Antibacterial electrospun poly(lactic acid) (PLA) nano fi brous webs incorporating triclosan/cyclodextrin inclusion complexes. J Agric Food Chem 61:3901–3908
Kiechel MA, Schauer CL (2013) Non-covalent crosslinkers for electrospun chitosan fibers. Carbohydr Polym 95:123–133
Kong L, Ziegler GR (2012) Role of molecular entanglements in starch fiber formation by electrospinning. Biomacromolecules 13:2247–2253
Kong L, Ziegler GR (2013) Quantitative relationship between electrospinning parameters and starch fiber diameter. Carbohydr Polym 92:1416–1422
Kong L, Ziegler GR (2014a) Rheological aspects in fabricating pullulan fibers by electro-wet-spinning. Food Hydrocoll 38:220–226
Kong L, Ziegler GR (2014b) Fabrication of pure starch fibers by electrospinning. Food Hydrocoll 36:20–25
Kong L, Ziegler GR (2014c) Formation of starch-guest inclusion complexes in electrospun starch fibers. Food Hydrocoll 38:211–219
Lachke A (2004) Xanthan—a versatile gum. Resonance 9:25–33

Laelorspoen N, Wongsasulak S, Yoovidhya T, Devahastin S (2014) Microencapsulation of Lactobacillus acidophilus in zein-alginate core-shell microcapsules via electrospraying. J Funct Foods 7:342–349

Lancuški A, Vasilyev G, Putaux JL, Zussman E (2015) Rheological properties and electrospinnability of high-amylose starch in formic acid. Biomacromolecules 16:2529–2536

Lancuški A, Abu Ammar A, Avrahami R et al (2017) Design of starch-formate compound fibers as encapsulation platform for biotherapeutics. Carbohydr Polym 158:68–76

Lee KY, Jeong L, Kang YOO et al (2009) Electrospinning of polysaccharides for regenerative medicine. Adv Drug Deliv Rev 61:1020–1032

Li L, Hsieh YL (2006) Chitosan bicomponent nanofibers and nanoporous fibers. Carbohydr Res 341:374–381

Li J, He A, Zheng J, Han CC (2006) Gelatin and gelatin—hyaluronic acid nanofibrous membranes produced by electrospinning of their aqueous solutions. Biomacromolecules 7:2243–2247

Li Y, Lim LT, Kakuda Y (2009) Electrospun zein fibers as carriers to stabilize (−)-epigallocatechin gallate. J Food Sci 74:C233–C240

Li J, Chotiko A, Narcisse DA, Sathivel S (2016a) Evaluation of alpha-tocopherol stability in soluble dietary fiber based nanofiber. LWT Food Sci Technol 68:485–490

Li H, Wang M, Williams GR, Wu J, Sun X, Lv Y, Zhu LM (2016b) Electrospun gelatin nanofibers loaded with vitamins A and E as antibacterial wound dressing materials. RSC Adv 6:50267–50277

Librán CM, Castro S, Lagaron JM (2016) Encapsulation by electrospray coating atomization of probiotic strains. Innovative Food Sci Emerg Technol 39:216–222

Lim LT (2015) Encapsulation of bioactive compounds using electrospinning and electrospraying technologies. In Nanotechnology and Functional Foods: Effective Delivery of Bioactive Ingredients, ed. CM Sabliov, H Chen, RY Yada, pp. 297–317. New York: Wiley.

Lindman B, Karlstrom G, Stigsson L (2010) On the mechanism of dissolution of cellulose. J Mol Liq 156:76–81

Lopez-Rubio A, Lagaron JM (2012) Whey protein capsules obtained through electrospraying for the encapsulation of bioactives. Innovative Food Sci Emerg Technol 13:200–206

López-Rubio A, Sanchez E, Wilkanowicz S et al (2012) Electrospinning as a useful technique for the encapsulation of living bifidobacteria in food hydrocolloids. Food Hydrocoll 28:159–167

Luo Y, Wang Q (2014) Recent development of chitosan-based polyelectrolyte complexes with natural polysaccharides for drug delivery. Int J Biol Macromol 64:353–367

Ma G, Liu Y, Fang D et al (2012) Hyaluronic acid/chitosan polyelectrolyte complexes nanofibers prepared by electrospinning. Mater Lett 74:78–80

McClements DJ (2015) Nanoparticle- and microparticle-based delivery systems—encapsulation, proteciton and release of active compounds. CRC Press, Boca Raton, pp 265–339

Mendes AC, Baran ET, Pereira RC et al (2012) Encapsulation and survival of a chondrocyte cell line within xanthan gum derivative. Macromol Biosci 12:350–359

Mendes AC, Gorzelanny C, Halter N et al (2016) Hybrid electrospun chitosan-phospholipids nanofibers for transdermal drug delivery. Int J Pharm 510:48–56

Mendes AC, Stephansen K, Chronakis IS (2017) Electrospinning of food proteins and polysaccharides. Food Hydrocoll 68:53–68

Meng Y, Cloutier S (2014) Gelatin and other proteins for microencapsulation. In: Gaonkar AG, Vasisht N, Khare AR (eds) Microencapsulation in the food industry. Academic Press, San Diego, pp 227–239

Mit-uppatham C, Nithitanakul M, Supaphol P (2004) Ultrafine electrospun polyamide-6 fibers: effect of solution conditions on morphology and average fiber diameter. Macromol Chem Phys 205:2327–2338

Moomand K, Lim LT (2014) Oxidative stability of encapsulated fish oil in electrospun zein fibres. Food Res Int 62:523–532

Moomand K, Lim LT (2015) Effects of solvent and n-3 rich fish oil on physicochemical properties of electrospun zein fibres. Food Hydrocoll 46:191–200

Neo YP, Ray S, Jin J, Gizdavic-Nikolaidis M, Nieuwoudt MK, Liu D (2013) Encapsulation food grade antioxidant in natural biopolymer by electrospinning technique: a physicochemical study based on zein-gallic acid system. Food Chem 136:1013–1021
Nie H, He A, Zheng J et al (2008) Effects of chain conformation and entanglement on the electrospinning of pure alginate. Biomacromolecules 9:1362–1365
Nieuwland M, Geerdink P, Brier P, van den Eijnden P, Henket JTM, Langelaan MLP, Stroeks N, van Deventer HC, Martin AH (2013) Food-grade electrospinning of proteins. Innovative Food Sci Emerg Technol 20:269–275
Ohkawa K, Cha D, Kim H et al (2004) Electrospinning of chitosan. Macromol Rapid Commun 25:1600–1605
Okutan N, Terzi P, Altay F (2014) Affecting parameters on electrospinning process and characterization of electrospun gelatin nanofibers. Food Hydrocoll 39:19–26
Pakravan M, Heuzey M-C, Ajji A (2012) Core-shell structured PEO-chitosan nanofibers by coaxial electrospinning. Biomacromolecules 13:412–421
Park WH, Jeong L, Il YD, Hudson S (2004) Effect of chitosan on morphology and conformation of electrospun silk fibroin nanofibers. Polymer (Guildf) 45:7151–7157
Pelayo A (2017) Production, characterization and oxidative stability of fish oil-loaded nanocapsules and yoghurt fortified with the nanocapsules. Master thesis, Technical University of Denmark
Pérez-Masiá R, Lagaron JM, López-Rubio A (2014a) Development and optimization of novel encapsulation structures of interest in functional foods through electrospraying. Food Bioprocess Technol 7:3236–3245
Pérez-Masiá R, Lagaron JM, Lopez-Rubio A (2014b) Morphology and stability of edible lycopene-containing micro- and nanocapsules produced through electrospraying and spray drying. Food Bioprocess Technol 8:459–470
Pérez-Masiá R, López-Nicolás R, Periago MJ et al (2015) Encapsulation of folic acid in food hydrocolloids through nanospray drying and electrospraying for nutraceutical applications. Food Chem 168:124–133
Pitigraisorn P, Srichaisupakit K, Wongpadungkiat N, Wongsasulak S (2017) Encapsulation of lactobacillus acidophilus in moist-heat-resistant multilayered microcapsules. J Food Eng 192:11–18
Reddy N, Yang Y (2015) Innovative biofibers from renewable resources. Springer, New York
Rezaei A, Nasirpour A, Fathi M (2015) Application of Cellulosic Nanofibers in Food Science Using Electrospinning and Its Potential Risk. Compr Rev Food Sci Food Saf 14:269–284
Ritcharoen W, Thaiying Y, Saejeng Y et al (2008) Electrospun dextran fibrous membranes. Cellulose 15:435–444
Rosalam S, England R (2006) Review of xanthan gum production from unmodified starches by Xanthomonas comprestris sp. Enzym Microb Technol 39:197–207
Sangsanoh P, Supaphol P (2006) Stability improvement of electrospun chitosan nanofibrous membranes in neutral or weak basic aqueous solutions. Biomacromolecules 7:2710–2714
Schiffman JD, Schauer CL (2007) One-step electrospinning of cross-linked chitosan fibers. Biomacromolecules 8:2665–2667
Shalumon KT, Anulekha KH, Girish CM et al (2010) Single step electrospinning of chitosan/poly(caprolactone) nanofibers using formic acid/acetone solvent mixture. Carbohydr Polym 80:414–420
Shekarforoush E, Mendes ACL, Baj V, Beeren SR, Chronakis IS (2017a) Electrospun phospholipid fibers as micro-encapsulation and antioxidant matrices. Molecules 22:1708
Shekarforoush E, Faralli A, Ndoni S et al (2017b) Electrospinning of xanthan polysaccharide. Macromol Mater Eng 201700067:1700067
Shukla R, Cheryan M (2001) Zein: the industrial protein from corn. Ind Crop Prod 13:171–192
Sobel R, Versic R, Gaonkar AG (2014) Introduction to microencapsulation and controlled delivery in foods. In: Gaonkar AG, Vasisht N, Khare AR (eds) Microencapsulation in the food industry. Academic Press, San Diego, pp 3–12

Songchotikunpan P, Tattiyakul J, Supaphol P (2008) Extraction and electrospinning of gelatin from fish skin. Int J Biol Macromol 42:247–255

Songsurang K, Praphairaksit N, Siraleartmukul K, Muangsin N (2011) Electrospray fabrication of doxorubicin-chitosan-tripolyphosphate nanoparticles for delivery of doxorubicin. Arch Pharm Res 34:583–592

Spano F, Massaro A (2012) Electrospun dextran-based nanofibers for biosensing and biomedical applications. Acad Res J 1:23–30

Sreekumar S, Lemke P, Moerschbacher BM et al (2017) Preparation and optimization of submicron chitosan capsules by water-based electrospraying for food and bioactive packaging applications. Food Addit Contam Part A Chem Anal Control Expo Risk Assess 34:1795–1806

Stephansen K, Chronakis IS, Jessen F (2014) Bioactive electrospun fish sarcoplasmic proteins as a drug delivery system. Colloids Surf B Biointerfaces 122:158–165

Stephansen K, García-Díaz M, Jessen F, Chronakis IS, Nielsen H (2015) Bioactive protein-based nanofibers interact with intestinal biological components resulting in transepithelial permeation of a therapeutic protein. Int J Pharm 495:58–66

Stephansen K, García-Díaz M, Jessen F, Chronakis IS, Nielsen HM (2016) Interactions between surfactants in solution and electrospun protein fibers: effects on release behavior and fiber properties. Mol Pharm 13:748–755

Stijnman AC, Bodnar I, Hans Tromp R (2011) Electrospinning of food-grade polysaccharides. Food Hydrocoll 25:1393–1398

Suárez G, Gutiérrez TJ (2017) Recent advances in the development of biodegadable films and foams from cassava starch. In: Klein C (ed) Handbook on cassava: production, potential uses and recent advances. Nova Science Publishers, New York, pp 297–312. ISBN: 978-1-53610-307-6

Sullivan ST, Tang C, Kennedy A, Talwar S, Khan SA (2014) Electrospinning and heat treatment of whey protein nanofibers. Food Hydrocoll 35:36–50

Sun K, Li ZH (2011) Preparations, properties and applications of chitosan based nanofibers fabricated by electrospinning. Express Polym Lett 5:342–361

Sun XB, Jia D, Kang WM et al (2013) Research on electrospinning process of pullulan nanofibers. Appl Mech Mater 268–270:198–201

Taepaiboon P, Rungsardthong U, Supaphol P (2007) Vitamin-loaded electrospun cellulose acetate nanofiber mats as transdermal and dermal therapeutic agents of vitamin A acid and vitamin E. Eur J Pharm Biopharm 67:387–397

Thien DVH, Hsiao SW, Ho MH (2012) Synthesis of electrosprayed chitosan nanoparticles for drug sustained release. Nano Life 2:1250003

Tomasula PM, Sousa AMM, Liou SC, Li R, Bonnaillie LM, Liu LS (2016) Short communication: electrospinning of casein/pullulan blends for food-grade application. J Dairy Sci 99:1837–1845

Torres-Giner S, Ocio MJ, Lagaron JM (2008) Development of active antimicrobial fiber-based chitosan polysaccharide nanostructures using electrospinning. Eng Life Sci 8:303–314

Torres-Giner S, Ocio MJ, Lagaron JM (2009) Novel antimicrobial ultrathin structures of zein/chitosan blends obtained by electrospinning. Carbohydr Polym 77:261–266

Torres-Giner S, Martinez-Abad A, Ocio MJ, Lagaron JM (2010) Stabilization of a nutraceutical omega-3 fatty acid by encapsulation in ultrathin electrosprayed zein prolamine. J Food Sci 75:N69–N79

Ungeheuer S, Bewersdorff H, Singh RP (1989) Turbulent drag effectiveness and shear stability of xanthan-gum-based graft copolymers. J Appl Polym Sci 37:2933–2948

Vega-Lugo AC, Lim LT (2009) Controlled release of allyl isothiocyanate using soy protein and poly(lactic acid) electrospun fibers. Food Res Int 42:933–940

Verdugo M, Lim LT, Rubilar M (2014) Electrospun protein concentrate fibers from microalgae residual biomass. J Polym Environ 22:373–383

Wang S, Bai J, Li C, Zhang J (2012) Functionalization of electrospun B-cyclodextrin/polyacrylonitrile (PAN) with silver nanoparticles: Broad-spectrum antibacterial property. Appl Surf Sci 261:499–503

Weiss J, Kanjanapongkul K, Wongsasulak S, Yoovidhya T (2012) Electrospun fibers: fabrication, functionalities and potential food industry applications. In: Huang Q (ed) Nanotechnology in the food, beverage and nutraceutical industries. Woodhead Publishing, Cambridge, pp 362–397

Wu X, Wang L, Yu H, Huang Y (2005) Effect of solvent on morphology of electrospinning ethyl cellulose fibers. J Appl Polym Sci 97:1292–1297
Xie JB, Hsieh YL (2003) Ultra-high surface fibrous membranes from electrospinning of natural proteins: casein and lipase enzyme. J Mater Sci 38:2125–2133
Xu W, Yang W, Yang Y (2009) Electrospun starch acetate nanofibers: development, properties, and potential application in drug delivery. Biotechnol Prog 25:1788–1795
Yang DZ, Li YN, Nie J (2007) Preparation of gelatin/PVA nanofibers and their potential application in controlled release of drugs. Carbohydr Polym 69:538–543
Yang H, Wen P, Feng K, Zong MH, Lou WY, Wu H (2017) Encapsulation of fish oil in a coaxial electrospun nanofibrous mat and its properties. RSC Adv 7:14939–14946
Zeleny J (1914) The electrical discharge from liquid points, and a hydrostatic method of measuring the electric intensity at their surfaces. Phys Rev 3:69–91
Zirnsak MA, Boger DV, Tirtaatmadja V (1999) Steady shear and dynamic rheological properties of xanthan gum solutions in viscous solvents. J Rheol (N Y N Y) 43:627

Chapter 18
Food Gel Emulsions: Structural Characteristics and Viscoelastic Behavior

Gabriel Lorenzo, Noemí Zaritzky, and Alicia Califano

Abstract If the continuous phase of an emulsion or foam is a semisolid system, these systems can be described as 'filled gels' or 'composite solids'. Gel emulsions are widely used in different industries like cosmetic, pharmaceutical, and food, among others. Typical examples are cheese, many desserts, sausages, low-fat mayonnaises and bakery products. The aggregation and cross-linking of protein and polysaccharides molecules into three-dimensional solid-like networks ('gels') is one of the most important mechanisms for developing microstructure with desirable textural attributes. Due to their elastic characteristics, oil droplets can be kept in suspension avoiding creaming. The structure and the rheological properties of gel emulsions are dependent on the nature of the interactions between the emulsifiers adsorbed on the surface of the droplets that fill the emulsion and the biopolymeric network formed in the aqueous phase. The present chapter deals with the viscoelastic behavior of o/w gel emulsions containing either polysaccharides or proteins in the aqueous phase. Two case studies are discussed, i.e., emulsions with low lipid content, stabilized with bovine gelatin of different molecular weights and heat-induced gel emulsions containing high acyl gellan gum. Small amplitude oscillatory shear tests (stress and frequency sweeps) and transient studies (creep-recovery) were performed over the different matrices and modeled to interpret the structural characteristics of the gel emulsions. The Broadened Baumgaertel-Schausberger-Winter spectrum was used to represent the linear viscoelastic behavior of the continuous phase and the emulsified system. Relaxation spectra were validated using creep experiments.

G. Lorenzo · N. Zaritzky (✉)
Centro de Investigación y Desarrollo en Criotecnología de Alimentos (CIDCA), Facultad de Cs. Exactas, UNLP—CICPBA—CONICET, La Plata, Argentina

Depto. Ingeniería Química, Facultad de Ingeniería, Universidad Nacional de La Plata (UNLP), La Plata, Argentina
e-mail: zaritzky@ing.unlp.edu.ar

A. Califano
Centro de Investigación y Desarrollo en Criotecnología de Alimentos (CIDCA), Facultad de Cs. Exactas, UNLP—CICPBA—CONICET, La Plata, Argentina

T. J. Gutiérrez (ed.), *Polymers for Food Applications*,
https://doi.org/10.1007/978-3-319-94625-2_18

Keywords Bovine gelatin · Creep and recovery tests · High acyl gellan gum · Small amplitude oscillatory shear tests

18.1 Introduction

Different perspectives can be adopted to study food structuring, from nanoscience and formulation engineering to the perspectives of soft matter physics and material science. The common denominator is the fact that food properties are determined by the distribution of the essential structural components during food preparation and subsequent rearrangements during storage (Dickinson 2012). To understand the underlying mechanisms, principles colloid science must be incorporated.

Simple emulsions consist of droplets of one liquid dispersed in a second immiscible liquid. As the droplet concentration increases, they exhibit a transition from a viscous fluid to an elastic solid at dispersed phase concentrations near the random close-packing concentration (Koenig et al. 2002). In principle, the type of emulsion is determined by the oil and water ratio, so that it is possible to obtain phase inversion by changing the ratio from oil-in-water (o/w) to water-in-oil (w/o) emulsions.

Emulsions are thermodynamically unstable systems that will tend towards the lowest possible energy level; normally, emulsions would separate into two stable phases: oil and water. Droplet stabilization can be achieved in different ways: steric stabilization, stabilization by solid particles, electrostatic stabilization, or by increasing the viscosity of the o/w emulsions by the addition of polymers such as starches, hydrocolloids, and proteins with gelling activity, which will retard the coalescence of aggregates and creaming of the oil droplets (Dickinson 2010).

If the continuous phase of an emulsion or foam is a semisolid, these systems can be described as 'filled gels' or 'composite solids'. Typical examples are cheese, many desserts, sausages, low-fat mayonnaises and bakery products. The aggregation and cross-linking of protein and polysaccharides molecules into three-dimensional solid-like networks ('gels') is one of the most important mechanisms for developing microstructure with desirable textural attributes. Several studies have found that the mechanical behavior of a filled gel is affected by the volume fraction, the shape, and the deformability of the filler (Manoi and Rizvi 2009). "Emulsion gel" designates a class of complex colloidal soft-solid material that exists as both an emulsion and a gel. They are widely used in cosmetic, food and pharmaceutical industry. Droplets suspended in a viscoelastic matrix may lead to the formation of solid like emulsions, and hence to materials with original textures. The viscoelastic properties of these materials depend not only on that of the gel matrix and on the droplet concentration but also on the type of the droplet-matrix interactions (Koenig et al. 2002). Due to their elastic characteristics, oil droplets can be kept in suspension avoiding creaming. They can be considered as complex mixtures of various components, in which particulate inclusions (fillers) are embedded in a polymeric gel matrix. Food emulsions are compositionally complex. Their droplets are stabilized to differing extents by proteins, small-molecule surfactants (emulsifiers), and, in certain cases, polysaccharides (hydrocolloids. In Dickinson's own words (2006) "in the presence of added polysaccharides or certain proteins, the kinetics of phase separation is controlled in the short/medium term by the rheological behavior of the interconnected oil droplet regions. That is, the gravitationally unstable liquid-like emulsion has become transformed into a stable gel-like emulsion containing trapped 'blobs' of hydrocolloid-structured water".

The rheological properties and the breakdown behavior of gels filled with emulsions droplets can be varied by changing the interactions between oil droplets and gel matrix, the oil content and the oil droplet size (Kim et al. 2001). Since the bipolymers are used to modify textural attributes, the study of their rheological behavior is essential as it is recognized that rheological properties play an important role in process design, evaluation and modeling. Furthermore, many of the sensory attributes of food emulsions are directly related to their rheological properties (e.g., creaminess, thickness, smoothness, spreadability, pourability, and flowability). The shelf-life of many food emulsions depends on the rheological characteristics of the component phases (e.g., the creaming of oil droplets depends on the viscosity of the aqueous phase). Information about the rheology of food products is used as an analytical tool to provide fundamental insights about the structural organization and interactions of the components within emulsions (McClements 1999). Thus, the correlation between microstructure information and rheology is useful to understand the macroscopic behavior in terms of the microstructure organization.

The viscoelastic properties of gel emulsions can be studied by measuring time-dependent rheological properties. In dynamic measurements, at low values of deformation the gels behave as an elastic solid, but at high values viscous flow processes occur. In transient measurements, a step function shear strain is applied to the gel and the shear stress is measured as a function of time.

To obtain an emulsion with good mechanical and stabilizing properties, it is necessary to consider many aspects and, consequently, the choice of the constituent components should be made carefully. The present chapter deals with the viscoelastic behavior of o/w gel emulsions containing either polysaccharides or proteins in the aqueous phase. Two case studies are discussed, i.e. emulsions with low lipid content, stabilized with bovine gelatin of different molecular weights (Lorenzo et al. 2011) and heat-induced gel emulsions containing high acyl gellan gum (Lorenzo et al. 2013).

18.2 Case Studies

18.2.1 Cold Gel Emulsions Stabilized with Bovine Gelatin

In protein-based systems two structural arrangements are distinguished: (a) the protein-stabilized emulsion gel, and (b) the emulsion-filled protein gel. The protein-stabilized emulsion gel is a type of particulate gel, and the properties of the network of aggregated emulsion droplets mainly determine its rheological properties. The

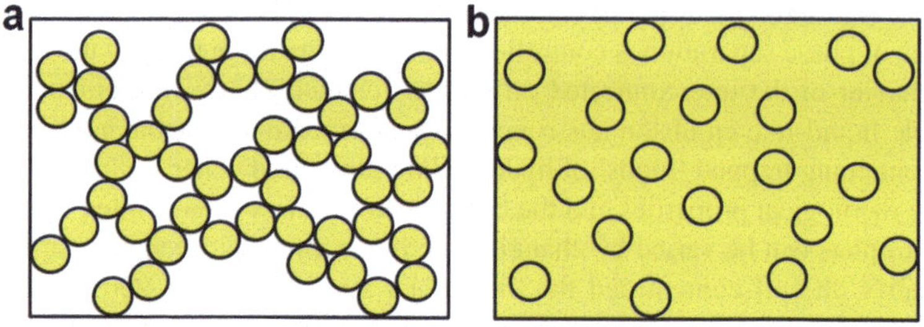

Fig. 18.1 Schematic presentation of two idealized structured arrangements: (**a**) protein-stabilized emulsion gel, (**b**) emulsion-filled protein gel

emulsion-filled protein gel is a protein gel matrix within which emulsion droplets are embedded and its solid-like rheological properties are determined predominantly by the network properties of the spatially continuous matrix (Fig. 18.1).

These two structural arrangements are idealized representations, the structural state of a particular protein-based emulsion gel is a hybrid network made up of a combination of cross-linked biopolymer molecules and partially aggregated droplets (Reiffers-Magnani et al. 1999).

The use of proteins in emulsified systems presents a growing trend in order to replace synthetic emulsifiers (Dickinson and Lopez 2001; Garti 1999). Proteins, and particularly gelatins, can be used as emulsifiers in foods because of their ability to facilitate the formation of an emulsion, improve the stability, and produce desirable physicochemical properties in oil-in-water (o/w) emulsions (Lobo 2002; Surh et al. 2006).

Gelatin is a relatively high molecular weight protein obtained by partial hydrolysis of animal collagen (Keenan 1998). Collagen is present in bovine hides and pig and fish skins. It acts as extracellular, structural protein in bone, tendon, skin, and the connective tissue of various organs. The characteristic features of collagen are the exceptional amino acid composition (33% glycine and 22% proline) and structure: the (rigid) triple extended helix. The triple-helix structure is characterized by three extended left-handed polyproline II-like helical chains that are supercoiled into a right-handed triple helix. The three chains are staggered by one residue with respect to each other, and are linked through interchain hydrogen bonds. The triple-helical conformation is associated with a distinctive amino acid sequence with glycine as every third residue and a high content of imino acids (de Wolf 2003; Harrington and Rao 1970). Type A gelatin is produced by acid processing of collagenus raw material; type B is produced by alkaline or lime processing.

Among commercial hydrocolloids used in the food industry, gelatin has been regarded as special and unique, serving multiple functions with a wide range of applications in various industries. Uses of gelatin are based on its combination of properties, reversible gel-sol transition of aqueous solution; ability to act as a protective colloid; viscosity of warm aqueous solutions; and insolubility in cold water but complete solubility in hot water.

Table 18.1 Main specifications of the tested gelatin samples

Gelatin sample	A	B	C
Weight-average molecular weight (Mw, kDa)	60	80	120
Bloom (g)	182	217	265
Moisture (g/100 g)	10.4	10.0	9.8
Ashes (g/100 g)	≤2.0	≤2.0	≤2.0
pH of a 1 g/100 g solution	5.5	5.8	5.3
Granulometry (sieve size, mm)	0.6	0.6	0.6

The main constituents of gelatin are large and complex polypeptide molecules of the same amino acid composition as the parent collagen, covering a broad molecular weight distribution range (20,000–250,000).

A solid-like emulsion gel may be generated from a stable liquid-like emulsion by gelling the continuous phase (Fig. 18.1b) and/or aggregating the emulsion droplets (Fig. 18.1a). For an emulsion kept at a constant low or moderate oil content, the transformation from liquid state to soft solid is normally brought about by some kind of processing operation (Dickinson 2012). The most popular methods of protein gelation are heat treatment, acidification, and enzyme treatment (van Vliet et al. 2004). However, a fourth possibility involves the gelification of the emulsion induced by shear as a consequence of the high energy involved during the emulsification process. In all cases protein structure is first altered (denatured) and secondly gel formation is due to a partial renaturation of the collagen molecule ("collagen flod"). Those parts of gelatin that are rich in proline and hydroxyproline regain some of their structure, following which they can interact. When many molecules are involved, a three dimensional structure is produced which is responsible for the gel at low temperatures. Usually, the rigidity of a gel increases on further cooling or, sometimes, on standing. In the latter case, the gel loses water and shrinks in a syneresis process (Shakuntala and Manay 2001).

18.2.1.1 Cold Formation of Protein-Based Emulsion Gels

Low-in-fat o/w emulsions (15 g/100 g) were prepared by adding sunflower oil to aqueous phases containing food grade bovine gelatin with different molecular weights (60 kDa, 80 kDa, and 120 kDa, PB Leiner, Argentina). Main specifications of these products were supplied by the producer and summarized in Table 18.1. Final concentration of gelatins in the studied emulsions was 2.55 g/100 g. Continuous phases were prepared by dispersing gelatin in a buffered aqueous solution (pH = 7) at 25 °C and stirred overnight to ensure a complete hydration of the protein.

The procedure to prepare o/w protein based gel emulsions involved two steps. In the first step the whole sample was pre-emulsified for 4 min at 11,500 rpm with the Ultra-Turrax T25 (IKA Labortechnik, Germany) homogenizer (rotor stator principle), equipped with a dispersing tool (520-25-NK196). In the second stage the pre-emulsified system was passed through a high pressure valve homogenizer (Stansted

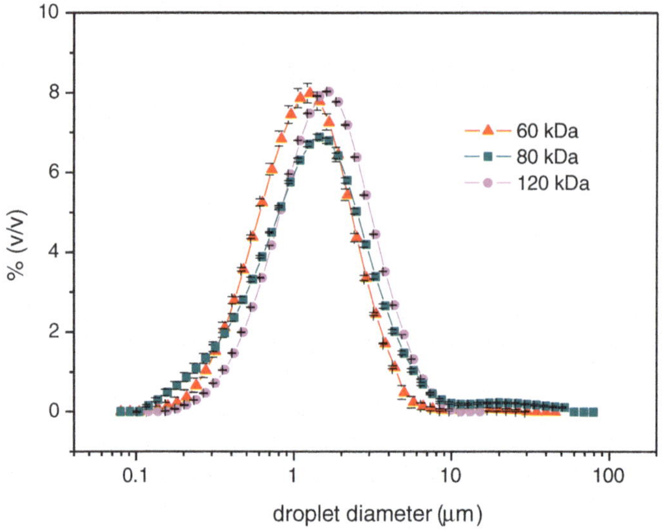

Fig. 18.2 Droplet size distribution of the gel-like emulsions stabilized with gelatin of 60 (red triangles), 80 (green squares), and 120 (pink circles) kDa. Error bars denote standard errors of the means

Fluid Power FPG 7400, Essex, UK). The pressure was set at 40 MPa and 4 MPa for the first and second valve, respectively, and the emulsion was recycled four times through the homogenizer to achieve a monodispersed system. The temperature during the high pressure homogenization was monitored to ensure that the emulsions never exceed 30 °C.

18.2.1.2 Droplet Size Distribution and Emulsion Stability

Oil droplet size distributions of model emulsions were determined by static light scattering using a Mastersizer 2000 (Malvern Instruments Ltd., Malvern, Worcester, UK); all the evaluated emulsions presented monomodal distributions as can be seen in Fig. 18.2. Values of the Sauter mean diameter, $d_{3,2}$, which is inversely proportional to the specific surface area of droplets, were obtained: 0.972 µm, 0.980 µm, and 1.289 µm for Mw 60 kDa, 80 kDa, and 120 kDa, respectively. Low polydispersity for all emulsions were observed as a result of the several passes through the high pressure valve homogenizer (standard deviations between 0.013 µm and 0.023 µm).

Stability analysis conducted in 50 mL glass graduated cylinders and quiescently stored at 20 °C showed that 60 kDa emulsions exhibited an incipient interface on the eighth day of storage, while the 80 kDa emulsions did not destabilize until the 20th day. On the contrary, emulsions formulated with 120 kDa gelatin (with greater Sauter diameter), remained stable for more than 6 weeks. This result clearly

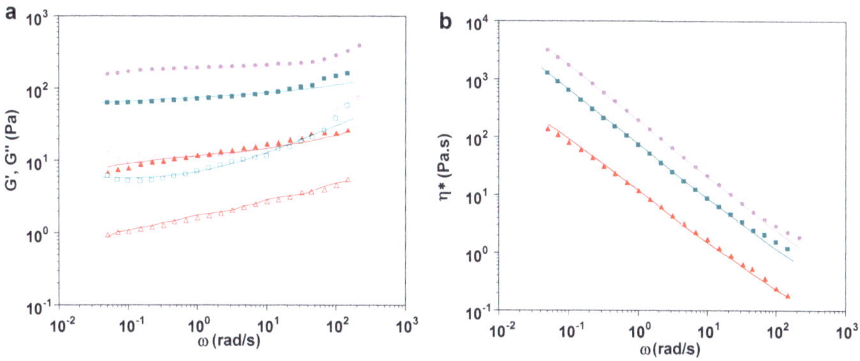

Fig. 18.3 Effect of the weight-average molecular weight of the gelatin used as stabilizer on the linear viscoelastic parameters. (**a**) Storage (G′; filled symbols) and loss (G″; empty symbols) moduli for of 60 (filled red triangles, open red triangles), 80 (filled green squares, open green squares), and 120 (filled pink circles, open pink circles) kDa. Optimized fitting with BSW generalized (continuous line), (**b**) complex viscosity (η^*), of 60 (filled red triangles), 80 (filled green squares), and 120 (filled pink circles) kDa. Optimized model according to Friedrich and Heymann (1988) (straight lines)

evidenced that the emulsion stability is related to the protein molecular weight rather than droplet diameter. When gelatin is composed of relatively small molecules (Mw 60 kDa), emulsions destabilize quickly. As Mw increases, stability also increases, because the steric protection provided by the interfacial film between oil droplets and the bulk of the aqueous continuous phase (Muller and Hermel 1994).

18.2.1.3 Effect of Molecular Weight of Gelatin on the Rheological Behavior of Emulsions

Dynamic oscillatory shear tests were performed in a controlled stress rheometer (Haake RS600, Thermo Fisher Sc., Germany) at 25 °C. The linear viscoelastic range (LVR) was determined for all samples through stress sweep test at a fixed frequency of 1 Hz.

Results of the dynamic measurements were expressed in terms of the elastic modulus (G′) and loss modulus (G″) as a function of the angular frequency (ω).The curves were qualitatively similar for all the formulations assayed. G′ was always greater than G″ in the frequency range measured and the increase of the two moduli with frequency was small. As G′ >> G″, the material exhibited a solid behavior (i.e. deformation in the linear range will be essentially elastic or recoverable). The slight dependence of the storage modulus on the oscillation frequency is known as the "plateau region". The plateau region is an intermediate zone of the mechanical spectra between the "terminal" and the "transition" zones (Ferry 1980). It is characterized by a decrease in the slope of both moduli (lower than 1) and a possible minimum in the loss modulus (G″). Figure 18.3a shows the frequency sweeps curves for emulsions stabilized with 60, 80 and 120 kDa gelatin. As the molecular weight of the gelatin increased, G′ showed higher values leading to more stable systems.

Spectra show the characteristic behavior of viscoelastic solids where the cross-links between macromolecules are not permanent but a dynamic equilibrium between formation and rupture of intermolecular interactions can exist contributing to the preservation of the structure during long observation times. In the range of small deformations, polymeric materials are expected to be characterized by a unique relaxation time spectrum, H(λ). Since this spectrum cannot be measured directly, there have been developed several theories to predict H(λ) from observable material functions such as the storage modulus, G′(ω),and the loss modulus, G″(ω) (Mours and Winter 2000).

The following equations relate the dynamic moduli with H(λ):

$$G'(\omega) = G_e + \int_0^\infty H(\lambda) \frac{\omega^2\lambda^2}{1+\omega^2\lambda^2} \frac{d\lambda}{\lambda} \tag{18.1}$$

$$G''(\omega) = \int_0^\infty H(\lambda) \frac{\omega\lambda}{1+\omega^2\lambda^2} \frac{d\lambda}{\lambda} \tag{18.2}$$

Using an appropriate representation of the relaxation time spectrum, it is possible to model the dynamic moduli. Based on a detailed analysis of dynamic mechanical data of linear model polymers, Baumgaertel and Winter (1992) proposed a specific form of the relaxation time spectrum for broadly distributed linear flexible polymers (Baumgaertel-Schausberger-Winter generalized spectrum or BSW spectrum). The spectrum has the following form:

$$H(\lambda) = G_N^0 \left[H_g \left(\frac{\lambda}{\lambda_0}\right)^{-n_0} + n_e \left(\frac{\lambda}{\lambda_e}\right)^{n_e} \right] \exp\left(-\left(\frac{\lambda}{\lambda_{\max}}\right)^{\beta}\right) \tag{18.3}$$

In this empirical model, G_N^0 is the plateau modulus, n_e and n_0 are the slopes of the spectrum in the entanglement and high frequency glass transition regimes respectively, H_g is the glass-transition front factor, λ_e the relaxation time corresponding to polymer chains with entanglement molar mass. The exponent β controls the sharpness of the cut-off of the spectrum, λ_{max} is the longest relaxation time and λ_0 is the crossover time to the glass transition.

The application of this theory to gel like emulsions stabilized with gelatin was made using the IRIS Rheo-Hub software (Winter and Mours 2006), and its satisfactory fitting to dynamic data is shown in Fig. 18.3a. Table 18.2 shows the predicted parameters of the BSW model. The displacement of λ_e to shorter relaxation times stands for an increasing molecular mobility. The spacing between the crossover time to the glass transition (λ_0) and the terminal relaxation time (λ_e) is a measure of the width of the plateau zone on the logarithmic time or frequency scale. It showed a significant increment with the molecular weight of the gelatin as was reported for other polymers (Ferry 1980).

Table 18.2 Predicted parameters of the Baumgaertel e Schausberger e Winter model for gel-like emulsions stabilized with gelatin of different molecular weights

M_W (kDa)	60	80	120
G_N^0 (Pa)	15.4	41.0	135
λ_e (s)	9.10	40.0	80.0
λ_0 (s)	3.6 10^{-3}	1.7 10^{-2}	1.7 10^{-2}
λ_{max} (s)	56.0	115	114
n_e	0.14	0.20	0.20
n_0	0.27	0.41	0.64
η_0 (Pa.s)	3.20 10^5	9.24 10^6	2.51 10^8

Physically, the response times are discrete and the continuous representation is a convenient mathematical device to facilitate the handling of the distributions of response times (Tschoegl 1997). The time dependence of a material is thus revealed in a finite discrete set of response times and their associated spectral strength $\{\lambda_I, G_i\}$. The relation between the continuous and the discrete spectra was calculated according to (Mours and Winter 2000):

$$G_i = H(\lambda_i)\ln\frac{\lambda_i}{\lambda_{i+1}} \tag{18.4}$$

Using the discrete spectrum obtained from the BSW model for the three tested emulsions it was possible to calculate other viscoelastic properties of interest such as zero-shear viscosity, η_0 (Jackson et al. 1994)

$$\eta_0 = \int_0^{\lambda_{max}} H(\lambda)d\lambda = \sum_{i=1}^{N} G_i\lambda_i \tag{18.5}$$

The zero-shear viscosity was related to the polymer molecular weight following a power law dependence as was previously express by Izuka et al. (1992). From the experimental data (Lorenzo et al. 2011) it was found that the incidence of Mw on the viscosity of gel like emulsions stabilized with gelatin was $\eta_0 \propto Mw^{9.2}$. Friedrich and Heymann (1988) demonstrated that in the linear viscoelastic regime, in the high frequency range or near the gel point, G′ and G″ are given by:

$$G'(\omega) = G_{\infty,\alpha} + \sqrt{\frac{2}{\pi}}S_\alpha^* \cos\left(\frac{\pi}{2}\alpha\right)\omega^\alpha \tag{18.6}$$

$$G''(\omega) = \sqrt{\frac{2}{\pi}}S_\alpha^* \sin\left(\frac{\pi}{2}\alpha\right)\omega^\alpha \tag{18.7}$$

where α is the order of the relaxation function, $G_{\infty,\alpha}$ is the equilibrium shear modulus and S_α^* is a material parameter related to the material strength. At the gel state, the complex viscosity (G^*) could be expressed as (Lorenzo et al. 2008):

$$\eta^* = \frac{G^*}{\omega} = \frac{\sqrt{G'^2 + G''^2}}{\omega} \approx A_\alpha \omega^{(\alpha-1)} \tag{18.8}$$

where A_α measures the "strength" of the cross-linking protein network in simple shear $\left(A_\alpha = \left(\sqrt{2/\pi}\right) S_\alpha^*\right)$.

The fitting was satisfactory in all the cases as is shown in Fig. 18.3b. The order of the relaxation function (α) did not present significant differences ($P > 0.05$) between the studied emulsions. The α values were lower than 0.1 for all formulations indicating the pronounced elastic character of the emulsions, typically observed in gel-like samples (Doublier et al. 1992; Steffe 1996).

As was expected from the results in Fig. 18.3a, η^* increased as the molecular weight of gelatin increased. Thus, A_α presented values of 10.8, 77.4 and 190.8 Pa.s$^{(2-\alpha)}$ for emulsions containing gelatin of 60, 80, and 120 kDa, respectively. This marked increment ($P < 0.05$) of A_α with the molecular weight could explain the increasing stability observed. In this kind of dispersions the thickness of the interfacial film was found to be a critical variable in stability. When the polymeric chains of the protein become larger (i.e. increment of the average molecular weight), the layer between droplets and the aqueous continuous phase become thicker and that is reflected in both a greater stability and a stronger polymeric network (large A_α value).

All the different analyses of the dynamic oscillatory experiments led to the conclusion that the complex viscosity of the system and its elastic characteristics increased as the molecular weight of the gelatin increased; changes in the material parameters G_0, λ_e, η_0, and A_α indicated that there was a positive correlation between the degree of crosslinking and the molecular weight of the protein.

18.2.1.4 Influence of the Storage Time on the Emulsions Rheology

Considering the higher stability exhibited by the emulsions containing gelatin of 120 kDa, this system was chosen to evaluate rheological evolution during storage time. Samples were also tested using small amplitude oscillatory shear analysis.

G' and G'' showed a marked increase with storage time, exhibiting an independent behavior of the storage modulus with frequency in all the curves (Fig. 18.4).

Taking into account that the linear viscoelastic functions always showed qualitatively similar frequency dependence, an empirical time-concentration superposition method has been applied, using the plateau modulus, G_N^0, as the normalization factor. The plateau modulus (G_N^0) is a viscoelastic parameter defined for polymers as the extrapolation of the entanglement contribution to the viscoelastic functions at high frequencies. This parameter can be considered as a characteristic parameter of this region and may be easily estimated from the minimum in the loss tangent ($\tan \delta = G''/G'$) as follows (Bais et al. 2005):

$$G_N^0 = \left[G'\right]_{\tan\delta \to \min} \tag{18.9}$$

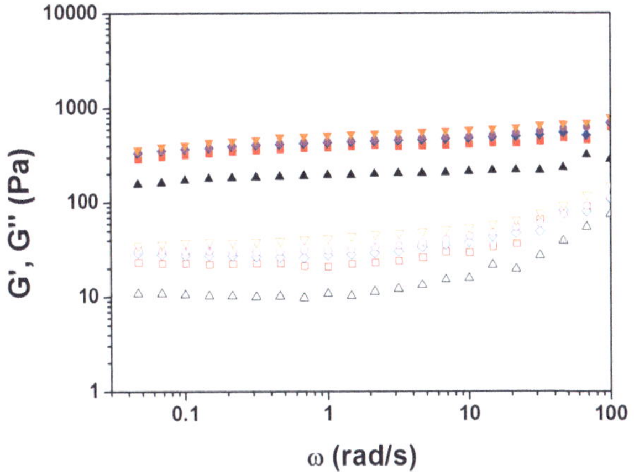

Fig. 18.4 Effect of storage time on the elastic (filled symbols) and viscous (voided symbols) moduli of emulsions formulated with aqueous phases containing 120 kDa gelatin: 1 (filled black triangles, open black triangles), 7 (filled red squares, open red squares), 14 (filled blue diamonds, open blue diamonds), 28 (filled pink circles, open pink circles), and 42 (filled inverted orange triangles, open inverted orange triangles) days of storage

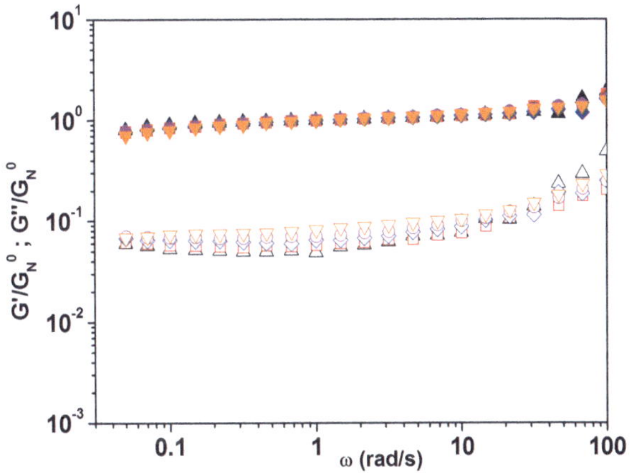

Fig. 18.5 Normalized dynamic master curves of the elastic (G' / G_N^0 , filled symbols) and viscous (G'' / G_N^0 , voided symbols) moduli for emulsions stabilized with 120 kDa gelatin at different storage times. Normalization factors (G_N^0) for each day: (day 1, 193.6 Pa); (day 7, 344.1 Pa); (day 14, 433.0 Pa); (day 28, 441.8 Pa); (day 42, 496.1 Pa)

Figure 18.5 shows the normalized dynamic master curves of the storage (G' / G_N^0) and the loss (G'' / G_N^0) moduli for emulsions stabilized with 120 kDa gelatin at different storage times.

There is quite enough evidence that gelatin gels obey the theory of rubber elasticity (Mitchell 1976). This theory has been very successful in explaining the properties of polymer networks. The change in free energy when a rubbery network is deformed

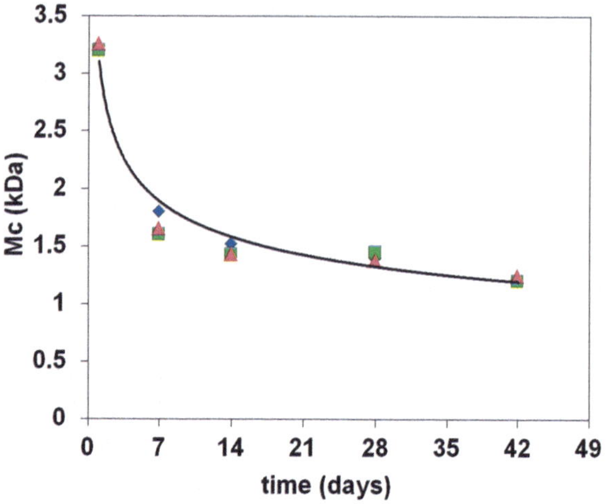

Fig. 18.6 Effect of storage time on the average molecular weight of the portion of chain between two consecutive cross-links (Mc) according to the rubber elasticity theory for emulsions stabilized with 120 kDa gelatin. Different symbols (green squares, pink triangles, blue diamonds) corresponds to different replicates

is made up of the sum of the entropy changes of the individual network chains (chains in joining adjacent crosslinks). By referring to the classical theory of rubber elasticity, Segeren et al. (1974), as well as Mitchell and Blanshard (1976), were able to estimate the number-average molecular mass (M_c) of the portion of the chain between two consecutive cross-links by using the equation valid for a network of Gaussian chains (Flory 1953; Moresi et al. 2004; Rao 2007):

$$G' = \frac{cRT}{M_c} \tag{18.10}$$

where c is the concentration of the polymer in the network (in g/m^3), T is the absolute temperature, and R, the ideal gas-law constant (8.31 J/mol K). The rubber elasticity theory was applied to the studied system assuming that the structure of the emulsions is given by a gelatin network where the oil droplets act as inactive fillers retained by this matrix. Within the plateau region, G_N^0 is the value of the elastic modulus G′. Considering this condition in Eq. (18.10), the effect of storage time on the average molecular weight of the portion of chain between two consecutive cross-links is shown in Fig. 18.6.

This result is reasonable when the gelatin tendency to collagen regeneration is taken into account. Considering that the temperature of the samples was maintained below 30 °C throughout the entire preparation process, emulsion gelification was shear induced as a consequence of the high energy involved during the emulsification.

The rearrangements that occurred during the gelification of gelatin were a consequence of both chemical and physical cross-links. Below the gel point temperature, gelatin chains reverted partially to the tropocollagenic triple helix, from disordered to rather ordered states. This phenomenon occurs usually at around 28 °C, where the coil to helix transition is observed (the transition temperature depends on species sources and protein concentration). The result is that parts of either three chains or two chains in a physical "hairpin" bond type are stabilized mainly through hydrogen bonds (Ledward 1986; Olivares et al. 2006), forming thus a basic physical network coupled to the covalent one.

During storage of the emulsions this slow reforming of the collagen molecules was still occurring, where the growth of junction zones was more important in relation to nucleation (Bot et al. 1996; Normand and Ravey 1997). This was a consequence of the fact that the small junction zones formed at short maturation times, started to growth at higher times introducing an harmonicity in the links (Ottone and Deiber 2005). Due to steric restraints imposed on the chains being held within the gel network it seems more reasonable that the growth of existing junctions will predominate over the formation of new ones. The decrease in the number-average molecular mass (M_c) of the portion of the chain between two consecutive cross-links during storage time could be related to these structural changes occurred in the emulsions, which induced an increment of the elasticity modulus based on a tighter network which expelled the droplets leading to its final destabilization.

18.2.2 Use of Gellan Gum to Produce Gel Emulsions

Polysaccharides are complex carbohydrates composed of long chains of monosaccharide units bound together by glycosidic linkages, arranged in chains. The cyclic monomer imparts considerable stiffness to polysaccharide chains, not only because of the inherent rigidity of the monomer unit, but also because its bulkiness results in a limited number of possible spatial arrangements of adjacent monomer residues. The flexibility of polysaccharide chains depends on the ease of rotation around the anomeric links. An exception is provided by (1,6)-linkages that have three-bond glycosidic linkages rather than the usual two; the extra bond allows much greater conformational freedom (Gidley and Nishinari 2009). Certain structural characteristics such as chain conformation and intermolecular associations will influence the physicochemical properties of polysaccharides. Interactions with the aqueous solvent may determine the preferred conformation by disrupting intramolecular hydrogen bonding (Kirschner and Woods 2001).

At low polysaccharide content few polymer chains are involved and only isolated n-fold helical arrangements are formed, and the rheology of these systems is essentially ruled by chain stiffness as in the cases of xanthan, rhamsan, welan, and various glucans (Grassi et al. 1996; Lapasin and Pricl 1995; Morris et al. 1996). At sufficiently high polymer concentration, chain–chain associations often lead to the formation of gel microdomains or span the whole system, with the consequent development of a three-dimensional network. In the latter case, the rheological properties sensibly differ from those of concentrated polymer solutions, where temporary networks are formed by topological interchain interactions (Bais et al. 2005).

Polysaccharides are frequently employed to stabilize aqueous suspensions or o/w emulsions (Manca et al. 2001; Whistler 1993). There are different mechanisms by which these polymers can stabilize an emulsion. The formation of an extended hydrogel network reflects into high viscosity of the continuous phase at low shear, thus slowing down the droplet motion (Lapasin and Pricl 1995, Manca et al. 2001, Whistler 1993); such a polymeric structure surrounds the oil droplets, ensuring effective steric hindrance of their coalescence (McClements 1999; Whistler 1993). Another contribution to the stabilization is provided by nonadsorbing depletion mechanism, due to the pronounced hydrophilicity, low flexibility, and low surface activity of these polymers (Garti et al. 1999; McClements 1999). Finally, due to the presence of some impurities, such as hydrophobic groups or proteinic moieties, an additional stabilizing effect (Garti et al. 1999) can derive from the formation of a viscoelastic adsorbed layer (Wilde 2000).

Gellan (Sworn 2000) is an extracellular anionic polysaccharide produced by the bacterium *Sphingomonas elodea* (ATCC31461) formerly known as *Auromonas elodea* or *Pseudomonas elodea*) on aerobic fermentation. It has a linear tetrasaccharide repeating sequence of: →3) -β-D-glucose - (1→4) - β-D-glucuronic acid - (1→4) - β-D-glucose - (1→4) -α -L-rhamnose - (1→(Jansson et al. 1983; O'Neill et al. 1983). The native polysaccharide, as biosynthesized, has an L-glyceryl substituent on O2 of the 3-linked glucose and about half the repeat units have an acetyl group on O6 of the same residue (Kuo et al. 1986). Its industrial usefulness is due to its various functional properties. This gelling agent has the ability to form a variety of gel textures depending on combinations of polymer concentration, degree of substitution, type of cation, ionic strength, and temperature (Goh et al. 2006; Ikeda et al. 2004). Gellan gum has been a subject of interest since its discovery in 1980 due to its ability to form transparent gels even at low concentrations (Rodríguez-Hernández et al. 2003) its thermal stability, and its high efficiency as gelling agent. Gellan gum can differ by the proportion of units with acyl substituents attached to the glucose molecule (degree of acylation) and the distribution of these acyl groups along the polysaccharide chain. Commercial samples are usually described as exhibiting a 'high' or 'low' degree of acylation. In low acyl gellan gum the acetyl groups were removed by alkaline treatment. The amount of acyl substituents strongly affect the properties of systems that contain this polysaccharide, in particular their rheological properties (Vilela and Lopes da Cunha 2016).

Traditional gelling agents such as agarose and carrageenan show reduced gelling capacity at low pH, whereas gellan gum can form strengthened gels (Moritaka et al. 1995; Yamamoto and Cunha 2007). These characteristics make emulsion-filled gels one of the potential fields of application of gellan gum.

The structure and the rheological properties of heat-set emulsion gels are dependent on the nature of the interactions between the emulsifiers adsorbed on the surface of the droplets that fill the emulsion and the biopolymeric network formed in the aqueous phase. Based on these interactions, mainly two types of filler particles can be distinguished: active fillers and inactive fillers (Chen and Dickinson 1999).

In the former case, a strong interaction exists between individual filler particles and the gel matrix; the elastic modulus of the gel at small deformation increases with the increase of the filler volume fraction. In the latter case, with little or no interaction between filler particles and gel matrix, the modulus decreases with increasing the volume fraction of the filler (Anton et al. 2001). On this basis, it is suggested that a gel matrix containing such non-interacting particles should behave at small deformations as it were filled with particles having the mechanical properties of a low viscosity liquid like water. Increasing the volume fraction of the filler two situations may be distinguished: if the filler is inactive the modulus of the emulsion gel decreases while an increase of the modulus occurs in the case of active fillers. (Chen and Dickinson 1999)

For food emulsion gels, the precise role of filler particles is not always easy to predict due to the complex composition and unknown properties of the monolayer. Thus, basic knowledge relevant for the physical properties of emulsion filled gels needs to be developed. This includes aspects such as rheological and stability properties of the emulsion filled gel as a whole, as well as the underlying properties of the gelled polymer network (Kokini and van Aken 2006).

The relaxation of polymeric and complex materials reveals the existence of a broad distribution of relaxation times. Knowledge of an expression for this relaxation time spectrum is advantageous because of two reasons: first, it provides analytical expressions for the material functions which are quantitative (as G', G'' data would yield) and, second, it reduces the description of the linear viscoelastic behavior of any complex material to the determination of only a few material-specific parameters. The resulting pattern is expected to be universally valid in the linear region (Baumgaertel and Winter 1989).

18.2.2.1 Rheological Analysis of High and Low Acyl Gellan Gels

Gellan gels were prepared in acetate buffer (pH 4.5) at 2. Three gellan concentrations were studied: 0.1 g/100 g, 0.3 g/100 g, and 0.5 g/100 g for low acyl and high acyl gellan. To ensure gel formation, samples were heated up to 90 °C at a constant rate (2 °C/min), then 5 mM $CaCl_2$ were added followed by rapid cooling down to 20 °C (20 °C/min).

Mechanical spectrum of gellan gum solutions subjected to the gelation process described above as well as gel emulsions were obtained in order to compare the viscoelastic behavior of emulsions and continuous phases; both low acyl and high acyl gellan gums were studied.

As expected rising hydrocolloid concentration produced a significant increase in G' and G'' for both types of gellan solutions. However depending on the degree of substitution of the hydrocolloid, the systems showed qualitatively different behaviors. Figure 18.7a shows the gels with low acyl gellan gum, where regardless of the content of hydrocolloid all the systems exhibited a "true gel" behavior. G' and G'' curves were almost parallel independent of the oscillation frequency and the storage modulus was practically a decade higher than the loss modulus over the entire frequency range. This behavior is in agreement with low acyl gellan gum gels, typical of rigid systems with hard and brittle structure previously described by several

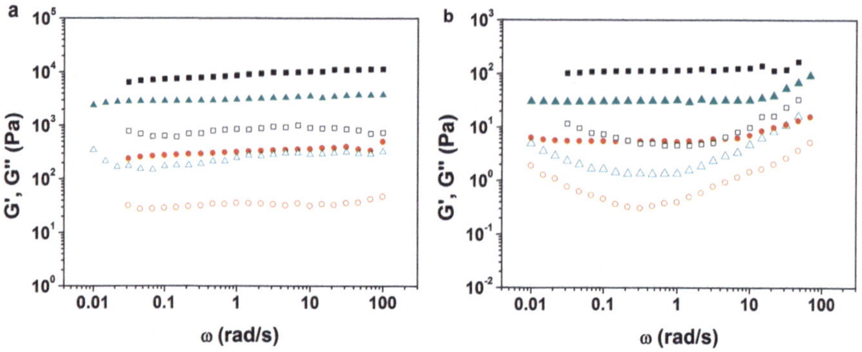

Fig. 18.7 Storage (G′; filled symbols) and loss (G″; open symbols) moduli as a function of frequency (ω) for gels obtained with (**a**) low acyl and (**b**) high acyl gellan gum and different hydrocolloid concentration. (filled red circles, open red circles) 0.1, (filled blue triangles, open blue triangles) 0.3, (filled black squares, open black squares), and 0.5 g/100 g

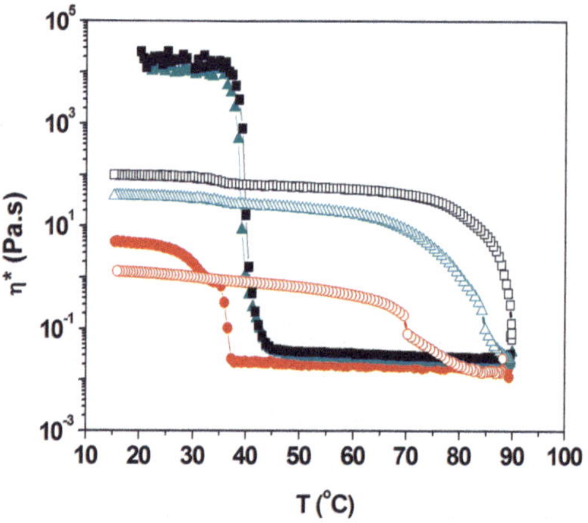

Fig. 18.8 Temperature dependence of the complex viscosity (η^*) during cooling for solutions containing high (open symbols) and low (filled symbols) acyl gellan gum at different concentrations: (filled red circles, open red circles) 0.1, (filled blue triangles, open blue triangles) 0.3, (filled black squares, open black squares), and 0.5 g/100 g

authors (Huang et al. 2003; Rodríguez-Hernández et al. 2003). In contrast, aqueous solutions of high acyl gellan gum showed lower values of both moduli and a minimum on the viscous modulus (G″) around 10^{-1}–10^{0} rad/s (Fig. 18.7b), because as the degree of acylation increases the matrix became more soft and flexible (Sworn 2000).

Temperature sweeps (90 °C to 15 °C at 0.5 °C/min) were performed on gellan gum dispersions within the LVR to determine the gel point. Figure 18.8 shows the complex viscosity as a function of temperature for all the tested continuous phases. During the cooling of the aqueous dispersions it was observed that the complex viscosity ($\eta^* = ((G'^2 + G''^2)^{1/2}/\omega)$) remained almost invariant down to a certain temperature value (T_{gel}, sol-gel transition) from which there was an abrupt increase in all dispersions considered. Both T_{gel} and the gel strength increased with hydrocolloid content in solution (Fig. 18.8).

Fig. 18.9 O/W gel emulsions (20 g oil/100 g) using 0.3 g/100 g of high acyl (left) or low acyl (right) gellan gum

The proposed gelation mechanism of gellan gum is based on the *domain model* (Morris et al. 1980; Sworn 2000). As a hot solution cools gellan gum undergoes a disorder–order transition. This transition has been attributed to a coil-helix transition (Grasdalen and Smidsrod 1987). In the case of low acyl gellan gum, gel promoting cations such as Na^{+} and Ca^{+2} promote aggregation of the gellan double helices to form a three-dimensional network. The acyl substituents have a profound effect on the structure of the gellan gum gels. The high acyl gellan gum undergoes a similar disorder to order transition as the solution is cooled, but further aggregation of the helices is limited by the presence of the acetyl group (Morris et al. 1996). The double helix structure of high acyl gellan is inherently more stable than that of the deacylated material, but it has little, if any, capacity for cation-mediated aggregation. The subsequent gels are, therefore, soft and elastic.

18.2.2.2 Stability of Gel Emulsions with High and Low Acyl Gellan Gum

To study the stability of gel emulsions appropriate quantities of sunflower oil were mixed with the hot gellan gum solutions to give three different concentrations of disperse phase in the final emulsion: 10 g/100 g, 20 g/100 g, and 30 g/100 g and homogenized.

As an example Fig. 18.9 shows o/w emulsions (oil = 20 g/100 g) using 0.3 g/100 g of high and low acyl gellan gum. The latter showed a marked creaming destabilization, while the high acyl emulsion showed a uniform distribution of the dispersed phase throughout the volume. Sauter mean diameters (d_{32}) of oil droplets ranged from 10.0 to 14.3 mm (standard deviation = 2.4), however, not significant differences were observed between d_{32} values. Consequently, the differences in emulsion stability could not be explained in terms of the differences in droplet size; creaming stability depends on the capacity of the continuous phase to gel and immobilize oil droplets.

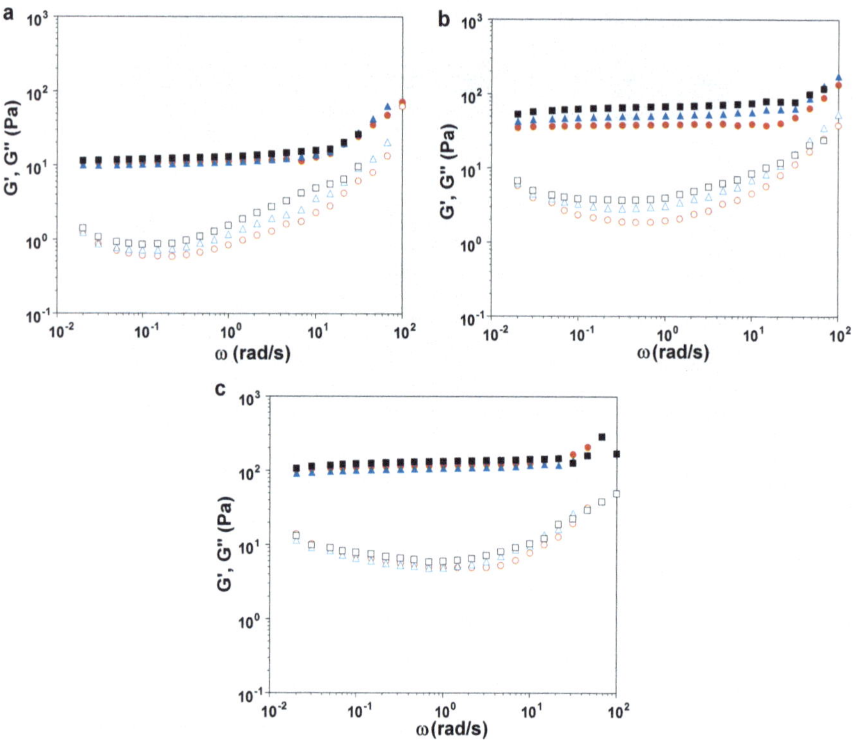

Fig. 18.10 G′ (filled symbols) and G″ (open symbols) as a function of frequency (ω) for emulsions containing (filled red circles, open red circles) 10, (filled blue triangles, open blue triangles) 20, (filled black squares, open black squares), and 30 g sunflower oil/100 g, stabilized with con (**a**) 0.1, (**b**) 0.3, and (**c**) 0.5 g high acyl gellan gum/100 g in the continuous phase

Regardless of hydrocolloid concentration it could be observed a marked difference in the sol-gel transition temperatures (Tgel) of high and low acyl gellan gum solutions; the former showed $T_{gel} > 70$ °C, while the low acyl gellan gum solutions gelled below 50 °C. As the high acyl gellan gum continuous phase gelled above 70 °C, the droplets in the emulsion were rapidly entrapped by the high acyl gel matrix being unable to move upwards. On the other hand, in spite of the rapid cooling rate (2 °C/min) used, the low acyl continuous phase remained fluid for longer times allowing oil droplets to move upwards, creaming.

18.2.2.3 Dynamic Rheology on Emulsion-Filled Gels with High Acyl Gellan Gum

Figure 18.10 shows frequency sweep curves for emulsion-filled gels obtained using different concentrations of high acyl gellan gum and sunflower oil. Curves were qualitatively similar for all the formulations assayed, and showed the same

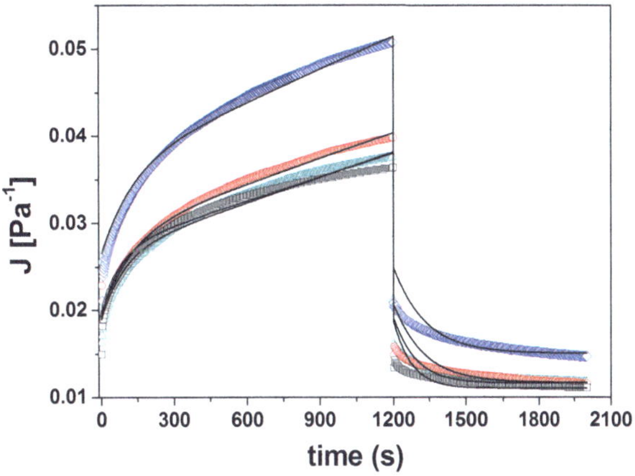

Fig. 18.11 Creep compliance curves for emulsion-filled gels containing 0 (blue diamonds), 10 (red circles), 20 (blue triangles), and 30 g/100 g (black squares) with a lipid phase using 0.3 g/100 g of high acyl gellan gum in the continuous phase. Four parameters Burgers Model (continuous line)

characteristic behavior already described for gelatin (Fig. 18.3a). Although increasing dispersed phase concentration slightly increased storage and loss moduli, the most significant change was observed when gellan gum concentration increased (Fig. 18.10).

18.2.2.4 Creep and Recovery Analysis

Within the linear viscoelastic range a creep-recovery analysis was done on both, the emulsion-filled gels and the corresponding high acyl gellan gels. Creep curves, (compliance, J (Pa^{-1}), vs. time) were fitted to the Burgers model of four elements consisting of a Maxwell element connected in series to a Kelvin-Voigt element (Steffe 1996). To obtain a more accurate group of parameters for each sample, experimental data corresponding to both creep and recovery experiments were simultaneously used to obtain the parameters of the Burgers model (Lorenzo et al. 2011; Steffe 1996). Thus, according to Boltzmann superposition principle, the complete model is represented by the following equations:

$$J(t)=\begin{cases} J_0+J_1\left(1-\exp\left(\dfrac{-t}{(\lambda_{ret})_1}\right)\right)+\dfrac{t}{\eta_0} & t\le t_1 \\ J_1\exp\left(\dfrac{-t}{(\lambda_{ret})_1}\right)\left(\exp\left(\dfrac{-t_1}{(\lambda_{ret})_1}\right)-1\right)+\dfrac{t_1}{\eta_0} & t>t_1 \end{cases} \tag{18.11}$$

where J_0 is the instantaneous compliance (Pa^{-1}), η_0 the viscosity of the Maxwell dashpot (Pa s), J_1 (Pa^{-1}) and $(\lambda_{ret})_1$ (s) are the compliance and the retardation time associated with the Kelvin-Voigt element, respectively, and t_1 is the time at which the stress was removed. J_0 represents the instantaneous elastic response of the system at $t = 0$; the lower the value of J_0, the greater is the elasticity (Kaschta and Schwarzl 1994).

As an example Fig. 18.11 shows the creep and recovery curves for formulations containing 0.3 g/100 g of high acyl gellan gum in the continuous phase. In all cases the Burgers model showed a satisfactory fitting to experimental data ($r^2 > 0.89$).

While all formulations showed qualitatively the same behavior, increasing gellan gum content significantly decreased ($P < 0.05$) the instantaneous compliance (J_0). Thus, when high acyl gellan gum concentration increased, the stiffness of the network increased resulting in less deformation at constant shear stress.

Emulsified systems showed a considerable decrease in creep compliance compared with their respective gels. Oil droplets act as active filler on the emulsions leading to a reinforcement of the structure when a dispersed phase is included in the gellan matrix. If a shear stress is applied to an emulsion gel containing interacting filler, the gel will deform to a certain degree in response. Because the filler particles are an integrated part of the gel, both the gel matrix and the filler particles make their contributions towards resisting the applied shear (Chen and Dickinson 1999). It is important to notice that a further increase in the oil content (between 10 and 30 g/100 g) did not produce significant changes ($P < 0.05$) in creep compliance curves.

The retarded compliance J_1 showed the same trend observed in J_0 but with values almost an order of magnitude lower, a behavior characteristic of highly flexible systems, where the material needs to deform initially before a flow is established, and deform again later (Jiménez-Avalos et al. 2005).

Although significant changes in the Burgers model parameters with the emulsion composition were observed the percentage contribution of each element remained constant (Lorenzo et al. 2011). Instantaneous compliance (J_0) represented the largest contribution, with an average 54% of the total deformation of the system, while J_1 only explained 21% of the total compliance. All the emulsion-filled gels exhibited a recovery percentage above 70% of the total deformation. This is in agreement with results in dynamic tests where, although a marked increase in G' and G'' was observed, the overall characteristics of the mechanical spectrum remained practically invariant.

18.2.2.5 Relaxation Spectrum Determination

During evaluation of mechanical properties, a broad spectrum of relaxation times generating relaxation spectra has been observed for macromolecules (Winter 1997). This relaxation spectrum is a fundamental quantity in the linear theory of viscoelastic materials (Honerkamp and Weese 1993) and its shape is often correlated with specific molecular architectures (Winter 1997). According to Malkin (2006), the spectrum correctly reflects main characteristics of viscoelastic behavior of real polymeric materials. Further, the molecular mobility of polymeric liquids and solids expresses itself in a relaxation-time spectrum (Baumgaertel and Winter 1992). However, direct measurement of relaxation spectra is impossible, as it can only be calculated on the basis of experimental data (Malkin 2006); dynamic mechanical

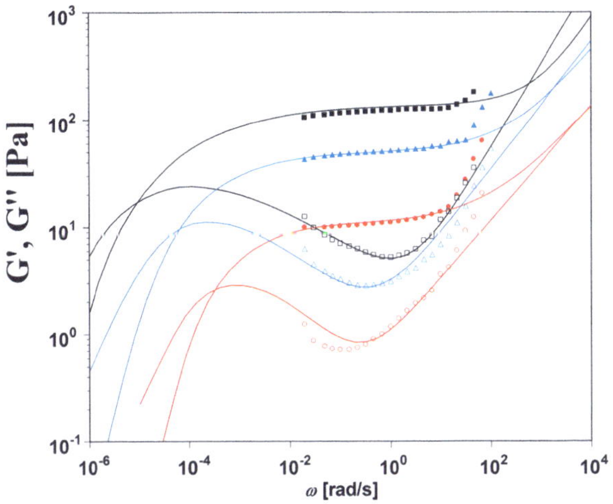

Fig. 18.12 Frequency sweep curves: experimental (symbols) and predicted with the BSW model (lines), G′ (filled symbols) and G″ (open symbols) of emulsion-filled gels containing 20 g/100 g oil with different hydrocolloids concentration: (filled red circles, open red circles) 0.1, (filled blue triangles, open blue triangles) 0.3, (filled black squares, open black squares), and 0.5 g/100 g

experiments are the most effective for measuring the relaxation modes of polymeric liquids and solids (Baumgaertel and Winter 1989). The complex viscoelastic modulus, G^*, consists of a real part (G') and an imaginary part (G''), i.e. $G^*(\omega) = G'(\omega) + iG''(\omega)$. From Eqs. (18.1) and (18.2), it could be expressed in terms of the relaxation spectrum as:

$$G^* = G_e + \int_0^\infty H(\lambda)\frac{(\lambda\omega)^2}{1+(\lambda\omega)^2}d\lambda + i\int_0^\infty H(\lambda)\frac{\lambda\omega}{1+(\lambda\omega)^2}d\lambda \qquad (18.12)$$

where, G_e represents the viscoelastic solid state of an additional spring attached in series to a Maxwell model that is called the equilibrium elasticity modulus (for viscoelastic liquids like the studied emulsions this parameter can be neglected).

In the response of a linearly viscoelastic material to a strain or stress excitation, the relaxation spectrum, $H(\lambda)$, contains complete information on the time-dependent part of the response. Thus, once the spectrum and the viscoelastic constants are known, it is possible, in theory, to generate the response to any desired type of excitation.

Using an appropriate representation of the relaxation time spectrum, it is possible to model the dynamic moduli. A frequently used representation of the relaxation time spectrum is the already mentioned Baumgaertel-Schausberger-Winter spectrum. In the present case, the satisfactory agreement between experimental oscillatory data and predictions calculated using the BSW model is shown in Fig. 18.12 for emulsion-filled gels containing 20 g/100 g of sunflower oil.

Increasing gellan concentration significantly increased the plateau modulus, while the dispersed phase content did not significantly affect G_N^0. This could be explained considering that the rheological behavior of the systems is controlled by the highly structured continuous phase rather than the contribution of filler lipid droplet in the emulsion, as was previously pointed out. It was observed a relationship between G_N^0 and gellan content "c" ($G_N^0 \propto c^2$) similar to the tendency found in other polymers (Ferry 1980; Graessley 1974).

As it was previously stated in the present chapter, the viscoelastic material functions of biopolymeric systems could be defined using an infinite number of Maxwell elements, by the relaxation time spectrum, H(λ). For the transient part of the relaxation modulus (G(t)), the expression would be:

$$G(t) = G_e + \int_{-\infty}^{+\infty} H(\lambda) e^{-\frac{t}{\lambda}} d\ln\lambda \tag{18.13}$$

where G_e is the equilibrium modulus. G_e is finite for viscoelastic solids and is zero for the liquids and for the materials at the gel point (de Rosa and Winter 1994).

In an entirely analogous manner, creep behavior can be described by the Generalized Voigt model. A group of Voigt elements in series represents a discrete spectrum of retardation times, each time λ_i being associated with a spectral compliance magnitude (J_i). Thus, creep compliance (J(t)) can be calculated as:

$$J(t) = J_0 + \sum_{i=1}^{N} J_i \left(1 - e^{-\frac{t}{\lambda_i}}\right) + \frac{t}{\eta_0} \tag{18.14}$$

where J_0 is the instantaneous compliance and η_0 is the zero-shear viscosity.

If the Voigt model is made infinite in extent, the continuous retardation spectrum, L(λ), is defined as the sum of infinitesimal contributions to the material behavior. Thus, creep compliance is expressed as:

$$J(t) = J_0 + \int_{-\infty}^{+\infty} L(\lambda) \left(1 - e^{-\frac{t}{\lambda}}\right) d\ln\lambda + \frac{t}{\eta_0} \tag{18.15}$$

The two spectra (H and L) are of the nature of distribution functions, although they have the dimensions of a modulus (H) and a compliance (L) respectively, rather than the dimensionless character of the usual distribution function. Moreover, the spectra H and L are useful qualitatively in gauging the distribution of relaxation or retardation mechanisms in different regions of the timescale (Ferry 1980).

There is an interrelation between the relaxation (H(λ)) and retardation (L(λ)) spectra, so if one spectrum is known over the entire range of time scale, the other spectrum can be calculated as follows (Ferry 1980):

$$L(\lambda) = \frac{H(\lambda)}{\left[G_e - \int_{-\infty}^{\infty} \frac{H(u)}{(\lambda/u) - 1} d\ln u\right]^2 + \pi^2 H(\lambda)^2} \tag{18.16}$$

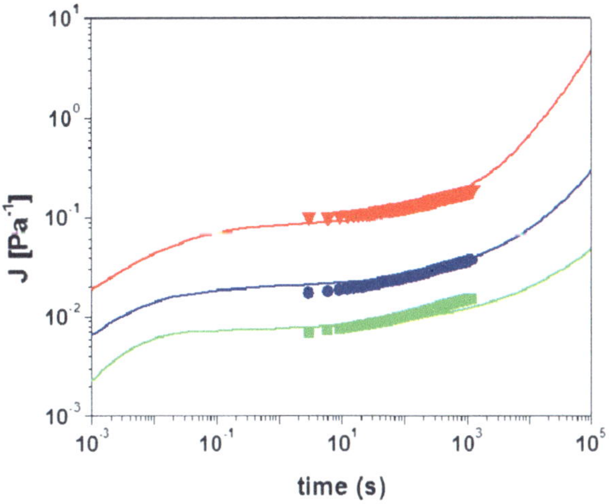

Fig. 18.13 Comparison between experimental creep compliance data and the prediction from broadened BSW model. Data corresponded to emulsion-filled gels containing 20 g/100 g oil with different hydrocolloids concentration: 0.1 g/100 g (red inverted triangles); 0.3 g/100 g (blue circles); 0.5 g/100 g (green squares)

Thus, the retardation spectrum was evaluated using Eq. (18.16) from the expression of H(λ) obtained for the gel emulsion (Eq. 18.3). Once the retardation spectrum was known, the creep-compliance behavior was predicted according to Eq. (18.15). Figure 18.13 shows the predicted compliance for emulsion-filled gels as well as the experimental data.

Any dynamic data measured in this work have been successfully converted into the time domain by the application of BSW model. It is a useful tool, especially for establishing a rheological data bank and analyzing viscoelastic experiments.

18.3 Conclusions

Most of the systems prepared in cold with bovine gelatin exhibited gel-like properties which delayed phase-separation for more than 1 month. The high shear stresses to which the samples were submitted during homogenization produced shear induced denaturation of the gelatin and gelation of the whole system. Thus, this type of gel emulsions could be an adequate alternative to deliver thermolabile hydrophobic compounds.

Emulsion stability was related to the protein molecular weight rather than droplet diameter. The viscosity of the gelatin systems and its elastic characteristics increased as the molecular weight of the gelatin increased; there was a positive correlation between the degree of crosslinking and the molecular weight of the protein.

By applying the theory of rubber elasticity, the number-average molecular mass of the gelatin chain portions between two consecutive cross-links (Mc) were calculated at different storage times. The decrease in the Mc values during storage could be related to the structural changes occurred in the emulsions, which induced an increment of the elasticity modulus based on a tighter network, which expelled the droplets, leading to its final destabilization.

Rheological behavior of gellan gum aqueous solutions was extremely dependent on the acylation degree of the biopolymer. Low acyl gellan gum gels exhibited almost parallel and frequency independent curves of storage (G′) and loss (G″) moduli, whereas high acyl gellan gum gels showed a minimum in G″ and both moduli presented lower values.

High acyl gellan gum gel emulsions showed qualitatively the same behavior as the continuous phase, with a storage modulus (G′) independent of the frequency, even in systems with low hydrocolloid content. All samples showed a viscoelastic solid behavior with the storage modulus (G′) dominating the viscoelastic response. Oil concentration slightly affected the elastic behavior of the emulsions, and it became less significant with the increase of gum content.

Creep and recovery data was analyzed using the Burgers model and a satisfactory fitting was obtained. Burgers parameters showed that regardless hydrocolloids concentration, emulsified systems presented a considerable decrease in creep compliance compared with their respective gels which suggests that oil droplets act as active filler particles reinforcing the structure and the gel strength of the gellan matrix.

Viscoelastic behavior of both gelatin and gellan systems was satisfactorily modeled using Baumgaertel-Schausberger-Winter equation. The use of the BSW spectrum is advantageous due to different reasons, i.e. it provides analytical expressions for the material functions (as G′, G″ data) and, it reduces the description of the linear viscoelastic behavior to the determination of only a few material-specific parameters. Validation of the predicted spectra was carried out through creep compliance data for emulsion-filled gels.

Acknowledgements The authors are grateful to CP Kelco (San Diego, CA) and PB Leiner Argentina, who kindly provided the gellan gum and gelatin used for this study. The financial support of the Consejo Nacional de Investigaciones Científicas y Técnicas (CONICET), Agencia Nacional de Promoción Científica y Tecnológica, and Universidad Nacional de La Plata are also acknowledged.

References

Anton M, Le Denmat M, Beaumal V (2001) Filler effects of oil droplets on the rheology of heat-set emulsion gels prepared with egg yolk and egg yolk fractions. Colloids Surf B Biointerfaces 21:137–147

Bais D, Trevisan A, Lapasin R et al (2005) Rheological characterization of polysaccharide-surfactant matrices for cosmetic o/w emulsions. J Colloid Interface Sci 290:546–556

Baumgaertel M, Winter HH (1989) Determination of discrete relaxation and retardation time spectra from dynamic mechanical data. Rheol Acta 28:511–519

Baumgaertel M, Winter HH (1992) Interrelation between continuous and discrete relaxation time spectra. J Non-Newtonian Fluid Mech 44:15–36

Bot A, van Amerongen IA, Groot RD et al (1996) Large deformation rheology of gelatin gels. Polym Gels Netw 4:189–227

Chen J, Dickinson E (1999) Effect of surface character of filler particles on rheology of heat-set whey protein emulsion gels. Colloids Surf B Biointerfaces 12:373–381
Dickinson E (2006) Colloid science of mixed ingredients. Soft Matter 2:642–652
Dickinson E (2010) Food emulsions and foams: stabilization by particles. Curr Opin Colloid Interface Sci 15:40–49
Dickinson E (2012) Emulsion gels: the structuring of soft solids with protein-stabilized oil droplets. Food Hydrocoll 28:224–241
Dickinson E, Lopez G (2001) Comparison of the emulsifying properties of fish gelatin and commercial milk proteins. J Food Sci 66:118–123
Doublier JL, Launay B, Cuvelier G (1992) Viscoelastic properties of food gels. In: Rao MA, Steffe JF (eds) Viscoelastic properties of foods. Elsevier Science, Barking, pp 371–434
Ferry JD (1980) Viscoelastic properties of polymers. Wiley, New York, pp 56–79, 366–403
Flory PJ (1953) Principles of polymer chemistry. Cornell University Press, New York
Friedrich CHR, Heymann L (1988) Extension of a model for crosslinking polymer at the gel point. J Rheol 32:235–241
Garti N (1999) What can nature offer from an emulsifier point of view: trends and progress? Colloids Surf A Physicochem Eng Asp 152:125–146
Garti N, Slavin Y, Aserin A (1999) Surface and emulsification properties of a new gum extracted from *Portulaca oleracea* L. Food Hydrocoll 13:145–155
Gidley MJ, Nishinari K (2009) Physico-chemistry of (1,3)-β-glucans. In: Bacic A, Fincher JB, Stone BA (eds) Chemistry, biochemistry, and biology of 1–3 beta glucans and related polysaccharides. Academic Press, San Diego, pp 47–118
Goh KKT, Hainsman DR, Singh H (2006) Characterization of a high acyl gellan polysaccharide using light scattering and rheological techniques. Food Hydrocoll 20:176–183
Graessley WW (1974) The entanglement concept in polymer rheology. Adv Polym Sci 16:1–179
Grasdalen H, Smidsrod O (1987) Gelation of gellan gum. Carbohydr Polym 7(5):371–393
Grassi M, Lapasin R, Pricl S (1996) A study of the rheological behavior of scleroglucan weak gel systems. Carbohydr Polym 29:169–181
Harrington WF, Rao NV (1970) Collagen structure in solution. I. Kinetics of helix regeneration in single-chain gelatins. Biochemistry 19:3714–3724
Honerkamp J, Weese J (1993) A nonlinear regularization method for the calculation of relaxation spectra. Rheol Acta 32:65–73
Huang Y, Tang J, Swanson BG et al (2003) Effect of calcium concentration on textural properties of high and low acyl mixed gellan gels. Carbohydr Polym 54(4):517–522
Ikeda S, Nitta Y, Temsiripong T et al (2004) Atomic force microscopy studies on cation-induced network formation of gellan. Food Hydrocoll 18:727–735
Izuka A, Winter HH, Hashimoto T (1992) Molecular weight dependence of viscoelasticity of polycaprolactone critical gels. Macromolecules 25:2422–2428
Jackson JK, De Rosa ME, Winter HH (1994) Molecular weight dependence of relaxation time spectra for the entanglement and flow behavior of monodisperse linear flexible polymers. Macromolecules 27:2426–2431
Jansson PE, Lindberg B, Sandford PA (1983) Structural studies of gellan gum, an extracellular polysaccharide elaborated by *Pseudomonas elodea*. Carbohydr Res 124:135–139
Jiménez-Avalos HA, Ramos-Ramírez EG, Salazar-Montoya JA (2005) Viscoelastic characterization of gum arabic and maize starch mixture using the Maxwell model. Carbohydr Polym 62:11–18
Kaschta J, Schwarzl ER (1994) Calculation of discrete retardation spectra from creep data-II. Analysis of measured creep curves. Rheol Acta 33:530–541
Keenan TR (1998) Gelatin. In: Domb AJ, Kost J, Wiseman D (eds) Handbook of biodegradable Polymers. CRC Press, pp 307–317
Kim KH, Renkema JMS, Vliet TV (2001) Rheological properties of soybean protein isolate gels containing emulsion droplets. Food Hydrocoll 15:295–302

Kirschner KN, Woods RJ (2001) Solvent interactions determine carbohydrate conformation. Proc Natl Acad Sci USA 98:10541–10545

Koenig A, Hébraud P, Perrin P (2002) Preparation and rheological properties of emulsion gels. Langmuir 18:6458–6461

Kokini J, van Aken G (2006) Discussion session on food emulsions and foams. Food Hydrocoll 20:438–445

Kuo MS, Mort AJ, Dell A (1986) Identification and location of L-glycerate, an unusual acyl substituent in gellan gum. Carbohydr Res 156:173–187

Lapasin R, Pricl S (1995) Rheology of industrial polysaccharides: theory and applications. Blackie Academic & Professional, Glasgow, pp 63–83, 373–423

Ledward DA (1986) Gelation of gelatin. In: Mitchell JR, Ledward DA (eds) Functional properties of food macromolecules. Elsevier, New York, pp 171–201

Lobo L (2002) Coalescence during emulsification. 3. Effect of gelatin on rupture and coalescence. J Colloid Interface Sci 254:165–174

Lorenzo G, Checmarev G, Zaritzky N et al (2011) Linear viscoelastic assessment of cold gel-like emulsions stabilized with bovine gelatin. LWT Food Sci Technol 44:457–464

Lorenzo G, Zaritzky NE, Califano A (2008) Optimization of non-fermented gluten-free dough composition based on rheological behavior for industrial production of "empanadas and pie-crusts". J Cereal Sci 48:224–231

Lorenzo G, Zaritzky NE, Califano A (2013) Rheological analysis of emulsion-filled gels based on high acyl gellan gum. Food Hydrocoll 30:672–680

Malkin AY (2006) Continuous relaxation spectrum—its advantages and methods of calculation. Int J Applied Mech Eng 11:235–243

Manca S, Lapasin R, Partal P, Gallegos C (2001) Influence of surfactant addition on the rheological properties of aqueous Welan matrices. Rheol Acta 40:128–134

Manoi K, Rizvi SSH (2009) Emulsification mechanisms and characterizations of cold, gel-like emulsions produced from texturized whey protein concentrate. Food Hydrocoll 23:1837–1847

McClements DJ (1999) Food emulsions: principles, practice and techniques. CRC Press, Boca Raton, pp 185–265

Mitchell JR (1976) Rheology of gels. J Texture Stud 7:313–339

Mitchell JR, Blanshard JMV (1976) Rheological properties of alginate gels. J Texture Stud 7:219–234

Moresi M, Bruno M, Parente E (2004) Viscoelastic properties of microbial alginate gels by oscillatory dynamic tests. J Food Eng 64:179–186

Moritaka H, Nishinari K, Taki M et al (1995) Effects of pH, potassium chloride, and sodium chloride on the thermal and rheological properties of gellan gum gels. J Agric Food Chem 43:1685–1689

Morris ER, Gothard MGE, Hember MWN et al (1996) Conformational and rheological transitions of welan, rhamsan and acylated gellan. Carbohydr Polym 30:165–175

Morris ER, Rees DA, Robinson G (1980) Cation-specific aggregation of carrageenan helices: domain model of polymer gel structure. J Mol Biol 138(2):349–362

Mours M, Winter HH (2000) Mechanical spectroscopy. In: Tanaka T (ed) Methods in polymer science: modern methods in polymer research and technology. Academic Press, San Diego, pp 495–546

Muller HJ, Hermel H (1994) On the relation between the molecular mass distribution of gelatin and its ability to stabilize emulsions. Colloid Polym Sci 272:433–439

Normand V, Ravey JC (1997) Dynamic study of gelatin gels by creep measurements. Rheol Acta 36:610–617

Olivares ML, Peirotti MB, Deiber JA (2006) Analysis of gelatin chain aggregation in dilute aqueous solutions through viscosity data. Food Hydrocoll 20:1039–1049

O'Neill MA, Selvendran RR, Morris VJ (1983) Structure of the acidic extracellular gelling polysaccharide produced by *Pseudomonas elodea*. Carbohydr Res 124:123–133

Ottone ML, Deiber JA (2005) Modeling the rheology of gelatin gels for finite deformations. Part 1. Elastic rheological model. Polymer 46:4928–4937
Rao MA (2007) Rheology of fluids, semisolids, and solid foods. Principles and applications. Springer, New York, pp 331–390
Reiffers-Magnani CK, Cuq JL, Watzke HJ (1999) Composite structure formation in whey protein stabilized O/W emulsions. I. Influence of the dispersed phase on viscoelastic properties. Food Hydrocoll 13:303–316
Rodríguez-Hernández AI, Durand S, Garnier C et al (2003) Rheology-structure properties of gellan systems: Evidence of network formation at low gellan concentrations. Food Hydrocoll 17:621–628
de Rosa ME, Winter HH (1994) The effect of entanglements on the rheological behavior of polybutadiene critical gels. Rheol Acta 33:220–237
Segeren AJM, Boskamp JV, van den Tempel M (1974) Rheological and swelling behaviour of alginate gels. Faraday Discuss 57:255–262
Shakuntala N, Manay O (2001) Food: facts and principles. New Age International Ltd, New Delhi, pp 376–394
Steffe JF (1996) Rheological methods in food process engineering. Freeman Press, East Lansing, pp 294–348
Surh J, Decker EA, McClements J (2006) Properties and stability of oil-in water emulsions stabilized by fish gelatin. Food Hydrocoll 20:596–606
Sworn G (2000) Gellan gum. In: Phillips GO, Williams PA (eds) Handbook of hydrocolloids. Woodhead Publishing Limited, Cambridge, pp 117–135
Tschoegl NW (1997) Time dependence in material properties: an overview. Mech Time-Depend Mater 1:3–31
de Wolf FA (2003) Collagen and gelatin. In: Aalbersberg WY et al (eds) Progress in biotechnology, vol 23. Elsevier Science BV, Amsterdam, pp 133–218
van Vliet T, Lakemond CMM, Visschers RW (2004) Rheology and structure of milk protein gels. Curr Opin Colloid Interface Sci 9:298–304
Vilela JAP, Lopes da Cunha R (2016) High acyl gellan as an emulsion stabilizer. Carbohydr Polym 139:115–124
Whistler RL (1993) Introduction to industrial gums. In: Whistler RL, JN BM (eds) Industrial gums: polysaccharides and their derivatives. Academic Press, San Diego, pp 1–20
Wilde PJ (2000) Interfaces: their role in foam and emulsion behaviour. Curr Opin Colloid Interface Sci 5:176–181
Winter HH (1997) Analysis of dynamic mechanical data: inversion into a relaxation time spectrum and consistency check. J Non-Newtonian Fluid Mech 68:225–239
Winter HH, Mours M (2006) The cyber infrastructure initiative for rheology. Rheol Acta 45:331–338
Yamamoto F, Cunha RL (2007) Acid gelation of gellan: effect of final pH and heat treatment conditions. Carbohydr Polym 68:517–527

Chapter 19
Polymers for Structure Design of Dairy Foods

Haotian Zheng

Abstract Dairy foods ranging from liquids, semi-solids to solids are considered as complex viscoelastic materials. Maintaining the physical stability over the shelf life and delivering superior mouthfeel of foods after series of treatments have always been challenging for the dairy industry. During the manufacturing process, both high shear stress and temperature changing history can significantly affect the macro- and microstructure of dairy food systems. Therefore, their physical stability and sensory attributes are altered as consequences. Food polymers as stabilizer and texturizer are used in different dairy systems for eliminating negative impacts of intensive processing treatments and for manipulating texture for meeting the specific sensory preference for a targeted group of consumers. As kinetically metastable systems, the optimum structure of dairy foods may be engineered by following a universal two steps principle: (1) Apply the proper mix of food polymers in the dairy food formulation; (2) Process the formula with dedicated parameters and procedures. Although the principle is simple, the implementation is complicated. Such existence of challenge is due to the detailed interaction mechanisms between non-dairy polymers and dairy components in various physicochemical environments are not entirely understood. In this chapter, the nondairy polymers induced destabilization/stabilization of dairy systems are explained, the technical challenges of stabilization of dairy systems are discussed. It focuses on three major topics regarding dairy food structure design: (1) Formulation strategy of thickening dairy matrices; (2) Formulation strategy of increasing perception of the creaminess of dairy matrices; (3) The current updates about the synergetic functionality of food polymers.

Keywords Food hydrocolloids · Dairy Foods · Stability · Texture

H. Zheng, PhD (✉)
Leprino Foods Dairy Innovation Institute, Animal Science Department, California Polytechnic State University, 1 Grand Avenue, San Luis Obispo 93407-0251, CALIFORNIA, United States
e-mail: hazheng@calpoly.edu

T. J. Gutiérrez (ed.), *Polymers for Food Applications*,
https://doi.org/10.1007/978-3-319-94625-2_19

19.1 Overview

Specific polymers as food hydrocolloids are often found in the ingredient list of dairy food products such as starch, pectin, carrageenan, gellan gum, β-glucan, inulin, or gelatin. The presence of these items in dairy systems is not expected from consumers' perspective, and most of the consumers have little knowledge about the significance of using these non-dairy materials in dairy foods. These materials are traditionally recognized as stabilizers which are responsible for maintaining the physical stability of dairy foods, preventing the dairy dispersion particles or gel systems are visibly separated from the aqueous phase. The US Food and Drug Administration (FDA) and European Food Safety Authority (EFSA) classified such food hydrocolloids as emulsifiers, stabilizers, gelling agents and thickeners as part of the larger family of food additives. FDA provides a Food Additives Status List which includes short notations on use limitations for each additive. The related regulations of usage of the additives are promulgated under the Federal Food, Drug, and Cosmetic Act (FD&C Act) under section 401 (food standards) and 409 (food additives) (FDA 2017). The detailed application guidelines of the most stabilizers can be found from the FDA Code of Federal Regulations Title 21, part 172—food additives permitted for direct addition to food for human consumption. According to EFSA, all stabilizers as food additives are identified by specific E numbers (EFSA 2017b). The European Union is carrying out a re-evaluation programme accessing the safety usage of all food additives. By 2020, EFSA's Expert Panel on Food Additives and Nutrient Sources Added to Food (ANS) is going to provide updated conclusions on the safety of the intended uses of the food additive for the consumers (EFSA 2017a).

The classical reason for applying food polymers in dairy products are: (1) constructing physically stable dairy liquid/gel systems preventing phase separation over the shelf-life or preventing the growth of ice crystals caused by temperature abuse. (2) Establishing and maintaining the texture of dairy foods for delivering body and specific mouthfeel for attracting consumers (Hansen 1993). In light of the development of food technology and the demand of nutritional foods, food polymers are also used as wall materials for encapsulating sensitive nutrients or probiotics in dairy material based food systems (Li et al. 2017; Anal and Singh 2007). Microencapsulation can reduce the reactivity of the core material with environmental reactants due to the low permeability of the barrier wall. Therefore, for instance, undesirable oil oxidation may be prevented or slowed down to an acceptable oxidation rate according to the shelf life; the release of specific nutrients or probiotics can be engineered for the targeted delivery; the unfavorable taste or flavor of the beneficial health materials are masked during consumption (Shahidi and Han 1993).

All in all, the aforementioned three areas for food polymer application in dairy foods can be summarized as the dairy structure and texture design. Based on the three core purposes of application of food polymers in dairy foods, this chapter is aiming to provide technical knowledge to dairy food technologists who use food polymers to formulate different dairy foods for achieving physical stability, and desired mouthfeel. One can use this chapter for acquiring knowledge about the working mechanisms of different food polymers in different types of dairy matrices and use it as guidelines for constructing ideal structure and texture of dairy foods which meet the various markets' needs. The term of "polymer" is extensively used the following texts, it represents non-dairy polymer molecules distinguishing from dairy proteins.

19.2 Milk System and Polymer-Protein Interactions

19.2.1 Milk Components

Milk is a complex oil-in-water (o/w) emulsion system containing oil droplets, dissolved whey protein, colloidal casein protein micelles, lactose and a small amount of minerals. In raw milk system, the milk fat presents as globular shape droplets named as milk fat globule (MFG). The size of MFGs ranges from 0.2 to 15 μm with a volume weighted mean diameter (d43) between 3.5 and 5.3 μm. The MFG is stabilized by a tri-layer membrane system namely the milk fat globule membrane (MFGM) which composed of phospholipids and larger molecular weight proteins (Huppertz and Kelly 2006)(Zheng et al. 2014). The MFGM system stabilizes the native MFGs against aggregation and flocculation. However, the MFGM cannot prevent creaming. Casein as the major type protein in milk accounts for around 80% protein proportion; it is a mixture of four proteins: α_{S1}-, α_{S2}-, β-, and κ-casein have molecular masses around 20 kDa (Dalgleish 1997; Walstra et al. 2006). Casein presents in milk aqueous phase in different states from dissolved macromolecules to stable colloidal larger proteinaceous particle namely casein micelle (Dalgleish 1997). The typical volume-average radius of casein micelle is of ~100 nm (Kruif and Holt 2003). About 20% bovine milk protein proteins are whey/serum proteins including α-lactalbumin (~19% w/w of total whey protein), β-lactoglobulin (~52% w/w of total whey protein), bovine serum albumin (BSA, ~6% w/w of total whey protein), Immunoglobulins (~13% w/w of total whey protein), proteose peptone (~13% w/w of total whey protein). Most of the serum proteins are globular proteins; unlike casein micelles, whey proteins present in a dissolved form in bovine milk. Both casein and whey proteins are hydrophobic having hydrophobic side groups ranging from 22 to 29%; they are all negatively charged at neutral pH of bovine milk (~pH 6.5) (Walstra 2006). To be able to engineer dairy structure and texture using nondairy food polymers three types of interaction mechanisms need to be elaborated (see 19.2.2).

19.2.2 Nondairy Polymers and Dairy Protein Interactions

Nondairy polymers together with whey protein and casein micelles in the aqueous phase of milk form a pseudoternary "protein-polysaccharide-water" polyelectrolyte dispersion system. Such ternary dispersion system is highly complicated due to the presences of polyelectrolytes and irregular distribution of charged groups along the nondairy polymer/dairy protein (Syrbe et al. 1998). For instance, α_{S1}-, α_{S2}-, β-, and κ-caseins have an uneven distribution of charges; such complexity indicates site-specific interactions among different nondairy polymers. Moreover, several different nondairy polymers (>3 types) are often mixed forming a "stabilizer mix" and applied in dairy systems. Therefore, the commercial dairy foods are more than a ternary system. For better understanding the nature of the pseudoternary system, it is important to point out that the physical behavior of the system is controlled by enthalpic effects (Syrbe et al. 1998).

Syrbe and co-authors (1998) summarized three equilibrium conditions of polymer-protein-solvent ternary systems; these conditions may apply to dairy matrices containing multi-polymers: (1) Incompatibility: the different polymers (dairy protein and polysaccharide) are concentrated in separated domains in the system, which is also termed as "segregative phase separation." For instance, an increase of the concentration of high methoxyl pectin in skim milk at pH 6.7 induces depletion flocculation of casein micelles forming casein-rich domains (Acero-Lopez et al. 2010). Such interaction mechanism is explained and understood as deletion (Tuinier et al. 2003); (2) Complex formation (coacervation): such interaction among polymers are initiated by electrostatic attraction, hydrogen binding and hydrophobic attraction, the protein-polymer complex results in a polymer-rich phase in the food matrix (Corredig et al. 2011; Thies 2003). The formation of protein-polymer complex induces a series of changes such as modification of a rheological property of the system (Wang et al. 2007) and cause of precipitation of protein (Niederauer and Glatz 1994). At relative lower pH (pH 5.3) both high and low methoxyl pectin interact with casein micelles via electrostatic force driven bridging flocculation mechanism and can form a physically stable system (Maroziene and de Kruif 2000). Using quasi-elastic light scattering and fluorescence spectroscopy techniques, it was found that globular whey protein (human serum albumin) can interact with polyethylene glycol (used antifoaming agent or plasticizers in aqueous film coatings) forming an intrapolymer complex via hydrogen bonding (Azegami et al. 1999). Polyelectrolytes may regulate protein/peptide drug delivery (Vasir et al. 2003), and it was found that both the concentration of soluble whey protein (free BSA) and the hydrophobicity of polyacrylates control the formation of the soluble protein-polymer complex (Porcar et al. 1999); (3) Miscibility: in such scenario, different polymer species are homogenously co-distributed in a system via only physical contact without chemical and physicochemical interactions. The true interaction mechanism between dairy protein and starch has been extensively studied, but it is not yet fully understood. The interactions between dairy protein ingredients and gelatinized starch paste (Kumar et al. 2017, 2018) and the interactions between starch granules and different dairy food systems (Considine et al. 2011) are studied and reviewed. In these studies, the authors demonstrated that although protein-polymer complexes are formed, protein/protein aggregates may be homogenously miscible in starch paste system (Fig. 19.1) or gelatinized starch granules may be homogenously distributed in acidified dairy protein gel systems (as shown in Fig. 19.2). The miscible system may become incompatible system under specific physical stress in a certain timescale. The miscible system is considered as a stable

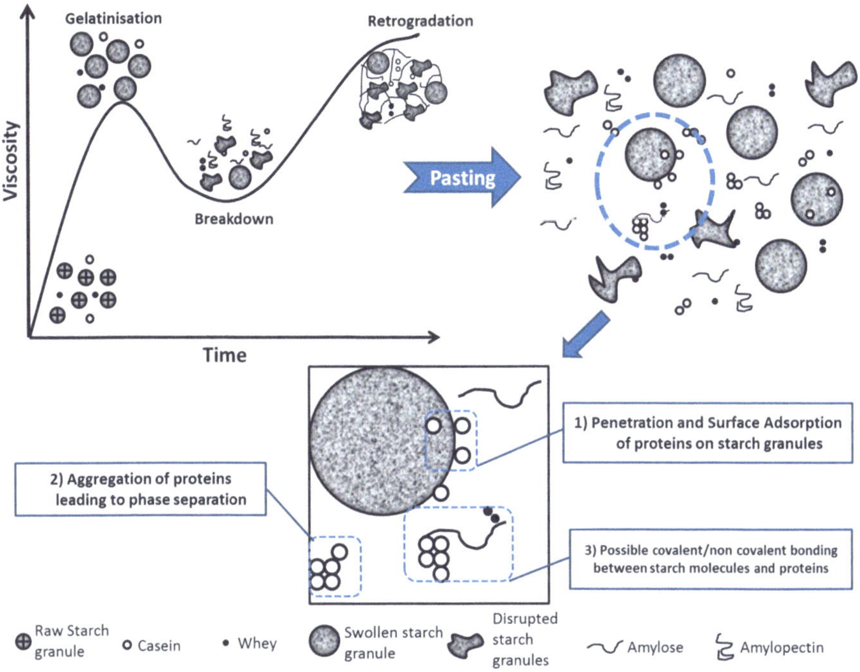

Fig. 19.1 Starch-milk proteins interactions during heat treatment under continuous shear (Kumar et al. 2017)

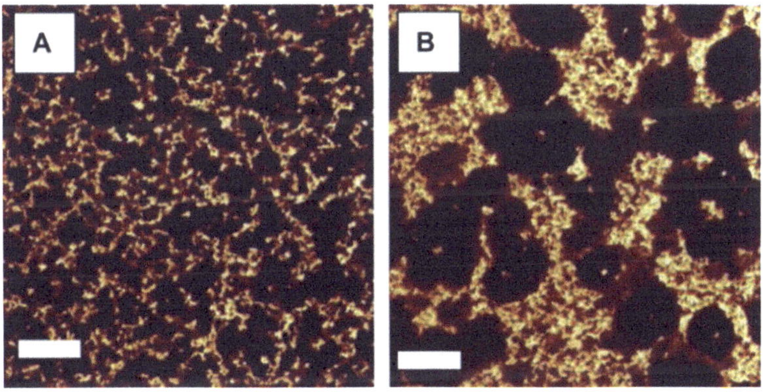

Fig. 19.2 Confocal laser scanning micrographs show homogenous distribution of gelatinized starch granules in the acidified milk gels containing (**b**) 2% w/w unmodified rice starch granules. Figure (**a**) is the control sample containing 0% w/w starch. Protein network appears in yellow, and dark spherical regions in (**b**) are the swollen starch granules. White scale bar: 20 μm (Considine et al. 2011)

system. However, such stability is only a relative condition subjected to mechanic treatment and time. One can utilize different polymers to increase the relative stability and therefore, improve the shelf life stability.

19.3 Destabilization/Stabilization Mechanisms

It is important to the point out that the complex formation does not necessarily result in a stable or unstable system. Moreover, physical stability is one of the primary concerns in dairy food manufacturing. Therefore, for the food technology research works investigated rheological behavior of protein-polymer mixtures (Lizarraga et al. 2006; Rodd et al. 2000), the physical stability and structural features must need to be reported (e.g., both rheological properties and stability phase diagrams are reported (Langendorff et al. 2000)). Rheological behavior is time, structure and shear force-dependent, without elaborating the stability and structure features, the published information of rheological parameters have little significance in terms of guiding the real-life application of food formulation design for the food industry.

Although the functionality of food polymer in multiphasic systems is not fully clear, the basic destabilization mechanisms of protein-polymer containing model food systems were illustrated in several studies (Corredig et al. 2011; Doublier et al. 2000; Syrbe et al. 1998). Entropic effect is greater than the enthalpic effect in polymer mixture systems. The fate of multiphasic system is separation. The physical stable food matrix means the homogenous distribution of incompatible materials cross the whole food system and nonappearance of undesirable (visible) phase separation. Microscale phase separation is negligible if non-negative impacts on the appearance and mouthfeel of the product are detected. Based on the affinities between protein, polymer, and solvent, two types phase separation mechanisms are introduced: (1) segregative phase separation (for describing non-adsorbing polymers containing systems); (2) associative phase separation (for describing adsorbing polymers containing systems). In summary, both types of phase separation depend on absolute concentrations of dairy protein and nondairy polymer; relative ratio of concentration between dairy protein and nondairy polymer, species of added polymer, pH and ionic strength of the mixture system.

19.3.1 Segregative Phase Separation (Non-adsorptive Stabilization)

Milk proteins are negatively charged at native milk pH (~pH 6.5–6.7). Neutral or anionic polymer/polysaccharide is incompatible with milk proteins at the natural pH range of milk, thus, pH and ionic strength determine the phase separation. Such polymers are recognized as non-adsorbing or non-interacting polymers. In a dairy protein dispersion system containing non-adsorbing polymers, its stability depends on the concentration of the free polymer in the aqueous phase (Corredig et al. 2011; Syrbe et al. 1998). (a) At extremely low concentration (0–1.5% w/w) of non-adsorbing polymer (amylopectin), the dairy protein (casein) dispersion system is stable (De Bont et al. 2002); (b) At relatively higher concentration level of non-adsorbing polymer (>0.3% w/w, κ-carrageenan) the casein dispersion system (0–5% w/w casein content) becomes unstable (Schorsch et al. 2000). This is due to the increased concentration of polymer induced an osmotic pressure gradient, consequently caused the depletion flocculation of the casein micelles (Suresh et al. 2006); (c) Keep increasing the non-adsorbing polymer in liquid milk system may overcome the depletion flocculation effect and result a stable milk-polymer mixture system (Fig. 19.3 a3). Konjac glucomannan (KGM) is a natural polysaccharide extracted from konjac which may be used as thickening and gelling agent in dairy food systems. Dai and co-workers (2017) found that KGM and milk components in a mixture system follow the segregative phase separation mechanism; also, the authors used binodal curve and showed that the KGM-milk mixtures may be stabilised from phase separation by using either low or high dosage of KGM. For instance, in diluted milk system (70% liquid milk), adding <0.2% or >0.7% KGM may result stable polymer-milk mixture systems; however, phase separation is induced by adding 0.25–0.6% KGM. In the same research work, the authors also observed the formation of aggregate structures when the concentration of KGM is higher than 0.5%. Such observations suggested that at higher concentration levels of KGM in milk system, the KGM polymers form non-adsorption self-packing structure/network at the aqueous phase, therefore, increase the viscosity of the continuous phase in the mixture system. The network of polymer (including gelation) in aqueous phase results increase of system viscosity (Hemar et al. 2001b). According to the stokes's law, the viscosity of the aqueous phase of dispersion system determines the phase separation rate (Huppertz and Kelly 2006). Therefore, the relatively higher volume of non-adsorbing polymer (higher than the critical concentration of depletion effect) may be considered in dairy food formulation for improving physical stability and thickness (Fig. 19.3a).

19.3.2 Associative Phase Separation (Adsorptive Stabilization)

Electrostatic attraction is the driving force of association between dairy protein and polysaccharides. Most of the formation of protein-polymer complex happens under the pI of dairy proteins due to both protein and polysaccharides for food applications are negatively charged at native pH of milk. If the protein-polymer complex is needed for stabilizing a system from phase separation, for the benefit of protein-polymer complex formation, the polysaccharides are expected to have relatively lower pI comparing with dairy proteins ($pI_{\text{casein micelle}}$: pH 4.6; $pI_{\beta\text{-Lactoglobulin}}$: pH 5.2; $pI_{\alpha\text{-Lactalbumin}}$: pH ~4.3) so that they remain negatively charged at acidic pH when dairy proteins have zero or positive net charges. Consequently, polysaccharides can

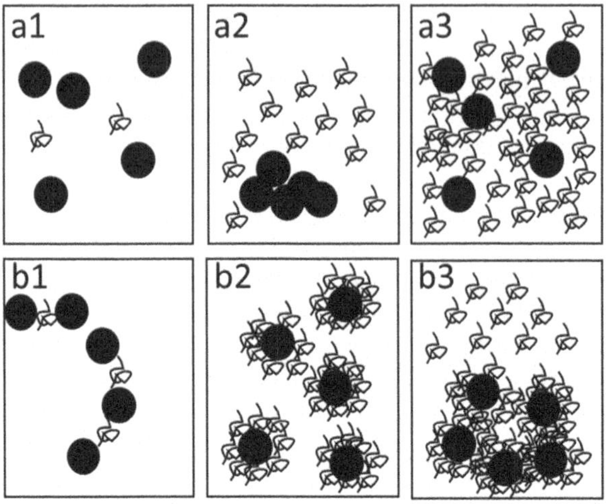

Fig. 19.3 (**a**) Non-adsorbing polymers in dairy protein dispersions. (a1) Stable. (a2) Segregative phase separation (depletion flocculation). (a3) Stable (polymer networks increase viscosity of aqueous phase). (**b**) Adsorbing polymer in dairy protein dispersions. (b1) Bridging flocculation. (b2) Adsorptive stabilization. (b3) Depletion flocculation. *Linear lines* nondairy polymers. *Solid dots* dairy (micellar) proteins. The figure is re-drawn according to (Dickinson 1998)

adsorb onto dairy proteins via electrostatic attraction. Moreover, weak coulombic complex formation between whey protein and polymer (pectin) was observed at region of pH 4–5 when the protein concentration is relatively low where pectin is negatively charged (Cape et al. 1974) and whey proteins have nearly zero net charge (Zaleska et al. 2000; Bystrický et al. 1990). This phenomenon suggests that not only counter charges initiate coulombic attraction, zero net charge molecules may also induce protein-polymer interaction. In general, under acidic condition, upon increasing the concentration of anionic polymer in dairy protein dispersion/solution system, it undergoes three stages transition. (a) Bridging flocculation, this is an unstable state in which polymers bridge dairy proteins forming flocculated complex particles (Gancz et al. 2006; Everett and Mcleod 2005; Langendorff et al. 1999). Bridging flocculation of milk proteins (e.g., casein micelles) are caused by insufficient addition volume of counter charged polymer. (b) Optimum adsorption, this is a stable state in which the amount of polymer molecules are just enough to encapsulate the individual milk protein molecules or micelles (Dickinson 1998; Syrbe et al. 1998). The newly formed protein-polymer complex particles have identical surface charge and being repulsive to each other. The adsorbing polymer on the surface of colloidal protein particles are saturated, and steric repulsion effect is generated between the outer layers formed by nondairy polymers (soybean & soluble polysaccharide) (Nakamura et al. 2006; Nobuhara et al. 2014). The presence of small quantity of free polymer molecules does not necessarily disrupt the stabilized system (Syrbe et al. 1998). (c) Depletion flocculation, if excessive amount of polymers are added into stable protein-polymer colloidal system where nondairy polymer has been already saturated on dairy protein particles, at a certain concentration, the free, non-adsorbed polymers induce depletion flocculation of protein-polymer complexes (Repin et al. 2012; Mession et al. 2012; Rohart and Michon 2014) (Fig. 19.3b).

19.3.3 Challenges of Applying Theories of Stabilization Mechanism

Even though the general destabilization/stabilization mechanisms of a polymer containing dairy system are understood, strong uncertainty exists when choosing proper polysaccharides and applying them for stabilizing a specific dairy colloidal/emulsion system. For example, carrageenan is one of the special stabilizers being used in dairy systems as its stabilization mechanism is still not fully clear. It is still controversial that whether carrageenan interacts with micellar casein forming a polymer-protein complex in milk systems or carrageenan only forms a self-supporting gel which holds up other colloidal particles responsible for the stabilization. Carrageenans are linear, negatively charged (at neutral pH), sulphated polysaccharides containing D-galactose and 3,6-anhydro-D-galactose extracted from red seaweed (*Rhodophyceae*). Three major types of carrageenan as kappa (κ), iota (ι), and lambda (λ)-carrageenans are widely applied in dairy systems (Spagnuolo et al. 2005; Lynch and Mulvihill 1994; Lin and Hansen 1970; Bayarri et al. 2010; Camacho et al. 1998); and they differ in number/position of sulphate groups and the content of 3,6-anhydrogalactosyl ring per disaccharide (Damodaran et al. 2007). k and ι-carrageenans are able to form self-supported gel network in the presence of cations (Drohan et al. 1997; Langendorff et al. 1997), their sulfate groups and the 3,6-anhydro-D-galactopyranosyl ring may undergo coil (disordered) to helix (ordered) transition as the response to temperature change; nevertheless, due to lack of 3,6-anhydro-D-galactopyranosyl ring, λ-carrageenan is not able to gel (Rees et al. 1969). Spagnuolo et al. (2005) summarized two theories which explain the stabilization of micellar casein using k-carrageenan. In the first theory, researchers believe negatively charged k-carrageenan may adsorb onto casein micelles via the interaction with a positively charged region of κ-casein (residues 97–112) (Dalgleish and Morris 1988). However, the second theory states k-carrageenan forms self-supporting gel system with the presence of cations (e.g., Ca^{2+}) and hold up casein micelles or dairy protein stabilized emulsion droplets without phase separation rather than to interact with casein proteins (Drohan et al. 1997; Vega et al. 2005).

A nondairy polymer can be dairy protein non adsorbing and protein adsorbing at different pH levels. Therefore, different stabilizing mechanisms need to be considered when pH, temperature, protein composition and concentration are all formulation variables even the same type of polysaccharide is used for stabilizing a dairy food (Gu et al. 2005; Langendorff et al. 1999; Corredig et al. 2011). For instance, in β-lactoglobulin stabilized emulsion system, at pH 3 (below the pI of whey protein), ι- and λ-carrageenans as adsorbing agents at different concentration levels caused both stabilization (at concentration range: 0% < x < 0.08% w/w) and creaming effects (at concentration range: 0.08% < x < 0.15% w/w) to the same emulsion system, moreover, ι-carrageenan as non-adsorptive agent is able to stabilize the same emulsion system from creaming at pH 6 (above the pI of whey protein) at concentration range 0–0.15% w/w (Gu et al. 2005).

The authors believed that at pH 3, the aggregation of oil droplets is induced by the droplet-carrageenan-droplet bridging effect moreover, they observed a greater change of ζ-potential in the system containing a higher level of carrageenan comparing the emulsion systems containing a lower concentration of carrageenan (Gu et al. 2005). However, according to the aforementioned mechanism of "optimum adsorption" in adsorptive stabilization and the gelling mechanism of carrageenan, it is reasonable to re-interpret the experimental results from this research using an alternative reaction mechanism which is explained as follows. At pH 3, at lower concentration levels of carrageenan (0% < x < 0.08% w/w), weak gel is formed by carrageenan–carrageenan cross-linkages (carrageenan is able to form gel at concentration as low as 0.018% w/w) (Drohan et al. 1997). The weak gel hinders oil droplets interaction and the relatively high surface charge results in electrostatic repulsion between oil droplets. The emulsion system is thus stable. At higher concentration levels of carrageenan (0.08% < x < 0.15% w/w), the carrageenan molecules adsorb onto the surface of emulsion droplets via the interaction with the emulsifying agent, β-lactoglobulin, therefore neutralizing the surface charge of oil droplets. The excessive free carrageenan interacts free β-lactoglobulin forming polymer-protein complex and induces depletion flocculation of oil droplets. Consequently, the emulsion system is destabilized. The interaction mechanism between the nondairy polymer and dairy protein or dairy protein stabilized emulsion droplets is complicated. It is just shown in this paragraph that different speculative mechanisms may be used to explain the rationale of the same set results. Such uncertainties in stabilizing dairy food systems are the driving force for initiating more systemic research so that one can draw a clearer picture of the functionality of a nondairy polymer in a colloidal/emulsion systems containing milk proteins or milk protein stabilized oil droplets.

The significant suggestion to dairy technologists is that one should hypothesize a stabilization mechanism of the targeted product before formulating and processing the dairy matrix. Subsequently, the microstructure and stability can be engineered by selecting proper nondairy polymers, applying them at proper concentrations, adjusting system temperature, pH, and ionic strength, etc. for achieving the non-adsorptive or adsorptive stabilization. Sometimes, probably most cases of commercial product applications, both non-adsorptive and adsorptive reactions may occur in one system for achieving the physical stability. For instance, in the stabilized system containing gelatinized starch granules, leached starch polymers (amylose and amylopectin) and dairy proteins (whey and casein), the starch-casein association presents in the dairy protein dispersion system. Moreover, starch self-associated networks also present in the dairy protein dispersion system. These starch-protein and starch-starch networks are responsible for increasing complex viscosity (Kumar et al. 2017).

19.4 Texture Establishment

Besides stabilization, nondairy polymers are applied as texturizers to manipulate the rheological, tribological properties of dairy foods so that the market preferred sensory attributes may be achieved (Foegeding et al. 2010; Marshall and Rawson 1999; Foegeding 2007; Van den Berg et al. 2007). The heterogeneous group of long-chain polymers mainly polysaccharides and proteins/polypeptides are able to engage water molecules forming viscous dispersion system in the aqueous phase of dairy dispersion/emulsion/gel systems. The engagement of water molecules is due to the presence of large quantity of hydroxyl groups in the nondairy polymers. The capability of forming higher viscous dispersion system makes the hydrocolloid as a "thickening agent" which is used for providing body and increasing creaminess in dairy food products. A gel as tangled and interconnected molecular network may also be formed in the aqueous phase of dairy food system through crosslinking of polymers (Oakenfull and Glicksman 1987). The gel system may provide mechanical rigidity, and it may be flowable liquid-like (injectable, low yield stress) gel or non-flowable (highly elastic, high yield stress) solid-like gel (Piron and Tholin 2001; Le et al. 2017; Brenner et al. 2015). In general, the polymer-polymer and polymer-dairy protein interactions defer to the aforementioned of non-adsorptive and adsorptive interaction mechanisms.

19.4.1 Thickening Dairy Matrixes

Polymers as thickeners in dispersion system presented as polymer random coils; its concentration determines the physical feature of the dispersion system (Daoud et al. 1975). Three concentration domains are defined as separated chains, overlapping chains, and concentrated solution regime (Berry et al. 1979). The thickening effect derives from polymer-solvent interactions, and it is strongly correlated with the restricted freedom of movement of individual chains of the polymer which is determined by the degree of overlapping chain (Morris 1994). Therefore, it is logical to state that an ideal concentration of polymer is needed for initiating functional thickening effect. The ideal concentration is named as "critical coil overlap concentration" (c^*), the polymer chain starts overlapping and entangling between each other. Consequently, the viscosity of the dispersion/emulsion systems significantly increases (Baines and Morris 1987). Therefore, for thickening and improving month feel of dairy food using nondairy hydrocolloids, one needs to know the c^* of the polymer candidate. Although the physical features and perception of taste intensity of polysaccharides fortified solution systems have been studied, more research through engineering approaches is required for constructing mathematical models containing c^* as one of the key predictors which may be used in predictions of sensory attributes.

c^* is the critical parameter which determines the transition of rheology nature of dispersion system between the Newtonian fluid and non-Newtonian fluid (Phillips and Williams 2009). The c^* may be measured and calculated via using experimental approaches (Morris et al. 1981; Cook et al. 2002), and may also be estimated from prediction models (Ying and Chu 1987). In the experimental approach, zero-shear specific viscosity (η_0) and intrinsic viscosity (η) are measured and estimated using Huggins and Kraemer models (Eqs. (19.1) and (19.2), η: intrinsic viscosity, η_{sp}: specific viscosity, c: concentration). Then c^* is calculated using the model log($c \times \eta$) (x-axis)-log (η_0) (y-axis), c^* is the c which results the intersect point on the log-log model (Morris et al. 1981, Cook et al. 2002). In a recent study, the author stated intrinsic viscosity is used for estimating c^* using the equation $c^* = 1/\eta$ (Van der Sman 2015); intrinsic viscosity can be expressed by Fiery-Fox relationship (Flory 1953).

$$\frac{\eta_{sp}}{c} = \eta + k'\eta^2 c \tag{19.1}$$

$$\frac{\ln \eta_{rel}}{c} = \eta + k''\eta^2 c \tag{19.2}$$

Starch as its native or modified forms has been widely used in dairy food formulations as a thickening agent (Gutiérrez et al. 2017; Gutiérrez 2018). This may due to its relatively lower cost comparing with other gum stabilizers and relatively clean taste (Saha and Bhattacharya 2010). In dairy pudding dessert system containing milk proteins, carrageenan, and unmodified starch granules, it was found that the gelatinized starch granules do not contribute the gel structure formation (Verbeken et al. 2004). Continuously increasing the concentration of starch results exclusion effect. This effect causes a concentration of gelling agent in the aqueous phase of the dairy food system, therefore, strengthens the gel structure. Exclusion effect dominates the starch enriched dairy food system containing gelling agents, and such effect is more pronounced when the effects caused from dairy protein and gelling polysaccharide (carrageenan) (Verbeken et al. 2004). Such starch induced exclusion effect may be utilized in the dairy formulation strategy when one attempts to enhance the dairy gel structure without using additional relatively expensive dairy proteins and gelling gums.

Xanthan gum as a thickening agent is used for increasing viscosity of the different dairy liquid or semi-solid systems at different pH levels (Hemar et al. 2001a, b; El-Sayed et al. 2002). Its viscosity dominated the viscosity of protein dispersion systems made from different dairy protein ingredients at neutral pH (incl. skim milk powder, milk protein concentrate and sodium caseinate) (Hemar et al. 2001b). However, phase separation appeared in the dispersion systems when xanthan gum is mixed with skim milk powder and milk protein concentrate (Hemar et al. 2001b). The author attributed the phase separation to depletion flocculation. Low concentration of xanthan gum ($\leq$0.2% w/w) induced visual creaming of milk protein stabilized emulsion droplets at neutral pH (Hemar et al. 2001a). In yogurt (acidified dairy protein gel system), the addition of a small volume of xanthan gum (at concentrations of 0.01% and 0.05%, w/w) resulted in a dramatic increase of gel curd tension and a significant decrease of syneresis rate (El-Sayed et al. 2002). Xanthan gum and locust bean gum (LBG) are co-used in food gel systems for improving sensory and rheological attributes due to the proven synergetic effects (Juszczak et al. 2003). However, such synergy does not necessarily appear in an emulsion system; it found that xanthan gum-LBG combination did not result in higher viscosity in a mayonnaise-like emulsion comparing the system containing the same amount of sole type of polymer (Dolz et al. 2007).

Other polysaccharides such as carboxymethyl cellulose (CMC), methyl cellulose (MC), hydroxypropylmethyl cellulose (HPMC), gum Arabic (GA), guar gum (GG), tara gum (TG), konjac mannan (KM), gum tragacanth (GT) are all know as thickening agents in dairy food systems at different pH levels (Bayarri et al. 2009; Zhao et al. 2009; Arboleya and Wilde 2005; Ibanoğlu 2002; Mudgil et al. 2014; Bourriot et al. 1999; Saha and Bhattacharya 2010; Tobin et al. 2011; Azarikia and Abbasi 2010). Moreover, some of these aforementioned polymers may also be used as emulsifiers for stabilizing emulsion oil droplets, for instance, MC, HPMC, GA (Arboleya and Wilde 2005; Mcnamee et al. 1998). Bayarri et al. (2009) found that when a higher concentration of CMC (>1%, w/w) is applied in skim milk, a weak gel rheology feature is identified. Therefore, it is important to point out that the concepts about thickening agent, gelling agent, emulsifying agent are an empirical description of the functionality of polymers rather than absolute definition. Dairy technologists need to know the broad functionalities of individual polysaccharides in different dairy systems rather than simply group a polymer as a thickening agent or a gelling agent.

19.4.2 Smoothing Dairy Matrixes

Creaminess is one of the key sensory attributes of dairy foods which strongly correlated with consumers' hedonic response (Folkenberg and Martens 2003). It is generally understood that creaminess perception in dairy products are determined by both flavor and texture, and the fat content plays a crucial role (Mela 1988). Reducing fat content from dairy matrix results in undesirable sensory properties and less consumer acceptance (Cardello 1994; Tuorila et al. 1994). Polysaccharides are used as fat replacers in fat-reduced dairy foods for compensating thickness, rheological behavior and creaminess perception (Bayarri et al. 2010). However, it is important to point out that the detailed mechanism of creaminess perception is not fully understood. It might be combined sensations of flavor, texture, and psychology (Drake 1989; Antmann et al. 2011; Frøst and Janhøj 2007; Kilcast and Clegg 2002; Elmore et al. 1999). The general agreement according to a series of research states that structure, texture, and smoothness are highly correlated with creaminess perception (Akhtar et al. 2006; Elmore et al. 1999; Kilcast and Clegg 2002). Extensive knowledge and research findings are available for manipulating structure and texture (mainly rheological features) properties of dairy foods. However, techniques about how to engineer dairy food structure using non-dairy polymers with enhanced smoothness are relatively scarce. Clearly, creaminess improvement is far beyond manipulating rheological properties. Classically, viscosity at 50 s^{-1} shear rate is used as an indication of mouthfeel perception (Wood 1968). However, it was noted that such parameter is insufficient for describing perceived thickness or creaminess of dairy-based emulsion systems containing nondairy polymers used as texturizer (Akhtar et al. 2006). Recently, tribology techniques are developed for quantitatively characterizing the smoothness of dairy systems (Sonne et al. 2014; Nguyen et al. 2016; Meyer et al. 2011b; Laguna et al. 2017; Dresselhuis et al. 2008). Such technical developments open a gate for screening tribological functionalities of food polymers in dairy food matrices.

Regarding modifying creaminess perception, maltodextrin (13.5% w/w) and xanthan gum (0.17%), respectively, are able to enhance creaminess perception of dairy protein stabilized emulsion system, and maltodextrin is relatively more functional comparing with xanthan gum in terms of elevation of creaminess level (Akhtar et al. 2006). The authors controlled the viscosities of the emulsion systems which contain the two different polymers at the same level at a reference shear rate (50 s^{-1}). They noted that the sensory creaminess perception might be altered by nondairy polymer without changing viscosity (Akhtar et al. 2006). The same research group also found that low methoxyl pectin is more functional than xanthan gum for increasing creaminess of dairy emulsion at the relatively thinner system (viscosity 50 mPa.s at 50 s^{-1}). Such discrimination of functionality vanished at the thicker system (viscosity 100 mPa.s at 50 s^{-1}) (Akhtar et al. 2005). It is important to reveal that although pectin and xanthan gum were able to increase creaminess perception, their capability of creaminess enhancement is not comparable with fat content. It is interesting to note that at the same viscosity level (50 mPa.s at 50 s^{-1}), the creaminess perception of pectin fortified emulsion is still lower than the plain emulsion system, even though the fat content of the pectin-containing emulsion (22% vol/vol) is slightly higher than the plain emulsion (20% vol/vol) (Akhtar et al. 2005). Therefore such polymers cannot be used as fat replacers for maintaining the intrinsic creaminess.

In the dairy protein gel structured system, nondairy polymers may be used for improving creaminess. λ-carrageenan as a non-gelling agent only increases the viscosity of aqueous phase in milk or dairy gel system; it was reported that λ-carrageenan at 0.06% (w/w) increased creaminess of gelled dairy dessert (Tarrega and Costell 2006). Application of long-chain chicory inulin (4% w/w) in no fat yogurt was found to be able to mimic the rheological features of full-fat yogurt. However, no sensory data is available in that research (Paseephol et al. 2008). Mimicking the rheological parameters does not guarantee matching of creaminess (Szczesniak 2002). 8% (w/w) of long-chain inulin was recommended to be added to skim milk for compensating the creaminess of whole fat milk (Villegas et al. 2007). Meyer and co-authors (2011a) elaborated the working mechanism of inulin in dairy products as texture modifier for enhancing creaminess, it was noticed that the smoothing effect of long-chain inulin is hindered by higher concentration of starch (4% w/w). For increasing the creaminess of low fat yogurt, inulin must be applied before fermentation and be part of the protein gel structure (Kip et al. 2003). Inulin can interfere with the extracellular polysaccharides, therefore, reduce the "brush friction" resulting elevating creaminess level (Kip et al. 2006; Marle et al. 1999).

19.5 Conclusion

Application of nondairy polymers offers dairy technologists a great opportunity to construct sophisticated dispersion/gel structures which are essential for maintaining the physical stability of innovative; value-added dairy products. The nondairy polymers are capable enough to modify the texture of dairy foods resulting desirable changes of sensory attributes. The general interaction mechanisms of different type of polymers in liquid/gel based dairy systems are explained in this chapter. Such information as a powerful guideline provides strategic approaches for future formulation of dairy foods, for instance, one would know how to prevent segregative phase separation and associative phase separation when thickening agents and gelling agents are needed in what type of dairy system. However, the technical boundaries of the functionality of individual polymers and detailed mechanism of polymer-dairy system interaction are not fully clear. For the interest of specific application, further systemic research is needed for mapping the functionalities of individual polymers in different dairy systems.

References

Acero-Lopez A, Alexander M, Corredig M (2010) Diffusing wave spectroscopy and rheological studies of rennet-induced gelation of skim milk in the presence of pectin and κ-carrageenan. Int Dairy J 20:328–335

Akhtar M, Stenzel J, Murray BS, Dickinson E (2005) Factors affecting the perception of creaminess of oil-in-water emulsions. Food Hydrocoll 19:521–526

Akhtar M, Murray BS, Dickinson E (2006) Perception of creaminess of model oil-in-water dairy emulsions: Influence of the shear-thinning nature of a viscosity-controlling hydrocolloid. Food Hydrocoll 20:839–847

Allen Foegeding E, Çakir E, Koç H (2010) Using dairy ingredients to alter texture of foods: implications based on oral processing considerations. Int Dairy J 20:562–570

Anal AK, Singh H (2007) Recent advances in microencapsulation of probiotics for industrial applications and targeted delivery. Trends Food Sci Technol 18:240–251

Antmann G, Ares G, Salvador ANA, Varela P, Fiszman SM (2011) Exploring and explaining creaminess perception: consumers' underlying concepts. J Sens Stud 26:40–47

Arboleya J-C, Wilde PJ (2005) Competitive adsorption of proteins with methylcellulose and hydroxypropyl methylcellulose. Food Hydrocoll 19:485–491

Azarikia F, Abbasi S (2010) On the stabilization mechanism of Doogh (Iranian yoghurt drink) by gum tragacanth. Food Hydrocoll 24:358–363

Azegami S, Tsuboi A, Izumi T, Hirata M, Dubin PL, Wang B, Kokufuta E (1999) Formation of an intrapolymer complex from human serum albumin and poly(ethylene glycol). Langmuir 15:940–947

Baines ZV, Morris ER (1987) Flavour/taste perception in thickened systems: the effect of guar gum above and below c*. Food Hydrocoll 1:197–205

Bayarri S, González-Tomás L, Costell E (2009) Viscoelastic properties of aqueous and milk systems with carboxymethyl cellulose. Food Hydrocoll 23:441–450

Bayarri S, Chuliá I, Costell E (2010) Comparing λ-carrageenan and an inulin blend as fat replacers in carboxymethyl cellulose dairy desserts. Rheological and sensory aspects. Food Hydrocoll 24:578–587

Berry GC, Nakayasu H, Fox TG (1979) Viscosity of poly(vinyl acetate) and its concentrated solutions. J Polym Sci Polym Phys Ed 17:1825–1844

Bourriot S, Garnier C, Doublier J-L (1999) Phase separation, rheology and structure of micellar casein-galactomannan mixtures. Int Dairy J 9:353–357

Brenner T, Tuvikene R, Fang Y, Matsukawa S, Nishinari K (2015) Rheology of highly elastic iota-carrageenan/kappa-carrageenan/xanthan/konjac glucomannan gels. Food Hydrocoll 44:136–144

Bystrický S, Malovíková A, Sticzay T (1990) Interaction of alginates and pectins with cationic polypeptides. Carbohydr Polym 13:283–294

Camacho M, Martínez-Navarrete N, Chiralt A (1998) Influence of locust bean gum/λ-carrageenan mixtures on whipping and mechanical properties and stability of dairy creams. Food Res Int 31:653–658

Cape JN, Cook DH, Williams DR (1974) Thermodynamic considerations in co-ordination. Part XIX. In vitro studies of complexing equilibria involved in oral iron(II) therapy. J Chem Soc Dalton Trans:1849–1852

Cardello AV (1994) Consumer expectations and their role in food acceptance. In: Macfie HJH, Thomson DMH (eds) Measurement of food preferences. Springer, Boston, MA

Considine T, Noisuwan A, Hemar Y, Wilkinson B, Bronlund J, Kasapis S (2011) Rheological investigations of the interactions between starch and milk proteins in model dairy systems: a review. Food Hydrocoll 25:2008–2017

Cook DJ, Hollowood TA, Linforth RST, Taylor AJ (2002) Perception of taste intensity in solutions of random-coil polysaccharides above and below c∗. Food Qual Prefer 13:473–480

Corredig M, Sharafbafi N, Kristo E (2011) Polysaccharide–protein interactions in dairy matrices, control and design of structures. Food Hydrocoll 25:1833–1841

Dai S, Jiang F, Shah NP, Corke H (2017) Stability and phase behavior of konjac glucomannan-milk systems. Food Hydrocolloids 73:30–40

Dalgleish DG (1997) Structure-function relationships of caseins. In: Damodaran S (ed) Food proteins and their applications. Taylor & Francis

Dalgleish DG, Morris ER (1988) Interactions between carrageenans and casein micelles: electrophoretic and hydrodynamic properties of the particles. Food Hydrocoll 2:311–320

Damodaran S, Parkin KL, Fennema OR (2007) Fennema's food chemistry, 4th edn. CRC Press, Boca Raton

Daoud M, Cotton JP, Farnoux B, Jannink G, Sarma G, Benoit H, Duplessix C, Picot C, de Gennes PG (1975) Solutions of flexible polymers. Neutron experiments and interpretation. Macromolecules 8:804–818

De Bont PW, van Kempen GMP, Vreeker R (2002) Phase separation in milk protein and amylopectin mixtures. Food Hydrocoll 16:127–138

Dickinson E (1998) Stability and rheological implications of electrostatic milk protein–polysaccharide interactions. Trends Food Sci Technol 9:347–354

Dolz M, Hernández MJ, Delegido J, Alfaro MC, Muñoz J (2007) Influence of xanthan gum and locust bean gum upon flow and thixotropic behaviour of food emulsions containing modified starch. J Food Eng 81:179–186

Doublier JL, Garnier C, Renard D, Sanchez C (2000) Protein–polysaccharide interactions. Curr Opin Colloid Interface Sci 5:202–214

Drake B (1989) Sensory textural/rheological properties—a polyglot list. J Texture Stud 20:1–27
Dresselhuis DM, de Hoog EHA, Cohen Stuart MA, Van Aken GA (2008) Application of oral tissue in tribological measurements in an emulsion perception context. Food Hydrocoll 22:323–335
Drohan DD, Tziboula A, Mcnulty D, Horne DS (1997) Milk protein-carrageenan interactions. Food Hydrocoll 11:101–107
EFSA (2017a) Food additives [Online]. European Food Safety Authority. http://www.efsa.europa.eu/en/topics/topic/food-additives
EFSA (2017b) Food additive re-evaluations [Online]. European Food Safety Authority. http://www.efsa.europa.eu/en/topics/topic/food-additive re-evaluations Accessed 26 Dec 2017
Elmore JR, Heymann H, Johnson J, Hewett JE (1999) Preference mapping: relating acceptance of "creaminess" to a descriptive sensory map of a semi-solid. Food Qual Prefer 10:465–475
El-Sayed E, Abd El-Gawad I, Murad H, Salah S (2002) Utilization of laboratory-produced xanthan gum in the manufacture of yogurt and soy yogurt. Eur Food Res Technol 215:298–304
Everett DW, Mcleod RE (2005) Interactions of polysaccharide stabilisers with casein aggregates in stirred skim-milk yoghurt. Int Dairy J 15:1175–1183
FDA (2017) Food Additives Status List [Online]. U.S. Food and Drug Administration, Spring, MD. https://www.fda.gov/Food/IngredientsPackagingLabeling/FoodAdditivesIngredients/ucm091048.htm#collapse2014. Accessed 26 Dec 2017
Flory PJ (1953) Principles of polymer chemistry. Cornell University Press, Ithaca, NY
Foegeding EA (2007) Rheology and sensory texture of biopolymer gels. Curr Opin Colloid Interface Sci 12:242–250
Folkenberg DM, Martens M (2003) Sensory properties of low fat yoghurts. Part B: hedonic evaluations of plain yoghurts by consumers correlated to fat content, sensory profile and consumer attitudes. Milchwissenschaft 58:154–157
Frøst MB, Janhøj T (2007) Understanding creaminess. Int Dairy J 17:1298–1311
Gancz K, Alexander M, Corredig M (2006) In situ study of flocculation of whey protein-stabilized emulsions caused by addition of high methoxyl pectin. Food Hydrocoll 20:293–298
Gu YS, Decker EA, Mcclements DJ (2005) Influence of pH and carrageenan type on properties of β-lactoglobulin stabilized oil-in-water emulsions. Food Hydrocoll 19:83–91
Gutiérrez TJ (2018) Characterization and in vitro digestibility of non-conventional starches from guinea arrowroot and La Armuña lentils as potential food sources for special diet regimens. Starch-Stärke 70(1–2). https://doi.org/10.1002/star.201700124
Gutiérrez TJ, González Seligra P, Medina Jaramillo C, Famá L, Goyanes S (2017) Effect of filler properties on the antioxidant response of thermoplastic starch composites. In: Thakur VK, Thakur MK, Kessler MR (eds) Handbook of composites from renewable materials. Wiley-Scrivener Publisher, pp 337–370. https://doi.org/10.1002/9781119441632.ch14. ISBN: 978-1-119-22362-7
Hansen PMT (1993) Food hydrocolloids in the dairy industry. In: Nishinari K, Doi E (eds) Food hydrocolloids: structures, properties, and functions. Springer US, Boston, MA
Hemar Y, Tamehana M, Munro PA, Singh H (2001a) Influence of xanthan gum on the formation and stability of sodium caseinate oil-in-water emulsions. Food Hydrocoll 15:513–519
Hemar Y, Tamehana M, Munro PA, Singh H (2001b) Viscosity, microstructure and phase behavior of aqueous mixtures of commercial milk protein products and xanthan gum. Food Hydrocoll 15:565–574
Huppertz T, Kelly AL (2006) Physical Chemistry of Milk Fat Globules. In: Fox PF, Mcsweeney PLH (eds) Advanced dairy chemistry volume 2: lipids. Springer, Boston, MA
Ibanoğlu E (2002) Rheological behaviour of whey protein stabilized emulsions in the presence of gum arabic. J Food Eng 52:273–277
Juszczak L, Fortuna T, Kośla A (2003) Sensory and rheological properties of Polish commercial mayonnaise. Food Nahrung 47:232–235
Kilcast D, Clegg S (2002) Sensory perception of creaminess and its relationship with food structure. Food Qual Prefer 13:609–623
Kip P, Peters B, Meyer D (2003) Improving mouthfeel and texture of stirred yoghurt by the addition of inulin [P37]. In Abstract of third NIZO dairy conference, Dynamics of texture, process & perception, NIZO, Papendal, The Netherlands

Kip P, Meyer D, Jellema RH (2006) Inulins improve sensoric and textural properties of low-fat yoghurts. Int Dairy J 16:1098–1103
Kruif CG, Holt C (2003) Casein micelle structure, functions and interactions. In: Fox PF, Mcsweeney PLH (eds) Advanced dairy chemistry—1 proteins: Part A/Part B. Springer, Boston, MA
Kumar L, Brennan MA, Mason SL, Zheng H, Brennan CS (2017) Rheological, pasting and microstructural studies of dairy protein–starch interactions and their application in extrusion-based products: a review. Starch Stärke 69:1600273
Kumar L, Brennan M, Zheng H, Brennan C (2018) The effects of dairy ingredients on the pasting, textural, rheological, freeze-thaw properties and swelling behaviour of oat starch. Food Chem 245:518–524
Laguna L, Farrell G, Bryant M, Morina A, Sarkar A (2017) Relating rheology and tribology of commercial dairy colloids to sensory perception. Food Funct 8:563–573
Langendorff V, Cuvelier G, Launay B, Parker A (1997) Gelation and flocculation of casein micelle/carrageenan mixtures. Food Hydrocoll 11:35–40
Langendorff V, Cuvelier G, Launay B, Michon C, Parker A, de Kruif CG (1999) Casein micelle/iota carrageenan interactions in milk: influence of temperature. Food Hydrocoll 13:211–218
Langendorff V, Cuvelier G, Michon C, Launay B, Parker A, de Kruif CG (2000) Effects of carrageenan type on the behaviour of carrageenan/milk mixtures. Food Hydrocoll 14:273–280
Le XT, Rioux L-E, Turgeon SL (2017) Formation and functional properties of protein–polysaccharide electrostatic hydrogels in comparison to protein or polysaccharide hydrogels. Adv Colloid Interf Sci 239:127–135
Li K, Woo MW, Patel H, Selomulya C (2017) Enhancing the stability of protein-polysaccharides emulsions via Maillard reaction for better oil encapsulation in spray-dried powders by pH adjustment. Food Hydrocoll 69:121–131
Lin C, Hansen P (1970) Stabilization of casein micelles by carrageenan. Macromolecules 3:269–274
Lizarraga MS, Piante Vicin DD, González R, Rubiolo A, Santiago LG (2006) Rheological behaviour of whey protein concentrate and λ-carrageenan aqueous mixtures. Food Hydrocoll 20:740–748
Lynch MG, Mulvihill DM (1994) The influence of caseins on the rheology of ι-carrageenan gels. Food Hydrocoll 8:317–329
Marle MEV, Ende DVD, Kruif CGD, Mellema J (1999) Steady-shear viscosity of stirred yogurts with varying ropiness. J Rheol 43:1643–1662
Maroziene A, de Kruif CG (2000) Interaction of pectin and casein micelles. Food Hydrocoll 14:391–394
Marshall VM, Rawson H (1999) Effects of exopolysaccharide-producing strains of thermophilic lactic acid bacteria on the texture of stirred yoghurt. Int J Food Sci Technol 34:137–143
Mcnamee BF, O'Riorda ED, O'Sullivan M (1998) Emulsification and microencapsulation properties of gum arabic. J Agric Food Chem 46:4551–4555
Mela DJ (1988) Sensory assessment of fat content in fluid dairy products. Appetite 10:37–44
Mession JL, Assifaoui A, Lafarge C, Saurel R, Cayot P (2012) Protein aggregation induced by phase separation in a pea proteins–sodium alginate–water ternary system. Food Hydrocoll 28:333–343
Meyer D, Bayarri S, Tárrega A, Costell E (2011a) Inulin as texture modifier in dairy products. Food Hydrocoll 25:1881–1890
Meyer D, Vermulst J, Tromp RH, de Hoog EHA (2011b) The effect of inulin on tribology and sensory profiles of skimmed milk. J Texture Stud 42:387–393
Morris ER (1994) Rheological and organoleptic properties of food hydrocolloids. Food hydrocolloids. Springer, Boston, MA
Morris ER, Cutler AN, Ross-Murphy SB, Rees DA, Price J (1981) Concentration and shear rate dependence of viscosity in random coil polysaccharide solutions. Carbohydr Polym 1:5–21
Mudgil D, Barak S, Khatkar BS (2014) Guar gum: processing, properties and food applications—a Review. J Food Sci Technol 51:409–418

Nakamura A, Yoshida R, Maeda H, Corredig M (2006) The stabilizing behaviour of soybean soluble polysaccharide and pectin in acidified milk beverages. Int Dairy J 16:361–369

Nguyen PTM, Bhandari B, Prakash S (2016) Tribological method to measure lubricating properties of dairy products. J Food Eng 168:27–34

Niederauer MQ, Glatz CE (1994) Model of the polyelectrolyte precipitation of genetically engineered enzymes possessing charged polypeptide tails. J Macromol Sci A 31:127–153

Nobuhara T, Matsumiya K, Nambu Y, Nakamura A, Fujii N, Matsumura Y (2014) Stabilization of milk protein dispersion by soybean soluble polysaccharide under acidic pH conditions. Food Hydrocoll 34:39–45

Oakenfull D, Glicksman M (1987) Gelling agents. CRC Crit Rev Food Sci Nutr 26:1–25

Paseephol T, Small DM, Sherkat F (2008) Rheology and texture of set yogurt as affected by inulin addition. J Texture Stud 39:617–634

Phillips GO, Williams PA (2009) Handbook of hydrocolloids. Elsevier Science, Amsterdam

Piron E, Tholin R (2001) Polysaccharide crosslinking, hydrogel preparation, resulting polysaccharide(s) and hydrogel(s), uses thereof. Google Patents

Porcar I, Cottet H, Gareil P, Tribet C (1999) Association between protein particles and long amphiphilic polymers: effect of the polymer hydrophobicity on binding isotherms. Macromolecules 32:3922–3929

Rees D, Steele I, Williamson F (1969) Conformational analysis of polysaccharides. III. The relation between stereochemistry and properties of some natural polysaccharide sulfates (1). J Polym Sci Polym Symp 28(1):261–276

Repin N, Scanlon MG, Gary Fulcher R (2012) Phase behaviour of casein micelles and barley beta-glucan polymer molecules in dietary fibre-enriched dairy systems. J Colloid Interface Sci 377:7–12

Rodd AB, Davis CR, Dunstan DE, Forrest BA, Boger DV (2000) Rheological characterisation of 'weak gel' carrageenan stabilised milks. Food Hydrocoll 14:445–454

Rohart A, Michon C (2014) Designing microstructure into xanthan gum-enriched acid milk gels. Innovative Food Sci Emerg Technol 25:53–57

Saha D, Bhattacharya S (2010) Hydrocolloids as thickening and gelling agents in food: a critical review. J Food Sci Technol 47:587–597

Schorsch C, Jones MG, Norton IT (2000) Phase behaviour of pure micellar casein/κ-carrageenan systems in milk salt ultrafiltrate. Food Hydrocoll 14:347–358

Shahidi F, Han XQ (1993) Encapsulation of food ingredients. Crit Rev Food Sci Nutr 33:501–547

Sonne A, Busch-Stockfisch M, Weiss J, Hinrichs J (2014) Improved mapping of in-mouth creaminess of semi-solid dairy products by combining rheology, particle size, and tribology data. LWT Food Sci Technol 59:342–347

Spagnuolo PA, Dalgleish DG, Goff HD, Morris ER (2005) Kappa-carrageenan interactions in systems containing casein micelles and polysaccharide stabilizers. Food Hydrocoll 19:371–377

Suresh B, Remco T, Peter S (2006) Spinodal decomposition in a food colloid–biopolymer mixture: evidence for a linear regime. J Phys Condens Matter 18:L339

Syrbe A, Bauer WJ, Klostermeyer H (1998) Polymer science concepts in dairy systems—an overview of milk protein and food hydrocolloid interaction. Int Dairy J 8:179–193

Szczesniak AS (2002) Texture is a sensory property. Food Qual Prefer 13:215–225

Tarrega A, Costell E (2006) Effect of composition on the rheological behaviour and sensory properties of semisolid dairy dessert. Food Hydrocoll 20:914–922

Thies C (2003) MICROCAPSULES A2—caballero, benjamin. encyclopedia of food sciences and nutrition, 2nd edn. Academic Press, Oxford

Tobin JT, Fitzsimons SM, Kelly AL, Fenelon MA (2011) The effect of native and modified konjac on the physical attributes of pasteurized and UHT-treated skim milk. Int Dairy J 21:790–797

Tuinier R, Rieger J, de Kruif CG (2003) Depletion-induced phase separation in colloid–polymer mixtures. Adv Colloid Interf Sci 103:1–31

Tuorila H, Cardello AV, Lesher LL (1994) Antecedents and consequences of expectations related to fat-free and regular-fat foods. Appetite 23:247–263

Van den Berg L, van Vliet T, van der Linden E, van Boekel M, van de Velde F (2007) Breakdown properties and sensory perception of whey proteins/polysaccharide mixed gels as a function of microstructure. Food Hydrocoll 21:961–976

Van der Sman RGM (2015) Biopolymer gel swelling analysed with scaling laws and Flory–Rehner theory. Food Hydrocoll 48:94–101

Vasir JK, Tambwekar K, Garg S (2003) Bioadhesive microspheres as a controlled drug delivery system. Int J Pharm 255:13–32

Vega C, Dalgleish DG, Goff HD (2005) Effect of κ-carrageenan addition to dairy emulsions containing sodium caseinate and locust bean gum. Food Hydrocoll 19:187–195

Verbeken D, Thas O, Dewettinck K (2004) Textural properties of gelled dairy desserts containing κ-carrageenan and starch. Food Hydrocoll 18:817–823

Villegas B, Carbonell I, Costell E (2007) Inulin milk beverages: sensory differences in thickness and creaminess using R-index analysis of the ranking data. J Sens Stud 22:377–393

Walstra P (2006) Dairy science and technology. CRC Press/Taylor & Francis, Boca Raton

Walstra P, Wouters JTM, Geurts TJ (2006) Dairy science and technology. CRC/Taylor & Francis, Boca Raton, FL

Wang X, Lee J, Wang Y-W, Huang Q (2007) Composition and Rheological Properties of β-Lactoglobulin/Pectin Coacervates: Effects of Salt Concentration and Initial Protein/Polysaccharide Ratio. Biomacromolecules 8:992–997

Wood FW (1968) Psychophysical studies on the consistency of liquid foods. Rheology and texture of foodstuffs. Society of Chemical Industry, London

Ying Q, Chu B (1987) Overlap concentration of macromolecules in solution. Macromolecules 20:362–366

Zaleska H, Ring SG, Tomasik P (2000) Apple pectin complexes with whey protein isolate. Food Hydrocoll 14:377–382

Zhao Q, Zhao M, Li J, Yang B, Su G, Cui C, Jiang Y (2009) Effect of hydroxypropyl methylcellulose on the textural and whipping properties of whipped cream. Food Hydrocoll 23:2168–2173

Zheng H, Jiménez-Flores R, Everett DW (2014) Lateral lipid organization of the bovine milk fat globule membrane is revealed by washing processes. Journal of Dairy Science 97(10):5964–5974

Chapter 20
Partially Hydrolyzed Guar Gum: Preparation and Properties

Deepak Mudgil

Abstract Guar gum is a galactomannan obtained from the endosperm portion of seed of the plant *Cyamopsis tetragonolobus*. Guar gum is used as thickener and stabilizer in food industry due to its high viscosity in aqueous solution. The partial hydrolysis of the guar galactomannan leads to production of partially hydrolyzed guar gum (PHGG) which is similar in structure to native guar galactomannan. PHGG can be produced by enzymatic hydrolysis, acid hydrolysis, irradiation, microwave and ultrasonication techniques. Enzymatic hydrolysis of guar gum is preferred for food processing applications. In enzymatic hydrolysis, enzymes such as mannanse, pectinase, cellulase cut the linkages between mannose units in the main chain. Partial hydrolysis of guar gum leads to the reduction in molecular weight of native guar gum. Partially hydrolyzed guar gum shows low viscosity in aqueous solutions. X-ray diffraction analysis of partially hydrolyzed guar gum revealed that it is little crystalline in nature as compared to native guar galactomannan which is amorphous in nature. Partially hydrolyzed guar gum produced from enzymatic hydrolysis can be used mainly for nutritional purpose i.e. for development of fiber enriched processed food products such as cookies, bread, noodles, yoghurt etc. Its specific physicochemical properties make it possible to improve the quality of food products. Partially hydrolyzed guar gum as soluble dietary fibre is beneficial in diabetes, heart disease and digestive problems.

Keywords Cardiovascular diseases · Diabetes · Dietary fiber · Galactomannan · Prebiotic

D. Mudgil (✉)
Department of Dairy and Food Technology, Mansinhbhai Institute of Dairy and Food Technology, Mehsana, India

T. J. Gutiérrez (ed.), *Polymers for Food Applications*,
https://doi.org/10.1007/978-3-319-94625-2_20

20.1 Introduction

Guar gum is a water soluble galactomannan obtained from seed endosperm of guar plant i.e. *Cyamopsis tetragonolobus*, a leguminous plant grown mainly in India and Pakistan and shares about 90% of world guar production. India's share in world guar production is 80% which makes it a world leader in guar production (Mudgil et al. 2014a). Guar gum is basically present in endosperm fraction of guar seeds as galactomannan. Its main property is its water solubility at room temperature and formation of highly viscous solution in water at low temperature as well as at high temperature. This high viscosity in aqueous solution makes guar gum commercially valuable. Most of the food products have water phase in them and guar gum has capability of water phase management in these food products hence can be used as a stabilizing agent, thickening agent and improvers in numerous food products such as beverages, ice cream, sauces, cakes, bread, noodles, fruit crush, squashes, etc. (Mudgil et al. 2011). The high viscosity of guar gum becomes a problem when it has to be utilized as dietary fiber (Greenberg and Sellman 1998). Dietary fibers are the complex carbohydrates which resist the human digestive secretion and resist their digestion and absorption in human alimentary canal. Dietary fibers are considered very useful in prevention of certain disease such as cardiovascular, diabetes and digestive problems. Human digestive enzymes cannot digest guar gum hence it can also be utilized as dietary fiber (Mudgil and Barak 2013; Butt et al. 2007). Unfortunately, native guar gum cannot be utilized as dietary fiber at physiological beneficial concentration as it gives very high viscosity at higher concentration and causes undesirable textural changes from consumer point of view. This restricts its application in food products especially liquid food products. This high viscosity of native guar galactomannan also interferes with the digestion and absorption of certain nutrients which are essential for good health (Mudgil et al. 2012a).

Partially hydrolyzed guar gum attracted interest of many food scientists and nutritionist because of its health benefits. Moreover, it has very low viscosity, no color, no taste at physiologically beneficial concentration which makes it acceptable from consumer point of view. Partially hydrolyzed guar gum can be utilized as dietary fiber source in various food products without interfering with their textural and sensory properties (Greenberg and Sellman 1998). Partially hydrolyzed guar gum can be produced by enzymatic hydrolysis, acid hydrolysis, irradiation, microwave and ultrasonication techniques (Tayal and Khan 2000; Cheng et al. 2002; Gupta et al. 2009; Singh and Tiwari 2009). Enzymatic hydrolysis of guar gum is preferred for food processing applications. In enzymatic hydrolysis, enzymes such as mannanse, pectinase, cellulase etc. are used for processing (Yoon et al. 2008; Shobha et al. 2005; Mudgil et al. 2014b). These enzymes or hydrolyzing agents used in hydrolysis process, act on the linkage between the mannose backbone and breaks these linkages. The acid hydrolysis, microwave and ultrasonication hydrolysis processes use chemicals or acids as processing aids which helps in the hydrolysis of native guar gum and leads to production of partially hydrolyzed guar gum. Irradiation can also be used for the preparation of depolymerized or partially hydrolyzed guar gum. Co-60 isotope is generally used as a source of gamma radiations for hydrolysis of native guar gum (Gupta et al. 2009). Microwave and ultrasonic waves are also capable of inducing hydrolysis in native guar galactomannan molecular chain. Partially hydrolyzed guar gum is similar in structure and function as compared to native guar gum as evidenced from various characterization techniques such as Fourier Transform Infrared Spectroscopy, Differential Scanning Calorimetry and X-ray Diffraction (Mudgil et al. 2012b). The only difference between the two is viscosity and molecular weight as evidenced from rheology and viscsometry. Partially hydrolyzed guar gum is colorless, tasteless, odorless and less viscous in aqueous phase. Various studies have been reported regarding the physicochemical and structural properties of partially hydrolyzed guar gum (Mudgil et al. 2012c). Enzymatically hydrolyzed guar gum is also reported to be safe for consumption based on the results from subchronic feeding study and mutagenecity tests. Hence, it can be used for fiber fortification of various food products without changing their sensory properties. Partially hydrolyzed guar gum as soluble dietary fibre is beneficial in various diseases such as diabetes, heart disease and digestive problems because of its resistance towards human digestive secretion and interaction with other nutrients and metabolites (Yoon et al. 2008).

20.2 Structure

Guar galactomannan is a high molecular weight heteropolysaccharide composed of linear backbone chain of β(1→4)-linked mannose units to which α(1→6)-linked galactose units are attached as side chains at alternate position. The galactose to mannose ratio in guar galactomannan is about 1:1.6 to 1:2. Side chain galactose is substituted at mannose chain as random distribution (Mudgil et al. 2014a). Partially hydrolyzed guar gum obtained from enzymatic hydrolysis by endo-β-D-mannanase have structure similar to native guar gum as it cleave β(1→4) linkages between mannose in backbone chain without disturbing the galactose side chain. Hence the structure of partially hydrolyzed guar gum remains unchanged in spite of the molecular weight. Enzymatic treatment of native guar gum with pectinase from *Aspergillus niger* not only causes the depolymerization but also causes debranching of galactose unit from mannose chain. Hence the molecular weight of resultant hydrolyzed product is reduced and the galactose to mannose ratio ranges from 1:2.1 to 1:2.8 (Shobha et al. 2005). It is also reported that pullulanase from *Bacillus acidopullulyticus* can also cause hydrolysis and debranching in native guar galactomannan which leads to reduction in molecular weight. Debranching of galactose unit leads to changes in galactose to mannose ratio in the resultant product which ranges between 1:2.4 to 1:3.8 (Shobha and Tharanathan 2009).

20.3 Processing

Native guar gum can be transformed into partially hydrolyzed guar gum via numerous processes such as enzymatic hydrolysis, thermal hydrolysis, acid hydrolysis, gamma irradiation, microwave and ultrasonication techniques. Among all these techniques, enzymatic hydrolysis is the considered as best for food processing points of view because of less use of hazardous chemicals and inactivation of enzymes can be carried after hydrolysis process.

20.3.1 Enzymatic Hydrolysis

Enzymatic hydrolysis can be performed using number of enzymes such as mannanase, cellulase, pectinase, pullulanase etc. Hydrolysis of native guar galactomannan via use of mannanase enzyme is most popular. Mannanse enzyme acts on the linkage which holds together the mannose molecules in backbone chain. It hydrolyzes this linkage and leads to production of partially hydrolyzed guar gum. Mannanse treated guar galactomannan has very reduced molecular weight and viscosity as compared to native guar galactomannan. It has similar galactose to mannose ratio as in native guar because mannanase do not affect the linkages which hold galactose side chains on the mannose chain (Yoon et al. 2008). Similar to mannanase, cellulase can also be used for the preparation of partially hydrolyzed guar gum which is similar to native guar gum except molecular weight and viscosity (Mudgil et al. 2014b). Pectinase and pullulanase enzymes are also reported to have depolymeration capacity against native guar galactomannan. Both these enzymes also cause debranching of galactose from the mannose chain and the resultant product from such hydrolysis has modified galactose to mannose ratio (Shobha et al. 2005; Shobha and Tharanathan 2009). All these enzymes require certain specific conditions during hydrolysis processes such as temperature, pH, time, enzyme to substrate ratio etc., which should be fulfilled to achieve the desire degree of hydrolysis. Hence enzymatic hydrolysis processes are generally carried out at optimum conditions. Native guar gum is partially hydrolyzed during these enzymatic hydrolysis processes hence these are referred to as partial hydrolysis processes.

20.3.2 Acid Hydrolysis

Controlled acid hydrolysis process of native guar galactomannan is also reported for the production of partially hydrolyzed guar gum. In acid hydrolysis, guar galactomannan is completely hydrolyzed into its monomer units if it is not carried out in controlled conditions (Wang et al. 2000). The process parameters required to be controlled are temperature, time, amount of guar gum and acid. The main acid used in this hydrolysis process in HCl. This acid hydrolysis reaction of guar gum is reported to be carried at room temperature (25 °C) as well as at elevated temperature (70 °C). The resultant product obtained after acid hydrolysis varied in their molecular weight depending on the conditions of the hydrolysis process (Cheng et al. 2002). Neutralization of acid is very essential after hydrolysis process as it affects the pH of the resultant product. The acid hydrolysis of native guar gum can be carried out in solid-state medium or in dilute solution.

20.3.3 Thermal Hydrolysis

Native guar gum can also be hydrolyzed via noncatalytic hydrothermal process. In this process, hydrothermal conditions are maintained to carry out degradation of guar gum. Temperature of 180–240 °C and reaction time of 3–60 min is required for the hydrothermal degradation of guar gum which leads to resultant product having low molecular weight and degree of polymerization. It is reported that at optimum conditions, i.e. 200 °C temperature and reaction time of 7 min leads to production of oligosaccharide with degree of polymerization of up to 20 with a yield of about 94%. Increase in reaction time leads to lower production of oligosaccharide and higher concentration of monosaccharides. Hence partially hydrolyzed guar gum can be prepared via hydrothermal process at optimum conditions of temperature and reaction time (Miyazawa and Funazukuri 2006). Effect of pH and addition of antioxidant rheological properties of guar galactomannan at elevated temperature have also been reported (Kok 2010). Kok (2010) reported that heat treatment (at 100 °C) within pH range of 5–9 showed remarkable variations in the viscosity of the resultant product. Kok (2010) reported also that lowest viscosity and maximum degradation can be obtained at pH 7, which means that this pH is optimum for the free radical activity. Addition of antioxidants provides stability to gums against free radical degradation. Hence, it also validated the hypotheses that the degradation in gum at higher temperature is caused due to the action of free radicals.

20.3.4 Irradiation Hydrolysis

Irradiation is reported as a useful technique for decrease the molecular weight and viscosity of native guar gum. It affects the molecular integrity of native guar galactomannan molecule. It is reported that molecular mass and viscosity of final irradiated guar galactomannan depends on the irradiation dose level. Hence, partially hydrolyzed guar gum can also be prepared via controlling irradiation dose level. Irradiation does not cause significant change in gross conformation of guar galactomannan. The proposed mechanism of breakdown/hydrolysis of guar galactomannan in this process is random fracture. Intrinsic viscosity value of hydrolyzed guar gum molecules confirms their lower molecular weight with increasing radiation dose. Intrinsic viscosity of hydrolyzed guar gum shows rapid decrease at lower dose levels as compared to higher dose levels. Irradiation of native guar galactomannan causes a significant change in rheological behavior of resultant irradiated product. Irradiation at higher dose level leads to shift in rheological behavior towards Newtonian behavior as compared to shear thinning behavior of native guar gum (Jumel et al. 1996). Native guar gum can be irradiated in powder form as well as in solution-form using Co-60 gamma irradiator at ambient temperature condition. Gupta et al. (2009) reported the effect of initial apparent viscosity and moisture content on degree of hydrolysis and rheological properties of irradiated guar gum. They reported that rate of degradation of guar gum during irradiation is dependent on initial moisture content of native guar gum before irradiation. They also reported that there were no considerable chemical functional group conversions in irradiated guar gum as compared to native guar gum as evidenced from FTIR spectroscopy. Radiation processing of native guar gum also leads to decrease in its apparent viscosity and consistency index. They also reported that at higher initial moisture content of guar gum the free radical formed are lost due to cross-linking and leads to reduced degradation rates (Gupta et al. 2009).

20.3.5 Microwave Hydrolysis

Microwave hydrolysis is an alternate and efficient method for the preparation of partially hydrolyzed guar gum. This process requires certain hydrolyzing agent or catalyst (potassium persulphate) for the hydrolysis of guar polymer. These catalysts are more effective when used on solid substrate (gum) as compared to the aqueous solution of gum. This method do not involved the use of acids for hydrolysis (Singh et al. 2006). Reddy and Tammishetti (2004) also reported microwave mediated free radical degradation of guar gum for the preparation of lower molecular weight fragments. In this method guar gum powder is mixed with fixed quantity of water containing potassium persulphate or hydrogen peroxide and kneaded well followed by grinding which leads to formation of spongy material. This spongy material subjected to microwave treatment followed by washing and drying operations. The resultant hydrolyzed gum obtained from this process is referred to as partially hydrolyzed guar gum as evidenced from its molecular weight which is much more higher than the monomer unit of guar galactomannan molecule. Partially hydrolyzed guar gum obtained from this microwave hydrolysis show less intrinsic viscosity and molecular weight than native guar gum and improved film forming properties. Thermo gravimetric analysis of the resultant hydrolysis product reveals that there is no major change in chemical structure of the gum (or cross-linking) as compared to native guar gum. It is considered to be more effective than conventional thermal hydrolysis process (Singh and Tiwari 2009).

20.3.6 Ultrasound Hydrolysis

Depolymerization of guar gum can also be achieved using ultrasound irradiation or hydrolysis for preparation of partially hydrolyzed guar gum. Ultrasound alone or in combination with enzyme or microwave has been reported for partial hydrolysis of guar gum. Ultrasonic hydrolysis of guar gum can be achieved using high intensity probe type ultrasonic processor. Guar gum solution at low concentration is prepared and subjected to ultrasound treatment which leads to production of partially hydrolyzed guar gum with reduced molecular weight (Ansari et al. 2013). Ultrasound treatment becomes more effective when used in combination with microwave irradiation. Microwave hydrolysis of native guar gum followed by ultrasonication treatment has been reported more effective than ultrasound alone for depolymerization of guar gum. Ultrasound hydrolysis when used in combination with initiator (potassium persulphate) is most effective to achieve depolymerization of native guar gum (Prajapat and Gogate 2015). Ultrasound assisted enzymatic depolymerization of aqueous solutions of native guar gum has also been reported in which the effect of ultrasound treatment along with enzyme (cellulase) at different temperature was studied. Combination of ultrasound treatment with cellulase enzyme exhibits superior competence than cellulase alone which is due to the modifications in the spatial conformation of enzyme contributing to the intensification. The combination of ultrasound and cellulase enzyme hydrolysis of guar gum was more beneficial at 50 °C as compared to the singular action of ultrasonication and enzyme hydrolysis. At this temperature enzyme activity is higher and the cavitational activity is optimum (Prajapat et al. 2016).

20.4 Composition

Partially hydrolyzed guar gum is most suitable for food applications because of its high solubility, low viscosity, colorless, tasteless, and odorless behavior. It gets slight sweet taste as a result of partial enzymatic hydrolysis. It forms transparent aqueous solution when dissolved in water. In dry form it is a fine white powder. In powder form it contains around 8% moisture, 2% protein, 2.5% ash and 1% fat (Yoon et al. 2008). Partially hydrolyzed guar gum is a rich source of dietary fiber. It contains about 83% of total dietary fiber. It has a very low molecular weight and viscosity value as compared to native guar gum (Mudgil et al. 2016a).

20.5 Physicochemical Properties

20.5.1 Intrinsic Viscosity

Intrinsic viscosity represents the characteristic of an isolated individual macromolecule present in solution. Its value depends on the shape and specific volume of the individual molecule and is associated with its molecular weight and gyration radius. Intrinsic viscosity of polymers is estimated by capillary viscsometry. For measurement of intrinsic viscosity, first relative viscosity is determined which is the ratio of flow time of solution and flow time of solvent. This relative viscosity is then converted into specific viscosity by deducting one from it and extrapolation of this specific viscosity to zero concentration of polymer. Intercept of extrapolation to zero concentration gives the value of intrinsic viscosity. Partially hydrolyzed guar gum shows intrinsic viscosity value ranges between 2.72 and 0.28 (Cheng et al. 2002; Mudgil et al. 2012c; Reddy and Tammishetti 2004).

20.5.2 Apparent Viscosity

Viscosity is generally considered as resistance to flow of liquid. It is a single viscosity measured at constant speed in a viscometer. Native guar gum in aqueous solution exhibits very high apparent viscosity as compared to partially hydrolyzed guar gum. Apparent viscosity is generally determined using spindle type viscometer at a specific temperature, with specific spindle and at a fixed spindle rotational speed. Low value of apparent viscosity of partially hydrolyzed guar gum is required for its utilization as a source of dietary fiber in formulation of fiber fortified or enriched food products. A very low value, i.e. 4 cps has been reported for 1% aqueous solution of partially hydrolyzed guar gum (Mudgil et al. 2012b).

20.5.3 Viscosity Stability

Aqueous solution of partially hydrolyzed guar gum shows more viscosity stability as compared to native guar gum solution. Viscosity of partially hydrolyzed guar gum shows stability at very high temperature, i.e. 130 °C and broad range of pH, i.e. 3.0–7.0. Partially hydrolyze guar gum solution is stable at high salt concentration, i.e. up to 10%. Digestive enzymes such as amylases cannot cause changes in the viscosity of the partially hydrolyzed guar gum solutions (Yoon et al. 2008).

20.5.4 Shear Viscosity

It is measured as the friction between the fluid and moving boundaries. Controlled stress rheometers are generally used of measuring shear viscosity of aqueous gum solutions using cone-plate and concentric cylinders geometry. Partially hydrolyzed guar gum has different shear viscosity as compared to native guar gum. Native guar gum shows pseudoplastic behavior or more specifically shears thinning behavior, whereas partially hydrolyzed guar gum shows Newtonian behavior as evidenced from the rheological analysis reported in literature. It is also reported that viscosity values at different shear rates for partially hydrolyzed guar gum are lower as compared to native guar gum. Gamma irradiated guar gum samples at low irradiation dose are reported to have large shear rate dependence on viscosity while the guar gum samples irradiated at higher dose levels (above 5 kGy) exhibits Newtonian behavior (Jumel et al. 1996).

20.5.5 Molecular Weight

Molecular weight of polymers or oligomers is an important inherent characteristic which is associated with many of its functional and physiological properties (Siracusa et al. 2008). Viscosity of guar gum is very much dependent on its molecular weight (Robinson et al. 1982). Partially hydrolyzed guar gum has very low molecular weight as compared to native guar gum. Molecular weight of native and partially hydrolyzed guar gum can be measured using techniques such as size exclusion chromatography (SEC), dilute solution viscsometry, ultracentrifugation, etc. Recent publications reported that molecular weight of guar gum and partially hydrolyzed guar gum is about 900 kDa and 8 kDa, respectively (Mudgil et al. 2012b). Enzymatic hydrolysis breaks the bonds between mannose molecules in the backbone chain and leads to reduction in chain length and ultimately the reduction in molecular weight of partially hydrolyzed guar gum. Average degree of polymerization of partially hydrolyzed guar gum after enzymatic hydrolysis is about 29 (Mudgil et al. 2012c).

20.5.6 Sedimentation Coefficient

The rate at which molecule moves in a centrifugal field is related to its molecular weight and viscous drag which oppose this movement of molecule in solution. Sedimentation coefficient thus has an importance as it give knowledge regarding the conformation of molecules in the solution. Sedimentation velocity experiments on partially hydrolyzed guar gum especially via irradiation reveal the monodisperse nature (Jumel et al. 1996).

20.5.7 Flow Behavior Index and Consistency Index

Flow behavior index is a measure of flow ability of a substance whereas consistency index is a measure of consistency of a substance. Flow behavior index and consistency index of gum samples can be determined using spindle viscometer or rheometer. Viscosity of gum samples are measured at different shear rates and the data obtained from viscosity measurement is fitted in Power Law model to get the value of flow behavior index and consistency index. Flow behavior index value of guar gum and partially hydrolyzed guar gum are reported as 0.31 and 1.70, respectively. The value of flow behavior index for any substance below one represents non-Newtonian behavior and above one represents Newtonian behavior. This is the reason why native guar gum has a shear thinning behavior, whereas partially hydrolyzed guar gum shows a Newtonian behavior. The reported value of consistency index of guar gum and partially hydrolyzed guar gum are about 4.0 and 0.07, respectively

(Mudgil et al. 2012c). The low consistency index value of partially hydrolyzed guar gum enables it for fortification of several food products including beverages without disturbing their original texture and consistency.

20.5.8 Crystallinity Index

Crystallinity index is measure of per cent crystalline behavior of molecule. Crystallinity index value describes the molecular alignment in particle structure of a substance. X-ray diffraction analysis is used to describe the crystalline or amorphous nature of the substance. X-ray diffraction analyse results reveal that native guar gum has amorphous structure. Empirical formula proposed by Segal et al. (1959) can be used for calculating the crystallinity index of guar gum and partially hydrolyzed guar gum which uses difference in intensity at two specific Bragg angles, i.e. 18° and 22.7°. Percent crystallinity index reported for guar gum and partially hydrolyzed guar gum are 3.86 and 13.2%, respectively. The higher crystallinity index value describes the higher crystallinity or molecular alignment in partially hydrolyzed guar gum as compared to native guar gum. Enzymatic hydrolysis of native guar gum reduces its molecular weight and chain length due to which they are more align in particle structure as compared with native guar gum molecules which are long polymer chain of high molecular weight. Shorter chain length of polymers leads to more alignment and higher crystallinity index.

20.5.9 Solubility

Partially hydrolyzed guar gum is water soluble and alcohol insoluble in nature. It is soluble in cold and hot water like native guar gum. The pH has no significant on the solubility of partially hydrolyzed guar gum. This water solubility and low viscosity of partially hydrolyzed guar gum at broad temperature and pH range makes it a significant dietary fiber source which can be utilized for the dietary fiber fortification of the solid and semi-solid foods as well as the beverages. In aqueous systems, it forms stable solution which shows longer stability during their period of storage.

20.5.10 Tensile Strength

Films prepared from partially hydrolyzed guar gum shows good tensile strength as compared to films obtained from native guar gum. Films prepared from native guar gum are brittle in nature due to high molecular weight of guar gum which leads to production of colloidal solution (Álvarez et al. 2017). True solutions of the material are considered to have better film-forming properties as compared to colloidal solutions. Films obtained from partially hydrolyzed guar gum have better properties because it forms true solution due to its enhanced solubility and low viscosity and molecular weight. Partial hydrolysis of guar gum improves the tensile strength and elongation at break values which is desirable for coating applications (Reddy and Tammishetti 2004).

20.5.11 Thermal Properties

Thermal properties of partially hydrolyzed guar gum can be analyzed using differential scanning calorimetry. Mudgil et al. (2012b) reported change in heat flow of partially hydrolyzed guar gum over a temperature range of 50–500 °C at a heating rate of 10 °C/min. They reported that partially hydrolyzed guar gum shows endothermic peaks at 223 and 274 °C, and exothermic peak at 295 °C. Endothermic peak at 223 °C is associated with initial disintegration, whereas endothermic peak at 274 °C is associated with commencement of combustion. Exothermic peak at 295 °C is associated with continuing thermal and oxidative decomposition and vaporization and elimination of volatile substances. DSC results revealed that enzymatic hydrolysis of guar gum leads to reduction in decomposition temperature as evidenced from the lower exothermic peak temperature. It is associated with shorter chain length and molecular weight of partially hydrolyzed guar gum.

20.6 Food Applications

Partially hydrolyzed guar gum does not interact with other ingredients in food products or we can say that it remains stable when used in food for fortification. It does not cause viscosity change in protein solutions, precipitation of soluble material and destabilization of emulsions.

Partially hydrolyzed guar gum is colorless, tasteless, odorless hence it does not interfere with the color and taste and odor of the food products which makes it a unique functional ingredient for development of dietary fiber fortified food products. Moreover, it is considered as shelf life enhancer of high-starch foods as it decreases dextrin turbidity at low temperature. PHGG has been reported for its application in various food products such as bread, cookies, noodles, yoghurt, beverages, etc. Partially hydrolyzed guar gum also improves the products characteristics besides fortification and enrichment of dietary fiber in such products. It also aids in the processing of food products such as cereals flow ability, improvement in dough rheology, stabilization and mellowing of flavor in beverages, texture improvement in yoghurt as well as in bakery products. Low viscosity of partially hydrolyzed guar gum makes it a suitable ingredient for dietary fiber fortification of food products including beverages.

20.6.1 Bread

Mudgil et al. (2016b) reported utilization of partially hydrolyzed guar gum for the fiber fortification in wheat bread. Wheat flour of bread was replaced with partially hydrolyzed guar gum at the levels of 1–5%. The effect of partially hydrolyzed guar gum fortification in bread on firmness, specific loaf volume and sensory acceptability of bread has been reported. Partially hydrolyzed guar gum addition in bread at level of up to 3% decreases the bread firmness or in other words softer breads can be prepared using partially hydrolyzed guar gum as fortificant at certain level which means that partially hydrolyzed guar gum in bread provides strength to the gluten network and helps in retaining the spongy structure which results in softer bread. Optimum concentration of partially hydrolyzed guar gum reported for bread fortification is 1.59%. At this concentration of PHGG, bread with lower firmness and higher specific loaf volume is obtained which has similar sensory acceptability as compared to bread containing no partially hydrolyzed guar gum (i.e. control bread). At this level of PHGG, total dietary fiber content increased from 2.52% (control bread) to 3.78% (fortified bread) and soluble dietary fiber content increased from 0.53% (control bread) to 1.76% (fortified bread). Hence it is reported that PHGG not only improves the dietary fiber content of bread but also improves the bread characteristics.

20.6.2 Cookies

Partially hydrolyzed guar gum has been reported for dietary fiber fortification in wheat flour cookies (Mudgil et al. 2012d). Wheat flour of cookies dough was replaced with partially hydrolyzed guar gum at the levels of 1–5%. The effect of partially hydrolyzed guar gum addition in cookie dough along with different water levels and baking time on spread ratio, cookie texture and sensory acceptability of cookies has been reported. Partially hydrolyzed guar gum causes significant changes in physical and sensory characteristics of cookies. It is reported that when cookies fortified with partially hydrolyzed guar gum at concentration of 2.21%, it results in decrease in the hardness, spread ratio and overall acceptability as compared to control cookies. This reduction in cookie hardness upon partially hydrolyzed guar gum fortification is caused by lower level of gluten formation as the portion of wheat flour was replaced with partially hydrolyzed guar gum which causes reduction in concentration of gluten forming proteins in partially hydrolyzed guar gum fortified cookie dough. The optimum level of partially hydrolyzed guar gum in cookie dough increases its total dietary fiber content approximately up to 4% as compared to control cookies. PHGG in cookies increases dietary fiber content along with softness in cookies (Mudgil et al. 2017a). Partially hydrolyzed guar gum also reported to have prebiotic effects on selected human colonic microflora when utilized for fortification of biscuits. The increase in number of bifidobacteria is more pronounced in volunteers having low initial population levels of bifidobacteria (Tuohy et al. 2001).

20.6.3 Yoghurt

Dairy products including yoghurt are naturally devoid of dietary fiber. Hence dietary fiber fortification of dairy products is of current interest. Partially hydrolyzed guar gum has been reported for dietary fiber fortification in yoghurt (Kondo et al. 2004; Mudgil et al. 2016c). Kondo et al. (2004) reported the effect of partially hydrolyzed guar gum in yoghurt on the rise of postprandial serum lipid levels. The results of this study suggested that partially hydrolyzed guar gum has potential for the reduction of hyperlipaemia. Mudgil et al. (2016c) reported the effect of partially hydrolyzed guar gum levels, culture levels and incubation time on physicochemical properties such as viscosity, water holding capacity, acidity and sensory acceptability of yoghurt. On the basis of desirable physicochemical properties the optimum value of PHGG for yoghurt fortification has been reported as 2%. This optimum level of PHGG when used for the fortification of yoghurt it results in improved physicochemical properties and slightly lower sensory acceptability. The effect of partially hydrolyzed guar gum fortification on textural characteristics of yoghurt has also been reported (Mudgil et al. 2017b). Results of this study reveal that partially hydrolyzed guar gum fortification in yoghurt is negatively correlated with firmness and gumminess of yoghurt. The results also reveal that adhesiveness, cohesiveness and springiness are positively correlated with the levels of partially hydrolyzed guar gum fortified in yoghurt. The optimum level of partially hydrolyzed guar gum reported for desired textural characteristics on the basis of response surface methodology is 3.37%. Another study also reported the stability of partially hydrolyzed guar gum in yoghurt when added at 3% level. Partially hydrolyzed guar gum is more stable as compared to other dietary fiber (such as polydextrose, inulin and resistant dextrin) in yoghurt. PHGG in yoghurt also show more lactic acid bacteria count after 1 week of storage as compared to polydextrose, inulin and resistant dextrin (Kapoor and Juneja 2009).

20.6.4 Beverage

Beverage segment is one of the very important and fast growing segments of processed food industry. Beverages are liquid products generally consumed for their refreshing, thirst quenching nature. They are instant source of energy hence also utilized by sport persons. The ingredients used in beverages should have certain specific characteristics such as solubility in water and it should not impart color and flavor to beverages. Partially hydrolyzed guar gum possesses all these characteristics hence suitable for dietary fiber fortification in beverages. Minekus et al. (2005)

reported the effect of partially hydrolyzed guar gum fortified yoghurt drink on bioaccessibility of fat and cholesterol in healthy volunteers. Partially hydrolyzed guar gum reduces the bioaccessibility of fat and cholesterol when used at 3 and 6% level in yoghurt drink. Mudgil and Barak (2016) also reported development of functional buttermilk by fortification with partially hydrolyzed guar gum. They also reported the effect of partially hydrolyzed guar gum on physicochemical properties such as acidity, pH, viscosity, whey separation. Partially hydrolyzed guar gum enhances the viscosity of buttermilk whereas it reduces the whey separation of buttermilk which is desirable from consumer point of view. Both these characteristics are very important parameter for any beverage quality. Higher viscosity of the beverage up to certain levels are generally liked by consumers as it increases the mouthfeel of the beverage. Phase separation in beverages is considered as a technical problem and is not accepted by consumers. Partially hydrolyzed guar gum interacts with milk constituents particularly proteins and reduces the phase separation in buttermilk during storage. Results reveal that fortification of buttermilk with 4% level of partially hydrolyzed guar gum gives desirable sensory characteristics.

20.6.5 Noodles

Noodles are very famous among natives of Asian countries. Noodles constitutes important portion of the daily diet in Asian continent. Noodles possess little quantity of dietary fiber as it is prepared from refined wheat flour hence dietary fiber fortification of noodles is essential as it is a staple food in several Asian countries and spreading its popularity in other countries of the world. Mudgil et al. (2016a) reported effect of partially hydrolyzed guar gum, water level and dough mixing time on textural characteristics of white salted noodles such as hardness, adhesiveness, cohesiveness, chewiness and resilience. Wheat flour for noodle making was replaced with partially hydrolyzed guar gum at the levels of 1–5 g/100 g of flour. Partially hydrolyzed guar gum fortification in noodles was positively correlated with hardness, adhesiveness, chewiness and resilience. Partially hydrolyzed guar gum improves the textural characteristics of white salted noodles by hampering the gluten protein network in noodles. Mudgil et al. (2017c) reported the effect of partially hydrolyzed guar gum fortification on cooking yield, cooking loss and sensory acceptability of white salted noodles. Partially hydrolyzed guar gum fortification in noodles decreases cooking yield and increases cooking loss due to its water soluble nature which leads to its loss in water and ultimate reduction in cooking yield and increase in cooking loss. Noodles fortified with optimum level of partially hydrolyzed guar gum are more acceptable to consumer due to their improved textural characteristics. Kapoor and Juneja (2009) also reported that partially hydrolyzed guar gum can also be utilized as additive in noodles as it reduces noodles tangling when added to seasoning sauce at a concentration level of 5% after 2 h of hydration.

20.7 Health Benefits

20.7.1 Cardiovascular Diseases

Higher levels of triglycerides and cholesterol are recognised as responsible factors for cardiovascular diseases. In last decade, lot of research has been carried out on dietary fibers which demonstrated that dietary fibers are associated with improved lipid metabolism and reduced risk of cardiovascular diseases. It is reported that soluble dietary fibers have lipid and cholesterol lowering effect (Kay 1982). Partially hydrolyzed guar gum as functional soluble fiber has ability to improve the blood lipid profile as evidenced from numerous animal and human studies. The effect of native guar gum (molecular weight 300 kDa) and partially hydrolyzed guar gum (molecular weight 24 kDa) have been reported on total and plasma cholesterol and triglyceride levels in rats (Takeno et al. 1990). In this study, rats were fed hypercholesterolemic diet having native guar gum or partially hydrolyzed guar gum for 21 days. Native guar gum effectively decreases the rise of total cholesterol and triglyceride levels, whereas partially hydrolyzed guar gum was effective in decreasing the elevation of plasma cholesterol and triglyceride levels. It suggests that native and partially hydrolyzed guar gum both are effective in reducing the plasma cholesterol. It proves the functionality of low viscosity and molecular weight partially hydrolyzed guar gum in plasma cholesterol lowering. Another study reported the effect of diet containing insoluble dietary fiber (cellulose) and soluble dietary fiber (pectin, native and partially hydrolyzed guar gum) on serum lipids. It reported that soluble fibers reduce total cholesterol and triglyceride levels (Yamada et al. 1999). Ide et al. (1991) also reported the cholesterol lowering effects of native and partially hydrolyzed guar gum (at 8% level) in rats. They reported that both native and partially hydrolyzed guar gum reduces serum lipid levels which validate the functionality of partially hydrolyzed guar gum in attenuation of blood cholesterol levels. The similar effect of partially hydrolyzed guar gum has also been reported in numerous human volunteer studies (Takahashi et al. 1993; Yamatoya et al. 1997; Alam et al. 2005). All these studies suggest that partially hydrolyzed guar gum can be effectively utilized as hypocholesterolemic agent as it binds the bile acids in the intestine and increases fecal excretion of bile acids which ultimately leads to increase in the cholesterol oxidation to bile acids and reduces the cholesterol levels.

20.7.2 Diabetes

Guar gum is reported very potent in postprandial glycemia in patients having diabetes. It reduces the increase in postprandial glucose level and insulin concentration and leads to control of diabetes. Enrichment of liquid enteral formulas at a concentration level of soluble fibre which is effective physiologically leads to its viscosity enhancement and restricts its oral intake. Partially hydrolyzed guar gum solves this problem as it do not show high viscosity as compared to native guar gum hence apt for oral intake. It is reported that intake of formulas containing partially hydrolyzed guar gum significantly reduces the blood glucose and insulin concentration in diabetes patients as compared to formula not containing PHGG (Golay et al. 1995; Gu et al. 2003). Some reports observed no considerable effects on glucose and insulin plasma concentrations. It is proposed that partially hydrolyzed guar gum reduces the postprandial serum glucose level by combined action of three different mechanisms. In first mechanism, partially hydrolyzed guar gum enhances the viscosity of small intestine juice and makes it viscous which hampers the diffusion of glucose. According to second mechanism, partially hydrolyzed guar gum binds the glucose molecules and leads to decrease in the concentration of glucose in small intestine. According to third mechanism partially hydrolyzed guar gum reduces α-amylase action via encapsulating starch molecules and makes them unavailable or least available for amylase action. The combined action of all three mechanisms leads to reduction in the glucose absorption rate and the postprandial serum glucose concentration. Reported results revealed that partially hydrolyzed guar gum is much lesser viscous as compared to native guar gum, but it has similar effect on the reduction of postprandial blood glucose level as compared to native guar gum (Jenkins et al. 1978; Yamatoya et al. 1993).

20.7.3 Irritable Bowel Syndrome

Irritable bowel syndrome is considered as typical gastrointestinal functional disorder in which abdominal uneasiness is caused by some factors such as defecation, change in bowel pattern and abnormal defecation (Brenner et al. 2009). Irritable bowel syndrome is not a life threatening disorder but it certainly diminishes the quality of life (Drossman et al. 2002). Dietary fiber is considered to play important role in treatment of irritable bowel syndrome. Water soluble and non-gelling types of dietary fiber are more prominent for the treatment of irritable bowel syndrome. Increased fiber intake is recommended for patient suffering from irritable bowel syndrome (Bijkerk et al. 2004). High amount of fiber intake in diet leads to softer and bulkier stools, improved colonic peristalsis, reduced gut transit time which ultimately results in ease in defecation. Partially hydrolyzed guar gum is reported very effective in these physiological conditions and helps in reducing the abdominal pain as well as symptoms in both forms of irritable bowel syndrome (i.e. constipation and diarrhea-predominant). In small bowel, partially hydrolyzed guar gum is hydrated by incorporation of water. In large bowel, hydrated partially hydrolyzed guar gum regularizes fecal moisture and boosts the volume of fecal output. Partially hydrolyzed guar gum in presence of greatly hydrated milieu enhances water retention and upgrades the stool consistency (Giannini et al. 2006).

20.7.4 Prebiotic Effects

Soluble types of dietary fibers are reported to have favourable effects on improvement of microbial gut health. It is reported that guar galactomannan is promptly undergo fermentation by fecal microflora (Salyers et al. 1977). The resultant product of guar galactomannan fermentation leads to reduction in colonic pH and enhanced production of short chain fatty acids. This reduction in pH improves the intestinal health via maintaining the absolute environment which is essential for the survival and growth of beneficial bacteria. These conditions are unfavourable for the formation of unhealthy microbial metabolites (Goldin and Gorbach 1976; Hood and Zottola 1988). Partially hydrolyzed guar gum has been reported to exhibit similar effects on gut health and immunological functions. Carlson et al. (2016) studied the effect of partially hydrolyzed guar gum fermentation on gut microbiota. They reported that partially hydrolyzed guar gum stimulates the growth of bacteria, i.e. *Bacteroides* and *Parabacteroides* which are negatively correlated with irritable bowel syndrome and ulcerative colitis. Hence it is concluded that partially hydrolyzed guar gum is a readily fermentable dietary fiber having prebiotic activity (Carlson et al. 2016).

20.7.5 Laxation Effect

Diarrhea is one of the gastrointestinal disorders associated with tube fed patients. Patient conditions such as stress, surgical procedure, bacterial contamination of gut, antibiotic medication and protein energy malnutrition attributed to diarrhea (Bliss et al. 1992). Enteral nutrition is suspended in conditions like diarrhea because it interfere with fluid and electrolyte balance in patients. In tube fed patients the incidence of diarrhea is between 2 and 63%. There are no standard factors which can describe diarrhea condition however three main factors such as stool consistency, stool frequency and stool quantity are generally used to define diarrhea condition. Reduction in diarrhea can be achieved by partially hydrolyzed guar gum addition to tube feeding formulas. Partially hydrolyzed guar gum can be added to various food and beverages due to its low viscosity and it reduces the quantity of laxative agents used in aged people. It is reported that partially hydrolyzed guar gum can suppress transitory diarrhea caused by ingestion of maltitol and lactitol. Alam et al. (2015) reported the efficacy of partially hydrolyzed guar gum supplemented oral rehydration solution (ORS) for the treatment of watery diarrhea in severely malnourished children. They concluded that partially hydrolyzed guar gum added to ORS effectively reduces duration of diarrhea, stool output and increases the weight gain in malnourished children. Suggested mechanism for suppression of diarrhea by fiber such as partially hydrolyzed guar gum is related to its bacterial fermentation leading to production of short chain fatty acids which expedite the absorption of sodium chloride and water in the colon (Roediger and Moore 1981; Binder and Mehta 1989). It also stimulates the epithelial cell proliferation and causes recovery of mucosal atrophy and damage (Slavin and Greenberg 2003). Hence partially hydrolyzed guar gum can be utilized as functional fiber due to its ability to furnish useful physiological effects on laxation and gastrointestinal disorders.

20.7.6 Appetite Control and Weight Management

Uncontrolled appetite and intake of extra calories leads to obesity type health problems hence control of appetite and fewer intakes of extra calories are considered as rational way for obesity control or weight management. Viscous types of dietary fiber are very adequate in appetite control but due to their high viscosity these cannot be supplemented at physiologically beneficial higher doses in diet. Partially hydrolyzed guar gum is low viscosity and water soluble type of dietary fiber hence it has certain health benefits. One of these benefits is its potential for use in appetite control. The main mechanism suggested for appetite control by partially hydrolyzed guar gum are extension of colonic transit time, and incitement of satiety hormone cholecystokinin which leads to extended feeling of post meal satiety. It is reported that 2 g per serving dose level of partially hydrolyzed guar gum provide the uninterrupted post meal satiety and it also minimizes the calorie intake (20%) in between two meals in normal subjects. Satiety effects of partially hydrolyzed guar gum are more pronounced when it is consumed along with protein. Hence partially hydrolyzed guar gum can be used as a novel ingredient for satiety control (Rao 2016).

20.8 Conclusions

Partially hydrolyzed guar gum is water soluble gum obtained after partial hydrolysis of native guar gum via enzymatic hydrolysis, acid hydrolysis, irradiation, microwave and ultrasonication techniques. It is similar in structure and functions to native guar galactomannan except molecular weight and viscosity. It has approximately ten times lesser molecular weight than native guar gum. Enzymatically hydrolyzed guar gum has been extensively studied for its food applications as dietary fiber. Partially hydrolyzed guar gum can be used for fiber fortification of food products due to its less viscosity, water soluble, tasteless, colorless and odorless behavior. Hence, it can also be used in liquid products such as beverages. Its fortification in food products not only increases the dietary fiber content but also improves their physicochemical and sensory characteristics. Partial hydrolysis of guar gum do not interfere with its functional nature hence its use in food products leads to certain health benefits such as diabetes, cardiovascular disease, prebiotic effects, satiety control, digestive problems, etc. similar to native guar gum.

References

Alam NH, Ashraf H, Kamruzzaman M, Ahmed T, Islam S, Olesen MK, Gyr N, Meier R (2015) Efficacy of partially hydrolyzed guar gum (PHGG) supplemented modified oral rehydration solution in the treatment of severely malnourished children with watery diarrhoea: a randomised double-blind controlled trial. J Health Popul Nutr 34(3):1–9

Alam NH, Meier R, Sarker SA, Bardhan PK, Schneider H, Gyr N (2005) Partially hydrolysed guar gum supplemented comminuted chicken diet in persistent diarrhoea: a randomized controlled trial. Arch Dis Child 90(2):195–199

Álvarez K, Famá L, Gutiérrez TJ (2017) Physicochemical, antimicrobial and mechanical properties of thermoplastic materials based on biopolymers with application in the food industry. In: Masuelli M, Renard D (eds) Advances in physicochemical properties of biopolymers: part 1. Bentham Science Publishers, pp 358–400. https://doi.org/10.2174/9781681084534117010015. ISBN: 978-1-68108-454-1. eISBN: 978-1-68108-453-4

Ansari SA, Matricardi P, Cencetti C, Di Meo C, Carafa M, Mazzuca C, Palleschi A, Capitani D, Alhaique F, Coviello T (2013) Sonication-based improvement of the physicochemical properties of guar gum as a potential substrate for modified drug delivery systems. Biomed Res Int 2013:985259. https://doi.org/10.1155/2013/985259

Bijkerk CJ, Muris JW, Knottnerus JA, Hoes AW, De Wit NJ (2004) Systematic review: the role of different types of fibre in the treatment of irritable bowel syndrome. Aliment Pharmacol Ther 19(3):245–251

Binder HJ, Mehta P (1989) Short-chain fatty acids stimulate active sodium and chloride absorption in vitro in the rat distal colon. Gastroenterology 96(4):989–996

Bliss DZ, Guenter PA, Settle RG (1992) Defining and reporting diarrhea in tube-fed patients—what a mess! Am J Clin Nutr 55(3):753–759

Brenner DM, Moeller MJ, Chey WD, Schoenfeld PS (2009) The utility of probiotics in the treatment of irritable bowel syndrome: a systematic review. Am J Gastroenterol 104(4):1033–1049

Butt MS, Shahzadi N, Sharif MK, Nasir M (2007) Guar gum: a miracle therapy for hypercholesterolemia, hyperglycemia and obesity. Crit Rev Food Sci Nutr 47(4):389–396

Carlson J, Gould T, Slavin J (2016) In vitro analysis of partially hydrolyzed guar gum fermentation on identified gut microbiota. Anaerobe 42:60–66

Cheng Y, Brown KM, Prud'homme RK (2002) Preparation and characterization of molecular weight fractions of guar galactomannans using acid and enzymatic hydrolysis. Int J Biol Macromol 31(1):29–35

Drossman DA, Camilleri M, Mayer EA, Whitehead WE (2002) AGA technical review on irritable bowel syndrome. Gastroenterology 123(6):2108–2131

Giannini EG, Mansi C, Dulbecco P, Savarino V (2006) Role of partially hydrolyzed guar gum in the treatment of irritable bowel syndrome. Nutrition 22(3):334–342

Golay A, Schneider H, Bloise D, Vadas L (1995) The effect of a liquid supplement containing guar gum and fructose on glucose tolerance in non-insulin-dependent diabetic patients. Nutr Metab Cardiovasc Dis 5(2):141–148

Goldin BR, Gorbach SL (1976) The relationship between diet and rat fecal bacterial enzymes implicated in colon cancer. J Natl Cancer Inst 57(2):371–375

Greenberg NA, Sellman D (1998) Partially hydrolyzed guar gum as a source of fiber. Cereal Food World 43(9):703–707

Gu Y, Yamashita T, Suzuki I, Juneja LR, Yokawa T (2003) Effect of enzyme hydrolyzed guar gum on the elevation of blood glucose levels after meal. Med Biol 147:19–24

Gupta S, Shah B, Sanyal B, Variyar PS, Sharma A (2009) Role of initial apparent viscosity and moisture content on post irradiation rheological properties of guar gum. Food Hydrocoll 23(7):1785–1791

Hood SK, Zottola EA (1988) Effect of low pH on the ability of *Lactobacillus acidophilus* to survive and adhere to human intestinal cells. J Food Sci 53(5):1514–1516

Ide T, Moriuchi H, Nihimoto K (1991) Hypolipidemic effects of guar gum and its enzyme hydrolysate in rats fed highly saturated fat diets. Ann Nutr Metab 35(1):34–44
Jenkins DJ, Wolever TM, Leeds AR, Gassull MA, Haisman P, Dilawari J, Goff DV, Metz GL, Alberti KG (1978) Dietary fiber, fiber analogues and glucose tolerance: importance of viscosity. Br Med J 1(6124):1392–1394
Jumel K, Harding SE, Mitchell JR (1996) Effect of gamma irradiation on macromolecular integrity of guar gum. Carbohydr Res 282(2):223–236
Kapoor MP, Juneja LR (2009) Partially hydrolyzed guar gum dietary fiber. In: Susan SC, Priscilla S (eds) Fiber ingredients: food applications and health benefits, 1st edn. CRC Press, Boca Raton, pp 79–112
Kay RM (1982) Dietary fiber. J Lipid Res 23(2):221–242
Kok MS (2010) Rheological study of galactomannan depolymerisation at elevated temperatures: effect of varying pH and addition of antioxidants. Carbohydr Polym 81(3):567–571
Kondo S, Xiao JZ, Takahashi N, Miyaji K, Iwatsuki K, Kokubo S (2004) Suppressive effects of dietary fiber in yogurt on the postprandial serum lipid levels in healthy adult male volunteers. Biosci Biotechnol Biochem 68(5):1135–1138
Minekus M, Jelier M, Xiao JZ, Kondo S, Iwatsuki K, Kokubo S, Bos M, Dunnewind B, Havenaar R (2005) Effect of partially hydrolyzed guar gum (PHGG) on the bioaccessibility of fat and cholesterol. Biosci Biotechnol Biochem 69(5):932–938
Miyazawa T, Funazukuri T (2006) Noncatalytic hydrolysis of guar gum under hydrothermal conditions. Carbohydr Res 341(7):870–877
Mudgil D, Barak S (2013) Composition, properties and health benefits of indigestible carbohydrate polymers as dietary fiber: a review. Int J Biol Macromol 61:1–6
Mudgil D, Barak S (2016) Development of functional buttermilk by soluble fibre fortification. Agro FOOD Industry Hi Tech 27(2):44–47
Mudgil D, Barak S, Khatkar BS (2011) Effect of hydrocolloids on the quality characteristics of tomato ketchup. Carpath J Food Sci Technol 3(1):39–43
Mudgil D, Barak S, Khatkar BS (2012a) Process optimization of partially hydrolyzed guar gum using response surface methodology. Agro FOOD Industry Hi Tech 23(1):13–15
Mudgil D, Barak S, Khatkar BS (2012b) X-ray diffraction, IR spectroscopy and thermal characterization of partially hydrolyzed guar gum. Int J Biol Macromol 50(4):1035–1039
Mudgil D, Barak S, Khatkar BS (2012c) Effect of enzymatic depolymerization on physicochemical and rheological properties of guar gum. Carbohydr Polym 90(1):224–228
Mudgil D, Barak S, Khatkar BS (2012d) Soluble fibre and cookie quality. Agro FOOD Industry Hi Tech 23(3):15–17
Mudgil D, Barak S, Khatkar BS (2014a) Guar gum: processing, properties and food applications—a review. J Food Sci Technol 51(3):409–418
Mudgil D, Barak S, Khatkar BS (2014b) Optimization of enzymatic hydrolysis of guar gum using response surface methodology. J Food Sci Technol 51(8):1600–1605
Mudgil D, Barak S, Khatkar BS (2016a) Optimization of textural properties of noodles with soluble fiber, dough mixing time and different water levels. J Cereal Sci 69:104–110
Mudgil D, Barak S, Khatkar BS (2016b) Optimization of bread firmness, specific loaf volume and sensory acceptability of bread with soluble fiber and different water levels. J Cereal Sci 70:186–191
Mudgil D, Barak S, Khatkar BS (2016c) Development of functional yoghurt via soluble fiber fortification utilizing enzymatically hydrolyzed guar gum. Food Biosci 14:28–33
Mudgil D, Barak S, Khatkar BS (2017a) Cookie texture, spread ratio and sensory acceptability of cookies as a function of soluble dietary fiber, baking time and different water levels. LWT Food Sci Technol 80:537–542
Mudgil D, Barak S, Khatkar BS (2017b) Texture profile analysis of yogurt as influenced by partially hydrolyzed guar gum and process variables. J Food Sci Technol 54(12):3810–3817
Mudgil D, Barak S, Khatkar BS (2017c) Development and characterization of soluble fiber enriched noodles via fortification with partially hydrolyzed guar gum. J Food Meas Charact 12(1):156–163. https://doi.org/10.1007/s116940179626y

Prajapat AL, Gogate PR (2015) Depolymerization of guar gum solution using different approaches based on ultrasound and microwave irradiations. Chem Eng Process 88:1–9

Prajapat AL, Subhedar PB, Gogate PR (2016) Ultrasound assisted enzymatic depolymerization of aqueous guar gum solution. Ultrason Sonochem 29:84–92

Rao TP (2016) Role of guar fiber in appetite control. Physiol Behav 164:277–283

Reddy TT, Tammishetti S (2004) Free radical degradation of guar gum. Polym Degrad Stab 86(3):455–459

Robinson G, Ross-Murphy SB, Morris ER (1982) Viscosity-molecular weight relationships, intrinsic chain flexibility, and dynamic solution properties of guar galactomannan. Carbohydr Res 107(1):17–32

Roediger WE, Moore A (1981) Effect of short chain fatty acids on sodium absorption in isolated human colon perfused through vascular bed. Dig Dis Sci 26(2):100–106

Salyers AA, West SHE, Vercellotti JR, Wilkins TD (1977) Fermentation of mucins and plant polysaccharides by anaerobic bacteria from the human colon. Appl Environ Microbiol 34(5):529–533

Segal LG, Creely JJ, Martin AE Jr, Conrad CM (1959) An empirical method for estimating the degree of crystallinity of native cellulose using the X-ray diffractometer. Text Res J 29(10):786–794

Shobha MS, Kumar AV, Tharanathan RN, Koka R, Gaonkar AK (2005) Modification of guar galactomannan with the aid of Aspergillus niger pectinase. Carbohydr Polym 62(3):267–273

Shobha MS, Tharanathan RN (2009) Rheological behaviour of pullulanase-treated guar galactomannan on co-gelation with xanthan. Food Hydrocoll 23(3):749–754

Singh V, Tiwari A (2009) Hydrolytic fragmentation of seed gums under microwave irradiation. Int J Biol Macromol 44(2):186–189

Singh V, Tiwari A, Kumari P, Tiwari S (2006) Microwave-promoted hydrolysis of plant seed gums on alumina support. Carbohydr Res 341(13):2270–2274

Siracusa V, Rocculi P, Romani S, Dalla Rosa M (2008) Biodegradable polymers for food packaging: a review. Trends Food Sci Technol 19(12):634–643

Slavin JA, Greenberg N (2003) Partially hydrolyzed guar gum: clinical nutrition uses. Nutrition 19(6):549–552

Takahashi H, Yang SI, Hayashi C, Kim M, Yamanaka J, Yamamoto T (1993) Effect of partially hydrolyzed guar gum on fecal output in human volunteers. Nutr Res 13(6):649–657

Takeno F, Yamada H, Sekiya K, Fujitani B, Ohtsu K (1990) Effect of partially decomposed guar gum on high-cholesterol-fed rats and non-dietary fiber-fed rats. J Jpn Nutr Food Sci 43:421–425

Tayal A, Khan SA (2000) Degradation of a water-soluble polymer: molecular weight changes and chain scission characteristics. Macromolecules 33(26):9488–9493

Tuohy KM, Kolida S, Lustenberger AM, Gibson GR (2001) The prebiotic effects of biscuits containing partially hydrolysed guar gum and fructo-oligosaccharides–a human volunteer study. Br J Nutr 86(3):341–348

Wang Q, Ellis PR, Ross-Murphy SB (2000) The stability of guar gum in an aqueous system under acidic conditions. Food Hydrocoll 14(2):129–134

Yamada K, Tokunaga Y, Ikeda A, Ohkura KI, Mamiya S, Kaku S, Sugano M, Tachibana H (1999) Dietary effect of guar gum and its partially hydrolyzed product on the lipid metabolism and immune function of Sprague-Dawley rats. Biosci Biotechnol Biochem 63(12):2163–2167

Yamatoya K, Kuwano K, Suzuki J (1997) Effects of hydrolyzed guar gum on cholesterol and glucose in humans. Food Hydrocoll 11(2):239–242

Yamatoya K, Sekiya K, Yamada H, Ichikawa T (1993) Effects of partially hydrolyzed guar gum on postprandial plasma glucose and lipid levels in humans. J Jpn Soc Nutr Food Sci 46:199–203

Yoon SJ, Chu DC, Juneja LR (2008) Chemical and physical properties, safety and application of partially hydrolyzed guar gum as dietary fiber. J Clin Biochem Nutr 42(1):1–7

Chapter 21
Development of Hydrogels from Edible Polymers

Akbar Ali and Shakeel Ahmed

Abstract Hydrogels are three dimensional crosslinked macromolecular network structures that can absorb and retain a significant amount of water without dissolving in it. This chapter reflects a detailed account on hydrogel developed from edible polymers and their potential applications highly emphasizing on food industries as packaging material. Hydrogels based on edibles polymers offers many valuable properties compared to their synthetic counterparts. Edibles polymers can contribute to the reduction of environmental contamination; advances recyclability, provides sustainability and there by increases its applicability along with providing environmentally benign products. The application of edible polymer hydrogel covers many areas including drug delivery to tissue engineering in biomedical fields, food industries; providing more safe and attractive products, and pharmaceuticals etc. The main objective of this chapter is to provide a brief idea and description about edible polymers based hydrogels, its synthesis, properties and applications.

Keywords Biopolymers · Cross-linking · Food packaging

21.1 Introduction

Hydrogels are three dimensionally crosslinked network structure having great affinity for water without disturbing their dimensional stability. The water absorbing capacity of the hydrogel depends on the presence and number of hydrophilic groups and domains in the network structure. The dimensional stability was maintained by either physical or chemical linkages between the polymer networks. Chemical linkages includes permanent bonding while physical linking comprises many transitory junctions like, chain entanglements, ionic interactions, hydrogen bonding and

A. Ali
Department of Chemistry, Jamia Millia Islamia, New Delhi, India

S. Ahmed (✉)
Department of Chemistry, Government Degree College Mendhar, Poonch, Jammu and Kashmir, India

T. J. Gutiérrez (ed.), *Polymers for Food Applications*,
https://doi.org/10.1007/978-3-319-94625-2_21

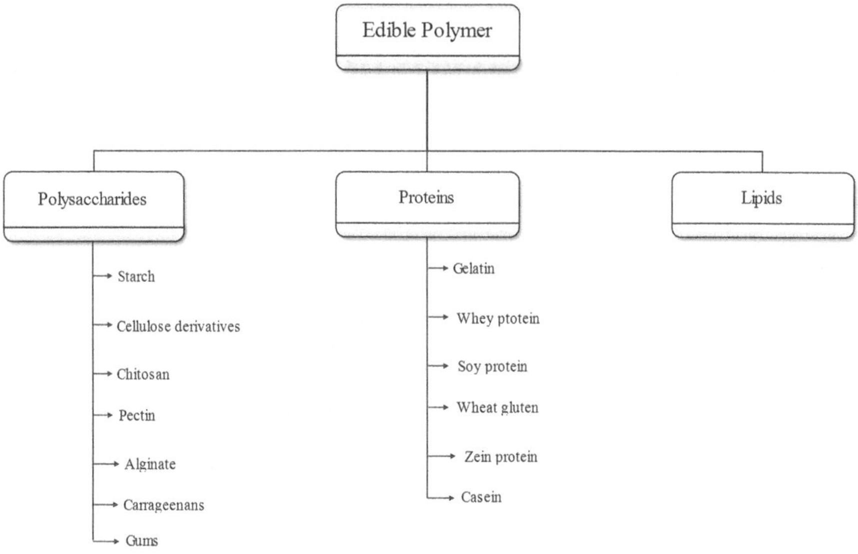

Fig. 21.1 Types of edible polymer

hydrophobic interaction etc. The structures of hydrogel were also influenced by the change in concentration, structure, monomer functionality and crosslinker.

On the basis of their origin hydrogels can be divided into two broad categories; the synthetic and the natural hydrogels. Synthetic hydrogels such polyvinyl alcohol, polyacrylamide, polyethylene glycol, etc. are known for their excellent mechanical strength. On contrary, the natural hydrogels lack enough mechanical strength. However, both of them have their own drawbacks and advantageous properties. Which mark them very important for different applications, subject to the proper choice of materials as per applications required? Both synthetic and natural polymer based hydrogels have been projected as scaffolding materials in tissue engineering for more than a decade. Dried hydrogels in the forms of films have being now reported as biodegradable packaging materials for food, pharmaceuticals products and cosmetics (Langmaier et al. 2008). Edible films and coatings have delivered an interesting and often crucial complementary means for regulating the quality and stability of various food products (Guilbert et al. 1995). Coatings can apply directly on the food product, while films can be applied or used after prepared separately.

21.2 Edible Polymer Based Hydrogels

Edible polymers mainly consist of polysaccharides, proteins and lipids (Fig. 21.1). As being totally biodegradable, edible polymer based packaging materials could contribute to an alternative solutions in reducing the amount of non-biodegradable plastic waste.

Table 21.1 Edible polymer based hydrogel films for food packaging

Edible polymers	Plasticizer	Additives	Method	Physical form	Mechanical strength (MPa)	Applications	References
Starch	–	Nano clay (C30B)	Casting	Films	–	Antimicrobial food packaging	Abreu et al. (2015)
Starch	Glycerol	TiO_2	Casting	Films	5.27–5.74	UV protective food packaging	Goudarzi et al. (2017)
Starch	Glycerol/Sorbitol	–	Casting	Films	–	–	García et al. (2000)
Starch	Glycerol	Carvacrol, thymol, linalool	Pressed under laboratory press	Films	–	Antimicrobial food packaging	Kuorwel et al. (2013)
Starch	Glycerol	Lignin	Casting	Films	3.48–4.20	Food packaging purpose	Bhat et al. (2013)
Starch-KGM	Glycerol		Solvent evaporation	Films	68.1	Edible coating and films	Chen et al. (2008)
Hydroxypropyl methylcellulose	–	Silver nanoparticles	Casting	Films	28.3–50.0	Active food packaging	de Moura et al. (2012)
Carboxymethyl cellulose (CMC)	Sodium benzoate/ glutaraldehyde/ gelatin	–	UV/chemical crosslinking	Films	14.2–46.8	Food packaging	Shahbazi et al. (2016)
Methylcellulose	Glutaraldehyde	Murta fruit	Air drying casting	Films	4.5–14.5	Antioxidant and antimicrobial food packaging	López de Dicastillo et al. (2016)
CMC	Glycerol	Potassium sorbate	Casting	Films	8.65–20.66	Food applications	Sayanjali et al. (2011)
CMC/PVA	Glycerol	Clove oil	Casting	Films	43–100	Packaging of ground chicken meat	Muppalla et al. (2014)

(continued)

Table 21.1 (continued)

Edible polymers	Plasticizer	Additives	Method	Physical form	Mechanical strength (MPa)	Applications	References
CMC/PVP	PEG, glycerine, agar	–	Dry air casting	Films	42.1–45.7	Food packaging	Gregorova et al. (2015)
Konjac Glucomannan (KGM)-gelatin	Glycerine	–	Casting	Films	59.7	Inner packaging materials in food industry	Li et al. (2006)
KGM-curdlan	Glycerine	–	Casting	Films	40.09–42.93	Edible food films and coatings	Wu et al. (2012)
Pullulan	Sorbitol/sucrose fatty acid ester	–	Casting	Films/ coating	–	Fruit preservation	Diab et al. (2001)
Chitosan	Glycerol	Thyme, clove and cinnamon oil	Casting	Films	12.2–21.35	Antimicrobial films and coatings	Hosseini et al. (2009)
Chitosan	–	Fatty acid	–	Coating	–	Oil barrier packaging	Ham-Pichavant et al. (2005)
Chitosan	–	Graphene, Cinnamaldehyde	Casting	Films	40–90	Food packaging	Demitri et al. (2016)
Chitosan	–	Propolis	Casting	Films	5.3–20.3	Active food with antioxidant and antimicrobial packaging	Siripatrawan and Vitchayakitti (2016)
Chitosan	Glycerol	Green tea extract	Casting	Films	23.66–28.35	Active food packaging	Siripatrawan and Harte (2010)
Pectin	Glycerol	Lysozyme	Casting	Films	24.3	Antimicrobial packaging films against lysozyme sensitive microorganisms	Bayarri et al. (2014)
Pectin-gelatin	Glycerol	–	Crosslinking than air drying	Films	19–48	Biomedical product	Gupta et al. (2014)

Edible polymers	Plasticizer	Additives	Method	Physical form	Mechanical strength (MPa)	Applications	References
Pectin-fish gelatin	Glycerol	Glutaraldehyde	Casting	Films	17.0–71.8	Packaging or coating of food or drugs	LiuLiu et al. (2007)
Pullulan	–	Silica	Roll coating	Coating	–	Food packaging	Farris et al. (2012)
Pullulan	–	Montmorillonite	Automatic film applicator	Coating	–	Food packaging	Introzzi et al. (2012)
Sodium alginate		Calcium chloride	TLC sprayer	Coating		Food protection	Kobašlija and McQuade (2006)
Sodium alginate	Glycerol	Garlic oil	Casting	Films	38.67–66.12	Antibacterial food applications	Pranoto et al. (2005b)
Alginate/pectin	Glycerol	Natamycin	Casting	Films	80.87–94.53	Antimicrobial packaging	Bierhalz et al. (2012)
Gelatin	Glycerol	Natural oxidants	Casting	Films	14.64–23.42	Active food packaging	Li et al. (2014)
Gelatin	Glycerol	Zataria oil	Casting	Films	2.7–4.4	Food packaging	Kavoosi et al. (2014)
Gelatin/chitosan	Sorbitol and glycerol	Oregano oil	Casting	Films	30.69–45.91	Fish preservation	Wu et al. (2014)
Whey protein isolate (WPI)	Glycerol	Nano clays	Casting	Films	1.55–3.40	Food packaging	Sothornvit et al. (2009)
WPI	Glycerol	SiO_2, TiO_2	Casting	Films	1.03–1.18	Active food packaging	Kadam et al. (2013)
WPI	Glycerol	Sorbic or p-Aminobenzoic acid	Casting	Films	2.60–6.00	Food protection	Cagri et al. (2001)
Soy protein	Glycerol	Ferulic acid	Casting	Films	1.47–2.60	Food packaging	Ou et al. (2005)

(continued)

Table 21.1 (continued)

Edible polymers	Plasticizer	Additives	Method	Physical form	Mechanical strength (MPa)	Applications	References
Soy Protein Isolate (SPI)	Glycerol	Montmorillonite	Melt extrusion	Films	1.86–6.28	Commercial application	Kumar et al. (2010)
SPI	Glycerine	Dialdehyde starch	Casting	Films	6.19–7.84	Packaging and mulching applications	Rhim et al. (1998)
SPI	Glycerol	Cellulose fibers	Casting	Films	1.60–10.83	Food applications	Jensen et al. (2015)
SPI	–	Flaxseed oil	Casting	Films	2.35–5.35	Packaging	Hopkins et al. (2015)
Whey/zein protein	Glycerol, olive oil	–	Casting	Films	3–12		Ghanbarzadeh and Oromiehi (2008)

Although there are some issues regarding its uses in industrial applications due to poor mechanical properties; which can be overwhelmed by various chemical modification. Ribeiro et al. thoroughly investigated the effect of polysaccharide based coatings (starch, carrageenan and chitosan) to extend the shelf-life of strawberry fruit particularly for industrial applications (Ribeiro et al. 2007). Table 21.1 provides a detailed account of edible polymers based hydrogel films, its preparation, properties and application in food industries.

21.3 Polysaccharides

Polysaccharides are complex carbohydrate molecules comprising of many monosaccharides units linked via glycosidic linkages. Polysaccharide can be obtained from animals, plants, microorganisms and seaweeds etc. Depending on its compositions, it could be either homo-polysaccharide or hetero-polysaccharide. Among edible polymers, polysaccharides in particular have some excellent properties and have been explored in every fields to their full potential (Pasqui et al. 2012). Due to their biocompatibility and biodegradability polysaccharide based materials found prospective applications in food industries, pharmaceuticals, biomedical, agricultural, and green chemistry et., (García-González et al. 2011; Prabaharan and Mano 2006; Shewan and Stokes 2013). Some polysaccharides were already commercially available as gelling and thickening agents, stabilizer, emulsifier and encapsulating agents in food and non-food industries. Polysaccharide based dry hydrogels (films) have suitable mechanical and organoleptic properties for food packaging (Guilbert et al. 1995). There are numbers of edible polysaccharides; in this chapter few of them have been discussed in details along with its hydrogel preparation.

21.3.1 Starch

Starch is the most abundant and extensively studied biopolymer in food industries due to its biocompatibility and biodegradability (Gutiérrez et al. 2015b, c, 2017; Suárez and Gutiérrez 2017). Starch consists of two homopolymers of D-glucose units, amylose (10–20%) the linear α-D-glucan (α-1,4 linkage) and branched amylopectin (80–90%) having α-1,4 as well as α-1,6 linkages (Fig. 21.2) (Lu et al. 2009). The relative proportion of these components differs depending upon the sources (e.g., potato, corn, wheat etc.) and also effects the crystallinity and molecular order of the polysaccharide (Ellis et al. 1998; Gutiérrez 2018a). The amorphous part mainly contains amylose along with some amylopectin, whereas the crystalline parts were dominated by amylopectin (Sajilata et al. 2006). Amylose fraction in starch act as an important parameter in determining the mechanical properties, water vapor sorption, stress strain behaviour and also affects the films forming capability of starch (Lourdin et al. 1995; Bader and Göritz 1994a, b, c; Gutiérrez et al. 2015d).

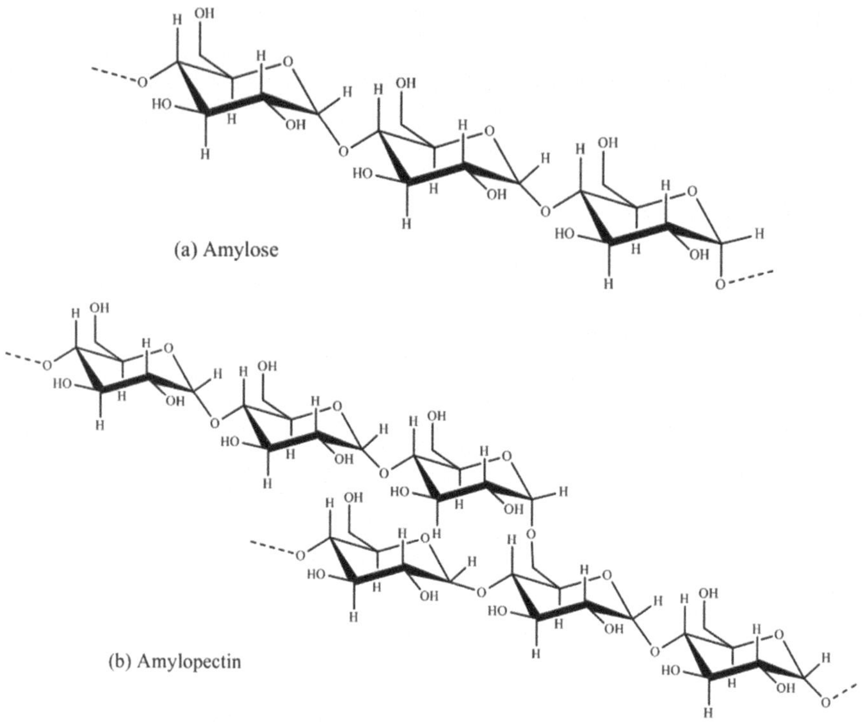

Fig. 21.2 Chemical components of starch (a) amylose and (b) amylopectin

Starch contains large number hydroxyl functionality on the backbone (C-2, C-3 and C-6 position of each glucose unit), which provides hydrophilic character to starch and also acts as sites for further chemical modification (Gutiérrez and Alvarez 2017a, b, c). In food industries starch based food packaging and edible films are two major applications. The organoleptic, nutritional, functional, mechanical and other properties can be modified by the addition of non-toxic chemicals in small amounts (Romero-Bastida et al. 2005). Starch exhibits physical properties similar to convention polymer plastic such as tasteless, odourless, nontoxic, resistant to oxygen passage, biologically absorbable and semipermeable to carbon dioxide.

The starch gel was usually regarded as a composite, in which swollen gelatinized starch granules are embedded in and reinforce an interpenetrating amylose gel matrix (Ring 1985). The gelation of starch involves thermally assisted three steps, hydration-plasticization of polymeric network (Ismail et al. 2013). First step, involves swelling of hydrophilic starch granules via water absorption, followed by heat induced gelatinization in the second step, resulting in leaching of the amylose component, destruction of granule structure and irreversible physical changes. The third step is the retrogradation step, in which hydrogel network formed upon cooling and aging. Partial reorganization and recrystallization in the network structure of polysaccharide take place during retrogradation (Matignon and Tecante 2017; Morris 1990). The gelatinization temperature and amylase content highly effect the gel formation (Miles et al. 1985; García-González et al. 2011).

Starch based hydrogels exhibits numerous advantageous properties, however there are some serious disadvantages also such as highly water sensitivity and poor mechanical properties making them inadequate for packaging applications (John and Thomas 2008; Xiong et al. 2008). Therefore chemical modification or blending with other polymer improves both of these drawbacks. Ghanbarzadeh et al. reported chemically modified starch based composite with enhanced barrier and mechanical property (Ghanbarzadeh et al. 2011). Modification of starch with citric acid (0–10%) and its blending with carboxymethyl cellulose (CMC) improves its water vapour barrier property as well as tensile strength. The improvement in barrier property were attributed to the substitution of hydrophilic hydroxyl (-OH) groups with hydrophobic ester groups. However, excess of citric acid increases the water vapor transmission through the films. Addition of CMC also showed positive effect on both moisture resistance and tensile strength. The films with CMC (15%, w/w) showed the lowest water vapour permeability values. Addition of plasticizer like sorbitol and glycerol improves heat sealability as well as seal strength (375 N/m at 3:1, sorbitol-glycerol) of starch based films (Abdorreza et al. 2011). Sorbitol plasticized films however possessed better heat sealability as compared to glycerol. The seal strength value become lower than synthetic polymers (≥730 N/m), however it become comparable to other polysaccharide films, e.g., carrageenan films (130 N/m for sorbitol and 137 N/m for glycerol), lactic acid-casein films plasticized with sorbitol (153–247 N/m) and protein isolate/lipid emulsion films (301–323 N/m) (Kim and Ustunol 2001).

For food preservation, researchers have considered the incorporation of food components and additives into starch based biodegradable films to improve food shelf life (Gutiérrez et al. 2015a; Fama et al. 2005). However, in non-food usage starch were modified through various chemical functionalities to achieve products with suitable properties for different applications. Esterification and grafting are two highly reported methods for the preparation of starch based hydrogel through chemical means (Chu et al. 2011; Hong et al. 2016; Ackar et al. 2015; Athawale and Lele 1998; Zhai et al. 2002). Athawale et al. reported preparation of acrylamide grafted maize starch based hydrogel initiated by ceric ions ($Ce4^{+}$) in an aqueous medium (Athawale and Vidyagauri 1998). The maximum water absorption achieved by this hydrogels was 170 g/g. The grafting of acrylamide on to starch takes place via a free radical mechanism. The Ce^{4+} form complex with –OH groups of starch at C-2 and C-3 position, which leads to decomposition generating the free radicals and Ce^{4+} gets reduced to Ce^{3+} releasing a proton. The grafting of acrylamide onto starch thus takes place at the generated active sites (Dragan and Apopei 2011). The grafting of vinyl monomers enhances the stability of starch (Athawale and Lele 2000). Acryl modified starch based hydrogels acts as excellent dyes sorbents materials (Dragan and Apopei 2011).

Now, a day's polysaccharide based nanocomposites hydrogels were under tremendous investigation due to its advanced property as compared to conventional hydrogels. Starch based nanocomposites films exhibit with excellent antimicrobial property as well as improved mechanical and barrier property (Gutiérrez 2018b).

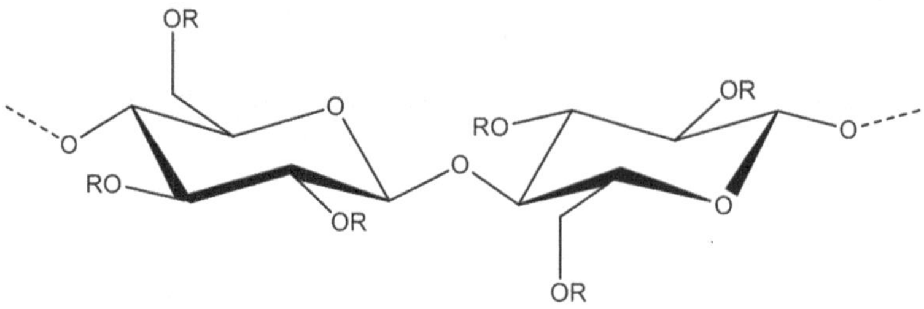

		R (groups)
Cellulose	→	H
Methylcellulose	→	H, CH_3
Ethylcellulose	→	H, CH_2CH_3
Hydroxyethylcellulose	→	H, $[CH_2CH_2O]_nH$
Hydroxyethylmethylcellulose	→	H,CH_3, $[CH_2CH_2O]_nH$
Carboxymethylcellulose	→	H, CH_2COONa

Fig. 21.3 Cellulose and cellulose derivatives (R—new functional group in different derivatives) repeating unit

Abreu et al. prepared a starch based nanocomposite films containing montmorillonite, C30B (Mnt) and silver nanoparticle (AgNP) for packaging applications (Abreu et al. 2015). The nanocomposite films exhibit good antimicrobial activity against *Staphylococcus aureus, E. coli* and *Candida albicans*. Contact test using food simulants also indicate safe applicability of such films for food packaging. Nanocomposites containing nano clay displayed significant improvement in tensile strength and modulus with effecting in elongation at break (Chung et al. 2010; Majdzadeh-Ardakani et al. 2010). Bozanic et al. reported sago starch based Ag composite films with good bioactivity against *S. aureus*, *E. coli* and *C. albicans* (Bozanic et al. 2011).

21.3.2 Cellulose and Cellulose Derivatives

Cellulose is the most abundant nature based polysaccharide, exists as the chief component of plants and also produced by some microorganism (Somerville 2006; Römling 2002). It is composed of glucose monomers in linear fashion linked by β-1,4-glucosidic bonds and every alternate glucose unit being flipped related to the next one, which gives high rigidity and crystallinity to cellulose (Fig. 21.3) (Gutiérrez and Alvarez 2017d; Varshney and Naithani 2011; Kamel et al. 2008). Therefore, become insoluble in water and most organic solvents which limit its applications. These drawback can be exalted through various chemical modifications such as esterification and etherification (Roy et al. 2009; Pang et al. 2016). Various cellulose derivatives such as methylcellulose (MC), ethylcellulose (EC), hydroxyethyl cellulose (HEC), hydroxypropyl cellulose (HPC), hydroxyethyl methyl cellulose (HEMC) and sodium carboxymethyl cellulose (NaCMC) etc. are water soluble and are widely used in different applications (Fig. 21.3) (Álvarez et al. 2017). Among the different cellulose derivatives NaCMC is a polyelectrolyte and displays sensitivity to pH and ionic strength. This polyelectrolyte nature of NaCMC makes it ideal candidate for the development of smart superabsorbent hydrogels (Qiu and Hu 2013; Eldin et al. 2011).

Cellulose derivatives such as MC, EC, HMPC, and CMC etc., turn out to be the first choice raw materials for the hydrogels preparation in different applications due to their abundant availability, low cost and nontoxicity. The excellent biocompatibility and cell mediated biodegradation promoted the large scale usage of cellulose based hydrogel materials for biomedical applications (Märtson et al. 1999; Entcheva et al. 2004). Other applications of cellulose based hydrogels includes superabsorbent, water reservoirs, drug delivery, tissue engineering and wound dressing etc. (Sannino et al. 2004, 2000, 2009; Yu et al. 2017; Trojani et al. 2005; Nguyen and West 2002; Pandey et al. 2017). Demitri et al. prepared sodium carboxymethyl cellulose (CMCNa) and hydroxyethyl cellulose (HEC) based superabsorbent hydrogel using citric acid as crosslinking agent, which combine good swelling properties with biodegradability (Demitri et al. 2008). The crosslinking reaction takes place via an anhydride intermediate formation. Gregorova et al. reported carboxymethyl cellulose (CMC) and polyvinylpyrrolidone (PVP) based hydrogel for food packaging (Gregorova et al. 2015). The hydrogels were prepared using casting method in the presence of other ingredients polyethylene glycol (1%), agar (2%) and glycerine (1%). Both CMC and PVP hydrogels exhibits comparable water retention and uptake capacity. The mechanical strength of CMC hydrogel ranging between 43.7–45.7 MPa, the addition of PVP further improves the mechanical strength of hydrogels. In another report the water vapour permeability (WVP) and mechanical properties of MC and MC-whey protein (WP) composite films using glycerol as plasticizer were assessed by Erdohan et al. (Erdohan and Turhan 2005). The increasing concentration of MC in the hydrogel decreases WVP, whereas tensile strength and percent elongation significantly increased. The maximum and minimum WVP values found at a concentration level in the ratio of 0.3 and 0.8 for MC:WP, respectively. Therefore such type of hydrogels could find potential applications in food packaging. Pal et al. prepared a pH sensitive hydrogel membrane via esterification of aqueous NaCMC with acryloyl chloride added dropwise in methyl ethyl ketone (Pal et al. 2006). Other method includes grafting of NaCMC with polyacrylic acid to develop smart pH-sensitive hydrogels (Eldin et al. 2011; El-Mohdy 2014; Pourjavadi et al. 2009).

21.3.3 Chitosan

Chitosan (CS) is a poly-cationic naturally occurring polysaccharide derived as a deacetylated form of chitin which is the second most abundant polysaccharide after cellulose in nature (Gutiérrez 2017). It contains more than 5000 glucosamine units

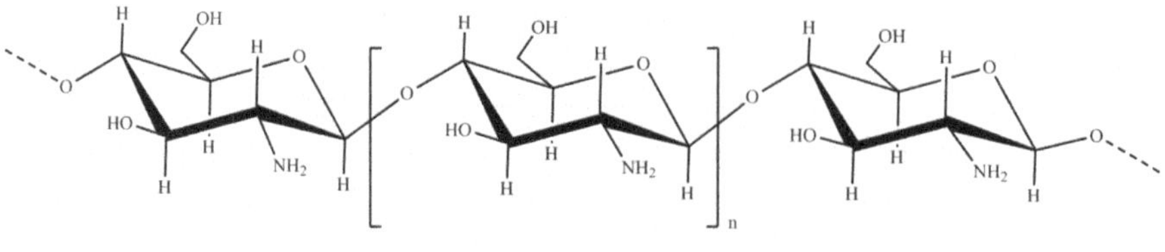

Fig. 21.4 Chemical structure of chitosan

and is obtained commercially from shrimp and crab shell containing chitin (*N*-acetyl glucosamine). It is commercially produced by partial chemical de-acetylation of chitin extracted from crustaceans (insects, crabs, shrimp shell or squid pen wastes) by deproteination, decalcification and decolorization and from fungi by enzymatic extraction. CS is a copolymer of *N*-acetyl-D-glucosamine and D-glucosamine units with one amino (NH_2) and two hydroxyl (-OH) functionality in each repeating glycosidic units (Fig. 21.4) (Ali and Ahmed 2018). Two important parameters which strongly influence the characteristic of CS are its molar mass and degree of acetylation (DA). CS hydrogels exhibit good biocompatibility, biodegradability, film forming ability, nontoxic, bioadhesive, blood anticoagulants and bioactivity. CS and chitosan derivatives have been extensively investigated in food industries as antimicrobials agents, coating and packaging, additives, dietary fibers, food processing, and encapsulating materials (Zargar et al. 2015; Wang et al. 2017a; Lagaron et al. 2007). Other applications include pharmaceuticals, biomedical, wound healing, tissue engineering, cosmetics, agricultural and water treatment etc. (Dodane and Vilivalam 1998; Dash et al. 2011; Islam et al. 2017; Bui et al. 2017; Boonlertnirun et al. 2017; Oladoja et al. 2017).

CS has demonstrated adequate ability to forms hydrogels, it can forms hydrogel in its native state, i.e. without crosslinker (Physical hydrogel) or in the presence of crosslinking agent (Chemical hydrogel). CS get dissolved under slight acidic environment, the acidic solution of CS when exposed to alkaline pH results in the formation of physical gels which is due the decrease in the apparent charge density of polymer (Fiamingo et al. 2016). Hydrogel bonding along with hydrophobic interaction plays important role in gel formation. These type of simple gel preparation are important in the fabrication of nontoxic hydrogels particularly for biomedical and food applications. CS films can be prepared by direct evaporation of CS solution which is known to be the simplest technique. It also forms stable films under basic pH or neutral conditions because of its pH dependent solubility (Racine et al. 2017). Chitosan molar mass, degree of acetylation (DA) and solvent have a strong effect on the physical and mechanical properties of the films (Park et al. 2002; Peh et al. 2000). Higher molar mass polysaccharide shown higher flexibility and high tensile strength, but high DA gives more brittle films (Tomihata and Ikada 1997; Rong Huei and Hwa 1996). Therefore, to improve its processability, brittleness and elastic properties, additives or plasticizers such as glycerol, PEG etc. were added to initial formulation, which increase polymer chains mobility via reducing polymer-polymer interaction (Rudyardjo and Wijayanto 2017).

CS based chemically crosslinked hydrogels offer enhanced control over physiological stability and mechanical strength. CS possesses two functionality $-NH_2$ and $-OH$ groups, which can be modified or crosslinked with various different crosslinkers or other polymers. CS based hydrogels have been researched for years in food industries due to their excellent performance. Pranoto et al. reported chitosan films incorporated garlic oil (GO), potassium sorbate (PS) and nisin (N) to enhance its antimicrobial activity (Pranoto et al. 2005a). The concentrations of additives GO at a minimum level of 100 μL/g, PS at 100 mg/g and N at 51000 IU/g of chitosan were found to have anti-microbial activity against *S. aureus, L. monocytogenes and B. cereus*. The addition of additives did not shown pronounced effect on the physical and mechanical properties of films. Thus, it can act as a good physical and antimicrobial barrier to food contamination. Bordenave et al. investigated the effect of chitosan coating on the liquid water and moisture barrier properties of packaging papers (Bordenave et al. 2007). Chitosan coating increases the moisture barrier properties of the paper, although not sufficient for packaging purpose as the surface hydrophilicity remains high. Pereira Jr. et al. reported anthocyanins (natural pH indicators) doped polyvinyl alcohol (PVA)/chitosan films as time-temperature indicator sensor based intelligent food packaging material (Pereira et al. 2015). The hydrogel was prepared by mixing PVA (1%) and CS (1%) solution in the ratio of 3:7 (v/v) respectively, through casting technique after incorporating anthocyanins (25%). To promote crosslinking 1.5% sodium tripolyphosphate solution were added to the mixture and pH maintained at 6.1 (using NaOH). The prepared hydrogel changes its colour in response to pH changing, thereby providing a simple and cheap way to detect the changes happening in the food chemical composition. However, the mechanical properties were diverged from those of commercially available polymers. Genipin, a naturally derived compound, has been extensively used as a crosslinker for chitosan based hydrogels due to its lower acute toxicity and higher biocompatibility as compared to synthetic crosslinker (Dimida et al. 2015; Mi et al. 2000, 2001). Klein et al. reported genipin crosslinked chitosan particles as a novel alternative for food processing applications (Klein et al. 2016). The toxicity effect of genipin is 5000–10,000 times less as compared to conventional crosslinker glutaraldehyde (Sung et al. 1999). However, the mechanical properties of hydrogel become almost comparable with both crosslinker. A graphical description of crosslinking between genipin and chitosan is given in Fig. 21.5. The crosslinking mechanism of genipin differs at different pH. Under neutral and acidic conditions, genipin reacts with primary amino groups of chitosan to form heterocyclic amines. The heterocyclic amines were further associated to form short chains crosslinked network. Further, in the presence of an acid catalyst a nucleophilic substitution reaction of the ester group on genipin by the primary amine group on chitosan occurred (Mi et al. 2005; Butler et al. 2003). Under basic condition, a ring opening reaction of genipin occurs through a nucleophilic attack by hydroxyl group ($-OH$), prior to crosslinking with chitosan. The ring opening leads to formation of intermediate aldehyde groups which undergo aldol condensation. The cross-linking reactions were than occur via Schiff reaction between terminal aldehyde group on genipin and

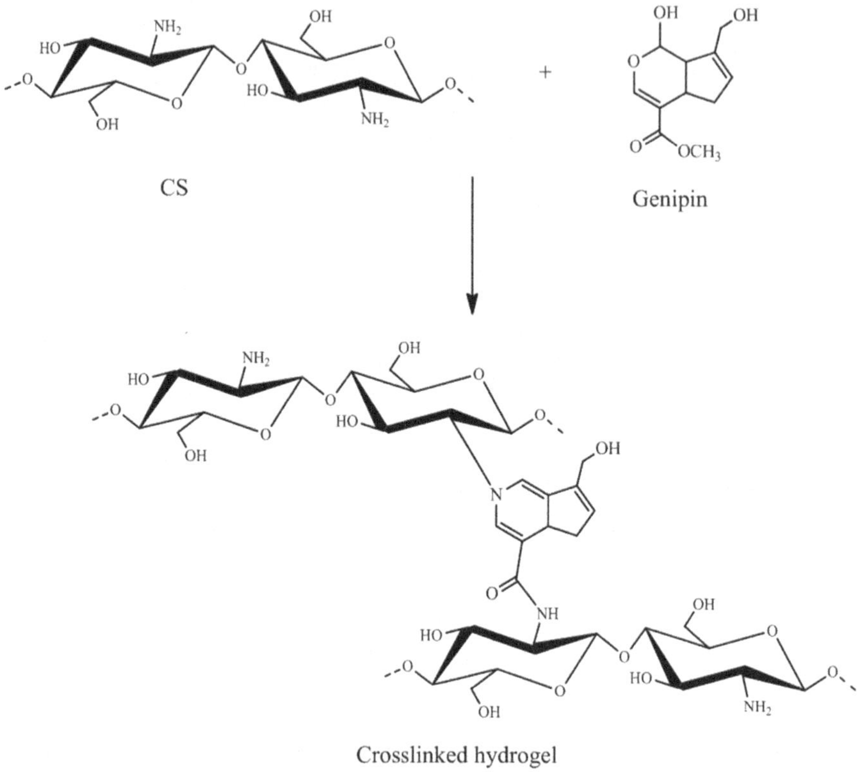

Fig. 21.5 Crosslinking of chitosan with genipin

amino groups on chitosan (Muzzarelli 2009; Sung et al. 1999). It can, thus, be concluded that pH conditions has a very strong influence on the crosslinking reaction.

Jin et al. reported chitosan and chitosan/polyethylene oxide (PEO) based films crosslinked by genipin (Jin et al. 2004). The crosslinked films demonstrated better mechanical properties, elasticity, stability and more hydrophobicity. Swelling characteristic of the films was time and pH dependent and the swelling extent can be reduced by higher concentration of genipin. Therefore such type of chitosan based hydrogel could be appropriate candidate in food-based applications.

21.3.4 Pectin

Pectin is a structural hetero-polysaccharide present in the primary cell wall of terrestrial plants. Pectins are polysaccharides bearing homopolymeric methyl-esterified poly-α-(1→4)-D-galacturonic acid and heteropolymeric α-(1→2)-L-rhamnosy-α-(1→4)-D-galacturonosyl units Fig. 21.6 (Liu et al. 2012). The carboxylate groups in pectin tends to expand the structure, while the methylated groups impart

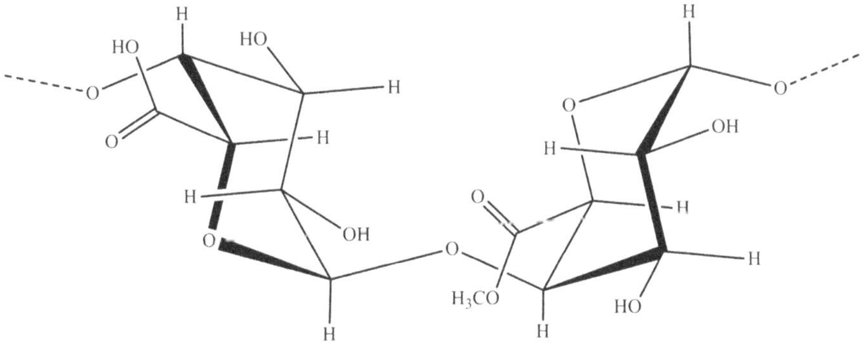

Fig. 21.6 Chemical structure of pectin

hydrophobicity to structure (Willats et al. 2006). Pectin has being used in food industries as thickening and gelling agent, which gives the desired firmness and modifies the texture of the gels (Yapo 2011; Willats et al. 2006).

Espitia et al. thoroughly elaborated the synthesis, techniques and applications of pectin based edible hydrogels films as active packaging material (Espitia et al. 2014). Pectin can be classified as high methoxyl (HMP, ≥50% of carboxyl groups esterified) and low methoxyl (LMP, ≤50% carboxyl groups esterified) pectin. HMP forms gels in acidic media in the presence of different sugars (glucose, sucrose, etc.), while LMP forms gels in the presence of multivalent ions (Mishra et al. 2012; Videcoq et al. 2011). Gelation of pectin involves various intermolecular interactions; LMP involves electrostatic interactions between negatively charged polymer and positively charged cation, while in HMP gelation occurs at low pH in the presence of co-solutes where the electrostatic repulsion and water cavity are reduced (Alonso-Mougán et al. 2002). Bayarri et al. investigated lysozyme/LMP properties to develop an edible antimicrobial film (Bayarri et al. 2014). The films were prepared by gradually increasing LMP concentration (0–2 g/L) at constant lysozyme concentration (0.14 g/L) up to a threshold LMP concentration. The developed film releases lysozyme and can be used against lysozyme sensitive microorganisms particularly pectinolytic enzyme producers (*Bacillus and Clostridium spp.*) as edible films.

21.3.5 Alginates

Alginate is a naturally occurring anionic hydrophilic, biocompatible, biodegradable, and relatively economical biopolymer. Structurally, alginate consists of two forms being a family of copolymers, i.e. (1→4) linked β-D-mannuronic acid (M) and α-L-guluronic acid (G) monomers (Fig. 21.7) (Narayanan et al. 2012). The structure of alginate has no regular repeating unit and it contains all the four possible glycosidic linkages. Alginate has been approved by U.S. Food and Drug Administration (FDA), and become one of the most important biomaterials having

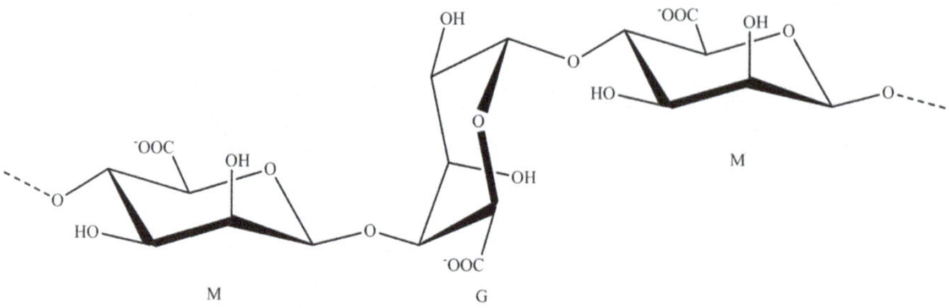

Fig. 21.7 Chemical structure of alginate (representing M and G block)

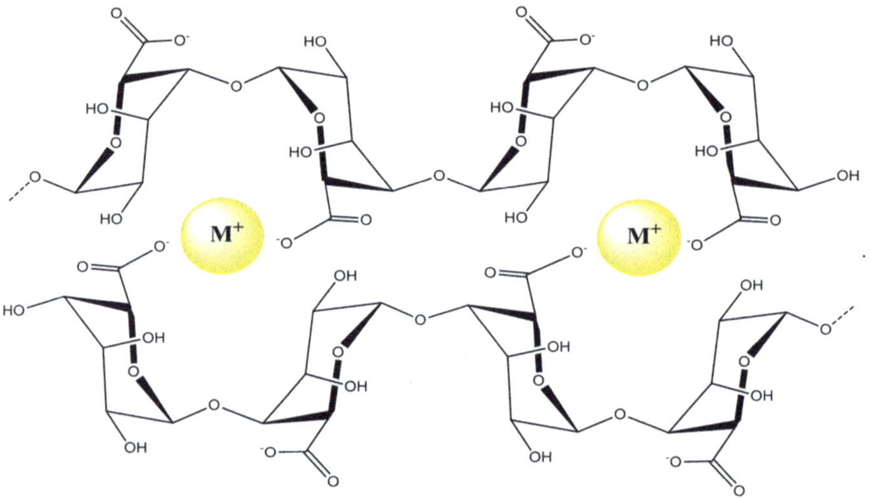

Fig. 21.8 Schematic representation of crosslinking in alginate by metal cation

diverse applications in medicine, pharmaceutics, nutrition supplement and semipermeable separation etc. (Sun and Tan 2013; Balakrishnan and Jayakrishnan 2005; Kong et al. 2004).

The aqueous solution of alginic acid can be converted to ionotropic hydrogels via crosslinking with various metal cations, such as Ca^{2+}, Fe^{3+}, Sr^{2+} and Ba^{2+} etc. (Remya et al. 2012; Morch et al. 2006; Sreeram et al. 2004). The crosslinking reaction was take place through chelation between metal cations (M^+) and carboxylate ions of alginate (Fig. 21.8) (Narayanan et al. 2012). In case of calcium alginate hydrogel the "egg-box" model structure were established in which four G subunits are bound by one Ca^{2+} ions in a 2/1 helical conformation (Li et al. 2007; Sikorski et al. 2007). The binding capacities of metal ions differ with the sub units (G, M or GM) or block structure of alginate. Ca^{2+} was found to bind to G- and MG-blocks, Ba^{2+} to G- and M-blocks, and Sr^{2+} to G-blocks (Mørch et al. 2006). In general, the affinity of alginate towards different metal cations was reported to decrease in the following order: $Pb^{2+} > Cu^{2+} > Cd^{2+} > Ba^{2+} > Sr^{2+} > Ca^{2+} > Co^{2+} = Ni^{2+} = Zn^{2+} > Mn^{2+}$ (Papageorgiou et al. 2010;

Mørch et al. 2012). Wang et al. reported a ternary blend hydrogel films composed of agar/alginate/collagen along with silver nanoparticles (AgNPs) and grapefruit seed extract (GSE).

21.3.6 Carrageenans

Carrageenans are naturally occurring sulphated polysaccharides extracted from many species of red edible seaweeds of the Rhodophyceae family. The structure is primarily based on linearly interconnected copolymers of 1→3-linked β-D-galactose and 1→4-linked α-D-galactose monomers with varying degree of sulfatation (Santo et al. 2009; Ficko-Blean et al. 2017). The disaccharide repeating units was formed by alternating α-1→3 and β-1→4 glycosidic linkages between the monomers (Fig. 21.9) (Abad et al. 2003). The negatively charged sulfate groups (OSO_3^-) imparts strongly anionic character to carrageenan polysaccharide. Carrageenan has been classified in different types, depending up on the number of sulfate groups in the disaccharide repeating units. This includes kappa (κ), iota (ι), lambda (λ), Mu (μ), Nu (υ) and theta (ϴ) carrageenan. Three very commonly used carrageenans are κ-carrageenan (one, $-OSO_3^-$ group), ι-carrageenan (two, $-OSO_3^-$ group) and λ-carrageenan (three, $-OSO_3^-$ groups) (Fig. 21.9) (Langendorff et al. 2000). The application of carrageenan is increasing in pharmaceuticals, food industry, cosmetics, biomedical, agriculture etc., due to biocompatibility and low toxicity (Prajapati et al. 2014; Rinaudo 2008; Khalil et al. 2017; Majee et al. 2017).

Carrageenans have been extensively engaged in pharmaceuticals and food industries as stabilizers, gelling agents, emulsifiers as well as base material for packaging films. They also exhibit excellent gelling competence. Carrageenans undergo gelation forming thermotropic and ionotropic gels, under appropriate salt conditions upon cooling via coil-helix conformational transitions (Chronakis et al. 2000). κ-carrageenan forms strong gels in the presence of potassium salt ions (K) and ι-carrageenan by calcium (Ca) ions (Mangione et al. 2005; Hermansson 1989; Hermansson et al. 1991). However, λ-carrageenan does not form gels and used as thickening agent in dairy products. The mechanism of gel formation in carrageenan depends on the temperature and gelling agent. At high temperature (>80 °C), carrageenans structure becomes random coil due to electrostatic repulsion between neighbouring chains, which changes to helical structure upon reducing the temperature (Tavassoli-Kafrani et al. 2016). Further cooling in the presence of gelling agent (metal cations) promote intermolecular interactions among polymer chains leading to aggregation of helical dimers and finally forming three dimensional stable network (Fig. 21.10).

Farhan et al. developed semi refined κ-carrageenan based edible packaging films plasticized with glycerol and sorbitol (Farhan and Hani 2017). Sorbitol as a plasticizer imparts more effective oxygen barrier to films as compared to glycerol. The mechanical properties and heat seal strength were also improved by the addition of glycerol and sorbitol (20–30%). Therefore, such hydrogels could be excellent packaging films for oxygen sensitive food products. Padhi et al. developed highly biocompatible composite

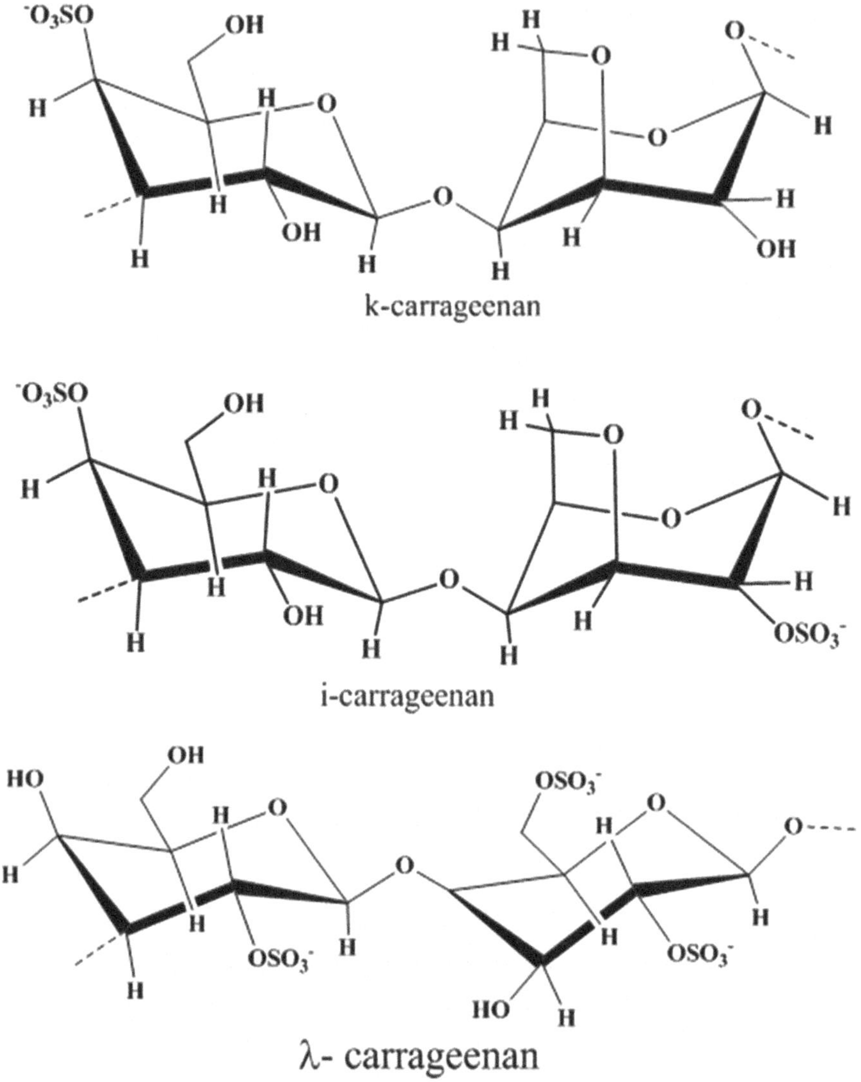

Fig. 21.9 Chemical structure of k, ι, and λ carrageenan

hydrogel composed of gelatin and i-carrageenan for biomedical applications (Padhi et al. 2016). Hambleton et al. investigated the aroma barrier property of ι-carrageenan emulsion based films for the encapsulation of active food component (Hambleton et al. 2009). The emulsion was formed with GBS (fat) an acetic acid ester of mono and diglycerides blended with 20% w/w beeswax and glycerol monostearate as emulsifier mixed to GBS. ĺ-carrageenan based films further demonstrate better mechanical properties, stabilization of emulsion and reduction in oxygen transfer. They concluded that

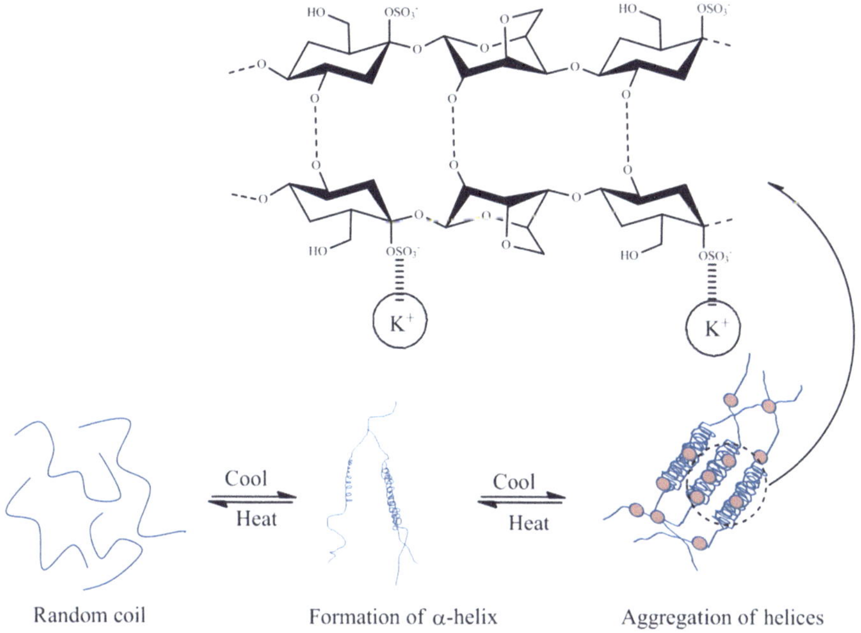

Fig. 21.10 The mechanism of gel formation in κ-carrageenan in the presence of K^+ ions (Rhein-Knudsen et al. 2015)

these type of hydrogel with ι-carrageenan as matrix leads to promising prospective for flavour encapsulation applications and coating for food surfaces. Oun et al. prepared carrageenan based bio-nanocomposite hydrogels and films containing zinc oxide (ZnO) and copper oxide (CuO) nanoparticles with strong antibacterial activity (Oun and Rhim 2017). Potassium chloride (KCl) was used as crosslinking agent to increase gel strength. The incorporation of nanoparticles in the films changes the transparent films to translucent, which also decrease UV light transmission. Biological activity of bio-nanocomposite hydrogel films showed strong antibacterial activity against food borne pathogenic bacteria *E. coli* and *Listeria monocytogenes*. Over all the effect of ZnONPs on various properties such as mechanical, swelling, UV transmission, thermal stability etc., become more pronounced than CuONPs. These types of bio-nanocomposite hydrogel films could find potential application in packaging areas. It can, overall, be concluded that carrageenan based hydrogels has potential scope for food based industries.

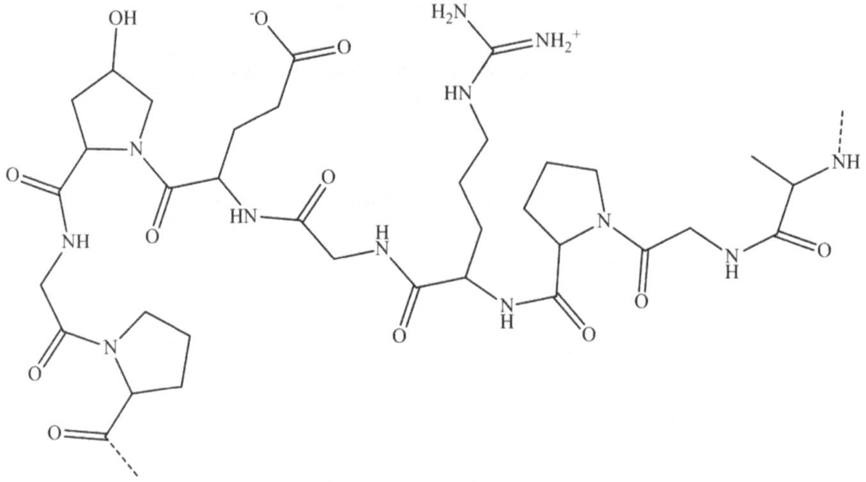

Fig. 21.11 Representative structure of gelatin

21.4 Proteins

Proteins are the second category of edible polymers which possess many constructive properties and have under tremendous investigation and exploration for hydrogel preparation. Proteins are large polypeptides composed of different amino acids with repeating amide bond in the backbone structure. Proteins hydrogels have directed renewed attention of scientist as degradable and renewable edible polymer. Proteins were employed as adhesives and edible films/coatings for food industry (Krochta 2002). The wide range of functionality and properties possessed by proteins attracted chemist attention towards the development of new biodegradable materials. Protein based materials (films) displayed excellent gas barrier properties, as well some are water resistant but not entirely hydrophobic (Miller and Krochta 1997). Unlike polysaccharide, there is less extent of research data available on protein-based hydrogels. However, different animals and plants proteins are generally used as biodegradable polymers, such as gelatin, soy protein, corn zein, whey protein, gluten, and casein etc. Here, we have discussed few proteins based hydrogels in details.

21.4.1 Gelatin

Gelatin is a colourless, translucent, flavourless fibrous protein derived from collagen by partial degradation which was present in various animal body parts as structural protein. The amino acid composition of gelatin is not defined clearly and defers from author to author. Muyonga et al. reported about 30% of total amino acids in mammalian gelatins were proline and hydroxyproline. However, Farris et al. reported 23% for the same (Muyonga et al. 2004; Farris et al. 2009). Figure 21.11 shows the representative structure of gelatin: Ala-Gly-Pro-Arg-Gy-Glu-4Hyp-Gly-Pro- (Nur Hanani et al. 2014). Gelatin exhibit excellent biological and physiochemical property, therefore has being extensively employed in food industry, pharmaceuticals, biomedical, tissue engineering and environmental recycling (Karim and Bhat 2008; Djagny et al. 2001; Su and Wang 2015; Wang et al. 2017b). Gelatin possess thermo-reversible rheological property, transforming between sol and gel (Zakaria and Bakar 2015). Gelatin also exhibit anti-oxidant and anti-microbial property (Gómez-Guillén et al. 2011). Generally, the properties of final gelatin product are mostly influenced by two key factors: initial collagen characteristic and extraction processes involved. The amino acid composition and molecular weight distribution of gelatin play also an important role in determining the mechanical and barrier properties of gelatin films. Wang et al. evaluated film-forming abilities of six different types of proteins (0–16% conc.) and polysaccharides (0–4% conc.) by changing their concentration and heating temperature (60–80 °C) using glycerol as plasticizer (Wang et al. 2007). The proteins includes gelatin, sodium caseinate and whey protein, while polysaccharide consists of CMC, sodium alginate and potato starch. Films produced from proteins showed more resistance to solvent as compared to polysaccharide. Sodium alginate films achieved higher tensile strength and barrier to water vapour and oxygen permeability, while gelatin gives higher flexible films. The overall results demonstrated gelatin, whey protein and sodium alginate films possess more desirable properties as compared to others.

Although gelatin exhibit excellent properties in food industries, but the inadequate mechanical and thermal stability limits its potential applications. Gelatin is also highly hygroscopic, due to which can be dissolved or swelled when comes in contact with high moisture content. Therefore to overcome such drawback researchers were trying to improve these properties by the addition of different chemical agents, such as plasticizer, crosslinker, additives with antimicrobial and antioxidant activity, strengthening agents and by complexation with other polysaccharide, etc. (De Carvalho and Grosso 2004; Zhang et al. 2010; Mao et al. 2003).

The crosslinking of gelatin with glutaraldehyde were reported by Farris et al. (Farris et al. 2009). The crosslinking reactions were taking place between aldehyde group in glutaraldehyde and free amino groups in gelatin via a nucleophilic addition type reaction. The reaction processed through an unstable carbinolamine intermediate formation followed by loss of water molecule yielding conjugated Schiff bases. However, due to cytotoxicity effect of glutaraldehyde, such types of crosslinker were avoided in food formulation and packaging applications. Therefore, there is a need of biocompatible low toxic crosslinker which can full fill the needs of researchers.

The use polysaccharide based on biocompatible and biodegradable crosslinker are emerging as new type of crosslinker particularly for food and biomedical applications. Polysaccharide chemically modified through chemical reaction (mainly periodate oxidation) to incorporate dialdehyde functional groups which

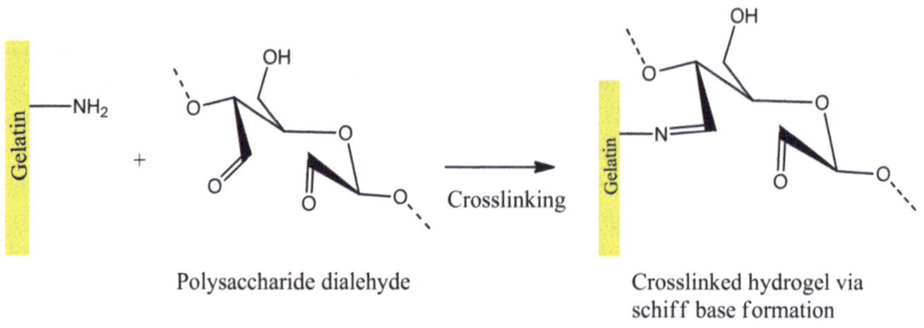

Fig. 21.12 Graphical representation showing crosslinking of gelatin with polysaccharide based crosslinker

react with amino functional group via Schiff base formation, resulting in crosslinking (Fig. 21.12).

Mu et al. prepared gelatin based on dialdehyde carboxymethyl cellulose (DCMC) crosslinked edible films (Mu et al. 2012). DCMC act as good crosslinking agent which also greatly reduce the WVP (1.5×10^{-10} gm/m^{2}s Pa) and equilibrium swelling ratio (150%) of the films. Thus DCMC has the potential to reduce the water sensitivity of gelatin based films. The addition of plasticizer (glycerol) in the films increase the elongation break and WVP, but the thermal stability decreases. Such types of films can be useful in food industries and biomedical fields.

Similarly, Zhuang et al. reported gelatin based on hydrogel using chemically modified cellulose as crosslinker (Zhuang et al. 2017). Microcrystalline cellulose were modified by chemical oxidation using 2,2,6,6-tetramethylpiperidine-1-oxyl (TEMPO) as oxidizing agent, which oxidize primary hydroxyl group at C-6 to carboxyl group followed by reacting with *N*-hydroxysuccinimide to give cellulose ester (TMN). The crosslinking reactions take place between the NH_2 group of gelatin and ester groups of TMN. The resulted film shows good mechanical strength, flexible, elasticity, increased moisture absorption and greater swelling capability. Overall such types of films act as safe, eco-friendly, stable and biorefractory gelatin based materials for food packaging purposes. Polyion-complex hydrogels based on gelatin–pectin was reported by Farris et al. (Farris et al. 2011). Ionic interactions between opposite charges in the polymers produce reversible hydrogels having homogenous molecular arrangement, resulting in improvement water resistance and mechanical property, but the thermal stability remains same relative to individual polymer gels.

21.4.2 Whey Protein

Whey is a yellow-green liquid separated from the curd during casein and cheese manufacturing and has been considered as waste by-product (Smithers et al. 1996). Whey represents a heterogeneous mixture with the main constituents as β-lactoglobulin and α-lactalbumin which accounts for around 70–80% of total whey protein. Whey protein has received considerable attention due to its films and coating forming ability with excellent aroma, oxygen and oil barrier properties (Sothornvit and Krochta 2000a). Hydrolysis of whey protein increases its solubility and forms better films hydrogel at low plasticizer concentration without effecting its WVP (Sothornvit and Krochta 2000b). Whey protein based films and coating possess the required physicochemical properties for packaging applications such as barrier, optical, WVP, mechanical and surface properties etc. (Ramos et al. 2012). Gunasekaran et al. prepared two sets of heat induced whey protein-based hydrogels (Gunasekaran et al. 2006). One sets prepared at constant concentration of whey protein (15% w/v) while varying the pH (5.1–10.0) and the other set at constant pH (10.0) by changing the protein concentration (12, 15 and 18%). At a particular protein concentration the properties of hydrogel like gelation time and mechanical behaviour are shown to be highly dependent on pH. The gels exhibit minimum equilibrium swelling ratio when the pH of swelling medium was close to isoelectric point of whey protein and become increase as the pH moves far from the isoelectric point values. As the pH further increase for the isoelectric point value the swelling behaviour becomes highly pH sensitive. Lacroix et al. used γ-irradiation to prepare hydrogel films form whey, casein and soya protein and investigated their structural and functional characteristics (Lacroix et al. 2002). γ-irradiation caused crosslinking and also the conformation of the protein changed to some extent, which results in more ordered an stable structure. The resulted films showed enhanced puncture strength and physico-chemical properties. Whey protein nanofibrils form cold set hydrogels in the presence of different divalent (Ca^{2+}, Mn^{2+}, Zn^{2+}) cations at lower concentration as compared to parent proteins (Mohammadian and Madadlou 2016).

21.4.3 Soy Protein

Soy protein (SP) is the most abundant plant based protein in nature. Soy protein isolate (SPI) have been extensively used as hydrogels, adhesives, plastics, films, emulsifiers and coatings (Liu et al. 2017). It can also be useful as food packaging materials due to its low toxicity and biodegradability (Kumar et al. 2002). Various properties of soy proteins such as film formation, tensile strength and WVP were affected by pH. The films formation will only occurred between pH 1–3 and 6–12, there is no formation of films happened between pH 4 and 5 (Gennadios et al. 1993). Cold gelation of proteins is an important technique reported by many researchers and proven to useful in explaining the gel forming mechanism (Glibowski et al.

2006; Remondetto and Subirade 2003). The physicochemical characteristics of the hydrogel, such as water holding capacity, opacity and mechanical properties can be controlled by varying the pH, gelation temperature, salt concentration and types (Lakemond et al. 2003). Maltais et al. investigated the effect of protein and salt concentration (calcium) on the cold-set gelation of SPI (Maltais et al. 2005). The gel formations were carried out at SPI concentration 6–9% with salt concentration ranging from 10 to 20 mM. The increasing concentration of SPI and salt improves elastic modulus of gel, but the water holding capacity only improves with SPI concentration. The gel opacity increases with increasing salt concentration (from 10 to 20 mM); however the opposite effect was showed on increasing SPI concentration. Tansaz et al. developed a composite hydrogel microcapsule and films containing SPI and alginate (Tansaz et al. 2017). The microcapsules showed higher water uptake capability as compared to films of the same composition due to its high surface area.

21.4.4 Wheat Gluten

Wheat gluten protein is a renewable resource having great potential of hydrogel formation in the form of films with excellent strength, plasticity and elasticity with glycerol as plasticizer. Wheat gluten mainly consists of fractions roughly present in equal amounts: gliadins consist of heterogeneous monomeric proteins and glutenins are composed of a number of subunits crosslinked by disulphide bridges, classified as per their solubility in alcohol (Lagrain et al. 2010). Disulphide bond play a significant role in determining the structure and properties of gluten proteins (Wieser 2007). Kontogiorgos thoroughly reviewed the microstructure of hydrated gluten network and divided into four structural levels (Kontogiorgos 2011). At molecular level discrete gluten components interact through various physical and covalent forces, leads to a transaction in morphology to a continuous sheet-like structure (secondary structural level), which may possibly considered as the fundamental units of gluten network. Further, the excessive association of gluten sheets in a disordered fashion concludes the third level of structural level. Before these some nanoporous ultrastructures were also forms which get distorted afterwards due to entrapment of aqueous phase. Finally, the arrangement of microstructural elements finishes the formation of fourth structural level displaying various morphologies and mechanical properties. Hydrogel films based on wheat gluten possess unique cohesive and elastic properties, due to which it has received great interest from researchers (Kayserilioğlu et al. 2003). Wheat gluten based bioplastics are fully biodegradable within 36 days under aerobic fermentation and 50 days in farmland soil (Domenek et al. 2004). Edible films from wheat gluten were prepared via compression molding technique using glycerol as plasticizer (Zubeldía et al. 2015). The concentration and plasticizer type found to be a dominating factor effecting the WVP and mechanical properties of films. Pressing temperature also influence the final film properties, but temperature higher than 100 °C is not appropriate, because it leads to darker films which would not good enough for packaging applications.

21.4.5 Zein Proteins

Zein is the key storage protein of corn kernel accounting for about 44–79% of the total protein content mostly exists in the endosperm (Shukla and Cheryan 2001). The protein is hydrophobic because of high proportion of apolar amino acids (approximately 55%), although some polar groups were also present but insoluble in water. Zein is generally divided in to four fractions: α, β, γ, and δ-zein, based on its solubility in aqueous alcoholic solution. Out of these four, α-zein accounts the highest proportion and is the main constituent of commercial zein. Zein exhibit promising properties such as biodegradability, non-toxicity, biocompatibility, self-assembly capacity and has been under tremendous investigation focusing on their utilization for pharmaceuticals, food and biomedical applications (Ni and Dumont 2017). Due to its self-assembly ability, zein based hydrogels act as a potential candidate for functional coating, drug delivery, encapsulation material and tissue engineering etc. (Hurtado-Lopez and Murdan 2005; Luo and Wang 2014; Landers et al. 2002). Zein based hydrogel were also explored as potential absorbent materials for the removal of heavy metals and oils (Ni et al. 2017, 2018). The films prepared from zein without plasticizers are more brittle and less flexible and thus are of low value (Lawton 2002). Therefore, low molecular weight compounds (as plasticizer) and chemical modification were applied to improve it overall physico-mechanical properties. Shi et al. chemically modified zein with lauryl chloride to improve the brittleness of films via acylation reaction (Shi et al. 2011). The new material shows sevenfold improvement in elongation at break with slight loss of mechanical strength, uniform surface with more hydrophobicity and low glass transition temperature. Thus, this new biomaterials can be useful in the development of biodegradable packaging material for food and deliver system.

21.4.6 Milk Proteins

Milk proteins (MP) are natural, inexpensive and widely available GRAS raw materials with high nutritional value (Livney 2010). MP are generally considered as good choice materials for microencapsulation of nutriceuticals and probiotics (Abd El-Salam and El-Shibiny 2015; Tavares et al. 2014). MP exhibit excellent gelation properties, for example rennet or acid curd formation of caseins, heat induced gelation of MP. The rennet gelation is centred on the proteolytic cleavage of κ-casein, resulting in micelle aggregation (Heidebach et al. 2009a), while the acid gelation is based on isoelectric precipitation. This technique has been applied for encapsulation of probiotic bacteria. Transglutaminase catalysed gelation of casein is another technique of MP gelation (Heidebach et al. 2009b). Song et al. reported genipin crosslinked casein based hydrogel for controlled drug delivery (Song et al. 2009). Genipin concentration and temperature has a direct effect on gelation time of casein.

Moreover, the mechanical property of hydrogel can be tuned by varying the concentration of genipin.

21.5 Lipids

Lipid compounds are generally hydrophobic in nature and mainly consist of acetylated monoglycerides, natural waxes and surfactants (Shit and Shah 2014). The most functioning lipids are paraffin wax and beeswax As compared to polysaccharides and proteins, the proportion of lipids based materials under investigation is not in abundant. Lipids based hydrogels were usually prepared accompanied by some polysaccharide to provide mechanical strength. Most lipids have the ability to stretch up to 102% of their original length in solid state before rupturing, with exception acetylated glycerol monostearate stretch up to 800%. Lipid based films exhibit enhanced water vapour barrier properties due to their hydrophobic nature (Guilbert et al. 1995). Lipids are frequently added to polysaccharide based hydrogels to reduce their hydrophobicity. Acetylated monoglyceride have been used as coating material on poultry and meat cuts to hinder moisture loss during storage (Bourtoom 2008). Waxes have been utilized to advance surface appearance and barrier for moisture and gases. Lipids particles dispersed in polysaccharide hydrogels demonstrated to have potential application for pharmaceuticals and cosmetic industries (Kulkarni et al. 2015). It also improves drug stability in the hydrogel which was either lost in native hydrogels due to possible interactions with hydrogel networks (Pardeike et al. 2009).

21.6 Properties of a Good of Packaging Material

Different materials have being used for packing purposes such as papers, clothes which are light and flexible, metals and glasses which are strong and corrosion free respectively, and polymers. However, the polymer materials, i.e. plastics are among the more demanding materials for packaging applications (Roy et al. 2011). Polymers give more advantageous properties like, softness, lightweight, heat sealing ability, transparency and good strength etc. (Mahalik and Nambiar 2010; Akelah 2013).

The main objective of food packaging includes covering and retaining the integrity of the food component, maintenance of food freshness, enhancing organoleptic characteristics of foods such as taste, appearance, aroma and prevention of food from environmental hazards (Zhao et al. 2008). The function of packaging materials is listed in Fig. 21.13. The quality of a good packaging material in food industries depends on the various properties associated with it such as gas barrier, vapour and aroma barrier, mechanical, thermal, antimicrobial, optical and biodegradable properties.

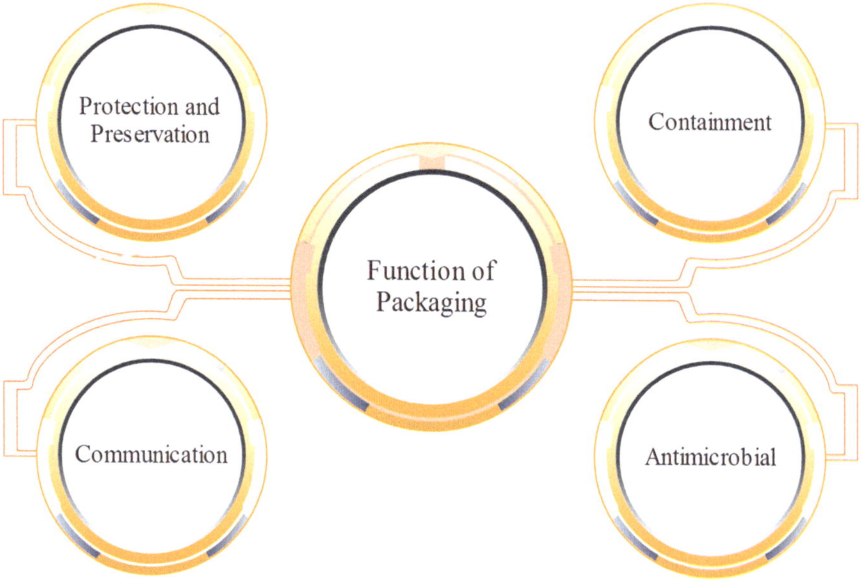

Fig. 21.13 Function of packaging materials

21.7 Conclusions

The increasing demand over the safety issues and applications of eco-friendly, non-toxic and biodegradable materials in food industries enforced scientist and industrialist to think towards an alternative to conventional nondegradable plastic materials. Edible polymers in this situation found as best options due to its various remarkable properties associated with them. Edible polymers mainly consist of polysaccharide, proteins and lipids, each of them having its on specific properties. This chapter highlights the importance of edible polymers based on hydrogels, synthesis, properties and applications with particularly emphasizing on its application in food industries. Edible polymers exhibit many advantageous properties such as available in abundant, renewable, eco-friendly, nontoxic, biocompatible and the most important is biodegradable. In food industries these hydrogels can be used as either coating or films and for packaging and protection purposes of food materials. The properties of theses hydrogels can be further enhanced through composite formations either with inter polymer composite formation or by the addition of additives like natural extracts nanoparticles etc. It can, thus, be concluded that edible polymers based hydrogels could be a potential alternative to conventional plastic materials preferring its nontoxicity and biodegradable properties.

References

Abad LV, Relleve LS, Aranilla CT, Dela Rosa AM (2003) Properties of radiation synthesized PVP-kappa carrageenan hydrogel blends. Radiat Phys Chem 68:901–908

Abd El-Salam MH, El-Shibiny S (2015) Preparation and properties of milk proteins-based encapsulated probiotics: a review. Dairy Sci Technol 95:393–412. https://doi.org/10.1007/s13594-015-0223-8

Abdorreza MN, Cheng LH, Karim AA (2011) Effects of plasticizers on thermal properties and heat sealability of sago starch films. Food Hydrocoll 25:56–60. https://doi.org/10.1016/j.foodhyd.2010.05.005

Abreu AS, Oliveira M, de Sá A et al (2015) Antimicrobial nanostructured starch based films for packaging. Carbohydr Polym 129:127–134. https://doi.org/10.1016/j.carbpol.2015.04.021

Ackar D, Babic J, Jozinovic A et al (2015) Starch modification by organic acids and their derivatives: a review. Molecules 20:19554–19570. https://doi.org/10.3390/molecules201019554

Akelah A (2013) Polymers in food packaging and protection. In: Functionalized polymeric materials in agriculture and the food industry. Springer, Boston, MA, pp 293–347

Ali A, Ahmed S (2018) A review on chitosan and its nanocomposites in drug delivery. Int J Biol Macromol 109:273–286. https://doi.org/10.1016/j.ijbiomac.2017.12.078

Alonso-Mougán M, Meijide F, Jover A et al (2002) Rheological behaviour of an amide pectin. J Food Eng 55:123–129

Álvarez K, Famá L, Gutiérrez TJ (2017) Physicochemical, antimicrobial and mechanical properties of thermoplastic materials based on biopolymers with application in the food industry. In: Masuelli M, Renard D (eds) Advances in physicochemical properties of biopolymers: part 1. Bentham Science Publishers, pp 358–400. https://doi.org/10.2174/9781681084534117010015. ISBN: 978-1-68108-454-1. eISBN: 978-1-68108-453-4

Athawale VD, Lele V (1998) Graft copolymerization onto starch. II. Grafting of acrylic acid and preparation of it's hydrogels. Carbohydr Polym 35:21–27. https://doi.org/10.1016/S0144-8617(97)00138-0

Athawale VD, Lele V (2000) Thermal studies on granular maize starch and its graft copolymers with vinyl monomers. Starch Stärke 52:205–213. https://doi.org/10.1002/1521-379X(200007)52:6/7<205::AID-STAR205>3.0.CO;2-3

Athawale VD, Vidyagauri VL (1998) Graft copolymerization onto starch. 3: Grafting of acrylamide using ceric ion initiation and preparation of its hydrogels. Starch Stärke 50:426–431

Bader HG, Göritz D (1994a) Investigations on high amylose corn starch films. Part 1: wide angle X-ray scattering (WAXS). Starch Stärke 46:229–232. https://doi.org/10.1002/star.19940460606

Bader HG, Göritz D (1994b) Investigations on high amylose corn starch films. Part 3: stress strain behaviour. Starch Stärke 46:435–439. https://doi.org/10.1002/star.19940461106

Bader HG, Göritz D (1994c) Investigations on high amylose corn starch films. Part 2: water vapor sorption. Starch Stärke 46:249–252. https://doi.org/10.1002/star.19940460704

Balakrishnan B, Jayakrishnan A (2005) Self-cross-linking biopolymers as injectable in situ forming biodegradable scaffolds. Biomaterials 26:3941–3951. https://doi.org/10.1016/j.biomaterials.2004.10.005

Bayarri M, Oulahal N, Degraeve P, Gharsallaoui A (2014) Properties of lysozyme/low methoxyl (LM) pectin complexes for antimicrobial edible food packaging. J Food Eng 131:18–25. https://doi.org/10.1016/j.jfoodeng.2014.01.013

Bhat R, Abdullah N, Din RH, Tay G-S (2013) Producing novel sago starch based food packaging films by incorporating lignin isolated from oil palm black liquor waste. J Food Eng 119:707–713. https://doi.org/10.1016/j.jfoodeng.2013.06.043

Bierhalz ACK, da Silva MA, Kieckbusch TG (2012) Natamycin release from alginate/pectin films for food packaging applications. J Food Eng 110:18–25. https://doi.org/10.1016/j.jfoodeng.2011.12.016

Boonlertnirun S, Boonraung C, Suvanasara R (2017) Application of chitosan in rice production

Bordenave N, Grelier S, Pichavant F, Coma V (2007) Water and moisture susceptibility of chitosan and paper-based materials: structure–property relationships. J Agric Food Chem 55:9479–9488. https://doi.org/10.1021/jf070595i

Bourtoom T (2008) Edible films and coatings: characteristics and properties. Int Food Res J 15:237–248

Bozanic DK, Djokovic V, Dimitrijevic-Brankovic S et al (2011) Inhibition of microbial growth by silver-starch nanocomposite thin films. J Biomater Sci Polym Ed 22:2343–2355. https://doi.org/10.1163/092050610X539532

Bui VKH, Park D, Lee Y-C (2017) Chitosan combined with ZnO, TiO2 and Ag nanoparticles for antimicrobial wound healing applications: a mini review of the research trends. Polymers (Basel) 9:21

Butler MF, Ng Y-F, Pudney PDA (2003) Mechanism and kinetics of the crosslinking reaction between biopolymers containing primary amine groups and genipin. J Polym Sci Part A Polym Chem 41:3941–3953. https://doi.org/10.1002/pola.10960

Cagri A, Ustunol Z, Ryser ET (2001) Antimicrobial, mechanical, and moisture barrier properties of low pH whey protein-based edible films containing p-aminobenzoic or sorbic acids. J Food Sci 66:865–870. https://doi.org/10.1111/j.1365-2621.2001.tb15188.x

Chen J, Liu C, Chen Y et al (2008) Structural characterization and properties of starch/konjac glucomannan blend films. Carbohydr Polym 74:946–952. https://doi.org/10.1016/j.carbpol.2008.05.021

Chronakis IS, Doublier J-L, Piculell L (2000) Viscoelastic properties for kappa- and iota-carrageenan in aqueous NaI from the liquid-like to the solid-like behaviour. Int J Biol Macromol 28:1–14. https://doi.org/10.1016/S0141-8130(00)00141-0

Chu H, Wei H, Zhu J, Hu S (2011) Preparation of starch esters with crosslinking structures derived from dianhydride. Front Chem Sci Eng 5:51–54. https://doi.org/10.1007/s11705-010-0534-5

Chung Y-L, Ansari S, Estevez L et al (2010) Preparation and properties of biodegradable starch–clay nanocomposites. Carbohydr Polym 79:391–396

Dash M, Chiellini F, Ottenbrite RM, Chiellini E (2011) Chitosan—a versatile semi-synthetic polymer in biomedical applications. Prog Polym Sci 36:981–1014

De Carvalho RA, Grosso CRF (2004) Characterization of gelatin based films modified with transglutaminase, glyoxal and formaldehyde. Food Hydrocoll 18:717–726

de Moura MR, Mattoso LHC, Zucolotto V (2012) Development of cellulose-based bactericidal nanocomposites containing silver nanoparticles and their use as active food packaging. J Food Eng 109:520–524. https://doi.org/10.1016/j.jfoodeng.2011.10.030

Demitri C, Sole RD, Scalera F et al (2008) Novel superabsorbent cellulose-based hydrogels crosslinked with citric acid. J Appl Polym Sci 110:2453–2460. https://doi.org/10.1002/app.28660

Demitri C, De Benedictis VM, Madaghiele M et al (2016) Nanostructured active chitosan-based films for food packaging applications: effect of graphene stacks on mechanical properties. Measurement 90:418–423. https://doi.org/10.1016/j.measurement.2016.05.012

Diab T, Biliaderis CG, Gerasopoulos D, Sfakiotakis E (2001) Physicochemical properties and application of pullulan edible films and coatings in fruit preservation. J Sci Food Agric 81:988–1000. https://doi.org/10.1002/jsfa.883

Dimida S, Demitri C, De Benedictis VM et al (2015) Genipin-cross-linked chitosan-based hydrogels: reaction kinetics and structure-related characteristics. J Appl Polym Sci 132:2804–2814

Djagny KB, Wang Z, Xu S (2001) Gelatin: a valuable protein for food and pharmaceutical industries: review. Crit Rev Food Sci Nutr 41:481–492. https://doi.org/10.1080/20014091091904

Dodane V, Vilivalam VD (1998) Pharmaceutical applications of chitosan. Pharm Sci Technol Today 1:246–253

Domenek S, Feuilloley P, Gratraud J et al (2004) Biodegradability of wheat gluten based bioplastics. Chemosphere 54:551–559. https://doi.org/10.1016/S0045-6535(03)00760-4

Dragan ES, Apopei DF (2011) Synthesis and swelling behavior of pH-sensitive semi-interpenetrating polymer network composite hydrogels based on native and modified potatoes starch as potential sorbent for cationic dyes. Chem Eng J 178:252–263

Eldin MSM, El-Sherif HM, Soliman EA et al (2011) Polyacrylamide-grafted carboxymethyl cellulose: smart pH-sensitive hydrogel for protein concentration. J Appl Polym Sci 122:469–479. https://doi.org/10.1002/app.33283

Ellis RP, Cochrane MP, Dale MFB et al (1998) Starch production and industrial use. J Sci Food Agric 77:289–311. https://doi.org/10.1002/(SICI)1097-0010(199807)77:3<289::AID-JSFA38>3.0.CO;2-D

El-Mohdy HLA (2014) Radiation initiated synthesis of 2-acrylamidoglycolic acid grafted carboxymethyl cellulose as pH-sensitive hydrogel. Polym Eng Sci 54:2753–2761. https://doi.org/10.1002/pen.23831

Entcheva E, Bien H, Yin L et al (2004) Functional cardiac cell constructs on cellulose-based scaffolding. Biomaterials 25:5753–5762

Erdohan ZÖ, Turhan KN (2005) Barrier and mechanical properties of methylcellulose–whey protein films. Packag Technol Sci 18:295–302. https://doi.org/10.1002/pts.700

Espitia PJP, Du W-X, de Jesús Avena-Bustillos R et al (2014) Edible films from pectin: physical-mechanical and antimicrobial properties—a review. Food Hydrocoll 35:287–296. https://doi.org/10.1016/j.foodhyd.2013.06.005

Fama L, Rojas AM, Goyanes S, Gerschenson L (2005) Mechanical properties of tapioca-starch edible films containing sorbates. LWT Food Sci Technol 38:631–639

Farhan A, Hani NM (2017) Characterization of edible packaging films based on semi-refined kappa-carrageenan plasticized with glycerol and sorbitol. Food Hydrocoll 64:48–58. https://doi.org/10.1016/j.foodhyd.2016.10.034

Farris S, Song J, Huang Q (2009) Alternative reaction mechanism for the cross-linking of gelatin with glutaraldehyde. J Agric Food Chem 58:998–1003

Farris S, Schaich KM, Liu L et al (2011) Gelatin–pectin composite films from polyion-complex hydrogels. Food Hydrocoll 25:61–70. https://doi.org/10.1016/j.foodhyd.2010.05.006

Farris S, Introzzi L, Fuentes-Alventosa JM et al (2012) Self-assembled pullulan–silica oxygen barrier hybrid coatings for food packaging applications. J Agric Food Chem 60:782–790

Fiamingo A, Montembault A, Boitard S-E et al (2016) Chitosan hydrogels for the regeneration of infarcted myocardium: preparation, physicochemical characterization, and biological evaluation. Biomacromolecules 17:1662–1672. https://doi.org/10.1021/acs.biomac.6b00075

Ficko-Blean E, Préchoux A, Thomas F et al (2017) Carrageenan catabolism is encoded by a complex regulon in marine heterotrophic bacteria. Nat Commun 8:1685

García MA, Martino MN, Zaritzky NE (2000) Microstructural Characterization of Plasticized Starch-Based Films. Starch Stärke 52:118–124. https://doi.org/10.1002/1521-379X(200006)52:4<118::AID-STAR118>3.0.CO;2-0

García-González CA, Alnaief M, Smirnova I (2011) Polysaccharide-based aerogels—promising biodegradable carriers for drug delivery systems. Carbohydr Polym 86:1425–1438. https://doi.org/10.1016/j.carbpol.2011.06.066

Gennadios A, Brandenburg AH, Weller CL, Testin RF (1993) Effect of pH on properties of wheat gluten and soy protein isolate films. J Agric Food Chem 41:1835–1839

Ghanbarzadeh B, Oromiehi AR (2008) Biodegradable biocomposite films based on whey protein and zein: barrier, mechanical properties and AFM analysis. Int J Biol Macromol 43:209–215. https://doi.org/10.1016/j.ijbiomac.2008.05.006

Ghanbarzadeh B, Almasi H, Entezami AA (2011) Improving the barrier and mechanical properties of corn starch-based edible films: effect of citric acid and carboxymethyl cellulose. Ind Crop Prod 33:229–235. https://doi.org/10.1016/j.indcrop.2010.10.016

Glibowski P, Mleko S, Wesolowska-Trojanowska M (2006) Gelation of single heated vs. double heated whey protein isolate. Int Dairy J 16:1113–1118. https://doi.org/10.1016/j.idairyj.2005.10.024

Gómez-Guillén MC, Giménez B, López-Caballero ME, Montero MP (2011) Functional and bioactive properties of collagen and gelatin from alternative sources: a review. Food Hydrocoll 25:1813–1827. https://doi.org/10.1016/j.foodhyd.2011.02.007

Goudarzi V, Shahabi-Ghahfarrokhi I, Babaei-Ghazvini A (2017) Preparation of ecofriendly UV-protective food packaging material by starch/TiO2 bio-nanocomposite: characterization. Int J Biol Macromol 95:306–313. https://doi.org/10.1016/j.ijbiomac.2016.11.065

Gregorova A, Saha N, Kitano T, Saha P (2015) Hydrothermal effect and mechanical stress properties of carboxymethylcellulose based hydrogel food packaging. Carbohydr Polym 117:559–568. https://doi.org/10.1016/j.carbpol.2014.10.009

Guilbert S, Gontard N, Cuq B (1995) Technology and applications of edible protective films. Packag Technol Sci 8:339–346. https://doi.org/10.1002/pts.2770080607

Gunasekaran S, Xiao L, Ould Eleya MM (2006) Whey protein concentrate hydrogels as bioactive carriers. J Appl Polym Sci 99:2470–2476

Gupta B, Tummalapalli M, Deopura BL, Alam MS (2014) Preparation and characterization of in-situ crosslinked pectin–gelatin hydrogels. Carbohydr Polym 106:312–318. https://doi.org/10.1016/j.carbpol.2014.02.019

Gutiérrez TJ (2017) Chitosan applications for the food industry. In: Ahmed S, Ikram S (eds) Chitosan: derivatives, composites and applications. Wiley-Scrivener Publisher, pp 183–232. https://doi.org/10.1002/9781119364849.ch8. ISBN: 978-1-119-36350-7

Gutiérrez TJ (2018a) Characterization and *in vitro* digestibility of non-conventional starches from guinea arrowroot and La Armuña lentils as potential food sources for special diet regimens. Starch Stärke 70(1–2). https://doi.org/10.1002/star.201700124

Gutiérrez TJ (2018b) Active and intelligent films made from starchy sources/blackberry pulp. J Polym Environ 15:445–448. https://doi.org/10.1007/s10924-017-1134-y

Gutiérrez TJ, Alvarez VA (2017a) Properties of native and oxidized corn starch/polystyrene blends under conditions of reactive extrusion using zinc octanoate as a catalyst. React Funct Polym 112:33–44. https://doi.org/10.1016/j.reactfunctpolym.2017.01.002

Gutiérrez TJ, Alvarez VA (2017b) Eco-friendly films prepared from plantain flour/PCL blends under reactive extrusion conditions using zirconium octanoate as a catalyst. Carbohydr Polym 178:260–269. https://doi.org/10.1016/j.carbpol.2017.09.026

Gutiérrez TJ, Alvarez VA (2017c) Data on physicochemical properties of active films derived from plantain flour/PCL blends developed under reactive extrusion conditions. Data Brief 15:445–448. https://doi.org/10.1016/j.dib.2017.09.071

Gutiérrez TJ, Alvarez VA (2017d) Cellulosic materials as natural fillers in starch-containing matrix-based films: a review. Polym Bull 74(6):2401–2430. https://doi.org/10.1007/s00289-016-1814-0

Gutiérrez TJ, Morales NJ, Tapia MS et al (2015a) Corn starch 80: 20 "waxy": regular,"native" and phosphated, as bio-matrixes for edible films. Procedia Mater Sci 8:304–310. https://doi.org/10.1016/j.mspro.2015.04.077

Gutiérrez TJ, Morales NJ, Pérez E, Tapia MS, Famá L (2015b) Physico-chemical study of edible films based on native and phosphating cush-cush yam and cassava starches. Food Packaging Shelf Life 3:1–8. https://doi.org/10.1016/j.fpsl.2014.09.002

Gutiérrez TJ, Tapia MS, Pérez E, Famá L (2015c) Edible films based on native and phosphated 80:20 waxy:normal corn starch. Starch Stärke 67(1–2):90–97. https://doi.org/10.1002/star.201400164

Gutiérrez TJ, Tapia MS, Pérez E, Famá L (2015d) Structural and mechanical properties of native and modified cush-cush yam and cassava starch edible films. Food Hydrocoll 45:211–217. https://doi.org/10.1016/j.foodhyd.2014.11.017

Gutiérrez TJ, González Seligra P, Medina Jaramillo C, Famá L, Goyanes S (2017) Effect of filler properties on the antioxidant response of thermoplastic starch composites. In: Thakur VK, Thakur MK, Kessler MR (eds) Handbook of composites from renewable materials. Wiley-Scrivener Publisher, pp 337–370. https://doi.org/10.1002/9781119441632.ch14. ISBN: 978-1-119-22362-7

Hambleton A, Fabra M-J, Debeaufort F et al (2009) Interface and aroma barrier properties of iota-carrageenan emulsion–based films used for encapsulation of active food compounds. J Food Eng 93:80–88. https://doi.org/10.1016/j.jfoodeng.2009.01.001

Ham-Pichavant F, Sèbe G, Pardon P, Coma V (2005) Fat resistance properties of chitosan-based paper packaging for food applications. Carbohydr Polym 61:259–265. https://doi.org/10.1016/j.carbpol.2005.01.020

Heidebach T, Först P, Kulozik U (2009a) Microencapsulation of probiotic cells by means of rennet-gelation of milk proteins. Food Hydrocoll 23:1670–1677. https://doi.org/10.1016/j.foodhyd.2009.01.006

Heidebach T, Först P, Kulozik U (2009b) Transglutaminase-induced caseinate gelation for the microencapsulation of probiotic cells. Int Dairy J 19:77–84. https://doi.org/10.1016/j.idairyj.2008.08.003

Hermansson A-M (1989) Rheological and microstructural evidence for transient states during gelation of kappa-carrageenan in the presence of potassium. Carbohydr Polym 10:163–181

Hermansson A-M, Eriksson E, Jordansson E (1991) Effects of potassium, sodium and calcium on the microstructure and rheological behaviour of kappa-carrageenan gels. Carbohydr Polym 16:297–320

Hong J, Zeng X-A, Brennan CS et al (2016) Recent advances in techniques for starch esters and the applications: a review. Foods 5:50. https://doi.org/10.3390/foods5030050

Hopkins EJ, Chang C, Lam RSH, Nickerson MT (2015) Effects of flaxseed oil concentration on the performance of a soy protein isolate-based emulsion-type film. Food Res Int 67:418–425. https://doi.org/10.1016/j.foodres.2014.11.040

Hosseini MH, Razavi SH, Mousavi MA (2009) Antimicrobial, physical and mechanical properties of chitosan-based films incorporated with thyme, clove and cinnamon essential oils. J Food Process Preserv 33:727–743. https://doi.org/10.1111/j.1745-4549.2008.00307.x

Hurtado-Lopez P, Murdan S (2005) Formulation and characterisation of zein microspheres as delivery vehicles. J Drug Deliv Sci Technol 15:267–272

Introzzi L, Blomfeldt TOJ, Trabattoni S et al (2012) Ultrasound-assisted pullulan/montmorillonite bionanocomposite coating with high oxygen barrier properties. Langmuir 28:11206–11214

Islam S, Bhuiyan MAR, Islam MN (2017) Chitin and chitosan: structure, properties and applications in biomedical engineering. J Polym Environ 25:854–866. https://doi.org/10.1007/s10924-016-0865-5

Ismail H, Irani M, Ahmad Z (2013) Starch-based hydrogels: present status and applications. Int J Polym Mater Polym Biomater 62:411–420. https://doi.org/10.1080/00914037.2012.719141

Jensen A, Lim L-T, Barbut S, Marcone M (2015) Development and characterization of soy protein films incorporated with cellulose fibers using a hot surface casting technique. LWT Food Sci Technol 60:162–170. https://doi.org/10.1016/j.lwt.2014.09.027

Jin J, Song M, Hourston DJ (2004) Novel chitosan-based films cross-linked by genipin with improved physical properties. Biomacromolecules 5:162–168. https://doi.org/10.1021/bm034286m

John MJ, Thomas S (2008) Biofibres and biocomposites. Carbohydr Polym 71:343–364

Kadam DM, Thunga M, Wang S et al (2013) Preparation and characterization of whey protein isolate films reinforced with porous silica coated titania nanoparticles. J Food Eng 117:133–140. https://doi.org/10.1016/j.jfoodeng.2013.01.046

Kamel S, Ali N, Jahangir K et al (2008) Pharmaceutical significance of cellulose: a review. Express Polym Lett 2:758–778

Karim AA, Bhat R (2008) Gelatin alternatives for the food industry: recent developments, challenges and prospects. Trends Food Sci Technol 19:644–656. https://doi.org/10.1016/j.tifs.2008.08.001

Kavoosi G, Rahmatollahi A, Mohammad Mahdi Dadfar S, Mohammadi Purfard A (2014) Effects of essential oil on the water binding capacity, physico-mechanical properties, antioxidant and antibacterial activity of gelatin films. LWT Food Sci Technol 57:556–561. https://doi.org/10.1016/j.lwt.2014.02.008

Kayserilioğlu BŞ, Bakir U, Yilmaz L, Akkaş N (2003) Use of xylan, an agricultural by-product, in wheat gluten based biodegradable films: mechanical, solubility and water vapor transfer rate properties. Bioresour Technol 87:239–246. https://doi.org/10.1016/S0960-8524(02)00258-4

Khalil HPSA, Saurabh CK, Tye YY et al (2017) Seaweed based sustainable films and composites for food and pharmaceutical applications: a review. Renew Sust Energ Rev 77:353–362

Kim S-J, Ustunol Z (2001) Thermal properties, heat sealability and seal attributes of whey protein isolate/lipid emulsion edible films. J Food Sci 66:985–990. https://doi.org/10.1111/j.1365-2621.2001.tb08223.x

Klein MP, Hackenhaar CR, Lorenzoni ASG et al (2016) Chitosan crosslinked with genipin as support matrix for application in food process: support characterization and β-d-galactosidase immobilization. Carbohydr Polym 137:184–190. https://doi.org/10.1016/j.carbpol.2015.10.069

Kobašlija M, McQuade DT (2006) Removable colored coatings based on calcium alginate hydrogels. Biomacromolecules 7:2357–2361

Kong HJ, Alsberg E, Kaigler D et al (2004) Controlling degradation of hydrogels via the size of crosslinked junctions. Adv Mater 16:1917–1921. https://doi.org/10.1002/adma.200400014

Kontogiorgos V (2011) Microstructure of hydrated gluten network. Food Res Int 44:2582–2586. https://doi.org/10.1016/j.foodres.2011.06.021

Krochta JM (2002) Proteins as raw materials for films and coatings: definitions, current status, and opportunities. In: Protein-based films and coatings. CRC Press, Boca Raton, FL, pp 1–41

Kulkarni CV, Moinuddin Z, Patil-Sen Y et al (2015) Lipid-hydrogel films for sustained drug release. Int J Pharm 479:416–421. https://doi.org/10.1016/j.ijpharm.2015.01.013

Kumar R, Choudhary V, Mishra S et al (2002) Adhesives and plastics based on soy protein products. Ind Crop Prod 16:155–172. https://doi.org/10.1016/S0926-6690(02)00007-9

Kumar P, Sandeep KP, Alavi S et al (2010) Preparation and characterization of bio-nanocomposite films based on soy protein isolate and montmorillonite using melt extrusion. J Food Eng 100:480–489. https://doi.org/10.1016/j.jfoodeng.2010.04.035

Kuorwel KK, Cran MJ, Sonneveld K et al (2013) Migration of antimicrobial agents from starch-based films into a food simulant. LWT Food Sci Technol 50:432–438. https://doi.org/10.1016/j.lwt.2012.08.023

Lacroix M, Le TC, Ouattara B et al (2002) Use of γ-irradiation to produce films from whey, casein and soya proteins: structure and functionals characteristics. Radiat Phys Chem 63:827–832. https://doi.org/10.1016/S0969-806X(01)00574-6

Lagaron JM, Fernandez-Saiz P, Ocio MJ (2007) Using ATR-FTIR spectroscopy to design active antimicrobial food packaging structures based on high molecular weight chitosan polysaccharide. J Agric Food Chem 55:2554–2562. https://doi.org/10.1021/jf063110j

Lagrain B, Goderis B, Brijs K, Delcour JA (2010) Molecular basis of processing wheat gluten toward biobased materials. Biomacromolecules 11:533–541

Lakemond CMM, de Jongh HHJ, Paques M et al (2003) Gelation of soy glycinin; influence of pH and ionic strength on network structure in relation to protein conformation. Food Hydrocoll 17:365–377. https://doi.org/10.1016/S0268-005X(02)00100-5

Landers R, Hübner U, Schmelzeisen R, Mülhaupt R (2002) Rapid prototyping of scaffolds derived from thermoreversible hydrogels and tailored for applications in tissue engineering. Biomaterials 23:4437–4447

Langendorff V, Cuvelier G, Michon C et al (2000) Effects of carrageenan type on the behaviour of carrageenan/milk mixtures. Food Hydrocoll 14:273–280

Langmaier F, Mokrejs P, Kolomaznik K, Mladek M (2008) Biodegradable packing materials from hydrolysates of collagen waste proteins. Waste Manag 28:549–556. https://doi.org/10.1016/j.wasman.2007.02.003

Lawton JW (2002) Zein: a history of processing and use. Cereal Chem 79:1–18

Li B, Kennedy JF, Jiang QG, Xie BJ (2006) Quick dissolvable, edible and heatsealable blend films based on konjac glucomannan—gelatin. Food Res Int 39:544–549. https://doi.org/10.1016/j.foodres.2005.10.015

Li L, Fang Y, Vreeker R et al (2007) Reexamining the egg-box model in calcium-alginate gels with X-ray diffraction. Biomacromolecules 8:464–468. https://doi.org/10.1021/bm060550a

Li J-H, Miao J, Wu J-L et al (2014) Preparation and characterization of active gelatin-based films incorporated with natural antioxidants. Food Hydrocoll 37:166–173. https://doi.org/10.1016/j.foodhyd.2013.10.015

Liu LS, Kost J, Yan F, Spiro RC (2012) Hydrogels from biopolymer hybrid for biomedical, food, and functional food applications. Polymers (Basel) 4:997–1011

Liu J, Su D, Yao J et al (2017) Soy protein-based polyethylenimine hydrogel and its high selectivity for copper ion removal in wastewater treatment. J Mater Chem A 5:4163–4171. https://doi.org/10.1039/C6TA10814H

LiuLiu LC-K, Fishman ML, Hicks KB (2007) Composite films from pectin and fish skin gelatin or soybean flour protein. J Agric Food Chem 55:2349–2355. https://doi.org/10.1021/jf062612u

Livney YD (2010) Milk proteins as vehicles for bioactives. Curr Opin Colloid Interface Sci 15:73–83. https://doi.org/10.1016/j.cocis.2009.11.002

López de Dicastillo C, Bustos F, Guarda A, Galotto MJ (2016) Cross-linked methyl cellulose films with murta fruit extract for antioxidant and antimicrobial active food packaging. Food Hydrocoll 60:335–344. https://doi.org/10.1016/j.foodhyd.2016.03.020

Lourdin D, Della VG, Colonna P (1995) Influence of amylose content on starch films and foams. Carbohydr Polym 27:261–270. https://doi.org/10.1016/0144-8617(95)00071-2

Lu DR, Xiao CM, Xu SJ (2009) Starch-based completely biodegradable polymer materials. Express Polym Lett 3:366–375

Luo Y, Wang Q (2014) Zein-based micro-and nano-particles for drug and nutrient delivery: a review. J Appl Polym Sci 131:40696

Mahalik NP, Nambiar AN (2010) Trends in food packaging and manufacturing systems and technology. Trends Food Sci Technol 21:117–128. https://doi.org/10.1016/j.tifs.2009.12.006

Majdzadeh-Ardakani K, Navarchian AH, Sadeghi F (2010) Optimization of mechanical properties of thermoplastic starch/clay nanocomposites. Carbohydr Polym 79:547–554. https://doi.org/10.1016/j.carbpol.2009.09.001

Majee SB, Avlani D, Biswas GR (2017) Pharmacological, pharmaceutical, cosmetic and diagnostic applications of sulfated polysaccharides from marine algae and bacteria. African J Pharm Pharmacol 11:68–77

Maltais A, Remondetto GE, Gonzalez R, Subirade M (2005) Formation of soy protein isolate cold-set gels: protein and salt effects. J Food Sci 70:C67–C73. https://doi.org/10.1111/j.1365-2621.2005.tb09023.x

Mangione MR, Giacomazza D, Bulone D et al (2005) K(+) and Na(+) effects on the gelation properties of kappa-Carrageenan. Biophys Chem 113:129–135. https://doi.org/10.1016/j.bpc.2004.08.005

Mao JS, Yin YJ, De Yao K (2003) The properties of chitosan–gelatin membranes and scaffolds modified with hyaluronic acid by different methods. Biomaterials 24:1621–1629

Märtson M, Viljanto J, Hurme T et al (1999) Is cellulose sponge degradable or stable as implantation material? An in vivo subcutaneous study in the rat. Biomaterials 20:1989–1995. https://doi.org/10.1016/S0142-9612(99)00094-0

Matignon A, Tecante A (2017) Starch retrogradation: from starch components to cereal products. Food Hydrocoll 68:43–52. https://doi.org/10.1016/j.foodhyd.2016.10.032

Mi F, Sung H, Shyu S (2000) Synthesis and characterization of a novel chitosan-based network prepared using naturally occurring crosslinker. J Polym Sci Part A Polym Chem 38:2804–2814

Mi F-L, Tan Y-C, Liang H-C et al (2001) In vitro evaluation of a chitosan membrane cross-linked with genipin. J Biomater Sci Polym Ed 12:835–850. https://doi.org/10.1163/156856201753113051

Mi F-L, Shyu S-S, Peng C-K (2005) Characterization of ring-opening polymerization of genipin and pH-dependent cross-linking reactions between chitosan and genipin. J Polym Sci Part A Polym Chem 43:1985–2000. https://doi.org/10.1002/pola.20669

Miles MJ, Morris VJ, Orford PD, Ring SG (1985) The roles of amylose and amylopectin in the gelation and retrogradation of starch. Carbohydr Res 135:271–281. https://doi.org/10.1016/S0008-6215(00)90778-X

Miller KS, Krochta JM (1997) Oxygen and aroma barrier properties of edible films: a review. Trends Food Sci Technol 8:228–237. https://doi.org/10.1016/S0924-2244(97)01051-0

Mishra RK, Banthia AK, Majeed ABA (2012) Pectin based formulations for biomedical applications: a review. Asian J Pharm Clin Res 5:1–7

Mohammadian M, Madadlou A (2016) Cold-set hydrogels made of whey protein nanofibrils with different divalent cations. Int J Biol Macromol 89:499–506. https://doi.org/10.1016/j.ijbiomac.2016.05.009

Mørch ÝA, Donati I, Strand BL, Skjåk-Braek G (2006) Effect of Ca2+, Ba2+, and Sr2+ on alginate microbeads. Biomacromolecules 7:1471–1480. https://doi.org/10.1021/bm060010d

Mørch YA, Qi M, Gundersen POM et al (2012) Binding and leakage of barium in alginate microbeads. J Biomed Mater Res A 100:2939–2947. https://doi.org/10.1002/jbm.a.34237

Morris VJ (1990) Starch gelation and retrogradation. Trends Food Sci Technol 1:2–6. https://doi.org/10.1016/0924-2244(90)90002-G

Mu C, Guo J, Li X et al (2012) Preparation and properties of dialdehyde carboxymethyl cellulose crosslinked gelatin edible films. Food Hydrocoll 27:22–29. https://doi.org/10.1016/j.foodhyd.2011.09.005

Muppalla SR, Kanatt SR, Chawla SP, Sharma A (2014) Carboxymethyl cellulose–polyvinyl alcohol films with clove oil for active packaging of ground chicken meat. Food Packaging Shelf Life 2:51–58. https://doi.org/10.1016/j.fpsl.2014.07.002

Muyonga JH, Cole CGB, Duodu KG (2004) Extraction and physico-chemical characterisation of Nile perch (Lates niloticus) skin and bone gelatin. Food Hydrocoll 18:581–592

Muzzarelli RAA (2009) Genipin-crosslinked chitosan hydrogels as biomedical and pharmaceutical aids. Carbohydr Polym 77:1–9. https://doi.org/10.1016/j.carbpol.2009.01.016

Narayanan RP, Melman G, Letourneau NJ et al (2012) Photodegradable Iron(III) cross-linked alginate gels. Biomacromolecules 13:2465–2471. https://doi.org/10.1021/bm300707a

Nguyen KT, West JL (2002) Photopolymerizable hydrogels for tissue engineering applications. Biomaterials 23:4307–4314

Ni N, Dumont M-J (2017) Protein-based hydrogels derived from industrial byproducts containing collagen, keratin, zein and soy. Waste Biomass Valori 8:285–300. https://doi.org/10.1007/s12649-016-9684-0

Ni N, Duquette D, Dumont M-J (2017) Synthesis and characterization of zein-based cryogels and their potential as diesel fuel absorbent. Eur Polym J 91:420–428. https://doi.org/10.1016/j.eurpolymj.2017.04.031

Ni N, Zhang D, Dumont M-J (2018) Synthesis and characterization of zein-based superabsorbent hydrogels and their potential as heavy metal ion chelators. Polym Bull 75:31–45. https://doi.org/10.1007/s00289-017-2017-z

Nur Hanani ZA, Roos YH, Kerry JP (2014) Use and application of gelatin as potential biodegradable packaging materials for food products. Int J Biol Macromol 71:94–102. https://doi.org/10.1016/j.ijbiomac.2014.04.027

Oladoja NA, Unuabonah EI, Amuda OS, Kolawole OM (2017) Progress and prospects of polysaccharide composites as adsorbents for water and wastewater treatment. In: Polysaccharides as a green and sustainable resources for water and wastewater treatment. Springer, pp 65–90

Ou S, Wang Y, Tang S et al (2005) Role of ferulic acid in preparing edible films from soy protein isolate. J Food Eng 70:205–210. https://doi.org/10.1016/j.jfoodeng.2004.09.025

Oun AA, Rhim J-W (2017) Carrageenan-based hydrogels and films: effect of ZnO and CuO nanoparticles on the physical, mechanical, and antimicrobial properties. Food Hydrocoll 67:45–53. https://doi.org/10.1016/j.foodhyd.2016.12.040

Padhi JR, Nayak D, Nanda A et al (2016) Development of highly biocompatible Gelatin & i-Carrageenan based composite hydrogels: in depth physiochemical analysis for biomedical applications. Carbohydr Polym 153:292–301. https://doi.org/10.1016/j.carbpol.2016.07.098

Pal K, Banthia AK, Majumdar DK (2006) Development of carboxymethyl cellulose acrylate for various biomedical applications. Biomed Mater 1:85

Pandey M, Mohamad N, Low W-L et al (2017) Microwaved bacterial cellulose-based hydrogel microparticles for the healing of partial thickness burn wounds. Drug Deliv Transl Res 7:89–99

Pang J, Liu X, Yang J et al (2016) Synthesis of highly polymerized water-soluble cellulose acetate by the side reaction in carboxylate ionic liquid 1-ethyl-3-methylimidazolium acetate. Sci Rep 6:33725

Papageorgiou SK, Kouvelos EP, Favvas EP et al (2010) Metal–carboxylate interactions in metal–alginate complexes studied with FTIR spectroscopy. Carbohydr Res 345:469–473

Pardeike J, Hommoss A, Müller RH (2009) Lipid nanoparticles (SLN, NLC) in cosmetic and pharmaceutical dermal products. Int J Pharm 366:170–184. https://doi.org/10.1016/j.ijpharm.2008.10.003

Park SY, Marsh KS, Rhim JW (2002) Characteristics of different molecular weight chitosan films affected by the type of organic solvents. J Food Sci 67:194–197

Pasqui D, De Cagna M, Barbucci R (2012) Polysaccharide-based hydrogels: the key role of water in affecting mechanical properties. Polymers (Basel) 4:1517–1534

Peh K, Khan T, Ch'ng H (2000) Mechanical, bioadhesive strength and biological evaluations of chitosan films for wound dressing. J Pharm Pharm Sci 3:303–311

Pereira VA, de Arruda INQ, Stefani R (2015) Active chitosan/PVA films with anthocyanins from Brassica oleraceae (Red Cabbage) as time–temperature indicators for application in intelligent food packaging. Food Hydrocoll 43:180–188. https://doi.org/10.1016/j.foodhyd.2014.05.014

Pourjavadi A, Ghasemzadeh H, Mojahedi F (2009) Swelling properties of CMC-g-poly (AAm-co-AMPS) superabsorbent hydrogel. J Appl Polym Sci 113:3442–3449. https://doi.org/10.1002/app.30094

Prabaharan M, Mano JF (2006) Stimuli-responsive hydrogels based on polysaccharides incorporated with thermo-responsive polymers as novel biomaterials. Macromol Biosci 6:991–1008. https://doi.org/10.1002/mabi.200600164

Prajapati VD, Maheriya PM, Jani GK, Solanki HK (2014) Carrageenan: a natural seaweed polysaccharide and its applications. Carbohydr Polym 105:97–112. https://doi.org/10.1016/j.carbpol.2014.01.067

Pranoto Y, Rakshit SK, Salokhe VM (2005a) Enhancing antimicrobial activity of chitosan films by incorporating garlic oil, potassium sorbate and nisin. LWT Food Sci Technol 38:859–865. https://doi.org/10.1016/j.lwt.2004.09.014

Pranoto Y, Salokhe VM, Rakshit SK (2005b) Physical and antibacte rial properties of alginate-based edible film incorporated with garlic oil. Food Res Int 38:267–272. https://doi.org/10.1016/j.foodres.2004.04.009

Qiu X, Hu S (2013) "Smart" materials based on cellulose: a review of the preparations, properties, and applications. Materials (Basel) 6:738–781. https://doi.org/10.3390/ma6030738

Racine L, Texier I, Auzély-Velty R (2017) Chitosan-based hydrogels: recent design concepts to tailor properties and functions. Polym Int 66:981–998. https://doi.org/10.1002/pi.5331

Ramos ÓL, Fernandes JC, Silva SI et al (2012) Edible films and coatings from whey proteins: a review on formulation, and on mechanical and bioactive properties. Crit Rev Food Sci Nutr 52:533–552. https://doi.org/10.1080/10408398.2010.500528

Remondetto GE, Subirade M (2003) Molecular mechanisms of Fe2+–induced beta-lactoglobulin cold gelation. Biopolymers 69:461–469. https://doi.org/10.1002/bip.10423

Remya P, Nicolas J, Nicole L (2012) Photodegradable iron (III) cross-linked alginate gels. Biomacromolecules 13:2465–2471

Rhein-Knudsen N, Ale MT, Meyer AS (2015) Seaweed hydrocolloid production: an update on enzyme assisted extraction and modification technologies. Mar Drugs 13:3340–3359. https://doi.org/10.3390/md13063340

Rhim J-W, Gennadios A, Weller CL et al (1998) Soy protein isolate–dialdehyde starch films1Journal Series No. 12010, Agricultural Research Division, Institute of Agriculture and Natural Resources, University of Nebraska-Lincoln. This study was conducted at the Industrial Agricultural Products Center. Ind Crop Prod 8:195–203. https://doi.org/10.1016/S0926-6690(98)00003-X

Ribeiro C, Vicente AA, Teixeira JA, Miranda C (2007) Optimization of edible coating composition to retard strawberry fruit senescence. Postharvest Biol Technol 44:63–70. https://doi.org/10.1016/j.postharvbio.2006.11.015

Rinaudo M (2008) Main properties and current applications of some polysaccharides as biomaterials. Polym Int 57:397–430. https://doi.org/10.1002/pi.2378

Ring SG (1985) Some studies on starch gelation. Starch Stärke 37:80–83. https://doi.org/10.1002/star.19850370303

Romero-Bastida CA, Bello-Pérez LA, García MA et al (2005) Physicochemical and microstructural characterization of films prepared by thermal and cold gelatinization from non-conventional sources of starches. Carbohydr Polym 60:235–244. https://doi.org/10.1016/j.carbpol.2005.01.004

Römling U (2002) Molecular biology of cellulose production in bacteria. Res Microbiol 153:205–212

Rong Huei C, Hwa H-D (1996) Effect of molecular weight of chitosan with the same degree of deacetylation on the thermal, mechanical, and permeability properties of the prepared membrane. Carbohydr Polym 29:353–358. https://doi.org/10.1016/S0144-8617(96)00007-0

Roy D, Semsarilar M, Guthrie JT, Perrier S (2009) Cellulose modification by polymer grafting: a review. Chem Soc Rev 38:2046–2064

Roy N, Saha N, Sáha P (2011) Biodegradable hydrogel film for food packaging. In: Proceedings of the 4th WSEAS international conference on engineering mechanics, structures, engineering geology (EMESEG'11), Corfu Islands, Greece. pp 329–334

Rudyardjo DI, Wijayanto S (2017) The synthesis and characterization of hydrogel chitosan-alginate with the addition of plasticizer lauric acid for wound dressing application. In: Journal of physics: conference series. IOP Publishing, p 12042

Sajilata MG, Singhal RS, Kulkarni PR (2006) Resistant starch—a review. Compr Rev Food Sci Food Saf 5:1–17. https://doi.org/10.1111/j.1541-4337.2006.tb00076.x

Sannino A, Esposito A, Nicolais L et al (2000) Cellulose-based hydrogels as body water retainers. J Mater Sci Mater Med 11:247–253

Sannino A, Mensitieri G, Nicolais L (2004) Water and synthetic urine sorption capacity of cellulose-based hydrogels under a compressive stress field. J Appl Polym Sci 91:3791–3796. https://doi.org/10.1002/app.13540

Sannino A, Demitri C, Madaghiele M (2009) Biodegradable cellulose-based hydrogels: design and applications. Materials (Basel) 2:353–373. https://doi.org/10.3390/ma2020353

Santo VE, Frias AM, Carida M et al (2009) Carrageenan-based hydrogels for the controlled delivery of PDGF-BB in bone tissue engineering applications. Biomacromolecules 10:1392–1401. https://doi.org/10.1021/bm8014973

Sayanjali S, Ghanbarzadeh B, Ghiassifar S (2011) Evaluation of antimicrobial and physical properties of edible film based on carboxymethyl cellulose containing potassium sorbate on some mycotoxigenic Aspergillus species in fresh pistachios. LWT Food Sci Technol 44:1133–1138. https://doi.org/10.1016/j.lwt.2010.12.017

Shahbazi M, Ahmadi SJ, Seif A, Rajabzadeh G (2016) Carboxymethyl cellulose film modification through surface photo-crosslinking and chemical crosslinking for food packaging applications. Food Hydrocoll 61:378–389. https://doi.org/10.1016/j.foodhyd.2016.04.021

Shewan HM, Stokes JR (2013) Review of techniques to manufacture micro-hydrogel particles for the food industry and their applications. J Food Eng 119:781–792. https://doi.org/10.1016/j.jfoodeng.2013.06.046

Shi K, Huang Y, Yu H et al (2011) Reducing the brittleness of zein films through chemical modification. J Agric Food Chem 59:56–61. https://doi.org/10.1021/jf103164r

Shit SC, Shah PM (2014) Edible polymers: challenges and opportunities. J Polym 2014:1–13

Shukla R, Cheryan M (2001) Zein: the industrial protein from corn. Ind Crop Prod 13:171–192

Sikorski P, Mo F, Skjak-Braek G, Stokke BT (2007) Evidence for egg-box-compatible interactions in calcium-alginate gels from fiber X-ray diffraction. Biomacromolecules 8:2098–2103. https://doi.org/10.1021/bm0701503

Siripatrawan U, Harte BR (2010) Physical properties and antioxidant activity of an active film from chitosan incorporated with green tea extract. Food Hydrocoll 24:770–775. https://doi.org/10.1016/j.foodhyd.2010.04.003

Siripatrawan U, Vitchayakitti W (2016) Improving functional properties of chitosan films as active food packaging by incorporating with propolis. Food Hydrocoll 61:695–702. https://doi.org/10.1016/j.foodhyd.2016.06.001

Smithers GW, John Ballard F, Copeland AD et al (1996) New opportunities from the isolation and utilization of whey proteins. J Dairy Sci 79:1454–1459. https://doi.org/10.3168/jds.S0022-0302(96)76504-9

Somerville C (2006) Cellulose synthesis in higher plants. Annu Rev Cell Dev Biol 22:53–78

Song F, Zhang L-M, Yang C, Yan L (2009) Genipin-crosslinked casein hydrogels for controlled drug delivery. Int J Pharm 373:41–47. https://doi.org/10.1016/j.ijpharm.2009.02.005

Sothornvit R, Krochta JM (2000a) Oxygen permeability and mechanical properties of films from hydrolyzed whey protein. J Agric Food Chem 48:3913–3916

Sothornvit R, Krochta JM (2000b) Water vapor permeability and solubility of films from hydrolyzed whey protein. J Food Sci 65:700–703. https://doi.org/10.1111/j.1365-2621.2000.tb16075.x

Sothornvit R, Rhim J-W, Hong S-I (2009) Effect of nano-clay type on the physical and antimicrobial properties of whey protein isolate/clay composite films. J Food Eng 91:468–473. https://doi.org/10.1016/j.jfoodeng.2008.09.026

Sreeram KJ, Yamini Shrivastava H, Nair BU (2004) Studies on the nature of interaction of iron(III) with alginates. Biochim Biophys Acta Gen Subj 1670:121–125. https://doi.org/10.1016/j.bbagen.2003.11.001

Su K, Wang C (2015) Recent advances in the use of gelatin in biomedical research. Biotechnol Lett 37:2139–2145. https://doi.org/10.1007/s10529-015-1907-0

Suárez G, Gutiérrez TJ (2017) Recent advances in the development of biodegadable films and foams from cassava starch. In: Klein C (ed) Handbook on cassava: production, potential uses and recent advances. Nova Science Publishers, Inc., New York, pp 297–312. ISBN: 978-1-53610-307-6

Sun J, Tan H (2013) Alginate-based biomaterials for regenerative medicine applications. Materials (Basel) 6:1285–1309. https://doi.org/10.3390/ma6041285

Sung HW, Huang RN, Huang LL, Tsai CC (1999) In vitro evaluation of cytotoxicity of a naturally occurring cross-linking reagent for biological tissue fixation. J Biomater Sci Polym Ed 10:63–78

Tansaz S, Durmann A, Detsch R, Boccaccini AR (2017) Hydrogel films and microcapsules based on soy protein isolate combined with alginate. J Appl Polym Sci 134:44358

Tavares GM, Croguennec T, Carvalho AF, Bouhallab S (2014) Milk proteins as encapsulation devices and delivery vehicles: applications and trends. Trends Food Sci Technol 37:5–20

Tavassoli-Kafrani E, Shekarchizadeh H, Masoudpour-Behabadi M (2016) Development of edible films and coatings from alginates and carrageenans. Carbohydr Polym 137:360–374. https://doi.org/10.1016/j.carbpol.2015.10.074

Tomihata K, Ikada Y (1997) In vitro and in vivo degradation of films of chitin and its deacetylated derivatives. Biomaterials 18:567–575. https://doi.org/10.1016/S0142-9612(96)00167-6

Trojani C, Weiss P, Michiels J-F et al (2005) Three-dimensional culture and differentiation of human osteogenic cells in an injectable hydroxypropylmethylcellulose hydrogel. Biomaterials 26:5509–5517

Varshney VK, Naithani S (2011) Chemical functionalization of cellulose derived from nonconventional sources. In: Cellulose fibers: bio-and nano-polymer composites. Springer, pp 43–60

Videcoq P, Garnier C, Robert P, Bonnin E (2011) Influence of calcium on pectin methylesterase behaviour in the presence of medium methylated pectins. Carbohydr Polym 86:1657–1664

Wang LZ, Liu L, Holmes J et al (2007) Assessment of film-forming potential and properties of protein and polysaccharide-based biopolymer films. Int J Food Sci Technol 42:1128–1138

Wang H, Qian J, Ding F (2017a) Emerging chitosan-based films for food packaging applications. J Agric Food Chem 66(2):395–413. https://doi.org/10.1021/acs.jafc.7b04528
Wang X, Yu T, Chen G et al (2017b) Preparation and characterization of a chitosan/gelatin/extracellular matrix scaffold and its application in tissue engineering. Tissue Eng Part C Methods 23:169–179. https://doi.org/10.1089/ten.TEC.2016.0511
Wieser H (2007) Chemistry of gluten proteins. Food Microbiol 24:115–119. https://doi.org/10.1016/j.fm.2006.07.004
Willats WGT, Knox JP, Mikkelsen JD (2006) Pectin: new insights into an old polymer are starting to gel. Trends Food Sci Technol 17:97–104. https://doi.org/10.1016/j.tifs.2005.10.008
Wu C, Peng S, Wen C et al (2012) Structural characterization and properties of konjac glucomannan/curdlan blend films. Carbohydr Polym 89:497–503. https://doi.org/10.1016/j.carbpol.2012.03.034
Wu J, Ge S, Liu H et al (2014) Properties and antimicrobial activity of silver carp (Hypophthalmichthys molitrix) skin gelatin-chitosan films incorporated with oregano essential oil for fish preservation. Food Packaging Shelf Life 2:7–16. https://doi.org/10.1016/j.fpsl.2014.04.004
Xiong H, Tang S, Tang H, Zou P (2008) The structure and properties of a starch-based biodegradable film. Carbohydr Polym 71:263–268
Yapo BM (2011) Pectic substances: from simple pectic polysaccharides to complex pectins—a new hypothetical model. Carbohydr Polym 86:373–385
Yu S, Zhang X, Tan G et al (2017) A novel pH-induced thermosensitive hydrogel composed of carboxymethyl chitosan and poloxamer cross-linked by glutaraldehyde for ophthalmic drug delivery. Carbohydr Polym 155:208–217
Zakaria S, Bakar NHA (2015) Extraction and characterization of gelatin from black tilapia (Oreochromis niloticus) scales and bones. In: International Conference on Advances in Science, Engineering, Technology & Natural Resources (ICASETNR-15). pp 27–28
Zargar V, Asghari M, Dashti A (2015) A review on chitin and chitosan polymers: structure, chemistry, solubility, derivatives, and applications. ChemBioEng Rev 2:204–226. https://doi.org/10.1002/cben.201400025
Zhai M, Yoshii F, Kume T, Hashim K (2002) Syntheses of PVA/starch grafted hydrogels by irradiation. Carbohydr Polym 50:295–303. https://doi.org/10.1016/S0144-8617(02)00031-0
Zhang X, Do MD, Casey P et al (2010) Chemical modification of gelatin by a natural phenolic cross-linker, tannic acid. J Agric Food Chem 58:6809–6815. https://doi.org/10.1021/jf1004226
Zhao R, Torley P, Halley PJ (2008) Emerging biodegradable materials: starch- and protein-based bio-nanocomposites. J Mater Sci 43:3058–3071. https://doi.org/10.1007/s10853-007-2434-8
Zhuang C, Tao F, Cui Y (2017) Eco-friendly biorefractory films of gelatin and TEMPO-oxidized cellulose ester for food packaging application. J Sci Food Agric 97:3384–3395. https://doi.org/10.1002/jsfa.8189
Zubeldía F, Ansorena MR, Marcovich NE (2015) Wheat gluten films obtained by compression molding. Polym Test 43:68–77. https://doi.org/10.1016/j.polymertesting.2015.02.001

Chapter 22
Food Grade Polymers for the Gelation of Edible Oils Envisioning Food Applications

A. J. Martins, L. M. Pastrana, A. A. Vicente, and M. A. Cerqueira

Abstract Oleogels are systems traditionally produced by the self-assembly of materials called gelators, which are responsible for inducing viscosity and solid-like capabilities to oil-based systems. The emergent interest concerning oil structuring strategies in food applications is related to oleogels' capacity to undergo structural and textural tailoring, and the possibility of their use in delivery of bioactive compounds.

The selection of the materials and the methodology used to produce oleogels are definitely the key aspects in this kind of systems. The existing, obligatory demand of using food grade ingredients, combined with the increasing requirements for bio-based solutions towards the total replacement of petroleum-based materials narrows the possibilities to a few set of materials. Overall, with the exception of ethylcellulose, a linear polysaccharide derived from cellulose with high hydrophobicity and semi-crystalline nature, polymeric gelators with ability to structure oil are scarce. It is known that the hydrophilic nature of biopolymers does not make them the first choice for oil structuring purposes. Despite of that, some biopolymers are amphiphilic and can interact with apolar phases; this seems to be the case of some polysaccharides and protein combinations (e.g. xanthan gum, gelatin, whey protein and chitin).

This chapter will emphasize food grade polymeric gelators with the capacity of oil structuring, using different methodologies. Mainly polysaccharides and proteins able to produce oleogels will be presented, and their distinct chemical structures will be related to their gelation capacity and final oleogel properties. Besides, different structuring methodologies and gelator combinations able to produce oleogels will also be addressed, together with the main advantages and drawbacks of the different methodologies. Finally, the main food applications will be showed, discussing their possible exploitation by the food industry.

Keywords Biopolymers · Gelator · Gels · Organogel

A. J. Martins · A. A. Vicente
Centre of Biological Engineering, University of Minho, Braga, Portugal

L. M. Pastrana · M. A. Cerqueira (✉)
International Iberian Nanotechnology Laboratory, Braga, Portugal
e-mail: miguelcerqueira@inl.int

T. J. Gutiérrez (ed.), *Polymers for Food Applications*,
https://doi.org/10.1007/978-3-319-94625-2_22

22.1 Introduction

Recent legislative updates in worldwide governmental initiatives fostered new developments aiming at banning trans-fats and decreasing the use of saturated fats in food products. These developments and the consumer demands for clean label products led to evolving discussions regarding the replacement of fats in foodstuffs and food processes for healthy and more acceptable alternatives. Such facts have increased the general interest in the use of healthier oils and innovative oil-structuring agents. Oil structuring technology presents as main advantages the use of healthy fat sources (mono- and poly-unsaturated fatty acids), the ability to modify and personalize food products (by means of structural and mechanical properties) and simplify the delivery of lipophilic bioactive compounds. All these factors make this technology extremely relevant and well perceived by the scientific community. This value is already acknowledged by the industry, as demonstrated by a considerable number of patents filled in the last decade or so (around 200 are found upon searching by "oleogel" and "food" at Google Patents), upholding the effectiveness of the oleogels' for distinct applications. With increased focus on improving health through nutrition, stated at the worldwide level through governmental framework projects, both scientists and industry are joining efforts towards conveying to the general public/consumers the health benefits that result from a more enriched and healthy diet. The introduction of oleogels in the food chain will surely be an outcome of these efforts. A number of edible gelators and gelation methodologies have been studied, generating very interesting products. Being so, different approaches regarding oil structuring can be pointed out as convenient and interesting. Both single-step and direct polymer gelation result in the formation of a supramolecular network that occurs by means of physical or chemical crosslinking of polymer strands (O'Sullivan et al. 2016; Suzuki and Hanabusa 2010).

The most common direct oil gelation methods comprise the use of low molecular weight gelators that reveal high affinity towards non-polar solvents. These gelators are commonly waxes (low and high melting ones), sterol-based gelators (combinations of gamma-oryzanol and phytosterols), lecithin and fatty acid derivatives. As single-component gelators, waxes remain probably as the most successful molecules; due to their unique (and complex) structure, these gelators are able to structure oil at very low concentrations (~3%) (Martins et al. 2017). Monoacylglycerides also fall in this same category. It is important to stress that legislation is, of course, one of the main aspects in this matter due to all legal restrictions imposed to the introduction of new food ingredients and additives. For the multi-component strategy, self-assembled fibrilar networks are the ones that present the most promising results when sterol-based molecules (e.g. phytosterols and gamma-oryzanol) are combined to develop a 3D network that is capable to impart a structure able to entrap oils (Matheson et al. 2017). Given this state-of-the-art, one of the most commonly mentioned challenges is to find bio-based functional molecules able to produce oleogels. In this field, the alternative that is still underexplored is the polymer gelation

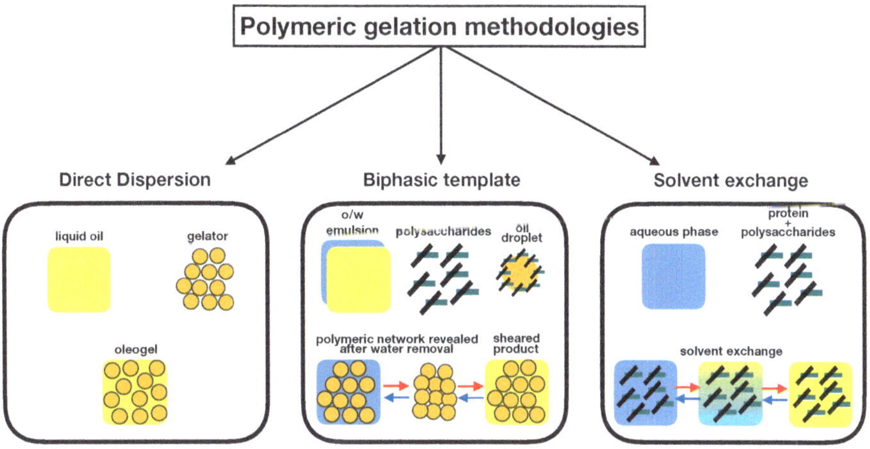

Fig. 22.1 Direct and indirect strategies of oil gelation using biopolymers

approach; on this regard research has been focused on the synthesis of novel amphiphilic compounds or multi-component mixtures capable of providing that functionality, being polymers very promising candidates. In fact, there is a great number of food grade polymers at our disposal, however, due their hydrophilic nature, very few of these can actually be used to form oleogels (Davidovich-Pinhas 2016).

Polymer oleogels can provide an interesting complement to hydrogels (aqueous polymer-gelled systems) towards the delivery of lipid-soluble molecules, as hydrogels are frequently used in delivery applications for hydrophilic molecules. Different approaches that use water as the continuous phase can be used as templates to oil gelation (Patel and Dewettinck 2016). Recently, some reports demonstrated the potential of using new sources of polysaccharides and proteins and their combinations to induce oil structuring. These compounds have a larger space for development because of the diversity of existing polymeric compounds and the benefit that such polymers can add to the final food products. The possible applications of polymeric oleogels are targeted to the replacement of a substantial amount of animal fat in food composition without compromising the functional and textural properties that it imparts to foods. In addition, the possibility of carrying and protecting bioactives by incorporating them in the oleogel matrix can be used to further increase the nutritional value of the food product. With that in mind, undesirable bioactive molecules' polymorphic transformations, that regularly occur in triacylglycerols' (TAG's) systems, can be prevented by the capability of polymeric gels to induce increased stability by slowing the bioactive compound mobility within the gel lipid-matrix (Suzuki and Hanabusa 2010). The aim of this chapter is to specifically establish a comparative discussion regarding oleogels' production methodologies within the field of polymer gelation of edible oils. Figure 22.1 shows the different polymeric gelation methodologies for the production of oleogels.

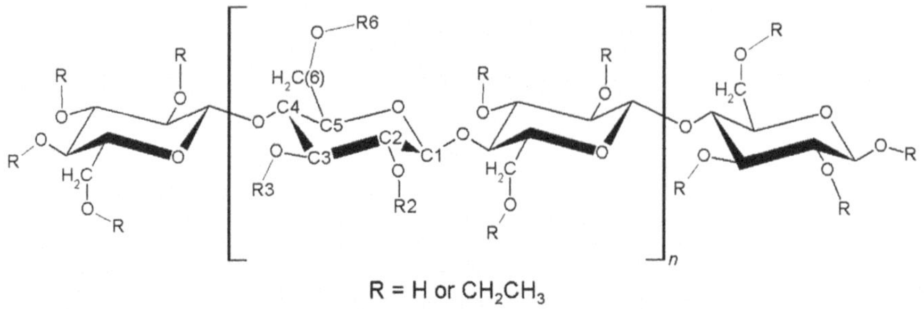

Fig. 22.2 Ethylcellulose structure. Reproduced from Stortz et al. (2014) with permission of The Royal Society of Chemistry (RSC) on behalf of the Centre National de la Recherche Scientifique (CNRS) and the RSC

22.2 Direct Dispersion Methodology

22.2.1 Ethylcellulose Used as Agent for Edible Oil Structuring

Ethylcellulose is a linear polysaccharide derived from cellulose (Fig. 22.2) that is produced from the ethoxylation of the hydroxyl groups, forming an ether bond. The molecule presents three hydroxyl groups on each monomer (excluding terminal monomers) which are available for the ethoxylation process. Because of the level of ethoxylation that can be performed, the substitution degree (SD) of the molecule will determine the compound's solubility, (i.e., $1.0 < SD < 1.5$ soluble in water; $2.4 < SD < 2.5$ soluble in organic solvent). It is common that ethylcellulose presents a SD of ~2.5 (Singh et al. 2017). The molecule configuration can produce eight possible monomers, in an ethylcellulose polymer, due to the three hydroxyl groups that are available for ethoxylation (Stortz et al. 2014). Consequently, it is quite a tailor-friendly polymeric compound due to the SD by the addition of ethylene groups and the length of the polymer backbone (Davidovich-Pinhas et al. 2015a). There are several studies on the application of ethylcellulose in cosmetics, plastics, ceramics and agricultural herbicides (Dubernet et al. 1991; Mukherjee et al. 2005; Melzer et al. 2003; Yamada et al. 2001; Stortz and Marangoni 2013; Okuda et al. 2012; Flores-Cespedes et al. 2018).

The production of ethylcellulose oleogels occurs after a complete polymer solubilization (at least 4% of polymer mass is needed to form a network capable to retain the solvent) in an oil at high temperatures. The ethylcellulose and oil mixture must be heated above ethylcellulose glass transition temperature (~140 °C) in order to achieve the softening of the polymer; if this is not done, the risk of an incomplete gelation is high. After ethylcellulose softening and mixture of the biopolymer in oil, the mixture must undergo the necessary cooling period to ensure that the polymer returns to its rigid form, forming hydrogen intermolecular interactions that will lead to the formation of an entangled polymeric network. This 3D network will be responsible for the entrapment of the solvent and, depending on the viscosity grade of the polymer, gelation confers higher or lower viscoelastic properties (solid-like characteristics) to the oleogel obtained (Zetzl et al. 2014). Reports on the gelation capability of ethylcellulose demonstrated a particular thermal transition, composed by a polymer glass transition followed by a solvent interaction with the polymer chain, which is responsible for the polymer dissolution in the oil. As a consequence of exposed polymer chains, this interaction between oil and existent polymeric chains will be responsible for the physical transition of the gel from a more slippery or glassy state to a rough, rubbery state (Davidovich-Pinhas et al. 2015a). Tailoring properties of ethylcellulose-based oleogels are extremely important, in the sense that distinct setting temperatures alongside the gelation process will determine different final morphological characteristics, namely gel strength. If thermoreversibility of the oleogel is induced by manipulating environmental conditions (e.g. temperature), the consequent gel-sol or sol-gel transitional behavior will be responsible for a rearrangement of the structure of the polymer network. This fact will consequently influence oleogel's final mechanical and optical properties (Davidovich-Pinhas et al. 2015b). A model to predict the amounts of distinct components for ethylcellulose oleogels production was developed. The model consists in two-dimensional fits of the experimental data of oil-surfactant-ethylcellulose system, using gel strength as the variable to be optimized as a function of mass fraction of ethylcellulose and surfactant (with three fixed ethylcellulose/surfactant ratios). According to this model, for all the tested conditions the final gel strength increases with mass fraction of ethylcellulose in a power law. Such model should be useful to understand the consequences of the different molecular weights of ethylcellulose regarding manufacturing conditions and final gel characteristics (e.g. gelling temperature and gel strength) (Gravelle et al. 2014). Studies on the microstructure of ethylcellulose oleogels showed how the 3D network of the polymer surrounds the oil, in an oil droplet configuration. Surely, the grade of ethylcellulose will induce significant differences in terms of average pore size of the oil droplets, changing the final structure of the oleogel and consequently its mechanical properties, which are important in terms of product acceptance by the consumers. However, factors other than pore size are contributing for the mechanical strength of the gels; these factors can be the strength of the walls surrounding the pores, the interconnections on the polymeric network, the number of junction zones per polymer bundle and interactions between the polymer, oil, and/or surfactant, if this is present (Zetzl et al. 2014). Figure 22.3 shows the polymeric microstructure of ethylcellulose that is able to impart solid structure to oils. Probably, the main issue regarding this type of oleogels is the polymer high melting temperature and consequent gelation temperature interval, because of that, stability issues (e.g. oil oxidation) can be a drawback.

Sullivan et al. reported on *in-vitro* digestion of beta-carotene loaded EC-oleogels, demonstrating bioactive delivery capability and how gel strength influences the behavior of gels during the digestive process (O'Sullivan et al. 2017). In this work, slow or incomplete lipolysis was observed in harder oleogels. This is relevant regarding the protection and consequent bioavailability of the desired compound under the harsh environmental conditions of the stomach. However, extremely high dosages of beta-carotene (as mentioned by the authors) are not the ideal.

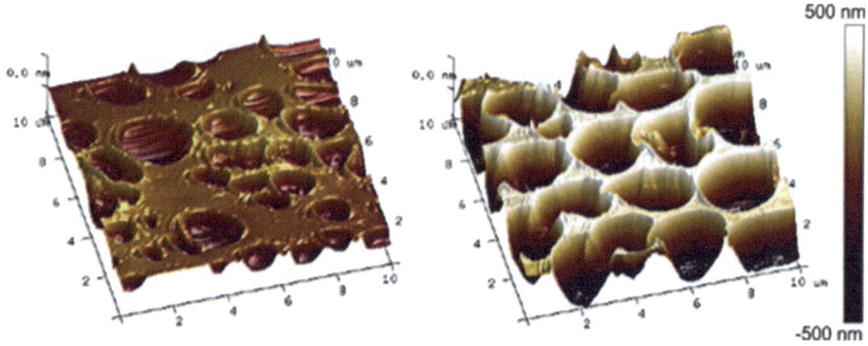

Fig. 22.3 AFM of ethylcellulose polymer microstructure in canola oil oleogel environment. (Reprinted from Food Structure, 2/(1–2), Zetzl, A. K., Gravelle, A. J., Kurylowicz, M., Dutcher, J., Barbut, S., & Marangoni, A. G., Microstructure of ethylcellulose oleogels and its relationship to mechanical properties, 27–40, Copyright (2014), with permission from Elsevier)

Despite that fact, this study revealed important information on oleogel structural behavior, when another compound is present in its internal matrix, and how these oleogels can act as a vehicle of lipophilic molecules. Significant outcomes can result from the use of ethylcellulose oleogels when comparing with liquid or solid-lipid matrices. Namely, improved shelf life stability and defense concerning recrystallization behavior and degradation of the encapsulated bioactive compound (Yu et al. 2012; O'Sullivan et al. 2017). Going forward, the incorporation of ethylcellulose oleogels in an aqueous system is pointed as one of the most interesting developments to target even more applications for this kind of oleogels. With that in mind the multi-component approach, with combination of different structuring materials aiming at certain synergic behaviors, can be a positive answer to overcome some of the existent problems (Pernetti et al. 2007; Wang et al. 2016b).

22.3 Biphasic Emulsion-Based Methodologies

22.3.1 Chitin Oleogels

Chitin is one of the most abundant natural aminopolysaccharide polymers, and can be obtain from several bio-based sources, (e.g. crustaceans, insects, and the cell walls of fungi) (Gutiérrez 2017). It has been used in an extensive way in the development of polymeric scaffolds (through enzymatic or chemical deacetylation, as also happens with its derivative, chitosan) (Elieh-Ali-Komi and Hamblin 2016). Chitin and chitosan have been also under study on the development of thickeners for oils, mainly by means of introducing reactive isocyanate groups into chitin or chitin-based molecules, and thus inducing the formation of gel-like structures that could act as biodegradable lubricants (Gallego et al. 2013; Sánchez et al. 2011). In the framework of oleogel development with these polymeric strands, they perform

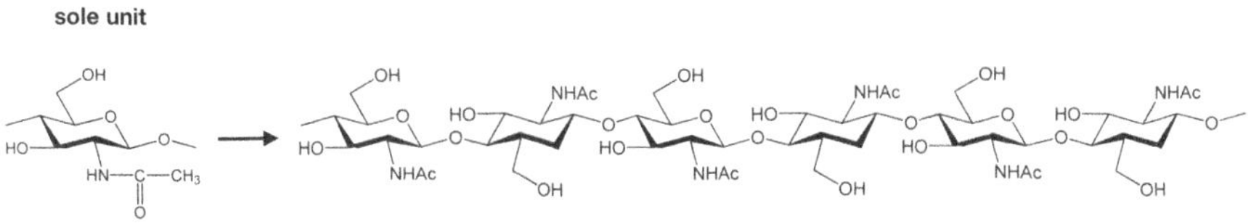

Fig. 22.4 Sole unit of 2-acetamido-2-deoxy-D-glucopyranose and chitin structure. Adapted from (Nikiforidis and Scholten 2015)—Published by The Royal Society of Chemistry

interactions by means of physical forces, such as van der Waals, and electrostatic interactions. These interactions are the ones responsible for the polymer–polymer and polymer–solvent synergies that will impart solid-like viscoelastic properties to the organic solvents used in the process (Laredo et al. 2011). Chitin is a strong hydrophobic compound with very poor solubility in water. The resemblance of the chitin molecular structure with the ethylcellulose structure (presented earlier in this chapter) is notorious in Fig. 22.4 (when compared with Fig. 22.2). In the C-2 position, chitin contains an acetamido group in contrast to the hydroxyl or ethyl ester groups in the case of ethylcellulose (Nikiforidis and Scholten 2015).

Still very little is known, regarding the use of chitin as an oil structuring agent. It was reported that chitin as sole gelator component has not been able to shown oil structuring ability. Testing only chitin solubilized in oil (e.g. soybean and castor oil) resulted in very weak stranded gels and precipitation of chitin occurred after a few days due to inability to develop a sufficiently strong polymeric network (Sánchez et al. 2011). However, it has been reported that chitin-based oleogels can be developed by the use of chitin (in an emulsion-based method) in combination with surfactants or derived chitin nanoparticles that will create building blocks with oil structuring capabilities. Figure 22.5 portrays both the chitin-based oleogels as well as their microstructural arrangements in the presence of phosphatidylcholine and nanocrystals of chitin. These combinations showed promising results towards oil gelation and interesting oleogel viscoelastic properties (Nikiforidis and Scholten 2015). The presence of surfactants helps preventing the aggregation of crude chitin in the nonpolar solvent; this process can overturn the chitin aggregation induced by hydrophilic chitin–chitin interactions, thus improving the stability and dispersion of crude chitin in the oil medium. Therefore, the minimization of chitin aggregation will lead to the subsequent formation of a network that can trap liquid oil into a gel structure.

Nikiforidis and Scholten (2015) showed that chitin combined with a non-ionic surfactant (Span 60), a zwitter-ionic surfactant (phosphatidylcholine) or an anionic enzymatically modified phosphatidylcholine, produced oleogels with different characteristics. Oleogels with enzymatically modified phosphodicholine showed rheological thermal stability up until 90 °C without modification of viscoelastic values.

Significant changes in oleogel's structure can be performed using different chitin compositions, by changing the length and the flexibility of the chitin strands. Temperatures above 70 °C, will be equally determinant regarding polymeric interactions, which will suffer changes in polymer-polymer, polymer-solvent and polymer-surfactant interactions and affect the final properties of the oleogels.

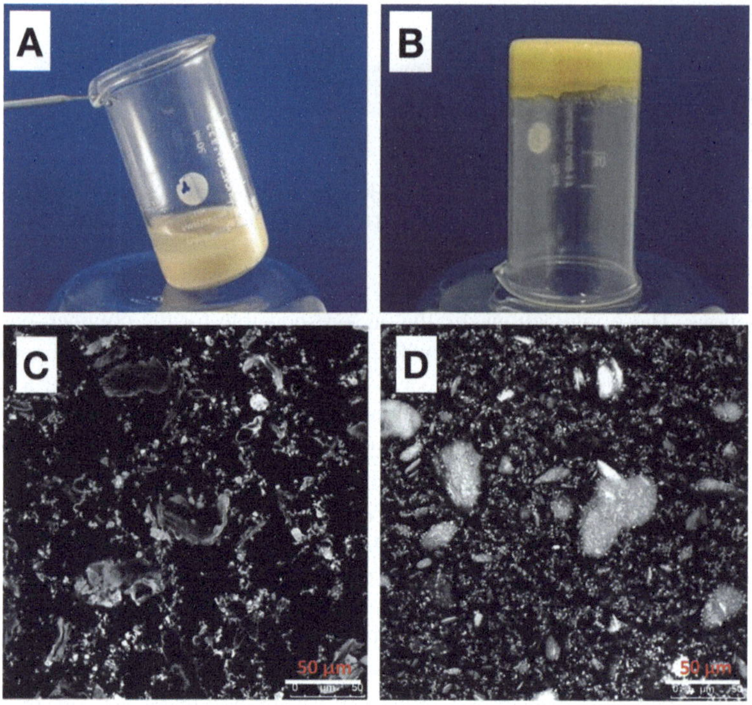

Fig. 22.5 (**a**) Phase separation of crude chitin in sunflower oil; (**b**) chitin concentration of 20 wt% in sunflower oil; (**c**) microstructure of oleogels with crude chitin and high phosphatidylcholine content (70 wt%); (**d**) microstructure of oleogels with chitin nanocrystals and phosphatidylcholine. Adapted from Nikiforidis and Scholten (2015)—Published by The Royal Society of Chemistry

22.3.2 Cellulose Derivatives-Based Oleogels

Like ethylcellulose, hydroxyl-propyl-methylcellulose (HPMC) is a cellulose derivative (Gutiérrez and Alvarez 2017). It is a polymeric strand that has shown to be suitable for oil structuring purposes. HPMC is non-toxic, biodegradable, biocompatible and presents good mechanical properties. Because of these characteristics, HPMC has numerous functionalities as regards its use by the food industry and can be subjected to fine-tuning processes (Patel and Dewettinck 2015). HPMC is used as an emulsifier to control dispersions' texture and rheological properties, once it allows controlling ice crystallization (Bell et al. 2006). This polymeric strand acts as a hydrophilic foaming agent, able to create a colloid with air–water interface, that will make possible to stabilize, at low temperatures, an hydrophilic foam that after drying (and consequent water removal) allows the formation of a porous template. This porous template, with great oil affinity, is then used to develop an oleogel (Patel et al. 2013; Patel and Dewettinck 2015).

A recent study on the gelling behavior of combinations of re-generated cellulose (RC) and carboxymethyl cellulose (CMC) demonstrated a feasible way to develop

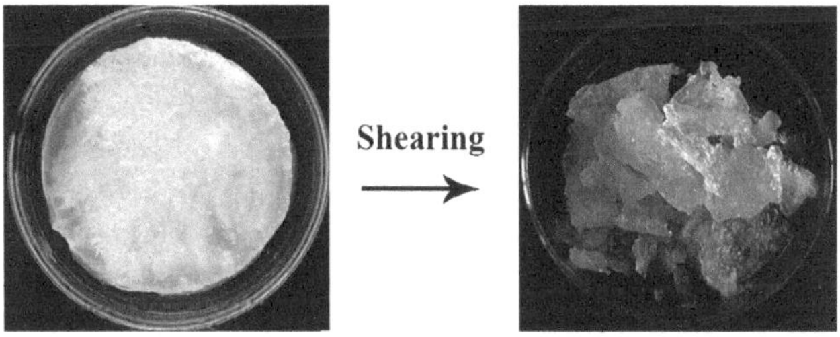

Fig. 22.6 Photographs of dried product (1.4 wt% RC and 1.4 wt% CMC) and oleogel prepared by simply shearing the dried product. (Reprinted from Food Hydrocolloids, Jiang, Y., Liu, L., Wang, B., Sui, X., Zhong, Y., Zhang, L., Mao, Z., Xu, H., Cellulose-rich oleogels prepared with an emulsion-templated approach, in press, Copyright (2017), with permission from Elsevier)

an indirect methodology to produce oleogels with oil concentrations up to ~97% (Jiang et al. 2017). The anti-solvent regeneration method to produce cellulose is quite effective in improving its role as dispersion and emulsion stabilizer agent. The mentioned methodology consists in developing water-continuous emulsions, that are stabilized by RC and CMC using a high-energy ultra-homogenization at 15,000 rpm for 3 min. Oleogels were obtained after removing water using freeze-drying and shearing the dried product. Oleogels revealed increased viscoelastic properties alongside with the increase of RC in the primary emulsion composition. Figure 22.6 shows the final cellulose-based oleogel appearance after shearing.

The possibility to develop linear amphiphilic block copolymers that are composed of hydrophilic HPMC and hydrophobic poly(l-lactide) (PLLA) has also been reported. These copolymers exhibit thermoresponsive behavior and microscopy revealed that they are able to self-assemble into distinct architectural conformations like spherical micelles, rod-like micelles, worm-like micelles, or filomicelles. These building blocks can possibly serve as templates for oil structuring purposes (Wang et al. 2017). However, due to the non-food grade status of PLLA, suitable alternatives should be investigated. One of the possibilities is polycaprolactone (PCL), which is a biodegradable and biocompatible polymer with current FDA approval for food applications and has been used for scaffold development with drug delivery purposes, as well as namely green tea polyphenols through PCL implants.

22.3.3 Protein and Polysaccharide Combinations

Proteins in combination with other compounds, such as polysaccharides, can form a dense building block layer, which can be used as an oleogelator. This is a process based in oil-in-water systems that are capable to ensure protein adsorption at their interface, by means of chemical crosslinking or just driven by physical interactions. Romoscanu and Mezzenga (2006) reported on a similar approach where in this case, after emulsion formation, the resultant dry "foamed" polymer showed increased capability of oil retention. Here, the original emulsion could be restored by rehydrating the final (dryed) solidified product (Romoscanu and Mezzenga 2006). This approach requires removal of the hydrophilic phase, after which the exposed polymeric stranded structure will imprison the oil content of the primary emulsion, forming an oleogel (de Vries et al. 2015). De Vries et al. (2015) reported on emulsion stabilization by means of a mixture of protein and polysaccharide, namely gelatin and xanthan gum. More than their hydrophilic nature, one of the important reasons to use both gelatin and xanthan gum is their affinity to interact with each other, even more than to interact with just water. This not only results from hydrophobic interactions, but also from non-coulombic ones, that happen with the involvement of NH and OH groups. However, the ratio of the components will contribute to the success of this combination; if this interaction is performed with an increased fraction of gelatin, the reaction yield will decrease (Lii et al. 2002). This also evidences that the interactions between these biopolymers can be tuned by controlling the protein/polysaccharide ratio, together with the pH and the ionic strength (Wang et al. 2016a). Gelatin and xanthan gum are used to provide structure to this oil-in-water emulsion due to their hydrophilic nature. Both polymeric substances are not directly mixed for this purpose; instead, each one is used to obtain hydrophilic solutions which are subsequently added to oil under high energy mixing (e.g. using an Ultra-Turrax), as described elsewhere (Patel et al. 2015a). After this, a complete dehydration process is needed to eliminate the entire water content, exposing the biopolymer structure that will serve as building block for the structuring process. The drying can be achieved by high or low-temperature processes, namely oven drying (e.g. at 70 °C for 48 h) or lyophilization. The resultant dried structure that is used to entrap the oil phase, exhibits very interesting features, in the sense that its microstructural morphology, which is configured by a polymer layer that would act as oil droplet coalescence preventer, will be responsible for the packing of the oil droplets. The strength of this tight packing would influence the gel strength and the solid-like characteristics of the oleogels formed. As the polymer concentration rises, the easier the drying procedure will be. If the amount of polymer is relatively low, oil leakage will be noticeable leading to phase separation. It has been reported by Patel et al. (2015a) that for formulations containing >97 wt% liquid oil, this emulsion-based method using gelatin and xanthan gum is quite successful and promising. As Patel et al. (2014) demonstrated, it is also possible to produce oil continuous emulsions and then promote the gelling of the water droplets, providing a strong enough interconnected network capable of imparting adequate structural sustainment to such emulsions. A successful gelation will produce a self-standing elastic gel-like structure, with sufficient strength to retain oil within its structure. This gelation of high internal phase emulsions (HIPEs) can be produced using a combination of different polysaccharides. Some options are galactomannans; these natural polymers are known to develop synergistic interactions with other food-grade hydrocolloids (e.g. carrageenan and xanthan gum). Also because of its higher mannose:galactose ratio (4:1), locust bean gum (LBG) displays a strong interaction with other non-gelling colloidal solutions. These interactions will form gels at interesting low polymer concentrations. The hydrocolloids used should present thermoreversibility, because after all, this is a heat-triggered water droplet gelation mechanism that forms a gel after cooling. Also, the combination of different hydrocolloids is presented as being more attractive in order to widen the tailoring properties of these gels. Reports on this type of HIPE gelling mechanism, demonstrated such capabilities by emulsifying the hydrocolloid solutions in oil, using high temperatures (>70 °C) followed by a cooling process that leads to the gelation of the solutions (Patel et al. 2014). The stabilization of this system relies probably in the use of non-surface-active hydrocolloids that will maintain the oil continuous property of the system. Using the same LBG and carrageenan combination, Patel et al. (2015b) also explored gel-in-oil-in-gel type structured emulsions, that allowed producing different types of microstructured gels with mutable rheological properties. These gels can possibly undergo processes aiming at increasing the final oil intake and serving as templates for the development of oleogels.

22.4 Solvent Exchange Methodologies

22.4.1 Proteins and Polysaccharides as Building Blocks for Oleogel Development

Producing hydrogels using proteins is a widely studied process; from the several possibilities can be pinpoint the use of collagen, gelatin, elastin, silk fibroin and milk proteins (Juncosa-Melvin et al. 2006; Kuijpers et al. 2000; Jonker et al. 2012; Altman et al. 2003; Kinsella 1984). The main processing step that will confer protein networks the ability of incorporating a nonpolar solvent as continuous phase, therefore forming oleogels, is based in a solvent substitution/replacement approach. It is well known that proteins display great affinity towards apolar solvents, being poorly capable of oil dispersion due to their hydrophilic character. One of the possibilities to change this behavior is to induce a solvent-exchange process in order to allow performing oleogelation, widening the field of applications of proteins towards oil structuring strategies, while also showing the applicability of different conceptual tools in colloid science. The solvent exchange approach is commonly used to develop aerogels by means of water removal by solvent addition (e.g. acetone or/and alcohol). One example has been presented by Yang et al. (2017) where these solvents were removed by supercritical CO_2. Whey protein was evaluated as a polymeric structural backbone for solvent exchange experiments, by first producing hydrogels with whey protein isolate (WPI) after ionic strength adjustments with NaCl. The heat denaturation of the proteins was performed at 80 °C during 30 min and, after cooling and stabilization at low temperatures ~4 °C, the hydrogels were subjected to solvent exchange procedures (de Vries et al. 2015). Solvents like acetone and tetrahydrofuran that are capable of mixing with both water and oils were used; being so, a two-step procedure was performed to substitute the water by the oil (sunflower oil in this case). The immersion of the gels in fresh 100% (v/v)

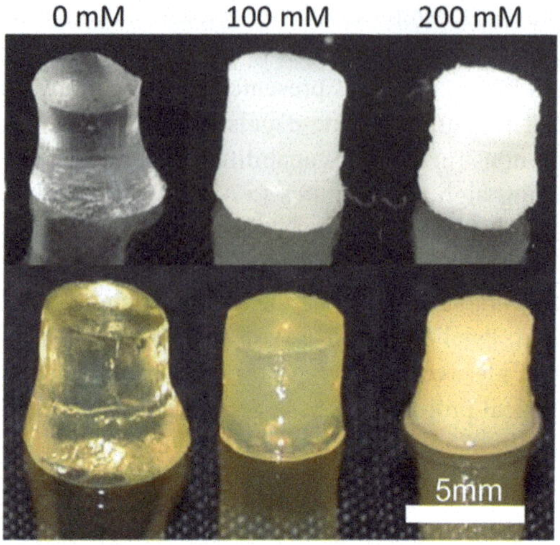

Fig. 22.7 WPI hydrogels on top and oleogels in the bottom after the solvent exchange procedure. The composition of the hydrogels was 15% WPI with 0, 100, and 200 mM NaCl. (Reprinted with permission from de Vries et al. (2015). Copyright (2015) American Chemical Society)

acetone (or tetrahydrofuran) led to water removal, after which the gels were immersed in sunflower oil to remove the acetone and incorporate the oil within the polymeric strands (de Vries et al. 2015). Figure 22.7 shows the hydrogels that led to the formation of oleogels after solvent exchange.

The tailoring ability of these protein oleogels was demonstrated by the different levels of oil holding capacity of the protein network. This customization can be made by choosing the polarity of the solvents used in the intermediate step, and also by controlling the kinetics of the solvent exchange methodology. Regarding mechanical properties, protein oleogels showed increased hardness (increased modulus by approximately two orders of magnitude in comparison to the primary hydrogels) and revealed to be more brittle as well (decrease in fracture strain). Still in Fig. 22.7, the upper row hydrogel samples resulted from a solution with higher ionic strength, helping proteins to aggregate in larger conformations (Yan and Pochan 2010). Because of that, a wiry or coarser conformation of protein structures will confer increased turbidity to protein gels (de Vries et al. 2015). The swelling behavior is a characteristic of protein-based gels; this is a feature that will help considerably this exchanging solvent approach in terms of improved oil intake. Despite the network that is responsible for the oil entrapment after solvent exchange being the same that primarily formed the hydrogels, the mechanical properties of the final oleogels will be distinct from their hydrogel counterpart. As showed by the authors, when comparing the oleogels with the corresponding hydrogels, the former showed a strong increase in modulus and fracture stress accompanied by a lower strain at fracture (Fig. 22.8).

The solvent exchange approach described above will also be responsible for some other modified properties. The polarity of the solvent will decrease alongside with the exchange of solvent, being this factor responsible for a conformational change of the gel, leading to its shrinkage. This change will influence the oil intake

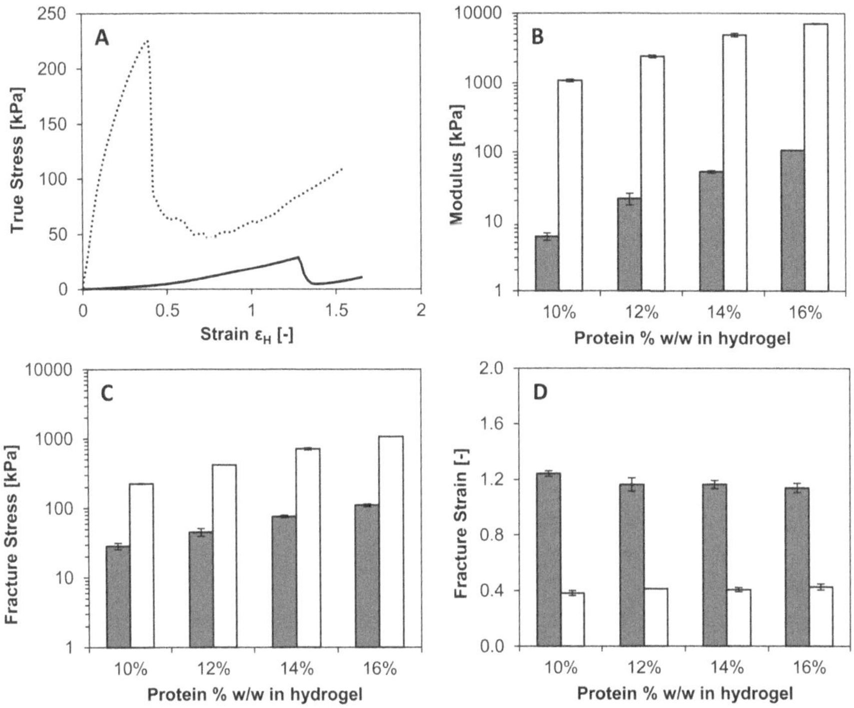

Fig. 22.8 Large deformation properties of protein-based hydrogels and oleogels after the solvent exchange procedure using acetone. (**a**) Stress-strain curve for a representative 10% of 50 mM NaCl hydrogel (solid line) and the resulting oleogel after the solvent exchange (dotted line). (**b–d**) Modulus, fracture stress, and fracture strain of hydrogels (gray bars) prepared at 50 mM NaCl with different protein concentrations and the resulting oleogels (white bars) after the solvent exchange. Error bars represent standard deviation of duplicate measurements. (Reprinted with permission from de Vries et al. (2015) Copyright (2015) American Chemical Society)

capabilities of the oleogels. Depending on the solvent used in the intermediate step, gel shrinkage and consequent oil holding capacity will vary.

A similar procedure was described by Manzocco et al. (2017) when reporting the development of oleogels by means of solvent exchange in polysaccharide aerogels. In this work, a single component polysaccharide (κ-carrageenan) was used to develop hydrogels, with increasing carrageenan concentrations. Afterwards, the hydrogels were submitted to immersion in alcohol solutions at increasing concentrations (25, 50, 75% v/v) at a constant 1:8 (v/v) hydrogel:ethanol ratio. Then, to prevent hydrogel architectural collapse, supercritical CO_2 drying was employed. Different CO_2 flow rates were tested allowing to minimize the drying time, so that the structural integrity of the gels remained unaltered, in particular without surface cracks. The final procedure consisted in introducing the aerogels in 125 mL of sunflower oil until a constant weight was recorded. The maximum observed oil intake was of ~80%. Table 22.1 shows parameters values for increasing concentrations of κ-carrageenan in hydrogel formulation and consequent oleogel production.

Table 22.1 Oil content, firmness and oil holding capacity (OHC) values of oleogels obtained from hydrogels produced with increasing κ-carrageenan concentration

κ-Carrageenan in hydrogel (% w/w)	κ-Carrageenan in oleogel (% w/w)	Oil in oleogel (% w/w)	Firmness (N)	OHC (% w/w)
0.4	27.58 ± 0.83a	72.42 ± 0.83a	158.33 ± 9.16c	83.44 ± 1.35a
1.0	26.76 ± 0.33a	73.24 ± 0.33a	311.70 ± 11.78a	82.18 ± 1.11a
2.0	18.72 ± 0.31b	81.28 ± 0.31b	216.40 ± 6.79b	62.21 ± 1.31b

Different letters in the same column are significantly different ($p < 0.05$)

Also within the solvent exchange methodology approach, another possible solution to use protein building blocks as gelators to endure oleogel formation consists in the usage of whey protein aggregates that result from hydrogel structure breakdown. De Vries et al. demonstrated that protein aggregates with very small sizes can be successfully used in the formation of a network in a nonpolar environment, which resembles the ones usually developed in aqueous media by larger conformations (de Vries et al. 2017). In this approach, contrary to the one referred before, the primary backbone for oleogels is not the protein architectural structure, but the resultant protein dispersion pellet obtained after hydrogel breakdown, homogenization and centrifugation processes. One important feature of is the ability of this protein to build blocks, forming gels with solid-like behavior (i.e. $G' > G''$), at a very low total mass fraction of gelator (~3%). Macro and microstructures of the protein gels at different stages of oleogels' development are shown in Fig. 22.9. The hydrophobic protein aggregated pellet, after the solvent exchange procedure (with acetone for example) and consequent solvent oil polarity changes with sunflower oil, is responsible for the formation of oleogels. This is materialized by the protein building blocks, with approx. diameter of 200 nm (measured for both water and oil), that act as gelator particles (de Vries et al. 2017). Also, for increasing amounts of protein, the rheological data shows modifications towards higher resemblance to an elastic covalent gel.

22.5 Food Applications and Final Remarks

Polymeric oil gelation strategies are still under development but it is possible to state that there are interesting options concerning the usage of different polymer strands that are usually more associated to water gelling and less with oil structuring. Therefore, it is safe to say that there are some unexplored options for oil-structuring that can be definitely extended to other biopolymers of interest.

In fact, and taking into account the already proposed applications for non-polymeric oleogels, such as chocolates (Marangoni 2014; Stortz and Marangoni 2011, 2013; Rogers et al. 2014), pastries (Mert and Demirkesen 2016; Hwang et al. 2016), spreads (Patel 2015; Patel and Dewettinck 2016), processed meats (Panagiotopoulou et al. 2016), and ice-cream (Zulim Botega et al. 2013; Moriano

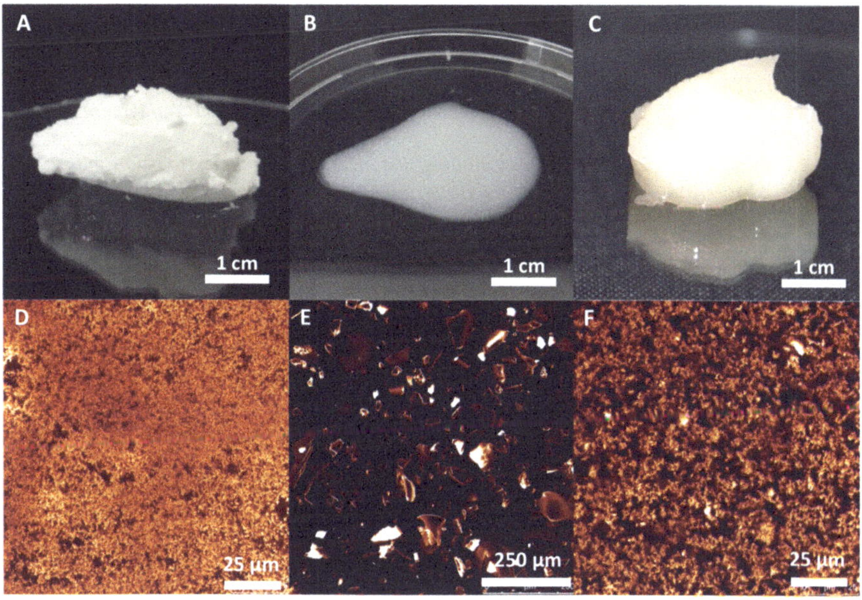

Fig. 22.9 (**a**) Appearance of heat-set WPI aggregates after centrifugation, (**b**) dispersion of freeze-dried WPI aggregates in sunflower oil and (**c**) WPI aggregates in sunflower oil via a solvent exchange with corresponding CLSM micrographs (**d–f**). (Reprinted from Journal of Colloid and Interface Science, 486, de Vries, A., Wesseling, A., van der Linden, E., & Scholten, E., Protein oleogels from heat-set whey protein aggregates, 75–83, Copyright (2017), with permission from Elsevier)

and Alamprese 2017), the use of polymeric-based oleogels opens an even wider span of possibilities. Considering the challenge of mimicking the mouthfeel and texture of common fats using non-polymeric oleogels, the introduction of polymeric oleogels can widen the possibilities of shortening the existing gap. Nevertheless, several challenges still need to be overcome if the focus is the application in the food industry. Mainly, the price of the biopolymers and difficulties in up scaling the methodologies used, that in some cases could condition the final price of oleogels and thus their economic viability.

Acknowledgments This research was supported by Norte Regional Operational Program 2014–2020 (Norte2020) through the European Regional Development Fund (ERDF) Nanotechnology based functional solutions (NORTE-01-0145-FEDER-000019). This study was supported by the Portuguese Foundation for Science and Technology (FCT) under the scope of the strategic funding of UID/BIO/04469/2013 unit and COMPETE 2020 (POCI-01-0145-FEDER-006684) and BioTecNorte operation (NORTE-01-0145-FEDER-000004) funded by the European Regional Development Fund under the scope of Norte2020—Programa Operacional Regional do Norte. The authors acknowledge the Project RECI/BBB-EBI/0179/2012 (FCOMP-01-0124-FEDER-027462). Artur Martins is recipient of a fellowship supported by a doctoral advanced training (call NORTE-69-2015-15) funded by the European Social Fund under the scope of Norte2020—Programa Operacional Regional do Norte.

References

Altman GH, Diaz F, Jakuba C, Calabro T, Horan RL, Chen J, Helen L, Richmond J, Kaplan DL (2003) Silk-based biomaterials. Biomaterials 24(3):401–416. https://doi.org/10.1016/s0142-9612(02)00353-8

Bell D, Coffey D, Henderson A (2006) Cellulose and cellulose derivatives:147–179. https://doi.org/10.1201/9781420015164.ch5

O'Sullivan CM, Davidovich-Pinhas M, Wright AJ, Barbut S, Marangoni AG (2017) Ethylcellulose oleogels for lipophilic bioactive delivery—effect of oleogelation on in vitro bioaccessibility and stability of beta-carotene. Food Funct 8(4):1438–1451. https://doi.org/10.1039/c6fo01805j

Davidovich-Pinhas M (2016) Oleogels: a promising tool for delivery of hydrophobic bioactive molecules. Ther Deliv 7(1):1–3. https://doi.org/10.4155/tde.15.83

Davidovich-Pinhas M, Barbut S, Marangoni AG (2015a) The gelation of oil using ethyl cellulose. Carbohydr Polym 117:869–878. https://doi.org/10.1016/j.carbpol.2014.10.035

Davidovich-Pinhas M, Gravelle AJ, Barbut S, Marangoni AG (2015b) Temperature effects on the gelation of ethylcellulose oleogels. Food Hydrocoll 46:76–83. https://doi.org/10.1016/j.foodhyd.2014.12.030

de Vries A, Hendriks J, van der Linden E, Scholten E (2015) Protein oleogels from protein hydrogels via a stepwise solvent exchange route. Langmuir 31(51):13850–13859. https://doi.org/10.1021/acs.langmuir.5b03993

de Vries A, Wesseling A, van der Linden E, Scholten E (2017) Protein oleogels from heat-set whey protein aggregates. J Colloid Interface Sci 486:75–83. https://doi.org/10.1016/j.jcis.2016.09.043

Dubernet C, Rouland JC, Benoit JP (1991) Ibuprofen-loaded ethylcellulose microspheres: analysis of the matrix structure by thermal analysis. J Pharm Sci 80(11):1029–1033. https://doi.org/10.1002/jps.2600801106

Elieh-Ali-Komi D, Hamblin MR (2016) Chitin and Chitosan: Production and Application of Versatile Biomedical Nanomaterials. Int J Adv Res 4(3):411–427

Flores-Cespedes F, Daza-Fernandez I, Villafranca-Sanchez M, Fernandez-Perez M, Morillo E, Undabeytia T (2018) Lignin and ethylcellulose in controlled release formulations to reduce leaching of chloridazon and metribuzin in light-textured soils. J Hazard Mater 343:227–234. https://doi.org/10.1016/j.jhazmat.2017.09.012

Gallego R, Arteaga JF, Valencia C, Franco JM (2013) Isocyanate-functionalized chitin and chitosan as gelling agents of castor oil. Molecules 18(6):6532–6549. https://doi.org/10.3390/molecules18066532

Gravelle AJ, Barbut S, Quinton M, Marangoni AG (2014) Towards the development of a predictive model of the formulation-dependent mechanical behaviour of edible oil-based ethylcellulose oleogels. J Food Eng 143:114–122. https://doi.org/10.1016/j.jfoodeng.2014.06.036

Gutiérrez TJ (2017) Chitosan applications for the food industry. In: Ahmed S, Ikram S (eds) Chitosan: derivatives, composites and applications. Wiley-Scrivener Publisher, pp 183–232. https://doi.org/10.1002/9781119364849.ch8. ISBN: 978-1-119-36350-7

Gutiérrez TJ, Alvarez VA (2017) Cellulosic materials as natural fillers in starch-containing matrix-based films: a review. Polym Bull 74(6):2401–2430. https://doi.org/10.1007/s00289-016-1814-0

Hwang HS, Singh M, Lee S (2016) Properties of cookies made with natural wax-vegetable oil organogels. J Food Sci 81(5):C1045–C1054. https://doi.org/10.1111/1750-3841.13279

Jiang Y, Liu L, Wang B, Sui X, Zhong Y, Zhang L, Mao Z, Xu H (2017) Cellulose-rich oleogels prepared with an emulsion-templated approach. Food Hydrocoll. https://doi.org/10.1016/j.foodhyd.2017.10.023

Jonker AM, Löwik DWPM, van Hest JCM (2012) Peptide- and protein-based hydrogels. Chem Mater 24(5):759–773. https://doi.org/10.1021/cm202640w

Juncosa-Melvin N, Shearn JT, Boivin GP, Gooch C, Galloway MT, West JR, Nirmalanandhan VS, Bradica G, Butler DL (2006) Effects of mechanical stimulation on the biomechanics and histology of stem cell-collagen sponge constructs for rabbit patellar tendon repair. Tissue Eng 12(8):2291–2300. https://doi.org/10.1089/ten.2006.12.2291

Kinsella JE (1984) Milk proteins: physicochemical and functional properties. Crit Rev Food Sci Nutr 21(3):197–262. https://doi.org/10.1080/10408398409527401
Kuijpers AJ, Engbers GHM, Krijgsveld J, Zaat SAJ, Dankert J, Feijen J (2000) Cross-linking and characterisation of gelatin matrices for biomedical applications. J Biomater Sci Polym Ed 11(3):225–243. https://doi.org/10.1163/156856200743670
Laredo T, Barbut S, Marangoni AG (2011) Molecular interactions of polymer oleogelation. Soft Matter 7(6):2734–2743. https://doi.org/10.1039/C0SM00885K
Lii CY, Liaw SC, Lai VMF, Tomasik P (2002) Xanthan gum-gelatin complexes. Eur Polym J 38(7):1377–1381. https://doi.org/10.1016/S0014-3057(02)00008-3
Manzocco L, Valoppi F, Calligaris S, Andreatta F, Spilimbergo S, Nicoli MC (2017) Exploitation of κ-carrageenan aerogels as template for edible oleogel preparation. Food Hydrocoll 71:68–75. https://doi.org/10.1016/j.foodhyd.2017.04.021
Marangoni AG (2014) Chocolate compositions containing ethylcellulose oleogel and methods for preparing. Google Patents
Martins AJ, Cerqueira MA, Cunha RL, Vicente AA (2017) Fortified beeswax oleogels: effect of β-carotene on gel structure and oxidative stability. Food Funct. https://doi.org/10.1039/c7fo00953d
Matheson AB, Koutsos V, Dalkas G, Euston S, Clegg P (2017) Microstructure of beta-sitosterol: gamma-oryzanol edible organogels. Langmuir 33(18):4537–4542. https://doi.org/10.1021/acs.langmuir.7b00040
Melzer E, Kreuter J, Daniels R (2003) Ethylcellulose: a new type of emulsion stabilizer. Eur J Pharm Biopharm 56(1):23–27. https://doi.org/10.1016/s0939-6411(03)00025-0
Mert B, Demirkesen I (2016) Reducing saturated fat with oleogel/shortening blends in a baked product. Food Chem 199:809–816. https://doi.org/10.1016/j.foodchem.2015.12.087
Moriano ME, Alamprese C (2017) Organogels as novel ingredients for low saturated fat ice creams. LWT Food Sci Technol 86:371–376. https://doi.org/10.1016/j.lwt.2017.07.034
Mukherjee B, Mahapatra S, Gupta R, Patra B, Tiwari A, Arora P (2005) A comparison between povidone-ethylcellulose and povidone-eudragit transdermal dexamethasone matrix patches based on in vitro skin permeation. Eur J Pharm Biopharm 59(3):475–483. https://doi.org/10.1016/j.ejpb.2004.09.009
Nikiforidis CV, Scholten E (2015) Polymer organogelation with chitin and chitin nanocrystals. RSC Adv 5(47):37789–37799. https://doi.org/10.1039/c5ra06451a
O'Sullivan CM, Barbut S, Marangoni AG (2016) Edible oleogels for the oral delivery of lipid soluble molecules: composition and structural design considerations. Trends Food Sci Technol 57:59–73. https://doi.org/10.1016/j.tifs.2016.08.018
Okuda Y, Irisawa Y, Okimoto K, Osawa T, Yamashita S (2012) Further improvement of orally disintegrating tablets using micronized ethylcellulose. Int J Pharm 423(2):351–359. https://doi.org/10.1016/j.ijpharm.2011.10.050
Panagiotopoulou E, Moschakis T, Katsanidis E (2016) Sunflower oil organogels and organogel-in-water emulsions (part II): implementation in frankfurter sausages. LWT Food Sci Technol 73:351–356. https://doi.org/10.1016/j.lwt.2016.06.006
Patel AR (2015) Alternative routes to oil structuring. Springer International Publishing
Patel AR, Dewettinck K (2015) Comparative evaluation of structured oil systems: shellac oleogel, HPMC oleogel, and HIPE gel. Eur J Lipid Sci Technol 117(11):1772–1781. https://doi.org/10.1002/ejlt.201400553
Patel AR, Dewettinck K (2016) Edible oil structuring: an overview and recent updates. Food Funct 7(1):20–29. https://doi.org/10.1039/c5fo01006c
Patel AR, Rajarethinem PS, Cludts N, Lewille B, De Vos WH, Lesaffer A, Dewettinck K (2015a) Biopolymer-based structuring of liquid oil into soft solids and oleogels using water-continuous emulsions as templates. Langmuir 31(7):2065–2073. https://doi.org/10.1021/la502829u
Patel AR, Dumlu P, Vermeir L, Lewille B, Lesaffer A, Dewettinck K (2015b) Rheological characterization of gel-in-oil-in-gel type structured emulsions. Food Hydrocoll 46:84–92. https://doi.org/10.1016/j.foodhyd.2014.12.029

Patel AR, Rodriguez Y, Lesaffer A, Dewettinck K (2014) High internal phase emulsion gels (HIPE-gels) prepared using food-grade components. RSC Adv 4(35):18136–18140. https://doi.org/10.1039/c4ra02119c

Patel AR, Schatteman D, Lesaffer A, Dewettinck K (2013) A foam-templated approach for fabricating organogels using a water-soluble polymer. RSC Adv 3(45):22900–22903. https://doi.org/10.1039/c3ra44763d

Pernetti M, Vanmalssen K, Kalnin D, Floter E (2007) Structuring edible oil with lecithin and sorbitan tri-stearate. Food Hydrocoll 21(5–6):855–861. https://doi.org/10.1016/j.foodhyd.2006.10.023

Rogers MA, Strober T, Bot A, Toro-Vazquez JF, Stortz T, Marangoni AG (2014) Edible oleogels in molecular gastronomy. Int J Gastron Food Sci 2(1):22–31. https://doi.org/10.1016/j.ijgfs.2014.05.001

Romoscanu AI, Mezzenga R (2006) Emulsion-templated fully reversible protein-in-oil gels. Langmuir 22(18):7812–7818. https://doi.org/10.1021/la060878p

Sánchez R, Stringari GB, Franco JM, Valencia C, Gallegos C (2011) Use of chitin, chitosan and acylated derivatives as thickener agents of vegetable oils for bio-lubricant applications. Carbohydr Polym 85(3):705–714. https://doi.org/10.1016/j.carbpol.2011.03.049

Singh A, Auzanneau FI, Rogers MA (2017) Advances in edible oleogel technologies—a decade in review. Food Res Int 97:307–317. https://doi.org/10.1016/j.foodres.2017.04.022

Stortz TA, De Moura DC, Laredo T, Marangoni AG (2014) Molecular interactions of ethylcellulose with sucrose particles. RSC Adv 4(98):55048–55061. https://doi.org/10.1039/c4ra12010h

Stortz TA, Marangoni AG (2011) Heat resistant chocolate. Trends Food Sci Technol 22(5):201–214. https://doi.org/10.1016/j.tifs.2011.02.001

Stortz TA, Marangoni AG (2013) Ethylcellulose solvent substitution method of preparing heat resistant chocolate. Food Res Int 51(2):797–803. https://doi.org/10.1016/j.foodres.2013.01.059

Suzuki M, Hanabusa K (2010) Polymer organogelators that make supramolecular organogels through physical cross-linking and self-assembly. Chem Soc Rev 39(2):455–463. https://doi.org/10.1039/b910604a

Wang C-S, Natale G, Virgilio N, Heuzey M-C (2016a) Synergistic gelation of gelatin B with xanthan gum. Food Hydrocoll 60:374–383. https://doi.org/10.1016/j.foodhyd.2016.03.043

Wang FC, Peyronel F, Marangoni AG (2016b) Phase diagram of glycerol monostearate and sodium stearoyl lactylate. Cryst Growth Des 16(1):297–306. https://doi.org/10.1021/acs.cgd.5b01241

Wang J, Caceres M, Li S, Deratani A (2017) Synthesis and self-assembly of amphiphilic block copolymers from biobased hydroxypropyl methyl cellulose and poly(l-lactide). Macromol Chem Phys 218(10):1600558. https://doi.org/10.1002/macp.201600558

Yamada T, Onishi H, Machida Y (2001) Sustained release ketoprofen microparticles with ethylcellulose and carboxymethylethylcellulose. J Control Release 75(3):271–282. https://doi.org/10.1016/s0168-3659(01)00399-6

Yan C, Pochan DJ (2010) Rheological properties of peptide-based hydrogels for biomedical and other applications. Chem Soc Rev 39(9):3528–3540. https://doi.org/10.1039/b919449p

Yang Y, Yu J, Xu H, Sun B (2017) Porous lightweight composites reinforced with fibrous structures. Springer, Berlin/Heidelberg. https://doi.org/10.1007/978-3-662-53804-3

Yu H, Ke S, Liu D, Huang Q (2012) Development of a food-grade organogel with high bioaccessibility and loading of curcuminoids. Food Chem 131(1):48–54. https://doi.org/10.1016/j.foodchem.2011.08.027

Zetzl AK, Gravelle AJ, Kurylowicz M, Dutcher J, Barbut S, Marangoni AG (2014) Microstructure of ethylcellulose oleogels and its relationship to mechanical properties. Food Struct 2(1–2):27–40. https://doi.org/10.1016/j.foostr.2014.07.002

Zulim Botega DC, Marangoni AG, Smith AK, Goff HD (2013) The potential application of rice bran wax oleogel to replace solid fat and enhance unsaturated fat content in ice cream. J Food Sci 78(9):C1334–C1339. https://doi.org/10.1111/1750-3841.12175

Chapter 23
Current Applications in Food Preservation Based on Marine Biopolymers

Mohamed E. I. Badawy and Entsar I. Rabea

Abstract Marine biopolymers, including polysaccharides such as alginate, carrageenan, chitin, chitosan and gelatin, are biocompatible, biodegradable and non-toxic to mammals and are widely used in a variety of industrial applications. In food, these biopolymers perform a number of functions including gelling and thickening aqueous solutions, as well as stabilizing foams, emulsions and dispersions, inhibiting ice and sugar crystal formation, preventing spoilage and control the release of additive materials. These food biopolymers play an important role in food structure, food functional properties, food processing and shelf life. They are generally hydrophilic due to the large number of hydroxyl groups, which confer high affinity for binding water molecules, so that they can be dispersed in water in the colloidal state. In this chapter, we provide recent collaborative studies of the application of some important biopolymers in food preservation. In addition, the chapter provides the latest technological applications and prospects of these products in food applications. It provides a better understanding of the food systems, improve food qualities, and make better use of food macromolecules.

Keywords Biopolymers · Food preservation · Technological applications

23.1 Introduction

Marine biopolymers are polymers produced from the living organisms in the marines and oceans such as microorganisms, insects, crustaceous and other marine environment (Weiner 1997). Some of biopolymers can directly replace synthetic polymers in traditional applications (Álvarez et al. 2017). The characteristics of these effective materials such as compatibility with biological system,

M. E. I. Badawy (✉)
Department of Pesticide Chemistry and Technology, Faculty of Agriculture, Alexandria University, Alexandria, Egypt

E. I. Rabea
Department of Plant Protection, Faculty of Agriculture, Damanhour University, Damanhour, Egypt

T. J. Gutiérrez (ed.), *Polymers for Food Applications*,
https://doi.org/10.1007/978-3-319-94625-2_23

Table 23.1 List of some marine biopolymers and their sources

Biopolymers	Source
Phycocolloids (agar, alginates and carrageenans)	Seaweeds
Chitin and chitosan	Shells of marine animals
Gelatin	Animal skins and bones and fish scales

biodegradability, non-toxicity to mammals, and potential antimicrobial activity make these polymers widely used in food and biomedical applications (Shimizu and Kamiya 1983). The isolation, characterization, and application of marine biomaterials show a significant role in developing the marine biotechnology industries (Scheuer 2013). A recent advance in technologies helps to isolate bioactive materials from the marine source in an effective manner. Recently, marine biopolymers have acquired a strong market position, which attracts various marine researchers and consumers as well (Hou et al. 2016; Nollet 2016; Centella et al. 2017). The main objective of this chapter is to highlight the current trends in marine biopolymers, particularly with the major implications in food preservation. The literature and information in this chapter promises to the concept of marine-derived biopolymers. In this context, the chapter chiefly deals with the latest approaches of new applications and new products related to marine biopolymers that safely used in food preservation.

23.2 Marine Biopolymers: Structure, Classification, Source and Characterizations

Marine biopolymers are polysaccharides, which are polymeric carbohydrate molecules composed of long chains of monosaccharide units bound composed by glycosidic bonds (Thakur and Thakur 2016; Ferreira et al. 2016). Their structure is heterogeneous and varies from linear to highly branched molecules. Depending on the structure, these macromolecules may have distinct properties from their monosaccharide building blocks and may be amorphous or even water-insoluble (Varki et al. 1999).

Marine biopolymers, such as agar, alginate, carrageenan, chitin, chitosan, collagen, fucoidan and gelatin, are highly abundant but underexploited renewable biomasses. In addition to their natural biological and structural functions, biopolymers can be adapted to new biomaterials with new functionalities. Control of the molecular structure of the biopolymers is fundamental to understanding their functional properties and developing targeted functions. The marine biopolymers can be grouped according to their method of formation as addition and condensation polymers. Most biopolymers are condensation polymers, which are formed because of monomer units linking to form a small molecule (usually water) as a by-product.

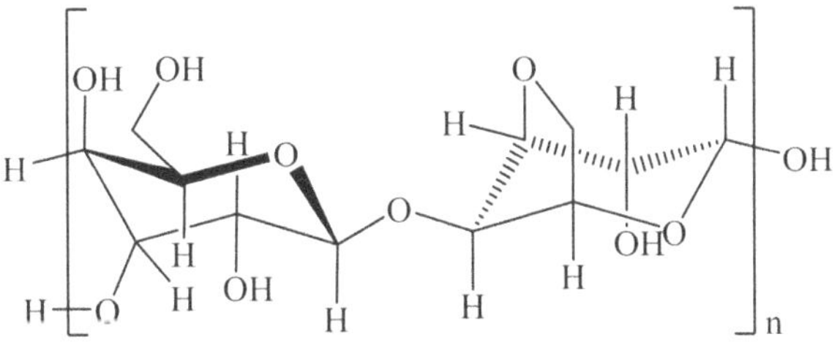

Fig. 23.1 The chemical structure of an agar or agarose polymer [(1→3)-β-D-galactopyranosyl-(1→4)-3,6-anhydro-α-L-galactopyranose units]

The additional biopolymers are those formed by direct combination of the monomer units constituting the polymer without any by-products. The biopolymers can be grouped into six main classifications with respect to their sources: proteins, polysaccharide polynucleotides, polyisoprenes, polyesters and lignin (Olatunji 2016). In Table 23.1, a list of polysaccharides from various sources is provided.

23.2.1 Agar

Agar is the phycocolloid (polysaccharides of high molecular weight from marine algae) of most ancient origin and derived from the polysaccharide agarose, which forms the support structure in the cell walls of certain algal species and which is released at the boil. These algae are known as agarophytes and belong to the phylum of Rhodophytes (red algae) (Balfour 1871). The main commercial genera of agarophytes are *Gelidium*, *Pterocladiella*, *Gelidiella*, and *Gracilaria*. Agar has also been found in species of *Ahnfeltia*, *Acanthopeltis*, *Campylaephora*, *Ceramium*, *Gracilariopsis* and *Phyllophora* (Pereira et al. 2013). In fact, agar is the subsequent combination of two components of polysaccharides: the linear agarose and a heterogeneous mixture of smaller molecules called agaropectin whose basic monomer is galactose (Fig. 23.1) (Phillips and Williams 2009). Generally, agarose is the predominant fraction of agar (50–90%) (Araki 1956) and is also responsible for its gelling properties (Nussinovitch and Gershon 1997). These polysaccharides can be sulfated to varying degrees, but to a lesser degree than in carrageenan (linear sulfated polysaccharides). Therefore, agar contains galactose, 3,6-anhydro-galactose (Hands and Peat 1938; Percival et al. 1938) and inorganic sulfate bonded to the carbohydrate (Samec and Isajevič 1922). Although some variations may occur, depending on factors such as algal species and environmental conditions (Armisen and Galatas 2000). The ash content is lower than that of carrageenan, furcelleran (Danish agar) and others. A maximum ash content of 5.0% is acceptable for the agar, although normally maintained between 2.5 and 4.0%. About 90% of the agar produced in the world is for food applications. The origin of agar as a food ingredient was in Asia, where it was expended for numerous centuries (Pereira 2011). Its surprising qualities as a thickening, stabilizing and gelling agent it make an essential ingredient for the preparation of processed foods. It has been classified as commonly documented as safe by the US Food and Drug Administration (FDA), which has set maximum use levels for specific applications.

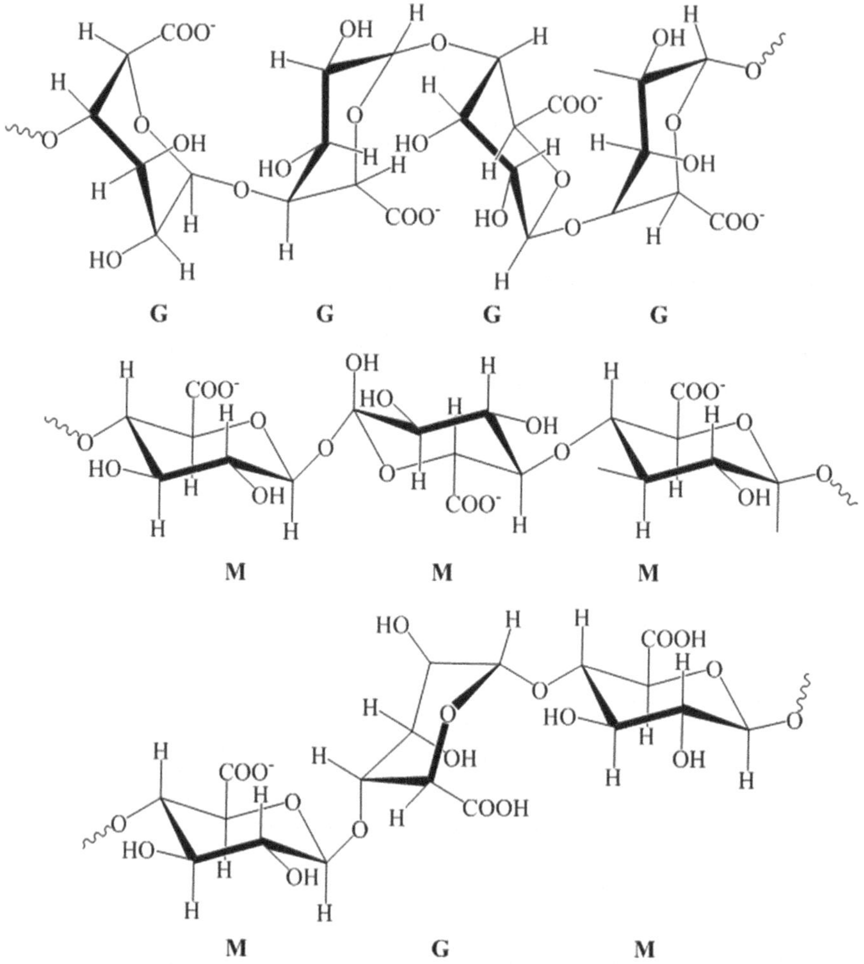

Fig. 23.2 Chemical structures of alginate [(1→4)-ß-D-mannuronate (Block M) and (1→4)-α-L-guluronate (Block G) units]

23.2.2 Alginate

Alginate or alginic acid is another marine biopolymer, naturally present in all species of brown algae and some bacteria; it exists within the cell walls providing flexibility and strength. Alginate is a long linear chain polymer, composed of (1→4)-linked-ß-D-mannuronate (Block M) unit and (1→4)-linked-α-L-guluronate (Block G) unit with different proportions and sequential arrangements (Clark and Green 1936; Lee and Mooney 2012; Venkatesan et al. 2015). The chemical structure is composed of repeated G residues (GGGGGG), M residues (MMMMMM) and alternating M and G residues (GMGMGM) (Fig. 23.2). It has been in use as food as far back as 600 B.C. However, it was not until 1896 that the purified form of alginate was extracted from seaweed by Krefting method (Krefting 1903). In 1929, alginate became a commercial product, with Kelco company being the first to market it as a stabilizer in ice cream (Sabra and Deckwer 2005). It is usually associated with other cations, such as sodium and calcium, which form sodium alginate and calcium alginate, respectively. The cations bound to the alginate affect its properties. The properties of alginate also depend on the type of algae, which depends mainly on the species of algae, which is mainly *Ascophyllum nodosum, Laminaria hyperborean, Laminaria digitata* and *Macrocystis pyrifera*. Bacteria of the species Pseudomonas and Azotobacter also produce alginate-like polymeric materials (Sabra and Deckwer 2005). Alginic acid serves various biological functions and industrial applications as a stabilizing agent, drug carrier, viscosifying agent and binding agent (Ogaji et al. 2012; Sabra and Deckwer 2005). It is also used in combination with other polymers such as chitosan and hyaluronic acid to serve more varied functions (Williams et al. 2017; George and Abraham 2006; Paul and Sharma 2004). When the alginate reacts with divalent or polyvalent cations, such as chitosan, three-dimensional structures are created in the form of an "egg-box" (Bajpai and Sharma 2004; Paula et al. 2006; Badawy et al. 2017b).

The molecular weight of commercially available alginates ranges from 32,000 to 400,000 g/mol. The parameters of the Mark-Houwink relationship ($[\eta] = KM_v^a$) for sodium alginate in 0.1 M NaCl solution at 25 °C are $K = 2 \times 10^{-3}$ and $a = 0.97$, where $[\eta]$ is intrinsic viscosity (mL/g) and M_v is the viscosity-average molecular weight (g/mol) (Rinaudo 1992). The viscosity of the alginate solutions increases as the pH decreases and reaches a maximum at pH 3.0–3.5 when the carboxylate groups of the alginate molecule become protonated and form hydrogen bonds. Increasing the molecular weight of the alginate can improve the physical properties of the resulting gels. However, an alginate solution formed from a high molecular weight polymer becomes very viscous, which is often undesirable in the treatment (Lee and Mooney 2012).

23.2.3 Carrageenan

Carrageenans, as hydrophilic colloids, are a family of linear sulfated polysaccharide galactans that are extracted from certain species of red algae (Rhodophyta). They are the third most important hydrocolloids in the food industry, after gelatin and starch (van de Velde and De Ruiter 2005). The industrial extraction of carrageenan started in 1930 in New England, from red seaweed species *Chondrus crispus* and *Mastocarpus stellatus*, for the preparation of chocolate milk. The greatest universally used marketable carrageenans are extracted from *Eucheuma denticulatum* and *Kappaphycus alvarezii* (McHugh 2003). Mostly wild-harvested genera such as *Chondrus*, *Chondracanthus*, *Furcellaria*, *Gigartina*, *Iridaea*, *Mazzaella*, *Mastocarpus*, *Sarcothalia* and *Tichocarpus* are also mainly grown as carrageenan raw materials. The annual world production of carrageenans is 45,000 tons, of which 70% is used in the food industry and the rest in the cosmetics and pharmaceutical industry (Campo et al. 2009). They are extensively used in the food industry,

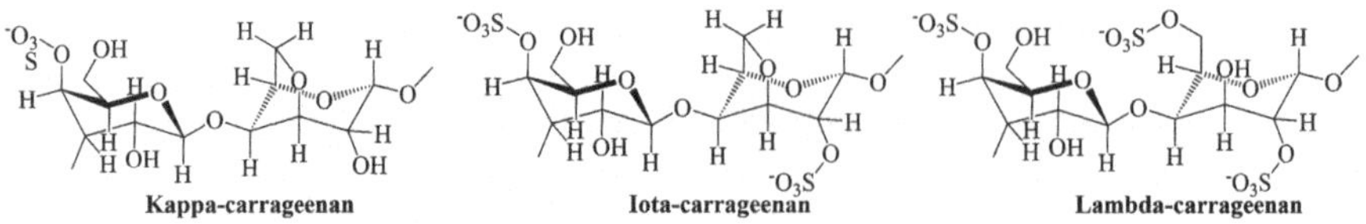

Fig. 23.3 Chemical structures of carrageenans (Kappa (κ)-carrageenan, Iota (ι)-carrageenan and Lambda (λ)-carrageenan)

for their gelling, thickening and stabilizing properties. However, the main application is in dairy and meat products, because of their strong binding to dietary proteins.

Carrageenan has a repeating disaccharides unit of ß-D-galactose and α-D-galactose or α-D-3,6-anhydrogalactose. There are three main types of carrageenans depending on the degree of substitution (DS) occurs on their free hydroxyl groups. Substitutions are usually either the addition of ester sulfate or the presence of the residues of 3,6-anhydride (Campo et al. 2009; Nanaki et al. 2010). Kappa (κ)-carrageenan has one sulphate group per disaccharide, Iota (ι)-carrageenan has two, and Lambda (λ)-carrageenan has three, whose chemical names according to International Union of Pure and Applied Chemistry (IUPAC) and to the letter code are 2,4′-disulfate (G4S-DA2S), carrageenose 4′-sulfate (G4S-DA), and carrageenose 2,6,2′-trisulfate (G2S-D2S,6S), respectively (Van de Velde et al. 2002; Knutsen et al. 1994) (Fig. 23.3). In addition to D-galactose and 3,6-anhydro-D-galactose as the main sugar residues and sulphate as the main substituent, other carbohydrate residues typically exist in carrageenans such as glucose, uronic acids and xylose (Prajapati et al. 2014). Accordingly, the structure of the different types of carrageenans is defined by the number and position of the sulfate groups, the presence of 3,6-anhydro-D-galactose and the conformation of the pyranose ring (Pereira et al. 2013). The carrageenans undergo a conformational transition from the random coils to the ordered double helix structure under appropriate conditions (decreasing temperature or increasing salt concentration). The branched polymer chains could then develop into a macromolecular network by crosslinking the double helixes, thus forming a gel or aggregates. Carrageenans have the characteristic solubility of hydrophilic colloids: they are soluble in water and insoluble in most organic solvents (Therkelsen 1993). Alcohols and ketones, although miscible with water, do not represent carrageenan solvents; however, they can be tolerated in a mixture of carrageenan solutions greater than 40%.

23.2.4 Chitin

Chitin is a naturally abundant polysaccharide and the supporting material of crustaceans, insects and fungi (Gutiérrez 2017). It is consists of β-(1-4)-2-acetamido-2-deoxy-β-D-glucose (GlcNAc) units through a β-(1→4) linkage

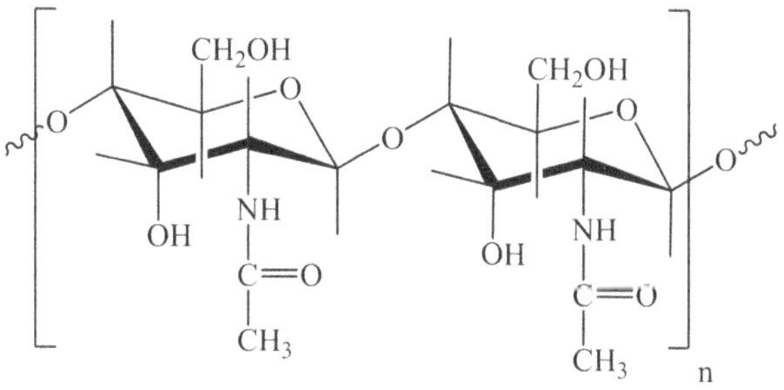

Fig. 23.4 Chemical structure of chitin (GlcNAc units)

(Fig. 23.4). The name chitin is derived from Greek, meaning "tunic" or "envelope" and it was first discovered in mushrooms in 1811 by Henri Braconnot (Braconnot 1811). In 1823, Odier found the same material in insects and plants and named it chitin (Odier 1823). The concept of chitin was more well-known by Lassaigne, who established the presence of nitrogen in its structure during the purification of the elytra beetles and *Bombyx mori* exuviae and then treated the residues with potassium in hot so obtaining potassium cyanide which has confirmed the presence of nitrogen in chitin (Lassaigne 1843). In 1859, Rouget treated chitin with strong alkali, which resulted in a substance that could, unlike chitin itself, be dissolved in acidic aqueous solutions that named as chitosan (Rouget 1859).

23.2.5 Chitosan

Chitosan is a linear biopolymer consists of higher than 70% of β-(1-4)-2-deoxy-β-D-glucopyranose (GlcN) and lower than 30% of (GlcNAc) units linked by β-1,4-glucosidic bonds (Fig. 23.5). It can be obtained through a deacetylation process of purified chitin (Dutta et al. 2004; Badawy et al. 2016b). Chitosan is of commercial interest due to its high percentage of nitrogen (~6.0–7.0%) compared to synthetically substituted cellulose (1.25%) that makes it as a good chelating agent for metal ions (Muzzarelli 1973). It has unique characteristics such as biocompatibility, biodegradability, low toxicity to mammals and possesses reactive functional groups. These important characteristics make chitosan useful in different fields of applications related to (1) agriculture (antimicrobial, seed-coating, postharvest and controlled release of agrochemicals) (Badawy and Rabea 2011; El Hadrami et al. 2010; Kong et al. 2010), (2) food industry and nutrition (anticholesterolemic dietary products and antimicrobial coatings) (Dutta et al. 2004; Badawy and Rabea 2016, 2017; Badawy et al. 2017a), (3) biotechnology (spinning, a dye-binder for textiles, strengthening additive in paper, enzyme and cell immobilization, protein separation, chromatography and cosmetics) (Liu et al. 2014; Mi et al. 2001; Kumar et al. 2014); (4) biomedicine (drug and gene delivery, blood coagulation, wound healing, bone regeneration, immunoadjuvant activity, pharmaceutics and ophthalmology) (Dodane and Vilivalam 1998; Krajewska 2004) and (5) combinations of chitosan

Fig. 23.5 Chemical structure of chitosan (GlcN and remaining of GlcNAc units). DA is a degree of acetylation

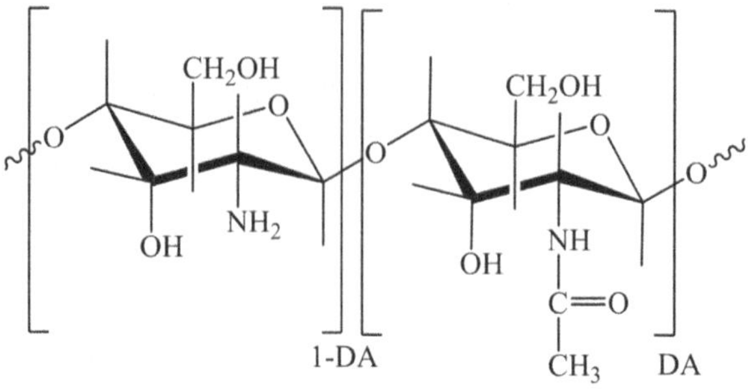

with other natural or synthetic polymers (grafting, polyelectrolyte complexation, blends, and coatings) (Honarkar and Barikani 2009). Chitosan, however, exhibits a limitation in its solubility and reactivity, therefore, many studies has been paid to modify its chemical structure (Badawy and Rabea 2011; Badawy et al. 2017b).

23.2.6 Gelatin

Gelatin is a translucent, lugubrious, weak (when dry) polymer, tasteless, obtained from collagen derived from various animal by-products. It contains proline, hydroxyproline, glycine, glutathione and catechin in its polypeptide chain (Fig. 23.6). Glycine is responsible for close packing of the chains, however, the presence of proline restricts the conformation (Street 2012). Gelatin is very attractive biopolymers due to its abundance, biodegradability, low cost and functional and excellent film-forming properties (Pereda et al. 2011). Nevertheless, they have low mechanical strength and dissolves relatively quickly in aqueous solution, thus limiting their practical applications.

Gelatin is of two types, type A which is the gelatin obtained by acid hydrolysis and type B obtained by basic hydrolysis (alkaline) (Mariod and Fadul 2013). The properties of gelatin depend on the source and method of extraction. For example, studies have shown that insect gelatin has different properties from commercial gelatin (Mariod and Fadul 2013, 2015). Studies also show that gelatin from different fish and their parts differs (Koli et al. 2012). The gelatin dissolves easily in hot water forming a gel and also soluble in most polar solvents. However, it does not dissolve well when added directly to cold water (Reinhard and Herbert 2007). Gelatin solutions show viscoelastic flow and continuous birefringence. Typically, the gelatin can be dispersed in a relatively concentrated acid. These dispersions are stable for 10–15 days with little or no chemical modification and are suitable for coating or for extrusion in a precipitation bath. The mechanical properties of gelatin gels are very sensitive to temperature variations and gels exist only over a small temperature range, typically less than 35 °C. The higher melting point is lower than the human

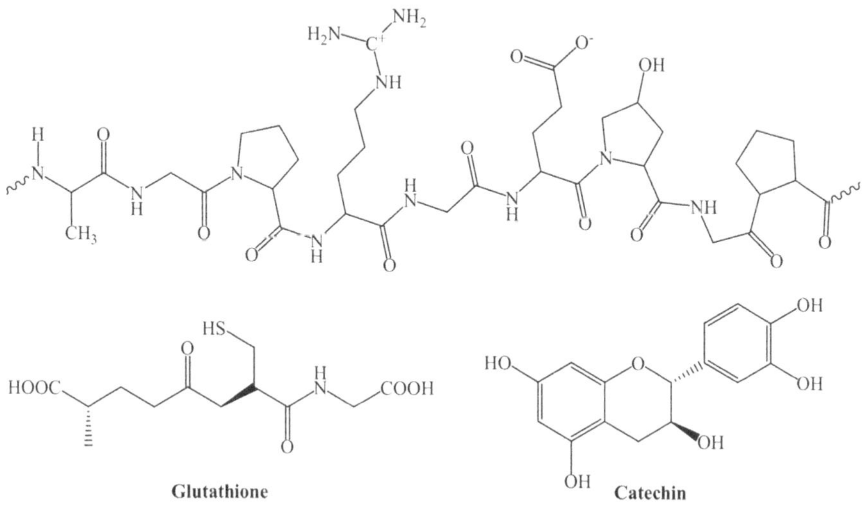

Fig. 23.6 Chemical structure of gelatin showing abundant amino acids glycine, proline and hydroxy proline in sphere and cylindrical model (-Gly-Pro-Hyp-), glutathione and catechin

body temperature, important factor for the mouth feel of foods produced with gelatin (Cole 2000). The viscosity of the gelatin solution is highest when the concentration is high and the mixture is kept cool at 4 °C. The strength of the gel is quantified using the Bloom test.

23.3 Marine Biopolymers-Based Packaging Materials

Marine biopolymer-based packaging is defined as packaging that contains raw materials originating from marine sources. The marketing of ecologically approachable packaging that uses such biodegradable materials has the greatest potential in countries where landfill is the core waste management tool. In this concept marine biopolymer-based packaging materials include both edible films and coatings, along with primary and secondary packaging materials in food preservation (Etxabide et al. 2017). Edible coatings are a thin layer of edible material that forms a shielding film or coating layer on the surface of fruits, vegetables and fresh food products to prolong their shelf life (Fig. 23.7). It is simple, convenient, economical, environment friendly and relatively inexpensive technology. It is widely used to reduce the deterioration of products and more acceptable by consumers to other conservative methods (Mohebbi et al. 2012).

A film of alginate gel with high salt concentration was used as an edible subsector. Better heating and shorter cooking times were demonstrated (Albert et al. 2012). Edible films have also been prepared in the form of emulsifiable alginate with novel barrier possessions and respectable mechanical properties which also protect encap-

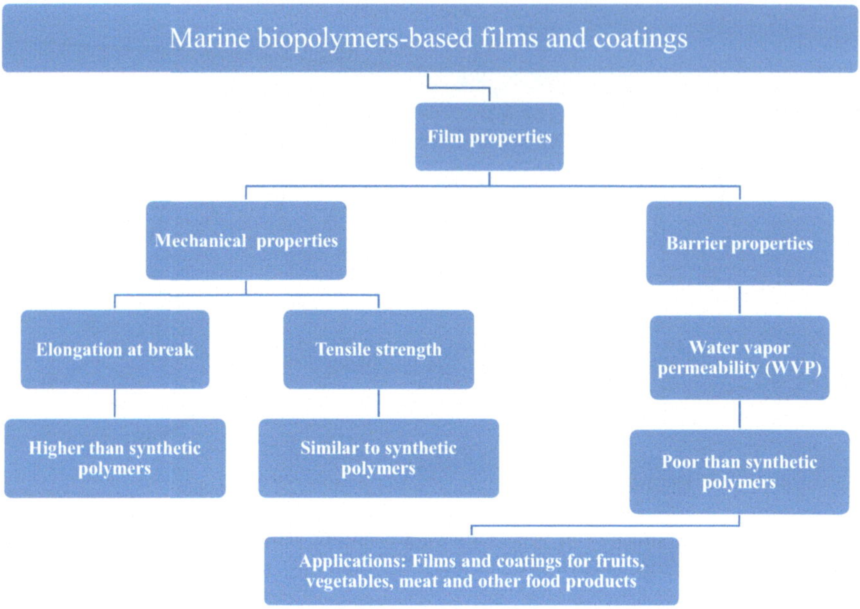

Fig. 23.7 Characterizations of marine biopolymers-based films and coatings as packaging materials

sulated active components (Hambleton et al. 2009). Vegetables and fruits preservation was also favored by incorporation of stable silver nanoparticles into alginate film (Mohammed Fayaz et al. 2009). Over the last decades, it has become increasingly interesting to develop and use bioactive chitosan films and coatings for food packaging, because of its particular physicochemical properties, its short-term biodegradability, its biocompatibility with human tissue and its antimicrobial activities (No et al. 2002; Dutta et al. 2009; Kong et al. 2010; Gómez-Estaca et al. 2010). Chitosan films were prepared with many organic acids and aromatic oils and showed high inhibitory measures against the growth of indigenous (lactic acid bacteria and Enterobacteriaceae) or inoculated bacteria (*Lactobacillus sakei* and *Serratia liquefaciens*) on the surfaces of meat products (Quintavalla and Vicini 2002). It was observed that the action of the bio-based films studied did not affect the growth of lactic acid bacteria, however, the growing of Enterobacteriaceae and *S. liquefaciens* was delayed or totally inhibited after storage for 21 days at 4 °C. The strongest inhibition was observed on surfaces with lower water activity values (bologna) on which acid release was slower and with films containing cinnamaldehyde, because of its superior antimicrobial action under these environments. Dutta et al. (2009) prepared a chitosan-starch film using a microwave treatment and proved that this product can be used in food packaging technology. Chitosan antimicrobial activity varies considerably with the type of chitosan; particularly the DS, molecular weight, the target organism and the conditions of the medium in which it is applied; in particular the pH, the ionic strength and the presence of solutes capable of reacting with chitosan by electrostatic interaction and/or covalent bonding (Rabea et al. 2003; Badawy and Rabea 2011). Many research studies proved that the chitosan with a molecular weight of less than 10 kDa has higher antimicrobial activity than high molecular weight. Moreover, the antimicrobial activity of chitosan and chitosan films has been shown to increase by decreasing the pH value. This effect can be considered as a synergistic action for the reasons of the obstacle effect of acid stress on bacterial cells (Kong et al. 2010). The antimicrobial action of chitosan-based edible coating solutions and chitosan-tapioca starch mixtures with and without the addition of potassium sorbate was investigated (Vásconez et al. 2009). Authors showed that the addition of chitosan reduced water vapor permeability (WVP) and solubility of starch films. One of the most perspective active biofilms is that based on chitosan, which is associated with different materials such as plant and animal proteins, polysaccharides and antimicrobial peptides such as nisin and divergicin, a novel bacteriocin from *Carnobacterium divergens* (Tahiri et al. 2004, 2009).

Generally, marine biopolymers forms coatings by breaking down interactions among long-chain polymer segments, thereby forming new intermolecular hydrogen and hydrophilic bonds upon solvent evaporation to generate a coating matrix. High molecular weight polymers contain large chain polymeric structures, which are required for creating polymer matrices having the suitable cohesive strength that provides mechanical strength to the coating (Moncayo et al. 2013).

23.3.1 Active and Intelligent Packaging of Foods

Among all living organisms, marine biopolymers derived from various resources such as algae (green, brown and red algae), crustaceans, and microorganisms forms one of the major components and considered as the greatest abundant source of products used in multiple food packaging applications because of their availability, eco-friendly connotation and singular properties (Manivasagan and Oh 2016; Bhatia 2016). These products extending the shelf life of foodstuffs include the application of many approaches such as temperature control, humidity control, addition of chemicals (salt, sugar, carbon dioxide or natural acids), elimination of oxygen, or a combination thereof with an effective packaging. Recently, the active packaging system is used to delay or eradicate the microbial, enzymatic and oxidative alterations, minimize contamination, weight loss and ensure the stability of the color and integrity of the products during the storage period (Imran et al. 2010; Al-Naamani et al. 2016). Active packaging is a state-of-the-art packaging technology that permits the product and its environment to interact to outspread the shelf life of the product and/or maintain its microbial safety while preserving the quality of the packaged food (Ahvenainen 2003; Imran et al. 2010; Realini and Marcos 2014; Bracone et al. 2016; Gutiérrez et al. 2016, 2018). It protects food from infection or degradation by providing a barrier to external environments while networking with the internal environment to control the atmosphere in the packaging. This technology is based on the theory of integrating certain constituents into packaging systems

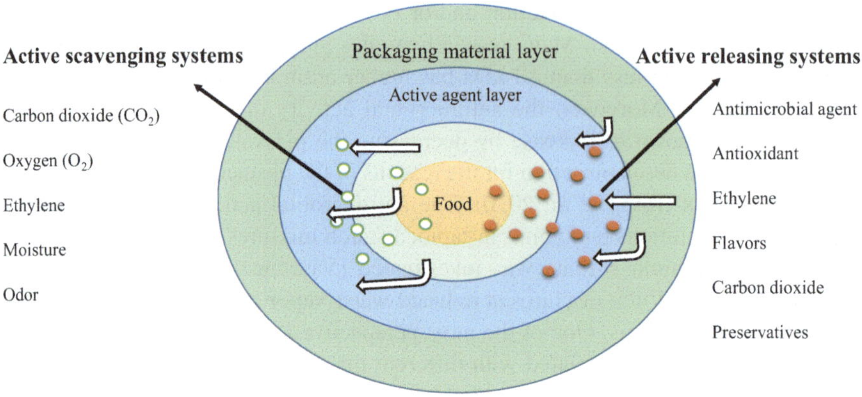

Fig. 23.8 Active and intelligent packaging of foods based on marine biopolymers

that release or absorb ingredients from or into the packaged food or into the surrounding environment to extend shelf life and maintain quality, safety and sensory properties (Camo et al. 2008; Vermeiren et al. 1999).

The progress of a whole range of active packaging systems is quite new (Kerry et al. 2006). Most important functions of active and intelligent packaging technologies include moisture control, oxygen permeability and scavenging, carbon dioxide (CO_2) controllers, odor controllers, flavor enhancement, ethylene removal and antimicrobial functions (Fig. 23.8) (Brody 2009; Restuccia et al. 2010; Kerry et al. 2006; Dainelli et al. 2008; Pereira de Abreu et al. 2012).

The existence of oxygen in food packages can trigger many food deterioration reactions. Oxygen can cause off-taste progress (rancidity due to lipid oxidation), color changes (pigment oxidation), nutrient loss (vitamin C oxidation), and microbial development (Brody et al. 2001). It also has a significant impact on the respiration rate and ethylene production of respiring foods such as fruits and vegetables. Oxygen scavenging is an effective way to prevent the growth of aerobic bacteria and mold in dairy and baked goods. Oxygen concentrations of 0.01% or less in the headspace are required for this purpose. There are three main modes of action for free radical scavenging activity by CH_2. The hydroxyl groups in the polysaccharide unit of marine biopolymers can react with radicals by; (1) the H-abstraction reaction, (2) the active hydrogen ions in free amino groups can react with radicals to form stable macromolecule radicals and (3) the amino groups can form ammonium (NH^{3+}) groups by absorbing hydrion from the solution. Then reacting with radicals through addition reaction (Xie et al. 2001). The use of oxygen scavengers, which absorb residual O_2 after packaging, may minimize the quality changes undergone by foods. Oxygen scavengers are reducing agents, i.e. substances able to react with oxygen and thus reduce its concentration within the package. These substances can be blended or dispersed in the polymeric materials rendering a smart polymer technology, namely oxygen scavenging films. One active packaging approach that is applicable to meat systems is the use of oxygen scavengers in the form of either oxygen absorbing materials in small sachets that are placed into the package or packaging films incorporated with oxygen scavenging materials (Cruz et al. 2012). Interest in healthier foods by today's consumers has led to the production by meat technologists of naturally preserved meat products. Edible coatings with natural, antimicrobial and antioxidant characteristics have become increasingly popular in the area of meat technology due to their effectiveness on the surface of the product for a long time, resulting in the delay or inhibition of chemical and microbial deteriorations (Devlieghere et al. 2004a). Chitosan coating was successfully used in reducing lipid oxidation (traditional dry fermented sausage) (Krkić et al. 2013), in extending the shelf-life of fresh shrimp (Aşik and Candoğan 2014), and ready-to-cook meat products (chicken balls, chicken seekh kababs, and mutton seekh kababs) (Kanatt et al. 2013), and in controlling *Salmonella* on fresh chicken breasts (Olaimat and Holley 2015). Wu et al. (2000) reported that wrapping with chitosan, wheat gluten or soy protein film was not effective in controlling lipid oxidation of precooked beef patties. These authors explained that a higher oxygen permeability might contribute to the higher thiobarbituric acid-reactive substances from these treatments. After being applied on patties and during the 3 days of storage, the wrappings absorbed moisture from meat samples and swelled. The plasticizing action of water molecules may have changed the barrier properties of the edible films, and resulted in higher oxygen permeability or even loss of film integrity. The composition of the chitosan film may also have donated to its higher oxygen permeability (Caner et al. 1998).

Another active antimicrobial packaging solution depends on CO_2 generators. The preservation of meat and poultry requires increased levels of CO_2 (10–80%) as they can effectively reduce the microbial growth of the surface and thus prolong the shelf life of the product (Vermeiren et al. 1999; Kerry et al. 2006). In addition, ethylene (C_2H_4) is a natural growth-stimulating plant hormone that accelerates ripening and senescence by increasing the rate of respiration of produce, thereby decreasing its shelf life. Thus, elimination of ethylene from the package headspace is of high importance. Removal of ethylene can be done by potassium permanganate ($KMnO_4$) imbedded in silica or other inert substrates at a level of 5–6%. The use of ethylene scavengers and absorbers is often combined with modified atmosphere in order to achieve maximal shelf-life extension. The combination of heat treatment plus chitosan coating of the fruits displayed the lowest respiratory rate, malondialdehyde levels, membrane leakage, ethylene evolution, and the highest strength and consumer acceptance among the treatments (Shao et al. 2012).

Aroma scavengers/absorbers are designed to eliminate undesirable flavors, aromas and odors present in the package headspace. The formation of these off-flavors and off-odors in food is mainly due to the oxidation of fats and oils, which leads to the formation of aldehydes and ketones or the breakdown of proteins from fish muscles to amines (Siro 2012; Akelah 2013). Although these malodors can be removed from the package headspace by using active aroma scavengers, these systems may mask or absorb off-flavors and off-odors that are indicative of spoilage, and therefore their application may pose a health risk to consumers. Packaging resources could similarly generate undesirable odors that must be minimized for consumer acceptance (Bonilla et al. 2012).

23.3.2 Active Releasers/Emitters

Release-type active packaging systems are designed to release one or more active agents or ingredients into the food product or into the package headspace at a desired time and at a specific rate (Madene et al. 2006; Siro 2012; Ahvenainen 2003). The aim of controlled-release packaging is to maintain effective levels of active compounds over a period by continual replenishment of inhibitory substances. There have been some approaches to ensure longer protection by merely decreasing the release rate of active compounds from the packaging. One of the most diffuse techniques is represented by the use of multilayer films, which include an outer barrier layer, a matrix layer containing the active agent and a control layer (Buonocore et al. 2005). Depending on the physical form of active packaging systems, absorbers and releasers can be a sachet, label or film type. As oxidation reactions and microbial activity are the main processes limiting shelf life, antioxidative and antimicrobial packaging systems are the most relevant representatives of such active packaging mechanism (Ahvenainen 2003). The others that are least important are carbon dioxide emitters and flavor releasers. The main types of active releasers/emitters and some controlled release issues are discussed in detail in the following topic of the active antimicrobial packaging based on marine biopolymers.

23.3.3 Active Antimicrobial Packaging

One favorable type of active packaging is the combination of antimicrobial ingredients in food packaging resources to regulate unwanted development of microorganisms on food surface (Han 2005; Silberbauer and Schmid 2017). Antimicrobial packaging is excellent technology that can prolong shelf life and increase food safety for synthetic polymers and edible films (Appendini and Hotchkiss 2002; Han 2003). Films or materials having antimicrobial properties can be classified into two types. In the first type, an active substance emits or migrates into the headspace of the package or on the food surface. In this type, the system need not be in direct contact with the food, but in the second case, it must be in contact. The second type are those which are effective against microbial growth without emitting or migrating the active agents into the headspace of the package or food. In this case, the material must be in direct contact with the food (Han 2003). Since both concepts require direct contact between the product and the packaging, vacuum-packed products can benefit from this technique. However, when volatile antimicrobials are used, no direct contact is needed thus; such mechanisms can be applied, for example, to the packaging of fruits and vegetables.

An antimicrobial film may be made by various techniques, including incorporation of the antimicrobial ingredients in a sachet associated with the package from which the bioactive constituent is released during further storage. Direct incorporation of the antimicrobial into the packaging film when it applied in a hot extrusion material, thermo-resistance and shearing resistance of the antimicrobial must be considered, coating of the packaging with a material that acts as a carrier for the additive. The material will not be introduced to high temperatures or shear forces; moreover, it can be applied as a subsequent step and antimicrobial molecules with film forming properties (Cooksey 2005; Falguera et al. 2011; Ramos et al. 2017). Antimicrobial food packaging acts to reduce or inhibit the growth of microorganisms that maybe present in the packed food or packaging material itself.

23.3.3.1 Chitosan Films and Coatings

Chitosan has been widely studied as an antimicrobial agent in food packaging because of its ability to form a suitable film or coating instead of non-biodegradable and non-renewable polymers (Park et al. 2005; Wang and Gao 2013; Gol et al. 2013; Devlieghere et al. 2004b; No et al. 2007; Elsabee and Abdou 2013; Dutta et al. 2009; Fernandez-Saiz et al. 2009). Chitosan coating, soaking or spraying of fresh, frozen and manufactured foods can improve the shelf life of perishable foods by semi-permeable surface formation and thus delaying breathing, reducing moisture and weight loss, and maintain the overall quality (Pereda et al. 2011; Zhang et al. 2011). Chitosan films have the advantage over other bio-based food packaging resources that they can contain functional substances such as minerals or vitamins and can serve as carriers for the delivery of antimicrobial agents (Avila-Sosa et al. 2012; Acevedo-Fani et al. 2015).

Edible chitosan coatings consisting solely of chitosan or a mixture of chitosan with other biopolymers such as sodium caseinate, gelatin and starch. The sodium caseinate/chitosan films inhibited the growth of bacteria and yeast and can be possibly applied to several food matrices (Moreira et al. 2011). Since the primary function of food packaging and coatings is often to avoid or reduce moisture transmission between the food and the surrounding atmosphere, or between two components of a heterogeneous food product, the water vapor transmission rate (ETV) should be as low as possible (Gontard et al. 1993). The efficacy of chitosan-coated plastic films including five commonly documented as safe antimicrobials (nisin, sodium lactate, sodium diacetate, potassium sorbate and sodium benzoate) against *Listeria monocytogenes* on cold-smoked salmon was evaluated. Chitosan-coated plastic films with 4.5 mg/cm^2 sodium lactate, 4.5 mg/cm^2 sodium lactate, 0.6 mg/cm^2 potassium sorbate and 2.3 mg/cm^2 sodium lactate −500 IU/cm^2 nisin were the most effective (Ye et al. 2008). In addition, the acetic acid chitosan coating was more effective in reducing *L. monocytogenes* counts on the surface of ready-to-eat roast beef than the lactic acid chitosan coating (Beverlya et al. 2008). Most foods are a mixture of different compounds, such as carbohydrates, proteins, fats, minerals, vitamins, salts and others, and many of them can interact with the chitosan resulting in loss or enhancement of antimicrobial activity. For example, Devlieghere et al. (2004b) extensively studied the influence of different food components include starch, protein, oil, and NaCl on the antimicrobial effect of chitosan. They reported that starch, whey protein and sodium chloride have negative effects on the antimicrobial action of chitosan, but oil has no influence. Chitosan coatings designed to slow the release of antimicrobials to the surface, so a smaller amount of antimicrobials would be needed. As a result, many studies have confirmed that chitosan coatings can be used as a vehicle for integrating functional components, such as antimicrobials or nutraceutical compounds that could enhance the effects of chitosan coatings in food preservation (Vu et al. 2011; Perdones et al. 2012; Shen and Kamdem 2015). Application of chitosan (0.5 and 1%) added singly or in combination with nitrites (150 mg/L) on the sausage surface by immersion protected the fresh pork sausages from microbial decay and extended the shelf life (Bostan and Mahan 2011). Lysozyme-chitosan composite films were developed for enhancing the antimicrobial properties of chitosan films. The 60% lysozyme incorporation films increased the inhibitory potency of chitosan films against both *Streptococcus faecalis* and *E. coli*, achieving 3.8 log cycles reduction in *S. faecalis* and 2.7 log cycles reduction in *E. coli* (Park et al. 2004b). Antibacterial activity of chitosan—guar gum films against *E. coli* and *S. aureus* was tested. The results showed that chitosan films were effective against both the organisms. Addition of guar gum to chitosan film at 15% (v/v) led to additional reduction in log CFU/ml of *E. coli* (Rao et al. 2010). Rosemary essential oil was used to develop an active film from chitosan. These films showed more antibacterial activity and total phenol content and can be used as active film in food preservation (Abdollahi et al. 2012a).

The antimicrobial activity of chitosan is influenced by several intrinsic and extrinsic factors, related to the species of the target organism, chemical properties (density of positive charges, molecular weight, concentration, hydrophilic/hydrophobic characteristics and chelating potential), water solubility and environmental factors (ionic strength of the medium, pH, temperature and pathogen exposure time pathogen) (Kong et al. 2010; Badawy and Rabea 2011; Rabea et al. 2003). Recently, nano-chitosan has been shown to break up well in biopolymers, having the potential to be used as antimicrobial agents in edible food packaging structures and decomposition. Several studies have concentrated on antimicrobial properties of nanostructures that combine chitosan and other antimicrobials. Chitosan nanoparticle-based can be efficiently used in the food industry because they offer different benefits, counting good edibility, biocompatibility with human tissues, aesthetically attractive appearance, barrier properties alongside pathogenic microorganisms, environmentally friendly and made of cheap material. Rhim et al. (2006, 2013) have established that a nanocomposite film of chitosan prepared with organic clay (Cloisite 30B) has a strong antimicrobial action against food poisoning bacteria, including Gram-positive, recommended that the antimicrobial activity was attributable to the quaternary ammonium groups in the modified films. Hong and Rhim (2008) evidenced that the same clay of Cloisite 30B has a durable antibacterial action to Gram-positive and Gram-negative bacteria, caused by the quaternary ammonium groups. These authors reported that the retarded biodegradability of this nanocomposite which previously observed by Lee et al. (2002) was responsible to the antimicrobial action in addition to the quaternary ammonium groups. Therefore, the biodegradability of bio-nanocomposites can be used for the expansion of biodegradable nanocomposite packaging materials with controlled biodegradation function (Souza and Fernando 2016).

Edible marine biopolymer coatings can improve the quality of fresh, frozen and processed meat and poultry products by delaying moisture loss, reducing lipid oxidation and discoloration, improving the appearance of the product, and improving the quality of the product. The shelf life of spicy beef treated with chitosan was remarkably improved by decreasing the bacterial cell counts and preventing lipid oxidation throughout storage for 10 days at 4 °C (Youn et al. 2004). Abdallah et al. (2017) applied chitosan as an edible coating for one of the best popular traditional dried meat products (pastirma) in order to improve its sensory, physicochemical and microbiological characteristics. The results revealed improvement of sensory attributes, lower moisture loss, and antioxidant effect with lower shear strength values in the chitosan coated samples. Approximately one log CFU/g reduction in aerobic plate count was recorded in chitosan coated samples, while psychrotrophic, anaerobic, yeast and mold counts were below detectable levels of direct plating (2-log CFU/g). The chitosan monomethyl fumaric acid derivative considerably reduced the total number of viable bacteria, enterobacteria, lactic bacteria, psychrotrophic bacteria and mold yeasts on beef during refrigeration storage, extending the shelf life of beef about 8 days (Khan et al. 2017).

23.3.3.2 Gelatin Films and Coatings

Gelatin is known for its excellent ability to form a film, and its use for coating or packaging can preserve food quality during storage due to its oxygen and light barrier, dehydration prevention and lipid oxidation (Hoque et al. 2010; Jongjareonrak et al. 2008; Ahmad et al. 2012; Shankar et al. 2016). Some properties of the gelatin film including mechanical properties, permeability, light absorption, transparency, antimicrobial activity and antioxidant ability, are affected by the addition of active substances (Tongnuanchan et al. 2012, 2014). Because gelatin is a protein, it can be completely degraded by protease and components are recycled to the environment. In addition, gelatin can be used as fully biodegradable and environmentally friendly food packaging films in combination with other biopolymers or nanofillers. Pereda et al. (2011) indicated that both *E. coli* and *Listeria monocytogenes* showed high sensitivity to all films that formed gelatin-chitosan-based solutions, however, the results obtained with edible films showed that only *E. coli* was sensitive to gelatin-chitosan composite. Nowzari et al. (2013) prepared bilayer and gelatin composite and chitosan coating and film to study their effect on the quality of rainbow trout. The coating was found to be better than the film in reducing the lipid oxidation of fillets.

The incorporation of nano metal or metal oxides, such as silver, titanium dioxide (TiO_2), copper oxide (CuO) and zinc oxide (ZnO), as reinforcing filler into gelatin-based nanocomposite films has emerged due to their high thermal stability, strong antimicrobial activity, antioxidant activity and UV-screening properties. For example, gelatin-based nanocomposite films blended with nanosilver (AgNPs), nanocopper (CuNPs), zinc oxide nanoparticles (ZnO NPs), and TiO_2 were prepared and showed reflective antibacterial action against both Gram-positive and Gram-negative foodborne pathogenic bacteria (Shankar et al. 2015, 2016). Active nanocomposite of AgNPs films have shown strong antibacterial activity against pathogens in food (*E. coli* and *L. monocytogenes*) and are said to have high potential as an active food packaging system for maintaining food safety and prolonging the shelf life of packaged food (Kanmani and Rhim 2014b). Nafchi et al. (2014) prepared gelatin/ZnO-nanorod nanocomposite film by incorporating various concentrations of ZnO-nanorod (0.01, 0.02, 0.03, and 0.05 g/g dried gelatin) into bovine gelatin matrices. They reported that the inhibition zones of the nano-incorporated films against *S. aureus* increased significantly with an increase in ZnO-nanorod content. A bovine gelatin/Cu(II)-exchanged montmorillonite nanocomposite film showed *in vitro* antibacterial activity against *E. coli* O157:H7 (Gram-negative) and *L. monocytogenes* (Gram-positive) as by the agar-disc diffusion assay was shown (Martucci and Ruseckaite 2017). The film showed antibacterial action against both pathogens that were tested under the same conditions and showed a stronger effect on *L. monocytes* than *E. coli*, as the cell wall of the latter differed significantly and this difference in susceptibility and responsiveness to active films. The authors reported that the incorporation of low levels as a vehicle for copper ions into the gelatin matrix is a good way to develop functional materials that can potentially be used in the design of food contact items. Some of the nanocomposite systems offer additional functional properties such as antioxidant and antimicrobial functions. Specially enriched with antimicrobial properties that expand the scope of the films in the food packaging industry. They reduce the risk of outbreaks of food pathogens, improving food safety and quality and extending the food shelf life (Llorens et al. 2012).

Moreover, several natural antimicrobial agents, especially plant essential oils have been added to the gelatin films to provide their preventive role in food technology (Gómez-Estaca et al. 2009b; Martucci et al. 2015; Atarés and Chiralt 2016). Gelatin-chitosan-based edible films incorporated with clove essential oil were elaborated and their antimicrobial activity tested against *E. coli*, *Lactobacillus acidophilus*, *L. innocua*, *Photobacterium phosphoreum*, *Pseudomonas fluorescens* and *Shewanella putrefaciens* (Gómez-Estaca et al. 2010). When the gelatin-chitosan film including clove oil was applied on the fish during refrigerated storage, the growth of microorganisms was significantly reduced in Gram-negative bacteria, especially enterobacteria, while the lactic acid bacteria remained virtually constant for many storage periods. The process of inserting oryganum oil into gelatin-catfish coating showed greater antimicrobial activity against *Salmonella typhimurium* compared to *E. coli* during storage at 4 and 10 °C for 12 days (Min and Oh 2009). The antimicrobial action of edible *Gelidium corneum*-gelatin (GCG) mixture films comprising grapefruit seed extract (GFSE) or green tea extract (GTE) against *E. coli* O157:H7 and *L. monocytogenes* improved with increasing concentration, resulting in a decrease in the populations of bacteria by 0.77–2.08 and 0.91–3.30 log CFU/g, respectively (Hong et al. 2009). The samples of pork loins packed with the GCG film containing GFSE (0.08%) or GTE (2.80%) had a decrease in the populations of *E. coli* O157:H7 and *L. monocytogenes* of 0.69–1.11 and 1.05–1.14 log CFU/g, respectively, compared to the control after 4 days of storage. Gelatin in the form of an enriched coating with cinnamon oil is suitable for keeping new rainbow trout slices and efficiently maintaining quality attributes to an acceptable level during storage (Andevari and Rezaei 2011). Films incorporated with lemongrass oils in various concentrations exhibited inhibitory activity in a concentration-dependent manner against *E. coli*, *L. monocytogenes*, *S. aureus* and *S. typhimurium*, but the bergamot oil film inhibited only *L. monocytogenes* and *S. aureus* (Ahmad et al. 2012). Composite films based on fish protein isolate and fish skin gelatin blend film which combined with 100% leaf essential oil, especially in combination with ZnO nanoparticles, showed strong antibacterial activity against pathogenic and perishable bacteria in food and thus could be used as an active food packaging material to ensure safety and to extending the shelf-life of packaged food (Arfat et al. 2014). Kavoosi et al. (2014) prepared 10% (w/v) gelatin film incorporated with Zataria multiflora essential oil that exhibited profound antioxidant and antimicrobial activity against both Gram-positive and Gram-negative bacteria. Silver carp skin gelatin-chitosan films combined with several concentrations of oregano essential oil showed the best antimicrobial action among oregano, cinnamon and anise essential oils against *Bacillus subtilis*, *B. enteritidis*, *E. coli*, *S. aureus*, and *Shiga bacillus* (Wu et al. 2014). In addition, gelatin films containing cinnamon essential oil nanoliposomes were prepared by thin film ultrasonic dispersion technique to enhance the antimicrobial stability (Wu et al. 2015). Further, fish skin gelatin films were prepared to which various concentrations of cinnamon essential oil were added and the films showed good inhibitory effects against bacteria of *E. coli* and *S. aureus* and fungi of *Aspergillus niger*, *Paecilomyces variotii* and *Rhizopus oryzae*, and their antifungal activity appeared to be more effective as the resistance to bacterial growth (Wu et al. 2017).

23.3.3.3 Alginate Films and Coatings

Alginate is an attractive film formation due to non-toxicity, biodegradability, compatibility with vital life, and low price (Kim et al. 2011; Yang et al. 2011). The antibacterial effect of alginate-based edible films has been studied by incorporating garlic oil up to 0.4% (v/v) against some pathogenic bacteria, suggesting their use for the preservation of various foods (Pranoto et al. 2005). The presence of 0.1% garlic oil in the nutrient broth reduced the live cell counts for *E. coli*, *S. typhimurium*, *S. aureus* and *B. cereus* by 2.28, 1.24, 4.31 and 5.61 log cycles, respectively 24 h incubation. The edible film showed high antibacterial activity against *S. aureus* and *B. cereus* as compared with *E. coli* and *S. typhimurium* using an agar diffusion test (Pranoto et al. 2005). Zactiti and Kieckbusch (2006) investigated potassium sorbate permeability behavior in sodium alginate films crosslinked at various Ca^{2+} concentrations using a diffusion cell. In addition, alginate films containing 1% (w/v) *Spanish oregano*, *Chinese cinnamon* or savory essential oils were added in 2% (w/v) or 20% (w/v) $CaCl_2$ solution dipped and then applied to bovine muscle slices to control the growth of *E. coli* O157: H7 and *S. typhimurium* on the beef surfaces

(Oussalah et al. 2006). After 5 days of storage, films containing oregano or cinnamon essential oils were most effective against *S. typhimurium*, regardless of the type of pretreatment used (2 or 20% $CaCl_2$). In the same period, meat inoculated with *E. coli* O157: H7 and coated with films treated with 2% $CaCl_2$ had significantly less bacteria when using oregano-based films than when cinnamon-based films were used than when cinnamon- and savory-based films were used. Edible films from a mixture of alginate and partially hydrolyzed sago starch containing lemongrass oil (0.1–0.4%, v/w) and glycerol (20%, w/w) were designed for antimicrobial activity (Maizura et al. 2007). The film containing essential oil was effective against *E. coli* O157:H7 at all levels of concentrations. Solutions of different potassium sorbate concentrations (150–1050 mg/L) in contact with the films increased the permeability constant of the preservative, reflecting adaptations in the polymeric film structure. In addition, the efficiency of alginate-based coating containing lactate and diacetate to control the growth of *L. monocytogenes* and to promote microbiological safety of smoked salmon slices and fillets has been well-known. It has been shown that alginate coating is the most effective carrier of various antimicrobial treatments in inhibiting *L. monocytogenes* growth (Neetoo et al. 2010). Alginate coatings improved with sodium lactate and potassium sorbate postponed the growth of *L. monocytogenes* with counts achieving 4.3 log CFU/g (home-style poached turkey) and 6.5 log CFU/g (roasted deli turkey), respectively (Juck et al. 2010). While the counts in their untreated equivalents were considerably higher reaching 9.9 and 7.9 log CFU/g, respectively. Alginate films combined with oregano essential oil were more active against Gram-positive bacteria (*Staphylococcus* and *L. monocytogenes*) than Gram-negative bacteria (*E. coli* and *S. enteritidis*) and a minimum concentration of 1.0% film was required to confirm their antibacterial effectiveness (Benavides et al. 2012). Bierhalz et al. (2012) prepared single and composite films based on alginate and pectin with natamycin as active ingredient and the release behavior in water and the diffusion coefficients were also evaluated. The result proved that the individual alginate films showed more suitable packaging properties than the single-pectin and composite films. The efficacy of a developed multilayered alginate-based antimicrobial edible coating in the freshly cut watermelon without compromising its quality properties was evaluated (Sipahi et al. 2013). A set of solutions enclosing sodium alginate (0.5, 1.2 g/100 g), beta-cyclodextrin and microencapsulated trans-cinnamaldehyde, pectin and calcium lactate were used as coating systems and prepared using the layer-by-layer technique. Both the 1 and 2 g/100 g alginate coatings prolonged the shelf life of freshly cut watermelons from 7 to 15 days. Application of the 1 g/100 g alginate multilayer edible coating will maintain the quality and sensory acceptability of freshly cut watermelon while extending shelf life (Sipahi et al. 2013). Edible coatings based on sodium alginate, galbanum oleo-resin gum and the composite of alginate and galbanum encompassing *Ziziphora* essential oil were *in vitro* evaluated and showed promising effect on the quality and shelf life of chicken fillet during cold storage (Hamedi et al. 2017). Moreover, nanocomposite alginate films were also prepared for packaging and preservation of foods (Huq et al. 2012; Alboofetileh et al. 2014; Rhim et al. 2006; Rhim and Ng 2007).

23.3.4 Carrageenan Films and Coatings

Carrageenans are water-soluble galactose polymers extracted from red algae that are commonly used in the food industry as gelling and stabilizing agents, and have been used for the formation of cohesive and transparent films (Choi et al. 2005; Alves et al. 2010). Seol et al. (2009) reported that the j-carrageenan-based film containing ovotransferrin combined with ethylene diamine tetra acetic acid (EDTA) displayed a slight antimicrobial action against *E. coli, S. typhimurium, S. aureus*, and *Candida albicans* but the effect was improved in the presence of 5 mM EDTA. Antimicrobial films based on carrageenan were prepared by combining grape seed extract in varying concentrations into the polymer using a solvent casting method. The films appeared yellowish in color due to the polyphenolic compounds in the grape extract and showed pronounced activity against food-borne bacteria. These results suggest that the carrageenan films have high potential application as antimicrobial or active food packaging (Kanmani and Rhim 2014a). In addition, κ-carrageenan and Na-alginate-based antimicrobial films were prepared alone and in combination with lysozyme, nisin, grape fruit seed extract and EDTA as antimicrobial agents (Cha et al. 2002). The biofilms exhibited good inhibitory zones and both Na-alginate- and κ-carrageenan-based films containing grape fruit seed extract-EDTA displayed inhibitory effect against all indicator microorganisms tested. Shojaee-Aliabadi revealed that the *Zataria multiflora Boiss* (ZEO) and *Mentha pulegium* (MEO) essential oils have good possibilities for inclusion in κ-carrageenan to make antioxidant and antimicrobial films for food applications (Shojaee-Aliabadi et al. 2014). Antimicrobial activities of the films prepared by incorporating essential oils were enhanced and *S. aureus* was found the most sensitive, followed by *B. cereus* and *E. coli*. The highest inhibition area of 544.05 mm^2 for *S. aureus* was observed around the films that were combined with 3% ZEO. The total inhibitory region of 3% MEO films was 20.43 for *S. typhimurium* and 10.15 mm^2 for *P. aeruginosa*. These results revealed that ZEO and MEO have good potential to be integrated into κ-carrageenan to make antioxidant and antimicrobial films in food technology (Shojaee-Aliabadi et al. 2014).

23.3.5 Active Antioxidative Packaging

Autoxidation in food and biological systems is responsible for a multitude of adverse effects and implications in human health as well as in food stability and preservation. Antioxidants play a major role in preventing autoxidation and have much attention as food stabilizers, dietary supplements and natural health products (Simic and Karel 2013; Ganiari et al. 2017). There has been a broad range of substances identified with antioxidant activity, which exert their oxidation-inhibitory effect via different mechanisms, with varied efficiencies, depending on their physical and chemical characteristics as well as the system environment involved

(Pokorny 2007). The antioxidants may act as free radical scavengers, singlet oxygen quenchers, inactivators of peroxides, metal ion chelators, quenchers of secondary oxidation products and inhibitors of pro-oxidative enzymes, among others (Ganiari et al. 2017).

23.3.5.1 Chitosan Films and Coatings

Edible film and coating technology based on marine biopolymers are close to active packaging technology and can also be a way to reduce oxidative deterioration of food. It can been noticed that chitosan films have been studied most widely. Chitosan film and coating possesses excellent antioxidant activity that allow for free radical scavenging and widely used in food packaging and preservation (Xie et al. 2001; Schreiber et al. 2013; Wang et al. 2016). However, the antioxidant action mechanism of chitosan is still debatable. While some researchers showed considerable antioxidant activities, *in vitro*, for a native chitosan molecule (Park et al. 2004a; Yen et al. 2007, 2008; Badawy et al. 2016a), many studies have clearly indicated low antioxidant activity of native chitosan, although the activity significantly increased with appropriate chemical modifications (Schreiber et al. 2013; Casettari et al. 2012; Xie et al. 2001).

The antioxidant activity of chitosan has been used to preserve salmon during frozen storage (Sathivel et al. 2007). Chitosan forms a physical barrier around the fruit, and therefore the darkening is reduced. In addition, the positive charges present in the coating can stabilize anthocyanin pigments, which contributes to the maintenance of fruit color, sensory properties and antioxidant properties (Tezotto-Uliana et al. 2014). The antioxidant activity of chitosan and its derivatives has attracted the most attention because they exert strong antioxidant activity and their effects are similar to those of phenolic antioxidants (Park et al. 2004a; Yen et al. 2008; Ngo and Kim 2014; Liu et al. 2012). Among the various reactive oxygen species, the chemical activity of the hydroxyl radical •OH is strongest, which can simply react with biomolecules such as amino acids, proteins, and DNA (Cacciuttolo et al. 1993). Two types of fungal chitosan were prepared by alkaline *N*-deacetylation of raw chitin for different periods of 60, 90 and 120 min and their antioxidant properties were investigated (Yen et al. 2007). The results showed that the chitosan has antioxidant activities of 61.6–82.4% at 1 mg/mL and showed reducing powers of 0.42–0.57 for 1,1-diphenyl-2-picrylhydrazyl (DPPH) radicals at 10000 mg/L. In addition, no significant difference in the antioxidant properties between both types of chitosan has been observed and the EC_{50} values of scavenging abilities on hydroxyl radicals were below 100 mg/L whereas those of chelating abilities on ferrous ions were 580–690 mg/L. In another study by the same authors (Yen et al. 2008), who found that the chitosan prepared from crab chitin and deacetylated for 60, 90 and 120 min exhibited antioxidant activities of 58.3–70.2% at 1000 mg/L and showed reducing powers of 0.32–0.44 at 10,000 mg/L. At this concentration, the scavenging ability of chitosan C60 on DPPH radicals was 28.4% whereas those of other chitosans were 46.4–52.3%. All EC_{50} values of antioxidant activity were below 1500 mg/L. Prabu and Natarajan (2012) reported that the hydroxyl radicals scavenging activity of chitosan isolated from *Podophthalmus vigil* was found to be higher than the ascorbic acid and the range was from 23 to 93%. Superoxide anion radical was ranging from 5.21 to 65% for the concentration between 125 and 2000 mg/L. In another study by Kim and Thomas (2007), who found that three different MWs chitosans (30, 90, and 120 kDa) exhibited antioxidative activities in salmon (*Salmo salar*). Incorporation of 0.2, 0.5, and 1% chitosan, employing 2-thiobarbituric acid-reactive substances and 2,2-diphenyl-1picrylhydrazyl scavenging assays, caused reduction of lipid oxidation in salmon for 7 days of storage. The free radical-scavenging activity of the 0.2 mM DPPH solution was saturated by 30 kDa chitosan at 7000 mg/L, resulting in a strong antioxidant activity of approximately 85% (Kim and Thomas 2007).

Hydroxypropyl chitosan and carboxymethyl chitosan (CMC) were reacted with maleic acid sodium to form water-soluble derivatives and their scavenging activities against hydroxyl radical •OH were investigated by chemiluminescence technique (Xie et al. 2001). They exhibited IC_{50} values ranging from 246 to 498 mg/L, depend on the type of substituent and the content of hydroxyl and amino groups (Garcıa et al. 2015; Pasanphan et al. 2015). Chitosan was irradiated in acetic acid solution (1%) with different doses (2–20 kGy) of Co-60 γ rays to investigate the enhancement of the antioxidant activity (Feng et al. 2008). Radical mediated inhibition of lipid peroxidation, reducing power, superoxide anion and hydroxyl radical quenching assays have been used for the evaluation of the antioxidant activity of irradiated chitosan. The result indicated that the 2.1 kDa chitosan exhibited high reducing capacity and expressed good inhibition of linoleic acid peroxidation. At a concentration of 100 mg/L could trap 74.2% of the superoxide radical. *N*-CMC was studied in the prevention of warmed-over flavor (WOF) (Angelo and Vercellotti 1989) and the results reported that this product was highly effective in controlling of the WOF at different temperatures tested. It was also reported that the *N*-CMC at 5000 mg/L caused 93% inhibition of the thiobarbituric acid (TBA) in the ground beef and 99% reduction in the hexanal content in the products. These results were further confirmed by Li et al. (1996), who found that the addition of 3000 mg/L of *N*-CMC to cooked pork was sufficient to prevent the oxidative rancidity of the product. However, the *N,O*-CMC and its lactate, acetate and pyrrolidine carboxylate salts showed high effect in controlling the oxidation and flavor deterioration of cooked meat over a 9 day storage of cold storage with inhibition percentages of 46.7, 69.9, 43.4, and 66.3% for the four products, respectively as reflected in their TBA values (Shahidi 1996).

The incorporation of antioxidants into chitosan films as packaging materials has become common because oxidation is a key problem affecting food quality and could affect the features of the film itself and also the packaged food. Therefore, before adding antioxidants to the films and coatings, it is required to evaluate firstly their antioxidant capacity, but also how they influence the properties of the packaging material and the characteristics of the food product (Ganiari et al. 2017; Bonilla et al. 2013; Abdollahi et al. 2012b). In this context, several studies reported that the antioxidant activity of chitosan was enhanced by incorporation of other compounds such as sugars, essential oils and gelatin (Altiok et al. 2010). Now, the most commonly used antioxidants in the active packaging are butylated hydroxyanisole (BHA), butylated hydroxytoluene (BHT) and other natural antioxidants such as phenolic compounds and plant essential oils and extracts as alternatives to synthetic antioxidants (Siripatrawan and Harte 2010; Talón et al. 2017). For example, Siripatrawan and Harte (2010) combined aqueous green tea extract in concentrations of 2, 5, 10 and 20% (w/v) to a chitosan film as an active packaging material in food with promising antioxidant activity. Peng et al. (2013) reported that the addition of tea extracts significantly improved the DPPH radical scavenging activity of chitosan films in all food simulants (0, 20, 75 and 95% ethanol). The higher the green tea extracts and black tea extracts concentration, the faster the release of the tea antioxidants from the film. This study revealed that chitosan film could be obtained through active incorporation of tea extracts, which may provide new formulation options for the development of active antioxidant packaging. The inclusion of essential rosemary oil at concentrations of 0.5, 1.0 and 1.5% showed antibacterial activity, and the total phenol content of chitosan film could be used as an active ingredient in food preservation and packaging (Abdollahi et al. 2012a). As expected in this study, the total phenol content of the chitosan film improved considerably by the addition of rosemary essential oil, which was consistent with other reported results (Gómez-Estaca et al. 2010, 2017; Norajit et al. 2010). Martins et al. (2012) reported that the antioxidant capacity of chitosan-based films incorporated with α-tocopherol was improved and suggested to use it to improve quality and shelf-life of food products.

The packaging release antioxidant is a type of food preservation system, wherein an antioxidant or mixtures are incorporated in the package, which contributes to extending the shelf life of food. Some studies have indicated that the antioxidant mechanism of chitosan may be a chelating action of metal ions such as copper and iron found in enzyme active sites thus to inactivate the oxidizing enzymes and/or binding with lipid such enzymes (Badawy and Rabea 2009). In addition, chitosan films have been reported to inhibit both primary and secondary lipid oxidation in herring and Atlantic cod (Jeon et al. 2002). In all studies, it can be concluded that chitosan has a good antioxidant activity, hydroxyl radical scavenging ability, and chelating ability on ferrous ions and can be used as a source of antioxidants, as a possible supplement or ingredient in the pharmaceutical industry.

Generally, the antioxidant abilities of the films are relative to the concentration of the active compound in the film. The loss of activity during the formation and conditioning of the film is not remarkable in most cases (Akhtar et al. 2012; Leon and Rojas 2007; Pastor et al. 2013; Pereira de Abreu et al. 2012), although significant losses have been reported in the case where an extrusion process was followed at 170 °C or higher (Manzanarez-López et al. 2011; Soto-Valdez et al. 2011; Eça et al. 2014). However, few studies have evaluated the stability of the active compounds during storage of the enriched film (Jiménez et al. 2013; Leon and Rojas 2007).

23.3.5.2 Gelatin Films and Coatings

Gelatin film is widely used for the protection of food quality. Gelatin contains amino acids or biologically active peptides which may acts as electron donors that indicate antioxidant activity of gelatin composite edible films (Yang et al. 2008; Alemán et al. 2011; Gómez-Estaca et al. 2009a; Martucci et al. 2015; Jridi et al. 2014). Edible films can improve the preservation of food, principally because of their ability to act as barriers to water, prevent dehydration and to oxygen and light, reducing the oxidation of lipids (Baldwin et al. 2011; Dutta et al. 2009; Aider 2010; Cheng et al. 2015). To improve the antioxidant activity of gelatin films, natural tea polyphenol-loaded antioxidant chitosan nanoparticles (TPCN) were prepared and incorporated into a gelatin film (Bao et al. 2009). The antioxidant effect of TPCN against the oxidation of fish oil was studied, as well as the release of tea polyphenols from nanoparticles in the film. The peroxide value of the fish oil packaged with films incorporated with TPCN was lower than that of the fish oil packed with control films during the incubation period. The radical-scavenging activity of films incorporated with TPCN was higher than that of the control films, and an increasing activity of radical trapping of the former was observed during the storage period (Bao et al. 2009). The effect of ginger, turmeric and plai essential oils was studied in 25, 50 and 100% based on protein content, on the properties and antioxidant activity of the gelatin-based skin (Tongnuanchan et al. 2013). The authors reported that the films incorporated with plai and turmeric essential oils showed the higher DPPH and ABTS radical scavenging activity, respectively, compared with the control film and ginger essential oil added film. The potential of gelatin films incorporated with thymol as natural antioxidant and antimicrobial nano film were also demonstrated (Kavoosi et al. 2013). A series of active gelatin-based films containing natural antioxidants (green tea extract—GTE, proanthocyanidins—OPC, grape seed polyphenols—GSP, ginger extract—GE, gingko leaf extract—GBE) were used as food packaging (Li et al. 2014). The results suggested that addition of GBE made the gelatin-based film possess the highest DPPH radical scavenging activity at 1.0 mg/mL. Gelatin-based films added with *Origanum vulgare* and *Lavandula officinalis* essential oils were prepared by casting (2000–6000 mg/L) and exhibited the highest antimicrobial against *E. coli* and *S. aureus* and good antioxidant properties (Martucci et al. 2015). Active composite chitosan-gelatin films membranes with improvement the antioxidant action were organized by molding the composite polymers of 2% chitosan and gelatin (2 and 5%) with changed concentrations of glycerol and sorbitol (1, 2, 5, and 10%, w/w) as plasticizers (Badawy et al. 2016a). Of the natural plant extracts, rosemary and oregano extracts are rich in phenolic substances and effective when incorporated into edible or inedible films, although the various extraction procedures, the film matrix used to incorporate the antioxidants or the food itself somehow different results. Gómez-Estaca et al. (2007) examined the transfer of total phenolics from gelatin films enriched with oregano or rosemary extracts and cold-smoked sardines from the beginning of storage. In the case of oregano, phenols did not show a significant increase over time, while an increase in phenol was found in samples packed with rosemary-enriched films in the following days of storage. This proves that the antioxidant capacity of rosemary is gradually released. Comparison of thiobarbituric acid reactive (TBAR) assay results for ore and rosemary enriched film-coated samples revealed that rosemary extract lowered the oxidation rate more than Oregano, despite that rosemary had lower CPTs (Gómez-Estaca et al. 2007).

Recently, development and characterization of active films based on gelatin, gelatin-sodium caseinate (G-Cs) and gelatin-chitosan (G-Ch) blends, applying active compounds (α-tocopherol, garlic essential oil and cinnamaldehyde) nano-emulsified in water were examined (Córdoba and Sobral 2017). The data proved that the films based on the gelatin-chitosan blend containing active compounds showed the best antioxidant activity, highlighting its potential use as active packaging for shelf life extension of foodstuffs.

23.3.5.3 Alginate Films and Coatings

Composite films based on alginate and pectin containing natamycin as active agent (Bierhalz et al. 2012), starch-alginate coatings with incorporated or coated tocopherols (Wu et al. 2001), alginate/clay nanocomposite films enriched with essential oils (Alboofetileh et al. 2014), and different alginate-based coatings carrying antimicrobials or antioxidants (Raybaudi-Massilia et al. 2008) have been studied for food packaging applications. Norajit et al. (2010) revealed that the alginate biodegradable films incorporating white, red and extruded white ginseng extracts showed good potential to make antioxidant biodegradable film or coating for various food applications. Alginate film comprising extruded white ginseng extract at a temperature of 130 °C showed the highest (61.12%) radical scavenging activity in contradiction of DPPH in methanol, followed by a film containing white ginseng extruded at 115 °C.

23.3.5.4 Carrageenan Films and Coatings

In food industry, carrageenan is generally used because of its excellent physical and functional properties including antioxidant activity, stabilizing ability, gelling, emulsifying and thickening agent (Rhim 2012; Tavassoli-Kafrani et al. 2016; Khalil et al. 2017). They can also be used in water-based foods, meat products (as an oxygen barrier for delayed fat oxidation (Varela and Fiszman 2011). Importantly, it was reported that oligosaccharides carrageenan and its derivatives showed an antioxidant-related activity both *in vitro* and in the cell system (Yuan et al. 2006). Sun et al. (2015) also noted that the antioxidant activity of carrageenan oligosaccharides was significantly affected by the sugar content limit, corresponding to the removal of polymerization from kaapa carrageenan. However, the preparation of edible films using carrageenan is not abundant in literature because carrageenan is mostly used as a coating (Tavassoli-Kafrani et al. 2016).

The *in vitro* antioxidant activities of ι, κ and λ carrageenans were investigated (de Souza et al. 2007). Inhibition of superoxide radical formation, IC_{50} was for kappa, and carrageenans were 0.112, 0.332 and 0.046 mg/mL, respectively, and all samples had an inhibitory effect on the formation of hydroxyl radicals. The results of the peroxidation tests showed that κ, ι and λ carrageenans had an IC_{50} of 2.753 and 2.338 and 0.323 mg/mL, respectively. Some derivatives of oligosaccharides carrageenan have shown the highest antioxidant activity of sugars and oligosaccharides in some antioxidant systems (Yuan et al. 2005). The over-sulphated and acetylated derivatives, which scavenge superoxide radicals, the low degree of substitution phosphorylated and acetylated carrageenan derivatives, which scavenge hydroxyl radicals, and the phosphorylated derivatives, which scavenge DPPH radicals all showed significant antioxidant activity in the systems under study. Further, Yuan et al. (2006) also investigated the antioxidant activity of κ-carrageenan oligosaccharides and their modified derivatives *in vitro* systems, such as reducing power, iron ion chelation, and total antioxidant activity using β-carotene-linoleic acid system. Ferric reducing antioxidant action of carrageenans and their inhibitory properties on hydroxyl radicals and superoxide anion radicals were also investigated *in vitro* (Sokolova et al. 2011). Fresh pork sausage patties containing carrageenan, without or with soy protein and an antioxidant were packaged with or without vacuum (Ho et al. 1995). In fat-control products, rosemary extract as an antioxidant in combination with vacuum packs providing optimal protection against rancidity and reduced-fat products maintained acceptable levels (TBA-reactive substances and sensory properties) during frozen storage for 16 weeks (Ho et al. 1995). Shojaee-Aliabadi et al. (2013) characterized the biodegradable composite κ-carrageenan films incorporated with *Satureja hortensis* (SEO) for potentially use in packaging a wide range of food products, particularly those that are highly oxidative and microbial sensitive. The authors pointed out that the carrageenan films containing the SEO have antioxidant properties. This effect was significantly improved when the SEO ratio was added 3%. Furthermore, antioxidant activity of κ-carrageenan-based films containing different concentrations of *Zataria multiflora Boiss* (ZEO) and *Mentha pulegium* (MEO) were evaluated as essential oils (Shojaee-Aliabadi et al. 2014). ZEO affected the antioxidant activities of the films more clearly than MEO, e.g. B. ZEO-containing films showed DPPH radical scavengers of 80.6%, twofold higher than those with MEO.

Chitosan, starch, cellulose, chitin, alginate/carrageenan composites and blends have been broadly used in biomedical applications of drug delivery, tissue engineering and many other applications (Derkach et al. 2015; Olaimat and Holley 2015). Homogeneous carrageenan-starch films with antioxidant extracts of Cuban red propolis and yerba mate were prepared by casting technique (Chang-Bravo et al. 2014). After 6 months of storage at 75% RH and 23 °C, the samples presented an increase in their DPPH scavenging activity. In addition, the release behavior of the phenolic compounds from the films in an aqueous medium was assayed finding significant differences between release rates of both extracts. Recently, the effects of κ-carrageenan on myofibrillar protein oxidation in peeled shrimp (*Litopenaeus vannamei*) were investigated during frozen storage by assessing the content of the carbonyls, total sulfhydryl, dityrosine, and surface hydrophobicity (Zhang et al. 2018). The obtained data revealed that *in vitro* carrageenan oligosaccharides exhibited good capacities of OH., O_2^-., and DPPH scavenging and Fe^{2+}-chelating activity. Chemical analysis showed that frozen storage improved the vulnerability of myofibrillar proteins to frozen oxidation. Commonly, the effect of antioxidant activities has played an important role in the effects of cryoprotective effects of carrageenan oligosaccharides on frozen shrimp.

23.3.6 Active Modified Atmosphere

Modified atmosphere packaging was rapidly grown in the food packaging market. It increases product quality, freshness and increases shelf life, provides convenience to the consumer and increases the value of the product. It allows fresh products or perishable products to be packaged when they are fresh, and then keep them in that condition, reducing distribution costs and improving taste and nutritional value. It also does not slow down the ongoing life processes by changing the product, but by adapting its environment (Jayas and Jeyamkondan 2002; Zagory and Kader 1988; Wilson et al. 2017). Nitrogen, oxygen, carbon dioxide and water vapor are the main gases used in modified atmosphere packaging. Nitrogen is an inert and tasteless gas, without any antimicrobial activity and it is primarily used to displace oxygen and prevent package collapse. Oxygen inhibits the growth of anaerobic microorganisms. It is responsible for several undesirable reactions in foods, including the oxidation of fats and oils, rapid ripening and senescence of fruits and vegetables, stale baked goods and color changes. Finally, carbon dioxide is soluble in both water and lipids. It has a bacteriostatic effect and slows down the respiration of many products (Rodriguez-Aguilera and Oliveira 2009). The packaging of food products in polymer films can produce modified atmospheres generated by commodities. In addition, the edible coatings can also provide protection for fresh products and give the same effect as modified atmosphere storage with respect to modifying internal gas composition (Park 1999; Zeuthen and Bøgh-Sørensen 2003; Raybaudi-Massilia et al. 2016). The modification of the atmosphere within such packages depends on the permeability of the film, the rate of respiration and the diffusion characteristics of the gas and the initial free volume and the atmospheric composition in the package. Temperature, relative humidity, and movement of air around the package can influence the permeability of the film. The temperature also affects the metabolic activity of the commodity and therefore the rate of attack of the desired modified atmospheres. All of these factors must be considered in developing a mathematical model to choose the most suitable film and coating for each product (Kader et al. 1989). For example, if a coating or film is too thick, harmful effects may result because the internal oxygen concentration is below a desirable and beneficial level and there is an increased carbon dioxide concentration that is greater than a critical tolerable level. These conditions lead to anaerobic fermentation (Zeuthen and Bøgh-Sørensen 2003). There are several possible edible coatings for food products, including gelatin, cellulose and chitosan, because they have the desirable characteristics of being generally odorless, tasteless and transparent. The biodegradable laminate of a chitosan-cellulose and polycaprolactone as a film was found suitable as a packaging material for storage of fresh produce (Makino and Hirata 1997). Temperature estimation of O_2, CO_2 and N_2 permeability parameters for biodegradable plates was studied. The efficacy of chitosan immobilized in edible coatings has been studied on the microbiological shelf life of the carrot sticks under two different passive atmospheric modification packaging conditions (Simões et al. 2009). The joint application of edible coatings containing chitosan and moderate O_2 and CO_2 levels maintain quality and promote phenolic content in carrot sticks. Campaniello et al. (2008) evaluated the possibility of prolonging the microbiological and physiochemical microorganisms of cut strawberries treated with 1% chitosan solution and packing them in passive and active atmosphere packaging. A chitosan coating inhibited the growth of microorganisms and affected significantly and positively the microbiological stability of the products. Besides, the presence of high percentage of oxygen, combined with a chitosan coating, seemed to affect positively the color. The addition of chitosan into cheese making, combined with an active coating (lysozyme and EDTA disodium salt) and modified atmosphere packaging (30% CO_2, 5% O_2 and 65% N_2), was used to prolong the microbiological shelf life of 'Fior di latte' cheese (Del Nobile et al. 2009). This combination represents a strategic solution to prolong the shelf life of cheese for more than 3 days, compared with the control sample in conventional saline that registers storage less than 1 day. Xing et al. (2010) investigated the effect of chitosan coating containing anti-browning agents and modified atmosphere packaging on browning and the shelf life of fresh lotus cuttings stored at 4 °C for 10 days. The atmosphere in the packages was evaluated for O_2 and CO_2 concentrations. At the end of storage, the edible coated and modified-atmosphere packaging samples modify the composition of the atmosphere and respiratory rate of the lotus root sections and have the lowest polyphenol oxidase and malondialdehyde content. Chitosan solutions (3%) incorporating 20% krill oil with or without the addition of 0.1 μL/mL cinnamon leaf essential oil, vacuum or modified atmosphere (60% CO_2 and 40% N_2) increased total lipid and omega-3 fatty acid contents of the lingcod (*Ophiodon elongates*) fillets by about twofold up to 21 days at 2 °C (Duan et al. 2010). Moreover, this treatment reduced lipid oxidation as represented in thiobarbituric acid reactive substance and microbiological spoilage as reported in total plate count (2.22–4.25 Log reductions during storage). Consumers preferred the overall quality of chitosan-coated, cooked lingcod samples to the control, based on their firm texture and less fishy aroma.

23.4 Concluding Remarks

Marine biopolymers have already found important applications in food preservation and other important areas, where cost is much less important than their function. It seems very unlikely to be biodegradable, edible and biologically compatible, and widely displaced from its current role in the application of food packaging, where cost is more important for the consumer market than environmental acceptability. Marine biopolymers have good film-forming properties along with desirable packaging properties such as low gas barrier and heat-sealing ability. However, there are some problems associated with films based on biopolymers, such as moisture sensitivity, rather than low mechanical strength and low process capacity. These problems have limited the industrial application of these films. Therefore, the recent development of the integration of other products or the use of composite nanomaterials has led to new opportunities for the use of films and coatings based on marine biopolymers in the food packaging industry. This type of packaging materials then requires more research and more added value such as introduction of smart and intelligent molecules (nanotechnology) capable of providing information about the characteristics of the food within the package (quality, shelf life and microbiological safety) and nutritional values. It is necessary to develop this type of materials to improve barrier properties, ensure the safety of food characteristics, integrate intelligent labeling and give the consumer the opportunity to obtain more detailed product information than the current technology.

Conflict of Interests The authors declare that there is no conflict of interests regarding the publication of this chapter.

References

Abdallah MR, Mohmaed MA, Mohamed HM, Emara MM (2017) Improving the sensory, physicochemical and microbiological quality of pastirma (a traditional dry cured meat product) using chitosan coating. LWT Food Sci Technol 86:247–253

Abdollahi M, Rezaei M, Farzi G (2012a) Improvement of active chitosan film properties with rosemary essential oil for food packaging. Int J Food Sci Technol 47(4):847–853

Abdollahi M, Rezaei M, Farzi G (2012b) A novel active bionanocomposite film incorporating rosemary essential oil and nanoclay into chitosan. J Food Eng 111(2):343–350

Acevedo-Fani A, Salvia-Trujillo L, Rojas-Graü MA, Martín-Belloso O (2015) Edible films from essential-oil-loaded nanoemulsions: physicochemical characterization and antimicrobial properties. Food Hydrocoll 47:168–177

Ahmad M, Benjakul S, Prodpran T, Agustini TW (2012) Physico-mechanical and antimicrobial properties of gelatin film from the skin of unicorn leatherjacket incorporated with essential oils. Food Hydrocoll 28(1):189–199

Ahvenainen R (2003) Active and intelligent packaging: An introduction. Novel Food Packag Tech:5–21

Aider M (2010) Chitosan application for active bio-based films production and potential in the food industry: review. LWT Food Sci Technol 43(6):837–842

Akelah A (2013) Polymers in food packaging and protection. In: Functionalized polymeric materials in agriculture and the food industry. Springer, pp 293–347

Akhtar MJ, Jacquot M, Jasniewski J, Jacquot C, Imran M, Jamshidian M, Paris C, Desobry S (2012) Antioxidant capacity and light-aging study of HPMC films functionalized with natural plant extract. Carbohydr Polym 89(4):1150–1158

Albert A, Salvador A, Fiszman S (2012) A film of alginate plus salt as an edible susceptor in microwaveable food. Food Hydrocoll 27(2):421–426

Alboofetileh M, Rezaei M, Hosseini H, Abdollahi M (2014) Antimicrobial activity of alginate/clay nanocomposite films enriched with essential oils against three common foodborne pathogens. Food Control 36(1):1–7

Alemán A, Giménez B, Montero P, Gómez-Guillén M (2011) Antioxidant activity of several marine skin gelatins. LWT Food Sci Technol 44(2):407–413

Al-Naamani L, Dobretsov S, Dutta J (2016) Chitosan-zinc oxide nanoparticle composite coating for active food packaging applications. Innov Food Sci Emerg Technol 38:231–237

Altiok D, Altiok E, Tihminlioglu F (2010) Physical, antibacterial and antioxidant properties of chitosan films incorporated with thyme oil for potential wound healing applications. J Mater Sci Mater Med 21(7):2227–2236

Álvarez K, Famá L, Gutiérrez TJ (2017) Physicochemical, antimicrobial and mechanical properties of thermoplastic materials based on biopolymers with application in the food industry. In: Masuelli M, Renard D (eds) Advances in physicochemical properties of biopolymers: part 1. Bentham Science Publishers, pp 358–400. https://doi.org/10.2174/97816810845341170100l5. ISBN: 978-1-68108-454-1. eISBN: 978-1-68108-453-4

Alves VD, Costa N, Coelhoso IM (2010) Barrier properties of biodegradable composite films based on kappa-carrageenan/pectin blends and mica flakes. Carbohydr Polym 79(2):269–276

Andevari GT, Rezaei M (2011) Effect of gelatin coating incorporated with cinnamon oil on the quality of fresh rainbow trout in cold storage. Int J Food Sci Technol 46(11):2305–2311

Angelo AJS, Vercellotti JR (1989) Inhibition of warmed-over flavor and preserving of uncured meat containing materials. Google Patents

Appendini P, Hotchkiss JH (2002) Review of antimicrobial food packaging. Innov Food Sci Emerg Technol 3(2):113–126

Araki C (1956) Structure of the agarose constituent of agar-agar. Bull Chem Soc Jpn 29(4):543–544

Arfat YA, Benjakul S, Prodpran T, Sumpavapol P, Songtipya P (2014) Properties and antimicrobial activity of fish protein isolate/fish skin gelatin film containing basil leaf essential oil and zinc oxide nanoparticles. Food Hydrocoll 41:265–273

Armisen R, Galatas F (2000) Extraction of agar. In: Phillips GO, Williams PA (eds) Handbook of hydrocolloids. CRC press, Cambridge

Aşik E, Candoğan K (2014) Effects of chitosan coatings incorporated with garlic oil on quality characteristics of shrimp. J Food Qual 37(4):237–246

Atarés L, Chiralt A (2016) Essential oils as additives in biodegradable films and coatings for active food packaging. Trends Food Sci Technol 48:51–62

Avila-Sosa R, Palou E, Munguía MTJ, Nevárez-Moorillón GV, Cruz ARN, López-Malo A (2012) Antifungal activity by vapor contact of essential oils added to amaranth, chitosan, or starch edible films. Int J Food Microbiol 153(1):66–72

Badawy MEI, Rabea EI (2009) Potential of the biopolymer chitosan with different molecular weights to control postharvest gray mold of tomato fruit. Postharvest Biol Technol 51(1):110–117

Badawy MEI, Rabea EI (2011) A biopolymer chitosan and its derivatives as promising antimicrobial agents against plant pathogens and their applications in crop protection. Int J Carbohydr Chem 2011:460381

Badawy MEI, Rabea EI (2016) Chitosan and Its derivatives as active ingredients against plant pests and diseases. In: Bautista-Baños S, Romanazzi G, Jiménez-Aparicio A (eds) Chitosan in the preservation of agricultural commodities. Academic Press, pp 179–219

Badawy MEI, Rabea EI (2017) Chitosan and its modifications as biologically active compounds in different applications. In: Masuelli M, Renard D (eds) Advances in physicochemical properties of biopolymers (part 2). Bentham Science, p 1

Badawy MEI, Rabea EI, El-Nouby MA (2016a) Preparation, physicochemical characterizations, and the antioxidant activity of the biopolymer films based on modified chitosan with starch, gelatin, and plasticizers. J Polym Mater 33(1):17

Badawy MEI, Rabea EI, Taktak NE, El Nouby MA (2016b) Production and properties of different molecular weights of chitosan from marine shrimp shells. J Chitin Chitosan Sci 4(1):46–54

Badawy MEI, Rabea EI, AM El-Nouby M, Ismail RIA, Taktak NEM (2017a) Strawberry shelf life, composition, and enzymes activity in response to edible chitosan coatings. Int J Fruit Sci 17(2):117–136

Badawy MEI, Taktak NEM, Awad OM, Elfiki SA, El-Ela NEA (2017b) Preparation and characterization of biopolymers chitosan/alginate/gelatin gel spheres crosslinked by glutaraldehyde. J Macromol Sci B 56(6):359–372

Bajpai SK, Sharma S (2004) Investigation of swelling/degradation behaviour of alginate beads crosslinked with Ca 2+ and Ba 2+ ions. React Funct Polym 59(2):129–140

Baldwin EA, Hagenmaier R, Bai J (2011) Edible coatings and films to improve food quality. CRC Press

Balfour E (1871) On the ethnology of hyderabad in the dekhan. Madras

Bao S, Xu S, Wang Z (2009) Antioxidant activity and properties of gelatin films incorporated with tea polyphenol-loaded chitosan nanoparticles. J Sci Food Agric 89(15):2692–2700

Benavides S, Villalobos-Carvajal R, Reyes J (2012) Physical, mechanical and antibacterial properties of alginate film: effect of the crosslinking degree and oregano essential oil concentration. J Food Eng 110(2):232–239

Beverlya RL, Janes ME, Prinyawiwatkula W, No HK (2008) Edible chitosan films on ready-to-eat roast beef for the control of Listeria monocytogenes. Food Microbiol 25(3):534–537

Bhatia S (2016) Marine bolysaccharides based nano-materials and Its applications. In: Natural polymer drug delivery systems. Springer, pp 185–225

Bierhalz ACK, da Silva MA, Kieckbusch TG (2012) Natamycin release from alginate/pectin films for food packaging applications. J Food Eng 110(1):18–25

Bonilla J, Atarés L, Vargas M, Chiralt A (2012) Edible films and coatings to prevent the detrimental effect of oxygen on food quality: possibilities and limitations. J Food Eng 110(2):208–213

Bonilla J, Talón E, Atarés L, Vargas M, Chiralt A (2013) Effect of the incorporation of antioxidants on physicochemical and antioxidant properties of wheat starch–chitosan films. J Food Eng 118(3):271–278

Bostan K, Mahan FI (2011) Microbiological quality and shelf-life of sausage treated with chitosan. İstanbul Üniversitesi Veteriner Fakültesi Dergisi 37(2):117–126

Bracone M, Merino D, González J, Alvarez VA, Gutiérrez TJ (2016) Nanopackaging from natural fillers and biopolymers for the development of active and intelligent films. In: Ikram S, Ahmed S (eds) Natural polymers: derivatives, blends and composites. Nova Science Publishers, New York, pp 119–155. ISBN: 978-1-63485-831-1

Braconnot H (1811) Sur la nature des champignons. Ann Chim Phys 79:265–304

Brody AL (2009) Innovations in fresh prepared meal delivery systems. Food Technol

Brody AL, Strupinsky E, Kline LR (2001) Active packaging for food applications. CRC press

Buonocore G, Conte A, Corbo M, Sinigaglia M, Del Nobile M (2005) Mono-and multilayer active films containing lysozyme as antimicrobial agent. Innov Food Sci Emerg Technol 6(4):459–464

Cacciuttolo MA, Trinh L, Lumpkin JA, Rao G (1993) Hyperoxia induces DNA damage in mammalian cells. Free Radic Biol Med 14(3):267–276

Camo J, Beltrán JA, Roncalés P (2008) Extension of the display life of lamb with an antioxidant active packaging. Meat Sci 80(4):1086–1091

Campaniello D, Bevilacqua A, Sinigaglia M, Corbo M (2008) Chitosan: Antimicrobial activity and potential applications for preserving minimally processed strawberries. Food Microbiol 25(8):992–1000

Campo VL, Kawano DF, da Silva DB, Carvalho I (2009) Carrageenans: biological properties, chemical modifications and structural analysis—a review. Carbohydr Polym 77(2):167–180

Caner C, Vergano P, Wiles J (1998) Chitosan film mechanical and permeation properties as affected by acid, plasticizer, and storage. J Food Sci 63(6):1049–1053

Casettari L, Gennari L, Angelino D, Ninfali P, Castagnino E (2012) ORAC of chitosan and its derivatives. Food Hydrocoll 28(2):243–247

Centella MH, Arévalo-Gallegos A, Parra-Saldivar R, Iqbal HM (2017) Marine-derived bioactive compounds for value-added applications in bio-and non-bio sectors. J Clean Prod 168:1559–1565

Cha DS, Choi JH, Chinnan MS, Park HJ (2002) Antimicrobial films based on Na-alginate and κ-carrageenan. LWT Food Sci Technol 35(8):715–719

Chang-Bravo L, López-Córdoba A, Martino M (2014) Biopolymeric matrices made of carrageenan and corn starch for the antioxidant extracts delivery of Cuban red propolis and yerba mate. React Funct Polym 85:11–19

Cheng S-Y, Wang B-J, Weng Y-M (2015) Antioxidant and antimicrobial edible zein/chitosan composite films fabricated by incorporation of phenolic compounds and dicarboxylic acids. LWT Food Sci Technol 63(1):115–121

Choi J, Choi W, Cha D, Chinnan M, Park H, Lee D, Park J (2005) Diffusivity of potassium sorbate in κ-carrageenan based antimicrobial film. LWT Food Sci Technol 38(4):417–423

Clark DE, Green HC (1936) Alginic acid and process of making same. Google Patents

Cole C (2000) Gelatin. In: Francis FJ (ed) Encyclopedia of food science and technology, vol 4. Wiley, New York, pp 1183–1188

Cooksey K (2005) Effectiveness of antimicrobial food packaging materials. Food Addit Contam 22(10):980–987

Córdoba LJP, Sobral PJ (2017) Physical and antioxidant properties of films based on gelatin, gelatin-chitosan or gelatin-sodium caseinate blends loaded with nanoemulsified active compounds. J Food Eng 213:47–53

Cruz RS, Camilloto GP, dos Santos Pires AC (2012) Oxygen scavengers: an approach on food preservation. In: Eissa, A.A. (ed), Structure and function of food engineering. InTech

Dainelli D, Gontard N, Spyropoulos D, Zondervan-van den Beuken E, Tobback P (2008) Active and intelligent food packaging: legal aspects and safety concerns. Trends Food Sci Technol 19:S103–S112

de Souza MCR, Marques CT, Dore CMG, da Silva FRF, Rocha HAO, Leite EL (2007) Antioxidant activities of sulfated polysaccharides from brown and red seaweeds. J Appl Phycol 19(2):153–160

Del Nobile MA, Gammariello D, Conte A, Attanasio M (2009) A combination of chitosan, coating and modified atmosphere packaging for prolonging Fior di latte cheese shelf life. Carbohydr Polym 78(1):151–156

Derkach SR, Ilyin SO, Maklakova AA, Kulichikhin VG, Malkin AY (2015) The rheology of gelatin hydrogels modified by κ-carrageenan. LWT Food Sci Technol 63(1):612–619

Devlieghere F, Vermeiren L, Debevere J (2004a) New preservation technologies: possibilities and limitations. Int Dairy J 14(4):273–285

Devlieghere F, Vermeulen A, Debevere J (2004b) Chitosan: antimicrobial activity, interactions with food components and applicability as a coating on fruit and vegetables. Food Microbiol 21(6):703–714

Dodane V, Vilivalam VD (1998) Pharmaceutical applications of chitosan. Pharm Sci Technol Today 1(6):246–253

Duan J, Jiang Y, Cherian G, Zhao Y (2010) Effect of combined chitosan-krill oil coating and modified atmosphere packaging on the storability of cold-stored lingcod (*Ophiodon elongates*) fillets. Food Chem 122(4):1035–1042

Dutta PK, Dutta J, Tripathi V (2004) Chitin and chitosan: chemistry, properties and applications. J Sci Ind Res 63(1):20–31

Dutta P, Tripathi S, Mehrotra G, Dutta J (2009) Perspectives for chitosan based antimicrobial films in food applications. Food Chem 114(4):1173–1182

Eça KS, Sartori T, Menegalli FC (2014) Films and edible coatings containing antioxidants—a review. Braz J Food Technol 17(2):98–112

El Hadrami A, Adam LR, El Hadrami I, Daayf F (2010) Chitosan in plant protection. Mar Drugs 8(4):968–987

Elsabee MZ, Abdou ES (2013) Chitosan based edible films and coatings: a review. Mater Sci Eng C 33(4):1819–1841

Etxabide A, Uranga J, Guerrero P, de la Caba K (2017) Development of active gelatin films by means of valorisation of food processing waste: a review. Food Hydrocoll 68:192–198

Falguera V, Quintero JP, Jiménez A, Muñoz JA, Ibarz A (2011) Edible films and coatings: Structures, active functions and trends in their use. Trends Food Sci Technol 22(6):292–303
Feng T, Du Y, Li J, Hu Y, Kennedy JF (2008) Enhancement of antioxidant activity of chitosan by irradiation. Carbohydr Polym 73(1):126–132
Fernandez-Saiz P, Lagaron J, Ocio M (2009) Optimization of the biocide properties of chitosan for its application in the design of active films of interest in the food area. Food Hydrocoll 23(3):913–921
Ferreira AR, Alves VD, Coelhoso IM (2016) Polysaccharide-based membranes in food packaging applications. Membranes 6(2):22
Ganiari S, Choulitoudi E, Oreopoulou V (2017) Edible and active films and coatings as carriers of natural antioxidants for lipid food. Trends Food Sci Technol 68:70–82
Garcıa MA, de la Paz N, Castro C, Rodrıguez JL, Rapado M, Zuluaga R, Ganán P, Casariego A (2015) Effect of molecular weight reduction by gamma irradiation on the antioxidant capacity of chitosan from lobster shells. J Radiat Res Appl Sci 8(2):190–200
George M, Abraham TE (2006) Polyionic hydrocolloids for the intestinal delivery of protein drugs: alginate and chitosan—a review. J Control Release 114(1):1–14
Gol NB, Patel PR, Rao TR (2013) Improvement of quality and shelf-life of strawberries with edible coatings enriched with chitosan. Postharvest Biol Technol 85:185–195
Gómez-Estaca J, Montero P, Giménez B, Gómez-Guillén M (2007) Effect of functional edible films and high pressure processing on microbial and oxidative spoilage in cold-smoked sardine (*Sardina pilchardus*). Food Chem 105(2):511–520
Gómez-Estaca J, Bravo L, Gómez-Guillén M, Alemán A, Montero P (2009a) Antioxidant properties of tuna-skin and bovine-hide gelatin films induced by the addition of oregano and rosemary extracts. Food Chem 112(1):18–25
Gómez-Estaca J, López de Lacey A, Gómez-Guillén M, López-Caballero M, Montero P (2009b) Antimicrobial activity of composite edible films based on fish gelatin and chitosan incorporated with clove essential oil. J Aquat Food Product Technol 18(1–2):46–52
Gómez-Estaca J, De Lacey AL, López-Caballero M, Gómez-Guillén M, Montero P (2010) Biodegradable gelatin–chitosan films incorporated with essential oils as antimicrobial agents for fish preservation. Food Microbiol 27(7):889–896
Gómez-Estaca J, Balaguer M, López-Carballo G, Gavara R, Hernández-Muñoz P (2017) Improving antioxidant and antimicrobial properties of curcumin by means of encapsulation in gelatin through electrohydrodynamic atomization. Food Hydrocoll 70:313–320
Gontard N, Guilbert S, Cuq JL (1993) Water and glycerol as plasticizers affect mechanical and water vapor barrier properties of an edible wheat gluten film. J Food Sci 58(1):206–211
Gutiérrez TJ (2017) Chitosan applications for the food industry. In: Ahmed S, Ikram S (eds) Chitosan: derivatives, composites and applications. Wiley-Scrivener Publisher, pp 183–232. https://doi.org/10.1002/9781119364849.ch8
Gutiérrez TJ, Guzmán R, Medina Jaramillo C, Famá L (2016) Effect of beet flour on films made from biological macromolecules: native and modified plantain flour. Int J Biol Macromol 82:395–403. https://doi.org/10.1016/j.ijbiomac.2015.10.020
Gutiérrez TJ, Herniou-Julien C, Álvarez K, Alvarez VA (2018) Structural properties and in vitro digestibility of edible and pH-sensitive films made from guinea arrowroot starch and wastes from wine manufacture. Carbohydr Polym 184:135–143. https://doi.org/10.1016/j.carbpol.2017.12.039
Hambleton A, Debeaufort F, Bonnotte A, Voilley A (2009) Influence of alginate emulsion-based films structure on its barrier properties and on the protection of microencapsulated aroma compound. Food Hydrocoll 23(8):2116–2124
Hamedi H, Kargozari M, Shotorbani PM, Mogadam NB, Fahimdanesh M (2017) A novel bioactive edible coating based on sodium alginate and galbanum gum incorporated with essential oil of *Ziziphora persica*: the antioxidant and antimicrobial activity, and application in food model. Food Hydrocoll 72:35–46
Han JH (2003) Antimicrobial food packaging. Novel Food Packag Tech 8:50–70

Han JH (2005) Innovations in food packaging. Academic Press
Hands S, Peat S (1938) Isolation of an anhydro-L-galactose derivative from agar. Nature 142:797
Ho C, Huffman D, Bradford D, Egbert W, Mikel W, Jones W (1995) Storage stability of vacuum packaged frozen pork sausage containing soy protein concentrate, carrageenan or antioxidants. J Food Sci 60(2):257–261
Honarkar H, Barikani M (2009) Applications of biopolymers I: chitosan. Monatshefte für Chemie Chem Month 140(12):1403–1420
Hong S-I, Rhim J-W (2008) Antimicrobial activity of organically modified nano-clays. J Nanosci Nanotechnol 8(11):5818–5824
Hong YH, Lim GO, Song K (2009) Physical properties of gelidium corneum–gelatin blend films containing grapefruit seed extract or green tea extract and its application in the packaging of pork loins. J Food Sci 74(1):C6–C10
Hoque MS, Benjakul S, Prodpran T (2010) Effect of heat treatment of film-forming solution on the properties of film from cuttlefish (*Sepia pharaonis*) skin gelatin. J Food Eng 96(1):66–73
Hou Y, Shavandi A, Carne A, Bekhit AA, Ng TB, Cheung RCF, Bekhit AE-dA (2016) Marine shells: potential opportunities for extraction of functional and health-promoting materials. Crit Rev Environ Sci Technol 46(11–12):1047–1116
Huq T, Salmieri S, Khan A, Khan RA, Le Tien C, Riedl B, Fraschini C, Bouchard J, Uribe-Calderon J, Kamal MR (2012) Nanocrystalline cellulose (NCC) reinforced alginate based biodegradable nanocomposite film. Carbohydr Polym 90(4):1757–1763
Imran M, Revol-Junelles A-M, Martyn A, Tehrany EA, Jacquot M, Linder M, Desobry S (2010) Active food packaging evolution: transformation from micro-to nanotechnology. Crit Rev Food Sci Nutr 50(9):799–821
Jayas DS, Jeyamkondan S (2002) PH—postharvest technology: Modified atmosphere storage of grains meats fruits and vegetables. Biosyst Eng 82(3):235–251
Jeon Y-J, Kamil JY, Shahidi F (2002) Chitosan as an edible invisible film for quality preservation of herring and Atlantic cod. J Agric Food Chem 50(18):5167–5178
Jiménez A, Fabra MJ, Talens P, Chiralt A (2013) Physical properties and antioxidant capacity of starch–sodium caseinate films containing lipids. J Food Eng 116(3):695–702
Jongjareonrak A, Benjakul S, Visessanguan W, Tanaka M (2008) Antioxidative activity and properties of fish skin gelatin films incorporated with BHT and α-tocopherol. Food Hydrocoll 22(3):449–458
Jridi M, Hajji S, Ayed HB, Lassoued I, Mbarek A, Kammoun M, Souissi N, Nasri M (2014) Physical, structural, antioxidant and antimicrobial properties of gelatin–chitosan composite edible films. Int J Biol Macromol 67:373–379
Juck G, Neetoo H, Chen H (2010) Application of an active alginate coating to control the growth of Listeria monocytogenes on poached and deli turkey products. Int J Food Microbiol 142(3):302–308
Kader AA, Zagory D, Kerbel EL, Wang CY (1989) Modified atmosphere packaging of fruits and vegetables. Crit Rev Food Sci Nutr 28(1):1–30
Kanatt SR, Rao M, Chawla S, Sharma A (2013) Effects of chitosan coating on shelf-life of ready-to-cook meat products during chilled storage. LWT Food Sci Technol 53(1):321–326
Kanmani P, Rhim J-W (2014a) Development and characterization of carrageenan/grapefruit seed extract composite films for active packaging. Int J Biol Macromol 68:258–266
Kanmani P, Rhim J-W (2014b) Physical, mechanical and antimicrobial properties of gelatin based active nanocomposite films containing AgNPs and nanoclay. Food Hydrocoll 35:644–652
Kavoosi G, Dadfar SMM, Purfard AM (2013) Mechanical, physical, antioxidant, and antimicrobial properties of gelatin films incorporated with thymol for potential use as nano wound dressing. J Food Sci 78(2):E244–E250
Kavoosi G, Rahmatollahi A, Dadfar SMM, Purfard AM (2014) Effects of essential oil on the water binding capacity, physico-mechanical properties, antioxidant and antibacterial activity of gelatin films. LWT Food Sci Technol 57(2):556–561
Kerry J, O'grady M, Hogan S (2006) Past, current and potential utilisation of active and intelligent packaging systems for meat and muscle-based products: a review. Meat Sci 74(1):113–130

Khalil HA, Saurabh CK, Tye Y, Lai T, Easa A, Rosamah E, Fazita M, Syakir M, Adnan A, Fizree H (2017) Seaweed based sustainable films and composites for food and pharmaceutical applications: a review. Renew Sust Energ Rev 77:353–362
Khan I, Tango CN, Oh DH (2017) Development and evaluation of chitosan and its derivative for the shelf life extension of beef meat under refrigeration storage. Int J Food Sci Technol 52(5):1111–1121
Kim KW, Thomas R (2007) Antioxidative activity of chitosans with varying molecular weights. Food Chem 101(1):308–313
Kim HS, Lee C-G, Lee EY (2011) Alginate lyase: structure, property, and application. Biotechnol Bioprocess Eng 16(5):843
Knutsen S, Myslabodski D, Larsen B, Usov A (1994) A modified system of nomenclature for red algal galactans. Bot Mar 37(2):163–170
Koli JM, Basu S, Nayak BB, Patange SB, Pagarkar AU, Gudipati V (2012) Functional characteristics of gelatin extracted from skin and bone of Tiger-toothed croaker (Otolithes ruber) and Pink perch (*Nemipterus japonicus*). Food Bioprod Process 90(3):555–562
Kong M, Chen XG, Xing K, Park HJ (2010) Antimicrobial properties of chitosan and mode of action: a state of the art review. Int J Food Microbiol 144(1):51–63
Krajewska B (2004) Application of chitin-and chitosan-based materials for enzyme immobilizations: a review. Enzym Microb Technol 35(2):126–139
Krefting A (1903) Process of extracting glutinous substances from seaweed. Google Patents
Krkić N, Šojić B, Lazić V, Petrović L, Mandić A, Sedej I, Tomović V (2013) Lipid oxidative changes in chitosan-oregano coated traditional dry fermented sausage *Petrovská klobása*. Meat Sci 93(3):767–770
Kumar M, McGlade D, Lawler J (2014) Functionalized chitosan derived novel positively charged organic–inorganic hybrid ultrafiltration membranes for protein separation. RSC Adv 4(42):21699–21711
Lassaigne JL (1843) C. r. acad. sci., Paris 16, 387 (1843). Ann Chem 48
Lee KY, Mooney DJ (2012) Alginate: properties and biomedical applications. Prog Polym Sci 37(1):106–126
Lee S-R, Park H-M, Lim H, Kang T, Li X, Cho W-J, Ha C-S (2002) Microstructure, tensile properties, and biodegradability of aliphatic polyester/clay nanocomposites. Polymer 43(8):2495–2500
Leon PG, Rojas AM (2007) Gellan gum films as carriers of L-(+)-ascorbic acid. Food Res Int 40(5):565–575
Li K, Hwang Y, Tsai T, Chi S (1996) Chelation of iron ion and antioxidative effect on cooked salted ground pork by *N*-carboxymethylchitosan (NCMC). Food Sci Taiwan 23:608–616
Li J-H, Miao J, Wu J-L, Chen S-F, Zhang Q-Q (2014) Preparation and characterization of active gelatin-based films incorporated with natural antioxidants. Food Hydrocoll 37:166–173
Liu X, Zeng A, Song T, Li L, Yang F, Wang Q, Wu B, Liu Y, Zhi X (2012) Hypocholesterolemic effects of *N*-[(2-Hydroxy-3-*N*,*N*-Dimethylhexadecyl Ammonium) propyl] chitosan chloride in high-fat-diet-induced rats. J Biomater Sci Polym Ed 23(8):1107–1114
Liu Q, Wu C, Cai H, Hu N, Zhou J, Wang P (2014) Cell-based biosensors and their application in biomedicine. Chem Rev 114(12):6423–6461
Llorens A, Lloret E, Picouet PA, Trbojevich R, Fernandez A (2012) Metallic-based micro and nanocomposites in food contact materials and active food packaging. Trends Food Sci Technol 24(1):19–29
Madene A, Jacquot M, Scher J, Desobry S (2006) Flavour encapsulation and controlled release—a review. Int J Food Sci Technol 41(1):1–21
Maizura M, Fazilah A, Norziah M, Karim A (2007) Antibacterial activity and mechanical properties of partially hydrolyzed sago starch–alginate edible film containing lemongrass oil. J Food Sci 72(6):C324–C330
Makino Y, Hirata T (1997) Modified atmosphere packaging of fresh produce with a biodegradable laminate of chitosan-cellulose and polycaprolactone. Postharvest Biol Technol 10(3):247–254

Manivasagan P, Oh J (2016) Marine polysaccharide-based nanomaterials as a novel source of nanobiotechnological applications. Int J Biol Macromol 82:315–327
Manzanarez-López F, Soto-Valdez H, Auras R, Peralta E (2011) Release of α-tocopherol from poly (lactic acid) films, and its effect on the oxidative stability of soybean oil. J Food Eng 104(4):508–517
Mariod AA, Fadul H (2013) Gelatin, source, extraction and industrial applications. Acta Sci Pol Technol Aliment 12(2):135–147
Mariod AA, Fadul H (2015) Extraction and characterization of gelatin from two edible Sudanese insects and its applications in ice cream making. Revista de Agaroquimica y Tecnologia de Alimentos 21(5):380–391
Martins JT, Cerqueira MA, Vicente AA (2012) Influence of α-tocopherol on physicochemical properties of chitosan-based films. Food Hydrocoll 27(1):220–227
Martucci JF, Ruseckaite RA (2017) Antibacterial activity of gelatin/copper (II)-exchanged montmorillonite films. Food Hydrocoll 64:70–77
Martucci J, Gende L, Neira L, Ruseckaite R (2015) Oregano and lavender essential oils as antioxidant and antimicrobial additives of biogenic gelatin films. Ind Crop Prod 71:205–213
McHugh D (2003) Chapter 7: Carrageenan. A guide to the seaweed industry: FAO fisheries technical paper 441
Mi F-L, Shyu S-S, Wu Y-B, Lee S-T, Shyong J-Y, Huang R-N (2001) Fabrication and characterization of a sponge-like asymmetric chitosan membrane as a wound dressing. Biomaterials 22(2):165–173
Min B, Oh JH (2009) Antimicrobial activity of catfish gelatin coating containing origanum (*Thymus capitatus*) oil against Gram-negative pathogenic bacteria. J Food Sci 74(4):M143–M148
Mohammed Fayaz A, Balaji K, Girilal M, Kalaichelvan P, Venkatesan R (2009) Mycobased synthesis of silver nanoparticles and their incorporation into sodium alginate films for vegetable and fruit preservation. J Agric Food Chem 57(14):6246–6252
Mohebbi M, Ansarifar E, Hasanpour N, Amiryousefi MR (2012) Suitability of *Aloe vera* and gum tragacanth as edible coatings for extending the shelf life of button mushroom. Food Bioprocess Technol 5(8):3193–3202
Moncayo D, Buitrago G, Algecira N (2013) The surface properties of biopolymer-coated fruit: a review. Ingeniería e Investigación 33(3):11–16
Moreira MR, Pereda M, Marcovich NE, Roura SI (2011) Antimicrobial effectiveness of bioactive packaging materials from edible chitosan and casein polymers: assessment on carrot, cheese, and salami. J Food Sci 76(1):M54–M63
Muzzarelli RAA (1973) Natural chelating polymers: alginic acid, chitin and chitosan. Pergamon Press, Oxford, New York
Nafchi AM, Moradpour M, Saeidi M, Alias AK (2014) Effects of nanorod-rich ZnO on rheological, sorption isotherm, and physicochemical properties of bovine gelatin films. LWT Food Sci Technol 58(1):142–149
Nanaki S, Karavas E, Kalantzi L, Bikiaris D (2010) Miscibility study of carrageenan blends and evaluation of their effectiveness as sustained release carriers. Carbohydr Polym 79(4):1157–1167
Neetoo H, Ye M, Chen H (2010) Bioactive alginate coatings to control *Listeria monocytogenes* on cold-smoked salmon slices and fillets. Int J Food Microbiol 136(3):326–331
Ngo D-H, Kim S-K (2014) Antioxidant effects of chitin, chitosan and their derivatives. Elsevier, Oxford, UK
No HK, Park NY, Lee SH, Meyers SP (2002) Antibacterial activity of chitosans and chitosan oligomers with different molecular weights. Int J Food Microbiol 74(1):65–72
No H, Meyers S, Prinyawiwatkul W, Xu Z (2007) Applications of chitosan for improvement of quality and shelf life of foods: a review. J Food Sci 72(5):R87–R100
Nollet LM (2016) Marine microorganisms: extraction and analysis of bioactive compounds. CRC Press, Boca Raton
Norajit K, Kim KM, Ryu GH (2010) Comparative studies on the characterization and antioxidant properties of biodegradable alginate films containing ginseng extract. J Food Eng 98(3):377–384

Nowzari F, Shábanpour B, Ojagh SM (2013) Comparison of chitosan–gelatin composite and bilayer coating and film effect on the quality of refrigerated rainbow trout. Food Chem 141(3):1667–1672

Nussinovitch A, Gershon Z (1997) Physical characteristics of agar—yeast sponges. Food Hydrocoll 11(2):231–237

Odier A (1823) Mémoire sur la composition chimique des parties cornées des insectes

Ogaji IJ, Nep EI, Audu-Peter JD (2012) Advances in natural polymers as pharmaceutical excipients. Pharm Anal Acta 3:146.

Olaimat AN, Holley RA (2015) Control of Salmonella on fresh chicken breasts by κ-carrageenan/chitosan-based coatings containing allyl isothiocyanate or deodorized Oriental mustard extract plus EDTA. Food Microbiol 48:83–88

Olatunji O (2016) Classification of natural polymers. In: Olatunji O (ed) Natural polymers. Springer, Basel, pp 1–17

Oussalah M, Caillet S, Salmieri S, Saucier L, Lacroix M (2006) Antimicrobial effects of alginate-based film containing essential oils for the preservation of whole beef muscle. J Food Prot 69(10):2364–2369

Park HJ (1999) Development of advanced edible coatings for fruits. Trends Food Sci Technol 10(8):254–260

Park P-J, Je J-Y, Kim S-K (2004a) Free radical scavenging activities of differently deacetylated chitosans using an ESR spectrometer. Carbohydr Polym 55(1):17–22

Park SI, Daeschel M, Zhao Y (2004b) Functional properties of antimicrobial lysozyme-chitosan composite films. J Food Sci 69(8):M215–M221

Park SI, Stan SD, Daeschel MA, Zhao Y (2005) Antifungal coatings on fresh strawberries (Fragaria× ananassa) to control mold growth during cold storage. J Food Sci 70(4):M202–M207

Pasanphan W, Rattanawongwiboon T, Choofong S, Güven O, Katti KK (2015) Irradiated chitosan nanoparticle as a water-based antioxidant and reducing agent for a green synthesis of gold nanoplatforms. Radiat Phys Chem 106:360–370

Pastor C, Sánchez-González L, Chiralt A, Cháfer M, González-Martínez C (2013) Physical and antioxidant properties of chitosan and methylcellulose based films containing resveratrol. Food Hydrocoll 30(1):272–280

Paul W, Sharma CP (2004) Chitosan and alginate wound dressings: a short review. Trends Biomater Artif Organs 18(1):18–23

Paula HCB, de Paula RCM, Bezerral SKF (2006) Swelling and release kinetics of larvicide-containing chitosan/cashew gum beads. J Appl Polym Sci 102(1):395–400.

Peng Y, Wu Y, Li Y (2013) Development of tea extracts and chitosan composite films for active packaging materials. Int J Biol Macromol 59:282–289

Percival E, Somerville J, Forbes I (1938) Isolation of an anhydro-sugar derivative from agar. Nature 142:797–798

Perdones A, Sánchez-González L, Chiralt A, Vargas M (2012) Effect of chitosan–lemon essential oil coatings on storage-keeping quality of strawberry. Postharvest Biol Technol 70:32–41

Pereda M, Ponce A, Marcovich N, Ruseckaite R, Martucci J (2011) Chitosan-gelatin composites and bi-layer films with potential antimicrobial activity. Food Hydrocoll 25(5):1372–1381

Pereira L (2011) A review of the nutrient composition of selected edible seaweeds. In: Pomin VH (ed) Seaweed: ecology, nutrient composition and medicinal uses. Nova Science, pp 15–47

Pereira de Abreu D, Cruz J, Paseiro Losada P (2012) Active and intelligent packaging for the food industry. Food Rev Intl 28(2):146–187

Pereira L, Gheda SF, Ribeiro-Claro PJ (2013) Analysis by vibrational spectroscopy of seaweed polysaccharides with potential use in food, pharmaceutical, and cosmetic industries. Int J Carbohydr Chem 2013:537202

Phillips GO, Williams PA (2009) Handbook of hydrocolloids. Elsevier

Pokorny J (2007) Antioxidants in food preservation. In: Handbook of food preservation, vol 2, pp 259–286

Prabu K, Natarajan E (2012) *In vitro* antimicrobial and antioxidant activity of chitosan isolated from *Podophthalmus vigil*. J Appl Pharm Sci 2(9):75–82

Prajapati VD, Maheriya PM, Jani GK, Solanki HK (2014) Carrageenan: a natural seaweed polysaccharide and its applications. Carbohydr Polym 105:97–112
Pranoto Y, Salokhe VM, Rakshit SK (2005) Physical and antibacte rial properties of alginate-based edible film incorporated with garlic oil. Food Res Int 38(3):267–272
Quintavalla S, Vicini L (2002) Antimicrobial food packaging in meat industry. Meat Sci 62(3):373–380
Rabea EI, Badawy ME-T, Stevens CV, Smagghe G, Steurbaut W (2003) Chitosan as antimicrobial agent: applications and mode of action. Biomacromolecules 4(6):1457–1465
Ramos M, Jiménez A, Garrigós MC (2017) Active nanocomposites in food contact materials. In: Nanoscience in food and agriculture 4. Springer, pp 1–44
Rao M, Kanatt S, Chawla S, Sharma A (2010) Chitosan and guar gum composite films: preparation, physical, mechanical and antimicrobial properties. Carbohydr Polym 82(4):1243–1247
Raybaudi-Massilia RM, Mosqueda-Melgar J, Martín-Belloso O (2008) Edible alginate-based coating as carrier of antimicrobials to improve shelf-life and safety of fresh-cut melon. Int J Food Microbiol 121(3):313–327
Raybaudi-Massilia R, Mosqueda-Melgar J, Soliva-Fortuny R, Martín-Belloso O (2016) Combinational edible antimicrobial films and coatings. In: Antimicrobial food packaging. Elsevier, pp 633–646
Realini CE, Marcos B (2014) Active and intelligent packaging systems for a modern society. Meat Sci 98(3):404–419
Reinhard S, Herbert G (2007) Gelatin handbook: theory and industrial practice. Wiley-VCH Press, Weinheim
Restuccia D, Spizzirri UG, Parisi OI, Cirillo G, Curcio M, Iemma F, Puoci F, Vinci G, Picci N (2010) New EU regulation aspects and global market of active and intelligent packaging for food industry applications. Food Control 21(11):1425–1435
Rhim JW (2012) Physical-mechanical properties of agar/κ-carrageenan blend film and derived clay nanocomposite film. J Food Sci 77(12):N66–N73
Rhim J-W, Ng PK (2007) Natural biopolymer-based nanocomposite films for packaging applications. Crit Rev Food Sci Nutr 47(4):411–433
Rhim J-W, Hong S-I, Park H-M, Ng PK (2006) Preparation and characterization of chitosan-based nanocomposite films with antimicrobial activity. J Agric Food Chem 54(16):5814–5822
Rhim J-W, Park H-M, Ha C-S (2013) Bio-nanocomposites for food packaging applications. Prog Polym Sci 38(10):1629–1652
Rinaudo M (1992) On the abnormal exponents a ν and a D in Mark Houwink type equations for wormlike chain polysaccharides. Polym Bull 27(5):585–589
Rodriguez-Aguilera R, Oliveira JC (2009) Review of design engineering methods and applications of active and modified atmosphere packaging systems. Food Eng Rev 1(1):66–83
Rouget C (1859) Des substances amylacees dans le tissue des animux, specialement les Articules (Chitine). Compt Rend 48:792–795
Sabra W, Deckwer W-D (2005) Alginate—a polysaccharide of industrial interest and diverse biological functions. Polysacharides Struct Divers Funct Versat 2:515–533
Samec M, Isajevič V (1922) Studien über Pflanzenkolloide, XIV. Fortschrittsberichte über Kolloide und Polymere 16(5):285–300
Sathivel S, Liu Q, Huang J, Prinyawiwatkul W (2007) The influence of chitosan glazing on the quality of skinless pink salmon (*Oncorhynchus gorbuscha*) fillets during frozen storage. J Food Eng 83(3):366–373
Scheuer PJ (2013) Marine natural products: chemical and biological perspectives. Academic Press, New York
Schreiber SB, Bozell JJ, Hayes DG, Zivanovic S (2013) Introduction of primary antioxidant activity to chitosan for application as a multifunctional food packaging material. Food Hydrocoll 33(2):207–214
Seol K-H, Lim D-G, Jang A, Jo C, Lee M (2009) Antimicrobial effect of κ-carrageenan-based edible film containing ovotransferrin in fresh chicken breast stored at 5 C. Meat Sci 83(3):479–483

Shahidi F (1996) Role of chemistry and biotechnology in value-added utilization of shellfish processing discards. ChemInform 27(15)
Shankar S, Teng X, Li G, Rhim J-W (2015) Preparation, characterization, and antimicrobial activity of gelatin/ZnO nanocomposite films. Food Hydrocoll 45:264–271
Shankar S, Jaiswal L, Rhim J-W (2016) Gelatin-based nanocomposite films: potential use in antimicrobial active packaging. In: Antimicrobial food packaging. Elsevier, pp 339–348
Shao X, Tu K, Tu S, Tu J (2012) A combination of heat treatment and chitosan coating delays ripening and reduces decay in “Gala” apple fruit. J Food Qual 35(2):83–92
Shen Z, Kamdem DP (2015) Development and characterization of biodegradable chitosan films containing two essential oils. Int J Biol Macromol 74:289–296
Shimizu Y, Kamiya H (1983) Bioactive marine biopolymers. In: Scheuer PJ (ed) Marine natural products: chemical and biological perspectives, vol 5. Academic Press, p 391
Shojaee-Aliabadi S, Hosseini H, Mohammadifar MA, Mohammadi A, Ghasemlou M, Ojagh SM, Hosseini SM, Khaksar R (2013) Characterization of antioxidant-antimicrobial κ-carrageenan films containing *Satureja hortensis* essential oil. Int J Biol Macromol 52:116–124
Shojaee-Aliabadi S, Hosseini H, Mohammadifar MA, Mohammadi A, Ghasemlou M, Hosseini SM, Khaksar R (2014) Characterization of κ-carrageenan films incorporated plant essential oils with improved antimicrobial activity. Carbohydr Polym 101:582–591
Silberbauer A, Schmid M (2017) Packaging concepts for ready-to-eat food: recent progress. J Packag Technol Res 1(3):113–126
Simic MG, Karel M (2013) Autoxidation in food and biological systems. Springer
Simões AD, Tudela JA, Allende A, Puschmann R, Gil MI (2009) Edible coatings containing chitosan and moderate modified atmospheres maintain quality and enhance phytochemicals of carrot sticks. Postharvest Biol Technol 51(3):364–370
Sipahi R, Castell-Perez M, Moreira R, Gomes C, Castillo A (2013) Improved multilayered antimicrobial alginate-based edible coating extends the shelf life of fresh-cut watermelon (*Citrullus lanatus*). LWT Food Sci Technol 51(1):9–15
Siripatrawan U, Harte BR (2010) Physical properties and antioxidant activity of an active film from chitosan incorporated with green tea extract. Food Hydrocoll 24(8):770–775
Siro I (2012) Active and intelligent packaging of food. In: Bhat R, Alias AK, Paliyath G (eds) Progress in food preservation. Wiley, Chichester, p 23
Sokolova E, Barabanova A, Homenko V, Solov’eva T, Bogdanovich R, Yermak I (2011) *In vitro* and *ex vivo* studies of antioxidant activity of carrageenans, sulfated polysaccharides from red algae. Bull Exp Biol Med 150(4):426–428
Soto-Valdez H, Auras R, Peralta E (2011) Fabrication of poly (lactic acid) films with resveratrol and the diffusion of resveratrol into ethanol. J Appl Polym Sci 121(2):970–978
Souza VGL, Fernando AL (2016) Nanoparticles in food packaging: biodegradability and potential migration to food—a review. Food Packag Shelf Life 8:63–70
Street H (2012) Gelatin handbook. Gelatin Handb 25
Sun Y, Yang B, Wu Y, Liu Y, Gu X, Zhang H, Wang C, Cao H, Huang L, Wang Z (2015) Structural characterization and antioxidant activities of κ-carrageenan oligosaccharides degraded by different methods. Food Chem 178:311–318
Tahiri I, Desbiens M, Benech R, Kheadr E, Lacroix C, Thibault S, Ouellet D, Fliss I (2004) Purification, characterization and amino acid sequencing of divergicin M35: a novel class IIa bacteriocin produced by Carnobacterium divergens M35. Int J Food Microbiol 97(2):123–136
Tahiri I, Desbiens M, Lacroix C, Kheadr E, Fliss I (2009) Growth of Sarnobacterium divergens M35 and production of Divergicin M35 in snow crab by-product, a natural-grade medium. LWT Food Sci Technol 42(2):624–632
Talón E, Trifkovic KT, Nedovic VA, Bugarski BM, Vargas M, Chiralt A, González-Martínez C (2017) Antioxidant edible films based on chitosan and starch containing polyphenols from thyme extracts. Carbohydr Polym 157:1153–1161
Tavassoli-Kafrani E, Shekarchizadeh H, Masoudpour-Behabadi M (2016) Development of edible films and coatings from alginates and carrageenans. Carbohydr Polym 137:360–374

Tezotto-Uliana JV, Fargoni GP, Geerdink GM, Kluge RA (2014) Chitosan applications pre-or postharvest prolong raspberry shelf-life quality. Postharvest Biol Technol 91:72–77
Thakur VK, Thakur MK (2016) Handbook of sustainable polymers: processing and applications. CRC Press
Therkelsen GH (1993) Carrageenan. In: Whistler RL, BeMiller JN (eds) Industrial gums, polysaccharides and their derivatives. Academic Press, New York, pp 145–180
Tongnuanchan P, Benjakul S, Prodpran T (2012) Properties and antioxidant activity of fish skin gelatin film incorporated with citrus essential oils. Food Chem 134(3):1571–1579
Tongnuanchan P, Benjakul S, Prodpran T (2013) Physico-chemical properties, morphology and antioxidant activity of film from fish skin gelatin incorporated with root essential oils. J Food Eng 117(3):350–360
Tongnuanchan P, Benjakul S, Prodpran T (2014) Structural, morphological and thermal behaviour characterisations of fish gelatin film incorporated with basil and citronella essential oils as affected by surfactants. Food Hydrocoll 41:33–43
Van de Velde F, De Ruiter GA (2005) Carrageenan. Biopolymers Online
Van de Velde F, Knutsen S, Usov A, Rollema H, Cerezo A (2002) ^{1}H and ^{13}C high resolution NMR spectroscopy of carrageenans: application in research and industry. Trends Food Sci Technol 13(3):73–92
Varela P, Fiszman S (2011) Hydrocolloids in fried foods. A review. Food Hydrocoll 25(8):1801–1812
Varki A, Cummings R, Esko J, Freeze H, Hart G, Marth J (1999) Essentials of glycobiology. Cold Spring Harbor Laboratory Press, New York, pp 66–84
Vásconez MB, Flores SK, Campos CA, Alvarado J, Gerschenson LN (2009) Antimicrobial activity and physical properties of chitosan–tapioca starch based edible films and coatings. Food Res Int 42(7):762–769
Venkatesan J, Bhatnagar I, Manivasagan P, Kang K-H, Kim S-K (2015) Alginate composites for bone tissue engineering: a review. Int J Biol Macromol 72:269–281
Vermeiren L, Devlieghere F, Van Beest M, De Kruijf N, Debevere J (1999) Developments in the active packaging of foods. Trends Food Sci Technol 10(3):77–86
Vu K, Hollingsworth R, Leroux E, Salmieri S, Lacroix M (2011) Development of edible bioactive coating based on modified chitosan for increasing the shelf life of strawberries. Food Res Int 44(1):198–203
Wang SY, Gao H (2013) Effect of chitosan-based edible coating on antioxidants, antioxidant enzyme system, and postharvest fruit quality of strawberries (Fragaria x aranassa Duch.). LWT Food Sci Technol 52(2):71–79
Wang J, Hu S, Nie S, Yu Q, Xie M (2016) Reviews on mechanisms of *in vitro* antioxidant activity of polysaccharides. Oxidative Med Cell Longev 2016:5692852
Weiner RM (1997) Biopolymers from marine prokaryotes. Trends Biotechnol 15(10):390–394
Williams PA, Campbell KT, Gharaviram H, Madrigal JL, Silva EA (2017) Alginate-chitosan hydrogels provide a sustained gradient of sphingosine-1-phosphate for *Therapeutic Angiogenesis*. Ann Biomed Eng 45(4):1003–1014
Wilson MD, Stanley RA, Eyles A, Ross T (2017) Innovative processes and technologies for modified atmosphere packaging of fresh and fresh-cut fruits and vegetables. Crit Rev Food Sci Nutr:1–12
Wu Y, Rhim J, Weller C, Hamouz F, Cuppett S, Schnepf M (2000) Moisture loss and lipid oxidation for precooked beef patties stored in edible coatings and films. J Food Sci 65(2):300–304
Wu Y, Weller C, Hamouz F, Cuppett S, Schnepf M (2001) Moisture loss and lipid oxidation for precooked ground-beef patties packaged in edible starch-alginate-based composite films. J Food Sci 66(3):486–493
Wu J, Ge S, Liu H, Wang S, Chen S, Wang J, Li J, Zhang Q (2014) Properties and antimicrobial activity of silver carp (Hypophthalmichthys molitrix) skin gelatin-chitosan films incorporated with oregano essential oil for fish preservation. Food Packag Shelf Life 2(1):7–16
Wu J, Liu H, Ge S, Wang S, Qin Z, Chen L, Zheng Q, Liu Q, Zhang Q (2015) The preparation, characterization, antimicrobial stability and *in vitro* release evaluation of fish gelatin films incorporated with cinnamon essential oil nanoliposomes. Food Hydrocoll 43:427–435

Wu J, Sun X, Guo X, Ge S, Zhang Q (2017) Physicochemical properties, antimicrobial activity and oil release of fish gelatin films incorporated with cinnamon essential oil. Aquac Fish 2(4):185–192

Xie W, Xu P, Liu Q (2001) Antioxidant activity of water-soluble chitosan derivatives. Bioorg Med Chem Lett 11(13):1699–1701

Xing Y, Li X, Xu Q, Jiang Y, Yun J, Li W (2010) Effects of chitosan-based coating and modified atmosphere packaging (MAP) on browning and shelf life of fresh-cut lotus root (*Nelumbo nucifera Gaerth*). Innov Food Sci Emerg Technol 11(4):684–689

Yang J-I, Ho H-Y, Chu Y-J, Chow C-J (2008) Characteristic and antioxidant activity of retorted gelatin hydrolysates from cobia (*Rachycentron canadum*) skin. Food Chem 110(1):128–136

Yang J-S, Xie Y-J, He W (2011) Research progress on chemical modification of alginate: a review. Carbohydr Polym 84(1):33–39

Ye M, Neetoo H, Chen H (2008) Effectiveness of chitosan-coated plastic films incorporating antimicrobials in inhibition of *Listeria monocytogenes* on cold-smoked salmon. Int J Food Microbiol 127(3):235–240

Yen M-T, Tseng Y-H, Li R-C, Mau J-L (2007) Antioxidant properties of fungal chitosan from shiitake stipes. LWT Food Sci Technol 40(2):255–261

Yen M-T, Yang J-H, Mau J-L (2008) Antioxidant properties of chitosan from crab shells. Carbohydr Polym 74(4):840–844

Youn S, Her J, Park S, Ahn D, Kim Y, Choi J (2004) Studies on the improvement of shelf-life in spicy beef meat using chitosan. J Korean Soc Food Sci Nutr 33(1):207–211

Yuan H, Zhang W, Li X, Lü X, Li N, Gao X, Song J (2005) Preparation and *in vitro* antioxidant activity of κ-carrageenan oligosaccharides and their oversulfated, acetylated, and phosphorylated derivatives. Carbohydr Res 340(4):685–692

Yuan H, Song J, Zhang W, Li X, Li N, Gao X (2006) Antioxidant activity and cytoprotective effect of κ-carrageenan oligosaccharides and their different derivatives. Bioorg Med Chem Lett 16(5):1329–1334

Zactiti E, Kieckbusch T (2006) Potassium sorbate permeability in biodegradable alginate films: effect of the antimicrobial agent concentration and crosslinking degree. J Food Eng 77(3):462–467

Zagory D, Kader AA (1988) Modified atmosphere packaging of fresh produce. Food Technol 42(9):70–77

Zeuthen P, Bøgh-Sørensen L (2003) Food preservation techniques. Elsevier

Zhang H, Wang M-L, Mao Y-T (2011) Advances in the application of chitosan coating in fresh-keeping of fruits and vegetables. J Guizhou Agric Sci 10:040

Zhang B, Fang C-D, Hao G-J, Zhang Y-Y (2018) Effect of kappa-carrageenan oligosaccharides on myofibrillar protein oxidation in peeled shrimp (*Litopenaeus vannamei*) during long-term frozen storage. Food Chem 245:254–261

Chapter 24
Functional Carbohydrate Polymers: Prebiotics

Jun Yang and Yixiang Xu

Abstract Increasing scientific evidence has identified the correlation among dietary intake, the gut microbiome, and human health. Controlling the microbiome within the human gut through dietary modifications sheds light on novel nutritional strategies and clinical practices in reducing some chronic diseases. The emerging field of prebiotics, probiotics, and synbiotics is associated with the development of nutritional interventions, gut microbiome with positively impact health outcomes. Although there is strong evidence to demonstrate the complex link between gut microbiota and human health, substantial challenges still remain in delivering effective, stable and cost efficient foods with positive health outcomes, building personalized diets based on the gut microbiome profile, and standardizing clinical practices and establishing regulation. Dietary intervention, as a strong applicator, on microbiota and consequently on physiology and immune system, could play significant role in reducing the risk and progression of some chronic diseases including cancer and obesity. In this chapter, the authors focus on prebiotics as functional carbohydrate polymers, including traditional ones of human milk oligosaccharides (HMOS), fructooligosaccharides (FOS), and galactooligosaccharides (GOS), as well as potential ones of pectin oligosaccharides (POS), xylooligosaccharides (XOS), arabinoxylan oligosaccharides (AXOS), and glucomannan oligosaccharides (GMOS). To better understand the complex interplay of diet, nutrition and the microbiome in food development, as well as the effects of diet on the diversity of human microbiome, the contents of source, chemical structure, processing, physiological functionalities for each prebiotic will be covered.

Keywords Functionality · Health · Plant polysaccharides · Production · Structure · Traditional and emerging prebiotics

Jun Yang is employee of PepsiCo Inc. The views expressed in this manuscript are those of the authors and do not necessarily reflect the position or policy of PepsiCo Inc.

J. Yang (✉)
Measurement Science, PepsiCo R&D, Plano, TX, USA
e-mail: Jun.Yang@pepsico.com

Y. Xu
Agricultural Research Station, Virginia State University, Petersburg, VA, USA

T. J. Gutiérrez (ed.), *Polymers for Food Applications*,
https://doi.org/10.1007/978-3-319-94625-2_24

Abbreviations

AX Arabinoxylan
AXOS Arabinoxylan oligosaccharides
DF Dietary fiber
DP Degree of polymerization
FOS Fructooligosaccharides
GI tract Gastrointestinal tract
GMOS Glucomannan oligosaccharides
GOS Galactooligosaccharides
HMOS Human milk oligosaccharides
MW Molecular weight
OS Oligosaccharides
PI Prebiotic index
POS Pectin oligosaccharides
PS Polysaccharides
SCFAs Short-chain fatty acids
XOS Xylooligosaccharides

24.1 Introduction

The human colonic microbiota, microbes in the human GI tract, include an enormous biomass of approximately 100 trillion (10^{14}) microorganisms consisting of a vast diversity of microbial species (Míguez et al. 2016). Microbiomes are their collective genome, which contains more than 100 times as many genes as the human genome. Accumulating evidence has shown that microbes in the human GI tract play a critical role in human health, functioning to be capable of harvesting nutrients, producing additional energy otherwise inaccessible to the host, generating vitamins, enhancing the absorption of minerals, metabolizing carcinogens, preventing colonization by pathogens, assisting in the development of a mature immune system, reducing gut infection level, and suppressing colon cancer initiation. There is increasing evidence to unveil that the gut microbiota has a profound effect on human illnesses, including obesity, type 2 diabetes and other metabolic disorders, which comprise of a rapidly increasing health problem worldwide (Nicholson et al. 2012). Moreover, interactions between gut microbiota and the brain affect autism, depression, and Parkinson's disease (Jiang et al. 2015; Scheperjans et al. 2015). Therefore, probiotics, beneficial gut microorganisms, are attracting increasing attention in the scientific community and dietary supplement industry.

Prebiotics are undigested by the enzymes of the upper GI tract, but are instead selectively fermented by various types of intestinal bacteria in the large intestine, as shown in humans (Gibson and Roberfroid 1995) and animals (Patterson and Burkholder 2003). As selectively fermented ingredients, prebiotics lead to specific changes in the composition and/or the activity of gastrointestinal flora (microbiota), with consequent health effects for the host. In general, the traditional and potential prebiotics include, but are not limited to, inulin, HMOS, FOS, GOS, KGMO, POS, AXOS, XOS, pyrodextrins, soy-derived OS, and isomaltooligosaccharides (Olveira and González-Molero 2016). This definition partially overlaps with that of DF, but includes the selectivity of prebiotics for some specific microorganisms. DFs are various carbohydrates and lignin that resist hydrolysis by digestive enzymes in the human small intestine, but may be completely or partially fermented by colonic (large intestine) microflora and/or partly excreted in feces. They contain non-starch polysaccharides (celluloses, hemicelluloses, pectins, gums, and mucilages), resistant starch, inulin, FOS, and GOS. Prebiotic compounds are able to reach the lower gut, being available for the intestinal bacteria, and modulate intestinal (ileum and colon) mobility (Roberfroid et al. 2010; Havenaar 2011; Al-Sheraji et al. 2013). There is convincing data to demonstrate the prebiotic effects of HMOS, FOS, GOS, inulin, β-glucan, and resistant starch in human clinical trials (Davis et al. 2010). Similar effects have been found for POS, GMOS, AXOS and XOS, and thus these constitute a group of emerging prebiotics (Broekaert et al. 2011). Although AXOS and XOS cannot be degraded under conditions similar to the human stomach (Courtin et al. 2009), they can be completely fermented by fecal microbiota *in vitro* cultivation (Kabel et al. 2002). The prebiotic activity of AX, AXOS, and XOS has been studied in pure culture *in vitro* experiments, and thus, *in vivo* and human clinical studies are further required to demonstrate a prebiotic role of those carbohydrates. In addition, the consumption of a combination of FOS and resistant starch prebiotics in rats has been shown to have a synergistic activity (Rodriguez-Cabezas et al. 2010). Insoluble fibers such as cellulose and lignin are generally not fermented by the microbiota in the colon. However, soluble fibers such as FOS and GOS tend to be fermented by colonic microbiota and confer health benefits through the products of their fermentation, including SCFAs like acetic acid, propionic acid, and butyric acid, which serve as energy sources for colonic epithelial cells. These SCFAs lower the pH in the colon, promoting beneficial microorganisms such as Bifidobacteria and Lactobacilli and suppressing pathogenic species. Therefore, these soluble fibers as potential prebiotics in promoting health benefits have been drawing much attention in both academic and industry research. The promotion of beneficial microorganisms in GI is the latest frontier of this growing field. A prebiotic index (PI) has been defined in order to quantify the prebiotic effects of dietary OS (Palframan et al. 2003). A positive PI suggests selective stimulation and growth of bacteria like Bifidobacterium and/or Lactobacillus, while a negative index is influenced by Bacteroides and/or Clostridium.

How can the microbiome influence the response to dietary components? McLaughlin et al. (2015) have assessed the capabilities of 20 carbohydrates including several commercially available prebiotics to promote the growth of 32 human-derived isolates in the genera Bifidobacterium and Lactobacillus in order to distinguish carbohydrates that beneficially modulate growth and the metabolic activity of these bacteria. The results revealed that bifidobacterial strains possessed more diverse carbohydrate utilization profiles as compared to the Lactobacillus species tested, proving that several bifidobacterial strains have the ability to metabolize

AX, XOS, maltodextrin, galactan and FOS components. In contrast, maltodextrin, galactan, arabinogalactan and galactomannan did not support robust growth of any of the Lactobacillus strains evaluated.

The distribution of gut microbiota is affected by various factors, including antibiotics administrated, stress, dietary patterns and ages (Claesson et al. 2012), although those microbes remain stable for individuals during life time. Diet shapes diversity in gut microbiome. It shifts the microbial community and affects exposure to dietary metabolites. Species such as *B. longum*, *B. adolescentis*, *L. rhamnosus* and *L. casei/paracasei* are predominant in adults, whereas *B. breve*, *B. longum subsp. infantis*, *L. gasseri* and *L. salivarius* are usually detected in the infant gut (Wall et al. 2007). Short-term feeding of plant- and animal-based diets alters gut microbiota (David et al. 2014). A DF-deprived microbiota in GI degraded the colonic mucus barrier and enhanced pathogen susceptibility (Desai et al. 2016). Mice fed a high fiber diet had more fiber-degrading bacteria, a thicker mucus lining, and were protected from Citrobacter rodentium infection. Sialyated milk oligosaccharides promoted growth of animals colonized with infant microbiota (Charbonneau et al. 2016). Bifidobacteria and lactobacilli are major probiotic bacteria resident in the human intestine, performing different fermentation capacities and molecular mechanisms upon utilizing carbohydrates. Watson et al. (2013) ascertained that the growth of bifidobacterial strains is promoted by a broader spectrum of carbohydrates than lactobacilli, which may be ascribed to over 8% of the bifidobacterial genome contributed to carbohydrate metabolism (Pokusaeva et al. 2011).

Probiotics such as Bifidobacteria and Lactobacilli are intervened by prebiotics as they have been documented to confer various health benefits, including, but not limited to, immune-modulation (Turroni et al. 2014), restriction of pathogenic bacteria through competition and production of SCFAs and modulation of mucosal barrier function (Miyauchi et al. 2012). Prebiotics are non-digestible food ingredients, which selectively stimulate the growth of gut bacteria. Synbiotics is defined as a combination of probiotics and prebiotics that beneficially affect the host by improving the survival and implantation of live microbial dietary supplements in the GI tract. Synbiotics selectively stimulate the growth and/or activate the metabolism of a limited number of health-promoting bacteria, thus improving host welfare (Pandey et al. 2015). There exists a synergistic relationship between probiotics and prebiotics, since it favors competitive advantage for probiotic rapid adaptation in the GI tract. Synbiotics are available as mixtures of bifidobacteria and/or lactobacillus with FOS and/or inulins.

Potential novel prebiotics could be developed from different sources. They can be found naturally in foods created by chemical and/or enzymatic hydrolysis from OS or other substrates. For example, by using β-galactosidase from various microbial sources, lactulose can be hydrolyzed and subsequent transgalactosylated to produce new prebiotic OS (Villamiel et al. 2014). OS including sucrose, raffinose and stachyose in Tofu whey permeate can be enzymatically synthesized to mixtures of fructosylated α-galactosides and FOS (Corzo-Martínez et al. 2016). Lactose-derived OS such as trisaccharides (4-galactosyl-kojibiose, lactulosucrose, and lactosucrose) can be potential prebiotics through enzymatic conversion from lactose obtained from cheese whey permeate and sucrose (García-Cayuela et al. 2014; Corzo-Martínez et al. 2015). Seaweeds and marine microalgae are alternative sources for the finding of potential prebiotic candidates (Moreno et al. 2017). The production of prebiotics can be hindered by carbohydrate polymers' structure complexity and associated costs. There is an urgent requirement for efficient, sustainable, and less costly processes for prebiotic oligosaccharides application on a large scale in food industry.

24.2 Traditional Prebiotics

24.2.1 Human Milk Oligosaccharides (HMOS)

HMOS are a group of bioactive carbohydrates that generally compose of 3–10 monosaccharide units (Bode 2012), representing the third most abundant solid component of human milk following lactose and lipids (Goehring et al. 2014). The average HMOS concentrations ranges from 5 to 23 g/L, with the highest levels of 20–30 g/L in colostrum and 10–15 g/L in maturate (Marx et al. 2014). HMOS are considered the first example of prebiotics discovered 100 years ago and are growth factors for bifidus flora in breast-fed infants. Recently, they have received much attention because of their potential benefits for the breast-fed neonate (Kunz et al. 2000; Jantscher-Krenn and Bode 2012; Venema 2012).

24.2.1.1 Source and Chemical Structure

HMOS are unique to human beings and are abundant in human milk. They possess an extremely complex structure comprised of at least 1000 different components (Smilowitz et al. 2014). Only trace amounts and structurally less complex oligosaccharides are found in mature bovine milk or bovine milk-based formula (Bode 2009). Currently, approximately 100–200 HMOS have been isolated and identified (Asakuma et al. 2011), and their levels and patterns vary significantly with different individuals and over the course of lactation (Goehring et al. 2014). Regardless of the structural complexity and variation, five common monosaccharides, namely D-glucose (Glu), D-galactose (Gal), *N*-acetylglucosamine (GluNAc), L-fucose (Fuc), and *N*-acetylneuraminic acid (NeuAc), are the building blocks that make up all HMOS (Oliveira et al. 2015). With a few exceptions, almost all HMOS share a basic core structure that includes a lactose (Galβ1-4Glu) at the reducing end with extended chains by glycosyltransferases addition of lacto-*N*-biose units (Gal β1-3GluNAc, LNB, type 1) or lactosamine units (Gal β1-4GluNAc, LacNAc, type 2) (Zivkovic et al. 2011; Yang et al. 2012; Ito et al. 2013). The core structure can further be modified by fucosylation, sialylation or sulphation through either adding fucose in α-(1–2), α-(1–3) and α-(1–4) linkages, sialic acids in α-(2–3) or α-(2–6) linkages or sulphate groups at the terminal positions (Petschacher and Nidetzky 2016; Smith-Brown et al. 2016; Lis-Kuberka et al. 2017). The modification processes not only lead to structural variations of HMOS, but also divide them into two groups: neutral and acid oligosaccharides. The former contains fucosylated carbohydrates and is a higher concentration in the human milk than the latter which contains either sialic acids or sulphate groups (Thurl et al. 2010).

HMOS composition depends on the woman's Secretor and Lewis blood group statuses which express certain glycosyltransferase (Espinosa et al. 2007). More than 70% of Caucasian women are believed to be Secretors and actively express the α-(1–2) fucosyltransferase FUT2 gene. Therefore, HMOS from those women are characterized by the presence of characteristic 2-fucosyllactose (2'FL), lacto-*N*-fucopentaose 1 (LNFP1) and other α-(1–2) fucosylated HMOS, which cannot be detected in the milk from non-Secretor women who do not express FUT2 gene (Morrow et al. 2004; Bode 2015). In addition, the milk from the women with Lewis positive blood group status who express the α-(1–3/4) fucosyltransferase FUT3 gene contains characteristic α-(1–4) fucosylated HMOS (LNFP2) (Jantscher-Krenn and Bode 2012; Marx et al. 2014).

Qualitative and quantitative identification of HMOS structure and composite will help better understanding their biological functions. Chromatography has become a classic technique to elucidate HMOS structures. Various chromatographic methods, including normal-phase, reversed-phase, gel permeation, high-performance anion-exchange, lectin affinity chromatography, porous graphitized carbon, hydrophilic-interaction liquid chromatography, gas chromatography, and capillary electrophoresis, have been used to analyze their composition and structure (Galeotti et al. 2014; Mantovanl et al. 2016; Yan et al. 2017). In addition, other advanced and sensitive analytical techniques such as mass spectrometry and nuclear magnetic resonance spectroscopy have also been utilized (Maliniak and Widmalm 2014; Wu et al. 2017). Structural complexity of HMOS presents two-sided effects. On one hand, the complexity makes the large-scale commercial production of HMOS impossible. But on the other hand, structural complexity is the basis for a multitude of biological functions.

24.2.1.2 Functionality and Health Effects

Although HMOS are resistant to gastrointestinal digestion and are considered to have no nutritional values, they could provide health benefits for breast-fed infants in unique ways.

Prebiotics Effect

HMOS exert prebiotic function to shape a healthy gut microbiota by selectively promoting the growth of beneficial bacteria, while suppressing that of pathogens (Musilova et al. 2014). The first recognition of the prebiotic effects of HMOS is found in the work of Schönfeld who described HMOS as a bifidus factor that stimulates the growth of specific *Bifidobacterium* and *Bacteroides* species (Kunz 2012). *Bifidobacterium longum* subsp. *infantis*, *Bacteroides fragilis* and *Bacteroides vulgatus* strains are found to effectively metabolize HMOS (Marcobal et al. 2010). HMOS are structure-specific and those containing GluNAc appear to be the most active in stimulating growth of bifidobacteria (Coppa et al. 2006). *Bifidobacterium* species are one of the first bacteria to colonize the human gastrointestinal tract, and certain species such as *Bifidobacterium longum* subsp. *infantis* (*B. infantis*) and *B. breve* are more common in the infant gut (Thomson et al. 2017). *B. infantis* is unique among gut bacteria because it can digest any HMOS structure and utilizes them as its sole carbon source (Ruiz-Moyano et al. 2013; Underwood et al. 2015; Wickramasinghe et al. 2015). However, it is worthy note that although *B. infantis* can grow very well on any type of HMOS, it prefers certain short-chain and fucosylated HMOS over their high MW counterparts (Garrido et al. 2015).

Antiadhesive Antimicrobials

HMOS possess anti-infective properties against many pathogenic microorganisms in the infant gastrointestinal tract, including *Salmonella*, *Listeria*, and *Campylobacter*, which generally bind or adhere to the host's cultured epithelial surface (Martin et al. 2016). The antiadhesive antimicrobial effects of HMOS are based on their capacity to act as a soluble decoy receptor analog to compete with epithelial ligands for bacterial binding and to prevent the attachment of pathogenic microorganisms to epithelial surfaces (Smilowitz et al. 2014). The similarities between HMOS and carbohydrates of epithelial cell surface lead the former to mimic the later to form specific interactions with pathogenic microorganisms (Houghteling and Walker 2015). Antiadhesive antimicrobial effects of HMOS are not only limited to bacteria, they also affect certain protozoan parasites and viruses. HMOS are found to effectively inhibit Entamoeba histolytica, a parasite that causes amoebic dysentery or amoebic liver abscess (Jantscher-Krenn et al. 2012; Kutty 2016). Furthermore, HMOS are also reported to have potential protective effects against HIV-1 mother-to-child transmission during breastfeeding (Bode 2012). The antiadhesive antimicrobial effects of HMOS contributes to a lower incidence of intestinal, upper respiratory and urinary tract infections in breast-fed infants (Bode 2012; Smilowitz et al. 2014).

Immune Modulators and Brain Development

HMOS possess immunomodulatory properties that modulate epithelial and immune cell responses (Bode 2012). Specific glycan-binding proteins expressed by nearly all epithelial and immune cells interact with HMOS, which directly modulate immune system response and development (Goehring et al. 2014). HMOS are found to act either locally on cells of the mucosa-associated lymphoid tissues or on a systemic level to reduce excessive mucosal leukocyte infiltration and activation. Through these interactions, they protect against necrotizing enterocolitis, one of the most common fatal disorders in pre-term infants.

HMOS are also reported to facilitate infant's brain development by providing essential nutrients such as galactose and sialic acid (Wang 2009; Ruhaak et al. 2014). Galactose, a major component in HMOS, is involved in the formation of galactolipids, an essential component for rapid brain development (Jantscher-Krenn and Bode 2012). In addition, sialic acid contains NeuAc, a component of gangliosides that is covalently linked to neural cell adhesion molecules in order to mediate cell-cell interactions for memory (Charbonneau et al. 2016). The role of the HMOS is to ensure that galactose and sialic acid levels do not become depleted during pre- and post-natal stages of brain development.

Overall, the beneficial effects of HMSO are not only based on total amount, but on their distinct structural composition (Marx et al. 2014). Currently, a lack of availability of these structures in sufficient quantities and of purity to clinically test their biological functions is the main barrier to discern the biological role of HMOS. Exclusively assigning some observed effects to HMOS is a challenge, since other components in the milk might also elicit the particular response when studying the physiology and clinical behaviors of breast-fed versus formula-fed infants (Espinosa et al. 2007).

24.2.2 *Fructooligosaccharides (FOS)*

Fructooligosaccharides (FOS) are non-digestible carbohydrates with plant origins, and represent one of the major bifidogenic oligosaccharides (Slavin 2013). Similar to HMOS, they also serve as prebiotic compounds that improve the well-being and health of the host. The average daily intake varies by region, with 2–4 g for North Americans and 2–12 g for Europeans, respectively (Sousa et al. 2011). Because of their functional properties and economic potential, FOS have been increasingly used in food and health industries. The global FOS market is expected to reach US$ 3.52 billion by 2024 (GVR 2016a).

24.2.2.1 Source and Chemical Structure

FOS are found in varying concentrations with milligram quantities in many foods, including (a) some vegetables and fruits, such as onion, garlic, leek, tomato, asparagus, artichoke, lettuce, beetroot, burdock, yacon, apples, and bananas; (b) grains, such as barley, wheat, rye, triticale, and oat; and (c) other food products, such as brown sugar, beer, and honey (Díez-Municio et al. 2013). Onions have the highest FOS content, ranging from 25 to 40% of dry matter basis. The daily intake of FOS from those natural resources is approximately 1 g in the western diet (Bornet 2001).

Beyond their natural presence in some common foods, FOS can be produced commercially through two ways: (1) synthesis from sucrose, and (2) hydrolysis of inulin (Fontana et al. 2011; Surin et al. 2012; Lorenzoni et al. 2014; Wang and Wang 2016). In the former route, the fructose moiety of substrate sucrose are either polymerized to sucrose or are transferred to other carbohydrates through transfructosylation reaction by enzymes such as fructansucrase or fructosyltransferase from different fungi and bacteria (Díez-Municio et al. 2013; Arthee and Vijila 2014). Sucrose plays a dual role during those reactions: as a fructose acceptor during polymerization and as a fructose donor during transfructosylation (Chuankhayan et al. 2010). FOS generated from this route have been considered as short-chain FOS which consist only of 1–3 fructose units (Bornet 2001). On the other hand, FOS produced in the second route are found to have longer chains with a DP ranging from 3 to 9 (Borromei et al. 2009). In spite of the production of short-chain FOS, the first route is to be economically advantageous, since substrate sucrose is cheaper than inulin (Hang et al. 2013).

The chemical structure of FOS consists of a chain of fructose units linked by β-(2–1) glycosidic bonds terminating with a glucose unit. Three most important FOS include: (1) 1-kestose which has two fructose units attached to the glucose unit; (2) nystose which attaches three fructose units to the glucose and (3) fructosylnystose has four fructose units (Flores-Maltos et al. 2016). Human digestive enzymes could not hydrolyze these β-(2–1) glycosidic bonds since they are specific on α-glycosidic bonds (Sabater-Molina et al. 2009). The difference between FOS and inulin molecules is the polymer chain length characterized by DP, defined by the number of individual fructose units. FOS have a DP value less than 10, while the DP values of inulin molecules range from 2 to 60 (van de Wiele et al. 2007). Similar to HMOS, chromatography has become a popular technique to characterize the structures and compositions of FOS, especially high-performance liquid chromatography (HPLC) with different detectors (charged aerosol and evaporative light scattering) (Li et al. 2013, 2014). Furthermore, high-performance anion-exchange chromatography (HPAEC) coupled with pulsed electrochemical detection (PED) has been routinely used to qualitatively analyze FOS. However, a lack of standardization makes it impossible to identify each component. Matrix-assisted laser desorption/ionization time-off light mass spectrometry (MALDI-TOF-MS) opens the door not only to determine chain length distribution but also to qualitatively and quantitatively analyze FOS (Borromei et al. 2009).

24.2.2.2 Functionality and Health Effects

Because of various chemical and structural conformations, FOS are recognized as a flexible and appealing health-promoting ingredient for different food applications. Several authors have conducted comprehensive reviews about their functionality and health benefits (Sabater-Molina et al. 2009; Dominguez et al. 2014; Sridevi et al. 2014).

Functional Ingredients in the Food Industry

A number of unique functional properties make FOS important food ingredients. There include high water solubility, high water-holing capacity and viscosity, thermal stability that improves texture, refined taste, and a longer shelf life (Sabater-Molina et al. 2009). Furthermore, less sweetness and small amount of energy make FOS a good candidate as a sugar substitute for development of various food products in which sucrose is prohibited. Examples of food products that contain FOS include light jam products, ice cream and confection such as hard candies, gums and marshmallows (Sangeetha et al. 2005).

Prebiotic Effect

As all non-digestible oligosaccharides, FOS resist gastric acidity, escape digestion by enzymatic hydrolysis in the small intestine and eventually reach the colon where they are thoroughly fermented by bacteria into the short-chain fatty acids (SCFA), lactate, CO_2 and H_2. The SCFA including acetate, propionate and butyrate create an acidic environment in the colon that changes the gut microflora population (Morrison and Preston 2016). In such environments, some beneficial bacteria such as *Bifidobacteria, Lactobacilli*, and *Eubacterium*, are favorably developed, while the growth of some pathogenic species such as *Escherichia coli, Clostridium, Streptococcus faecalis* and *Proteus* are suppressed (Zhang et al. 2015b; O'Callaghan and van Sinderen 2016). The prebiotic effect of FOS is based on their capacity to promote a healthy balance of intestinal microbiota and decrease gastrointestinal infections. The functions of SCFA generated from FOS are not only limited to improving intestinal environment; they also exert multiple beneficial effects to directly and indirectly influence human bodies.

Improving Mineral Absorption

FOS have positive effects on mineral absorption, especially for magnesium, calcium, iron and zinc (van den Heuvela et al. 2009; Wang et al. 2010). Those minerals play an important role in the development and growth of the body, and mineral deficiencies remain a critical nutritional issue worldwide. Mineral absorption generally occurs in small intestine. However, when fed with a FOS-supplemented diet, FOS bind with the minerals to reduce their absorption in the small intestine and to push them to the colon where their bioavailability increases (Mussatto and Mancilha 2007; Scholz-Ahrens et al. 2007). The increased mineral bioavailability in the colon is attributed to a fact that SCFA produced from FOS fermentation build an acid environment that favors mineral solubility (Wang et al. 2010; Sheridan et al. 2014). In addition, the presence of FOS increases the water content and the absorptive capacity of the colon and improves gut health, which also contributes to enhancing mineral absorption (Sabater-Molina et al. 2009).

Reducing Lipids and Cholesterol Levels

FOS have hypotriglyceridemia and hypocholesterolemic effects that reduce the levels of serum lipids and cholesterol (Costa et al. 2015; Cronin et al. 2016). These effects can be ascribed to three mechanisms. First, some SCFA produced through FOS fermentation, specifically butyrate and propionate, inhibit cholesterogenesis and lipogenesis pathways. Butyrate is well known to inhibit liver cholesterol synthesis. Propionate reduces the hepatic cholesterol synthesis rate, thereby lowering plasma cholesterol levels. It also may inhibit the synthesis of fatty acids in the liver, which could lead to lower rates of triacylglycerol secretion rate (Ooi and Liong 2010). Second, FOS significantly reduce the production of glucose and insulin, main stimulators in the control of lipogenesis (Costa et al. 2012). The decrease in sugar and insulin levels leads to a reduction of lipogenesis (Dehghan et al. 2013). Finally, FOS alter the intestinal environment and enhance the growth of *Bifidobacterium* and *Lactobacillus*. These beneficial bacteria facilitate the precipitation and excretion of fecal bile acid, thereby decreasing their intestinal concentration, which results in the reduction of lipase activity and the levels of serum lipid and cholesterol (Xu et al. 2002).

Preventing Colon Cancer

FOS have inhibitory effects on colon cancer, which are mainly linked to FOS prebiotic effects and SCFA production during FOS fermentation in the colon (Sabater-Molina et al. 2009). First, FOS prebiotic effects inhibit the growth of pathogenic bacteria and reduce the production of the carcinogenic substances. SCFA, especially butyrate and propionate, are able to induce the differentiation of T-regulatory cells to assist in the control of intestinal inflammation (Ríos-Covián et al. 2016). They also are found to protect against the development of colorectal cancer. Among various SCFA, butyrate has received the most attention and can be both preventive and inhibitory in colon carcinogenesis by playing multiple roles: (1) providing a source of energy for human colon epithelial cells, (2) promoting normal cell proliferation and inhibiting carcinogenic cell, and (3) presenting a chemo preventive effect in carcinogenesis (Gill and Dudeja 2011). Finally, the presence of FOS increases stool mass and decreases colonic transit time, thereby reducing the exposure time of the colonic microbials to potential carcinogenic agents (Sabater-Molina et al. 2009).

24.2.3 Galactooligosaccharides (GOS)

Like HMOS and FOS, GOS refer to a group of non-digestible carbohydrates, which serve as prebiotic compounds that enhance health-related physiological activities (Sangwan et al. 2011). Because of their similarity to HMOS, GOS are one of only three prebiotics recognized by experts for its breadth of clinical evidence and have drawn a great of attention worldwide (Torres et al. 2010; Sangwan et al. 2011). The global GOS market size was estimated at US$643.3 million in 2015 and is predicted to grow rapidly within the next 10 years (GVR 2016b).

24.2.3.1 Source and Chemical Structure

Although GOS occur naturally in human milk with the total GOS concentration in human milk approximately 1 g/L (Angus et al. 2005), most available GOS are synthesized from lactose through a classic transgalactosylation reactions (Rodriguez-Colinas et al. 2012; Vera et al. 2012). This reaction was first observed in the early 1950s. Since then, it has become an important process to commercially produce GOS, with Japan and Europe as pioneers and leaders in the production and development of GOS worldwide (GVR 2016b). During the reaction, many parameters, including the source and concentration of substrate, enzyme type, pH, time, and temperature, influence GOS synthesis. Highly concentrated lactose, either in the refined form or in concentrated lactose syrup, has been used as the substrate. Lactose-rich cheese whey permeate during whey ultrafiltration which has the lactose concentration of 75% (g/g) is an ideal substrate for GOS production (Barile et al. 2009). In terms of enzymes, both glycosyltransferases and glycosidases, which are responsible for the transfer of glycosyl moieties from a donor sugar to an acceptor, can be used to convert lactose to GOS. Although the former is highly selective and can produce a high yield of GOS, a high cost and the requirement of specific sugar nucleotides as substrates rule it out as a good candidate for the commercial production of GOS (Torres et al. 2010). On the other hand, β-galactohydrolases, one type of glycosidase that can simultaneously hydrolyze lactose to D-galactose and D-glucose and catalyze the transgalactosylation reaction to form GOS, is much more readily available, although it is less stero-selective and might result in low yield (Lamsal 2012). Converting lactose into GOS by β-galactosidases is a kinetically controlled reaction, and the ratio between lactose hydrolysis and GOS synthesis is dependent on the reaction conditions, such as lactose concentration, the reaction temperature and the intrinsic enzyme properties (Torres et al. 2010; Rodriguez-Colinas et al. 2016).

The chemical structure of GOS consists of a chain of galactose units polymerized to a terminal glucose unit with different glycosidic linkages, namely β-(1–1), β-(1–2), β-(1–3), β-(1–4) and β-(1–6), to form a generic structure Glu α-(1–4) (β-(Gal1–6)n, where Glu refers to glucose molecule, Gal refers to galactose, and n is the number of galactose units and varies from 2–10 (Lamsal 2012; Belorkar and Gupta 2016). Chemical structures, including the saccharide composition, type of glucoside linkage, and DP, greatly influence their chemical and physiological characteristics, functionality and application in food and health industries (Gosling et al. 2010). Based on their nature, GOS are classified as dietary fiber. An analytical method based on enzymatic treatment with β-galactosidase followed by a HPAEC, and coupled with pulsed amperometric detection (PAD) has been developed and approved as AOAC Method 2001.2 to analyze GOS in the food and feed products (AOAC 2005). However, the major shortcoming of this method is that it is not suitable for the analysis of the products containing high levels of lactose. Recent efforts have made use of liquid chromatography to analyze broad samples including infant formula (Austin et al. 2014).

24.2.3.2 Functionality and Health Effects

Differences in GOS structure and composition provide them with a number of unique physicochemical and biological characteristics as functional ingredients for food and health applications (Lamsal 2012).

Functional Ingredients in the Food Industry

As a food ingredient, GOS possess functional benefits including: (1) high solubility that results in a transparent aqueous solution and easy dissolution in milk and other dairy products; (2) a clean taste and lessened sweetness that do not affect the taste of products being applied; and (3) the ability to withstand high temperatures and acid stability that allows them to be incorporated into a wide variety of foods (Otieno 2010). Because of these unique properties, GOS are widely used in various food systems, including infant formula, dairy product, beverage, bakery product, pet feed, and sugar substitutes (Sangwan et al. 2011). Foods and beverages are reported to dominate the demand of GOS with a revenue share of over 90% in 2015, with dairy products, ice cream, and cereals contributing to an increase in demand (GVR 2016b). GOS is an ingredient used in food products that offers double benefits: improved organoleptic quality and better-balanced nutrition.

Beneficial Health Effects

GOS easily mimic HMOS in terms of providing health benefits. Although HMOS exhibit a wide range of biological activities, they are not commercially available due to the complexity of their structures. GOS share similarities with the core structure as HMOS and are widely used in infant formula as an inexpensive alternative to HMOS. This is especially true of galactose-containing hetero-oligosaccharides, which have an even closer structure to that of HMOS (Intanon et al. 2014; Guo et al. 2017). Various strategies, including microbial and fermentative methods, have been developed to produce HMOS that more accurately and structurally- mimic GOS by employing single, appropriately engineered microorganisms or by sialylation of GOS using an enzymatic approach (Wang et al. 2015b).

Similar to other non-digestible dietary oligosaccharides, GOS are useful dietary tools that provide health benefits. These benefits include:

1. Improving intestinal health by modifying intestinal microflora and barrier functions. The proliferation of bifidobacteria is selectively promoted, while deteriorating bacteria are inhibited. In addition, GOS also regulate mucosal transport and host defense against adherence and invasion of harmful bacteria and viruses to enhance the intestinal barrier. One study indicated that compared with FOS and other dietary oligosaccharides, GOS have the greatest adherence inhibition of *E. coli* E2348/69 cells on HEp-2 and Caco-2 (Shoaf et al. 2006).
2. Improving mineral absorption including iron and calcium. One study found that calcium was absorbed more efficiently and bone ash weight was significantly higher in rats fed with a diet containing GOS than their control counterparts (Chonan et al. 1995).
3. Other health benefits, including reducing constipation, preventing cancer and allergies, and modulating the immune system, are well documented (Kukkonen et al. 2007; Niittynen et al. 2007; Macfarlane et al. 2008). Similarly to FOS, the benefits of GOS are also associated with the production of SCFA during fermentation in the colon.

24.3 Potential Prebiotics

24.3.1 Arabinoxylan Oligosaccharides (AXOS) and Xylooligosaccharides (XOS)

As the second most abundant carbohydrate in terms of plant biomass after cellulose, heteroxylans are localized in plant cell walls and are composed of a backbone of β-1,4-linked D-xylopyranoside units. The DP of xylan backbone, type, and relative abundance, as well as monosaccharide unit of the side branches results in the structural versatility of those glycan polymers, and further result in compositional, functional, and health benefits discrepancies. Prebiotic and other health-associated effects of cereal-derived AXs, AXOS, and XOS have been extensively and comprehensively reviewed and summarized by Broekaert et al. (2011). Here, we focus on some recent research in terms of the potential prebiotic effects of AX, AXOS, and XOS.

24.3.1.1 Source

Xylan, a major component of hemicellulose and lignocellulosic biomass, is available broadly and abundantly in nature. Cereal and agricultural wastes including brans, stalk, straw, cob, hull, husk, bagasse and wood pulp are principal sources of xylan. As polysaccharides residing in the cell wall matrix of cereals and other plants, AXs can be obtained from corn (Rose 2011), wheat (Wang et al. 2004), rye (Ragaee et al. 2001), barley and triticale (Dervilly-Pinel et al. 2001), barley, oat, and sorghum (Izydorczyk and Dexter 2008). AXs are mainly present in the bran and aleurone fractions, and represent 50% of dietary fibers in wheat (Neyrinck and Delzenne 2010). For example, approximately 2.4 million tons of corns is dry-milled annually, leading to the production of 350,000 tons of corn bran.

As a representative of a new class of potential prebiotics, AXs can be selectively broken down in the colon by intestinal bacterial enzymes including xylanases and arabinofuranosidases (Hughes et al. 2007; Grootaert et al. 2007) to become low MW AXOS. XOS is the hydrolysis product of xylan, and can serve as a carbon source for the colonic commensal bacterial population, acting as a potential prebiotic. In addition, there are some phenolics such as ferulic and coumaric acids bound to the XOS, which exert antioxidant and immunomodulatory activities.

24.3.1.2 Chemical Structure

AX is a major type of hemicellulose in cereals with structural variations in different species and plants. Its backbone contains a linear β-1,4-D-xylopyranosyl (Xyl) xylose, where α-L-arabinofuranosyl (Ara) branches (substituents) are attached to the Xyl at the *O-2* and/or *O-3* positions (Izydorczyk and Biliaderis 2007). Additional substituents can be other hexoses, pentoses, uronic acids, acetic acid, or ferulic acid (Andersson and Åman 2008). AXs obtained from various sources vary in the linkage point of the Ara branches on the Xyl backbone, the presence of the Ara branches in different arabinose/xylose (A/X) ratios, and the abundance of the types of branches. A/X ratios exhibit wide variability even with the same source. For instance, the A/X ratios of AXs fractionated from wet-milled corn fiber for coarse fiber, fine fiber, and spent flake are 0.63, 0.82, and 0.97, respectively, suggesting varied degrees of branching (Doner et al. 2001). Variability in molecular property leads to different rheological characteristics, as revealed by Andrewartha et al. (1979). The rheological properties of wheat AXs and enzymatically modified AX solutions indicate the degree of branching and the MW distribution. Hence, it is possible for AXs as potential prebiotics through enzymatic hydrolysis by modifying sugar composition, branch linkage points, and the structure of the branches including Xyl galactose (Gal), glucuronic acid (GlcA), and the branch distributions on the Xyl backbone (Coda et al. 2014). Corn bran heteroxylan possesses about 75 ferulic acid esters and is connected by about 30 diferulic crosslinks (Saulnier and Thibault 1999). Structurally, the cell wall of cereal pericarp in corn is a three-dimensional network of cellulose (fibrils) and hemicellulose bundles connected by insoluble AXs and a gel-type matrix made of soluble AXs, where diferulic acids are esterified to the xylans as bridges and structural proteins are embedded in the networks. Each corn bran heteroxylan includes about 75 ferulic acid esters and is connected by about 30 diferulic crosslinks. Therefore, fibers in cereal grains are commonly insoluble, which cannot be well fermentable.

AXs are generally classified as water extractable (WE-AX) or water unextractable (WU-AX) (Courtin and Delcour 2002). In wheat, the inner endosperm are 2–3% AX, comprising of about two thirds WU-AX and one third WE-AX. The wheat bran fraction consists of more than 25% AX with major WU-AX (Maes and Delcour 2002). In addition, AX in the pericarp contains an A/X ratio of about one, whereas the AX in the aleurone layer contains a much lower A/X ratio ranging from 0.2 to 0.4 (Benamrouche et al. 2002). AXs from crude corn fiber with alkaline extraction consist of 60% of total bran weight with some glucuronic acid and trace amounts of D and L-galactose. Hydrogen peroxide is usually added into the extraction medium to remove lignin and bleach brown color (Doner et al. 2000; Gelroth and Ranhotra 2001). To fragment AX, chemical extraction of AX from cereal grains or bran is followed by enzymatic hydrolysis of the extracted AX by specific AX-degrading enzymes to yield AXOS with different DP and A/X ratios (DS) (Swennen et al. 2006; Grootaert et al. 2007). Therefore, AXOS can be structurally characterized by both the average DP (x) and DS (y), represented as AXOS-x-y. The DP, type and degree of substituents of XOS have effects on the rate of fermentation as indicated by XOS from Eucalyptus wood during *in vitro* fermentation (Kabel et al. 2002). Physicochemical and fermentation characteristics (Monobe et al. 2008) as well as immunological modulation in different organisms (Geraylou et al. 2013) depend mainly on the structural properties of AXOS. Although the chemical structure of XOS varies with the xylan source, it commonly possesses chains of xylose linked by β-(1–4) bonds, with a DP ranging from 2 to 10. The structural characteristics of AXOS such as A/X (ara/xyl) branching ratio, average DP, acylation, and other molecular architecture may affect AXOS fermentation through sterically impeding the formation of xylanase-substrate intermediates, and thereby influencing nutrient digestibility and ultimately human or animal performance (van Craeyveld et al. 2010). Therefore, the prebiotic effects of AX-derived OS are highly affected by their structures and MW sizes (Damen et al. 2011; Broekaert et al. 2011).

24.3.1.3 Production of AXs, AXOS, and XOS

Industrially, AXS, AXOS, and XOS are produced by chemical or enzymatic hydrolysis from xylan in lignocellulosic materials. A traditional method to create soluble AXs was to extract AXs with sodium hydroxide followed by decantation, neutralization and precipitation of insoluble material from the liquid phase with sulfuric acid, centrifugal separation and microfiltration or ultrafiltration of the supernatant followed by spray drying of the retentate (Bataillon et al. 1998). In order to break the network crosslinks in cereal bran polysaccharides, and further solubilize AXs, an alkali treatment of the bran is applied to cleave the crosslinks followed by a sequential precipitation using ethanol or ammonium sulphate along with the isolation of different AX fractions (Doner et al. 2001). Swennen et al. (2006) utilized different enzymes including amylases, proteinases and endoxylanases to hydrolyze wheat bran and yield soluble AXs. Recently, autohydrolysis technology was used to solubilize AXs to produce hydrolyzates of a DP of over 2 (Rose 2011). This hydrothermal process with temperatures of between 160 and 220 °C under acidic conditions hydrolyzes bran to release soluble AXs. Compared to enzymatic hydrolysis, this autohydrolysis is energy demanding and costly. Various technologies have been applied to extract xylan from wheat bran, including steam, chemicals such as barium and calcium hydroxide (Bergmans et al. 1996), and a combination of both (Maes and Delcour 2002). Xylan has been purified, characterized, and enzymatically hydrolyzed to produce XOS in different studies (Swennen et al. 2006; Zhang et al. 2011). Furthermore, XOS has been discovered to have a prebiotic effect. Recently, xylan was extracted from wheat bran using either an autoclave or a microwave oven, and then was enzymatically hydrolyzed by thermostable xylanase RmXyn10A to produce XOS (Immerzeel et al. 2014). Human gut bacteria including *L. brevis* (DSMZ 1269), *B. adolescentis* (ATCC 15703) and the species pair *Weissella cibaria* were utilized to assess the potential prebiotic properties of the XOS. The results revealed that xylobiose from XOS hydrolysate as carbon source was used by all tested bacterial species while xylotriose could be consumed by *B. adolescentis* and the Weissella strains. *L. brevis*, *B. adolescentis* and the *Weissella spp.* all exhibited growth on XOS with increases in cell density, lactic acid and acetic acid production after 48 h of *in vitro* incubation.

Various types of AX-degrading enzymes with respective substrate specificities are useful tools in converting AX into OS with different DP. Specially, endo-xylanases are of critical due to their breakdown of the main chain, and arabinofuranosidases are responsible for removing arabinose substituents. Endo-acting xylanases hydrolyze and solubilize AX by cleaving the β-1,4-linked xylopyranose backbone of the AX heteropolymer into AXOS. The commonly used xylanases are in the glycoside hydrolase families 10 or 11, which possess different tolerance for substituents in AX. They are active on unsubstituted groups in the xylan chain, but have limited tolerance for arabinose substituents. Morgan et al. (2017) have evaluated the influence of xylanases from families 10 and 11 on the production of small chain AXOS (X2–X4) in wheat AXs. Xylanase from family 10 has been shown to have a larger hydrolytic efficiency in AXOS production as compared with xylanase from family 11 or combination of family 10 and 11. The maximal conversion of AXs into AXOS was seen at pH 2.5. Xylanases from *Clostridium thermocellum* belong to glycoside hydrolase family 5, and are active through arabinose substituents (Correia et al. 2011). Arabinoxylanases in the glycoside hydrolase family 5 are categorized under subfamily 34 to display similar substrate specificity as xylanases (Labourel et al. 2016). Recently, Falck et al. (2018) conducted a study of utilizing an arabinose-specific xylanase from glycoside hydrolase family 5 (GH5) to hydrolyze wheat and rye AX, thus generating arabinose substituted oligosaccharides with 2–10 xylose residues in the main chain but no unsubstituted XOS. Molecular modelling has further revealed that the hydroxyl groups of xylose residues in the active site were exposed to solvent, suggesting that arabinose substituents can be attached with both glycone and aglycone groups. In addition, those AX hydrolysates could stimulate the growth of *B adolescentis* but not of *L. brevis*, indicating that this GH5 enzyme is a useful tool in selection of prebiotics. Xylanases (EC 3.2.1.8) in a pilot scale were used to cleave internal β-1,4-xylosidic bonds in AX to yield AXOS with a low avDP in wheat bran (Swennen et al. 2006). XOS (xylobiose and xylotriose) from hardwood xylan were produced by using an immobilized endo-xylanase of *Clostridium* strain BOH3 (Rajagopalan et al. 2016).

Probiotics and prebiotics may be an alternative solution in antibiotic-free diets to stabilize the intestinal health of monogastric animals. The prebiotic potential of a commercial feed GH11 xylanase was evaluated *in vitro* through degradation of AX from wheat bran cell walls (Ravn et al. 2017). The AXOS produced by the enzyme were monitored using immuno-microscopy, and analyzed by non-starch polysaccharide (NSP), mass spectrometry (MS) and chromatography to assess the effect of AXOS glycan complexity and enzyme dosage on fermentation patterns in a wheat-based diet. The results showed that butyrate-producing bacterial genera Faecalibacterium and Intestinimonas significantly increased in the fermentation of wheat bran with GH11 xylanase. The GH11 xylanase was capable of solubilizing and degrading wheat bran AX to produce low DP AXOS, which can be fermented by cecal microbiota, leading to microbiota shifts and beneficial effects on *in vitro* transepithelial resistance.

AXs can be WE- or WU-AX depending on the ratio of arabinose, xylose, MW, and degree of cross-linking with other polymers. The influences of three commercial xylanases and a peptidase on WU-AX were studied in Brewer's spent grain (BSG) (Severini et al. 2015). The results showed that xylanase alone was capable of solubilizing 23.7% of WU-AX, while the peptidase exhibited no effect. Furthermore, the combination of xylanase and peptidase boosted WU-AX solubilization over 1.6 times, demonstrating that access to xylan backbone has a synergistic, enhancing effect with proteolytic activities. WU-AXs have been displayed to exert several benefits on human health. For instance, hydrolysis of WU-AX yields AXOS small molecules, which performed prebiotic effects in the colon. Damen et al. (2012) investigated the *in situ* production of AXOS by 3 mesophilic xylanases obtained from *Bacillus subtilis*, *Aspergillus niger* and *Hypocrea jecorina*, and one thermophilic xylanase from *H. jecorina* (HjXynA) during bread-making. The results showed that HjXynA maximally solubilized the AX fraction among 4 enzymes, generating 2.1% (dry basis) of AXOS with an avDP of 9 after delivering the enzyme amounts without impairment of dough manageability. In addition, rye or wheat bran fortified breads applied HjXynA produced high quality breads of over 2% AXOS with an avDP of 26 and 19, respectively.

24.3.1.4 Functionality and Health Effects

AXs are chemically and/or enzymatically hydrolyzed into AXOS and XOS, which have been shown to exert prebiotic activities in the colon of humans and animals through selective stimulation of beneficial intestinal microbiota. Some *in vitro* studies and *in vivo* intervention trials on animals and humans have demonstrated health-related effects from the dietary intake of AX, AXOS, or XOS, as potential prebiotics.

AXs have gained considerable attention due to their potential positive effect in the human GI tract (Redgwell and Fischer 2005). The consumption of AX has been linked to metabolic improvement in diabetic individuals (Garcia et al. 2007). There are several forms of AXs including soluble, insoluble, high MW, and enzymatically modified short-chain fractions, with high MW possessing the most activity (Monobe et al. 2008). Varying with DP, while AXs play some role in lipid and glucose metabolism, the physiological influences of AXs are not completely understood. AXs with prebiotic properties from wheat play a role in the prevention of obesity. Neyrinck et al. (2011) reported water-extractable high MW AXs were capable of modulating gut microbiota and lipid metabolism in high-fat (HF) diet-induced obese mice. The results showed that a HF diet supplemented with AXs (10% w/w) restored the number of bacteria including Bacteroides-*Prevotella spp.* and *Roseburia spp.*, and markedly elevated the amount of *B. animalis lactis* in caecal bifidobacteria, which was accompanied by the improvement of gut barrier function and by the lowering of circulating inflammatory marker. This study also revealed that the HF diet with AXs reduced adipocyte size, HF diet-induced expression of genes mediating differentiation, fatty acid uptake, fatty acid oxidation and inflammation, lipogenic enzyme activity in the subcutaneous adipose tissue, hepatic cholesterol accumulation, and insulin resistance. Data in literature is inconsistent with respect to the relationships between the immunomodulatory effects and the structure and source of AXs. Zhang et al. (2015a) comprehensively reviewed possible mechanisms between immune-enhancing effects of AXs and their structural heterogeneity through *in vitro*, *in vivo*, animal and human research.

As emerging prebiotics, AXOS and XOS are drawing attention as functional ingredients in the pharmaceutics, feed, and food industries (Aachary and Prapulla 2011; Broekaert et al. 2011). The prebiotic effects of AXOS on gut microbiota in promoting health and growth performance have been demonstrated in terrestrial animals, aquatic vertebrates (Gomez and Balcazar 2008), and humans. *In vitro* studies using a simulator of the human intestinal microbial ecosystem have unveiled that AXOS showed a higher potential than inulin to shift the beneficial carbohydrate fermentation toward the distal parts of the colon (Grootaert et al. 2009). In a rat trial, supplementation of AXOS with an avDP of 3 or 5 promoted bifidobacterial growth and butyrate production, while AXOS with an avDP of 61 presented no effect on bifidobacteria nor on butyrate production (van Craeyveld et al. 2008). In healthy humans, daily administration of 2.2 g of AXOS with an avDP and A/X of 15 and 0.26, respectively, beneficially modulated the colonic bacterial metabolism (Cloetens et al. 2008). *In vitro* studies using Simulator of Human Intestinal Microbial Ecosystem (SHIME) have revealed that an efficient fermentation of the AXOS resulted in the stimulation of certain bifidobacterial species and produced profiles with high propionate (van den Abbeele et al. 2009). Studies have shown that colonic microbial metabolism plays an important role in accumulating uremic retention solutes in patients with chronic kidney disease (CKD) (Meyer and Hostetter 2012). In order to assess the effect of prebiotics on lowering intestinal generation of microbial metabolites and improving insulin resistance. Poesen et al. (2016) delivered AXOS (10 g twice daily) and maltodextrin for 4 weeks in a randomized, placebo-controlled, double-blind, cross-over study in 40 CKD patients without dialysis. Interestingly, there was no significant effect of AXOS on serum *p*-cresyl sulfate, *p*-cresyl glucuronide, indoxyl sulfate, or phenylacetylglutamine. In addition, no significant change in the homeostatic model assessment (HOMA-IR) was found. Therefore, the influence of prebiotic AXOS on microbiota derived uremic retention solutes and insulin resistance in patients with CKD not yet on dialysis was not demonstrated. Studies performed in mammals have revealed that XOS promoted the growth of beneficial intestinal bacteria as a carbon source, especially Bifidobacterium and Lactobacillus species, giving rise to an increase of caeca SCFAs. These effects are associated with some health benefits of XOS, including the improvement of mineral absorption, modulation of bowel function, promotion of lipid and glucose metabolism, regulation of immunomodulatory activity, reduction of the risk of colon cancer, enhancement of antioxidant, as well as anti-inflammatory and anti-microbial activities (Aachary and Prapulla 2011).

Prebiotics as feed additives are considered to be a promising alternative for antibiotics in aquaculture, particularly due to the potential development of antibiotic-resistant bacteria. Several studies were conducted to assess the potential effects of XOS in fish (Li et al. 2008; Xu et al. 2009; Guerreiro et al. 2015). A diet supplemented with 400 mg kg^{-1} of XOS promoted growth performance and nonspecific immunity in juvenile turbot (*Scophthalmus maximus*) (Li et al. 2008). As compared with the control diet, allogynogenetic crucian carp (*Carassius auratus gibelio*) fed 100 mg XOS kg^{-1} showed greater growth performance and digestive enzyme activity (Xu et al. 2009). In the European sea bass (*Dicentrarchus labrax*) with fish meal diets, XOS and short-chain FOS prebiotics enhanced glycolytic activity (Guerreiro et al. 2015). AXOS as a potential prebiotic have been evaluated in the improvement of growth and gut microbiota of juvenile *Acipenser baerii* (Geraylou et al. 2013). Two trials of basic diets fortified with 2% or 4% AXOS-32-0.30 (high DP) (trial 1) and 2% AXOS-32-0.30 or AXOS-3-0.25 (low DP) (trial 2), respectively, were delivered to fish for 10 or 12 weeks. Microbiota was quantified using 454-pyrosequencing with barcoded primers targeting the V3 region of the 16S rRNA gene. The results revealed that the supplementation of 2% AXOS-32-0.30 enhanced the relative abundance of *Eubacteriaceae*, *Clostridiaceae*, *Streptococcaceae* and *Lactobacillaceae*, while 4% AXOS-32-0.30 and 2% AXOS-3-0.25 increased the amount of *Bacillaceae*. A significant clustering of the gut microbiota of individuals consuming an AXOS diet was demonstrated in redundancy analysis. In two trials of fish fed 2% AXOS-32-0.30, acetate, butyrate, and total SCFAs were boosted. AXOS consumption is also associated with multiple effects on health promotion in rat and chicken (Courtin et al. 2008; van Craeyveld et al. 2008).

24.3.2 *Pectin Oligosaccharides (POS)*

24.3.2.1 Source

Pectin is considered a functional food ingredient, commonly present in fruits and vegetables. PS in plant cell walls as by-products from food processing are potential raw materials for producing novel prebiotics. Byproducts rich in pectin, including citrus pulp, apple pomace, potato pulp, grape pomace, sugar beet pulp, and bergamot peel, have been investigated as sources of uronic acid-based OS and/or neural sugar-rich prebiotics originating from the homogalacturonan (HG) region of the cell wall (Hotchkiss et al. 2003; Chen et al. 2013; Gullón et al. 2013; Khodaei et al. 2016). It was shown that neutral sugar-rich OS and oligomers had higher prebiotic effects compared with those mainly containing uronic acid (Gullón et al. 2011). Pectin functions as a gelling agent, thickener, texturizer, emulsifier, stabilizer, or fat/sugar replacer to form gel at low concentrations, increase viscosity in beverages, and stabilize liquid system. Apples (pomace) and citrus fruits (peel) are the primary sources of commercial pectin. Waste streams from fruits and vegetables are potential sources of pectin for functional and nutritional use (Christiaens et al. 2015). Pectins can be controllably hydrolyzed to obtain a mixture of OS considered as pectic-oligosaccharides (POS), containing oligogalacturonides, glucooligosaccharides, arabinooligosaccharides, arabinogalactooligosaccharides, and rhamnogalacturonoligosaccharides (Corzo et al. 2015).

24.3.2.2 Chemical Structure

Pectin is a primary component in the middle lamella of cell walls in terrestrial plants. It acts as a cementing substance among adjacent cells in cell walls (Naqash et al. 2017). Pectin molecules are complex polysaccharides with high diversity in structure. They are comprised of a linear homopolymer of α-1,4-linked-D-galacturonic acid residues (GalA: 65%, main chain) that are partly esterified with methyl and acetyl groups. They also contain L-rhamnose (Rha), D-galactose (Gal), D-arabinose (Ara), and 13 different monosaccharides (Fissore et al. 2012). Pectin is an alternation of smooth and hairy regions along the main chain molecule. Smooth areas as linear backbones contain homogalacturonanas (HGs), whereas hairy regions contain rhamnogalacturonans (RG-I and RG-II) (Koubala et al. 2014). RG-I has the ramified structure (Sousa et al. 2015), while RG-II designates substituted galacturonans (GS) of the pectin molecule (Caffall and Mohnen 2009). The main chain backbone is covalently connected to RG-I and RG-II. The RG-I contains an alternating sequence of α-1,4-linked-D-galacturonosyl and α-1,2-linked-L-rhamnosyl residues. In this region, highly branched neutral sugar chains including rabinan, galactan and arabinogalactan, are attached to the backbone. The RG-II consists of a complex pectin polysaccharide with a low MW. Its backbone only possesses α-1,4-linked-D-galacturonosyl residues. RG-I and RG-II constitute the side chain region in the pectin molecule. Yapo (2011) proposed a new model of pectin structure, suggesting that HGs and RG-I constitute 65% and 20–35% of the pectin molecule, respectively; the rest is GS. GalA can be acetylated at positions O-2 and/or O-3. The carboxyl groups of the GalA can be esterified methyl groups. The degree of methyl esterification (DM) is the percentage of GalA units esterified by methanol. Pectins are thus classified as high methoxyl (HM) with DM >50%, and low methoxyl (LM) with DM <50%. In apple, HGs are almost completely esterified, whereas HGs have an average degree of methoxylation of 70% (de Vries et al. 1983). In addition, pectin can be cross-linked through ferulic acid, calcium ions, and/or borate ester dimerization (Oechslin et al. 2003). Potato pulp was identified as a pectin-rich by-product, composed of 56% (w/w) pectic OS (Khodaei and Karboune 2013). It is of great interest as a source of neutral sugar-based prebiotic OS/oligomers due to its high amount of galactan-rich RG-I.

24.3.2.3 Production of POS

The technologies to develop POS as a potential prebiotic primarily include acid hydrolysis (Hu et al. 2009), hydrothermal processing (Gómez et al. 2016), and enzymatic hydrolysis (Concha and Zúñiga 2012). Studies of POS generated from agro-wastes are associated with lemon peels (Gómez et al. 2013), orange peels (Martínez et al. 2010), olive by-products (Lama-Muñoz et al. 2012), apple pomaces and citrus peels (Wang et al. 2014) and sugar beet pulp. Although pectics are largely fermented in the human colon, the degradation mechanisms are not fully understood. Pectic enzymes and pectinolytic bacteria from the human intestine have been detected and characterized. Jensen and Canale-Parola (1986) reported that the major end product of lyase in pectin is unsaturated tri-GalA. An extract from human feces showed both pectate lyase and hydrolase activity (Matsuura 1991). *Bacteroides thetaiotaomicron* isolated from human gut flora was found to degrade pectins with different degrees of esterification (Dongowski et al. 2000).

Pectin is traditionally extracted by utilizing strong acids at high temperatures. For the sake of environmental sustainability, novel and/or eco-friendly technologies using Green Chemistry are employed in pectin extraction, including accelerated solvent extraction (ASE), microwave assisted extraction (MAE), supercritical fluid extraction (SFE), and ultrasound-assisted extraction (UAE) (González-Centeno et al. 2014). Conventional chemical and combined physical-enzymatic (CPE) methods were applied to extract pectin from Yuza (*Citrus junos*) pomace (Jongbin et al. 2012). The results revealed that chemically extracted pectin had fewer amounts of neutral sugar residues than that of CPE. Enzymes with pectin esterase activity generate blocks of carboxyl groups as compared with acids resulting in random de-esterification. Pectin extracted by different methods in gold kiwifruit (*Actinidia chinensis*) pomace has been shown to possess different characteristics and functional properties (Yuliarti et al. 2015). Pectin extracted by enzymes displayed the highest yield and amount of GalA, the lowest degree of branching, with the side chains interspersed after every 50th GalA residue. Pectin extracted by water exhibited a higher proportion of RG-I branching, indicating the retention of RG-I chains. In addition, pectin extracted by acid and water extracted has a higher degree of branching. On the other hand, pectin extracted by acid and enzymes yielded more linear structures than that extracted by water. Subcritical water extraction (SWE), also known as pressurized hot water extraction (PHWE), can improve the extraction rate at the subcritical water temperature of 160 °C. An increased yield of pectin extracted from soy hulls was observed in combination with ultrasonic treatment and phosphate-assisted subcritical water (Liu et al. 2016), suggesting the enhancement of yield after the combination of different technologies. MAE with rapidly heated sample solvent mixture showed short processing time and high extraction yield, low amounts of solvent usage, and a high purity of GalA in the extraction of grapefruit pectin (Bagherian et al. 2011). With an increased irradiation time, the amount of pectin released into medium and subsequently diffused out from the raw material enhanced (Hosseini et al. 2016). UAE was applied to yield pectin from sisal waste (Maran and Priya 2015). It was reported that UAE had more effects on side chains in the degradation of pectin, as indicated by a lowered ratio of Gal and Ara/Rha associated with the size of neutral sugar side chains (Liu et al. 2013). The influence on the pectin main chain is less, as evidenced by a slight change in the Rha/GalA ratio (Zhang et al. 2013). Apple pectin applied with ultrasound in extraction did not see changes in the main chain, but in the neutral sugar side chains of RG-I with a decreased DE. In addition, high intensity ultrasound-assisted heat extraction (UAHE) was employed in pectin extraction. Pectin in grapefruit peels was extracted using UAHE with improvement of bioactivity (Wang et al. 2015a). The combination of ultrasound and microwave extraction (UME) to generate a synergistic effect was utilized to obtain pectin from the albedo of pomelo peels using a Bronsted acidic ionic liquid as extraction medium (Liu et al. 2017). The yield from UME was significantly higher than from using conventional methods and solvents.

24.3.2.4 Functionality and Health Effects

The prebiotic potentials of pectin and POS have been extensively investigated (Guevara-Arauza et al. 2012; Chen et al. 2013; Leijdekkers et al. 2014; Gómez et al. 2016). In addition, POS have exhibited other biological effects such as immunomodulation and antibacterial activities (Lee et al. 2013), antiproliferative activity on a breast cancer cell line (Concha et al. 2013), and antitumor activity (Lee et al. 2014). Pectin and its OS exert physiological effects on the GI tract by increasing hypocholesterolemia effects, lowering glucose absorption, postponing gastric emptying, promoting beneficial bacteria as new generation prebiotics (Hotchkiss et al. 2003; Ho et al. 2017), protecting against pathogenic *E. coli* (Rhoades et al. 2008), preventing colonocytes from *E. coli* verotoxins, stimulating apoptosis in human colonic adenocarcinoma cells (Olano-Martin et al. 2003), and inhibiting invasion of Caco-2 cells from Campylobacter (Ganan et al. 2010). As a water-soluble DF, pectin contributes to lower blood pressure, improved serum lipid concentration, improved blood glucose control in diabetes, and weight loss in humans (Anderson et al. 2009). Pectin exerts anti-inflammatory activity partially through prebiotic effects (Tian et al. 2016).

POS modulates beneficial bacteria as a novel prebiotic. The fermentation of Valencia orange peels revealed bifidogenic effects with an increase of SCFAs such as acetate, butyrate, and propionate (Hotchkiss et al. 2003). An *in vitro* study has demonstrated the POS promoted *bifidobacteria* and *Eubacterium rectale* (Manderson et al. 2005). OS from bergamot peel enhanced populations of bifidobacteria and lactobacilli over clostridia. It also contained a higher PI than that of FOS (Mandalari et al. 2007). Parkar et al. (2010) examined the gut health benefits of kiwifruit pectin, which was extracted by the selective resolubilization of fruit fiber using chemical treatments. Using the *in vitro* Caco-2 model, kiwifruit pectin was shown to facilitate bacterial adhesion to intestinal epithelial cells, enhance the adhesion of *L. rhamnosus,* and prevent *S. typhimurium* from adhesion. Endo-β-galactanase catalyzed potato pulp polysaccharides to produce POS, which promotes the growth of *B. longum* and *L. acidophilus*, and inhibits the growth of *Clostridium perfringens* (Michalak et al. 2012). The potential prebiotic effect of pectin and its derived OS have been assessed by Gullón et al. (2013). POS from apple pomace was evaluated in substrate consumption, distribution of metabolic products, and effect on bacterial numbers. Dynamic high-pressure microfluidization (DHPM) was applied in apple pomace to yield POS, which enhanced the numbers of bifidobacteria and lactobacilli, as shown by *in vitro* fecal batch culture fermentation and increased SCFAs of acetic, lactic, and propionic acids as compared to the pectin from which they were derived. The supplementation of POS has been observed to drop the number of bacteroides and clostridia (Chen et al. 2013). Khodaei et al. (2016) have found that potato galactan-rich RG-I was enzymatically hydrolyzed by endo-β-1,4-galactanase and underwent Depol 670L multi-enzymatic preparation to yield galactose-rich OS/oligomers (oligo-RG-I). The *in vitro* digestibility of potato RG-I and oligo-RG-I was assessed using the TIM-1 model to monitor their ability to reach the colon intact. Furthermore, the corresponding fermentation was investigated using a continuous culture system inoculated with immobilized fecal microbiota to simulate and mimic the colonic environment. The results showed that both RG-I and oligo-RG-I promoted the growth of *Bifidobacterium spp.* and *Lactobacillus spp.*, with oligo-RG-I hydrolysates more selectively fermented by these beneficial bacteria through a continuous culture system inoculated with immobilized fecal microbiota. Total amounts of SCFAs produced from the fermentation of oligo-RG-I were higher than those obtained from its parent RG-I and the control (FOS). Citrus pectin with 60% esterification hydrolyzed by Peclyve CP to yield pectin enzyme hydrolysate (PEH) with various hydrolysis times (6, 12, 24, or 48 h) or concentrations (1%, 2%, and 4%) was applied for its growth stimulation effect on *B. bifidum* and *L. acidophilus* (Ho et al. 2017). This study indicated that PEH-derived from citrus could be an effective prebiotic to enhance the growth, fermentation, acid tolerance, and survival in nonfat milk for the probiotics tested. Enzymatically hydrolyzed Citrus pectin promoted the growth and acid tolerance of *B. bifidum* and *L. acidophilus*. The monosaccharides in hydrolyzates could be consumed by the bacteria (Ho et al. 2017).

24.3.3 Glucomannan Oligosaccharides (GMOS)

24.3.3.1 Source

Glucomannan (GM) is a mannan-type polysaccharide, widely distributed in nature, including softwood and hardwood plants (Al-Ghazzewi et al. 2007; Tester and Al-Ghazzewi 2010). *Amorphophallus konjac* has long been used in China, Japan and Southeast Asia as a food source and a traditional medicine. The *Amorphophallus konjac* tubers and roots are primary sources of GM (Alonso-Sande et al. 2009; Xu et al. 2013). Konjac glucomannan (KGM) majorly consists of D-mannose and D-glucose units connected by β-1,4-glycosidic bonds. It was reported that mannan derived OS could be obtained from by-products such as coconut meal (Khuwijitjaru et al. 2012) and fiberboard industry (*Pinus pinaster*) (Rivas et al. 2012) through subcritical water technologies.

24.3.3.2 Chemical Structure

GMs commonly contain mannose residues with glucose as the second sugar, consisting of glucuronic acid, acetyl substituents, and phosphoric acid groups (Nishinari et al. 2007). KGM, as a biodegradable polymer, is a water-soluble heteropolysaccharide. It is a linear random copolymer connected by 1,4-linked-β-D-mannopyranose and β-D-glucopyranose units in a molar ratio of 1.6:1 or 1.4:1, with about 1 in 19 sugar units being acetylated at the C-6 position. In addition, KGM is a slightly branched polysaccharide with branches containing 11–16 hexose residues of mannose, glucose, and galactose. Usually, there is 5–10% degree of acetylation in KGM, which facilitates dispersion and solubility by inhibiting intra-molecular hydrogen bonds (Liu et al. 2015; Zia et al. 2016).

24.3.3.3 Functionality and Health Effects

GMs have been shown to possess potential health benefits including reduction of serum cholesterol levels and risk of constipation, modulation of diabetes, prevention of certain cancers, stimulation in immune systems, inactivation of some food-poisoning bacteria, or inhibition of bacterial adhesion to epithelial cells, and generation of SCFAs through fermentation (Tester and Al-Ghazzewi 2010). KGM degradation products with various MWs have immunomodulatory, anti-cancer, and cytothesis effects (Liu et al. 2015). Particularly, the functional and nutritional limitations for the applications of native KGM can be resolved by depolymerization to produce OS with different DP (Al-Ghazzewi et al. 2007). Depolymerized KGM has health-promoting benefits including reduced risk of gut cancer, modulation of immune responses by the gut-associated lymphoid tissue system, restriction of aflatoxin toxicity, protection against oxidative stress in the human colon, and prevention of atopic diseases through suppression of IgE production in mice (Schley and Field 2002). Additionally, KGM is widely utilized in food, pharmaceutical, and cosmetic industries as a gelling agent, thickener, film-former, emulsifier, texturizer, and water binder (Liu et al. 2015). Gelation and high viscosity are the prominent properties of KGM. Specifically, KGM was used to alter the rheology and matrix structure of difference starches, acting as a barrier to prevent amylopectin chain association (Silva et al. 2013).

An *in vitro* assay of KGM in batch cultures inoculated with human feces was reported by Connolly et al. (2010). It was observed that the fermentation of KGM hydrolysates (KGMH) resulted in a favorable SCFA profile and selectively promoted the beneficial bacteria compared to the inulin prebiotics. The populations of *Bifidobacterium* genus, *Lactobacillus–Enterococcus* group and the *Atopobium* group all significantly increased after KGMH and inulin addition. The *Bacteroides–Prevotella* abundance decreased after KGMH fermentation and increased after inulin fermentation. As with inulin, KGMH stimulated beneficial gut microbiobes and resulted in a favorable SCFA profile. Additionally, KGM inclusion led to an increase in bifidobacteria and lactobacilli and a decrease in *Escherichia coli* and *Clostridium perfringens* in the lower gut. Rivas et al. (2012) revealed the bifidogenic potential of wood mannan hydrolysates in a study involving hydrothermal production, membrane fractionation, characterization, and *in vitro* fermentation assays with fecal cultures. Yang et al. (2017) recently studied the effect of KGMOS obtained by the combination of mechanical shear pre-treatment and enzymatic hydrolysis on eight lactobacilli and bifidobacteria from various sources. Two KGMH were developed by either enzymatic hydrolysis (KGMH I) or the combination of mechanical shear pre-treatment and enzymatic hydrolysis (KGMH II). Enzymatic treatment of native KGM reduced the average MW of the supernatant and pellet by 3-fold and 3.1-fold, respectively. Additional pre-treatment by mechanical shear further reduced the MW of the supernatant and pellet of hydrolyzed KGM by 5% and 35%, respectively. It was postulated that pulverized and depolymerized short-chain KGM promotes the growth of lactobacilli and bifidobacteria when compared to long-chain native KGM and other fermentables, such as glucose, mannose, and inulin. The results showed that lactobacilli, excluding *L. delbrueckii* subsp. *bulgaricus*, were KGM hydrolysate fermenters. More specifically, *L. acidophilus*, *L. plantarum* ATCC 14917 and ATCC 8014 fermented KGMH I and KGMH II at a faster pace than inulin. Overall, the bifidobacteria were not effective fermenters of the KGM hydrolysates. B. *longum* subsp. *infantis* is a moderate fermenter of the KGMH II but not as strongly as the lactobacilli. It was concluded that both pulverization and enzymatic depolymerization have a significant effect on the MW of KGM, suggesting that human GI bacteria are capable of utilizing molecules with reduced weights as carbon substrates.

An *in vivo* study of KGM hydrolysates on gut microbiota, blood glucose, and cholesterol levels in Old Wistar mice was conducted (Elamir et al. 2008). It was monitored that the hydrolysates promoted the growth of anaerobes and lactobacilli, and lowered the *Clostridium perfringens* and *Escherichia coli* counts. An *in vivo* study found that a diet supplemented with porang glucomannan decreased the abundance of *Escherichia coli*, enhanced the production of total SCFA, and reduced cecal pH (Harmayani et al. 2014). Recently, the effects of mannan oligosaccharides in striped catfish (*Pangasianodon hypophthalmus*) were reported by Akter et al. (2016). It unveiled that mannan OS improved the growth, feed utilization, digestive enzyme activities, gut morphology, and microbiota. While an increase in bacterial groups may correlate with the inclusion of prebiotics in growth media or product formulation, it may be difficult to demonstrate a direct relationship.

24.3.3.4 Production of GMOS

KGM polysaccharides can be degraded into OS with varying DP via different means, such as physical (γ-irradiation, shear force, microwave-assisted), chemical (acid and oxidation), and enzymatic hydrolysis (Albrecht et al. 2011). Especially in enzymatic hydrolysis, the action of β-mannanase, β-mannosidase, and β-glucosidase is required to completely hydrolyze the KGM backbone into monosaccharides. Catalysis of the random cleavage of β-D-1,4-mannopyranosyl linkages in KGM by β-mannanase yields mannobiose and mannotriose. The enzyme β-mannosidase catalyzes the removal of D-mannose residues from β-1,4-linked manno-oligosaccharides and leads to free D-mannose molecules. Finally, β-glucosidase removes glucose units from a terminus and halts catalysis at oxidized residues or mannose units (Alonso-Sande et al. 2009).

24.4 Conclusions

Increasing scientific evidence has identified the correlation among dietary intake, gut microbiome, and human health. Controlling microbiome that consume through dietary modifications and reside within the human gut sheds light on novel nutritional strategies and clinical practice in reducing some chronic diseases. As an emerging field, prebiotics, probiotics, and synbiotics are associated with the development of nutritional interventions and gut microbiota with positive health outcomes. Although there is strong evidence to demonstrate the complex link between gut microbiota and human health, large challenges still remain in delivering effective, stable, and cost efficient foods with positive health outcomes, building personalized diets based on gut microbiome profiles, standardizing clinical practices and establishing regulatory. Dietary intervention, as a strong applicator, on microbiota and consequently on physiology and the immune system, could play an important role in the reduction of risk and progression of some chronic diseases including cancer and obesity. Since the first definition of prebiotic was raised by Gibson and Roberfroid (1995), with new findings and continuously increasing knowledge, including the complexity of the gut microbiota and their interaction with the host, the discrepancies between DF and prebiotics, the metagenomics development approaches and other omic tools, the definition of prebiotics has been evolved and been modified over two decades (Hutkins et al. 2016). Basically, prebiotics are known as non-viable dietary components that are targeted by the probiotic to be fermented and to be utilized as carbons, which promote and modulate certain gut flora. Here, we focus on prebiotics as functional carbohydrate polymers, including traditional ones of HMOS, FOS, and GOS, as well as potential ones of POS, XOS, AXOS, and GMOS. To better understand the complex interplay of diet, nutrition and the microbiome in food development, as well as the effects of diet on the diversity of the human microbiome, the contents of resource, chemical structure, and physiological functionalities for each prebiotic have been included.

In vitro studies are useful tool in selecting polysaccharides with prebiotic potential, which will be validated in animal models, and human interventions. Pure cultures, mixed cultures, and human feces can be employed *in vitro* studies (Moon et al. 2016). The interaction between prebiotics and probiotics is widely studied in the fermentation model of the human gut microbiota using immobilized cell technologies (Cinquin et al. 2004). On the other hand, *in vivo* studies employ oral administration of prebiotic candidates to confirm the prebiotic activities on the colon dynamics and the intestinal microbiota effects in both animal and human clinical observations. The intestinal microbiota play a paramount role in homeostasis, gut morphology of hosts, nutrition, and pathogenesis of intestinal diseases, as well as the immune response, and thereby lead to beneficial effects on well-being and performance (Ducatelle et al. 2015). Although the mechanisms responsible for changes in the gut microbiota due to consumption of prebiotics are not fully elucidated, this chapter will not concentrate on those areas. There are further opportunities dedicated to expanding the existing knowledge on prebiotics, probiotics, synbiotics, colonic microbiota, and human physiopathology.

There is an increased interest to develop, identify, and characterize novel fibers as potential prebiotics, which remain a current trend in the development of new food products and the reformulation of current food products. Potential prebiotics, especially the usage of by-products (agro-industrial wastes) to yield added-value products, and attainable strategies to improve the competitiveness of the industry in the market are of great interest. Little information is currently available in utilizing POS, AXOS, XOS, or GMOS as prebiotics in either animal or human clinical trials, studies are essential in this direction. Moreover, research of the beneficial bacteria associated with dietary patterns and eating habits to deliver and choose the best prebiotics is also challenging. From future research and application perspectives, novel prebiotic candidates with health promotion properties need to be exploited, to be demonstrated their prebiotic potentiality in animals and humans, and further to be commercialized in industry.

In summary, microbiome research associated with probiotics and prebiotics is an emerging area of science, which has many research opportunities available. It presents a two-sided relationship between diets (prebiotics) and the microbiome. The microbiome is integral to human physiology, maintenance of health, and prevention of disease. The food industry should actively stay informed about advances in this field to develop potential prebiotics.

References

Aachary AA, Prapulla SG (2011) Xylooligosaccharides (XOS) as an emerging prebiotic: microbial synthesis, utilization, structural characterization, bioactive properties, and applications. Compr Rev Food Sci Food Saf 10:2–16

Akter MN, Sutriana A, Talpur AD, Hashim R (2016) Dietary supplementation with mannan oligosaccharide influences growth, digestive enzymes, gut morphology, and microbiota in juvenile striped catfish, Pangasianodon hypophthalmus. Aquac Int 24:127–144

Albrecht S, van Muiswinkel GC, Xu J, Schols HA, Voragen AG, Gruppen H (2011) Enzymatic production and characterization of konjac glu-comannan oligosaccharides. J Agric Food Chem 59:12658–12666

Al-Ghazzewi FH, Khanna S, Tester RF, Piggott J (2007) The potential use of hydrolysed konjac glucomannan as a prebiotic. J Sci Food Agric 87:1758–1766

Alonso-Sande M, Teijeiro-Osorio D, Remuñán-López C, Alonso MJ (2009) Glucomannan, a promising polysaccharide for biopharmaceutical purposes. Eur J Pharm Biopharm 72:453–462

Al-Sheraji SH, Ismail A, Manap MY, Mustafa S, Yusof RM, Hassan FA (2013) Prebiotics as functional foods: a review. J Funct Foods 5:1542–1553

Anderson JW, Baird P, Davis RH, Ferreri S, Knudtson M, Koraym A et al (2009) Health benefits of dietary fiber. Nutr Rev 67:188–205

Andersson R, Åman P (2008) Cereal arabinoxylan: Occurence, structure and properties. In: McCleary BV, Prosky L (eds) Advanced dietary fiber technology. Blackwell Science, pp 301–314

Andrewartha K, Phillips DR, Stone BA (1979) Solution properties of wheat-flour arabinoxylans and enzymically modified arabinoxylans. Carbohydr Res 77:191–204

Angus F, Smart S, Shortt C (2005) Prebiotic ingredients with emphasis on galacto-oligosaccharides and fructooligosaccharides. In: Tamine A (ed) Probiotic dairy products. Blackwell Publishing, Oxford, pp 120–137

Arthee R, Vijila K (2014) Study on fructosyltransferase enzyme from *Aspergillus* sp. in fructooligosaccharides production. Res J Recent Sci 3:147–153

Asakuma S, Hatakeyama E, Urashima T, Yoshida E, Katayama T, Yamamoto K, Kumagai H et al (2011) Physiology of consumption of human milk oligosaccharides by infant gut-associated bifidobacteria. J Biol Chem 286:34583–34592

Association of Official Analytical Chemists (AOAC) (2005) Determination of trans-galactooligosaccharides (TGOS) in selected food products. Method 2001.02. In: Latimer GW, Horwitz W (eds) Official methods of analysis of AOAC International, 18th edn. AOAC International, Gaithersburg, MD

Austin S, Bénet T, Michaud J, Cuany D, Rohfritsch P (2014) Determination of β-galactooligosaccharides by liquid chromatography. Int J Anal Chem 2014:768406. https://doi.org/10.1155/2014/768406

Bagherian H, Ashtiani FZ, Fouladitajar A, Mohtashamy M (2011) Comparisons between conventional, microwave- and ultrasound-assistedmethods for extraction of pectin from grapefruit. Chem Eng Process Process Intensif 50:1237–1243

Barile D, Tao N, Lebrilla CN, Coisson JD, Arlorio M, Germana JB (2009) Permeate from cheese whey ultrafiltration is a source of milk oligosaccharides. Int Dairy J 19:524–530

Bataillon M, Mathaly P, Nunes Cardinali AP, Duchiron F (1998) Extraction and Purification of arabinoxylans from destarched wheat bran in a pilot scale. Ind Crop Prod 8:37–43

Belorkar SA, Gupta AK (2016) Oligosaccharides: a boon from nature's desk. AMB Express 6:82

Benamrouche S, Cronier D, Debeire P, Chabbert B (2002) A chemical and histological study on the effect of (1,4)-beta-endo-xylanase treatment on wheat bran. J Cereal Sci 36:253–260

Bergmans MEF, Beldman G, Gruppen H, Voragen AGJ (1996) Optimisation of the selective extraction of (glucurono) arabinoxylans from wheat bran: use of barium and calcium hydroxide solution at elevated temperatures. J Cereal Sci 23:235–245

Bode L (2009) Human milk oligosaccharides: prebiotics and beyond. Nutr Rev 67:S183–S191

Bode L (2012) Human milk oligosaccharides: every baby needs a sugar mama. Glycobiology 22:1147–1162

Bode L (2015) The functional biology of human milk oligosaccharides. Early Hum Dev 91:619–622

Bornet F (2001) Fructo-oligosaccharides and other fructans: chemistry, structure and nutritional effects. In: McCleary BV, Prosky L (eds) Advanced dietary fiber technology. Blackwell Science, pp 480–493

Borromei C, Careri M, Cavazza A, Corradini C, Elviri L, Mangia A, Merusi C (2009) Evaluation of fructooligosaccharides and inulins as potentially health benefiting food ingredients by HPAEC-PED and MALDI-TOF MS. Int J Anal Chem 2009:530639. https://doi.org/10.1155/2009/530639

Broekaert WF, Courtin CM, Verbeke K, Van de Wiele T, Verstraete W, Delcour JA (2011) Prebiotic and other health-related effects of cereal-derived arabinoxylans, arabinoxylan-oligosaccharides, and xylooligosaccharides. Crit Rev Food Sci Nutr 51:178–194

Caffall KH, Mohnen D (2009) The structure, function, and biosynthesis ofplant cell wall pectic polysaccharides. Carbohydr Res 344:1879–1900

Charbonneau MR, O'Donnell D, Blanton LV, Totten SM, Davis JC, Barratt MJ et al (2016) Sialylated milk oligosaccharides promote microbiota-dependent growth in models of infant undernutrition. Cell 164:859–871

Chen J, Liang RH, Liu W, Li T, Liu CM, Wu SS et al (2013) Pectic-oligosaccharides prepared by dynamic high-pressure microfluidization and their in vitro fermentation properties. Carbohydr Polym 91:175–182

Chonan O, Matsumoto K, Watanuki M (1995) Effect of galactooligosaccharides on calcium absorption and preventing bone loss in ovariectomized rats. Biosci Biotech Biochem 59:236–239

Christiaens S, Uwibambe D, Uyttebroek M, Droogenbroeck BV, Loey AMV, Hendrickx ME (2015) Pectin characterisation in vegetable waste streams: a starting point for waste valorisation in the food industry. LWT Food Scie Technol 61:275–282

Chuankhayan P, Hsleh CY, Huang YC, Hsleh YY, Guan HH, Hsleh YC, Tlen YC et al (2010) Crystal structures of Aspergillus japonicus fructosyltransferase complex with donor/acceptor substrates reveal complete subsites in the active site for catalysis. J Biol Chem 285:23251–23264

Cinquin C, Le Blay G, Fliss I, Lacroix C (2004) Immobilization of infant fecal microbiota and utilization in an in vitro colonic fermentation model. Microb Ecol 48:128–138

Claesson MJ, Jeffery IB, Conde S, Power SE, O'Connor EM, Cusack S et al (2012) Gut microbiota composition correlates with diet and health in the elderly. Nature 488:178–184

Cloetens L, De Preter V, Swennen K, Brockaert WF, Courtin CM, Delcour JA et al (2008) Dose-response effect of arabinoxylooligosaccharides on gastrointestinal motility and on colonic bacterial metabolism in healthy volunteers. J Am Coll Nutr 27:512–518

Coda R, Kärki I, Nordlund E, Heiniö R-L, Poutanen K, Katina K (2014) Influence of particle size on bioprocess induced changes on technological functionality of wheat bran. Food Microbiol 37:69–77

Concha J, Zúñiga ME (2012) Enzymatic depolymerization of sugar beet pulp: production and characterization of pectin and pectic-oligosaccharides as apotential source for functional carbohydrates. Chem Eng J 192:29–36

Concha J, Weinstein C, Zúñiga ME (2013) Production of pectic extracts from sugar beet pulp with antiproliferative activity on a breast cancer cell line. Front Chem Sci Eng 7:482–489

Connolly ML, Lovegrove JA, Tuohy KM (2010) Konjac glucomannan hydrolysate beneficially modulates bacterial composition and activity within the faecal microbiota. J Funct Foods 2:219–224

Coppa GV, Zampini L, Galeazzi T, Facinelli B, Ferrante L, Capretti R, Orazio G (2006) Human milk oligosaccharides inhibit the adhesion to Caco-2 cells of diarrheal pathogens: *Escherichia coli, Vibrio cholerae*, and *Salmonella fyris*. Pediatr Res 59:377–382

Correia MAS, Mazumder K, Bras JLA, Firbank SJ, Zhu YP, Lewis RJ et al (2011) Structure and function of an arabinoxylan-specific xylanase. J Biol Chem 286:22510–22520

Corzo N, Alonso JL, Azpiroz F, Calvo MA, Cirici M, Leis R et al (2015) Prebióticos: concepto, propiedades y efectos beneficiosos. Nutr Hosp 31:99–118
Corzo-Martínez M, Luscher A, de las Rivas B, Muñoz R, Moreno FJ (2015) Valorization of cheese and tofu whey through enzymatic synthesis of lactosucrose. PLoS One 10:e0139035
Corzo-Martínez M, García-Campos G, Montilla A, Moreno FJ (2016) Tofu whey permeate is an efficient source to enzymatically produce prebiotic fructooligosaccharides and novel fructosylated a-galactosides. J Agric Food Chem 64:4346–4352
Costa GT, Guimaraes SB, de Carvalho Sampaio HA (2012) Fructo-oligosaccharide effects on blood glucose. An overview. Acta Cir Bras 27:279–282
Costa GT, Abreu GC, Guimarães AB, Vasconcelos PR, Guimarães SB (2015) Fructo-oligosaccharide effects on serum cholesterol levels. An overview. Acta Cir Bras 30:366–370
Courtin CM, Delcour JA (2002) Arabinoxylans and endoxylanases in wheat flour bread-making. J Cereal Sci 35:225–243
Courtin CM, Broekaert WF, Swennen K, Lescroart O, Onagbesan O, Buyse J et al (2008) Dietary inclusion of wheat bran arabinoxylooligosaccharides induces beneficial nutritional effects in chickens. Cereal Chem 85:607–613
Courtin CM, Swennen K, Verjans P, Delcour JA (2009) Heat and pH stability of prebiotic arabinoxylooligosaccharides, xylooligosaccharides and fructooligosaccharides. Food Chem 112:831–837
Cronin BE, Allsopp PJ, Slevin MM, Magee PJ, Livingstone MB, Strain JJ et al (2016) Effects of supplementation with a calcium-rich marine-derived multi-mineral supplement and short-chain fructo-oligosaccharides on serum lipids in postmenopausal women. Br J Nutr 115:658–665
Damen B, Verspreet J, Pollet A, Broekaert WF, Delcour JA, Courtin CM (2011) Prebiotic effects and intestinal fermentation of cereal arabinoxylans and arabinoxylan oligosaccharides in rats depend strongly on their structural properties and joint presence. Mol Nutr Food Res 55:1862–1874
Damen B, Pollet A, Dornez E, Broekaert AF, Van Haesendonck I et al (2012) Xylanase-mediated in situ production of arabinoxylan oligosaccharides with prebiotic potential in whole meal breads and breads enriched with arabinoxylan rich materials. Food Chem 131:111–118
David LA, Maurice CF, Carmody RN, Gootenberg DB, Button JE, Wolfe BE et al (2014) Diet rapidly and reproducibly alters the human gut microbiome. Nature 505:559–563
Davis LM, Martinez I, Walter J, Hutkins R (2010) A dose dependent impact of prebiotic galactooligosaccharides on the intestinal microbiota of healthy adults. Int J Food Microbiol 144:285–292
de Vries JA, Voragen AGJ, Rombouts FM, Pilnik W (1983) Distribution of methoxyl groups in apple pectic substances. Carbohydr Polym 3:245–258
Dehghan P, Gargari BP, Asgharijafarabadi M (2013) Effects of high performance inulin supplementation on glycemic status and lipid profile in women with type 2 diabetes: a randomized, placebo-controlled clinical trial. Health Promot Perspect 3:55–63
Dervilly-Pinel G, Rimsten L, Saulnier L, Andersson R, Aman P (2001) Water extractable arabinoxylan from pearled flours of wheat, barley, rye and triticale. Evidence for the presence of ferulic acid dimers and their involvement in gel formation. J Cereal Sci 34:207–214
Desai MS, Seekatz AM, Koropatkin NM, Kamada N, Hickey CA, Wolter M et al (2016) A dietary fiber-deprived gut microbiota degrades the colonic mucus barrier and enhances pathogen susceptibility. Cell 167:1339–1353
Díez-Municio M, DeLasRivas B, Jimeno ML, Munoz R, Moreno F, Herrero M (2013) Enzymatic synthesis and characterization of fructooligosaccharides and novel maltosylfructosides by inulosucrase from *Lactobacillus gasseri* DSM20604. Appl Environ Microbiol 79:4129–4140
Dominguez AL, Rodrigues LR, Lima NM, Teixeira JA (2014) An overview of the recent developments on fructooligosaccharide production and application. Food Bioprocess Technol 7:324–337
Doner, L. W., Sweeney, G. A., & Hicks, K. B. (2000). Isolation of hemicellulose from corn fiber, U.S. patent 6,147,206

Doner LW, Johnston DB, Singh V (2001) Analysis and properties of arabinoxylans from discrete corn wet-milling fiber fractions. J Agric Food Chem 49:1266–1269

Dongowski G, Lorenz A, Anger H (2000) Degradation of pectins with different degrees of esterification by Bacteroides thetaiotaomicron isolated from human gut flora. Appl Environ Microbiol 66:1321–1327

Ducatelle R, Eeckhaut V, Haesebrouck F, Van Immerseel F (2015) A review on prebiotics and probiotics for the control of dysbiosis: present status and future perspectives. Animal 9:43–48

Elamir AA, Tester RF, Al-Ghazzewi FH, Kaal HY, Ghalbon AA, Elmegrahai NA (2008) Effects of konjac glucomannan hydrolysates on the gut microflora of mice. Nutr Food Sci 38:422–429

Espinosa RM, Tamez M, Prieto P (2007) Efforts to emulate human milk oligosaccharides. Br J Nutr 98:S74–S79

Falck P, Linares-Pastén JA, Karlsson EN, Adlercreutz P (2018) Arabinoxylanase from glycoside hydrolase family 5 is a selective enzyme for production of specific arabinoxylooligosaccharides. Food Chem 242:579–584

Fissore EN, Rojas AM, Gerschenson LN (2012) Rheological performance ofpectin-enriched products isolated from red beet (Beta vulgaris L: var. *conditiva*) through alkaline and enzymatic treatments. Food Hydrocoll 26:249–260

Flores-Maltos DA, Mussatto SI, Contreras-Esquivel JC, Rodríguez-Herrera R, Teixeira JA, Aguilar CN (2016) Biotechnological production and application of fructooligosaccharides. Crit Rev Biotechnol 36:259–267

Fontana JD, Grzybowski A, Tiboni M, Passos M (2011) Fructo-oligosaccharide production from inulin through partial citric or phosphoric acid hydrolyses. J Med Food 14:1425–1430

Galeotti F, Coppa GV, Zampini L, Maccari F, Galeazzi T, Padella L et al (2014) Capillary electrophoresis separation of human milk neutral and acidic oligosaccharides derivatized with 2-aminoacridone. Electrophoresis 35:811–818

Ganan M, Collins M, Rastall R, Hotchkiss AT, Chau HK, Carrascosa AV et al (2010) Inhibition by pectic oligosaccharides of the invasion of undifferentiated and differentiated CaCo-2 cells by Campylobacter jejuni. Int J Food Microbiol 137:181–185

Garcia AL, Otto B, Reich S-C, Weickert MO, Steiniger J, Machowetz A et al (2007) Arabinoxylan consumption decreases postprandial serum glucose, serum insulin and plasma total ghrelin response in subjects with impaired glucose tolerance. Eur J Clin Nutr 61:334–341

García-Cayuela T, Díez-Municio M, Herrero M, Martínez-Cuesta MC, Peláez C, Requena T, Moreno FJ (2014) Selective fermentation of potential prebiotic lactose-derived oligosaccharides by probiotic bacteria. Int Dairy J 38:11–15

Garrido D, Ruiz-Moyano S, Lemay DG, Sela DA, German JB, Mills DA (2015) Comparative transcriptomics reveals key differences in the response to milk oligosaccharides of infant gut-associated bifidobacteria. Sci Rep 5:13517. https://doi.org/10.1038/srep13517

Gelroth J, Ranhotra GS (2001) Food uses of fiber. In: Cho SS, Dreher ML (eds) Handbook of dietary fiber. Marcel Dekker, New York, pp 435–451

Geraylou Z, Souffreau C, Rurangwa E, Maes GE, Spanier KI et al (2013) Prebiotic effects of arabinoxylan oligosaccharides on juvenile Siberian sturgeon (Acipenser baerii) with emphasis on the modulation of the gut microbiota using 454 pyrosequencing. FEMS Microbiol Ecol 86:357–371

Gibson R, Roberfroid MR (1995) Dietary modulation of the human colonic microbiota introducing the concept of prebiotics. J Nutr 125:1401–1412

Gill RK, Dudeja PK (2011) A novel facet to consider for the effects of butyrate on its target cells. Focus on "The short-chain fatty acid butyrate is a substrate of breast cancer resistance protein". Am J Physiol Cell Physiol 301(5):C977–C979

Goehring KC, Kennedy AD, Prieto PA, Buck RH (2014) Direct evidence for the presence of human milk oligosaccharides in the circulation of breastfed infants. PLoS One 9:e101692. https://doi.org/10.1371/journal.pone.0101692

Gomez GD, Balcazar JL (2008) A review on the interactions between gut microbiota and innate immunity of fish. FEMS Immunol Med Microbiol 52:145–154

Gómez B, Gullon B, Yáñez R, Parajo JC, Alonso JL (2013) Pectic oligosaccharides from lemon peel wastes: production, purification, and chemical characterization. J Agric Food Chem 61:10043–10053
Gómez B, Gullon B, Yanez R, Schols H, Alonso JL (2016) Prebiotic potential of pectins and pectic oligosaccharides derived from lemon peel wastes and sugar beet pulp: a comparative evaluation. J Funct Foods 20:108–121
González-Centeno MR, Knoerzer K, Sabarez H, Simal S, Rosselló C, Femenia A (2014) Effect of acoustic frequency and power density on the aqueous ultrasonic-assisted extraction of grape pomace (Vitis vinifera L.)—a response surface approach. Ultrason Sonochem 21:2176–2184
Gosling A, Stevens GW, Barber AR, Kentish SE, Gras SL (2010) Recent advances refining galactooligosaccharide production from lactose. Food Chem 121:307–318
Grand View Research (GVR) (2016a) Fructooligosaccharides (FOS) market analysis by source (inulin, sucrose) by application (food & beverages, infant formula, dietary supplements, animal Feed, pharmaceuticals) and segment forecasts to 2024. https://www.grandviewresearch.com/industry-analysis/fructooligosaccharides-market/toc. Accessed 1 Dec 2017
Grand View Research (GVR) (2016b) Galacto-oligosaccharide (GOS) market trend analysis by application (food & beverage, dietary supplements), by region (North America, Europe, Asia Pacific, Latin America, Middle East & Africa), by country, and segment forecasts, 2014–2025. https://www.grandviewresearch.com/industry-analysis/galacto-oligosaccharides-gos-market. Accessed 1 Dec 2017
Grootaert C, Delcour JA, Courtin CM, Broekaert WF, Verstraete W, Van de Wiele T (2007) Microbial metabolism and prebiotic potency of arabinoxylan oligosaccharides in the human intestine. Trends Food Sci Technol 18:64–71
Grootaert C, Van den Abbeele P, Marzorati M, Broekaert WF, Courtin CM, Delcour JA et al (2009) Comparison of prebiotic effects of arabinoxylan oligosaccharides and inulin in a simulator of the human intestinal microbial ecosystem. FEMS Microbiol Ecol 69:231–242
Guerreiro I, Oliva-Teles A, Enes P (2015) Improved glucose and lipid metabolism in European sea bass (Dicentrarchus labrax) fed short-chain fructooligosaccharides and xylooligosaccharides. Aquaculture 441:57–63
Guevara-Arauza JC, de Jesús Ornelas Paz J, Pimentel-González DJ, Rosales Mendoza S, Soria Guerra RE, Paz Maldonado LMT (2012) Prebiotic effect of mucilage and pectic derived oligosaccharides from nopal (Opuntia ficus-indica). Food Sci Biotechnol 21:997–1003
Gullón B, Gullón P, Sanz Y, Alonso JL, Parajó J (2011) Prebiotic potential of a refined product containing pectic oligosaccharides. LWT Food Sci Technol 44:1687–1696
Gullón B, Gómez B, Martínez-Sabajanes M, Yáñez R, Parajó J, Alonso J (2013) Pectic oligosaccharides: manufacture and functional properties. Trends Food Sci Technol 30:153–161
Guo Q, Goff HD, Cui SW (2017) Structural characterization of galacto-oligosaccharides (VITAGOS™) synthesized by transgalactosylation of lactose. Bioactive Carbohydrate. Dietary Fibre Available online July 21, 2017
Hang H, Miao M, Li Y, Jiang B, Mu W, Zhang T (2013) Difructosan anhydrides III preparation from sucrose by coupled enzyme reaction. Carbohydr Polym 92:1608–1611
Harmayani E, Aprilia V, Marsono Y (2014) Characterization of glucomannan from Amorphophallus oncophyllus and its prebiotic activity in vivo. Carbohydr Polym 112:475–479
Havenaar R (2011) Intestinal health functions of colonic microbial metabolites: a review. Benefic Microbes 2:103–114
Ho Y-Y, Lin C-M, Wu M-C (2017) Evaluation of the prebiotic effects of citrus pectin Hydrolysate. J Food Drug Anal 25:550–558
Hosseini SS, Khodaiyan F, Yarmand MS (2016) Optimization of microwave assisted extraction of pectin from sour orange peel and its physicochemical properties. Carbohydr Polym 140:59–65
Hotchkiss AT, Olano-Martin E, William EG, Gibson GR, Rastall RA (2003) Pectic oligosaccharides as prebiotics. In: Eggleston G, Cote GL (eds) Oligosaccharides in food and agriculture, ACS symposium series, vol 849. American Chemical Society, Washington, pp 54–62

Houghteling PD, Walker WA (2015) Why is initial bacterial colonization of the intestine important to the infant's and child's health? J Pediatr Gastroenterol Nutr 60:294–307

Hu K, Liu Q, Wang S, Ding K (2009) New oligosaccharides prepared by acidhydrolysis of the polysaccharides from Nerium indicum Mill and theirantiangiogenesis activities. Carbohydr Res 344:198–203

Hughes SA, Shewry PR, Li L, Gibson GR, Sanz ML et al (2007) In vitro fermentation by human fecal microflora of wheat arabinoxylans. J Agric Food Chem 55:4589–4595

Hutkins RW, Krumbeck JA, Bindels LB, Cani PD, Fahey G Jr, Goh YJ et al (2016) Prebiotics: why definitions matter. Curr Opin Biotechnol 37:1–7

Immerzeel P, Falck P, Galbe M, Adlercreutz P, Karlsson EN, Stålbrand H (2014) Extraction of water-soluble xylan from wheat bran and utilization of enzymatically produced xylooligosaccharides by Lactobacillus, Bifidobacterium and Weissella spp. LWT Food Sci Technol 56:321–327

Intanon M, Arreola SL, Pham NH, Kneifel W, Haltrich D, Nguyen TH (2014) Nature and biosynthesis of galacto-oligosaccharides related to oligosaccharides in human breast milk. FEMS Microbiol Lett 353:89–97

Ito T, Katayama T, Hattie M, Sakurama H, Wada J, Suzuki R, Ashisa H, Wakagi T, Yamamto K, Stubbs KA, Fushionobu S (2013) Crystal structures of a glycoside hydrolase family 20 lacto-n-biosidase from bifidobacterium bifidum. J Biol Chem 288:11795–11806

Izydorczyk MS, Biliaderis CG (2007) Arabinoxylans: technologically and nutritionally functional plant polysaccharides. In: Functional food carbohydrates. CRC Press Taylor & Francis Group, Boca Raton, FL, pp 249–290

Izydorczyk MS, Dexter JE (2008) Barley b-glucans and arabinoxylans: molecular structure, physicochemical properties, and uses in food products—a review. Food Res Int 41:850–868

Jantscher-Krenn E, Bode L (2012) Human milk oligosaccharides and their potential benefits for the breast-fed neonate. Minerva Pediatr 64:83–99

Jantscher-Krenn E, Lauwae T, Bliss LA, Reed SL, Gillin FD, Bode L (2012) Human milk oligosaccharides reduce Entamoeba histolytica attachment and cytotoxicity in vitro. Br J Nutr 108:1839–1846

Jensen NS, Canale-Parola E (1986) Bacteroides pectinophilus sp. nov. and Bacteroides galacturonicus sp. nov.: two pectinolytic bacteria from the human intestinal tract. Appl Environ Microbiol 52:880–887

Jiang H, Ling Z, Zhang Y, Mao H, Ma Z, Yin Y et al (2015) Altered fecal microbiota composition in patients with major depressive disorder. Brain Behav Immun 48:186–194

Jongbin L, Jiyoung Y, Ko S, Lee S (2012) Extraction and characterization ofpectin from Yuza (Citrus junos) pomace: a comparison of conventional-chemical and combined physical-enzymatic extractions. Food Hydrocoll 29:160–165

Kabel MA, Kortenoeven L, Schols HA, Voragen AG (2002) In vitro fermentability of differently substituted xylo-oligosaccharides. J Agric Food Chem 50:6205–6210

Khodaei N, Karboune S (2013) Extraction and structural characterisation of rhamnogalacturonan I-type pectic polysaccharides from potato cell wall. Food Chem 139:617–623

Khodaei N, Fernandez B, Fliss I, Karboune S (2016) Digestibility and prebiotic properties of potato rhamnogalacturonan Ipolysaccharide and its galactose-rich oligosaccharides/oligomers. Carbohydr Polym 136:1074–1084

Khuwijitjaru P, Watsanit K, Adachi S (2012) Carbohydrate content and composition of product from subcritical water treatment of coconut meal. J Ind Eng Chem 18:225–229

Koubala BB, Christiaens S, Kansci G, Loey AMV, Hendrickx ME (2014) Isolation and structural characterisation of papaya peel pectin. Food Res Int 55:215–221

Kukkonen K, Savilahti E, Haahtela T, Juntunen-Backman K, Korpela R, Poussa T, Tuure T, Kuitunen M (2007) Probiotics and prebiotic galactooligosaccharides in the prevention of allergic diseases: a randomized, double-blind, placebo-controlled trial. J Allergy Clin Immunol 119:192–198

Kunz C (2012) Historical aspects of human milk oligosaccharides. Adv Nutr 3(Suppl):430S–439S

Kunz C, Rudloff S, Baier W, Klein N, Strobel S (2000) Oligosaccharides in human milk: structural, functional, and metabolic aspects. Annu Rev Nutr 20:699–722
Kutty PK (2016) Breastfeeding and risk of parasitic infection—a review. Asian Pac J Trop Biomed 4:847–858
Labourel A, Crouch LI, Bras JLA, Jackson A, Rogowski A, Gray J et al (2016) The mechanism by which arabinoxylanases can recognize highly decorated xylans. J Biol Chem 291:22149–22159
Lama-Muñoz A, Rodríguez-Gutiérrez G, Rubio-Senent F, Fernández-Bolaños J (2012) Production, characterization and isolation of neutral and pectic oligosaccharides with low molecular weights from olive by-products thermally treated. Food Hydrocoll 28:92–104
Lamsal B (2012) Production, health aspects and potential food uses of dairy prebiotic galactooligosaccharides. J Sci Food Agric 92:2020–2028
Lee JS, Synytsya A, Kim HB, Choi DJ, Lee S, Lee J et al (2013) Purification, characterization and immune-modulating activity of a pectic polysaccharide isolated from Korean mulberry fruit Oddi (Morus alba L.). Int Immunopharmacol 17:858–866
Lee EH, Park H-R, Shin M-S, Cho SY, Choi H-J, Shin K-S (2014) Antitumor metastasis activity of pectic polysaccharide purified from the peels of Korean Citrus Hallabong. Carbohydr Polym 111:72–79
Leijdekkers AGM, Aguirre M, Venema K, Bosch G, Gruppen H, Schols HA (2014) In vitro fermentability of sugar beet pulp derived oligosaccharides using human and pig fecal inocula. J Agric Food Chem 62:1079–1087
Li Y, Wang YJ, Wang L, Jiang KY (2008) Influence of several non-nutrient additives on non-specific immunity and growth of juvenile turbot, Scophthalmus maximus L. Aquac Nutr 14:387–395
Li J, Liu X, Zhou B, Zhao J, Li S (2013) Determination of fructooligosaccharides in burdock using HPLC and microwave-assisted extraction. J Agric Food Chem 61:5888–5892
Li J, Hu D, Zong W, Lv G, Li S (2014) Determination of inulin-type fructooligosaccharides in edible plants by high-performance liquid chromatography with charged aerosol detector. J Agric Food Chem 62:7707–7713
Lis-Kuberka J, Berghausen-Mazurb M, Orczyk-Pawiłowicz M (2017) Alpha 2, 3- and alpha 2,6-sialylation of human skim milk glycoproteins during milk maturation. J Appl Biomed 15:196–203
Liu D, Zhang L, Xu Y, Zhang X (2013) The influence of ultrasound on the structure, rheological properties and degradation path of citrus pectin. Proc Meet Acoust 19:045092
Liu JH, Xu QH, Zhang JJ, Zhou XX et al (2015) Preparation, composition analysis and antioxidant activities of konjacoligo-glucomannan. Carbohydr Polym 130:398–404
Liu C, Cheng FF, Wang JM, Wan ZL, Sun YE, Yang XQ (2016) Preparation and characterization of surface-active pectin from soya hulls by phosphate-assisted subcritical water combined with ultrasonic treatment. Int J Food Sci Technol 51:61–68
Liu Z, Qiao L, Yang F, Gu H, Yang L (2017) Bronsted acidic ionic liquid based ultrasound-microwave synergistic extraction of pectin from pomelo peels. Int J Biol Macromol 94:309–318
Lorenzoni ASG, Aydos LF, Klein MP, Rodrigues RC, Hertz PF (2014) Fructooligosaccharides synthesis by highly stable immobilized β-fructofuranosidase from *Aspergillus aculeatus*. Carbohydr Polym 103:193–197
Macfarlane GT, Steed H, Macfarlane S (2008) Bacterial metabolism and health-related effects of galacto-oligosaccharides and other prebiotics. J Appl Microbiol 104:305–344
Maes C, Delcour JA (2002) Structural characterisation of water-extractable and water-unextractable arabinoxylans in wheat bran. J Cereal Sci 35:315–326
Maliniak A, Widmalm G (2014) Structural analysis of carbohydrates by nuclear magnetic resonance spectroscopy and molecular simulations: application to human milk ligosaccharides. In: Moreno FJ, Sanz ML (eds) Food oligosaccharides: production, analysis and bioactivity. Wiley, Chichester, UK
Mandalari G, Nueno Palop C, Tuohy K, Gibson GR, Bennett RN, Waldron KW (2007) In vitro evaluation of the prebiotic activity of a pectic oligosaccharide-rich extract enzymatically derived from bergamot peel. Appl Environ Microbiol 73:1173–1179

Manderson K, Pinart M, Tuohy KM, Grace WE, Hotchkiss AT, Widmer W et al (2005) In vitro determination of prebiotic properties of oligosaccharides derived from an orange juice manufacturing by-product stream. Appl Environ Microbiol 71:8383–8389

Mantovanl V, Galeotti F, Maccari F, Volpl N (2016) Recent advances on separation and characterization of human milk oligosaccharides. Electrophoresis 73:1514–1524

Maran JP, Priya B (2015) Ultrasound assisted extraction of pectin from sisal waste. Carbohydr Polym 115:732–738

Marcobal A, Barboza M, Froehlich JW, Block DE, German JB, Lebrilla CB (2010) Consumption of human milk oligosaccharides by gut-related microbes. J Agric Food Chem 58:5334–5340

Martin CR, Ling PR, Blackburn GL (2016) Review of infant feeding: key features of breast milk and infant formula. Nutrients 8:279. https://doi.org/10.3390/nu8050279

Martínez M, Yáñez R, Alonsó JL, Parajó JC (2010) Chemical production of pectic oligosaccharides from orange peel wastes. Ind Eng Chem Res 49:8470–8476

Marx C, Bridge R, Wolf AK, Rich W, Kim JH, Bode L (2014) Human milk oligosaccharide composition differs between donor milk and mother's own milk in the NICU. J Hum Lact 30:54–61

Matsuura Y (1991) Pectic acid degrading enzymes from human feces. Agric Biol Chem 55:885–886

McLaughlin HP, Motherway MO, Lakshminarayanan B, Stanton C, Ross RP et al (2015) Carbohydrate catabolic diversity of bifidobacteria and lactobacilli of human origin. Int J Food Microbiol 203:109–121

Meyer TW, Hostetter TH (2012) Uremic solutes from colon microbes. Kidney Int 81:949–954

Michalak M, Thomassen LV, Roytio H, Ouwehand AC, Meyer AS, Mikkelsen JD (2012) Expression and characterization of an endo-1,4-β-galactanase from Emericella nidulans and Pichia pastoris for enzymatic design of potentially prebiotic oligosaccharides from potato galactans. Enzym Microb Technol 50, 121–129

Míguez B, Gómez B, Gullón P, Gullón B, Alonso JL (2016) Pectic oligosaccharides and other emerging prebiotics. In: Rao V, Rao LG (eds) Prebiotics and probiotics in human nutrition and health. InTech, Rijeka

Miyauchi E, O'Callaghan J, Butto LF, Hurley G, Melgar S, Tanabe S et al (2012) Mechanism of protection of transepithelial barrier function by Lactobacillus salivarius: strain dependence and attenuation by bacteriocin production. Am J Physiol Gastrointest Liver Physiol 303:G1029–G1041

Monobe M, Maeda-Yamamoto M, Matsuoka Y, Kaneko A, Hiramoto S (2008) Immunostimulating activity and molecular weight dependence of an arabinoxylan derived from wheat bran. J Jpn Soc Food Sci Technol/Nippon Shokuhin Kagaku Kogaku Kaishi 55:245–249

Moon JS, Li L, Bang J, Soo Han N (2016) Application of in vitro gut fermentation models to food components: a review. Food Sci Biotechnol 25:1–7

Moreno FJ, Corzo N, Montilla A, Villamiel M, Olano A (2017) Current state and latest advances in the concept, production and functionality of prebiotic oligosaccharides. Curr Opin Food Sci 13:50–55

Morgan NK, Wallace A, Bedford MR, Choct M (2017) Efficiency of xylanases from families 10 and 11 in production of xylo-oligosaccharides from wheat arabinoxylans. Carbohydr Polym 167:290–296

Morrison DJ, Preston T (2016) Formation of short chain fatty acids by the gut microbiota and their impact on human metabolism. Gut Microbes 7:189–200

Morrow AL, Ruiz-Palacios GM, Altaye M, Jiang X, Guerrero ML, Meinzen-Derr JK et al (2004) Human milk oligosaccharides are associated with protection against diarrhea in breast-fed infants. J Pediatr 145:297–303

Musilova S, Rada V, Vlkova E, Bunesova V (2014) Beneficial effects of human milk oligosaccharides on gut microbiota. Benefic Microbes 5:273–283

Mussatto SI, Mancilha IM (2007) Nondigestible oligosaccharides: a review. Carbohydr Polym 68:587–597

Naqash F, Masoodi FA, Rather SA, Wani SM, Gani A (2017) Emerging concepts in the nutraceutical and functional properties of Pectin—a review. Carbohydr Polym 168:227–239

Neyrinck AM, Delzenne NM (2010) Potential interest of gut microbial changes induced by non-digestible carbohydrates of wheat in the management of obesity and related disorders. Curr Opin Clin Nutr Metab Care 13:722–728

Neyrinck AM, Possemiers S, Druart C, Van de Wiele T, De Backer F et al (2011) Prebiotic effects of wheat arabinoxylan related to the increase in bifidobacteria, roseburia and bacteroides/prevotella in diet-induced obese mice. PLoS One 6:e20944

Nicholson JK, Holmes E, Kinross J, Burcelin R, Gibson G, Jia W et al (2012) Host-gut microbiota metabolic interactions. Science 336:1262–1267

Niittynen L, Kajander K, Korpela R (2007) Galacto-oligosaccharides and bowel function. Scand J Food Nutr 51:62–66

Nishinari K, Takemasa M, Zhang H, Takahashi R (2007) Storage plant polysaccharides: xyloglucans, galactomannans, glucomannans. In: Kamerling JP (ed) Comprehensive glycoscience. Elsevier, pp 613–623

O'Callaghan A, van Sinderen D (2016) Bifidobacteria and their role as members of the human gut microbiota. Front Microbiol 7:925

Oechslin R, Lutz M, Amado R (2003) Pectic substances isolated from apple cellulosic residue: Structural characterization of a new type of rhamnogalacturonan I. Carbohydr Polym 51:301–310

Olano-Martin E, Williams MR, Gibson GR, Rastall RA (2003) Pectins and pectic oligosaccharides inhibit Escherichia coli O157:H7 Shiga toxin as directed towards the human colonic cell line HT29. FEMS Microbiol Lett 218:101–105

Oliveira D, Wilbey RA, Grandison AS, Roseiro LB (2015) Milk oligosaccharides: a review. Int J Dairy Technol 68:305–321

Olveira G, González-Molero I (2016) An update on probiotics, prebiotics and symbiotics in clinical nutrition. Endocrinol Nutr 63:482–494

Ooi LG, Liong M-T (2010) Cholesterol-lowering effects of probiotics and prebiotics: a review of *in vivo* and *in vitro* findings. Int J Mol Sci 11:2499–2522

Otieno DO (2010) Synthesis of β-galactooligosaccharides from lactose using microbial β-galactosidases. Compr Rev Food Sci Food Saf 9:471–482

Palframan R, Gibson GR, Rastall RA (2003) Development of a quantitative tool for the comparison of the prebiotic effect of dietary oligosaccharides. Lett Appl Microbiol 37(4):281–284

Pandey KR, Naik SR, Vakil BV (2015) Probiotics, prebiotics and synbiotics—a review. J Food Sci Technol 52:7577–7587

Parkar SG, Redgate EL, Wibisono R, Luo X, Koh ETH, Schroder R (2010) Gut health benefits of kiwifruit pectins: comparison with commercial functional polysaccharides. J Funct Foods 2:210–218

Patterson JA, Burkholder KM (2003) Application of prebiotics and probiotics in poultry production. Poult Sci 82:627–631

Petschacher B, Nidetzky B (2016) Biotechnological production of fucosylated human milk oligosaccharides: prokaryotic fucosyltransferases and their use in biocatalytic cascades or whole cell conversion systems. J Biotechnol 235:61–83

Poesen R, Evenepoel P, de Loor H, Delcour JA, Courtin CM et al (2016) The influence of prebiotic arabinoxylan oligosaccharides on microbiota derived uremic retention solutes in patients with chronic kidney disease: a randomized controlled trial. PLoS One 11(4):e0153893. https://doi.org/10.1371/journal.pone.0153893

Pokusaeva K, Fitzgerald GF, van Sinderen D (2011) Carbohydrate metabolism in bifidobacteria. Genes Nutr 6:285–306

Ragaee SM, Campbell GL, Scoles GJ, McLeod JG, Tyler RT (2001) Studies on rye (Secale cereale L.) lines exhibiting a range of extract viscosities. 2. Rheological and baking characteristics of rye and rye/wheat blends and feeding value for chicks of wholemeals and breads. J Agric Food Chem 49:2446–2453

Rajagopalan G, Shanmugavelu K, Yang K-L (2016) Production of xylooligosaccharides from hardwood xylan by using immobilized endoxylanase of Clostridium strain BOH3. RSC Adv 6:81818–81825

Ravn JL, Thøgersen JC, Eklöf J, Pettersson D, Ducatelle R, Immerseel FV, Pedersen NR (2017) GH11 xylanase increases prebiotic oligosaccharides from wheat bran favouring butyrate-producing bacteria in vitro. Anim Feed Sci Technol 226:113–123

Redgwell RJ, Fischer M (2005) Dietary fiber as a versatile food component: an industrial perspective. Mol Nutr Food Res 49:421–535

Rhoades J, Manderson K, Wells A, Hotchkiss AT, Gibson GR, Formentin K et al (2008) Oligosaccharide-mediated inhibition of the adhesion of pathogenic Escherichia coli strains to human gut epithelial cells in vitro. J Food Prot 71:2272–2277

Ríos-Covián D, Ruas-Madiedo P, Margolles A, Gueimonde M, de los Reyes-Gavilán CG, Salazar N (2016) Intestinal short chain fatty acids and their link with diet and human health. Front Microbiol 7:185

Rivas S, Gullón B, Gullón P, Parajó JC (2012) Manufacture and properties of bifidogenic saccharides derived from wood mannan. J Agric Food Chem 60:4296–4305

Roberfroid M, Gibson GR, Hoyles L, McCartney AL, Rastall R, Rowland I et al (2010) Prebiotic effects: metabolic and health benefits. Br J Nutr 104(Suppl 2):S1–S63

Rodriguez-Cabezas ME, Camuesco D, Arribas B, Garrido-Mesa N, Comalada M, Bailon E et al (2010) The combination of fructooligosaccharides and resistant starch shows prebiotic additive effects in rats. Clin Nutr 29:832–839

Rodriguez-Colinas B, Poveda A, Jimenez-Barbero J, Ballesteros AO, Plou FJ (2012) Galacto-oligosaccharide synthesis from lactose solution or skim milk using the β-galactosidase from Bacillus circulans. J Agric Food Chem 60:6391–6398

Rodriguez-Colinas B, Fernandez-Arrojo L, Santos-Moriano P, Ballesteros AO, Plou FJ (2016) Continuous packed bed reactor with immobilized β-Galactosidase for production of galactooligosaccharides (GOS). Catalysts 6:189. https://doi.org/10.3390/catal6120189

Rose DJ (2011) Autohydrolytic production of feruloylated arabinoxylan hydrolysates from cereal processing coproducts for food applications. In: Awika J, Piironen V, Bean S (eds) Advances in cereal science: implications to food processing and health promotion, vol 1089. America Chemical Society, Washington, DC, pp 111–130

Ruhaak LR, Stroble C, Underwoodm MA, Lebrilla CB (2014) Detection of milk oligosaccharides in plasma of infants. Anal Bioanal Chem 406:5775–5784

Ruiz-Moyano S, Totten SM, Garrido DA, Smilowitz JT, German JB, Lebrilla CB, Mills DA (2013) Variation in consumption of human milk oligosaccharides by infant gut-associated strains of Bifidobacterium breve. Appl Environ Microbiol 79:6040–6049

Sabater-Molina M, Larqué E, Torrella F, Zamora S (2009) Dietary fructooligosaccharides and potential benefits on health. J Physiol Biochem 65:315–328

Sangeetha PT, Ramesh MN, Prapulla SG (2005) Recent trends in the microbial production, analysis and application of fructooligosaccharides. Trends Food Sci Technol 16:442–457

Sangwan V, Tomar SK, Singh RRB, Singh AK, Ali B (2011) Galactooligosaccharides: novel components of designer foods. J Food Sci 76:R103–R111

Saulnier L, Thibault JF (1999) Ferulic acid and diferulic acids as components of sugar-beet pectins and maize bran heteroxylans. J Sci Food Agric 79:396–402

Scheperjans F, Aho V, Pereira PAB, Koskinen K, Paulin L, Pekkonen E et al (2015) Gut microbiota are related to Parkinson's disease and clinical phenotype. Mov Disord 30:350–358

Schley PD, Field CJ (2002) The immune-enhancing effects of dietary fibres and prebiotics. Br J Nutr 87(Suppl. 2):S221–S230

Scholz-Ahrens KE, Ade P, Marten B, Weber P, Timm WY, Gluer CC, Schrezenmeir J (2007) Prebiotics, probiotics, and synbiotics affect mineral absorption, bone mineral content, and bone structure. J Nutr 137:838S–846S

Severini C, Azzollini D, Jouppila K, Jussi L, Derossi A (2015) Effect of enzymatic and technological treatments on solubilisation of arabinoxylans from brewer's spent grain. J Cereal Sci 65:162–166
Sheridan PO, Bindels LB, Saulnier DM, Reid G, Nova E, Holmgren K et al (2014) Can prebiotics and probiotics improve therapeutic outcomes for undernourished individuals? Gut Microbes 5:74–82
Shoaf K, Mulvey GL, Armstrong GD, Hutkins RW (2006) Prebiotic galactooligosaccharides reduce adherence of enteropathogenic *Escherichia coli* to tissue culture cells. Infect Immun 74:6920–6928
Silva E, Birkenhake M, Scholten E, Sagis LMC, van der Linden E (2013) Controlling rheology and structure of sweet potato starch noodles with high broccoli powder content by hydrocolloids. Food Hydrocoll 30:42–52
Slavin J (2013) Fiber and prebiotics: mechanisms and health benefits. Nutrients 5:1417–1435
Smilowitz JT, Lebrilla CB, Mills DA, German JB, Freeman SL (2014) Breast milk oligosaccharides: structure-function relationships in the neonate. Annu Rev Nutr 34:14.1–14.27
Smith-Brown P, Morrison M, Krause L, Davies PS (2016) Mother's secretor status affects development of childrens microbiota composition and function: a pilot study. PLoS One 11:e0161211
Sousa V, Santos E, Sgarbieri V (2011) The importance of prebiotics in functional foods and clinical practice. Food Nutr Sci 2:133–144
Sousa AG, Nielsen HL, Armagan I, Larsen J, Sorensen SO (2015) The impact of rhamnogalacturonan-I side chain monosaccharides on the rheological properties of citrus pectin. Food Hydrocoll 47:130–139
Sridevi V, Sumathi V, Guru Prasad M, Satish Kumar M (2014) Fructooligosaccharides—type prebiotic: a Review. J Pharm Res 8:321–330
Surin S, Seesuriyachan P, Thakeow P, Phimolsiripol Y (2012) Optimization of enzymatic production of fructooligosaccharides. J Appl Sci 12:1118–1123
Swennen K, Courtin CM, Lindemans GCJE, Delcour JA (2006) Large-scale production and characterisation of wheat bran arabinoxylooligosaccharides. J Sci Food Agric 86:1722–1731
Tester RF, Al-Ghazzewi FH (2010) Utilisation of glucomannans for health. In: Hollingworth CS (ed) Food hydrocolloids: characteristics, properties and structures. Nova Science Publishers, Hauppauge, NY, pp 243–252
Thomson P, Medina DA, Garrido D (2017) Human milk oligosaccharides and infant gut bifidobacteria: molecular strategies for their utilization. Food Microbiol. https://doi.org/10.1016/j.fm.2017.09.001
Thurl S, Munzert M, Henker J, Boehm G, Muller-Werner B, Jelinek J, Stahl B (2010) Variation of human milk oligosaccharides in relation to milk groups and lactational periods. Br J Nutr 104:1261–1271
Tian L, Scholte J, Borewicz K, van den Bogert B, Smidt H, Scheurink AJW et al (2016) Effects of pectin supplementation on the fermentation patterns of different structural carbohydrates in rats. Mol Nutr Food Res 60:2256–2266
Torres DPM, Goncalves M d PF, Teixeira JA, Rodrigues LR (2010) Galacto-oligosaccharides: production, properties, applications, and significance as prebiotics. Compr Rev Food Sci Food Saf 9:438–454
Turroni F, Ventura M, Butto LF, Duranti S, O'Toole PW, Motherway MO, van Sinderen D (2014) Molecular dialogue between the human gut microbiota and the host: a Lactobacillus and Bifidobacterium perspective. Cell Mol Life Sci 71:183–203
Underwood M, German JB, Lebrilla CB, Mills DA (2015) Bifidobacterium longum subspecies infantis: champion colonizer of the infant gut. Pediatr Res 77:229–235
van Craeyveld V, Swennen K, Dornez E et al (2008) Structurally different wheat-derived arabinoxylooligosaccharides have different prebiotic and fermentation properties in rats. J Nutr 138:2348–2355

van Craeyveld V, Dornez E, Holopainen U, Selinheimo E, Poutanen K et al (2010) Wheat bran AX properties and choice of xylanase affect enzymic production of wheat bran derived arabinoxylan-oligosaccharidesaccharides. Cereal Chem 87:283–291

van de Wiele T, Boon N, Possemiers S, Jacob H, Verstrate W (2007) Inulin-type fructans of longer degree of polymerization exert more pronounced in vitro prebiotic effects. J Appl Microbiol 102:452–460

van den Abbeele P, Van de Wiele T, Grootaert C, Verstraete W, Gerard P et al (2009) Arabinoxylans and inulin modulate the luminal and mucosaassociated bacteria in vitro and in vivo. In: McCleary B, Jones JM, Topping D, van der Kamp JW (eds) Dietary fibre—new frontiers for food and health. Academic Publishers, Wageningen, pp 233–249

van den Heuvela EGHM, Muijsa T, Brouns F, Hendriksa HFJ (2009) Short-chain fructo-oligosaccharides improve magnesium absorption in adolescent girls with a low calcium intake. Nutr Res 29:229–237

Venema K (2012) Intestinal fermentation of lactose and prebiotic lactose derivatives, including human milk oligosaccharides. Int Dairy J 22:123–140

Vera C, Guerrero C, Conejeros R, Illanes A (2012) Synthesis of galactoligosaccharides by β-galactosidase from Aspergillus oryzae using partially dissolved and supersaturated solution of lactose. Enzym Microb Technol 50:188–194

Villamiel M, Montilla A, Olano A, Corzo N (2014) Production and bioactivity of oligosaccharides derived from lactose. In: Moreno FJ, Sanz ML (eds) Food oligosaccharides: production, analysis and bioactivity. Wiley-Blackwell, pp 137–167

Wall R, Fitzgerald G, Hussey S, Ryan T, Murphy B, Ross P, Stanton C (2007) Genomic diversity of cultivable Lactobacillus populations residing in the neonatal and adult gastrointestinal tract. FEMS Microbiol Ecol 59:127–137

Wang B (2009) Sialic acid is an essential nutrient for brain development and cognition. Annu Rev Nutr 29:177–222

Wang D, Wang S-A (2016) A one-step bioprocess for production of high-content fructo-oligosaccharides from inulin by yeast. Carbohydr Polym 151:1220–1226

Wang M, van Vliet T, Hamer RJ (2004) How gluten properties are affected by pentosans. J Cereal Sci 39:395–402

Wang Y, Zeng T, Wang SE, Wang Q, Yu HX (2010) Fructo-oligosaccharides enhance the mineral absorption and counteract the adverse effects of phytic acid in mice. Nutrition 26:305–311

Wang X, Chen Q, Lu¨ X (2014) Pectin extracted from apple pomace and citrus peel by subcritical water. Food Hydrocoll 38:129–137

Wang W, Ma X, Xu Y, Cao Y, Jiang Z, Ding T et al (2015a) Ultrasound-assisted heating extraction of pectin from grapefruit peel: optimization and comparison with the conventional method. Food Chem 178:106–114

Wang Y, Jiang K, Ma H, Zeng W, Wang PG, Yao N, Han W et al (2015b) Enzymatic production of HMO mimics by the sialylation of galacto-oligosaccharides. Food Chem 181:51–56

Watson D, O'Connell Motherway M, Schoterman MH, van Neerven RJ, Nauta A, van Sinderen D (2013) Selective carbohydrate utilization by lactobacilli and bifidobacteria. J Appl Microbiol 114:1132–1146

Wickramasinghe S, Pacheco AR, Lemay DG, Mills DA (2015) Bifidobacteria grown on human milk oligosaccharides downregulate the expression of inflammation-related genes in Caco-2 cells. BMC Microbiol 15:172

Wu LD, Ruhaak LR, Lebrilla CB (2017) Analysis of milk oligosaccharides by mass spectrometry. Methods Mol Biol 1503:121–129

Xu ZR, Zou XT, Hu CH, Xia MS, Zhan XA, Wang MQ (2002) Effects of dietary fructooligosaccharide on digestive enzyme activities, intestinal microflora and morphology of growing pigs. Asian Australas J Anim Sci 15:1784–1789

Xu BH, Wang YB, Li JR, Lin Q (2009) Effect of prebiotic xylooligosaccharides on growth performances and digestive enzyme activities of allogynogenetic crucian carp (Carassius auratus gibelio). Fish Physiol Biochem 35:351–357

Xu M, Li D-S, Li L, Wang C, Zhu Y-P, Lv W-P, Xie B-J (2013) Comparative study on molecular weight of konjac glucomannan by gel permeation chromatography-laser light scattering-refractive index and laser light-scattering methods. J Spectrosc 2013:685698
Yan J, Ding J, Liang X (2017) Chromatographic methods for the analysis of oligosaccharides in human milk. Anal Methods 9:1071–1077
Yang B, Chuang H, Chen RF (2012) Protection from viral infections by human milk oligosaccharides: direct blockade and indirect modulation of intestinal ecology and immune reactions. Open Glycosci 5:19–25
Yang J, Vittori N, Wang W-W, Shi Y-C, Hoeflinger JL, Miller MJ, Pan Y (2017) Molecular weight distribution and fermentation of mechanically pre-treated konjac enzymatic hydrolysates. Carbohydr Polym 159:58–65
Yapo BM (2011) Pectic substances: from simple pectic polysaccharides to complex pectins—a new hypothetical model. Carbohydr Polym 86:373–385
Yuliarti O, Goh KKT, Matia-Merino L, Mawson J, Brennan C (2015) Extraction and characterisation of pomace pectin from gold kiwifruit (Actinidiachinensis). Food Chem 187:290–296
Zhang Y, Pitkänen L, Douglade J, Tenkanen M, Remond C, Joly C (2011) Wheat bran arabinoxylans: chemical structure and film properties of three isolated fractions. Carbohydr Polym 86:852–859
Zhang L, Ye X, Ding T, Sun X, Xu Y, Liu D (2013) Ultrasound effects on thedegradation kinetics, structure, and rheological properties of apple pectin. Ultrason Sonochem 20:222–231
Zhang SY, Li W, Smith CJ, Musa H (2015a) Cereal-derived arabinoxylans as biological response modifiers: extraction, molecular features, and immune-stimulating properties. Crit Rev Food Sci Nutr 55:1035–1052
Zhang YJ, Li S, Gan R-Y, Zhou T, Xu D-P, Li H-B (2015b) Impacts of gut bacteria on human health and diseases. Int J Mol Sci 16:7493–7519
Zia F, Zia KM, Zuber M, Ahmad HB, Muneer M (2016) Glucomannan based polyurethanes: a critical short review of recent advances and future perspectives. Int J Biol Macromol 87:229–236
Zivkovic A, German JB, Lebrilla CB, Mills DA (2011) Human milk glycobiome and its impact on the infant gastrointestinal microbiota. Proc Natl Acad Sci U S A 108:4653–4658

Chapter 25
Role of Different Polymers on the Development of Gluten-Free Baked Goods

Manuel Gómez and Laura Román

Abstract Storage proteins in wheat are able to develop a network upon hydration and application of mechanical work, giving rise to unique properties of doughs made with wheat flour. Due to this gluten network, these doughs have a certain extensibility and elasticity. They are also capable of retaining the gases produced during fermentation and yielding spongy products. However, there is an increasing number of people choosing to consume gluten-free products. Both celiac and wheat allergic patients and people with non-celiac gluten sensitivity are included in this group. So far, no appropriate material has been found capable of replacing gluten and conferring all its properties to doughs. Therefore, the development of gluten-free products is based on different physicochemical principles than those used in the manufacture of gluten products. In these gluten-free developments, starch, a major component of bread, plays an essential role, and it must be carefully selected in order to obtain a high quality product. Conversely, the use of certain hydrocolloids has been shown to significantly enhance the quality of gluten-free breads by improving gas retention during fermentation and baking. In these products, it is also common to use proteins that help promote a more roasted color of the crust by increasing Maillard reactions. Likewise, it is possible to use celluloses and other fibers to improve water retention in the final product and increase the juiciness of the loaves. All these polymers play a fundamental role not only in the sensory quality of the obtained products but also in the nutritional quality. In this chapter, the role of different polymers used for the development of gluten-free products will be assessed, providing key insights for their correct choice, both sensorially and nutritionally.

Keywords Fibers · Hydrocolloids · Proteins · Starches

M. Gómez (✉) · L. Román
Food Technology Area, College of Agricultural Engineering, University of Valladolid, Palencia, Spain
e-mail: pallares@iaf.uva.es

T. J. Gutiérrez (ed.), *Polymers for Food Applications*,
https://doi.org/10.1007/978-3-319-94625-2_25

25.1 Introduction

In western countries wheat culture is broadly spread, wheat is the cereal predominantly utilized for the production of many food products that constitute the basis of the population's diet. However, there is a part of the population that cannot (under medical prescription), or is reluctant, to consume wheat based products. Among these people, the celiac group is prominent, composed of people who possess an intolerance to wheat proteins and other cereals such as barley or rye. This intolerance affects approximately 1% of the population (Catassi et al. 2014). There are also people allergic to wheat proteins, although the prevalence of this allergy is much lower (<0.5%) than that of celiacs. In addition, there is a group of people who, without being celiac or allergic to gluten, suffer from intestinal and extra-intestinal symptoms related to the ingestion of gluten-containing food (non-celiac gluten sensitivity, NCGS) and should restrict their intake of gluten products. The overall prevalence of NCGS in the general population is still unknown, mainly because many patients are currently self-diagnosed and start a gluten-free diet (GFD) without medical advice or consultation. Despite the lack of accurate data, NCGS prevalence is estimated to be slightly higher than that of celiac disease (Catassi et al. 2010). However, it is important to bear in mind that recent studies claim that the cause of digestive problems in patients with self-reported NCGS might not be gluten but instead fructans present in cereals forbidden to celiacs (Skodje et al. 2017). Last but not least, there is a large number of people who follow a GFD without suffering any celiac pathology [up to 25% in the USA (Reilly 2016)], as they believe in its benefits on health, weight loss, treating disease and/or minimising future risk of disease (Lis et al. 2015; Staudacher and Gibson 2015).

Since GFD is gaining importance due to the increasing number of the above mentioned groups, the food industry is trying to respond to the recent demand for high quality gluten-free products (GFPs). Production of good quality GFPs is challenging due to the lack of wheat gluten proteins and their unique functionality. Then, food scientists and technologists have attempted to meet industrial requirements by developing more suitable GFPs based on these approaches; (1) the study of the functionality of the different polymers of wheat flour in each elaboration (requirement or no requirement for a gluten network) in order to find alternative products that replace such functionality, (2) selection and thorough knowledge of appropriate gluten-free ingredients or additives that can mimic the desired functionality of the lacking wheat components and (3) improvement of the nutritional quality of the products obtained (Capriles et al. 2016).

In wheat flour, starch is the major component (~70–80% of the total content of the flour). The role of this polymer is different depending on the product, but basically starch acts as a source of sugars, after amylase hydrolysis (in those products in which fermentation occurs), and as a structural component of the dough. During kneading and fermentation of the dough only starch that has been damaged in previous stages, (i.e., milling), can be hydrolyzed by the amylases present in the grain (Struyf et al. 2016). Damaged starch is hydrolyzed to produce fermentable sugars, namely glucose and maltose, that are transformed by yeasts into ethanol and CO_2. If the CO_2 produced is properly retained by the dough network, this leads to its adequate expansion and an ultimate increase in product, yielding aerated products. Sugars also participate in Maillard and caramelization reactions, especially in the external parts of the products subjected to higher temperatures. These high temperature requiring reactions produce aromatic compounds and other colored compounds that are responsible for the color and aroma of the bread crusts. As structural element, the undamaged native starch granules, which are the major part of the total, contribute together with the protein to give the desired consistency to the dough/batter, forming the starch-protein continuous network in which gas cells are dispersed. During baking, starch granules are gelatinized as a result of the high moisture and temperature conditions. Gelatinization mainly comprises two stages: an initial phase in which starch granules absorb water and swell until they reach a certain point, and a second one in which the granule integrity is disrupted, the granular structure starts to break down and the starch components (mainly amylose) are solubilized. The nearly complete breakage is markedly pronounced in bread crumbs and after gelatinization and upon cooling there is a rapid settling of the structure, causing the formation of the crumb. Although the contour of some swollen and/or deformed starch granules is still visible in crumb walls, gelatinization phenomena is known to be complete. In contrast, due to the rapid drying of the surface, gelatinization phenomena are minimized in the crust and granule integrity is mostly retained (Vanin et al. 2009; Martinez et al. 2018). Upon cooling, the amorphous gelatinized starch undergoes a rearrangement of its molecules to a more ordered or crystalline state, process known as retrogradation. Initially, this phenomenon mainly affects amylose (rapid retrogradation) and is responsible for the initial texture of the crumb. Thus, double helices are rapidly formed between the solubilized amylose molecules, and a continuous network develops conferring amylose an essential structural role in crumb. Conversely, due to the architecture of amylopectin molecules, amylopectin side-chain recrystallisation is a much slower process which has been associated with bread staling (Kulp and Ponte Jr 1981). During storage, bread stales losing its freshness, but bread staling is complex and multiple mechanisms are involved such as crust toughening, crumb firmness and loss of its elasticity, moisture and flavor loss, etc. Amylopectin retrogradation is responsible, at least partially, for the firming of the crumb during storage. In addition, this retrogradation phenomena results in the release of part of the incorporated water and facilitates its migrations, and, desiccation of the breads (Goesaert et al. 2008; Ziobro et al. 2013a). Therefore, in the preparation of gluten-free products, it will be necessary to use a starch source which fulfills the functions of wheat starch in a similar way.

The second most abundant component of wheat flour is protein (8–18% of the total content of the flour). In wheat flour, gluten is the main protein (between 80 and 85% of total protein) composed of glutenins and gliadins. The degree of difficulty in producing gluten-free products is due to the technological/functional role of gluten. Gluten protein is responsible for the visco-elastic characteristics of the dough. Specifically, the wheat gluten protein, in the presence of water and under mechanical work, form a continuous phase known as gluten network. This gluten network provides unique cohesiveness, elasticity and reduced stickiness properties to the dough, while retains fermentation gases (Delcour et al. 2012; Khatkar and Schofield 1997). Therefore this protein network gives rise to a dough that can be handled properly (adequate consistency and stickiness), and after fermentation, aerated products are obtained, with a softer and more pleasant texture. The most challenging gluten-free products to formulate and produce are gluten-free breads (GFBs), due to the fact that gluten plays a key role in the dough. A gluten network is also formed in other fermented products, such as croissant, puff pastry or pizza. Unlike these products, there are products made with wheat flour in which the gluten network is not fully developed, either because the proteins are not hydrated or because the "doughs" do not undergo enough mechanical work during kneading/mixing. As clear examples, batters or "liquid type doughs" (such as cakes, crepes, waffles, tea pastes, batters) which exhibit low viscosity (partially caused by the higher ratio of liquid ingredients in the formula) that reduces friction and mechanical work during mixing or whipping. Added to that, the gluten network is not developed in some types of cookies containing high amounts of sugar and fat with low water content in their formulation, and whose mixing times are shortened to reduce the input of mechanical work. For these products, the preparation of their gluten-free counterparts is feasible and replacement of the wheat flour with an alternative gluten-free flour, such as rice or maize, is feasible without the need for additional ingredients in such a way that high quality gluten-free cakes (de la Hera et al. 2013b), and cookies (Mancebo et al. 2015b) are easily obtained. In fact, in the case of muffins made with rice flour, the inclusion of wheat proteins neither improved the volume nor the softness of the muffins (Matos et al. 2014). Regarding cookies, the creation of a gluten network has even been considered negative, as it increases the consistency of the doughs and reduces their expansion (Pareyt and Delcour 2008). In addition, Souza et al. (1994) stated that the amount of protein in flour is more important than its characteristics. Thus, in these products the functionality of the polymers is the same as in their gluten counterparts and, for that reason, they will not be further analyzed in this chapter. In contrast, in the products in which the gluten network is formed, it is necessary to find a substitute for gluten, in order to retain the fermentation gases, and, if possible, to obtain a dough that can be handled in a similar way to a wheat one. To this end, research has been carried out on proteins other than wheat, but the most commonly used alternative to gluten is hydrocolloids, as can be seen in the reviews of Taylor et al. (2016) and Mir et al. (2016), respectively. These gluten substitutes provide a certain cohesion to the dough while improving its gas retention capability. Other than that, gluten proteins also fulfill a nutritional function and are responsible for the color and aroma of the product crusts through the Maillard reactions with simple sugars (Smak 1972). Thus, the presence of proteins, and their composition in amino acids, must also be evaluated in the elaboration of products without gluten.

In this way, the common ingredients in gluten-free breads are flour and/or starch mainly from maize, rice, potato (or other tubers), with the addition of proteins or hydrocolloids that may partially act as substitutes of gluten. The role of each of these ingredients will be further discussed in this book chapter.

25.2 Starches

The raw materials used as substitutes of starchy material for wheat flour for gluten-free formulations comprise starches and flours from different origin, mainly gluten-free cereals, pseudo-cereals, tubers and pulses, among others. Rice flour and maize starch are among the preferred ingredients in the preparation of GFPs (Masure et al. 2016), since they are well known/widely studied and readily available. Maize starch allows higher in volume GFBs to be obtained compared to rice flour, although the sensory evaluation of maize starch based breads is worse, which is usually associated with its poor, excessively dry and not very cohesive, crumb texture (Mancebo et al. 2015a). For this reason, formulations based on starch in mixture with flour have been developed. The alternatives to maize starch include starches from tubers, such as potato, tapioca, and gluten-free formulations based on their blends with other starches or flours. More specifically, tapioca starch results in breads with lower specific volume and overly rubbery texture compared to maize starch (Lopez et al. 2004). Likewise, potato starch also produces breads with lower volume, which has been attributed to the larger size of its granules (less prone to pack), preventing the formation of a continuous starch-hydrocolloid matrix and lowering dough elasticity (compared to maize and wheat starches) (Martinez and Gomez 2017a). In fact, Zhang et al. (2017) observed that doughs made with potato starch presented higher than delta (lower dough elasticity) as well as lower creep compliance, followed successively by the doughs made with tapioca, tuber and wheat starches. The same order was found for the degree of dependence of G′ on frequency sweep, suggesting that the resistance to deformation depends on network structure stability. Despite the poor quality of the breads obtained with these starches alone, both are used in combination with maize starch or rice flour (Crockett et al. 2011a; Matos and Rosell 2013; Parra et al. 2015; Sciarini et al. 2012b) since they are able to minimize some of the defects of the breads obtained with them, such as the excessively dry texture and the poor cohesiveness. In this way, Sanchez et al. (2002), through a central composite design, proposed a mixture of maize starch, tapioca starch and rice flour to optimize the organoleptic quality of GFBs. The incorporation of these starches can also be due to cultural or traditional influences, in areas where the consumption of these tubers is common. In fact, in a study on GFPs in Brazil, the presence of tapioca starch was found to be superior to that of other starches, such as maize starch (do Nascimento et al. 2013).

Despite the controversy over its safety, wheat starch represents a highly suitable alternative to maize starch for GFPs. Wheat starch has been traditionally avoided for the formulations of GFPs, due to the risk of gluten contamination, and to the reluctance from consumers who adhere to a gluten-free diet. However, in recent years the improvement in the separation process of starch and gluten, as well as in the quality control for the absence of gluten, have enabled the starch industry to offer gluten-free wheat starch. Concerning its safety, Peräaho et al. (2003) indicated that gluten-free wheat starch shows a similar histological and clinical recovery in patients with newly detected celiac disease compared to a natural gluten-free diet. Recent studies

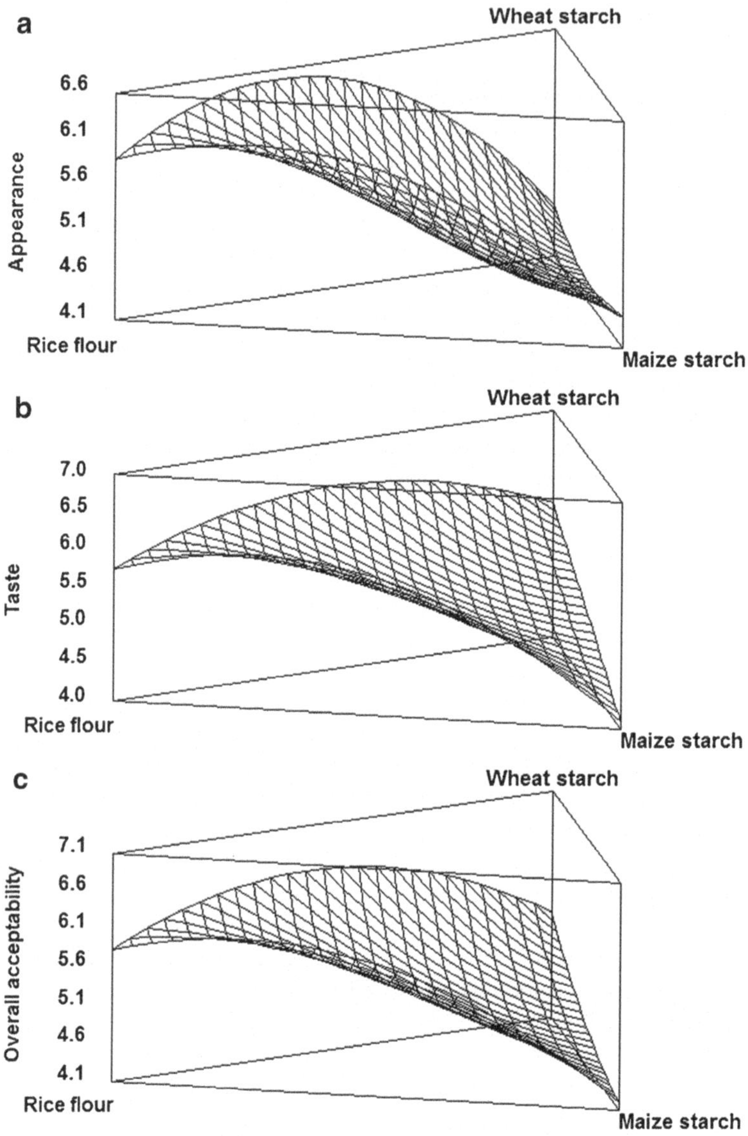

Fig. 25.1 Effect of rice flour, maize starch and wheat starch content on appearance (**a**), taste (**b**) and overall acceptability (**c**) of the breads (Mancebo et al. 2015a)

have demonstrated wheat starch has great potential in the production of GFBs (Fig. 25.1). In particular, Mancebo et al. (2015a) found that wheat starch yielded GFBs with better textural properties than those made with maize starch. These authors also reported a higher volume of wheat starch based breads and an improved acceptability related to the better flavor and aroma of these breads compared to maize ones. The potential of wheat starch as an ingredient for GFBs was further discussed by Martinez and Gomez (2017a). In the aforementioned work, Martinez and Gomez (2017a) highlighted that wheat starch had positive packing properties, i.e. small wheat starch granules filled the interstitial spaces of the big ones, increasing the continuity of the starch-hydrocolloid matrix of the dough, which reduced dough consistency and in turn, increased GFB volume compared to other starch sources (maize and rice). In wheat starch, two distinct types of granules exist that differ in size and shape, i.e. large, lenticular (A-type) and small, spherical (B-type) granules. These wheat starch granules also differ in composition and properties. Generally, A-type granules have higher amylose content and lower gelatinization temperature, whereas B-type granules have higher lipid content and water absorption capacity (Kim and Huber 2010; Maningat et al. 2009). Likewise, ratios of A- to B-type wheat starch granules were obtained to prepare GFBs (Román et al. 2018). In this work, the increase in the proportion of B-type granules enhanced the packing and uniformity of the continuous phase as well as the Pickering stabilization of bubbles. Interestingly, combinations of 75–25 and 25–75 of A to B starch granules gave rise to breads with the highest specific volume and best textural properties (higher cohesiveness and lower hardness). In addition, A- type based GFBs presented an open grain structure (fewer cells with larger size), while a more closed structure was visible in B-type based ones. Wheat starch contains approximately 30% B granules by weight, therefore usual distribution in commercial starches would be adequate to maximize the volume.

Rice starches have been less studied, maybe because of their lower supply in the market or their higher cost. However, recent research indicates that rice starch, provided it is obtained through methods that reduce damage to starch granules (damaged starch <5%), such as wet milling, have a high emulsifying capacity and, in particular, a stabilization capacity by the Pickering effect. Pickering is based on the potential of small granules to be adsorbed onto the air-liquid interface, stabilizing air bubbles. In this way, the preparation of GFBs with these low damaged rice starches has been proposed, with no need for gluten substitutes, such as hydrocolloids (Yano et al. 2017). In view of the above details, maize starch is the most commonly used starchy material, as it yields breads with higher specific volumes compared to those made by simply using gluten-free flours. However, due to its poor organoleptic characteristics, its combination with potato or tapioca starches or with gluten-free flours is advisable. The use of wheat and rice starches seems promising, but existing research is scarce and more in-depth studies that analyze these starches or their combination with other starch and flour sources are needed.

Regarding starches, a final possibility is the utilization of modified starches obtained either by physical, chemical or enzymatic methods. This approach is generally less common since these starches are more expensive and usually need to be specifically designated on the product label. So far, only chemically modified starches are considered as additives and must be indicated as "modified" on the label of the products in which they are utilized, but the denomination will depend on the legislation of each country. For this reason, food companies try to avoid the use of modified starches in order to meet consumer demands for more clean labeled products. Nonetheless, these starches can improve the shelf-life of GFBs by reducing the staling phenomenon. Thus, the incorporation of hydroxypropyl distarch phosphate was shown to minimize the hardness and loss of cohesiveness during storage of GFBs (Ziobro et al. 2012). However, these authors could not relate these events to changes in amylopectin retrogradation, suggesting that the improved texture attributes were the result of a lower moisture loss during storage. Physically modified starches such as tapioca pregelatinized starch have also been proposed to increase the volume of rice breads, reduce their initial hardness and promote darker crusts (Pongjaruvat et al. 2014). The substitution of pregelatinized starch in rice based breads slightly decreased the dough elasticity (storage modulus, G′) but increased dough resistance to deformation, thus permitting better expansion of gas cells during proofing and baking. It is known that the reduction in G′ is highly correlated with the volume increase of GFBs (Mancebo et al. 2017). Similarly, the increase in volume is associated with the lower hardness of the crumb (Gallagher et al. 2003; Martinez and Gomez 2017a). The fact that the crust of these breads is slightly darker may be due to the greater accessibility of pregelatinized starches to enzymatic degradation, and therefore the possible generation of reducing sugars that would participate in Maillard reactions. More research that confirms these results using starches with different types of modifications would be of great help for encouraging the utilization of these starch sources.

It is important to bear in mind that, in the processes for producing starch, a minimum amount of granules is damaged and amylases present in the grain are eliminated. Therefore, during baking of starch based breads the generation of sugars by enzymatic hydrolysis of the damaged granules will be minimal. For this reason, sugars should be incorporated into the formulations, usually in the form of sucrose or glucose. These sugars are fermentable and will contribute to the fermentation processes carried out by the yeasts. In the case that the sugars are not entirely consumed by the yeasts, they can participate in Maillard reactions, between sugars and amino acids, contributing to the final color of the crusts.

Another way to incorporate starch into GFPs is through gluten-free flours. Among gluten-free flours, rice flour is the most commonly used both in research (Masure et al. 2016) and in commercial products because of its lower price, higher availability, bland taste, and hypoallergenic properties. Rice, along with wheat and maize, is one of the most widely cultivated cereals worldwide. Most of rice production is consumed as pearled grain, in which the husk and the external parts (bran and germ) of the kernel are removed. During the process by which white rice is obtained, parts of the grains are broken. These grains are considered to be by-products from the rice industry, and once they are polished and separated from the rest, they can be used to make flours using different milling systems, with the hammer and rolling mills being frequently used. In general, rice flours bring about breads with lower specific volume than those made with maize starch and doughs with higher G′ and G″ values, and lower compliance values in both creep and recovery phases (Martinez and Gomez 2017a). However, as seen in Fig. 25.1, an increase in the percentage of rice flour in the mixtures with starch can improve the organoleptic quality of these breads, especially by improving taste and external appearance (Mancebo et al. 2015a). In the baking process, parts of the flour particles are broken, releasing the starch granules, while the harder fragments remain intact (de la Hera et al. 2013a). Therefore, the particle size and the breakage of the particles during processing can influence the compactness of the insoluble components, and, consequently, the rheology of the doughs, the increase of the volume during fermentation and the quality of the resulting bread. Thus one of the derived problems of the utilization of rice flours is the lack of regularity of the flours, since the milling conditions or the rice varieties can widely affect the characteristics of the bread obtained.

In case a finer particle size is desired the mechanical process of milling has to be forced, increasing the amount of damaged starch, and therefore its ability to absorb water. Another option would be the mechanical sieving of the flour after milling, separating the largest particles from the smallest and obtaining flours of different particle size. In general, it has been observed that coarser flours lead to breads of higher specific volume. This phenomenon has been reported both when producing flours with different milling systems (Kadan et al. 2008) and when obtaining different fractions by sieving (de la Hera et al. 2013a). Nonetheless, the particle size should not be excessively large, since otherwise the breads will present a sandy texture. De la Hera et al. (2013a) found that short-grain varieties, with a lower amylose content, generated breads with higher specific volume, with crumbs that were less firm and more cohesive and resilient. However, Sivaramakrishnan et al. (2004) obtained better results with a flour from long rice grain than from short grain, but in this case the long grain flour had a larger particle size, so this factor could have also influenced the final results. In fact, studies with a greater number of variables have not found a direct correlation between the amylose content of the flours and their quality for the preparation of gluten-free breads (Han et al. 2012; Nishita and Bean 1979). It has also been stated that rice varieties with more rounded starch granules may be more suitable for the preparation of GFBs, compared to those with the more usual polyhedral granules, by increasing the volume and reducing firmness of the breads (Kang et al. 2015). However, this study was conducted only with three varieties that differed in the shape of the starch granules as well as in the kernel hardness (and in turn, damaged starch), thus deeper studies are necessary in this regard. In a later study, the influence of rice flour particle size on the volume of breads obtained was confirmed, and it was observed that breads made with coarser flours had a higher content in slowly digestible starch and resistant starch when the dough hydration exceeded 90% (de la Hera et al. 2014). These differences were attributed to the lower gelatinization degree of the starch in the coarser flours during baking (albeit it was not measured), as the penetration of water inside the bigger particles becomes more difficult. These authors also mentioned that in coarse flours the lower surface area of the particles exposed to digestive enzymes could limit enzyme binding and absorption, and, then, decrease the rate of digestion. It is important to bear in mind that flour particles in flour-based GFBs are not completely disrupted during the kneading process (de la Hera et al. 2013c; Martinez and Gomez 2017a).

As an alternative to rice flours, maize or sorghum flours can be used, but their utilization is limited due to availability and organoleptic characteristics. In the case of maize, although there are floury varieties, the maize used for milling usually comes from hard varieties since it is the maize type mainly used for the production of grits and semolina, fundamental ingredients in some snack and ready to eat cereal industries. On the other hand, sorghum has been used mainly for the elaboration of feed, with minor use in the food industry. Both in maize and sorghum, there are colored varieties and more whitish varieties, with the latter more preferable for the preparation of GFPs, due to their similar appearance to wheat flours and bland flavor. However, in some countries with high consumption of maize, it may be interesting to use yellow maize flours for cultural reasons. Thus, in the southern area of USA it is common to make maize breads, as well as in Portugal (*Broa*). Regarding these types of flour, there is little information on how their characteristics affect the quality of GFB. However, it seems clear that both the variety (Brites et al. 2010; Garzon et al. 2017) and the particle size (de la Hera et al. 2013c) of maize flours influence the characteristics of GFBs, with coarser flours being more preferable, as in the case of rice. Regarding sorghum flours, Trappey et al. (2015) observed an opposite tendency, as there was a slight improvement of bread volumes when particle size was reduced, but in this case all the flours with low degree of extraction had an average size lower than 100 μm. In this same study, it was observed that by increasing the degree of extraction of the flours, and therefore increasing their content in bran, the volume of the breads was reduced, being these darker and harder. As in the case of bread made with other gluten-free flours, the replacement of these by starches improves the volume of the bread and reduces its firmness (Onyango et al. 2011a; Velazquez et al. 2012), this effect is more evident with native starches that with pregelatinized (Onyango et al. 2011b).

Apart from the aforementioned rice, maize and sorghum flours, which are the most dominant, studies about the utilization of other less common gluten-free flours from cereals such as millet (Taylor et al. 2006) or teff (Zhu 2018), pseudo-cereals (Alvarez-Jubete et al. 2010b; Gimenez-Bastida et al. 2015), legumes (Foschia et al. 2016; Melini et al. 2017) or chestnuts (Zhu 2017) are reported. The use of these flours in commercial breads is usually limited because of the lower availability, as well as the higher cost and unpleasant sensory attributes of the resulting breads (bitter and extraneous flavors). Nonetheless, in certain areas it may be convenient to include them for tradition or cultural reasons. However, studies on the influence of varieties, milling system, or other processing variables on these alternative flours and their effect on the quality of GFBs are scarce. As in the case of other flours, it seems that the incorporation of starches in the formulas can help increase the volume of the breads and improve their texture and organoleptic characteristics, so it is usual to use them in mixtures. The main advantage of these flours is usually their nutritional content, since many of them are used as whole meal flours. Thus, all these flours are characterized by a higher content in minerals, vitamins and dietary fiber than starches and rice flour, especially when they are whole flours. In addition, both pulses and some pseudo-cereals (quinoa and amaranth) are good sources of protein. These flours can also be a source of antioxidant compounds (Torres et al. 2017). These nutritional characteristics are of great interest since the gluten-free diet is deficient in fiber and micronutrients such as vitamins (B, folates and D) and minerals (iron, zinc, magnesium and calcium) (Theethira and Dennis 2015; Vici et al. 2016). Therefore, supplementation with some of these micronutrients has been proposed to improve the quality of the gluten-free diet (Caruso et al. 2013).

Furthermore, GFBs are characterized by a higher fat content and a lower protein content than gluten containing breads (Mazzeo et al. 2015; Miranda et al. 2014).

It is noteworthy that oats have been traditionally considered harmful for celiac patients. However, more recent research has shown that oat consumption is safe for the celiac population, provided that cross-contamination with other toxic cereals for this group is avoided (Hüttner and Arendt 2010; Pawlowska et al. 2012). However, it must be borne in mind that a small number of the celiacs does not tolerate the avenins (prolamin fraction of the oats). The incorporation of oats in gluten-free diets can offer important advantages, since breads made with oatmeal are characterized by a higher content of proteins and β-glucans, and better sensory perception than breads obtained from other gluten-free cereals (Hager et al. 2012). The protein in oats is neither able to give rise to doughs with a protein network similar to gluten, being also the addition of a gluten substitute. Due to their nutritional and sensory qualities, different countries and groups are changing their legislation and recommendations to facilitate the inclusion of products with oat, properly controlled, in gluten-free diets (Fric et al. 2011). Although there are more countries that admit the consumption of oats for people with celiac condition the situation in each country regarding oat consumption is different. In Finland, oat consumption is allowed and usual, while in other countries celiacs are advised to consult with their doctors in advance, and in others there is still a certain resistance to consumption.

A final point to consider about the role of starch in gluten-free products is the glycemic index. Most studies agree that GFBs have a higher postprandial glucose level than their counterparts made with wheat (Berti et al. 2004; Jenkins et al. 1987). This may be because the starch present in most commercial gluten-free breads is usually isolated maize starch, either alone or mixed with potato starch and rice flour, as observed in the Matos and Rosell (2011). Therefore, this starch does not have a protein network that protects the granules from the action of amylases, as could take place in wheat breads (Jenkins et al. 1987). However, when certain gluten-free flours are used, the glycemic index of these breads may be lower than that of wheat breads and clear differences are observed between the different flours (Wolter et al. 2013). In this study, it was found that breads made with teff or sorghum flours (gluten-free cereals) had lower glycemic indexes than those made with buckwheat or quinoa (pseudocereals), something that could be related to differences in the starch structure of these grains (i.e. starch granules size and shape, damaged starch, amylose content, etc.) and the interactions between protein and starch. However, more studies are needed evaluating different flours and different milling systems to reach more solid conclusions. On the other hand, the germination of the grains also allows for reduction in the estimated glycemic index (IGe) of the breads made with these flours (Cornejo et al. 2015; Xu et al. 2012). This event may be related to changes in the starch structure leading to a more crystalline and less accessible form for amylases, fostered by annealing conditions of moisture and temperature during soaking and drying of the grains. Xu et al. (2012) also highlighted the possibility that during germination the most amorphous part of the starch was consumed since it was more available, while the most crystalline part was less accessible to the enzymatic attack and more prominent in the final flour.

25.3 Proteins

In wheat, gluten storage proteins present unique properties once hydrated with enough mechanical energy input, forming a tridimensional network with singular rheological properties. Gluten proteins are transformed into a continuous cohesive, extensible, visco-elastic protein network that makes wheat flour suitable for breadmaking, obtaining a pliable dough. Furthermore, this gluten network enables gas retention produced during fermentation, thus giving raise to aerated, spongy products. These unique properties of wheat protein cannot even be found in closely related cereals such as barley or rye. In this way, obtaining elastic and resilient crumbs is still a major technological challenge in gluten-free breads due to the lack of the polymeric glutenins from gluten (the high molecular weight glutenin subunits, HMW-GS) which are mainly liable for elasticity and dough strength (Shewry et al. 2002). Based on this, attempts to modify proteins from different origin to achieve similar functionality to that of wheat have been made. However, these investigations have not been successful, and at the moment none of them have been included in commercial products, suggesting the necessity of utilizing other types of gluten substitutes, such as hydrocolloids.

Thus, to reinforce the bonds between the protein molecules, the use of transglutaminase has been proposed (Marco and Rosell 2008; Moore et al. 2006; Renzetti et al. 2008; Storck et al. 2013). However, in no case has a significant improvement in the specific volume of the breads been reported. Indeed, bread volume was very low when hydrocolloids were not used in the formulation, meaning that this enzyme, whatever the protein source is, did not help create a protein network similar to that formed by wheat proteins. What transglutaminase apparently achieves is a more closed cell grain and a greater crumb firmness, which may be related to a lower extensibility of the crumb cell walls (Moore et al. 2006; Renzetti et al. 2008). The incorporation of mesostructured whey protein has also been proposed. In this case, whey protein needs to be previously heated and subsequently mixed with locust bean gum accompanied by reduction of the pH of the protein-gum mixture. Mixtures of this protein-gum combination with starch gave rise to a cohesive dough that can be comparable to a wheat dough regarding their rheological behavior (Van Riemsdijk et al. 2011a). Furthermore, the rheology of these doughs can be improved by promoting the creation of disulfide bonds, since doughs tend to have less extensibility than wheat counterparts (Van Riemsdijk et al. 2011b). Breads produced with mixtures of these particles (2.4%) and maize starch were similar in volume and crumb grain to wheat breads, but to achieve this the baking process had to be slightly modified, by suppressing the sheeting step, since this negatively affected bread volume (Van Riemsdijk et al. 2011c). However, these proteins have not been studied in depth, and they do not seem to have been incorporated in commercial gluten-free products, perhaps because there are still some aspects to improve. The breads obtained in these studies had an excessively whitish color, which could be enhanced with the inclusion of other proteins in a greater percentage to promote Maillard reactions, and no data was reported on the textural properties, staling behavior or sensory quality of the breads.

One of the proteins that has attracted the attention of scientists as a possible gluten substitute is zein, which is the prolamin fraction of maize. This interest began with the work of Schober et al. (2008) in which the kneading of a mixture containing starch and zein above the glass transition temperature of zein (Tg ~40 °C), resulted in the formation of finer zein strands in the dough. These zein strands allowed the dough to be handled similarly to a wheat one, by providing them with cohesiveness and extensibility. However, if the dough was cooled down upon mechanical impact below Tg these fibers were broken and no further recovered, while this breakage was not observed if the dough did not suffer the mechanical impact. In this work, it was also observed that the incorporation of hydroxypropyl methyl cellulose (HPMC) improved bread quality by stabilizing the air bubbles in the dough, so a combination of both techniques was recommended. Attempts to work with this kind of doughs at lower temperatures have focused on the reduction of the glass transition temperature with the use of plasticizing agents, such as citric acid (Berta et al. 2015), but this did not demonstrate to be effective in obtaining baking processes similar to those of wheat. Another problem found in these doughs is based on the negative effect that salt exerts on them, since it reduces the strength of the network, prevents the formation of an extensible dough and, thus, lowers the specific volume of the breads obtained (Smith et al. 2017). This disadvantage is particularly important since salt is a common compound in the formulations of bread. Likewise, studies have also been carried out based on kafirin, a protein present in sorghum (Oom et al. 2008), but the results are not as promising as those with zein and the knowledge on the use of this protein for baking is more scarce.

Given the interest in the utilization of zein, more in depth studies have been conducted. In this sense, Schober et al. (2010) proved that the quality of zein containing breads could be improved if zein was subjected to a previous degreasing treatment, since viscoelastic moduli (G′ and G″ values) of the dough were increased. It is noteworthy that these works seek to develop doughs with a network similar to that of wheat, which can be handled and formed, as depicted in Fig. 25.2, and therefore, it must be understood that a greater dough consistency (viscoelasticity) is usually beneficial, unlike what happens in the rest of gluten-free doughs normally based on hydrocolloids, more similar to a cake batter than to a wheat dough.

In addition, this degreasing treatment enhanced the extensibility of zein strands and lowered the temperature at which protein cross-linking occurred by 2 °C, enabling an increased specific volume of the GFBs and preventing the flattening of the rolls of dough formed during processing. Added to that, it is known that protein functionality in GF products is more linked to hydrophobic interactions rather than disulfide bonds (as in gluten doughs). Thus, α-zeins are preferable to obtain structured fibers with extensibility properties to other types of zein (β- or γ-zeins), which concentration will depend on the isolation procedures (Schober et al. 2011). Similar to wheat doughs, the mechanical work applied during kneading of zein doughs is a key factor for the development of zein fibers. Thus, a finer fiber network may be formed by increasing shear through an addition of viscosity-increasing hydrocolloids, reducing water content in the dough or the use of appropriate mixing equipment (Andersson et al. 2011). These authors observed that doughs without

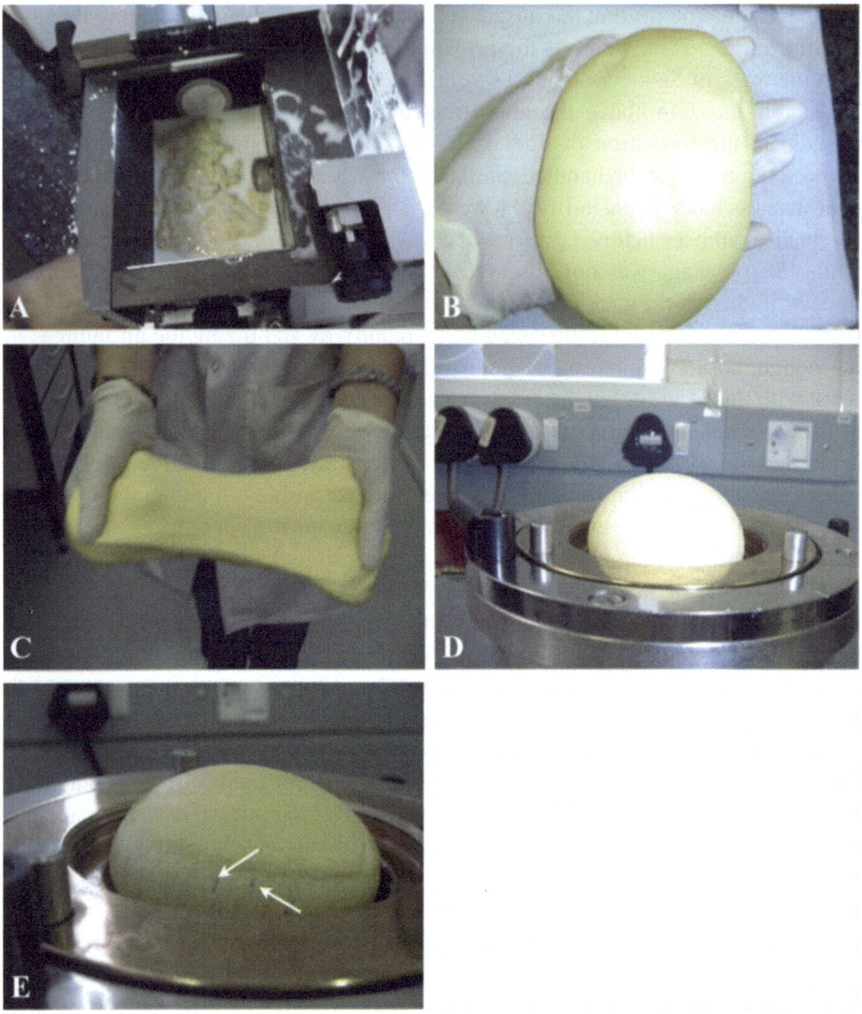

Fig. 25.2 Zein-maize starch doughs. (**a**) Dough prepared with water after kneading, showing phase separation; (**b**) dough prepared with 1.3% lactic acid after kneading; (**c**) malleability of dough prepared with 1.3% lactic acid; (**d**) alveography of dough prepared with 1.3% lactic acid; (**e**) loss of uniformity and tearing of Alveograph bubble of dough prepared with 5.4% lactic acid (Sly et al. 2014)

hydrocolloids (HPMC or oat bran with a high content in β-glucans) exhibited rapid age-related stiffening and poor bread-making performance and, among hydrocolloids, HPMC was preferable to obtain a finer crumb structure due to its surfactant effect. Doughs made with mixtures of zein and maize starch or rice flour can also be improved by the addition of diluted acetic or lactic acid, acids that are present in the sourdough. Sly et al. (2014) reported that doughs formed with distilled water did not present an adequate consistency, not being possible to develop the dough for alveographic analysis. Conversely, doughs made with diluted acids gave rise to alveographic results similar to a wheat dough (Fig. 25.2), which was related to the creation of a more uniform fibril network and to the increase of the proportion of α-helical conformation in the zein doughs, possibly as a result of deamination. Detailed information about the utilization of zein for GFB making is available in the reviews works of Erickson et al. (2012) and Taylor et al. (2016). Briefly, it can be stated that the use of structured zein (under application of mechanical work to the dough), allows GFBs similar in volume to wheat breads to be obtained (although this can also be achieved with other alternatives, such as hydrocolloids, see Sect. 25.4), and with dough rheological properties resembling those of gluten doughs. However, zein based makings require changes in the formulation that can negatively affect the sensory quality of the bread, as well as changes in the processing, that are difficult to apply in industrial processes.

On the other hand, apart from the structural function of these proteins as substitutes for the gluten network, it is also important to consider the amount of protein present in the products without gluten. Thus, protein content in GFPs tends to be lower than in their gluten containing counterparts (Matos and Rosell 2011; Mazzeo et al. 2015; Miranda et al. 2014). Undoubtedly, wheat flour is characterized by a protein level much higher than that of the most commonly used alternatives in the preparation of GFBs (for instance, starch or rice flour), but protein in wheat flour also accomplishes several functions other than the formation of a network with unique extensibility, elasticity able to retain gas. Proteins participate in Maillard reactions and are responsible, at least in part, for the color and aroma of the breads (Smak 1972). Furthermore, the nutritional importance of proteins should be considered (Friedman 1996). Indeed, the lower protein content associated with formulations of GFBs is responsible, not only for its lower nutritional value, but also for its paler crust color. Gluten-free flours with higher protein content, such as those of some pseudocereals or legumes (see Sect. 25.2), and protein isolates or concentrates (~ more than 80 and 50% of protein content) could be incorporated to alleviate these problems. In this sense, there are several studies that focus on protein incorporation in different GFB formulations based on starch (Horstmann et al. 2017; Kittisuban et al. 2014; Krupa-Kozak et al. 2013; Witczak et al. 2017; Ziobro et al. 2013b, 2016), rice flour (Marco and Rosell 2008; Nozawa et al. 2016; Phongthai et al. 2016; Shin et al. 2010; Storck et al. 2013) or a starch-flour mixture (Aprodu et al. 2016; Crockett et al. 2011a; Gallagher et al. 2003; Smerdel et al. 2012).

The color of bread crust is a factor that drastically influences its acceptability (Castro et al. 2017), and in general, gluten-free breads have an unduly whitish color, in such a way that darker breads usually have better acceptability (Carini et al. 2015). Regarding the influence of proteins on crust color, dairy proteins seem to be the most effective, since a decrease in brightness (L^*) is observed in all studies compared to a control bread with no protein addition, even with small doses (3%) (Gallagher et al. 2003; Krupa-Kozak et al. 2013). On the contrary, with the addition of plant proteins, Phongthai et al. (2016) observed a clear decrease in brightness with low amounts (2%) of rice bran protein concentrate, while Marco and Rosell (2008) did not find such effect with larger amounts (13%) of soy protein isolate. In the same way,

Phongthai et al. (2016) did not observe significant differences in the brightness of GFBs with 4% egg albumin addition. However, it is difficult to draw general conclusions because the studies are not comparable due to the use of different formulations, protein amounts and baking conditions. In fact, the brightness of the control bread is very different in each of the studies mentioned. Nonetheless, it seems that there is a clear influence of the type of protein on bread brightness (affecting this value to a different extent), which may be affected by factors such as protein solubility or its amino acid content.

Protein addition also seems to have a clear effect on bread specific volume, but in some cases the results are contradictory. Notably, Krupa-Kozak et al. (2013) reported a significant increase in specific volume when 12% dairy proteins were incorporated, which was confirmed by Aprodu et al. (2016) with 15% addition, although lower specific volume was obtained in most cases with 24% incorporation. Meanwhile, Gallagher et al. (2003) found scarce differences in volume or even a reduction, with protein levels that did not exceed 9%. As for egg proteins, there is general agreement that they improve the specific volume of breads. Nonetheless, Crockett et al. (2011a) needed to reach levels of 15% to perceive this effect, while Nozawa et al. (2016) noticed it with percentages close to 1%, which can be partially explained by the high variability in the formulations, and then, the different effect of ingredients on dough network interactions. Regarding plant proteins, there is also a great variability and even inconsistency in the results. Phongthai et al. (2016) and Witczak et al. (2017) observed an increased volume with rice bran protein and potato protein at levels of 2%, but significant declines when this level was exceeded. Added to that, Crockett et al. (2011a) did not find differences in bread volume with 1–2% soy protein, but a decrease in those of 3% addition, while Aprodu et al. (2016) highlighted an improvement in volume with 15% addition of either soy or lupine proteins. In a recent study, Horstmann et al. (2017) also observed differences between different plant proteins (potato, pea, carob, lupine and soy), but were not able to relate dough and bread properties to any inherent characteristic (solubility, emulsifying, foaming, water hydration properties) of the proteins. Therefore, it seems that a range or combination of properties contributes to the characteristics ultimately observed in bread, as no single property is able to explain bread attributes.

One of the difficulties in comparing the different studies is the lack of homogeneity in the bread formulas utilized. Differences were prevalent in terms of types of starchy materials, types of hydrocolloid (if used), and baking conditions, which help explain the huge variability found in the specific volume of the control breads. It is noteworthy that one of the most important factors affecting bread volume is the hydration of the dough. However, in most studies the hydration level is kept constant, and the different water absorption capacity of the proteins is not taken into account. In general, the greater water content in the dough, the higher the specific volume of the breads, up to a maximum at which point the dough collapses (too weak structure) in the fermentation or baking steps (Mancebo et al. 2017). Therefore, if a protein promotes an increase in dough consistency and the hydration level is not corrected, a decreased volume may be obtained, whereas if a protein reduces the consistency, it can lead to the opposite effect. In agreement with this, soluble proteins, such as milk proteins or egg proteins, usually reduce dough consistency when they replace part of the flour, while insoluble proteins, mainly plant proteins, tend to have the opposite effect. In this way, the negative results found for some proteins, which may reduce bread volume, can be reversed into positives by increasing the hydration level in the dough, since they probably withstand greater water addition. On this basis, Ziobro et al. (2013b) modified the amount of water in the formulation based on farinographic analysis and observed different absorption values for each protein, with higher values in plant proteins than in egg proteins. These authors reported that, among all protein sources, breads with egg protein gave rise to the highest specific volume. In addition, significant differences were found between the vegetable proteins used (pea, lupine and soy), with lupine being the only one increasing the volume of the breads.

Regarding textural attributes, the inclusion of insoluble plant proteins such as potato, soy, lupine or pea generally reduce, or do not modify, the cohesiveness, resilience or springiness of breads (Marco and Rosell 2008; Ziobro et al. 2013b; Witczak et al. 2017). Meanwhile egg protein increases elasticity (Phongthai et al. 2016) as well as cohesiveness and springiness (Storck et al. 2013; Ziobro et al. 2013b), which would be positive effects since GFBs are characterized by having less cohesive crumbs than those of wheat breads. This effect may be related to the soluble character of this protein and its ability to coagulate when heated. Egg protein also produces a closer crumb grain which should be associated with its foaming and air stabilization abilities. As for milk proteins, while casein seems to increase the cohesiveness and springiness of the crumb (Storck et al. 2013), serum proteins reduce these values (Krupa-Kozak et al. 2013).

25.4 Hydrocolloids

As mentioned before (see Sect. 25.3), gluten proteins in wheat flour are able to create a network when they are hydrated and mechanical energy is applied, giving rise to unique rheological and textural properties of the doughs and breads made with them. In order to obtain gluten-free breads with a crumb texture that is more similar to gluten counterparts, hydrocolloids and gums have been used as gluten replacement to attain visco-elasticity of the dough (Anton and Artfield 2008; Sciarini et al. 2010). Among the research published, HPMC is found to be the most used hydrocolloid, followed by xanthan gum, although other hydrocolloids such as guar gum, pectins, or their mixtures are also frequent (Masure et al. 2016; Mir et al. 2016). In contrast to research, Matos and Rosell (2011) observed that xanthan gum was the main hydrocolloid in commercially available gluten-free products. In general terms, hydrocolloids, hydrophilic in nature, improve dough structure, mouthfeel, acceptability and shelf-life of GF products.

Research works comparing different hydrocolloids are scarce and their results are not conclusive. Indeed, Lazaridou et al. (2007) showed that xanthan, carboxymethyl cellulose (CMC), pectin, agarose and β-glucans increased water binding capacity and elastic modulus (G′) of gluten-free doughs (based on maize starch and rice flour), but different results were found when assessing the quality of the GFBs. These authors found that xanthan gum led to breads with excessive hardness, while pectin and CMC yielded breads with the best properties in terms of higher specific volume and more elastic and porous crumbs. Conversely, Liu et al. (2018) showed that HPMC was the most suitable hydrocolloid in steamed bread based on potato flour, producing breads with the highest specific volume, followed by xanthan gum, CMC and pectins. These authors also pointed out that loaves with HPMC presented a less cohesive crumb. Differences in the results of these two studies can be associated with differences in their formula (although this was not reported) as well as to the hydration level used to make those GFBs. While Lazaridou et al. (2007) used a water content in the range of 130–140%, a remarkably lower water content (70%) was used by Liu et al. (2018). On the other hand, Akin and Miller (2017) studied chemically leavened sorghum breads and reported that, although hydrocolloids tended to enhance bread volume, the improvement on such property was dependent on the formula used, not finding great differences among HPMC, xanthan gum or locus bean gum. As in the above mentioned study, dough hydration was not adjusted based on the rheology, i.e. the same hydration level was chosen for the same dose of hydrocolloids regardless of the differences in dough rheology, although in some cases hydration was increased with the dose of hydrocolloid. Bearing in mind differences in dough rheology, Sabanis and Tzia (2011) opted for increasing water content in the doughs made with xanthan gum, whereas for HPMC, guar and carrageenans the hydration level was kept constant. As a result, breads with higher volume and lower firmness were reported using HPMC followed by guar gum, while the opposite trend was obtained for xanthan gum (lower volume and softness). However, a minimum hydrocolloid dose is possibly necessary for its effects to be evident, since Sciarini et al. (2012a) indicated that no improvement was visible in precooked breads with 0.5% incorporation of either CMC or xanthan gum compared to a control bread with no hydrocolloid. In general, hydrocolloid doses range between 1 and 3% in most studies.

In view of the different studies, it is difficult to obtain general conclusions and it seems that the choice of hydrocolloid will depend on the formulation, processing conditions and type of bread developed, but in any case it is important to adjust the hydration level since, as previously mentioned, dough consistency will determine the expansion of the loaves, and the presence of hydrocolloids will significantly modify the rheology. One way to adjust the amount of water needed in the formula is to design experiments through Response Surface Methodology. In this way, both Mancebo et al. (2015c) and Ylimaki et al. (1988) have observed that mixtures of HPMC with CMC or psyllium can improve the volume of the breads when the water content is increased, with a more pronounced effect when the CMC or psyllium content is increased rather than that of HPMC. In this respect, it seems that the hydrocolloids in GF breads, although they exert a thickening function, are not used for this purpose. Their role is mainly related to their ability to provide cohesion to the dough, and their greater thickener effect must be corrected with the addition of more water, since it is generally assumed that a lower dough consistency gives rise to loaves with higher volume.

Among the different hydrocolloids, gelling agents (pectins, agars, carrageenans) are believed to be less effective, partly because the gels they form after cooling are brittle and do not facilitate dough expansion, and otherwise partly because previous solubilization is required for many of them. Thus, for complete solubilization, it is necessary to heat the mixture in advance, something that is not usually done. Regarding the rest of hydrocolloids, it appears that HPMC works in a different way than other thickeners, such as xanthan, guar or locust bean gum. As an example, xanthan addition in mixtures containing xanthan and HPMC (maintaining constant water content) resulted in a reduction of GFB volume, unlike HPMC, whose addition increased or maintained the volume, depending on the flour used (Hager and Arendt 2013). The differential effect of HPMC has been linked to its ability to form a thermo-reversible gel which strengthens when heated and reverts back to a weak entanglement after cooling (Grover 1982). Thus, this temperature dependence stabilizes the gelatinizing crumb structure during baking but reduces crumb hardness of the final bread (Crockett et al. 2011b). Despite the advantages of using HPMC to improve bread volume, this hydrocolloid has a higher cost and results in poor cohesive crumbs. Thus, it is usually mixed with other hydrocolloids in order to minimize these drawbacks.

It is note to highlight that the term HPMC encompasses a large number of cellulose derivatives with hydroxypropyl and methyl groups in different proportions, and with higher or lower molecular weight, and therefore different thickening power. HPMC 4KM with low percentage of methoxyl groups and high thickening power is one of the most commonly used HPMC in GFBs. However, Crockett et al. (2011b) obtained loaves with a higher specific volume using HPMC E15, with a higher percentage of methoxyl groups. These authors attributed this outcome in part to the higher foaming ability of this HPMC and to the reinforced micelle integrity during gas expansion, preventing coalescing of the air cells and improving the specific loaf volume of the GFB. Another plausible explanation for the better performance of this HPMC E15 could be its lower molecular weight compared to 4KM and lower thickening ability (lower viscosity in cold water). In fact, doughs made with E15 had much lower G' values than those made with 4KM, something that could have been reversed by modifying the amount of water. In addition, Perez-Quirce et al. (2014) observed a higher volume for breads containing HPMC with a higher percentage of methoxyl groups when they compared two types with the same thickening power, which they attributed to the creation of firmer gels after heating. However, they also observed the creation of large cavities when this high methoxyl HPMC was used, maybe due to the rigidity of the gels. Morreale et al. (2017) proposed the use of HPMC types with greater thickening power, such as K15, would yield better results provided that the amount of water incorporated is optimized (i.e., higher water contents are required compared to other HPMC sources).

From a nutritional perspective, hydrocolloids are considered as soluble dietary fiber (Viebke et al. 2014). Therefore, with their use in GFPs, they could increase the dietary fiber content of these products. Although the levels of hydrocolloid added are usually not very high (generally less than 4%), for reasons associated with their cost and mainly with their significant effect on dough rheology, the study of their health benefits should not be dismissed. Indeed, some of the hydrocolloids broadly used for GFPs have interesting health properties. Most studies on the health properties of hydrocolloids have focused on aspects such as its role in cholesterol and glycemic index reduction, prebiotic effect, control of food intake due to its satiating effect, prevention of constipation and even the possible prevention of cancer, especially colon cancer (Li and Nie 2016). Likewise, there are some health claims relating to food hydrocolloids which have been approved by EFSA included under article 13.1 (Viebke et al. 2014). Among those claims, it is important to mention those related to maintenance of normal blood cholesterol concentrations (β-glucans, guar, konjac gum, pectins, HPMC), and the reduction of post-prandial glycaemic response (β-glucans, HPMC, pectins).

The interest in some hydrocolloids, such as HPMC or pectins, for the development of gluten-free breads has been previously discussed. Unlike those hydrocolloids, guar gum has not been very successful for gluten replacement. However, it can potentially be used in combination with other gums. Notably, a synergistic effect of this gum with xanthan gum has been shown (as occurs with locust bean and xanthan gum) in such a way that the addition of xanthan gum, even at a very low concentration, to a guar gum solution promoted the transition of the system from a macromolecular solution to a structured system displaying gel-like properties (Schorsch et al. 1997). In this sense, some works have suggested the combined use of these gums in the preparation of gluten-free breads, with better results than for each of the individual gums (Demirkesen et al. 2010a, b).

In the case of β-glucans, there is some research examining their effect on gluten-free formulations, either as the only hydrocolloid in the formulation (Lazaridou et al. 2007; Perez-Quirce et al. 2014), or in combination with other more commonly used hydrocolloids, like xanthan gum (Hager et al. 2011b) or HPMC (Kittisuban et al. 2014; Perez-Quirce et al. 2017; Ronda et al. 2015). Results from these articles agree in the fact that by itself β-glucan is not a good substitute for gluten, but it can be used in combination with other hydrocolloids to obtain good results. Nonetheless, it is necessary to increase the hydration level of the dough in order to obtain good volume and desired texture, and always bearing in mind that the effect of the addition of β-glucans will depend on their dose and molecular size. For instance, Perez-Quirce et al. (2017) reported that an increase of β-glucan molecular weight and level lowers the bread starch digestibility.

Another product of great interest for the preparation of GFPs is the husk from psyllium. Psyllium (*Plantago ovata*) is an original plant from India, also known as Ispaghula, whose husk is very rich in a water-soluble mucilage, mainly composed of arabinoxylans (Sandhu and Hudson 1981). Proven beneficial effects of psyllium include reduced cholesterol levels (Xing et al. 2017), risk of coronary heart disease (Bernstein et al. 2013) and constipation problems (Gelinas 2013). For the production of gluten-free breads, psyllium has great potential because it presents similar properties to those of xanthan gum (Haque et al. 1993). In addition, since it is a product simply obtained by grinding the husk of a seed it is not considered an additive, and, hence it can be used to obtain products with a cleaner label. Zandonadi et al. (2009) produced gluten-free breads with psyllium as a single substitute for gluten and obtained products with good organoleptic characteristics, but other technological characteristics such as volume or texture were not measured. Moreover, psyllium has also been included in GF formulations in combination with other hydrocolloids, such as carrageenans (Aprodu and Banu 2015), HPMC (Cappa et al. 2013; Haque and Morris 1994; Mancebo et al. 2015c), or inulin (Tubili et al. 2016). In these studies, a positive effect of the inclusion of psyllium on the specific volume of the loaves is generally reported, provided that hydration level is corrected, since as previously mentioned with β-glucans, psyllium also possesses a high thickening capability.

In general, it can be concluded that, so far, hydrocolloids represent the most suitable gluten substitutes for GFB making, but the correct choice will depend on the formula used and the manufacturing conditions (such as time and temperature of fermentation). Among these polymers, HPMC is the one that yields higher bread volumes, but at the same time, less cohesive crumbs are obtained. Thus, studying its use in a mixture with other hydrocolloids is of immense interest. In all cases, increasing the water content in the dough formulation may be required to compensate for the great thickening effects of adding hydrocolloids, since an increase in dough consistency will hinder dough expansion during fermentation and baking. Furthermore, not only the technological properties of gluten replacement with hydrocolloids should be analyzed but also the nutritional implications of these replacements.

25.5 Fibers

Health benefits of fiber are widely known and include enhanced control of blood glucose and cholesterol levels, prevention and protection against cardiovascular diseases, regulation of intestinal function, promotion of gut health and protection against colon cancer (Ktenioudaki and Gallagher 2012). Different studies have confirmed that patients who follow a strict gluten-free diet present a lower fiber intake than the rest of the population, with intake levels below those recommended by different international organizations (Hager et al. 2011a; Martin et al. 2013; Shepherd and Gibson 2013; Wild et al. 2010). It should be noted that the fiber content of gluten-free breads, although very varied, is not usually lower than those of wheat counterparts, with contents ranging between 1.3 and 7.2% for GFBs and regular values between 3 and 6% (Matos and Rosell 2011; Mazzeo et al. 2015). Since the fiber content of starches and flours commonly used for GFPs is low, the above mentioned fiber content found in GFBs would be partially explained by the fact that the hydrocolloids added for gluten replacement are considered as fiber from a nutritional point of view. One of the main problems faced by the celiac collective to increase their fiber intake is the shortage in supply of whole grain products. In this way, one of the possibilities to enhance the fiber content in the diet can rely on the elaboration of breads with whole grain gluten-free flours (Alvarez-Jubete et al. 2009, 2010a; Cornejo and Rosell 2015; Cornejo et al. 2015; Renzetti et al. 2008).

However, when it comes to the utilization of whole grain flours from the most abundant gluten-free cereals, there are some drawbacks that complicate or potentially impede the utilization of their brans. In the case of maize, the bran is excessively hard and, hence, more difficult to mill which may result in coarser particle sizes. Although there are no studies dealing with maize bran incorporation in GFPs, when added in wheat-based products, product acceptability is substantially reduced in terms of flavor, mouthfeel, texture and/or color (Rose et al. 2010). In addition, both maize germ and the by-products of rice milling, consisting of bran and germ, contain a significant amount of lipids that easily becomes rancid after milling, something undesirable that negatively affects consumer acceptance. To minimize lipid rancidity in rice, the application of thermal treatments has been proposed as reviewed by Sharif et al. (2014). A different approach to enrich fiber content of GFPs is the incorporation of defatted rice bran (Phimolsiripol et al. 2012), other gluten-free brans such as quinoa bran (Foste et al. 2014) or the addition of by-products rich in fiber (sources of cellulose, hemicellulose, pectin or lignin) from the food industry such as orange pomace (O'Shea et al. 2013, 2015), apple pomace (Parra et al. 2015) or tiger-nut by-product (Aguilar et al. 2015). Added to the advantage of high fiber content (up to 93.2 g of fiber per 100 g of solids in apple pomace) (Nawirska and Kwasniewska 2005), fruit and vegetable by-products are characterized by their content in other components of nutritional interest, such as polyphenols with antioxidant capacity. Despite the nutritional advantages, the utilization of these by-products in baked goods usually leads to the loss of acceptability and product quality. This is mostly caused by the presence of extraneous flavor and different color and the interference of these ingredients with dough structure that, in turn, affects its expansion, respectively. Therefore, in order to minimize these negative effects, by-products must be added in small quantities and physical properties (e.g., particle size, water holding capacity, gel-formation ability) must be taken into consideration (Martinez and Gomez 2017b). Thus, as previously explained, the advantage of such enriched products in the market is low. The use of fruit and vegetable by-products as ingredients in gluten-free baked goods are reviewed in more detail elsewhere (Martinez and Gomez 2017b) and will not be further discussed in this book chapter. Added to that, results from several studies have been published on the incorporation of fiber from different sources in GFBs. Most of these have focused on the incorporation of fructo-oligosaccharides (Capriles and Areas 2013; Korus et al. 2006; Ziobro et al. 2013a), resistant starch isolates (50–60% of resistant starch) (Korus et al. 2009; Sciarini et al. 2017) and cereal insoluble fibers (Sabanis et al. 2009a, b).

However, it is noteworthy that fiber should be added with careful consideration of its characteristics, especially in terms of water binding capacity, solubility, thickening power/viscosity or gel ability properties since differences in these properties may entail significant changes during manufacturing and on final product quality. In this sense, it is broadly assumed that soluble fibers with low thickening ability, such as inulin, help increase bread volume and reduce crumb firmness (Capriles and Areas 2013; Korus et al. 2006), which may be associated with a reduction in dough consistency and thereby allowing for easy dough expansion during fermentation and baking. In fact, Ziobro et al. (2013a) reported that doughs containing inulin (soluble fiber) needed a lower water content to achieve a similar consistency (as measured with a back extrusion rig) than the control. In contrast, when inulin incorporated in the formula had a high polymerization degree (i.e., higher molecular weight) its thickening power increased, and doughs with higher consistency could be obtained since higher water content was needed. Nonetheless, taking only dough consistency into consideration does not explain differences in volume because, although Ziobro et al. (2013a) optimized water content in the formula to unify dough rheology (same consistency), inulins of lower degree of polymerization yielded breads of higher volumes and remarkably lower firmness than those of higher polymerization degree and the control. Unlike inulins with low degree of polymerization, an increase in hardness and bread staling (rate of crumb hardening), was reported when inulins of higher degree of polymerization were added, but in this case no effect was visible on bread volume. Furthermore, addition of inulins resulted in a more open and less uniform crumb structure compared to the control. On the other hand, resistant starch incorporation was found to increase G′ and G″ (viscous and elastic moduli) values of the doughs leading to the subsequent decrease of specific volume, but it was surprising to find that addition of resistant starch seemed to decrease bread hardness, something that was associated with its effect on starch gelatinization (Korus et al. 2009).

In a broader study, where fibers from several sources were studied, the positive effect of soluble fibers was confirmed. Martinez et al. (2014) reported that soluble fibers with low viscosity (nutriose and polydextrose), which are similar to inulin in their low polymerization degree, reduced complex modulus (G^*) of the dough, and, in turn, greatly increased bread volume while reducing its hardness. These soluble fibers also yielded breads with a finer and more uniform cell grain (Fig. 25.3), and a darker crust (Martinez et al. 2014). This effect on crust color, already observed by Capriles and Areas (2013) and Korus et al. (2006) with inulin, was attributed to the partial hydrolysis of these compounds during baking, giving rise to mono- and disaccharides that could participate in Maillard reactions. In fact, Korus et al. (2006) also reported a higher monosaccharide content in the breads made with inulin. Sciarini et al. (2017) reported that incorporation of insoluble fibers resulted in bread with a low specific volume due to the higher firmness of the dough. However, staling rate was generally reduced after fiber addition. It is noteworthy that results shown by Martinez et al. (2014) also suggested that both shape and particle size of cellulose fibers (insoluble fiber) were remarkably important for final GFB quality (Fig. 25.3). Thus, elongated fibers with a finer particle size led to an increase in bread volume, while coarser and more irregular in shape fibers reduced bread volume. This is in agreement with the results reported by Sabanis et al. (2009b) and it can be related to the physical interaction between fiber particles and the rest of the dough components (mainly starch/flour particles). This physical interaction would be associated with differences in fiber microstructure in terms of particle size and deformability and compaction, which ultimately affect dough leavening during proofing.

To conclude, the positive results from the nutritional standpoint of fiber addition in GFB must be understood without forgetting the sensory quality of the final products. Attention should be paid not only to the type of fiber and its physical properties

Fig. 25.3 Crumb detail: images of finished breads. (**a**) Control, (**b**) pea fibre, (**c**) potato fibre, (**d**) fine bamboo fibre, (**e**) coarse bamboo fibre (**f**) Nutriose, (**g**) polydextrose (Martinez et al. 2014)

but also to its optimum level in the formula. Secondly, it should also be considered that nutritional benefits are dependent on the fiber source, and, more specifically, on its solubility and thickening ability (Padayachee et al. 2017).

25.6 Concluding Remarks

The interest in gluten-free cereal products has increased significantly over the last decades. Although considerable efforts have been made in recent years to produce high quality gluten-free baked goods, especially gluten-free bread, technological and sensory quality of gluten-free products in which a viscoelastic gluten-like network is developed is still far from wheat analogues, and then, consumer demands.

The main reason for this is that, at the moment, there are no raw materials (proteins, hydrocolloids, and enzymes) capable of completely mimicking gluten functionality, although a combination of different ingredients or additives could help mimic this functionality. Yet, further deeper study is still needed.

At the moment, hydrocolloids, alone or in combination with other fibers and hydrocolloids, are the most used gluten substitutes both in laboratory and industrial scales. In order to mimic gluten functionality, attempts have also been made to use protein from other cereal sources, sometimes involving physical, chemical or enzymatic modification, but for the time being none of these proteins have been commercially applied. Apart from hydrocolloids, starches, which are another major ingredient in gluten-free formulations, also play an important role in determining the rheological properties of gluten-free dough networks. Starches from different botanical sources contribute differently to the rheological properties of doughs due to variations in structural and physicochemical properties of starch granules. Among all starch sources, maize starch and rice flour are the most prominent starchy materials in gluten-free breads. The adequate mixture of flour and starch in the formulas can lead to the appropriate physical interaction among dough particles (i.e. starch-hydrocolloid interactions) that help increase the volume of breads as well as improve their texture and organoleptic characteristics.

In addition, features more related to the processing variables of the baked products should also be considered as key elements in order to minimize the loss of quality in gluten-free formulations. In this way, not only adaptation of fermentation and baking conditions is important, but also hydration level used in gluten-free formulations. Since it is known that water content markedly affects dough rheology, and in turn, bread volume, the appropriate dosage of water can help obtain more desired baked goods.

Other than polymer functionality in these products, a nutritional standpoint should also be considered, focusing on the utilization of starch ingredients that can limit the glycemic index of the product and enrich its contents of dietary fiber and protein, since gluten-free diets and breads, respectively, tend to be poor sources of these components. Given all this, food manufacturers should make strategic efforts to select gluten-free ingredients with optimal composition and physical properties that attenuate the negative effects of lack of gluten.

References

Aguilar N, Albanell E, Minarro B, Guamis B, Capellas M (2015) Effect of tiger nut-derived products in gluten-free batter and bread. Food Sci Technol Int 21:323–331

Akin PA, Miller RA (2017) Starch-hydrocolloid interaction in chemically leavened gluten-free sorghum bread. Cereal Chem 94:897–902

Alvarez-Jubete L, Arendt EK, Gallagher E (2009) Nutritive value and chemical composition of pseudocereals as gluten-free ingredients. Int J Food Sci Nutr 60:240–257

Alvarez-Jubete L, Arendt EK, Gallagher E (2010a) Nutritive value of pseudocereals and their increasing use as functional gluten-free ingredients. Trends Food Sci Technol 21:106–113

Alvarez-Jubete L, Auty M, Arendt EK, Gallagher E (2010b) Baking properties and microstructure of pseudocereal flours in gluten-free bread formulations. Eur Food Res Technol 230:437–445
Andersson H, Ohgren C, Johansson D, Kniola M, Stading M (2011) Extensional flow, viscoelasticity and baking performance of gluten-free zein-starch doughs supplemented with hydrocolloids. Food Hydrocoll 25:1587–1595
Anton AA, Artfield SD (2008) Hydrocolloids in gluten-free breads: a review. Int J Food Sci Nutr 59:11–23
Aprodu I, Badiu EA, Banu I (2016) Influence of protein and water addition on gluten-free dough properties and bread quality. Int J Food Eng 12:355–363
Aprodu I, Banu I (2015) Influence of dietary fiber, water, and glucose oxidase on rheological and baking properties of maize based gluten-free bread. Food Sci Biotechnol 24:1301–1307
Bernstein AM, Titgemeier B, Kirkpatrick K, Golubic M, Roizen MF (2013) Major cereal grain fibers and psyllium in relation to cardiovascular health. Nutrients 5:1471–1487
Berta M, Gmoser R, Krona A, Stading M (2015) Effect of viscoelasticity on foam development in zein-starch dough. LWT Food Sci Technol 63:1229–1235
Berti C, Riso P, Monti LD, Porrini M (2004) In vitro starch digestibility and in vivo glucose response of gluten-free foods and their gluten counterparts. Eur J Nutr 43:198–204
Brites C, Trigo MJ, Santos C, Collar C, Rosell CM (2010) Maize-based gluten-free bread: influence of processing parameters on sensory and instrumental quality. Food Bioprocess Technol 3:707–715
Cappa C, Lucisano M, Mariotti M (2013) Influence of Psyllium, sugar beet fibre and water on gluten-free dough properties and bread quality. Carbohydr Polym 98:1657–1666
Capriles VD, Areas JAG (2013) Effects of prebiotic inulin-type fructans on structure, quality, sensory acceptance and glycemic response of gluten-free breads. Food Funct 4:104–110
Capriles VD, dos Santos FG, Areas JAG (2016) Gluten-free breadmaking: improving nutritional and bioactive compounds. J Cereal Sci 67:83–91
Carini E, Scazzina F, Curti E, Fattori F, Mazzeo T, Vittadini E (2015) Physicochemical, sensory properties and starch in vitro digestion of gluten-free breads. Int J Food Sci Nutr 66:867–872
Caruso R, Pallone F, Stasi E, Romeo S, Monteleone G (2013) Appropriate nutrient supplementation in celiac disease. Ann Med 45:522–531
Castro W, Oblitas J, Chuquizuta T, Avila-George H (2017) Application of image analysis to optimization of the bread-making process based on the acceptability of the crust color. J Cereal Sci 74:194–199
Catassi C, Gatti S, Fasano A (2014) The new epidemiology of celiac disease. J Pediatr Gastroenterol Nutr 59:S7–S9
Catassi C et al (2010) Non-celiac gluten sensitivity: the new frontier of gluten related disorders. Nutrients 5:3839–3853
Cornejo F, Caceres PJ, Martinez-Villaluenga C, Rosell CM, Frias J (2015) Effects of germination on the nutritive value and bioactive compounds of brown rice breads. Food Chem 173:298–304
Cornejo F, Rosell CM (2015) Influence of germination time of brown rice in relation to flour and gluten free bread quality. J Food Sci Technol Mysore 52:6591–6598
Crockett R, Ie P, Vodovotz Y (2011a) Effects of soy protein isolate and egg white solids on the physicochemical properties of gluten-free bread. Food Chem 129:84–91
Crockett R, Ie P, Vodovotz Y (2011b) How do xanthan and hydroxypropyl methylcellulose individually affect the physicochemical properties in a model gluten-free dough? J Food Sci 76:E274–E282
de la Hera E, Martinez M, Gomez M (2013a) Influence of flour particle size on quality of gluten-free rice bread. LWT Food Sci Technol 54:199–206
de la Hera E, Martinez M, Oliete B, Gomez M (2013b) Influence of flour particle size on quality of gluten-free rice cakes. Food Bioprocess Technol 6:2280–2288
de la Hera E, Talegon M, Caballero P, Gomez M (2013c) Influence of maize flour particle size on gluten-free breadmaking. J Sci Food Agr 93:924–932
de la Hera E, Rosell CM, Gomez M (2014) Effect of water content and flour particle size on gluten-free bread quality and digestibility. Food Chem 151:526–531

Delcour JA, Joye IJ, Pareyt B, Wilderjans E, Brijs K, Lagrain B (2012) Wheat gluten functionality as a quality determinant in cereal-based food products. Annu Rev Food Sci Technol 3:469–492

Demirkesen I, Mert B, Sumnu G, Sahin S (2010a) Utilization of chestnut flour in gluten-free bread formulations. J Food Eng 101:329–336

Demirkesen I, Mert B, Sumnu G, Sahin S (2010b) Rheological properties of gluten-free bread formulations. J Food Eng 96:295–303

do Nascimento AB, Fiates GMR, dos Anjos A, Teixeira E (2013) Analysis of ingredient lists of commercially available gluten-free and gluten-containing food products using the text mining technique. Int J Food Sci Nutr 64:217–222

Erickson DP, Campanella OH, Hamaker BR (2012) Functionalizing maize zein in viscoelastic dough systems through fibrous, beta-sheet-rich protein networks: an alternative, physicochemical approach to gluten-free breadmaking. Trends Food Sci Technol 24:74–81

Foschia M, Horstmann SW, Arendt EK, Zannini E (2016) Legumes as functional ingredients in gluten-free bakery and pasta products. Annu Rev Food Sci Technol 8:4.1–4.22

Foste M, Nordlohne SD, Elgeti D, Linden MH, Heinz V, Jekle M, Becker T (2014) Impact of quinoa bran on gluten-free dough and bread characteristics. Eur Food Res Technol 239:767–775

Fric P, Gabrovska D, Nevoral J (2011) Celiac disease, gluten-free diet, and oats. Nutr Rev 69:107–115

Friedman M (1996) Nutritional value of proteins from different food sources. A review. J Agr Food Chem 44:6–29

Gallagher E, Gormley TR, Arendt EK (2003) Crust and crumb characteristics of gluten-free breads. J Food Eng 56:153–161

Garzon R, Rosell CM, Malvar RA, Revilla P (2017) Diversity among maize populations from Spain and the United States for dough rheology and gluten-free breadmaking performance. Int J Food Sci Technol 52:1000–1008

Gelinas P (2013) Preventing constipation: a review of the laxative potential of food ingredients. Int J Food Sci Technol 48:445–467

Gimenez-Bastida JA, Piskula M, Zielinski H (2015) Recent advances in development of gluten-free buckwheat products. Trends Food Sci Technol 44:58–65

Goesaert H, Leman P, Delcour JA (2008) Model approach to starch functionality in bread making. J Agr Food Chem 56:6423–6431

Grover JA (1982) Methylcellulose (MC) and hydroxypropyl methylcellulose (HPMC). In: Glicksmen M (ed) Food hydrocolloids. CRC Press, Boca Raton, pp 121–154

Hager AS, Arendt EK (2013) Influence of hydroxypropylmethylcellulose (HPMC), xanthan gum and their combination on loaf specific volume, crumb hardness and crumb grain characteristics of gluten-free breads based on rice, maize, teff and buckwheat. Food Hydrocoll 32:195–203

Hager AS, Axel C, Arendt EK (2011a) Status of carbohydrates and dietary fiber in gluten-free diets. Cereal Foods World 56:109–114

Hager AS, Ryan LAM, Schwab C, Ganzle MG, O'Doherty JV, Arendt EK (2011b) Influence of the soluble fibres inulin and oat beta-glucan on quality of dough and bread. Eur Food Res Technol 232:405–413

Hager A-S, Wolter A, Czerny M, Bez J, Zannini E, Arendt EK, Czerny M (2012) Investigation of product quality, sensory profile and ultrastructure of breads made from a range of commercial gluten-free flours compared to their wheat counterparts. Eur Food Res Technol 235:333–344

Han HM, Cho JH, Kang HW, Koh BK (2012) Rice varieties in relation to rice bread quality. J Sci Food Agr 92:1462–1467

Haque A, Morris ER (1994) Combined use of ispaghula and HPMC to replace or augment gluten in breadmaking. Food Res Int 27:379–393

Haque A, Richardson RK, Morris ER, Dea ICM (1993) Xanthan-like weak gel rheology from dispersions of ispaghula seed husk. Carbohydr Polym 22:223–232

Horstmann SW, Foschia M, Arendt EK (2017) Correlation analysis of protein quality characteristics with gluten-free bread properties. Food Funct 8:2465–2474

Hüttner EK, Arendt EK (2010) Recent advances in gluten-free baking and the current status of oats. Trends Food Sci Technol 21:303–312

Jenkins DJ, Thorne MJ, Wolever TM, Jenkins AL, Rao AV, Thompson LU (1987) The effect of starch-protein interaction in wheat on the glycemic response and rate of in vitro digestion. Am J Clin Nutr 45:946–951

Kadan RS, Bryant RJ, Miller JA (2008) Effects of milling on functional properties of rice flour. J Food Sci 3:151–154

Kang TY, Sohn KH, Yoon MR, Lee JS, Ko S (2015) Effect of the shape of rice starch granules on flour characteristics and gluten-free bread quality. Int J Food Sci Technol 50:1743–1749

Khatkar BS, Schofield JD (1997) Molecular and physico-chemical basis of breadmaking properties of wheat gluten proteins: a critical appraisal. J Food Sci Technol Mysore 34:85–102

Kim HS, Huber KC (2010) Physicochemical properties and amylopectin fine structures of A- and B-type granules of waxy and normal soft wheat starch. J Cereal Sci 51:256–264

Kittisuban P, Ritthiruangdej P, Suphantharika M (2014) Optimization of hydroxypropylmethylcellulose, yeast β-glucan, and whey protein levels based on physical properties of gluten-free rice bread using response surface methodology. LWT Food Sci Technol 57:738–748

Korus J, Grzelak K, Achremowicz K, Sabat R (2006) Influence of prebiotic additions on the quality of gluten-free bread and on the content of inulin and fructooligosaccharides. Food Sci Technol Int 12:489–495

Korus J, Witczak M, Ziobro R, Juszczak L (2009) The impact of resistant starch on characteristics of gluten-free dough and bread. Food Hydrocoll 23:988–995

Krupa-Kozak U, Baczek N, Rosell CM (2013) Application of dairy proteins as technological and nutritional improvers of calcium-supplemented gluten-free bread. Nutrients 5:4503–4520

Ktenioudaki A, Gallagher E (2012) Recent advances in the development of high-fibre baked products. Trends Food Sci Technol 28:4–14

Kulp K, Ponte JG Jr (1981) Staling of white pan bread: fundamental causes. Crit Rev Food Sci Nutr 15:1–48

Lazaridou A, Duta D, Papageorgiou M, Belc N, Biliaderis CG (2007) Effects of hydrocolloids on dough rheology and bread quality parameters in gluten-free formulations. J Food Eng 79:1033–1047

Li JM, Nie SP (2016) The functional and nutritional aspects of hydrocolloids in foods. Food Hydrocoll 53:46–61

Lis DM, Stellingwerff T, Shing CM, Ahuja KD, Fell JW (2015) Exploring the popularity, experiences, and beliefs surrounding gluten-free diets in nonceliac athletes. Int J Sport Nutr Exerc Metab 25:37–45

Liu XL, Mu TH, Sun HN, Zhang M, Chen JW, Fauconnier ML (2018) Influence of different hydrocolloids on dough thermo-mechanical properties and in vitro starch digestibility of gluten-free steamed bread based on potato flour. Food Chem 239:1064–1074

Lopez ACB, Pereira AJG, Junqueira RG (2004) Flour mixture of rice flour, corn and cassava starch in the production of gluten-free white bread. Braz Arch Biol Technol 47:63–70

Mancebo CM, Martínez MM, Merino C, de la Hera E, Gómez M (2017) Effect of oil and shortening in rice bread quality: relationship between dough rheology and quality characteristics. J Texture Stud 48:597–606

Mancebo CM, Merino C, Martínez MM, Gómez M (2015a) Mixture design of rice flour, maize starch and wheat starch for optimization of gluten free bread quality. J Food Sci Technol 52:6323–6333

Mancebo CM, Picon J, Gómez M (2015b) Effect of flour properties on the quality characteristics of gluten free sugar-snap cookies. LWT Food Sci Technol 64:264–269

Mancebo CM, San Miguel MA, Martínez MM, Gómez M (2015c) Optimisation of rheological properties of gluten-free doughs with HPMC, psyllium and different levels of water. J Cereal Sci 61:8–15

Maningat CC, Seib PA, Bassi SD, Woo KS, Lasater GD (2009) Wheat starch: production, properties, modification and uses. In: BeMiller J, Whistler R (eds) Starch. Chemistry and technology. Academic Press, New York, pp 441–510

Marco C, Rosell CM (2008) Breadmaking performance of protein enriched, gluten-free breads. Eur Food Res Technol 227:1205–1213

Martin J, Geisel T, Maresch C, Krieger K, Stein J (2013) Inadequate nutrient intake in patients with celiac disease: Results from a German dietary survey. Digestion 87:240–246
Martinez MM, Diaz A, Gomez M (2014) Effect of different microstructural features of soluble and insoluble fibres on gluten-free dough rheology and bread-making. J Food Eng 142:49–56
Martinez MM, Gomez M (2017a) Rheological and microstructural evolution of the most common gluten-free flours and starches during bread fermentation and baking. J Food Eng 197:78–86
Martinez MM, Gomez M (2017b) Fruit and vegetable by-products as novel ingredients to improve the nutritional quality of baking foods. Crit Rev Food Sci Nutr:1–17. https://doi.org/10.1080/10408398.2017.1305946
Martinez MM, Román L, Gómez M (2018) Implications of hydration depletion in the in vitro starch digestibility of white bread crumb and crust. Food Chem 239:295–303
Masure HG, Fierens E, Delcour JA (2016) Current and forward looking experimental approaches in gluten-free bread making research. J Cereal Sci 67:92–111
Matos ME, Rosell CM (2011) Chemical composition and starch digestibility of different gluten-free breads. Plant Food Hum Nutr 66:224–230
Matos ME, Rosell CM (2013) Quality indicators of rice-based gluten-free breadlike products: relationships between dough rheology and quality characteristics. Food Bioprocess Technol 6:2331–2341
Matos ME, Sanz T, Rosell CM (2014) Establishing the function of proteins on the rheological and quality properties of rice based gluten free muffins. Food Hydrocoll 35:150–158
Mazzeo T, Cauzzi S, Brighenti F, Pellegrini N (2015) The development of a composition database of gluten-free products. Public Health Nutr 18:1353–1357
Melini F, Melini V, Luziatelli F, Ruzzi M (2017) Current and forward-looking approaches to technological and nutritional improvements of gluten-free bread with legume flours: a critical review. Compr Rev Food Sci Food Saf 16:1101–1122
Mir SA, Shah MA, Naik HR, Zargar IA (2016) Influence of hydrocolloids on dough handling and technological properties of gluten-free breads. Trends Food Sci Technol 51:49–57
Miranda J, Lasa A, Bustamante MA, Churruca I, Simon E (2014) Nutritional differences between a gluten-free diet and a diet containing equivalent products with gluten. Plant Foods Hum Nutr 69:182–187
Moore MM, Heinbockel M, Dockery P, Ulmer HM, Arendt EK (2006) Network formation in gluten-free bread with application of transglutaminase. Cereal Chem 83:28–36
Morreale F, Garzón R, Rosell CM (2017) Understanding the role of hydrocolloids viscosity and hydration in developing gluten-free bread. A study with hydroxypropylmethylcellulose. Food Hydrocoll 77:629–635. https://doi.org/10.1016/j.foodhyd.2017.11.004.
Nawirska A, Kwasniewska M (2005) Dietary fibre fractions from fruit and vegetable processing waste. Food Chem 91:221–225
Nishita KD, Bean MM (1979) Physicochemical properties of rice in relation to rice bread. Cereal Chem 56:185–189
Nozawa M, Ito S, Arai E (2016) Effect of ovalbumin on the quality of gluten-free rice flour bread made with soymilk. LWT Food Sci Technol 66:598–605
Onyango C, Mutungi C, Unbehend G, Lindhauer MG (2011a) Modification of gluten-free sorghum batter and bread using maize, potato, cassava or rice starch. LWT Food Sci Technol 44:681–686
Onyango C, Mutungi C, Unbehend G, Lindhauer MG (2011b) Rheological and textural properties of sorghum-based formulations modified with variable amounts of native or pregelatinised cassava starch. LWT Food Sci Technol 44:687–693
Oom A, Pettersson A, Taylor JRN, Stading M (2008) Rheological properties of kafirin and zein prolamins. J Cereal Sci 47:109–116
O'Shea N, Doran L, Auty M, Arendt E, Gallagher E (2013) The rheology, microstructure and sensory characteristics of a gluten-free bread formulation enhanced with orange pomace. Food Funct 4:1856–1863
O'Shea N, Rossle C, Arendt E, Gallagher E (2015) Modelling the effects of orange pomace using response surface design for gluten-free bread baking. Food Chem 166:223–230

Padayachee A, Day L, Howella K, Gidley MJ (2017) Complexity and health functionality of plant cell wall fibers from fruits and vegetables. Crit Rev Food Sci Nutr 57:59–81
Pareyt B, Delcour JA (2008) The role of wheat flour constituents, sugar, and fat in low moisture cereal based products: a review on sugar-snap cookies. Crit Rev Food Sci Nutr 48:824–839
Parra AFR, Ribotta PD, Ferrero C (2015) Apple pomace in gluten-free formulations: effect on rheology and product quality. Int J Food Sci Technol 50:682–690
Pawlowska P, Diowksz A, Kordialik-Bogacka E (2012) State-of-the-art incorporation of oats into a gluten-free diet. Food Rev Int 28:330–342
Peräaho M, Kaukinen K, Paasikivi K, Sievanen H, Lohiniemi S, Maki M, Collin P (2003) Wheat-starch-based gluten-free products in the treatment of newly detected coeliac disease: prospective and randomized study. Aliment Pharmacol Ther 17:587–594
Perez-Quirce S, Collar C, Ronda F (2014) Significance of healthy viscous dietary fibres on the performance of gluten-free rice-based formulated breads. Int J Food Sci Technol 49:1375–1382
Perez-Quirce S, Lazaridou A, Biliaderis CG, Ronda F (2017) Effect of beta-glucan molecular weight on rice flour dough rheology, quality parameters of breads and in vitro starch digestibility. LWT Food Sci Technol 82:446–453
Phimolsiripol Y, Mukprasirt A, Schoenlechner R (2012) Quality improvement of rice-based gluten-free bread using different dietary fibre fractions of rice bran. J Cereal Sci 56:389–395
Phongthai S, D'Amico S, Schoenlechner R, Rawdkuen S (2016) Comparative study of rice bran protein concentrate and egg albumin on gluten-free bread properties. J Cereal Sci 72:38–45
Pongjaruvat W, Methacanon P, Seetapan N, Fuongfuchat A, Gamonpilas C (2014) Influence of pregelatinised tapioca starch and transglutaminase on dough rheology and quality of gluten-free jasmine rice breads. Food Hydrocoll 36:143–150
Reilly NR (2016) The gluten-free diet: recognizing fact, fiction, and fad. J Pediatr 175:206–2010
Renzetti S, Dal Bello F, Arendt EK (2008) Microstructure, fundamental rheology and baking characteristics of batters and breads from different gluten-free flours treated with a microbial transglutaminase. J Cereal Sci 48:33–45
Román L, de la Cal E, Gómez M, Martínez MM (2018) Specific ratio of A- to B-type wheat starch granules improves the quality of gluten-free breads: optimizing dough viscosity and Pickering stabilization. Food Hydrocoll (in press)
Ronda F, Perez-Quirce S, Lazaridou A, Biliaderis CG (2015) Effect of barley and oat beta-glucan concentrates on gluten-free rice-based doughs and bread characteristics. Food Hydrocoll 48:197–207
Rose DJ, Inglett GE, Liu SX (2010) Utilisation of corn (*Zea mays*) bran and corn fiber in the production of food components. J Sci Food Agr 90:915–924
Sabanis D, Lebesi D, Tzia C (2009a) Development of fibre-enriched gluten-free bread: a response surface methodology study. Int J Food Sci Nutr 60:174–190
Sabanis D, Lebesi D, Tzia C (2009b) Effect of dietary fibre enrichment on selected properties of gluten-free bread. LWT Food Sci Technol 42:1380–1389
Sabanis D, Tzia C (2011) Effect of hydrocolloids on selected properties of gluten-free dough and bread. Food Sci Tech Int 17:279–291
Sanchez HD, Osella CA, de la Torre MA (2002) Optimization of gluten-free bread prepared from cornstarch, rice flour, and cassava starch. J Food Sci 67:416–419
Sandhu JS, Hudson GH (1981) The gel nature and structure of the carbohydrate of ispaghula husk ex plantago ovata forsk. Carbohydr Res 93:247–259
Schober TJ, Bean SR, Boyle DL, Park SH (2008) Improved viscoelastic zein-starch doughs for leavened gluten-free breads: Their rheology and microstructure. J Cereal Sci 48:755–767
Schober TJ, Bean SR, Tilley M, Smith BM, Ioerger BP (2011) Impact of different isolation procedures on the functionality of zein and kafirin. J Cereal Sci 54:241–249
Schober TJ, Moreau RA, Bean SR, Boyle DL (2010) Removal of surface lipids improves the functionality of commercial zein in viscoelastic zein-starch dough for gluten-free breadmaking. J Cereal Sci 52:417–425
Schorsch C, Garnier C, Doublier JL (1997) Viscoelastic properties of xanthan/galactomannan mixtures: comparison of guar gum with locust bean gum. Carbohydr Polym 34:165–175

Sciarini LS, Bustos MC, Vignola MB, Paesani C, Salinas CN, Perez GT (2017) A study on fibre addition to gluten free bread: its effects on bread quality and in vitro digestibility. J Food Sci Technol Mysore 54:244–252
Sciarini LS, Perez GT, de Lamballerie M, Leon AE, Ribotta PD (2012a) Partial-baking process on gluten-free bread: Impact of hydrocolloid addition. Food Bioprocess Tech 5:1724–1732
Sciarini LS, Ribotta PD, Leon AE, Perez GT (2010) Effect of hydrocolloids on gluten-free batter properties and bread quality. Int J Food Sci Technol 45:2306–2312
Sciarini LS, Ribotta PD, Leon AE, Perez GT (2012b) Incorporation of several additives into gluten free breads: effect on dough properties and bread quality. J Food Eng 111:590–597
Sharif MK, Butt MS, Anjum FM, Khan SH (2014) Rice bran: a novel functional ingredient. Crit Rev Food Sci Nutr 54:807–816
Shepherd SJ, Gibson PR (2013) Nutritional inadequacies of the gluten-free diet in both recently-diagnosed and long-term patients with coeliac disease. J Hum Nutr Diet 26:349–358
Shewry PR, Halford NG, Belton PS, Tatham AS (2002) The structure and properties of gluten: an elastic protein from wheat grain. Philos Trans R Soc Lond B Biol Sci 357:133–142
Shin M, Gang DO, Song JY (2010) Effects of protein and transglutaminase on the preparation of Gluten-free rice bread. Food Sci Biotechnol 19:951–956
Sivaramakrishnan HP, Senge B, Chattopadhyay PK (2004) Rheological properties of rice dough for making rice bread. J Food Eng 62:37–45
Skodje GI, Sarna VK, Minelle IH, Rolfsen KL, Muir JG, Gibson PR, Veierød MB, Henriksen C, Lundin KEA (2017) Fructan, rather than gluten, induces symptoms in patients with self-reported non-celiac gluten sensitivity. Gastroenterology 154(3):529–539.e2. https://doi.org/10.1053/j.gastro.2017.10.040
Sly AC, Taylor J, Taylor JRN (2014) Improvement of zein dough characteristics using dilute organic acids. J Cereal Sci 60:157–163
Smak C (1972) New approach to determine the brownness of bread crust. Correlation between crust color and protein content. Cereal Chem 49:554–560
Smerdel B, Pollak L, Novotni D, Čukelj N, Benković M, Lušić D, Ćurić D (2012) Improvement of gluten-free bread quality using transglutaminase, various extruded flours and protein isolates. J Food Nutr Res 51:242–253
Smith BM, Bean SR, Selling G, Sessa D, Aramouni FM (2017) Effect of salt and ethanol addition on zein-starch dough and bread quality. J Food Sci 82:613–621
Souza E, Kruk M, Sunderman DW (1994) Association of sugar-snap cookie quality with high molecular weight glutenin alleles in soft white spring wheats. Cereal Chem 70:601–605
Staudacher HM, Gibson PR (2015) How healthy is a gluten-free diet? Br J Nutr 114:1539–1541
Storck CR, Zavareze ED, Gularte MA, Elias MC, Rosell CM, Dias ARG (2013) Protein enrichment and its effects on gluten-free bread characteristics. LWT Food Sci Technol 53:346–354
Struyf N, Verspreet J, Courtin CM (2016) The effect of amylolytic activity and substrate availability on sugar release in non-yeasted dough. J Cereal Sci 69:111–118
Taylor JRN, Schober TJ, Bean SR (2006) Novel food and non-food uses for sorghum and millets. J Cereal Sci 44:252–271
Taylor JRN, Taylor J, Campanella OH, Hamaker BR (2016) Functionality of the storage proteins in gluten-free cereals and pseudocereals in dough systems. J Cereal Sci 67:22–34
Theethira TG, Dennis M (2015) Celiac disease and the gluten-free diet: consequences and recommendations for improvement. Dig Dis 33:175–182
Torres MD, Arufe S, Chenlo F, Moreira R (2017) Coeliacs cannot live by gluten-free bread alone—every once in awhile they need antioxidants. Int J Food Sci Technol 52:81–90
Trappey EF, Khouryieh H, Aramouni F, Herald T (2015) Effect of sorghum flour composition and particle size on quality properties of gluten-free bread. Food Sci Technol Int 21:188–202
Tubili C, Di Folco U, Hassan OMS, Agrigento S, Carta G, Pandolfo MM, Nardone MR (2016) Fiber enriched protein-free pasta and bread: is it a useful tool in chronic kidney disease in type 2 diabetes? Med J Nutr Metab 9:95–99
van Riemsdijk LE, Peigrom PJM, van der Goot AJ, Boom RM, Hamer RJ (2011a) A novel method to prepare gluten-free dough using a meso-structured whey protein particle system. J Cereal Sci 53:133–138

van Riemsdijk LE, van der Goot AJ, Hamer RJ (2011b) The use of whey protein particles in gluten-free bread production, the effect of particle stability. Food Hydrocoll 25:1744–1750

van Riemsdijk LE, van der Goot AJ, Hamer RJ, Boom RM (2011c) Preparation of gluten-free bread using a meso-structured whey protein particle system. J Cereal Sci 53:355–361

Vanin FM, Lucas T, Trystram G (2009) Crust formation and its role during bread baking. Trends Food Sci Technol 20:333–343

Velazquez N, Sanchez H, Osella C, Santiago LG (2012) Using white sorghum flour for gluten-free breadmaking. Int J Food Sci Nutr 63:491–497

Vici G, Belli L, Biondi M, Polzonetti V (2016) Gluten free diet and nutrient deficiencies: a review. Clin Nutr 35:1236–1241

Viebke C, Al-Assaf S, Phillips GO (2014) Food hydrocolloids and health claims. Bioact Carbohydr Diet Fibre 4:101–114

Wild D, Robins GG, Burley VJ, Howdle PD (2010) Evidence of high sugar intake, and low fibre and mineral intake, in the gluten-free diet. Aliment Pharmacol Ther 32:573–581

Witczak T, Juszczak L, Ziobro R, Korus J (2017) Rheology of gluten-free dough and physical characteristics of bread with potato protein. J Food Process Eng 40:e12491

Wolter A, Hager A-S, Zannini E, Arendt EK (2013) In vitro starch digestibility and predicted glycaemic indexes of buckwheat, oat, quinoa, sorghum, teff and commercial gluten-free bread. J Cereal Sci 58:431–436

Xing LC, Santhi D, Shar AG, Saeed M, Arain MA, Shar AH, Bhutto ZA, Kakar MU, Manzoor R, El-Hack MEA, Alagawany M, Dhama K, Ling MC (2017) Psyllium husk (*Plantago ovata*) as a potent hypocholesterolemic agent in animal, human and poultry. Int J Pharmacol 13:690–697

Xu J, Zhang H, Guo XN, Qian HF (2012) The impact of germination on the characteristics of brown rice flour and starch. J Sci Food Agr 92:380–387

Yano H, Fukui A, Kajiwara K, Kobayashi I, Yoza KI, Satake A, Villeneuve M (2017) Development of gluten-free rice bread: Pickering stabilization as a possible batter-swelling mechanism. LWT Food Sci Technol 79:632–639

Ylimaki G, Hawrysh ZJ, Hardin RT, Thomson ABR (1988) Application of response-surface methodology to the development of rice flour yeast breads—objective measurements. J Food Sci 53:1800–1805

Zandonadi RP, Assunçao Botelho RB, Coelho Araújo WM (2009) Psyllium as a substitute for gluten in bread. J Am Dietetic Assoc 109:1781–1784

Zhang D, Mu T, Sun H (2017) Comparative study of the effect of starches from five different sources on the rheological properties of gluten-free model doughs. Carbohydr Polym 176:345–355

Zhu F (2017) Properties and food uses of chestnut flour and starch. Food Bioprocess Technol 10:1173–1191

Zhu F (2018) Chemical composition and food uses of teff (*Eragrostis tef*). Food Chem 239:402–415

Ziobro R, Juszczak L, Witczak M, Korus J (2016) Non-gluten proteins as structure forming agents in gluten free bread. J Food Sci Technol 53:571–580

Ziobro R, Korus J, Juszczak L, Witczak T (2013a) Influence of inulin on physical characteristics and staling rate of gluten-free bread. J Food Eng 116:21–27

Ziobro R, Korus J, Witczak M, Juszczak L (2012) Influence of modified starches on properties of gluten-free dough and bread. Part II: quality and staling of gluten-free bread. Food Hydrocoll 29:68–74

Ziobro R, Witczak T, Juszczak L, Korus J (2013b) Supplementation of gluten-free bread with non-gluten proteins. Effect on dough rheological properties and bread characteristic. Food Hydrocoll 32:213–220

Chapter 26
3D Food Printing: Perspectives

Jie Sun, Weibiao Zhou, Dejian Huang, and Liangkun Yan

Abstract Due to consumers' growing attention to personal health, food products that focus on personal care, healthy concepts and functional claims are emerging as a new trend. This motivates a growing market for personalized healthy food, which aims to tailor and fabricate diet specifically based on an individual's health condition. Traditional food preparation processes even with advanced processing technologies cannot meet such demands.

Three-dimensional (3D) food printing, also known as Food Layered Manufacture, can be one of the potential ways to bridge this gap. This is a digitally controlled, robotic construction process, which can build up complex 3D food products layer by layer. It aims to revolutionize food manufacturing with customized shape, color, flavor, texture, and even nutrition. Hence, food products can be designed and fabricated to meet individual needs through controlling the amount of printing material and nutrition content. Foods created by 3D printing have already entered the market. A range of 3D printing methods, platforms, materials and recipes have been utilized. This chapter reviews the common approaches and techniques used in food printing. The market challenges, technical difficulties and possible strategies along the pathway of commercialization are also discussed.

J. Sun (✉)
Xi'an Jiaotong-Liverpool University, Suzhou, People's Republic of China

National University of Singapore (Suzhou) Research Institute,
Suzhou, People's Republic of China
e-mail: Jie.sun@xjtlu.edu.cn

W. Zhou · D. Huang
National University of Singapore (Suzhou) Research Institute,
Suzhou, People's Republic of China

Food Science and Technology Programme, c/o Department of Chemistry, National University of Singapore, Singapore, Singapore
e-mail: chmzwb@nus.edu.sg; chmhdj@nus.edu.sg

L. Yan
Keio-NUS CUTE Center, Smart System Institute (SSI), National University of Singapore, Singapore, Singapore

T. J. Gutiérrez (ed.), *Polymers for Food Applications*,
https://doi.org/10.1007/978-3-319-94625-2_26

Keywords Customized food fabrication · Personalized nutrition · Platform design · Printing recipe

26.1 Introduction

Consumers' attitudes for food decisions are decided by four criteria: taste, cost, experience, convenience and nutrition (Deloitte 2015). Typical convenience foods enjoyed by a wide range of consumers in the market have high glycaemic index, but low price. Healthy concepts and functional claims are emerging as a new trend, due to consumers' growing attention to personal health and food products. According to the 2015 American Pantry Study, 47% of consumers described themselves as "health conscious", and 35% described themselves as "ingredient sensitive" (Deloitte 2015). It is noted that food ingredients and their effects on metabolism vary among individuals. This motivates a growing market for personalized healthy food, which aims to tailor and fabricate diet specifically based on an individual's health condition.

Meanwhile, available customized food products such as frosted patterns on biscuits and chocolates, letters carved into cookies, and logos painted onto food, have created an amazing sector in the personal gift market. Compared with foods manufactured in mass production, they are more nutrition controllable, but significantly more expensive and only available from very limited suppliers. Traditional mass food preparation processes, even those with advanced processing technologies, cannot meet such personalized demands (Zoran and Coelho 2011). Some food companies are exploring alternative food preparation methods to capture and maintain market share, in turn providing opportunities for food preparation methods such as three-dimensional (3D) food printing. 3D Food Printing, also known as Food Layered Manufacture (Wegrzyn et al. 2012), is a digitally-controlled, robotic construction process which can build up complex 3D food products layer by layer (Huang et al. 2013). It has started a revolution in cooking by precisely mixing, depositing and cooking layers of ingredients, so that users can easily and rapidly experiment with different material combinations.

Foods created by 3D printing have already entered the market, although most consumers may still be unaware. In the Netherlands, people have begun to use 3D printing in microwave pancake fabrication, and this suggests that there could be a rise in the popularity of 3D food printing machines, much like microwave ovens (Hadhazy 2013). In the below session, the history and motivation of food printing are discussed, followed by a comparison between food printing and robotics-based food manufacturing.

26.1.1 History

"Food synthesizer", a very primary food printer concept, was described in the movie Star Trek: The Original Series in the 1960s as a "replicating machine" that could synthesize meals based on user requirements. The concept reveals people's

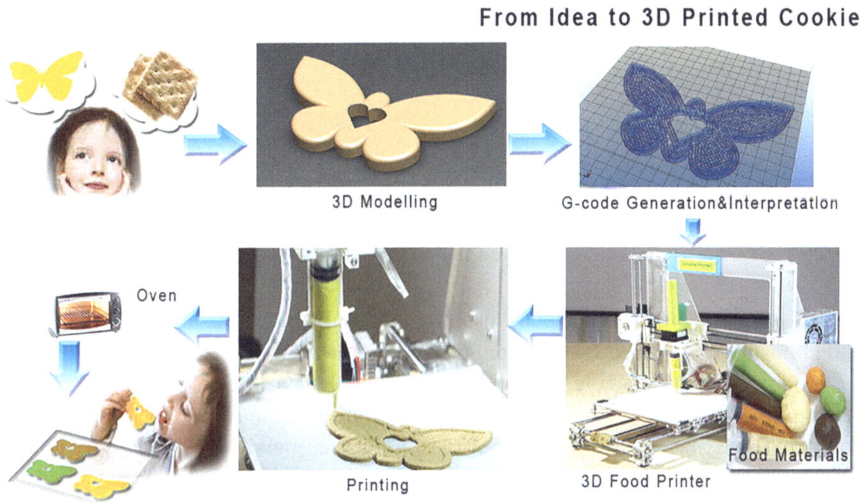

Fig. 26.1 Overview of 3D food printing process

desire of instantly making personalized meals and replicating exciting food designs. The first generation food printer concept designs was introduced to the general public more than 15 years. Nanotek Instruments Inc. patented a rapid prototyping and fabrication method for producing 3D food objects in 2001 (Yang et al. 2001), such as a customer-designed birthday cake; however, no physical prototype was built. Nico Kläber (Electrolux 2009) came out with a Moléculaire concept design in Electrolux Design Lab 2009 competition, which could print a multi-material customized meal using a small robotic arm. Philips Design (2010) proposed creating a custom-designed food product using food cartridges, and an interactive graphical user interface to select ingredients, quantities, shapes, textures and other properties. Massachusetts Institute of Technology (MIT) introduced a digital gastronomy concept into food printer design and presented three conceptual designs. Each conceptual design focused on different aspects of gastronomy from mixing, modelling, to transformation. These concepts seem more realistic compared to the previous conceptual designs, but are still far away to be technically feasible.

A few printing projects have been carried out (Cohen et al. 2009; Hao et al. 2010; Lipton et al. 2009). For example, NASA (National Aeronautics and Space Administration) has award a Small Business Innovation Research (SBIR) Phase I contract to System and Materials Research Consultancy of Austin, Texas to develop a 3D printed food system for long duration space missions (NASA 2013). To obtain fully grown healthy snack, Chloé Rutzerveld (2014) created the concept of Edible Growth, where a mixture of seeds, spores and yeast is printed in the shape, and this structure grows organically and naturally over the course of 3–5 days.

As shown in Fig. 26.1, the current food printing process starts with designing a virtual 3D model. Slicing software translates this model into individual layers and

Fig. 26.2 Customized food design and fabrication samples at the National University of Singapore

finally generates machine codes for printing. After uploading the codes into a printer and choosing a preferred food recipe, the food printing starts.

Food printing process creates food pieces in a layer-by-layer manner, which does not require a high energy source to completely remove liquid ingredients from food composition. Fabricated layers do not need to be completely solidified, but require sufficient rigidity and strength to support its own weight and the weight of subsequent layers without a significant deformation or shape change.

26.1.2 Motivation

Food printing can support customized food design and personalized nutrition, and provide sustainable food supply. It can also be applied as a prototyping tool in new product development. The printed food pieces with customized shapes can be used to teach shape, taste, color and design in early child education. Printed foods with personalized nutrition are also very helpful for elderly people with swallowing or chewing difficulties (Kira 2015). These motivations are further discussed as follows.

26.1.2.1 Customized Food Design

Food manufacturing techniques are mainly developed for mass production, while creativity on shapes, structures and flavors are usually compromised. Customized food involves specifically hand-made skills with low production rate and high cost. Food printing technologies target to replace such traditional operation by providing a platform to experiment food design on shapes, colors and flavors automatically. More design solutions are generated such as customized chocolate shaping (Causer 2009; Zoran and Coelho 2011), and personalized full-color images onto solid food formats (Golding et al. 2011). Figure 26.2 shows printed biscuit samples fabricated by our group at the National University of Singapore using materials such as flour, butter, sugar and egg white.

The quality of fabricated food products depend on the fabrication process rather than operator skills. Besides, complex food pieces can be produced in a single process.

As such, production can be easily synchronized with customer demands. The need for warehousing, transportation, and packaging can be reduced significantly.

26.1.2.2 Personalized Nutrition

Besides existing nutritional preferences, the concept of personalized nutrition care according to a person's dietary needs, allergies, or taste preferences is on the research agenda of food industries (Watzke and German 2010). Studies have shown that individuals respond differently to various nutrients, and they may benefit more or less for particular dietary components. Only personalized nutrition can meet the needs and preferences in terms of individual's health status and body-type requirement. Netherlands Organisation for Applied Scientific Research (TNO) has suggested printing customized meals for seniors, athletes, expectant mothers through varying food component levels like protein and fat (Gray 2010). Elderly people suffering from dysphagia have trouble chewing and swallowing food. These elderly people typically get their meals in the form of an unappealing milkshake of pureed chicken and broccoli, and lead to loss of appetite and malnourishment. Serizawa et al. (2014) developed a 3D edible gel printer using syringe pump and dispenser to make soft food for the elderly who cannot swallow food well.

Under the traditional food supply chain, foods with personalized nutrition are produced with additional cost. Marketing and distributing such foods may not be financially viable. From a technical perspective, foods with controlled ingredient formulation will be much more challenging to produce (de Roos 2013). Food printing can personalize nutrition through controlling the amount of food to be printed, and/or ingredients design. This can be achieved using food printers in house or service store, thus the additional cost for distribution is minimized.

26.1.2.3 Food Supply Sustainability

An increase of global population results in growing demand for food. Alternative ingredients extracted from algae, fungi, seaweed, lupine, and waste from the current agricultural and food production can be utilized as printing materials in the future. Using other advanced technologies, these food materials can be scaled down to a greater extent, and fed into food printers to fabricate novel taste and texture food pieces. This progress may also ease the growing demand for food production in an environmentally friendly and efficient manner.

In short, food printing can help move people toward alleviating starvation in some parts of the world and prompt healthy and natural side of manufactured food. However, some people are very conservative to the fabricated food pieces, and suspect that applying advanced manufacturing process may lead something unnatural and unhealthy.

26.1.2.4 Prototyping Tool for Small Batch Production

In the food industry, consumer demands on improving food safety, shelf life, nutritional value, and reducing wastage, create a complicated scenario for food product design. The food industry preferred to re-developing the existing products with incremental changes, rather than creating a radical change in products (Winger and Wall 2006). This apparently 'safe' approach perpetuates the problem of a high food product failure rate at around 75% (Stewart-Knox and Mitchell 2003). To improve the communications between food scientists, food engineers, marketing people, distributors and consumers during the product development stage, food producers need to explore ingredient combination and fabricate new design samples. However, it is always difficult to find suitable equipment with simple design and reliable performance for a small batch production. A promising solution is to further develop food printer as a prototyping tool to conduct small batch production in a cost effective and time efficient way. It can help to fully understand comprehensive technical requirements, explore ingredient combination, taste and mouthfeel prior to starting mass production. The fabricated food products may be used to verify consumer interest in a proposed design, and ingredient stability of specific designs. This could also help filter out a large number of design candidates that do not meet the requirements in a short time at acceptable cost.

26.1.3 Food Printing and Robotics-Based Food Manufacturing

Food processing often requires rapid and repetitive movements, thus applying automation in a food manufacturing process can improve its efficiency and the resultant food quality. Both food printing and robotics-based manufacturing can automate a food preparation process and reduce human workload, yet they create totally different user experience. The former places users' creativity and control at the center of the process by allowing the users to manipulate food forms and materials directly, and the latter aims to reduce human involvement and workload by automating various manual processes.

Robotics-based technologies have been designed to replace labor-intensive operations, to automate individual steps or replace manual operations in household, food catering service and food manufacturing industries. For example, baking cookies robots can locate ingredients, mix them in a correct order, and place the resulting dough in a baking tray (Bollini et al. 2011). Motion libraries embedded into these robots can perform cooking tasks, through basic actions such as picking up an object, putting it down or pouring (Beetz et al. 2011). These technologies changed the way of food producing, improve food preparation efficiency, but had very little relevance to nutrition control and customized fabrication.

Food printing is a digital food fabrication process integrating 3D printing and digital gastronomy technique to manufacture food pieces. It allows users to design

Table 26.1 Comparison of recipes in food printing and robotics-based food manufacturing

	Food printing	Robotics-based food manufacturing
Cookies	Printable sugar cookies: Flour, powdered sugar, egg yolk, unsalted butter (Lipton et al. 2010) Snowflake-shaped sugar cookies: All-purpose flour, granulated sugar, unsalted butter, egg and egg yolk, salt, vanilla extract (Bosker 2013)	Afghan biscuits: Flour, sugar, butter, cocoa powder, cornflakes (Bollini et al. 2011) Quick'N easy sugar cookies: all-purpose flour, sugar, eggs, vegetable oil, vanilla, granulated baking powder, salt (Bollini et al. 2011)
Chocolate	Cadbury milk chocolate (Hao et al. 2010)	Cocoa and cocoa butter, full cream milk, sugar, special flavoring, emulsifier (Gunstone and Padley 1997)
Sugar cubes	Sugar, sweet and sour flavored candies, and milk chocolates (3D Systems 2013)	Sugar, food colorings, aromatic herbs and spices (Labau 2014)

and fabricate food with customized color, shape, flavor, texture and even nutrition. As a result, our eating experiences can go beyond taste to encompass all aspects of gastronomy such as food preparation, culture, economy, physics and chemistry (van Bommel and Spicer 2011). While, the efficiency of current food printers is too slow, and cannot meet consumer requirements.

Engaging consumers is also emphasized in food printer design, i.e. a convenient and friendly interaction between consumers and machines. Burritobot (2014) extruded customizable amounts of Mexican ingredients onto a pre-made tortilla to a user's taste, which might become a perfect way for the fast food industry to save manpower cost, improve service efficiency and reduce consumer waiting time. Consumers in Japan may order chocolates made from a 3D scan of their face for Valentine's Day (Gorkin and Dodds 2013).

Table 26.1 gives a comparison of recipes for cookies, chocolates and sugar cubes between food printing and robotics-based food manufacturing. The basic ingredients of cookies are quite similar including flour, sugar, egg and butter. It is the same case for the sugar cube recipes. For robotics-based chocolate manufacturing, raw materials such as cocoa, cocoa butter, full cream milk, and sugar are used for processing. However, commercial chocolates are utilized in printing.

Substantial efforts have been made to pre-process materials suitable for printing, and improve their thermal stability for post-processing. Hence, the recipes used in printing would have to be slightly different from traditional recipes (Lipton et al. 2010). Ingredients even with well-known material properties needed be tailored for each printing application. In ChocALM machine, formulations of chocolate were modified to meet the rheological and post-deposition fusion requirements during the development (Hao et al. 2010). These modified recipes may have commercial value in the near future, hence not much details are released in publications. Meanwhile, researchers from TNO had started to explore more fundamental topics such as converting ingredients into tasty products for healthy and environmental concerns (van Bommel and Spicer 2011).

26.2 Materials and Recipes

26.2.1 Available Printing Materials

The available materials for food printing can be classified into three categories: natively printable materials, non-printable traditional food materials and alternative ingredients.

26.2.1.1 Natively Printable Materials

Natively printable materials like hydrogel, cake frosting, cheese, hummus and chocolate can be extruded smoothly from a syringe (Cohen et al. 2009). Final products are fabricated with diverse taste, nutritional value, and texture. However, none of them are served as a main course in meals.

Some of these natively printable materials are stable enough to hold the shape after deposition. For example, the mixture of sugars, starch, and mashed potato was used as powder materials in Z Corporation powder/binder 3D printers (Southerland et al. 2011) to fabricate sugar teeth. The fabricated teeth were strong enough without further post processing. Other composite formulations such as batters and protein pastes may require a post-cooking process (Lipton et al. 2010), resulting in fabricated structures difficult to retain their printed shapes.

26.2.1.2 Non-printable Traditional Food Material

Printability tests for traditional food materials were judged by viscosity, consistency and solidifying properties (Fabaroni 2007), and the most successful printable material was pasta dough. Food like rice, meat, fruit and vegetables, largely consumed by people every day, are not printable by nature. To enable their capability of extrusion, adding hydrocolloids in these solid materials has been utilized in many culinary fields. Although some solid and semi-solid foods have already been manipulated to become printable by gastronomic tricks, it is difficult to test and modify the whole list. One potential solution is to create an element set using a small group of ingredients which can generate a high degree of freedom on texture and flavor. Cohen et al. (2009) investigated on fine tuning concentration of hydrocolloids (xanthan gum and gelatin), and achieved a very wide range of textures (i.e. mouthfeels).

After printing process, the majority of traditional edibles need post-deposition cooking, such as baking, steaming or frying. These processes involve different levels of heat penetration and result in non-homogenous texture. Lipton et al. (2010) experimented on modifying cookie recipes for both printing and post-cooking. They managed to find one recipe which can print 3D models with complex internal geometries and retain their shape after deep frying.

26.2.1.3 Alternative Ingredients

Alternative ingredients extracted from algae, fungi, seaweed, lupine and insects, are novel sources for protein and fiber. In the 'Insects Au Gratin' project, insect powders mixed with extrudable icing and soft cheese were used as printing materials to shape food structures and make tasty pieces (Southerland et al. 2011). Residues from the current agricultural and food processing can be transformed to biologically active metabolites, enzymes, and food flavor compounds (Silva et al. 2007; Nikitina et al. 2007), as sustainable and eco-friendly printing material sources. Available food processing technologies can further scale down the size of alternative food material molecules, create more particles for an overall greater surface area, and improve food nutrition absorption and stability. Briefly, introducing alternative ingredients into food printing would aid in developing healthier (e.g. low-fat) food products.

26.2.2 Printing Recipes

Based on these available food materials, printing recipes can be categorized into element-based recipe printing and traditional recipe printing.

The element-based recipe printing uses a standard set of dispensing elements to control the taste, texture, flavor and nutrition of fabricated food pieces. The elements mentioned here refer to ingredients. This method was proposed in Massachusetts Institute of Technology (MIT) conceptual designs on Virtuoso Mixer, Digital Fabricator and Robotic Chef (Zoran and Coelho 2011), and a few other prototype designs. Cohen et al. (2009) used a small group of ingredients to print solid and semi-solid foods. In their study, a wide range of textures could be achieved by fine tuning hydrocolloid concentration and combination, and the flavor could be tuned by using concentrated flavoring additives. van Bommel and Spicer (2011) extracted basic carbohydrates, proteins, and other nutrients from algae or insects and mixed them together in varied proportion to print something resembling steak and chicken. Such printing recipes would be an eco-friendly and sustainable solution to deal with a growing demand for food.

The traditional recipe printing works on modification of existing recipes for customized food fabrication. Examples include fabricated food items with complex structure from Cornell University's Fab@Home 3D Printer, customized chocolate products from Exeter University's ChocALM machine, and edible 3D printing items from University of the West of England. However, the link between recipes and ingredient control has not been built in the current research activities.

A combination of element-based recipe printing and traditional recipe printing can support a food design workflow to experiment with diverse recipes. It can easily substitute ingredients based on nutritional contents as well as personal and social preferences. Leveraging on their highly customizable features and digital fabrication flexibilities would create an information-driven food culture for healthier life.

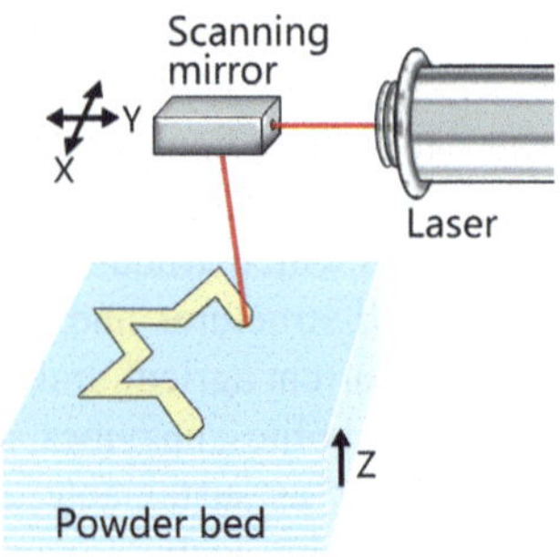

Fig. 26.3 Selective laser sintering

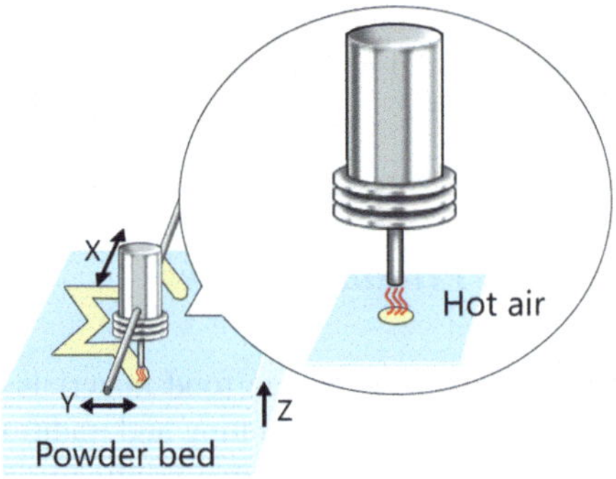

Fig. 26.4 Selective hot air sintering

26.3 Food Printing Technologies

Different food printing technologies are developed to process diverse food materials from powder form to polymers that vary in flowability. Below is a summary of applicable 3D food printing technologies.

26.3.1 Selective Laser Sintering/Hot Air Sintering

Both laser (as in Fig. 26.3) and hot air (as in Fig. 26.4) can be utilized as a sintering source to fuse powder particles and form a solid layer. TNO's Food Jetting Printer (Gray 2010) applied laser to sinter sugars and NesQuik powders to build solid 3D objects. The sintered material formed the product part whilst the un-sintered powder remained in place to support the structure. The CandyFab (2007) applied a selective low-velocity stream of hot air to sinter and melt a bed of sugar. The fabrication powder bed is heated to just below the material's melting point to minimize thermal distortion and facilitate fusion to the previous layer. The two sintering processes offer the freedom to quickly build complex food items in a short time without post curing. However, they are only suitable for sugar and fat based materials with

Table 26.2 Comparison of commercial extrusion-based food printing machines

Company	Choc edge	Natural machines	BeeHex	Barilla	Dovetailed
Machine	Choc creator	Foodini	BeeHex robot pizza printer	3D pasta printer	Nūfood 3D food printer
Technologies	HME	RTE	RTE	RTE	HFE
Materials	Melted chocolate	Semi-solid material such as dough	Dough, sauce and cheese	Durum wheat, semolina, water	Sodium solution, calcium chloride
Printhead number	Single printhead	Single printhead	Multi-printhead for multi-material	Dual-printhead with the same material	Single printhead
Extrusion mechanism	Syringe	Air pressure	Air pressure with multi-output	Air pressure with dual-output	Syringe
Fabricated products	Customized chocolate objects	Customized pizza and cookies	Customized pizza	Pasta	Fruit
Post-processing	N.A.	Baking	Baking	Boiling	N.A.
Reference	Choc creator (2014)	Foodini (2014)	BeeHex (2016)	Barilla (2015)	Dovetailed (2014)

HME hot-melt extrusion, *RTE* room temperature extrusion, *HFE* hydrogel-forming extrusion

relatively low melting point, and the fabrication processes as well as machine structure are complicated as many variables are involved.

26.3.2 Extrusion Based Food Printing

Extrusion is the action to force liquid/semisolid/solid materials through a die opening in order to create objects with desired cross sections. The extrusion process in food printing starts with loading of material, pushing the material out of the nozzle in a controlled manner, moving the material stream according to a predefined path, and eventually bonding the deposited layer to form a coherent solid structure. Although a range of 3D printing methods have been utilized for food printing, this method is the most widely adopted one in commercial machine designs. The ideal target of the extrusion-based 3D food printing is to achieve the output of the food extrusion cooking physically with digitalized design and personalized nutrition control.

Table 26.2 summarizes such commercial machine designs in terms of applicable materials, extrusion mechanism and printed products. They usually have a compact size, and low maintenance cost, but are greatly limited by material choices, long fabrication time and delamination caused by temperature fluctuation.

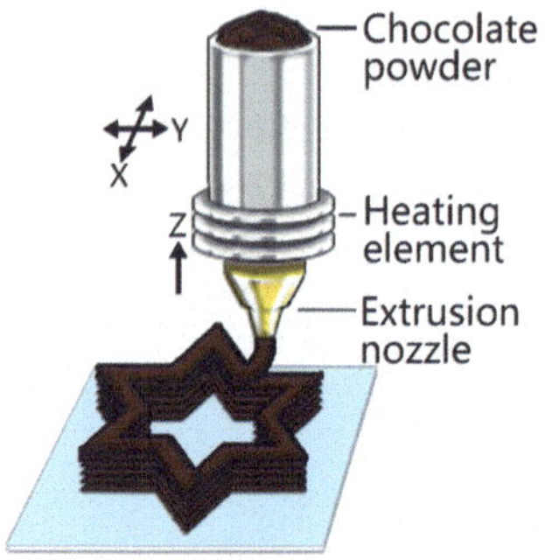

Fig. 26.5 Schematic diagram of HME

26.3.2.1 Hot Melt Extrusion (HME)

Hot-melt extrusion, also called fused deposition modeling (FDM), was firstly described in Crump's work (1991). In Fig. 26.5, melted semi-solid food polymer is extruded from a movable FDM head, solidifies almost immediately after extrusion and welds to the previous layers.

Hot-melt extrusion has been applied to create customized 3D chocolate products (Causer 2009; Hao et al. 2010). MIT Researchers used hot melt chocolate as a dispensing liquid and developed a functional prototype "digital chocolatier" (Zoran and Coelho 2011). In this project, compressed air was applied to push the melt chocolate out of chambers for customized candy fabrication. Using the hot-melt extrusion method, a '3D Food-Inks Printer' printed 3D color images on an extruded base (Golding et al. 2011), while a post-cooking step was required to fuse layers together. Over the years, Choc Edge (Choc Edge 2014), Natural Machines (2014) and TNO (van der Linden 2015) have utilized HME to build 3D chocolate objects. Among them, Foodini from Natural Machines is considered as "microwave oven of the future" and an equally important staple in the average kitchen.

26.3.2.2 Room Temperature Extrusion (RTE)

RTE refers to smoothly extruding natively printable materials like dough, cheese, frosting, creamy peanut butter, jelly, Nutella and hummus at room temperature (Cohen et al. 2009; Periard et al. 2007; Millen 2012). RTE has been applied to fabricate complex confections with high repeatability, which are difficult to make by hand (Periard et al. 2007). Essential carbohydrates, proteins, meat purees and other nutrients extracted from alternative sources (algae and insects) have been used as printing materials in RTE (van der Linden 2015). This extrusion method can fabricate complex confections using a single material with high repeatability, which were difficult to make by hand (Periard et al. 2007). The RTE can also be applied for pasta printing using classical recipes (durum wheat semolina and water), and surface filling on pizza, cookie and graphical decoration (van der Linden 2015).

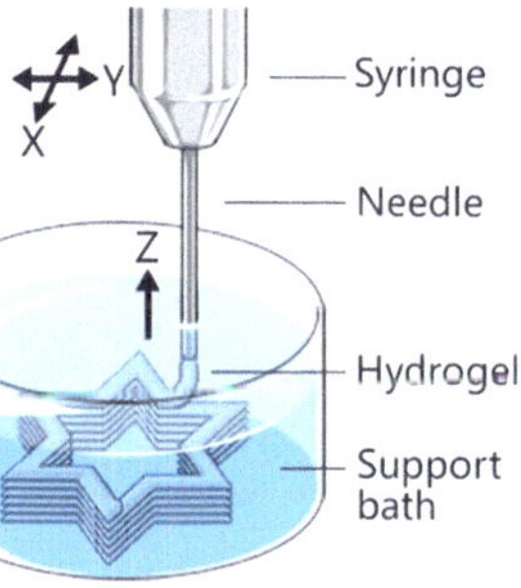

Fig. 26.6 Schematic diagram of HFE

26.3.2.3 Hydrogel-Forming Extrusion (HFE)

As shown in Fig. 26.6, HFE is the extrusion of hydrocolloid solutions or dispersion into a polymer/hardening/gel setting bath using syringe pipette, jet cutter, vibrating nozzle and similar apparatus. In general, gel droplets' diameter is about 0.2–5 mm, and solution temperature control is the key to forming stable shapes in HFE.

This extrusion is critically dependent on the rheological properties of polymer and the gel-formation mechanism. In other words, the polymer solution should present viscoelastic characteristic first, and then turn into self-supporting gels prior to the consecutive deposited layers. HFE has been used in commercial machine designs to print intricate food pieces. Serizawa et al. (2014) developed a 3D edible gel printer using a syringe pump and dispenser to produce soft foods for elderly people with swallowing problems. A UK firm, Dovetailed developed a 3D fruit printer which combined strawberry flavoring with a sodium rich gel, and deposited little spheres into a cold calcium chloride solution to create something like raspberry (Molitch-Hou 2014).

In short, many commercial machines are designed using temperature control: HME for chocolate printing (Choc Edge 2014), RTE for pizza and pasta printing (Alec 2015a; Molitch-Hou 2014), and HFE for fruit printing (Molitch-Hou 2014). Some even crossed the boundary between RTE, HME and HFE. For example, Bocusini plans to control the printer head's temperature between 20 and 70 °C to print more than 30 different pre-filled cartridges under six categories: confectionary products, bakery products, snack products, fruit and vegetable products, meat products and dairy products (Millsaps 2015). This breakthrough offers a wide range of printable food material using a single machine setup, in turn becoming more attractive for both professional chefs and home users. It can also apply interchangeable extruders and multi-printhead to deal with multiple-material food design and fabrication. This empowers consumers to take control of food design for the diversity of printing materials.

26.3.3 *Binder Jetting*

In binder jetting as Fig. 26.7, each powder layer is distributed evenly across the fabrication platform, and a liquid binder sprays to bind two consecutive powder layers (Sachs et al. 1992). Before fabrication, a layer of water mist is sprayed to stabilize

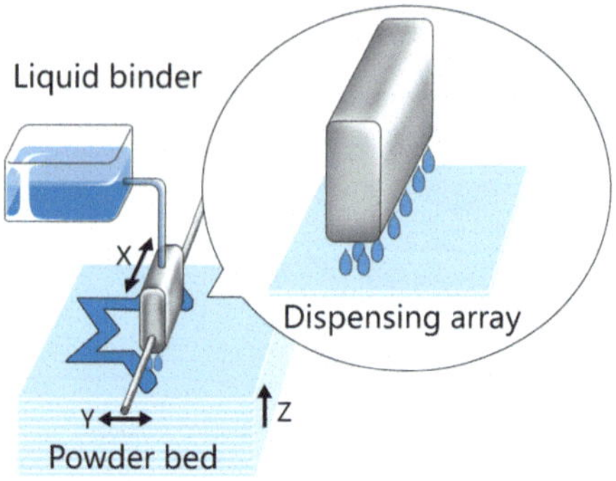

Fig. 26.7 Powder bed binder jetting

powder material and minimize disturbance caused by binder dispensing. In the edible 3D printing project, Southerland et al. (2011) utilized sugars and starch mixtures as the powder and a Z Corporation powder/binder 3D printer as the platform to fabricate customized shapes. Binder jetting offers advantages such as fast fabrication and low material cost, but suffers from rough surface finish and high machine cost. In 2013, Sugar Lab (3D Systems 2013) used sugar and different flavor binders to fabricate complex sculptural cakes for weddings and other special events. This fabrication adopted 3D Systems' Color Jet Printing technology, and the material and fabrication process met all food safety requirements. However, food items with high-sugar content and little nutritional value may not be attractive, which are often linked to obesity, Type-2 diabetes and heart disease. This greatly limits this technology's market potential.

26.3.4 Inkjet Printing

As shown in Fig. 26.8, inkjet food printing dispenses a stream of droplets from a syringe-type printhead in a drop-on-demand way for cookie, cake, or pastry fabrication. De Grood Innovations' FoodJet Printer (2012) used pneumatic membrane nozzle-jets to deposit drops onto pizza bases, biscuits and cupcakes. The drops fallen under gravity and formed a two and a half-dimensional digital image as decoration or surface fill on substrates.

The current drawbacks of 3D food printing technologies include limited use of ingredients and the public's general preference for traditionally made foods. Besides, food safety concerns have greatly limited the applying of technologies that involved laser, electron beam and unsafe food additives.

26.4 Platform Design and Configuration

In this section, two types of platform design are compared: specialized food printers and universal 3D printers with food printing function, followed by a discussion on printing stage configuration.

Fig. 26.8 Inkjet printing

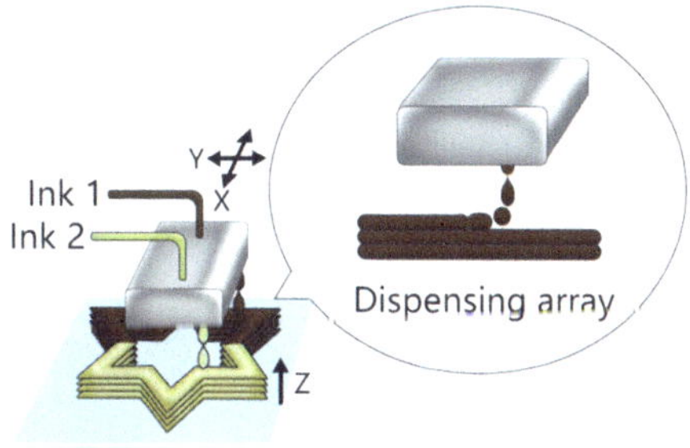

26.4.1 Specialized Food Printers and 3D Printers with Food Printing Function

Both specialized food printers and universal 3D printers with food printing function are available in the market.

To simplify a development process and shorten its development time, researchers have modified open source commercial 3D printers for food printing purpose. The Fab@Home system was one of them (Cohen et al. 2009). This machine is not specifically designed for food applications, but as a universal desktop fabricator compatible with food materials. MakerBot was also modified for food printing by installing Frostruder MK2 as a printhead (Millen 2012). These universal 3D printers with food printing function can help researchers quickly create 3D food shapes and investigate food materials' properties. However, they are applicable to limit materials and cannot support in-depth research.

These universal 3D printers with food printing function are relatively cheap, where the food printing is included as an additional feature. There are concerns about the general food safety for this type of printers, and it is difficult to certify them as safe. For example, bacteria can buildup in the extruder when food gets stuck in small cracks and spaces. Most of desktop printers are made of plastics, which may emit ultra-fine particles. Such toxic particles can release during the printing process and lead to adverse health effects.

The specialized food printers are usually built based on specific requirements to support fabrication related researches (Torrone 2007; Hao et al. 2010) such as dispensing mechanism, material property and so on. Such printers usually can give better fabrication performance, but are very expensive. For example, Choc Creator II even under a discounted price is about US$3800, and the retail price of Foodini is around $1000. The components of these machines are fabricated using food safe materials, and are considered safe when being in contact with food. A component cleaning process is also designed and implemented to avoid food contamination. For example, Foodini uses a re-usable extruder made of stainless steel, and the nozzle is easy to clean either in a dishwasher or by hand.

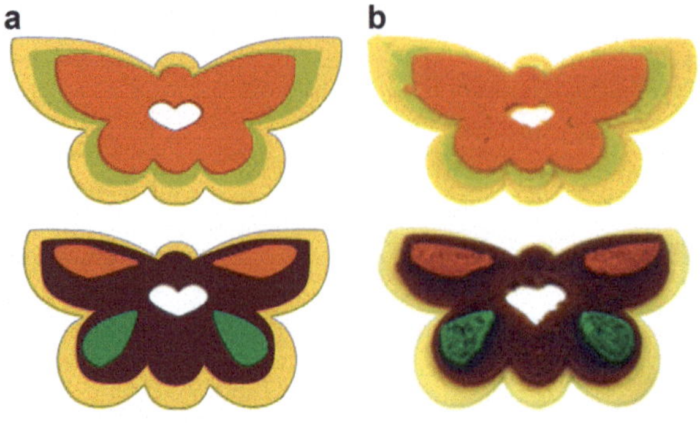

Fig. 26.9 Food design and fabrication samples at the National University of Singapore: (**a**) multiple material cookies 3D model, (**b**) actual printed cookies

26.4.2 Interchangeable Extruders and Multi-Printhead

Both multi-printhead and interchangeable extruders design have applied in food printing, which are capable of delivering multi-material fabrication with geometric complexity more easily than manual operation. The quality of fabricated food products depends more on the fabrication process rather than operator skills.

The universal 3D printers with food printing function apply interchangeable extruders which can be easily switched between polymer printing and food printing. For example, FOCUS 3D Printer from byFlow applies interchangeable extruders to print PLA, ABS filaments, paste food materials, and chocolate. This 3D printer can automatically recognise and adapt its settings to a set of interchangeable extruders (Grunewald 2015).

The specialized food printers also use interchangeable extruders to switch among food materials. As shown in Fig. 26.9, our group applied interchangeable extruders to fabricate multi-material butterfly cookies, where the same dough material with different food dyes was used to present multiple materials (Sun et al. 2015). This can also be utilized to replace an existing printhead with a continuous inkjet printhead loaded with food dyes to decorate food pieces, or with special nutrients to create functional food.

Multi-printhead design has been used in both specialized food printers and universal 3D printers with food printing function. To achieve controlled material deposition and distribution, multiple printheads are allocated to print supporting or fabrication materials. The data from each layer is directed to a platform controller, which activates the associated motors to move the corresponding dispensing head and control its feeding rate and deposition area. Printing multi-material from multi-printhead is a highly attractive feature which allows switching among material sources in a computer control way. The BeeHex 3D printer adapted this design to dispense pizza dough, sauce, and cheese individually via multiple nozzles (Garfield 2016) for personalized pizza making. It can also be applied to test various nutrition/ingredient combinations in a food product development process or tailor nutrition for individual preference.

Multi-printheads need not only multiple motors in the syringe-based extrusion or multi-channel output in the air pressure driven extrusion, but also the synchronization

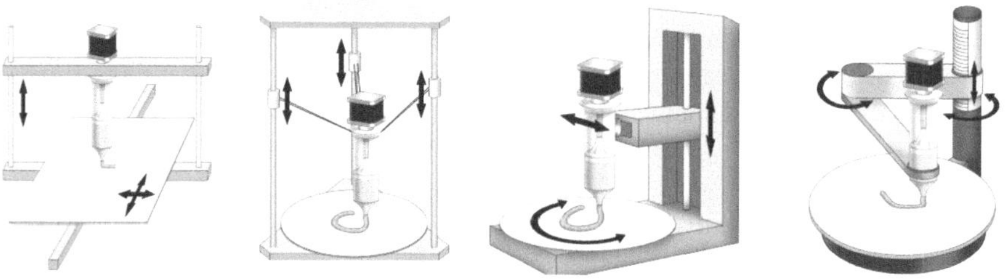

(a) Cartesian (b) Delta (c) Polar (d) Scara

Fig. 26.10 Food printer configuration

of these printheads' extrusion processes (Sun et al. 2018). This would greatly increase the system design complexity. Customers usually expect these printers to be compact and fit inside a bakery or home kitchen. Hence, multi-printhead and interchangeable extruders are usually combined together. For a dual-printhead printer, the interchangeable extruders design can bring in more flexibility and maximize the diversity of printing materials without increasing much on the machine's complexity.

26.4.3 Printing Stage Configuration

The multi-axis stages used in food printing include Cartesian, Delta, Polar and Scara configurations. The advantages and disadvantages of each configuration are discussed, including their applications in commercial machine designs. Some key factors and considerations on configuration selection are summarized after that.

26.4.3.1 Cartesian Configuration

As shown in Fig. 26.10a, the Cartesian configuration has X, Y, and Z axes for left to right, front to back and up and down motion, respectively. It may have a square stage moving along Z-axis and a printhead sitting on X-Y axis or a printhead moving along X-Z axis and a square stage sitting on Y-axis.

Many first generation food printers use this configuration, since a machine with this configuration is simpler to design, easier to maintain and calibrate. Examples of this Cartesian configuration include Choc Creator (Choc Edge 2014), Foodini (Natural Machines 2014), BeeHex Robot pizza printer (BeeHex 2016). Most importantly, comprehensive software and hardware resources are available to support the design and development work under this configuration, such as slicing software, printing path planning and dual printhead design. This configuration requires a large space for printing operation, thus it is not practical as a consumer end device. The moving printhead loaded with food material is also heavy, which compromises printing speed and results in a constant jerking motion when printing direction changes. It may result in collapse of 3D printed food pieces with large height. Last but not least, relatively slow printing speed in the Cartesian configuration is always a bottleneck to limit its application in commercial machine designs.

26.4.3.2 Delta Configuration

In Delta configuration as shown in Fig. 26.10b, a circular print stage is fixed, and the printhead is suspended above it by three arms in a triangular configuration. The number of components in this configuration is less, thus reducing maintenance and machine costs. Machines with Delta configuration such as Pinya3 food printer (Alcalde 2016) and Tytan 3D delta printer (Krassenstein 2014) are currently being developed towards commercialization. Compared with the Cartesian configuration, the Delta configuration printers are cheaper and faster, and can fabricate larger volume food piece in a shorter time period. This configuration is not so precise in position control but is sufficient for digitalized food fabrication, which is not so demanding on precision. When the printhead loaded with liquid material (such as melted chocolate) is moving under faster speed, the acceleration/deceleration may cause liquid vibration within the food material feeding unit. Thus, the printing process may become unstable. A modified Delta configuration is suggested for liquid material, i.e. a fixed printhead with a moving print stage.

26.4.3.3 Polar Configuration

Different from the Cartesian configuration, Polar configuration uses polar coordinates to describe points on a circular grid rather than a square. As shown in Fig. 26.10c, a Polar food printer usually has a spinning stage, plus a print head that can move up-down to cover Z axis, and left-right to cover X and Y tangentially.

This configuration can achieve a perfect circle, and equal performance for all the direction movement with minor mechanical errors and minimum calibration. Examples include the XOCO 3d-printer (Michiel Cornelissen Ontwerp n.d), consisting of a rotating build plate and a single pillar, and the TNO food printer (van der Linden 2015). This configuration is moving from conceptual stage to engineering designs. Machines with this configuration can build a greater volume in a higher speed and a smaller space than that of other configurations.

26.4.3.4 Scara Configuration

Selective Compliant Assembly Robot Arm (SCARA) has gained growing interest in food manufacturing industries greatly since the implementation of the FDA Food Safety Modernization Act (FSMA) in 2011 (Nowak 2015). This configuration is easy to build and has been re-purposed for 3D printing. Shown by Fig. 26.10d, it consists of a robot arm moving along the X-Y plane and an additional actuator to move along the Z-Axis. "Sanna: the food printer of 2020", a conceptual design from Columbia University (Creative Machines Lab 2016), has applied this configuration to convert unprocessed raw, frozen food purees into tasty, cooked and texturized dishes.

26.4.4 *Consideration in Configuration Selection*

Considering a limited kitchen space and busy lifestyle among urban populations, future kitchen devices should be enabled for fast cooking and operated within a compact space. Due to these requirements, factors such as the ratio of printed food piece volume to printer dimension, fabrication time, machine price, cleaning time and maintenance cost become crucial for developing commercial food printers.

An interest on Delta or Polar configuration is growing due to their larger ratio of printable food piece volume to printer dimension, faster fabrication time and lower cost. Although printing accuracy should be maintained for a reliable and repetitive fabrication, it is commonly less demanding in food printing than that in plastic material printing. This is quite different from conventional 3D printers, where the configuration should be selected to optimize printing accuracy. The flexibility of installing multiple-printhead should also be considered, due to the possible involvement of multiple food materials.

Most food printers are modified from the extrusion-based plastic material 3D printers. The printable food materials' density is relatively higher, which increases the payload of motors to drive printheads which were originally designed for plastic printing. Some designs even place a kitchen dish or oven plates on the printing stage to collect the printed food pieces directly. Hence, dynamic and kinematic performance should be studied in the configuration selection, which is lacking in the current projects.

26.5 Innovation on Food Piece Design

The overall perception of food design covers visual appearance, sense of touch (stickiness, roughness, hardness), first bite, chewing, swallowing (flow properties, roughness or smoothness), and residual effects on mouth. Traditional food processing technologies mainly use formulation and process parameters to alter food design. Food printing provides greater freedom to experiment with designs in a manner that was previously impossible. To change mouthfeel, people can print one material with different patterns or modify two materials during printing to create new patterns. This enables users to do things such as building back the texture of an existing food product after changing some key ingredients to upgrade nutritional profile, improving the texture of an existing product to become more desirable, and developing new food products with desired texture as part of the overall eating experience.

Food designers can also purposely feed macronutrients such as starch, protein and fat into the printers, and add appropriate flavors to form appealing textures. How to alter food design and fabrication from the perspectives of layer structure, appearance, nutrition and formulation is discussed as below.

26.5.1 Layer Structure and Unique Taste

The extrusion-based food printing builds 3D food pieces layer-by-layer, thus producing a staircase effect on oblique and curved surfaces. This staircase effect can be used to create decorative features on printed cookie and chocolate objects. Layer structure on pasta models (cubes, moons, and rose shapes) can decorate original designs to form unique styles (Alec 2015a). Along a very gradual slope, the layers with staircase effect create more artistic visual appearance. Layered structure from complicated hexagons and intricately laced patterns fabricated by CocoJet (3D Systems 2015), gives consumers endless possibilities for personalizing their chocolates.

This layered structure also creates a new form of chewing experience. One example is that, in addition to customized shape, colour and flavor, people chewing 3D printed chewing gum can feel its layered texture in their mouths (Alec 2015b).

In taste design, Nūfood Robot can print jelly-like liquid capsules that pack intense flavor into unexpected shapes and textures. For example, a jelly looks like a strawberry, but tastes like a raspberry (Molitch-Hou 2014). The capsules contain only natural ingredients, which can imitate existing fruits and foods, or combine them to create completely new tastes. Cohen et al. (2009) has successfully simulated a broad range of taste sensations using two types of hydrocolloids (xanthan gum and gelatine).

26.5.2 Formulation Innovation and Uplifted Nutrition Profile

Food recipes with varied formulation and ingredients can impart a variety of textural experiences. Fat based ingredients can provide lubricity or mouthcoating, starch based ingredients can provide viscosity, and protein based ingredients can provide gel and a range of textures and nutrition. Cookies loaded with large amounts of such ingredients enjoy by a wide range of consumers, especially kids.

Typically, a common shortbread cookie of 12 g composes of fat (¼, 3 g), sugar (1/3, or 4 g) and flours (~1/2) (Allrecipes n.d). With obesity and chronic diseases being the major health problems of modern society, there is a great need to alter such recipes and improve nutritional profiles. In digitalized food printing, the key is to upgrade the recipe of existing food products' nutritional profile by reducing or eliminating undesirable ingredients, or replacing them with healthy ingredients (i.e. fiber and micronutrients such as vitamins and minerals). Ingredients such as soluble fibre (from oats and barley) and plant proteins can be sourced to replace some portion of flours. Natural sweeteners can be introduced to replace sugar. Some vegetable oils rich in omega-3 fatty acids (e.g. Chia seed oil and flax seed oil) can be attempted to replace certain percentage of fats.

The above described reformulation creates a great challenge, and a compensation is needed by adding a combination of ingredients so as to maintain a desired rheological property suitable for printing, and build back the texture of an existing product. Hence, the printed food pieces' major nutritional profile will be enhanced by lowering glycemic load and glycemic index while increasing the contents of soluble fiber, protein, and omega-3 fatty acids. FoodJet 3D Printer has used both pureed ingredients and gelling agents to mimic dozens of classic meals on their taste, whilst creating soft, pureed texture that make it easy for elderly people to swallow (Kira 2015).

Furthermore, personalized minor ingredients can be added in the food printing such as vitamins, natural colorants, plant extracts with health promotion properties (anthocyanins, carotenoids, betanidins, plant polyphenolic compounds), and fat or protein based ingredients. These will give diverse colours, texture properties and nutritional values to the printed food pieces.

26.5.3 Digitalized Nutrition Control

The growing need for mass customisation and healthier options, as well as calls for more efficient use of food resources will drive the global market for 3D food printing. Biometric data about individual health condition and the corresponding complex nutritional recommendations will be available in the near future. Food printing is a promising way to bring the benefits of personalized data-driven health to people's kitchen tables by offering a healthy and personalized food choice. The EU Personalized Food for the Nutrition of Elderly Consumers (PERFORMANCE) project has applied food printing to fabricate customized soft foods with tailored nutritional contents for elderly with swallowing or chewing difficulties (Kira 2015). Each meal fabricated by FoodJet 3D Printer can deliver not only the specific nutrients and portions to patients on a weekly basis, but also the original dish in appearance and texture, allowing the elderly to enjoy gnocchi, peas, broccoli, carrots and so on. This endeavour enriches the 3D printed meal to ensure a well-balanced diet, which has achieved positive preliminary results in some EU care homes (Kira 2015).

The food printing mechanism has freedom to portion out materials to virtually fit any meal plan such as specific caloric and protein portions. Figure 26.11 displays how a food piece's nutrition may be personalized. The input of the food printer includes a 3D design file and a recipe. The volume of this design can be estimated using volume counting algorithms. With the information on density and volume, the material weight of the printed food pieces is obtained. Following this, the food piece's nutrition can be calculated using the weight, ingredients list and associated

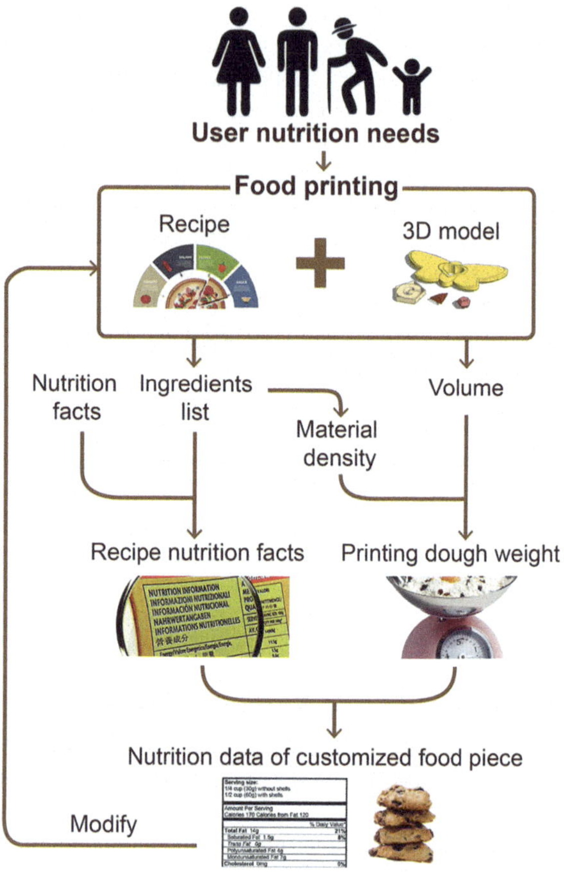

Fig. 26.11 Nutrition control in food piece design

nutritional details. Thus, the fabricated food pieces can be digitally controlled to adjust the design in terms of size and recipe. The distribution of ingredients and their impacts on the texture and construction, and even the accurate effect of cooking-related chemical reactions should also be modelled, so as to make foods specific to users' daily nutritional needs.

26.6 Potential Technologies, Challenges and Strategies along the Pathway of Commercialization

26.6.1 Potential Technologies

Besides the above-described 3D printing technologies, there is a need to bring in more established technologies to further enhance the printing process, such as electrospinning and microencapsulation. They have been embedded into bioprinter design for structural coating and microsphere fabrication (Xu et al. 2013; Yu et al. 2014). In food science, the applications of electrospinning and microencapsulation include extracting fibers and encapsulating nutrition, thus providing additional material sources for printing. The two technologies can also be directly integrated into the food printing process through multi-printhead platform, to control fibers and nutrition dispensing. This may be a potential way to fabricate on-demand food.

26.6.1.1 Electrospinning

Electrospinning is capable of producing thin, solid polymer strands ranging from 10 to 1000 nm in diameter. It can generate antimicrobial nanofibers from chitin (Kriegel et al. 2009) and biopolymer zein nanofibers to encapsulate beta-carotene (Fernandez et al. 2009) for bioactive food packaging.

Electrospinning can produce food materials with controlled size and structure, thus generate healthier foods (lower fat, lower salt) with desirable sensory properties and ingredients with improved properties (Neethirajan and Jayas 2011). It is also capable to shape non-traditional food materials under multi-scale into appealing edible structures.

An integration of electrospinning and food printing may offer a possible all-in-one solution to fabricate food products with personalized nutrition, i.e. extracting fibers out of materials, encapsulating nutrients, controlling their dispensing volume and constructing food structures with a controlled release of the nutrients. Gray (2010) proposed using electrospinning to produce multiple food sub-components at micro-scale and further assemble them into multi-component composite structures for a variety of materials. Micro-scale fibers can provide structure and texture to food products with a pleasant taste experience, such as muscle fibers in meat, cellulose fibers in vegetables, citrus fibers in low fat full taste mayonnaise. From technical perspective, the current challenge is to integrate and manipulate electrospinning process in food printing platform.

26.6.1.2 Microencapsulation

Simply adding ingredients to food products can improve nutritional value, but may compromise aroma, taste, color and texture. Also, the bioavailability of ingredients may suffer due to slow degradation, oxidation and reactions between ingredients and other food components. Microencapsulation can pack minerals, vitamins, flavors and essential oils within another material for the purpose of shielding active ingredients from the surrounding environment (Gutiérrez and Álvarez 2017; Gutiérrez 2018). One of the microencapsulation approaches, electrohydrodynamic atomization has been incorporated into bioprinter design to generate double-walled microspheres for bioactive drug delivery system (Xu et al. 2013). Integrating such technology into food printing can be achieved by using multi-printhead system, where at least one printhead generates and dispenses microcapsules in the fabricated food products. This would help fragile and sensitive materials survive in processing and packaging conditions, and stabilize the shelf life of active ingredients, and create appealing aroma release, and taste, odor and color masking. In other words, microcapsules containing flavor or nutritional elements would remain dormant in the food and will only be released when triggered by consumers (Dunn 2004). This method simplifies the current functional food manufacturing process, enhances functional ingredient stability (e.g. probiotics and bioactive ingredients) and realizes controlled release of flavorings and nutrients.

26.6.1.3 Digital Cooking

The current post-deposition cooking methods do not match with the digital food fabrication process. For example, we can print multilayer heterogeneous food pieces, but cook them using conventional heating methods. Thus, a digital cooking method is definitely required that can precisely and selectively control the food piece heating process so as to trigger chemical reactions. The evolution of cooked ingredients can change the formation of taste, aroma and color within the food pieces. This can subjectively tune flavors for personal preferences with desirable taste in a repeatable way. The reaction products and their spatial distribution in the food pieces can be achieved through controlling the intensity and speed of the digital cooking.

As the ingredients are organized in a digitally determined structure, they can precisely control the printed food's taste, flavor, and texture as part of the eating experience. However, the influence of post-deposition cooking on the essential constituents of food (carbohydrates, proteins and fat) either singly or in combination is not clear. Further studies on how to apply digital cooking to uplift the final products, to support health and to design novel textures are needed. This may lead to a more advanced concept, which can help develop functional foods and design rational food pieces to reduce diet-related diseases.

26.6.2 Challenges

More and more companies and research institutes are working on improving food printing technologies, so as to commercialize new digital cooking devices in kitchens, and promote innovative design and healthy life style. While, there are many challenges along the way of commercialization.

26.6.2.1 Gap between Consumer Needs and Current Functionalities

The two largest 3D printing companies invested on food printing commercialization. Stratasys filed patents towards developing a commercial chocolate printer, and 3D Systems released ChefJet series using the Z-Corp inkjet process to produce chocolate treats from powdered sugar and cocoa. Several startup companies including Chocedge and Foodini have pursued to commercialize either 3D printed foods or food printers. While, most of these companies or projects are not growing fast as people expected before. The low fabrication efficiency and lack of healthy food concepts place their efforts in an unfavorable circumstance.

Food printers may potentially be used at home, small scale food production at shops, restaurants, bakeries, and industrial scale food production. The purchase consideration for home users includes selling price, maintenance cost and time, functionalities and convenience. The current machines have limit functions, but require preparing additional food materials or purchasing pre-made food capsules, which actually increases time, cost and steps to make daily meal.

Food printers have been used in restaurants or high-end kitchens as chef assistant, to perform repetitive tasks, such as creating uniquely-shaped desserts or appetizers, and to help prepare meals and speed up the preparation process. Small bakeries may also benefit from these machines to produce decorating food items partially and fully in an automatic way. In this case, the printing efficiency and capabilities to produce geometric complex 3D shapes for artistic presentation and diverse textures become critical under a simple maintenance and adjustment procedure. Till now, no machines designed for industrial scale have officially been released. Such a printer is expected to have a large printhead array to produce hundreds of food pieces at the same time with low cost and high reliability.

Another challenge on the pathway of commercialization is the long fabrication time which includes printing time, material pre-process and product post-process time. Some food materials need longer waiting time to cool down for successive layer printing. With the target of approaching market service, the new generation printers should not only save time on styling, but also cut down the time taken during the printing process. For example, the commercial pizza printer BeeHex, launched in 2016, has an average fabrication time of 4 min, while it would take a human 9 min to perform the same task (Chew 2016).

Besides, the cost of 3D printed food products is the key factor to influence purchase. It includes expenses associated with printing platforms (hardware and software) and printing materials, labor cost, operation cost, and general overhead for maintaining the production facility. The current price of commercial food printing platforms is at least a few thousand dollars (3D Systems 2013; Choc Edge 2014), which is too high as a consumer product in terms of fabrication capabilities. In addition, consumers need to purchase printing materials from the platform companies which are more expensive than that of the similar materials in the market.

26.6.2.2 Issues from Food Material Property and Processing Technologies

For digitized recipes used in food printing, an accurate selection of materials with appropriate physical, chemical, rheological and mechanical properties are essential. The food printing process can be affected by specific gelation mechanisms (cross-linking) and thermal properties (melting point and glass transition temperature), material particle size distribution, bulk density, wettability and flowability. The food material properties vary among batches and suppliers. Besides, the rheology of the same dough material may vary when adding edible dyes. Food materials prepared for printing have an extremely limited shelf life, and their rheology may tend to shift with time. However, the current studies have not addressed these issues.

Even theoretical and experimental models can be established to explore the relationship among the food printing process, material composition, and printhead design, it is not cost effective to calibrate the diverse material choices. This issue becomes another bottleneck to develop food printing materials and recipes.

Either as a household device or industrial food processing machine, a future food printer should be able to replace current multi-step food processing technologies. While, conventional food processing technologies are unlikely to fit into a food printing process directly. Reformulating food manufacturing processes is essential, such as pre-conducting some processes (e.g. gluten formation and leavening) and replacing remaining processes (e.g. shaping and baking). Conceptual designs focusing on mixing, modelling and transformation processes may create new food processing technologies (Zoran and Coelho 2011). Besides, food materials' mechanical and thermal properties, biological variation, microbiological and biochemical properties place additional limitations on handling and processing. Hence, process reformulation and material processing will shape future printer design.

Ingredient formulations with varied combinations and manipulation conditions can generate various textures in products, which may go beyond a manageable level. A simulation model should be built to digitalize this process, including data quantification for each process (e.g. ingredient metering and mixing, printing, baking and so on), communication protocols between different functions or processes, and process predictions. Embedding such a model into food printer design can digitally visualize the printing process and increase consistency in the printed products. However, these topics have not been explored yet.

26.6.3 Strategies on User Interface Design

Various user interfaces have been developed consisting of 3D model design, material selection, path planning, processing parameters selection, and template library. Efforts have also been spent to develop an open-access web-based template library for food design innovation (Lipton et al. 2010; Malone and Lipson 2007). With such user interface, customers should be able to design their own food pieces, obtain design files online or share their designs on websites for others to download and modify.

The current consumer user interface for food printing is pretty straightforward. It takes input from Computer Aided Design (CAD), and relies on a CAD model to plan and construct food pieces. While most consumers are not capable of creating designs using professional CAD software. To prompt the usage of 3D Systems' ChefJet series, Digital Cookbook software has been designed for users who are not familiar with 3D modeling software (Quick 2014).

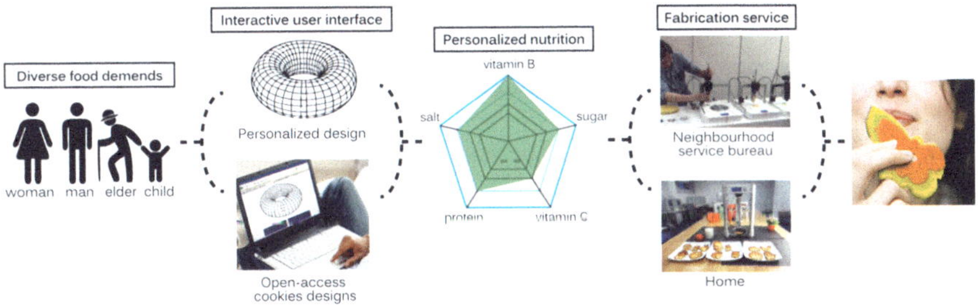

Fig. 26.12 Schematic diagram of food design and fabrication service

As shown in Fig. 26.12, this interface is expected to (1) transform designs into a digital 3D model with printing pathway planning and processing associated parameters, and (2) obtain or share design files online through a technology service provider. This interface should also be able to demonstrate the aesthetic style of the food piece, and assist fine-tuning the quantity of final products and distribution of ingredients prior to constructing. A growing number of makers, academics and startups have been playing with the idea of interactive user interface design for food printers. In the Burritobot printer (2014), consumers can play with sliding scales embedded in the app to calibrate the exact amount of each ingredient on the burrito via iOS app or a Ruby-based web app. In the near future, web/iOS interfaces will move to the next level by letting consumers create their exact meal via proxy.

Besides engaging consumers in food design and diet control, the user interface should support the value chain as shown in Fig. 26.12. The target is to facilitate ordinary users to transmit their designs to the service bureau, handle the ingredients sourcing from suppliers for printing preparation, and eventually arrange delivery services. This may radically alter food production practices by enabling companies to manage resources more responsibly and reduce waste across the food continuum, including food processors, distributors and consumers.

26.7 Conclusion

3D food printing has demonstrated its capability of making personalized chocolates or producing simple homogenous snacks. Currently, these applications are still primitive with limited internal structures and monotonous textures. It is necessary to develop a systematic way to investigate recipes, platform design, printing technologies and their influences on food fabrication. Food printing technology can manipulate food forms and materials in domestic cooking or catering service, which will allow efficient delivery of high quality, freshly prepared food items to consumers. It can also deliver personalized nutrition, new flavors, textures and shapes of food products.

With the development of an open web-based media interface, food printers may form an ecology of networked machines that can order new ingredients, prepare favorite food on demand and even collaborate with doctors to develop healthier diets.

Acknowledgement This research is partially supported by the National Research Foundation, Prime Minister's Office, Singapore under its International Research Centres in Singapore Funding Initiative.

References

3D Systems (2013) 3D systems acquires the Sugar Lab. http://www.3dsystems.com/de/press-releases/3d-systems-acquires-sugar-lab. Accessed 22 Dec 2014

3D Systems (2015) 3D Systems previews new chocolate 3d printer CocoJetTM at 2015 International CES. https://www.3dsystems.com/sites/default/files/2016/01062015_3d_systems_previews_new_chocolate_3d_printer_cocojettm_at_2015_international_ces.pdf. Accessed July 2017

Alcalde LR (2016) Walkthrough Pinya3, 3D food printing platform. 3Digital Cooks. http://3digitalcooks.com/2016/03/walkthrough-pinya3/. Accessed July 2017

Alec (2015a) Pasta maker Barilla to show off its 3D pasta printer at the Milan EXPO 2015. 3ders.org. http://www.3ders.org/articles/20150505-pasta-maker-barilla-to-show-off-its-3d-pasta-printer-at-the-milan-expo-2015.html. Accessed Dec 2016

Alec (2015b) Soon you'll be able to 3D print chewing gum with GumJet 3D printer. 3ders.org. http://www.3ders.org/articles/20150218-two-london-students-develop-3d-printed-chewing-gum.html. Accessed July 2017

Allrecipes (n.d.) Shortbread cookies II. http://allrecipes.com/recipe/10269/shortbread-cookies-ii/. Accessed Dec 2016

Barilla (2015) Pasta maker Barilla to show off its 3D pasta printer at the Milan EXPO 2015. 3ders.org. http://www.3ders.org/articles/20150505-pasta-maker-barilla-to-show-off-its-3d-pasta-printer-at-the-milan-expo-2015.html. Accessed Dec 2016

BeeHex (2016) BeeHex robots 3D-print pizza. http://beehex.com/. Accessed Aug 2016

Beetz M, Klank U, Kresse I (2011) Robotic roommates making pancakes. In: Proceedings of 11th IEEE-RAS International Conference on Humanoid Robots, Bled, Slovenia, 26–28 Oct 2011, p 529–536

Bollini M, Barry J, Rus D (2011) Bakebot: Baking cookies with the PR2. In: The PR2 Workshop: Challenges and Lessons Learned in Advancing Robots with a Common Platform, IEEE/RSJ International Conference on Intelligent Robots and Systems, San Francisco, 25–30 Sept 2011

Bosker B (2013) 3D printers could actually make donuts healthy. http://www.huffingtonpost.com/2013/04/24/3d-printed-food_n_3148598.html. Accessed 24 Apr 2013

Burritobot (2014) A 3-D printer that spits out burritos. https://www.pinterest.com/pin/115967759125319211/. Accessed Dec 2014

CandyFab (2007) The CandyFab Project. http://wiki.candyfab.org/Main_Page. Accessed Dec 2014

Causer C (2009) They've got a golden ticket. Potentials IEEE 28(4):42–44

Chloé Rutzerveld (2014) Edible growth. Chloé Rutzerveld Food & Concept Design. http://www.chloerutzerveld.com/edible-growth-2014/. Accessed 22 July 2017

Chew C (2016) Did Beehex just hit 'print' to make pizza at home? Huffpost Tech. http://www.huffingtonpost.co.uk/cohan-chew/did-beehex-just-hit-print_b_10108424.html. Accessed Sept 2016

Choc Edge (2014) Choc Creator. http://chocedge.com/. Accessed Dec 2014

Cohen DL, Jeffrey IL, Cutler M et al (2009) Hydrocolloid printing: A novel platform for customized food production. In: Proceedings of Solid Freeform Fabrication Symposium (SFF'09), Austin, TX, 3–5 Aug 2009

Creative Machines Lab (2016) Sanna: the food printer of 2020. Columbia University, School of Engineering. https://www.youtube.com/watch?v=yPS6FbNuY_I. Accessed July 2017
Crump SS (1991) Fast, precise, safe prototypes with FDM. In: Proceedings of ASME Annual Winter Conference, Atlanta, Dec 1991
de Roos B (2013) Personalised nutrition: ready for practice? Proc Nutr Soc 72(01):48–52
Deloitte (2015) The 2015 American pantry study: The call to re-connect with consumers. http://www2.deloitte.com/content/dam/Deloitte/us/Documents/consumer-business/us-cb-2015-american-pantry-study.pdf. Accessed Aug 2016
Dovetailed (2014) https://www.dovetailed.co/. Accessed Dec 2014
Dunn J (2004) A mini revolution-food manufacture. http://www.food.manufacture.co.uk/news/fullstory.php/aid/472/A%20mini%20revolution.htm. Accessed 23 Apr 2009
Electrolux (2009) Interview with Nico Kläber (Moléculaire) Electrolux design lab finalist. http://group.electrolux.com/en/interview-with-nico-klaber-moleculaire-electrolux-design-lab-finalist-2040/. Accessed Dec 2014
Fabaroni (2007) Fabaroni: A homemade 3D printer. http://fab.cba.mit.edu/classes/MIT/863.07/11.05/fabaroni/. Accessed Dec 2014
Fernandez A, Torres-Giner S, Lagaron JM (2009) Novel route to stabilization of bioactive antioxidants by encapsulation in electrospun fibers of zein prolamine. Food Hydrocoll 23(5):1427–1432. https://doi.org/10.1007/s11483-009-9119-6
Foodjet (2012) Foodjet. http://foodjet.nl/. Accessed Dec 2014
Garfield L (2016) This robot can 3D-print a pizza in under five minutes. Teck insider. http://www.techinsider.io/how-the-beehex-pizza-3d-printer-works-2016-6. Accessed Sept 2016
Golding M, Archer R, Gupta G et al (2011) Design and development of a 3-D food printer. In: Proceedings of NZIFST 2011 Conference, Rotorua, New Zealand, July 2011, p 10–12
Gorkin R, Dodds S (2013) The ultimate iron chef—when 3D printers invade the kitchen. The Conversation, 15 Oct 2013
Gray N (2010) Looking to the future: Creating novel foods using 3D printing. http://www.foodnavigator.com/Science-Nutrition/Looking-to-the-future-Creating-novel-foods-using-3D-printing. Accessed 23 Dec 2010
Grunewald SJ (2015) Multi-material, swappable extruder, portable 3d printer: The Focus launches on Kickstarter. https://3dprint.com/67155/focus-3d-printer-kickstarter/. Accessed July 2017
Gunstone FD, Padley FB (1997) Lipid technologies and applications. CRC press
Gutiérrez TJ (2018) Processing nano- and microcapsules for industrial applications. In: Hussain CM (ed) Handbook of nanomaterials for industrial applications. Elsevier, 989-1011. https://doi.org/10.1016/B978-0-12-813351-4.00057-2. ISBN: 978-0-12-813351-4
Gutiérrez TJ, Álvarez K (2017) Biopolymers as microencapsulation materials in the food industry. In: Masuelli M, Renard D (eds) Advances in physicochemical properties of biopolymers: part 2. Bentham Science Publishers, pp 296–322. https://doi.org/10.2174/978168108544911701000 9. ISBN: 978-1-68108-545-6. eISBN: 978-1-68108-544-9
Hadhazy A (2013) Will 3D printers manufacture your meals? Popular mechanics. http://www.popularmechanics.com/technology/gadgets/a8816/will-3d-printers-manufacture-yourmeals-15265101/. Accessed 27 Sept 2016
Hao L, Mellor S, Seaman O et al (2010) Material characterisation and process development for chocolate additive layer manufacturing. Virtual Physical Prototyping 5:57–64
Huang SH, Liu P, Mokasdar A (2013) Additive manufacturing and its societal impact: a literature review. Int J Adv Manuf Technol 67(5–8):1191–1203
Kira (2015) EU develops PERFORMANCE 3D printed food for elderly and patients with dysphagia. 3der.org. http://www.3ders.org/articles/20151026-eu-develops-performance-3dprinted-food-for-elderly-and-patients-with-dysphagia.html. Accessed Sept 2016
Krassenstein B (2014) Tytan 3D to launch a multi-material 3d printer which prints in salt, paper, food, adhesives & more. 3DPrint.com. https://3dprint.com/6893/tytan-3d-printer/. Accessed July 2017

Kriegel C, Kit KM, Mcclements DJ, Weiss J (2009) Influence of surfactant type and concentration on electrospinning of chitosan-poly (ethylene oxide) blend nanofibers. Food Biophys 4(3):213–228

Labau E (2014) Sugar cubes. http://candy.about.com/od/sugarcandy/r/Sugar-Cubes.htm. Accessed 22 Dec 2014

Lipton J, Cohen D, Heinz M, Lobovsky M (2009) Fab@Home model 2: towards ubiquitous personal fabrication devices. In: Solid Freeform Fabrication Symposium (SFF'09), Austin, TX, 3–5 Aug 2009, p 70–81

Lipton J, Arnold D, Nigl F (2010) Multi-material food printing with complex internal structure suitable for conventional post-processing. In: Proceedings of Solid Freeform Fabrication Symposium, Austin TX, Aug 2010, p 809–815

Malone E, Lipson H (2007) Fab@home: the personal desktop fabricator kit. Rapid Prototyp J 13(4):245–255

Michiel Cornelissen Ontwerp (n.d.) XOCO chocolate printer. http://www.michielcornelissen.com/portfolio_page/xoco-chocolate-printer/. Accessed July 2017

Millen CI (2012) The development of colour 3D food printing system. Master thesis. Massey University, New Zealand

Millsaps BB (2015) Bocusini to launch complete system for 3d printing food with pre-filled cartridges on kickstarter tomorrow. 3DPrint.com. https://3dprint.com/64431/embargoed-until-tuesmay-12-at-11-am-kickstarter-bocusini-3d-printing-food-cartridges/. Accessed Sept 2016

Molitch-Hou M (2014) The 3D fruit printer and the raspberry that tasted like a strawberry. 3D printing industry. https://3dprintingindustry.com/news/3d-fruit-printer-raspberry-tasted-likestrawberry-27713/. Accessed Sept 2016

NASA (2013) 3D Printing: Food in space. https://www.nasa.gov/directorates/spacetech/home/feature_3d_food.html. Accessed 22 July 2017

Natural Machines (2014) Foodini. http://www.naturalmachines.com/press-kit/. Accessed Dec 2014

Neethirajan S, Jayas DS (2011) Nanotechnology for the food and bioprocessing industries. Food Bioprocess Technol 4(1):39–47

Nikitina VE, Tsivileva OM, Pankratov AN (2007) Lentinula edodes biotechnology—from lentinan to lectins. Food Bioprocess Technol 45(3):230–237

Nowak J (2015) Robotics in the food and beverage industry. Bastian Solutions. https://www.bastiansolutions.com/blog/index.php/2015/05/28/robotics-in-the-food-and-beverage-industry/#.WYaZV4SGNFE. Accessed July 2017

Periard D, Schaal N, Schaal M (2007) Printing food. In: Proceedings of the 18th Solid Freeform Fabrication Symposium, Austin TX, p 564–574

Philips Design (2010) New design probe explores the future of food. www.design.philips.com/philips/sites/philipsdesign/about/design/designnews/pressreleases/foodprobes.page. Accessed 22 Mar 2010

Quick D (2014) 3D Systems cooks up ChefJet 3D printers to print sugary treats. http://newatlas.com/3d-systems-chefjet-sugar-candy-printer/30407/. Accessed July 2017

Sachs E, Cima M, Williams P et al (1992) Three-dimensional printing: rapid tooling and prototypes directly from a CAD model. J Manuf Sci Eng 114(4):481–488

Serizawa R, Shitara M, Gong J et al (2014) 3D jet printer of edible gels for food creation. In: Proceedings of SPIE Smart Structures and Materials+ Nondestructive Evaluation and Health Monitoring, San Diego, 9–13 Mar 2014

Silva ÉS, Cavallazzi JRP, Muller G (2007) Biotechnological applications of Lentinus edodes. J Food Agric Environ 5(3&4):403–407

Sun J, Zhou W, Huang D, Fuh JY, Hong GS (2015) An overview of 3D printing technologies for food fabrication. Food Bioprocess Technol 8(8):1605–1615. https://doi.org/10.1007/s11947-015-1528-6

Sun J, Zhou W, Yan L, Huang D, Lin LY (2018) Extrusion-based food printing for digitalized food design and nutrition control. J Food Eng 220:1–11

Southerland D, Walters P, Huson D (2011) Edible 3D printing. In: Proceeding of NIP & Digital Fabrication Conference, Society for Imaging Science and Technology, p 819–822

Stewart-Knox B, Mitchell P (2003) What separates the winners from the losers in new food product development? Trends Food Sci Technol 14(1):58–64

Torrone P (2007) Solid Freeform Fabrication: DIY, on the cheap, and made of pure sugar. http://www.evilmadscientist.com/2007/solid-freeform-fabrication-diy-on-the-cheap-and-made-of-pure-sugar/. Accessed 22 Dec 2014

van Bommel K, Spicer A (2011) Hail the snail: hegemonic struggles in the slow food movement. Organ Stud 32(12):1717–1744

van der Linden D (2015) 3D food printing. TNO. https://www.tno.nl/media/5517/3d_food_printing_march_2015.pdf. Accessed Sept 2016

Watzke H, German J (2010) Personalizing foods. In: Moskovitz H, Saguy I, Strauss T (eds) An integrated approach to new food product development. CRC Press, Boca Raton, pp 33–173

Wegrzyn TF, Golding M, Archer RH (2012) Food layered manufacture: a new process for constructing solid foods. Trends Food Sci Technol 27(2):66–72

Winger R, Wall G (2006) Food product innovation: a background paper. Food And Agric Organ UN, Rome

Xu Q, Qin H, Yin Z (2013) Coaxial electrohydrodynamic atomization process for production of polymeric composite microspheres. Chem Eng Sci 104:330–346

Yang J, Wu L, Liu J (2001) Rapid prototyping and fabrication method for 3-D food objects. US Patent 6280785, 28 August 2001

Yu YZ, Zheng LL, Chen HP (2014) Fabrication of hierarchical polycaprolactone/gel scaffolds via combined 3D bioprinting and electrospinning for tissue engineering. Adv Manuf 2(3):231–238

Zoran A, Coelho M (2011) Cornucopia: the concept of digital gastronomy. Leonardo 44(5):425–431

Chapter 27
Sensors Based on Conducting Polymers for the Analysis of Food Products

Constantin Apetrei, Mateus D. Maximino, Cibely S. Martin, and Priscila Alessio

Abstract Quality and safety have become important food attributes within the last years. With the expansion of food industry and food fabrication on a big scale is not infrequent to found adulterant and contaminant on food, i.e. of harmful chemicals and microorganisms, which can cause consumer illness. Increasingly, the public authorities are charging the food industry to develop quality management systems and restructure the food inspection system to improve food quality and safety. However, contamination can be taking place during all steps of food manufacturing. For instance, even during storage, the products can experience a significant degradation in quality due to a variety of physical, chemical, and biological interactions. Consequently, fast and inexpensive ways to control the food freshness, contamination or adulteration are requested to keep the quality control. Thus, there is an increasing demand for analytical methods easily operate with highly sensitive, selective and accurate for determining compounds in foods in general. We present here a review about conducting polymers and their application on the development of electrical and electrochemical sensors for detection of contamination or adulteration in food samples.

Keywords Biosensor · Conductive polymers · Electrochemical sensor · Electronic nose · Food samples

C. Apetrei
Department of Chemistry, Physics, and Environment, Faculty of Sciences and Environment, "Dunarea de Jos" University of Galati, Galati, Romania

M. D. Maximino · C. S. Martin · P. Alessio (✉)
School of Technology and Applied Sciences, São Paulo State University (UNESP), Presidente Prudente, SP, Brazil
e-mail: priscila.alessio@unesp.br

T. J. Gutiérrez (ed.), *Polymers for Food Applications*,
https://doi.org/10.1007/978-3-319-94625-2_27

27.1 Introduction

27.1.1 Conducting Polymers: A General Approach

Organic polymers are generally electrical insulators, but the conductive polymers behave in the exact opposite way, being able to displace a flow of electrons or ions in an ordered way to the length of its structure resulting in electronic or ionic conductivity. Conducting polymers are a class of materials relatively new but also of such importance that culminated in the Nobel Prize of Chemistry of the year 2000 for Alan J. Heeger, Alan MacDiarmid, and Hideki Shirakawa by the discovery and the development of electrically conductive polymers. The first report about conducting polymer was published in 1977 (Shirakawa et al. 1977a). They described "*When silvery films of the semiconducting polymer, trans 'polyacetylene', are exposed to chlorine, bromine, or iodine vapour, uptake of halogen occurs, and the conductivity increases markedly (over seven orders of magnitude in the case of iodine) to give, depending on the extent of halogenation, silvery or silvery-black films, some of which have a remarkably high conductivity at room temperature*". Thenceforth, a great deal of work has been done studying the synthesis and development of new conducting polymers, its characteristics, and applications in several areas. Nowadays, the main conducting polymers are the polyaniline, polypyrrole, PEDOT, and polythiophene. Other polymers play a minor role as compared with the polymers mentioned above are poly(*o*-phenylenediamine), polycarbazole, polyindole, polyfurane, and polyphenazine (Inzelt 2017). In addition, azines derivatives (dyes), like phenothiazines, phenoxazines and phenazines (Malinauskas 1999; Thiemann and Brett 2001; Schlereth and Karyakin 1995), have also been used as conducting polymers due to charge transfer properties.

Since the very beginning of this research field, a major point was the quest for useful applications of the conducting polymers. The advantages of using conducting polymers are based on the inexpensive, easy to synthesize and versatility due to their properties can be readily modulated by surface functionalization techniques and the use of a wide range of molecules that can be entrapped or used as dopants. Nowadays a wide range of application in biological areas as drug delivery systems, bioactuators, and tissue engineering has increased (Ravichandran et al. 2010). Besides, the research and development of high-performance energy devices based on conducting polymers have grown with the high demand for electricity, once one cannot imagine the everyday life without electricity which includes the electrochemical power sources (Inzelt 2017). As important as the applications cited above, the rapid growth of the human population claims for rapid techniques to the disease diagnosis and food quality assessment, which are essential role developed by sensors and biosensors (Kausar 2017).

In this chapter, we review and discuss the synthesis and the main techniques to the fabrication of electrochemical sensors and biosensors based on conducting polymers for the analysis of food products.

27.1.2 Classification of Conducting Polymers

It may be considered that there are two classes of polymeric materials that can conduct electricity:

- extrinsically conducting polymers
- intrinsically conducting polymers

As described above, the extrinsically conducting polymers are conventional insulating polymeric materials, whose conductivity increases by several orders by incorporating certain conductive particles in their mass (metallic powders or conductive fibers, carbon black, graphene, etc.). Conductive composite materials have recently been obtained, in which an intrinsically conducting polymer is dispersed over an insulating matrix (Dai 2004).

On the other hand, intrinsically conducting polymers can conduct electricity, a property that is inherent to the chemical structure of the material. The most commonly encountered polymers belonging to this category are polyacetylene, polypyrrole, polythiophene, and polyaniline, which have sp^2 hybridization carbons on the main chain. This hybridization creates conjugated double and simple covalent bonds between the component atoms of the polymer's main chain (Thanh-Hai et al. 2017).

This class of conducting polymers comprises the following types:

- redox-center polymers;
- electroactive ion-counter ion changing polymers;
- polymers with coordination centers for metals with redox properties;
- conjugated chain polymers.

In the first three categories of polymers, electricity is conducted by combining reactions of exchanging electrons between neighboring atoms that are oxidized or reduced and the side redox groups, by mechanisms of diffusive type. These mechanisms of electronic conductivity are due to the concentration gradients between the oxidized and reduced atoms as well as the electric potential gradients.

The fourth group contains the conjugated chain polymers, with alternate double and single bonds, which have been called organic conducting polymers or more simply conducting polymers, among which the previously mentioned polyacetylene, polypyrrole, polythiophene, polyaniline and their derivatives. In these materials, load transport is due to the delocalized defects on the main chain, generated by doping the polymer backbone (Taylor et al. 2012).

A classification of conducting polymers may also be performed according to their chemical composition. The conductive properties may be achieved in various manners, by conjugating the unsaturated carbon atom chains, the carbons, and the heteroatoms, or even the heteroatom chains (Bunting et al. 1997). Figure 27.1 shows a classification of the various types of conducting polymers.

The most common conducting polymers are the polyvinylenes, polyarylenes, and polyheterocycles. Polyvinylenes have good thermal stability and electric conductivity. Poly (p-phenylenes), polyazulenes and poly-(phenylene vinylenes) belong

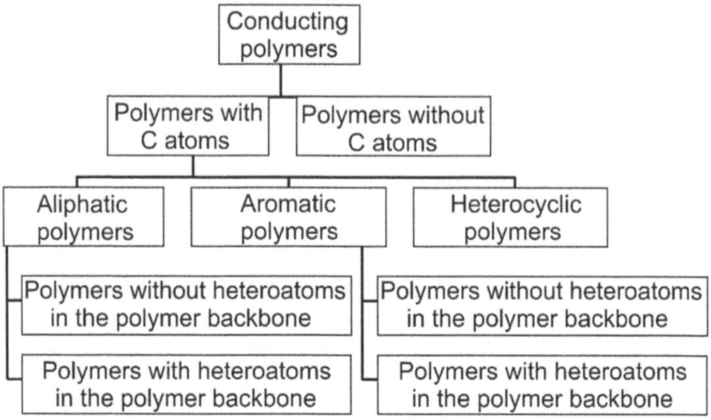

Fig. 27.1 General classification of conducting polymers

to the family of polyaromatic polymers. Poly(p-phenylene) was the first non-acetylenic polymer showing high conductivity in a doped state. Many polyheterocyclic polymers may be added on the list of organic conducting polymers, such as polypyrrole, polythiophene and polyfuran, all with a five-atom cycle, out of which one is a heteroatom, such as nitrogen, sulfur or oxygen (Khan et al. 2018)

Polypyrrole and its derivatives have received a lot of attention due to its relatively easy synthesis, good chemical and thermal stability, and high conductivity. Phene and its derivatives have a high chemical and electrochemical stability in both doped and undoped state. Polyaniline is also an organic polymer with good stability and excellent conductive properties, which may be controlled by doping it with hydrogen ions. The alteration of the physical and electrical properties was also achieved by the synthesis of numerous copolymers of these monomers (Sahmetlioglu et al. 2009).

27.1.3 Synthesis Methods of Conducting Polymers

The most common synthesis methods of conducting polymers are:

Directly catalyzed synthesis: the method was developed by Shirakawa in 1977. The inner wall of a glass recipient is coated with a Ziegler-Natta catalyst (derivatives of aluminum alkyl and titan halogenides), and then an acetylene flux is passed over it, leading to the formation of a film of bright silvery polyacetylene, due to a catalyst excess (Shirakawa et al. 1977b).

The chemical oxidation of the monomer: an oxidant is added in a monomer solution, whose potential corresponds to the monomer oxidation potential, such as Fe^{3+} in pyrrole solutions, resulting in the formation of a black polypyrrole precipitate, or ammonium persulphate in acid aniline solution, yielding polyaniline (Upadhyay and Ahmad 2010).

Electrochemical oxidation: it is a method similar to chemical oxidation, by a heterogeneous process, taking place on the surface of an electrode (Labaye et al. 2002).

Plasma oxidation: in this case, generating plasma initiates polymerization on the modifying surface (Berndt et al. 2017).

Out of precursor polymers: a precursor polymer is used, which is usually a soluble one, applied to the desired surface. It breaks down by heating, resulting in a gaseous molecule and an insoluble conducting polymer (Uemura et al. 2013).

There are also other less used methods, such as photoinitiated polymerization, condensation polymerizations, synthesis in the inverse emulsion, or oxidizing the monomer with a cathodically generated intermediary during the reduction of the dissolved O_2 (Wan 2009).

27.1.4 Doping of Conducting Polymers

The doping phenomenon is the only and most important characteristic that distinguishes conducting polymers from other types of polymers. During the doping process, an insulating or semiconductive organic polymer with low conductivity, usually in the range of 10^{-10}–10^{-5} S × cm^{-1}, turns into a polymer which is in a metallic conduction regime (approx. 1–10^4 S × cm^{-1}). The strictly controlled addition of nonstoichiometric amounts of chemical substances leads to major changes in the polymer's electric, magnetic, optical and structural properties. Doping is reversible, as the initial polymer can be recovered without or with minimal degrading of the polymer's backbone. Both the doping and the undoping process presuppose the use of counter ions stabilizing the doped stage of the polymer, and the processes may be performed chemically or electrochemically. There are also transitory doping methods which do not presuppose the introduction of doping ions (Macdiarmid and Epstein 1995).

By controlling the doping level, one may easily obtain polymer conductivity, between the undoped form (insulating or semiconducting polymer) and the completely doped form (highly conducting polymer). Various types of conductive wires and plates may be manufactured by mixing a conductive (doped) polymer with a conventional (insulating) polymer, and their conductivity may be adjusted by varying the relative proportions of each polymer in the mixture (Skotheim et al. 1995). The conductivity reached by a conducting polymer has raised an infinity of times when the superconductivity of regioregular poly-3-hexylthiophene was found, although this phenomenon was obtained only in a very thin polymer layer at very low temperatures (about 2K) (Bard and Rubenstein 1996).

Before discovering the protonic acid doping of polyaniline, during which the number of electrons associated to the polymer's backbone remains unchanged, all conducting polymers were doped by redox methods. The latter presupposes adding or removing electrons in the π system of the main backbone by processes of oxidation or reduction (Sapurina and Shishov 2012).

Fig. 27.2 Iodine doping of *trans*-polyacetylene

$$trans\text{-}(CH)_x + 1{,}5xy\ I_2 \longrightarrow \left[CH^{y+}(I_3)_y^-\right]_x$$

Ever since polyacetylene was discovered in 1977, which may be p or n doped by chemical or electrochemical methods up to the metallic state, the development of conducting polymer field has evolved at a fast pace, leading to the discovery of a wide range of conducting polymers and their derivatives.

In the doped state, the backbone of a conducting polymer is a delocalized π system. In the undoped state, the polymer may have a conjugated backbone, maintained in a modified form after doping, or an unconjugated backbone. An example is polyaniline in the basic form leucoesmeraldine, which only becomes conjugated after p-type doping, or an unconjugated structure as in the basic emeraldine form of polyaniline, which only becomes conjugated after proton doping.

27.2 Redox Doping

All conducting polymers may undergo transformations after redox doping of the p and n-type, by chemical and/or electrochemical processes during which the number of electrons associated to the polymer's backbone is changed (Shirakawa et al. 1977b).

27.2.1 The p-Type Chemical and Electrochemical Doping

The p-type doping, which is partial oxidation of the backbone with conjugated π bonds of an organic polymer, was first discovered by treating *trans*-polyacetylene (*trans*-$(CH)_x$) with an oxidizing agent such as iodine. The chemical process taking place, in this case, is shown in Fig. 27.2.
This process leads to increasing conductivity from about 10^{-5} S × cm^{-1} to about 10^3 S × cm^{-1} if the polymer, before doping, has a strictly unidirectional orderly structure. Under optimum circumstances, the conductivities that may be obtained in parallel to the ordering direction are higher than 10^5 S × cm^{-1}.

About 85% of the positive loads are delocalized more than 15 CH units, with the formation of a positive soliton. The delocalized structure of a positive soliton may be seen in Fig. 27.3.

Similarly, the p-type doping may also be performed by anodic electrochemical oxidation. The processes is based on the immersing of a *trans*-$(CH)_x$ film in an electrolyte solution, dissolved in an organic solvent, and attaching it to the positive pole of a continuous current source, the negative pole being attached to an inert electrode immersed in the same solution. The process taking place in the case of using $LiClO_4$ as electrolyte is seen in Fig. 27.4.

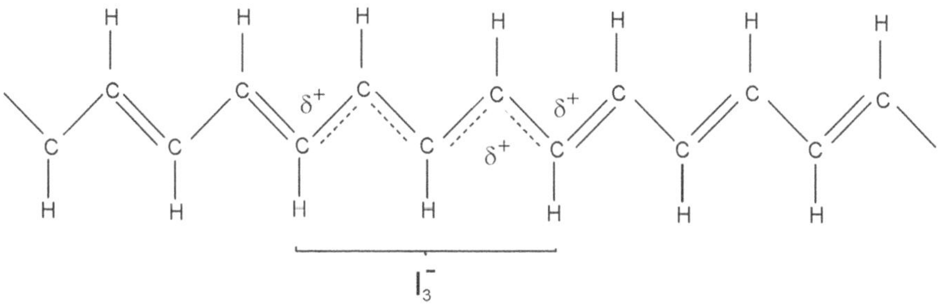

Fig. 27.3 Delocalized structure of a positive soliton

$$trans\text{-}(CH)_x + (xy)(ClO_4)^- \longrightarrow \left[CH^{y+}(ClO_4)_y^-\right]_x + (xy)e^-$$

Fig. 27.4 The electrochemical doping process of trans-polyacetylene with $LiClO_4$

$$trans\text{-}(CH)_x + (xy)Na^+(Naft)^- \longrightarrow \left[(Na^+)_y(CH)^{y-}\right]_x + Naft$$

Fig. 27.5 The n-type doping process of *trans*-polyacetylene with sodium naphthalene

$$trans\text{-}(CH)_x + (xy)Li^+ + (xy)e^- \longrightarrow \left[(Li^+)_y(CH)^{y-}\right]_x$$

Fig. 27.6 The n-type electrolytic doping process of *trans*-polyacetylene with $LiClO_4$ dissolved in tetrahydrofuran

27.2.2 The n-Type Chemical and Electrochemical Doping

The n-type doping, i.e. the partial reduction of the conjugated backbone of an organic polymer, was also discovered by treating *trans*-$(CH)_x$ with a reducing agent, like for instance sodium naphthalene. The process that takes place is seen in Fig. 27.5.

Through this process the π anti-bond system is partially populated with electrons, resulting in a conductivity increase up to approx. 10^3 S × cm^{-1}.

The n-type doping may also be performed by cathodic electrochemical reduction, by immersing a film of *trans*-$(CH)_x$ in a $LiClO_4$ solution, dissolved in tetrahydrofuran, and attaching to the negative pole of a continuous current source, while the positive pole is attached to an inert electrode immersed in the same solution. The process may be seen in Fig. 27.6.

All the n- or p-type doping processes achieved by chemical and electrochemical procedures discovered for polyacetylene and other conducting polymers involve doping counter-ions that are introduced in the polymer matrix to the purpose of stabilizing the load of the polymer backbone. The existence of the electrically

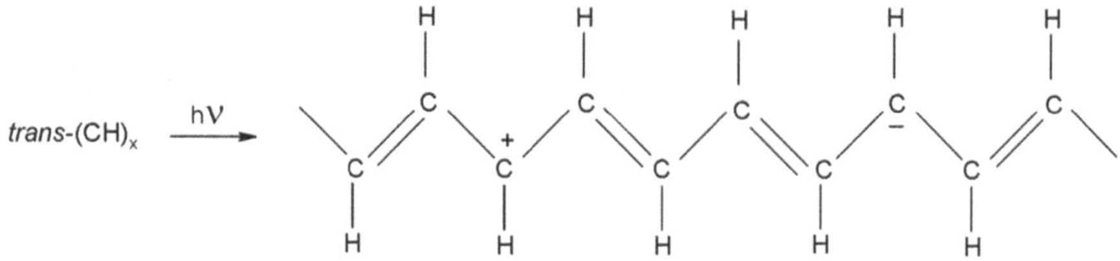

Fig. 27.7 Scheme of soliton photogeneration in *trans*-polyacetylene

charged polymer backbone was proved spectroscopically from the special characteristics of solitons, polarons, bipolarons, etc. (Feast et al. 1996; Shirakawa and Macdiarmid 1977; Wöll 2009).

Polymer doping procedures have continuously developed, including processes where no counter ions are involved, such as photodoping or doping by electrical charge injection.

27.2.3 Photo-Doping

If *trans*-$(CH)_x$ is exposed to energy radiation higher than the energy of the forbidden bad, electrons are transported on this band, and the polymer becomes photodoped. In optimum experimental conditions, solitons are generated, which may be evinced spectrometrically (Balint et al. 2014). Figure 27.7 shows the photodoping process of trans-acetylene.

The positive and negative solitons are shown in this figure as if belonging to only one CH unit. In fact, they are delocalized over approx. 15 CH units. But they disappear rapidly as the electrons, and the gaps recombine when irradiation is interrupted. If during the irradiation a potential is applied, then the electrons and the gaps are separated, and photoconductivity is observed (Su et al. 1979).

27.2.4 Doping by Injecting Charges

Doping by injecting charges is achieved by using a metal/insulator/semiconductor configuration, including a metal and a conducting polymer separated by a thin insulator layer. The application of the adequate potential along the structure may provide, for instance, a loaded superficial layer called accumulation layer which has been extensively studied for conducting polymers. The charges resulting in the polymer are present without being associated with a doping ion. The spectroscopic properties of the charged species formed in this manner may be examined in the absence of the doping ion. The spectroscopic studies of polyacetylene doped in this manner

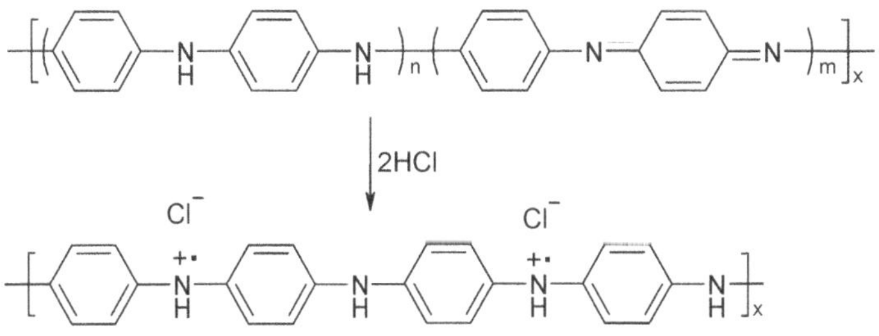

Fig. 27.8 Acid doping of polyaniline

show the characteristics of solitons as well as the mid-gap absorption band observed in chemically and electrochemically doped polymers (Hosseini et al. 2017).

27.3 Non-redox Doping

This type of doping type differs from the redox doping above by the fact that the number of electrons associated to the polymer's backbone is not altered during the doping process. However, the energy levels of the macromolecule are rearranged during doping. The form of emeraldine-base of polyaniline was the first example of doping an organic polymer towards a high conductivity regime by a process producing a stable polysemiquinonic radical-cation (Macdiarmid and Epstein 1995). It was obtained by treating the base emeraldine with aqueous acid solutions (Fig. 27.8), and this transformation determines an increase in conductivity of about 9–10 size orders (close to the value 3 $S \times cm^{-1}$) by obtaining the protonated base of emeraldine. Figure 27.8 shows the process of acid doping of aniline.

The mechanism of acid doping of polyaniline with hydrochloric acid is shown in Fig. 27.9.

Proton doping was extended to other conducting polymers, such as heteroaromatic polyvinylenes. In conclusion, irrespective of the manner of achieving the doping process, the number of polarons, bipolarons and solitons increases with the doping level (Fig. 27.10).

At high doping levels, localized polarons, bipolarons or solitons in the proximity of doping ions may overlap, leading to new energy bands or even the overlap of the valence and conduction bands, where electrons may move freely.

The global conductivity of conducting polymers includes contributions from the transport of intra-chain, inter-chain and inter-domain electrons (Dai 2004). The relative importance of the transport processes in global conductivity is not yet fully understood, but the main factors that influence conductivity were identified. The doping process is the most important factor influencing the conductivity of conducting polymers. The other factors include the orientation, crystallinity, and purity of conducting polymers (Kaur et al. 2015).

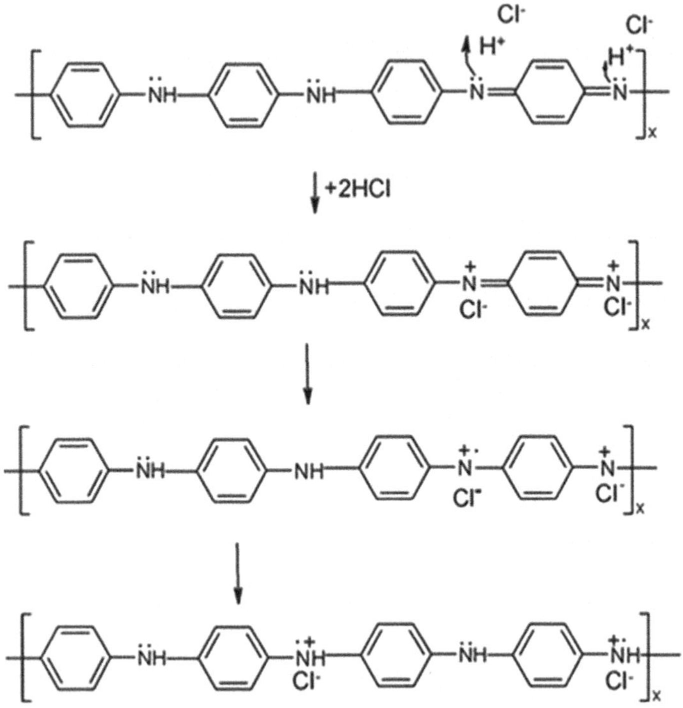

Fig. 27.9 Mechanism of acid doping of polyaniline

Fig. 27.10 Schematic description of polaron, bipolaron and soliton pair formation in a trans-polyacetylene chain by doping: (**a**) polyene; (**b**) polaron (radical-anion); (**c**) bipolaron (dianion); (**d**) soliton pair. Reprinted with permission (Dai 2004)

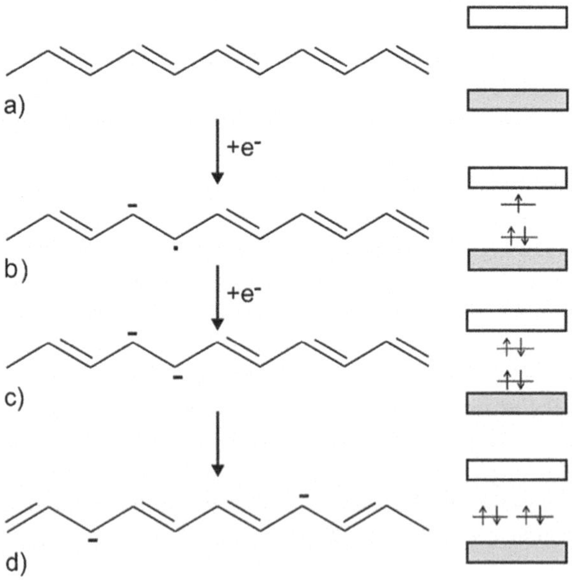

27.4 Properties of Conducting Polymers

Besides electrical conductivity, these materials have properties related to their electrochemical properties, among which the following may be mentioned:

Load storage capacity: the possibility to introduce or remove electrons in a reversible manner in the conducting polymers determines the formation of polarons and bipolarons, thus creating a load accumulation along the polymer's backbone, easily detectable by electrochemical techniques (Heeger 1985).

Electrochromic properties: these are based on the color change when the polymer's oxidation state changes. Thin polymer films have different colors when they are in an oxidized or reduced state, due to the formation and destruction of the polaron/bipolaron states, and the unoccupied electronic levels between the valence band and the conduction band of the polymer (Brooke et al. 2017).

Electrochemiomechanical properties: they consist of the reversible variation of the polymer volume accompanying the processes of oxidation and reduction. Due to this reversibility, a mechanical movement is obtained by the electrochemical control of the redox processes of the polymer. This phenomenon was observed in 1982 by Burgmayer et al. (1982). Okabayashi et al. studied the volume change during the doping process of polyaniline, obtaining an increase in the volume of 120%, with applications in the field of artificial muscles (Okabayashi et al. 1987).

Electrocatalytic properties and catalytic selectivity in certain reactions: in 2005, Isaacs et al. performed electropolymerization by the continuous scanning (cyclic voltammetry) of tetraaminophthalocyanine on vitreous carbon electrodes using as solvents dimethylformamide and dimethylsulfoxide (Isaacs et al. 2005).

Corrosion-inhibiting properties: D'Elia et al. (2001) studied the electro-coating of poly-(*o*-phenylenediamine) films on steel electrodes in potentiostatic conditions, from *o*-phenylenediamine solutions in phosphoric acid, resulting in thin films, adhering to the electrode surface, acting as an inhibitor of steel corrosion in an HCl/NaCl environment (D'Elia et al. 2001).

27.5 Chemical and Electrochemical Synthesis of Some Conducting Polymers

27.5.1 Polypyrrole

27.5.1.1 The Chemical Synthesis of Polypyrrole

Pyrrole is a heterocycle that has high reactivity with the tendency to form polymers in the presence of strong acids and/or oxidants. The structure of these pyrrole polymers consists of pyrrole units linked together by positions 2 and 5, as confirmed by the oxidative degradation of these compounds, which lead to the formation of the

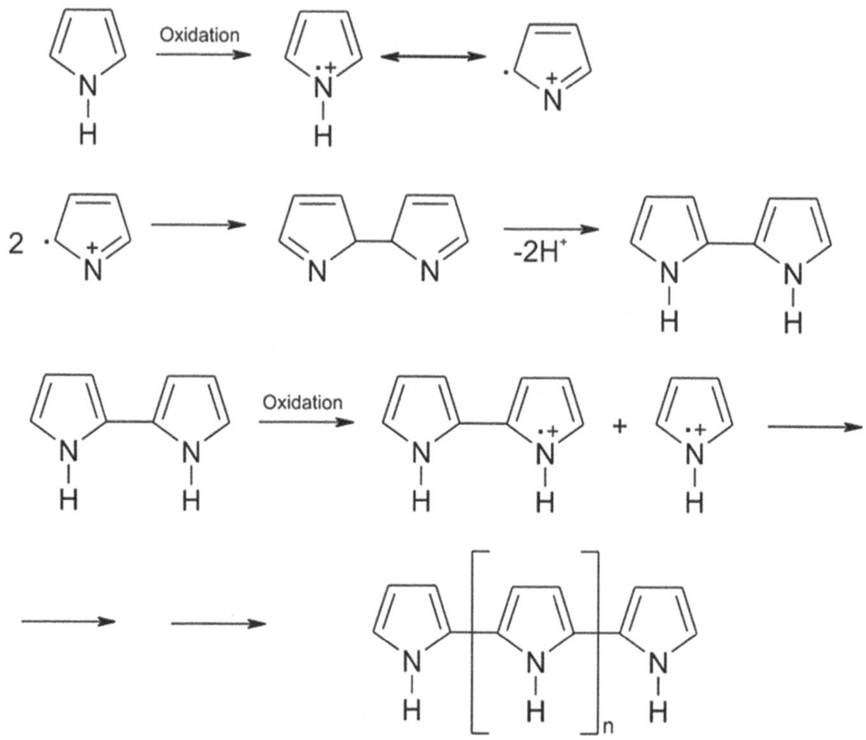

Fig. 27.11 Pyrrole polymerization mechanism in an oxidant environment

2,5-pyrroledicarboxylic acid. Besides, the 2,5-alkyl substituted derivatives of pyrrole do not form polymers (Ateh et al. 2006).

In acid solution and the absence of oxidants, pyrrole forms a polymer by addition, resulting in the mixture of several products, obtained under the form of a black precipitate. The mechanism of pyrrole polymerization is shown in Fig. 27.11.

The first stage is a redox reaction, in a homogeneous state, completely displaced to the right, when the condensation of two cationic radicals of pyrrole occurs. This condensation takes place according to a radical reaction of dimerization of the radical cation, Py^+, followed by aromatization, resulting in a chain of PPy which simultaneously is partially oxidized (doped), as its oxidation potential is considerably lower then pyrrole's. To maintain electroneutrality in the polymer chains of PPy, the anion of the salt used as an oxidant is incorporated in its structure (Fig. 27.12). As a result, the polypyrrole obtained is PPy-X (where X is the incorporated doping anion). The general reaction resulting in the doped state polymer is seen in Fig. 27.12.

In this reaction, two hydrogen ions are released for each pyrrole molecule that forms a polymer so that the acidity of the environment increases. Due to incorporating the anion of the salt used in the PPy obtained, the use of various salts of the same oxidation allows for obtaining polypyrrole with a different structure and

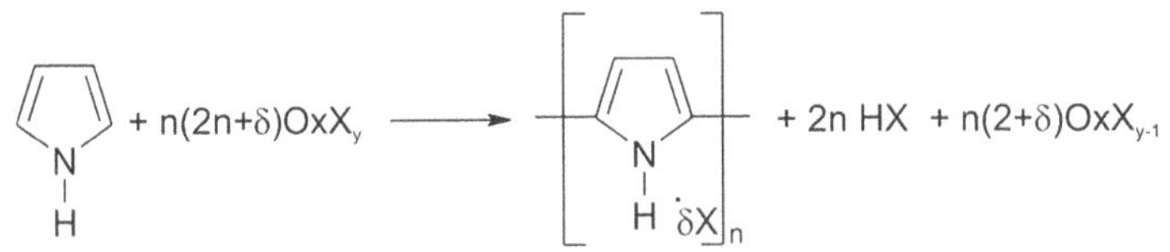

Fig. 27.12 The general reaction of obtaining PPy in doped state, where δ is the doping or oxidation degree of the polymer

properties (electrical conductivity, stability, mechanical properties, etc.). Polypyrrole is certainly the conducting polymer that has been studied the most, as it can incorporate a large number of counter-ions in its polymer matrix. Several salts may be mentioned, whose anions are incorporated in the polypyrrole polymer matrix, such as $K_4[Fe(CN)_6]$, sodium *para*-toluenesulfonate, tetrasulfonate nickel phthalocyanine, decane sulfonic acid, acid sodium sulfate, ferric chloride, etc. (Efimov and Chem 1997; Mason and Weber 2011).

27.5.1.2 Electrochemical Synthesis of Polypyrrole

The flux of an anodic current through a solution containing a solvent, an electrolyte, and monomer (pyrrole, thiophene, aniline, furan, etc.) produces a polymer on the anode. The polymer pellicles generated through electrosynthesis are characterized by high electric conductivity (up to 10^5 S × cm^{-1}), which is intrinsic electrical conductivity of polymer backbones. Electropolymerization (or electrochemical polymerization) is a fast process. A few seconds after beginning the anodic polarization or the anodic current flux, the electrode is covered in a black film. The practical requirements are not restrictive, as the aqueous solutions may be handled at ambient temperatures and pressures. The most restrictive requirement is the inert atmosphere in case the synthesis of polymer films is desired for specific applications.

The mechanism of the electrochemical polymerization may be seen in Fig. 27.13.

In the previous mechanism it may be noted that at first, an initial monomer oxidation process occurs (1), where, by extracting one electron, a cation radical of the monomer is generated. The next step is a coupling reaction, i.e. joining another cation radical monomer (2). This union between cation radicals is followed by deprotonation of the dimer to regenerate the aromatic cycles. The dimer gets oxidized, thus forming a new oligomeric radical cation, which may get coupled with another cation radical dimer forming a tetramer, or a monomer cation radical forming a trimer, and so on. By repeating these stages, a polymer chain is obtained, which is conductive in an oxidized state, with a system of delocalized π electrons (Gvozdenovi et al. 2014). The oxi-reduction processes of polypyrrole are shown in Fig. 27.14.

The application of an anodic potential to trigger this polymerization reaction involves a strict selection of the system components. First and foremost, the nature of the electrode material where the polymer is to form is of utmost importance. As

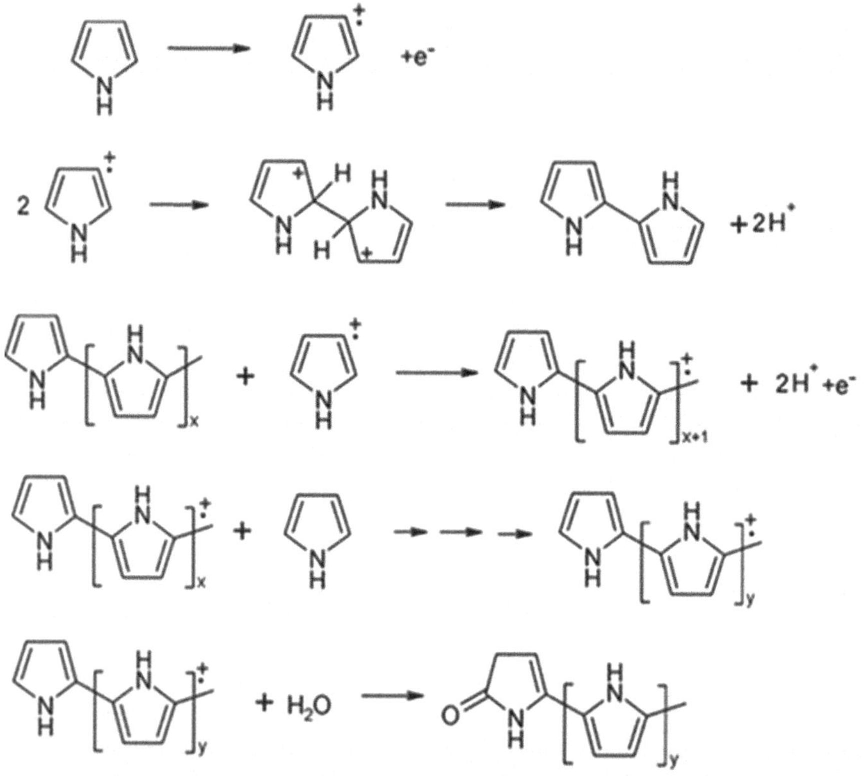

Fig. 27.13 Mechanism of electrochemical pyrrole polymerization

the polymer is obtained by an oxidation process, the working electrode should not get oxidized at the same time as the monomer. That is why the most research was conducted using gold, platinum or carbon electrodes.

Secondly, because polymerization is produced by oxidation and takes place using cationic species (cation-radical), it is very sensitive in the presence of nucleophiles close to the electrode surface. Thus, it was proved that when the e character of the solvent is important, the polymer pellicle is minimized, and that is why it is convenient to use aprotic solvents whose nucleophilic character is low.

Another aspect that should be taken into account regarding electropolymerization is the electrolytic salt used. The same as in the case of the solvent, it is necessary to use salts with a low nucleophilic character allowing for the formation of high-quality films. In any case, the use of various salts in the polymerization solution triggers important changes in the conductivity of the pellicles formed, that may reach up to 4 size orders (Bonifas and Mccreery 2012).

During the electro-generation of these polymers, a simultaneous phenomenon that occurs is the oxidation of the monomer, and also the newly formed polymer, so that positive charges are formed along the polymer chain. These charges make the

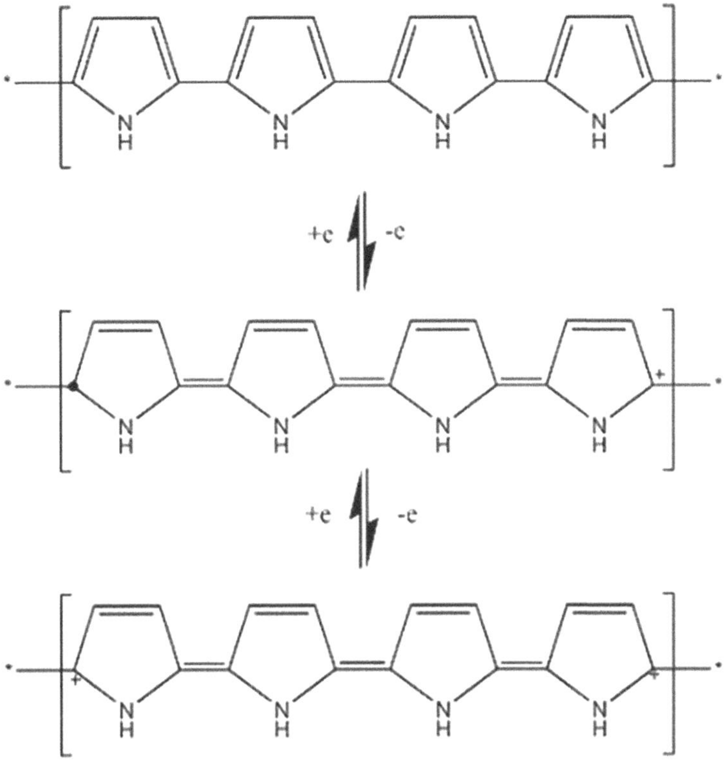

Fig. 27.14 Reversible redox processes of polypyrrole

anions in the reaction environment enter the polymer layer, to maintain the material electrically neutral during the entire length of the reaction, a principle used in the selective polymer doping. That is why, depending on the polymer oxidation level, the anion content varies. It was shown that this content might be up to 35% of the total mass of the deposit. The determination of these parameters was carried out using gravimetric techniques (crystal quartz microbalance), which is very useful in determining the movement of ionic species in and on the polymer film surface during the oxi-reduction processes (Torresi et al. 1995).

27.5.2 Polyaniline

Polyaniline may be synthesized both chemically (in powder form) and electrochemically (as film) on the surface of an electrode.

Chemical synthesis requires three reagents: aniline, an acid (aqueous or organic) environment, and an oxidant. The two most widely used acid are the

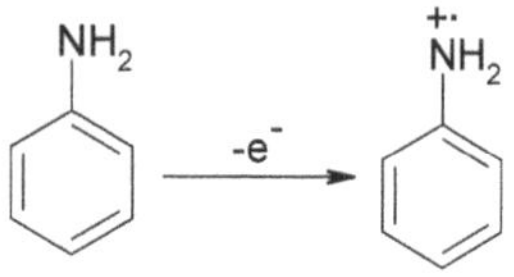

Fig. 27.15 Electrochemical oxidation of aniline

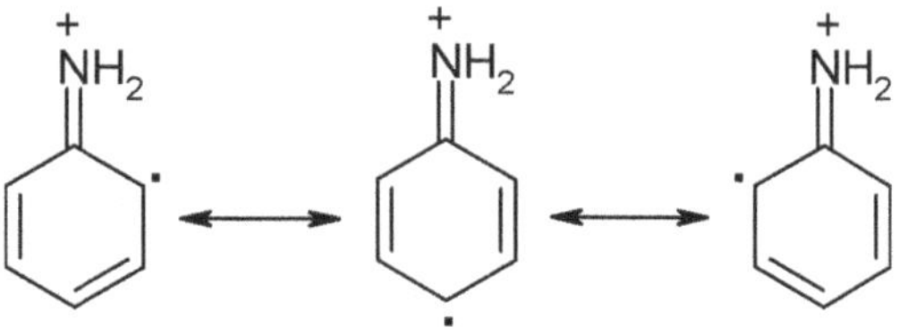

Fig. 27.16 Resonance structures of radical cation of polyaniline

sulphuric acid (H_2SO_4) and the hydrochloric acid (HCl). The oxidizing agents used may be the ammonium persulphate ($NH_4)_2S_2O_8$), potassium dichromate ($K_2Cr_2O_7$), cerium sulphate ($Ce(SO_4)_2$), sodium vanadate (Na_3VO_3), potassium ferrocyanide $K_3[(Fe(CN)_6)]$, potassium iodide (KIO_3), hydrogen peroxide (H_2O_2), etc. (Boeva and Sergeyev 2014).

Electrochemical synthesis uses an electrochemical cell with three electrodes, irrespective of the potential application technique: a working electrode on which the polymer is deposited, a counter electrode (platinum plate or wire) and a reference electrode (SCE or Ag/AgCl). The most common is the working electrode made of Pt. It may also be made of ITO, Cu, Au, carbon, etc. (Boeva and Sergeyev 2014).

The mechanism of aniline polymerization corresponds to a polycondensation. The first oxidation stage corresponds to the formation of a radical-cation by transferring an electron from the nitrogen atom of aniline, as shown in Fig. 27.15. Kinetically speaking, this stage sets the speed.

The radical cation of aniline has three forms of resonance, shown in Fig. 27.16. Out of these resonance forms, the *para* form is the most reactive, partially due to the important inductive effect of the substitute, and on the other hand, the absence of steric prevention.

The next stage (Fig. 27.17) is the radical-cation reaction and the *para* resonance form, the so-called head to tail reaction, favored in acid (aqueous or organic) environment, corresponding to the formation of a dimer. The dimer undergoes deprotonation reactions with the recovery of the aromatic structure.

Then, the dimer is oxidized, forming a new radical-cation (or dication-diradical), as shown in Fig. 27.18.

The radical-cation formed may react with either a radical-cation of the monomer, or a radical-cation of the dimer, forming a trimer or a tetramer, and so on, until a polymer is formed (Fig. 27.19).

The electrochemical synthesis of polyaniline allows for the control of the polymer film morphology, as the experimental factors may be controlled fairly easily, and they are the ones influencing the kinetics of polymerization, as well as the polymer's conductive or electrochemical properties. Thus, conductivity,

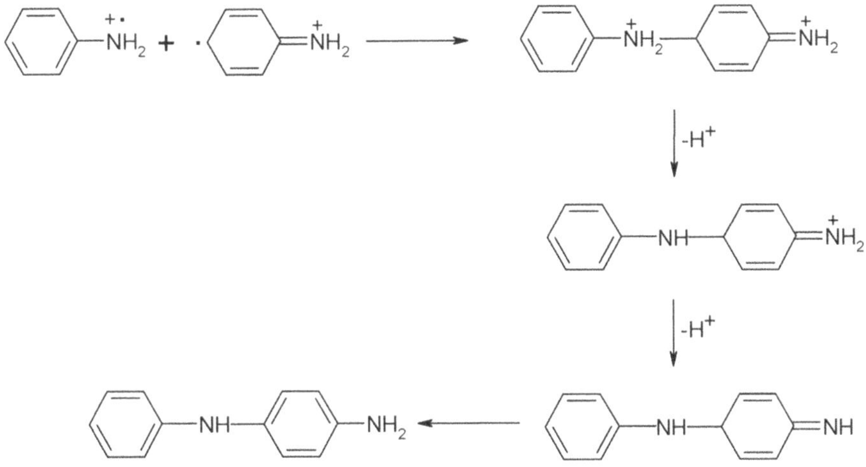

Fig. 27.17 The coupling mechanism in *para* position

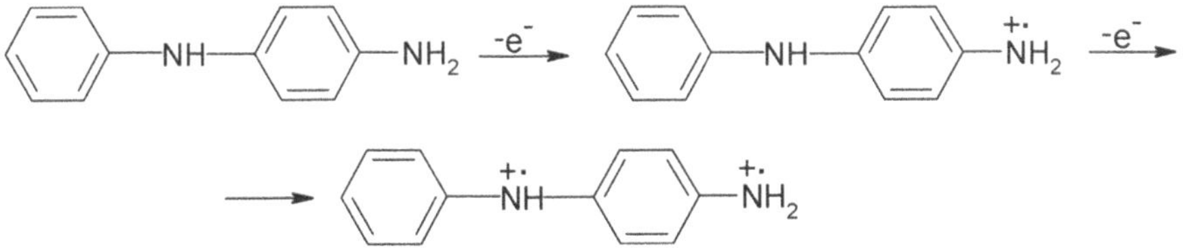

Fig. 27.18 Electrochemical oxidation of the aniline dimer

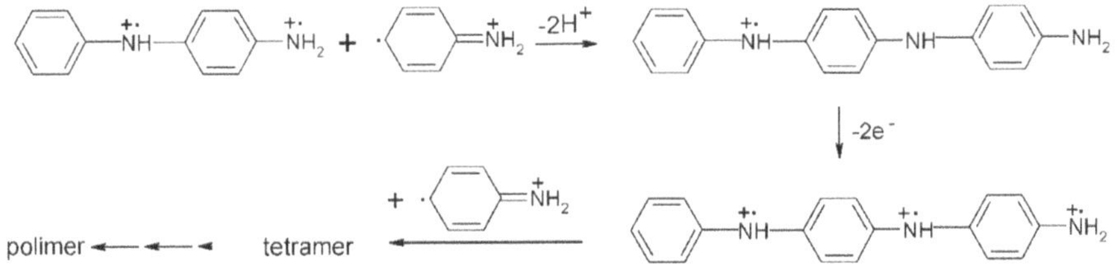

Fig. 27.19 General scheme of aniline polymerization

processability and thermal stability of PANI depend on the protonation degree and the nature of the doping ion. The size of the doping ion is important as the polymer is subjected to major volume changes during the doping and undoping processes, a phenomenon influencing the polymer's mechanical resistance. Also, it has been noticed that the electroactivity of polyaniline depends on the size of the doping anion. The properties of polyaniline depend on the strength of the acid used in the polymerization reaction, as its conductivity is higher when synthesized in stronger acids (Priyanka et al. 2015).

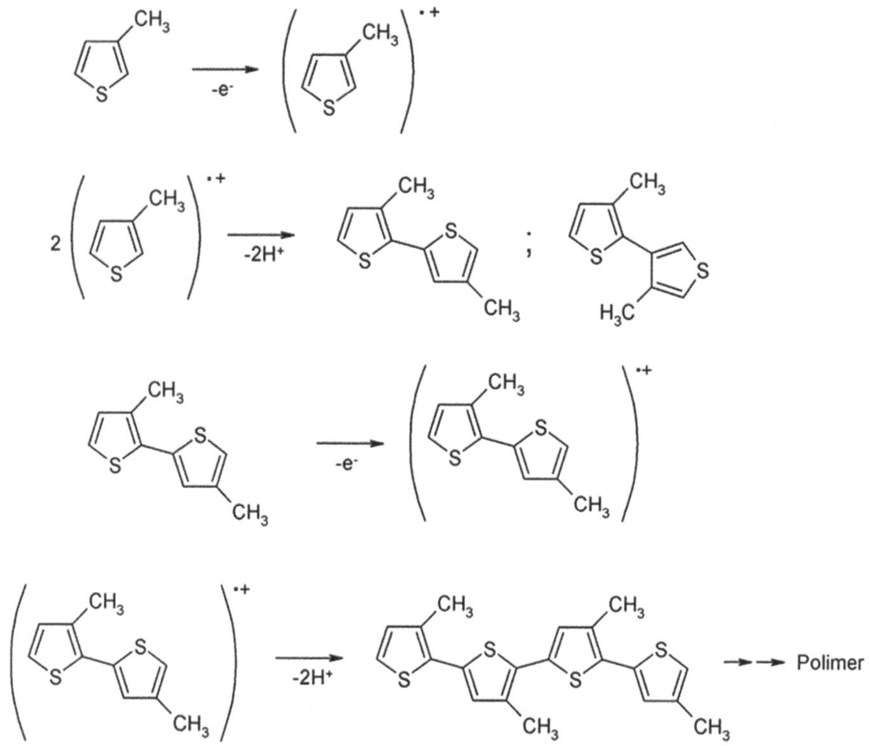

Fig. 27.20 The polymerization mechanism for 3-methylthiophene

27.5.3 Poly-3-Methylthiophene

Poly-3-methylthiophene may be synthesized both chemically and electrochemically. The chemical synthesis is based on monomer oxidation, triggering its polymerization, using an oxidizing agent acting in solution. The various oxidizing agents used were $FeCl_3$, Mg/THF, $Cu/PdCl_2$, etc. (Fichou 2008).

The electrochemical synthesis of poly-3-methylthiophene leads to a polymeric film on a metallic or semiconductive electrode, by passing an electrical current through a solution containing the monomer, a solvent, and an electrolyte. The pellicles obtained are good conductors due to the intense doping with the anions of the electrolyte. Due to the lack of solubility in an aqueous environment, it is necessary to use organic solvents to start the polymerization reaction. Among the most commonly used solvents, there are acetonitrile, propylene carbonate, nitrobenzene, etc. The electrolytes used in the electrochemical synthesis of poly-3-methylthiophene are $LiN(CF_3SO_2)_2$, KPF_6, $TBABF_4$, $LiCF_3SO_3$, etc. (Dubey et al. 1999). The mechanism of polymerization is seen in Fig. 27.20.

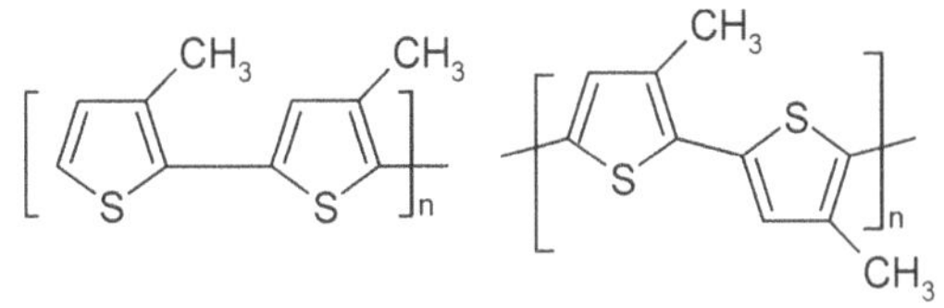

Fig. 27.21 Geometrical (*cis* and *trans*) isomers of poly-3-methylthiophene

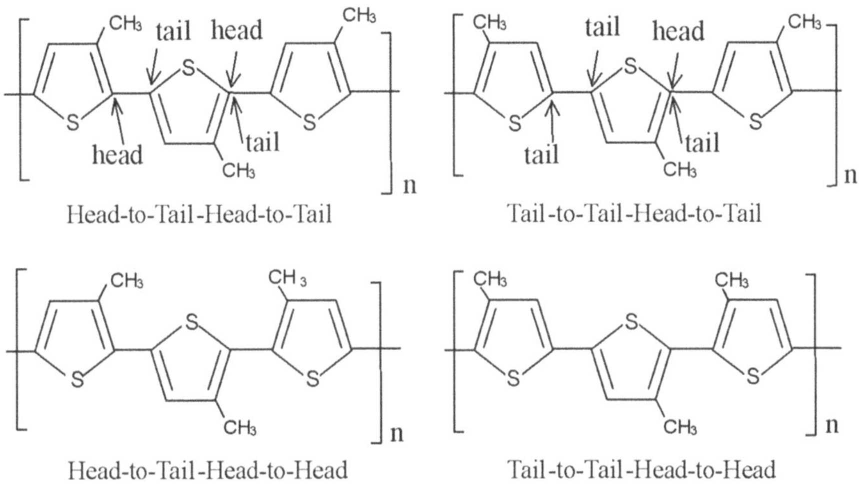

Fig. 27.22 Types of monomer couplings in poly-3-methylthiophene

The mechanism of polymerization is similar to pyrrole polymerization, including the same stages: forming a cation-radical out of the monomer by process of oxidation, followed by coupling two cation-radicals, eliminating two protons to recover the aromatic structure. The dimer obtained is oxidized to a radical-cation, and so on, until a polymer chain is obtained.

In this case, attention should be given to certain peculiarities, due to the presence of the methyl group in the monomer, triggering the regioselectivity of the polymerization reaction, and the existence of geometric isomers (Fig. 27.21).

Regarding the position of the methyl group in the polymer chain, it can be seen that there are different couplings between the thiophene rings: head-to-tail-head-to-tail (HT-HT) coupling; tail-to-tail-head-to-tail (TT-HT) coupling; head-to-tail-head-to-head (HT-HH) coupling; tail-to-tail-head-to-head (TT-HH) coupling (Fig. 27.22) (Skotheim 1997).

The HT-HT coupling is the most stable and abundant (>90%). The increase in the percentage of HT-HT couplings makes poly-3-methylthiophene have a higher order degree on longer distances, higher intrinsic conductivity, smaller forbidden band and higher wavelengths of the absorption maximum (Pern and Frank 1990).

27.5.4 Electrochemical Sensors Based on Conducting Polymers for Food Analysis

Among the fabrication techniques applied to the development of sensors and biosensors based on conducting polymer two has received special attention: Electropolymerization and Molecular Imprinted Polymers (MIP). The electropolymerization has the advantage of being simple and fast, and MIPs increase the sensitivity and specificity of the sensors. We will discuss these two techniques below.

27.5.4.1 Electropolymerization

One of most commonly used technique to fabricate sensors is the electropolymerization, through the deposition of different polymers combined with other material to enhance the sensibility and selectivity of the sensor. Usually, the film is deposited onto a conducting electrode, such as glassy carbon electrode (GCE), ITO (indium tin oxide) and other. The electropolymerization process leads to simple and reproducible formations of polymer films and consists basically in the application of a range of potential over the desirable monomer solution. In general, the electropolymerization process for the formation of the films can be carried out by a successive cyclic sweep at a more positive potential range than the oxidation potential of the functional group, forming radical-cations and preserving monomer structure (Steter et al. 2008). However, the conducting polymer can also be obtained from electropolymerization at a fixed potential or fixed current (potentiostatic or galvanostatic mode) (Apetrei and Apetrei 2013; Palmisano et al. 2000). Briefly, the electropolymerization mechanism is based on the interaction between the radical-cations (radical-cation coupling) formed from the oxidation of the monomer at high potential values (above 0.9 V) (Steter et al. 2008; Chen et al. 2008; Zhang et al. 2010). For example, Chen et al. (2008) described the electropolymerization of Nile Blue at glassy carbon electrode and their application as nitrite sensing in food samples. The Nile Blue conducting polymer was obtained by electropolymerization applying the cyclic potential sweeps from −0.60 V to +1.2 V using a 5.0×10^{-4} mol/L of Nile Blue solution (Fig. 27.23). The electropolymerization mechanism was ascribed to radical-cation coupling by the one-electron oxidation of Nile Blue monomer at ~0.9 V. The electropolymerized Nile Blue was evaluated as nitrite sensing, showing a linear range from 5.0×10^{-7} to 1.0×10^{-4} mol/L of nitrite and a limit of detection of 1.0×10^{-7} mol/L in standard solution. The application in real food samples (sausage samples) indicates effective and sensitive sensing when compared with spectrophotometric method (UV-vis absorption) (Chen et al. 2008).

The selectivity is a very important factor when it comes to detect or determine an analyte, this statement it is well represented by the work of Filik et al. (2017). They fabricated an electropolymerized sensor using Poly(alizarin red S) into a glassy carbon electrode. They performed cyclic voltammetry measurements with two analytes, caffeine, and vanillin (Fig. 27.24). In one of the measurements, they kept the

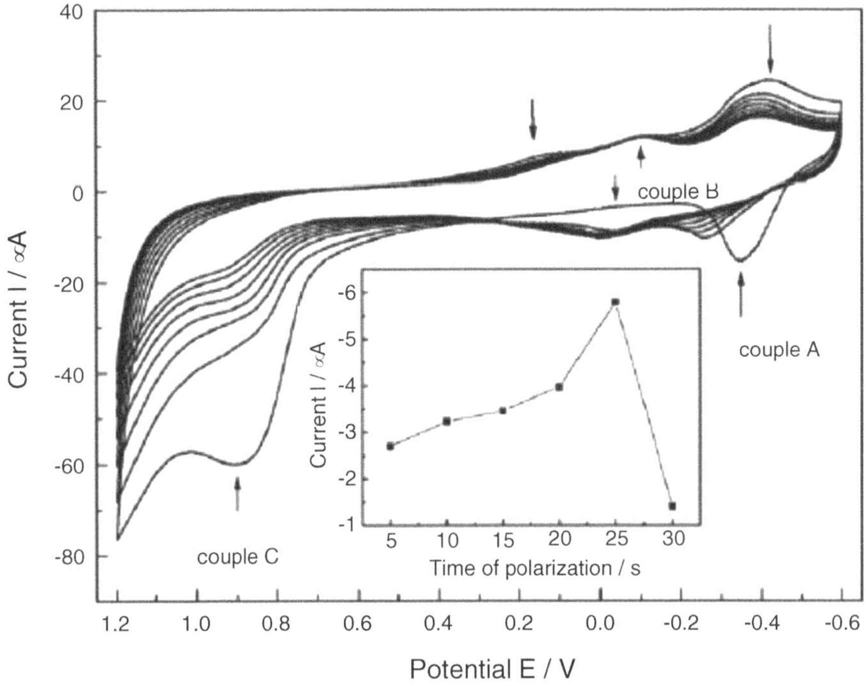

Fig. 27.23 Cyclic voltammograms of electropolymerization of 5.0×10^{-4} mol/L Nilo blue at glassy carbon electrode in 0.15 mol/L PBS (pH 5.3) at a scan rate of 50 mV/s. Reprinted with permission (Chen et al. 2008)

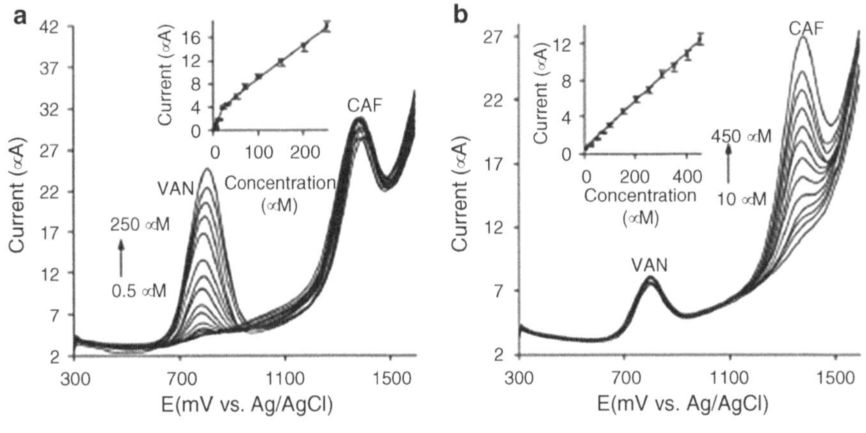

Fig. 27.24 Square wave votamograms of (**a**) a fixed caffenine concentration and in different concentrations of vanillin and (**b**) and square wave voltamograms of a fixed vanillin concentration and different caffeine concentrations. Reprinted with permission (Filik et al. 2017)

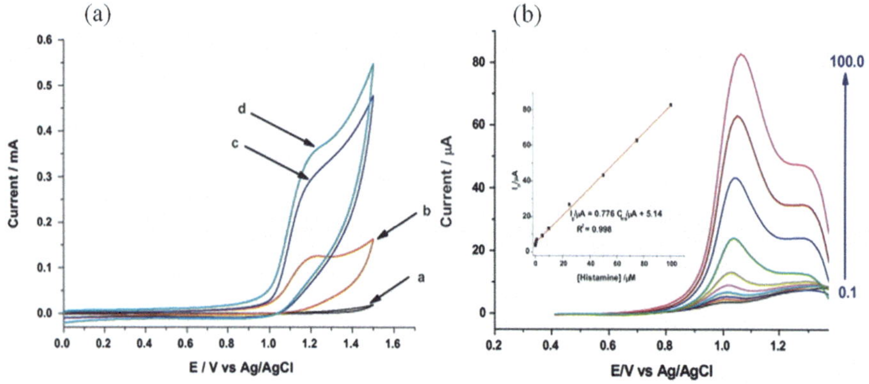

Fig. 27.25 (**a**) Cyclic voltamograms of different electrodes modified with the conducting polymer p-(AHNSA) and the MWCNTs in 1.0×10^{-3} mol/L of histimine in buffer solution. (**b**) Differential pulse voltamograms obtained for the p-(AHNSA)/MWCNT electrode in the range of 0.1–100.0 μmol/L of histidine. Reprinted with permission (Geto et al. 2014)

caffeine concentration constant and variate the vanillin concentration, and they obtained a constant peak for the caffeine and variable peak for vanillin. They performed opposite, keeping the vanillin concentration constant and the caffeine variating, and a similar result was obtained, showing a good selectivity of the sensor. They also performed the determination of caffeine and vanillin in real samples, achieving recovery values in the range of 100–102% of vanillin in energy drink samples, and 97.3–99.8% of the caffeine in vanilla sugar samples. Therefore, this is a very interesting result, because a very simple electropolymerized film with a single compound was enough to differentiate two analytes.

The presence of the amino acid histidine in fish muscle was evaluated in the work of Geto et al. (2014). They modified the GCE with electropolymerization of the polymer 4-amino-3-hydroxynaphthalene (p-(AHNSA)) sulfonic acid and then coated with the multiwalled carbon nanotubes (MWCNTs) suspension. This assembling presented a 34-fold the bare electrode response, and also showed a wide range in concentration during the calibration curve (Fig. 27.25). The histidine determination in real samples in fish muscles obtained recovery values between 96.6 and 102.9%. The interference study revealed that the presence of creatine, methionine, phenylalanine and uric acid did not affect much the histidine peak. However, the presence of glycine caused a considerable interference.

Saber-Tehrani et al. (2013) reported the use of the monomer 2-aminothiophenol deposited onto a gold electrode surface by self-assembly. Then, the conducting polymer was formed by potential cycling in acid media. The produced electrode was applied to the determination of nitrite in different samples of sausage. The DPV measurements respond linearly to the increase in the nitrite concentration. The obtained recovery values for the nitrite in the sausage samples are up to 93.9–102.6%.

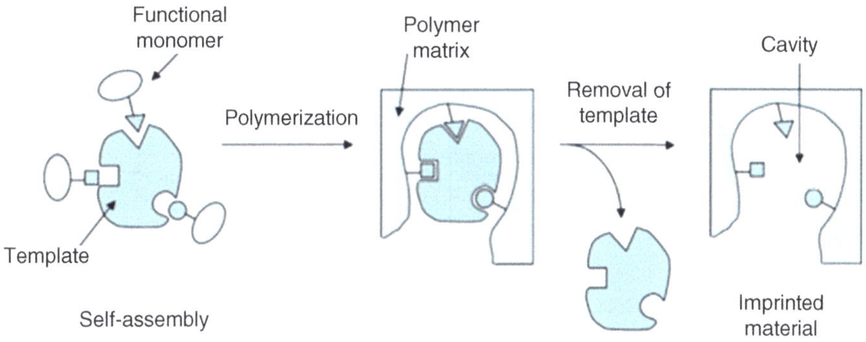

Fig. 27.26 Generalized diagram of MIP synthesis and selective analyte interaction. Reprinted with permission (Sarafraz-Yazdi and Razavi 2015)

27.5.4.2 Molecular Imprinted Polymers (MIP)

Molecular Imprinting Technology (MIT) is basically designed to mimic the natural ability that some compounds have to recognize some specific materials, such as enzymes and antibodies. However, MIP presents a robust design and higher stability, presenting the same capacity compared with enzyme and antibodies based sensors. Due to these properties, MIT has become an important and active area of research made in recent years. MIT is considered a promising technique, which can recognize both biological and chemical molecules including amino acids and proteins (Bossi et al. 2007; Morelli et al. 2010; Scorrano et al. 2011), nucleotide derivatives (Longo and Vasapollo 2008), pollutants (Chapuis-Hugon 2008; Tamayo et al. 2005) and drugs (Puoci et al. 2007). However, here we are going to discuss the recent work reported in food detection and determination.

MIPs is based on the formation of a complex between an analyte (template) and a functional monomer. In the presence of a large excess of a cross-linking agent, a three-dimensional polymer network (Ramström and Mosbach 1999) is formed. After polymerization process, the template is removed from the polymer leaving specific recognition sites complementary in shape, size and chemical functionality to the template molecule (Fig. 27.26).

The main advantages of molecularly imprinted polymers (MIPs) are their high selectivity and affinity for the target molecule used in the imprinting procedure. Imprinted polymers compared to biological systems such as proteins and nucleic acids, have higher physical robustness, strength, resistance to elevated temperature and pressure and inertness towards acids, bases, metal ions and organic solvents. In addition, they are also less expensive to be synthesized, and the storage life of the polymers can be very high, keeping their recognition capacity also for several years at room temperature.

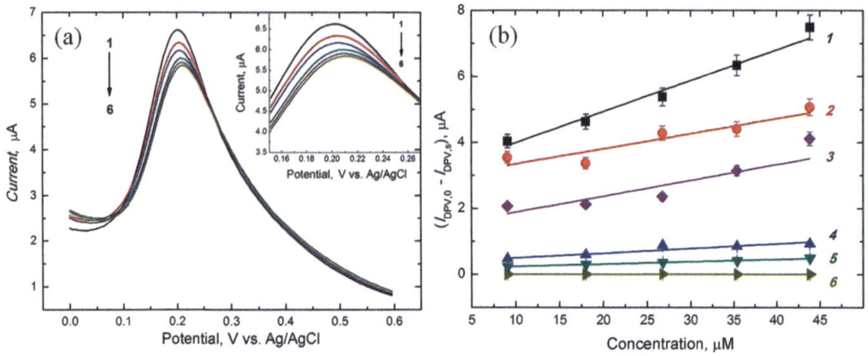

Fig. 27.27 (**a**) Response of the MIP electrode to the presence of the template analyte in different concentrations and (**b**) the calibration plots of different interferers. Reprinted with permission (Lach et al. 2017)

The concern with the quality of the food is very important, and knowing that, probably one of the most affected products that we eat daily is meat. Also, there are several studies about different compounds found in different kinds of meat such as fish, chickens, and pigs. Wang et al. (2017) performed one of these studies, they prepared a sensor based on a GCE, gold nanoparticles, MWCNTs and most importantly the MIP using olaquindox (which is a compound used to increase the gain of weight of pigs as a desired analyte and *o*-phenylenediamine as the monomer (Barber et al. 1979). They applied the sensor described before to detect the olaquindox in a real sample of pork fish meat using DPV measurements. They obtained recoveries values up to 80.7–115.8%, with the relative standard deviation (RSD) less than 11.4%, and also with no interference when applied with other similar materials, showing a good selectivity.

In another work involving pork meat, but this time using *N*-nitroso-l-proline as analyte and tetrabutylammonium perchlorate as a monomer to prepare the MIP sensor into the platinum disk and a gold film (Lach et al. 2017). The results showed recovery values in the range of 77–118%. The MIP sensor also presented selectivity if applied to other interference materials such as glucose and urea (Fig. 27.27). Although, other materials like creatine and adrenaline presented some interference, due to the similar size, shape, and functional groups. As it can notice some MIP sensor have a better response when applies to the interferers. However, if the material has some similar structure or functional groups, this material can fit inside the template just like the original analyte causing the interference to happen.

Just like solid food, beverages also suffer from the presence of some chemicals, when it comes to beverages determinations, some articles presented very interesting results. Such as the work of Pacheco et al. (2015), who fabricated a GCE modified with MWNTs and a MIP, using polypyrrole as the monomer to detect the analyte ochratoxin A in beers and wines. Ochratoxin A is one the most common toxins found in food and beverages; it is produced by a few species of fungus (Al-Anati and Petzinger 2006). Using DPV measurements they achieved recovery values in the range of 104–110% for beer, 84–108 for white wine and 93–98% for red wine.

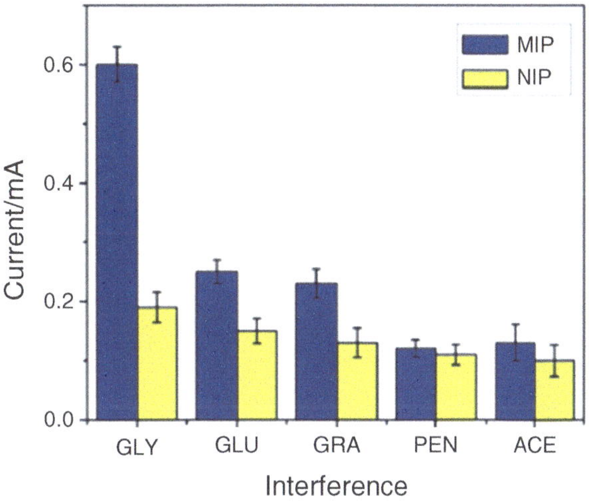

Fig. 27.28 Selective adsorption of the MIP electrode for detecting glyphosate and other pesticides (concentration 1000 ng/mL). Reprinted with permission (Xu et al. 2017)

Some commercial drinks have dyes to generate artificial colors or even chemicals who simulate the taste and smell of products, such sodas. In the work of Yu et al. (2017) they applied a glassy carbon electrode modified with graphene oxide, silver nanoparticles an a MIP based on sunset yellow (template). The results presented recovery values in the range of 87–111% for Fanta drink, Mirinda drink, orange, and mango juices. The sensor presented a selectivity when applied to a coexistent analyte such as tartrazine, brilliant blue G, and ascorbic acid. However, the presence of amaranth caused an interference, due to the similar chemical structure.

Farming products also have issues with materials, mainly with pesticides, due to the widespread use of these compounds to prevent crop loss. The evaluation of pesticides (glyphosate) in corn samples has been recently reported using MIP sensors. Xu et al. (2017) modified an ITO electrode with gold nanoparticles, Prussian blue, and polypyrrole as the monomer in the MIP glyphosate template. The determination of glyphosate in corn samples was performed by DPV measurements (Fig. 27.28), and achieved recovery values between 97.5 and 101%. The sensor also presented a satisfactory selectivity if applied with other pesticides (interferer agent).

27.5.5 Electrochemical Biosensors Based on Conducting Polymers for Food Analysis

A wide variety of biosensors has also been constructed during the last years (Gerard 2002). The main advantages of biosensors are the sensitivity and selectivity to the detection of the molecule of interest. In food analysis, the biosensors can be used for detection of glucose, fructose, ethanol, sucrose, lactate, malate, galactose, citrate, lactose, urea, starch, etc. in food industries (Gerard 2002). However, the biosensor can be applied not only to control of nutritional food quality but also to

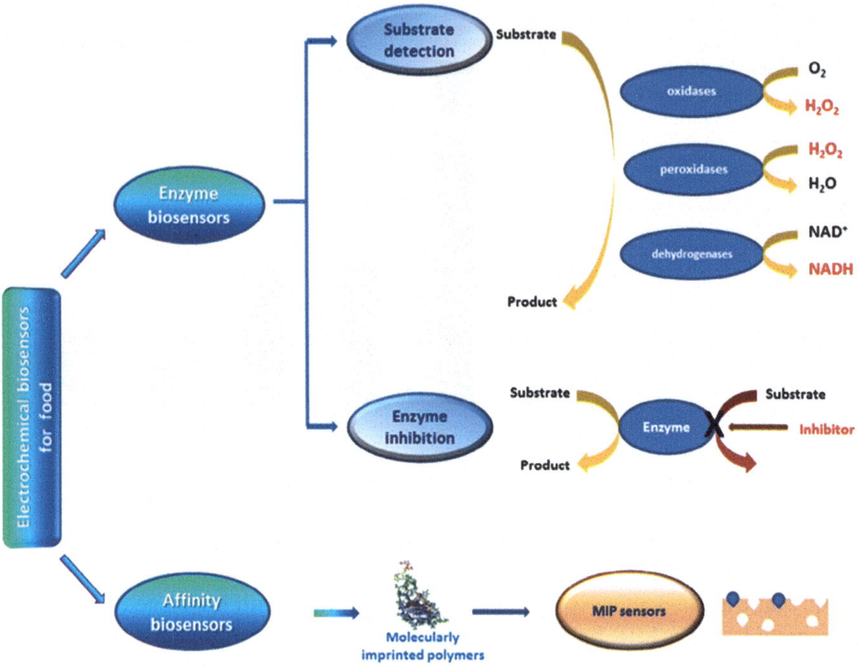

Fig. 27.29 Types of biosensors in food analysis. Reprinted with permission (Rotariu et al. 2016)

the determination of possible contaminants. The food biosensor can be a classified principal in two classes: enzymes biosensor and affinity biosensor (Fig. 27.29) (Rotariu et al. 2016), which are based on bioreceptor or transducer type. At the enzymes biosensor, the sensing reaction is based on reaction catalyzed by enzymes, which can be used directly or immobilized in solid substrates, which the conducting polymers are used as good matrices and provide an appropriate environment (Trojanowicz and Krawczyński vel Krawczyk 1995; Kranz et al. 1998; Cooper and Hall 1992).

Among the conducting polymers, the derivatives of polyindole, polyaniline, polyacetylene, polythiophene and polypyrrole (Apetrei and Apetrei 2013; Palmisano et al. 2000; Cooper and Hall 1992; Bartlett and Whitaker 1987), are widely used as enzyme immobilization matrices. The immobilization onto conducting polymer surface can involve physical adsorption, entrapment, cross-linking or covalent bonding (Gerard 2002).

In 1986, Umana and Waller (1986) reported the simple immobilization of glucose oxidase, by electropolymerization of pyrrole in the presence of the enzyme. The enzyme becomes entrapped by the polypyrrole film growing on the electrode surface. The electrodes can be used in the determination of glucose in aqueous solutions for a period of up to 7 days.

Ghosh (Hazra) et al. (1998) developed a biosensor system for qualitative measurement of fish freshness. Freshness is indicated by an index which depends on the amount of adenosine 5′-triphosphate degradation products present in fish meat. The metabolites in fish tissue, hypoxanthine, inosine and inosine monophosphate were efficiently measured quantitatively. The biosensor system was based on amperometric method, using three electrodes composed by ferrocene carboxylic acid mediator incorporated conducting polypyrrole with immobilized xanthine oxidase, nucleoside phosphorylase, or nucleotidase enzymes. The results suggest that fish freshness can be monitored quantitatively with this biosensor.

Apetrei and Apetrei (2013) described a biosensor based on Tyrosinase enzyme immobilized onto polypyrrole film to detect tyramine. Briefly, tyramine is one of the biogenic amines produced by microbial activity degradation commonly present in fermented foods and beverages, meat, fish, seafood and dairy products (Onal 2007). First, the polypyrrole film was electrochemically synthesized on a Pt electrode surface by applying a fixed current at 60 μA for 15 s. After that, the Tyrosinase enzyme solution was added onto polypyrrole film and exposed to 2.5% (v/v) glutaraldehyde solution. The amperometric biosensor was evaluated in tyramide standard solution showing a linear range of concentration from 4–80 × 10^{-6} mol/L and a limit of detection of 5.7 × 10^{-7} mol/L. The analysis in the real sample (salted sauerkraut samples) showed an average recovery of 99.3%.

Palmisano et al. (2000) also report the development of biosensor based on enzyme immobilization onto polypyrrole film. More, specifically, the authors developed a dual biosensor based on glucose and lactate oxidase onto the overoxidase polypyrrole film surface. Initially, the polypyrrole film was obtained by electropolymerization using a constant potential at +0.7 V. After, a glucose and lactate oxidase solution was added on polypyrrole film surface, keeping in room temperature to drying and washed before the sensing application. The dual biosensor showed a great response to simultaneous monitoring of glucose and lactate in tomatoes juice. The measurements were carried out in tomatoes juice without any pre-treatment with 95% of confidence level (Palmisano et al. 2000). Based on this aspect, the biosensor shows advantages compared with spectrometric or chromatographic methods, like selectivity, direct detection using real samples without any pre-treatment, fast measurements and results acquisition, and the possibility of miniaturization.

27.5.6 Electronic Nose-Type Systems Based on Conducting Polymers for Food Analysis

To analyze food, networks of sensors based on conducting polymers are used, which make up the detection system of various electronic nose-type systems. Routine analyses in food quality control are one of the most promising applications of the electronic nose.

Several successful applications of the electronic nose were published in monitoring synthetic or natural aromas, as well as aroma components in various production stages.

Many applications of the electronic nose mentioned in specialized literature referred to the shelf-life in the maturation process of fruits and vegetables, from harvesting to consumption, dairy products, and olive oil.

Freshness is another important property in the food industry, related to product quality. During storage, various volatile compounds are generated, and the electronic nose has shown its potential in forecasting the deterioration of certain foods, especially in the case of food items that give off volatile substances due to fast degradation through microbial contamination. The electronic nose was widely used to evaluate and classify extra virgin olive oils, according to their aromatic attributes. Similarly, it was used in product quality control, such as meat, oil, etc.

First papers discussed will include different aspects such as the control of the quality, authenticity, adulteration, falsification of different foods with electronic noses based on resistive sensors based on conducting polymers. In a study were developed to differentiate three artificial aromas: strawberry, grape, and apple. The sensors array was made from chemical sensors based on polyaniline films. The sensors were manufactured by polymerization of aniline on interdigitated graphite electrodes and doped with different acids (hydrochloric acid, camphor sulfonic acid, and dodecylbenzene sulfonic acid). The study was demonstrated by principal component analysis (PCA) that the sensor array was highly effective to discriminate artificial aromas, thereby being a promising tool for aroma quality analysis in various food industry sectors (Tiggemann et al. 2017).

Neely et al. (2001) used an electronic nose equipped with 14 semi-conducting polymer film sensors to discriminate between different types of meat. The data from the electronic nose was analyzed by linear discriminant analysis (LDA) method. The results showed that this electronic nose classified the meat types correctly, in agreement with the biological origin (Neely et al. 2001).

In another work, an array of sensors based on polyaniline and poly-3-methylthiophene were used to discriminate between olive oils of different qualities. The gas sensors were prepared using different electrochemical techniques (chronoamperometry, chronopotentiometry, and cyclic voltammetry) and to optimize the electrochemical conditions (polymerization potential, polymerization time, polymerization current, doping agent and type of substrate) to improve the performance characteristics. The performance characteristics of the sensors were determined by studying the response of the sensors to volatile organic compounds from the headspace of olive oils samples of different qualities. The results obtained by PCA show that the polymeric sensors array coupled with the appropriate data analysis is able to discriminate among olive oils of different qualities (Guadarrama et al. 2004).

The same electronic nose was able to discriminate among Spanish virgin olive oils obtained from different varieties of olives and even of different geographic origins (Guadarrama et al. 2001). The selection and test of an array of conducting polymer sensors with extra virgin olive oil samples were presented in another paper. Different polymeric sensors made by both electrochemical and chemical methods were developed. The performance characteristics of sensors were tested towards pure compounds usually found in the headspace of extra virgin olive oils. The compounds studied are related to overall sensorial characteristics of extra virgin olive oils. Different experimental set-ups

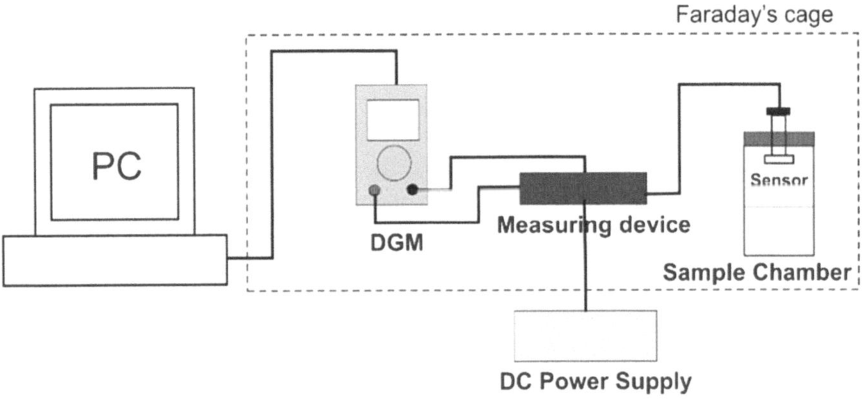

Fig. 27.30 The electronic nose experimental set-up. Reprinted with permission (Rañola et al. 2016)

and procedures for olive oil volatile sampling were tested and compared. The best procedure for the practical purposes in the olive oil industry was validated in real sample analysis. Three different extra virgin olive oils from Italy of were distinguished with an array of four polymeric sensors (Stella et al. 2000).

Rañola et al. (2016) developed an array of gas sensors based on conducting polymers for the discrimination of coconut oils. The sensors were fabricated through the electrodeposition of polyaniline, polypyrrole, and poly-3-methylthiophene on the gap separating two planar gold electrodes set on a Teflon substrate. The modification in electrical resistance of the gas sensors was observed and measured after exposing the array to the headspace of oil samples.

The scheme of the experimental set-up is illustrated in Fig. 27.30.

Different responses were obtained for each coconut oil sample, and multivariate data analysis were employed for the processing of the data. The electronic nose was able to distinguish virgin coconut oil from refined, bleached and deodorized coconut oil, flavored virgin coconut oil, homemade virgin coconut oil, and rancid virgin coconut oil (Rañola et al. 2016).

Another category of studies consists in the detection of gases related to the degradation of foods. Matindoust et al. (2017) developed a sensor for ammonia gas based on polyaniline. The fabrication procedure consists of several steps. Polyaniline was synthesized by oxidative polymerization technique. Then, a polyaniline film was deposited on a printed circuit board. Finally, polyaniline microdevice was included in an interdigitated electrode arrays to produce the sensor for ammonia gas detection.

The response time of this sensor and humidity impact were examined for NH_3 sensitivity and compared with a commercial gas sensor (Taguchi Model 826). The sensor was found to have a rapid and stable linear response to NH_3 in the concentration ranging from 50 to150 ppm, under room temperature operation condition. The sensor could be used in the detection of NH_3 without variation of the signal in the presence of environmental humidity. The low cost, flexibility, low power consumption and high sensitivity are the main benefits of this sensor. In real-time application conditions polyaniline based gas sensor exhibits good performance and accurate evaluation of food spoilage (Matindoust et al. 2017).

Siegmund and Pfannhauser (1999) used an electronic nose to study the modifications of the volatile compounds of heat-treated chicken meat during storage. The electronic nose detector system consists of a sensor array having 32 conducting polymer sensors. The signals of the sensors array were analyzed using PCA, and cluster analysis. The PCA data scores illustrate the differentiation of samples stored at different times. The results indicated a good correlation between the results obtained with the electronic nose with that obtained using the gas chromatography-mass spectrometry and gas chromatography-olfactometry (Siegmund and Pfannhauser 1999).

Tung et al. (2014) developed chemoresistive vapor sensors by the assembly of magnetic nanoparticles-decorated reduced graphene oxide with poly-(3,4-ethylene dioxythiophene) and poly(ionic liquid). This hybrid sensing material demonstrated improved sensitivity, selectivity, and reduced response time. The polymer nanocomposite sensor exhibited stable and reproducible signals for both polar (ethanol, acetone, water) and non-polar (chloroform, dichlorobenzene, toluene) volatile organic compounds. These compounds are recognized as food degradation biomarkers. The sensor responses are well defined even at the under ppm level, and the application is the field of smart packaging (Tung et al. 2014).

Sotzing et al. (2000) developed chemically-sensitive resistors for the detection of biogenic amines, at levels of 1–10 parts per trillion in ambient air. The sensitive materials were composites of emeraldine salt of polyaniline with particles of carbon black (Sotzing et al. 2000).

The main focus of the work of Hossain et al. (2012) was on the development of a carbon black polymer sensor array to detect stored grain model volatiles related to insect secretions (benzene derivatives) and fungi (aliphatic hydrocarbon derivatives). The statistical analysis was used to select polymer sensors for the array, which were optimum for distinguishing between the classes of compounds resulted from the degradation of grains (i.e., quinones, alcohols). The performance of the developed sensor array was suitable to demonstrate the discrimination of stored grain model volatiles at ambient conditions (Hossain et al. 2012).

In the work of Ridgway et al. (1999), several wheat samples, to which were added flour mites (*Acarus siro L.*), were studied, 1–2 weeks after preparation, by a conducting polymer sensor based electronic nose with transient flow sampling. The results demonstrate the capacity of the electronic nose to detect the mites in wheat at concentrations of relevance to the cereal trade. Undecane was identified as the main compound responsible for the electronic nose response (Ridgway et al. 1999).

Steffens et al. (2010) report the development of gas sensors based on polyaniline doped with dodecyl benzene sulphonic acid deposited onto interdigitated line patterns of graphite. The sensors exhibited different behavior related to moisture and also to the vapor-phase atmosphere of ethanol, acetone, n-hexane and ethyl acetate. Results demonstrate that sensors can detect the ripeness of banana. This technology can be used as an accurate method for monitoring the maturation of fruits (Steffens et al. 2010).

Spanier et al. (1999) developed a successful method to distinguish differences in and keeping-quality of the whole, fresh-cut, and minimally processed Gala variety apple using a commercial electronic nose based on 32 polymer sensors.

Abbey et al. (2004) developed an electronic nose conducting based on polymer sensor to classify onion headspace volatiles. Onion bulb quality discriminated by the electronic nose is related to N, S and soil type treatments, and related to their interactions. The chemical compounds in the onion headspace volatiles that interacted with the sensor polymer are correlated with the onion qualities including flavor (Abbey et al. 2004).

In conclusion, these works highlight the present-day tendency to use non-destructive analytical instruments to evaluate foods, proving that this technology (the electronic nose) has excellent potential for the quality control of foodstuffs.

The electronic nose is a useful analysis instrument in characterizing and monitoring products and processes in the food industry. The advantages of this technology are mobility, speed, reliability, ease of use and price. Using the electronic nose, the quality of a sample is assessed without collecting data on its chemical composition or the concentration of its compounds.

The manufacturing technology of the sensors and the use of adequate software for the analysis of the olfactory print of food products have made fast progress, and many industrial applications have been developed. The challenges for the research in this field are related to how to increase the selectivity, sensitivity, and reproducibility of the analyses.

References

Abbey BL, Joyce DC, Aked J et al (2004) Electronic nose evaluation of onion headspace volatiles and bulb quality as affected by nitrogen, sulphur and soil type. Ann Appl Biol 145(1):41–50

Al-Anati L, Petzinger E (2006) Immunotoxic activity of ochratoxin A. J Vet Pharmacol Ther 29:79–90

Apetrei IM, Apetrei C (2013) Amperometric biosensor based on polypyrrole and tyrosinase for the detection of tyramine in food samples. Sensors Actuators B Chem 178:40–46. https://doi.org/10.1016/j.snb.2012.12.064

Ateh DD, Navsaria HA, Vadgama P (2006) Polypyrrole-based conducting polymers and interactions with biological tissues. J R Soc Interface 3(11):741–752. https://doi.org/10.1098/rsif.2006.0141

Balint R, Cassidy NJ, Cartmell SH (2014) Acta biomaterialia conductive polymers: towards a smart biomaterial for tissue engineering. Acta Biomater 10:2341–2353. https://doi.org/10.1016/j.actbio.2014.02.015

Barber RS, Braude R, Hosking ZD, Mitchell KG (1979) Olaquindox as performance-promoting feed additive for growing pigs. Anim Feed Sci Technol 4:117–123. https://doi.org/10.1016/0377-8401(79)90036-1

Bard A, Rubenstein I (1996) Electroanalytical chemistry: a series of advances. CRC Press

Bartlett PN, Whitaker RG (1987) Strategies for the development of amperometric enzyme electrodes. Biosensors 3:359–379. https://doi.org/10.1016/0265-928X(87)80018-4

Berndt J, Pattyn C, Hussain S, Kovacevic E (2017) Plasma based synthesis of conductive polymers: experimental results and some remarks about general strategies for plasma based polymerization processes. ECS Trans 77:49–52

Boeva ZA, Sergeyev VG (2014) Polyaniline: synthesis, properties, and application. Polym Sci Ser C 56:144–153. https://doi.org/10.1134/S1811238214010032

Bonifas AP, Mccreery RL (2012) Solid state spectroelectrochemistry of redox reactions in polypyrrole/oxide molecular heterojunctions. Anal Chem 84(5):2459–2465. https://doi.org/10.1021/ac2032047

Bossi A, Bonini F, Turner APF, Piletsky SA (2007) Molecularly imprinted polymers for the recognition of proteins: the state of the art. Biosens Bioelectron 22:1131–1137. https://doi.org/10.1016/j.bios.2006.06.023

Brooke R, Mitraka E, Sardar S et al (2017) Infrared electrochromic conducting polymer devices. J Mater Chem C 5:5824–5830. https://doi.org/10.1039/c7tc00257b

Bunting RK, Swarat K, Yan D et al (1997) Synthesis and characterization of a conducting polymer: an electrochemical experiment for general chemistry. J Chem Educ 74:421–422

Burgmayer P, Murray RW, Hill C, Carolina N (1982) An ion gate membrane: electrochemical control of ion permeability through a membrane with an embedded electrode. J Am Chem Soc 104(6139–6140):6139–6140. https://doi.org/10.1021/ja00386a061

Chapuis-Hugon F (2008) Role of molecularly imprinted polymers for selective determination of environmental pollutants—a review. Anal Chim Acta 622(1–2):48–61. https://doi.org/10.1016/j.aca.2008.05.057

Chen X, Wang F, Chen Z (2008) An electropolymerized Nile Blue sensing film-based nitrite sensor and application in food analysis. Anal Chim Acta 623:213–220. https://doi.org/10.1016/j.aca.2008.06.021

Cooper JC, Hall EAH (1992) Electrochemical response of an enzyme-loaded polyaniline film. Biosens Bioelectron 7:473–485. https://doi.org/10.1016/0956-5663(92)80004-U

D'Elia LF, Ortíz RL, Márquez OP, Márquez J, Martínez Y (2001) Electrochemical deposition of poly(*o*-phenylenediamine) films on type 304 stainless steel. J Electrochem Soc 148(2):297–300. https://doi.org/10.1149/1.1354619

Dai L (2004) Intelligent macromolecules for smart devices: from materials synthesis to device applications. Springer, London

Dubey S, Prasad BM, Misra RA (1999) Electrochemical synthesis and properties of poly(3-methylthiophene): novel synthesis of poly(3-methylthiophene) with pentachlorostannate and hexachloroantimonate. J Appl Polym Sci 73:91–102

Efimov ON, Chem R (1997) Polypyrrole: a conducting polymer; its synthesis, properties and applications. Russ Chem Rev 66(5):443–457

Feast W, Tsibouklis J, Pouwer K et al (1996) Synthesis, processing and material properties of conjugated polymers. Polymer 37:5017–5047

Fichou D (2008) Handbook of oligo- and polythiophenes. Wiley, Weinheim

Filik H, Avan AA, Mümin Y (2017) Simultaneous electrochemical determination of caffeine and vanillin by using poly(Alizarin Red S) modified glassy carbon electrode. Food Anal Methods 10:31–40. https://doi.org/10.1007/s12161-016-0545-z

Gerard M (2002) Application of conducting polymers to biosensors. Biosens Bioelectron 17:345–359. https://doi.org/10.1016/S0956-5663(01)00312-8

Geto A, Tessema M, Admassie S (2014) Determination of histamine in fish muscle at multi-walled carbon nanotubes coated conducting polymer modified glassy carbon electrode. Synth Met 191:135–140. https://doi.org/10.1016/j.synthmet.2014.03.005

Ghosh (Hazra) S, Sarker D, Misra TN (1998) Development of an amperometric enzyme electrode biosensor for fish freshness detection. Sensors Actuators B Chem 53:58–62. https://doi.org/10.1016/S0925-4005(98)00285-8

Guadarrama A, Rodríguez-Méndez ML, Sanz C et al (2001) Electronic nose based on conducting polymers for the quality control of the olive oil aroma: discrimination of quality, variety of olive and geographic origin. Anal Chim Acta 432:283–292

Guadarrama A, Rodríguez-Méndez ML, De Saja JA (2004) Influence of electrochemical deposition parameters on the performance of poly-3-methyl thiophene and polyaniline sensors for virgin olive oils. Sensors Actuators B Chem 100:60–64. https://doi.org/10.1016/j.snb.2003.12.063

Gvozdenovi MM, Jugovi BZ, Stevanovi JS, Grgur BN (2014) Electrochemical synthesis of electroconducting polymers. Hem Ind 68(6):673–684. https://doi.org/10.2298/HEMIND131122008G

Heeger A (1985) Charge storage in conducting polymers—solitons, polarons, and bipolarons. Polym J 17:201–208

Hossain E, Rahman GMA, Freund MS et al (2012) Fabrication and optimization of a conducting polymer sensor array using stored grain model volatiles. J Agric Food Chem 60(11):2863–2873. https://doi.org/10.1021/jf204631q

Hosseini AR, Wong MH, Shen Y et al (2017) Charge injection in doped organic semiconductors Charge injection in doped organic semiconductors. J Appl Phys 97:23705. https://doi.org/10.1063/1.1835542

Inzelt G (2017) Recent advances in the field of conducting polymers. J Solid State Electrochem 21:1965–1975. https://doi.org/10.1007/s10008-017-3611-6

Isaacs M, Armijo F, Ram G et al (2005) Electrochemical reduction of CO 2 mediated by poly-M-aminophthalocyanines (M = Co, Ni, Fe): poly-Co-tetraaminophthalocyanine, a selective catalyst. J Mol Catal A Chem 229:249–257. https://doi.org/10.1016/j.molcata.2004.11.026

Kaur G, Adhikari R, Cass P et al (2015) Electrically conductive polymers and composites for biomedical applications. RSC Adv 5:37553–37567. https://doi.org/10.1039/C5RA01851J

Kausar A (2017) Overview on conducting polymer in energy storage and energy conversion system. J Macromol Sci Part A 54:640–653. https://doi.org/10.1080/10601325.2017.1317210

Khan A, Jawaid M, Khan A, Asiri A (2018) Electrically conductive polymers and polymer composites: from synthesis to biomedical applications. Wiley, Weinheim

Kranz C, Wohlschläger H, Schmidt H-L, Schuhmann W (1998) Controlled electrochemical preparation of amperometric biosensors based on conducting polymer multilayers. Electroanalysis 10(8):546–552. https://doi.org/10.1002/(SICI)1521-4109(199807)10:8<546::AID-ELAN546>3.0.CO;2-#

Labaye DE, Jérôme C, Geskin VM et al (2002) Full electrochemical synthesis of conducting polymer films chemically grafted to conducting surfaces. Langmuir 18:5222–5230. https://doi.org/10.1021/la011439n

Lach P, Sharma PS, Golebiewska K et al (2017) Molecularly imprinted polymer chemosensor for selective determination of an N-nitroso-l-proline food toxin. Chem Eur J 23:1942–1949. https://doi.org/10.1002/chem.201604799

Longo L, Vasapollo G (2008) Molecularly imprinted polymers as nucleotide receptors. Mini Rev Org Chem 5(3):163–170

Macdiarmid AG, Epstein AJ (1995) Secondary doping: a new concept in conducting polymers. Macromol Symp 98:835–842

Malinauskas A (1999) Electrocatalysis at conducting polymers. Synth Met 107:75–83. https://doi.org/10.1016/S0379-6779(99)00170-8

Mason E, Weber A (2011) Polypyrrole: properties, performance and applications. Nova Science Publisher, New York

Matindoust S, Farzi A, Baghaei M (2017) Ammonia gas sensor based on flexible polyaniline films for rapid detection of spoilage in protein-rich foods. J Mater Sci Mater Electron 28:7760–7768. https://doi.org/10.1007/s10854-017-6471-z

Morelli I, Chiono V, Vozzi G et al (2010) Sensors and actuators B: chemical molecularly imprinted submicronspheres for applications in a novel model biosensor-film. Sensors Actuators B Chem 150:394–401. https://doi.org/10.1016/j.snb.2010.06.046

Neely K, Taylor C, Prosser O, Hamlyn PF (2001) Assessment of cooked alpaca and llama meats from the statistical analysis of data collected using an 'electronic nose. Meat Sci 58:53–58

Okabayashi K, Goto F, Abe K, Yoshida T (1987) Electrochemical studies of polyaniline and its application. Synth Met 18:365–370. https://doi.org/10.1016/0379-6779(87)90906-4

Onal A (2007) A review: current analytical methods for the determination of biogenic amines in foods. Food Chem 103(4):1475–1486. https://doi.org/10.1016/j.foodchem.2006.08.028

Pacheco JG, Castro M, Machado S et al (2015) Molecularly imprinted electrochemical sensor for ochratoxin A detection in food samples. Sensors Actuators B Chem 215:107–112. https://doi.org/10.1016/j.snb.2015.03.046

Palmisano F, Rizzi R, Centonze D, Zambonin PG (2000) Simultaneous monitoring of glucose and lactate by an interference and cross-talk free dual electrode amperometric biosensor based on electropolymerized thin films. Biosens Bioelectron 15:531–539. https://doi.org/10.1016/S0956-5663(00)00107-X

Pern F, Frank AJ (1990) Electrochemical and optical characterization of poly(3-methylthiophene): effects of solvent, anion, and applied potential. J Electrochem Soc 137:2769–2777

Priyanka D, Venkatesh KS, Science M (2015) Synthesis, characterization and electrical properties of polyaniline doped with different acids. Int J Eng Res Appl 5:53–61

Puoci F, Cirillo G, Curcio M et al (2007) Molecularly imprinted solid phase extraction for the selective HPLC determination of α-tocopherol in bay leaves. Anal Chim Acta 593(2):164–170. https://doi.org/10.1016/j.aca.2007.04.053

Ramström O, Mosbach K (1999) Synthesis and catalysis by molecularly imprinted materials. Curr Opin Chem Biol 3(6):759–764

Rañola RAG, Santiago KS, Sevilla FB (2016) Use of array of conducting polymers for differentiation of coconut oil products. Talanta 146:75–82. https://doi.org/10.1016/j.talanta.2015.08.026

Ravichandran R, Sundarrajan S, Venugopal JR et al (2010) Applications of conducting polymers and their issues in biomedical engineering. J R Soc Interface 7:S559–S579. https://doi.org/10.1098/rsif.2010.0120.focus

Ridgway C, Chambers J, Portero-Larragueta E, Prosser O (1999) Detection of mite infestation in wheat by electronic nose with transient flow sampling. J Sci Food Agric 19(15):2067–2074

Rotariu L, Lagarde F, Jaffrezic-Renault N, Bala C (2016) Electrochemical biosensors for fast detection of food contaminants—trends and perspective. TrAC Trends Anal Chem 79:80–87. https://doi.org/10.1016/j.trac.2015.12.017

Saber-Tehrani M, Pourhabib A, Husain SW, Arvand M (2013) A simple and efficient electrochemical sensor for nitrite determination in food samples based on Pt nanoparticles distributed poly(2-aminothiophenol) modified electrode. Food Anal Methods 6:1300–1307. https://doi.org/10.1007/s12161-012-9543-y

Sahmetlioglu E, Yuruk H, Toppare L, Yagci Y (2009) Synthesis and characterization of conducting copolymers of bisphenol A-diglycidyl ether with thiophene side-groups and pyrrole. J Macromol Sci A 46(6):37–41. https://doi.org/10.1080/10601320902851819

Sapurina IY, Shishov M (2012) Oxidative polymerization of aniline: molecular synthesis of polyaniline and the formation of supramolecular structures. In: Gomes ADS (ed) New polymers for special applications. InTech

Sarafraz-Yazdi A, Razavi N (2015) Application of molecularly-imprinted polymers in solid-phase microextraction techniques. Trends Anal Chem 73:81–90. https://doi.org/10.1016/j.trac.2015.05.004

Schlereth DD, Karyakin AA (1995) Electropolymerization of phenothiazine, phenoxazine and phenazine derivatives: characterization of the polymers by UV-visible difference spectroelectrochemistry and Fourier transform IR spectroscopy. J Electroanal Chem 395:221–232. https://doi.org/10.1016/0022-0728(95)04127-A

Scorrano S, Mergola L, Del Sole R et al (2011) Synthesis of molecularly imprinted polymers for amino acid derivates by using different functional monomers. Int J Mol Sci 12(3):1735–1743. https://doi.org/10.3390/ijms12031735

Shirakawa H, Macdiarmid AG (1977) Electrical conductivity in doped polyacetylene. Phys Rev Lett 24(39):1098–1101

Shirakawa H, Louis EJ, MacDiarmid AG et al (1977a) Synthesis of electrically conducting organic polymers: halogen derivatives of polyacetylene, $(CH)_x$. J Chem Soc Chem Commun 16:578–580. https://doi.org/10.1039/c39770000578

Shirakawa H, Louis E, Macdiarmid A et al (1977b) Synthesis of electrically conducting organic polymers: halogen derivatives of polyacetylene. J Chem Soc Chem Commun 16:578–580

Siegmund B, Pfannhauser W (1999) Changes of the volatile fraction of cooked chicken meat during chill storing: results obtained by the electronic nose in comparison to GC-MS and GC olfactometry. Zeitschrift für Lebensmitteluntersuchung und Forschung A 208(5–6):336–341

Skotheim T (1997) Handbook of conducting polymers, 2nd edn. CRC Press, Boca Raton, FL

Skotheim T, Elsenbaumer R, Reynolds J (1995) Handbook of conducting polymers. Marcel Dekker, New York

Sotzing GA, Phend JN, Grubbs RH, Lewis NS (2000) Highly sensitive detection and discrimination of biogenic amines utilizing arrays of polyaniline/carbon black composite vapor detectors. Chem Mater 12(3):593–595. https://doi.org/10.1021/cm990694e

Spanier A, Beaulieu J, Bett K, Gross K (1999) Use of electronic nose technology to examine apple quality. In: Shahidi F, Ho CT (eds) Flavor chemistry of ethnic foods. Springer, Boston, MA

Steffens C, Franceschi E, Corazza FC et al (2010) Gas sensors development using supercritical fluid technology to detect the ripeness of bananas. J Food Eng 101:365–369. https://doi.org/10.1016/j.jfoodeng.2010.07.021

Stella R, Barisci JN, Serra G et al (2000) Characterisation of olive oil by an electronic nose based on conducting polymer sensors. Sensors Actuators B Chem 63(1–2):1–9

Steter JR, Pontólio JO, Campos MLAM, Romero JR (2008) Modified electrodes prepared with polyphenolic film containing ruthenium complex and metal ligand anchored by azo covalent bond. J Braz Chem Soc 19:660–666. https://doi.org/10.1590/S0103-50532008000400008

Su W, Schrieffer R, Heeger A (1979) Solitons in polyacetylene. Phys Rev Lett 42:1698–1701

Tamayo FG, Casillas JL, Martin-Esteban A (2005) Clean up of phenylurea herbicides in plant sample extracts using molecularly imprinted polymers. Anal Bioanal Chem 381:1234–1240. https://doi.org/10.1007/s00216-005-3071-1

Taylor P, Das TK, Prusty S (2012) Polymer-plastics technology and engineering review on conducting polymers and their applications. Polym Plast Technol Eng 51(14):37–41. https://doi.org/10.1080/03602559.2012.710697

Thanh-Hai L, Kim Y, Yoon H (2017) Electrical and electrochemical properties of conducting polymers. Polymers 9:1–32. https://doi.org/10.3390/polym9040150

Thiemann C, Brett CMA (2001) Electrosynthesis and properties of conducting polymers derived from aminobenzoic acids and from aminobenzoic acids and aniline. Synth Met 123:1–9. https://doi.org/10.1016/S0379-6779(00)00364-7

Tiggemann L, Ballen SC, Bocalon CM et al (2017) Electronic nose system based on polyaniline films sensor array with different dopants for discrimination of artificial aromas. Innovative Food Sci Emerg Technol 43:112–116. https://doi.org/10.1016/j.ifset.2017.08.003

Torresi RM, Córdoba de Torresi SI, Matencio T, de Paoli MA (1995) Ionic exchanges in dodecylbenzenesulfonate-doped polypyrrole Part II: Electrochemical quartz crystal microbalance study. Synth Met 72:283–287

Trojanowicz M, Krawczyński vel Krawczyk T (1995) Electrochemical biosensors based on enzymes immobilized in electropolymerized films. Mikrochim Acta 121:167–181. https://doi.org/10.1007/BF01248249

Tung T, Castroa M, Pillin I et al (2014) Graphene–Fe_3O_4/PIL–PEDOT for the design of sensitive and stable quantum chemo-resistive VOC sensors. Carbon 74:104–112. https://doi.org/10.1016/j.carbon.2014.03.009

Uemura T, Mamada M, Kumaki D, Tokito S (2013) Synthesis of semiconducting polymers through soluble precursor polymers with thermally removable groups and their application to organic transistors. ACS Macro Lett 2(9):830–833. https://doi.org/10.1021/mz400328r

Umana M, Waller J (1986) Protein-modified electrodes. The glucose oxidase/polypyrrole system. Anal Chem 58:2979–2983. https://doi.org/10.1021/ac00127a018

Upadhyay PK, Ahmad A (2010) Chemical synthesis, spectral characterization and stability of some electrically conducting polymers. Chin J Polym Sci 28:191–197. https://doi.org/10.1007/s10118-010-9004-2

Wan M (2009) Conducting polymers with micro or nanometer structure. Springer, Berlin

Wang H, Yao S, Liu Y et al (2017) Molecularly imprinted electrochemical sensor based on Au nanoparticles in carboxylated multi-walled carbon nanotubes for sensitive determination of olaquindox in food and feedstuffs. Biosens Bioelectron 87:417–421. https://doi.org/10.1016/j.bios.2016.08.092

Wöll C (2009) Physical and chemical aspects of organic electronics. Wiley, Weinheim
Xu J, Zhang Y, Wu K et al (2017) A molecularly imprinted polypyrrole for ultrasensitive voltammetric determination of glyphosate. Microchim Acta 184:1959–1967. https://doi.org/10.1007/s00604-017-2200-9
Yu L, Zheng H, Shi M et al (2017) A novel electrochemical sensor based on poly(diallyldimethylammonium chloride)-dispersed graphene supported palladium nanoparticles for simultaneous determination of sunset yellow and tartrazine in soft drinks. Food Anal Methods 10:200–209. https://doi.org/10.1007/s12161-016-0569-4
Zhang Y, Wang J, Xu M (2010) A sensitive DNA biosensor fabricated with gold nanoparticles/ploy (p-aminobenzoic acid)/carbon nanotubes modified electrode. Colloids Surf B 75:179–185. https://doi.org/10.1016/j.colsurfb.2009.08.030

Index

T. J. Gutiérrez (ed.), *Polymers for Food Applications*,
https://doi.org/10.1007/978-3-319-94625-2

U

V

The manufacturer's authorised representative in the EU is Springer Nature Customer Service Centre GmbH, Europaplatz 3, 69115 Heidelberg, Germany. If you have any concerns regarding our products, please contact ProductSafety@springernature.com

Printed and bound by CPI Group (UK) Ltd, Croydon, CR0 4YY
05/10/2026
02228141-0018